FOCUS ON THE BIG PICTURE

Built from the ground up for optimal learning, and refined to help students focus on the big picture.

Building on the research-proven instructional techniques introduced in Knight's *Physics for Scientists and Engineers, College Physics: A Strategic Approach* sets a new standard for algebra-based introductory physics—gaining widespread critical acclaim from professors and students alike.

New to the Tech Update

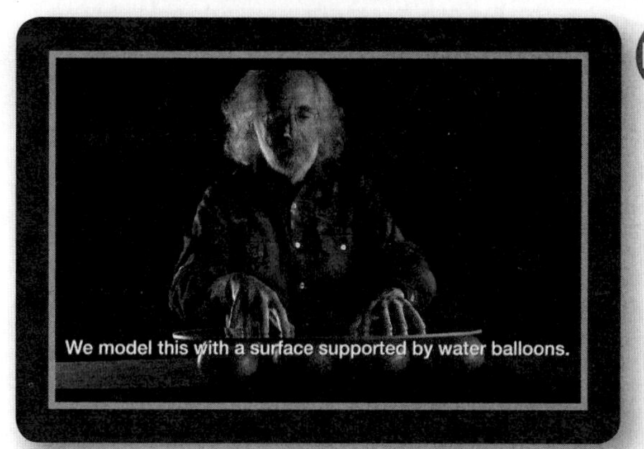

We model this with a surface supported by water balloons.

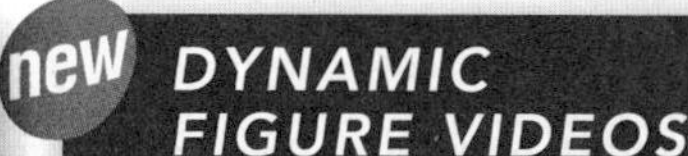

new DYNAMIC FIGURE VIDEOS

These one-minute videos are based on a key figure for each chapter that develops important, but often challenging, physics principles for students to grasp. By pairing a dynamic video with a static figure, students are better able to visualize the core concept of the figure, while text and graphic overlays highlight the physics coming to life.

new VIDEO TUTOR SOLUTIONS

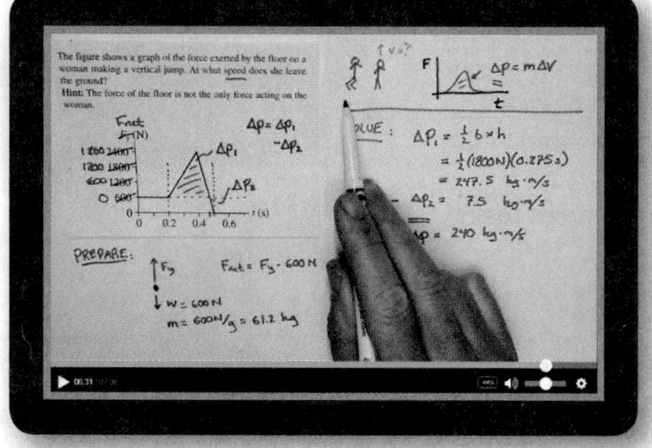

Created by co-authors Brian Jones and Stuart Field, these videos are an engaging and helpful walk-through of worked examples and typical End-of-Chapter (EOC). They are designed to help students solve key problems for each main topic through demos, animations, and guided solutions. Seven new videos per chapter provide targeted help for select EOC problems. Brian and Stuart carefully guide students through the steps needed to solve each problem using white-board style visuals. All VTS are accessible via QR code in the book and through MasteringPhysics.

The text, supplements, and MasteringPhysics® work together to help students see and understand the big picture, gain crucial problem-solving skills and confidence, and better prepare for lecture and their future. Now, more than 200 new QR codes appear throughout the textbook, enabling students to use their smartphone or tablet to instantly watch interactive videos about relevant demonstrations, new **Dynamic Figure Videos**, problem-solving strategies, and solutions explained by the authors.

FOCUS STUDENTS...

BEFORE:

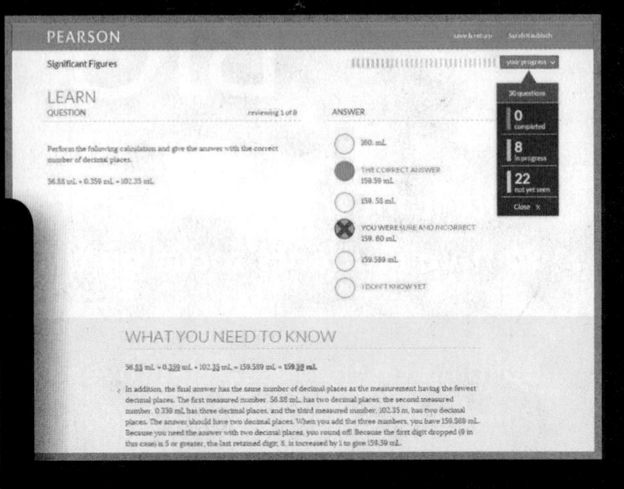

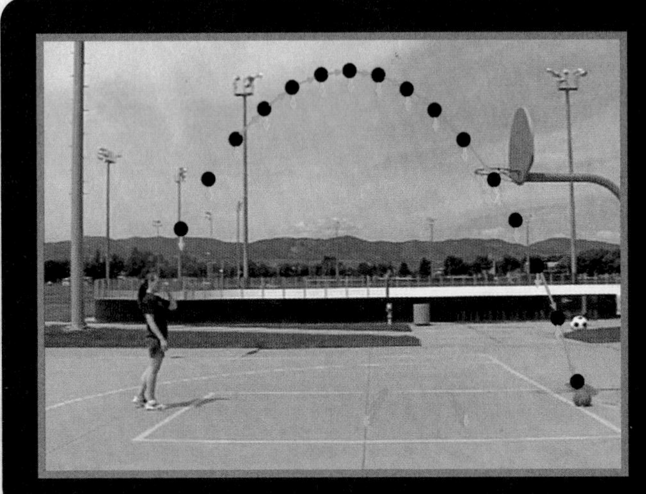

new PRELECTURE VIDEOS

Presented by co-author Brian Jones, these engaging videos are assignable through MasteringPhysics and expand on the ideas in the textbook's chapter previews, giving context, examples, and a chance for students to practice the concepts they are studying via short multiple-choice questions.

new DYNAMIC STUDY MODULES (DSMs)

DSMs help students study on their own by continuously assessing their activity and performance in real time. Students complete a set of questions with a unique answer format that also asks them to indicate their confidence level. Once completed, Dynamic Study Modules explain the concept for each correct and incorrect answer. Questions repeat until the student can answer them all correctly and confidently.

DURING:

LEARNING CATALYTICS™

With Learning Catalytics, a "bring your own device" student engagement, assessment, and classroom intelligence system, you can:

- Assess students in real time **using open-ended tasks to probe student understanding.**
- Understand immediately where students are **and adjust your lecture accordingly.**
- Add your own questions, choose from existing publisher questions, or find questions from your colleagues at other institutions **to make Learning Catalytics fit your course exactly.**
- Manage student interactions **with intelligent grouping and timing.**

BEFORE, DURING, AND AFTER CLASS

AFTER:

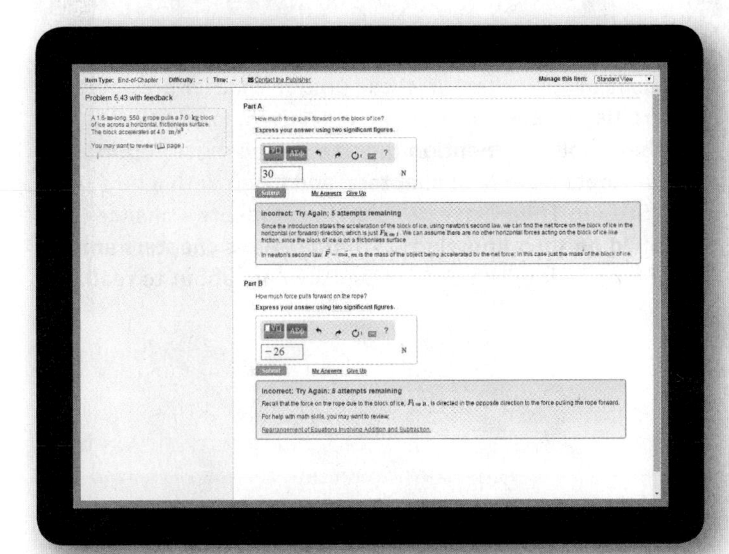

ENHANCED END-OF-CHAPTER QUESTIONS

Enhanced questions in MasteringPhysics offer students instructional support right when they need it, including wrong-answer specific feedback, links to the eText, and math remediation when completing homework assignments.

updated VIDEO TUTOR DEMONSTRATIONS

Video Tutor Demonstrations Feature "pause and predict" demonstrations of key physics concepts and incorporate assessment with answer-specific feedback on select videos to actively engage students in understanding key conceptual ideas.

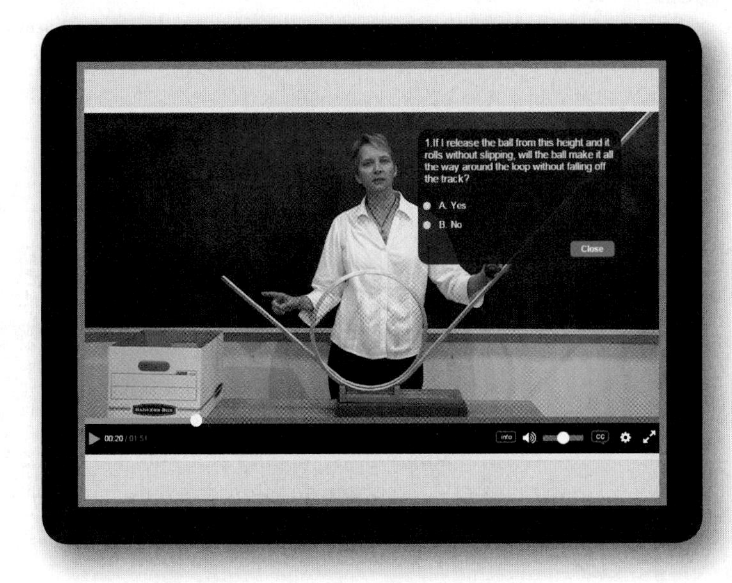

FOCUS STUDENTS...

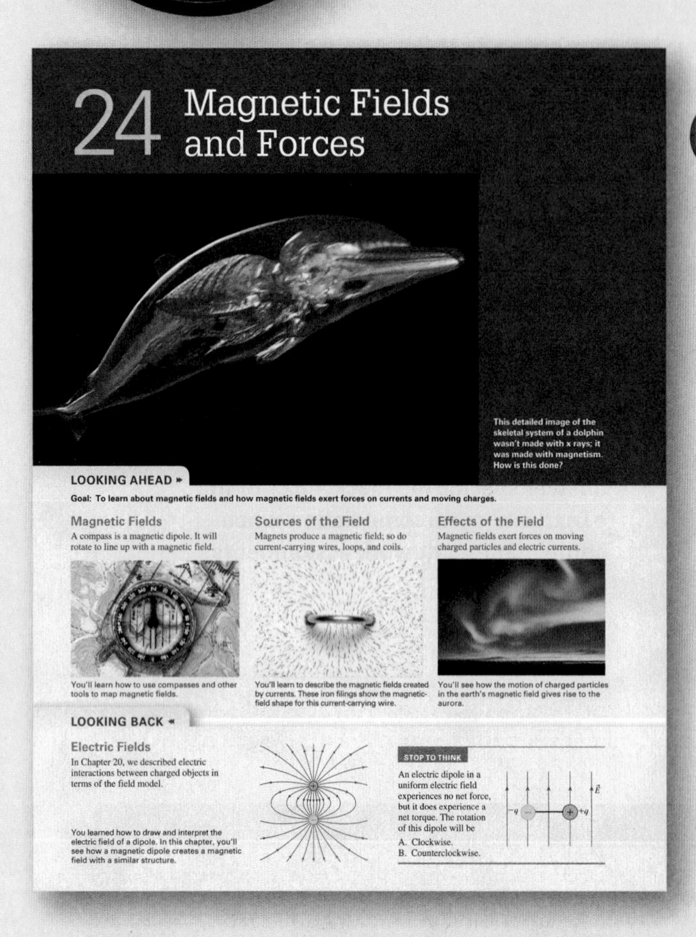

new ENHANCED CHAPTER PREVIEWS

Streamlined and focused on the three most important ideas in each chapter, these unique chapter previews are tied to specific learning objectives. In addition, they explicitly mention the one or two most important concepts from past chapters, and finish with a new "Stop to Think" question, giving students a chance to build on their knowledge from previous chapters and integrate it with new content they are about to read.

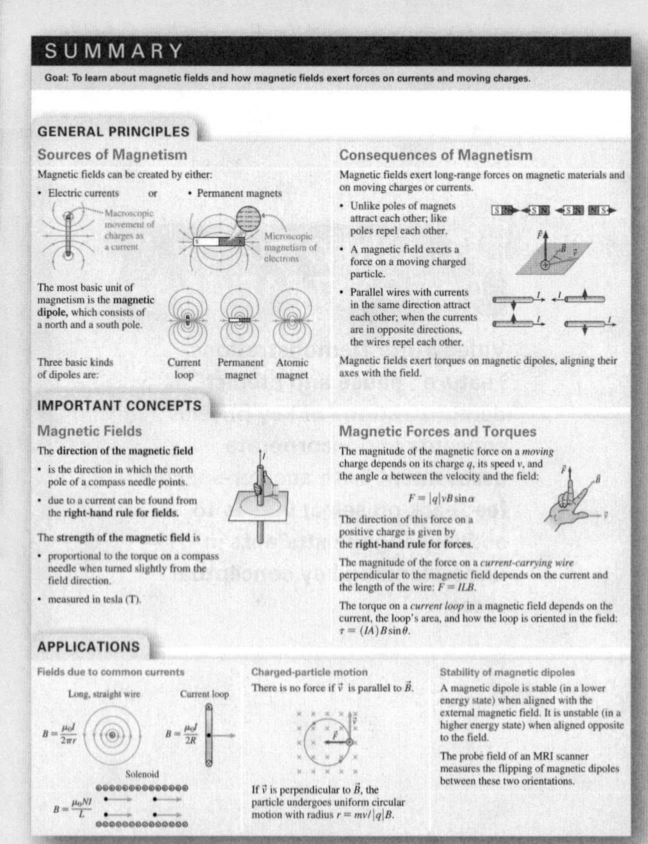

CHAPTER SUMMARIES

Chapter previews are mirrored by visual chapter summaries, helping students to review and organize what they've learned before moving ahead. They consolidate understanding by providing each concept in words, math, and figures, and organizing these into a coherent hierarchy—from General Principles to Applications.

ON THE BIG PICTURE

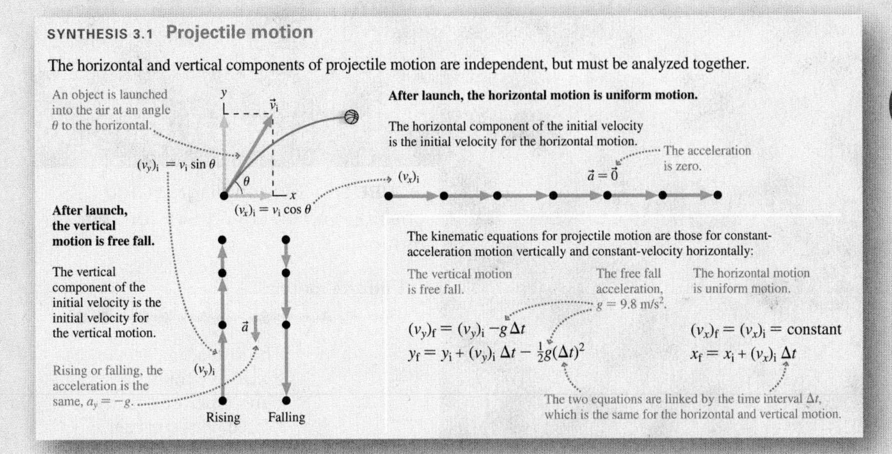

new SYNTHESIS BOXES

Bringing together key concepts, principles, and equations, this novel feature is designed to highlight connections and differences. More than a summary, they emphasize deeper relations and point out common or contrasting details.

new CONCEPT CHECK FIGURES

To encourage students to actively engage with a key or complex figure, they are asked to reason with a related "Stop to Think" question.

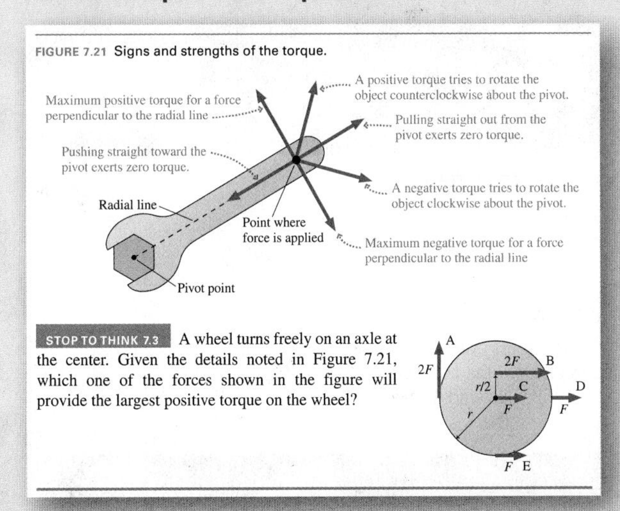

FIGURE 7.21 Signs and strengths of the torque.
Maximum positive torque for a force perpendicular to the radial line
A positive torque tries to rotate the object counterclockwise about the pivot.
Pulling straight out from the pivot exerts zero torque.
Pushing straight toward the pivot exerts zero torque.
A negative torque tries to rotate the object clockwise about the pivot.
Radial line
Point where force is applied
Maximum negative torque for a force perpendicular to the radial line
Pivot point
STOP TO THINK 7.3 A wheel turns freely on an axle at the center. Given the details noted in Figure 7.21, which one of the forces shown in the figure will provide the largest positive torque on the wheel?
A
B
C
D
E
2F
2F
F
F
F
r/2
r

PROBLEM-SOLVING STRATEGIES

Topic-specific Problem-Solving Strategies give students a framework and guidance for broad classes of problems. New overview statements now provide the "big picture," giving clear statements of what types of problems a strategy is intended for and/or how to use it.

PROBLEM-SOLVING STRATEGY 9.1 Conservation of momentum problems

new

We can use the law of conservation of momentum to relate the momenta and velocities of objects *after* an interaction to their values *before* the interaction.

PREPARE Clearly define the *system*.

- If possible, choose a system that is isolated ($\vec{F}_{net} = \vec{0}$) or within which the interactions are sufficiently short and intense that you can ignore external forces for the duration of the interaction (the impulse approximation). Momentum is then conserved.
- If it's not possible to choose an isolated system, try to divide the problem into parts such that momentum is conserved during one segment of the motion. Other segments of the motion can be analyzed using Newton's laws or, as you'll learn in Chapter 10, conservation of energy.

Following Tactics Box 9.1, draw a before-and-after visual overview. Define symbols that will be used in the problem, list known values, and identify what you're trying to find.

SOLVE The mathematical representation is based on the law of conservation of momentum, Equations 9.15. Because we generally want to solve for the velocities of objects, we usually use Equations 9.15 in the equivalent form

$$m_1(v_{1x})_f + m_2(v_{2x})_f + \cdots = m_1(v_{1x})_i + m_2(v_{2x})_i + \cdots$$
$$m_1(v_{1y})_f + m_2(v_{2y})_f + \cdots = m_1(v_{1y})_i + m_2(v_{2y})_i + \cdots$$

ASSESS Check that your result has the correct units, is reasonable, and answers the question.

Exercise 17

FOCUS ON STUDENTS...

Multiple Representations...

FIGURE 2.14 Motion diagram and position-versus-time graph for uniform motion.

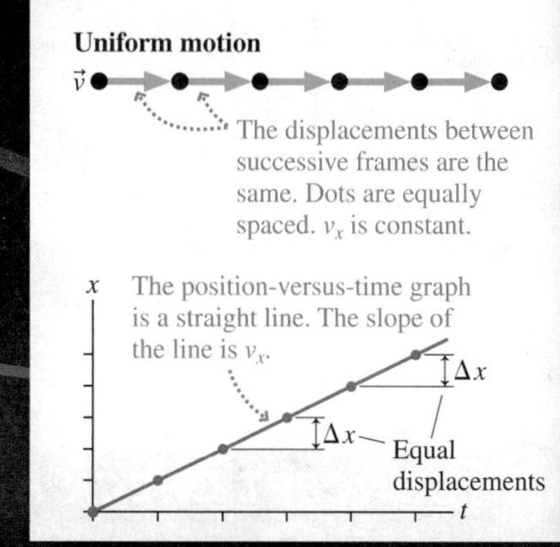

Balancing Qualitative and Quantitative Reasoning...

Figures that *Teach*...

FIGURE 11.18 The operation of a heat engine.

(a) 1. Heat energy Q_H is transferred from the hot reservoir to the system.

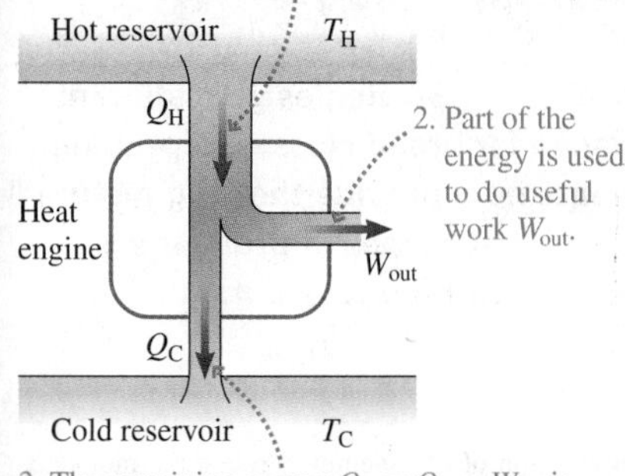

3. The remaining energy $Q_C = Q_H - W_{out}$ is exhausted to the cold reservoir as waste heat.

Addressing Misconceptions...

TACTICS BOX 24.1 **Right-hand rule for fields** MP

❶ Point your *right* thumb in the direction of the current.

❷ Wrap your fingers around the wire to indicate a circle.

❸ Your fingers curl in the direction of the magnetic field lines around the wire.

I

Exercises 6–11

An *Inductive* Approach...

Emphasis on Explicit Skills...

RESEARCH-BASED

Knight/Jones/Field was designed from the ground up with students in mind, and is based on a solid foundation of the latest physics education research . . .

AND HOW THEY LEARN

Guided Practice...

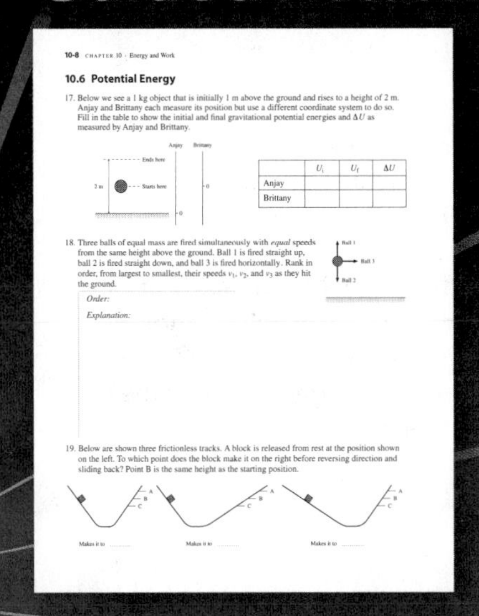

10-8 CHAPTER 10 · Energy and Work

10.6 Potential Energy

17. Below we see a 1 kg object that is initially 1 m above the ground and rises to a height of 2 m. Anjay and Brittany each measure its position but use a different coordinate system to do so. Fill in the table to show the initial and final gravitational potential energies and ΔU as measured by Anjay and Brittany.

Anjay Brittany

Ends here
2 m
Starts here
0
0

	U_i	U_f	ΔU
Anjay			
Brittany			

18. Three balls of equal mass are fired simultaneously with *equal* speeds from the same height above the ground. Ball 1 is fired straight up, ball 2 is fired straight down, and ball 3 is fired horizontally. Rank in order, from largest to smallest, their speeds v_1, v_2, and v_3 as they hit the ground.

Ball 1
Ball 3
Ball 2

Order:

Explanation:

19. Below are shown three frictionless tracks. A block is released from rest at the position shown on the left. To which point does the block make it on the right before reversing direction and sliding back? Point B is the same height as the starting position.

A B C A B C A B C

Makes it to ______ Makes it to ______ Makes it to ______

Annotated Equations...

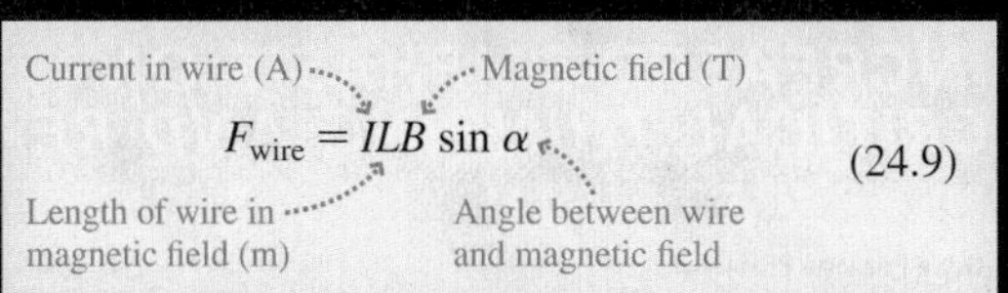

Visual Analogies...

FIGURE 22.7 Water in a pipe turns a turbine.

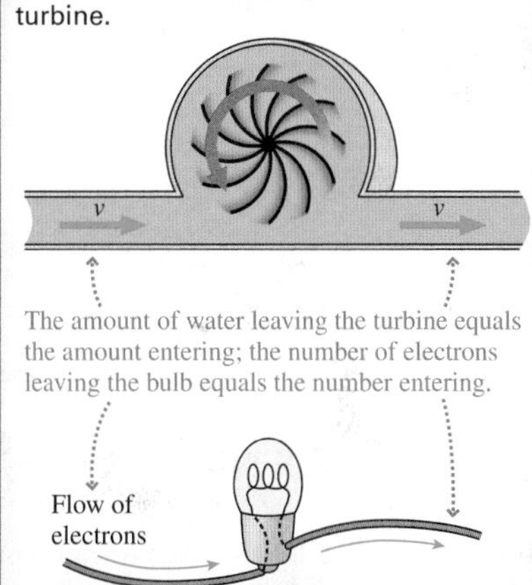

v
v
The amount of water leaving the turbine equals the amount entering; the number of electrons leaving the bulb equals the number entering.
Flow of electrons

Streamlining use of Color...

Active Learning...

STOP TO THINK 10.4 Rank in order, from largest to smallest, the gravitational potential energies of identical balls 1 through 4.

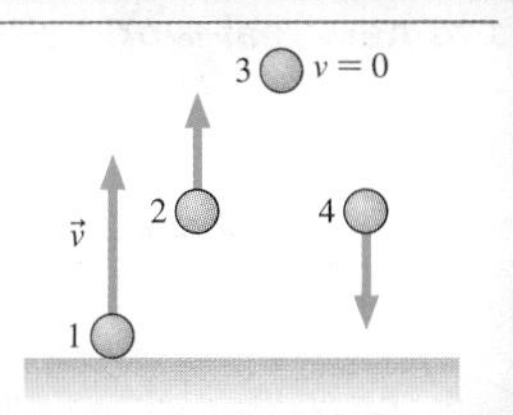

Structured Problem Solving...

PEDAGOGY

. . . Evident in the text, art, design, pedagogy, and technology, using ideas from multimedia learning theory, students are given the best tools to succeed in physics

FOCUS STUDENTS... ON THEIR GOALS

new INCREASED EMPHASIS ON CRITICAL THINKING AND REASONING

MCAT-Style Passage Problems

Kangaroo Locomotion BIO

Kangaroos have very stout tendons in their legs that can be used to store energy. When a kangaroo lands on its feet, the tendons stretch, transforming kinetic energy of motion to elastic potential energy. Much of this energy can be transformed back into kinetic energy as the kangaroo takes another hop. The kangaroo's peculiar hopping gait is not very efficient at low speeds but is quite efficient at high speeds.

Figure P11.68 shows the energy cost of human and kangaroo locomotion. The graph shows oxygen uptake (in mL/s) per kg of body mass, allowing a direct comparison between the two species.

Oxygen uptake (mL/kg · s)

2.0, 1.8, 1.6, 1.4, 1.2, 1.0, 0.8, 0.6, 0.4, 0.2, 0

Human, running

Red kangaroo, hopping

0, 2, 4, 6, 8, 10, 12 Speed (m/s)

FIGURE P11.68 Oxygen uptake (a measure of energy use per second) for a running human and a hopping kangaroo.

For humans, the energy used per second (i.e., power) is proportional to the speed. That is, the human curve nearly passes through the origin, so running twice as fast takes approximately twice as much power. For a hopping kangaroo, the graph of energy use has only a very small slope. In other words, the energy used per second changes very little with speed. Going faster requires very little additional power. Treadmill tests on kangaroos and observations in the wild have shown that they do not become winded at any speed at which they are able to hop. No matter how fast they hop, the necessary power is approximately the same.

68. | A person runs 1 km. How does his speed affect the total energy needed to cover this distance?
    A. A faster speed requires less total energy.
    B. A faster speed requires more total energy.
    C. The total energy is about the same for a fast speed and a slow speed.
69. | A kangaroo hops 1 km. How does its speed affect the total energy needed to cover this distance?
    A. A faster speed requires less total energy.
    B. A faster speed requires more total energy.
    C. The total energy is about the same for a fast speed and a slow speed.
70. | At a speed of 4 m/s,
    A. A running human is more efficient than an equal-mass hopping kangaroo.
    B. A running human is less efficient than an equal-mass hopping kangaroo.
    C. A running human and an equal-mass hopping kangaroo have about the same efficiency.
71. | At approximately what speed would a human use half the power of an equal-mass kangaroo moving at the same speed?
    A. 3 m/s B. 4 m/s C. 5 m/s D. 6 m/s
72. | At what speed does the hopping motion of the kangaroo become more efficient than the running gait of a human?
    A. 3 m/s B. 5 m/s C. 7 m/s D. 9 m/s

The MCAT has been restructured to test competencies. Students are required to reason, to do more than simply plug in numbers into equations. Of the hundreds of new end-of-chapter problems, many of these require students to reason using ratios and proportionality, to reason using real-world data, and to assess answers to see if they make physical sense.

EXPANDED LIFE-SCIENCE AND BIOMEDICAL APPLICATIONS

Building on the book's acclaimed real-world focus, even more applications from the living world have been added to text, worked examples, and end-of-chapter problems, giving students essential practice in applying core physical principles to new real-world situations.

CONCEPTUAL EXAMPLE 13.13 Blood pressure and cardiovascular disease BIO

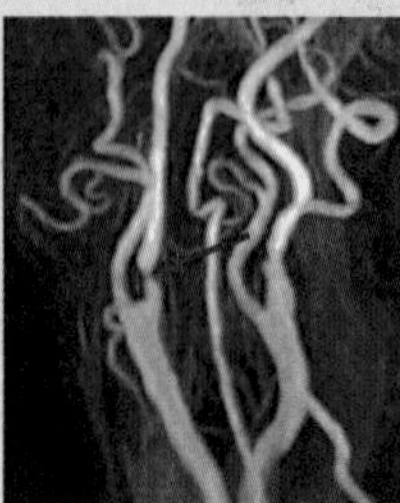

Cardiovascular disease is a narrowing of the arteries due to the buildup of plaque deposits on the interior walls. Magnetic resonance imaging, which you'll learn about in Chapter 24, can create exquisite three-dimensional images of the internal structure of the body. Shown are the carotid arteries that supply blood to the head, with a dangerous narrowing—a *stenosis*—indicated by the arrow.

If a section of an artery has narrowed by 8%, not nearly as much as the stenosis shown, by what percentage must the blood-pressure difference between the ends of the narrowed section increase to keep blood flowing at the same rate?

REASON According to Poiseuille's equation, the pressure difference Δp must increase to compensate for a decrease in the artery's radius R if the blood flow rate Q is to remain unchanged. If we write Poiseuille's equation as

$$R^4 \Delta p = \frac{8\eta L Q}{\pi}$$

we see that the product $R^4 \Delta p$ must remain unchanged if the artery is to deliver the same flow rate. Let the initial artery radius and pressure difference be R_i and Δp_i. Disease decreases the radius by 8%, meaning that $R_f = 0.92R_i$. The requirement

$$R_i^4 \Delta p_i = R_f^4 \Delta p_f$$

can be solved for the new pressure difference:

$$\Delta p_f = \frac{R_i^4}{R_f^4} \Delta p_i = \frac{R_i^4}{(0.92R_i)^4} \Delta p_i = 1.4\, \Delta p_i$$

The pressure difference must increase by 40% to maintain the flow.

ASSESS Because the flow rate depends on R^4, even a small change in radius requires a large change in Δp to compensate. Either the person's blood pressure must increase, which is dangerous, or he or she will suffer a significant reduction in blood flow. For the stenosis shown in the image, the reduction in radius is much greater than 8%, and the pressure difference will be large and very dangerous.

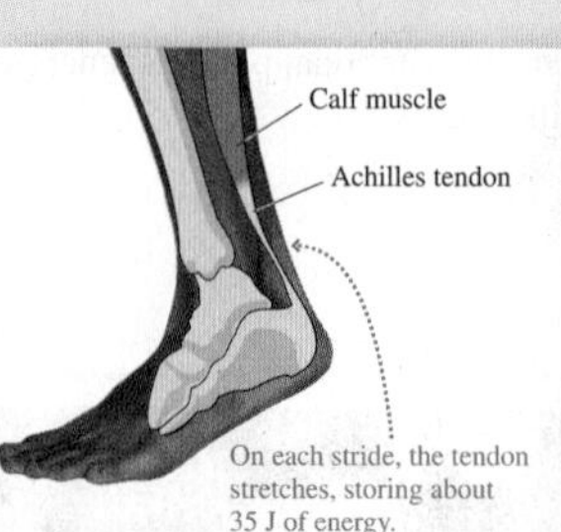

Calf muscle
Achilles tendon
On each stride, the tendon stretches, storing about 35 J of energy.

Spring in your step BIO As you run, you lose some of your mechanical energy each time your foot strikes the ground; this energy is transformed into unrecoverable thermal energy. Luckily, about 35% of the decrease of your mechanical energy when your foot lands is stored as elastic potential energy in the stretchable Achilles tendon of the lower leg. On each plant of the foot, the tendon is stretched, storing some energy. The tendon springs back as you push off the ground again, helping to propel you forward. This recovered energy reduces the amount of internal chemical energy you use, increasing your efficiency.

Table of Problem-Solving Strategies

Table of Synthesis Boxes

Table of Math Relationships

Note for users of the two-volume edition:
Volume 1 (pp. 1–533) includes Chapters 1–16.
Volume 2 (pp. 534–1011) includes Chapters 17–30.

Brief Contents

THIRD EDITION

VOLUME 2

college physics

a strategic approach

RANDALL D. KNIGHT

California Polytechnic State University, San Luis Obispo

BRIAN JONES

Colorado State University

STUART FIELD

Colorado State University

PEARSON

Editor-in-Chief:	Jeanne Zalesky
Acquisitions Editor:	Darien Estes
Project Manager:	Martha Steele
Program Manager:	Katie Conley
Senior Development Editor:	Alice Houston, Ph.D.
Art Development Editor:	Margot Otway
Development Manager:	Cathy Murphy
Program and Project Management Team Lead:	Kristen Flathman
Production Management:	Rose Kernan
Compositor:	Cenveo® Publisher Services
Design Manager:	Mark Ong
Cover Designer:	Tandem Creative, Inc.
Illustrators:	Rolin Graphics
Rights & Permissions Project Manager:	Maya Gomez
Rights & Permissions Management:	Rachel Youdelman
Photo Researcher:	Eric Schrader
Manufacturing Buyer:	Maura Zaldivar-Garcia
Executive Marketing Manager:	Christy Lesko
Marketing Manager:	Elizabeth Ellsworth
Cover Photo Credit:	Borut Trdina/Getty Images

Acknowledgements of third party content appear on page C-1, which constitutes an extension of this copyright page.

Library of Congress CIP data on file.

www.pearsonhighered.com

ISBN 10: 0-134-20194-9; ISBN 13: 978-0-134-20194-8 (Volume 2)

2 16

About the Authors

Randy Knight taught introductory physics for 32 years at Ohio State University and California Polytechnic State University, where he is Professor Emeritus of Physics. Professor Knight received a Ph.D. in physics from the University of California, Berkeley and was a post-doctoral fellow at the Harvard-Smithsonian Center for Astrophysics before joining the faculty at Ohio State University. It was at Ohio State that he began to learn about the research in physics education that, many years later, led to *Five Easy Lessons: Strategies for Successful Physics Teaching* and this book, as well as *College Physics: A Strategic Approach,* co-authored with Brian Jones and Stuart Field. Professor Knight's research interests are in the fields of laser spectroscopy and environmental science. When he's not in front of a computer, you can find Randy hiking, sea kayaking, playing the piano, or spending time with his wife Sally and their five cats.

Brian Jones has won several teaching awards at Colorado State University during his 25 years teaching in the Department of Physics. His teaching focus in recent years has been the College Physics class, including writing problems for the MCAT exam and helping students review for this test. In 2011, Brian was awarded the Robert A. Millikan Medal of the American Association of Physics Teachers for his work as director of the Little Shop of Physics, a hands-on science outreach program. He is actively exploring the effectiveness of methods of informal science education and how to extend these lessons to the college classroom. Brian has been invited to give workshops on techniques of science instruction throughout the United States and in Belize, Chile, Ethiopia, Azerbaijan, Mexico and Slovenia. Brian and his wife Carol have dozens of fruit trees and bushes in their yard, including an apple tree that was propagated from a tree in Isaac Newton's garden.

Stuart Field has been interested in science and technology his whole life. While in school he built telescopes, electronic circuits, and computers. After attending Stanford University, he earned a Ph.D. at the University of Chicago, where he studied the properties of materials at ultralow temperatures. After completing a postdoctoral position at the Massachusetts Institute of Technology, he held a faculty position at the University of Michigan. Currently at Colorado State University, Stuart teaches a variety of physics courses, including algebra-based introductory physics, and was an early and enthusiastic adopter of Knight's *Physics for Scientists and Engineers.* Stuart maintains an active research program in the area of superconductivity. Stuart enjoys Colorado's great outdoors, where he is an avid mountain biker; he also plays in local ice hockey leagues.

Preface to the Instructor

In 2006, we published *College Physics: A Strategic Approach,* a new algebra-based physics textbook for students majoring in the biological and life sciences, architecture, natural resources, and other disciplines. As the first such book built from the ground up on research into how students can more effectively learn physics, it quickly gained widespread critical acclaim from professors and students alike. For the second edition, and now for this third edition, we have continued to build on the research-proven instructional techniques introduced in the first edition and the extensive feedback from thousands of users to take student learning even further.

Objectives

Our primary goals in writing *College Physics: A Strategic Approach* have been:

- To provide students with a textbook that's a more manageable size, less encyclopedic in its coverage, and better designed for learning.
- To integrate proven techniques from physics education research into the classroom in a way that accommodates a range of teaching and learning styles.
- To help students develop both quantitative reasoning skills and solid conceptual understanding, with special focus on concepts well documented to cause learning difficulties.
- To help students develop problem-solving skills and confidence in a systematic manner using explicit and consistent tactics and strategies.
- To motivate students by integrating real-world examples relevant to their majors—especially from biology, sports, medicine, the animal world—and that build upon their everyday experiences.
- To utilize proven techniques of visual instruction and design from educational research and cognitive psychology that improve student learning and retention and address a range of learner styles.

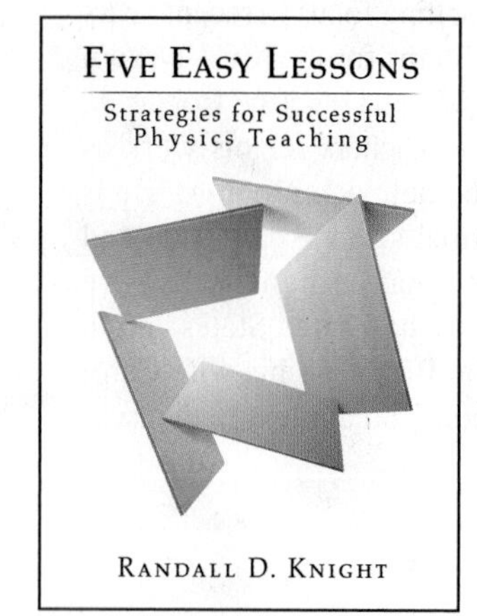

A more complete explanation of these goals and the rationale behind them can be found in Randy Knight's paperback book, *Five Easy Lessons: Strategies for Successful Physics Teaching*. Please request a copy from your local Pearson sales representative if it would be of interest to you (ISBN 978-0-805-38702-5).

What's New to This Edition

In revising the book for this third edition, we have renewed our basic focus on students and how they learn. We've considered extensive feedback from scores of instructors and thousands of students, including our student advisory panel, in order to enhance and improve the text, figures, and the end-of-chapter problems. Changes include the following:

- More focused **Chapter Previews** provide a brief, visual, and non-technical preview, proven to help students organize their thinking and improve their understanding of the upcoming material.
- New **Synthesis boxes** bring together key concepts, principles, and equations in order to highlight connections and differences.
- New **Concept Check figures** encourage students to actively engage with key or complex figures by asking them to reason with a related Stop To Think question.
- Additional **Stop To Think questions** provide students with more crucial practice and concept checks as they go through the chapters.

- New in-line **Looking Back pointers** encourage students to review important material from earlier chapters. These are given right at the moment they are needed, rather than at the start of the chapter (where they are often overlooked).
- New **Problem-Solving Strategy Overviews** give students the "big picture" of the strategy before delving into details, just as the chapter previews give the "big picture" of the chapter.
- **Streamlined text and figures** tighten and focus the presentation to more closely match student needs. We've scrutinized every figure, caption, discussion, and photo in order to enhance their clarity and focus their role.
- Expanded use of **annotated equations** helps students decipher what they "say," and what the variables and units are.
- Increased emphasis on **critical thinking and reasoning,** both in worked examples and end-of-chapter problems, promotes these key skills. These skills are especially important for those taking the MCAT exam.
- Expanded use of **realistic and real-world data** ensures students can make sense of answers that are grounded in the real world. Our examples and problems use real numbers and real data, and test different types of reasoning using equations, ratios, and graphs.
- **Enhanced end-of-chapter problems,** based on the wealth of data from MasteringPhysics, student advisory panel input, and a rigorous blind-solving and accuracy cross-checking process, optimize clarity, utility, and variety. We've added problems based on real-world and biomedical situations and problems that expand the range of reasoning skills students need to use in the solution.

We have made many small changes to the flow of the text throughout, streamlining derivations and discussions, providing more explanation for complex concepts and situations, and reordering and reorganizing material so that each section and each chapter has a clearer focus. There are small changes on nearly every page. The more significant content changes include:

- The circular motion material in Chapters 3, 6 and 7 has been reworked for a more natural progression of topics. Acceleration in circular motion is now introduced in Chapter 3, frequency and period are now introduced in Chapter 6, while angular position and angular velocity are now in Chapter 7. The treatment of circular motion in Chapter 3 emphasizes the use of vectors to understand the nature of centripetal acceleration. In Chapter 6, the focus is on dynamics, and in Chapter 7, we extend these ideas to rotational motion.
- The discussion of the law of conservation of energy in Section 10.6 has been updated to provide a more logical and coherent flow from the most general form of the law to more specialized versions for isolated systems and then to systems with only mechanical energy.
- The material in Chapter 11 making the microscopic connection between thermal energy and temperature for an ideal gas has been moved to Chapter 12, where it fits better with the atomic model of an ideal gas presented there.
- Minor topics that have been removed to focus the presentation include antinodal lines for sound waves in Chapter 16, maximum intensity of a diffraction grating's bright fringes in Chapter 17, exposure in Chapter 19, and elevation graphs in Chapter 21.
- The start of Chapter 21 has been revised to clarify the origin of electric potential energy by making a more concrete connection between electric potential energy and more familiar potential energies, such as gravitational and elastic potential energy.
- The treatment of electromagnetic waves in Chapter 25 was streamlined to focus on the nature of the waves, the meaning of polarization, and the application of these ideas to real-world situations.
- Chapters 29 and 30 have been significantly streamlined, improving the overall flow and removing some extraneous details so that students can better focus on the physics.

We know that students increasingly rely on sources of information beyond the text, and instructors are looking for quality resources that prepare students for engagement in lecture. The text will always be the central focus, but we have added additional media elements closely tied to the text that will enhance student understanding. In the Technology Update to the Second Edition, we added Class Videos, Video Tutor Solutions, and Video Tutor Demonstrations. In the Third Edition, we have added an exciting new supplement, **Prelecture Videos,** short videos with author Brian Jones that introduce the topics of each chapter with accompanying assessment questions. In the front of this book, you'll find an illustrated walkthrough of the new media available in this technology update for the third edition:

- **Dynamic Study Modules (DSMs)** help students study on their own by continuously assessing their activity and performance in real time. Students complete a set of questions with a unique answer format that repeats each question until students can answer them all correctly and confidently.
- **New! Dynamic Figure Videos in each chapter** are one-minute videos based on figures from the textbook that depict important, but often challenging, physics principles.
- **Video Tutor Solutions** created by co-author Brian Jones are an engaging and helpful walkthrough of worked examples and select End-of-Chapter (EOC) problems designed to help students solve problems for each main topic. Each chapter has seven Video Tutor Solutions.
- **Prep questions aligned with the new 2015 MCAT exam** are based on the Foundational Concepts and Content Categories outlined by the Association of American Medical Colleges. These 140 new problems are assignable in MasteringPhysics and available for self study in the Study Area.
- **Updated Video Tutor Demonstrations**, accessible via QR codes, feature "pause-and-predict" demonstrations of key physics concepts and incorporate assessment with answer-specific feedback.

  Please note: Users must download a QR code reader to their mobile phone. Data and roaming charges may also apply.

Textbook Organization

College Physics: A Strategic Approach is a 30-chapter text intended for use in a two-semester course. The textbook is divided into seven parts: Part I: *Force and Motion*, Part II: *Conservation Laws*, Part III: *Properties of Matter*, Part IV: *Oscillations and Waves*, Part V: *Optics*, Part VI: *Electricity and Magnetism*, and Part VII: *Modern Physics*.

Part I covers Newton's laws and their applications. The coverage of two fundamental conserved quantities, momentum and energy, is in Part II, for two reasons. First, the way that problems are solved using conservation laws—comparing an *after* situation to a *before* situation—differs fundamentally from the problem-solving strategies used in Newtonian dynamics. Second, the concept of energy has a significance far beyond mechanical (kinetic and potential) energies. In particular, the key idea in thermodynamics is energy, and moving from the study of energy in Part II into thermal physics in Part III allows the uninterrupted development of this important idea.

Optics (Part V) is covered directly after oscillations and waves (Part IV), but *before* electricity and magnetism (Part VI). Further, we treat wave optics before ray optics. Our motivations for this organization are twofold. First, wave optics is largely just an extension of the general ideas of waves; in a more traditional organization, students will have forgotten much of what they learned about waves by the time they get to wave optics. Second, optics as it is presented in introductory physics makes no use of the properties of electromagnetic fields. The documented difficulties that students have with optics are difficulties with waves, not difficulties with electricity and magnetism. There's little reason other than historical tradition to delay optics. However, the optics chapters are easily deferred until after Part VI for instructors who prefer that ordering of topics.

- **Complete edition,** with MasteringPhysics® and Student Workbook (ISBN 978-0-321-90255-9): Chapters 1–30.
- **Books a la Carte edition,** with MasteringPhysics® and Student Workbook (ISBN 978-0-321-90882-7)
- **Complete edition,** without MasteringPhysics® (ISBN 978-0-321-87972-1): Chapters 1–30.
- **Volume 1** without MasteringPhysics® (ISBN 978-0-321-90877-3): Chapters 1–16.
- **Volume 2** without MasteringPhysics® (ISBN 978-0-321-90878-0): Chapters 17–30.

Split your text the way you split your course! Log on to www.pearsoncustomlibrary.com and create your own splits of *College Physics: A Strategic Approach,* 3e, including *your* choice of chapters.

Instructional Package

College Physics: A Strategic Approach: Technology Update, third edition, provides an integrated teaching and learning package of support material for students and instructors.

NOTE For convenience, most instructor supplements can be downloaded from the "Instructor Resources" area of MasteringPhysics® and the Instructor Resource Center (www.pearsonhighered.com/educator).

Supplement	Print	Online	Instructor or Student Supplement	Description
MasteringPhysics with Pearson eText (ISBN 0321908813)		✓	Instructor and Student Supplement	This product features all of the resources of MasteringPhysics in addition to the new Pearson eText 2.0. Now available on smartphones and tablets, Pearson eText 2.0 comprises the full text, including videos and other rich media.
Instructor's Solutions Manual (ISBN 0321902971)		✓	Instructor Supplement	This comprehensive solutions manual contains complete solutions to all end-of-chapter questions and problems.
Instructor's Guide (ISBN 0321907264)		✓	Instructor Supplement	Written by the textbook authors, this resource provides chapter-by-chapter creative ideas for using active-learning techniques in your class.
TestGen Test Bank (ISBN 0321851412)		✓	Instructor Supplement	The Test Bank contains more than 2,000 high-quality problems, with a range of multiple-choice, true/false, short answer, and regular homework-type questions. Test files are provided in both TestGen® and Word format.
Instructor's Resource DVD (ISBN 0321907256)	✓	✓	Instructor Supplement	This cross-platform DVD includes all the text images, editable content for key features, all the instructor resources listed above, and solutions to the Student Workbook exercises.
Student's Workbook Standard (CH1–30) (ISBN 032166258X) Volume 1 (CH1–16) (ISBN 0321908864) Volume 2 (CH17–30) (ISBN 0321908872)	✓		Student Supplement	For a more detailed description of the *Student's Workbook*, see page viii.
Student's Solutions Manual Volume 1 (CH1–16) (ISBN 0321908848) Volume 2 (CH17–30) (ISBN 9321908856)	✓		Student Supplement	These solutions manuals contain detailed solutions to more than half of the odd-numbered end-of-chapter problems from the textbook.

The Student Workbook

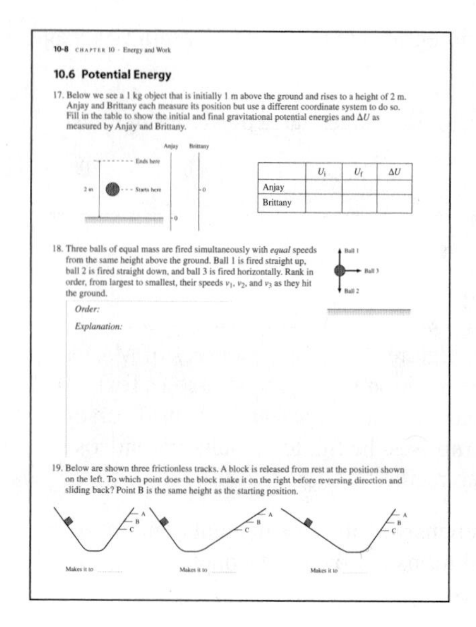
10-8 CHAPTER 10 · Energy and Work

10.6 Potential Energy

17. Below we see a 1 kg object that is initially 1 m above the ground and rises to a height of 2 m. Anjay and Brittany each measure its position but use a different coordinate system to do so. Fill in the table to show the initial and final gravitational potential energies and ΔU as measured by Anjay and Brittany.

	U_i	U_f	ΔU
Anjay			
Brittany			

18. Three balls of equal mass are fired simultaneously with *equal* speeds from the same height above the ground. Ball 1 is fired straight up, ball 2 is fired straight down, and ball 3 is fired horizontally. Rank in order, from largest to smallest, their speeds v_1, v_2, and v_3 as they hit the ground.

Order:

Explanation:

19. Below are shown three frictionless tracks. A block is released from rest at the position shown on the left. To which point does the block make it on the right before reversing direction and sliding back? Point B is the same height as the starting position.

A key component of *College Physics: A Strategic Approach* is the accompanying *Student Workbook*. The workbook bridges the gap between textbook and homework problems by providing students the opportunity to learn and practice skills prior to using those skills in quantitative end-of-chapter problems, much as a musician practices technique separately from performance pieces. The workbook exercises, which are keyed to each section of the textbook, focus on developing specific skills, ranging from identifying forces and drawing free-body diagrams to interpreting field diagrams.

The workbook exercises, which are generally qualitative and/or graphical, draw heavily upon the physics education research literature. The exercises deal with issues known to cause student difficulties and employ techniques that have proven to be effective at overcoming those difficulties. **New to the third edition workbook** are *jeopardy problems* that ask students to work backwards from equations to physical situations, enhancing their understanding and critical thinking skills. The workbook exercises can be used in-class as part of an active-learning teaching strategy, in recitation sections, or as assigned homework. More information about effective use of the *Student Workbook* can be found in the *Instructor's Guide*.

Available versions: Volume 1 (ISBN 978-0-321-90886-5): Chapters 1–16, and Volume 2 (978-0-321-90887-2): Chapters 17–30. A package of both volumes is also available (ISBN 978-0-321-90724-0).

Acknowledgments

We have relied upon conversations with and, especially, the written publications of many members of the physics education community. Those who may recognize their influence include Arnold Arons, Uri Ganiel, Fred Goldberg, Ibrahim Halloun, David Hestenes, Leonard Jossem, Jill Larkin, Priscilla Laws, John Mallinckrodt, Lillian McDermott and members of the Physics Education Research Group at the University of Washington, Edward "Joe" Redish, Fred Reif, John Rigden, Rachel Scherr, Bruce Sherwood, David Sokoloff, Ronald Thornton, Sheila Tobias, and Alan Van Heuleven.

We are very grateful to Larry Smith for the difficult task of writing the *Instructor Solutions Manual*; to Scott Nutter for writing out the Student Workbook answers; to Wayne Anderson, Jim Andrews, Nancy Beverly, David Cole, Karim Diff, Jim Dove, Marty Gelfand, Kathy Harper, Charlie Hibbard, Robert Lutz, Matt Moelter, Kandiah Manivannan, Ken Robinson, and Cindy Schwarz-Rachmilowitz for their contributions to the end-of-chapter questions and problems; to Wayne again for helping with the Test Bank questions; and to Steven Vogel for his careful review of the biological content of many chapters and for helpful suggestions.

We especially want to thank our editor Becky Ruden, development editor Alice Houston, project manager Martha Steele, and all the other staff at Pearson for their enthusiasm and hard work on this project. Having a diverse author team is one of the strengths of this book, but it has meant that we rely a great deal on Becky to help us keep to a single focus, on Martha to be certain that one of us attends to all details, and on Alice's tireless efforts and keen editorial eye as she helps us synthesize our visions into a coherent whole.

Rose Kernan and the team at Nesbitt Graphics/Cenveo, copy editor Carol Reitz, and photo researcher Eric Schrader get much credit for making this complex project all come together. In addition to the reviewers and classroom testers listed below, who gave invaluable feedback, we are particularly grateful to Charlie Hibbard for his close scrutiny of every word, symbol, number, and figure.

Randy Knight: I would like to thank my Cal Poly colleagues, especially Matt Moelter, for many valuable conversations and suggestions. I am endlessly grateful to my wife Sally for her love, encouragement, and patience, and to our many cats for nothing in particular other than being cats.

Brian Jones: I would like to thank my fellow AAPT and PIRA members for their insight and ideas, the creative students and colleagues who are my partners in the Little Shop of Physics, the students in my College Physics classes who help me become a better teacher, and, most of all, my wife Carol, my best friend and gentlest editor, whose love makes the journey worthwhile.

Stuart Field: I would like to thank my wife Julie and my children, Sam and Ellen, for their love, support, and encouragement.

Reviewers and Classroom Testers

Special thanks go to our third edition review panel: Taner Edis, Marty Gelfand, Jason Harlow, Charlie Hibbard, Jeff Loats, Amy Pope, and Bruce Schumm.

David Aaron, *South Dakota State University*
Susmita Acharya, *Cardinal Stritch University*
Ugur Akgun, *University of Iowa*
Ralph Alexander, *University of Missouri—Rolla*
Kyle Altmann, *Elon University*
Donald Anderson, *Ivy Tech*
Michael Anderson, *University of California—San Diego*
Steve Anderson, *Montana Tech*
James Andrews, *Youngstown State University*
Charles Ardary, *Edmond Community College*
Charles Bacon, *Ferris State University*
John Barry, *Houston Community College*
David H. Berman, *University of Northern Iowa*
Phillippe Binder, *University of Hawaii—Hilo*
Jeff Bodart, *Chipola College*
James Boger, *Flathead Valley Community College*
Richard Bone, *Florida International University*
James Borgardt, *Juniata College*
Daniela Bortoletto, *Purdue University*
Don Bowen, *Stephen F. Austin State University*
Asa Bradley, *Spokane Falls Community College*
Elena Brewer, *SUNY at Buffalo*
Dieter Brill, *University of Maryland*
Hauke Busch, *Augusta State University*
Kapila Castoldi, *Oakland University*
Raymond Chastain, *Louisiana State University*
Michael Cherney, *Creighton University*
Lee Chow, *University of Central Florida*
Song Chung, *William Paterson University*
Alice Churukian, *Concordia College*
Christopher M. Coffin, *Oregon State University*
John S. Colton, *Brigham Young University*
Kristi Concannon, *Kings College*
Teman Cooke, *Georgia Perimeter College at Lawrenceville*
Daniel J. Costantino, *The Pennsylvania State University*
Jesse Cude, *Hartnell College*
Melissa H. Dancy, *University of North Carolina at Charlotte*
Loretta Dauwe, *University of Michigan—Flint*
Mark Davenport, *San Antonio College*
Chad Davies, *Gordon College*
Lawrence Day, *Utica College*
Carlos Delgado, *Community College of Southern Nevada*
David Donovan, *Northern Michigan University*
James Dove, *Metropolitan State University of Denver*
Archana Dubey, *University of Central Florida*
Andrew Duffy, *Boston University*
Taner Edis, *Truman State University*
Ralph Edwards, *Lurleen B. Wallace Community College*
Steve Ellis, *University of Kentucky*
Paula Engelhardt, *Tennessee Technical University*
Davene Eryes, *North Seattle Community College*
Gerard Fasel, *Pepperdine University*
Luciano Fleischfresser, *OSSM Autry Tech*
Cynthia Galovich, *University of Northern Colorado*
Bertram Gamory, *Monroe Community College*
Sambandamurthy Ganapathy, *SUNY at Buffalo*
Delena Gatch, *Georgia Southern University*
Richard Gelderman, *Western Kentucky University*
Martin Gelfand, *Colorado State University*
Terry Golding, *University of North Texas*
Robert Gramer, *Lake City Community College*
William Gregg, *Louisiana State University*
Paul Gresser, *University of Maryland*
Robert Hagood, *Washtenaw Community College*
Jason Harlow, *University of Toronto*
Heath Hatch, *University of Massachusetts*
Carl Hayn, *Santa Clara University*
James Heath, *Austin Community College*
Zvonko Hlousek, *California State University Long Beach*
Greg Hood, *Tidewater Community College*
Sebastian Hui, *Florence-Darlington Technical College*
Eric Hudson, *The Pennsylvania State University*
Joey Huston, *Michigan State University*
David Iadevaia, *Pima Community College—East Campus*
Fred Jarka, *Stark State College*
Ana Jofre, *University of North Carolina—Charlotte*
Daniel Jones, *Georgia Tech*
Erik Jensen, *Chemeketa Community College*
Todd Kalisik, *Northern Illinois University*
Ju H. Kim, *University of North Dakota*
Armen Kocharian, *California State University Northridge*
J. M. Kowalski, *University of North Texas*
Laird Kramer, *Florida International University*
Christopher Kulp, *Eastern Kentucky University*
Richard Kurtz, *Louisiana State University*
Kenneth Lande, *University of Pennsylvania*
Tiffany Landry, *Folsom Lake College*
Todd Leif, *Cloud County Community College*
John Levin, *University of Tennessee—Knoxville*
John Lindberg, *Seattle Pacific University*
Jeff Loats, *Metropolitan State University of Denver*
Rafael López-Mobilia, *The University of Texas at San Antonio*
Robert W. Lutz, *Drake University*
Lloyd Makorowitz, *SUNY Farmingdale*
Colleen Marlow, *Rhode Island College*
Eric Martell, *Millikin University*
Mark Masters, *Indiana University—Purdue*
John McClain, *Temple College*
Denise Meeks, *Pima Community College*

Henry Merrill, *Fox Valley Technical College*
Mike Meyer, *Michigan Technological University*
Karie Meyers, *Pima Community College*
Tobias Moleski, *Nashville State Tech*
April Moore, *North Harris College*
Gary Morris, *Rice University*
Krishna Mukherjee, *Slippery Rock University*
Charley Myles, *Texas Tech University*
Meredith Newby, *Clemson University*
David Nice, *Bryn Mawr*
Fred Olness, *Southern Methodist University*
Charles Oliver Overstreet, *San Antonio College*
Paige Ouzts, *Lander University*
Russell Palma, *Minnesota State University—Mankato*
Richard Panek, *Florida Gulf Coast University*
Joshua Phiri, *Florence-Darling Technical College*
Iulia Podariu, *University of Nebraska at Omaha*
David Potter, *Austin Community College*
Promod Pratap, *University of North Carolina—Greensboro*
Michael Pravica, *University of Nevada, Las Vegas*
Earl Prohofsky, *Purdue University*
Marilyn Rands, *Lawrence Technological University*
Andrew Rex, *University of Puget Sound*
Andrew Richter, *Valparaiso University*
William Robinson, *North Carolina State University*
Phyliss Salmons, *Embry—Riddle Aeronautical University*
Michael Schaab, *Maine Maritime Academy*
Bruce Schumm, *University of California, Santa Cruz*
Mizuho Schwalm, *University of Minnesota Crookston*
Cindy Schwarz, *Vassar College*
Natalia Semushkhina, *Shippensburg University*
Khazhgery (Jerry) Shakov, *Tulane University*
Kathy Shan, *University of Toledo*
Anwar Sheikh, *Colorado Mesa University*
Bart Sheinberg, *Houston Community College*
Marllin Simon, *Auburn University*
Kenneth Smith, *Pennsylvania State University*
Michael Smutko, *Northwestern University*
Jon Son, *Boston University*
Noel Stanton, *Kansas State University*
Donna Stokes, *University of Houston*
Chuck Stone, *North Carolina A&T*
Chun Fu Su, *Mississippi State University*
Jeffrey Sudol, *West Chester University*
Scott Thompson, *Georgia Gwinnett College*
William Tireman, *Northern Michigan University*
Negussie Tirfessa, *Manchester Community College*
Rajive Tiwari, *Belmont Abbey College*
Herman Trivilino, *College of the Mainland*
Dmitri Tsybychev, *Stony Brook University*
Douglas Tussey, *Pennsylvania State University*
Stephen Van Hook, *Pennsylvania State University*
Manuel Valera, *Slippery Rocky University*
Christos Valiotis, *Antelope Valley College*
James Vesenka, *University of New England*
Stamatis Vokos, *Seattle Pacific University*
James Wanliss, *Embry—Riddle Aeronautical University*
Henry Weigel, *Arapahoe Community College*
Luc T. Wille, *Florida Atlantic University*
Courtney Willis, *University of Northern Colorado*
Katherine Wu, *University of Tampa*
Ali Yazdi, *Jefferson State Community College*
David Young, *Louisiana State University*
Hsiao-Ling Zhou, *Georgia State University*
Todd Zimmerman, *University of Wisconsin—Stout*
Ulrich Zurcher, *Cleveland State University*

Student Advisory Board for the Third Edition

Nathalia Alzate, *Auburn University*
Hannah Chapman, *Colorado State University*
Rachel Eckert, *California Polytechnic State University—San Luis Obispo*
Emily Garban, *Colorado State University*
Tyrel Heckendorf, *Georgia State University*
Alex Keifer, *California Polytechnic State University—San Luis Obispo*
Isaac Moore, *The Pennsylvania State University*
John Peter Polites, *Georgia State University*
Blair Porterfield, *The Pennsylvania State University*
Rebecca Rogers, *Auburn University*
Rashawn D. Simmons, *Georgia State University*
Mary-Catherin Skoulos, *Stony Brook University*
Brittany Swiderski, *Stony Brook University*
Jenna Tustin, *Colorado State University*
Aaron Vermeersch, *Michigan State University*
Philip E. Weinberg, *Michigan State University*
Keith Wood, *Auburn University*
Timothy Yuan, *Stony Brook University*
Andrew Zilavy, *Colorado State University*

Preface to the Student

The most incomprehensible thing about the universe is that it is comprehensible.
—Albert Einstein

If you are taking a course for which this book is assigned, you probably aren't a physics major or an engineering major. It's likely that you aren't majoring in a physical science. So why are you taking physics?

It's almost certain that you are taking physics because you are majoring in a discipline that requires it. Someone, somewhere, has decided that it's important for you to take this course. And they are right. There is a lot you can learn from physics, even if you don't plan to be a physicist. We regularly hear from doctors, physical therapists, biologists and others that physics was one of the most interesting and valuable courses they took in college.

So, what can you expect to learn in this course? Let's start by talking about what physics is. Physics is a way of thinking about the physical aspects of nature. Physics is not about "facts." It's far more focused on discovering *relationships* between facts and the *patterns* that exist in nature than on learning facts for their own sake. Our emphasis will be on thinking and reasoning. We are going to look for patterns and relationships in nature, develop the logic that relates different ideas, and search for the reasons *why* things happen as they do.

The concepts and techniques you will learn will have a wide application. In this text we have a special emphasis on applying physics to understanding the living world. You'll use your understanding of charges and electric potential to analyze the electric signal produced when your heart beats. You'll learn how sharks can detect this signal to locate prey and, further, how and why this electric sensitivity seems to allow hammerhead sharks to detect magnetic fields, aiding navigation in the open ocean.

Like any subject, physics is best learned by doing. "Doing physics" in this course means solving problems, applying what you have learned to answer questions at the end of the chapter. When you are given a homework assignment, you may find yourself tempted to simply solve the problems by thumbing through the text looking for a formula that seems like it will work. This isn't how to do physics; if it was, whoever required you to take this course wouldn't bother. The folks who designed your major want you to learn to *reason,* not to "plug and chug." Whatever you end up studying or doing for a career, this ability will serve you well.

How do you learn to reason in this way? There's no single strategy for studying physics that will work for all students, but we can make some suggestions that will certainly help:

- **Read each chapter *before* it is discussed in class.** Class attendance is much more effective if you have prepared.
- **Participate actively in class.** Take notes, ask and answer questions, take part in discussion groups. There is ample scientific evidence that *active participation* is far more effective for learning science than is passive listening.
- **After class, go back for a careful rereading of the chapter.** In your second reading, pay close attention to the details and the worked examples. Look for the *logic* behind each example, not just at what formula is being used.
- **Apply what you have learned to the homework problems at the end of each chapter.** By following the techniques of the worked examples, applying the tactics and problem-solving strategies, you'll learn how to apply the knowledge you are gaining.
- **Form a study group with two or three classmates.** There's good evidence that students who study regularly with a group do better than the rugged individualists who try to go it alone.

And we have one final suggestion. As you read the book, take part in class, and work through problems, step back every now and then to appreciate the big picture. You are going to study topics that range from motions in the solar system to the electrical signals in the nervous system that let you tell your hand to turn the pages of this book. It's a remarkable breadth of topics and techniques that is based on a very compact set of organizing principles.

Now, let's get down to work.

Studying for and Taking the MCAT Exam

If you are taking the College Physics course, there's a good chance that you are majoring in the biological sciences. There's also a good chance that you are preparing for a career in the health professions, and so might well be required to take the Medical College Admission Test, the MCAT exam.

The *Chemical and Physical Foundations of Biological Systems* section of the MCAT assesses your understanding of the concepts of this course by testing your ability to apply these concepts to living systems. You will be expected to use what you've learned to analyze situations you've never seen before, making simplified but realistic models of the world. Your reasoning skills will be just as important as your understanding of the universal laws of physics.

Structure of the MCAT Exam

Most of the test consists of a series of passages of technical information followed by a series of questions based on each passage, much like the passage problems at the end of each chapter in this book. Some details:

- **The passages and the questions are *always* integrated.** Understanding the passage and answering the questions will require you to use knowledge from several different areas of physics.
- **Passages will generally be about topics for which you do not have detailed knowledge.** But, if you read carefully, you'll see that the treatment of the passage is based on information you should know well.
- **The test assumes a basic level of background knowledge.** You'll need to have facility with central themes and major concepts, but you won't need detailed knowledge of any particular topic. Such detailed information, if needed, will be provided in the passage.
- **You can't use calculators on the test, so any math that you do will be reasonably simple.** Quickly estimating an answer with ratio reasoning or a knowledge of the scale of physical quantities will be a useful skill.
- **The answers to the questions are all designed to be plausible.** You can't generally weed out the "bad" answers with a quick inspection.
- **The test is given online.** Practicing with MasteringPhysics will help you get used to this format.

Preparing for the Test

Because you have used this book as a tool for learning physics, you should use it as a tool for reviewing for the MCAT exam. Several of the key features of the book will be useful for this, including some that were explicitly designed with the MCAT exam in mind.

As you review the chapters:

- Start with the *Chapter Previews,* which provide a "big picture" overview of the content. What are the major themes of each chapter?
- Look for the *Synthesis* boxes that bring together key concepts and equations. These show connections and highlight differences that you should understand and be ready to apply.
- Go through each chapter and review the *Stop to Think* exercises. These are a good way to test your understanding of the key concepts and techniques.
- Each chapter closes with a passage problem that is designed to be "MCAT-exam-like." They'll give you good practice with the "read a passage, answer questions" structure of the MCAT exam.

The passage problems are a good tool, but the passages usually don't integrate topics that span several chapters—a key feature of the MCAT exam. For integrated passages and problems, turn to the *Part Summaries:*

- For each Part Summary, read the *One Step Beyond* passage and answer the associated questions.
- After this, read the passages and answer the questions that end each Part Summary section. These passages and associated problems are—by design—very similar to the passages and questions you'll see on the actual MCAT exam.

Taking the Test: Reading the Passage

As you read each passage, you'll need to interpret the information presented and connect it with concepts you are familiar with, translating it into a form that makes sense based on your background.

The next page shows a passage that was written to very closely match the style and substance of an actual MCAT passage. Blue annotations highlight connections you should make as you read. The passage describes a situation (the mechanics and energetics of sled dogs) that you probably haven't seen before. But the basic physics (friction, energy conversion) are principles that you are familiar with, principles that you have seen applied to related situations. When you read the passage, think about the underlying physics concepts and how they apply to this case.

Translating the Passage
As you read the passage, do some translation. Connect the scenario to examples you've seen before, translate given information into forms you are familiar with, think about the basic physical principles that apply.

Passage X

For travel over snow, a sled with runners that slide on snow is the best way to get around. Snow is slippery, but there is still friction between runners and the ground; the forward force required to pull a sled at a constant speed might be 1/6 of the sled's weight.

As you read this part of the passage, think about the forces involved: For a sled moving at a constant speed, there is no net force. The downward weight force is equal to the upward normal force; the forward pulling force must be equal to the friction force, which is acting opposite the sled's motion. There are many problems like this in Chapter 5.

Part of translating is converting given information into a more usual or more useful form. This is really a statement about the coefficient of kinetic friction.

The pulling force might well come from a dog. In a typical sled, the rope that the dog uses to pull attaches at a slight angle, as in Figure 1. The pulling force is the horizontal component of the tension in the rope.

The force applied to the sled is the tension force in the rope, which is shown at an angle. The horizontal component is the pulling force; you're told this. There is a vertical component of the force as well.

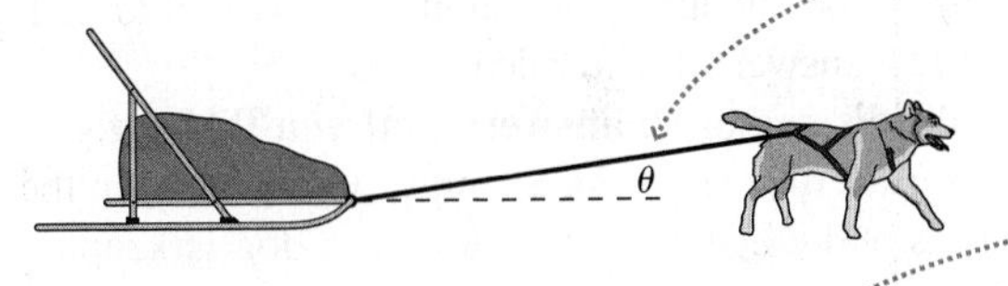

Figure 1

In the data given here, and the description given above, the sled moves at a constant speed—there is no mention of acceleration anywhere in this passage. In such cases, the net force is zero and the kinetic energy of the sled isn't changing.

Sled dogs have great aerobic capacity; a 40 kg dog can provide output power to pull with a 60 N force at 2.2 m/s for hours. The output power is related to force and velocity by $P = F \cdot v$, so they can pull lighter loads at higher speeds.

Notice that the key equation relating power, force and velocity is given to you. That's to be expected. Any specific information, including equations, constants and other such details, will generally be given in the passage. The MCAT is a test of reasoning, not recall.

Doing 100 J of work means that a dog must expend 400 J of metabolic energy. The difference must be exhausted as heat; given the excellent insulation provided by a dog's fur, this is mostly via evaporation as it pants. At a typical body temperature, the evaporation of 1.0 l of water carries away 240,000 J, so this is an effective means of cooling.

The concepts of metabolic energy and energy output are treated in Chapter 11. The details here match those in the chapter (as they should!); this corresponds to an efficiency of 25%. 400 J of energy is used by the body; 25% of this, 100 J, is the energy output. This means that 300 J is exhausted as heat.

Chapter 12 discusses means of heat transfer: conduction, convection, radiation, evaporation. This paragraph gives biological details about dogs that you can interpret as follows: A dog's fur limits transfer by conduction, convection and radiation; evaporation of water by a panting dog must take up the slack.

The specific data for energy required to evaporate water is given. If you need such information to answer questions, it will almost certainly be provided. As we noted above, this is a test of reasoning, not recall.

FIGURE MCAT-EXAM.1 Interpreting a passage.

Taking the Test: Answering the Questions

The passages on the MCAT exam seem complicated at first, but, as we've seen, they are about basic concepts and central themes that you know well. The same is true of the questions; they aren't as difficult as they may seem at first. As with the passage, you should start by translating the questions, identifying the physical concepts that apply in each case. You then proceed by reasoning, determining the solution to the question, using your understanding of these basic concepts. The practical suggestions below are followed by a detailed overview of the solutions to the questions based on the passage on the previous page.

You Can Answer the Questions in Any Order

The questions test a range of skills and have a range of difficulties. Many questions will involve simple reading comprehension; these are usually quite straightforward. Some require sophisticated reasoning and (slightly) complex mathematical manipulations. Start with the easy ones, ones that you can quickly solve. Save the more complex ones for later, and skip them if time is short.

Take Steps to Simplify or Eliminate Calculations

You won't be allowed to use a calculator on the exam, so any math that you do will be reasonably straightforward. To rapidly converge on a correct answer choice, there are some important "shortcuts" that you can take.

- **Use ratio reasoning.** What's the relationship between the variables involved in a question? You can use this to deduce the answer with only a very simple calculation, as we've seen many times in the book. For instance, suppose you are asked the following question:

  A model rocket is powered by chemical fuel. A student launches a rocket with a small engine containing 1.0 g of combustible fuel. The rocket reaches a speed of 10 m/s. The student then launches the rocket again, using an engine with 4.0 g of fuel. If all other parameters of the launch are kept the same, what final speed would you expect for this second trial?

  This is an energy conversion problem: Chemical energy of the fuel is converted to kinetic energy of the rocket. Kinetic energy is related to the speed by $K = \frac{1}{2}mv^2$. The chemical energy—and thus the kinetic energy—in the second trial is increased by a factor of 4. Since $K \sim v^2$, the speed must increase by a factor of 2, to 20 m/s.
- **Simplify calculations by liberally rounding numbers.** You can round off numbers to make calculations more straightforward. Your final result will probably be close enough to choose the correct answer from the list given. For instance, suppose you are asked the following question:

  A ball moving at 2.0 m/s rolls off edge of table that's 1.2 m high. How far from the edge of the table does the ball land?

  A. 2 m B. 1.5 m C. 1 m D. 0.5 m

  We know that the vertical motion of the ball is free fall; so the vertical distance fallen by the ball in a time Δt is $\Delta y = -\frac{1}{2}gt^2$. The time to fall 1.2 m is $\Delta t = \sqrt{2(1.2\text{ m})/g}$. Rather than complete this calculation, we estimate the results as follows: $\Delta t = \sqrt{2.4/9.8} \approx \sqrt{1/4} = 1/2 = 0.5\text{ s}$ During this free fall time, the horizontal motion is constant at 2.0 m/s, so we expect the ball to land about 1 m away. Our quick calculation shows us that the correct answer is choice C—no other answer is close.
- **For calculations using values in scientific notation, compute either the first digits or the exponents, not both.** In some cases, a quick calculation can tell you the correct leading digit, and that's all you need to figure out the correct answer. In other cases, you'll find possible answers with the same leading digit but very different exponents or decimal places. In this case, all you need is a simple order of magnitude estimate to decide on the right result.
- **Where possible, use your knowledge of the expected scale of physical quantities to quickly determine the correct answer.** For instance, suppose a question asks you to find the photon energy for green light of wavelength 550 nm. Visible light has photon energies of about 2 eV, or about 3×10^{-19} J, and that might be enough information to allow you to pick out the correct answer with no calculation.
- **Beware of "distractors", answers that you'll get if you make common mistakes.** For example, Question 4 on the next page is about energy conversion. The dog is keeping the sled in motion, so it's common for students to say that the dog is converting chemical energy in its body into kinetic energy. However, the kinetic energy isn't changing. The two answer choices that involve kinetic energy are common, but incorrect, choices. Be aware that the questions are constructed to bring out such misconceptions and that these tempting, but wrong, answer choices will be provided.

One Final Tip: Look at the Big Picture

The MCAT exam tests your ability to look at a technical passage about which you have some background knowledge and quickly get a sense of what it is saying, enough to answer questions about it. Keep this big picture in mind:

- **Don't get bogged down in technical details of the particular situation.** Focus on the basic physics.
- **Don't spend too much time on any one question.** If one question is taking too much time, make an educated guess and move on.
- **Don't get confused by details of notation or terminology.** For instance, different people use different symbols for physical variables; in this text we use the symbol K for kinetic energy; others use E_K.

Finally, don't forget the most important aspect of success on the MCAT exam: The best way to prepare for this or any test is simply to understand the subject. As you prepare for the test, focus your energy on reviewing and refining your knowledge of central topics and techniques, and practice applying your knowledge by solving problems like you'll see on the actual MCAT.

Translating

Look at the questions and think about the physics principles that apply, how they connect to concepts you know and understand.

Tips

- Numerical choices are presented in order; that's the usual practice on the test. Estimate the size of the answer, and think about where it falls.
- For questions with sentences as choices, decide on the solution before you look at the choices; this will save time reading.

Reasoning

Think about the question and the range of possible answers, and converge to a solution with as few steps as possible—time is limited!

This is a question about the size of the friction force. You are told that it takes a force that's about 1/6 of the sled's weight to pull it forward on snow. You can estimate the friction coefficient from this information.

1. What is the approximate coefficient of kinetic friction for a sled on snow?

A. 0.35
B. 0.25
C. 0.15
D. 0.05

For an object on level ground, the normal force equals the weight force. If the sled is moving at a constant speed, the pulling force equals the friction force. This implies that $\mu = f_k/n = f_{pull}/w = 1/6$. Two of the answer choices convert easily to fractions: 0.25 = 1/4; 0.05 = 1/20. 1/6 is between these, so C must be our choice. (Indeed, 1/6 = 0.167, so 0.15 is pretty close.)

If the speed is constant, there is no net force. We are told that the pulling force is the horizontal component of the tension force, not the tension force itself. Because there is no net force, this horizontal component is equal to the friction force, which is directed backward. So this is really a question about the friction force.

2. If a rope pulls at an angle, as in Figure 1, how will this affect the pulling force necessary to keep the sled moving at a constant speed?

A. This will reduce the pulling force.
B. This will not change the pulling force.
C. This will increase the pulling force.
D. It will increase or decrease the pulling force, depending on angle.

A vertical component of the tension force will reduce the normal force, reducing the friction force—and thus the pulling force.

We assume that the output power is the same for the two cases—this is implied in the passage.

3. A dog pulls a 40 kg sled at a maximum speed of 2 m/s. What is the maximum speed for an 80 kg sled?

A. 2 m/s
B. 1.5 m/s
C. 1.0 m/s
D. 0.5 m/s

Doubling the weight doubles the normal force, which doubles the friction force. This will double the necessary pulling force as well. Given the expression for power given in the passage, this means the maximum speed will be halved.

This is a question about energy transformation. For such questions, think about changes. What forms of energy are *changing*? We know that thermal energy is part of the picture because some of the chemical energy is converted to thermal energy in the dog's body.

4. As a dog pulls a sled at constant speed, chemical energy in the dog's body is converted to

A. kinetic energy
B. thermal energy
C. kinetic energy and thermal energy
D. kinetic energy and potential energy

Choice B is correct, but A and C are clever distractors. It's tempting to choose an answer that includes kinetic energy. The sled is in motion, after all! But don't be swayed. The kinetic energy isn't changing, and friction to the sled converts any energy the dog supplies into thermal energy.

Increasing speed increases power, as the passage told us. But the energy to pull the sled is not the *power*, it's the *work*, and we know that the work is $W = F\Delta x$. This is a question about work and energy, not about power.

5. A dog pulls a sled for a distance of 1.0 km at a speed of 1 m/s, requiring an energy output of 60,000 J. If the dog pulls the sled at 2 m/s, the necessary energy is

A. 240,000 J
B. 120,000 J
C. 60,000 J
D. 30,000 J

Doubling the speed doubles the power, but it doesn't change the force; that's fixed by friction. The distance is the same as well, and so is the work done, the energy required. Since the speed doubles, it's tempting to think the energy doubles, though. This "obvious" but incorrect solution is one of the choices—expect such situations on the actual MCAT.

The passage tells us that the dog uses 400 J of metabolic energy to do 100 J of work. 300 J, or 75%, must be exhausted to the environment. We can assume the same efficiency here.

6. A dog uses 100,000 J of metabolic energy pulling a sled. How much energy must the dog exhaust by panting?

A. 100,000 J
B. 75,000 J
C. 50,000 J
D. 25,000 J

If 75% of the energy must be exhausted to the environment, that's 75,000 J.

FIGURE MCAT-EXAM.2 Answering the questions for the passage of Figure MCAT-EXAM.1.

Real-World Applications

Applications of biological or medical interest are marked BIO in the list below, including MCAT-style Passage Problems. Other end-of-chapter problems of biological or medical interest are marked BIO in the chapter.

Part II Summary

Chapter 12

Chapter 13

Part III Summary

Chapter 14

Chapter 15

Chapter 16

Part IV Summary

Chapter 17

Chapter 18

Chapter 19

Part V Summary

Chapter 20

Detailed Contents

PART VII Modern Physics

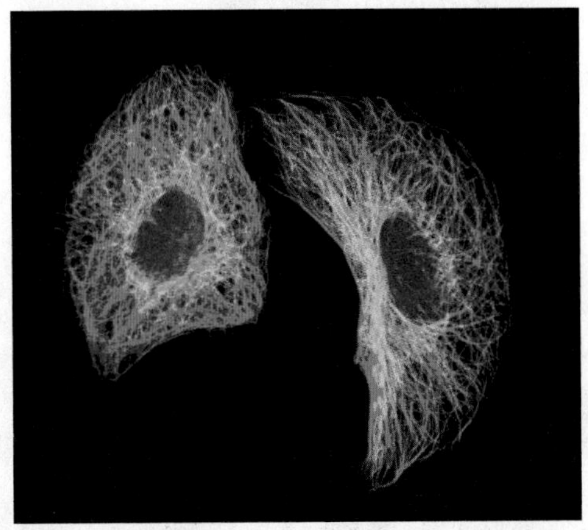

PART V

Optics

These images of animal eyes and light-sensing organs reveal the wide range of structures that can produce a visual sense. There is a certain similarity among eyes as well; did you spot the false eyes, patterns that are designed to mimic the appearance of an eye?

Light Is a Wave

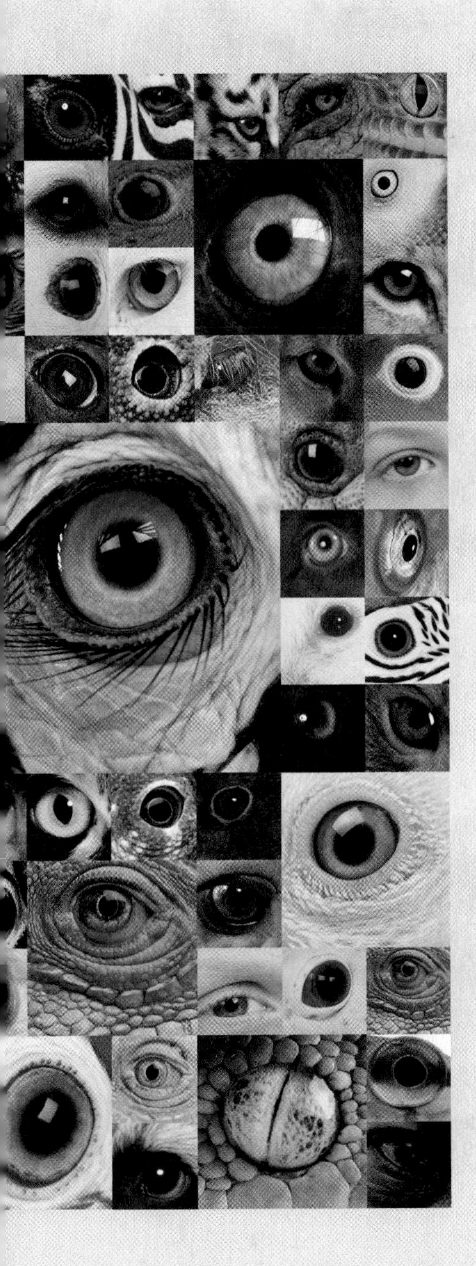

Isaac Newton is best known for his studies of mechanics and the three laws that bear his name, but he also did important early work on optics. He was the first person to carefully study how a prism breaks white light into colors. Newton was a strong proponent of the "corpuscle" theory of light, arguing that light consists of a stream of particles.

In fact, Newton wasn't quite right. As you will see, the beautiful colors of a peacock's feathers and the shimmery rainbow of a soap bubble both depend on the fact that light is a wave, not a particle. In particular, light is an electromagnetic wave, although these chapters depend on nothing more than the "waviness" of light waves for your understanding. The wave theory we developed in Part IV will be put to good use in Part V as we begin our investigation of light and optics with an analysis of the *wave model* of light.

The Ray Model

Yet Newton was correct in his observation that light seems to travel in straight lines, something we wouldn't expect a wave to do. Consequently, our investigations of how light works will be greatly aided by another model of light, the *ray model,* in which light travels in straight lines, bounces from mirrors, and is bent by lenses.

The ray model will be an excellent tool for analyzing many of the practical applications of optics. When you look in a mirror, you see an *image* of yourself that appears to be behind the mirror. We will use the ray model of light to determine just how it is that mirrors and lenses form images. At the same time, we will need to reconcile the wave and ray models, learning how they are related to each other and when it is appropriate to use each.

Working with Light

The nature of light is quite subtle and elusive. In Parts VI and VII, we will turn to the question of just what light is. As we will see, light has both wave-like *and* particle-like aspects. For now, however, we will set this question aside and work with the wave and ray models to develop a practical understanding of light. This will lead us, in Chapter 19, to an analysis of some common optical instruments. We will explore how a camera captures images and how telescopes and microscopes work.

Ultimately, the fact that you are reading this textbook is due to the optics of the first optical instrument you ever used, your eye! We will investigate the optics of the eye, learn how the cornea and lens bend light to create an image on your retina, and see how glasses or contact lenses can be used to correct the image should it be out of focus.

17 Wave Optics

The vivid colors of this hummingbird's feathers have a sheen unlike that of ordinary pigments, and they change strikingly depending on the angle at which they're viewed. How does light interact with the feathers to produce this bright display?

LOOKING AHEAD ▸▸

Goal: To understand and apply the wave model of light.

The Wave Model of Light

The varying colors reflected from this DVD can be understood using the **wave model** of light, which is the focus of this chapter.

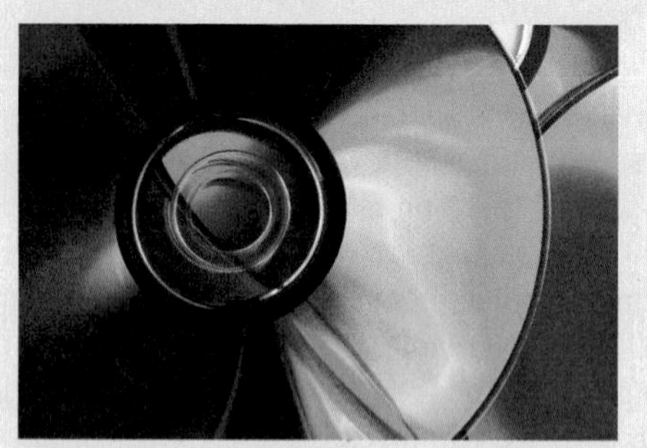

You'll learn that the wave model applies when light passes through small apertures or when waves from several small sources combine.

Interference

This beetle's colorful look is the result of **interference** between light waves reflecting off microscopic layers of its shell.

You'll learn how light waves from two or more sources can interfere constructively or destructively.

Diffraction

The light waves passing this sewing needle are slightly bent, or *diffracted*, causing the bands seen in the image.

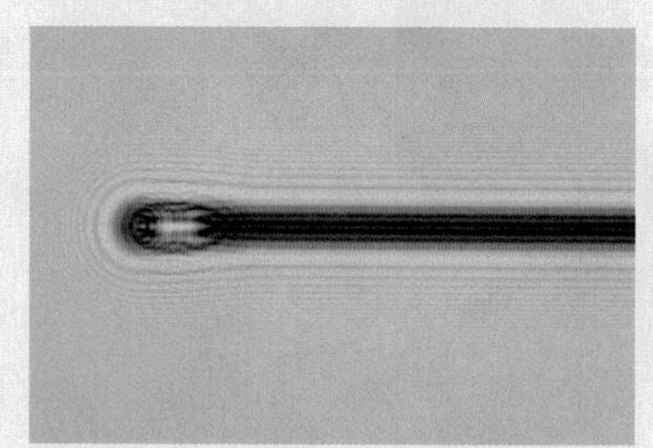

You'll learn how **diffraction** occurs for light passing through a narrow slit or a small, circular aperture.

LOOKING BACK ◂◂

Interference of Two Waves

In Section 16.6, you studied the interference of sound waves. We'll use these same ideas when we study the interference of light waves in this chapter.

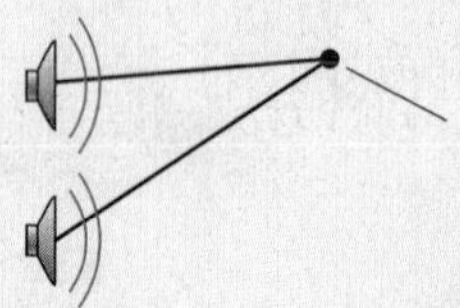

You learned that waves interfere constructively at a point if their path-length difference is an integer number of wavelengths.

STOP TO THINK

Sound waves spread out from two speakers; the circles represent crests of the spreading waves. The interference is

A. Constructive at both points 1 and 2.
B. Destructive at both points 1 and 2.
C. Constructive at 1, destructive at 2.
D. Destructive at 1, constructive at 2.

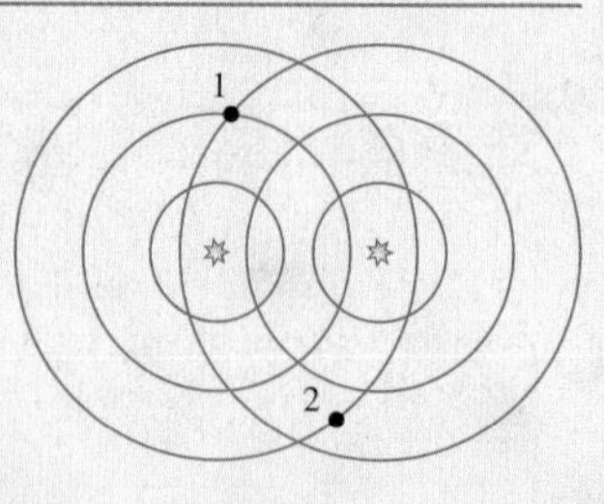

17.1 What Is Light?

The nature of light is elusive. Under some circumstances, light acts like particles traveling in straight lines. But change the circumstances, and light shows the same kinds of wave-like behavior as sound waves or water waves. Change the circumstances yet again, and light exhibits behavior that is neither wave-like nor particle-like but has characteristics of both.

Rather than an all-encompassing "theory of light," it will be better to develop three **models of light.** Each model successfully explains the behavior of light within a certain domain—that is, within a certain range of physical situations. We'll begin with a brief summary of all three models, so that you will have a road map of where we're headed. Each of these models will be developed in the coming chapters.

Three models of light

The Wave Model

The wave model of light is the most widely applicable model, responsible for the widely known "fact" that light is a wave. It is certainly true that, under many circumstances, light exhibits the same behavior as sound or water waves. Lasers and electro-optical devices, critical technologies of the 21st century, are best understood in terms of the wave model of light. Some aspects of the wave model of light were introduced in Chapters 15 and 16, and the wave model is the primary focus of this chapter. The study of light as a wave is called **wave optics.**

The Ray Model

An equally well-known "fact" is that light travels in a straight line. These straight-line paths are called *light rays.* The properties of prisms, mirrors, lenses, and optical instruments such as telescopes and microscopes are best understood in terms of light rays. Unfortunately, it's difficult to reconcile the statement "light travels in a straight line" with the statement "light is a wave." For the most part, waves and rays are mutually exclusive models of light. An important task will be to learn when each model is appropriate. The ray model of light, the basis of **ray optics,** is the subject of the next chapter.

The Photon Model

Modern technology is increasingly reliant on quantum physics. In the quantum world, light consists of *photons* that have both wave-like and particle-like properties. Photons are the *quanta* of light. Much of the quantum theory of light is beyond the scope of this textbook, but we will take a peek at the important ideas in Chapters 25 and 28 of this text.

The Propagation of Light Waves

FIGURE 17.1 shows an everyday observation about light. The sunlight passing through an opening makes a sharp-edged shadow as it falls upon the floor. This behavior is exactly what you would expect if light consisted of rays that travel in straight paths.

FIGURE 17.1 Light passing through an opening makes a sharp-edged shadow.

Is this behavior, where light seems to travel in straight lines, consistent with the motion of a wave? Light waves oscillate so rapidly—at some 10^{14} Hz—that we cannot hope to directly observe the crests and troughs of its electric or magnetic fields. Instead, let's use ordinary water waves to illustrate some basic properties of wave motion common to all waves, including light. **FIGURE 17.2** shows a water wave, as seen looking straight down on the water surface. The wave enters from the top of the picture, then passes though a window-like opening in a barrier. After passing through the opening, the wave *spreads out* to fill the space behind the opening. This spreading of a wave is the phenomenon called **diffraction.** Diffraction is a sure sign that whatever is passing through the opening is a wave.

FIGURE 17.2 A water wave passing through a narrow opening in a barrier.

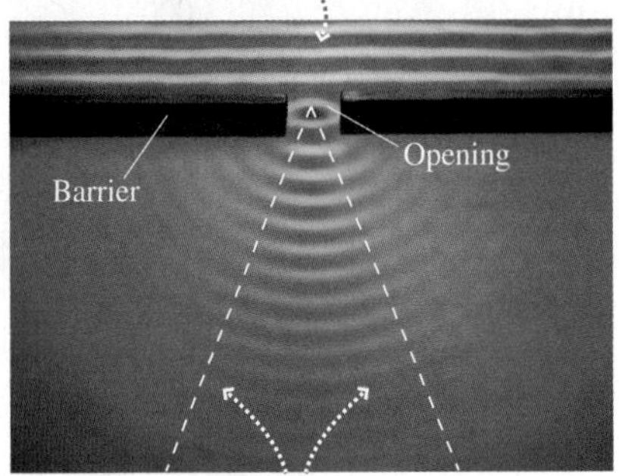

The straight-line travel of light appears to be incompatible with this spreading of a wave. Notice, however, that the width of the opening in Figure 17.2 is only slightly larger than the wavelength of the water wave. As **FIGURE 17.3** on the next page shows, something quite different occurs when we make the opening much wider. Rather than spreading out, now the wave continues to move straight forward, with a well-defined boundary between where the wave is moving and its "shadow," where there is no wave. This is just the behavior observed for light in Figure 17.1.

Whether a wave spreads out or travels straight ahead, with sharp shadows on both sides, evidently depends on the size of the objects with which the wave interacts. The spreading of diffraction becomes noticeable only when an opening or object is "narrow," comparable in size to the wavelength of the wave. Thus we would expect a light wave to spread out in this way only when it passes objects that are comparable in size to the wavelength of light.

FIGURE 17.3 A water wave passing through an opening that's many wavelengths wide.

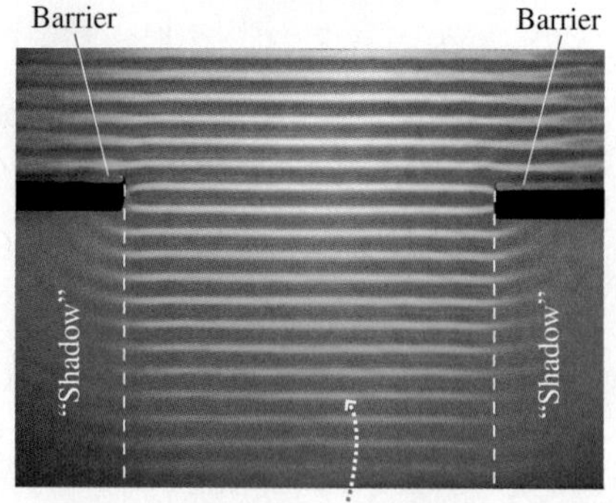

But the wavelength of light is *extremely* short. Later in this chapter we'll see how its wavelength can be measured, and we'll find that a typical wavelength of light is only about 0.5 μm. When light waves interact with everyday-sized objects, such as the opening in Figure 17.1, the situation is like that of the water wave in Figure 17.3. The wave travels straight ahead, and we'll be able to use the ray model of light. Only when the size of an object or an opening approaches the wavelength of light does the wave-like spreading become important.

Light Is an Electromagnetic Wave

If light is a wave, what is it that is waving? As we briefly noted in Chapter 15, light consists of very rapidly oscillating electric and magnetic fields: It is an *electromagnetic wave*. We will examine the nature of electromagnetic waves in more detail in Part VI after we introduce the ideas of electric and magnetic fields. For now we can say that light waves are a "self-sustaining oscillation of the electromagnetic field." Being self-sustaining means that electromagnetic waves require *no material medium* in order to travel; hence electromagnetic waves are not mechanical waves. Fortunately, we can learn about the wave properties of light without having to understand electromagnetic fields. In fact, the discovery that light propagates as a wave was made 60 years before it was realized that light is an electromagnetic wave. We, too, will be able to learn much about the wave nature of light without having to know just what it is that is waving.

Recall that all electromagnetic waves, including light waves, travel in a vacuum at the same speed, called the *speed of light*. Its value is

$$v_{\text{light}} = c = 3.00 \times 10^8 \text{ m/s}$$

where the symbol c is used to designate the speed of light.

Recall also that the wavelengths of light are extremely small, ranging from about 400 nm for violet light to 700 nm for red light. Electromagnetic waves with wavelengths outside this range are not visible to the human eye. A prism is able to spread the different wavelengths apart, from which we learn that "white light" is all the colors, or wavelengths, combined. The spread of colors seen with a prism, or seen in a rainbow, is called the *visible spectrum*.

If the wavelengths of light are incredibly small, the oscillation frequencies are unbelievably high. The frequency for a 600 nm wavelength of light is

$$f = \frac{v}{\lambda} = \frac{3.00 \times 10^8 \text{ m/s}}{600 \times 10^{-9} \text{ m}} = 5.00 \times 10^{14} \text{ Hz}$$

The frequencies of light waves are roughly a factor of a trillion (10^{12}) higher than sound frequencies.

The Index of Refraction

TABLE 17.1 Typical indices of refraction

Material	Index of refraction
Vacuum	1 exactly
Air	1.0003
Water	1.33
Glass	1.50
Diamond	2.42

Light waves travel with speed c in a vacuum, but they slow down as they pass through transparent materials such as water or glass or even, to a very slight extent, air. The slowdown is a consequence of interactions between the electromagnetic field of the wave and the electrons in the material. The speed of light in a material is characterized by the material's **index of refraction** n, defined as

$$n = \frac{\text{speed of light in a vacuum}}{\text{speed of light in the material}} = \frac{c}{v} \tag{17.1}$$

where v is the speed of light in the material. The index of refraction of a material is always greater than 1 because v is always less than c. A vacuum has $n = 1$ exactly. Table 17.1 lists the indices of refraction for several materials. Liquids and solids have higher indices of refraction than gases, simply because they have a much higher density of atoms for the light to interact with.

NOTE ▸ An accurate value for the index of refraction of air is relevant only in very precise measurements. We will assume $n_{\text{air}} = 1.00$ in this text. ◂

If the speed of a light wave changes as it enters into a transparent material, such as glass, what happens to the light's frequency and wavelength? Because $v = \lambda f$, either λ or f or both have to change when v changes.

As an analogy, think of a sound wave in the air as it impinges on the surface of a pool of water. As the air oscillates back and forth, it periodically pushes on the surface of the water. These pushes generate the compressions of the sound wave that continues on into the water. Because each push of the air causes one compression of the water, the frequency of the sound wave in the water must be *exactly the same* as the frequency of the sound wave in the air. In other words, **the frequency of a wave does not change as the wave moves from one medium to another.**

The same is true for electromagnetic waves, although the pushes are a bit more complex as the electric and magnetic fields of the wave interact with the atoms at the surface of the material. Nonetheless, the frequency does not change as the wave moves from one material to another.

FIGURE 17.4 shows a light wave passing through a transparent material with index of refraction n. As the wave travels through a vacuum it has wavelength λ_{vac} and frequency f_{vac} such that $\lambda_{vac} f_{vac} = c$. In the material, $\lambda_{mat} f_{mat} = v = c/n$. The frequency does not change as the wave enters ($f_{mat} = f_{vac}$), so the wavelength must change. The wavelength in the material is

$$\lambda_{mat} = \frac{v}{f_{mat}} = \frac{c}{n f_{mat}} = \frac{c}{n f_{vac}} = \frac{\lambda_{vac}}{n} \tag{17.2}$$

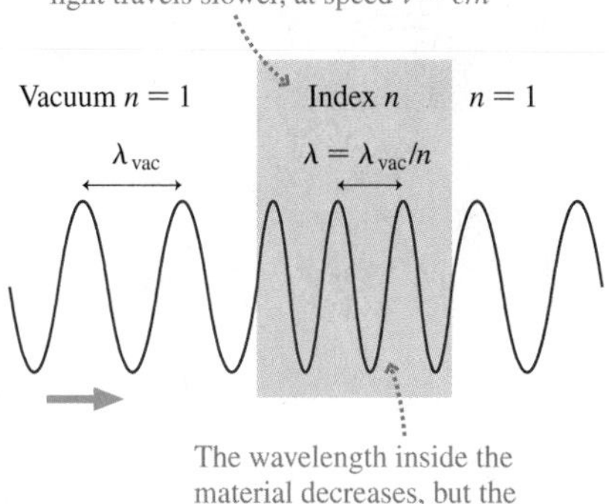

FIGURE 17.4 Light passing through a transparent material with index of refraction n.

The wavelength in the transparent material is shorter than the wavelength in a vacuum. This makes sense. Suppose a marching band is marching at one step per second at a speed of 1 m/s. Suddenly they slow their speed to $\frac{1}{2}$ m/s but maintain their march at one step per second. The only way to go slower while marching at the same pace is to take *smaller steps*. When a light wave enters a material, the only way it can go slower while oscillating at the same frequency is to have a *shorter wavelength*.

EXAMPLE 17.1 Analyzing light traveling through glass

Orange light with a wavelength of 600 nm is incident on a 1.00-mm-thick glass microscope slide.

a. What is the light speed in the glass?
b. How many wavelengths of the light are inside the slide?

SOLVE

a. From Table 17.1 we see that the index of refraction of glass is $n_{glass} = 1.50$. Thus the speed of light in glass is

$$v_{glass} = \frac{c}{n_{glass}} = \frac{3.00 \times 10^8 \text{ m/s}}{1.50} = 2.00 \times 10^8 \text{ m/s}$$

b. Because $n_{air} = 1.00$, the wavelength of the light is the same in air and vacuum: $\lambda_{vac} = \lambda_{air} = 600$ nm. Thus the wavelength inside the glass is

$$\lambda_{glass} = \frac{\lambda_{vac}}{n_{glass}} = \frac{600 \text{ nm}}{1.50} = 400 \text{ nm} = 4.00 \times 10^{-7} \text{ m}$$

N wavelengths span a distance $d = N\lambda$, so the number of wavelengths in $d = 1.00$ mm is

$$N = \frac{d}{\lambda} = \frac{1.00 \times 10^{-3} \text{ m}}{4.00 \times 10^{-7} \text{ m}} = 2500$$

ASSESS The fact that 2500 wavelengths fit within 1 mm shows how small the wavelengths of light are.

STOP TO THINK 17.1 A light wave travels through three transparent materials of equal thickness. Rank in order, from the highest to lowest, the indices of refraction n_1, n_2, and n_3.

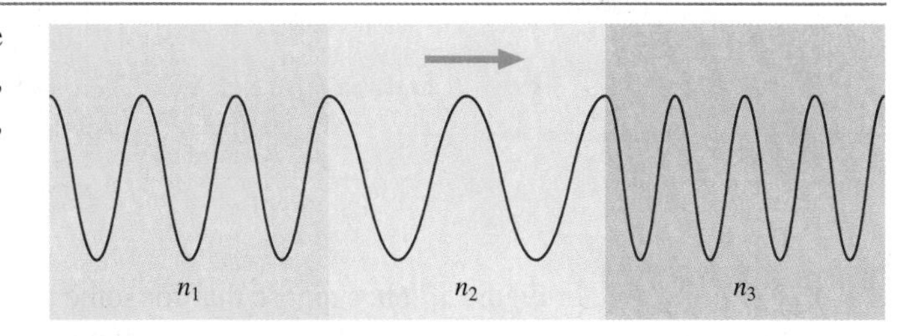

17.2 The Interference of Light

In ◀◀ SECTION 16.6 you learned that waves of equal wavelength emitted from two sources—such as the sound waves emitted from two loudspeakers—can overlap and *interfere*, leading to a large amplitude where the waves interfere constructively and a small or even zero amplitude where they interfere destructively. Interference is inherently a wave phenomenon, so if light acts like a wave we should be able to observe the interference of light waves.

To do so we need two sources of light waves with exactly the same wavelength. This is difficult to arrange with conventional light sources. Instead, consider the situation shown in FIGURE 17.5. Laser light passes through a narrow opening—a *slit*—that is only 0.1 mm wide, about twice the width of a human hair. This situation is similar to Figure 17.2, where water waves spread out after passing through a narrow opening. When light waves pass through a narrow slit, they too spread out behind the slit, just as the water wave did behind the opening in the barrier. The light is exhibiting *diffraction*, the sure sign of waviness. We will look at diffraction in more detail later in the chapter. For now, though, we see that a sufficiently narrow slit does act as a source of light waves that spread out behind it.

FIGURE 17.5 Light, just like a water wave, spreads out behind a slit in a screen if the slit is sufficiently small.

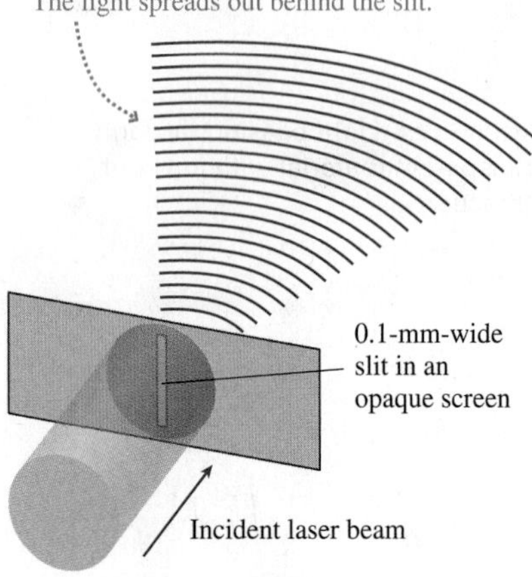

Young's Double-Slit Experiment

In order to observe interference, we need *two* light sources whose waves can overlap and interfere. FIGURE 17.6a shows an experiment in which a laser beam is aimed at an opaque screen containing two long, narrow slits that are very close together. This pair of slits is called a **double slit,** and in a typical experiment they are ≈0.1 mm wide and spaced ≈0.5 mm apart. We will assume that the laser beam illuminates both slits equally, and any light passing through the slits impinges on a viewing screen. Such a double-slit experiment was first performed by Thomas Young in 1801, using sunlight instead of a laser beam. It provided the first definitive evidence that light is a wave.

What should we expect to see on the viewing screen? FIGURE 17.6b is a view from above the experiment, looking down on the top ends of the slits and the top edge of the viewing screen. Because the slits are very narrow, **light spreads out behind each slit** as it did in Figure 17.5, and these two spreading waves overlap in the region between the slits and the screen.

FIGURE 17.6 A double-slit interference experiment.

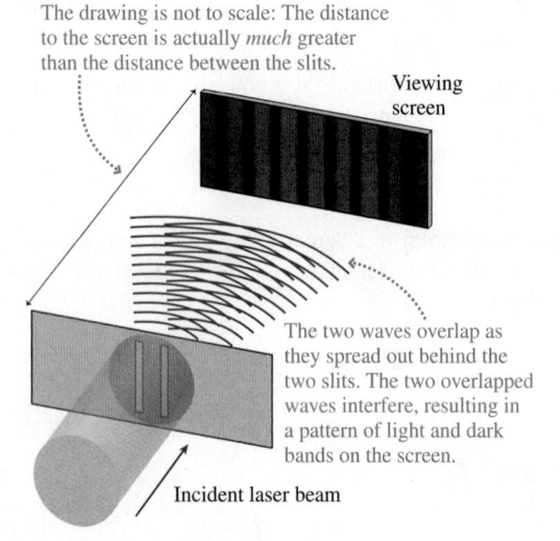

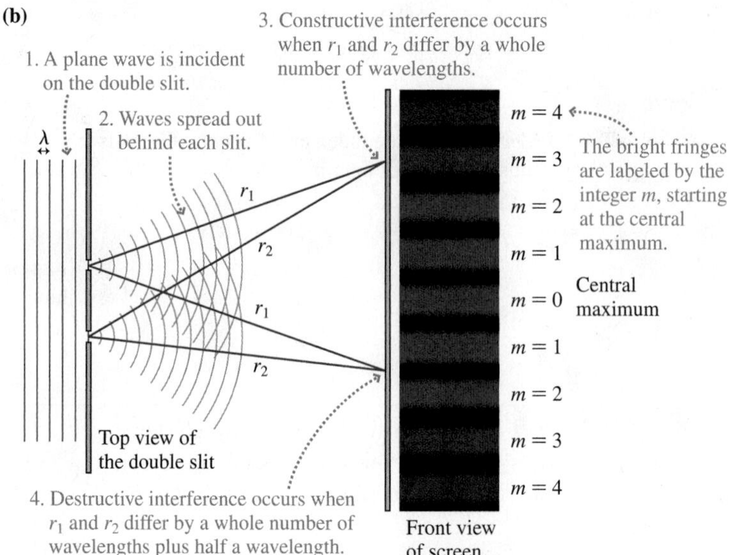

STOP TO THINK 17.2 In Figure 17.6b, suppose that for some point P on the screen $r_1 = 5{,}002{,}248.5\lambda$ and $r_2 = 5{,}002{,}251.5\lambda$, where λ is the wavelength of the light. The interference at point P is

A. Constructive. B. Destructive. C. Something in between.

In Chapter 16 we found that *constructive* interference occurs at a given point when the distances r_1 and r_2 from speakers 1 and 2 to that position differ by a *whole number* of wavelengths. For light waves, we thus expect constructive interference to occur, and the intensity of light on the screen to be high, when the distances r_1 and r_2 from the two slits to a point on the screen differ by a whole number of wavelengths. And, just as we found for sound waves, destructive interference for light waves occurs at positions on the screen for which r_1 and r_2 differ by a whole number of wavelengths plus half a wavelength. At these positions the screen will be dark.

The photograph in Figure 17.6b shows how the screen looks. As we move along the screen, the difference $\Delta r = r_2 - r_1$ alternates between being a whole number of wavelengths and a whole number of wavelengths plus half a wavelength, leading to a series of alternating bright and dark bands of light called **interference fringes.** The bright fringes are numbered by an integer $m = 1, 2, 3, \ldots$, going outward from the center. The brightest fringe, at the midpoint of the viewing screen, has $m = 0$ and is called the **central maximum.**

Analyzing Double-Slit Interference

Figure 17.6b showed qualitatively that interference is produced behind a double slit by the overlap of the light waves spreading out behind each opening. Now let's analyze the experiment more carefully. **FIGURE 17.7a** shows the geometry of a double-slit experiment in which the spacing between the two slits is d and the distance to the viewing screen is L. **We will assume that L is *very* much larger than d.**

Our goal is to determine if the interference at a particular point on the screen is constructive, destructive, or something in between. As we've just noted, constructive interference between two waves from identical sources occurs at points for which the path-length difference $\Delta r = r_2 - r_1$ is an integer number of wavelengths, which we can write as

$$\Delta r = m\lambda \qquad m = 0, 1, 2, 3, \ldots \tag{17.3}$$

Thus the interference at a particular point is constructive, producing a bright fringe, if $\Delta r = m\lambda$ at that point.

We need to find the specific positions on the screen where $\Delta r = m\lambda$. Point P in Figure 17.7a is a distance r_1 from one slit and r_2 from the other. We can specify point P either by its distance y from the center of the viewing screen or by the angle θ shown in Figure 17.7a; angle θ and distance y are related by

$$y = L \tan \theta \tag{17.4}$$

Because the screen is very far away compared to the spacing between the slits, the two paths to point P are virtually parallel, and thus the small triangle that is shaded green in the enlargement of **FIGURE 17.7b** is a right triangle whose angle is also θ. The path-length difference between the two waves is the short side of this triangle, so

$$\Delta r = d \sin \theta \tag{17.5}$$

Bright fringes due to constructive interference then occur at angles θ_m such that

$$\Delta r = d \sin \theta_m = m\lambda \qquad m = 0, 1, 2, 3, \ldots \tag{17.6}$$

We have added the subscript m to denote that θ_m is the angle of the mth bright fringe, starting with $m = 0$ at the center.

The center of the viewing screen at $y = 0$ is equally distant from both slits, so $\Delta r = 0$. This point of constructive interference, with $m = 0$, is the bright fringe identified as the central maximum in Figure 17.6b. The path-length difference increases as you move away from the center of the screen, and the $m = 1$ fringes occur at the positions where $\Delta r = 1\lambda$. That is, one wave has traveled exactly one wavelength farther than the other. In general, **the mth bright fringe occurs where one wave has traveled m wavelengths farther than the other and thus $\Delta r = m\lambda$.**

FIGURE 17.7 Geometry of the double-slit experiment.

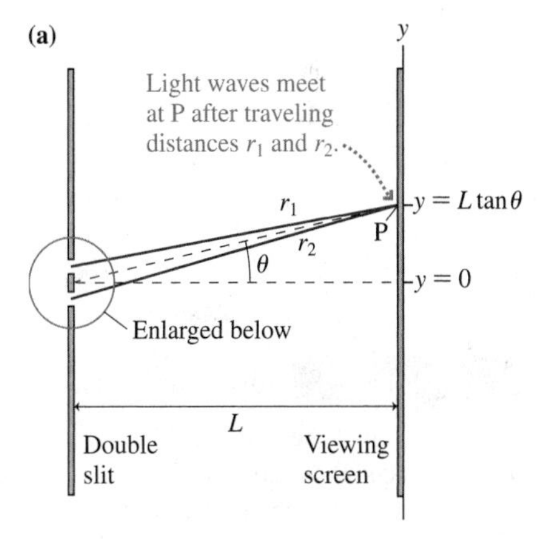

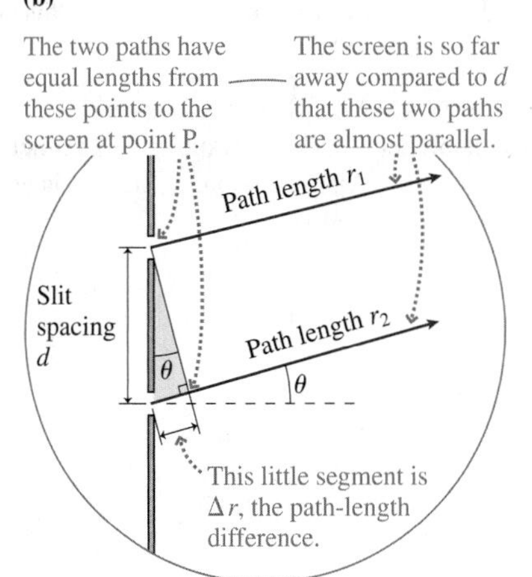

In practice, the angle θ in a double-slit experiment is almost always a very small angle ($<1°$). Recall, from ◀◀ SECTION 14.2, the *small-angle approximation* $\sin\theta \approx \theta$, where θ must be in *radians*. We can use this approximation to write Equation 17.6 as

$$\theta_m = m\frac{\lambda}{d} \qquad m = 0, 1, 2, 3, \ldots \tag{17.7}$$

Angles (in radians) of bright fringes for double-slit interference with slit spacing d

This gives the angular positions *in radians* of the bright fringes in the interference pattern.

It is usually more convenient to measure the *position* of the mth bright fringe, as measured from the center of the viewing screen. Using the small-angle approximation once again, this time in the form $\tan\theta \approx \theta$, we can substitute θ_m from Equation 17.7 for $\tan\theta_m$ in Equation 17.4 to find that the mth bright fringe occurs at position

$$y_m = \frac{m\lambda L}{d} \qquad m = 0, 1, 2, 3, \ldots \tag{17.8}$$

Positions of bright fringes for double-slit interference at screen distance L

The interference pattern is symmetrical, so there is an mth bright fringe at the same distance on both sides of the center. You can see this in Figure 17.6b.

NOTE ▶ Equations 17.7 and 17.8 do *not* apply to the interference of sound waves from two loudspeakers. The approximations we've used (small angles, $L \gg d$) are usually not valid for the much longer wavelengths of sound waves. ◀

EXAMPLE 17.2 How far do the waves travel?

Light from a helium-neon laser ($\lambda = 633$ nm) illuminates two slits spaced 0.40 mm apart. A viewing screen is 2.0 m behind the slits. A bright fringe is observed at a point 9.5 mm from the center of the screen. What is the fringe number m, and how much farther does the wave from one slit travel to this point than the wave from the other slit?

PREPARE A bright fringe is observed when one wave has traveled an integer number of wavelengths farther than the other. Thus we know that Δr must be $m\lambda$, where m is an integer. We can find m from Equation 17.8.

SOLVE Solving Equation 17.8 for m gives

$$m = \frac{y_m d}{\lambda L} = \frac{(9.5 \times 10^{-3}\text{ m})(0.40 \times 10^{-3}\text{ m})}{(633 \times 10^{-9}\text{ m})(2.0\text{ m})} = 3$$

Then the extra distance traveled by one wave compared to the other is

$$\Delta r = m\lambda = 3(633 \times 10^{-9}\text{ m}) = 1.9 \times 10^{-6}\text{ m}$$

ASSESS The path-length differences in two-slit interference are generally very small, just a few wavelengths of light. Here, Δr is only about one part in a million of the 2 m distance traveled by the waves!

Equation 17.8 predicts that **the interference pattern is a series of equally spaced bright lines** on the screen, exactly as shown in Figure 17.6b. How do we know the fringes are equally spaced? The **fringe spacing** between fringe m and fringe $m + 1$ is

$$\Delta y = y_{m+1} - y_m = \frac{(m+1)\lambda L}{d} - \frac{m\lambda L}{d}$$

which simplifies to

$$\Delta y = \frac{\lambda L}{d} \tag{17.9}$$

Spacing between any two adjacent bright fringes

Because Δy is independent of m, *any* two adjacent bright fringes have the same spacing.

The dark fringes in the photograph are bands of destructive interference. You learned in Chapter 16 that destructive interference occurs at positions where the path-length difference of the waves is a whole number of wavelengths plus half a wavelength:

$$\Delta r = \left(m + \frac{1}{2}\right)\lambda \qquad m = 0, 1, 2, 3, \ldots \tag{17.10}$$

We can use Equation 17.6 for Δr and the small-angle approximation to find that the dark fringes are located at positions

$$y'_m = \left(m + \frac{1}{2}\right)\frac{\lambda L}{d} \qquad m = 0, 1, 2, 3, \ldots \tag{17.11}$$

Positions of dark fringes for double-slit interference

We have used y'_m, with a prime, to distinguish the location of the mth minimum from the mth maximum at y_m. You can see from Equation 17.11 that **the dark fringes are located exactly halfway between the bright fringes.** FIGURE 17.8 summarizes the symbols we use to describe two-slit interference.

FIGURE 17.9 is a graph of the double-slit intensity versus y. Notice the unusual orientation of the graph, with the intensity increasing toward the left so that the y-axis can match the experimental layout. You can see that the intensity oscillates between dark fringes, where the intensity is zero, and equally spaced bright fringes of maximum intensity. The maxima occur at positions where $y_m = m\lambda L/d$.

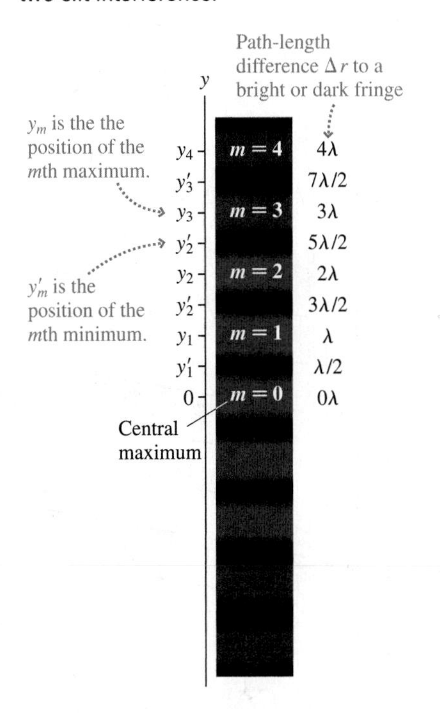

FIGURE 17.8 Symbols used to describe two-slit interference.

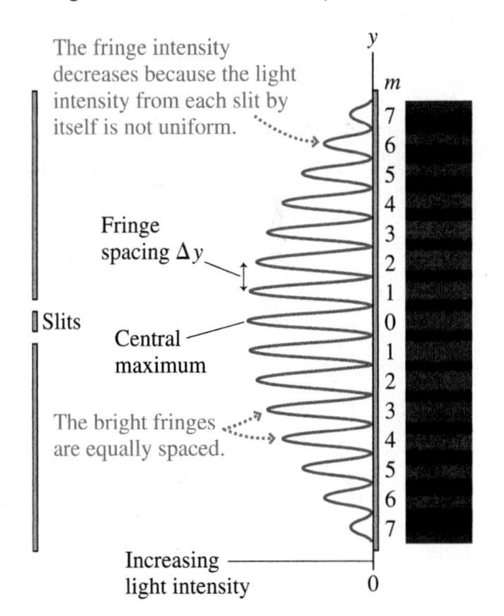

FIGURE 17.9 Intensity of the interference fringes in the double-slit experiment.

EXAMPLE 17.3 Measuring the wavelength of light

A double-slit interference pattern is observed on a screen 1.0 m behind two slits spaced 0.30 mm apart. From the center of one particular fringe to the center of the ninth bright fringe from this one is 1.6 cm. What is the wavelength of the light?

PREPARE It is not always obvious which fringe is the central maximum. Slight imperfections in the slits can make the interference fringe pattern less than ideal. However, you do not need to identify the $m = 0$ fringe because you can make use of the fact, expressed in Equation 17.9, that the fringe spacing Δy is uniform. The interference pattern looks like the photograph of Figure 17.6b.

SOLVE The fringe spacing is

$$\Delta y = \frac{1.6 \text{ cm}}{9} = 1.78 \times 10^{-3} \text{ m}$$

Using this fringe spacing in Equation 17.9, we find that the wavelength is

$$\lambda = \frac{d}{L}\Delta y = \frac{3.0 \times 10^{-4} \text{ m}}{1.0 \text{ m}}(1.78 \times 10^{-3} \text{ m})$$
$$= 5.3 \times 10^{-7} \text{ m} = 530 \text{ nm}$$

It is customary to express the wavelengths of visible light in nanometers. Be sure to do this as you solve problems.

ASSESS You learned in Chapter 15 that visible light spans the wavelength range 400–700 nm, so finding a wavelength in this range is reasonable. In fact, it's because of experiments like the double-slit experiment that we're able to measure the wavelengths of light.

STOP TO THINK 17.3 Light of wavelength λ_1 illuminates a double slit, and interference fringes are observed on a screen behind the slits. When the wavelength is changed to λ_2, the fringes get closer together. Is λ_2 larger or smaller than λ_1?

17.3 The Diffraction Grating

FIGURE 17.10 Top view of a diffraction grating with $N = 10$ slits.

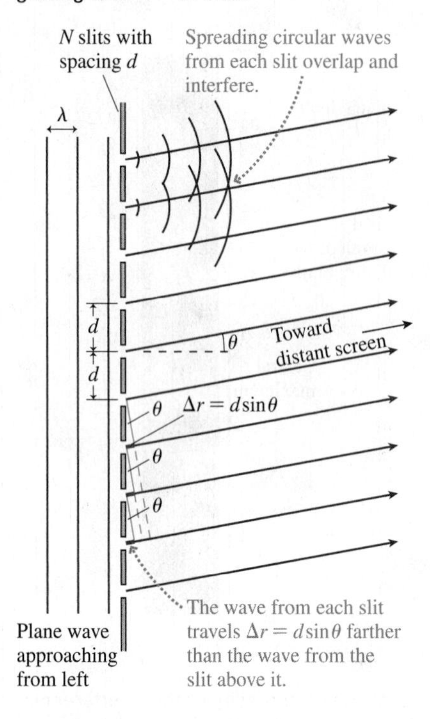

Suppose we were to replace the double slit with an opaque screen that has N closely spaced slits. When illuminated from one side, each of these slits becomes the source of a light wave that diffracts, or spreads out, behind the slit. Such a multi-slit device is called a **diffraction grating.** The light intensity pattern on a screen behind a diffraction grating is due to the interference of N overlapped waves.

FIGURE 17.10 shows a diffraction grating in which N slits are equally spaced a distance d apart. This is a top view of the grating, as we look down on the experiment, and the slits extend above and below the page. Only 10 slits are shown here, but a practical grating will have hundreds or even thousands of slits. Suppose a plane wave of wavelength λ approaches from the left. The crest of a plane wave arrives *simultaneously* at each of the slits, causing the wave emerging from each slit to be *in phase* with the wave emerging from every other slit—that is, all the emerging waves crest and trough simultaneously. Each of these emerging waves spreads out, just like the light wave in Figure 17.5, and after a short distance they all overlap with each other and interfere.

We want to know how the interference pattern will appear on a screen behind the grating. The light wave at the screen is the superposition of N waves, from N slits, as they spread and overlap. As we did with the double slit, we'll assume that the distance L to the screen is very large in comparison with the slit spacing d; hence the path followed by the light from one slit to a point on the screen is *very nearly* parallel to the path followed by the light from neighboring slits. You can see in Figure 17.10 that the wave from one slit travels distance $\Delta r = d\sin\theta$ farther than the wave from the slit above it and $\Delta r = d\sin\theta$ less than the wave below it. This is the same reasoning we used in Figure 17.7 to analyze the double-slit experiment.

FIGURE 17.11 Interference for a grating with five slits.

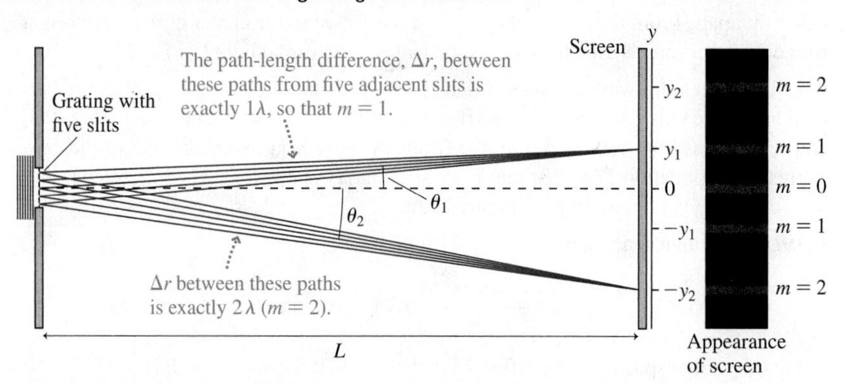

Figure 17.10 was a magnified view of the slits. FIGURE 17.11 steps back to where we can see the viewing screen, for a grating with five slits. If the angle θ is such that $\Delta r = d\sin\theta = m\lambda$, where m is an integer, then the light wave arriving at the screen from one slit will travel *exactly* m wavelengths more or less than light from the two slits next to it, so these waves will be *exactly in phase* with each other. But each of those waves is in phase with waves from the slits next to them, and so on until we reach the end of the grating. In other words, **N light waves, from N different slits, will *all* be in phase with each other when they arrive at a point on the screen at angle θ_m such that**

$$d\sin\theta_m = m\lambda \qquad m = 0, 1, 2, 3, \ldots \tag{17.12}$$

Angles of bright fringes due to a diffraction grating with slits distance d apart

The screen will have bright constructive-interference fringes at the values of θ_m given by Equation 17.12. When this happens, we say that the light is "diffracted at angle θ_m." Because it's usually easier to measure distances rather than angles, the position y_m of the mth maximum is

$$y_m = L \tan \theta_m \tag{17.13}$$

Positions of bright fringes due to a diffraction grating distance L from screen

The integer m is called the **order** of the diffraction. Practical gratings, with very small values for d, display only a few orders. Because d is usually very small, it is customary to characterize a grating by the number of *lines per millimeter.* Here "line" is synonymous with "slit," so the number of lines per millimeter is simply the inverse of the slit spacing d in millimeters.

NOTE ▶ The condition for constructive interference in a grating of N slits is identical to Equation 17.6 for just two slits. Equation 17.12 is simply the requirement that the path-length difference between adjacent slits, be they two or N, is $m\lambda$. But unlike the angles in double-slit interference, the angles of constructive interference from a diffraction grating are generally *not* small angles. The reason is that the slit spacing d in a diffraction grating is usually so small that λ/d is not a small number. Thus you *cannot* use the small-angle approximation to simplify Equations 17.12 and 17.13. ◀

There is an important difference between the intensity pattern of double-slit interference shown in Figure 17.8 and the intensity pattern of a multiple-slit diffraction grating shown in Figure 17.11: The bright fringes of a diffraction grating are *much* narrower. In general, as the number of slits N increases, the bright fringes get narrower and brighter. This trend is shown in **FIGURE 17.12** for gratings with two slits (double-slit interference), 10 slits, and 50 slits. For a practical diffraction grating, which may have thousands of closely spaced slits, the interference pattern consists of a small number of *very* bright and *very* narrow fringes while most of the screen remains dark.

FIGURE 17.12 The intensity on the screen due to three diffraction gratings. Notice that the intensity axes have different scales.

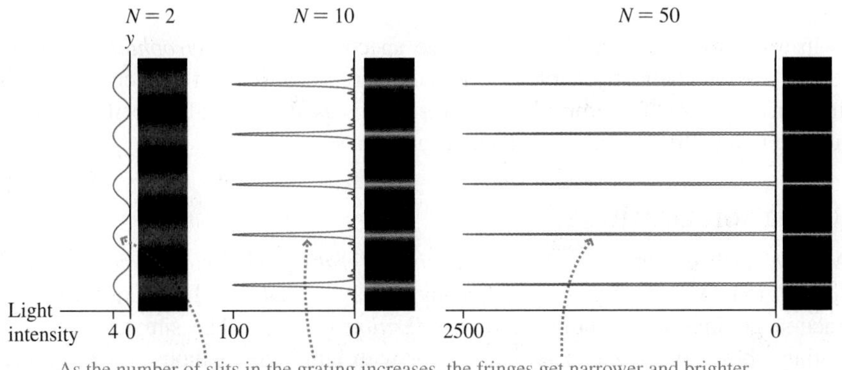

Spectroscopy

As we'll see in Chapter 29, each atomic element in the periodic table, if appropriately excited by light, electricity, or collisions with other atoms, emits light at only certain well-defined wavelengths. By accurately measuring these wavelengths, we can deduce the various elements in a sample of unknown composition. Molecules also emit light that is characteristic of their composition. The science of measuring the wavelengths of atomic and molecular emissions is called **spectroscopy.**

FIGURE 17.13 A diffraction grating can be used to measure wavelengths of light.

The two wavelengths have been separated.
Blue light has a longer wavelength than violet, and thus diffracts more.
Grating
All wavelengths overlap at $y = 0$.
Light intensity

Figure Video

Because their bright fringes are so distinct, diffraction gratings are an ideal tool for spectroscopy. Suppose the light incident on a grating consists of two slightly different wavelengths. According to Equation 17.12, each wavelength will diffract at a slightly different angle and, if N is sufficiently large, we'll see two distinct fringes on the screen. **FIGURE 17.13** illustrates this idea. By contrast, the fringes in a double-slit experiment are so broad that it would not be possible to distinguish the fringes of one wavelength from those of the other.

EXAMPLE 17.4 **Measuring wavelengths emitted by sodium atoms**

Light from a sodium lamp passes through a diffraction grating having 1000 slits per millimeter. The interference pattern is viewed on a screen 1.000 m behind the grating. Two bright yellow fringes are visible 72.88 cm and 73.00 cm from the central maximum. What are the wavelengths of these two fringes?

PREPARE This situation is similar to that in Figure 17.13. The two fringes are very close together, so we expect the wavelengths to be only slightly different. No other yellow fringes are mentioned, so we will assume these two fringes are the first-order diffraction ($m = 1$).

SOLVE The distance y_m of a bright fringe from the central maximum is related to the diffraction angle by $y_m = L\tan\theta_m$. Thus the diffraction angles of these two fringes are

$$\theta_1 = \tan^{-1}\left(\frac{y_1}{L}\right) = \begin{cases} 36.085^\circ & \text{fringe at 72.88 cm} \\ 36.129^\circ & \text{fringe at 73.00 cm} \end{cases}$$

These angles must satisfy the interference condition $d\sin\theta_1 = \lambda$, so the wavelengths are

$$\lambda = d\sin\theta_1$$

What is d? If a 1 mm length of the grating has 1000 slits, then the spacing from one slit to the next must be 1/1000 mm, or $d = 1.00 \times 10^{-6}$ m. Thus the wavelengths creating the two bright fringes are

$$\lambda = d\sin\theta_1 = \begin{cases} 589.0\text{ nm} & \text{fringe at 72.88 cm} \\ 589.6\text{ nm} & \text{fringe at 73.00 cm} \end{cases}$$

ASSESS In Chapter 15 you learned that yellow light has a wavelength of about 600 nm, so our answer is reasonable.

Class Video

Instruments that measure and analyze spectra, called *spectrophotometers,* are widely used in chemistry, biology, and medicine. Because each molecule has a distinct spectrum—a "fingerprint"—spectroscopy is used to identify specific biomolecules in tissue, drugs in urine, and chlorophyll in seawater.

Reflection Gratings

We have analyzed what is called a *transmission grating,* with many parallel slits. It's difficult to make such a grating with many closely spaced slits. In practice, most diffraction gratings are manufactured as *reflection gratings.* The simplest reflection grating, shown in **FIGURE 17.14**, is a mirror with hundreds or thousands of narrow, parallel grooves cut into the surface. The grooves divide the surface into many parallel reflective stripes, each of which, when illuminated, becomes the source of a spreading wave. Thus an incident light wave is divided into N overlapped waves. The interference pattern is exactly the same as the interference pattern of light transmitted through N parallel slits, and so **Equation 17.13 applies to reflection gratings as well as to transmission gratings.**

The rainbow of colors seen on the surface of a CD or DVD is an everyday display of interference. The surface of a DVD is smooth plastic with a mirror-like reflective coating. As shown in **FIGURE 17.15**, billions of microscopic holes, each about 320 nm in diameter, are "burned" into the surface with a laser. The presence or absence of a hole at a particular location on the disk is interpreted

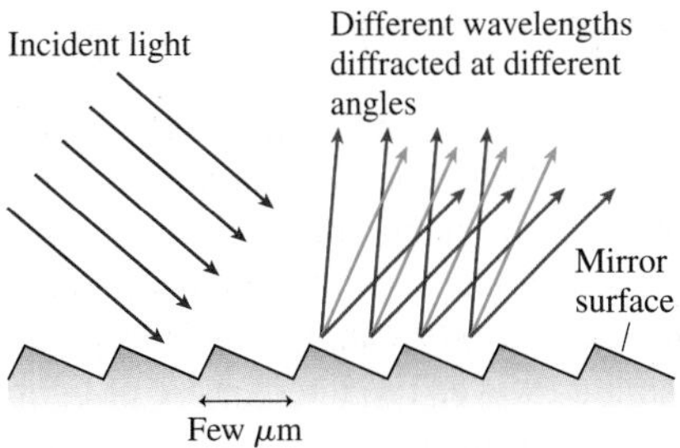

FIGURE 17.14 A reflection grating.

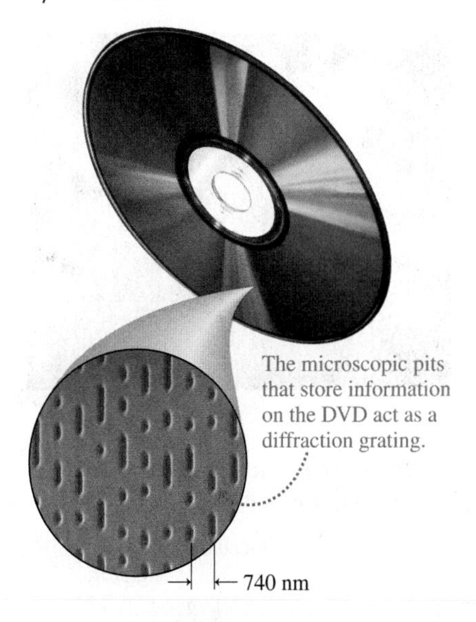

The microscopic pits that store information on the DVD act as a diffraction grating.
740 nm

FIGURE 17.15 A DVD's colors are caused by diffraction.

as the 0 or 1 of digitally encoded information. But from an optical perspective, the array of holes in a shiny surface is a two-dimensional version of the reflection grating shown in Figure 17.14. Less precise plastic reflection gratings can be manufactured at very low cost simply by stamping holes or grooves into a reflective surface, and these are widely sold as toys and novelty items. Rainbows of color are seen as each wavelength of white light is diffracted at a unique angle.

SYNTHESIS 17.1 Double-slit interference and diffraction gratings

The physical principles underlying double-slit interference and diffraction gratings are the same: A bright fringe occurs on a screen when the path-length difference between waves is an integer number $m = 0, 1, 2, 3...$ times the wavelength λ.

Double-slit interference

$$\theta_m = m\frac{\lambda}{d} \qquad y_m = \frac{m\lambda L}{d} \qquad y'_m = \left(m + \frac{1}{2}\right)\frac{\lambda L}{d}$$

Wavelength (m); Distance between slits (m); Distance to screen (m)

θ_m is the angle *in radians* of the *m*th bright fringe. y_m is the position of the *m*th bright fringe. y'_m is the position of the *m*th dark fringe.

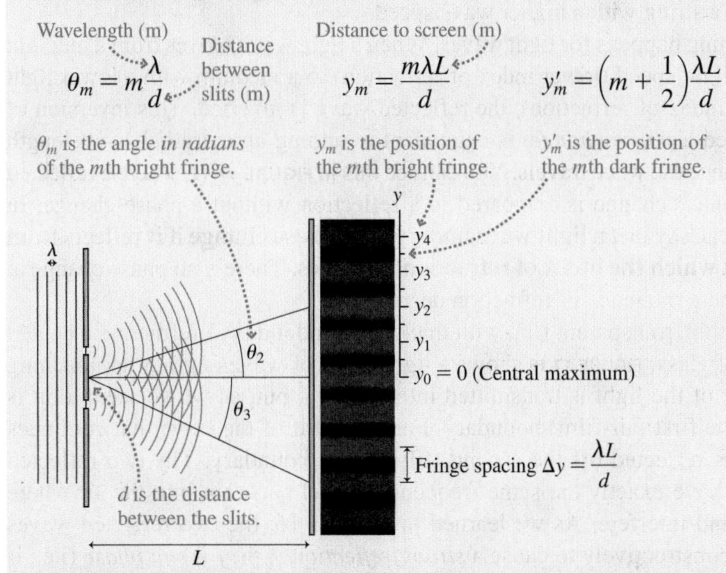

In double-slit interference, the angles θ are usually very small and we can use the small-angle approximation in the equations.

Diffraction gratings

$$d\sin\theta_m = m\lambda \qquad y_m = L\tan\theta_m$$

θ_m is the angle of the *m*th bright fringe. y_m is the position of the *m*th bright fringe.

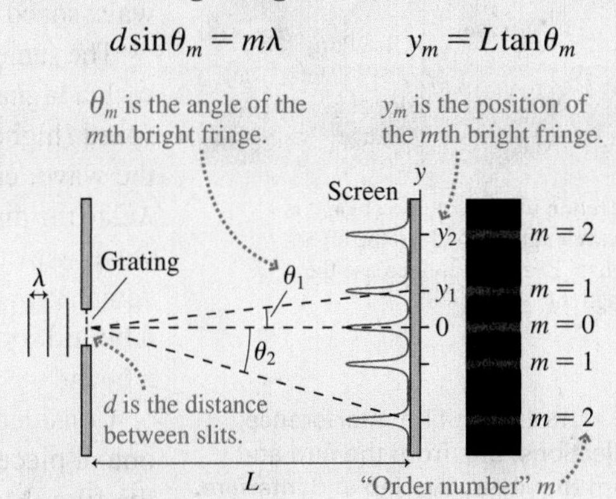

For a diffraction grating, the small-angle approximation is generally not valid. To find the bright-fringe positions y_m, you must first find the angle θ_m from $d\sin\theta_m = m\lambda$, then use $y_m = L\tan\theta_m$.

A grating with N slits or lines per mm has slit spacing $d = (1\text{ mm})/N$.

STOP TO THINK 17.4 White light passes through a diffraction grating and forms rainbow patterns on a screen behind the grating. For each rainbow,

A. The red side is on the right, the violet side on the left.
B. The red side is on the left, the violet side on the right.
C. The red side is closest to the center of the screen, the violet side is farthest from the center.
D. The red side is farthest from the center of the screen, the violet side is closest to the center.

FIGURE 17.16 Two reflections are visible in the window, one from each surface.

FIGURE 17.17 Reflected waves with and without a phase change.

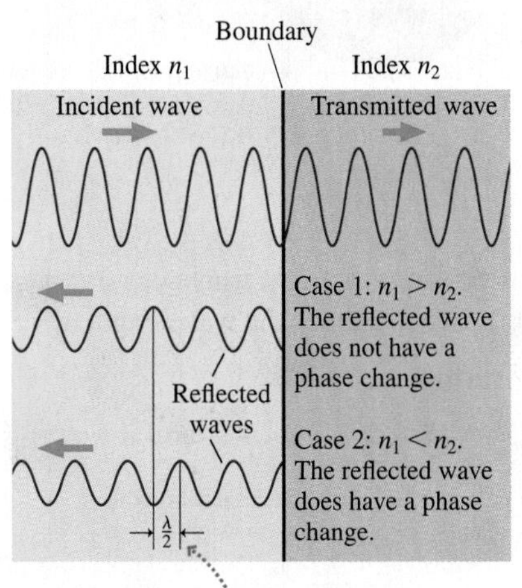

FIGURE 17.18 In thin-film interference, two reflections, one from the film and one from the glass, overlap and interfere.

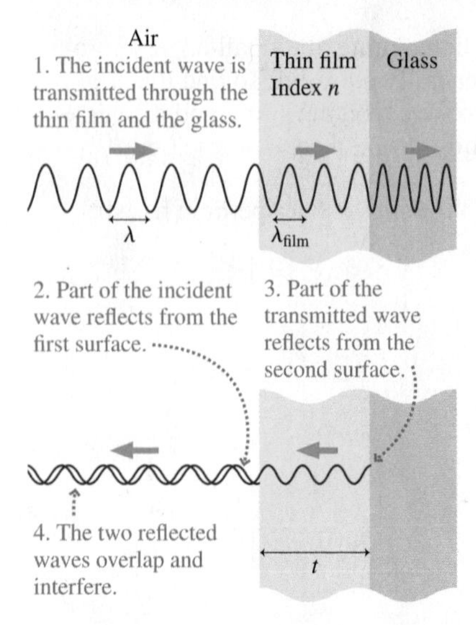

17.4 Thin-Film Interference

In ◀◀SECTION 16.6 you learned about the interference of sound waves in one dimension. Depending on whether they are in phase or out of phase, two sound waves of the same frequency, traveling in the same direction, can undergo constructive or destructive interference. Light waves can also interfere in this way. Equal-frequency light waves are produced when *partial reflection* at a boundary splits a light wave into a reflected wave and a transmitted wave. The interference of light waves reflected from the two boundaries of a thin film, such as the thin film of water that makes a soap bubble, is called **thin-film interference.**

Thin-film interference has important applications in the optics industry. Thin-film coatings, less than 1 μm (10^{-6} m) thick, are used for the antireflection coatings on the lenses in cameras, microscopes, and other optical equipment. The bright colors of oil slicks and soap bubbles are also due to thin-film interference.

Interference of Reflected Light Waves

As you know, and as we discussed in Chapter 16, a light wave encountering a piece of glass is partially transmitted and partially reflected. In fact, a light wave is partially reflected from *any* boundary between two transparent media with different indices of refraction. Thus light is partially reflected not only from the front surface of a sheet of glass, but from the back surface as well, as it exits from the glass into the air. This leads to the *two* reflections seen in FIGURE 17.16.

Another important aspect of wave reflections was shown for strings in Figure 16.10b of the last chapter. If a wave moves from a string with a higher wave speed to a string with a lower wave speed, the reflected wave is *inverted* with respect to the incoming wave. It is not inverted if the wave moves from a string with a lower wave speed to a string with a higher wave speed.

The same thing happens for light waves. When a light wave moves from a medium with a higher light speed (lower index of refraction) to a medium with a lower light speed (higher index of refraction), the reflected wave is inverted. This inversion of the wave, called a *phase change,* is equivalent to adding an extra half-wavelength $\lambda/2$ to the distance the wave travels. You can see this in FIGURE 17.17, where a reflected wave with a phase change is compared to a reflection without a phase change. In summary, we can say that **a light wave undergoes a phase change if it reflects from a boundary at which the index of refraction increases.** There's no phase change at a boundary where the index of refraction decreases.

Consider a thin, transparent film with thickness t and index of refraction n coated onto a piece of glass. FIGURE 17.18 shows a light wave of wavelength λ approaching the film. Most of the light is transmitted into the film, but, as we've seen, a bit is reflected off the first (air-film) boundary. Further, a bit of the wave that continues into the film is reflected off the second (film-glass) boundary. The two reflected waves, which have exactly the same frequency, travel back out into the air where they overlap and interfere. As we learned in Chapter 16, the two reflected waves will interfere constructively to cause a *strong reflection* if they are *in phase* (i.e., if their crests overlap). If the two reflected waves are *out of phase,* with the crests of one wave overlapping the troughs of the other, they will interfere destructively to cause a *weak reflection* or, if their amplitudes are equal, *no reflection* at all.

We found the interference of two sound waves to be constructive if their path-length difference is $\Delta d = m\lambda$ and destructive if $\Delta d = \left(m + \frac{1}{2}\right)\lambda$, where m is an integer. The same idea holds true for reflected light waves, for which the path-length difference is the extra distance traveled by the wave that reflects from the second surface. Because this wave travels twice through a film of thickness t, the path-length difference is $\Delta d = 2t$.

We noted above that the phase change when a light wave reflects from a boundary with a higher index of refraction is equivalent to adding an extra half-wavelength to the distance traveled. This leads to two situations:

1. If *neither* or *both* waves have a phase change due to reflection, the net addition to the path-length difference is zero. The *effective path-length difference* is $\Delta d_{\text{eff}} = 2t$.
2. If only *one* wave has a phase change due to reflection, the effective path-length difference is increased by one half-wavelength to $\Delta d_{\text{eff}} = 2t + \frac{1}{2}\lambda$.

The interference of the two reflected waves is then constructive if $\Delta d_{\text{eff}} = m\lambda_{\text{film}}$ and destructive if $\Delta d_{\text{eff}} = \left(m + \frac{1}{2}\right)\lambda_{\text{film}}$. Why λ_{film}? Because the extra distance is traveled inside the film, so we need to compare $2t$ to the wavelength in the film. Further, the film's index of refraction is n, so the wavelength in the film is $\lambda_{\text{film}} = \lambda/n$, where λ is the wavelength of the light in vacuum or air.

With this information, we can write the conditions for constructive and destructive interference of the light waves reflected by a thin film:

$$2t = m\frac{\lambda}{n} \qquad m = 0, 1, 2, \ldots \tag{17.14}$$

Condition for constructive interference with either 0 or 2 reflective phase changes
Condition for destructive interference with only 1 reflective phase change

$$2t = \left(m + \frac{1}{2}\right)\frac{\lambda}{n} \qquad m = 0, 1, 2, \ldots \tag{17.15}$$

Condition for destructive interference with either 0 or 2 reflective phase changes
Condition for constructive interference with only 1 reflective phase change

NOTE ► Equations 17.14 and 17.15 give the film thicknesses that yield constructive or destructive interference. At other thicknesses, the waves will interfere neither fully constructively nor fully destructively, and the reflected intensity will fall somewhere between these two extremes. ◄

These conditions are the basis of a procedure to analyze thin-film interference.

TACTICS BOX 17.1 Analyzing thin-film interference

Follow the light wave as it passes through the film. The wave reflecting from the second boundary travels an extra distance $2t$.

1. Note the indices of refraction of the three media: the medium before the film, the film itself, and the medium beyond the film. The first and third may be the same. There's a reflective phase change at any boundary where the index of refraction increases.
2. If *neither* or *both* reflected waves undergo a phase change, the phase changes cancel and the effective path-length difference is $\Delta d = 2t$. Use Equation 17.14 for constructive interference and 17.15 for destructive interference.
3. If *only one* wave undergoes a phase change, the effective path-length difference is $\Delta d = 2t + \frac{1}{2}\lambda$. Use Equation 17.14 for destructive interference and 17.15 for constructive interference.

Exercises 12, 13 

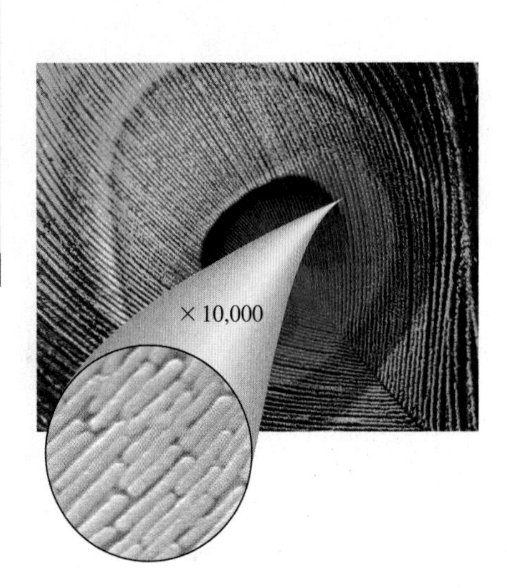

► **Iridescent feathers** BIO The gorgeous colors of the hummingbird shown at the beginning of this chapter are due not to pigments but to interference. This *iridescence,* present in some bird feathers and insect shells, arises from biological structures whose size is similar to the wavelength of light. The sheen of an insect, for instance, is due to thin-film interference from multiple thin layers in its shell. Peacock feathers are also a layered structure, but each layer itself consists of nearly parallel rods of melanin, as shown in the micrograph, that act as a diffraction grating. Thus a peacock feather combines thin-film interference and grating-like diffraction to produce its characteristic multicolored iridescent hues.

EXAMPLE 17.5 Designing an antireflection coating

The glasses on the top have an antireflection coating on them. Those on the bottom do not.

To keep unwanted light from reflecting from the surface of eyeglasses or other lenses, a thin film of a material with an index of refraction $n = 1.38$ is coated onto the plastic lens ($n = 1.55$). It is desired to have destructive interference for $\lambda = 550$ nm because that is the center of the visible spectrum. What is the thinnest film that will do this?

PREPARE We follow the steps of Tactics Box 17.1. As the light traverses the film, it first reflects at the front surface of the coating. Here, the index of refraction increases from that of air ($n = 1.00$) to that of the film ($n = 1.38$), so there will be a reflective phase change. The light then reflects from the rear surface of the coating. The index again increases from that of the film ($n = 1.38$) to that of the plastic ($n = 1.55$). With two phase changes, Tactics Box 17.1 tells us that we should use Equation 17.15 for destructive interference.

SOLVE We can solve Equation 17.15 for the thickness t that causes destructive interference:

$$t = \frac{\lambda}{2n}\left(m + \frac{1}{2}\right)$$

The thinnest film is the one for which $m = 0$, giving

$$t = \frac{550 \text{ nm}}{2(1.38)} \times \frac{1}{2} = 100 \text{ nm}$$

ASSESS Interference effects occur when path-length differences are on the order of a wavelength, so our answer of 100 nm seems reasonable.

Thin Films of Air

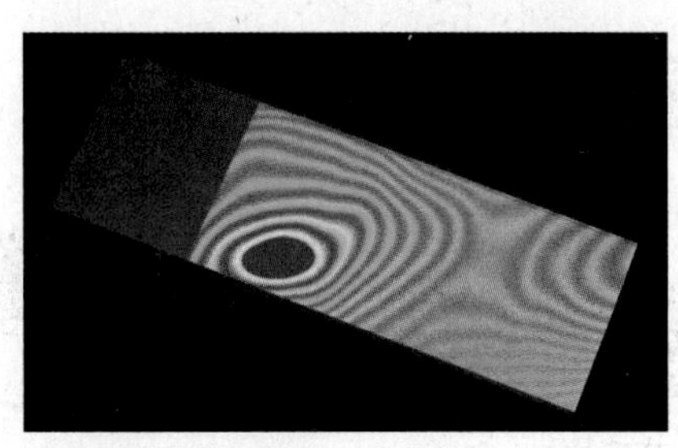

FIGURE 17.19 Light and dark fringes caused by thin-film interference due to the air layer between two microscope slides.

A film need not be of a solid material. A thin layer of air sandwiched between two glass surfaces also exhibits thin-film interference due to the waves that reflect off both interior air-glass boundaries. FIGURE 17.19 shows two microscope slides being pressed together. The light and dark "fringes" occur because the slides are not exactly flat and they touch each other only at a few points. Everywhere else there is a thin layer of air between them. At some points, the air layer's thickness is such as to give constructive interference (light fringes), while at other places its thickness gives destructive interference (dark fringes). These fringes can be used to accurately measure the flatness of two glass plates, as Example 17.6 will show.

EXAMPLE 17.6 Finding the fringe spacing from a wedge-shaped film of air

Two 15-cm-long flat glass plates are separated by a 10-μm-thick spacer at one end, leaving a thin wedge of air between them, as shown in FIGURE 17.20. The plates are illuminated by light from a sodium lamp with wavelength $\lambda = 589$ nm. Alternating bright and dark fringes are observed. What is the spacing between two bright fringes?

FIGURE 17.20 Two glass plates with an air wedge between them.

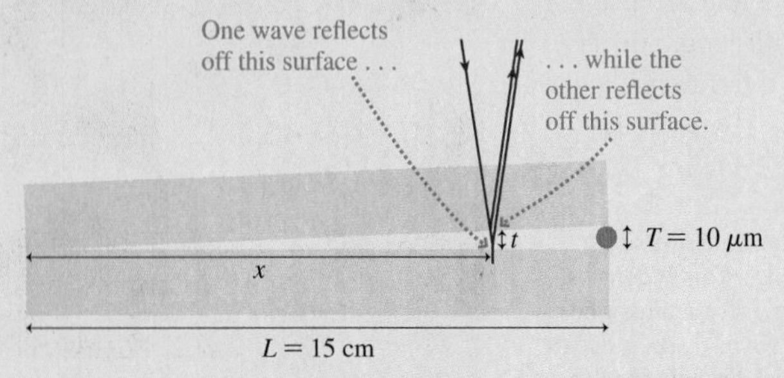
One wave reflects off this surface . . .
. . . while the other reflects off this surface.
t
T = 10 μm
x
L = 15 cm

PREPARE The wave reflected from the lower plate has a reflective phase change, but the top reflection does not because the index of refraction decreases at the glass-air boundary. According to Tactics Box 17.1, we should use Equation 17.15 for constructive interference:

$$2t = \left(m + \frac{1}{2}\right)\frac{\lambda}{n}$$

This is a film of air, so here n is the index of refraction of air. Each integer value of m corresponds to a wedge thickness t for which there is constructive interference and thus a bright fringe.

SOLVE Let x be the distance from the left end to a bright fringe. From Figure 17.20, by similar triangles we have

$$\frac{t}{x} = \frac{T}{L}$$

or $t = xT/L$. From the condition for constructive interference, we then have

$$2\frac{xT}{L} = \left(m + \frac{1}{2}\right)\frac{\lambda}{n}$$

There will be a bright fringe for any integer value of m, and so the position of the mth fringe, as measured from the left end, is

$$x_m = \frac{\lambda L}{2nT}\left(m + \frac{1}{2}\right)$$

We want to know the spacing between two adjacent fringes, m and $m + 1$, which is

$$\Delta x = x_{m+1} - x_m = \frac{\lambda L}{2nT}\left(m + 1 + \frac{1}{2}\right) - \frac{\lambda L}{2nT}\left(m + \frac{1}{2}\right) = \frac{\lambda L}{2nT}$$

Evaluating, we find

$$\Delta x = \frac{\lambda L}{2nT} = \frac{(5.89 \times 10^{-7}\ \text{m})(0.15\ \text{m})}{2(1.00)(10 \times 10^{-6}\ \text{m})} = 4.4\ \text{mm}$$

ASSESS As the photo shows, if the two plates are very flat, the fringes will appear as straight lines perpendicular to the direction of increasing air thickness. However, if the plates are not quite flat, the fringes will appear curved. The amount of curvature indicates the departure of the plates from perfect flatness.

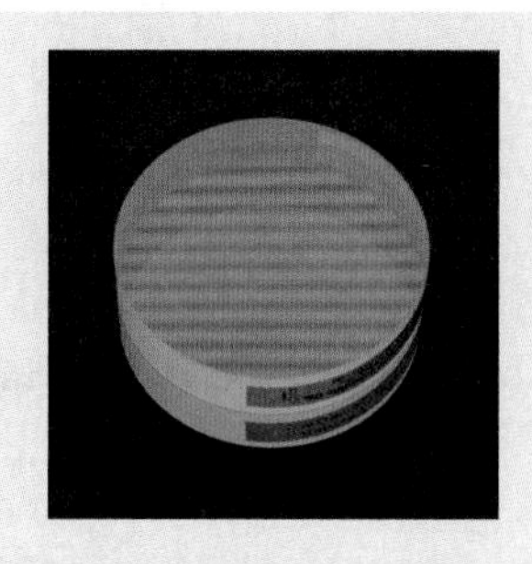

The Colors of Soap Bubbles and Oil Slicks

So far we have considered thin-film interference for only single wavelengths of light. The bright colors of soap bubbles and oil slicks on water are due to thin-film interference of white light, which, as we've seen, is a mixture of *all* wavelengths.

A soap bubble is a very thin, spherical film of soapy water ($n = 1.33$). Consider a soap film with thickness $t = 470$ nm. Light waves reflect from both surfaces of the film, and these reflected waves interfere. The light reflecting from the front (air-water) surface has a reflective phase change, but the back reflection does not. Thus Equation 17.15 describes constructive interference and Equation 17.14 is destructive interference. Table 17.2 shows wavelengths of constructive and destructive interference for three values of m.

Light near the red end (625 nm) and violet end (417) of the spectrum undergoes destructive interference; these colors are *not* reflected by the film. At the same time, light near 500 nm (green) interferes constructively and so is strongly reflected. Consequently, a soap film of this thickness will appear green.

Real soap bubbles and oil slicks have thicknesses that vary from point to point. At some thicknesses, green light is strongly reflected, while at others red or violet light is. **FIGURE 17.21** shows that the colors of a soap bubble are predominately greens and red/violets.

TABLE 17.2 Wavelengths for constructive and destructive interference from a 470-nm-thick soap bubble. Visible wavelengths are shown in **bold.**

Equation	$m = 1$	$m = 2$	$m = 3$
$\lambda_{\text{con}} = \dfrac{2nt}{m + \frac{1}{2}}$	833 nm	**500 nm**	357 nm
$\lambda_{\text{des}} = \dfrac{2nt}{m}$	1250 nm	**625 nm**	**417 nm**

FIGURE 17.21 The colors of a soap bubble.

CONCEPTUAL EXAMPLE 17.7 **Colors in a vertical soap film**

FIGURE 17.22 shows a soap film in a metal ring. The ring is held vertically. Explain the colors seen in the film.

FIGURE 17.22 A soap film in a metal ring.

REASON Because of gravity, the film is thicker near the bottom and thinner at the top. It thus has a wedge shape, and the interference pattern consists of lines of alternating constructive and destructive interference, just as for the air wedge of Example 17.6. Because this soap film is illuminated by white light, colors form as just discussed for any soap film.

Notice that the very top of the film, which is extremely thin, appears black. This means that it is reflecting no light at all. When the film is very thin—much thinner than the wavelength of light—there is almost no path-length difference between the two waves reflected off the front and the back of the film. However, the wave reflected off the back undergoes a reflective phase change and is out of phase with the wave reflected off the front. The two waves thus *always* interfere destructively, no matter what their wavelength.

ASSESS This simple experiment shows directly that the two reflected waves have different reflective phase changes.

STOP TO THINK 17.5 Reflections from a thin layer of air between two glass plates cause constructive interference for a particular wavelength of light λ. By how much must the thickness of this layer be increased for the interference to be destructive?

A. $\lambda/8$ B. $\lambda/4$ C. $\lambda/2$ D. λ

17.5 Single-Slit Diffraction

We opened this chapter with a photograph of a water wave passing through a hole in a barrier, then spreading out on the other side. The preview to this chapter showed how light, after passing a narrow needle, also spreads out on the other side. This phenomenon is called *diffraction.* We're now ready to look at the details of diffraction.

FIGURE 17.23 again shows the experimental arrangement for observing the diffraction of light through a narrow slit of width a. Diffraction through a tall, narrow slit of width a is known as **single-slit diffraction.** A viewing screen is placed a distance L behind the slit, and we will assume that $L \gg a$. The light pattern on the viewing screen consists of a *central maximum* flanked by a series of weaker **secondary maxima** and dark fringes. Notice that the central maximum is significantly broader than the secondary maxima. It is also significantly brighter than the secondary maxima, although that is hard to tell here because this photograph has been overexposed to make the secondary maxima show up better.

FIGURE 17.23 A single-slit diffraction experiment.

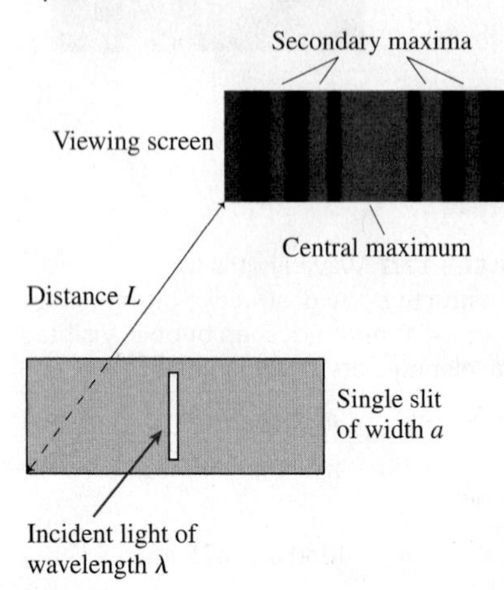
Secondary maxima
Viewing screen
Central maximum
Distance L
Single slit of width a
Incident light of wavelength λ

Huygens' Principle

Our analysis of the superposition of waves from distinct sources, such as two loudspeakers or the two slits in a double-slit experiment, has tacitly assumed that the sources are *point sources,* with no measurable extent. To understand diffraction, we need to think about the propagation of an *extended* wave front. This problem was first considered by the Dutch scientist Christiaan Huygens, a contemporary of Newton.

In ◄◄ SECTION 15.5 you learned how *wave fronts*—the "crests" of a wave—evolve with time for plane and spherical waves. Huygens developed a geometrical model to visualize how *any* wave, such as a wave passing through a narrow slit, evolves. **Huygens' principle** has two parts:

1. Each point on a wave front is the source of a spherical *wavelet* that spreads out at the wave speed.
2. At a later time, the shape of the wave front is the curve that is tangent to all the wavelets.

FIGURE 17.24 illustrates Huygens' principle for a plane wave and a spherical wave. As you can see, the curve tangent to the wavelets of a plane wave is a plane that has propagated to the right. The curve tangent to the wavelets of a spherical wave is a larger sphere.

FIGURE 17.24 Huygens' principle applied to the propagation of plane waves and spherical waves.

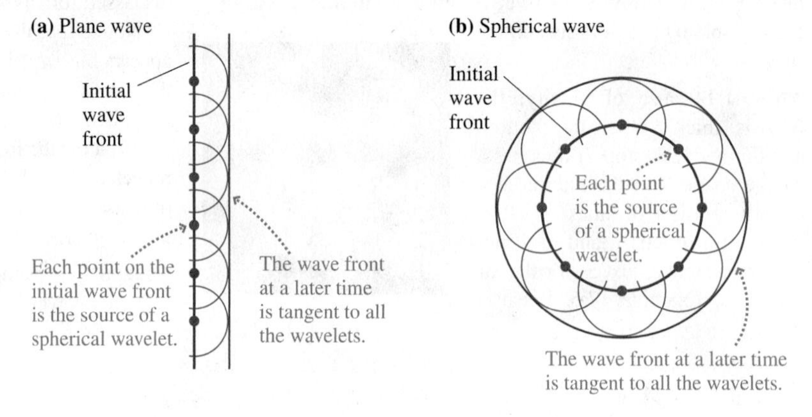
(a) Plane wave
Initial wave front
Each point on the initial wave front is the source of a spherical wavelet.
The wave front at a later time is tangent to all the wavelets.
(b) Spherical wave
Initial wave front
Each point is the source of a spherical wavelet.
The wave front at a later time is tangent to all the wavelets.

Analyzing Single-Slit Diffraction

FIGURE 17.25a shows a wave front passing through a narrow slit of width a. According to Huygens' principle, each point on the wave front can be thought of as the source of a spherical wavelet. These wavelets overlap and interfere, producing the diffraction

FIGURE 17.25 Each point on the wave front is a source of spherical wavelets. The superposition of these wavelets produces the diffraction pattern on the screen.

(a) Greatly magnified view of slit

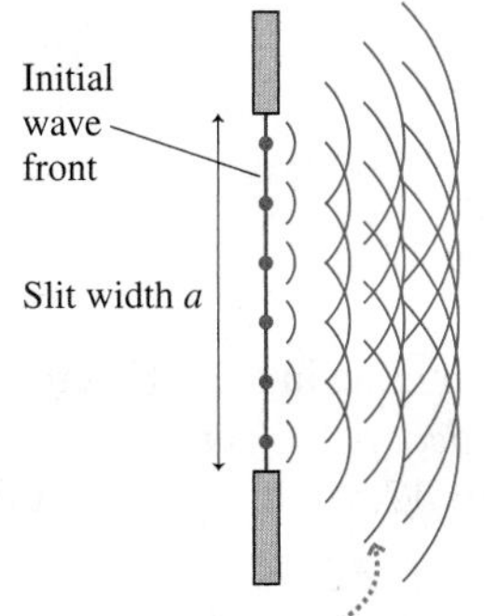

The wavelets from each point on the initial wave front overlap and interfere, creating a diffraction pattern on the screen.

(b)

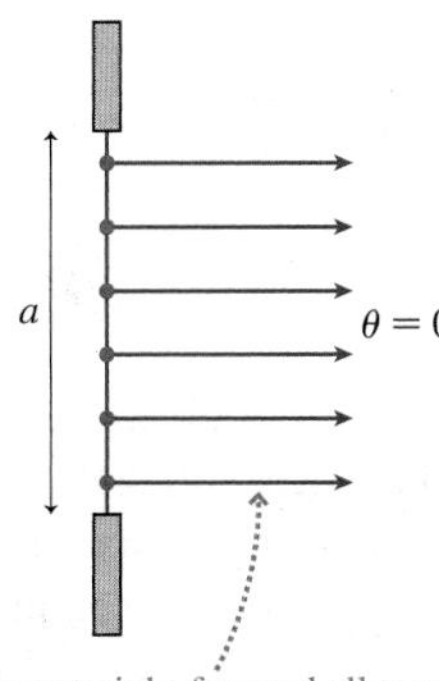

The wavelets going straight forward all travel the same distance to the screen. Thus they arrive in phase and interfere constructively to produce the central maximum.

(c)

Each point on the wave front is paired with another point distance $a/2$ away.

$\frac{a}{2}$ θ 1 3 5 2 4 6 Δr_{12}

These wavelets all meet on the screen at angle θ. Wavelet 2 travels distance $\Delta r_{12} = (a/2)\sin\theta$ farther than wavelet 1.

pattern seen on the viewing screen. The full mathematical analysis, using *every* point on the wave front, is a fairly difficult problem in calculus. We'll be satisfied with a geometrical analysis based on just a few wavelets.

FIGURE 17.25b shows the paths of several wavelets as they travel straight ahead to the central point on the screen. (The screen is *very* far to the right in this magnified view of the slit.) The paths to the screen are very nearly parallel to each other; thus all the wavelets travel the same distance and arrive at the screen *in phase* with each other. The *constructive interference* between these wavelets produces the central maximum of the diffraction pattern at $\theta = 0$.

Water waves can be seen diffracting behind the "slit" between the two breakwaters. The wave pattern can be understood using Huygens' principle.

The situation is different at points away from the center of the screen. Wavelets 1 and 2 in FIGURE 17.25c start from points that are distance $a/2$ apart. If the angle is such that Δr_{12}, the extra distance traveled by wavelet 2, happens to be $\lambda/2$, then wavelets 1 and 2 arrive out of phase and interfere destructively. But if Δr_{12} is $\lambda/2$, then the difference Δr_{34} between paths 3 and 4 and the difference Δr_{56} between paths 5 and 6 are also $\lambda/2$. Those pairs of wavelets also interfere destructively. The superposition of all the wavelets produces perfect destructive interference.

Figure 17.25c happens to show six wavelets, but our conclusion is valid for any number of wavelets. The key idea is that **every point on the wave front can be paired with another point that is distance *a*/2 away.** If the path-length difference is $\lambda/2$, the wavelets that originate at these two points will arrive at the screen out of phase and interfere destructively. When we sum the displacements of all N wavelets, they will—pair by pair—add to zero. The viewing screen at this position will be dark. This is the main idea of the analysis, one worth thinking about carefully.

You can see from Figure 17.25c that $\Delta r_{12} = (a/2)\sin\theta$. This path-length difference will be $\lambda/2$, the condition for destructive interference, if

$$\Delta r_{12} = \frac{a}{2}\sin\theta_1 = \frac{\lambda}{2} \qquad (17.16)$$

or, equivalently, $\sin\theta_1 = \lambda/a$.

We can extend this idea to find other angles of perfect destructive interference. Suppose each wavelet is paired with another wavelet from a point $a/4$ away. If Δr between these wavelets is $\lambda/2$, then all N wavelets will again cancel in pairs to give complete destructive interference. The angle θ_2 at which this occurs is found by replacing $a/2$ in Equation 17.16 with $a/4$, leading to the condition $a\sin\theta_2 = 2\lambda$. This process can be continued, and we find that the general condition for complete destructive interference is

$$a\sin\theta_p = p\lambda \qquad p = 1, 2, 3, \ldots \qquad (17.17)$$

When $\theta_p \ll 1$ rad, which is almost always true for light waves, we can use the small-angle approximation to write

$$\theta_p = p\frac{\lambda}{a} \qquad p = 1, 2, 3, \ldots \tag{17.18}$$

Angles (in radians) of *dark* fringes in single-slit diffraction with slit width a

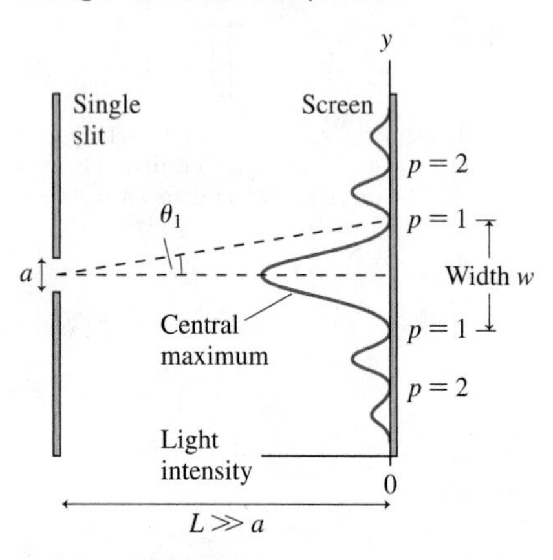

FIGURE 17.26 A graph of the intensity of a single-slit diffraction pattern.

Equation 17.18 gives the angles *in radians* to the dark minima in the diffraction pattern of a single slit. Notice that $p = 0$ is explicitly *excluded.* $p = 0$ corresponds to the straight-ahead position at $\theta = 0$, but you saw in Figures 17.5 and 17.25b that $\theta = 0$ is the central *maximum,* not a minimum.

NOTE ▸ Equations 17.17 and 17.18 are *mathematically* the same as the condition for the mth *maximum* of the double-slit interference pattern. But the physical meaning here is quite different. Equation 17.18 locates the *minima* (dark fringes) of the single-slit diffraction pattern. ◂

It is possible, although beyond the scope of this textbook, to calculate the entire light intensity pattern. The results of such a calculation are shown graphically in FIGURE 17.26. You can see the bright central maximum at $\theta = 0$, the weaker secondary maxima, and the dark points of destructive interference at the angles given by Equation 17.18. Compare this graph to the photograph of Figure 17.23 and make sure you see the agreement between the two.

The Width of a Single-Slit Diffraction Pattern

We'll find it useful, as we did for the double slit, to measure positions on the screen rather than angles. The position of the pth dark fringe, at angle θ_p, is $y_p = L\tan\theta_p$, where L is the distance from the slit to the viewing screen. Using Equation 17.18 for θ_p and the small-angle approximation $\tan\theta_p \approx \theta_p$, we find that the dark fringes in the single-slit diffraction pattern are located at

$$y_p = \frac{p\lambda L}{a} \qquad p = 1, 2, 3, \ldots \tag{17.19}$$

Positions of dark fringes for single-slit diffraction with screen distance L

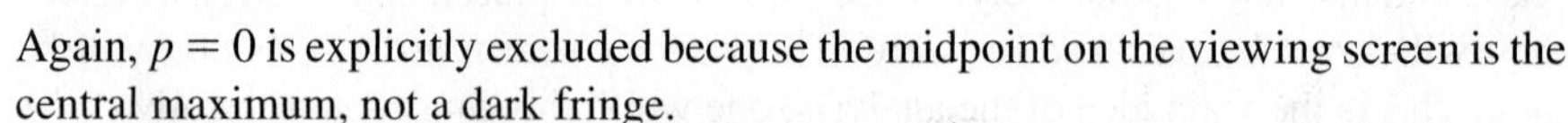

Again, $p = 0$ is explicitly excluded because the midpoint on the viewing screen is the central maximum, not a dark fringe.

A diffraction pattern is dominated by the central maximum, which is much brighter than the secondary maxima. The width w of the central maximum, shown in Figure 17.26, is defined as the distance between the two $p = 1$ minima on either side of the central maximum. Because the pattern is symmetrical, the width is simply $w = 2y_1$. This is

$$w = \frac{2\lambda L}{a} \tag{17.20}$$

Width of the central maximum for single-slit diffraction

The central maximum of this single-slit diffraction pattern appears white because it is overexposed. The width of the central maximum is clear.

An important implication of Equation 17.20 and one contrary to common sense, is that a narrower slit (smaller a) causes a *wider* diffraction pattern. **The smaller the opening a wave squeezes through, the *more* it spreads out on the other side.**

EXAMPLE 17.8 Finding the width of a slit

Light from a helium-neon laser ($\lambda = 633$ nm) passes through a narrow slit and is seen on a screen 2.0 m behind the slit. The first minimum in the diffraction pattern is 1.2 cm from the middle of the central maximum. How wide is the slit?

PREPARE The first minimum in a diffraction pattern corresponds to $p = 1$. The position of this minimum is given as $y_1 = 1.2$ cm. We can then use Equation 17.19 to find the slit width a.

SOLVE Equation 17.19 gives

$$a = \frac{p\lambda L}{y_p} = \frac{(1)(633 \times 10^{-9}\text{ m})(2.0\text{ m})}{0.012\text{ m}}$$
$$= 1.1 \times 10^{-4}\text{ m} = 0.11\text{ mm}$$

ASSESS This value is typical of the slit widths used to observe single-slit diffraction.

STOP TO THINK 17.6 The figure shows two single-slit diffraction patterns. The distance between the slit and the viewing screen is the same in both cases. Which of the following could be true?

A. The slits are the same for both; $\lambda_1 > \lambda_2$
B. The slits are the same for both; $\lambda_2 > \lambda_1$
C. The wavelengths are the same for both; $a_1 > a_2$
D. The wavelengths are the same for both; $a_2 > a_1$
E. The slits and the wavelengths are the same for both; $p_1 > p_2$
F. The slits and the wavelengths are the same for both; $p_2 > p_1$

17.6 Circular-Aperture Diffraction

Diffraction occurs if a wave passes through an opening of any shape. Diffraction by a single slit establishes the basic ideas of diffraction, but a common situation of practical importance is diffraction of a wave by a **circular aperture.** Circular diffraction is mathematically more complex than diffraction from a slit, and we will present results without derivation.

Consider some examples. A loudspeaker cone generates sound by the rapid oscillation of a diaphragm, but the sound wave must pass through the circular aperture defined by the outer edge of the speaker cone before it travels into the room beyond. This is diffraction by a circular aperture. Telescopes and microscopes are the reverse. Light waves from outside need to enter the instrument. To do so, they must pass through a circular lens. In fact, the performance limit of optical instruments is determined by the diffraction of the circular openings through which the waves must pass. This is an issue we'll look at more closely in Chapter 19.

FIGURE 17.27 shows a circular aperture of diameter D. Light waves passing through this aperture spread out to generate a *circular* diffraction pattern. You should compare this to Figure 17.23 for a single slit to note the similarities and differences. The diffraction pattern still has a *central maximum,* now circular, and it is surrounded by a series of secondary bright fringes. Most of the intensity is contained within the central maximum.

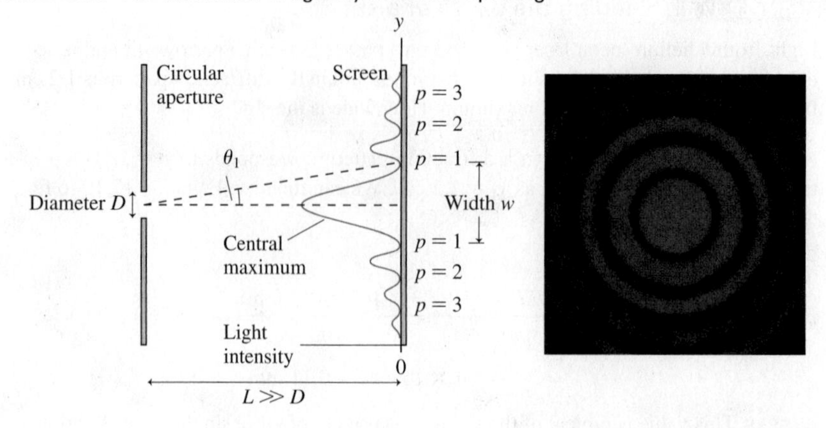

FIGURE 17.27 The diffraction of light by a circular opening.

Angle θ_1 locates the first minimum in the intensity, where there is perfect destructive interference. A mathematical analysis of circular diffraction finds that

$$\theta_1 = \frac{1.22\lambda}{D} \tag{17.21}$$

where D is the *diameter* of the circular opening. Equation 17.21 has assumed the small-angle approximation, which is almost always valid for the diffraction of light.

Within the small-angle approximation, the width of the central maximum on a screen a distance L from the aperture is

$$w = 2y_1 = 2L\tan\theta_1 \approx \frac{2.44\lambda L}{D} \tag{17.22}$$

Width of central maximum for diffraction from a circular aperture of diameter D

The diameter of the diffraction pattern increases with distance L, showing that light spreads out behind a circular aperture, but it decreases if the size D of the aperture is increased.

EXAMPLE 17.9 **Finding the right viewing distance**

Light from a helium-neon laser ($\lambda = 633$ nm) passes through a 0.50-mm-diameter hole. How far away should a viewing screen be placed to observe a diffraction pattern whose central maximum is 3.0 mm in diameter?

SOLVE Equation 17.22 gives us the appropriate screen distance:

$$L = \frac{wD}{2.44\lambda} = \frac{(3.0\times10^{-3}\text{ m})(5.0\times10^{-4}\text{ m})}{2.44(633\times10^{-9}\text{ m})} = 0.97\text{ m}$$

As we've seen, we need to use the wave model of light to understand the passage of light through narrow apertures, where "narrow" means comparable in size to the wavelength of the light. In the next two chapters, we'll consider the interaction of light with objects much larger than the wavelength. There, the ray model of light will be more appropriate for describing how light reflects from mirrors and refracts from lenses. But the wave model will reappear in Chapter 19 when we study telescopes and microscopes. We'll find that the *resolution* of these instruments has a fundamental limit set by the wave nature of light.

INTEGRATED EXAMPLE 17.10 Laser range finding

Scientists use *laser range finding* to measure the distance to the moon with great accuracy. A very brief (100 ps) laser pulse, with a wavelength of 532 nm, is fired at the moon, where it reflects off an array of 100 4.0-cm-diameter mirrors placed there by Apollo 14 astronauts in 1971. The reflected laser light returns to earth, where it is collected by a telescope and detected. The average earth-moon distance is 384,000 km.

The laser beam spreads out on its way to the moon because of diffraction, reaching the mirrors with an intensity of 300 W/m^2. The reflected beam spreads out even more on its way back because of diffraction due to the circular aperture of the mirrors.

a. What is the round-trip time for the laser pulse to travel to the moon and back?
b. If we want to measure the distance to the moon to an accuracy of 1.0 cm, how accurately must the arrival time of the returning pulse be measured?
c. Because of the spread of the beam due to diffraction, the light arriving at earth from one of the mirrors will be spread over a circular spot. Estimate the diameter of this spot.
d. What is the intensity of the laser beam when it arrives back at the earth?

PREPARE When the light reflects from one of the circular mirrors, diffraction causes it to spread out as shown in **FIGURE 17.28**. The width w of the central maximum for circular-aperture diffraction is given by Equation 17.22.

FIGURE 17.28 The geometry of the returning laser beam.

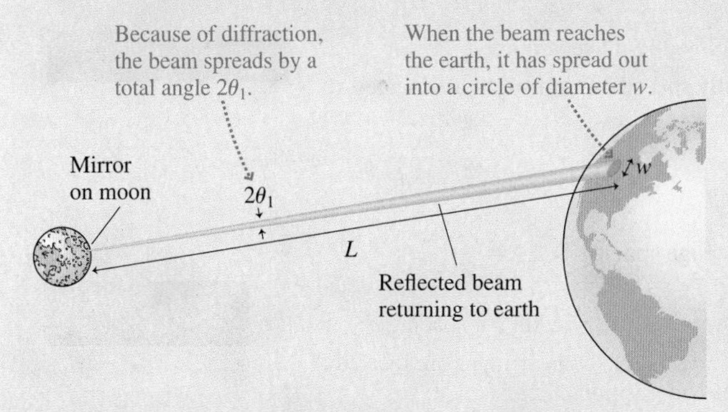
Because of diffraction, the beam spreads by a total angle $2\theta_1$.
When the beam reaches the earth, it has spread out into a circle of diameter w.
Mirror on moon
$2\theta_1$
w
L
Reflected beam returning to earth

Because we know the intensity of the laser beam as it strikes the mirrors, and we can easily find a mirror's area, we can use Equation 15.11 to calculate the power of the beam as it leaves a mirror. The intensity of the beam when it arrives back on earth will be much lower and can also be found from Equation 15.11.

SOLVE a. The round-trip distance is $2L$. Thus the round-trip travel time for the pulse, traveling at speed c, is

$$\Delta t = \frac{2L}{c} = \frac{2(3.84 \times 10^8 \text{ m})}{3.00 \times 10^8 \text{ m/s}} = 2.56 \text{ s}$$

b. If we wish to measure the moon's distance from the earth to an accuracy of ± 1.0 cm, then, because the laser beam travels both to and from the moon, we need to know the round-trip distance to an accuracy of ± 2.0 cm. The time it takes light to travel $\Delta x = 2.0$ cm is

$$\Delta t = \frac{\Delta x}{c} = \frac{0.020 \text{ m}}{3.00 \times 10^8 \text{ m/s}} = 6.6 \times 10^{-11} = 66 \text{ ps}$$

Thus the arrival time of the pulses must be timed to an accuracy of about 70 ps.

c. The light arriving at the moon reflects from circular mirrors of diameter D. Diffraction by these circular apertures causes the returning light to spread out with angular width $2\theta_1$, where θ_1 is the angle of the first minimum on either side of the central maximum. Equation 17.22 found that the width of the central maximum—the diameter of the circular spot of light when it reaches the earth—is

$$w = \frac{2.44\lambda L}{D} = \frac{2.44(532 \times 10^{-9} \text{ m})(3.84 \times 10^8 \text{ m})}{0.040 \text{ m}}$$
$$= 12{,}000 \text{ m}$$

d. The light is reflected from circular mirrors of radius $r = D/2$ and area $a = \pi r^2 = \pi D^2/4$. The power reflected from one mirror is, according to Equation 15.11,

$$P = Ia = I\pi\frac{D^2}{4} = (300 \text{ W/m}^2)\pi\frac{(0.040 \text{ m})^2}{4} = 0.38 \text{ W}$$

When the pulse returns to earth, it is now spread over the large area $\pi w^2/4$. Thus the intensity at the earth's surface from one mirror reflection is

$$I_1 = \frac{P}{\pi\frac{w^2}{4}} = \frac{0.38 \text{ W}}{\pi\frac{(12{,}000 \text{ m})^2}{4}} = 3.3 \times 10^{-9} \text{ W/m}^2$$

There are 100 reflecting mirrors in the array, so the total intensity reaching earth is $I = 100I_1 = 3.3 \times 10^{-7}$ W/m^2. A large telescope is needed to detect this very small intensity.

ASSESS A 1.0-m-diameter telescope collects only a few *photons,* or particles of light, in each returning pulse. Despite the difficult challenge of detecting this very weak signal, the accuracy of these measurements is astounding: The latest experiments can measure the instantaneous distance to the moon to ± 1 mm, a precision of 3 parts in a trillion!

SUMMARY

Goal: To understand and apply the wave model of light.

GENERAL PRINCIPLES

The Wave Model

The wave model considers light to be a wave propagating through space. Interference and diffraction are important. The wave model is appropriate when light interacts with objects whose size is comparable to the wavelength of light, or roughly less than about 0.1 mm.

Huygens' principle says that each point on a wave front is the source of a spherical wavelet. The wave front at a later time is tangent to all the wavelets.

IMPORTANT CONCEPTS

The **index of refraction** of a material determines the speed of light in that material: $v = c/n$. The index of refraction of a material is always greater than 1, so that v is always less than c.

The wavelength λ in a material with index of refraction n is *shorter* than the wavelength λ_{vac} in a vacuum: $\lambda = \lambda_{vac}/n$.

The *frequency* of light does not change as it moves from one material to another.

Diffraction is the spreading of a wave after it passes through an opening.

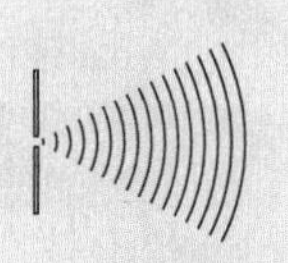

Constructive and destructive **interference** are due to the overlap of two or more waves as they spread behind openings.

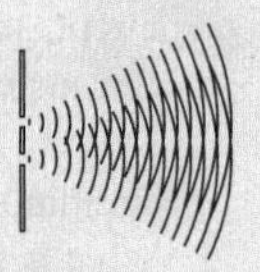

APPLICATIONS

Diffraction from a single slit

A single slit of width a has a bright **central maximum** of width

$$w = \frac{2\lambda L}{a}$$

that is flanked by weaker **secondary maxima.**

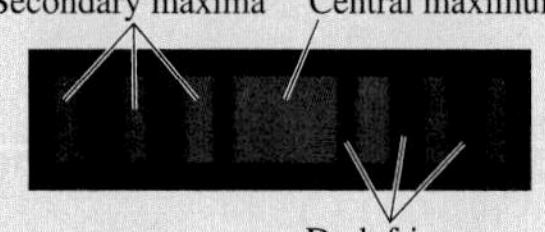

Dark fringes are located at angles such that

$$a\sin\theta_p = p\lambda \qquad p = 1, 2, 3, \ldots$$

If $\lambda/a \ll 1$, then from the small-angle approximation,

$$\theta_p = \frac{p\lambda}{a} \qquad y_p = \frac{p\lambda L}{a}$$

Circular aperture of diameter D

A bright central maximum of diameter

$$w = \frac{2.44\lambda L}{D}$$

is surrounded by circular secondary maxima. The first dark fringe is located at

$$\theta_1 = \frac{1.22\lambda}{D} \qquad y_1 = \frac{1.22\lambda L}{D}$$

For an aperture of any shape, a smaller opening causes a greater spreading of the wave behind the opening.

Interference from multiple slits

Waves overlap as they spread out behind slits. Bright fringes are seen on the viewing screen at positions where the path-length difference Δr between successive slits is equal to $m\lambda$, where m is an integer.

Double slit with separation d

Equally spaced bright fringes are located at

$$\theta_m = \frac{m\lambda}{d} \qquad y_m = \frac{m\lambda L}{d} \qquad m = 0, 1, 2, \ldots$$

The **fringe spacing** is $\Delta y = \dfrac{\lambda L}{d}$

Diffraction grating with slit spacing d

Very bright and narrow fringes are located at angles and positions

$$d\sin\theta_m = m\lambda \qquad y_m = L\tan\theta_m$$

Thin-film interference

Interference occurs between the waves reflected from the two surfaces of a thin film with index of refraction n. A wave that reflects from a surface at which the index of refraction increases has a phase change.

Interference	0 or 2 phase changes	1 phase change
Constructive	$2t = m\dfrac{\lambda}{n}$	$2t = \left(m + \dfrac{1}{2}\right)\dfrac{\lambda}{n}$
Destructive	$2t = \left(m + \dfrac{1}{2}\right)\dfrac{\lambda}{n}$	$2t = m\dfrac{\lambda}{n}$

Problem difficulty is labeled as | (straightforward) to ||||| (challenging). Problems labeled INT integrate significant material from earlier chapters; BIO are of biological or medical interest.

For assigned homework and other learning materials, go to MasteringPhysics®

Scan this QR code to launch a Video Tutor Solution that will help you solve problems for this chapter.

QUESTIONS

Conceptual Questions

1. The frequency of a light wave in air is 5.3×10^{14} Hz. Is the frequency of this wave higher, lower, or the same after the light enters a piece of glass?
2. Rank in order the following according to their speeds, from slowest to fastest: (i) 425-nm-wavelength light through a pane of glass, (ii) 500-nm-wavelength light through air, (iii) 540-nm-wavelength light through water, (iv) 670-nm-wavelength light through a diamond, and (v) 670-nm-wavelength light through a vacuum.
3. The wavelength of a light wave is 700 nm in air; this light appears red. If this wave enters a pool of water, its wavelength becomes $\lambda_{air}/n = 530$ nm. If you were swimming underwater, the light would still appear red. Given this, what property of a wave determines its color?
4. A double-slit interference experiment shows fringes on a screen. The entire experiment is then immersed in water. Do the fringes on the screen get closer together, farther apart, remain the same, or disappear entirely? Explain.
5. Figure Q17.5 shows the fringes observed in a double-slit interference experiment when the two slits are illuminated by white light. The central maximum is white, but as we move away from the central maximum, the fringes become less distinct and more colorful. What is special about the central maximum that makes it white? Explain the presence of colors in the outlying fringes.

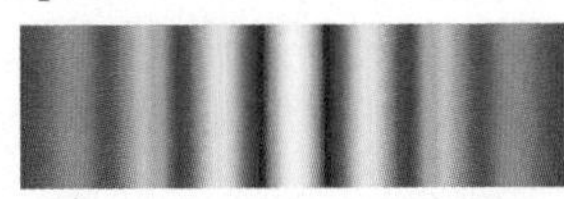

FIGURE Q17.5

6. In a double-slit interference experiment, interference fringes are observed on a distant screen. The width of both slits is then doubled without changing the distance between their centers.
   a. What happens to the spacing of the fringes? Explain.
   b. What happens to the intensity of the bright fringes? Explain.
7. Figure Q17.7 shows the viewing screen in a double-slit experiment with monochromatic light. Fringe C is the central maximum.
   a. What will happen to the fringe spacing if the wavelength of the light is decreased?
   b. What will happen to the fringe spacing if the spacing between the slits is decreased?
   c. What will happen to the fringe spacing if the distance to the screen is decreased?
   d. Suppose the wavelength of the light is 500 nm. How much farther is it from the dot on the screen in the center of fringe E to the left slit than it is from the dot to the right slit?

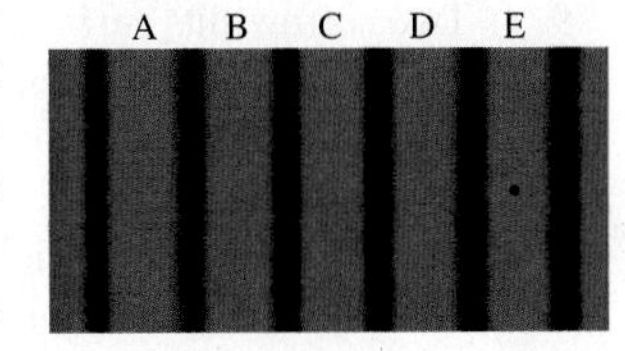

FIGURE Q17.7

8. Figure Q17.7 is the interference pattern seen on a viewing screen behind 2 slits. Suppose the 2 slits were replaced by 20 slits having the same spacing d between adjacent slits.
   a. Would the number of fringes on the screen increase, decrease, or stay the same?
   b. Would the fringe spacing increase, decrease, or stay the same?
   c. Would the width of each fringe increase, decrease, or stay the same?
   d. Would the brightness of each fringe increase, decrease, or stay the same?
9. Figure Q17.9 shows the light intensity on a viewing screen behind a single slit of width a. The light's wavelength is λ. Is $\lambda < a$, $\lambda = a$, $\lambda > a$, or is it not possible to tell? Explain.

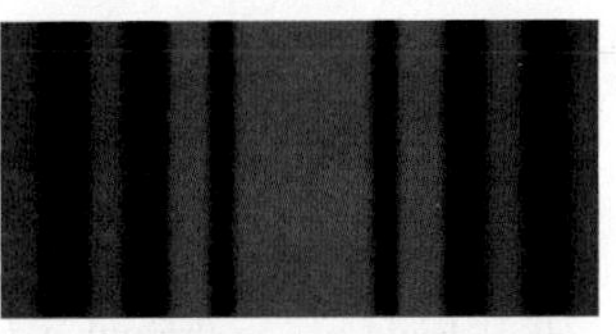

FIGURE Q17.9

10. Figure Q17.10 shows the light intensity on a viewing screen behind a circular aperture. What happens to the width of the central maximum if
    a. The wavelength is increased?
    b. The diameter of the aperture is increased?
    c. How will the screen appear if the aperture diameter is less than the light wavelength?

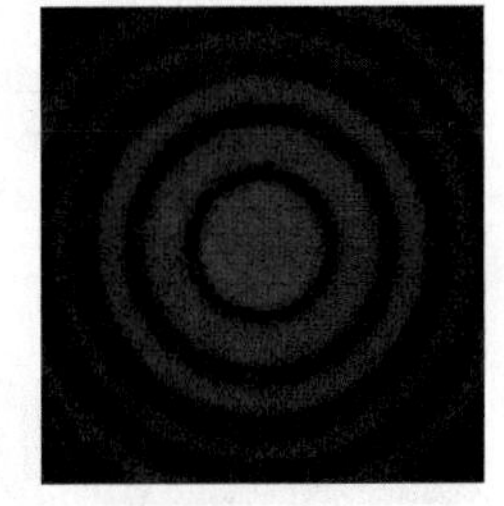

FIGURE Q17.10

11. Light with a wavelength of 600 nm is incident on a diffraction grating that has 100 slits. The first-order maximum is observed at a point P on a distant screen. How much farther does the light travel from the first slit of the grating to point P than does the light from the 100th slit?
12. White light is incident on a diffraction grating. What color is the central maximum of the interference pattern?
13. Figure Q17.13 shows a light wave incident on and passing through a thin soap film. Reflections from the front and back surfaces of the film create smaller waves (not shown in the figure) that travel to the left of the film, where they interfere. Is the interference constructive, destructive, or something in between? Explain.

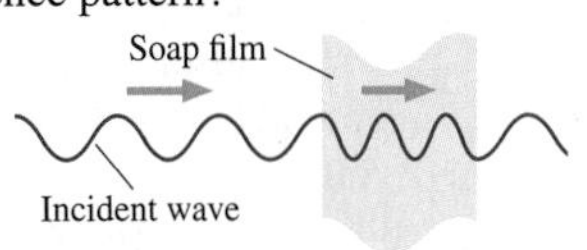

FIGURE Q17.13

14. A soap bubble usually pops because some part of it becomes too thin due to evaporation or drainage of fluid. The change in thickness also changes the color of light the bubble reflects. Why?
15. An oil film on top of water has one patch that is much thinner than the wavelength of visible light. The index of refraction of the oil is less than that of water. Will the reflection from that extremely thin part of the film be bright or dark? Explain.

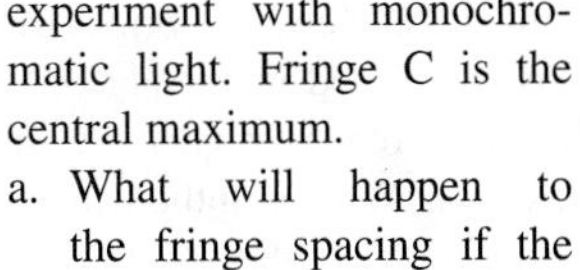

16. Should the antireflection coating of a microscope objective lens designed for use with ultraviolet light be thinner, thicker, or the same thickness as the coating on a lens designed for visible light?
17. Example 17.5 showed that a thin film whose thickness is one-quarter of the wavelength of light in the film serves as an antireflection coating when coated on glass. In Example 17.5, $n_{film} < n_{glass}$. If a quarter-wave thickness film with $n_{film} > n_{glass}$ were used instead, would the film still serve as an antireflection coating? Explain.
18. You are standing against the wall near a corner of a large building. A friend is standing against the wall that is around the corner from you. You can't see your friend. How is it that you can hear her when she talks to you?

Multiple-Choice Questions

19. | Light of wavelength 500 nm in air enters a glass block with index of refraction $n = 1.5$. When the light enters the block, which of the following properties of the light will not change?
    A. The speed of the light
    B. The frequency of the light
    C. The wavelength of the light
20. | The frequency of a light wave in air is 4.6×10^{14} Hz. What is the wavelength of this wave after it enters a pool of water?
    A. 300 nm B. 490 nm C. 650 nm D. 870 nm
21. | Light passes through a diffraction grating with a slit spacing of 0.001 mm. A viewing screen is 100 cm behind the grating. If the light is blue, with a wavelength of 450 nm, at about what distance from the center of the interference pattern will the first-order maximum appear?
    A. 5 cm B. 25 cm C. 50 cm D. 100 cm
22. || Blue light of wavelength 450 nm passes through a diffraction grating with a slit spacing of 0.001 mm and makes an interference pattern on the wall. How many bright fringes will be seen?
    A. 1 B. 3 C. 5 D. 7
23. | Yellow light of wavelength 590 nm passes through a diffraction grating and makes an interference pattern on a screen 80 cm away. The first bright fringes are 1.9 cm from the central maximum. How many lines per mm does this grating have?
    A. 20 B. 40 C. 80 D. 200
24. | Light passes through a 10-μm-wide slit and is viewed on a screen 1 m behind the slit. If the width of the slit is narrowed, the band of light on the screen will
    A. Become narrower.
    B. Become wider.
    C. Stay about the same.
25. || Reflected light from a thin film of oil gives constructive interference for light with a wavelength inside the film of λ_{film}. By how much would the film thickness need to be increased to give destructive interference?
    A. $2\lambda_{film}$ B. λ_{film} C. $\lambda_{film}/2$ D. $\lambda_{film}/4$
26. | You want to estimate the diameter of a very small circular pinhole that you've made in a piece of aluminum foil. To do so, you shine a red laser pointer ($\lambda = 632$ nm) at the hole and observe the diffraction pattern on a screen 3.5 m behind the foil. You measure the width of the central maximum to be 15 mm. What is the diameter of the hole?
    A. 0.18 mm B. 0.29 mm C. 0.36 mm D. 1.1 mm

PROBLEMS

Section 17.1 What Is Light?

1. ||| a. How long does it take light to travel through a 3.0-mm-thick piece of window glass?
   b. Through what thickness of water could light travel in the same amount of time?
2. | a. How long (in ns) does it take light to travel 1.0 m in a vacuum?
   b. What distance does light travel in water, glass, and diamond during the time that it travels 1.0 m in a vacuum?
3. |||| A 5.0-cm-thick layer of oil ($n = 1.46$) is sandwiched between a 1.0-cm-thick sheet of glass and a 2.0-cm-thick sheet of polystyrene plastic ($n = 1.59$). How long (in ns) does it take light incident perpendicular to the glass to pass through this 8.0-cm-thick sandwich?
4. || A light wave has a 670 nm wavelength in air. Its wavelength in a transparent solid is 420 nm.
   a. What is the speed of light in this solid?
   b. What is the light's frequency in the solid?
5. || How much time does it take a pulse of light to travel through 150 m of water?
6. || A helium-neon laser beam has a wavelength in air of 633 nm. It takes 1.38 ns for the light to travel through 30.0 cm of an unknown liquid. What is the wavelength of the laser beam in the liquid?

Section 17.2 The Interference of Light

7. || Two narrow slits 50 μm apart are illuminated with light of wavelength 500 nm. What is the angle of the $m = 2$ bright fringe in radians? In degrees?
8. ||| Light from a sodium lamp ($\lambda = 589$ nm) illuminates two narrow slits. The fringe spacing on a screen 150 cm behind the slits is 4.0 mm. What is the spacing (in mm) between the two slits?
9. || Two narrow slits are illuminated by light of wavelength λ. The slits are spaced 20 wavelengths apart. What is the angle, in radians, between the central maximum and the $m = 1$ bright fringe?
10. || A double-slit experiment is performed with light of wavelength 600 nm. The bright interference fringes are spaced 1.8 mm apart on the viewing screen. What will the fringe spacing be if the light is changed to a wavelength of 400 nm?
11. ||||| Light from a helium-neon laser ($\lambda = 633$ nm) is used to illuminate two narrow slits. The interference pattern is observed on a screen 3.0 m behind the slits. Eleven bright fringes are seen, spanning a distance of 52 mm. What is the spacing (in mm) between the slits?
12. || Two narrow slits are 0.12 mm apart. Light of wavelength 550 nm illuminates the slits, causing an interference pattern on a screen 1.0 m away. Light from each slit travels to the $m = 1$ maximum on the right side of the central maximum. How much farther did the light from the left slit travel than the light from the right slit?

13. ||| In a double-slit experiment, the distance from one slit to the $m = 3$ bright fringe is $2{,}000{,}198.2\lambda$. What is the distance from this bright fringe to the other, more distant slit?

Section 17.3 The Diffraction Grating

14. ||| A diffraction grating with 750 slits/mm is illuminated by light that gives a first-order diffraction angle of 34.0°. What is the wavelength of the light?
15. |||| A 1.0-cm-wide diffraction grating has 1000 slits. It is illuminated by light of wavelength 550 nm. What are the angles of the first two diffraction orders?
16. ||| Light of wavelength 600 nm illuminates a diffraction grating. The second-order maximum is at angle 39.5°. How many lines per millimeter does this grating have?
17. ||| A lab technician uses laser light with a wavelength of 670 nm to test a diffraction grating. When the grating is 40.0 cm from the screen, the first-order maxima appear 6.00 cm from the center of the pattern. How many lines per millimeter does this grating have?
18. || The human eye can readily detect wavelengths from about 400 nm to 700 nm. If white light illuminates a diffraction grating having 750 lines/mm, over what range of angles does the visible $m = 1$ spectrum extend?
19. ||| A diffraction grating with 600 lines/mm is illuminated with light of wavelength 500 nm. A very wide viewing screen is 2.0 m behind the grating.
    a. What is the distance between the two $m = 1$ fringes?
    b. How many bright fringes can be seen on the screen?
20. || A 500 line/mm diffraction grating is illuminated by light of wavelength 510 nm. How many diffraction orders are seen, and what is the angle of each?

Section 17.4 Thin-Film Interference

21. || What is the thinnest film of MgF_2 ($n = 1.38$) on glass that produces a strong reflection for orange light with a wavelength of 600 nm?
22. |||| A very thin oil film ($n = 1.25$) floats on water ($n = 1.33$). What is the thinnest film that produces a strong reflection for green light with a wavelength of 500 nm?
23. || A film with $n = 1.60$ is deposited on glass. What is the thinnest film that will produce constructive interference in the reflection of light with a wavelength of 550 nm?
24. || Antireflection coatings can be used on the *inner* surfaces of BIO eyeglasses to reduce the reflection of stray light into the eye, thus reducing eyestrain.
    a. A 90-nm-thick coating is applied to the lens. What must be the coating's index of refraction to be most effective at 480 nm? Assume that the coating's index of refraction is less than that of the lens.
    b. If the index of refraction of the coating is 1.38, what thickness should the coating be so as to be most effective at 480 nm? The thinnest possible coating is best.
25. || Solar cells are given antireflection coatings to maximize their efficiency. Consider a silicon solar cell ($n = 3.50$) coated with a layer of silicon dioxide ($n = 1.45$). What is the minimum coating thickness that will minimize the reflection at the wavelength of 700 nm, where solar cells are most efficient?
26. ||| A thin film of MgF_2 ($n = 1.38$) coats a piece of glass. Constructive interference is observed for the reflection of light with wavelengths of 500 nm and 625 nm. What is the thinnest film for which this can occur?
27. | Looking straight downward into a rain puddle whose surface is covered with a thin film of gasoline, you notice a swirling pattern of colors caused by interference inside the gasoline film. The point directly beneath you is colored a beautiful iridescent green. You happen to remember that the index of refraction of gasoline is 1.38 and that the wavelength of green light is about 540 nm. What is the minimum possible thickness of the gasoline layer directly beneath you?

Section 17.5 Single-Slit Diffraction

28. || A helium-neon laser ($\lambda = 633$ nm) illuminates a single slit and is observed on a screen 1.50 m behind the slit. The distance between the first and second minima in the diffraction pattern is 4.75 mm. What is the width (in mm) of the slit?
29. ||| For a demonstration, a professor uses a razor blade to cut a thin slit in a piece of aluminum foil. When she shines a laser pointer ($\lambda = 680$ nm) through the slit onto a screen 5.5 m away, a diffraction pattern appears. The bright band in the center of the pattern is 8.0 cm wide. What is the width of the slit?
30. || A 0.50-mm-wide slit is illuminated by light of wavelength 500 nm. What is the width of the central maximum on a screen 2.0 m behind the slit?
31. ||| The second minimum in the diffraction pattern of a 0.10-mm-wide slit occurs at 0.70°. What is the wavelength of the light?
32. || What is the width of a slit for which the first minimum is at 45° when the slit is illuminated by a helium-neon laser ($\lambda = 633$ nm)?
    Hint: The small-angle approximation is not valid at 45°.

Section 17.6 Circular-Aperture Diffraction

33. ||| A 0.50-mm-diameter hole is illuminated by light of wavelength 500 nm. What is the width of the central maximum on a screen 2.0 m behind the slit?
34. || Light from a helium-neon laser ($\lambda = 633$ nm) passes through a circular aperture and is observed on a screen 4.0 m behind the aperture. The width of the central maximum is 2.5 cm. What is the diameter (in mm) of the hole?
35. ||| You want to photograph a circular diffraction pattern whose central maximum has a diameter of 1.0 cm. You have a helium-neon laser ($\lambda = 633$ nm) and a 0.12-mm-diameter pinhole. How far behind the pinhole should you place the viewing screen?
36. || Infrared light of wavelength 2.5 μm illuminates a 0.20-mm-diameter hole. What is the angle of the first dark fringe in radians? In degrees?

General Problems

37. ||| An advanced computer sends information to its various parts via infrared light pulses traveling through silicon fibers ($n = 3.50$). To acquire data from memory, the central processing unit sends a light-pulse request to the memory unit. The memory unit processes the request, then sends a data pulse back to the central processing unit. The memory unit takes 0.50 ns to process a request. If the information has to be obtained from memory in 2.00 ns, what is the maximum distance the memory unit can be from the central processing unit?

Watch video solutions for problems: 17.39  17.53 

38. |||| Figure P17.38 shows the light intensity on a screen behind a double slit. The slit spacing is 0.20 mm and the wavelength of the light is 600 nm. What is the distance from the slits to the screen?

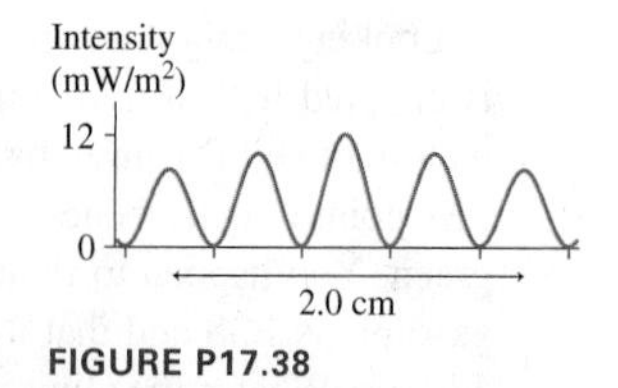

FIGURE P17.38

39. ||||| Figure P17.38 shows the light intensity on a screen behind a double slit. The slit spacing is 0.20 mm and the screen is 2.0 m behind the slits. What is the wavelength of the light?

40. || Your friend has been given a laser for her birthday. Unfortunately, she did not receive a manual with it and so she doesn't know the wavelength that it emits. You help her by performing a double-slit experiment, with slits separated by 0.18 mm. You find that two adjacent bright fringes are 5.5 mm apart on a screen 1.6 m from the slits. What is the wavelength the laser emits?

41. | A double slit is illuminated simultaneously with orange light of wavelength 600 nm and light of an unknown wavelength. The $m = 4$ bright fringe of the unknown wavelength overlaps the $m = 3$ bright orange fringe. What is the unknown wavelength?

42. || Figure P17.42 shows the light intensity on a screen 2.5 m behind a double slit. The wavelength of the light is 532 nm. What is the spacing between the slits?

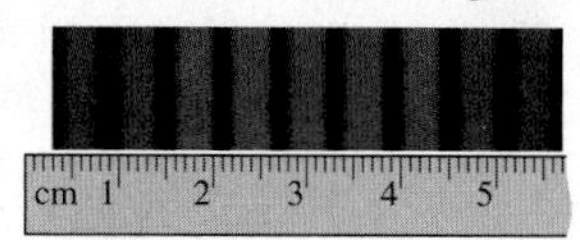

FIGURE P17.42

43. | A laser beam of wavelength 670 nm shines through a diffraction grating that has 750 lines/mm. Sketch the pattern that appears on a screen 1.0 m behind the grating, noting distances on your drawing and explaining where these numbers come from.

44. ||| The two most prominent wavelengths in the light emitted by a hydrogen discharge lamp are 656 nm (red) and 486 nm (blue). Light from a hydrogen lamp illuminates a diffraction grating with 500 lines/mm, and the light is observed on a screen 1.50 m behind the grating. What is the distance between the first-order red and blue fringes?

45. ||| A diffraction grating produces a first-order maximum at an angle of 20.0°. What is the angle of the second-order maximum?

46. | A diffraction grating is illuminated simultaneously with red light of wavelength 660 nm and light of an unknown wavelength. The fifth-order maximum of the unknown wavelength exactly overlaps the third-order maximum of the red light. What is the unknown wavelength?

47. |||| White light (400–700 nm) is incident on a 600 line/mm diffraction grating. What is the width of the first-order rainbow on a screen 2.0 m behind the grating?

48. |||| A miniature spectrometer used for chemical analysis has a diffraction grating with 800 slits/mm set 25.0 mm in front of the detector "screen." The detector can barely distinguish two bright lines that are 30 μm apart in the first-order spectrum. What is the *resolution* of the spectrometer at a wavelength of 600 nm? That is, if two distinct wavelengths can barely be distinguished, one of them being 600.0 nm, what is the wavelength difference $\Delta\lambda$ between the two?

49. ||| Figure P17.49 shows the interference pattern on a screen 1.0 m behind an 800 line/mm diffraction grating. What is the wavelength of the light?

Intensity

89.7 cm 43.6 cm 43.6 cm 89.7 cm

FIGURE P17.49

50. ||| Figure P17.49 shows the interference pattern on a screen 1.0 m behind a diffraction grating. The wavelength of the light is 600 nm. How many lines per millimeter does the grating have?

51. || Because sound is a wave, it is possible to make a diffraction INT grating for sound from a large board with several parallel slots for the sound to go through. When 10 kHz sound waves pass through such a grating, listeners 10 m from the grating report "loud spots" 1.4 m on both sides of center. What is the spacing between the slots? Use 340 m/s for the speed of sound.

52. |||| The shiny surface of a CD is imprinted with millions of tiny pits, arranged in a pattern of thousands of essentially concentric circles that act like a reflection grating when light shines on them. You decide to determine the distance between those circles by aiming a laser pointer (with $\lambda = 680$ nm) perpendicular to the disk and measuring the diffraction pattern reflected onto a screen 1.5 m from the disk. The central bright spot you expected to see is blocked by the laser pointer itself. You do find two other bright spots separated by 1.4 m, one on either side of the missing central spot. The rest of the pattern is apparently diffracted at angles too great to show on your screen. What is the distance between the circles on the CD's surface?

53. ||| If sunlight shines straight onto a peacock feather, the feather BIO appears bright blue when viewed from 15° on either side of the incident beam of sunlight. The blue color is due to diffraction from the melanin bands in the feather barbules, as was shown in the photograph on page 549. Blue light with a wavelength of 470 nm is diffracted at 15° by these bands (this is the first-order diffraction) while other wavelengths in the sunlight are diffracted at different angles. What is the spacing of the melanin bands in the feather?

54. ||| The wings of some beetles BIO have closely spaced parallel lines of melanin, causing the wing to act as a reflection grating. Suppose sunlight shines straight onto a beetle wing. If the melanin lines on the wing are spaced 2.0 μm apart, what is the first-order diffraction angle for green light ($\lambda = 550$ nm)?

55. |||| A diffraction grating having 500 lines/mm diffracts visible light at 30°. What is the light's wavelength?

56. ||| Light emitted by element X passes through a diffraction grating that has 1200 slits/mm. The interference pattern is observed on a screen 75.0 cm behind the grating. First-order maxima are observed at distances of 56.2 cm, 65.9 cm, and 93.5 cm from the central maximum. What are the wavelengths of light emitted by element X?

57. |||| Light of a single wavelength is incident on a diffraction grating with 500 slits/mm. Several bright fringes are observed on a screen behind the grating, including one at 45.7° and one next to it at 72.6°. What is the wavelength of the light?

58. || A sheet of glass is coated with a 500-nm-thick layer of oil ($n = 1.42$).
    a. For what *visible* wavelengths of light do the reflected waves interfere constructively?
    b. For what *visible* wavelengths of light do the reflected waves interfere destructively?
    c. What is the color of reflected light? What is the color of transmitted light?

59. | A soap bubble is essentially a thin film of water surrounded by air. The colors you see in soap bubbles are produced by interference. What visible wavelengths of light are strongly reflected from a 390-nm-thick soap bubble? What color would such a soap bubble appear to be?
60. |||| A laboratory dish, 20 cm in diameter, is half filled with INT water. One at a time, 0.50 μL drops of oil from a micropipette are dropped onto the surface of the water, where they spread out into a uniform thin film. After the first drop is added, the intensity of 600 nm light reflected from the surface is very low. As more drops are added, the reflected intensity increases, then decreases again to a minimum after a total of 13 drops have been added. What is the index of refraction of the oil?
61. ||| You need to use your cell phone, which broadcasts an 830 MHz signal, but you're in an alley between two massive, radio-wave-absorbing buildings that have only a 15 m space between them. What is the angular width, in degrees, of the electromagnetic wave after it emerges from between the buildings?
62. ||| Light from a sodium lamp ($\lambda = 589$ nm) illuminates a narrow slit and is observed on a screen 75 cm behind the slit. The distance between the first and third dark fringes is 7.5 mm. What is the width (in mm) of the slit?
63. |||| The opening to a cave is a tall, 30-cm-wide crack. A bat that INT is preparing to leave the cave emits a 30 kHz ultrasonic chirp. How wide is the "sound beam" 100 m outside the cave opening? Use $v_{sound} = 340$ m/s.
64. || A diffraction grating has 500 slits/mm. What is the longest wavelength of light for which there will be a third-order maximum?
65. ||| Figure P17.65 shows the light intensity on a screen behind a single slit. The wavelength of the light is 500 nm and the screen is 1.0 m behind the slit. What is the width (in mm) of the slit?

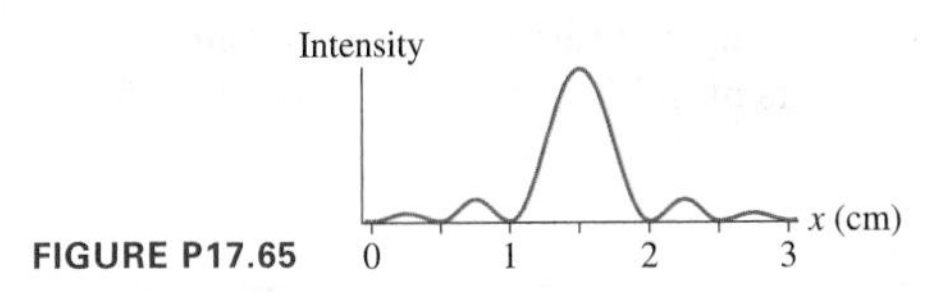

FIGURE P17.65

66. ||| Figure P17.65 shows the light intensity on a screen behind a single slit. The wavelength of the light is 600 nm and the slit width is 0.15 mm. What is the distance from the slit to the screen?
67. || Figure P17.67 shows the light intensity on a screen 2.5 m behind an aperture. The aperture is illuminated with light of wavelength 600 nm.
    a. Is the aperture a single slit or a double slit? Explain.
    b. If the aperture is a single slit, what is its width? If it is a double slit, what is the spacing between the slits?

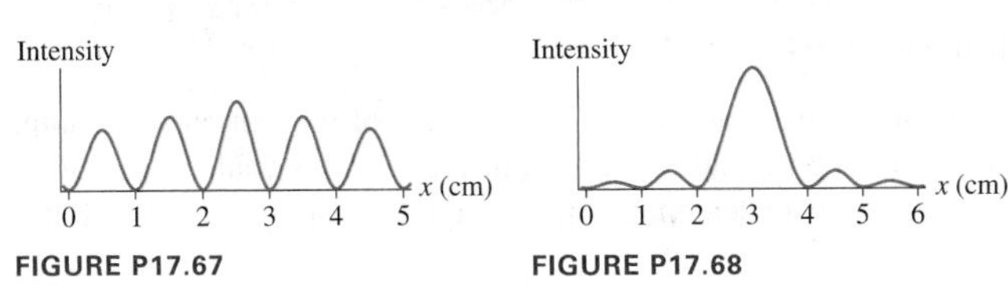

FIGURE P17.67 **FIGURE P17.68**

68. || Figure P17.68 shows the light intensity on a screen 2.5 m behind an aperture. The aperture is illuminated with light of wavelength 600 nm.
    a. Is the aperture a single slit or a double slit? Explain.
    b. If the aperture is a single slit, what is its width? If it is a double slit, what is the spacing between the slits?
69. ||| One day, after pulling down your window shade, you notice that sunlight is passing through a pinhole in the shade and making a small patch of light on the far wall. Having recently studied optics in your physics class, you're not too surprised to see that the patch of light seems to be a circular diffraction pattern. It appears that the central maximum is about 3 cm across, and you estimate that the distance from the window shade to the wall is about 3 m. Knowing that the average wavelength of sunlight is about 500 nm, estimate the diameter of the pinhole.
70. |||| A radar for tracking aircraft broadcasts a 12 GHz microwave INT beam from a 2.0-m-diameter circular radar antenna. From a wave perspective, the antenna is a circular aperture through which the microwaves diffract.
    a. What is the diameter of the radar beam at a distance of 30 km?
    b. If the antenna emits 100 kW of power, what is the average microwave intensity at 30 km?
71. ||||| A helium-neon laser ($\lambda = 633$ nm), shown in Figure P17.71, is built with a glass tube of inside diameter 1.0 mm. One mirror is partially transmitting to allow the laser beam out. An electrical discharge in the tube causes it to glow like a neon light. From an optical perspective, the laser beam is a light wave that diffracts out through a 1.0-mm-diameter circular opening.
    a. Explain why a laser beam can't be *perfectly* parallel, with no spreading.
    b. The angle θ_1 to the first minimum is called the *divergence angle* of a laser beam. What is the divergence angle of this laser beam?
    c. What is the diameter (in mm) of the laser beam after it travels 3.0 m?
    d. What is the diameter of the laser beam after it travels 1.0 km?

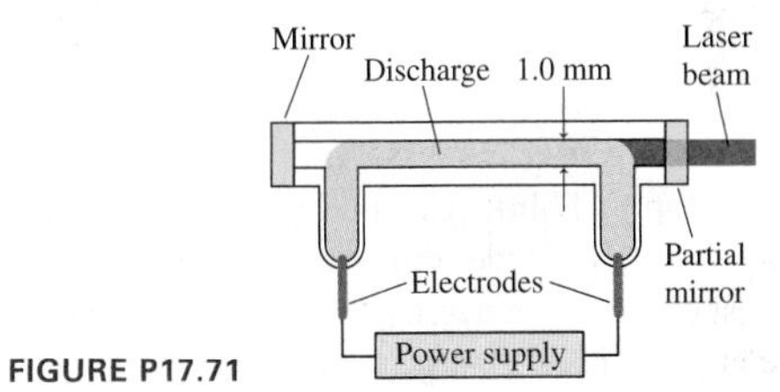

FIGURE P17.71

72. || In the laser range-finding experiments of Example 17.10, the laser beam fired toward the moon spreads out as it travels because it diffracts through a circular exit as it leaves the laser. In order for the reflected light to be bright enough to detect, the laser spot on the moon must be no more than 1 km in diameter. Staying within this diameter is accomplished by using a special large-diameter laser. If $\lambda = 532$ nm, what is the minimum diameter of the circular opening from which the laser beam emerges? The earth-moon distance is 384,000 km.

MCAT-Style Passage Problems

The Blue Morpho Butterfly BIO

The brilliant blue color of a blue morpho butterfly is, like the colors of peacock feathers, due to interference. Figure P17.73a shows an easy way to demonstrate this: If a drop of the clear solvent acetone is placed on the wing of a blue morpho butterfly, the color changes from a brilliant blue to an equally brilliant green—returning to blue once the acetone evaporates. There would be no change if the color were due to pigment.

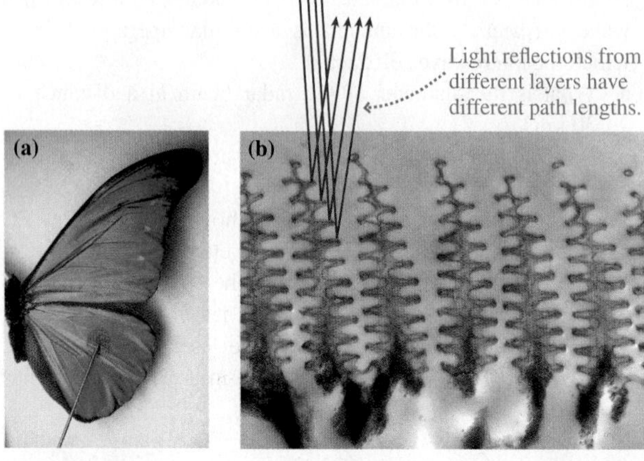

FIGURE P17.73

A cross section of a scale from the wing of a blue morpho butterfly reveals the source of the butterfly's color. As Figure P17.73b shows, the scales are covered with structures that look like small Christmas trees. Light striking the wings reflects from different layers of these structures, and the differing path lengths cause the reflected light to interfere constructively or destructively, depending on the wavelength. For light at normal incidence, blue light experiences constructive interference while other colors undergo destructive interference and cancel. Acetone fills the spaces in the scales with a fluid of index of refraction $n = 1.38$; this changes the conditions for constructive interference and results in a change in color.

73. | The coloring of the blue morpho butterfly is protective. As the butterfly flaps its wings, the angle at which light strikes the wings changes. This causes the butterfly's color to change and makes it difficult for a predator to follow. This color change is because
    A. A diffraction pattern appears only at certain angles.
    B. The index of refraction of the wing tissues changes as the wing flexes.
    C. The motion of the wings causes a Doppler shift in the reflected light.
    D. As the angle changes, the differences in paths among light reflected from different surfaces change, resulting in constructive interference for a different color.
74. | The change in color when acetone is placed on the wing is due to the difference between the indices of refraction of acetone and air. Consider light of some particular color. In acetone,
    A. The frequency of the light is less than in air.
    B. The frequency of the light is greater than in air.
    C. The wavelength of the light is less than in air.
    D. The wavelength of the light is greater than in air.
75. | The scales on the butterfly wings are actually made of a transparent material with index of refraction 1.56. Light reflects from the surface of the scales because
    A. The scales' index of refraction is different from that of air.
    B. The scales' index of refraction is similar to that of glass.
    C. The scales' density is different from that of air.
    D. Different colors of light have different wavelengths.

STOP TO THINK ANSWERS

Chapter Preview Stop to Think: C. At point 1, two crests overlap, leading to a large wave amplitude; this is constructive interference. At point 2, the crest of one wave overlaps a trough of the other wave, so the amplitude is zero; this is destructive interference.

Stop to Think 17.1: $n_3 > n_1 > n_2$. $\lambda = \lambda_{vac}/n$, so a shorter wavelength corresponds to a higher index of refraction.

Stop to Think 17.2: A. The type of interference observed depends on the path-length *difference* $\Delta r = r_2 - r_1$, not on the values of r_1 and r_2 alone. Here, $\Delta r = 3\lambda$, so the path-length difference is an integer number of wavelengths, giving constructive interference.

Stop to Think 17.3: Smaller. The fringe spacing Δy is directly proportional to the wavelength λ.

Stop to Think 17.4: D. Longer wavelengths have larger diffraction angles. Red light has a longer wavelength than violet light, so red light is diffracted farther from the center.

Stop to Think 17.5: B. An extra path difference of $\lambda/2$ must be added to change from constructive to destructive interference. In thin-film interference, one wave passes *twice* through the film. To increase the path length by $\lambda/2$, the thickness needs to be increased by only one-half this, or $\lambda/4$.

Stop to Think 17.6: B or C. The width of the central maximum, which is proportional to λ/a, has increased. This could occur either because the wavelength has increased or because the slit width has decreased.

18 Ray Optics

These thin beams of light at this laser show are well described using ray optics. How do these light beams behave when they reflect from shiny surfaces or pass through transparent materials?

LOOKING AHEAD ➤

Goal: To understand and apply the ray model of light.

Reflection

Light rays can bounce, or **reflect,** off a surface. Rays from the bird's head reflect from the water, forming an upside-down image.

You'll learn how the **law of reflection** can be used to understand image formation by mirrors.

Refraction

The two images of the turtle are due to **refraction,** the bending of light rays as they travel from one material into another.

You'll learn **Snell's law** for refraction and how images can be formed by refraction.

Lenses and Mirrors

Rays refracting at the surfaces of this lens form a magnified **image** of the girl behind it.

You'll learn how to locate and characterize the images formed by lenses and mirrors.

LOOKING BACK ◀

The Ray Model of Light

In Chapter 17, you learned that light spreads out as it passes through a narrow slit, but travels straight forward through wide openings.

Waves travel in a straight path—a *ray*—behind a wide slit.

In this chapter, we'll study the behavior of light in the *ray model,* applicable when light interacts with objects of everyday size such as mirrors or lenses.

STOP TO THINK

The dark screen has a 2-mm-diameter hole. The bulb is the only source of light. What do you see on the viewing screen?

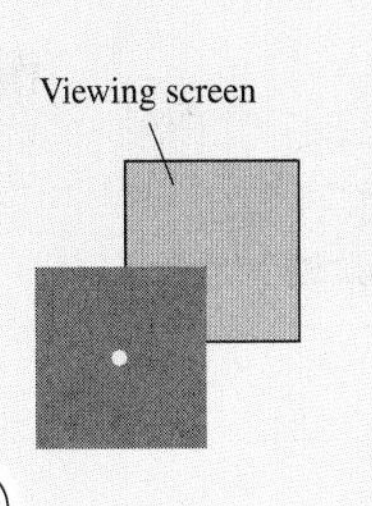

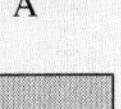

A

B

C

D

18.1 The Ray Model of Light

A flashlight makes a beam of light through the night's darkness, sunbeams stream into a darkened room through a small hole in the shade, and laser beams are even more well defined. Our everyday experience that light travels in straight lines is the basis of the ray model of light.

The ray model is an oversimplification of reality, but nonetheless is very useful within its range of validity. As we saw in Chapter 17, diffraction and other wave aspects of light are important only for apertures and objects comparable in size to the wavelength of light. Because the wavelength is so small, typically 0.5 μm, the wave nature of light is not apparent when light interacts with ordinary-sized objects. The ray model of light, which ignores diffraction, is valid as long as any apertures through which the light passes (lenses, mirrors, holes, and the like) are larger than about 1 mm.

To begin, let us define a **light ray** as a line in the direction along which light energy is flowing. A light ray is an abstract idea, not a physical entity or a "thing." Any narrow beam of light, such as the laser beam in FIGURE 18.1, is actually a bundle of many parallel light rays. You can think of a single light ray as the limiting case of a laser beam whose diameter approaches zero. Laser beams are good approximations of light rays, certainly adequate for demonstrating ray behavior, but any real laser beam is a bundle of many parallel rays.

FIGURE 18.1 A laser beam is a bundle of parallel light rays.

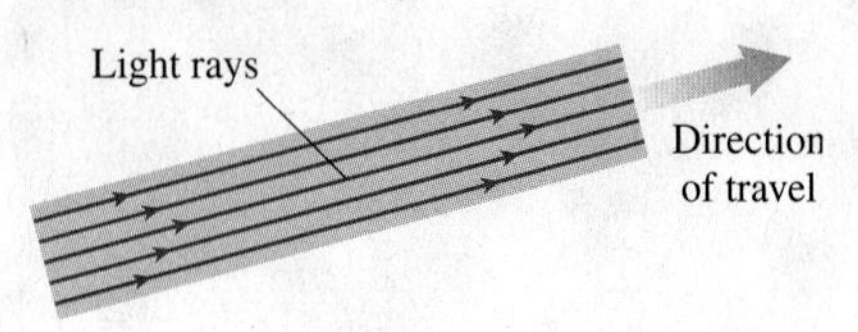

The following table outlines five basic ideas and assumptions of the ray model of light.

The ray model of light

Light rays travel in straight lines.

Light travels through a vacuum or a transparent material in straight lines called light rays. The speed of light in a material is $v = c/n$, where n is the index of refraction of the material.

Light rays can cross.

Light rays do not interact with each other. Two rays can cross without either being affected in any way.

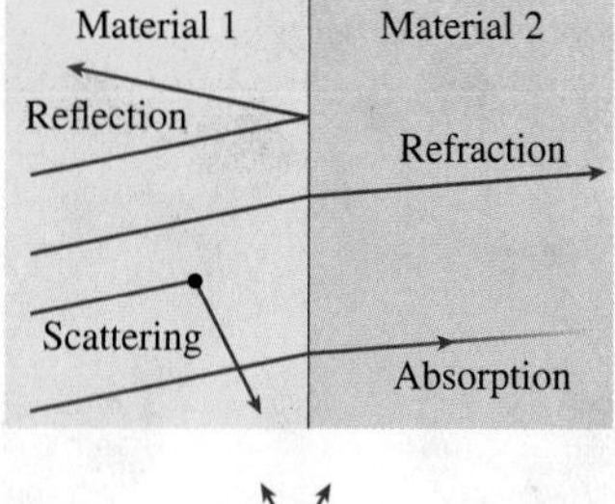

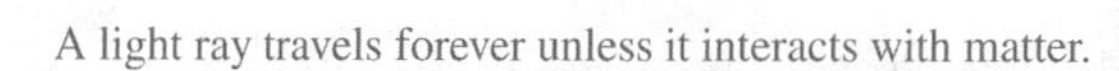

A light ray travels forever unless it interacts with matter.

A light ray continues forever unless it has an interaction with matter that causes the ray to change direction or to be absorbed. Light interacts with matter in four different ways:

- At an interface between two materials, light can be *reflected, refracted,* or both.
- Within a material, light can be either *scattered* or *absorbed.*

These interactions are discussed later in the chapter.

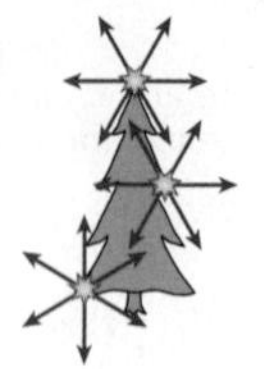

An object is a source of light rays.

An **object** is a source of light rays. Rays originate from *every* point on the object, and each point sends rays in *all* directions. Objects may be self-luminous—they create light rays—or they may be reflective objects that reflect only rays that originate elsewhere.

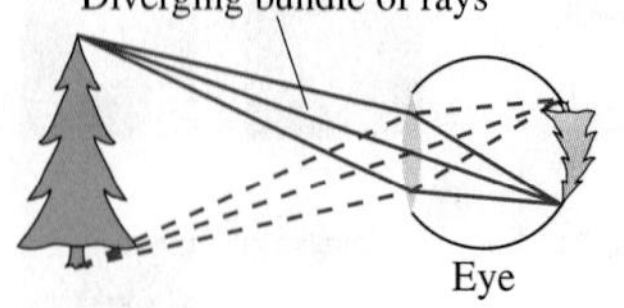

The eye sees by focusing a bundle of rays.

The eye sees an object when *diverging* bundles of rays from each point on the object enter the pupil and are focused to an image on the retina. Imaging is discussed later in the chapter, and the eye will be treated in much greater detail in Chapter 19.

Sources of Light Rays

In the ray model, there are two kinds of objects. **Self-luminous objects** (or *sources*) directly create light rays. Self-luminous objects include lightbulbs and the sun. Other objects, such as a piece of paper or a tree, are **reflective objects** that reflect rays originating in self-luminous objects. The table below shows four important kinds of self-luminous sources. Note that although ray and point sources are idealizations, they are useful in understanding the propagation of rays.

Self-luminous objects

A ray source	A point source	An extended source	A parallel-ray source
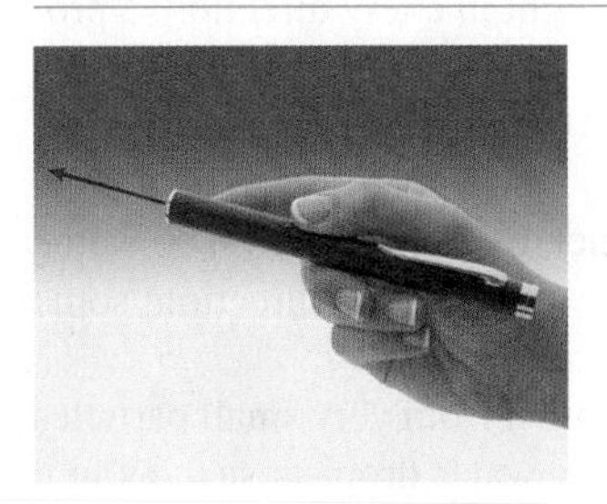			
Since a light ray is an idealization, there are no true ray sources. Still, the thin beam of a laser is often a good approximation of a single ray.	A point source is also an idealized source of light. It is infinitely small and emits light rays in every direction. The tiny filaments of these bulbs approximate point sources.	This is the most common light source. The *entire surface* of an extended source is luminous, so that **every point of an extended source acts as a point source.** Lightbulbs, flames, and the sun are extended sources.	Certain sources, such as flashlights and movie projectors, produce a bundle of parallel rays. Rays from a very distant object, such as a star, are very nearly parallel.

Reflective objects, such as a newspaper, a face, or a mirror, can also be considered as sources of light rays. However, the origin of these rays is not in the object itself. Instead, light rays from a self-luminous object strike a reflective object and "bounce" off of it. These rays can then illuminate other objects, or enter our eyes and form images of the reflective object, just as rays from self-luminous objects do.

Ray Diagrams

Rays originate from *every* point on an object and travel outward in *all* directions, but a diagram trying to show all these rays would be hopelessly messy and confusing. To simplify the picture, we usually use a **ray diagram** that shows only a few rays. For example, FIGURE 18.2 is a ray diagram showing only a few rays leaving the top and bottom points of the object and traveling to the right. These rays will be sufficient to show us how the object is imaged by lenses or mirrors.

FIGURE 18.2 A ray diagram simplifies the situation by showing only a few rays.

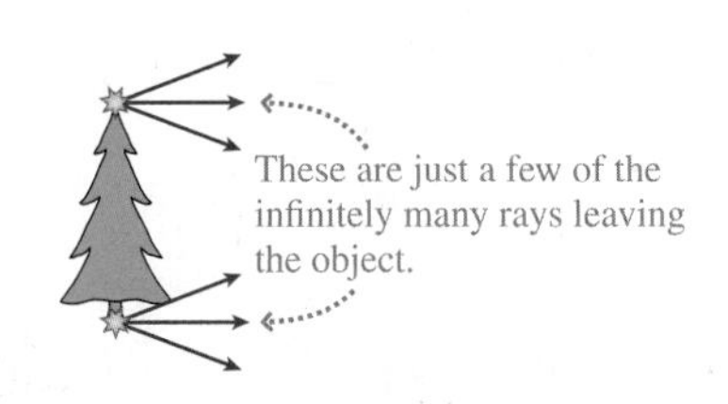

These are just a few of the infinitely many rays leaving the object.

NOTE ▸ Ray diagrams are the basis for a *visual overview* that we'll use throughout this chapter. Be careful not to think that a ray diagram shows all of the rays. The rays shown on the diagram are just a subset of the infinitely many rays leaving the object. ◂

Seeing Objects

How do we *see* an object? The eye works by focusing an image of an object on the retina, a process we'll examine in Chapter 19. For now, we'll ignore the details of image formation and instead make use of a simpler fact from the figure in the ray model table above: **In order for our eye to see an object, rays from that object must enter the eye.** This idea helps explain some subtle points about seeing.

Consider a ray source such as a laser. Can you see a laser beam traveling across the room? Under ordinary circumstances, the answer is no. As we've seen, a laser beam is a good approximation of a single ray. This ray travels in a straight line from the laser to whatever it eventually strikes. As FIGURE 18.3 shows, no light

FIGURE 18.3 A laser beam traveling through air is invisible.

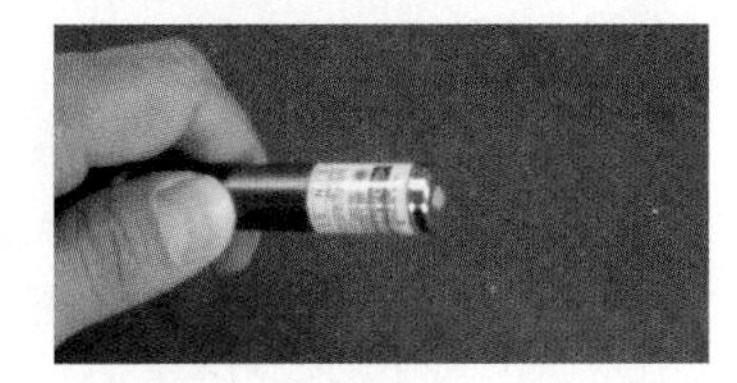

You can't see a laser beam crossing the room because no light ray enters your eye.

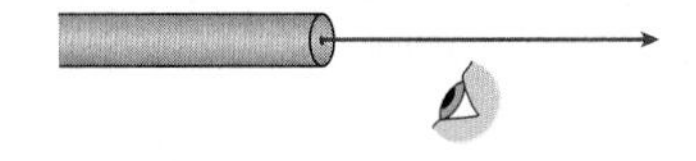

FIGURE 18.4 Point and extended sources can be seen by all observers.

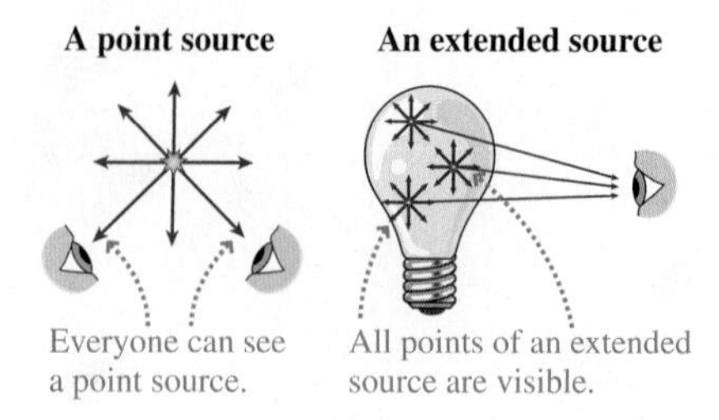

FIGURE 18.5 Reading a book by scattered light.

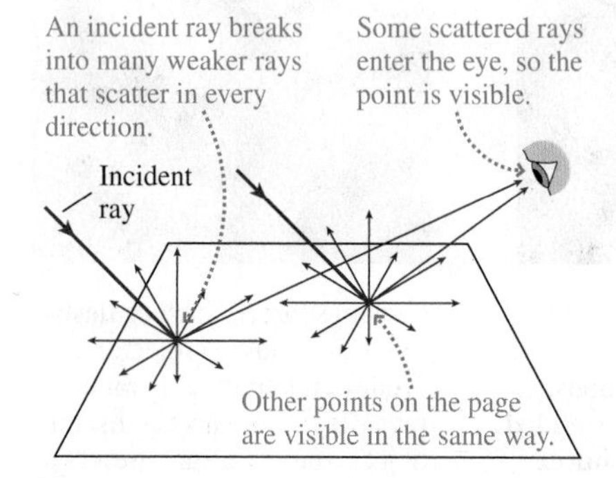

Class Video

ray enters the eye, so the beam is invisible. The same argument holds for a parallel-ray source.

A point source and an extended source behave differently, as shown in **FIGURE 18.4**. Because a point source emits rays in *every* direction, some of these rays will enter the eye no matter where it is located. Thus a point source is visible to everyone looking at it. And, since every point on the surface of an extended source is itself a point source, all parts of an extended source (not blocked by something else) can be viewed by all observers as well.

We can also use our simple model of seeing to explain how we see non-luminous objects. As we've already mentioned, such objects *reflect* rays that strike them. Most ordinary objects—paper, skin, grass—reflect incident light in every direction, a process called **diffuse reflection.** **FIGURE 18.5** illustrates the idea. Single rays are broken into many weaker rays that leave in all directions, a process called **scattering.**

Scattered light is what allows you to read a book by lamplight. As shown in Figure 18.5, every point on the surface of the page is struck by a ray (or rays) from the lamp. Then, because of diffuse reflection, these rays scatter in every direction; some of the scattered rays reach your eye, allowing you to see the page.

It is possible to make a laser beam visible by scattering it from very small particles suspended in air. These particles can be dust, smoke, or water droplets such as fog. **FIGURE 18.6** shows that as the beam strikes such a particle, it scatters rays in every direction. Some of these rays enter the eye, making each particle in the path of the beam visible and outlining the beam's path across the room.

FIGURE 18.6 A laser beam is visible if it travels through smoke or dust.

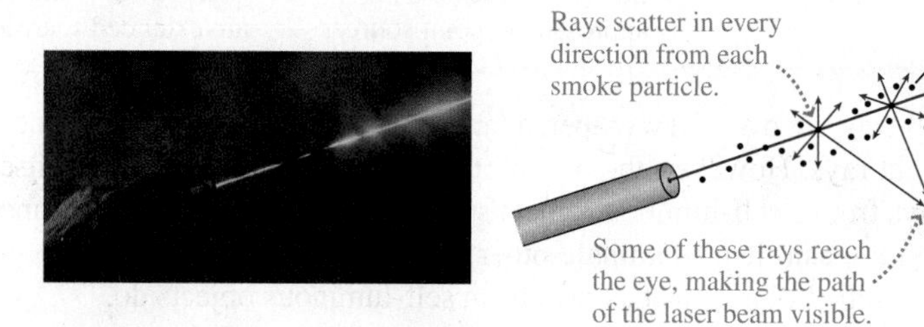

Shadows

Our ray model of light also explains the common phenomenon of *shadows.* Suppose an opaque object (such as a cardboard disk) is placed between a source of light and a screen. The object intercepts some of the rays, leaving a dark area behind it. Other rays travel on to a screen and illuminate it. The simplest shadows are those cast by a point source of light. This process is shown in **FIGURE 18.7a**. With a point source, the shadow is completely dark, and the edges of the shadow are sharp.

FIGURE 18.7 Shadows produced by point and extended sources of light.

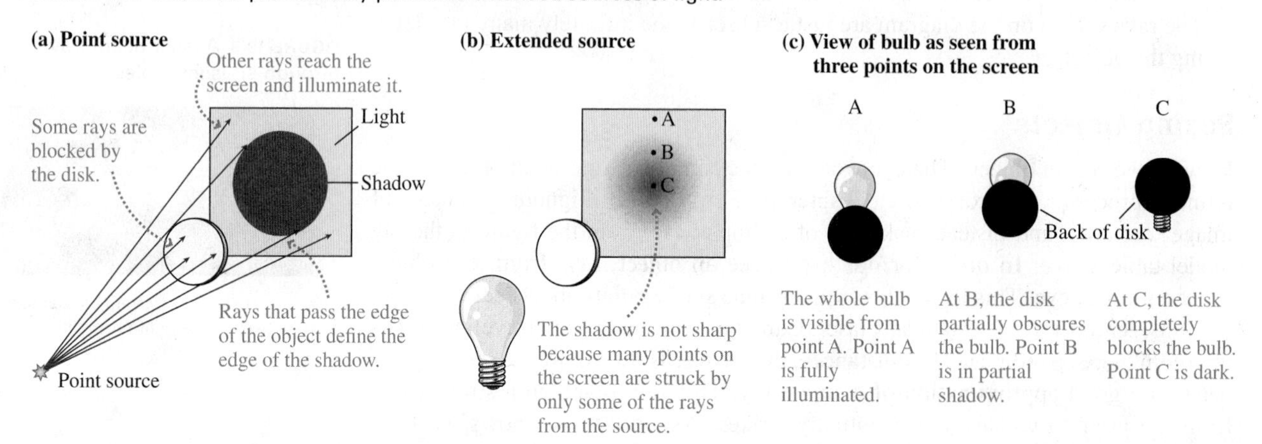

(a) Point source
Other rays reach the screen and illuminate it.
Some rays are blocked by the disk.
Light
Shadow
Rays that pass the edge of the object define the edge of the shadow.
Point source
(b) Extended source
A
B
C
The shadow is not sharp because many points on the screen are struck by only some of the rays from the source.
(c) View of bulb as seen from three points on the screen
A
B
C
Back of disk
The whole bulb is visible from point A. Point A is fully illuminated.
At B, the disk partially obscures the bulb. Point B is in partial shadow.
At C, the disk completely blocks the bulb. Point C is dark.

Shadows cast by extended sources are more complicated. An extended source is a collection of a large number of point sources, each of which casts its own shadow. However, as shown in **FIGURE 18.7b**, the patterns of shadow and light from each point overlap and thus the shadow region is no longer sharp. **FIGURE 18.7c** shows the view of the *bulb* as seen from three points on the screen. Depending on the size of the source, there is often a true shadow that no light reaches, surrounded by a fuzzy region of increasing brightness.

STOP TO THINK 18.1 The only source of light in a room is the lightbulb shown. An opaque disk is placed in front of the bulb. A screen is then placed successively at positions A, B, and C. At which screen position(s) does the disk cast a shadow that is completely dark at the center?

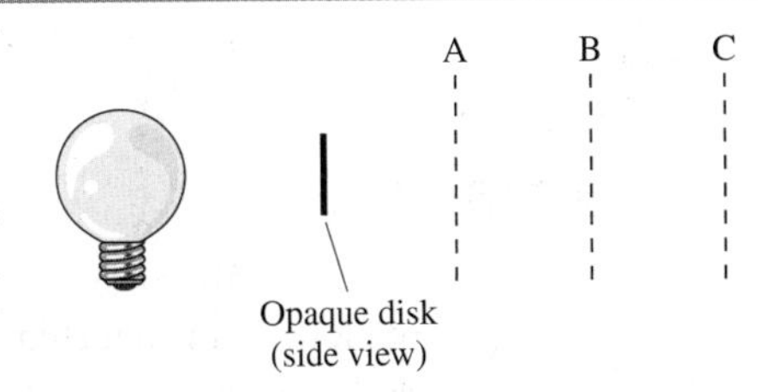

18.2 Reflection

Reflection is an everyday experience.

Reflection of light is a familiar, everyday experience. You see your reflection in the bathroom mirror first thing every morning, reflections in your car's rearview mirror as you drive to school, and the sky reflected in puddles of standing water. Reflection from a smooth, shiny surface, such as a mirror or a piece of polished metal, is called **specular reflection.**

FIGURE 18.8a shows a bundle of parallel light rays reflecting from a mirror-like surface. You can see that the incident and reflected rays are both in a plane that is normal, or perpendicular, to the reflective surface. A three-dimensional perspective accurately shows the relation between the light rays and the surface, but figures such as this are hard to draw by hand. Instead, it is customary to represent reflection with the simpler visual overview of **FIGURE 18.8b**. In this figure,

- The incident and reflected rays are in the plane of the page. The reflective surface extends into and out of the page.
- A *single* light ray represents the entire bundle of parallel rays. This is oversimplified, but it keeps the figure and the analysis clear.

FIGURE 18.8 Specular reflection of light.

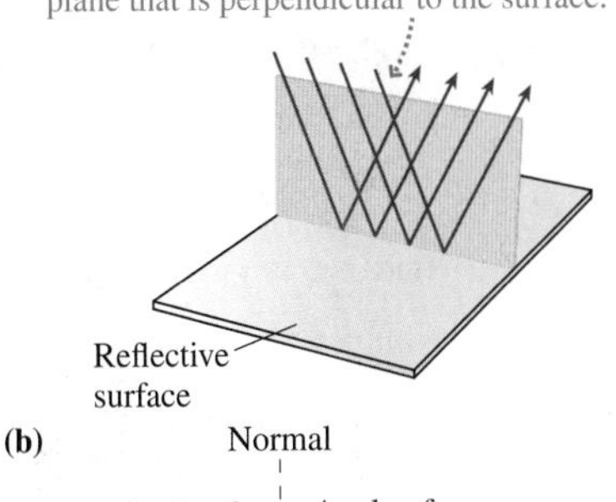

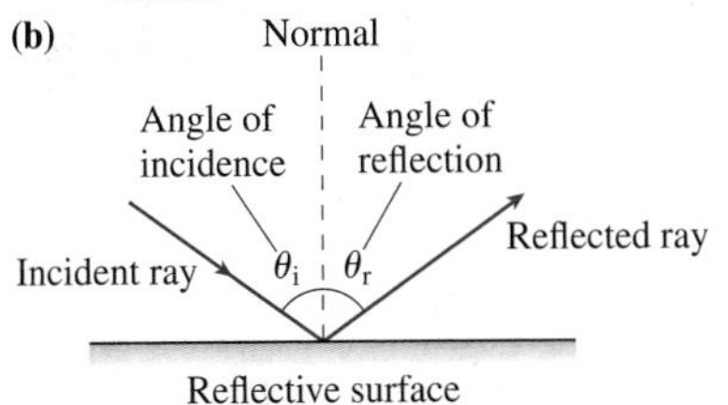

The angle θ_i between the incident ray and a line perpendicular to the surface—the *normal* to the surface—is called the **angle of incidence.** Similarly, the **angle of reflection** θ_r is the angle between the reflected ray and the normal to the surface. The **law of reflection,** easily demonstrated with simple experiments, states that

1. The incident ray and the reflected ray are both in the same plane, which is perpendicular to the surface, and
2. The angle of reflection equals the angle of incidence: $\theta_r = \theta_i$.

NOTE ► Optics calculations *always* use the angle measured from the normal, not the angle between the ray and the surface. ◄

EXAMPLE 18.1 Light reflecting from a mirror

A full-length mirror on a closet door is 2.0 m tall. The bottom touches the floor. A bare lightbulb hangs 1.0 m from the closet door, 0.5 m above the top of the mirror. How long is the streak of reflected light across the floor?

PREPARE Treat the lightbulb as a point source and use the ray model of light. **FIGURE 18.9** is a visual overview of the light rays. We need to consider only the two rays that strike the edges of the mirror. All other reflected rays will fall between these two.

FIGURE 18.9 Visual overview of light rays reflecting from a mirror.

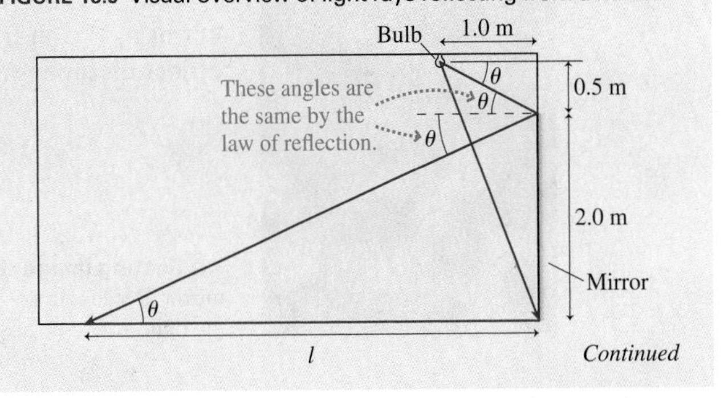

Continued

SOLVE The ray that strikes the bottom of the mirror reflects from it and hits the floor just where the mirror meets the floor. For the top ray, Figure 18.9 has used the law of reflection to set the angle of reflection equal to the angle of incidence; we call both θ. By simple geometry, the other angles shown are also equal to θ. From the small triangle at the upper right,

$$\theta = \tan^{-1}\left(\frac{0.5\text{ m}}{1.0\text{ m}}\right) = 26.6°$$

But we also have $\tan\theta = (2.0\text{ m})/l$, or

$$l = \frac{2.0\text{ m}}{\tan\theta} = \frac{2.0\text{ m}}{\tan 26.6°} = 4.0\text{ m}$$

Since the lower ray struck right at the mirror's base, the total length of the reflected streak is 4.0 m.

Diffuse Reflection

FIGURE 18.10 Diffuse reflection from an irregular surface.

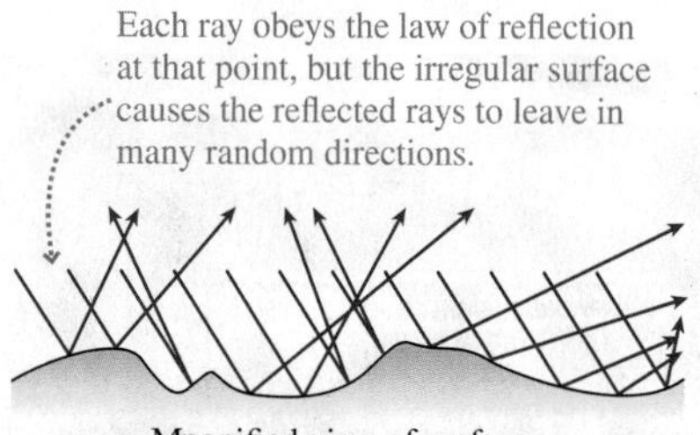

We've already discussed diffuse reflection, the reflection of light rays off of a surface such as paper or cloth that is not shiny like a mirror. If you magnify the surface of a diffuse reflector, you'll find that on the microscopic scale it is quite rough. The law of reflection $\theta_r = \theta_i$ is still obeyed at each point, but the irregularities of the surface cause the reflected rays to leave in many random directions. This situation is shown in FIGURE 18.10. Diffuse reflection is actually much more common than the mirror-like specular reflection.

The Plane Mirror

One of the most commonplace observations is that you can see yourself in a mirror. How? FIGURE 18.11a shows rays from point source P reflecting from a flat mirror, called a **plane mirror.** Consider the two particular rays shown in FIGURE 18.11b. The horizontal ray strikes the mirror at a zero angle of incidence, so it reflects straight back with a zero angle of reflection. The second, angled ray reflects with $\theta_r = \theta_i$. As indicated by the dashed lines, both reflected rays appear to have come from a point P′. But because our argument applies to any incoming ray, *all* reflected rays appear to be coming from point P′, as FIGURE 18.11c shows. We call P′, the point from which the reflected rays diverge, the **virtual image** of P. The image is "virtual" in the sense that no rays actually leave P′, which is in darkness behind the mirror. But as far as your eye is concerned, the light rays act exactly *as if* the light really originated at P′. So while you may say "I see P in the mirror," what you are actually seeing is the virtual image of P.

FIGURE 18.11 The light rays reflecting from a plane mirror.

(a)

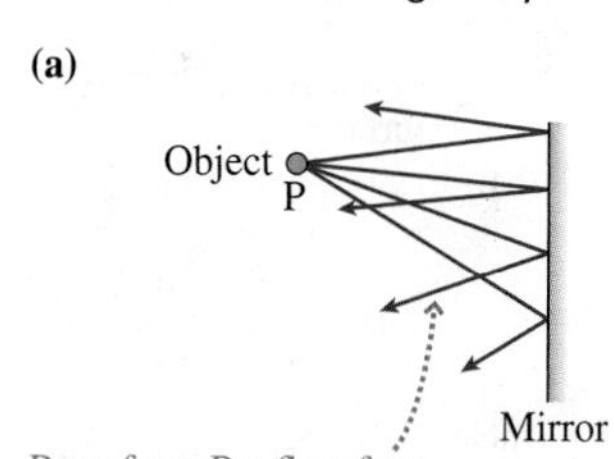

(b)

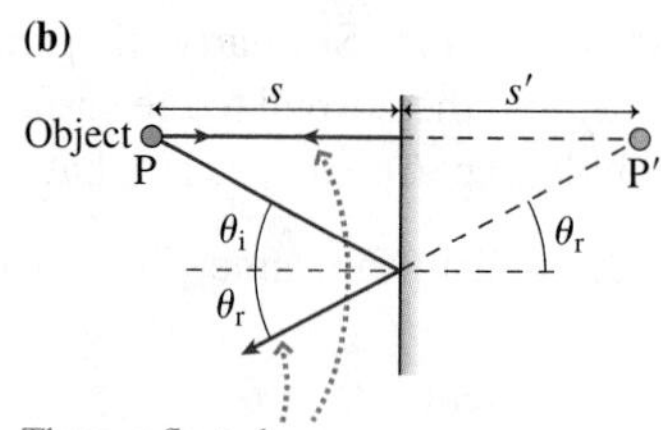

(c) Object distance Image distance

s s′

Object P

Virtual image P′

Eye

The reflected rays *all* diverge from P′, which appears to be the source of the reflected rays. Your eye collects the bundle of diverging rays and "sees" the light coming from P′.

In Figure 18.11b, simple geometry dictates that P′ is the same distance behind the mirror as P is in front of the mirror. That is, the **image distance** s' is equal to the **object distance** s:

$$s' = s \quad \text{(plane mirror)} \qquad (18.1)$$

◀ **A floating image** The student stands on her leg that is hidden behind the mirror. Because every point of her body to the left of the mirror has its image to the right, when she raises her other leg she appears to float above the floor.

For an extended object, such as the one in **FIGURE 18.12**, each point on the object has a corresponding image point an equal distance on the opposite side of the mirror. The eye captures and focuses diverging bundles of rays from each point of the image in order to see the full image in the mirror. Two facts are worth noting:

1. Rays from each point on the object spread out in all directions and strike *every point* on the mirror. Only a very few of these rays enter your eye, but the other rays are very real and might be seen by other observers.
2. Rays from points P and Q enter your eye after reflecting from *different* areas of the mirror. This is why you can't always see the full image of an object in a very small mirror.

FIGURE 18.12 Viewing an extended object.

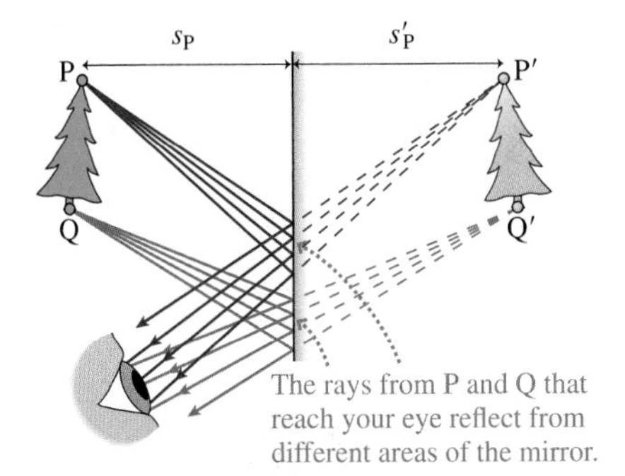

EXAMPLE 18.2 How high is the mirror?

If your height is h, what is the shortest mirror on the wall in which you can see your full image? Where must the top of the mirror be hung?

PREPARE Use the ray model of light. **FIGURE 18.13** is a visual overview of the light rays. We need to consider only the two rays that leave the top of your head and your feet and reflect into your eye.

FIGURE 18.13 Visual overview of light rays from your head and feet reflecting into your eye.

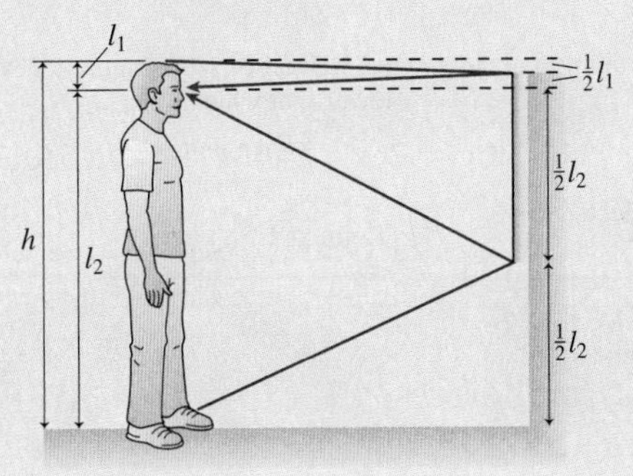

SOLVE Let the distance from your eyes to the top of your head be l_1 and the distance to your feet be l_2. Your height is $h = l_1 + l_2$. A light ray from the top of your head that reflects from the mirror at $\theta_r = \theta_i$ and enters your eye must, by congruent triangles, strike the mirror a distance $\frac{1}{2}l_1$ above your eyes. Similarly, a ray from your foot to your eye strikes the mirror a distance $\frac{1}{2}l_2$ below your eyes. The distance between these two points on the mirror is $\frac{1}{2}l_1 + \frac{1}{2}l_2 = \frac{1}{2}h$. A ray from anywhere else on your body will reach your eye if it strikes the mirror between these two points. Pieces of the mirror outside these two points are irrelevant, not because rays don't strike them but because the reflected rays don't reach your eye. Thus the shortest mirror in which you can see your full reflection is $\frac{1}{2}h$. But this will work only if the top of the mirror is hung midway between your eyes and the top of your head.

ASSESS It is interesting that the answer does not depend on how far you are from the mirror.

STOP TO THINK 18.2 An object is placed in front of a mirror. The observer is positioned as shown. Which of the points, A, B, or C, best indicates where the observer would perceive the image to be located?

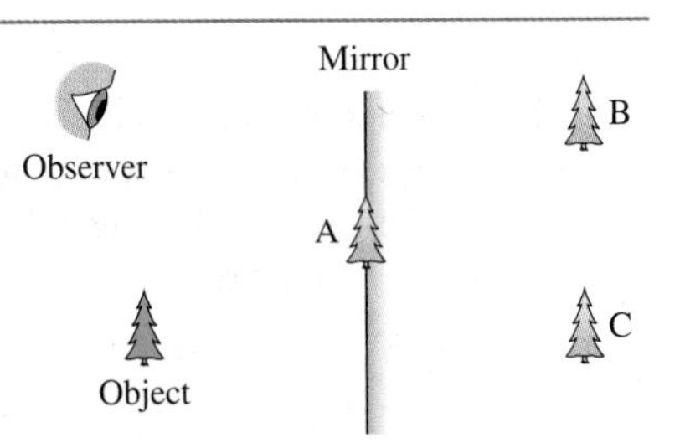

18.3 Refraction

In **FIGURE 18.14**, two things happen when a light ray crosses the boundary between the air and the glass:

1. Part of the light *reflects* from the boundary, obeying the law of reflection. This is how you see reflections from pools of water or storefront windows, even though water and glass are transparent.
2. Part of the light continues into the second medium. It is *transmitted* rather than reflected, but the transmitted ray changes direction as it crosses the boundary. The transmission of light from one medium to another, but with a change in direction, is called **refraction**.

FIGURE 18.14 A light beam refracts twice in passing through a glass prism.

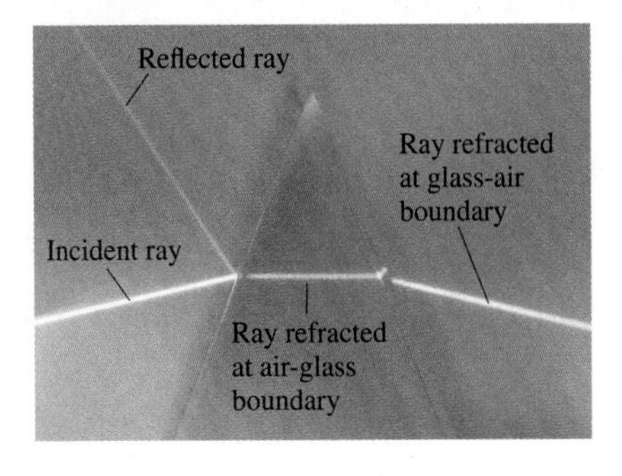

In Figure 18.14, notice the refraction of the light beam as it passes through the prism. Notice also that the ray direction changes as the light enters and leaves the glass. You can also see a weak reflection leaving the first surface of the prism.

Reflection from the boundary between transparent media is usually weak. Typically 95% of the light is transmitted and only 5% is reflected. Our goal in this section is to understand refraction, so we will usually ignore the weak reflection and focus on the transmitted light.

NOTE ▸ The transparent material through which light travels is called the *medium* (plural *media*). ◂

FIGURE 18.15a shows the refraction of light rays from a parallel beam of light, such as a laser beam, and rays from a point source. These pictures remind us that an infinite number of rays are incident on the boundary, but our analysis will be simplified if we focus on a single light ray. FIGURE 18.15b is a ray diagram showing the refraction of a single ray at a boundary between medium 1 and medium 2. Let the angle between the ray and the normal be θ_1 in medium 1 and θ_2 in medium 2. Just as for reflection, the angle between the incident ray and the normal is the *angle of incidence*. The angle on the transmitted side, *measured from the normal,* is called the **angle of refraction.** Notice that θ_1 is the angle of incidence in Figure 18.15b but is the angle of refraction in FIGURE 18.15c, where the ray is traveling in the opposite direction.

Class Video

FIGURE 18.15 Refraction of light rays.

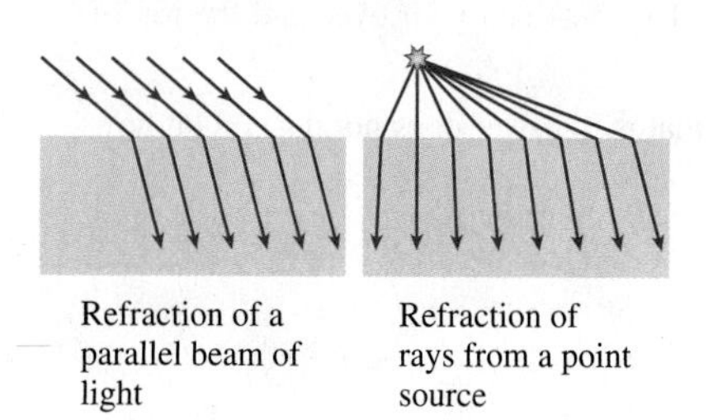

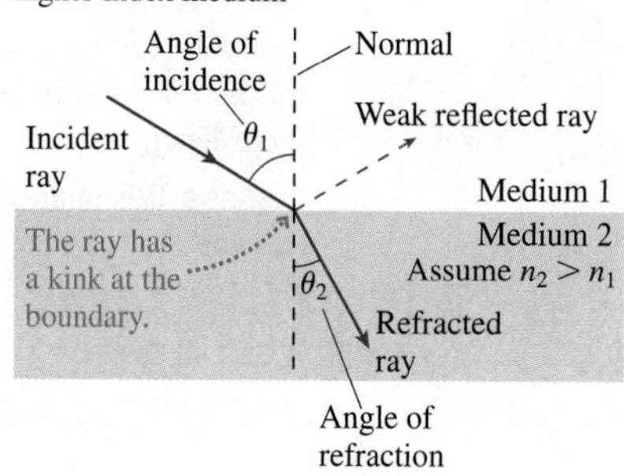

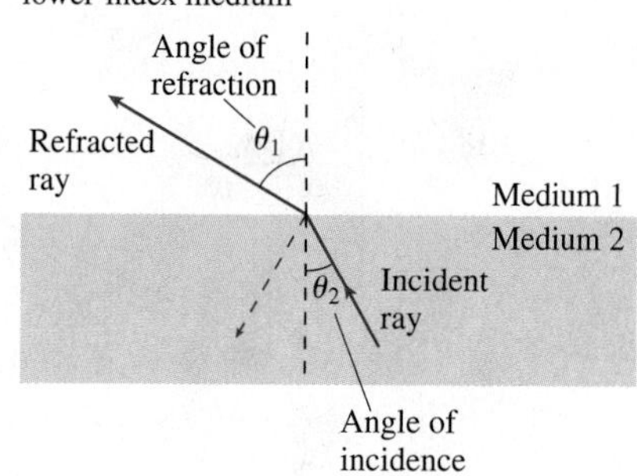

In 1621, Dutch scientist Willebrord Snell proposed a mathematical statement of the "law of refraction" or, as we know it today, Snell's law. If a ray refracts between medium 1 and medium 2, having indices of refraction n_1 and n_2, the ray angles θ_1 and θ_2 in the two media are related by

$$n_1 \sin\theta_1 = n_2 \sin\theta_2 \qquad (18.2)$$

Snell's law for refraction between two media

Notice that Snell's law does not mention which is the incident angle and which the refracted angle.

Table 18.1 lists the indices of refraction for several media. It is interesting to note that the n in the law of refraction, Equation 18.2, is the same index of refraction n we studied in ◂◂SECTION 17.1. There we found that the index of refraction determines the speed of a light wave in a medium according to $v = c/n$. Here, it appears to play a different role, determining by how much a light ray is bent when crossing the boundary between two different media. Although we won't do so here, it is possible to use Huygens' principle to show that Snell's law is a *consequence* of the change in the speed of light as it moves across a boundary between media.

TABLE 18.1 Indices of refraction

Medium	n
Vacuum	1 exactly
Air (actual)	1.0003
Air (accepted)*	1.00
Water	1.33
Ethyl alcohol	1.36
Oil	1.46
Glass (typical)	1.50
Polystyrene plastic	1.59
Cubic zirconia	2.18
Diamond	2.42
Silicon (infrared)	3.50

*Use this value in problems.

Examples of Refraction

Look back at Figure 18.15. As the ray in Figure 18.15b moves from medium 1 to medium 2, where $n_2 > n_1$, it bends closer to the normal. In Figure 18.15c, where the ray moves from medium 2 to medium 1, it bends away from the normal. This is a general conclusion that follows from Snell's law:

- When a ray is transmitted into a material with a higher index of refraction, it bends to make a smaller angle with the normal.
- When a ray is transmitted into a material with a lower index of refraction, it bends to make a larger angle with the normal.

This rule becomes a central idea in a procedure for analyzing refraction problems.

TACTICS BOX 18.1 Analyzing refraction

1. **Draw a ray diagram.** Represent the light beam with one ray.
2. **Draw a line normal (perpendicular) to the boundary.** Do this at each point where the ray intersects a boundary.
3. **Show the ray bending in the correct direction.** The angle is larger on the side with the smaller index of refraction. This is the qualitative application of Snell's law.
4. **Label angles of incidence and refraction.** Measure all angles from the normal.
5. **Use Snell's law.** Calculate the unknown angle or unknown index of refraction.

Exercises 10–13

EXAMPLE 18.3 Deflecting a laser beam

A laser beam is aimed at a 1.0-cm-thick sheet of glass at an angle 30° above the glass.

a. What is the laser beam's direction of travel in the glass?

b. What is its direction in the air on the other side of the glass?

PREPARE Represent the laser beam with a single ray and use the ray model of light. **FIGURE 18.16** is a visual overview in which the first four steps of Tactics Box 18.1 have been identified. Notice that the angle of incidence must be measured from the normal, so $\theta_1 = 60°$, not the 30° value given in the problem. The index of refraction of glass was taken from Table 18.1.

FIGURE 18.16 The ray diagram of a laser beam passing through a sheet of glass.

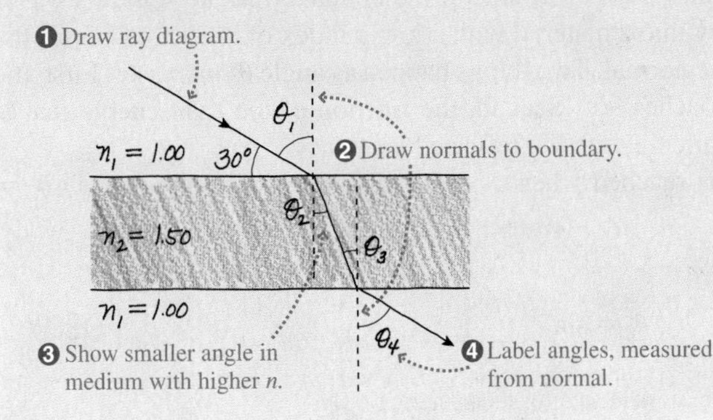

SOLVE

a. Snell's law, the final step in the Tactics Box, is $n_1 \sin\theta_1 = n_2 \sin\theta_2$. Using $\theta_1 = 60°$, we find that the direction of travel in the glass is

$$\theta_2 = \sin^{-1}\left(\frac{n_1 \sin\theta_1}{n_2}\right) = \sin^{-1}\left(\frac{\sin 60°}{1.5}\right)$$
$$= \sin^{-1}(0.577) = 35.3°$$

or 35° to two significant figures.

b. Snell's law at the second boundary is $n_2 \sin\theta_3 = n_1 \sin\theta_4$. You can see from Figure 18.16 that the interior angles are equal: $\theta_3 = \theta_2 = 35.3°$. Thus the ray emerges back into the air traveling at angle

$$\theta_4 = \sin^{-1}\left(\frac{n_2 \sin\theta_3}{n_1}\right) = \sin^{-1}(1.5 \sin 35.3°)$$
$$= \sin^{-1}(0.867) = 60°$$

This is the same as θ_1, the original angle of incidence.

ASSESS As expected, the laser beam bends toward the normal as it moves into the higher-index glass, and away from the normal as it moves back into air. The beam exits the glass still traveling in the same direction as it entered, but its path is *displaced.* This is a general result for light traveling through a medium with parallel sides. As the glass becomes thinner, the displacement becomes less; there is no displacement as the glass thickness becomes zero. This will be an important observation when we later study lenses.

EXAMPLE 18.4 **Measuring the index of refraction**

FIGURE 18.17 shows a laser beam deflected by a 30°-60°-90° prism. What is the prism's index of refraction?

FIGURE 18.17 A prism deflects a laser beam.

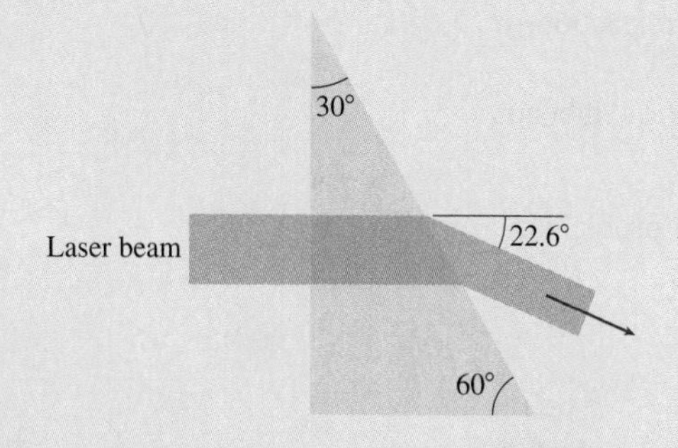

PREPARE Represent the laser beam with a single ray and use the ray model of light. **FIGURE 18.18** uses the steps of Tactics Box 18.1 to draw a ray diagram. The ray is incident perpendicular to the front face of the prism ($\theta_i = 0°$); thus it is transmitted through the first boundary without deflection. At the second boundary it is especially important to *draw the normal to the surface* at the point of incidence and to *measure angles from the normal.*

FIGURE 18.18 Visual overview of a laser beam passing through the prism.

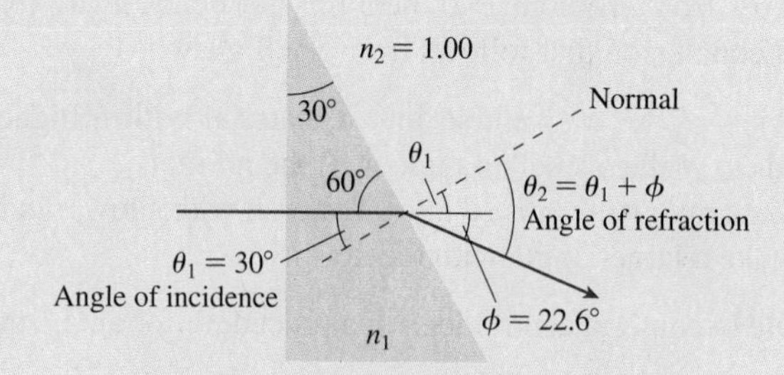

SOLVE From the geometry of the triangle you can find that the laser's angle of incidence on the hypotenuse of the prism is $\theta_1 = 30°$, the same as the apex angle of the prism. The ray exits the prism at angle θ_2 such that the deflection is $\phi = \theta_2 - \theta_1 = 22.6°$. Thus $\theta_2 = 52.6°$. Knowing both angles and $n_2 = 1.00$ for air, we can use Snell's law to find n_1:

$$n_1 = \frac{n_2 \sin\theta_2}{\sin\theta_1} = \frac{1.00 \sin 52.6°}{\sin 30°} = 1.59$$

ASSESS Referring to the indices of refraction in Table 18.1, we see that the prism is made of polystyrene plastic.

Total Internal Reflection

What would have happened in Example 18.4 if the prism angle had been 45° rather than 30°? The light rays would approach the rear surface of the prism at an angle of incidence $\theta_1 = 45°$. When we try to calculate the angle of refraction at which the ray emerges into the air, we find

$$\sin\theta_2 = \frac{n_1}{n_2}\sin\theta_1 = \frac{1.59}{1.00}\sin 45° = 1.12$$

$$\theta_2 = \sin^{-1}(1.12) = ???$$

FIGURE 18.19 One of the three beams of light undergoes total internal reflection.

Angle θ_2 cannot be computed because the sine of an angle can't be greater than 1. The ray is unable to refract through the boundary. Instead, 100% of the light *reflects* from the boundary back into the prism. This process is called **total internal reflection,** often abbreviated TIR. That it really happens is illustrated in **FIGURE 18.19.** Here, three light beams strike the surface of the water at increasing angles of incidence. The two beams with the smallest angles of incidence refract out of the water, but the beam with the largest angle of incidence undergoes total internal reflection at the water's surface.

FIGURE 18.20 Refraction and reflection of rays as the angle of incidence increases.

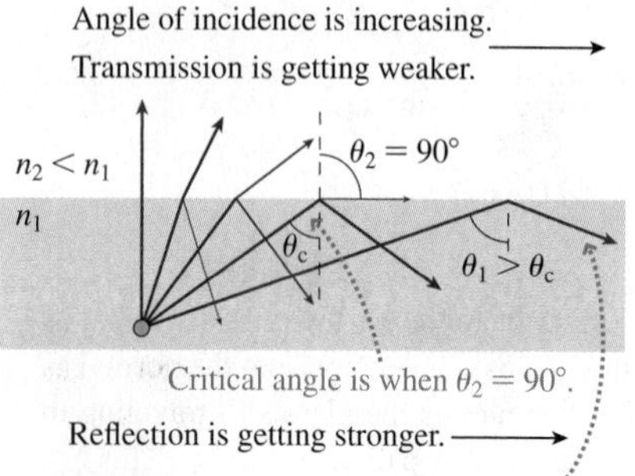

Figure Video

FIGURE 18.20 shows several rays leaving a point source in a medium with index of refraction n_1. The medium on the other side of the boundary has $n_2 < n_1$. As we've seen, crossing a boundary into a material with a lower index of refraction causes the ray to bend away from the normal. Two things happen as angle θ_1 increases. First, the refraction angle θ_2 approaches 90°. Second, the fraction of the light energy that is transmitted decreases while the fraction reflected increases.

A **critical angle** θ_c is reached when $\theta_2 = 90°$. Snell's law becomes $n_1 \sin\theta_c = n_2 \sin 90°$, or

$$\theta_c = \sin^{-1}\left(\frac{n_2}{n_1}\right) \qquad (18.3)$$

Critical angle of incidence for total internal reflection

The refracted light vanishes at the critical angle and the reflection becomes 100% for any angle $\theta_1 \geq \theta_c$. The critical angle is well defined because of our assumption that $n_2 < n_1$. **There is no critical angle and no total internal reflection if $n_2 > n_1$.**

We can compute the critical angle in a typical piece of glass at the glass-air boundary as

$$\theta_{c\ glass} = \sin^{-1}\left(\frac{1.00}{1.50}\right) = 42°$$

The fact that the critical angle is smaller than 45° has important applications. For example, **FIGURE 18.21** shows a pair of binoculars. The lenses are much farther apart than your eyes, so the light rays need to be brought together before exiting the eyepieces. Rather than using mirrors, which get dirty, are easily scratched, and require alignment, binoculars use a pair of prisms on each side. Thus the light undergoes two TIRs and emerges from the eyepiece. (The actual prism arrangement in binoculars is a bit more complex, but this illustrates the basic idea.)

FIGURE 18.21 Binoculars and other optical instruments make use of total internal reflection (TIR).

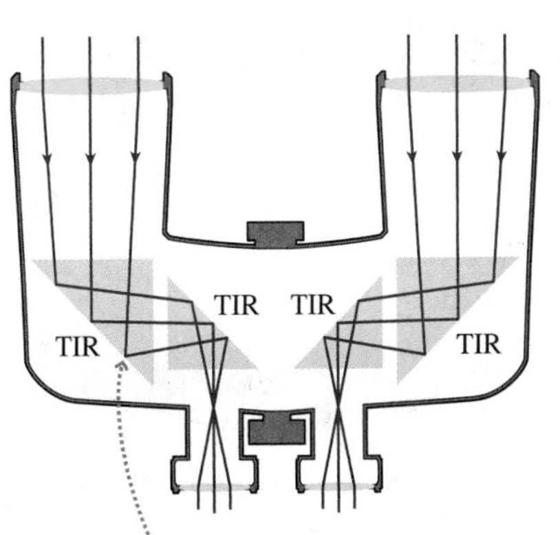

EXAMPLE 18.5 Seeing a submerged light

A lightbulb is set in the bottom of a 3.0-m-deep swimming pool. What is the diameter of the circle inside which a duck swimming on the surface could see the bulb?

PREPARE Represent the lightbulb as a point source and use the ray model of light. **FIGURE 18.22** is a visual overview of the light rays. The lightbulb emits rays at all angles, but only some of the rays refract into the air where they can be seen from above. Rays striking the surface at greater than the critical angle undergo TIR back down into the water. The diameter of the circle of light is the distance D between the two points at which rays strike the surface at the critical angle.

FIGURE 18.22 Visual overview of the rays leaving a lightbulb at the bottom of a swimming pool.

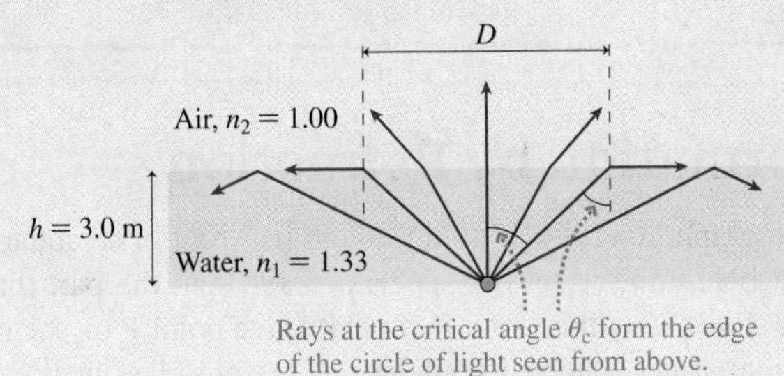

SOLVE From trigonometry, the circle diameter is $D = 2h\tan\theta_c$, where h is the depth of the water. The critical angle for a water-air boundary is $\theta_c = \sin^{-1}(1.00/1.33) = 48.7°$. Thus

$$D = 2(3.0\text{ m})\tan 48.7° = 6.8\text{ m}$$

ASSESS Light rays emerging at the edge of the circle actually skim the surface of the water. By reversing the direction of the rays, we can understand what a diver sees when she's underwater. This idea is explored further in the discussion below.

▶ **Snell's window** We can understand what a diver sees when she's underwater by reversing the direction of all the rays in Figure 18.22 in the preceding example. The drawing shows that she can see the sun overhead and clouds at larger angles. She can even see objects sitting at the waterline—but they appear at the edge of a circle as she looks up. Anything outside of this circle is a reflection of something in the water. The photo shows what she sees: a bright circle from the sky above—*Snell's window*—surrounded by the dark reflection of the water below.

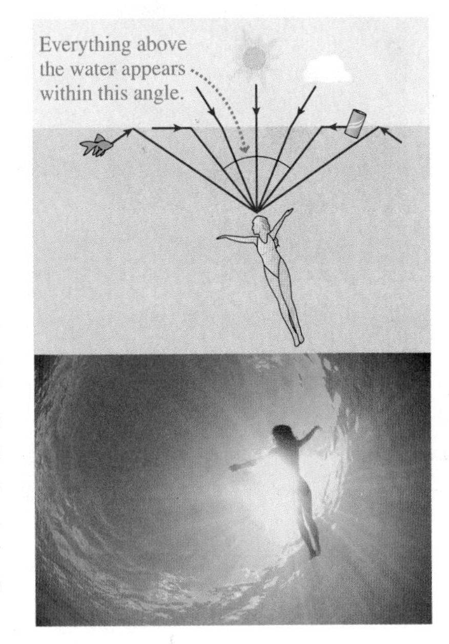

Everything above the water appears within this angle.

Fiber Optics

The most important modern application of total internal reflection is the transmission of light through optical fibers. **FIGURE 18.23a** on the next page shows a laser beam shining into the end of a long, narrow-diameter glass fiber. The light rays pass easily from the air into the glass, but they then strike the inside wall of the fiber at an angle of incidence θ_1 approaching 90°. This is much larger than the critical angle, so the laser beam undergoes TIR and remains inside the glass. The laser beam continues to "bounce" its way down the fiber as if the light were inside a pipe. Indeed, optical fibers are sometimes called "light pipes." The rays have an angle of incidence *smaller* than the critical angle ($\theta_1 \approx 0$) when they finally reach the flat end of the fiber; thus they refract out without difficulty and can be detected.

FIGURE 18.23 Light rays are confined within an optical fiber by TIR.

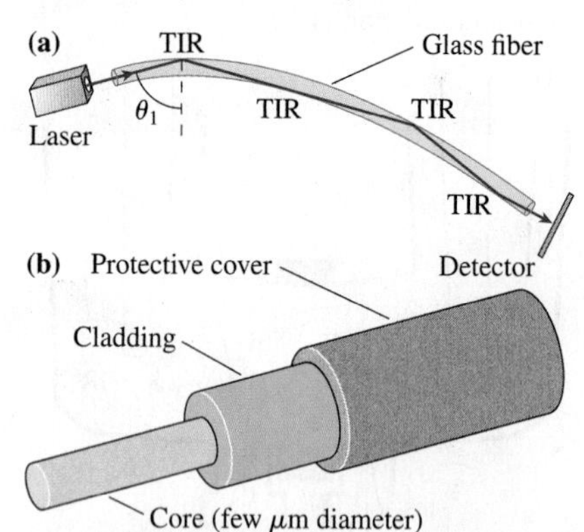

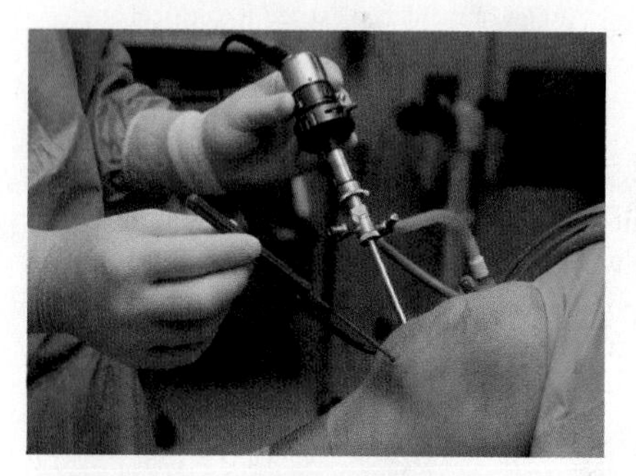

While a simple glass fiber can transmit light, reliance on a glass-air boundary is not sufficiently reliable for commercial use. Any small scratch on the side of the fiber alters the rays' angle of incidence and allows leakage of light. FIGURE 18.23b shows the construction of a practical optical fiber. A small-diameter glass *core* is surrounded by a layer of glass *cladding.* The glasses used for the core and the cladding have $n_{core} > n_{cladding}$. Thus, light undergoes TIR at the core-cladding boundary and remains confined within the core. This boundary is not exposed to the environment and hence retains its integrity even under adverse conditions.

Optical fibers have found important applications in medical diagnosis and treatment. Thousands of small fibers can be fused together to make an *endoscope,* a flexible bundle capable of transmitting high-resolution images along its length. During *arthroscopic surgery,* operations on injured joints are performed using an endoscope inserted through a small incision. The endoscope allows the surgeon to observe the procedure, which is performed with instruments inserted through another incision. The recovery time for such surgery is usually much shorter than for conventional operations requiring a full incision to expose the interior of the joint.

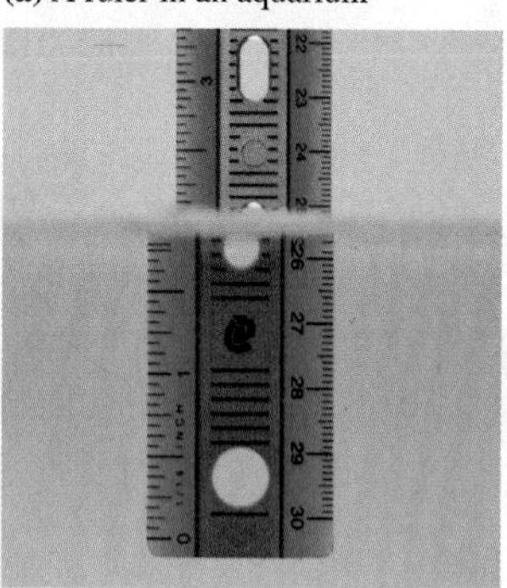

Arthroscopic surgery using an endoscope. BIO

STOP TO THINK 18.3 A light ray travels from medium 1 to medium 3 as shown. For these media,

A. $n_3 > n_1$
B. $n_3 = n_1$
C. $n_3 < n_1$
D. We can't compare n_1 to n_3 without knowing n_2.

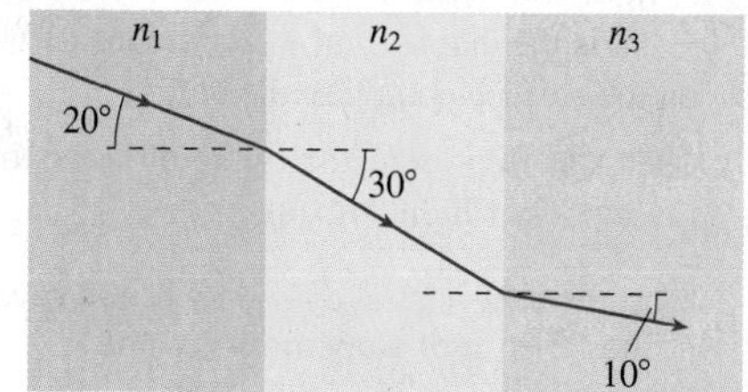

18.4 Image Formation by Refraction

FIGURE 18.24a shows a photograph of a ruler as seen through the front of an aquarium tank. The part of the ruler below the waterline appears *closer* than the part that is above water. FIGURE 18.24b shows why this is so. Rays that leave point P on the ruler refract away from the normal at the water-air boundary. (The thin glass wall of the aquarium has little effect on the refraction of the rays and can be ignored.) To your eye, outside the aquarium, these rays appear to diverge not from the object at point P, but instead from point P′ that is *closer* to the boundary. The same argument holds for every point on the ruler, so that **the ruler appears closer than it really is because of refraction of light at the boundary.**

FIGURE 18.24 Refraction causes an object in an aquarium to appear closer than it really is.

(a) A ruler in an aquarium

(b) Finding the image of the ruler

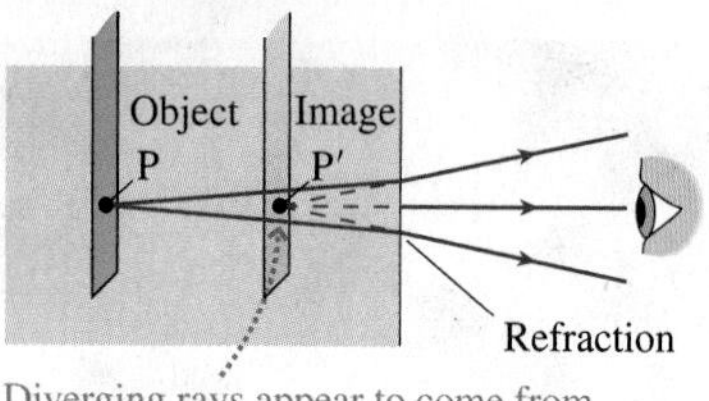

Diverging rays appear to come from this point. This is a virtual image.

We found that the rays reflected from a mirror diverge from a point that is not the object point. We called that point a *virtual image.* Similarly, if rays from an object point P refract at a boundary between two media such that the rays then diverge from a point P′ and *appear* to come from P′, we call P′ a virtual image of point P. The virtual image of the ruler is what you see.

Let's examine this image formation a bit more carefully. FIGURE 18.25 shows a boundary between two transparent media having indices of refraction n_1 and n_2. Point P, a source of light rays, is the object. Point P′, from which the rays *appear* to diverge, is the virtual image of P. The figure assumes $n_1 > n_2$, but this assumption isn't necessary. Distance s, measured from the boundary, is the object distance. Our goal is to determine the image distance s'.

The line through the object and perpendicular to the boundary is called the **optical axis.** Consider a ray that leaves the object at angle θ_1 with respect to the optical axis. θ_1 is also the angle of incidence at the boundary, where the ray refracts into the second medium at angle θ_2. By tracing the refracted ray backward, you can see that θ_2 is also the angle between the refracted ray and the optical axis at point P′.

The distance l is common to both the incident and the refracted rays, and you can see that $l = s\tan\theta_1 = s'\tan\theta_2$. Thus

$$s' = \frac{\tan\theta_1}{\tan\theta_2}s \qquad (18.4)$$

Snell's law relates the sines of angles θ_1 and θ_2; that is,

$$\frac{\sin\theta_1}{\sin\theta_2} = \frac{n_2}{n_1} \qquad (18.5)$$

In practice, the angle between any of these rays and the optical axis is very small because the pupil of your eye is very much smaller than the distance between the object and your eye. (The angles in the figure have been greatly exaggerated.) The small-angle approximation $\sin\theta \approx \tan\theta \approx \theta$, where θ is in radians, is therefore applicable. Consequently,

$$\frac{\tan\theta_1}{\tan\theta_2} \approx \frac{\sin\theta_1}{\sin\theta_2} = \frac{n_2}{n_1} \qquad (18.6)$$

Using this result in Equation 18.4, we find that the image distance is

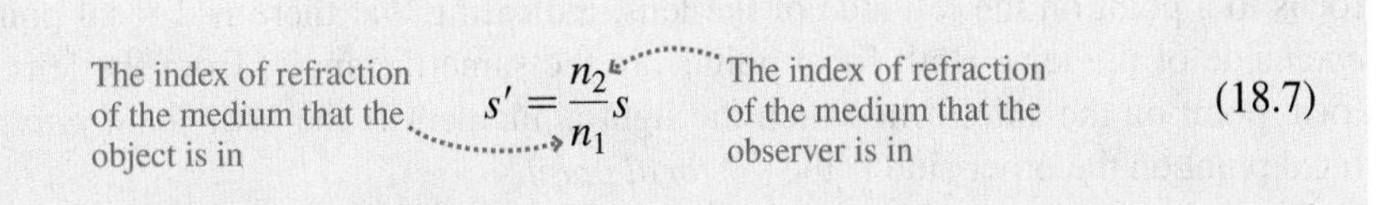

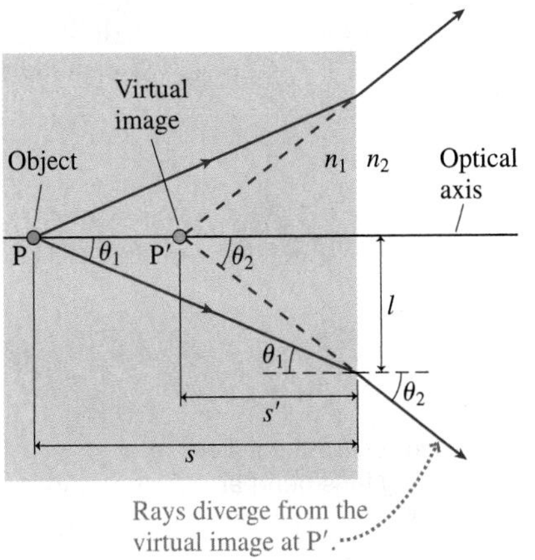

FIGURE 18.25 Finding the virtual image P′ of an object at P.

NOTE ► The fact that the result for s' is independent of θ_1 implies that *all* rays appear to diverge from the same point P′. This property of the diverging rays is essential in order to have a well-defined image. ◄

This section has given us a first look at image formation via refraction. We will extend this idea to image formation with lenses in the next section.

EXAMPLE 18.6 **An air bubble in a window**

A fish and a sailor look at each other through a 5.0-cm-thick glass porthole in a submarine. There happens to be a small air bubble right in the center of the glass. How far behind the surface of the glass does the air bubble appear to the fish? To the sailor?

PREPARE Represent the air bubble as a point source and use the ray model of light. Light rays from the bubble refract into the air on one side and into the water on the other. The ray diagram looks like Figure 18.25.

SOLVE The index of refraction of the glass is $n_1 = 1.50$. The bubble is in the center of the window, so the object distance from either side of the window is $s = 2.5$ cm. From the water side, the fish sees the bubble at an image distance

$$s' = \frac{n_2}{n_1}s = \frac{1.33}{1.50}(2.5\text{ cm}) = 2.2\text{ cm}$$

(1.33: The observer (the fish) is in water. 1.50: The object (the bubble) is in glass.)

This is the apparent depth of the bubble. The sailor, in air, sees the bubble at an image distance

$$s' = \frac{n_2}{n_1}s = \frac{1.00}{1.50}(2.5\text{ cm}) = 1.7\text{ cm}$$

ASSESS The image distance is *shorter* for the sailor because of the *larger* difference between the two indices of refraction.

18.5 Thin Lenses: Ray Tracing

A **lens** is a transparent material that uses refraction of light rays at *curved* surfaces to form an image. In this section we want to establish a pictorial method of understanding image formation. This method is called **ray tracing.** We will defer a mathematical analysis of the image formation by lenses until the next section.

Class Video

FIGURE 18.26 Converging and diverging lenses.

(a) Converging lenses, which are thicker in the center than at the edges, refract parallel rays toward the optical axis.

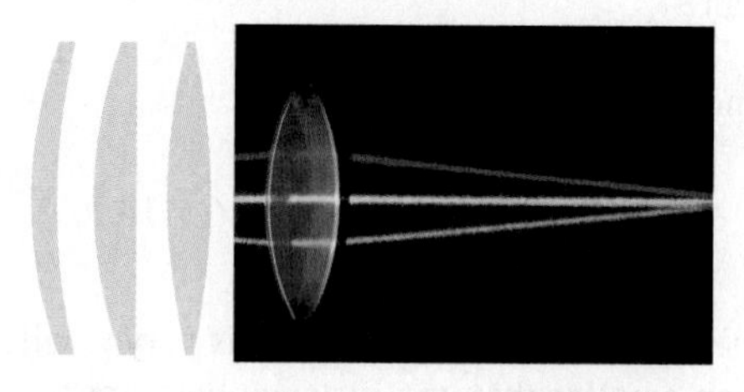

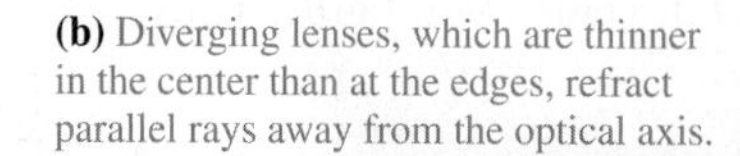

(b) Diverging lenses, which are thinner in the center than at the edges, refract parallel rays away from the optical axis.

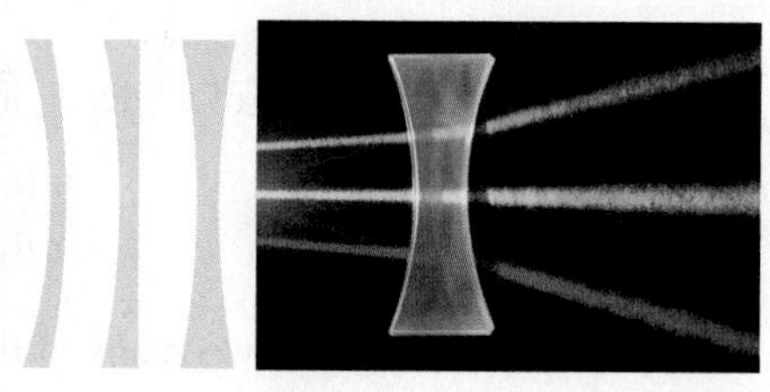

FIGURE 18.27 Both surfaces of a converging lens bend an incident ray toward the optical axis.

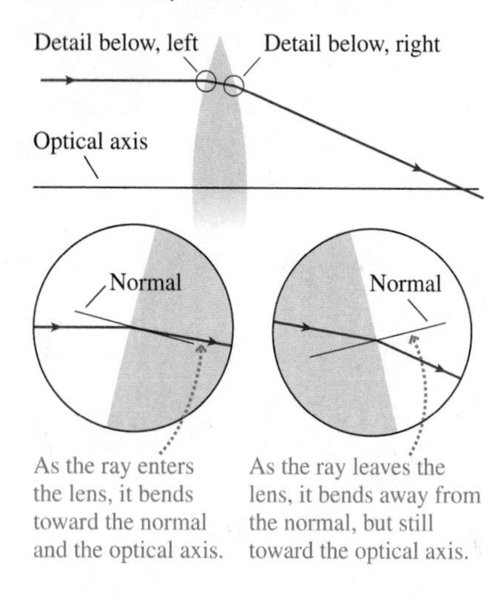
Detail below, left
Detail below, right
Optical axis
Normal
Normal
As the ray enters the lens, it bends toward the normal and the optical axis.
As the ray leaves the lens, it bends away from the normal, but still toward the optical axis.

FIGURE 18.26 shows parallel light rays entering two different lenses. The lens in Figure 18.26a, called a **converging lens,** causes the rays to refract *toward* the optical axis. FIGURE 18.27 shows how this works. An incoming ray refracts toward the optical axis at both the first, air-to-glass boundary *and* the second, glass-to-air boundary. FIGURE 18.28a shows that *all* incoming rays initially parallel to the optical axis converge at the *same* point, the **focal point** of the lens. The distance of the focal point from the lens is called the **focal length** f of the lens.

If the parallel rays approached the lens from the right side instead, they would focus to a point on the left side of the lens, indicating that there is a focal point on *each* side of the lens. Both focal points are the same distance f from the lens. The focal point on the side from which the light is incident is the *near focal point;* the focal point on the other side is the *far focal point.*

The lens in Figure 18.26b, called a **diverging lens,** causes the rays to refract *away* from the axis. A diverging lens also has two focal points, although these are not as obvious as for a converging lens. Figure 18.28b clarifies the situation. A backward projection of the diverging rays shows that they all *appear* to have started from the same point. This is the near focal point of a diverging lens, and its distance from the lens is the focal length of the lens. For both types of lenses, **the focal length is the distance from the lens to the point at which rays parallel to the optical axis converge or from which they appear to diverge.**

It is important to note that the focal points and focal length are properties of the particular lens itself. The curvature of its surfaces and its index of refraction determine its focal length. Further, for a converging lens, only rays initially parallel to the optical axis converge at the focal point. Rays not initially parallel to the axis may also converge at a different point, but this point is not the focal point. Likewise, for a diverging lens, only rays initially parallel to the optical axis appear to diverge from the focal point.

FIGURE 18.28 The focal point and focal length of converging and diverging lenses.

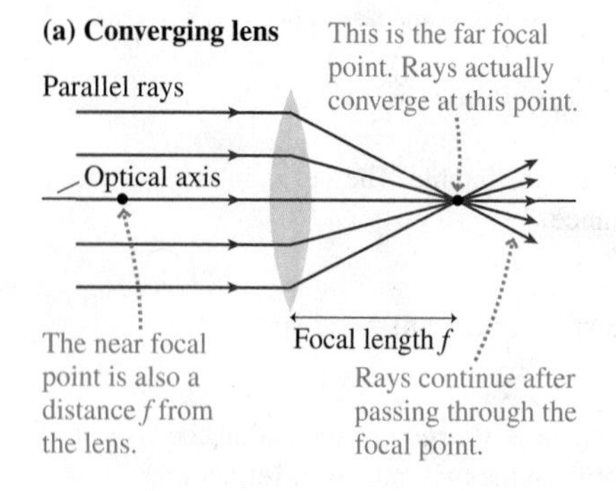
(a) Converging lens
Parallel rays
This is the far focal point. Rays actually converge at this point.
Optical axis
The near focal point is also a distance f from the lens.
Focal length f
Rays continue after passing through the focal point.

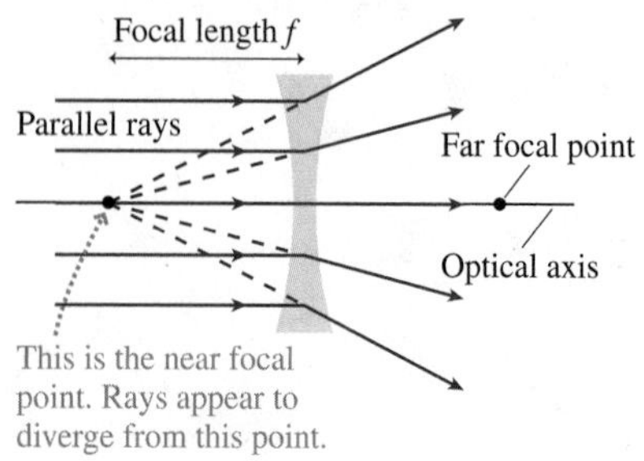
(b) Diverging lens
Focal length f
Parallel rays
Far focal point
Optical axis
This is the near focal point. Rays appear to diverge from this point.

NOTE ▸ For now, we take the focal length f to be a positive quantity. Later, when we introduce the *thin-lens equation,* we'll find that the focal length is sometimes negative. ◂

Converging Lenses

These basic observations about lenses are enough for us to understand image formation by a **thin lens,** an idealized lens whose thickness is zero and that lies entirely in a plane called the **lens plane.** Within this *thin-lens approximation,* **all refraction occurs as the rays cross the lens plane, and all distances are measured from the lens plane.** Fortunately, the thin-lens approximation is quite good for most practical applications of lenses.

NOTE ▸ We'll *draw* lenses as if they have a thickness, because that is how we expect lenses to look, but our analysis will not depend on the shape or thickness of a lens. ◂

FIGURE 18.29 shows three important sets of light rays passing through a thin, converging lens. Part (a) is familiar from Figure 18.28. If the direction of each of the rays in Figure 18.29a is reversed, Snell's law tells us that each ray will exactly retrace its path and emerge from the lens parallel to the optical axis. This leads to Figure 18.29b, which is the "mirror image" of part (a).

FIGURE 18.29 Three important sets of rays passing through a thin, converging lens.

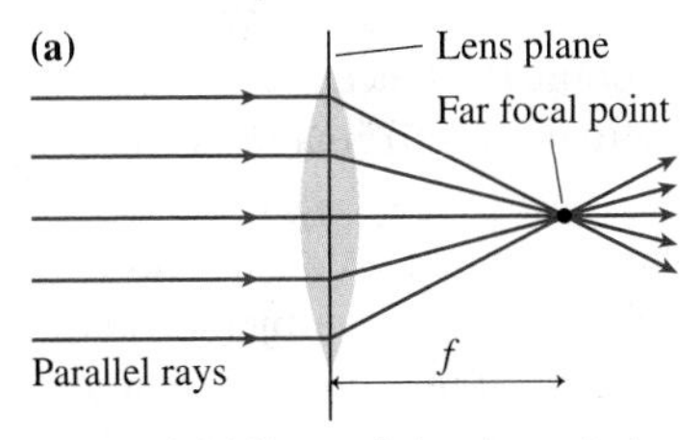

Any ray initially parallel to the optical axis will refract through the focal point on the far side of the lens.

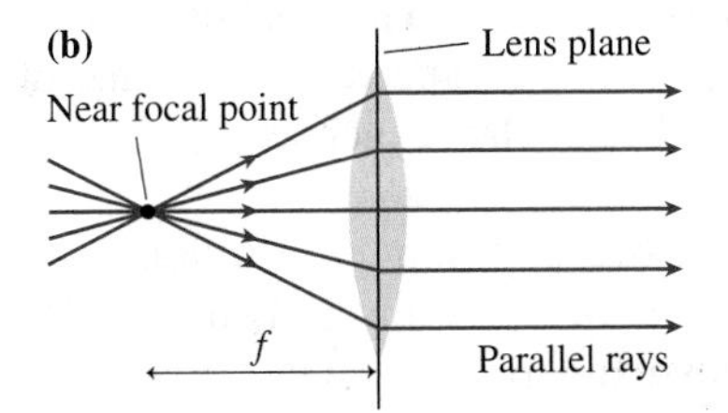

Any ray passing through the near focal point emerges from the lens parallel to the optical axis.

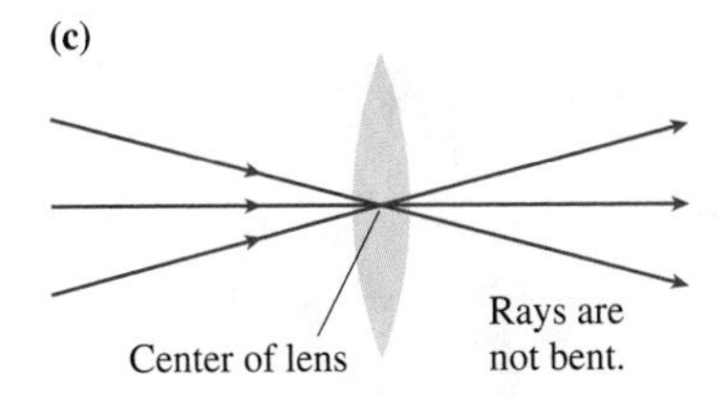

Any ray directed at the center of the lens passes through in a straight line.

Figure 18.29c shows three rays passing through the *center* of the lens. At the center, the two sides of a lens are very nearly parallel to each other. Earlier, in Example 18.3, we found that a ray passing through a piece of glass with parallel sides is *displaced* but *not bent* and that the displacement becomes zero as the thickness approaches zero. Consequently, a ray through the center of a thin lens, which has zero thickness, is neither bent nor displaced but travels in a straight line.

These three situations form the basis for ray tracing.

Real Images

FIGURE 18.30 shows a lens and an object whose distance s from the lens is larger than the focal length. Rays from point P on the object are refracted by the lens so as to converge at point P′ on the opposite side of the lens, at a distance s' from the lens. If rays diverge from an object point P and interact with a lens such that the refracted rays *converge* at point P′, actually meeting at P′, then we call P′ a **real image** of point P. Contrast this with our prior definition of a *virtual image* as a point from which rays appear to *diverge*, but through which no rays actually pass.

FIGURE 18.30 Rays from an object point P are refracted by the lens and converge to a real image at point P′.

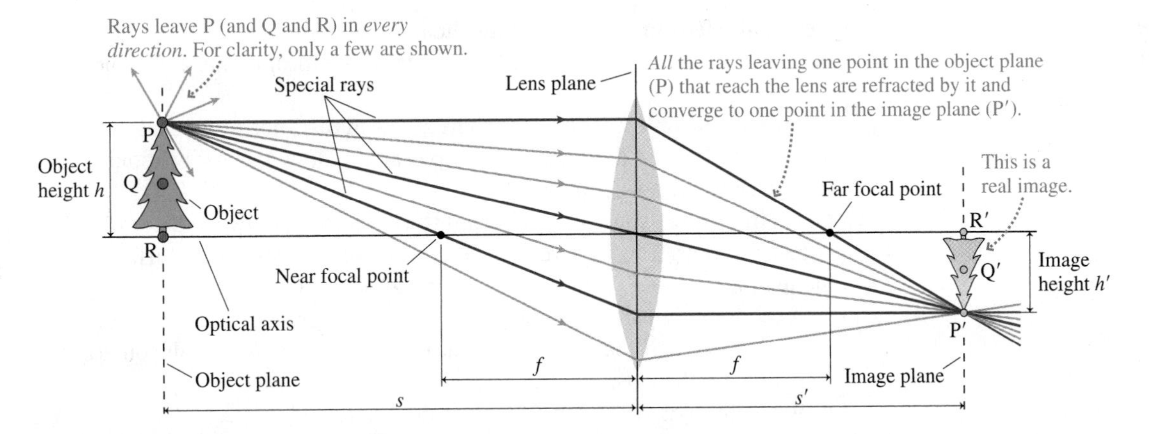

STOP TO THINK 18.4 An object and lens are positioned to form a well-focused, inverted image on a viewing screen. Then a piece of cardboard is lowered just in front of the lens to cover the top half of the lens. Using what you've learned from Figure 18.30, what happens to the image on the screen?

A. Nothing
B. The upper half of the image vanishes.
C. The lower half of the image vanishes.
D. The image becomes fuzzy and out of focus.
E. The image becomes dimmer but remains in focus.

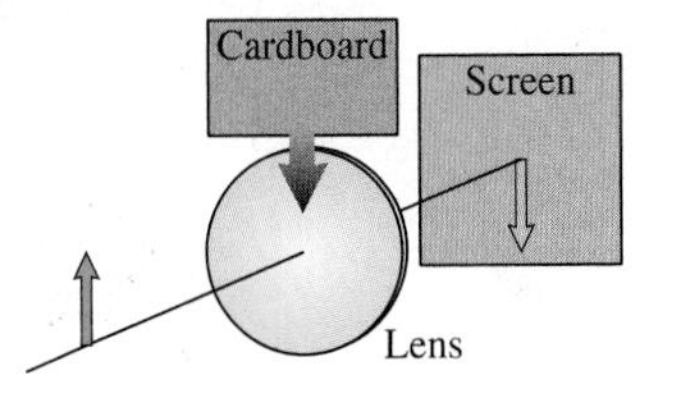

FIGURE 18.31 The lamp's image is upside down.

FIGURE 18.32 A close-up look at the rays and images near the image plane.

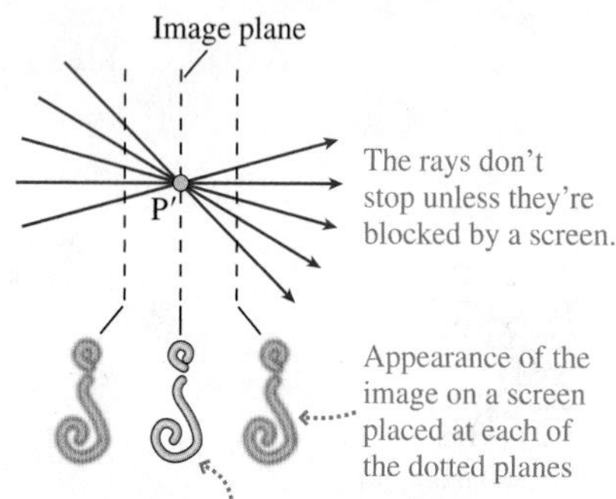
Image plane
P′
The rays don't stop unless they're blocked by a screen.
Appearance of the image on a screen placed at each of the dotted planes
A sharp, well-focused image is seen on a screen placed in the image plane.

All points on the object that are in the same plane, the **object plane,** converge to image points in the **image plane.** Points Q and R in the object plane of Figure 18.30 have image points Q′ and R′ in the same plane as point P′. Once we locate *one* point in the image plane, such as point P′, we know that the full image lies in the same plane.

There are two important observations to make about Figure 18.30. First, as also seen in **FIGURE 18.31**, the image is upside down with respect to the object. This is called an **inverted image,** and it is a standard characteristic of real-image formation. Second, rays from point P *fill* the entire lens surface, so that all portions of the lens contribute to the image. A larger lens will "collect" more rays and thus make a brighter image.

FIGURE 18.32 is a close-up view of the rays and images very near the image plane. The rays don't stop at P′ unless we place a screen in the image plane. When we do so, we see a sharp, well-focused image on the screen. If a screen is placed other than in the image plane, an image is produced on the screen, but it's blurry and out of focus.

NOTE ▸ Our ability to see a real image on a screen sets real images apart from virtual images. But keep in mind that we need not *see* a real image in order to *have* an image. A real image exists at a point in space where the rays converge even if there's no viewing screen in the image plane. ◂

Figure 18.30 highlights the three "special rays" that are based on the three situations of Figure 18.29. Notice that these three rays alone are sufficient to locate the image point P′. That is, we don't need to draw all the rays shown in Figure 18.30. The procedure known as *ray tracing* consists of locating the image by the use of just these three rays.

TACTICS BOX 18.2 Ray tracing for a converging lens

MP

1. **Draw an optical axis.** Use graph paper or a ruler! Establish an appropriate scale.
2. **Center the lens on the axis.** Draw the lens plane perpendicular to the axis through the center of the lens. Mark and label the focal points at distance f on either side.
3. **Represent the object with an upright arrow at distance s.** It's usually best to place the base of the arrow on the axis and to draw the arrow about half the radius of the lens.
4. **Draw the three "special rays" from the tip of the arrow.** Use a straightedge or a ruler. The rays refract at the lens plane, *not* at the surfaces of the lens.
   a. A ray initially parallel to the axis refracts through the far focal point.
   b. A ray that enters the lens along a line through the near focal point emerges parallel to the axis.
   c. A ray through the center of the lens does not bend.
5. **Extend the rays until they converge.** The rays converge at the image point. Draw the rest of the image in the image plane. If the base of the object is on the axis, then the base of the image will also be on the axis.
6. **Measure the image distance s'.** Also, if needed, measure the image height relative to the object height. The magnification can be found from Equation 18.8.

Exercise 19

EXAMPLE 18.7 Finding the image of a flower

A 4.0-cm-diameter flower is 200 cm from the 50-cm-focal-length lens of a camera. How far should the plane of the camera's light detector be placed behind the lens to record a well-focused image? What is the diameter of the image on the detector?

PREPARE The flower is in the object plane. Use ray tracing to locate the image.

SOLVE FIGURE 18.33 shows the ray-tracing diagram and the steps of Tactics Box 18.2. The image has been drawn in the plane where the three special rays converge. You can see *from the drawing* that the image distance is $s' \approx 65$ cm. This is where the detector needs to be placed to record a focused image. The heights of the object and image are labeled h and h'. The ray through the center of the lens is a straight line; thus the object and image both subtend the same angle θ. From similar triangles,

$$\frac{h'}{s'} = \frac{h}{s}$$

Solving for h' gives

$$h' = h\frac{s'}{s} = (4.0\text{ cm})\frac{65\text{ cm}}{200\text{ cm}} = 1.3\text{ cm}$$

The flower's image has a diameter of 1.3 cm.

ASSESS We've been able to learn a great deal about the image from a simple geometric procedure.

FIGURE 18.33 Ray-tracing diagram for the image of a flower.

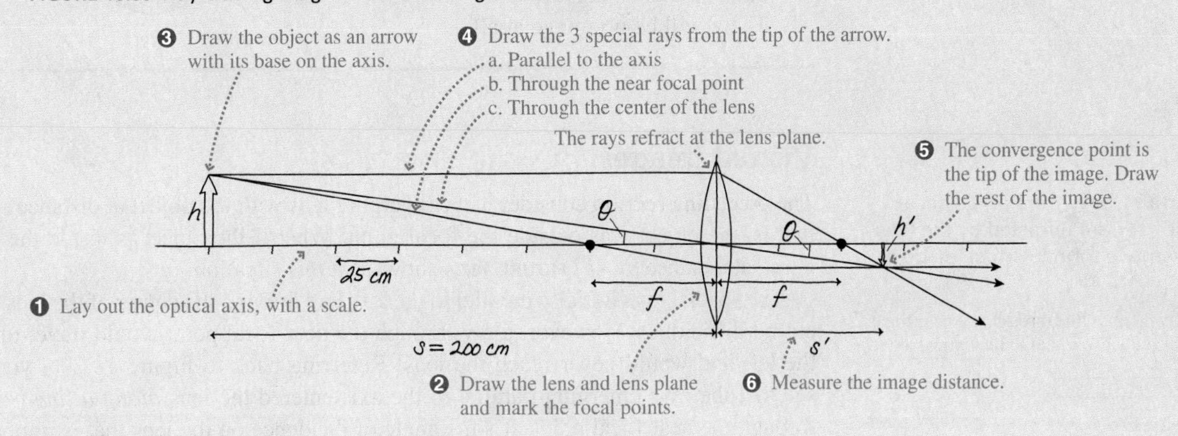
❸ Draw the object as an arrow with its base on the axis.
❹ Draw the 3 special rays from the tip of the arrow.
a. Parallel to the axis
b. Through the near focal point
c. Through the center of the lens
The rays refract at the lens plane.
❺ The convergence point is the tip of the image. Draw the rest of the image.
h
θ
θ
h′
25 cm
f
f
❶ Lay out the optical axis, with a scale.
s = 200 cm
s′
❷ Draw the lens and lens plane and mark the focal points.
❻ Measure the image distance.

Magnification

The image can be either larger or smaller than the object, depending on the location and focal length of the lens. Because the image height scales with that of the object, we're usually interested in the *ratio* h'/h of the image height to the object height. This ratio is greater than 1 when the image is taller than the object, and less than 1 when the image is shorter than the object.

But there's more to a description of the image than just its size. We also want to know its *orientation* relative to the object; that is, is the image upright or inverted? It is customary to combine image size and orientation information in a single number, the **magnification** m, defined as

$$m = -\frac{s'}{s} \tag{18.8}$$

Magnification of a lens or mirror

You saw in Example 18.7 that $s'/s = h'/h$. Consequently, we interpret the magnification m as follows:

1. The absolute value of m gives the ratio of image height to object height: $h'/h = |m|$.
2. A positive value of m indicates that the image is upright relative to the object. A negative value of m indicates that the image is inverted relative to the object.

The magnification in Example 18.7 would be

$$m = -\frac{s'}{s} = -\frac{65\text{ cm}}{200\text{ cm}} = -0.33$$

indicating that the image is 33% the size of the object and, because of the minus sign, is inverted.

NOTE ▶ Equation 18.8 applies to real or virtual images produced by both lenses and mirrors. Although s and s' are both positive in Example 18.7, leading to a negative magnification, we'll see that this is *not* the case for virtual images. ◀

STOP TO THINK 18.5 A lens produces a sharply focused, inverted image on a screen. What will you see on the screen if the lens is removed?

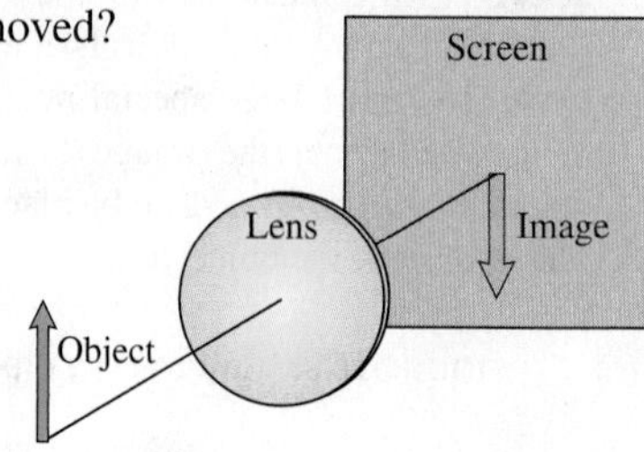

A. The image will be inverted and blurry.
B. The image will be upright and sharp.
C. The image will be upright and blurry.
D. The image will be much dimmer but otherwise unchanged.
E. There will be no image at all.

Virtual Images

The preceding section considered a converging lens with the object at distance $s > f$; that is, the object was outside the focal point. What if the object is inside the focal point, at distance $s < f$? FIGURE 18.34 shows just this situation.

FIGURE 18.34 Rays from an object at distance $s < f$ are refracted by the lens and diverge to form a virtual image at point P′.

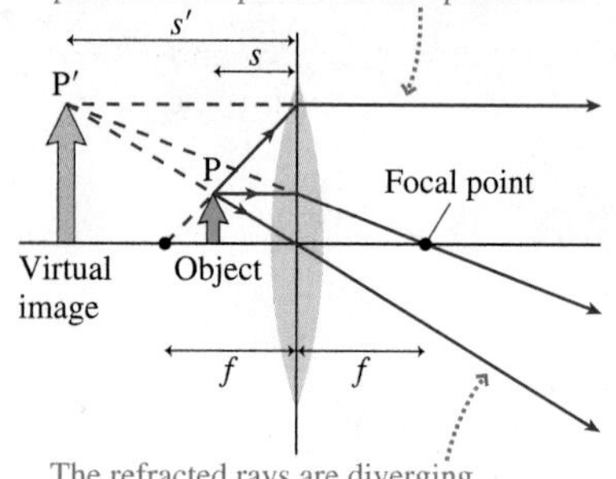

The special rays initially parallel to the axis and through the center of the lens present no difficulties. However, a ray through the near focal point would travel toward the left and would never reach the lens! Referring back to Figure 18.29b, you can see that the rays emerging parallel to the axis entered the lens *along a line* passing through the near focal point. It's the angle of incidence on the lens that is important, not whether the light ray actually passes through the focal point. This was the basis for the wording of step 4b in Tactics Box 18.2 and is the third special ray shown in Figure 18.34.

The three refracted rays don't converge. Instead, all three rays appear to *diverge* from point P′. This is the situation we found for rays reflecting from a mirror and for the rays refracting out of an aquarium. Point P′ is a *virtual image* of the object point P. Furthermore, it is an **upright image,** having the same orientation as the object.

The refracted rays, which are all to the right of the lens, *appear* to come from P′, but none of the rays were ever at that point. No image would appear on a screen placed in the image plane at P′. So what good is a virtual image?

Your eye collects and focuses bundles of diverging rays. Thus, as FIGURE 18.35a shows, you can "see" a virtual image by looking through the lens. This is exactly what you do with a magnifying glass, producing a scene like the one in FIGURE 18.35b. In fact, you view a virtual image any time you look through the eyepiece of an optical instrument such as a microscope or binoculars.

FIGURE 18.35 A converging lens is a magnifying glass when the object distance is $< f$.

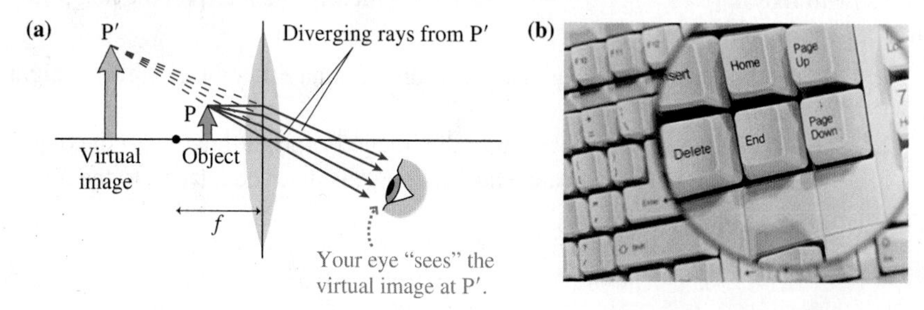

NOTE ► Recall that a lens thicker in the middle than at the edges is classified as a converging lens. The light rays from an object *can* converge to form a real image after passing through such a lens, but only if the object distance is greater than the focal length of the lens: $s > f$. If $s < f$, the rays leaving a converging lens diverge to produce a virtual image. ◄

Because a virtual image is upright, the magnification $m = -s'/s$ is positive. This means that the ratio s'/s must be *negative*. We can ensure this if **we define the image distance s' to be negative for a virtual image,** indicating that the image is on the *same* side of the lens as the object. This is our first example of a **sign convention** for the various distances that appear in understanding image formation from lenses and mirrors. We'll have more to say about sign conventions when we study the thin-lens equation in a later section.

EXAMPLE 18.8 Magnifying a flower

To see a flower better, you hold a 6.0-cm-focal-length magnifying glass 4.0 cm from the flower. What is the magnification?

PREPARE The flower is in the object plane. Use ray tracing to locate the image. Once the image distance is known, Equation 18.8 can be used to find the magnification.

SOLVE FIGURE 18.36 shows the ray-tracing diagram. The three special rays diverge from the lens, but we can use a straightedge to extend the rays backward to the point from which they diverge. This point, the image point, is seen to be 12 cm to the left of the lens. Because this is a virtual image, the image distance is $s' = -12$ cm. From Equation 18.8 the magnification is

$$m = -\frac{s'}{s} = -\frac{-12\text{ cm}}{4.0\text{ cm}} = 3.0$$

FIGURE 18.36 Ray-tracing diagram for a magnifying glass.

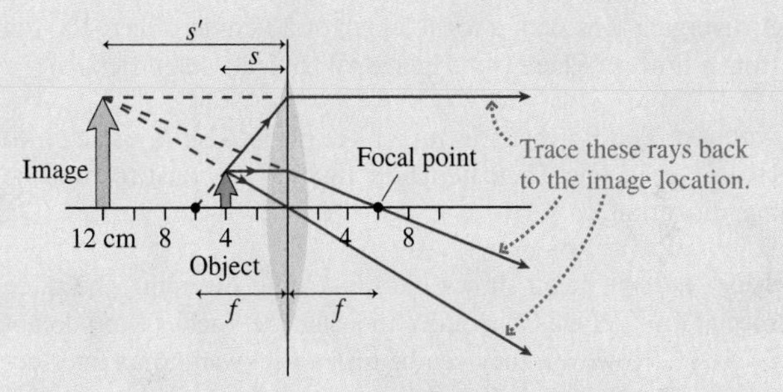

ASSESS The image is three times as large as the object and, as we see from the ray-tracing diagram and the fact that $m > 0$, upright.

Diverging Lenses

As Figure 18.26b showed, a *diverging lens* is one that is thinner at its center than at its edge. FIGURE 18.37 shows three important sets of rays passing through a diverging lens. These are based on Figures 18.26 and 18.28, where you saw that rays initially parallel to the axis diverge after passing through a diverging lens.

FIGURE 18.37 Three important sets of rays passing through a thin, diverging lens.

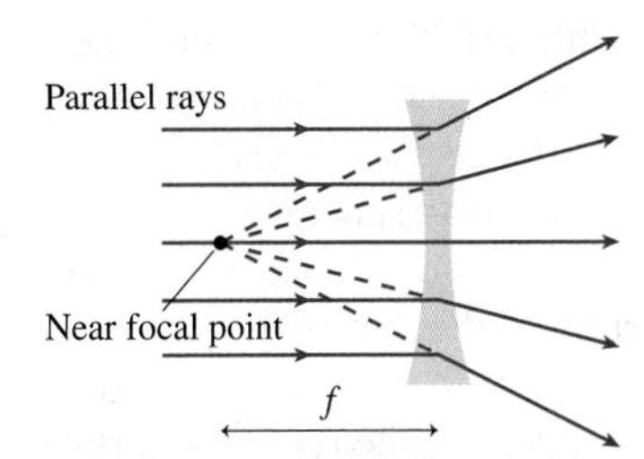

Any ray initially parallel to the optical axis diverges along a line through the near focal point.

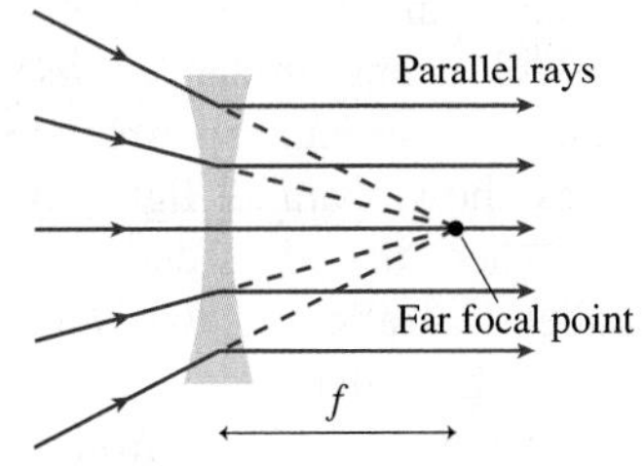

Any ray directed along a line toward the far focal point emerges from the lens parallel to the optical axis.

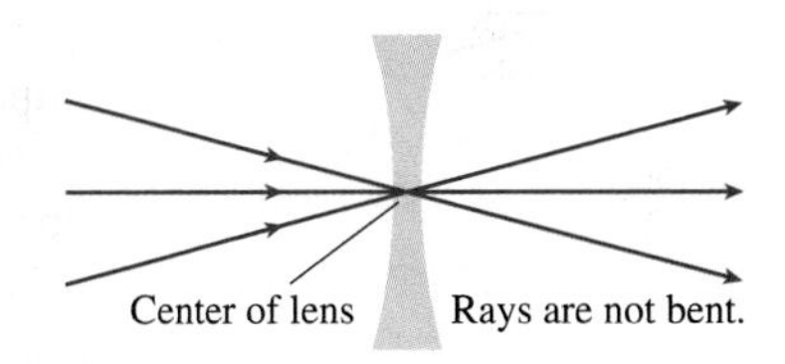

Any ray directed at the center of the lens passes through in a straight line.

Ray tracing follows the steps of Tactics Box 18.2 for a converging lens *except* that two of the three special rays in step 4 are different.

TACTICS BOX 18.3 **Ray tracing for a diverging lens**

❶–❸ **Follow steps 1 through 3 of Tactics Box 18.2.**

❹ **Draw the three "special rays" from the tip of the arrow.** Use a straightedge or a ruler. The rays refract at the lens plane.
  a. A ray parallel to the axis diverges along a line through the near focal point.
  b. A ray along a line toward the far focal point emerges parallel to the axis.
  c. A ray through the center of the lens does not bend.

❺ **Trace the diverging rays backward.** The point from which they are diverging is the image point, which is always a virtual image.

❻ **Measure the image distance s',** which, because the image is virtual, we will take as a negative number. Also, if needed, measure the image height relative to the object height. The magnification can be found from Equation 18.8.

EXAMPLE 18.9 **Demagnifying a flower**

A diverging lens with a focal length of 50 cm is placed 100 cm from a flower. Where is the image? What is its magnification?

PREPARE The flower is in the object plane. Use ray tracing to locate the image. Then Equation 18.8 can be used to find the magnification.

SOLVE FIGURE 18.38 shows the ray-tracing diagram. The three special rays (labeled a, b, and c to match the Tactics Box) do not converge. However, they can be traced backward to an intersection ≈ 33 cm to the left of the lens. Because the rays appear to diverge from the image, this is a virtual image and s' is < 0. The magnification is

$$m = -\frac{s'}{s} = -\frac{-33 \text{ cm}}{100 \text{ cm}} = 0.33$$

FIGURE 18.38 Ray-tracing diagram for demagnifying.

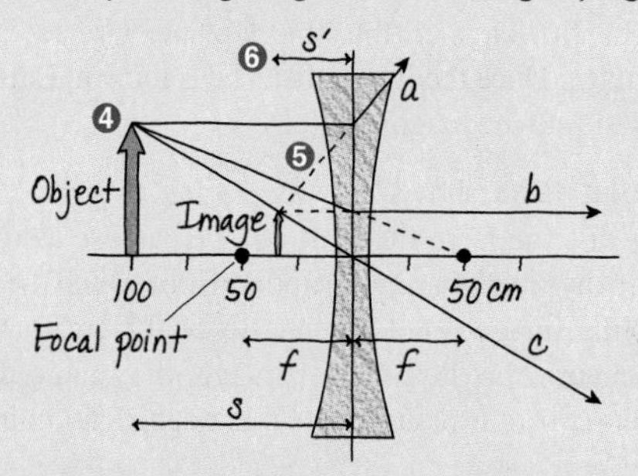

The image, which can be seen by looking *through* the lens, is one-third the size of the object and upright.

ASSESS Ray tracing with a diverging lens is somewhat trickier than with a converging lens, so this example is worth careful study.

Diverging lenses *always* make virtual images and, for this reason, are rarely used alone. However, they have important applications when used in combination with other lenses. Cameras, eyepieces, and eyeglasses often incorporate diverging lenses.

FIGURE 18.39 The focal point and focal length of concave and convex mirrors.

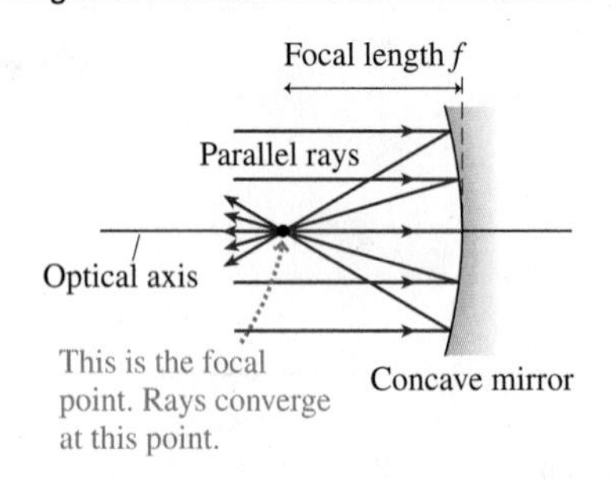

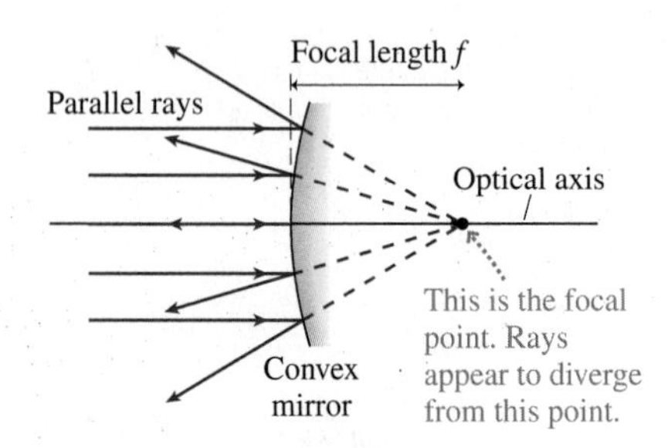

18.6 Image Formation with Spherical Mirrors

Curved mirrors can also be used to form images. Such mirrors are commonly used in telescopes, security and rearview mirrors, and searchlights. Their images can be analyzed with ray diagrams similar to those used with lenses. We'll consider only the important case of **spherical mirrors,** whose surface is a section of a sphere.

FIGURE 18.39 shows parallel light rays approaching two spherical mirrors. The upper mirror, where the edges curve toward the light source, is called a **concave mirror.** Parallel rays reflect off the shiny front surface of the mirror and pass through a single point on the optical axis. This is the focal point of the mirror. The lower mirror, where the edges curve away from the light source, is called a **convex mirror.** Parallel rays that reflect off its surface appear to have come from a point behind the mirror. This is the focal point for a convex mirror. For both mirrors, the focal length is the distance from the mirror surface to the focal point.

Concave Mirrors

To understand image formation by a concave mirror, consider the three special rays shown in FIGURE 18.40. These rays are closely related to those used for ray tracing with lenses. Figure 18.40a shows two incoming rays parallel to the optical axis. As Figure 18.39 showed, these rays reflect off the mirror and pass through the focal point.

Figure 18.40b is the same as Figure 18.40a, but with the directions of the rays reversed. Here we see that rays passing through the focal point emerge parallel to the axis. Finally, Figure 18.40c shows what happens to a ray that is directed toward the center of the mirror. Right at its center, the surface of the mirror is perpendicular to the optical axis. The law of reflection then tells us that the incoming ray will reflect at the same angle, but on the opposite side of the optical axis.

Let's begin by considering the case where the object's distance s from the mirror is greater than the focal length ($s > f$), as shown in **FIGURE 18.41**. The three special rays just discussed are enough to locate the position and size of the image. Recall that when ray tracing a thin lens, although we drew the lens as having an actual thickness, the rays refracted at an imaginary plane centered on the lens. Similarly, when ray tracing mirrors, the incoming rays reflect off the **mirror plane** as shown in Figure 18.41, not off the curved surface of the mirror. We see that the image is *real* because rays converge at the image point P′. Further, the image is *inverted.*

FIGURE 18.40 Three special rays for a concave mirror.

(a)

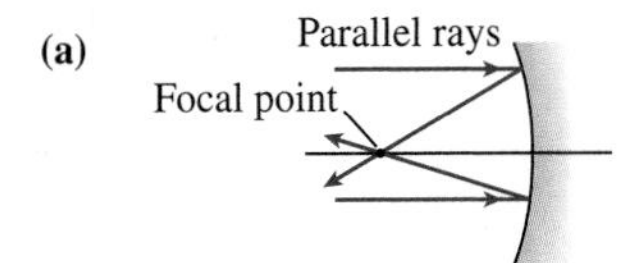

Any ray initially parallel to the optical axis will reflect through the focal point.

(b)

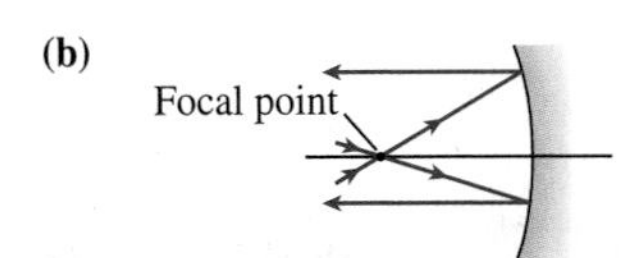

Any ray passing through the focal point will, after reflection, emerge parallel to the optical axis.

(c)

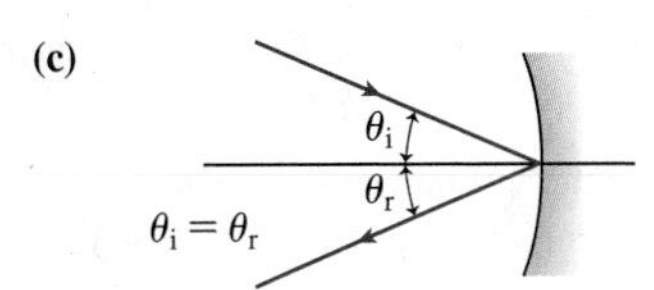

Any ray directed at the center of the mirror will reflect at an equal angle on the opposite side of the optical axis.

FIGURE 18.41 A real image formed by a concave mirror.

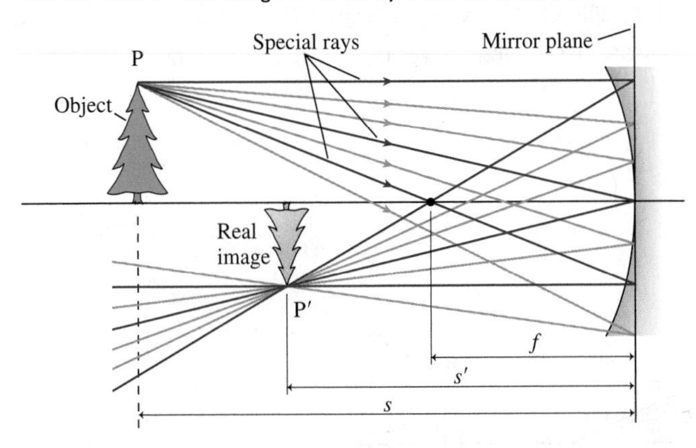

Figure 18.41 suggests the following Tactics Box for using ray tracing with a concave mirror:

TACTICS BOX 18.4 Ray tracing for a concave mirror

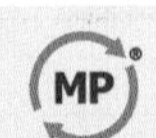

1. **Draw an optical axis.** Use graph paper or a ruler! Establish an appropriate scale.
2. **Center the mirror on the axis.** Mark and label the focal point at distance f from the mirror's surface. Draw the mirror plane through the mirror's center, perpendicular to the axis.
3. **Represent the object with an upright arrow at distance s.** It's usually best to place the base of the arrow on the axis and to draw the arrow about half the radius of the mirror.
4. **Draw the three "special rays" from the tip of the arrow.** Use a straightedge or a ruler. The rays should reflect off the mirror plane.
   a. A ray parallel to the axis reflects through the focal point.
   b. An incoming ray that passes through the focal point reflects back parallel to the axis.
   c. A ray that strikes the center of the mirror reflects at an equal angle on the opposite side of the optical axis.
5. **Extend the rays until they converge.** The rays converge at the image point. Draw the rest of the image in the image plane. If the base of the object is on the axis, then the base of the image will also be on the axis.
6. **Measure the image distance s'.** Also, if needed, measure the image height relative to the object height. The magnification can be found from Equation 18.8.

Exercises 21a, 22

▶ **Look into my eyes** BIO The eyes of most animals use lenses to focus an image. The *gigantocypris,* a deep-sea crustacean, is unusual in that it uses two concave mirrors to focus light onto its retina. Because it lives at depths where no sunlight penetrates, it is believed that gigantocypris uses its mirror eyes to hunt bioluminescent animals.

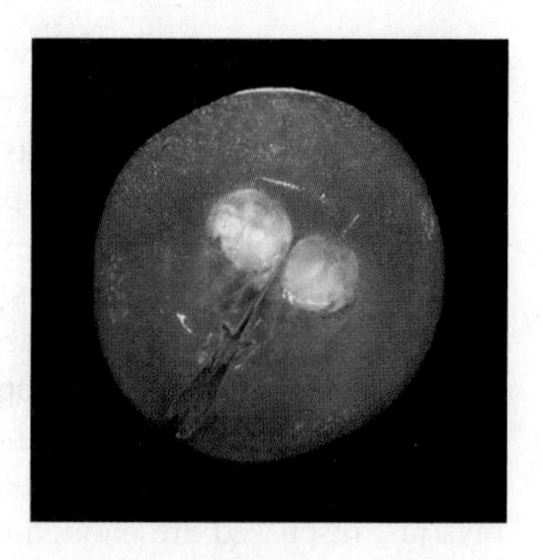

EXAMPLE 18.10 Analyzing a concave mirror

A 3.0-cm-high object is located 60 cm from a concave mirror. The mirror's focal length is 40 cm. Use ray tracing to find the position, height, and magnification of the image.

PREPARE FIGURE 18.42 shows the ray-tracing diagram and the steps of Tactics Box 18.4.

SOLVE After preparing a careful drawing, we can use a ruler to find that the image position is $s' \approx 120$ cm. The magnification is thus

$$m = -\frac{s'}{s} \approx -\frac{120\text{ cm}}{60\text{ cm}} = -2.0$$

The negative sign indicates that the image is inverted. The image height is thus twice the object height, or $h' \approx 6$ cm.

ASSESS The image is a *real* image because light rays converge at the image point.

FIGURE 18.42 Ray-tracing diagram for a concave mirror.

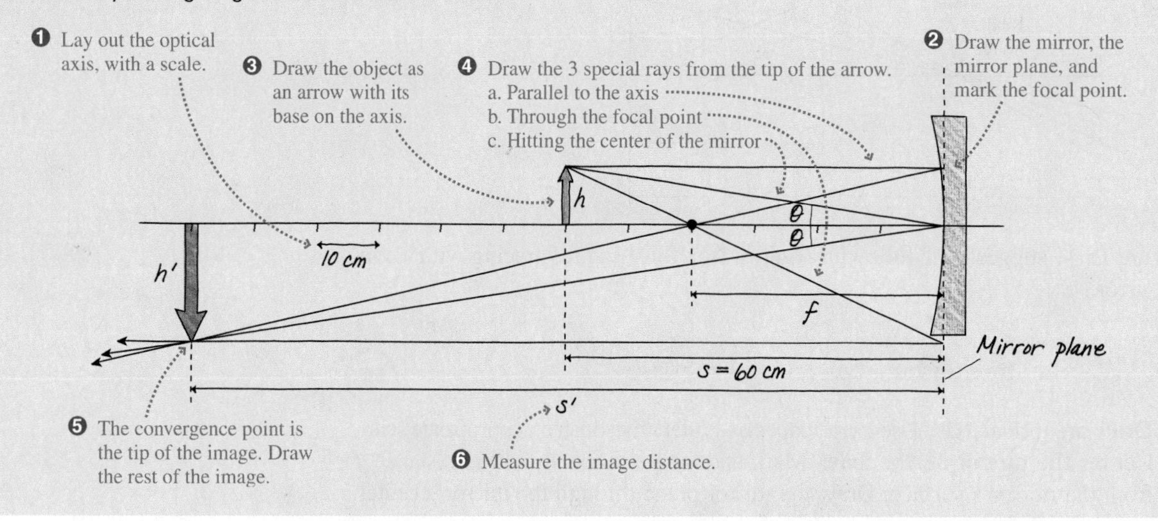
❶ Lay out the optical axis, with a scale.
❷ Draw the mirror, the mirror plane, and mark the focal point.
❸ Draw the object as an arrow with its base on the axis.
❹ Draw the 3 special rays from the tip of the arrow.
a. Parallel to the axis
b. Through the focal point
c. Hitting the center of the mirror
h
θ
θ
10 cm
h'
f
Mirror plane
s = 60 cm
s'
❺ The convergence point is the tip of the image. Draw the rest of the image.
❻ Measure the image distance.

If the object is inside the focal point ($s < f$), ray tracing can be used to show that the image is a virtual image. This situation is analogous to the formation of a virtual image by a lens when the object is inside the focal point.

FIGURE 18.43 Reflection from a silvered ball.

Convex Mirrors

A common example of a convex mirror is a silvered ball, such as a tree ornament. You may have noticed that if you look at your reflection in such a ball, your image appears right side up but is quite small, as shown in FIGURE 18.43. Let's use ray tracing to see why the image appears this way.

Once more, there are three special rays we can use to find the location of the image. These rays are shown in FIGURE 18.44; they are similar to the special rays we've already studied in other situations.

FIGURE 18.44 Three special rays for a convex mirror.

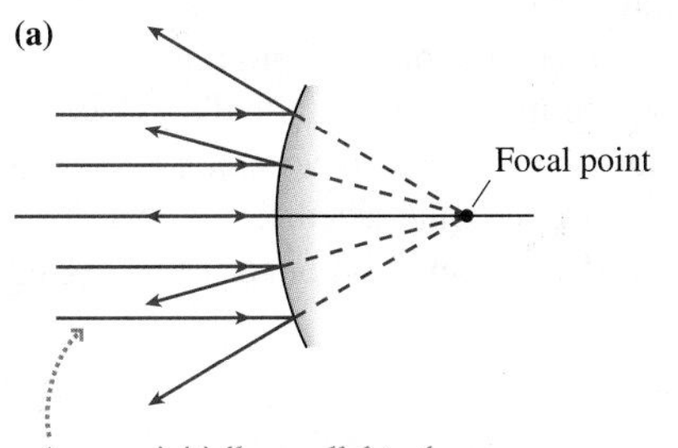

Any ray initially parallel to the optical axis will reflect as though it came from the focal point.

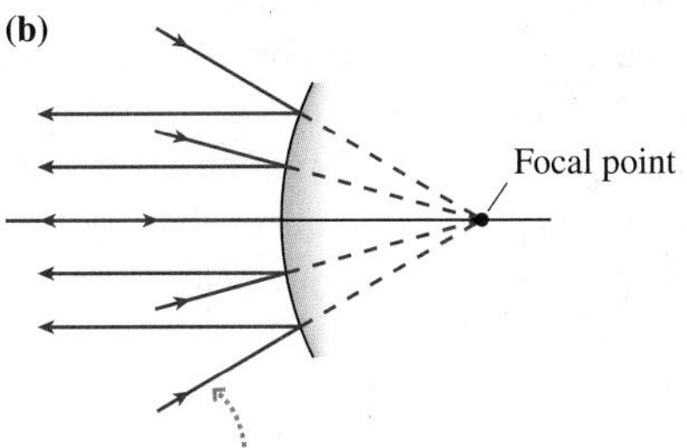

Any ray initially directed toward the focal point will reflect parallel to the optical axis.

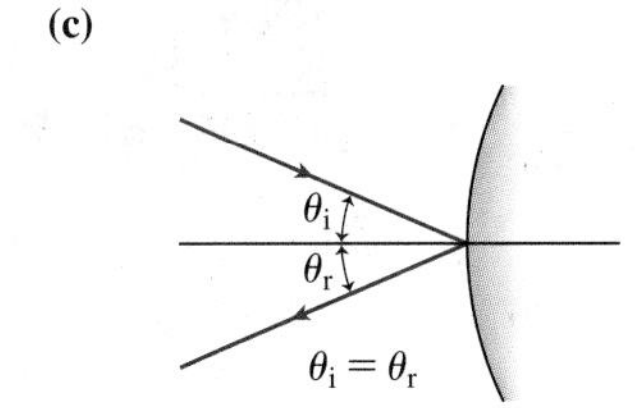

Any ray directed at the center of the mirror will reflect at an equal angle on the opposite side of the optical axis.

We can use these three special rays to find the image of an object, as shown in FIGURE 18.45. We see that the image is virtual—no actual rays converge at the image point P′. Instead, diverging rays *appear* to have come from this point. The image is also upright and much smaller than the object, in accord with our experience and the drawing of Figure 18.43.

FIGURE 18.45 Rays from point P reflect from the mirror and appear to have come from point P′.

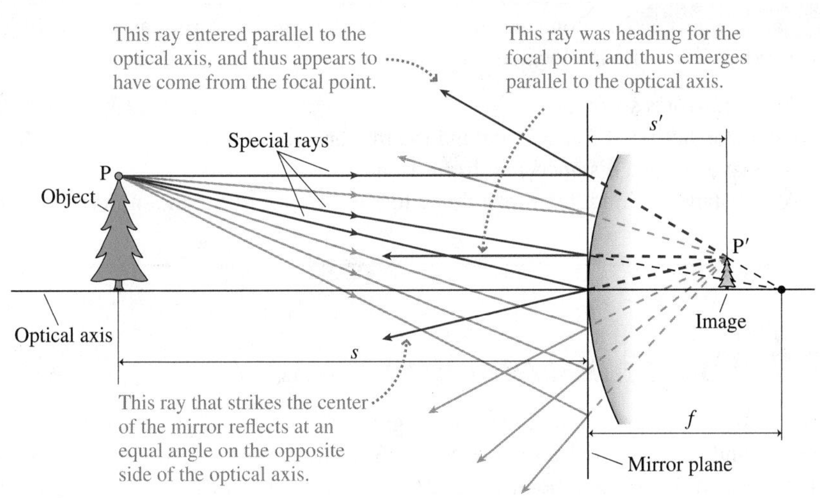

These observations form the basis of the following Tactics Box.

TACTICS BOX 18.5 Ray tracing for a convex mirror

❶–❸ **Follow steps 1 through 3 of Tactics Box 18.4.**

❹ **Draw the three "special rays" from the tip of the arrow.** Use a straightedge or a ruler. The rays should reflect off the mirror plane.

a. A ray parallel to the axis reflects as though it came from the focal point.
b. A ray initially directed toward the focal point reflects parallel to the axis.
c. A ray that strikes the center of the mirror reflects at an equal angle on the opposite side of the optical axis.

❺ **Extend the emerging rays *behind the mirror* until they converge.** The point of convergence is the image point. Draw the rest of the image in the image plane. If the base of the object is on the axis, then the base of the image will also be on the axis.

❻ **Measure the image distance s'.** Also, if needed, measure the image height relative to the object height. The magnification can be found from Equation 18.8.

Exercises 21b, 23

The small image in a convex mirror allows a wide-angle view of the store to be visible.

Convex mirrors are used for a variety of safety and monitoring applications, such as passenger-side rearview mirrors and the round mirrors used in stores to keep an eye on the customers. The idea behind such mirrors can be understood from Figure 18.45. When an object is reflected in a convex mirror, the image appears smaller. Because the image is, in a sense, a miniature version of the object, you can *see much more of it* within the edges of the mirror than you could with an equal-sized flat mirror. This wide-angle view is clearly useful for checking traffic behind you or for checking up on your store.

CONCEPTUAL EXAMPLE 18.11 Driver and passenger mirrors

The rearview mirror on the driver's side of a car is a plane (flat) mirror, while the mirror on the passenger's side is convex. Why is this?

REASON It is important for the driver to have a wide field of view from either mirror. He sits close to the driver-side mirror, so it appears large and can reflect a fairly wide view of what's behind. The passenger-side mirror is quite far from the driver, so it appears relatively small. If it were flat, it would offer only a narrow view of what's behind. Making it convex, like the security mirror discussed above, provides a wider field of view, but the trade-off is a smaller image. That's why the passenger-side mirror usually contains a warning: Objects in mirror are closer than they appear!

STOP TO THINK 18.6 A concave mirror of focal length f forms an image of the moon. Where is the image located?

A. At the mirror's surface
B. Almost exactly a distance f behind the mirror
C. Almost exactly a distance f in front of the mirror
D. At a distance behind the mirror equal to the distance of the moon in front of the mirror

18.7 The Thin-Lens Equation

Ray tracing is an important tool for quickly grasping the overall positions and sizes of an object and its image. For more precise work, however, we would like a mathematical expression that relates the three fundamental quantities of an optical system: the focal length f of the lens or mirror, the object distance s, and the image distance s'. We can find such an expression by considering the converging lens in **FIGURE 18.46**. Two of the special rays are shown: one initially parallel to the optical axis that then passes through the far focal point, and the other traveling undeviated through the center of the lens.

FIGURE 18.46 Deriving the thin-lens equation.

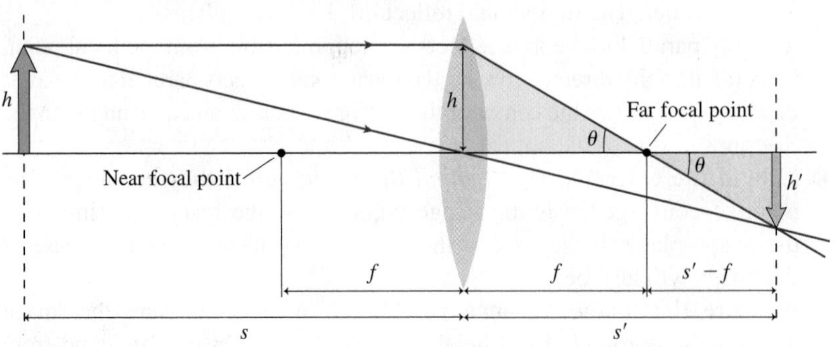

Consider the two right triangles highlighted in green and pink. Because they both have one 90° angle and a second angle θ that is the same for both, the two triangles are *similar*. This means that they have the same shape, although their sizes may be different. For similar triangles, the ratios of any two similar sides are the same. Thus we have

$$\frac{h'}{h} = \frac{s' - f}{f} \tag{18.9}$$

Further, we found in Example 18.7 that

$$\frac{h'}{h} = \frac{s'}{s} \tag{18.10}$$

Combining Equations 18.9 and 18.10 gives

$$\frac{s'}{s} = \frac{s' - f}{f}$$

Dividing both sides by s' gives

$$\frac{1}{s} = \frac{s' - f}{s'f} = \frac{1}{f} - \frac{1}{s'}$$

which we can write as

$$\frac{1}{s} + \frac{1}{s'} = \frac{1}{f} \tag{18.11}$$

Thin-lens equation (also works for mirrors) relating object and image distances to focal length

This equation, the **thin-lens equation,** relates the three important quantities f, s, and s'. In particular, if we know the focal length of a lens and the object's distance from the lens, we can use the thin-lens equation to find the position of the image.

NOTE ▸ Although we derived the thin-lens equation for a converging lens that produced a real image, it works equally well for *any* image—real or virtual—produced by both converging and diverging lenses. And, in spite of its name, the thin-lens equation also describes the images formed by *mirrors.* ◂

It's worth checking that the thin-lens equation describes what we already know about lenses. In Figure 18.29a, we saw that rays initially parallel to the optical axis are focused at the focal point of a converging lens. Initially parallel rays come from an object extremely far away, with $s \rightarrow \infty$. Because $1/\infty = 0$, the thin-lens equation tells us that the image distance is $s' = f$, as we expected. Or suppose an object is located right at the focal point, with $s = f$. Then, according to Equation 18.11, $1/s' = 1/f - 1/s = 0$. This implies that the image distance is infinitely far away ($s' = \infty$), so the rays leave the lens parallel to the optical axis. Indeed, this is what Figure 18.29b showed. Now, it's true that no real object or image can be at infinity. But if either the object or image is more than several focal lengths from the lens ($s \gg f$ or $s' \gg f$), then it's an excellent approximation to consider the distance to be infinite, the rays to be parallel to the axis, and the reciprocal ($1/s$ or $1/s'$) in the thin-lens equation to be zero.

Sign Conventions for Lenses and Mirrors

We've already noted that the image distance s' is positive for real images and negative for virtual images. In the thin-lens equation, the sign of the focal length can also be either positive or negative, depending on the type of lens or mirror. Synthesis 18.1 shows the sign conventions that we will use in this text and brings together what we've learned about lenses, mirrors, and the images they create.

SYNTHESIS 18.1 Lenses, mirrors, and their sign conventions

There are six distinct image situations that can occur for lenses and mirrors. We outline these here, and give the sign of each quantity. Note that the object distance s is always positive.

Converging lens or concave mirror

The focal length f of a converging lens or concave mirror is *positive*.

Object outside focal point

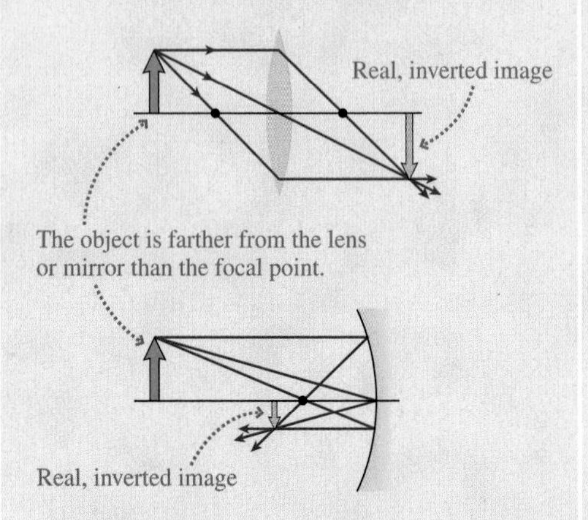

Object inside focal point

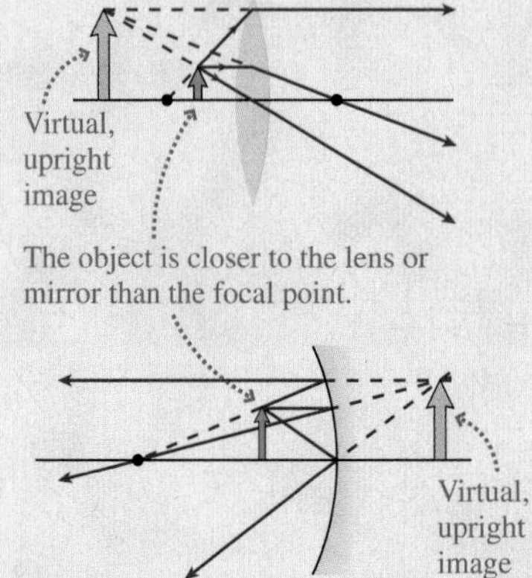

These two cases give *real* images that are *inverted* (the magnification m is *negative*). The image distance s' is *positive*.

Diverging lens or convex mirror

The focal length f of a diverging lens or convex mirror is *negative*.

For diverging lenses and convex mirrors, there is no distinction between an object inside or outside the focal point.

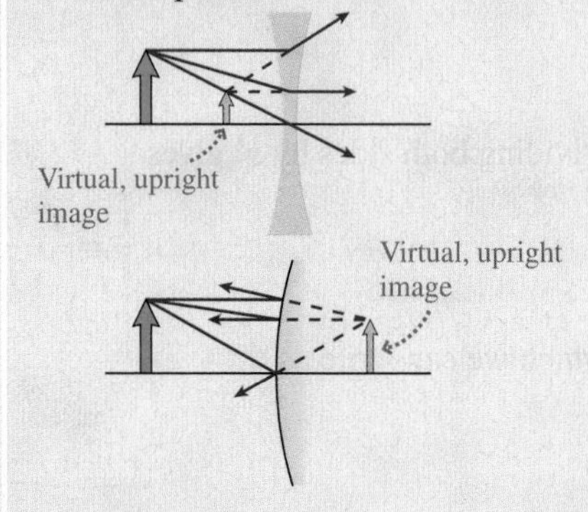

These four cases give *virtual* images that are *upright* (the magnification m is *positive*). The image distance s' is *negative*.

NOTE ▶ When the thin-lens equation is used the focal length must be taken as positive or negative according to Synthesis 18.1. In the pictorial method of ray tracing, however, focal lengths are just *distances* and are therefore always positive. ◀

EXAMPLE 18.12 Analyzing a magnifying lens

A stamp collector uses a magnifying lens that sits 2.0 cm above the stamp. The magnification is 4. What is the focal length of the lens?

FIGURE 18.47 Ray-tracing diagram of a magnifying lens.

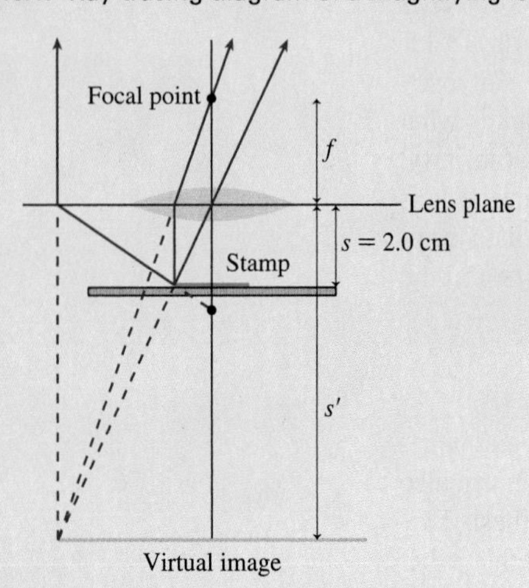

PREPARE A magnifying lens is a converging lens with the object distance less than the focal length ($s < f$). Assume it is a thin lens. The user looks *through* the lens and sees a virtual image. FIGURE 18.47 shows the lens and a ray-tracing diagram.

SOLVE A virtual image is upright, so $m = +4$. The magnification is $m = -s'/s$; thus

$$s' = -4s = -4(2.0\text{ cm}) = -8.0\text{ cm}$$

We can use s and s' in the thin-lens equation to find the focal length:

$$\frac{1}{f} = \frac{1}{s} + \frac{1}{s'} = \frac{1}{2.0\text{ cm}} + \frac{1}{-8.0\text{ cm}} = 0.375\text{ cm}^{-1}$$

Thus

$$f = \frac{1}{0.375\text{ cm}^{-1}} = 2.7\text{ cm}$$

ASSESS $f > 2$ cm, as expected because the object has to be inside the focal point.

EXAMPLE 18.13 **What is the focal length?**

An object is 38.0 cm to the left of a lens. Its image is found to be 22.0 cm from the lens on the same side as the object. What is the focal length of the lens? Draw a ray diagram for the lens and object.

PREPARE We know the object distance is $s = 38.0$ cm. According to Synthesis 18.1, the image distance is a negative number because the image is on the same side of the lens as the object, so $s' = -22.0$ cm.

SOLVE The thin-lens equation, Equation 18.11, is

$$\frac{1}{f} = \frac{1}{s} + \frac{1}{s'} = \frac{1}{38.0\text{ cm}} + \frac{1}{-22.0\text{ cm}} = -0.0191\text{ cm}^{-1}$$

from which we find

$$f = \frac{1}{-0.0191\text{ cm}^{-1}} = -52.3\text{ cm}$$

Because f is negative, Synthesis 18.1 indicates that the lens is a diverging lens. We can use the techniques learned in Section 18.5 to draw the ray diagram shown in **FIGURE 18.48**. Notice that the negative value of s' corresponds to a virtual image.

FIGURE 18.48 Ray-tracing diagram of a diverging lens.

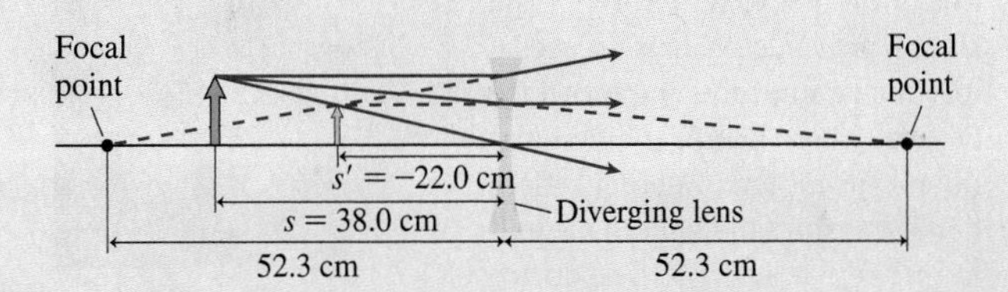

ASSESS In Synthesis 18.1, you can see that the only case for which the image is closer to the lens than the object, but on the same side, is in fact a diverging lens.

This problem would be difficult to solve graphically because you need to locate the focal points in order to draw a ray diagram. Using the thin-lens equation gives rapid and accurate results. Nonetheless, sketching a ray diagram once f is known is very helpful in fully understanding the situation.

EXAMPLE 18.14 **Finding the mirror image of a candle**

A concave mirror has a focal length of 2.4 m. Where should a candle be placed so that its image is inverted and twice as large as the object?

PREPARE Even though this is a mirror, we can use the thin-lens equation. We have $f = +2.4$ m because, according to Synthesis 18.1 a concave mirror has a positive focal length. We also know that the magnification is $m = -2$ because an inverted image implies a negative magnification.

SOLVE Equation 18.8, $m = -s'/s$, relates s' and s:

$$s' = -ms$$

We can insert this expression for s' into the thin-lens equation:

$$\frac{1}{f} = \frac{1}{s} + \frac{1}{s'} = \frac{1}{s} + \frac{1}{-ms} = \frac{1}{s} + \frac{1}{-(-2)s} = \frac{1}{s} + \frac{1}{2s} = \frac{3}{2s}$$

from which we get

$$s = \frac{3}{2}f = \frac{3}{2}(2.4\text{ m}) = 3.6\text{ m}$$

The candle should be placed 3.6 m in front of the mirror.

ASSESS The magnification equation tells us that $s' = 2s = 7.2$ m. This is positive, so the image is real and on the same side of the mirror as the object.

STOP TO THINK 18.7 A candle is placed in front of a converging lens. A well-focused image of the flame is seen on a screen on the opposite side of the lens. If the candle is moved farther away from the lens, how must the screen be adjusted to keep showing a well-focused image?

A. The screen must be moved closer to the lens.
B. The screen must be moved farther away from the lens.
C. The screen does not need to be moved.

INTEGRATED EXAMPLE 18.15 Optical fiber imaging BIO

An *endoscope* is a narrow bundle of optical fibers that can be inserted through a bodily opening or a small incision to view the interior of the body. As **FIGURE 18.49** shows, an *objective* lens focuses a real image onto the entrance face of the fiber bundle. Individual fibers—using total internal reflection—transport the light to the exit face, where it emerges. The doctor observes a magnified image of the exit face by viewing it through an *eyepiece* lens.

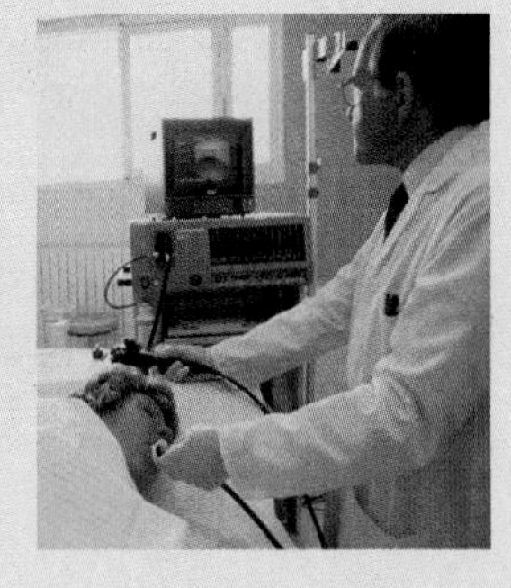

FIGURE 18.49 An endoscope.

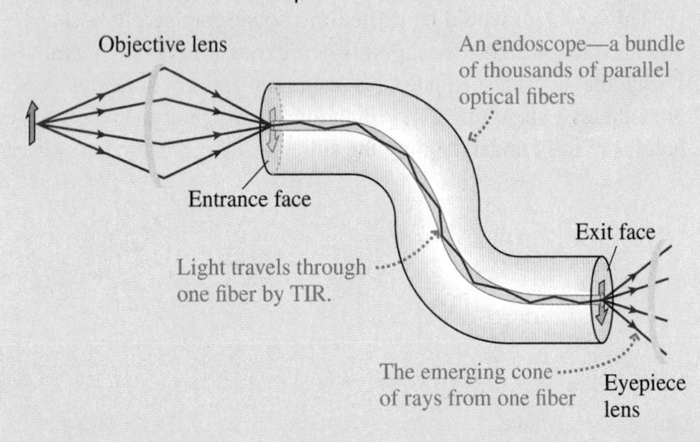

A single optical fiber, as discussed in Section 18.3 and shown in **FIGURE 18.50**, consists of a higher-index glass core surrounded by a lower-index cladding layer. To remain in the fiber, light rays propagating through the core must strike the cladding boundary at angles of incidence greater than the critical angle θ_c. As Figure 18.50 shows, this means that rays that enter the core at angles of incidence smaller than θ_{max} are totally internally reflected down the fiber; those that enter at angles larger than θ_{max} are not totally internally reflected, and escape from the fiber.

FIGURE 18.50 Cross section of an optical fiber (greatly magnified).

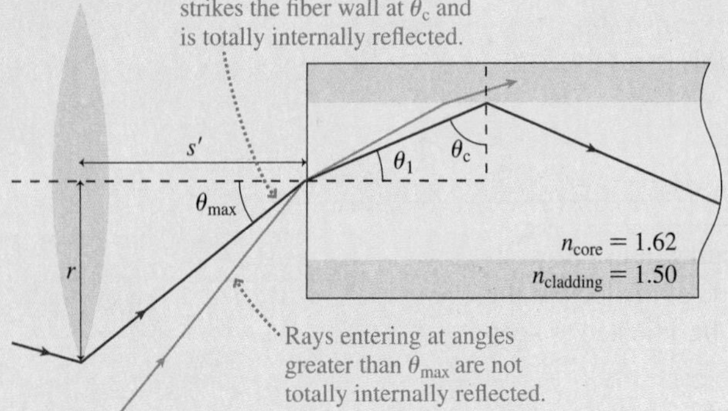

The objective lens of an endoscope must be carefully matched to the fiber. Ideally, the lens diameter is such that rays from the edge of the lens enter the fiber at θ_{max}, as shown in Figure 18.50. A lens larger than this is not useful because rays from its outer regions will enter the fiber at an angle larger than θ_{max} and will thus escape from the fiber. A lens smaller than this will suffer from a reduced light-gathering power.

a. What is θ_{max} for the fiber shown in Figure 18.50?
b. A typical objective lens is 3.0 mm in diameter and can focus on an object 3.0 mm in front of it. What focal length should the lens have so that rays from its edge just enter the fiber at angle θ_{max}?
c. What is the magnification of this lens?

PREPARE From Figure 18.50 we can find the critical angle θ_c and then use geometry to find θ_1. Snell's law can then be used to find θ_{max}. Once θ_{max} is known, we can find the image distance s', and, because we know that the object distance is $s = 3$ mm, we can use the thin-lens equation to solve for f.

SOLVE

a. The critical angle for total internal reflection is given by Equation 18.3:

$$\theta_c = \sin^{-1}\left(\frac{n_2}{n_1}\right) = \sin^{-1}\left(\frac{n_{cladding}}{n_{core}}\right) = \sin^{-1}\left(\frac{1.50}{1.62}\right) = 67.8^\circ$$

Because of the right triangle, this ray's angle of refraction from the fiber's face is $\theta_1 = 90^\circ - 67.8^\circ = 22.2^\circ$. Then, by Snell's law, this ray enters from the air at angle θ_{max} such that

$$n_{air}\sin\theta_{max} = 1.00\sin\theta_{max} = n_{core}\sin\theta_1$$
$$= 1.62(\sin 22.2^\circ) = 0.612$$

Thus

$$\theta_{max} = \sin^{-1}(0.612) = 37.7^\circ$$

b. As Figure 18.50 shows, the distance of the lens from the fiber's face—the image distance s'—is related to the lens radius $r = 1.5$ mm by

$$\frac{r}{s'} = \tan\theta_{max}$$

Thus the image distance is

$$s' = \frac{r}{\tan\theta_{max}} = \frac{1.5\text{ mm}}{\tan 37.7^\circ} = 1.94\text{ mm}$$

Then the thin-lens equation gives

$$\frac{1}{f} = \frac{1}{s} + \frac{1}{s'} = \frac{1}{3.0\text{ mm}} + \frac{1}{1.94\text{ mm}} = 0.849\text{ mm}^{-1}$$

so that

$$f = \frac{1}{0.849\text{ mm}^{-1}} = 1.18\text{ mm}$$

c. The magnification is

$$m = -\frac{s'}{s} = -\frac{1.94\text{ mm}}{3.0\text{ mm}} = -0.65$$

ASSESS The object distance of 3.0 mm is greater than the 1.2 mm focal length we calculated, as must be the case when a converging lens produces a real image.

SUMMARY

Goal: To understand and apply the ray model of light.

GENERAL PRINCIPLES

Reflection

Law of reflection: $\theta_r = \theta_i$

Reflection can be **specular** (mirror-like) or **diffuse** (from rough surfaces).

Plane mirrors: A virtual image is formed at P′ with $s' = s$, where s is the **object distance** and s' is the **image distance.**

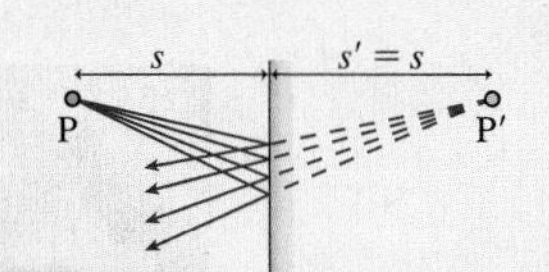

Refraction

Snell's law of refraction:

$$n_1 \sin\theta_1 = n_2 \sin\theta_2$$

Index of refraction is $n = c/v$. The ray is closer to the normal on the side with the larger index of refraction.

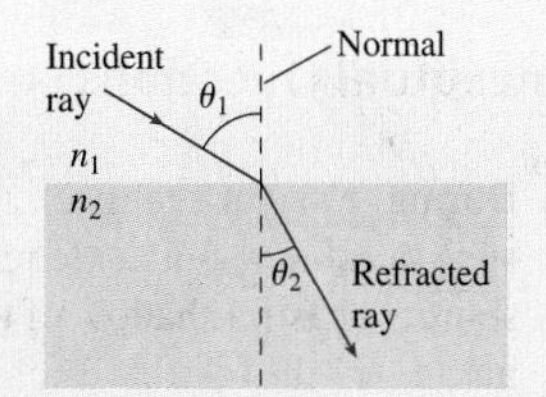

If $n_2 < n_1$, **total internal reflection** (TIR) occurs when the angle of incidence θ_1 is greater than $\theta_c = \sin^{-1}(n_2/n_1)$.

IMPORTANT CONCEPTS

The ray model of light

Light travels along straight lines, called **light rays,** at speed $v = c/n$.

A light ray continues forever unless an interaction with matter causes it to reflect, refract, scatter, or be absorbed.

Light rays come from self-luminous or reflective **objects.** Each point on the object sends rays in all directions.

Ray diagrams use only a few select rays to represent all the rays emitted by an object.

In order for the eye to see an object (or image), rays from the object or image must enter the eye.

Image formation

If rays diverge from P and, after interacting with a lens or mirror, *appear* to diverge from P′ without actually passing through P′, then P′ is a **virtual image** of P.

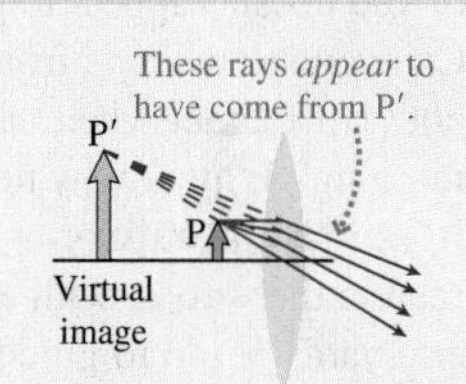

If rays diverge from P and interact with a lens or mirror so that the refracted rays *converge* at P′, then P′ is a **real image** of P. Rays actually pass through a real image.

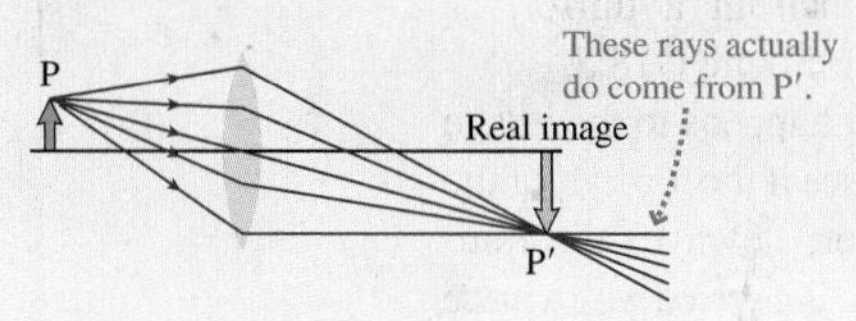

APPLICATIONS

Ray tracing for lenses

Three special rays in three basic situations:

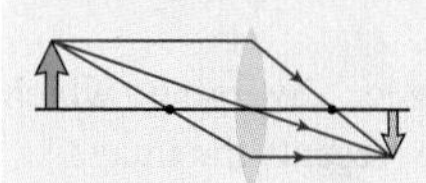
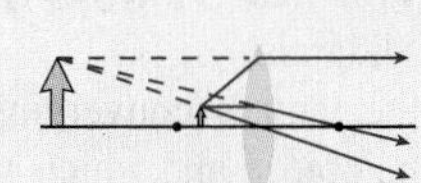
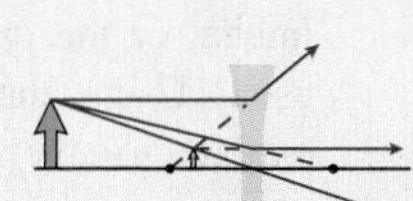

Converging lens Real image — Converging lens Virtual image — Diverging lens Virtual image

Ray tracing for mirrors

Three special rays in three basic situations:

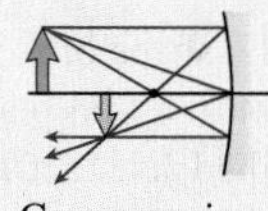
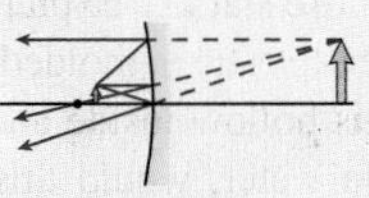
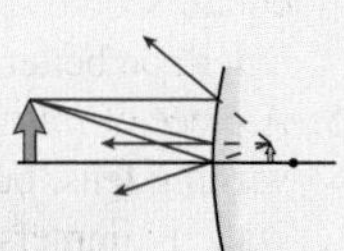

Concave mirror Real image — Concave mirror Virtual image — Convex mirror Virtual image

The thin-lens equation

For a lens or curved mirror, the object distance s, the image distance s', and the focal length f are related by the thin-lens equation:

$$\frac{1}{s} + \frac{1}{s'} = \frac{1}{f}$$

The **magnification** of a lens or mirror is $m = -s'/s$.

Sign conventions for the thin-lens equation:

Quantity	Positive when	Negative when
s	Always	Not treated here
s'	*Real* image; on opposite side of a lens from object, or in front of a mirror	*Virtual* image; on same side of a lens as object, or behind a mirror
f	Converging lens or concave mirror	Diverging lens or convex mirror
m	Image is upright.	Image is inverted.

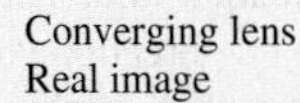

Problem difficulty is labeled as | (straightforward) to ||||| (challenging). Problems labeled INT integrate significant material from earlier chapters; BIO are of biological or medical interest.

For assigned homework and other learning materials, go to MasteringPhysics®

Scan this QR code to launch a Video Tutor Solution that will help you solve problems for this chapter.

QUESTIONS

Conceptual Questions

1. During a solar eclipse, the sun—a small but extended source—casts a shadow of the moon on the earth. Explain why the moon's shadow has a dark center surrounded by a region of increasing brightness.

2. If you turn on your car headlights during the day, the road ahead of you doesn't appear to get brighter. Why not?
3. Can you see the rays from the sun on a clear day? Why or why not? How about when they stream through a forest on a foggy morning? Why or why not?
4. If you take a walk on a summer night along a dark, unpaved road in the woods, with a flashlight pointing at the ground several yards ahead to guide your steps, any water-filled potholes are noticeable because they appear much darker than the surrounding dry road. Explain why.
5. You are looking at the image of a pencil in a mirror, as shown in Figure Q18.5.
   a. What happens to the image you see if the top half of the mirror, down to the midpoint, is covered with a piece of cardboard? Explain.
   b. What happens to the image you see if the bottom half of the mirror is covered with a piece of cardboard?

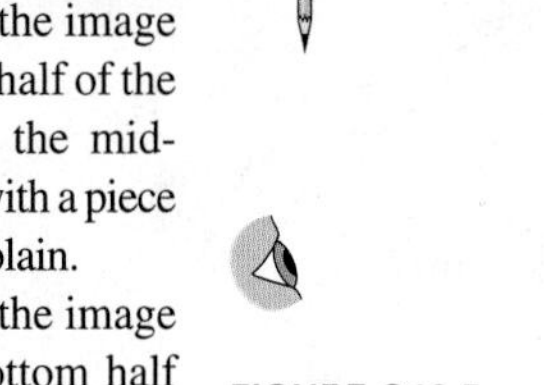

FIGURE Q18.5

6. In *The Toilet of Venus* by Velázquez (see Figure Q18.6), we can see the face of Venus in the mirror. Can she see her own face in the mirror, when the mirror is held as shown in the picture? If yes, explain why; if not, what does she see instead?

FIGURE Q18.6

Diego de Silva Velazquez (1599-1660), "Venus and Cupid," 1650. Oil on canvas. National Gallery, London. Erich Lessing/Art Resource, N.Y.

7. In Manet's *A Bar at the Folies-Bergère* (see Figure Q18.7) the reflection of the barmaid is visible in the mirror behind her. Is this the reflection you would expect if the mirror's surface is parallel to the bar? Where is the man seen facing her in the mirror actually standing?

FIGURE Q18.7

Edouard Manet 1832-1883, "Bar at the Folies-Bergere". 1881/82. Oil on Canvas. 37 13/16″ × 51″ (90 × 130 cm). Courtauld Institute Galleries, London. AKG-Images.

8. Explain why ambulances have the word "AMBULANCE" written backward on the front of them.
9. a. Consider *one* point on an object near a lens. What is the minimum number of rays needed to locate its image point?
   b. For each point on the object, how many rays from this point actually strike the lens and refract to the image point?
10. When you look at your reflection in the bowl of a spoon, it is upside down. Why is this?
11. A concave mirror brings the sun's rays to a focus at a distance of 30 cm from the mirror. If the mirror were submerged in a swimming pool, would the sun's rays be focused nearer to, farther from, or at the same distance from the mirror?
12. A student draws the ray diagram shown in Figure Q18.12 but forgets to label the object, the image, or the type of lens used. Using the diagram, explain whether the lens is converging or diverging, which arrow represents the object, and which represents the image.

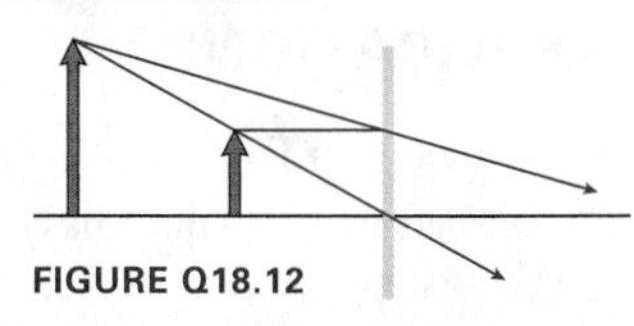

FIGURE Q18.12

13. You are looking straight into the front of an aquarium. You see a fish off to your right. Is the fish actually in the direction you're looking, farther to the right, or farther to the left? Explain.
14. A lens can be used to start a fire by focusing an image of the sun onto a piece of flammable material. All other things being equal, would a lens with a short focal length or a long focal length be better as a fire starter? Explain.
15. A piece of transparent plastic is molded into the shape of a converging lens, but it is hollow inside and has only a thin plastic wall. If immersed in water, would this air-filled lens act as a converging lens, act as a diverging lens, or not act as a lens at all?

16. From where you stand one night, you see the moon directly over a nearby streetlamp. You use a converging lens to get the moon's image sharply focused on a card; when you do so, the image of the lamp is a little out of focus. To bring the lamp's image into sharp focus, do you need to move the card closer to the lens or farther from the lens?

Multiple-Choice Questions

Questions 17 through 19 are concerned with the situation sketched in Figure Q18.17, in which a beam of light in the air encounters a transparent block with index of refraction $n = 1.53$. Some of the light is reflected and some is refracted.

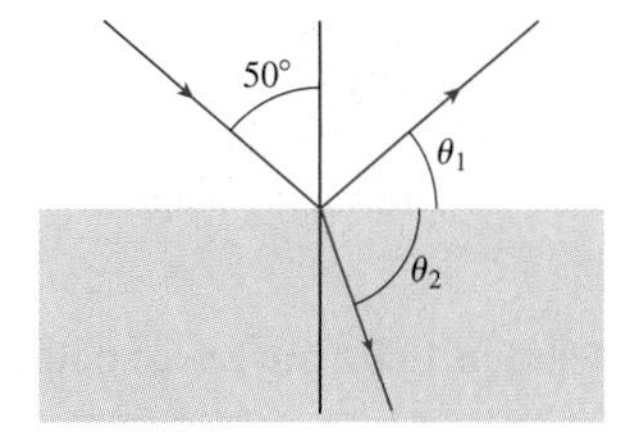

FIGURE Q18.17

17. | What is θ_1?
A. 40° B. 45°
C. 50° D. 90°
18. | What is θ_2?
A. 20° B. 30°
C. 50° D. 60°
19. | Is there an angle of incidence between 0° and 90° such that all of the light will be reflected?
A. Yes, at an angle greater than 50°
B. Yes, at an angle less than 50°
C. No
20. | A 2.0-m-tall man is 5.0 m from the converging lens of a camera. His image appears on a detector that is 50 mm behind the lens. How tall is his image on the detector?
A. 10 mm B. 20 mm
C. 25 mm D. 50 mm
21. || You are 2.4 m from a plane mirror, and you would like to take a picture of yourself in the mirror. You need to manually adjust the focus of the camera by dialing in the distance to what you are photographing. What distance do you dial in?
A. 1.2 m B. 2.4 m
C. 3.6 m D. 4.8 m
22. | As shown in Figure Q18.22, an object is placed in front of a convex mirror. At what position is the image located?

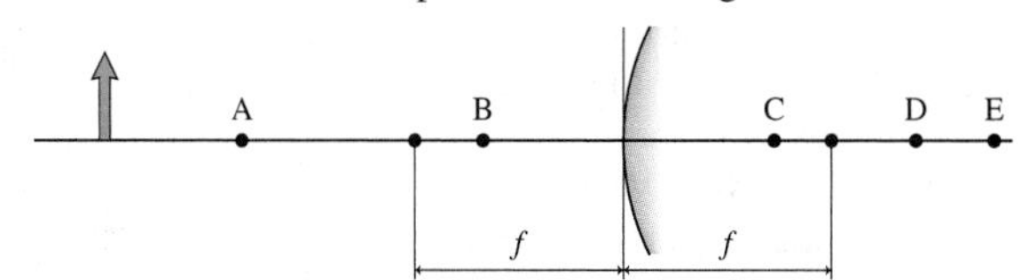

FIGURE Q18.22

23. || A real image of an object can be formed by
A. A converging lens. B. A plane mirror.
C. A convex mirror. D. Any of the above.
24. | An object is 40 cm from a converging lens with a focal length of 30 cm. A real image is formed on the other side of the lens, 120 cm from the lens. What is the magnification?
A. 2.0 B. 3.0 C. 4.0 D. 1.33 E. 0.33
25. | The lens in Figure Q18.25 is used to produce a real image of a candle flame. What is the focal length of the lens?
A. 9.0 cm
B. 12 cm
C. 24 cm
D. 36 cm
E. 48 cm

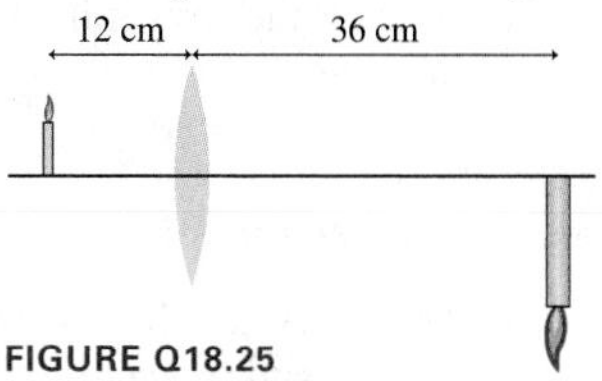

12 cm 36 cm

FIGURE Q18.25

26. | A converging lens of focal length 20 cm is used to form a real image 1.0 m away from the lens. How far from the lens is the object?
A. 20 cm B. 25 cm C. 50 cm D. 100 cm
27. | You look at yourself in a convex mirror. Your image is
A. Upright. B. Inverted.
C. It's impossible to tell without knowing how far you are from the mirror and its focal length.
28. || An object is 50 cm from a diverging lens with a focal length of −20 cm. How far from the lens is the image, and on which side of the lens is it?
A. 14 cm, on the same side as the object
B. 14 cm, on the opposite side from the object
C. 30 cm, on the same side as the object
D. 33 cm, on the same side as the object
E. 33 cm, on the opposite side from the object

PROBLEMS

Section 18.1 The Ray Model of Light

1. || A 5.0-ft-tall girl stands on level ground. The sun is 25° above the horizon. How long is her shadow?
2. |||| A 10-cm-diameter disk emits light uniformly from its surface. 20 cm from this disk, along its axis, is an 8.0-cm-diameter opaque black disk; the faces of the two disks are parallel. 20 cm beyond the black disk is a white viewing screen. The lighted disk illuminates the screen, but there's a shadow in the center due to the black disk. What is the diameter of the *completely dark* part of this shadow?
3. |||| A point source of light illuminates an aperture 2.00 m away. A 12.0-cm-wide bright patch of light appears on a screen 1.00 m behind the aperture. How wide is the aperture?

Section 18.2 Reflection

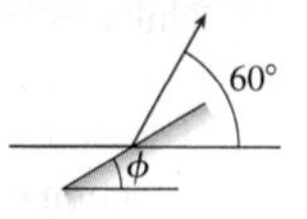

FIGURE P18.4

4. | The mirror in Figure P18.4 deflects a horizontal laser beam by 60°. What is the angle ϕ?
5. | Figure P18.5 shows an object O in front of a plane mirror. Use ray tracing to determine from which locations A–D the object's image is visible.

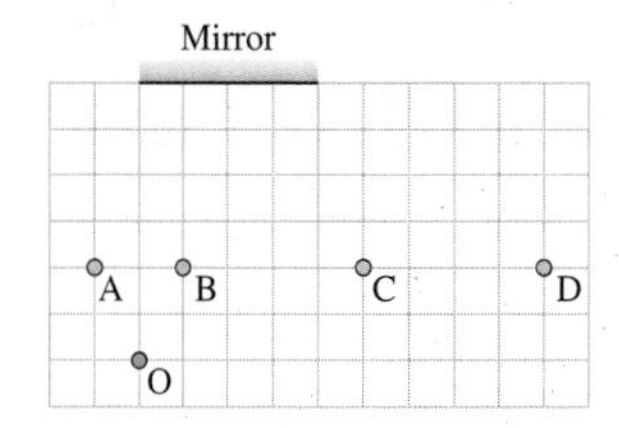

FIGURE P18.5

Watch video solutions for problems: 18.9 18.19

6. ||| A light ray leaves point A in Figure P18.6, reflects from the mirror, and reaches point B. How far below the top edge does the ray strike the mirror?

5 cm
10 cm
A
15 cm
Mirror
B
15 cm

FIGURE P18.6

7. |||| It is 165 cm from your eyes to your toes. You're standing 200 cm in front of a tall mirror. How far is it from your eyes to the image of your toes?

8. ||| A ray of light impinges on a mirror as shown in Figure P18.8. A second mirror is fastened at 90° to the first.
   a. After striking both mirrors, at what angle relative to the incoming ray does the outgoing ray emerge?
   b. What is the answer if the incoming angle is 30°?

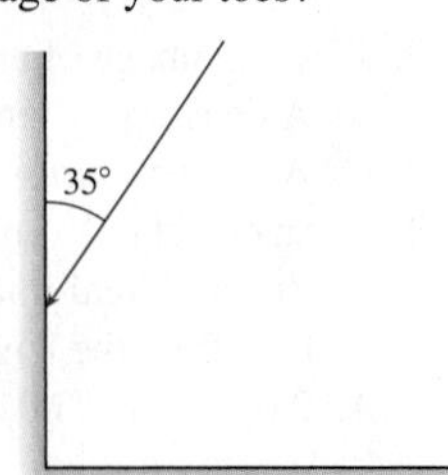

FIGURE P18.8

Section 18.3 Refraction

9. ||| An underwater diver sees the sun 50° above horizontal. How high is the sun above the horizon to a fisherman in a boat above the diver?

10. || A laser beam in air is incident on a liquid at an angle of 37° with respect to the normal. The laser beam's angle in the liquid is 26°. What is the liquid's index of refraction?

11. ||| A 1.0-cm-thick layer of water stands on a horizontal slab of glass. A light ray in the air is incident on the water 60° from the normal. After entering the glass, what is the ray's angle from the normal?

12. ||| A 4.0-m-wide swimming pool is filled to the top. The bottom of the pool becomes completely shaded in the afternoon when the sun is 20° above the horizon. How deep is the pool?

13. || A diamond is underwater. A light ray enters one face of the diamond, then travels at an angle of 30° with respect to the normal. What was the ray's angle of incidence on the diamond?

14. || A thin glass rod is submerged in oil. What is the critical angle for light traveling inside the rod?

15. || A light ray travels inside a horizontal plate of glass, striking its upper surface at an angle of incidence of 60°. This ray is totally internally reflected at the glass-air boundary. A liquid is then poured on top of the glass. What is the largest index of refraction that the liquid could have such that the ray is still totally internally reflected?

16. || A light ray travels inside a block of sodium fluoride that has index of refraction $n = 1.33$ as shown in Figure P18.16. The ray strikes the vertical wall at the critical angle, totally reflects, and then emerges into the air above the block. What is the angle θ_2 at which the ray emerges?

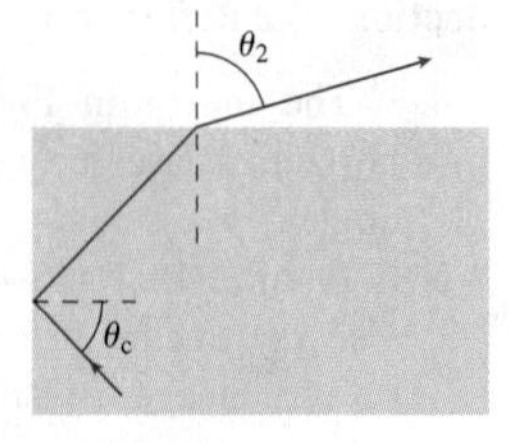

FIGURE P18.16

Section 18.4 Image Formation by Refraction

17. ||| A biologist keeps a specimen of his favorite beetle embedded in a cube of polystyrene plastic. The hapless bug appears to be 2.0 cm within the plastic. What is the beetle's actual distance beneath the surface?

18. || A fish in a flat-sided aquarium sees a can of fish food on the counter. To the fish's eye, the can looks to be 30 cm outside the aquarium. What is the actual distance between the can and the aquarium? (You can ignore the thin glass wall of the aquarium.)

19. | A swim mask has a pocket of air between your eyes and the flat glass front.

   a. If you look at a fish while swimming underwater with a swim mask on, does the fish appear closer or farther than it really is? Draw a ray diagram to explain.
   b. Does the fish see your face closer or farther than it really is? Draw a ray diagram to explain.

Section 18.5 Thin Lenses: Ray Tracing

20. | An object is 30 cm in front of a converging lens with a focal length of 10 cm. Use ray tracing to determine the location of the image. Is the image upright or inverted? Is it real or virtual?

21. | An object is 6.0 cm in front of a converging lens with a focal length of 10 cm. Use ray tracing to determine the location of the image. Is the image upright or inverted? Is it real or virtual?

22. || An object is 20 cm in front of a diverging lens with a focal length of 10 cm. Use ray tracing to determine the location of the image. Is the image upright or inverted? Is it real or virtual?

23. | An object is 15 cm in front of a diverging lens with a focal length of 10 cm. Use ray tracing to determine the location of the image. Is the image upright or inverted? Is it real or virtual?

Section 18.6 Image Formation with Spherical Mirrors

24. | A concave cosmetic mirror has a focal length of 40 cm. A 5-cm-long mascara brush is held upright 20 cm from the mirror. Use ray tracing to determine the location and height of its image. Is the image upright or inverted? Is it real or virtual?

25. | A light bulb is 60 cm from a concave mirror with a focal length of 20 cm. Use ray tracing to determine the location of its image. Is the image upright or inverted? Is it real or virtual?

26. | BIO The illumination lights in an operating room use a concave mirror to focus an image of a bright lamp onto the surgical site. One such light has a mirror with a focal length of 15.0 cm. Use ray tracing to find the position of its lamp when the patient is positioned 1.0 m from the mirror (you'll need a careful drawing to get a good answer).

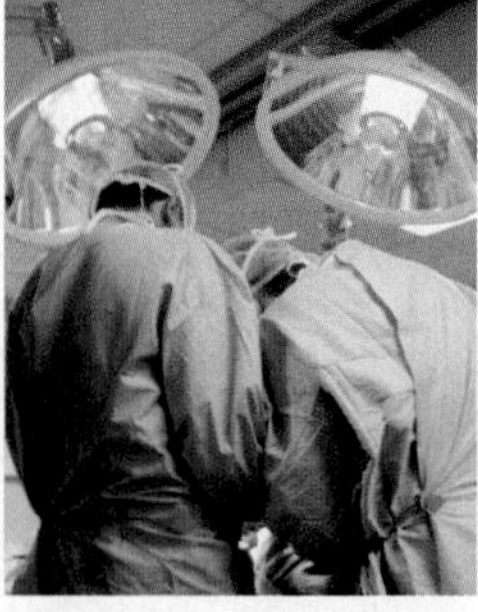

27. || BIO A dentist uses a curved mirror to view the back side of teeth on the upper jaw. Suppose she wants an erect image with a magnification of 2.0 when the mirror is 1.2 cm from a tooth. (Treat this problem as though the object and image lie along a straight line.) Use ray tracing to decide whether a concave or convex mirror is needed, and to estimate its focal length.

28. | A convex mirror, like the passenger-side rearview mirror on a car, has a focal length of 2.0 m. An object is 4.0 m from the mirror. Use ray tracing to determine the location of its image. Is the image upright or inverted? Is it real or virtual?

29. ||| An object is 12 cm in front of a convex mirror. The mirror creates an image that is 75% as tall as the object. Use ray tracing to find the distance of the focal point from the mirror.

Section 18.7 The Thin-Lens Equation

For Problems 30 through 39, calculate the image position and height.

30. | A 2.0-cm-tall object is 40 cm in front of a converging lens that has a 20 cm focal length.
31. || A 1.0-cm-tall object is 10 cm in front of a converging lens that has a 30 cm focal length.
32. || A 2.0-cm-tall object is 15 cm in front of a converging lens that has a 20 cm focal length.
33. | A 1.0-cm-tall object is 75 cm in front of a converging lens that has a 30 cm focal length.
34. | A 2.0-cm-tall object is 15 cm in front of a diverging lens that has a −20 cm focal length.
35. | A 1.0-cm-tall object is 60 cm in front of a diverging lens that has a −30 cm focal length.
36. | A 3.0-cm-tall object is 15 cm in front of a convex mirror that has a −25 cm focal length.
37. || A 3.0-cm-tall object is 45 cm in front of a convex mirror that has a −25 cm focal length.
38. | A 3.0-cm-tall object is 15 cm in front of a concave mirror that has a 25 cm focal length.
39. | A 3.0-cm-tall object is 45 cm in front of a concave mirror that has a 25 cm focal length.
40. | At what distance from a concave mirror with a 35 cm focal length should an object be placed so that its image is the same distance from the mirror as the object?

General Problems

41. ||| Starting 3.5 m from a department store mirror, Suzanne walks INT toward the mirror at 1.5 m/s for 2.0 s. How far is Suzanne from her image in the mirror after 2.0 s?
42. || You slowly back away from a plane mirror at a speed of INT 0.10 m/s. With what speed does your image appear to be moving away from you?
43. || At what angle ϕ should the laser beam in Figure P18.43 be aimed at the mirrored ceiling in order to hit the midpoint of the far wall?

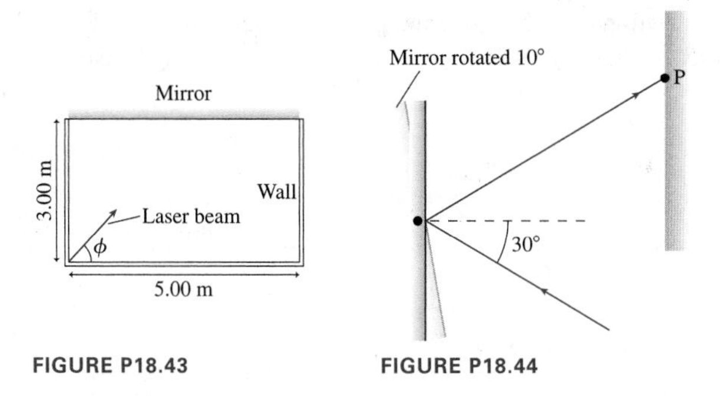

FIGURE P18.43

FIGURE P18.44

44. |||| A laser beam is incident on a mirror at an angle of 30°, as shown in Figure P18.44. It reflects off the mirror and strikes a wall 2.0 m away at point P. By what distance does the laser spot on the wall move if the mirror is rotated by 10°?

45. || Figure P18.45 shows a light ray incident on a polished metal cylinder. At what angle θ will the ray be reflected?

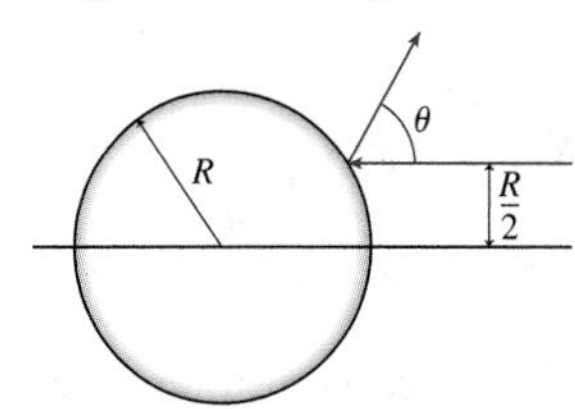

FIGURE P18.45

46. ||| The place you get your hair cut has two nearly parallel mirrors 5.0 m apart. As you sit in the chair, your head is 2.0 m from the nearer mirror. Looking toward this mirror, you first see your face and then, farther away, the back of your head. (The mirrors need to be slightly nonparallel for you to be able to see the back of your head, but you can treat them as parallel in this problem.) How far away does the back of your head appear to be? Neglect the thickness of your head.
47. | You shine your laser pointer through the flat glass side of a rectangular aquarium at an angle of incidence of 45°. The index of refraction of this type of glass is 1.55.
    a. At what angle from the normal does the beam from the laser pointer enter the water inside the aquarium?
    b. Does your answer to part a depend on the index of refraction of the glass?
48. || A ray of light traveling through air encounters a 1.2-cm-thick sheet of glass at a 35° angle of incidence. How far does the light ray travel in the glass before emerging on the far side?
49. || What is the angle of incidence in air of a light ray whose angle of refraction in glass is half the angle of incidence?
50. |||| Figure P18.50 shows a light ray incident on a glass cylinder. What is the angle α of the ray after it has entered the cylinder?

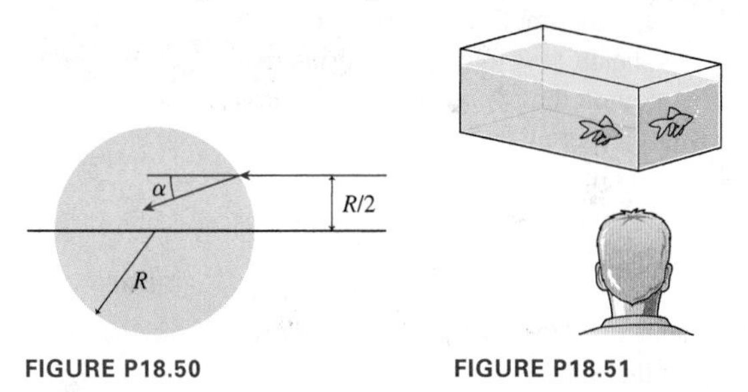

FIGURE P18.50

FIGURE P18.51

51. | If you look at a fish through the corner of a rectangular aquarium you sometimes see two fish, one on each side of the corner, as shown in Figure P18.51. Sketch some of the light rays that reach your eye from the fish to show how this can happen.
52. ||||| It's nighttime, and you've dropped your goggles into a swimming pool that is 3.0 m deep. If you hold a laser pointer 1.0 m directly above the edge of the pool, you can illuminate the goggles if the laser beam enters the water 2.0 m from the edge. How far are the goggles from the edge of the pool?
53. |||| One of the contests at the school carnival is to throw a spear at an underwater target lying flat on the bottom of a pool. The water is 1.0 m deep. You're standing on a small stool that places your eyes 3.0 m above the bottom of the pool. As you look at the target, your gaze is 30° below horizontal. At what angle below horizontal should you throw the spear in order to hit the target? Your raised arm brings the spear point to the level of your eyes as you throw it, and over this short distance you can assume that the spear travels in a straight line rather than a parabolic trajectory.

Watch video solutions for problems: 18.59 18.65

54. ||| Figure P18.54 shows a meter stick lying on the bottom of a 100-cm-long tank with its zero mark against the left edge. You look into the tank at a 30° angle, with your line of sight just grazing the upper left edge of the tank. What mark do you see on the meter stick if the tank is (a) empty, (b) half full of water, and (c) completely full of water?

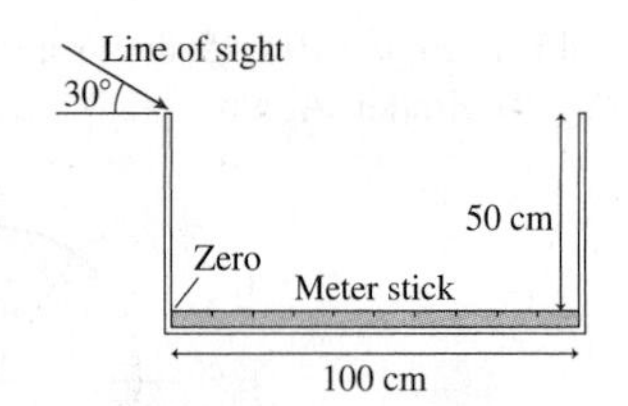

FIGURE P18.54

55. ||| You are standing in a 1.5-m-deep swimming pool at night. The water is very still. You hold a laser pointer just above the water's surface and shine it nearly parallel to the surface, but tilted slightly down so that the beam enters the water 5.0 m from you. How far from you does the beam strike the bottom of the pool?

56. ||| What is the exit angle θ from the glass prism in Figure P18.56?

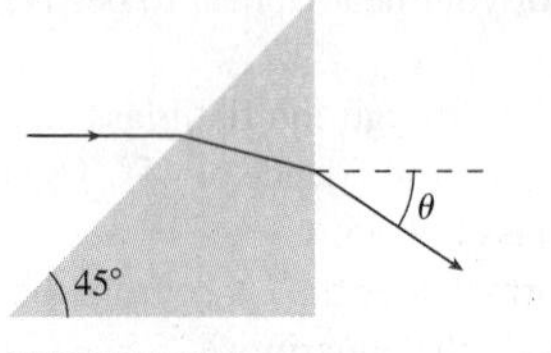

FIGURE P18.56

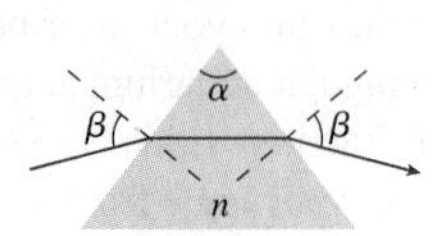

FIGURE P18.57

57. ||| There is just one angle of incidence β onto a prism for which the light inside an isosceles prism travels parallel to the base and emerges at that same angle β, as shown in Figure P18.57.
   a. Find an expression for β in terms of the prism's apex angle α and index of refraction n.
   b. A laboratory measurement finds that $\beta = 52.2°$ for a prism that is shaped as an equilateral triangle. What is the prism's index of refraction?

58. |||| What is the smallest angle θ_1 for which a laser beam will undergo total internal reflection on the hypotenuse of the glass prism in Figure P18.58?

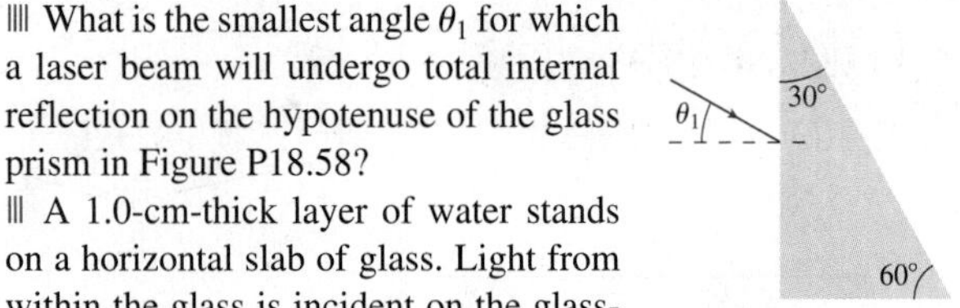

FIGURE P18.58

59. ||| A 1.0-cm-thick layer of water stands on a horizontal slab of glass. Light from within the glass is incident on the glass-water boundary. What is the maximum angle of incidence for which a light ray can emerge into the air above the water?

60. ||||| The glass core of an optical fiber has index of refraction 1.60. The index of refraction of the cladding is 1.48. What is the maximum angle between a light ray and the wall of the core if the ray is to remain inside the core?

61. |||| A 150-cm-tall diver is standing completely submerged on the bottom of a swimming pool full of water. You are sitting on the end of the diving board, almost directly over her. How tall does the diver appear to be?

62. || To a fish, the 4.00-mm-thick aquarium walls appear only 3.50 mm thick. What is the index of refraction of the walls?

63. || A microscope is focused on an amoeba. When a 0.15-mm-thick cover glass ($n = 1.50$) is placed over the amoeba, by how far must the microscope objective be moved to bring the organism back into focus? Must it be raised or lowered? BIO

64. || A ray diagram can be used to find the location of an object if you are given the location of its image and the focal length of the mirror. Draw a ray diagram to find the height and position of an object that makes a 2.0-cm-high upright virtual image that appears 8.0 cm behind a convex mirror of focal length 20 cm.

65. | A 2.0-cm-tall object is located 8.0 cm in front of a converging lens with a focal length of 10 cm. Use ray tracing to determine the location and height of the image. Is the image upright or inverted? Is it real or virtual?

66. || You need to use a 24-cm-focal-length lens to produce an inverted image twice the height of an object. At what distance from the object should the lens be placed?

67. |||| A near-sighted person might correct his vision by wearing diverging lenses with focal length $f = -50$ cm. When wearing his glasses, he looks not at actual objects but at the virtual images of those objects formed by his glasses. Suppose he looks at a 12-cm-long pencil held vertically 2.0 m from his glasses. Use ray tracing to determine the location and height of the image. BIO

68. | A 1.0-cm-tall object is 20 cm in front of a converging lens that has a 10 cm focal length. Use ray tracing to find the position and height of the image. To do this accurately, use a ruler or paper with a grid. Determine the image distance and image height by making measurements on your diagram.

69. || A 2.0-cm-tall object is 20 cm in front of a converging lens that has a 60 cm focal length. Use ray tracing to find the position and height of the image. To do this accurately, use a ruler or paper with a grid. Determine the image distance and image height by making measurements on your diagram.

70. || A 1.0-cm-tall object is 7.5 cm in front of a diverging lens that has a 10 cm focal length. Use ray tracing to find the position and height of the image. To do this accurately, use a ruler or paper with a grid. Determine the image distance and image height by making measurements on your diagram.

71. || A 1.5-cm-tall object is 90 cm in front of a diverging lens that has a 45 cm focal length. Use ray tracing to find the position and height of the image. To do this accurately, use a ruler or paper with a grid. Determine the image distance and image height by making measurements on your diagram.

72. || The moon is 3.5×10^6 m in diameter and 3.8×10^8 m from the earth's surface. The 1.2-m-focal-length concave mirror of a telescope focuses an image of the moon onto a detector. What is the diameter of the moon's image?

73. ||| A 2.0-cm-tall candle flame is 2.0 m from a wall. You happen to have a lens with a focal length of 32 cm. How many places can you put the lens to form a well-focused image of the candle flame on the wall? For each location, what are the height and orientation of the image?

74. || A 2.0-cm-diameter spider is 2.0 m from a wall. Determine the focal length and position (measured from the wall) of a lens that will make a half-size image of the spider on the wall.

75. ||| Figure P18.75 shows a meter stick held lengthwise along the optical axis of a concave mirror. How long is the image of the meter stick?

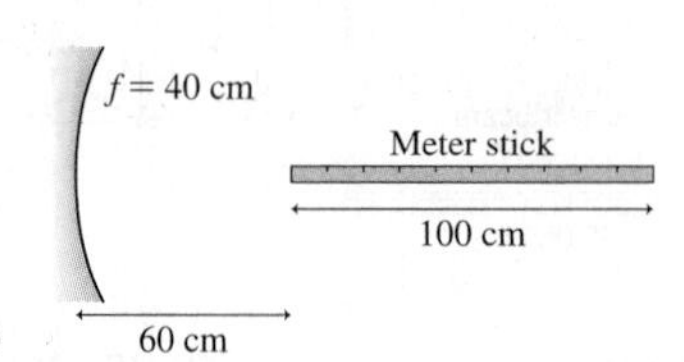

FIGURE P18.75

76. || A slide projector needs to create a 98-cm-high image of a 2.0-cm-tall slide. The screen is 300 cm from the slide.
   a. What focal length does the lens need? Assume that it is a thin lens.
   b. How far should you place the lens from the slide?

77. ||| The writing on the passenger-side mirror of your car says "Warning! Objects in mirror are closer than they appear." There is no such warning on the driver's mirror. Consider a typical convex passenger-side mirror with a focal length of −80 cm. A 1.5-m-tall cyclist on a bicycle is 25 m from the mirror. You are 1.0 m from the mirror, and suppose, for simplicity, that the mirror, you, and the cyclist all lie along a line.

a. How far are you from the image of the cyclist?
b. How far would you have been from the image if the mirror were flat?
c. What is the image height?
d. What would the image height have been if the mirror were flat?
e. Why is there a label on the passenger-side mirror?

MCAT-Style Passage Problems

Mirages

There is an interesting optical effect you have likely noticed while driving along a flat stretch of road on a sunny day. A small, distant dip in the road appears to be filled with water. You may even see the reflection of an oncoming car. But, as you get closer, you find no puddle of water after all; the shimmering surface vanishes, and you see nothing but empty road. It was only a *mirage,* the name for this phenomenon.

The mirage is due to the different index of refraction of hot and cool air. The actual bending of the light rays that produces the mirage is subtle, but we can make a simple model as follows. When air is heated, its density decreases and so does its index of refraction. Consequently, a pocket of hot air in a dip in a road has a lower index of refraction than the cooler air above it. Incident light rays with large angles of incidence (that is, nearly parallel to the road, as shown in Figure P18.78) experience total internal reflection. The mirage that you see is due to this reflection. As you get nearer, the angle goes below the critical angle and there is no more total internal reflection; the "water" disappears!

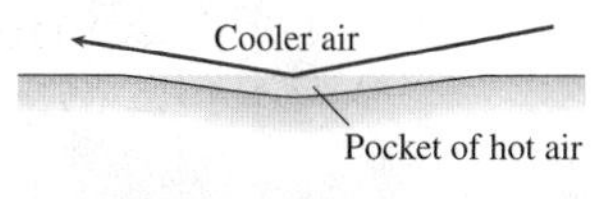

FIGURE P18.78

78. | The pocket of hot air appears to be a pool of water because
A. Light reflects at the boundary between the hot and cool air.
B. Its density is close to that of water.
C. Light refracts at the boundary between the hot and cool air.
D. The hot air emits blue light that is the same color as the daytime sky.

79. | Which of these changes would allow you to get closer to the mirage before it vanishes?
A. Making the pocket of hot air nearer in temperature to the air above it
B. Looking for the mirage on a windy day, which mixes the air layers
C. Increasing the difference in temperature between the pocket of hot air and the air above it
D. Looking at it from a greater height above the ground

80. | If you could clearly see the image of an object that was reflected by a mirage, the image would appear
A. Magnified.
B. With up and down reversed.
C. Farther away than the object.
D. With right and left reversed.

STOP TO THINK ANSWERS

Chapter Preview Stop to Think: B. The diameter of the hole is much greater than the wavelength of light, so the ray model is applicable and light will travel in rays from the bulb to the screen. Light rays from the top of the bulb will go through the hole and hit near the bottom of the screen, while rays from the bottom of the bulb will go through the hole and hit near the top of the screen. When all the rays are taken into account, the image of the bulb on the screen will be inverted.

Stop to Think 18.1: A and B. Rays from the very top and bottom of the bulb that go through the top and bottom of the disk define the regions with no light (shadow) and those with light. The shadow is complete at only positions A and B.

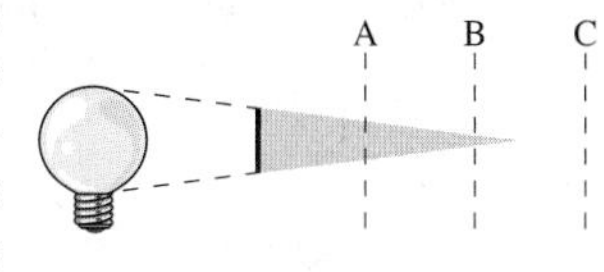

Stop to Think 18.2: C. The image due to a plane mirror is always located on the opposite side of the mirror as the object, along a line passing through the object and perpendicular to the mirror, and the same distance from the mirror as the object. The position of the observer has no bearing on the position of the image.

Stop to Think 18.3: A. The ray travels closer to the normal in both media 1 and 3 than in medium 2, so n_1 and n_3 are both greater than n_2. The angle is smaller in medium 3 than in medium 1, so $n_3 > n_1$.

Stop to Think 18.4: E. From Figure 18.30, the image at P′ is created by many rays from P that go through *all* parts of the lens. If the upper half of the lens is obscured, all the rays that go through the lower half still focus at P′. The image remains sharp but, because of the fewer rays reaching it, becomes dimmer.

Stop to Think 18.5: E. The rays from the object are diverging. Without a lens, the rays cannot converge to form any kind of image on the screen.

Stop to Think 18.6: C. For a converging mirror, the focal length f is the distance from the mirror at which incoming parallel rays meet. The moon is so distant that rays from any point on the moon are very nearly parallel. Thus the image of the moon would be very nearly at a distance f in front of the mirror.

Stop to Think 18.7: A. The thin-lens equation is $1/s + 1/s' = 1/f$. The focal length of the lens is fixed. Because $1/s$ gets smaller as s is increased, $1/s'$ must get larger to compensate. Thus s' must decrease.

19 Optical Instruments

This anableps is called the "four-eyed fish." How must the upper half of its eye differ from the lower half so that it has clear vision both above and below the waterline at the same time?

LOOKING AHEAD ▸▸

Goal: To understand how common optical instruments work.

The Human Eye

Our most important optical instruments are our own eyes, which use a lens to focus light onto the light-sensitive retina.

You'll learn how near- and farsightedness can be corrected using eyeglasses or contact lenses.

Optical Instruments

A converging lens is the simplest magnifier. We'll also study microscopes and telescopes.

You'll learn how optical instruments can be designed to magnify objects up to a thousand times.

Optical Resolution

This magnified image of chromosomes is slightly blurry because of a limit to the microscope's **resolution** due to diffraction.

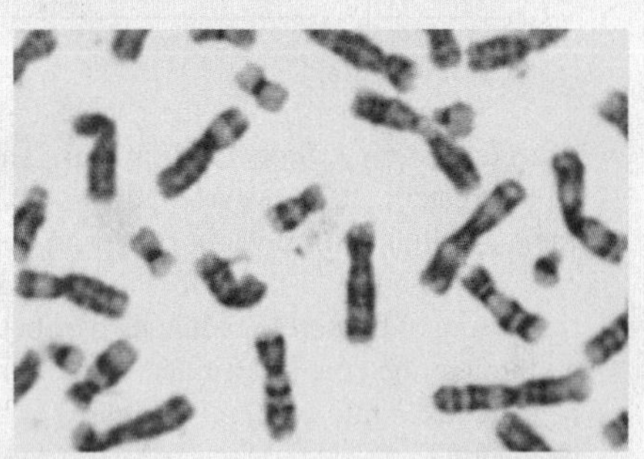

You'll learn that a microscope cannot resolve features much smaller than the wavelength of light.

LOOKING BACK ◂◂

Image Formation by Lenses

In Section 18.5, you learned how a single lens can form a real image of an object. In this chapter, we'll study how *combinations* of two lenses can form highly magnified images, such as those in microscopes and telescopes.

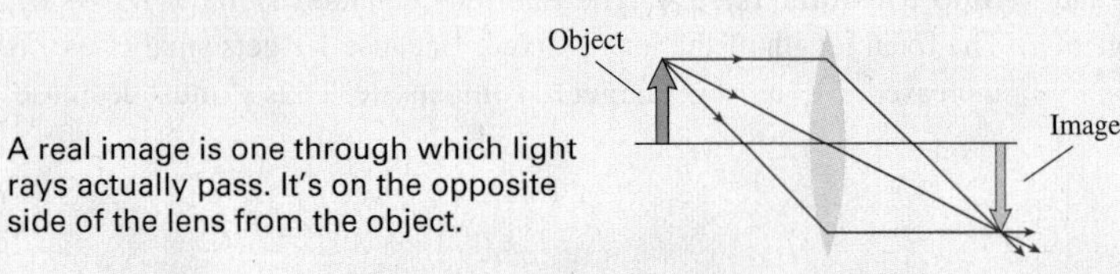

A real image is one through which light rays actually pass. It's on the opposite side of the lens from the object.

STOP TO THINK

A converging lens creates a real, inverted image. For this to occur, the object must be

A. Closer to the lens than the focal point.
B. Farther from the lens than the focal point.
C. At the focal point.

19.1 The Camera

This chapter will investigate a number of optical instruments in which a combination of lenses and mirrors performs a useful function. We'll start with an instrument familiar to everyone: the camera. A **camera** is a device that projects a real image onto a plane surface, where the image can be recorded onto film or, in today's digital cameras, an electronic detector. Although modern cameras are marvels of optical engineering, it is possible to produce decent images using only a light-proof box with a small hole punched in it. Such a **pinhole camera** is shown in FIGURE 19.1a. FIGURE 19.1b uses the ray model of light passing through a small hole to illustrate how the pinhole camera works. Each point on an object emits light rays in all directions, but, ideally, only one of these rays passes through the hole and reaches the film. Each point on the object thus illuminates just one point on the film, forming the image. As the figure illustrates, the geometry of the rays causes the image to be upside down.

A pinhole eye BIO The chambered nautilus is the only animal with a true pinhole "camera" as an eye. Light rays passing through the small opening form a crude image on the back surface of the eye, where the rays strike light-sensitive cells. The image may be poor, but it's sufficient to allow the nautilus to catch prey and escape predators.

FIGURE 19.1 A pinhole camera.

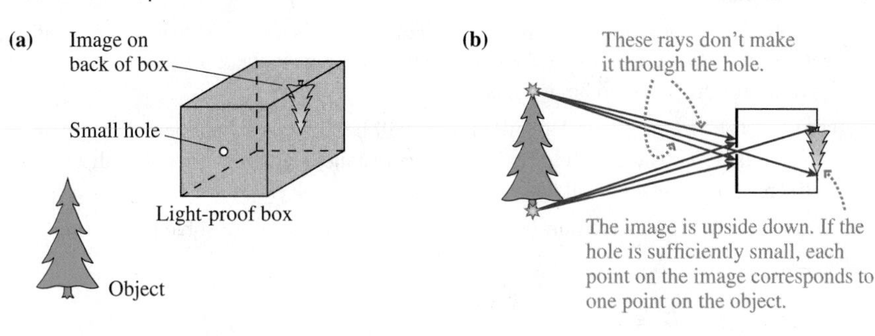

Actually, as you may have realized, each *point* on the object illuminates a small but finite *patch* on the film. This is because the finite size of the hole allows several rays from each point on the object to pass through at slightly different angles. As a result, the image is slightly blurred. Maximum sharpness is achieved by making the hole smaller and smaller, which makes the image dimmer and dimmer. (Diffraction also becomes an issue if the hole gets too small.) A real pinhole camera has to accept a small amount of blurring as the trade-off for having an image bright enough to be practical.

Figure Video

In a standard camera, a dramatic improvement is made possible by using a *lens* in place of a pinhole. FIGURE 19.2 shows how a camera's converging lens projects an inverted real image onto its electronic detector, just as a pinhole camera does. Unlike a pinhole, however, a lens can be large, letting in plenty of light while still giving a sharply focused image. Not shown are the *shutter,* an opaque barrier that is briefly moved out of the way in order for light to pass through the lens, and the *diaphragm,* a set of leaves that can move in from the outside of the lens to effectively reduce its diameter. The diaphragm helps control the amount of light that reaches the camera's detector.

FIGURE 19.2 A camera.

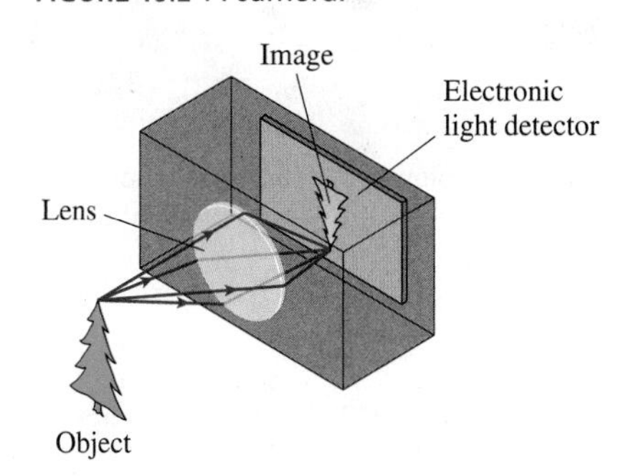

FIGURE 19.3 shows light rays from an object passing through the lens and converging at the image plane B. Here, a single point P on the object focuses to a single point P′

FIGURE 19.3 Focusing a camera.

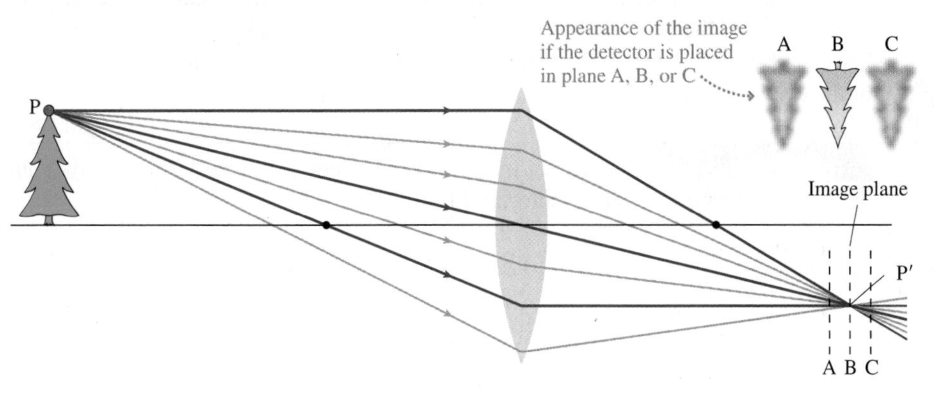

in the image plane. If the electronic detector is located in this plane, a sharp image will form on it. If, however, the detector had been located a bit in *front* of the image plane, at position A, rays from point P would not yet have completely converged and would form a small blurry *circle* on the detector instead of a sharp point. Thus the image would appear blurred, as shown. Similarly, if the detector is placed at C, *behind* the image plane, the rays will be diverging from their perfect focus and again form a blurry image.

Thus to get a sharp image, the detector must be accurately located in the image plane. Figure 19.3 shows that one way to do this is to move the detector until it coincides with the image plane. More commonly, however, a camera is **focused** by moving the *lens* either toward or away from the detector plane until the image is sharp. In either case, the lens-detector distance is varied.

EXAMPLE 19.1 Focusing a camera

A digital camera whose lens has a focal length of 8.0 mm is used to take a picture of an object 30 cm away. What must be the distance from the lens to the light-sensitive detector in order for the image to be in focus?

PREPARE As shown in Figure 19.3, the image will be in focus when the detector is in the image plane. Thus we need to find the image distance, knowing the object distance $s = 30$ cm and the lens's focal length $f = 8.0$ mm.

SOLVE We can rearrange the thin-lens equation, Equation 18.11, to solve for the image distance s':

$$\frac{1}{s'} = \frac{1}{f} - \frac{1}{s} = \frac{1}{0.0080\text{ m}} - \frac{1}{0.30\text{ m}} = 122\text{ m}^{-1}$$

Thus $s' = 1/122\text{ m}^{-1} = 0.0082\text{ m} = 8.2\text{ mm}$. The lens-detector distance has to be 8.2 mm.

ASSESS When the object is infinitely far away, the image, by definition, is at the focal length: $s' = f = 8.0$ mm. If the object is brought to 30 cm, the lens has to move forward a distance of only 8.2 mm $-$ 8.0 mm $=$ 0.2 mm to bring the object into focus. In general, camera lenses don't need to move far.

If a digital picture is magnified enough, you can see the individual pixels that make it up.

To record the image, digital cameras use an electronic light-sensitive detector called a *charge-coupled device* or **CCD.** A CCD consists of a rectangular array of many millions of small detectors called *pixels.* When light hits one of these pixels, it generates an electric charge proportional to the light intensity. Thus an image is recorded on the CCD in terms of little packets of charge. After the CCD has been exposed, the charges are read out, the signal levels are digitized, and the picture is stored in the digital memory of the camera.

FIGURE 19.4 A CCD chip used in a digital camera.

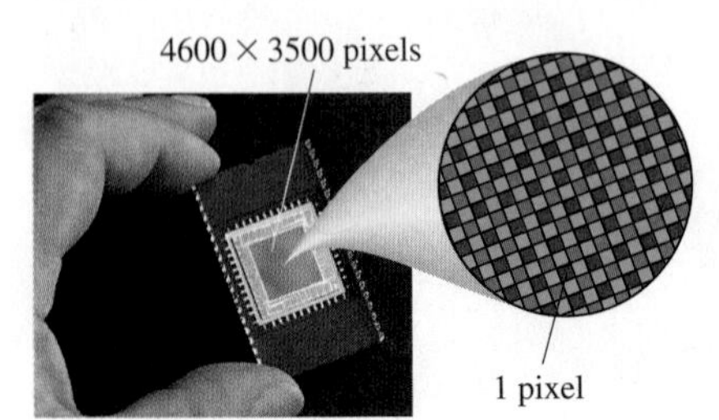

FIGURE 19.4 shows a CCD "chip" and, schematically, the magnified appearance of the pixels on its surface. To record color information, different pixels are covered by red, green, or blue filters; a pixel covered by a green filter, for instance, records only the intensity of the green light hitting it. Later, the camera's microprocessor interpolates nearby colors to give each pixel an overall true color. The structure of the retina of the eye is remarkably similar, as we'll see in Chapter 25.

STOP TO THINK 19.1 The screen in a pinhole camera is moved farther away from the pinhole. The image on the screen will

A. Become larger.
B. Become smaller.
C. Remain the same size.

19.2 The Human Eye BIO

The human eye functions much like a camera. Like the camera, it has three main functional groups: an optical system to focus the incoming light, a diaphragm to adjust the amount of light entering the eye, and a light-sensitive surface to detect the resulting image. The parts of the eye making up these three groups are shown in **FIGURE 19.5**. The *cornea,* the *aqueous humor,* and the *lens* are together responsible for refracting incoming light rays and producing an image. The adjustable *iris* determines how much light enters the eye, in much the same way as does the diaphragm of a camera. And the *retina* is the light-sensitive surface on which the image is formed. The retina is the biological equivalent of the CCD in a digital camera.

FIGURE 19.5 The human eye.

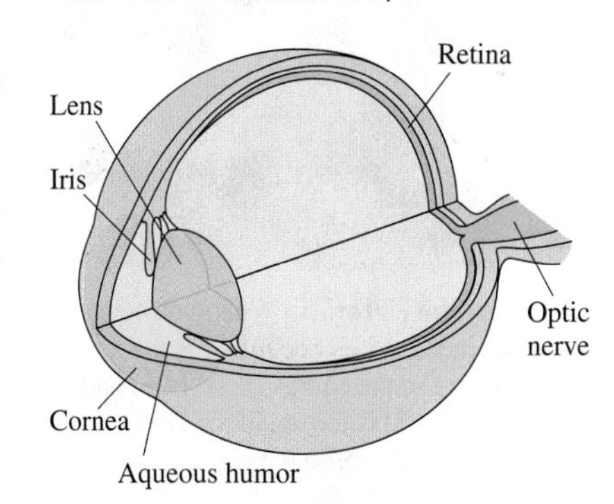
Retina
Lens
Iris
Optic nerve
Cornea
Aqueous humor

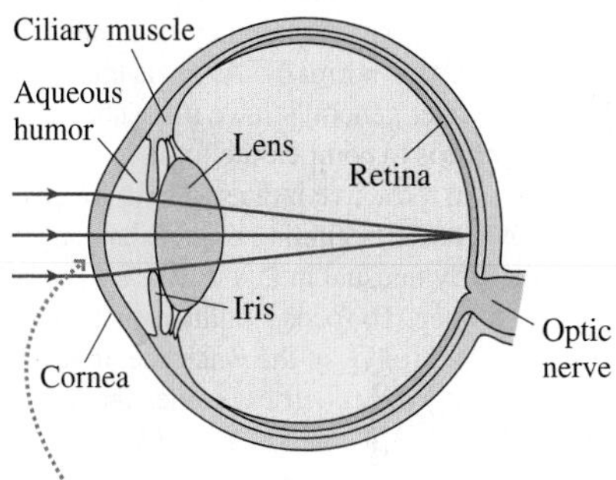
Ciliary muscle
Aqueous humor
Lens
Retina
Iris
Optic nerve
Cornea

Most of the refraction occurs at the cornea's surface, where Δn is the largest.

Focusing and Accommodation

Like a camera, the eye works by focusing incoming rays onto a light-sensitive surface, here the retina. To do so, light is refracted by, in turn, the cornea, the aqueous humor, and the lens, as shown in Figure 19.5. The indices of refraction (n) of these parts of the eye vary somewhat, but average around 1.4. Perhaps surprisingly, most of the eye's refraction occurs not in the lens but at the surface of the cornea. This is due both to the strong curvature of the cornea and to the large difference between the indices of refraction on either side of the surface. Light refracts less as it passes through the lens because the lens's index of refraction doesn't differ much from that of the fluid in which it is embedded. If the lens is surgically removed, which it often is for people with cataracts (a clouding of the lens), the cornea alone still provides a marginal level of vision.

The eye must constantly refocus as it views distant objects, then closer ones. It does this so automatically that we're not normally aware of the process. A camera focuses by changing the distance between the lens and the light detector, but your eye focuses in a different way: by changing the focal length of the lens itself. As shown in **FIGURE 19.6**, it does so by using the ciliary muscles to *change the shape* of the lens. This process of changing the lens shape as the eye focuses at different distances is called **accommodation.**

Class Video

FIGURE 19.6 Accommodation by the eye.

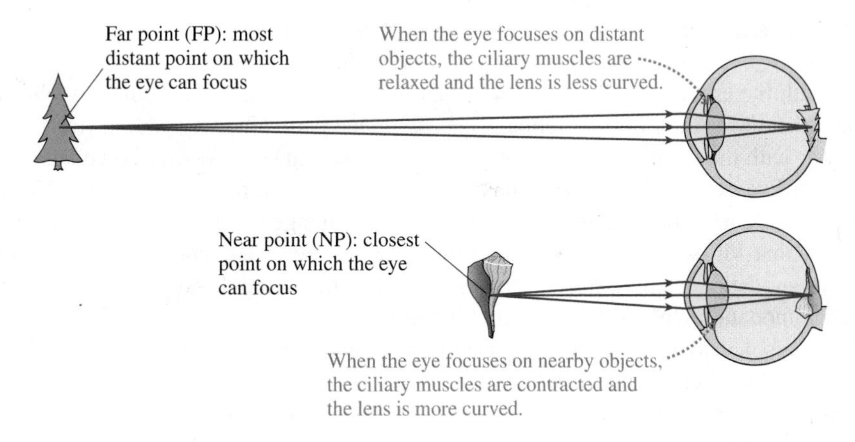
Far point (FP): most distant point on which the eye can focus
When the eye focuses on distant objects, the ciliary muscles are relaxed and the lens is less curved.
Near point (NP): closest point on which the eye can focus
When the eye focuses on nearby objects, the ciliary muscles are contracted and the lens is more curved.

The most distant point on which the completely relaxed eye can focus is called the eye's **far point** (FP). For normal vision, the far point is at infinity. The closest point on which the eye can focus, with the ciliary muscles fully contracted, is called the **near point** (NP). Objects closer than the near point cannot be brought into sharp focus.

Seeing underwater BIO When you swim underwater, the difference in refractive indices between the cornea ($n = 1.38$) and water ($n = 1.33$) is too small to allow significant refraction, so the eye cannot focus. If you wear goggles, the surface of the cornea is in contact with air, not water, and the eye can focus normally. Animals that live underwater generally have more sharply curved corneas to compensate for the small difference in refractive indices. The *anableps* fish shown at the beginning of this chapter is particularly unusual in that it lives at the water's surface. To focus simultaneously on objects on both sides of the waterline, it has evolved a very asymmetrical cornea that is more strongly curved below the waterline.

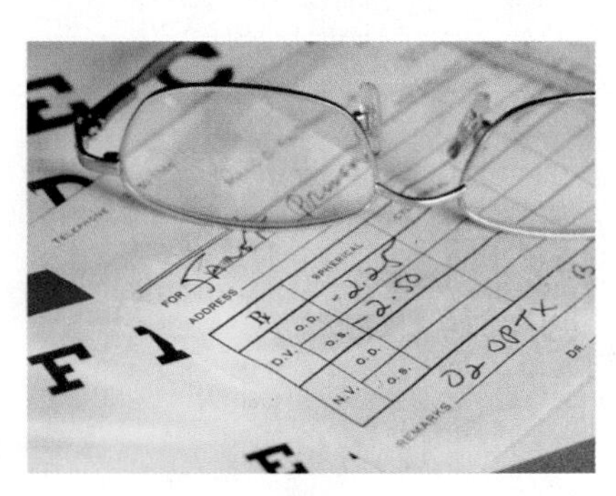

The optometrist's prescription is -2.25 D for the right eye (top) and -2.50 D for the left (bottom).

Class Video

Vision Defects and Their Correction

The near point of normal vision is considered to be 25 cm, but the near point of an individual changes with age. The near point of young children can be as little as 10 cm. The "normal" 25 cm near point is characteristic of young adults, but the near point of most individuals begins to move outward by age 40 or 45 and can reach 200 cm by age 60. This loss of accommodation, which arises because the lens loses flexibility, is called **presbyopia.** Even if their vision is otherwise normal, individuals with presbyopia need reading glasses to bring their near point back to 25 or 30 cm, a comfortable distance for reading.

Presbyopia is known as a *refractive error* of the eye. Two other common refractive errors are *hyperopia* and *myopia.* All three can be corrected with lenses—either eyeglasses or contact lenses—that assist the eye's focusing. Corrective lenses are prescribed not by their focal length but by their **refractive power.** The refractive power of a lens is the inverse of its focal length:

$$P = \frac{1}{f} \qquad (19.1)$$

Refractive power of a lens with focal length f

INVERSE p. 108

A lens with higher refractive power (shorter focal length) causes light rays to refract through a larger angle. The SI unit of lens refractive power is the **diopter,** abbreviated D, defined as $1\text{ D} = 1\text{ m}^{-1}$. Thus a lens with $f = 50\text{ cm} = 0.50\text{ m}$ has refractive power $P = 2.0\text{ D}$.

When writing prescriptions, optometrists don't write the D because the lens maker already knows that prescriptions are in diopters. If you look at your eyeglass prescription next time you visit the optometrist, it will look something like $+2.5/+2.7$. This says that your right eye needs a corrective lens with $P = +2.5$ D, the + indicating a converging lens with a positive focal length. Your left eye needs a lens with $P = +2.7$ D. Most people's eyes are not the same, so each eye usually gets a slightly different lens. Prescriptions with negative numbers indicate diverging lenses with negative focal lengths.

A person who is *farsighted* can see faraway objects (but even then must use some accommodation rather than a relaxed eye), but his near point is larger than 25 cm, often much larger, so he cannot focus on nearby objects. The cause of farsightedness—called **hyperopia**—is an eyeball that is too short for the refractive power of the cornea and lens. As FIGURE 19.7a and b on the next page show, no amount of accommodation allows the eye to focus on an object 25 cm away, the normal near point.

With hyperopia, the eye needs assistance to focus the rays from a nearby object onto the closer-than-normal retina. This assistance is obtained by adding refractive power with the positive (i.e., converging) lens shown in FIGURE 19.7c. To understand why this works, recall that the goal is to allow the person to focus on an object 25 cm away. If a corrective lens forms an upright, virtual image at the person's actual near point, that virtual image acts as an object for the eye itself and, with maximum accommodation, the eye can focus these rays onto the retina. Presbyopia, the loss of accommodation with age, is corrected in the same way.

A person who is *nearsighted* can clearly see nearby objects when the eye is relaxed (and extremely close objects by using accommodation), but no amount of relaxation allows her to see distant objects. Nearsightedness—called **myopia**—is caused by an eyeball that is too long. As FIGURE 19.8a on the next page shows, rays from a distant object come to a focus in front of the retina and have begun to diverge by the time they reach the retina. The eye's far point, shown in FIGURE 19.8b, is less than infinity.

To correct myopia, we need a diverging lens, as shown in FIGURE 19.8c, to slightly defocus the rays and move the image point back to the retina. To focus on a very distant object, the person needs a corrective lens that forms an upright, virtual image at her actual far point. That virtual image acts as an object for the eye itself and, when fully relaxed, the eye can focus these rays onto the retina.

FIGURE 19.7 Hyperopia.

(a)

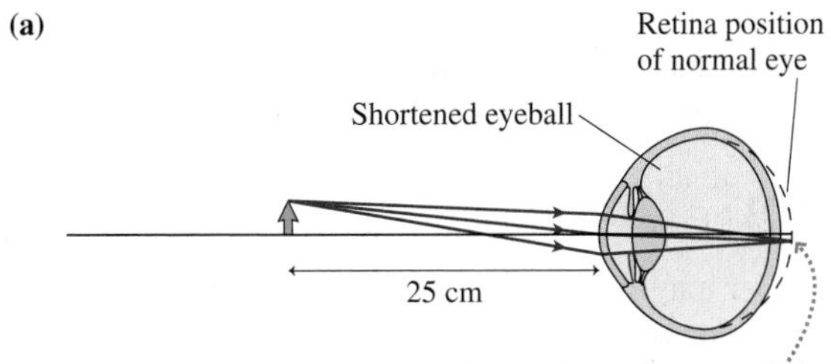
Retina position
of normal eye
Shortened eyeball
25 cm
Even with maximum accommodation,
the image is focused behind the retina.
Thus the image is blurry.

(b)

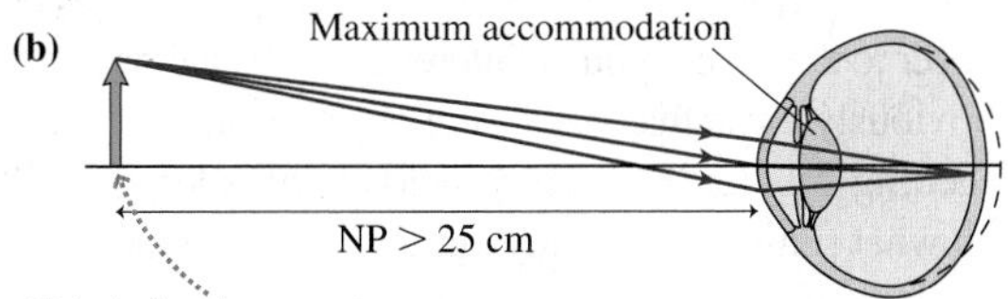
Maximum accommodation
NP > 25 cm
This is the closest point at
which the eye can focus.

(c)

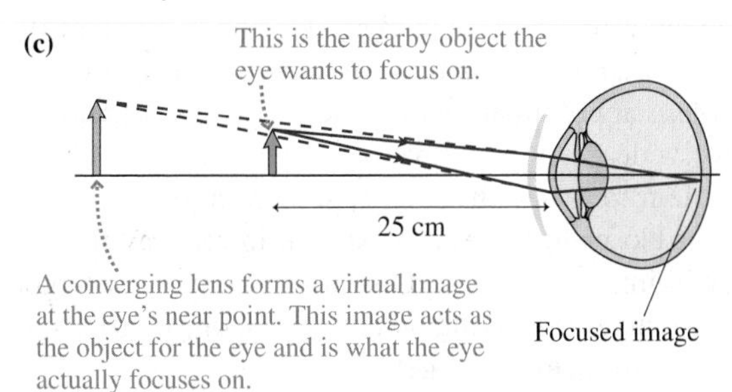
This is the nearby object the
eye wants to focus on.
25 cm
A converging lens forms a virtual image
at the eye's near point. This image acts as
the object for the eye and is what the eye
actually focuses on.
Focused image

FIGURE 19.8 Myopia.

(a)

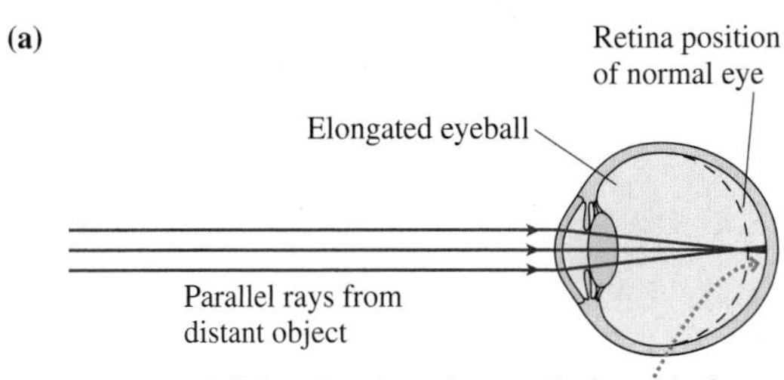
Retina position
of normal eye
Elongated eyeball
Parallel rays from
distant object
A fully relaxed eye focuses the image in front
of the actual retina. The image is blurry.

(b)

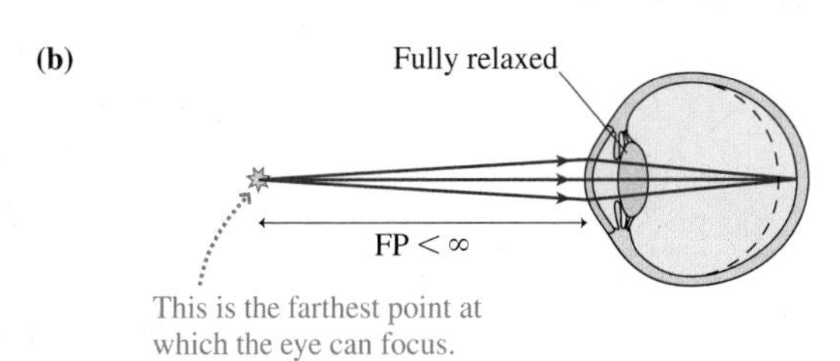
Fully relaxed
FP < ∞
This is the farthest point at
which the eye can focus.

(c)

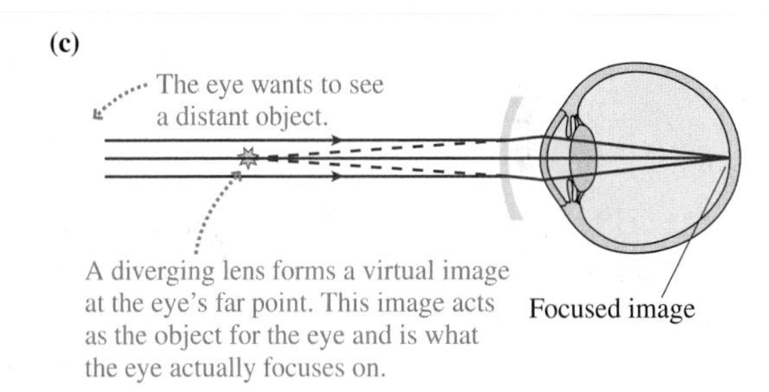
The eye wants to see
a distant object.
A diverging lens forms a virtual image
at the eye's far point. This image acts
as the object for the eye and is what
the eye actually focuses on.
Focused image

STOP TO THINK 19.2 Kara has a near-point distance of 40 cm. So that she can focus on a book 25 cm away, her corrective lenses should create

A. A virtual image of the book 40 cm from her eye.
B. A real image of the book 40 cm from her eye.
C. A virtual image of the book 25 cm from her eye.
D. A real image of the book 25 cm from her eye.

EXAMPLE 19.2 **Correcting hyperopia**

Sanjay has hyperopia, The near point of his left eye is 150 cm. What prescription lens will restore normal vision?

PREPARE Normal vision will allow Sanjay to focus on an object 25 cm away. In measuring distances, we'll ignore the small space between the lens and his eye.

SOLVE Because Sanjay can see objects at 150 cm, using maximum accommodation, we want a lens that creates a virtual image at position $s' = -150$ cm (negative because it's a virtual image) of an object held at $s = 25$ cm. From the thin-lens equation,

$$\frac{1}{f} = \frac{1}{s} + \frac{1}{s'} = \frac{1}{0.25\text{ m}} + \frac{1}{-1.50\text{ m}} = 3.3\text{ m}^{-1}$$

$1/f$ is the lens power, and m^{-1} are diopters. Thus the prescription is for a lens with power $P = 3.3$ D.

ASSESS Hyperopia is always corrected with a converging lens.

EXAMPLE 19.3 **Correcting myopia**

Martina has myopia. The far point of her left eye is 200 cm. What prescription lens will restore normal vision?

PREPARE Normal vision will allow Martina to focus on a very distant object. In measuring distances, we'll ignore the small space between the lens and her eye.

SOLVE Because Martina can see objects at 200 cm with a fully relaxed eye, we want a lens that will create a virtual image at position $s' = -200$ cm (negative because it's a virtual image) of a distant object at $s = \infty$ cm. From the thin-lens equation,

$$\frac{1}{f} = \frac{1}{s} + \frac{1}{s'} = \frac{1}{\infty\text{ m}} + \frac{1}{-2.0\text{ m}} = -0.5\text{ m}^{-1}$$

Thus the prescription is for a lens with power $P = -0.5$ D.

ASSESS Myopia is always corrected with a diverging lens.

STOP TO THINK 19.3 With her right eye, Maria can focus on a vase 0.5 m away, but not on a tree 10 m away. What could be the eyeglass prescription for her right eye?

A. +3.0 D B. +10 D C. −5.0 D D. −1.5 D

19.3 The Magnifier

You've no doubt used a magnifier, or magnifying glass, to get a better look at a small object such as an insect or a coin. As we saw in ◀◀ SECTION 18.5, a magnifier is a simple converging lens, but why objects appear larger when viewed through such a lens is actually rather subtle.

Let's begin by considering the simplest way to magnify an object, one that requires no extra optics at all. You simply get closer to the object you're interested in. The closer you get, the bigger the object appears. Obviously the actual size of the object is unchanged as you approach it, so what exactly is getting "bigger"? A penny, held at arm's length, will more than cover the distant moon. In what sense is the penny "larger" than the moon?

Angular Size and Apparent Size

Consider an object such as the green arrow in FIGURE 19.9a. To find the size of its image on the retina, we can trace the two rays shown that go through the center of the eye lens. (Here, we'll use the thin-lens approximation to reduce the eye's entire optical system to one thin lens.) As we've learned, such rays are undeviated as they pass through the lens, so we can use them to locate the image, shown in green, on the retina. If the arrow is then brought closer to the eye, as shown in red, ray tracing reveals that the size of the arrow's image is *larger.* Our brain interprets a larger image on the retina as representing a larger-appearing *object.* As the object is moved closer, its size doesn't change, but its **apparent size** gets larger.

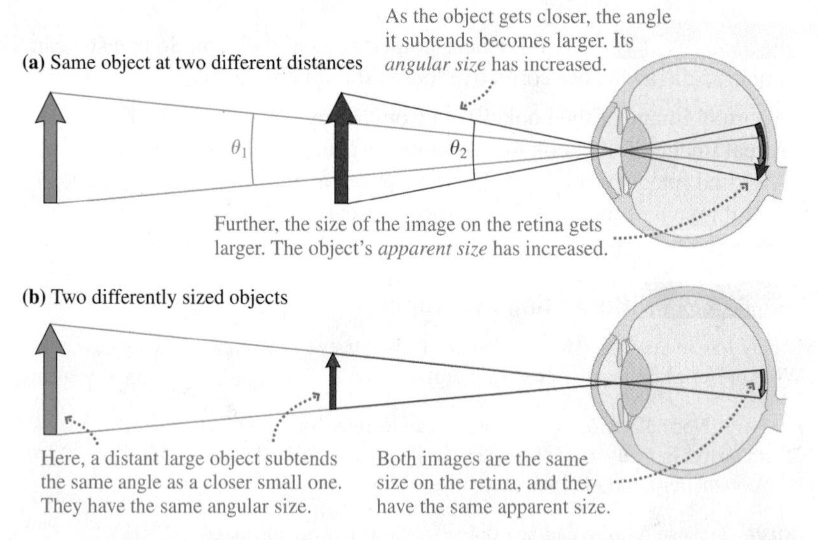

▶ FIGURE 19.9 How the apparent size of an object is determined.

Angles θ_1 and θ_2 in Figure 19.9a are the angles *subtended* by the green and red arrows. The angle subtended by an object is called its **angular size.** As you can see from the figure, **objects that subtend a larger angle appear larger to the eye.** It's also possible, as FIGURE 19.9b shows, for objects with different actual sizes to have the same angular size and thus the same apparent size.

Using a Magnifier

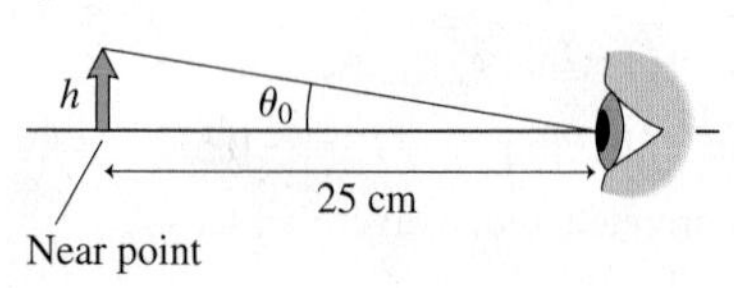

FIGURE 19.10 Angular size without a magnifier.

From our discussion of apparent size, it's clear that the easiest way to magnify an object is to just get closer to it. But there's a limit to this approach. Recall that the eye cannot focus on an object closer than its near point, which we conventionally take as 25 cm. Thus an in-focus object has its maximum angular size θ_0 when the object is at the near point, as shown in FIGURE 19.10. The geometry of this figure shows that

$\tan\theta_0 = h/25$ cm. If this angle is fairly small, which it usually is, we can use the small-angle approximation $\tan\theta_0 \approx \theta_0$ to write

$$\theta_0 \approx \frac{h}{25\text{ cm}}$$

How does a magnifier lead to an angular size larger than this? In the way that a magnifier is usually used, the lens is held such that the object is at or just inside the lens's focal point. As shown graphically in **FIGURE 19.11**, this produces a virtual image that is quite far from the lens. Your eye, looking through the lens, "sees" the virtual image. This is a convenient image location, because your eye's muscles are fully relaxed when looking at a distant image. Thus you can use the magnifier in this way for a long time without eye strain.

FIGURE 19.11 The magnifier.

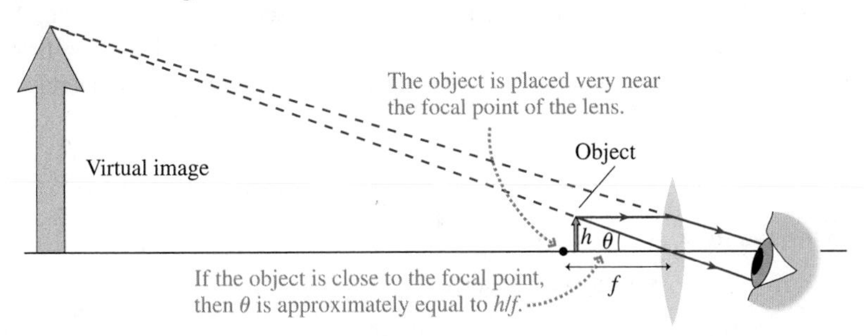
The object is placed very near the focal point of the lens.
Object
Virtual image
h
θ
f
If the object is close to the focal point, then θ is approximately equal to h/f.

We can find the angular size of the image using Figure 19.11. Suppose that the object is almost exactly at the focal point, a distance f from the lens. Then, tracing the ray that goes through the lens's center, we can see that the angular size θ of the image is such that $\tan\theta = h/f$ or, again using the small-angle approximation,

$$\theta \approx \frac{h}{f} \tag{19.2}$$

Thus the angular size θ when using the magnifier is larger than that without the magnifier by a factor of

$$M = \frac{\theta}{\theta_0} = \frac{h/f}{h/25\text{ cm}} = \frac{25\text{ cm}}{f} \tag{19.3}$$

where, for this calculation, the focal length f must be in cm. M is called the **angular magnification** of the magnifier. With a lens of short focal length it is possible to get magnifications as high as about 20.

Movie magic? Even if the more distant of two equally sized objects *appears* smaller, we don't usually believe it actually *is* smaller because there are abundant visual clues that tell our brain that it's farther away. If those clues are removed, however, the brain readily accepts the illusion that the farther object is smaller. The technique of *forced perspective* is a special effect used in movies to give this illusion. Here, Camelia, who is actually closer to the camera, looks like a giant compared to Kevin. The lower photo shows how the trick was done.

CONCEPTUAL EXAMPLE 19.4 **The angular size of a magnified image**

An object is placed right at the focal point of a magnifier. How does the apparent size of the image depend on where the *eye* is placed relative to the lens?

REASON When the object is precisely at the focal point of the lens, Equation 19.2 holds exactly. The angular size is equal to h/f *independent* of the position of the eye. Thus the object's apparent size is independent of the eye's position as well. **FIGURE 19.12** shows a calculator at the focal point of a magnifier. The apparent size of the COS button is the same whether the camera taking the picture is close to or far from the lens.

ASSESS When the object is at the magnifier's focus, we've seen that the image is at infinity. The situation is similar to observing any "infinitely" distant object, such as the moon. If you walk closer to or farther from the moon, its apparent size doesn't change at all. The same holds for a virtual *image* at infinity: Its apparent size is independent of the point from which you observe it.

FIGURE 19.12 Viewing a magnifier with the object at its focus.

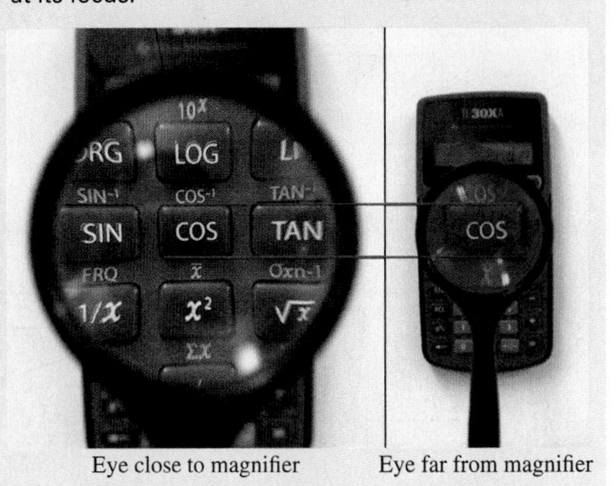
LOG
SIN
COS
TAN
COS
Eye close to magnifier
Eye far from magnifier

STOP TO THINK 19.4 A student tries to use a *diverging* lens as a magnifier. She observes a coin placed at the focal point of the lens. She sees

A. An upright image, smaller than the object.
B. An upright image, larger than the object.
C. An inverted image, smaller than the object.
D. An inverted image, larger than the object.
E. A blurry image.

19.4 The Microscope

To get higher magnifications than are possible using a simple magnifier, a *combination* of lenses must be used. This is how microscopes and telescopes are constructed. A simple rule governs how two lenses work in combination: **The image from the first lens acts as the object for the second lens.** The following example illustrates this rule.

EXAMPLE 19.5 Finding the image for two lenses in combination

A 5.0-cm-focal-length converging lens is 16.0 cm in front of a 10.0-cm-focal-length diverging lens. A 4.0-cm-tall object is placed 11.0 cm in front of the converging lens. What are the position and size of the final image?

PREPARE Let's start with ray tracing. A ray-tracing diagram helps us understand the situation and tells us what to expect for an answer. A diagram can often alert you to a calculation error. **FIGURE 19.13a** first uses the three special rays of the converging lens to locate its image. We see that the image of the first lens is a real image falling between the two lenses. According to the rule, we then use this image as the object for the second lens. This is done in **FIGURE 19.13b**, where we see that the final image is inverted, virtual, and roughly 4 cm to the left of the diverging lens.

Mathematically, we can use the thin-lens equation to find the image location and size due to the first lens, then use this as the object for the second lens in a second use of the lens equation.

FIGURE 19.13 Two lenses in combination.

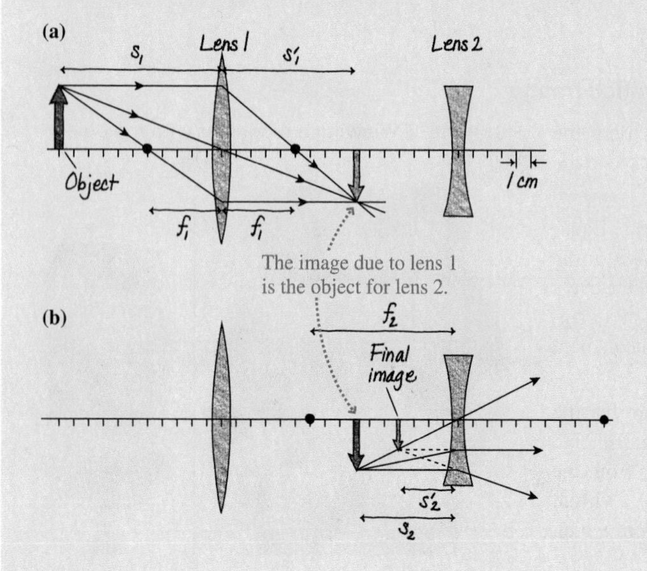

SOLVE We first solve for the image due to lens 1. We have

$$\frac{1}{s_1'} = \frac{1}{f_1} - \frac{1}{s_1} = \frac{1}{5.0\text{ cm}} - \frac{1}{11.0\text{ cm}}$$

from which $s_1' = 9.17$ cm. Because this is a positive image distance, the image is real and located to the right of the first lens. The magnification of the first lens is $m_1 = -s_1'/s_1 = -(9.17\text{ cm})/(11.0\text{ cm}) = -0.833$.

This image is the object for the second lens. Because it is 9.17 cm to the right of the first lens, and the lenses are 16.0 cm apart, it is 16.0 cm − 9.17 cm = 6.83 cm in front of the second lens. Thus $s_2 = 6.83$ cm.

Applying the thin-lens equation again, we have

$$\frac{1}{s_2'} = \frac{1}{f_2} - \frac{1}{s_2} = \frac{1}{-10.0\text{ cm}} - \frac{1}{6.83\text{ cm}}$$

from which we find $s_2' = -4.06$ cm, or -4.1 cm to two significant figures. Because this image distance is negative, the image is virtual and located to the left of lens 2, as shown in Figure 19.13b. The magnification of the second lens is $m_2 = -s_2'/s_2 = -(-4.06\text{ cm})/(6.83\text{ cm}) = 0.594$.

Thus the final image size is

$$h_2' = m_2 h_2 = m_2 h_1' = m_2(m_1 h_1) = m_1 m_2 h_1$$
$$= (-0.833)(0.594)(4.0\text{ cm}) = -2.0\text{ cm}$$

Here we used the fact that the object height h_2 of the second lens is equal to the image height h_1' of the first lens.

ASSESS When calculating the final image size, we found that $h_2' = m_1 m_2 h_1$. This shows the important fact that **the total magnification of a combination of lenses is the *product* of the magnifications for each lens alone.**

A microscope, whose major parts are shown in **FIGURE 19.14**, attains a magnification of up to 1000 by using two lenses in combination. A specimen to be observed is placed on the *stage* of the microscope, directly beneath the **objective lens** (or simply the **objective**), a converging lens with a relatively short focal length. The objective creates a magnified real image that is further enlarged by the **eyepiece,** a lens used as an ordinary magnifier. In most modern microscopes a prism bends the path of the rays from the object so that the eyepiece can be held at a comfortable angle. However, we'll consider a simplified version of a microscope without a prism. The light then travels along a straight tube.

Let's examine the magnification process in more detail. In **FIGURE 19.15** we draw a microscope tilted horizontally. The object distance is just slightly greater than the focal length f_o of the objective lens, so the objective forms a highly magnified real image of the object at a distance $s' = L$. This distance, known as the **tube length,** has been standardized for most biological microscopes at $L = 160$ mm. Most microscopes, such as the one shown in Figure 19.14, are focused by moving the sample stage up and down, using the focusing knob, until the object distance is correct for placing the image at L.

FIGURE 19.14 A microscope.

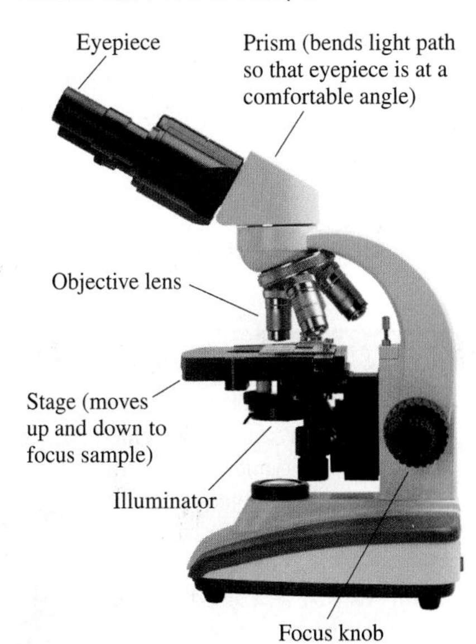

FIGURE 19.15 A horizontal view of the optics in a microscope.

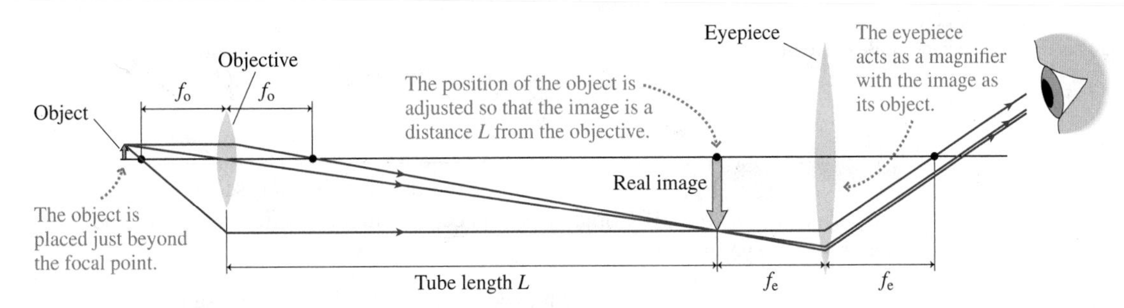

From the magnification equation, Equation 18.8, the magnification of the objective lens is

$$m_o = -\frac{s'}{s} \approx -\frac{L}{f_o} \tag{19.4}$$

Here we used the fact that the image distance s' is equal to the tube length L and the object distance s is very close to the focal length f_o of the objective. The minus sign tells us that the image is inverted with respect to the object.

The image of the objective acts as the object for the eyepiece, which functions as a simple magnifier. The angular magnification of the eyepiece is given by Equation 19.3: $M_e = (25\text{ cm})/f_e$. Together, the objective and eyepiece produce a total angular magnification

$$M = m_o M_e = -\frac{L}{f_o}\frac{25\text{ cm}}{f_e} \tag{19.5}$$

The minus sign shows that the image seen in a microscope is inverted.

EXAMPLE 19.6 Finding the focal length of a microscope objective

A biological microscope objective is labeled "20×." What is its focal length?

PREPARE The "20×" means that the objective has a magnification m_o of -20. We can use Equation 19.4 with L as 160 mm, which we've seen is the standard length for a biological microscope.

SOLVE From Equation 19.4 we have

$$f_o = -\frac{L}{m_o} = -\frac{160\text{ mm}}{-20} = 8.0\text{ mm}$$

ASSESS This focal length is significantly shorter than the tube length, in agreement with Figure 19.15.

The magnification of a microscope objective is called its "power." Thus we would say that the 20× objective of Example 19.6 has a power of 20. (The power or magnification of a microscope objective should not be confused with the refractive power of a lens.) Many microscopes have a set of objectives that can be pivoted into place to change the overall magnification. A complete set of objectives might include 5×, 10×, 20×, 40×, and 100×. Eyepieces are also specified by their magnification, or "power," and are available with magnifications in the range of 10× to 20×. With these lenses, the lowest magnification available would be $5 \times 10 = 50\times$, while the highest magnification would be $100 \times 20 = 2000\times$.

EXAMPLE 19.7 Viewing blood cells BIO

A pathologist inspects a sample of 7-μm-diameter human blood cells under a microscope. She selects a 40× objective and a 10× eyepiece. What size object, viewed from 25 cm, has the same apparent size as a blood cell seen through the microscope?

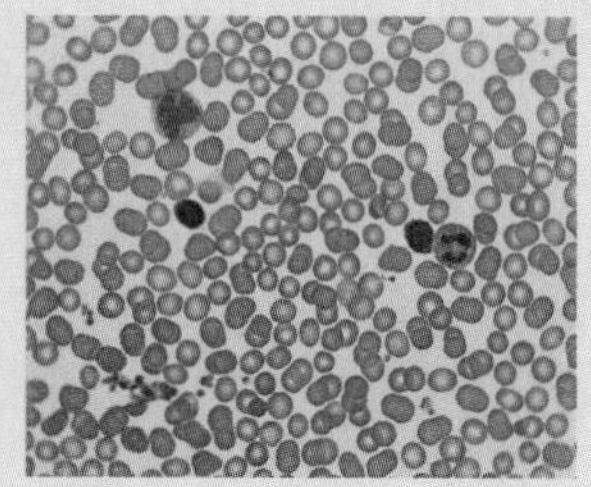

PREPARE Angular magnification compares the magnified angular size to the angular size seen at the near-point distance of 25 cm.

SOLVE The microscope's angular magnification is $M = -(40) \times (10) = -400$. The magnified cells will have the same apparent size as an object $400 \times 7\ \mu\text{m} \approx 3$ mm in diameter seen from a distance of 25 cm.

ASSESS 3 mm is about the size of a capital O in this textbook, so a blood cell seen through the microscope will have about the same apparent size as an O seen from a comfortable reading distance.

STOP TO THINK 19.5 A biologist rotates the turret of a microscope to replace the 20× objective with a 10× objective. To keep the magnification the same, the focal length of the eyepiece must

A. Be doubled. B. Be halved. C. Remain the same.
D. The magnification cannot stay the same if the objective power is changed.

19.5 The Telescope

The microscope magnifies small objects that can be placed near its objective lens. A *telescope* is used to magnify distant objects. The two-lens arrangement shown in **FIGURE 19.16** is similar to that of the microscope, but the objective lens has a long focal length instead of the very short focal length of a microscope objective. Because the object is very far away ($s \approx \infty$), the converging objective lens forms a real image of the distant object at the lens's focal point. A second lens, the eyepiece, is then used as a simple magnifier to enlarge this real image for final viewing by the eye.

FIGURE 19.16 The telescope.

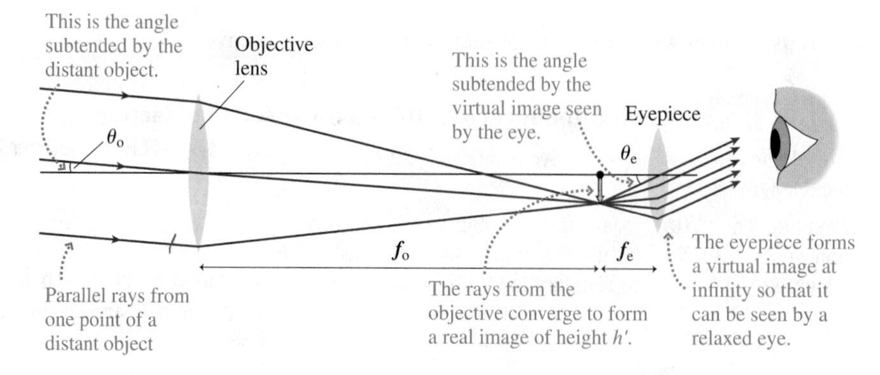

We can use Figure 19.16 to find the magnification of a telescope. The original object subtends an angle θ_o. Because the object is distant, its image is formed in the focal plane of the objective lens, a distance f_o from the objective. From the geometry of Figure 19.16, the height of this image (negative because it's inverted) is

$$h' \approx -f_o\theta_o$$

where we have used the small-angle approximation $\tan\theta_o \approx \theta_o$. This image is now the object for the eyepiece lens, which functions as a magnifier. The height of the "object" it views is h', so, from Equation 19.2, the angular size θ_e of the virtual image formed by the eyepiece is

$$\theta_e = \frac{h'}{f_e} = \frac{-f_o\theta_o}{f_e}$$

where f_e is the focal length of the eyepiece. The telescope's angular magnification is the ratio of the angular size seen when looking through the telescope to that seen without the telescope, so we have

$$M = \frac{\theta_e}{\theta_o} = -\frac{f_o}{f_e} \tag{19.6}$$

The minus sign indicates that you see an upside-down image when looking through a simple telescope. This is not a problem when looking at astronomical objects, but it could be disconcerting to a bird watcher. More sophisticated telescope designs produce an upright image.

To get a high magnification with a telescope, the focal length of the objective should be large and that of the eyepiece small. Contrast this with the magnification of a microscope, Equation 19.5, which is high when the focal lengths of both objective and eyepiece are small.

The telescope shown in Figure 19.16 uses a lens as its objective and hence is known as a *refracting telescope*. It is also possible to make a *reflecting telescope* with a concave mirror instead of a lens, as shown in **FIGURE 19.17**. One problem with this arrangement, for small telescopes, is that the image is formed in front of the mirror where it's hard to magnify with an eyepiece without getting one's head in the way. Newton, who built the first such telescope, used a small angled plane mirror, called a *secondary mirror,* to deflect the image to an eyepiece on the side of the telescope.

For large telescopes, such as those used in astronomy, mirrors have two important advantages over lenses. First, objectives of astronomical telescopes must be quite large in order to gather as much light from faint objects as possible. The Subaru Telescope in Hawaii is the world's largest single-mirror telescope; its mirror has a diameter of 8.3 m (27 ft)! A giant lens of this diameter would sag under its own weight. A mirror, on the other hand, can be supported along its entire back surface. Second, mirrors are free from chromatic aberration, the tendency that lenses have of splitting light into its constituent colors, as we'll see in a later section.

A clearer view The performance of telescopes on the earth is limited by the atmosphere. Even at night, the atmosphere glows faintly, interfering with the long exposures needed to photograph faint astronomical objects. Further, atmospheric turbulence—visible to the naked eye as the twinkling of stars—obscures the finest details of the object being observed. Because of this, the Hubble Space Telescope orbits the earth high above the atmosphere. With its 2.4-m-diameter mirror, it has produced some of the most spectacular images of astronomical objects, such as this gas cloud surrounding the star V838 Monocerotis.

FIGURE 19.17 A reflecting telescope.

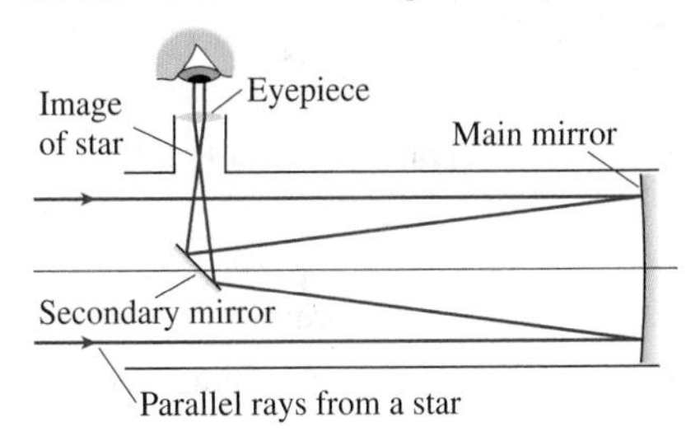

19.6 Color and Dispersion

I procured me a triangular glass prism to try therewith the celebrated phenomena of colors.
Isaac Newton

One of the most obvious visual aspects of light is the phenomenon of color. Yet color, for all its vivid sensation, is not inherent in the light itself. Color is a *perception,* not a physical quantity. Color is associated with the wavelength of light, but the fact that we see light with a wavelength of 650 nm as "red" tells us how our visual system responds to electromagnetic waves of this wavelength. There is no "redness" associated with the light wave itself.

Most of the results of optics do not depend on color. We generally don't need to know the color of light—or, to be more precise, its wavelength—to use the laws of reflection and refraction. Nonetheless, color is an interesting subject, one worthy of a short digression.

FIGURE 19.18 Newton used prisms to study color.

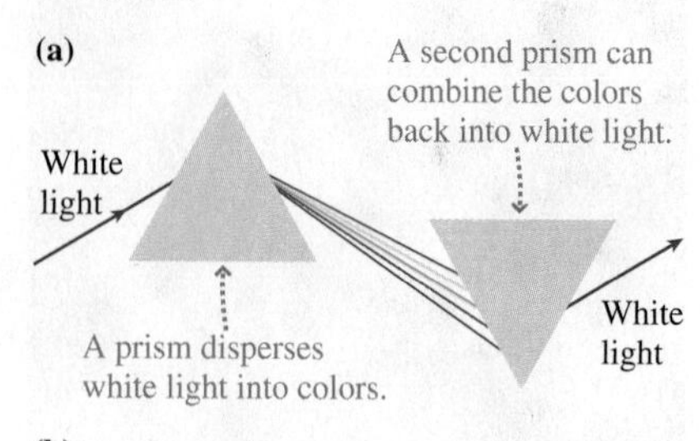
(a)
A second prism can combine the colors back into white light.
White light
White light
A prism disperses white light into colors.

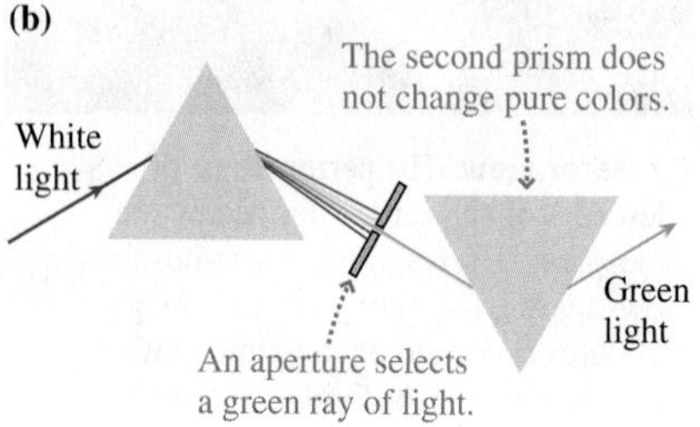
(b)
The second prism does not change pure colors.
White light
Green light
An aperture selects a green ray of light.

Color

It has been known since antiquity that irregularly shaped glass and crystals cause sunlight to be broken into various colors. A common idea was that the glass or crystal somehow altered the properties of the light by *adding* color to the light. Newton suggested a different explanation. He first passed a sunbeam through a prism, producing the familiar rainbow of light. We say that the prism *disperses* the light. Newton's novel idea, shown in FIGURE 19.18a, was to use a second prism, inverted with respect to the first, to "reassemble" the colors. He found that the light emerging from the second prism was a beam of pure white light.

But the emerging light beam is white only if *all* the rays are allowed to move between the two prisms. Blocking some of the rays with small obstacles, as in FIGURE 19.18b, causes the emerging light beam to have color. This suggests that color is associated with the light itself, not with anything that the prism is "doing" to the light. Newton tested this idea by inserting a small aperture between the prisms to pass only the rays of a particular color, such as green. If the prism alters the properties of light, then the second prism should change the green light to other colors. Instead, the light emerging from the second prism is unchanged from the green light entering the prism.

These and similar experiments show that:

1. What we perceive as white light is a mixture of all colors. White light can be dispersed into its various colors and, equally important, mixing all the colors produces white light.
2. The index of refraction of a transparent material differs slightly for different colors of light. Glass has a slightly higher index of refraction for violet light than for green light or red light. Consequently, different colors of light refract at slightly different angles. A prism does not alter the light or add anything to the light; it simply causes the different colors that are inherent in white light to follow slightly different trajectories.

Dispersion

TABLE 19.1 A brief summary of the visible spectrum of light

Color	Approximate wavelength
Deepest red	700 nm
Red	650 nm
Yellow	600 nm
Green	550 nm
Blue	450 nm
Deepest violet	400 nm

In 1801 Thomas Young performed his two-slit interference experiment, which first showed that different colors are associated with light of different wavelengths. The longest wavelengths are perceived as red light and the shortest wavelengths are perceived as violet light. Table 19.1 is a brief summary of the *visible spectrum* of light. Visible-light wavelengths are used so frequently that it is well worth committing this short table to memory.

The slight variation of index of refraction with wavelength is known as **dispersion.** FIGURE 19.19 shows the *dispersion curves* of two common glasses. Notice that **n is *higher* when the wavelength is *shorter;*** thus violet light refracts more than red light.

FIGURE 19.19 Dispersion curves show how the index of refraction varies with wavelength.

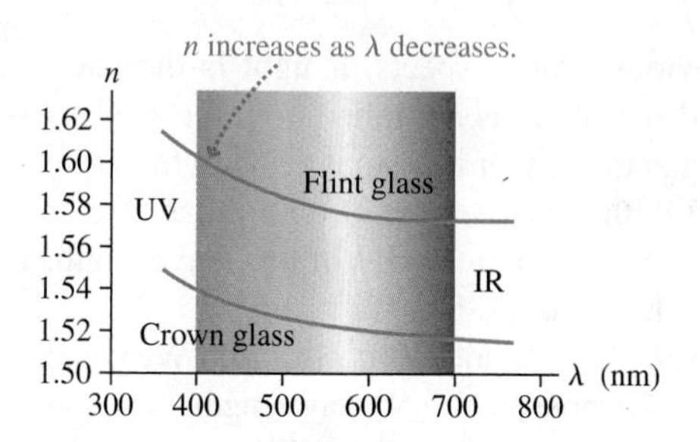
n increases as λ decreases.
n
1.62
1.60
1.58
1.56
1.54
1.52
1.50
UV
Flint glass
IR
Crown glass
300 400 500 600 700 800
λ (nm)

EXAMPLE 19.8 Dispersing light with a prism

Example 18.4 in Chapter 18 found that a ray incident on a 30° prism is deflected by 22.6° if the prism's index of refraction is 1.59. Suppose this is the index of refraction of deep violet light, and that deep red light has an index of refraction of 1.54.

a. What is the deflection angle for deep red light?
b. If a beam of white light is dispersed by this prism, how wide is the rainbow spectrum on a screen 2.0 m away?

PREPARE Figure 18.18 in Example 18.4 showed the geometry. A ray is incident on the hypotenuse of the prism at $\theta_1 = 30°$.

SOLVE a. If $n_1 = 1.54$ for deep red light, the refraction angle is

$$\theta_2 = \sin^{-1}\left(\frac{n_1 \sin\theta_1}{n_2}\right) = \sin^{-1}\left(\frac{1.54 \sin 30°}{1.00}\right) = 50.35°$$

Example 18.4 showed that the deflection angle is $\phi = \theta_2 - \theta_1$, so deep red light is deflected by $\phi_{red} = 20.35°$. This angle is slightly smaller than the deflection angle for violet light, $\phi_{violet} = 22.60°$.

b. The entire spectrum is spread between $\phi_{red} = 20.35°$ and $\phi_{violet} = 22.60°$. The angular spread is

$$\delta = \phi_{violet} - \phi_{red} = 2.25° = 0.0393 \text{ rad}$$

At distance r, the spectrum spans an arc length

$$s = r\delta = (2.0 \text{ m})(0.0393 \text{ rad}) = 0.0785 \text{ m} = 7.9 \text{ cm}$$

ASSESS Notice that we needed three significant figures for ϕ_{red} and ϕ_{violet} in order to determine δ, the *difference* between the two angles, to two significant figures. The angle is so small that there's no appreciable difference between arc length and a straight line. The spectrum will be 7.9 cm wide at a distance of 2.0 m.

Rainbows

One of the most interesting sources of color in nature is the rainbow. The details get somewhat complicated, but **FIGURE 19.20a** shows that the basic cause of the rainbow is a combination of refraction, reflection, and dispersion.

FIGURE 19.20 Light seen in a rainbow has undergone refraction + reflection + refraction in a raindrop.

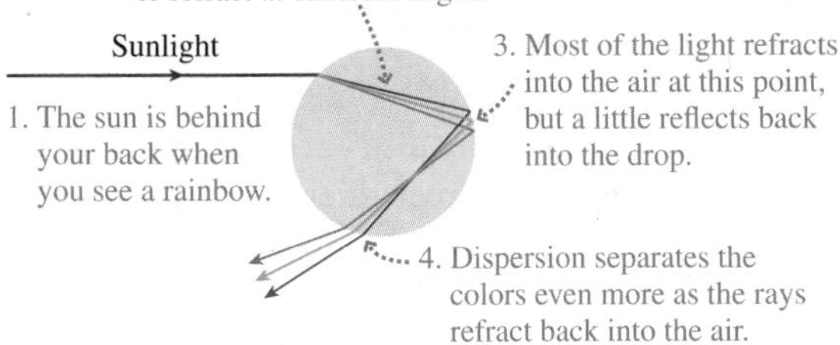
(a)
2. Dispersion causes different colors to refract at different angles.
Sunlight
3. Most of the light refracts into the air at this point, but a little reflects back into the drop.
1. The sun is behind your back when you see a rainbow.
4. Dispersion separates the colors even more as the rays refract back into the air.
(b)
Sunlight
42.5°
40.8°
Eye
You see a rainbow with red on the top, violet on the bottom.
Red light is refracted predominantly at 42.5°. The red light reaching your eye comes from drops higher in the sky.
Violet light is refracted predominantly at 40.8°. The violet light reaching your eye comes from drops lower in the sky.

Figure 19.20a might lead you to think that the top edge of a rainbow is violet. In fact, the top edge is red, and violet is on the bottom. The rays leaving the drop in Figure 19.20a are spreading apart, so they can't all reach your eye. As **FIGURE 19.20b** shows, a ray of red light reaching your eye comes from a drop *higher* in the sky than a ray of violet light. In other words, the colors you see in a rainbow refract toward your eye from different raindrops, not from the same drop. You have to look higher in the sky to see the red light than to see the violet light.

Colored Filters and Colored Objects

White light passing through a piece of green glass emerges as green light. A possible explanation would be that the green glass *adds* "greenness" to the white light, but Newton found otherwise. Green glass is green because it *removes* any light that is "not green." More precisely, a piece of colored glass *absorbs* all wavelengths except those of one color, and that color is transmitted through the glass without hindrance. We can think of a piece of colored glass or plastic as a *filter* that removes all wavelengths except a chosen few.

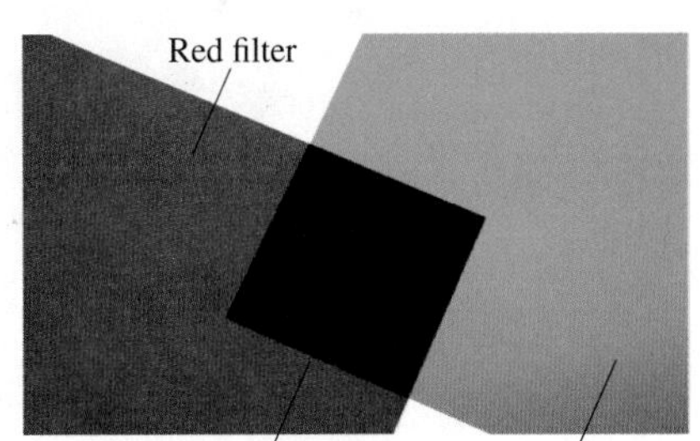
Red filter
Black where filters overlap
Green filter

No light at all passes through both a green and a red filter.

CONCEPTUAL EXAMPLE 19.9 Filtering light

White light passes through a green filter and is observed on a screen. Describe how the screen will look if a second green filter is placed between the first filter and the screen. Describe how the screen will look if a red filter is placed between the green filter and the screen.

REASON The first filter removes all light except for wavelengths near 550 nm that we perceive as green light. A second green filter doesn't have anything to do. The nongreen wavelengths have already been removed, and the green light emerging from the first filter will pass through the second filter without difficulty. The screen will continue to be green and its intensity will not change. A red filter, by contrast, absorbs all wavelengths except those near 650 nm. The red filter will absorb the green light, and *no* light will reach the screen. The screen will be dark.

FIGURE 19.21 The absorption curve of chlorophyll.

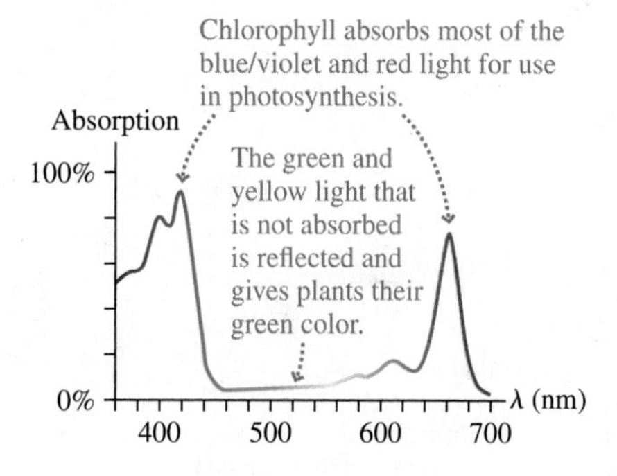

Opaque objects appear colored by virtue of *pigments* that absorb light of some wavelengths but *reflect* light of other wavelengths. For example, red paint contains pigments that reflect light of wavelengths near 650 nm while absorbing all other wavelengths. Pigments in paints, inks, and natural objects are responsible for most of the color we observe in the world, from the red of lipstick to the bright yellow of a goldfinch's feathers.

As an example, FIGURE 19.21 shows the absorption curve of *chlorophyll.* Chlorophyll is essential for photosynthesis in green plants. The chemical reactions of photosynthesis are able to use red light and blue/violet light; thus chlorophyll has evolved to absorb red light and blue/violet light from sunlight and put it to use. But green and yellow light are not absorbed. Instead, these wavelengths are mostly *reflected* to give the object a greenish-yellow color. When you look at the green leaves on a tree, you're seeing the light that was reflected because it *wasn't* needed for photosynthesis.

STOP TO THINK 19.6 A red apple is viewed through a green filter. The apple appears

A. Red. B. Green. C. Yellow. D. Black.

19.7 Resolution of Optical Instruments

Suppose you wanted to study the *E. coli* bacterium. It's quite small, about 2 μm long and 0.5 μm wide. You might imagine that you could pair a 150× objective (the highest magnification available) with a 25× eyepiece to get a total magnification of 3750! At that magnification, the *E. coli* would appear about 8 mm across—about the size of Lincoln's head on a penny—with much fine detail revealed. But if you tried this, you'd be disappointed. Although you would see the general shape of a bacterium, you wouldn't be able to make out any real details. All real optical instruments are limited in the details they can observe. Some limits are practical: Lenses are never perfect, suffering from **aberrations.** But even a perfect lens would have a fundamental limit to the smallest details that could be seen. As we'll see, this limit is set by the diffraction of light, and so is intimately related to the wave nature of light itself. Together, lens aberrations and diffraction set a limit on an optical system's **resolution**—its ability to make out the fine details of an object.

Aberrations

Consider the simple lens shown in **FIGURE 19.22** imaging an object located at infinity, so that the incoming rays are parallel. An ideal lens would focus all the rays to a single point. However, for a real lens with spherical surfaces, the rays that pass near the lens's center come to a focus a bit farther from the lens than those that pass near its edge. There is no single focal point; even at the "best" focus the image is a bit blurred. This inability of a real lens to focus perfectly is called **spherical aberration.**

A careful examination of Figure 19.22 shows that the outer rays are most responsible for the poor focus. Consequently, the effects of spherical aberration can be minimized by using a diaphragm to pass only rays near the optical axis. This "stopping down" of a lens improves its imaging characteristics at the expense of its light-gathering capabilities. Part of the function of the iris of the human eye is to improve vision in this way. Our vision is poorer at night with the iris wide open—but our ancestors probably avoided many a predator with this poor but sensitive night vision.

FIGURE 19.22 Spherical aberration.

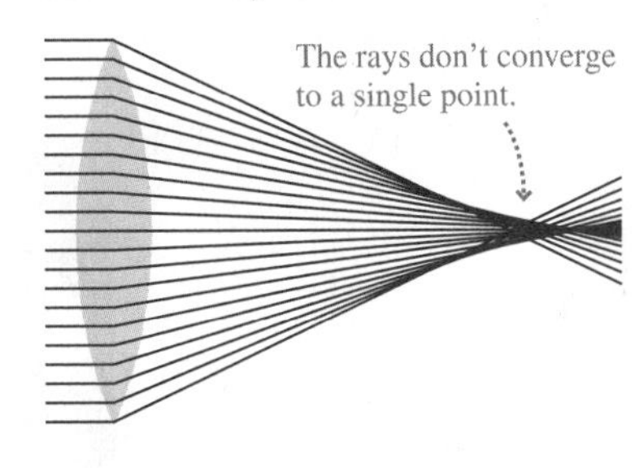
The rays don't converge to a single point.

As we learned in the previous section, glass has *dispersion;* that is, the index of refraction of glass varies slightly with wavelength. The higher a lens's index of refraction, the more it bends incoming light rays. Because the index of refraction for violet light is higher than that for red light, a lens's focal length is slightly shorter for violet light than for red light. Consequently, different colors of light come to a focus at slightly different distances from the lens. If red light is sharply focused on a viewing screen, then blue and violet wavelengths are not well focused. This imaging error, illustrated in **FIGURE 19.23**, is called **chromatic aberration.**

FIGURE 19.23 Chromatic aberration.

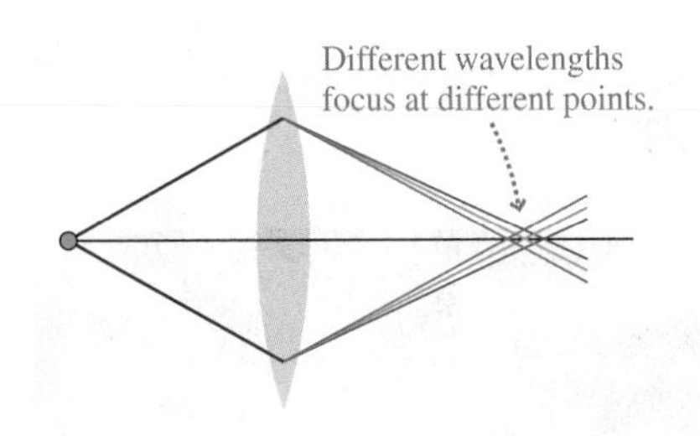
Different wavelengths focus at different points.

Correcting Aberrations

Single lenses always have aberrations of some kind. For high-quality optics, such as those used in microscopes or telescopes, the aberrations are minimized by using a careful *combination* of lenses. An important example is the **achromatic doublet** (achromatic = "without color"), two lenses used in combination to greatly reduce chromatic aberration. **FIGURE 19.24** shows how this works. A converging lens is paired with a weaker diverging lens; the combination has an overall positive refractive power and so is converging. However, the glasses are chosen so that the weaker diverging lens has a greater dispersion than the stronger converging lens. In this way the colors of white light, separated at first by the converging lens, are brought back together by the diverging lens. Achromatic doublets also minimize spherical aberration. Real microscope objectives are even more complex, but are based on the same principle as the achromatic doublet.

FIGURE 19.24 An achromatic lens.

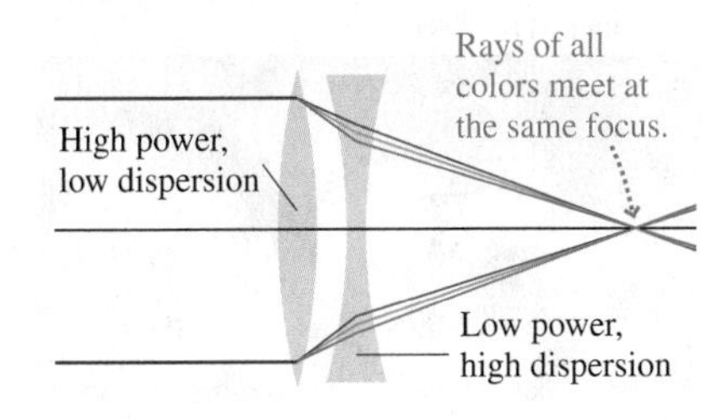
Rays of all colors meet at the same focus.
High power, low dispersion
Low power, high dispersion

Resolution and the Wave Nature of Light

Modern lenses can be well corrected for aberrations, so they might be expected to focus perfectly. According to the ray model of light, a perfect lens should focus parallel rays to a single point in the focal plane. However, we've already hinted that there's a more fundamental limit to the performance of an optical instrument, a limit set by the wave nature of light.

► **Eyeglasses in space** After launch, it was discovered that the mirror of the Hubble Space Telescope had been ground to the wrong shape, giving it severe spherical aberration. In a later service mission, corrective optics—in essence, very high-tech glasses—were put in place to correct for this spherical aberration. The photos to the right show an image of a galaxy before and after the corrective optics were added.

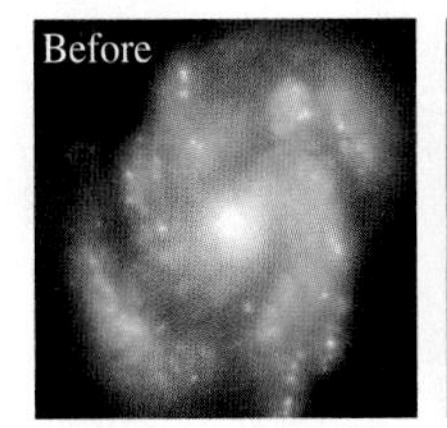
Before

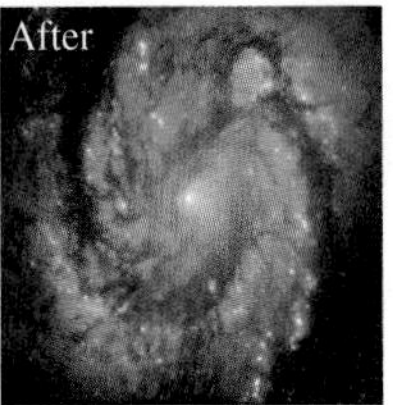
After

FIGURE 19.25 The image of a distant point source is a circular diffraction pattern.

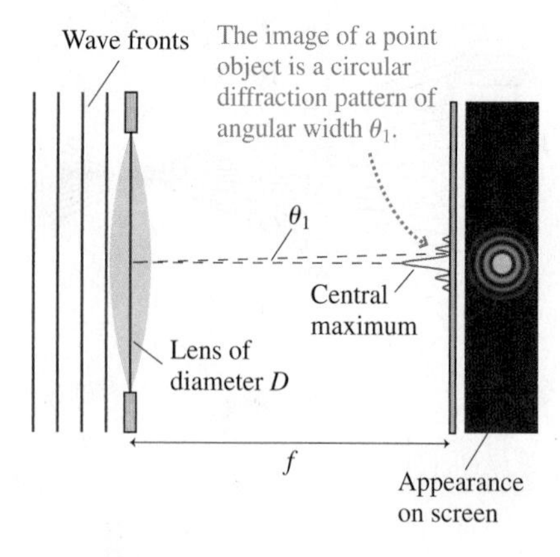

FIGURE 19.25 shows a plane wave from a distant point source such as a star being focused by a lens of diameter D. Only those waves passing *through* the lens can be focused, so the lens acts like a circular aperture of diameter D in an opaque barrier. In other words, the lens both focuses *and diffracts* light waves.

You learned in ◀◀ SECTION 17.6 that a circular aperture produces a diffraction pattern with a bright central maximum surrounded by dimmer circular fringes. Consequently, as Figure 19.25 shows, light from a distant point source focuses not to a perfect point but, instead, to a small circular diffraction pattern. Equation 17.21 in Chapter 17 gave the angle θ_1 of the outer edge of the central maximum as

$$\theta_1 = \frac{1.22\lambda}{D} \tag{19.7}$$

Because the wavelength of light λ is so much shorter than the lens diameter D of an ordinary lens, the angular size of the central maximum is very small—but it is not zero.

The fact that light is focused to a small spot, not a perfect point, has important consequences for how well a telescope can resolve two stars separated by only a small angle in the sky. FIGURE 19.26a shows how two nearby stars would appear in a telescope. Instead of perfect points, they appear as two diffraction images. Nonetheless, because they're clearly two separate stars, we say they are *resolved*. FIGURE 19.26b shows two stars that are closer together. Here the two diffraction patterns overlap, and it is becoming difficult to see them as two independent stars: They are barely resolved. The two very nearby stars in FIGURE 19.26c are so close together that we can't resolve them at all.

FIGURE 19.26 The resolution of a telescope.

(a) Stars resolved

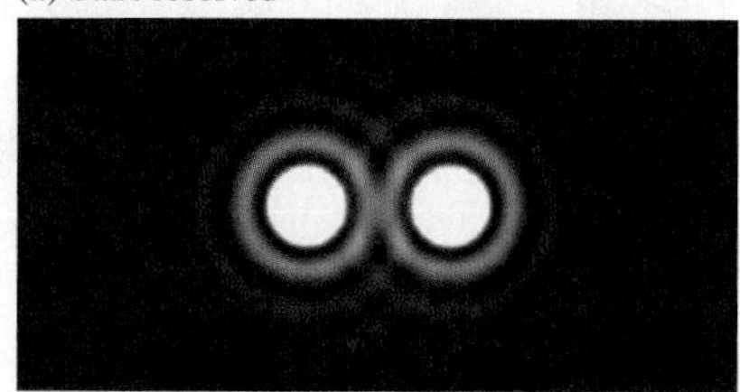

(b) Stars just resolved

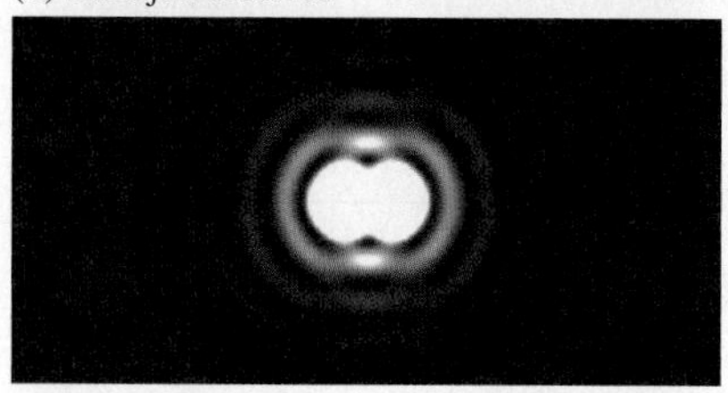

(c) Stars not resolved

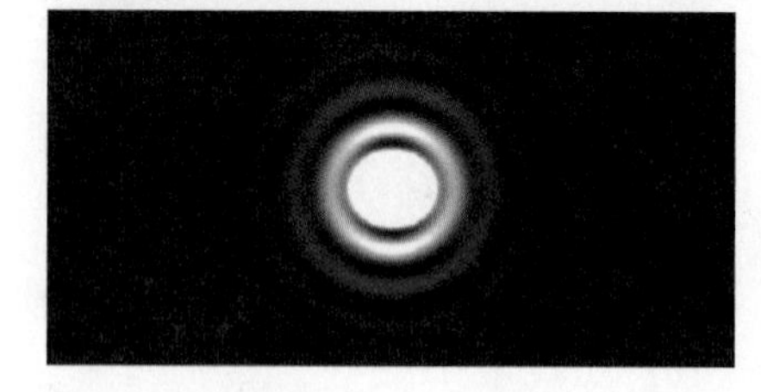

How close can the two diffraction patterns be before you can no longer resolve them? One of the major scientists of the 19th century, Lord Rayleigh, studied this problem and suggested a reasonable rule that today is called **Rayleigh's criterion.** In Figure 19.26b, where the two stars are just resolved, *the central maximum of the diffraction pattern of one star lies on top of the first dark fringe of the diffraction pattern of the other star.* Because the angle between the central maximum and the first dark fringe is θ_1, this means that the centers of the two stars are separated by angle $\theta_1 = 1.22\lambda/D$. Thus Rayleigh's criterion is:

> Two objects are resolvable if they are separated by an angle θ that is greater than $\theta_1 = 1.22\lambda/D$. If their angular separation is less than θ_1, then they are not resolvable. If their separation is equal to θ_1, then they are just barely resolvable.

For telescopes, the angle $\theta_1 = 1.22\lambda/D$ is called the *angular resolution* of the telescope. The angular resolution depends only on the lens diameter and the wavelength; the magnification is not a factor. Two overlapped, unresolved images will remain overlapped and unresolved no matter what the magnification. For visible light, where λ is pretty much fixed, the only parameter over which the astronomer has any control is the diameter of the lens or mirror of the telescope. The urge to build ever-larger telescopes is motivated, in part, by a desire to improve the angular resolution. (Another important motivation is to increase the light-gathering power so as to see objects farther away.)

The Resolution of a Microscope

A microscope differs from a telescope in that it magnifies objects that are very close to the lens, not far away. Nonetheless, the wave nature of light still sets a limit on the ultimate resolution of a microscope. FIGURE 19.27 shows the objective lens of a microscope that is observing two small objects. An analysis based on Rayleigh's criterion finds that the smallest resolvable separation between the two objects is

$$d_{min} = \frac{0.61\lambda}{n\sin\phi_0} \tag{19.8}$$

Here, ϕ_0, defined in Figure 19.27, is the angular size of the objective lens and n is the index of refraction of the medium between the objective lens and the specimen being observed. Usually this medium is air, so that $n = 1$, but biologists often use an *oil-immersion microscope* in which this space is filled with oil having $n \approx 1.5$. From Equation 19.8, you can see that this higher value of n reduces d_{min}, allowing objects that are closer together to be resolved.

The quantity $n \sin \phi_0$ is called the **numerical aperture** NA of the objective when immersed in a fluid of index n. The minimum resolvable distance of a microscope, also called its **resolving power** RP, is given by

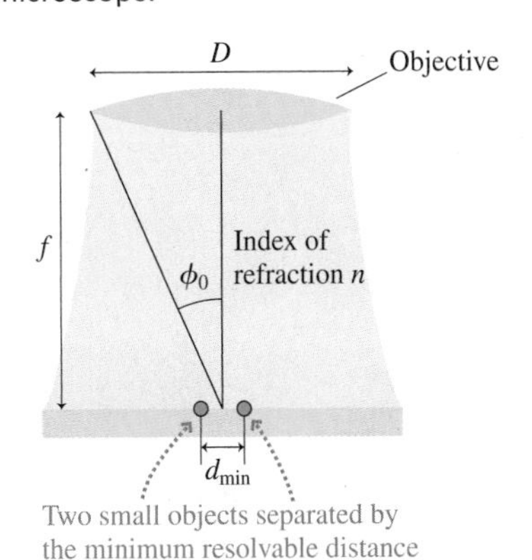

FIGURE 19.27 The resolution of a microscope.

$$\text{RP} = d_{min} = \frac{0.61\lambda_0}{\text{NA}} \tag{19.9}$$

The resolving power . . . is the smallest resolvable separation between two objects. Wavelength of light in vacuum. Numerical aperture of objective lens.

The lower the resolving power, the *better* the objective is at seeing small details.

In principle, it would appear from Equation 19.9 that the resolving power of a microscope could be made as low as desired simply by increasing the numerical aperture. But there are rather severe practical limits on how high the numerical aperture can be made. The highest possible numerical aperture for a high-magnification 100× objective used in air is about 0.95. For an oil-immersion objective, the numerical aperture might be as high as 1.3. With such an objective, the resolving power would be

$$\text{RP} \approx 0.5\lambda_0$$

This illustrates the fundamental fact that **the minimum resolving power of a microscope, and thus the size of the smallest detail observable, is about half the wavelength of light.** This is a *fundamental limit* set by the wave nature of light. For $\lambda \approx 400$ nm, at the short-wavelength edge of the visible spectrum, the maximum possible resolving power is RP ≈ 200 nm.

FIGURE 19.28 shows an actual micrograph of the bacillus *E. coli*. The width is about equal to the wavelength of light in the middle of the spectrum, about 500 nm, and the smallest resolved features are about half this. A higher magnification would not reveal any more detail because this micrograph is at the resolution limit set by the diffraction of light.

In contrast, the **electron microscope** micrograph of *E. coli* shows a wealth of detail unobservable in the optical picture. In Chapter 28 we'll find out why the resolving power of an electron microscope is so much lower than that of an optical microscope.

FIGURE 19.28 Optical and electron micrographs of *E. coli*. BIO

Optical microscope

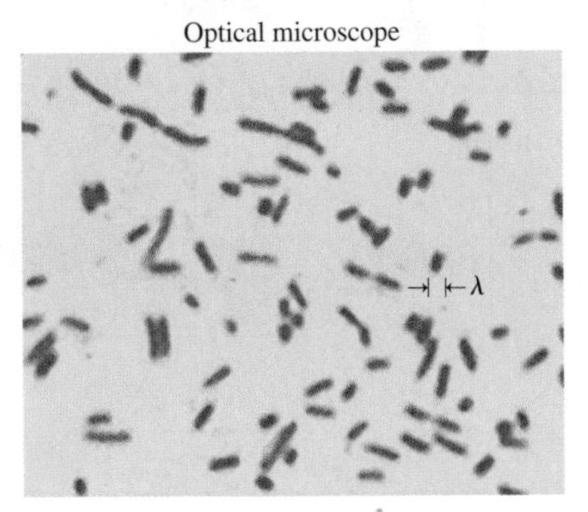

Electron microscope

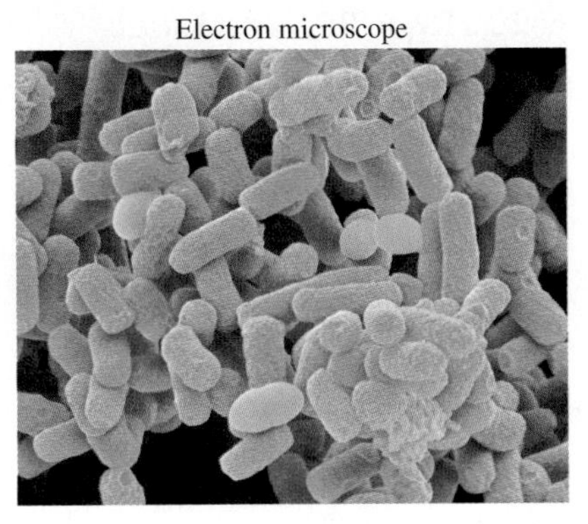

▶ The anatomy of a microscope objective The two most important specifications of a microscope objective, its magnification and its numerical aperture, are prominently displayed on its barrel. Other important information is shown as well. This objective is designed to project its real image at a tube length of infinity (∞), instead of at the standard distance of 160 mm. Extra lenses in the microscope move this image to just in front of the eyepiece. Many biological studies are conducted through a cover glass. This cover glass can introduce spherical aberration, blurring the image. By turning the correction collar, you can adjust this objective to correct for the exact thickness of the cover glass used.

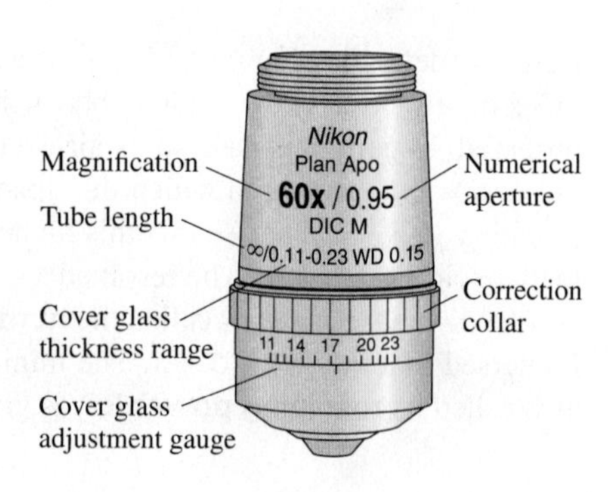

EXAMPLE 19.10 Finding the resolving power of a microscope

A microscope objective lens has a diameter of 6.8 mm and a focal length of 4.0 mm. For a sample viewed in air, what is the resolving power of this objective in red light? In blue light?

PREPARE We can use Equation 19.9 to find the resolving power. We'll need the numerical aperture of the objective, given as $\text{NA} = n \sin\phi_0$.

SOLVE From the geometry of Figure 19.27,

$$\tan\phi_0 = \frac{D/2}{f} = \frac{3.4\text{ mm}}{4.0\text{ mm}} = 0.85$$

from which $\phi_0 = \tan^{-1}0.85 = 40.4°$ and $\sin\phi_0 = \sin 40.4° = 0.65$. Hence the numerical aperture is (since $n = 1$ in air)

$$\text{NA} = n\sin\phi_0 = 1 \times 0.65 = 0.65$$

Then, from Equation 19.9, the resolving power is

$$\text{RP} = \frac{0.61\lambda_0}{0.65} = 0.94\lambda_0$$

Wavelengths of different colors of light were listed in Table 19.1. For red light, with $\lambda_0 = 650$ nm, RP $= 610$ nm, while blue light, with $\lambda_0 = 450$ nm, has RP $= 420$ nm.

ASSESS We see that shorter-wavelength light yields a higher resolution (lower RP). Unfortunately, wavelengths much shorter than 400 nm are invisible, and glass lenses are opaque to light of very short wavelength.

STOP TO THINK 19.7 Four lenses are used as microscope objectives, all for light with the same wavelength λ. Rank in order, from highest to lowest, the resolving powers RP_1 to RP_4 of the lenses.

$f = 10$ mm, 1, 2 mm
$f = 5$ mm, 2, 2 mm
$f = 10$ mm, 3, 4 mm
$f = 24$ mm, 4, 8 mm

INTEGRATED EXAMPLE 19.11 **The visual acuity of a kestrel** BIO

Like most birds of prey, the American kestrel has excellent eyesight. The smallest angular separation between two objects that an eye can resolve is called its *visual acuity;* a smaller visual acuity means better eyesight because objects closer together can be resolved. The eye of a particular kestrel has a pupil diameter of 3.0 mm. The fixed distance from its lens to the retina is 9.0 mm, and the space within the eye is filled with a clear liquid whose index of refraction is 1.31. Assume that the kestrel's optical system can be adequately modeled as a thin lens and a detector.

a. As the bird focuses on an insect sitting 0.80 m away, what is the focal length of its lens?
b. Laboratory measurements indicate that the kestrel can just resolve two small objects that have an angular separation of only 0.013°. How does this result compare with the visual acuity predicted by Rayleigh's criterion? Take the wavelength of light in air to be 550 nm.
c. What is the distance on the retina between the images of two small objects that can just be resolved? How does this distance compare to the 2.0 μm distance between two photoreceptors, the light-sensitive cells on the retina? Does this comparison make sense from the standpoint of vision?

PREPARE a. Recall that an eye focuses by changing the focal length of its lens. The image distance s' from the lens to the image plane (at the retina) is unchanged as the bird focuses from a distant object to a nearby one, so, as **FIGURE 19.29** shows, $s' = 9.0$ mm. We can then use the thin-lens equation to find f.

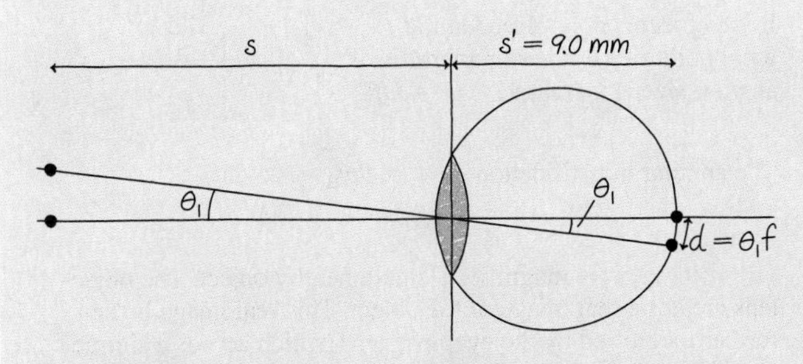

FIGURE 19.29 The kestrel's eye observing two closely spaced objects.

b. By Rayleigh's criterion, visual acuity—the smallest resolvable angle of an ideal lens—is proportional to the wavelength of light used. Inside the eye, however, the wavelength of light is shorter than its wavelength λ_0 in air because of the index of refraction of the liquid within the eye. Thus we must use $\lambda = \lambda_0/n = (550 \text{ nm})/(1.31) = 420$ nm in Rayleigh's criterion.

c. Figure 19.29 shows that, within the small-angle approximation, the distance d on the retina between the images of two small objects subtending an angle θ_1 is simply $d = \theta_1 f$, where θ_1 is in radians.

SOLVE a. From the thin-lens equation, Equation 18.11, we have

$$\frac{1}{f} = \frac{1}{s} + \frac{1}{s'} = \frac{1}{800 \text{ mm}} + \frac{1}{9.0 \text{ mm}} = 0.112 \text{ mm}^{-1}$$

Thus $f = 1/0.112 \text{ mm}^{-1} = 8.9$ mm.

b. For a perfect lens, with no aberrations, the smallest resolvable angular separation between two objects is given by Rayleigh's criterion as

$$\theta_1 = \frac{1.22\lambda}{D} = \frac{1.22(420 \times 10^{-9} \text{ m})}{3.0 \times 10^{-3} \text{ m}} = 1.7 \times 10^{-4} \text{ rad}$$

Recalling that there are 360° in 2π rad, this angle is

$$\theta_1 = (1.7 \times 10^{-4} \text{ rad}) \times \frac{360°}{2\pi \text{ rad}} = 0.0098°$$

The observed visual acuity of 0.013° is about 30% greater than this theoretical value. Presumably this is due to aberrations in the optical system of the eye.

c. The angle of 0.013° corresponds to 2.3×10^{-4} rad. Thus the distance between the images of the two small objects is

$$d = \theta_1 f = (2.3 \times 10^{-4} \text{ rad})(9.0 \text{ mm})$$
$$= 2.1 \times 10^{-3} \text{ mm} = 2.1 \text{ } \mu\text{m}$$

This is just about the same as the photoreceptor distance. This makes sense. If d were significantly greater than the photoreceptor distance, then many receptors would be wasted. If d were smaller, the eye's resolution would be determined by the photoreceptor spacing and would not reach the visual acuity predicted by Rayleigh's criterion.

ASSESS An object far from a converging lens gives an image distance close to the lens's focal length. Thus our focal length of 8.9 mm—close to the image distance of 9.0 mm—is reasonable.

Although the visual acuity of the kestrel is impressive—equivalent to being able to resolve a mouse at a distance of 600 feet—it turns out not to be significantly greater than that for the human eye. Raptors are also aided in hunting their prey by a highly evolved cerebral function that allows them to pick out small movements that humans would miss.

SUMMARY

Goal: To understand how common optical instruments work.

IMPORTANT CONCEPTS

Color and dispersion

The eye perceives light of different wavelengths as having different colors.

Dispersion is the dependence of the index of refraction n of a transparent medium on the wavelength of light: Long wavelengths have the lowest n, short wavelengths the highest n.

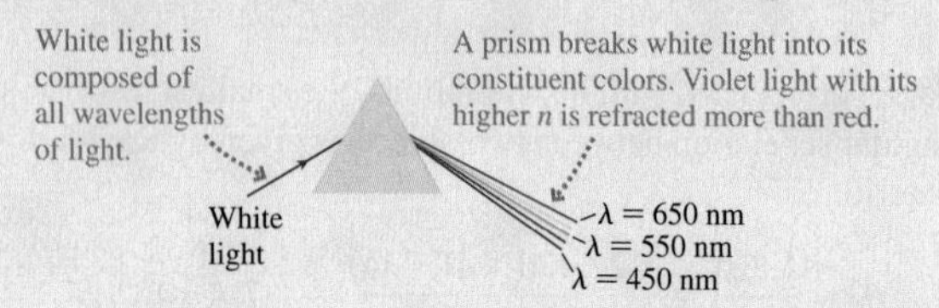

White light is composed of all wavelengths of light.
A prism breaks white light into its constituent colors. Violet light with its higher n is refracted more than red.
White light
λ = 650 nm
λ = 550 nm
λ = 450 nm

Lenses in combination

When two lenses are used in combination, the image from the first lens serves as the object for the second.

The **refractive power** P of a lens is the inverse of its focal length: $P = 1/f$. Refractive power is measured in diopters:

$$1\text{ D} = 1\text{ m}^{-1}$$

Resolution of optical instruments

The **resolution** of a telescope or microscope is limited by imperfections, or **aberrations,** in the optical elements, and by the more fundamental limits imposed by diffraction.

For a *microscope,* the minimum resolvable distance between two objects is

$$d_{\text{min}} = \frac{0.61\lambda}{\text{NA}}$$

For a *telescope,* the minimum resolvable angular separation between two objects is

$$\theta_1 = \frac{1.22\lambda}{D}$$

Angular and apparent size

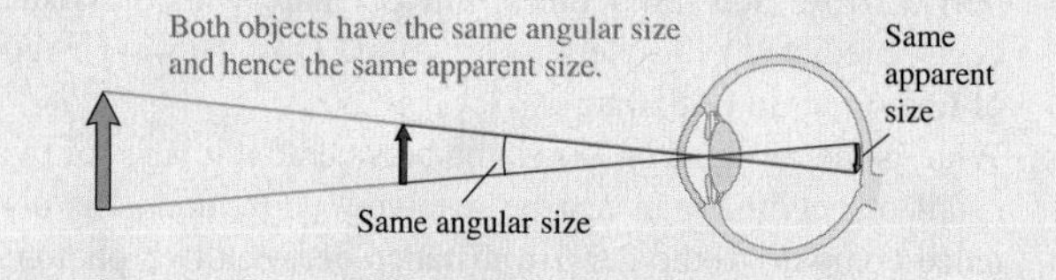

APPLICATIONS

The camera and the eye

Both the camera and the eye work by focusing an image on a light-sensitive surface.

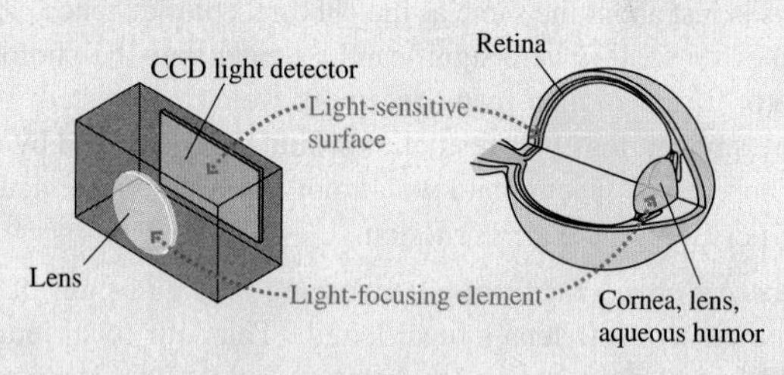

The camera focuses by changing the lens-detector distance, while the eye focuses by changing the focal length of its lens.

The telescope magnifies distant objects. The objective lens creates a real image of the distant object. This real image is then magnified by the eyepiece lens, which acts as a simple magnifier. The angular magnification is $M = -f_o/f_e$.

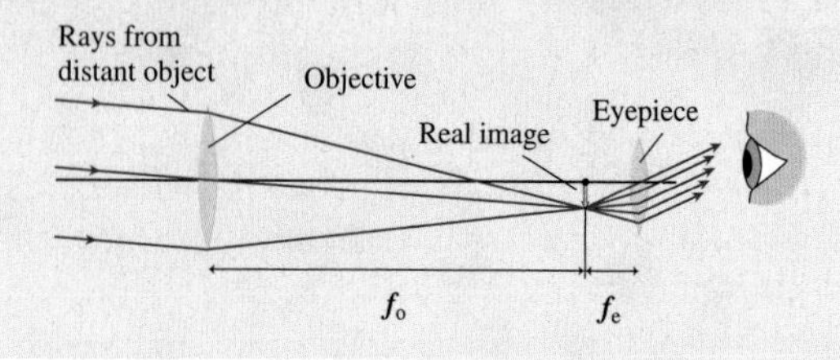

The magnifier

Without a lens, an object cannot be viewed closer than the eye's near point of ≈25 cm. Its angular size θ_0 is $h/25$ cm.

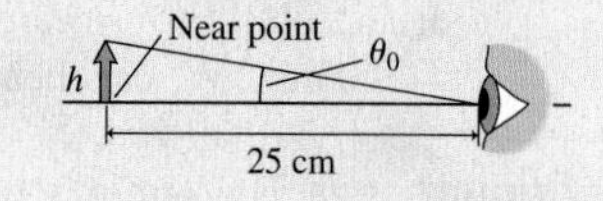

If the object is now placed at the focal point of a converging lens, its angular size is increased to $\theta = h/f$.

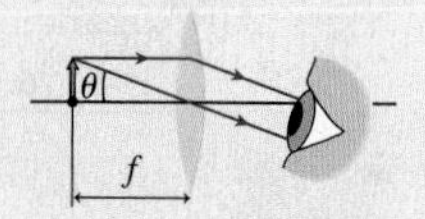

The angular magnification is $M = \theta/\theta_0 = 25\text{ cm}/f$.

The microscope magnifies a small, nearby object. The objective lens creates a real image of the object. This real image is then further magnified by the eyepiece lens, which acts as a simple magnifier. The angular magnification is

$$M = -\frac{L \times 25\text{ cm}}{f_o f_e}$$

Problem difficulty is labeled as | (straightforward) to ||||| (challenging). Problems labeled INT integrate significant material from earlier chapters; BIO are of biological or medical interest.

For assigned homework and other learning materials, go to MasteringPhysics®

Scan this QR code to launch a Video Tutor Solution that will help you solve problems for this chapter.

QUESTIONS

Conceptual Questions

1. On a sunny summer day, with the sun overhead, you can stand under a tree and look on the ground at the pattern of light that has passed through gaps between the leaves. You may see illuminated circles of varying brightness. Why are there circles, when the gaps between the leaves have irregular shapes?
2. Suppose you have two pinhole cameras. The first has a small round hole in the front of the camera. The second is identical in every regard, except that it has a square hole of the same area as the round hole in the first camera. Would the pictures taken by these two cameras, under the same conditions, be different in any obvious way? Explain.
3. A photographer focuses his camera on his subject. The subject then moves closer to the camera. To refocus, should the lens be moved closer to or farther from the detector? Explain.
4. The object for a magnifier is usually placed very close to the focal point of the lens, creating a virtual image very far away where it can be viewed with the relaxed eye. But the object could be placed so that the image is at the eye's near-point distance. In this case, the image can be viewed only by using the full accommodation of the eye. When used this way, is the angular magnification greater than, equal to, or less than the magnification when the image is far away? Explain.
5. A nature photographer taking a close-up shot of an insect replaces the standard lens on his camera with a lens that has a shorter focal length and is positioned farther from the CCD detector. Explain why he does this.
6. The CCD detector in a certain camera has a width of 8 mm. The photographer realizes that with the lens she is currently using, she can't fit the entire landscape she is trying to photograph into her picture. Should she switch to a lens with a longer or shorter focal length? Explain.
7. BIO All humans have what is known as a *blind spot,* where the optic nerve exits the eye and no light-sensitive cells exist. To locate your blind spot, look at the figure of the cross. Close your left eye and place your index finger on the cross. Slowly move your finger to the left while following it with your right eye. At a certain point the cross will disappear. Is your right eye's blind spot on the right or left side of your retina? Explain.
8. Suppose you wanted special glasses designed to wear underwater, without a face mask. Should the glasses use a converging or diverging lens in order for you to be able to focus under water? Explain.
9. You have lenses with the following focal lengths: $f = 25$ mm, 50 mm, 100 mm, and 200 mm. Which lens or pair of lenses would you use, and in what arrangement, to get the highest-power magnifier, microscope, and telescope? Explain.
10. BIO An 8-year-old child and a 75-year-old man both use the same magnifier to observe a bug. For whom does the magnifier more likely have the higher magnification? Explain.
11. A friend lends you the eyepiece of his microscope to use on your own microscope. He claims that since his eyepiece has the same diameter as yours but twice the focal length, the resolving power of your microscope will be doubled. Is his claim valid? Explain.
12. An astronomer is using a telescope to observe two distant stars. The stars are marginally resolved when she looks at them through a filter that passes green light near 550 nm. Which of the following actions would improve the resolution? Assume that the resolution is not limited by the atmosphere.
    a. Changing the filter to a different wavelength? If so, should she use a shorter or a longer wavelength?
    b. Using a telescope with an objective lens of the same diameter but a different focal length? If so, should she select a shorter or a longer focal length?
    c. Using a telescope with an objective lens of the same focal length but a different diameter? If so, should she select a larger or a smaller diameter?
    d. Using an eyepiece with a different magnification? If so, should she select an eyepiece with more or less magnification?
13. A student makes a microscope using an objective lens and an eyepiece. If she moves the lenses closer together, does the microscope's magnification increase or decrease? Explain.
14. BIO Is the wearer of the glasses in Figure Q19.14 nearsighted or farsighted? How can you tell?
15. A red card is illuminated by red light. What color does it appear to be? What if it's illuminated by blue light?

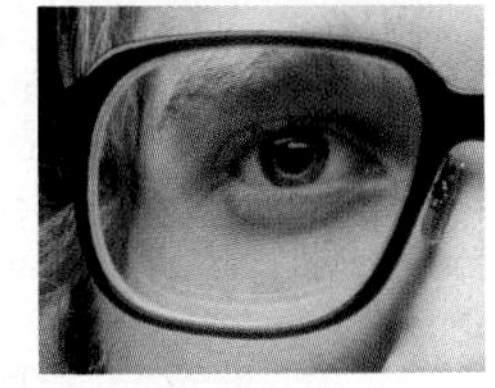

FIGURE Q19.14

Multiple-Choice Questions

16. | A collector notices a rare beetle on a tree 1.0 m away. From this vantage point, the beetle has an angular size of 0.69°. What would its angular size be if the collector approached it and looked at it with a magnifier having a 5.0 cm focal length?
    A. 0.69° B. 3.5° C. 6.9° D. 14°
17. | A microscope has a tube length of 20 cm. What combination of objective and eyepiece focal lengths will give an overall magnification of 100?
    A. 1.5 cm, 3 cm B. 2 cm, 2 cm
    C. 1 cm, 5 cm D. 3 cm, 8 cm

Watch video solution for problem: 19.5

18. || The distance between the objective and eyepiece of a telescope is 55 cm. The focal length of the eyepiece is 5.0 cm. What is the angular magnification of this telescope?
A. −10 B. −11 C. −50 D. −275
19. | A nearsighted person has a near point of 20 cm and a far point of 40 cm. When he is wearing glasses to correct his distant vision, what is his near point?
BIO
A. 10 cm B. 20 cm C. 40 cm D. 1.0 m
20. | A nearsighted person has a near point of 20 cm and a far point of 40 cm. What refractive power lens is necessary to correct this person's vision to allow her to see distant objects?
BIO
A. −5.0 D B. −2.5 D C. +2.5 D D. +5.0 D
21. | A 60-year-old man has a near point of 100 cm, making it impossible to read. What refractive power reading glasses would he need to focus on a newspaper held at a comfortable distance of 40 cm?
BIO
A. −2.5 D B. −1.5 D C. +1.5 D D. +2.5 D
22. | A person looking through a −10 D lens sees an image that appears 8.0 cm from the lens. How far from the lens is the object?
A. 10 cm B. 20 cm C. 25 cm D. 40 cm
23. | In a darkened room, red light shines on a red cup, a white card, and a blue toy. The cup, card, and toy will appear, respectively,
A. Red, red, blue.
B. Red, white, blue.
C. Red, red, black.
D. Red, black, blue.
24. || An amateur astronomer looks at the moon through a telescope with a 15-cm-diameter objective. What is the minimum separation between two objects on the moon that she can resolve with this telescope? Assume her eye is most sensitive to light with a wavelength of 550 nm.
A. 120 m B. 1.7 km C. 26 km D. 520 km

PROBLEMS

Section 19.1 The Camera

1. | The human eye has a lot in common with a pinhole camera, being essentially a small box with a hole in the front (the pupil) and a detector at the back (the retina). The distance from the pupil to the retina is approximately 24 mm.
BIO
   a. Suppose you look at a 180-cm-tall friend who is standing 7.4 m in front of you. Assuming your eye functions like a pinhole camera, what will be the height, in mm, of your friend's image on your retina?
   b. Suppose your friend's image begins to get bigger. How does your brain interpret this information?
2. | A student has built a 20-cm-long pinhole camera for a science fair project. She wants to photograph the Washington Monument, which is 167 m (550 ft) tall, and to have the image on the detector be 5.0 cm high. How far should she stand from the Washington Monument?
3. || A pinhole camera is made from an 80-cm-long box with a small hole in one end. If the hole is 5.0 m from a 1.8-m-tall person, how tall will the image of the person on the detector be?
4. || A photographer uses his camera, whose lens has a 50 mm focal length, to focus on an object 2.0 m away. He then wants to take a picture of an object that is 40 cm away. How far, and in which direction, must the lens move to focus on this second object?
5. || An older camera has a lens with a focal length of 50 mm and uses 36-mm-wide film to record its images. Using this camera, a photographer takes a picture of the Golden Gate Bridge that completely spans the width of the film. Now he wants to take a picture of the bridge using his digital camera with its 12-mm-wide CCD detector. What focal length should this camera's lens have for the image of the bridge to cover the entire detector?
6. ||| In Figure P19.6 the camera lens has a 50 mm focal length. How high is the man's well-focused image on the CCD detector?

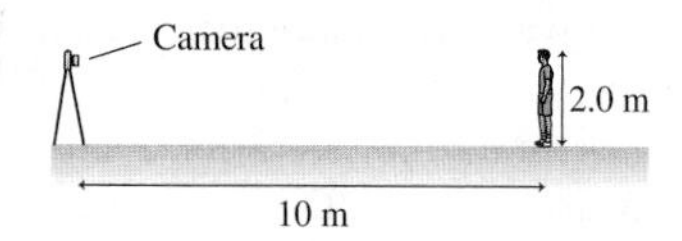

FIGURE P19.6

Section 19.2 The Human Eye

7. | a. Estimate the diameter of your eyeball.
BIO
   b. Bring this page up to the closest distance at which the text is sharp—not the closest at which you can still read it, but the closest at which the letters remain sharp. If you wear glasses or contact lenses, leave them on. This distance is called the *near point* of your (possibly corrected) eye. Record it.
   c. Estimate the effective focal length of your eye. The effective focal length includes the focusing due to the lens, the curvature of the cornea, and any corrections you wear. Ignore the effects of the fluid in your eye.
8. || A farsighted person has a near point of 50 cm rather than the normal 25 cm. What strength lens, in diopters, should be prescribed to correct this vision problem?
BIO
9. | A nearsighted woman has a far point of 300 cm. What kind of lens, converging or diverging, should be prescribed for her to see distant objects more clearly? What refractive power should the lens have?
BIO
10. || Martin has severe myopia, with a far point of only 17 cm. He wants to get glasses that he'll wear while using his computer, whose screen is 65 cm away. What refractive power will these glasses require?
BIO
11. | Mary, like many older people, has lost all ability to accommodate and can focus only on distant objects. She'd like to get reading glasses so that she can read a book held at a comfortable distance of 50 cm. What strength lenses, in diopters, does Mary need?
BIO
12. | The near point for your myopic uncle is 10 cm. Your own vision is normal; that is, your near point is 25 cm. Suppose you and your uncle hold dimes (which are 1.7 cm in diameter) at your respective near points.
   a. What is the dime's angular size, in radians, according to you?
   b. What is the dime's angular size, in radians, according to your uncle?
   c. Do these calculations suggest any benefit to near-sightedness?

13. || For a patient, a doctor prescribes glasses with a converging
BIO lens having a refractive power of 4.0 D.
   a. Is the patient nearsighted or farsighted?
   b. If the patient is nearsighted, what is the location of her eye's far point? If she is farsighted, what is the location of her near point?
14. ||| Rank the following people from the most nearsighted to the
BIO most farsighted, indicating any ties:
   A. Bernie has a prescription of +2.0 D.
   B. Carol needs diverging lenses with a focal length of −0.35 m.
   C. Maria Elena wears converging lenses with a focal length of 0.50 m.
   D. Janet has a prescription of +2.5 D.
   E. Warren's prescription is −3.2 D.

Section 19.3 The Magnifier

15. | The diameter of a penny is 19 mm. How far from your eye must it be held so that it has the same apparent size as the moon? (Use the astronomical data inside the back cover.)
16. ||| A magnifier has a magnification of 4× for someone with a normal near-point distance of 25 cm. What is the magnification of this lens when used by a person with a near-point distance of 15 cm?
17. | A magnifier has a magnification of 5×. How far from the lens should an object be placed so that its (virtual) image is at the near-point distance of 25 cm?
18. ||| A farsighted man has a near point of 40 cm. What power lens should he use as a magnifier to see clearly at a distance of 10 cm without wearing his glasses?

Section 19.4 The Microscope

19. || An inexpensive microscope has a tube length of 12.0 cm, and its objective lens is labeled with a magnification of 10×.
   a. Calculate the focal length of the objective lens.
   b. What focal length eyepiece lens should the microscope have to give an overall magnification of 150×?
20. ||| A standard biological microscope is required to have a magnification of 200×.
   a. When paired with a 10× eyepiece, what power objective is needed to get this magnification?
   b. What is the focal length of the objective?
21. ||| A forensic scientist is using a standard biological microscope with a 15× objective and a 5× eyepiece to examine a hair from a crime scene. How far from the objective is the hair?
22. ||| A microscope with an 8.0-mm-focal-length objective has a tube length of 16.0 cm. For the microscope to be in focus, how far should the objective lens be from the specimen?
23. ||| The distance between the objective and eyepiece lenses in a microscope is 20 cm. The objective lens has a focal length of 5.0 mm. What eyepiece focal length will give the microscope an overall angular magnification of 350?

Section 19.5 The Telescope

24. || For the combination of two identical lenses shown in Figure P19.24, find the position, size, and orientation of the final image of the 2.0-cm-tall object.

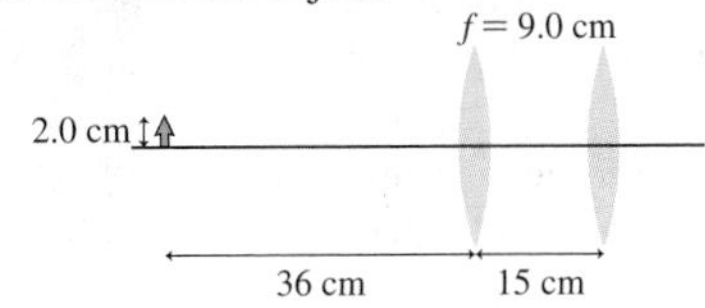

FIGURE P19.24

25. || For the combination of two lenses shown in Figure P19.25, find the position, size, and orientation of the final image of the 1.0-cm-tall object.

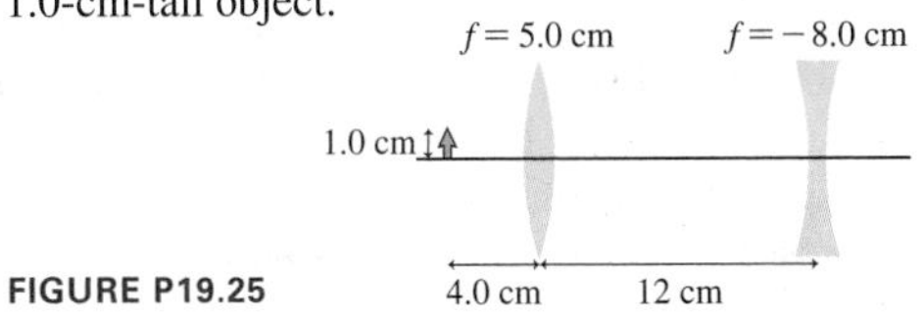

FIGURE P19.25

26. || A researcher is trying to shoot a tranquilizer dart at a 2.0-m-tall rhino that is 150 m away. Its angular size as seen through the rifle telescope is 9.1°. What is the magnification of the telescope?
27. || The objective lens of the refracting telescope at the Lick Observatory in California has a focal length of 57 ft.
   a. What is the refractive power of this lens?
   b. What focal length (mm) eyepiece would give a magnification of 1000× for this telescope?
28. ||| You use your 8× binoculars to focus on a yellow-rumped warbler (length 14 cm) in a tree 18 m away from you. What angle (in degrees) does the image of the warbler subtend on your retina?
29. || Your telescope has a 700-mm-focal-length objective and a 26-mm-focal-length eyepiece. One evening you decide to look at the full moon, which has an angular size of 0.52° when viewed with the naked eye.

   a. What angle (in degrees) does the image of the moon subtend when you look at it through your telescope?
   b. Suppose you decide to take a photograph of the moon using your telescope. You position a CCD detector so that it captures the image produced by the objective lens. What is the diameter of that image?

Section 19.6 Color and Dispersion

30. |||| A narrow beam of light with wavelengths from 450 nm to 700 nm is incident perpendicular to one face of a 40.00° prism made of crown glass, for which the index of refraction ranges from $n = 1.533$ to $n = 1.517$ for those wavelengths. What is the angular spread of the beam after passing through the prism?
31. ||| A ray of white light strikes the surface of a 4.0-cm-thick slab of flint glass as shown in Figure P19.31. As the ray enters the glass, it is dispersed into its constituent colors. Estimate how far apart the rays of deepest red and deepest violet light are as they exit the bottom surface. Which exiting ray is closer to point P?

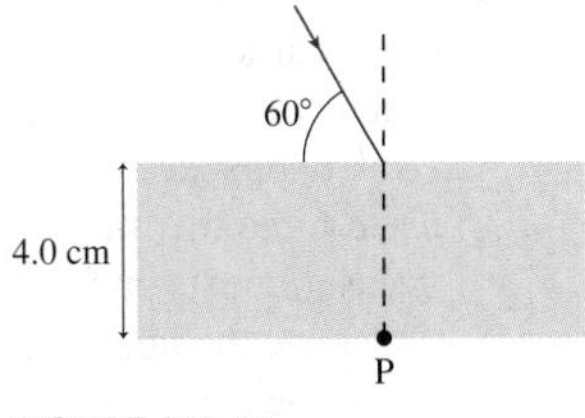

FIGURE P19.31

32. |||| A ray of red light, for which $n = 1.54$, and a ray of violet light, for which $n = 1.59$, travel through a piece of glass. They meet right at the boundary between the glass and the air, and emerge into the air as one ray with an angle of refraction of 22.5°. What is the angle between the two rays in the glass?

Watch video solutions for problems: 19.35 19.37 19.47 19.49

Section 19.7 Resolution of Optical Instruments

33. ||| Two lightbulbs are 1.0 m apart. From what distance can these lightbulbs be marginally resolved by a small telescope with a 4.0-cm-diameter objective lens? Assume that the lens is limited only by diffraction and $\lambda = 600$ nm.
34. | A 1.0-cm-diameter microscope objective has a focal length of 2.8 mm. It is used in visible light with a wavelength of 550 nm.
    a. What is the objective's resolving power if used in air?
    b. What is the resolving power of the objective if it is used in an oil-immersion microscope with $n_{oil} = 1.45$?
35. || A microscope with an objective of focal length 1.6 mm is used to inspect the tiny features of a computer chip. It is desired to resolve two objects only 400 nm apart. What diameter objective is needed if the microscope is used in air with light of wavelength 550 nm?

General Problems

36. | Suppose you point a pinhole camera at a 15-m-tall tree that is 75 m away.
    a. If the detector is 22 cm behind the pinhole, what will be the size of the tree's image on the detector?
    b. If you would like the image to be larger, should you get closer to the tree or farther from the tree? Explain.
    c. If you had time, you could make the image larger by rebuilding the camera, changing the length or the pinhole size. What one change would give a larger image?
37. |||| Jason uses a lens with a focal length of 10.0 cm as a magnifier by holding it right up to his eye. He is observing an object that is 8.0 cm from the lens. What is the angular magnification of the lens used this way if Jason's near-point distance is 25 cm?
38. | A magnifier is labeled "5×." What would its magnification be if used by a person with a near-point distance of 50 cm?
39. || A 20× microscope objective is designed for use in a microscope with a 16 cm tube length. The objective is marked NA = 0.40. What is the diameter of the objective lens?
40. |||| Two converging lenses with focal lengths of 40 cm and 20 cm are 10 cm apart. A 2.0-cm-tall object is 15 cm in front of the 40-cm-focal-length lens. INT
    a. Use ray tracing to find the position and height of the image. To do this accurately use a ruler or paper with a grid. Determine the image distance and image height by making measurements on your diagram.
    b. Calculate the image height and image position relative to the second lens. Compare with your ray-tracing answers in part a.
41. ||| A converging lens with a focal length of 40 cm and a diverging lens with a focal length of −40 cm are 160 cm apart. A 2.0-cm-tall object is 60 cm in front of the converging lens. INT
    a. Use ray tracing to find the position and height of the image. To do this, accurately use a ruler or paper with a grid. Determine the image distance and image height by making measurements on your diagram.
    b. Calculate the image height and image position relative to the second lens. Compare with your ray-tracing answers in part a.
42. || A lens with a focal length of 25 cm is placed 40 cm in front of a lens with a focal length of 5.0 cm. How far from the second lens is the final image of an object infinitely far from the first lens? Is this image in front of or behind the second lens?
43. || A microscope with a 5× objective lens images a 1.0-mm-diameter specimen. What is the diameter of the real image of this specimen formed by the objective lens?
44. || Your task in physics lab is to make a microscope from two lenses. One lens has a focal length of 10 cm, the other a focal length of 3.0 cm. You plan to use the more powerful lens as the objective, and you want its image to be 16 cm from the lens, as in a standard biological microscope.
    a. How far should the objective lens be from the object to produce a real image 16 cm from the objective?
    b. What will be the magnification of your microscope?
45. || A 20× objective and 10× eyepiece give an angular magnification of 200× when used in a microscope with a 160 mm tube length. What magnification would this objective and eyepiece give if used in a microscope with a 200 mm tube length?
46. || The objective lens and the eyepiece lens of a telescope are 1.0 m apart. The telescope has an angular magnification of 50. Find the focal lengths of the eyepiece and the objective.
47. ||| Your telescope has an objective lens with a focal length of 1.0 m. You point the telescope at the moon, only to realize that the eyepiece is missing. Even so, you can still see the real image of the moon formed by the objective lens if you place your eye a little past the image so as to view the rays diverging from the image plane, just as rays would diverge from an object at that location. What is the angular magnification of the moon if you view its real image from 25 cm away, your near-point distance?
48. || Martha is viewing a distant mountain with a telescope that has a 120-cm-focal-length objective lens and an eyepiece with a 2.0 cm focal length. She sees a bird that's 60 m distant and wants to observe it. To do so, she has to refocus the telescope. By how far and in which direction (toward or away from the objective) must she move the eyepiece in order to focus on the bird?
49. || Susan is quite nearsighted; without her glasses, her far point is 35 cm and her near point is 15 cm. Her glasses allow her to view distant objects with her eye relaxed. With her glasses on, what is the closest object on which she can focus?
50. ||| A spy satellite uses a telescope with a 2.0-m-diameter mirror. It orbits the earth at a height of 220 km. What minimum spacing must there be between two objects on the earth's surface if they are to be resolved as distinct objects by this telescope? Assume the telescope's resolution is limited only by diffraction and that it is recording light with a wavelength of 500 nm.
51. ||||| Two stars have an angular separation of 3.3×10^{-6} rad. What diameter telescope objective is necessary to just resolve these two stars, using light with a wavelength of 650 nm?
52. ||| Frank is nearsighted and his glasses require a prescription of −1.5 D. One day he can't find his glasses, but he does find an older pair with a prescription of −1.0 D. What is the most distant object that Frank can focus on while wearing this older pair of glasses?
53. ||||| What is the angular resolution of the Hubble Space Telescope's 2.4-m-diameter mirror when viewing light with a wavelength of 550 nm? The resolution of a reflecting telescope is calculated exactly the same as for a refracting telescope.

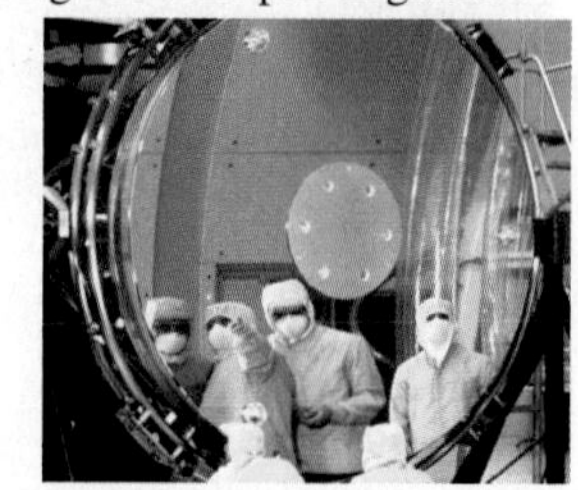

54. | The Hubble Space Telescope has a mirror diameter of 2.4 m. Suppose the telescope was used to photograph the surface of the moon from a distance of 3.8×10^8 m. What is the distance between two objects that the telescope can barely resolve? Assume the wavelength of light is 600 nm.

55. ||| Once dark adapted, the pupil of your eye is approximately BIO 7 mm in diameter. The headlights of an oncoming car are 120 cm apart. If the lens of your eye is limited only by diffraction, at what distance are the two headlights marginally resolved? Assume the light's wavelength in air is 600 nm and the index of refraction inside the eye is 1.33. (Your eye is not really good enough to resolve headlights at this distance, due both to aberrations in the lens and to the size of the receptors in your retina, but it comes reasonably close.)

56. |||| The normal human eye has maximum visual acuity with a BIO pupil size of about 3 mm. For larger pupils, acuity decreases due to increasing aberrations; for smaller pupils, acuity decreases due to increasing effects of diffraction. If your pupil diameter is 2.0 mm, as it would be in fairly bright light, what is the smallest diameter circle that you can barely see as a circle, rather than just a dot, if the circle is at your near point, 25 cm from your eye? Assume the light's wavelength in air is 600 nm and the index of refraction inside the eye is 1.33.

57. ||| Microtubules are filamentous structures in cells that maintain BIO cell shape and facilitate the movement of molecules within the cell. They are long, hollow cylinders with a diameter of about 25 nm. It is possible to incorporate fluorescent molecules into microtubules; when illuminated by an ultraviolet light, the fluorescent molecules emit visible light that can be imaged by the optical system of a microscope. If the emitted light has a wavelength of 500 nm and the NA of the microscope objective is 1.4, can a biologist looking through the microscope tell whether she is looking at a single microtubule or at two microtubules lying side by side?

MCAT-Style Passage Problems

Surgical Vision Correction BIO

Light that enters your eyes is focused to form an image on your retina. The optics of your visual system have a total refractive power of about +60 D—about +20 D from the lens in your eye and +40 D from the curved shape of your cornea. Surgical procedures to correct vision generally do not work on the lens; they work to reshape the cornea. In the most common procedure, a laser is used to remove tissue from the center of the cornea, reducing its curvature. This change in shape can correct certain kinds of vision problems.

58. | Flattening the cornea would be a good solution for someone who was
A. Nearsighted. B. Farsighted.
C. Either nearsighted or farsighted.

59. | Suppose a woman has a far point of 50 cm. How much should the refractive power of her cornea be changed to correct her vision?
A. −2.0 D B. −1.0 D C. +1.0 D D. +2.0 D

60. | A *cataract* is a clouding or opacity that develops in the eye's lens, often in older people. In extreme cases, the lens of the eye may need to be removed. This would have the effect of leaving a person
A. Nearsighted. B. Farsighted.
C. Neither nearsighted nor farsighted.

61. | The length of your eye decreases slightly as you age, making the lens a bit closer to the retina. Suppose a man had his vision surgically corrected at age 30. At age 70, once his eyes had decreased slightly in length, he would be
A. Nearsighted. B. Farsighted.
C. Neither nearsighted nor farsighted.

STOP TO THINK ANSWERS

Chapter Preview Stop to Think: B. A real image is formed only if the object is beyond the focal point. See Synthesis 18.1 for details.

Stop to Think 19.1: A. The rays from the top and bottom of the tree in Figure 19.1b will get farther apart as the screen is moved back, resulting in a larger image.

Stop to Think 19.2: A. Kara can focus on an object no closer to her eyes than 40 cm, so the image needs to be at 40 cm and no closer. Because the image is on the same side of the lens as the object, it will be virtual.

Stop to Think 19.3: D. Because Maria can focus on an object 0.5 m away, but not on one 10 m away, her far point must lie between these two distances. Following Example 19.3, we see that the prescription for her lens must then lie between $1/(-10\text{ m}) = -0.1$ D and $1/(-0.5\text{ m}) = -2$ D. Only the −1.5 D prescription falls in this range.

Stop to Think 19.4: A. Ray tracing shows why:

Stop to Think 19.5: B. The total magnification is the product of the objective magnification m_o and the eyepiece angular magnification M_e. If m_o is halved, from 20× to 10×, M_e must be doubled. Because M_e is inversely proportional to the eyepiece focal length, the focal length of the eyepiece must be halved.

Stop to Think 19.6: D. A green filter lets through only green light, so it blocks the red light from the apple. No light from the apple can pass through the filter, so it appears black.

Stop to Think 19.7: $RP_1 > RP_4 > RP_2 = RP_3$. The resolving power is $RP = 0.61\lambda/\sin\phi_0$ for objectives used in air ($n = 1$), so the resolving power is higher (worse resolution) when the angle ϕ_0 is smaller. From Figure 19.27 you can see that ϕ_0 is smaller when the ratio D/f is smaller. These ratios are $(D/f)_1 = 1/5$, $(D/f)_2 = 2/5$, $(D/f)_3 = 2/5$, and $(D/f)_4 = 1/3$.

Optics

Light is an elusive entity. It is everywhere around us, but exactly what *is* it? One of the more curious aspects of light is that its basic properties depend on the circumstances under which it's studied. Thus it's difficult to develop a single theory of light that applies under all circumstances. Because of this, we have developed two *models* of light in Part V, the wave model and the ray model. We found that each model has its particular realm of applicability.

Many experiments show that light has distinct wave-like properties. Light waves exhibit interference and diffraction, just as water and sound waves do. However, we're usually not aware of the wave aspects of light because the wavelengths of visible light are so short. Wave phenomena become apparent only when light interacts with objects or holes whose size is less than about 0.1 mm.

We can usually ignore the wave nature of light when we consider the propagation of light on larger length scales. In this case, we model light as traveling outward in straight lines, or *rays,* from its source. Light rays change direction at an interface between two media with different indices of refraction (different light speeds). At this interface the rays both reflect, heading back into their original medium, and refract, moving into the new medium but in a new direction. These processes are governed by the laws of reflection and refraction.

Despite light's subtle nature, the practical applications of optics are crucial to many of today's technologies. Cameras, telescopes, and microscopes all employ basic ideas of image formation with lenses and mirrors. We found that the ultimate resolution of an optical instrument is set by the wave nature of light, bringing our study of optics full circle.

KNOWLEDGE STRUCTURE V Optics

BASIC GOALS What are the consequences of the wave nature of light? In the ray model, how do light rays refract and reflect to form images?

GENERAL PRINCIPLES Light is understood using two models, the **wave model,** in which light exhibits wave properties such as interference and diffraction, and the **ray model,** in which light travels in straight lines until it reflects or refracts.

Wave model

- Light spreads out when passing through a narrow opening. This is **diffraction.**

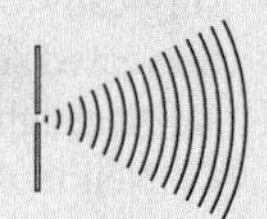

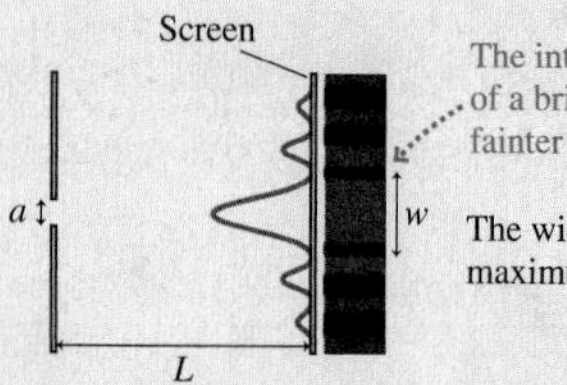

The intensity on a screen consists of a bright central maximum and fainter secondary maxima.

The width of the central maximum is

$$w = \frac{2\lambda L}{a}$$

- Light waves from multiple slits in a screen **interfere** where they overlap. The light intensity is large where the interfering waves are in phase, and small where they are out of phase.

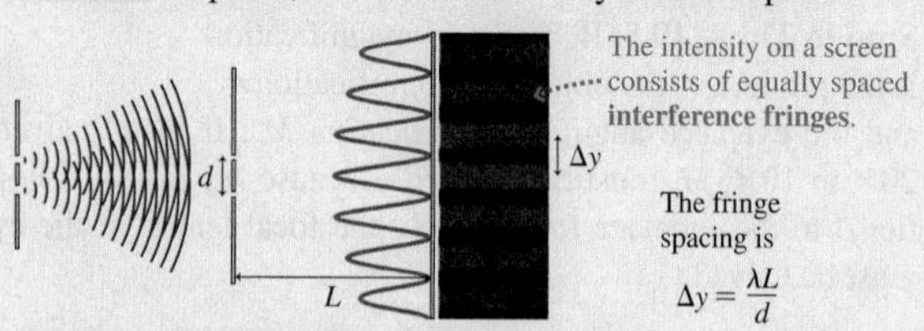

The intensity on a screen consists of equally spaced **interference fringes.**

The fringe spacing is

$$\Delta y = \frac{\lambda L}{d}$$

- Light waves reflected from the two surfaces of a thin transparent film also interfere.

The resolution of optical instruments

Diffraction limits how close together two point objects can be and still be resolved.

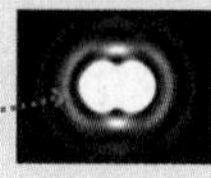

For a *telescope,* the minimum resolvable angular separation between two objects is

$$\theta_1 = \frac{1.22\lambda}{D}$$

For a *microscope,* the minimum resolvable distance between two objects is

$$d_{min} = \frac{0.61\lambda}{\text{NA}}$$

where the *numerical aperture* NA is a characteristic of the microscope objective.

Ray model

- Light travels out from its source in straight lines, called **rays.**
- Rays reflect off a surface between two media, obeying the **law of reflection,** $\theta_i = \theta_r$.
- Light rays change direction as they cross the surface between two media. The angles of incidence and refraction are related by **Snell's law:**

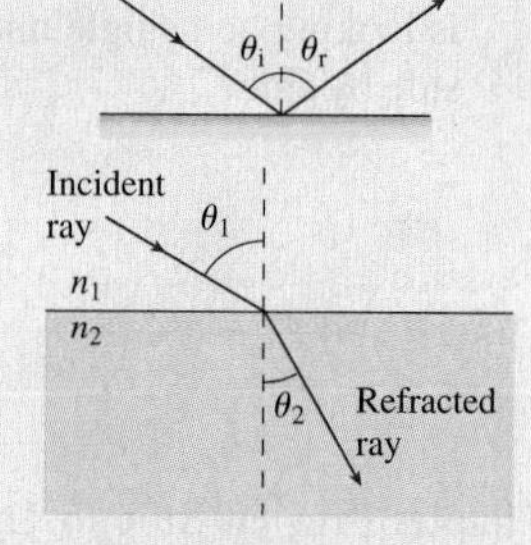

$$n_1 \sin\theta_1 = n_2 \sin\theta_2$$

where n is the **index of refraction.** The speed of light in a transparent material is $v = c/n$.

Image formation by lenses and mirrors

A lens or mirror has a characteristic **focal length** f. Rays parallel to the optical axis come to focus a distance f from the lens or mirror.

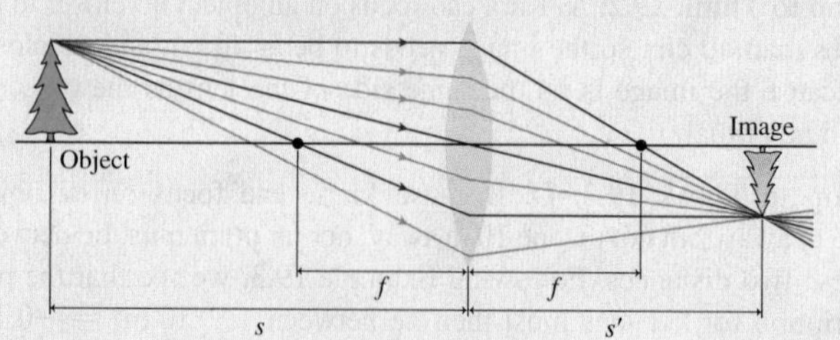

The **object distance** s, the **image distance** s', and the focal length are related by the *thin-lens equation,* which also works for mirrors:

$$\frac{1}{s} + \frac{1}{s'} = \frac{1}{f}$$

Scanning Confocal Microscopy

Although modern microscopes are marvels of optical engineering, their basic design is not too different from the 1665 compound microscope of Robert Hooke. Recently, advances in optics, lasers, and computer technology have made practical a new kind of optical microscope, the *scanning confocal microscope*. This microscope is capable of taking images of breathtaking clarity.

The figure shows the microscope's basic principle of operation. The left part of the figure shows how the translucent specimen is illuminated by light from a laser. The laser beam is converted to a diverging bundle of rays by suitable optics, reflected off a mirror, then directed through a microscope objective lens to a focus within the sample. The microscope objective focuses the laser beam to a very small ($\approx 0.5\ \mu m$) spot. Note that light from the laser passes through other regions of the specimen but, because the rays are not focused in those regions, they are not as intensely illuminated as is the point at the focus. This is the first important aspect of the design: Very intensely illuminate one very small volume of the sample while leaving other regions only weakly illuminated.

As shown in the right half of the figure, light is reflected from all illuminated points in the sample and passes back through the objective lens. The mirror that had reflected the laser light downward is actually a *partially transparent mirror* that reflects 50% of the light and transmits 50%. Thus half of the light reflected upward from the sample passes through the mirror and is focused on a screen containing a small hole. Because of the hole, only light rays that emanate from the brightly illuminated volume in the sample can completely pass through the hole and reach the light detector behind it. Rays from other points in the sample either miss the hole completely or are out of focus when they reach the screen, so that only a small fraction of them pass through the hole. This second key design aspect limits the detected light to only those rays that are emitted from the point in the sample at which the laser light was originally focused.

So we see that (a) the point in the sample that is at the focus of the objective is much more intensely illuminated than any other point, so it reflects more rays than any other point, and (b) the hole serves to further limit the detected rays to only those that emanate from the focus. Taken together, these design aspects ensure the detected light comes from a very small, very well-defined volume in the sample.

The microscope as shown would only be useful for examining one small point in the sample. To make an actual *image*, the objective is *scanned* across the sample while the intensity is recorded by a computer. This procedure builds up an image of the sample one *scan line* at a time. The final result is a picture of the sample in the very narrow plane in which the laser beam is focused. Different planes within the sample can be imaged by moving the objective up or down before scanning. It is actually possible to make three-dimensional images of a specimen in this way.

The improvement in contrast and resolution over conventional microscopy can be striking. The images show a section of a mouse kidney taken using conventional and confocal microscopy. Because light reflected from all parts of the specimen reaches the camera in a conventional microscope, that image appears blurred and has low contrast. The confocal microscope image represents a single plane or slice of the sample, and many details become apparent that are invisible in the conventional image.

The screen blocks out light reflected from points other than the focus of the laser beam.
Light detector
Hole in screen
Partially transparent mirror reflects 50%, transmits 50%.
Laser source
Objective lens
The intensity at the focus of the laser beam is very high.
$\approx 0.5\ \mu m$
Specimen
The intensity at other points is much lower.
Bright light reflected from point at focus
Faint light reflected from some other point

A confocal microscope.

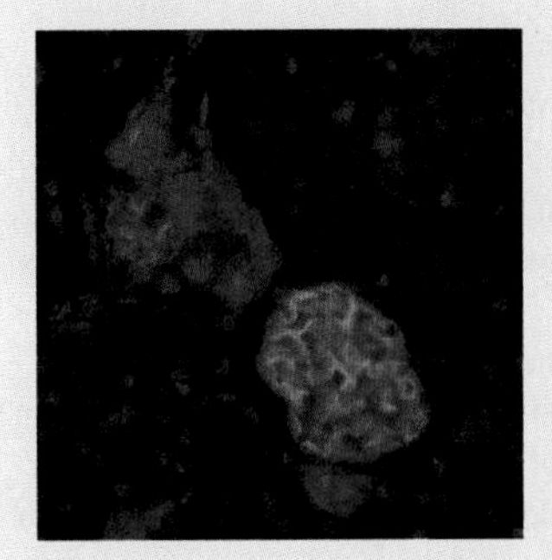

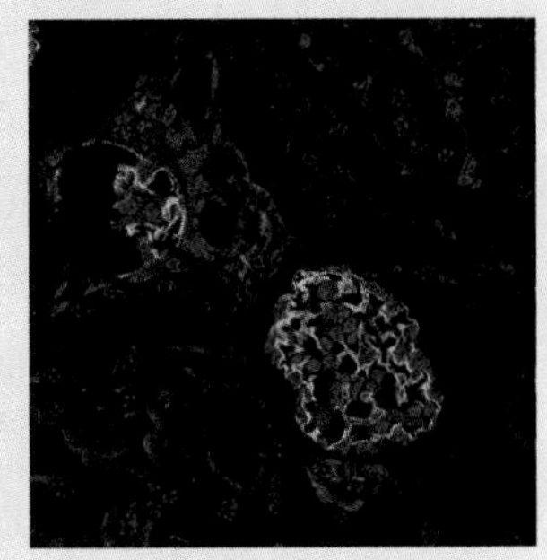

A section of fluorescently stained mouse kidney imaged using standard optical microscopy (left) and scanning confocal microscopy (right).

The following questions are related to the passage "Scanning Confocal Microscopy" on the previous page.

1. A laser beam consists of parallel rays of light. To convert this light to the diverging rays required for a scanning confocal microscope requires
   A. A converging lens.
   B. A diverging lens.
   C. Either a converging or a diverging lens.
2. If, because of a poor-quality objective, the light from the laser illuminating the sample in a scanning confocal microscope is focused to a larger spot,
   A. The image would be dimmer because the light illuminating the point imaged would be dimmer.
   B. The image would be blurry because light from more than one point would reach the detector.
   C. The image would be dimmer and blurry—both of the above problems would exist.
3. The resolution of a scanning confocal microscope is limited by diffraction, just as for a regular microscope. In principle, switching to a laser with a shorter wavelength would provide
   A. Greater resolution.
   B. Lesser resolution.
   C. The same resolution.
4. In the optical system shown in the passage, the distance from the source of the diverging light rays to the sample is ________ the distance from the sample to the screen.
   A. greater than
   B. the same as
   C. less than

The following passages and associated questions are based on the material of Part V.

Horse Sense BIO

The ciliary muscles in a horse's eye can make only small changes to the shape of the lens, so a horse can't change the shape of the lens to focus on objects at different distances as humans do. Instead, a horse relies on the fact that its eyes aren't spherical. As Figure V.1 shows, different points at the back of the eye are at somewhat different distances from the front of the eye. We say that the eye has a "ramped retina"; images that form on the top of the retina are farther from the cornea and lens than those that form at lower positions. The horse uses this ramped retina to focus on objects at different distances, tipping its head so that light from an object forms an image at a vertical location on the retina that is at the correct distance for sharp focus.

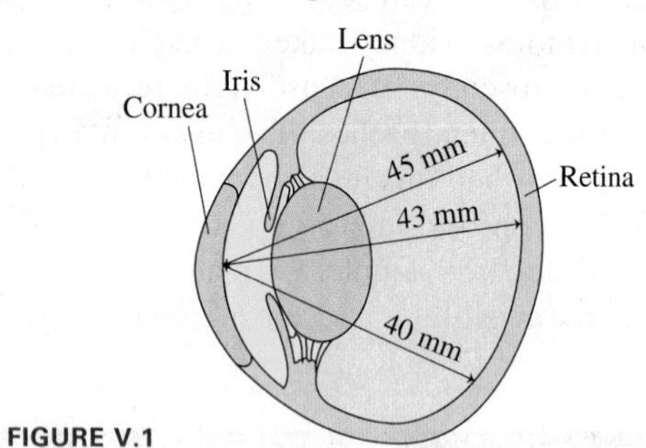

FIGURE V.1

5. In a horse's eye, the image of a close object will be in focus
   A. At the top of the retina.
   B. At the bottom of the retina.
6. In a horse's eye, the image of a distant object will be in focus
   A. At the top of the retina.
   B. At the bottom of the retina.
7. A horse is looking straight ahead at a person who is standing quite close. The image of the person spans much of the vertical extent of the retina. What can we say about the image on the retina?
   A. The person's head is in focus; the feet are out of focus.
   B. The person's feet are in focus; the head is out of focus.
   C. The person's head and feet are both in focus.
   D. The person's head and feet are both out of focus.
8. Certain medical conditions can change the shape of a horse's eyeball; these changes can affect vision. If the lens and cornea are not changed but all of the distances in Figure V.1 are increased slightly, then the horse will be
   A. Nearsighted.
   B. Farsighted.
   C. Unable to focus clearly at any distance.

The Fire in the Eye BIO

You have certainly seen the reflected light from the eyes of a cat or a dog at night. This "eye shine" is the reflection of light from a layer at the back of the eye called the *tapetum lucidum* (Latin for "bright carpet"). The tapetum is a common structure in the eyes of animals that must see in low light. Light that passes through the retina is reflected by the tapetum back through the cells of the retina, giving them a second chance to detect the light.

Sharks and related fish have a very well-developed tapetum. Figure V.2a shows a camera flash reflected from a shark's eye back toward the camera. This reflected light is much brighter than the diffuse reflection from the body of the shark. How is this bright reflection created?

Figure V.2b shows a typical tapetum structure for a fish. (The tapetum in land animals such as cats, dogs, and deer uses similar principles but has a different structure.) The reflection comes from the interfaces between two layers of nearly transparent cells (whose index of refraction is essentially that of water) and a stack of guanine crystals sandwiched between. Light is reflected from the interface at both sides of the stack of crystals. For certain wavelengths, constructive interference leads to an especially strong reflection.

Bright light from a distant source is focused by the lens of a shark's eye to a point on the retina, as shown in Figure V.2c. The tapetum reflects these rays back through the lens, where refraction bends them into parallel rays traveling back toward the source of the light. Because the reflected light from the tapetum is directional, it is much brighter than the diffuse reflection from the shark's body. But the bright reflection is seen by an observer—or a camera—only at or near the source of the flash that produced the reflection.

(a) "Eye shine" in a flash photo of a shark

Class Video

(b) Structure of the tapetum

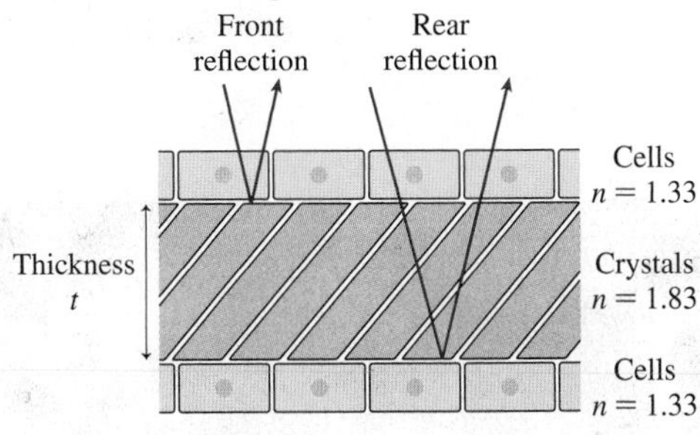
Front reflection
Rear reflection
Cells
$n = 1.33$
Thickness
t
Crystals
$n = 1.83$
Cells
$n = 1.33$

(c) Reflection of light rays

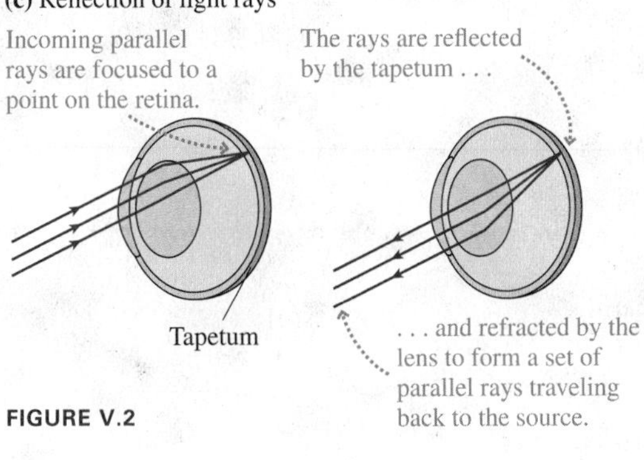
Incoming parallel rays are focused to a point on the retina.
The rays are reflected by the tapetum . . .
Tapetum
. . . and refracted by the lens to form a set of parallel rays traveling back to the source.

FIGURE V.2

9. Light of wavelength 600 nm in air passes into the layer of guanine crystals. What is the wavelength of the light in this layer?
   A. 1100 nm B. 600 nm
   C. 450 nm D. 330 nm
10. Figure V.2b shows rays that reflect from the two interfaces in the tapetum. Given the indices of refraction of the cells and the crystals, there will be a phase shift on reflection for
    A. The front reflection and the rear reflection.
    B. The front reflection only.
    C. The rear reflection only.
    D. Neither the front nor the rear reflection.
11. What is the (approximate) smallest thickness of the crystal layer that would lead to constructive interference between the front reflection and the rear reflection for light of wavelength 600 nm?
    A. 80 nm B. 160 nm
    C. 240 nm D. 320 nm
12. In human vision, the curvature of the cornea provides much of the power of the visual system. This is not the case in fish; in Figure V.2c, the light rays are bent by the lens but are not bent when they enter the cornea. This is because
    A. Fish eyes work in water, and the index of refraction of the fluids in the eye is similar to that of water.
    B. Fish eyes have a much smaller curvature of the cornea.
    C. Most fish have eyes that are more sensitive to light than the eyes of typical land animals.
    D. The reflection of the tapetum interferes with the refraction of the cornea.
13. Flash photographs of cats will generally show the tapetum reflection unless you are careful to avoid it. If you want to take a flash photograph of your cat while minimizing the "eye shine," which of the following strategies will *not* work?
    A. Take the photographs in dim light so that the irises of your cat's eyes are wide open.
    B. Use a flash on a stand at some distance from the camera.
    C. Use a diffuser so that the light from the flash is spread over a wide area.
    D. Use multiple flashes at different positions around the room.
14. Figure V.2c shows the lens of the eye bringing parallel rays together right at the retina. The retina is located
    A. In front of the focal point of the lens.
    B. At the focal point of the lens.
    C. Behind the focal point of the lens.

Additional Integrated Problems

15. BIO The pupil of your eye is smaller in bright light than in dim light. Explain how this makes images seen in bright light appear sharper than images seen in dim light.
16. BIO People with good vision can make out an 8.8-mm-tall letter on an eye chart at a distance of 6.1 m. Approximately how large is the image of the letter on the retina? Assume that the distance from the lens to the retina is 24 mm.
17. A photographer uses a lens with $f = 50$ mm to form an image of a distant object on the CCD detector in a digital camera. The image is 1.2 mm high, and the intensity of light on the detector is 2.5 W/m^2. She then switches to a lens with $f = 300$ mm that is the same diameter as the first lens. What are the height of the image and the intensity now?
18. Sound and other waves undergo diffraction just as light does. Suppose a loudspeaker in a 20°C room is emitting a steady tone of 1200 Hz. A 1.0-m-wide doorway in front of the speaker diffracts the sound wave. A person on the other side walks parallel to the wall in which the door is set, staying 12 m from the wall. When he is directly in front of the doorway, he can hear the sound clearly and loudly. As he continues walking, the sound intensity decreases. How far must he walk from the point where he was directly in front of the door until he reaches the first quiet spot?

PART

VI

Electricity and Magnetism

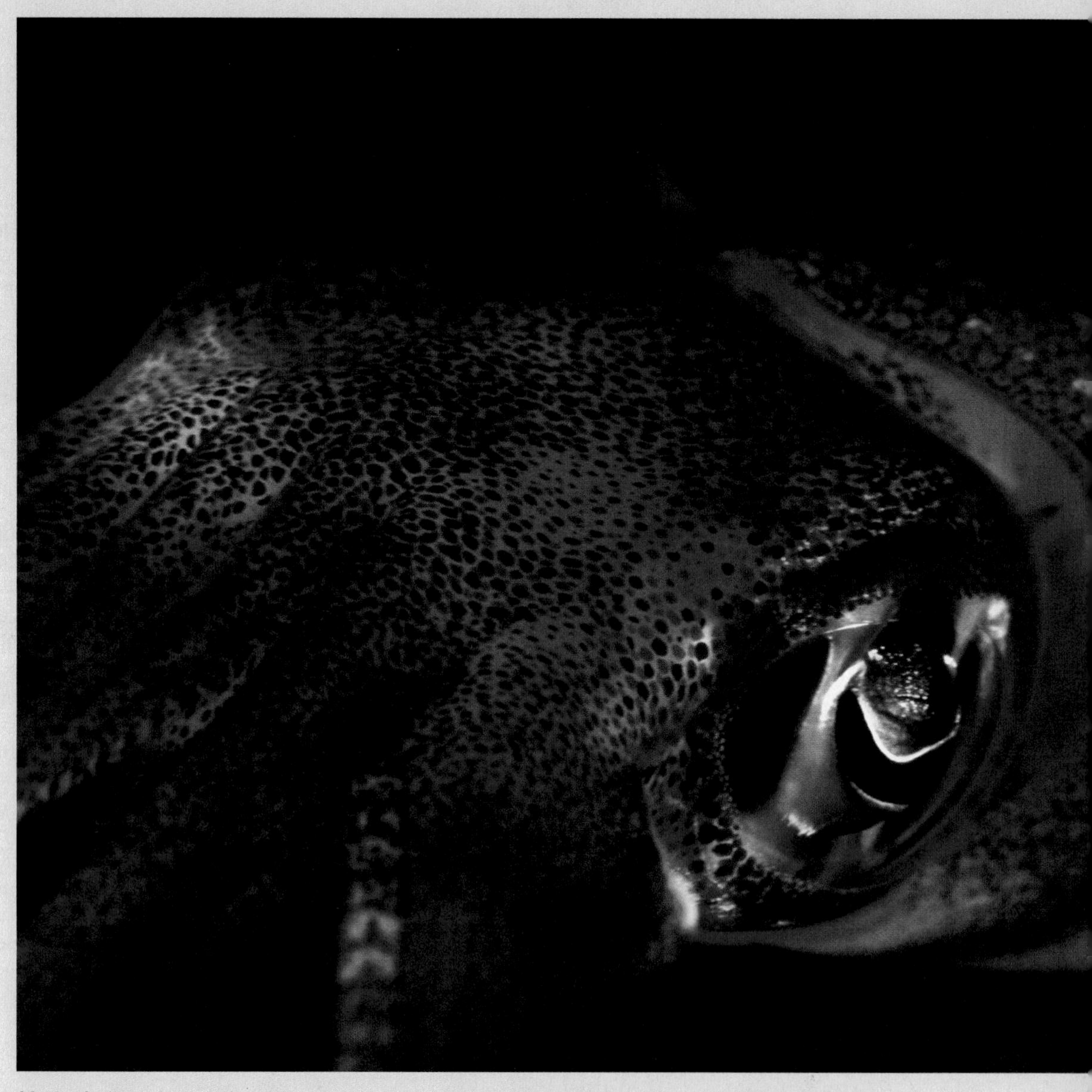

Much of what is known about your nervous system comes from the study of an animal that seems quite different from humans—the squid. Nerve fibers conduct electrical signals along their length, allowing the brain to direct the actions of the body. How is an electrical signal generated and transmitted in the nervous system of a human or a squid?

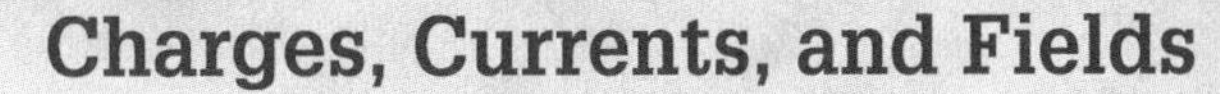

Charges, Currents, and Fields

The early Greeks discovered that a piece of amber that has been rubbed briskly can attract feathers or small pieces of straw. They also found that certain stones from the region they called *Magnesia* can pick up pieces of iron. These first experiences with the forces of electricity and magnetism began a chain of investigations that has led to today's high-speed computers, lasers, fiber-optic communications, and magnetic resonance imaging, as well as mundane modern-day miracles such as the lightbulb.

The development of a successful electromagnetic theory, which occupied the leading physicists of Europe for most of the 19th century, led to sweeping revolutions in both science and technology. The complete formulation of the theory of the electromagnetic field has been called by no less than Einstein "the most important event in physics since Newton's time."

The basic phenomena of electricity and magnetism are not as familiar to most people as those of mechanics. We will deal with this lack of experience by placing a large emphasis on these basic phenomena. We will begin where the Greeks did, by looking at the forces between objects that have been briskly rubbed, exploring the concept of *electric charge.* It is easy to make systematic observations of how charges behave, and we will be led to consider the forces between charges and how charges behave in different materials. *Electric current,* whether it be for lighting a lightbulb or changing the state of a computer memory element, is simply a controlled motion of charges through conducting materials. One of our goals will be to understand how charges move through electric circuits.

When we turn to magnetic behavior, we will again start where the Greeks did, noting how magnets stick to some metals. Magnets also affect compass needles. And, as we will see, an electric current can affect a compass needle in exactly the same way as a magnet. This observation shows the close connection between electricity and magnetism, which leads us to the phenomenon of *electromagnetic waves.*

Our theory of electricity and magnetism will introduce the entirely new concept of a *field.* Electricity and magnetism are about the long-range interactions of charges, both static charges and moving charges, and the field concept will help us understand how these interactions take place.

Microscopic Models

The field theory provides a macroscopic perspective on the phenomena of electricity and magnetism, but we can also take a microscopic view. At the microscopic level, we want to know what charges are, how they are related to atoms and molecules, and how they move through various kinds of materials. Electromagnetic waves are composed of electric and magnetic fields. The interaction of electromagnetic waves with matter can be analyzed in terms of the interactions of these fields with the charges in matter. When you heat food in a microwave oven, you are using the interactions of electric and magnetic fields with charges in a very fundamental way.

20 Electric Fields and Forces

DNA analysis is often done using gel electrophoresis. A solution of DNA segments is placed in a well at one end of a plate of gel. Different segments migrate through the gel at different rates, leading to the lines in the photo. What force causes the DNA segments to move through the gel?

LOOKING AHEAD ▸

Goal: To develop a basic understanding of electric phenomena in terms of charges, forces, and fields.

Charges and Coulomb's Law

A comb rubbed through your hair attracts a thin stream of water. The **charge model** of electricity explains this force.

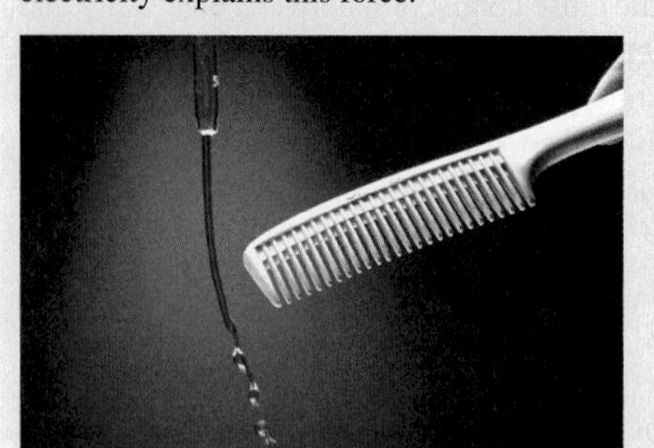

You'll learn to use **Coulomb's law** to calculate the force between two charged particles.

The Electric Field

Charges create an **electric field** around them. In thunderclouds, the field can be strong enough to ionize air, causing lightning.

You'll learn how to calculate the electric field for several important arrangements of charges.

Forces in Electric Fields

The electric field inside this smoke detector exerts a force on charged smoke particles, moving them toward a detecting electrode.

You'll learn how electric fields exert forces and torques on charged particles.

LOOKING BACK ◂

Vectors and Components

In Sections 3.1–3.3 you studied how a vector could be resolved into its component vectors.

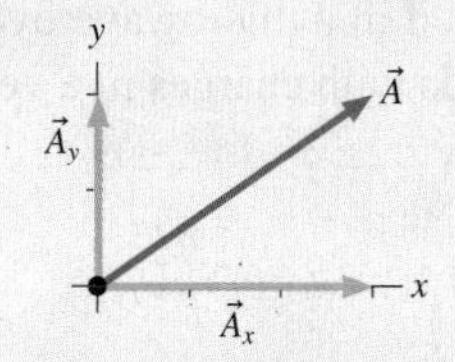

You learned that a vector $\vec{A}$ can be represented as the sum of its component vectors $\vec{A}_x$ and $\vec{A}_y$.

Electric forces and electric fields are vectors, so you will need to use vector components to solve electric force and field problems.

STOP TO THINK

The tension in the rope is 100 N. Given that $\sin 30° = 0.50$ and $\cos 30° = 0.87$, the x- and y-components of the tension are

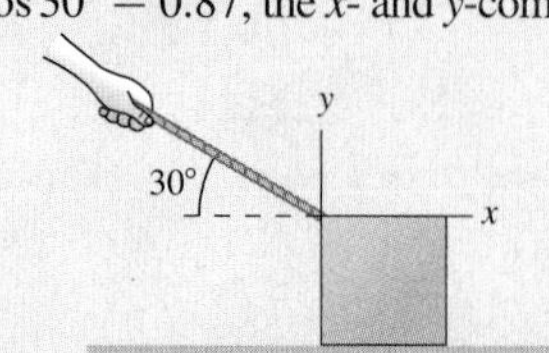

A. −87 N, 50 N
B. 87 N, 50 N
C. −50 N, 87 N
D. 50 N, −87 N
E. 87 N, −50 N

20.1 Charges and Forces

You can receive a mildly unpleasant shock and produce a little spark if you touch a metal doorknob after scuffing your shoes across a carpet. A plastic comb that you've run through your hair will pick up bits of paper and other small objects. In both of these cases, two objects are *rubbed* together. Why should rubbing an object cause forces and sparks? What kind of forces are these? These are the questions with which we begin our study of electricity.

Our first goal is to develop a model for understanding electric phenomena in terms of *charges* and *forces.* We will later use our contemporary knowledge of atoms to understand electricity on a microscopic level, but the basic concepts of electricity make *no* reference to atoms or electrons. The theory of electricity was well established long before the electron was discovered.

The ancient Greeks first noted the electrical nature of matter by observing amber, a form of fossilized tree resin. When rubbed with fur, amber buttons would attract bits of feather, hair, or straw. The Greek word for amber, *elektron*, is the source of our words "electric," "electricity," and—of course—"electron."

Experimenting with Charges

Let us enter a laboratory where we can make observations of electric phenomena. This is a modest laboratory, much like one you would have found in the year 1800. The major tools in the lab are:

- A number of plastic and glass rods, each several inches long. These can be held in your hand or suspended by threads from a support.
- Pieces of wool and silk.
- Small metal spheres, an inch or two in diameter, on wood stands.

We will manipulate and use these tools with the goal of developing a theory to explain the phenomena we see. The experiments and observations described below are very much like those of early investigators of electric phenomena.

Discovering electricity I

Experiment 1

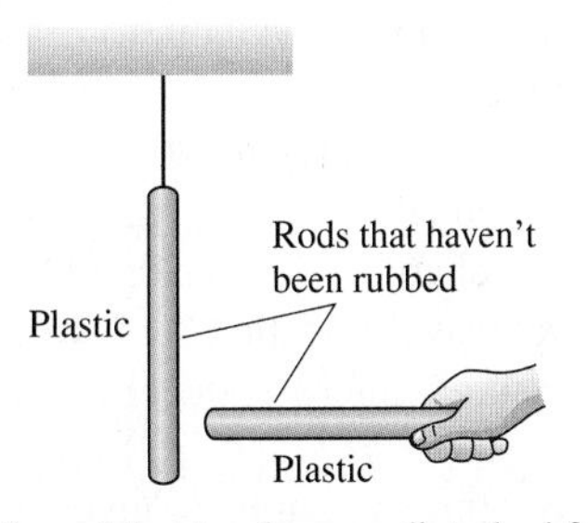

Take a plastic rod that has been undisturbed for a long period of time and hang it by a thread. Pick up another undisturbed plastic rod and bring it close to the hanging rod. Nothing happens to either rod.

Interpretation: There are no special electrical properties to these undisturbed rods. We say that they are **neutral.**

Experiment 2

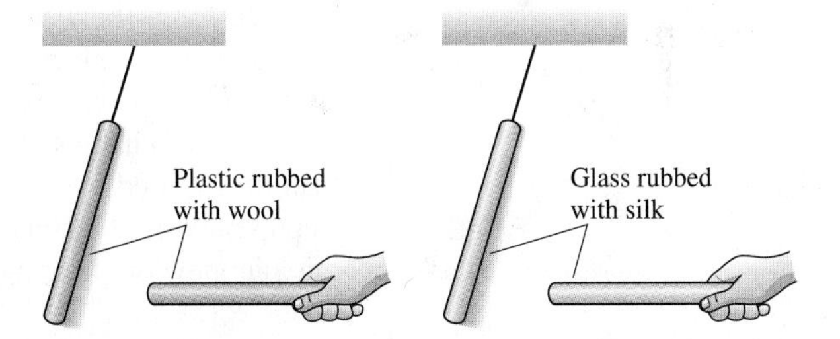

Vigorously rub both the hanging plastic rod and the handheld plastic rod with wool. Now the hanging rod *moves away* from the handheld rod when you bring the two close together. Rubbing two glass rods with silk produces the same result: The two rods repel each other.

Interpretation: Rubbing a rod somehow changes its properties so that forces now act between two such rods. We call this process of rubbing **charging** and say that the rubbed rod is *charged*, or that it has *acquired a* ***charge.***

Experiment 2 shows that there is a *long-range repulsive force* (i.e., a force requiring no contact) between two identical objects that have been charged in the *same* way, such as two plastic rods both rubbed with wool or two glass rods rubbed with silk. The force between charged objects is called the **electric force.** We have seen a long-range force before, gravity, but the gravitational force is always attractive. This is the first time we've observed a repulsive long-range force. However, the electric force is not always repulsive, as the next experiment shows.

Discovering electricity II

Experiment 3

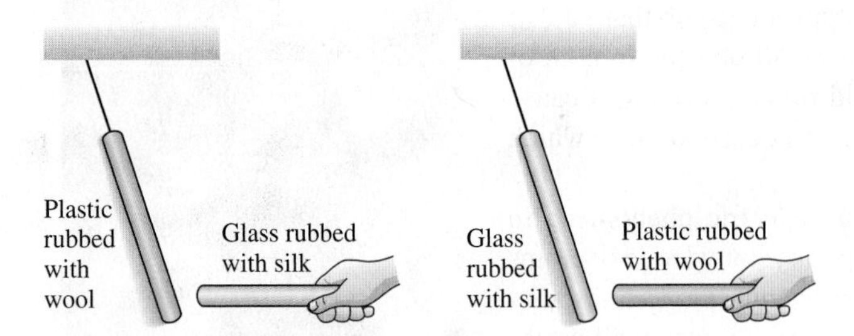

Bring a glass rod that has been rubbed with silk close to a hanging plastic rod that has been rubbed with wool. These two rods *attract* each other.

Interpretation: We can explain this experiment as well as Experiment 2 by assuming that there are two *different* kinds of charge that a material can acquire. We *define* the kind of charge acquired by a glass rod as *positive* charge, and that acquired by a plastic rod as *negative* charge. Then these two experiments can be summarized as **like charges** (positive/positive or negative/negative) exert repulsive forces on each other, while **opposite charges** (positive/negative) exert attractive forces on each other.

Experiment 4

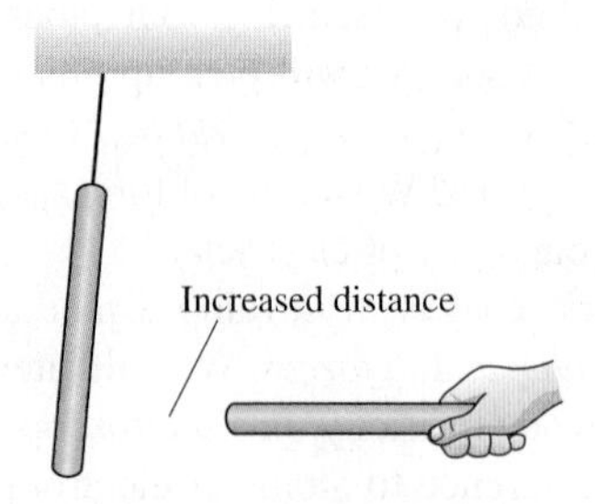

- If the two rods are held farther from each other, the force between them decreases.
- The strength of the force is greater for rods that have been rubbed more vigorously.

Interpretation: Like the gravitational force, the electric force decreases with the distance between the charged objects. And, the greater the charge on the two objects, the greater the force between them.

Although we showed experimental results for only plastic rods rubbed with wool and glass rods rubbed with silk, further experiments show that there are *only* two kinds of charge, positive and negative. For instance, when you rub a balloon on your hair, the balloon becomes negatively charged, while nylon rubbed with a polyester cloth becomes positively charged.

Visualizing Charge

Diagrams are going to be an important tool for understanding and explaining charges and the forces between charged objects. FIGURE 20.1 shows how to draw a *charge diagram,* which gives a schematic picture of the distribution of charge on an object. It's important to realize that the + and − signs drawn in Figure 20.1 do not represent "individual" charges. At this point, we are thinking of charge only as something that can be acquired by an object by rubbing, so in charge diagrams the + and − signs represent where the charge is only in a general way. In Section 20.2 we'll look at an atomic view of charging and learn about the single microscopic charges of protons and electrons.

FIGURE 20.1 Visualizing charge.

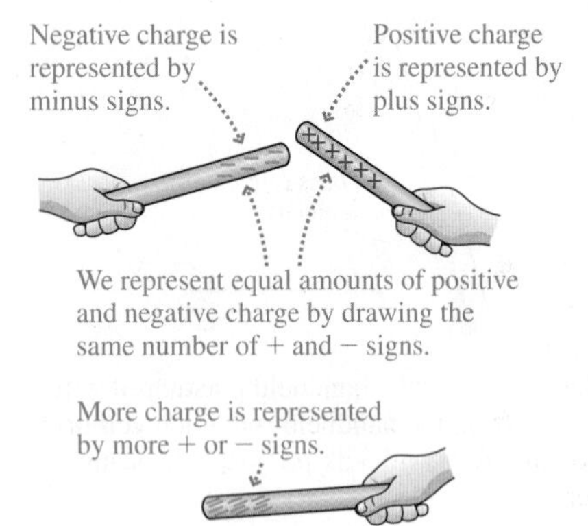

Negative charge is represented by minus signs.
Positive charge is represented by plus signs.
We represent equal amounts of positive and negative charge by drawing the same number of + and − signs.
More charge is represented by more + or − signs.

We can gain an important insight into the nature of charge by investigating what happens when we bring together a plastic rod and the wool used to charge it, as the following experiment shows.

Discovering electricity III

Experiment 5

Start with a neutral, uncharged hanging plastic rod and a piece of wool. Rub the plastic rod with the wool, then hold the wool close to the rod. The rod is *attracted* to the wool.

Interpretation: From Experiment 3 we know that the plastic rod has a negative charge. Because the wool attracts the rod, the wool must have a *positive* charge.

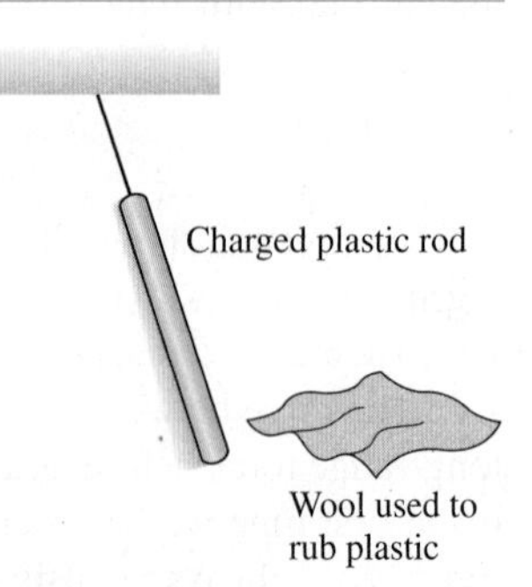

Experiment 5 shows that when a plastic rod is rubbed by wool, not only does the plastic rod acquire a negative charge, but also the wool used to rub it acquires a positive charge. This observation can be explained if we postulate that a neutral object is not one that has no charge at all; rather, **a neutral object contains** *equal amounts* **of positive and negative charge.** Just as in ordinary addition, where $2 + (-2) = 0$, equal amounts of opposite charge "cancel," leaving no overall or *net* charge.

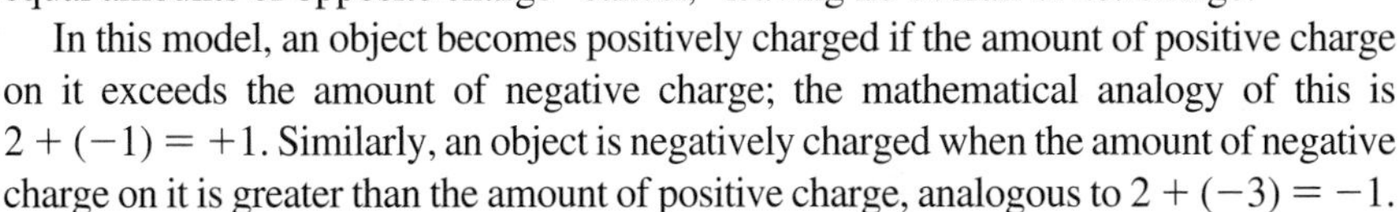

In this model, an object becomes positively charged if the amount of positive charge on it exceeds the amount of negative charge; the mathematical analogy of this is $2 + (-1) = +1$. Similarly, an object is negatively charged when the amount of negative charge on it is greater than the amount of positive charge, analogous to $2 + (-3) = -1$.

As **FIGURE 20.2** shows, the rubbing process works by *transferring* charge from one object to the other. When the two are rubbed together, negative charge is transferred from the wool to the rod. This clearly leaves the rod with an excess of negative charge. But the wool, having lost some of its negative charge to the rod, now has an *excess* of positive charge, leaving it positively charged. (We'll see in Section 20.2 why it is usually negative charge that moves.)

FIGURE 20.2 How a plastic rod and wool acquire charge during the rubbing process.

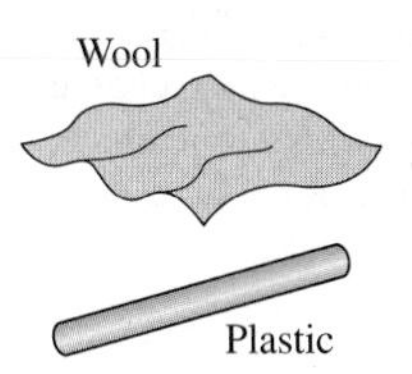

1. The rod and wool are initially neutral.

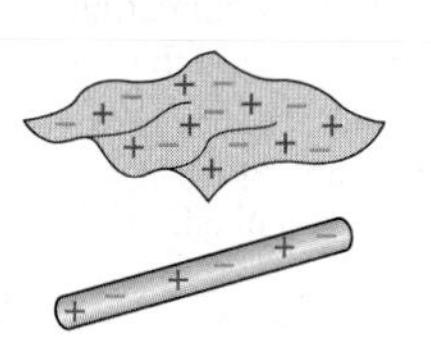

2. "Neutral" actually means that they each have equal amounts of positive and negative charge.

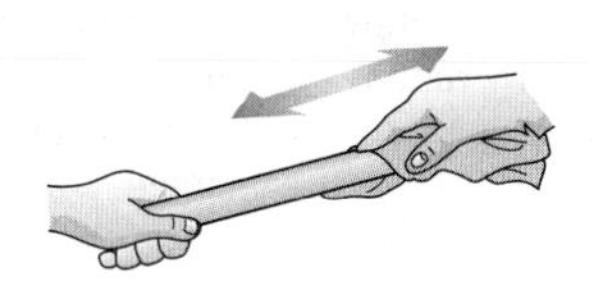

3. Now we rub the plastic and wool together. As we do so, negative charge moves from the wool to the rod.

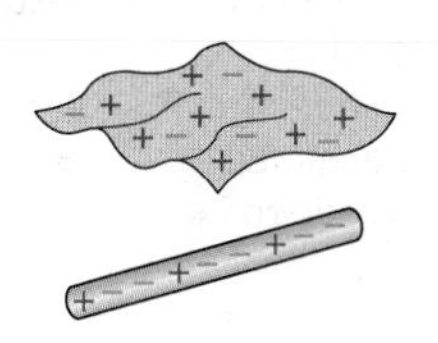

4. This leaves the rod with extra negative charge. The wool is left with more positive charge than negative.

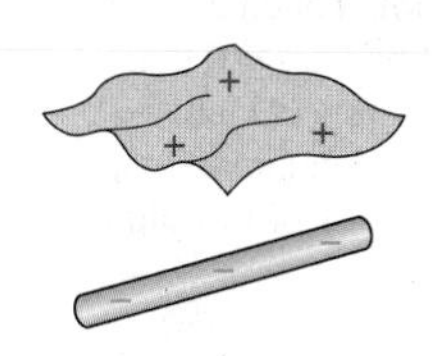

5. In a charge diagram, we draw only the *excess* charge. This shows how the rod and wool acquire their charge.

There is another crucial fact about charge implicit in Figure 20.2: Nowhere in the rubbing process was charge either created or destroyed. Charge was merely transferred from one place to another. It turns out that this fact is a fundamental law of nature, the **law of conservation of charge.** If a certain amount of positive charge appears somewhere, an equal amount of negative charge must appear elsewhere so that the net charge doesn't change.

The results of our experiments, and our interpretation of them in terms of positive and negative charge, can be summarized in the following **charge model.**

Charge model, part I The basic postulates of our model are:

1. Frictional forces, such as rubbing, add something called *charge* to an object or remove it from the object. The process itself is called *charging.* More vigorous rubbing produces a larger quantity of charge.
2. There are two kinds of charge, positive and negative.
3. Two objects with *like charge* (positive/positive or negative/negative) exert repulsive forces on each other. Two objects with *opposite charge* (positive/negative) exert attractive forces on each other. We call these *electric forces.*
4. The force between two charged objects is a long-range force. The magnitude of the force increases as the quantity of charge increases and decreases as the distance between the charges increases.
5. *Neutral* objects have an *equal mixture* of positive and negative charge.
6. The rubbing process charges the objects by *transferring* charge (usually negative) from one to the other. The objects acquire equal but opposite charges.
7. Charge is conserved: It cannot be created or destroyed.

Insulators and Conductors

Experiments 2, 3, and 5 involved a transfer of charge from one object to another. Let's do some more experiments with charge to look at how charge *moves* on different materials.

Discovering electricity IV

Experiment 6

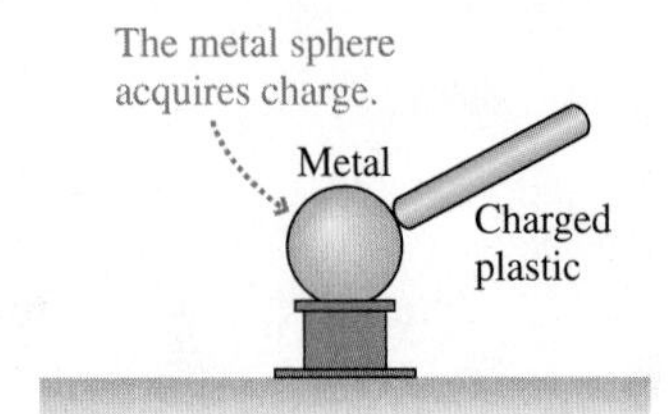

Charge a plastic rod by rubbing it with wool. Touch a neutral metal sphere with the rubbed area of the rod. The metal sphere then repels a charged, hanging plastic rod. The metal sphere appears to have acquired a charge of the same sign as the plastic rod.

Experiment 7

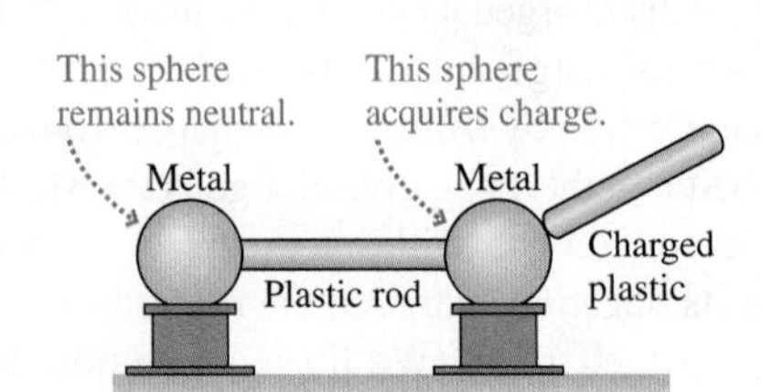

Place two metal spheres close together with a plastic rod connecting them. Charge a second plastic rod, by rubbing, and touch it to one of the metal spheres. Afterward, the metal sphere that was touched repels a charged, hanging plastic rod. The other metal sphere does not.

Experiment 8

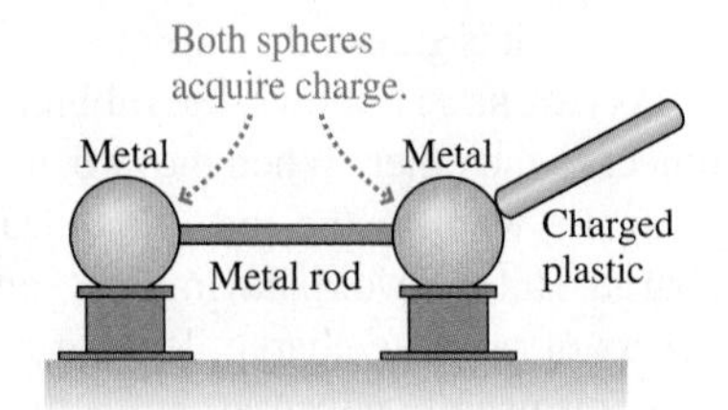

Repeat Experiment 7 with a metal rod connecting the two metal spheres. Touch one metal sphere with a charged plastic rod. Afterward, *both* metal spheres repel a charged, hanging plastic rod.

Our final set of experiments has shown that charge can be transferred from one object to another only when the objects *touch*. Contact is required. Removing charge from an object, which you can do by touching it, is called **discharging**.

In Experiments 7 and 8, charge is transferred from the charged rod to the metal sphere as the two are touched together. In Experiment 7, the other sphere remains neutral, indicating that no charge moved along the plastic rod connecting the two spheres. In Experiment 8, by contrast, the other sphere is found to be charged; evidently charge has moved along the metal rod connecting the spheres, transferring some charge from the first sphere to the second. We define **conductors** as those materials through or along which charge easily moves and **insulators** as those materials on or in which charges remain immobile. Glass and plastic are insulators; metal is a conductor.

This new information allows us to add more postulates to our charge model:

Charge model, part II

8. There are two types of materials. Conductors are materials through or along which charge easily moves. Insulators are materials on or in which charges remain fixed in place.
9. Charge can be transferred from one object to another by contact.

NOTE ▸ Both insulators and conductors can be charged. They differ in the ability of charge to *move*. ◂

◂ A dry day, a plastic slide, and a child with clothes of the right fabric lead to a startling demonstration of electric charges and forces. The rubbing of the child's clothes on the slide has made her build up charge. The body is a good conductor, so the charges spread across her body and her hair. The resulting repulsion produces a dramatic result!

CONCEPTUAL EXAMPLE 20.1 **Transferring charge**

In Experiment 8, touching a metal sphere with a charged plastic rod caused a second metal sphere, connected by a metal rod to the first, to become charged with the same type of charge as the rod. Use the postulates of the charge model to construct a charge diagram for the process.

REASON We need the following ideas from the charge model:

1. **Charge is transferred upon contact.** The plastic rod was charged by rubbing with wool, giving it a negative charge. The charge doesn't move around on the rod, an insulator, but some of the charge is transferred to the metal upon contact.
2. **Metal is a conductor.** Once in the metal, which is a conductor, the charges are free to move around.
3. **Like charges repel.** Because like charges repel, these negative charges quickly move as far apart as they possibly can. Some move through the connecting metal rod to the second sphere. Consequently, the second sphere acquires a net negative charge. The repulsive forces drive the negative charges as far apart as they can possibly get, causing them to end up on the *surfaces* of the conductors.

The charge diagram in **FIGURE 20.3** illustrates these three steps.

FIGURE 20.3 A charge diagram for Experiment 8.

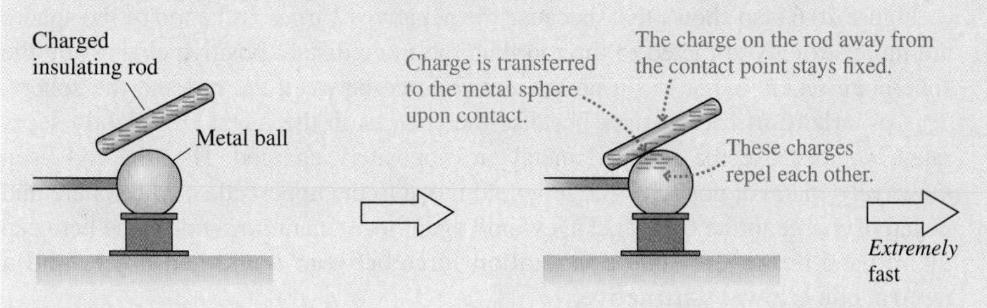
Charged insulating rod
Metal ball
Charge is transferred to the metal sphere upon contact.
The charge on the rod away from the contact point stays fixed.
These charges repel each other.
Extremely fast

The repulsive forces cause the charge to spread out over the surfaces of the conductors.

In Conceptual Example 20.1, once the charge is placed on the conductor it rapidly distributes itself over the conductor's surface. This movement of charge is *extremely* fast. Other than this very brief interval during which the charges are adjusting, the charges on an isolated conductor are in static equilibrium with the charges at rest. This condition is called **electrostatic equilibrium.**

CONCEPTUAL EXAMPLE 20.2 **Drawing a charge diagram for an electroscope**

Many electricity demonstrations are carried out with the help of an *electroscope* like the one shown in **FIGURE 20.4**. Touching the sphere at the top of an electroscope with a charged plastic rod causes the leaves to fly apart and remain hanging at an angle. Use charge diagrams to explain why.

REASON We will use the charge model and our understanding of insulators and conductors to make a series of charge diagrams in **FIGURE 20.5** that show the charging of the electroscope.

ASSESS The charges move around, but, because charge is conserved, the total number of negative charges doesn't change from picture to picture.

FIGURE 20.4 A charged electroscope.

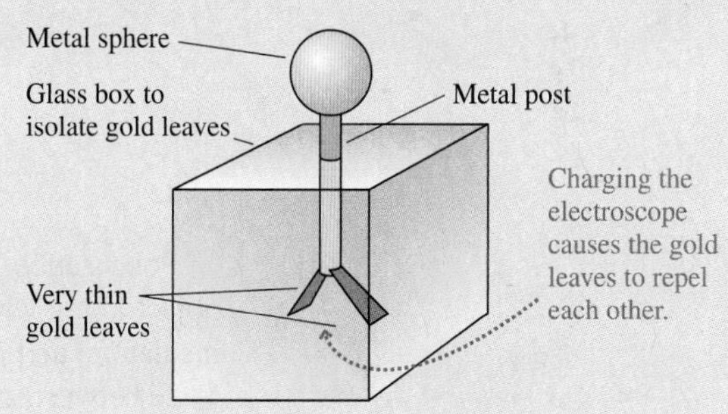
Metal sphere
Glass box to isolate gold leaves
Metal post
Charging the electroscope causes the gold leaves to repel each other.
Very thin gold leaves

FIGURE 20.5 Charging an electroscope.

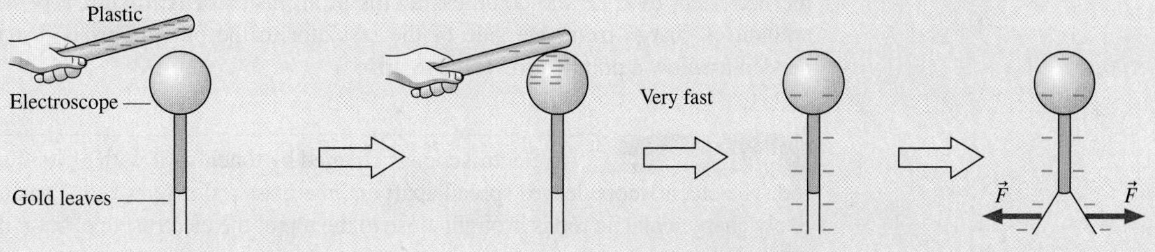
Plastic
Electroscope
Gold leaves
Very fast
$\vec{F}$
$\vec{F}$
1. Negative charges are transferred from the rod to the metal sphere upon contact.
2. Metal is a conductor. Therefore charge spreads (very rapidly) throughout the entire electroscope. The leaves become negatively charged.
3. Like charges repel. The negatively charged leaves exert repulsive forces on each other, causing them to spread apart.

Picking up pollen BIO Rubbing a rod with a cloth gives the rod an electric charge. In a similar fashion, the rapid motion of a bee's wings through the air gives the bee a small positive electric charge. As small pieces of paper are attracted to a charged rod, so are tiny grains of pollen attracted to the charged bee, helping it collect and hold the pollen.

Polarization

At the beginning of this chapter we showed a picture of a small feather being picked up by a piece of amber that had been rubbed with fur. The amber was charged by rubbing, but the feather had not been rubbed—it was *neutral.* How can our charge model explain the attraction of a neutral object toward a charged one?

Although a feather is an insulator, it's easiest to understand this phenomenon by first considering how a neutral *conductor* is attracted to a charged object. **FIGURE 20.6** shows how this works. Because the charged rod doesn't touch the sphere, no charge is added to or removed from the sphere. Instead, the rod attracts some of the sphere's negative charge to the side of the sphere near the rod. This leaves a deficit of negative charge on the opposite side of the sphere, so that side is now positively charged. This slight *separation* of the positive and negative charge in a neutral object when a charged object is brought near is called **charge polarization.**

Figure 20.6 also shows that, because the negative charges at the top of the sphere are more strongly attracted to the rod than the more distant positive charges on the sphere are repelled, there is a net *attractive* force between the rod and the sphere. This **polarization force** arises because the charges in the metal are slightly separated, *not* because the rod and metal are oppositely charged. Had the rod been *negatively* charged, positive charge would move to the upper side of the sphere and negative charge to the bottom. This would again lead to an *attractive* force between the rod and the sphere. **The polarization force between a charged object and a neutral one is always attractive.**

FIGURE 20.6 Why a neutral metal object is attracted to a charged object.

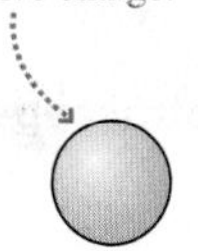
The neutral sphere contains equal amounts of positive and negative charge.

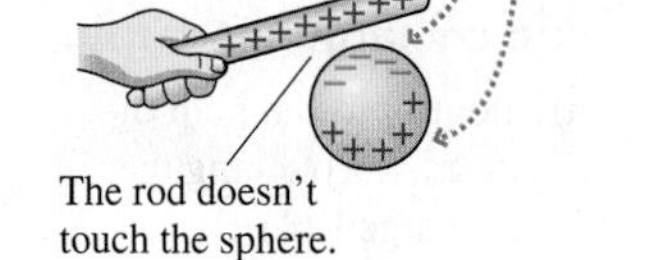
Negative charge is attracted to the positive rod. This leaves behind positive charge on the other side of the sphere.
The rod doesn't touch the sphere.

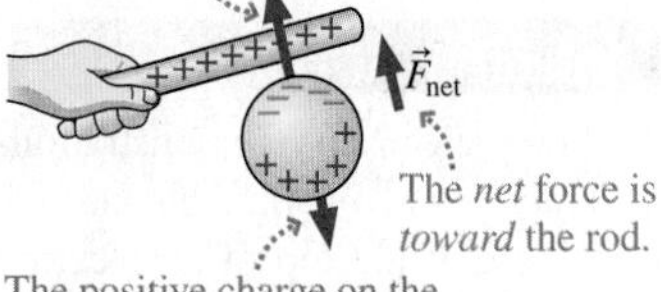
The negative charge on the sphere is close to the rod, so it is strongly attracted to the rod.
$\vec{F}_{net}$
The *net* force is *toward* the rod.
The positive charge on the sphere is far from the rod, so it is weakly repelled by the rod.

Polarization explains why forces arise between a charged object and a metal object along which charge can freely move. But the feathers attracted to amber are insulators, and charge can't move through an insulator. Nevertheless, the attractive force between a charged object and an insulator is also a polarization force. As we'll learn in the next section, the charge in each *atom* that makes up an insulator can be slightly polarized. Although the charge separation in one atom is exceedingly small, the net effect over all the countless atoms in an insulator is to shift a perceptible amount of charge from one side of the insulator to the other. This is just what's needed to allow a polarization force to arise.

STOP TO THINK 20.1 An electroscope is charged by touching it with a positive glass rod. The electroscope leaves spread apart and the glass rod is removed. Then a negatively charged plastic rod is brought close to the top of the electroscope, but it doesn't touch. What happens to the leaves?

A. The leaves move closer together.
B. The leaves spread farther apart.
C. The leaves do not change their position.

20.2 Charges, Atoms, and Molecules

We have been speaking about giving objects positive or negative charge without explaining what is happening at an atomic level. You already know that the basic constituents of atoms—the nucleus and the electrons surrounding it—are charged. In this section we will connect our observations of the previous section with our understanding of the atomic nature of matter.

Our current model of the atom is that it is made up of a very small and dense positively charged *nucleus,* containing positively charged *protons* as well as neutral particles called *neutrons,* surrounded by much-less-massive orbiting negatively charged *electrons* that form an **electron cloud** surrounding the nucleus, as illustrated in FIGURE 20.7. The atom is held together by the attractive electric force between the positive nucleus and the negative electrons.

FIGURE 20.7 Our modern view of the atom.

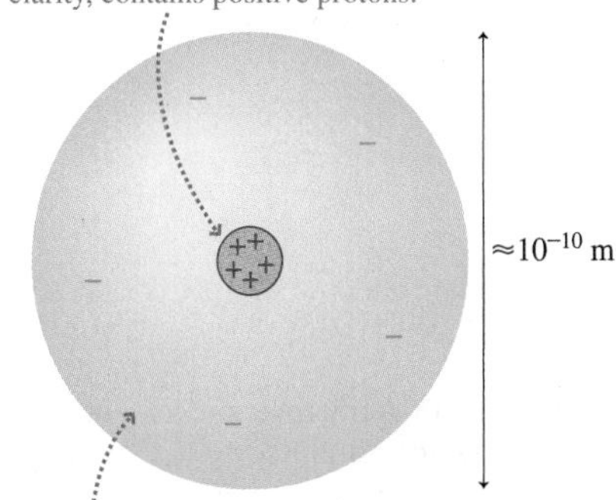

The nucleus, exaggerated in size for clarity, contains positive protons.
≈10^−10 m
The electron cloud is negatively charged.

Experiments show that **charge, like mass, is an inherent property of electrons and protons.** It's no more possible to have an electron without charge than it is to have an electron without mass.

An Atomic View of Charging

Electrons and protons are the basic charges in ordinary matter. **There are no other sources of charge.** Consequently, the various observations we made in Section 20.1 need to be explained in terms of electrons and protons.

Experimentally, it's found that electrons and protons have charges of opposite sign but *exactly* equal magnitude. Thus, because charge is due to electrons and protons, **an object is charged if it has an unequal number of electrons and protons.** An object with a negative charge has more electrons than protons; an object with a positive charge has more protons than electrons. Most macroscopic objects have an *equal number* of protons and electrons. Such an object has no *net* charge; we say it is *electrically neutral.*

FIGURE 20.8 Positive and negative ions.

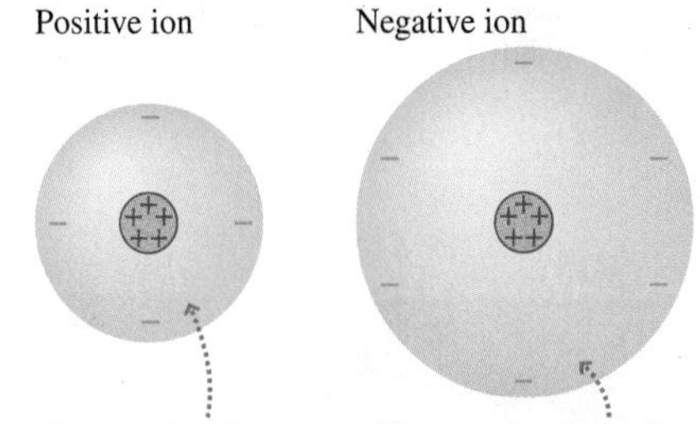

In practice, objects acquire a positive charge not by gaining protons but by losing electrons. Protons are *extremely* tightly bound within the nucleus and cannot be added to or removed from atoms. Electrons, on the other hand, are bound much more loosely than the protons and can be removed with little effort.

The process of removing an electron from the electron cloud of an atom is called **ionization.** An atom that is missing an electron is called a *positive ion.* Some atoms can accommodate an *extra* electron and thus become a *negative ion.* FIGURE 20.8 shows positive and negative ions.

The charging processes we observed in Section 20.1 involved rubbing and friction. The forces of friction often cause molecular bonds at the surface to break as two materials slide past each other. Molecules are electrically neutral, but FIGURE 20.9 shows that *molecular ions* can be created when one of the bonds in a large molecule is broken. If the positive molecular ions remain on one material and the negative ions on the other, one of the objects being rubbed ends up with a net positive charge and the other with a net negative charge. This is the way in which a plastic rod is charged by rubbing with wool or a comb is charged by passing through your hair.

FIGURE 20.9 Charging by friction may result from molecular ions produced as bonds are broken.

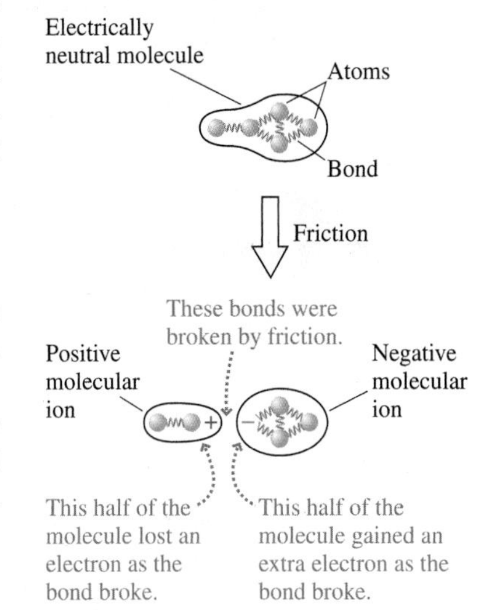

Charge Conservation

Charge is represented by the symbol q (or sometimes Q). The SI unit of charge is the **coulomb** (C), named for French scientist Charles Coulomb, one of many scientists investigating electricity in the late 18th century.

Protons and electrons, the charged particles in ordinary matter, have the same amount of charge, but of opposite signs. We use the symbol e for the **fundamental charge,** the magnitude of the charge of an electron or a proton. The fundamental charge e has been measured to have the value

$$e = 1.60 \times 10^{-19}\ \text{C}$$

TABLE 20.1 Protons and electrons

Particle	Mass (kg)	Charge (C)
Proton	1.67×10^{-27}	$+e = 1.60 \times 10^{-19}$
Electron	9.11×10^{-31}	$-e = -1.60 \times 10^{-19}$

Table 20.1 lists the masses and charges of protons and electrons.

NOTE ► The amount of charge produced by rubbing plastic or glass rods is typically in the range 1 nC (10^{-9} C) to 100 nC (10^{-7} C). This corresponds to an excess or deficit of 10^{10} to 10^{12} electrons. But, because of the enormous number of atoms in a macroscopic object, this represents an excess or deficit of only perhaps 1 electron in 10^{13}. ◄

That charge is associated with electrons and protons explains why charge is conserved. Because electrons and protons are neither created nor destroyed in ordinary processes, their associated charge is conserved as well.

Insulators and Conductors

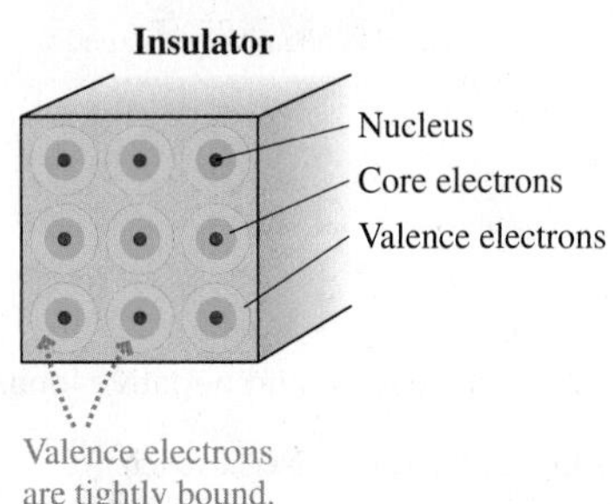

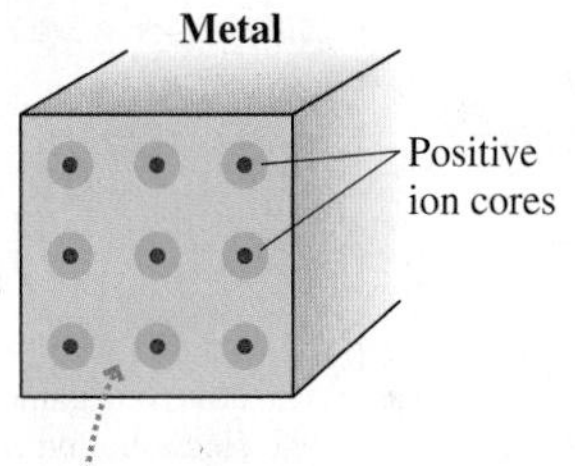

FIGURE 20.10 A microscopic look at insulators and conductors.

FIGURE 20.10 looks inside an insulator and a metallic conductor. The electrons in the insulator are all tightly bound to the positive nuclei and not free to move around. Charging an insulator by friction leaves patches of molecular ions on the surface, but these patches are immobile.

In metals, the outer atomic electrons (called the *valence electrons* in chemistry) are only weakly bound to the nuclei. As the atoms come together to form a solid, these outer electrons become detached from their parent nuclei and are free to wander about through the entire solid. The solid *as a whole* remains electrically neutral, because we have not added or removed any electrons, but the electrons are now rather like a negatively charged gas or liquid—what physicists like to call a **sea of electrons**—permeating an array of positively charged **ion cores.** However, although the electrons are highly mobile *within* the metal, they are still weakly bound to the ion cores and will not leave the metal.

Electric Dipoles

In the last section we noted that an insulator, such as paper, becomes polarized when brought near a charged object. We can use an atomic description of matter to see why.

Consider what happens if we bring a positive charge near a neutral atom. As FIGURE 20.11 shows, the charge polarizes the atom by attracting the electron cloud while repelling the nucleus. The polarization of just one atom is a very small effect, but there are an enormous number of atoms in an insulator. Added together, their net polarization—and the resulting polarization force—can be quite significant. This is how the rubbed amber picks up a feather, exerting an upward polarization force on it larger than the downward force of gravity.

FIGURE 20.11 An induced electric dipole.

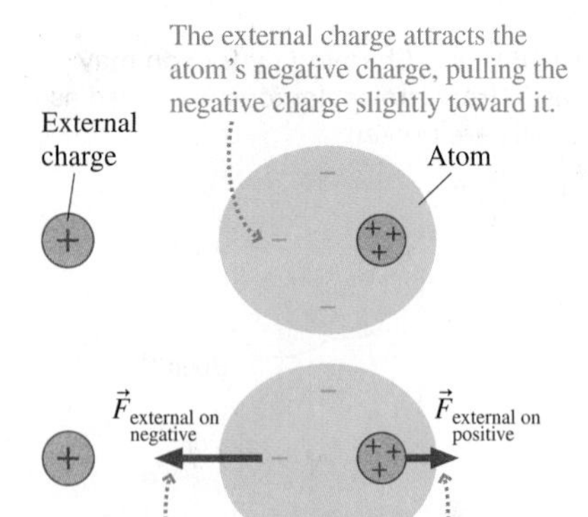

Two equal but opposite charges with a separation between them are called an **electric dipole.** In this case, where the polarization is caused by the external charge, the atom has become an *induced electric dipole.* Because the negative end of the dipole is slightly closer to the positive charge, the attractive force on the negative end slightly exceeds the repulsive force on the positive end, and there is a net force toward the external charge. If a charged rod causes all the atoms in a piece of paper to become induced electric dipoles, the net force is enough to lift the paper to the rod.

Hydrogen Bonding

Some molecules have an asymmetry in their charge distribution that makes them *permanent electric dipoles.* An important example is the water molecule. Bonding between the hydrogen and oxygen atoms results in an unequal sharing of charge that, as shown in FIGURE 20.12, leaves the hydrogen atoms with a small positive charge and the oxygen atom with a small negative charge.

When two water molecules are close, the attractive electric force between the positive hydrogen atom of one molecule and the negative oxygen atom of the second molecule can form a weak bond, called a **hydrogen bond,** as illustrated in FIGURE 20.13. These weak bonds result in a certain "stickiness" between water molecules that is

FIGURE 20.12 A water molecule is a permanent electric dipole.

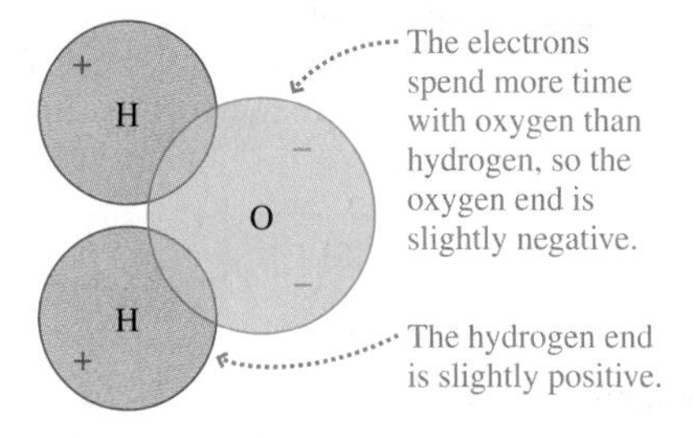

H
O
H
The electrons spend more time with oxygen than hydrogen, so the oxygen end is slightly negative.
The hydrogen end is slightly positive.

FIGURE 20.13 Hydrogen bonds between water molecules.

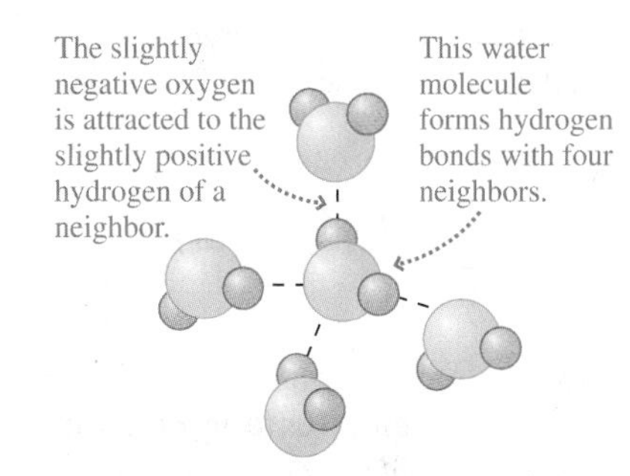

The slightly negative oxygen is attracted to the slightly positive hydrogen of a neighbor.
This water molecule forms hydrogen bonds with four neighbors.

responsible for many of water's special properties, including its expansion on freezing, the wide range of temperatures over which it is liquid, and its high heat of vaporization.

Hydrogen bonds are extremely important in biological systems. As you know, the DNA molecule has the structure of a double helix. Information in DNA is coded in the *nucleotides,* the four molecules guanine, thymine, adenine, and cytosine. The nucleotides on one strand of the DNA helix form hydrogen bonds with the nucleotides on the opposite strand.

The nucleotides bond only in certain pairs: Cytosine always forms a bond with guanine, adenine with thymine. This preferential bonding is crucial to DNA replication. When the two strands of DNA are taken apart, each separate strand of the DNA forms a template on which another complementary strand can form, creating two identical copies of the original DNA molecule.

The preferential bonding of nucleotide base pairs in DNA is explained by hydrogen bonding. In each of the nucleotides, the hydrogen atoms have a small positive charge, oxygen and nitrogen a small negative charge. The positive hydrogen atoms on one nucleotide attract the negative oxygen or nitrogen atoms on another. As the detail in **FIGURE 20.14** shows, the geometry of the nucleotides allows cytosine to form a hydrogen bond only with guanine, adenine only with thymine.

FIGURE 20.14 BIO Hydrogen bonds in DNA base pairs.

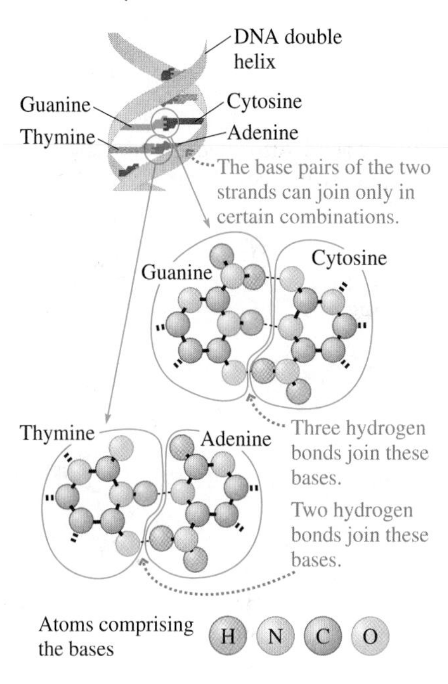

STOP TO THINK 20.2 Rank in order, from most positive to most negative, the charges q_A to q_E of these five systems.

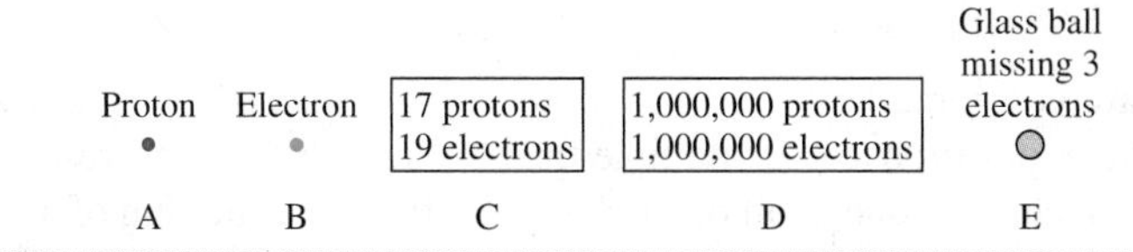

20.3 Coulomb's Law

The last two sections established a *model* of charges and electric forces. This model is very good at explaining electric phenomena and providing a general understanding of electricity. Now we need to become quantitative. Experiment 4 in Section 20.1 found that the electric force increases for objects with more charge and decreases as charged objects are moved farther apart. The force law that describes this behavior is known as *Coulomb's law.*

In the mathematical formulation of Coulomb's law, we will use the magnitude of the charge only, not the sign. We show this by using the absolute value notation we used earlier in the textbook. $|q|$ therefore represents the magnitude of the charge. It is always a positive number, whether the charge is positive or negative.

Video Tutor Demo

Class Video

Coulomb's law

Magnitude: If two charged particles having charges q_1 and q_2 are a distance r apart, the particles exert forces on each other of magnitude

$$F_{1\,\text{on}\,2} = F_{2\,\text{on}\,1} = \frac{K|q_1||q_2|}{r^2} \qquad (20.1)$$

p. 176 INVERSE-SQUARE

where the charges are in coulombs (C), and $K = 8.99 \times 10^9\ \text{N} \cdot \text{m}^2/\text{C}^2$ is called the **electrostatic constant.** These forces are an action/reaction pair, equal in magnitude and opposite in direction. It is customary to round K to $9.0 \times 10^9\ \text{N} \cdot \text{m}^2/\text{C}^2$ for all but extremely precise calculations, and we will do so.

Direction: The forces are directed along the line joining the two particles. The forces are *repulsive* for two like charges and *attractive* for two opposite charges.

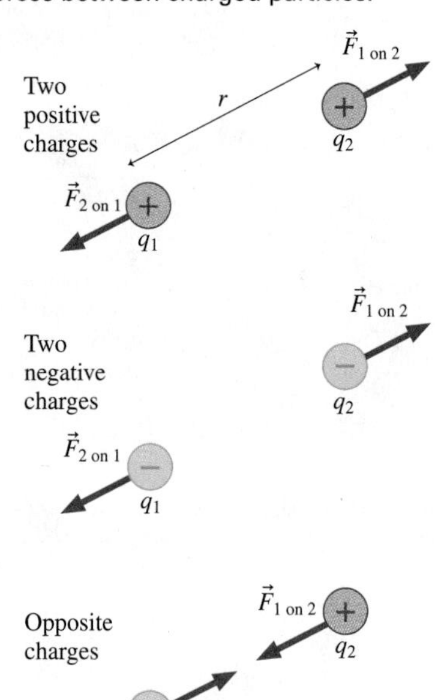

FIGURE 20.15 Attractive and repulsive forces between charged particles.

We sometimes speak of the "force between charge q_1 and charge q_2," but keep in mind that we are really dealing with charged *objects* that also have a mass, a size, and other properties. Charge is not some disembodied entity that exists apart from matter. Coulomb's law describes the force between charged *particles.*

NOTE ▶ Coulomb's law applies only to *point charges.* A point charge is an idealized material object with charge and mass but with no size or extension. For practical purposes, two charged objects can be modeled as point charges if they are much smaller than the separation between them. ◀

Coulomb's law looks much like Newton's law of gravity, which you studied in ◀◀ SECTION 6.5, but there is a key difference: The charge q can be either positive or negative, so the forces can be attractive or repulsive. Consequently, the absolute value signs in Equation 20.1 are especially important. The first part of Coulomb's law gives only the *magnitude* of the force, which is always positive. The direction must be determined from the second part of the law. FIGURE 20.15 shows the forces between different combinations of positive and negative charges.

Using Coulomb's Law

Coulomb's law is a force law, and forces are vectors. **Electric forces, like other forces, can be superimposed.** If multiple charges 1, 2, 3, ... are present, the *net* electric force on charge j due to all other charges is therefore the sum of all the individual forces due to each charge; that is,

$$\vec{F}_{\text{net}} = \vec{F}_{1\,\text{on}\,j} + \vec{F}_{2\,\text{on}\,j} + \vec{F}_{3\,\text{on}\,j} + \cdots \qquad (20.2)$$

where each of the forces $\vec{F}_{i\,\text{on}\,j}$ is given by Equation 20.1.

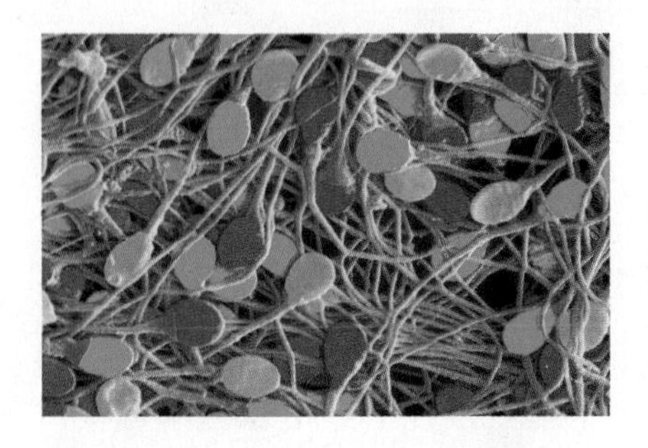

◀ **Separating the girls from the boys** BIO Sperm cells can be sorted according to whether they contain an X or a Y chromosome. The cells are put into solution, and the solution is forced through a nozzle, which breaks the solution into droplets. Suppose a droplet contains a sperm cell. An optical test measures which type of chromosome, X or Y, the cell has. The droplet is then given a positive charge if it has an X sperm cell, negative if it contains a Y. The droplets fall between two oppositely charged plates where they are pushed left or right depending on their charge—separating the X from the Y.

PROBLEM-SOLVING STRATEGY 20.1

Electric forces and Coulomb's law

MP

We can use Coulomb's law to find the electric force on a charged particle due to one or more other charged particles.

PREPARE Identify point charges or objects that can be modeled as point charges. Create a visual overview in which you establish a coordinate system, show the positions of the charges, show the force vectors on the charges, define distances and angles, and identify what the problem is trying to find.

SOLVE The magnitude of the force between point charges is given by Coulomb's law:

$$F_{1\text{ on }2} = F_{2\text{ on }1} = \frac{K|q_1||q_2|}{r^2}$$

Use your visual overview as a guide to the use of this law:

- Show the directions of the forces—repulsive for like charges, attractive for opposite charges—on the visual overview.
- Draw the lengths of your force vectors according to Coulomb's law: A particle with a greater charge, or one that is closer, will lead to a longer force vector.
- When possible, do graphical vector addition on the visual overview. While not exact, it tells you the type of answer you should expect.
- Write each force vector in terms of its x- and y-components, then add the components to find the net force. Use the visual overview to determine which components are positive and which are negative.

ASSESS Check that your result has the correct units, is reasonable, and answers the question.

Exercise 21

EXAMPLE 20.3

Adding electric forces in one dimension

Two +10 nC charged particles are 2.0 cm apart on the x-axis. What is the net force on a +1.0 nC charge midway between them? What is the net force if the charged particle on the right is replaced by a −10 nC charge?

PREPARE We proceed using the steps of Problem-Solving Strategy 20.1. We model the charged particles as point charges. The visual overview of **FIGURE 20.16** establishes a coordinate system and shows the forces $\vec{F}_{1\text{ on }3}$ and $\vec{F}_{2\text{ on }3}$. Figure 20.16a shows a +10 nC charge on the right; Figure 20.16b shows a −10 nC charge.

FIGURE 20.16 A visual overview of the forces for the two cases.

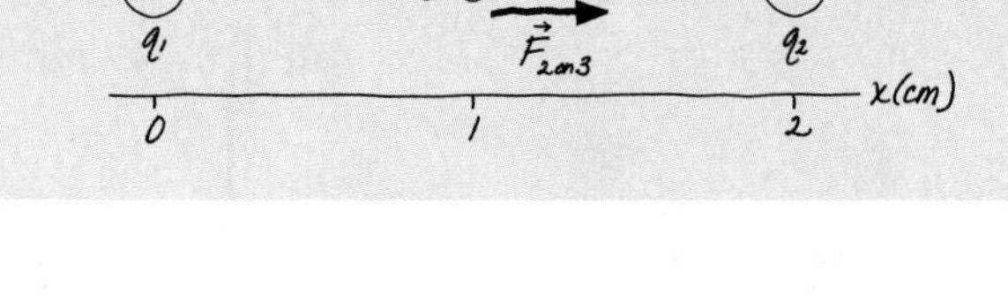

SOLVE Electric forces are vectors, and the net force on q_3 is the *vector* sum $\vec{F}_{\text{net}} = \vec{F}_{1\text{ on }3} + \vec{F}_{2\text{ on }3}$. Charges q_1 and q_2 each exert a repulsive force on q_3, but these forces are equal in magnitude and opposite in direction. Consequently, $\vec{F}_{\text{net}} = \vec{0}$. The situation changes if q_2 is negative, as in Figure 20.16b. In this case, the two forces are equal in magnitude but in the *same* direction, so $\vec{F}_{\text{net}} = 2\vec{F}_{1\text{ on }3}$. The magnitude of the force is given by Coulomb's law. The force due to q_1 is

$$F_{1\text{ on }3} = \frac{K|q_1||q_3|}{r_{13}^2}$$

$$= \frac{(9.0 \times 10^9\ \text{N}\cdot\text{m}^2/\text{C}^2)(10 \times 10^{-9}\ \text{C})(1.0 \times 10^{-9}\ \text{C})}{(0.010\ \text{m})^2}$$

$$= 9.0 \times 10^{-4}\ \text{N}$$

There is an equal force due to q_2, so the net force on the 1.0 nC charge is $\vec{F}_{\text{net}} = (1.8 \times 10^{-3}\ \text{N, to the right})$.

ASSESS This example illustrates the important idea that electric forces are *vectors*. An important part of assessing our answer is to see if it is "reasonable." In the second case, the net force on the charge is approximately 1 mN. Generally, charges of a few nC separated by a few cm experience forces in the range from a fraction of a mN to several mN. With this guideline, the answer appears to be reasonable.

EXAMPLE 20.4 Adding electric forces in two dimensions

Three charged particles with $q_1 = -50$ nC, $q_2 = +50$ nC, and $q_3 = +30$ nC are placed as shown in FIGURE 20.17. What is the net force on charge q_3 due to the other two charges?

FIGURE 20.17 The arrangement of the charges.

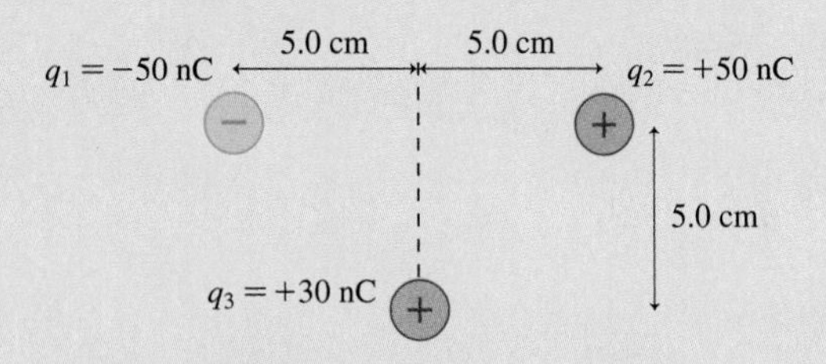

PREPARE We solve for the net force using the steps of Problem-Solving Strategy 20.1, beginning with the visual overview shown in FIGURE 20.18a. We have defined a coordinate system, with charge q_3 at the origin. We have drawn the forces on charge q_3, with directions determined by the signs of the charges. We can see from the geometry that the forces $\vec{F}_{1\text{ on }3}$ and $\vec{F}_{2\text{ on }3}$ are at the angles noted in the figure. The vector addition in FIGURE 20.18b shows the anticipated direction of the net force; this will be a good check on our final result. The distance between charges q_1 and q_3 is the same as that between charges q_2 and q_3; this distance is $r = \sqrt{(5.0\text{ cm})^2 + (5.0\text{ cm})^2} = 7.07$ cm. Because the magnitudes of q_1 and q_2 are equal and they are equidistant from q_3, we expect the magnitudes of the forces they exert to be equal, as we have drawn in Figure 20.18.

FIGURE 20.18 A visual overview of the charges and forces.

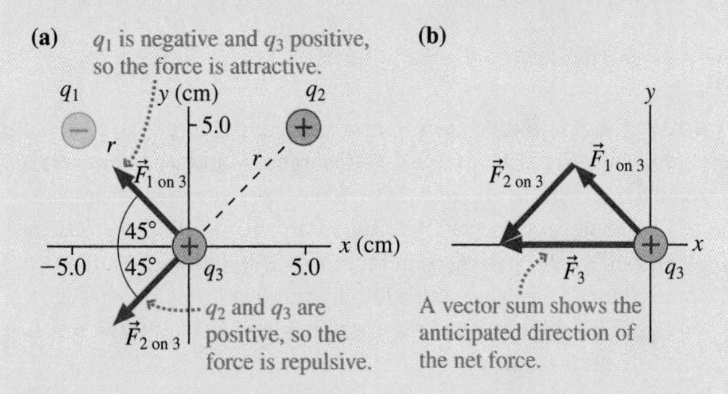

SOLVE We are interested in the net force on charge q_3. Let's start by using Coulomb's law to compute the magnitudes of the two forces on charge q_3:

$$F_{1\text{ on }3} = \frac{K|q_1||q_3|}{r^2}$$

$$= \frac{(9.0\times10^9\ \text{N}\cdot\text{m}^2/\text{C}^2)(50\times10^{-9}\ \text{C})(30\times10^{-9}\ \text{C})}{(0.0707\ \text{m})^2}$$

$$= 2.7\times10^{-3}\ \text{N}$$

The magnitudes of the charges and the distance are the same for $F_{2\text{ on }3}$, so

$$F_{2\text{ on }3} = \frac{K|q_2||q_3|}{r^2} = 2.7\times10^{-3}\ \text{N}$$

The components of these forces are illustrated in FIGURE 20.19a.

FIGURE 20.19 The net force on q_3 is to the left.

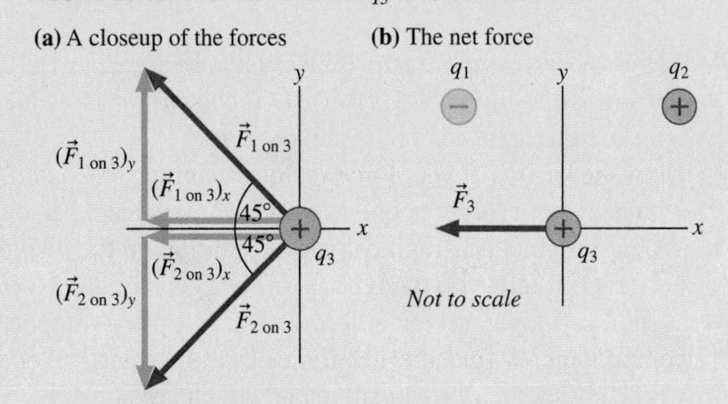

Computing values for the components, we find

$$(F_{1\text{on}3})_x = -(2.7\times10^{-3}\ \text{N})\cos45° = -1.9\times10^{-3}\ \text{N}$$

$$(F_{1\text{on}3})_y = (2.7\times10^{-3}\ \text{N})\sin45° = 1.9\times10^{-3}\ \text{N}$$

$$(F_{2\text{on}3})_x = -(2.7\times10^{-3}\ \text{N})\cos45° = -1.9\times10^{-3}\ \text{N}$$

$$(F_{2\text{on}3})_y = -(2.7\times10^{-3}\ \text{N})\sin45° = -1.9\times10^{-3}\ \text{N}$$

Next, we add components of the net force:

$$F_{3x} = (F_{1\text{on}3})_x + (F_{2\text{on}3})_x = -1.9\times10^{-3}\ \text{N} - 1.9\times10^{-3}\ \text{N}$$

$$= -3.8\times10^{-3}\ \text{N}$$

$$F_{3y} = (F_{1\text{on}3})_y + (F_{2\text{on}3})_y$$

$$= +1.9\times10^{-3}\ \text{N} - 1.9\times10^{-3}\ \text{N} = 0$$

Thus the net force, as shown in FIGURE 20.19b, is

$$\vec{F}_3 = (3.8\times10^{-3}\ \text{N},\ -x\text{-direction})$$

ASSESS The net force is directed to the left, as we anticipated. The magnitude of the net force, a few mN, seems reasonable as well.

EXAMPLE 20.5 Comparing electric and gravitational forces

A small plastic sphere is charged to −10 nC. It is held 1.0 cm above a small glass bead at rest on a table. The bead has a mass of 15 mg and a charge of +10 nC. Will the glass bead "leap up" to the plastic sphere?

PREPARE We model the plastic sphere and glass bead as point charges. FIGURE 20.20 establishes a y-axis, identifies the plastic sphere as q_1 and the glass bead as q_2, and shows a free-body diagram. The glass bead will rise if $F_{1\text{ on }2} > w$; if $F_{1\text{ on }2} < w$, the bead will remain at rest on the table, which then exerts a normal force $\vec{n}$ on the bead.

FIGURE 20.20 A visual overview showing the charges and forces.

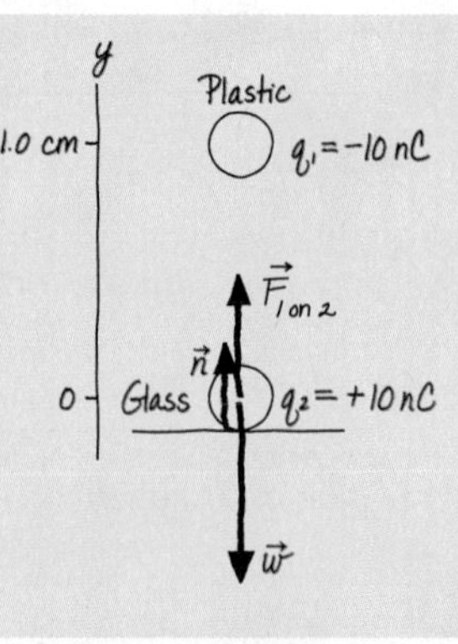

SOLVE Using the values provided, we have

$$F_{1\text{ on }2} = \frac{K|q_1||q_2|}{r^2} = 9.0 \times 10^{-3}\text{ N}$$
$$w = m_2 g = 1.5 \times 10^{-4}\text{ N}$$

$F_{1\text{ on }2}$ exceeds the bead's weight by a factor of 60, so the glass bead will leap upward.

ASSESS The values used in this example are realistic for spheres ≈ 2 mm in diameter. In general, as in this example, electric forces are *significantly* larger than weight forces. Consequently, we can ignore weight forces when working electric-force problems unless the particles are fairly massive.

STOP TO THINK 20.3 Charges 1 and 2 exert repulsive forces on each other. $q_1 = 4q_2$. Which statement is true?

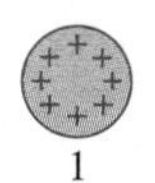

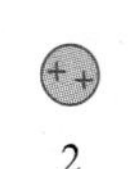

A. $F_{1\text{ on }2} > F_{2\text{ on }1}$ B. $F_{1\text{ on }2} = F_{2\text{ on }1}$ C. $F_{1\text{ on }2} < F_{2\text{ on }1}$

Figure Video

20.4 The Concept of the Electric Field

Coulomb's law is the basic law of electrostatics. We can use Coulomb's law to calculate the force a positive charge exerts on a nearby negative charge. But there is an unanswered question: How does the negative charge "know" that the positive charge is there? Coulomb's law tells us how to calculate the magnitude and direction of the force, but it doesn't tell us how the force is transmitted through empty space from one charge to the other. To answer this question, we will introduce the *field model,* first suggested in the early 19th century by Michael Faraday, a British investigator of electricity and magnetism.

FIGURE 20.21 Visualizing the electric field.

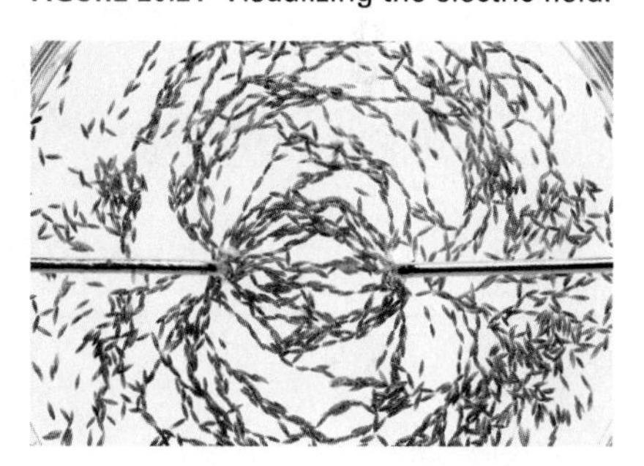

FIGURE 20.21 shows a photograph of the surface of a shallow pan of oil with tiny grass seeds floating on it. When charged wires, one positive and one negative, touch the surface of the oil, the grass seeds line up to form a regular pattern. The pattern suggests that some kind of electric influence from the charges *fills the space* around the charges. Perhaps the grass seeds are reacting to this influence, creating the pattern that we see. This alteration of the space around the charges could be the *mechanism* by which the long-range Coulomb's law force is exerted.

This is the essence of the field model. Consider the attractive force between a positive charge A and a negative charge B. **FIGURE 20.22** shows the difference between the force model, which we have been using, and the field model. **In the field model, it is the alteration of space around charge A that is the *agent* that exerts a force on charge B.** This alteration of space is what we call a **field.** The charge makes an alteration *everywhere* in space. Other charges then respond to the alteration at their position.

The field model applies to many branches of physics. The space around a charge is altered to create the **electric field.** The alteration of the space around a mass is called the **gravitational field.** The alteration of the space around a magnet is called the **magnetic field,** which we will consider in Chapter 24.

FIGURE 20.22 The force and field models for the interaction between two charges.

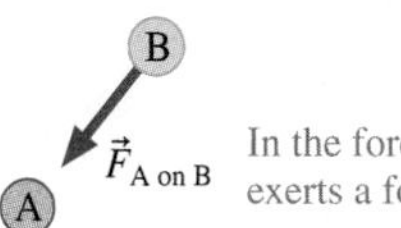

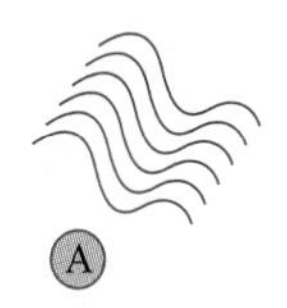

A
In the field model, A alters the space around it. (The wavy lines are poetic license. We'll soon learn a better representation.)

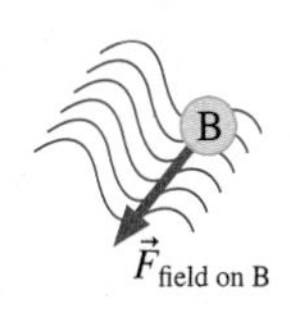

The Field Model

The field model might seem arbitrary and abstract. We can already calculate forces between charges; why introduce another way of looking at things? We will find that the field model is a very useful tool for visualizing and calculating forces for complex arrangements of charges. However, there's a more fundamental reason for introducing the electric field. When we begin to study fields that change with time, we'll find phenomena that can be understood *only* in terms of fields.

We begin our investigation of electric fields by postulating a **field model** that describes how charges interact:

1. A group of charges, which we will call the **source charges,** alter the space around them by creating an *electric field* $\vec{E}$.
2. If another charge is then placed in this electric field, it experiences a force $\vec{F}$ exerted *by the field.*

TABLE 20.2 Typical electric field strengths

Field	Field strength (N/C)
Inside a current-carrying wire	10^{-2}
Earth's field, near the earth's surface	10^{2}
Near objects charged by rubbing	10^{3} to 10^{6}
Needed to cause a spark in air	10^{6}
Inside a cell membrane	10^{7}
Inside an atom	10^{11}

Suppose charge q experiences an electric force $\vec{F}_{\text{on}\,q}$ due to other charges. The strength and direction of this force vary as q is moved from point to point in space. This suggests that "something" is present at each point in space to cause the force that charge q experiences. We define the electric field $\vec{E}$ at the point (x, y, z) as

$$\vec{E} \text{ at } (x, y, z) = \frac{\vec{F}_{\text{on}\,q} \text{ at } (x, y, z)}{q} \tag{20.3}$$

Electric field at a point defined by the force on charge q

We're *defining* the electric field as a force-to-charge ratio; hence the units of the electric field are newtons per coulomb, or N/C. The magnitude E of the electric field is called the **electric field strength.** Typical electric field strengths are given in Table 20.2.

You can think of using charge q as a *probe* to determine whether an electric field is present at a point in space. If charge q experiences an electric force at a point in space, as FIGURE 20.23a shows, then there is an electric field at that point causing the force. Further, we *define* the electric field at that point to be the vector given by Equation 20.3. FIGURE 20.23b shows the electric field at two points, but you can imagine "mapping out" the electric field by moving the charge q throughout all space.

FIGURE 20.23 Charge q is a probe of the electric field.

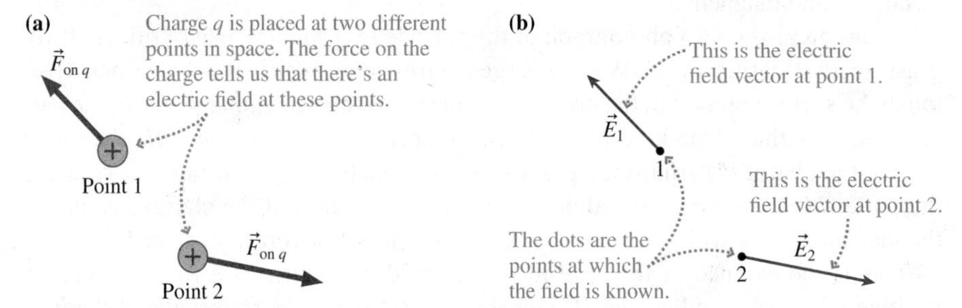

The basic idea of the field model is that **the field is the agent that exerts an electric force on a particle with charge *q*.** Notice three important things about the field:

1. The electric field, a vector, exists at every point in space. Electric field diagrams will show a sample of the vectors, but there is an electric field vector at every point whether one is shown or not.
2. If the probe charge q is positive, the electric field vector points in the same direction as the force on the charge; if negative, the electric field vector points opposite the force.
3. Because q appears in Equation 20.3, it may seem that the electric field depends on the magnitude of the charge used to probe the field. It doesn't. We know from Coulomb's law that the force $\vec{F}_{\text{on}\,q}$ is proportional to q. Thus the electric field defined in Equation 20.3 is *independent* of the charge q that probes the field. The electric field depends on only the source charges that create the field.

The Electric Field of a Point Charge

FIGURE 20.24a shows a point source charge q that creates an electric field at all points in space. We can use a second charge, shown as q' in FIGURE 20.24b, to serve as a probe of the electric field created by charge q.

For the moment, assume both charges are positive. The force on q', which is repulsive and points directly away from q, is given by Coulomb's law:

$$\vec{F}_{\text{on } q'} = \left(\frac{Kqq'}{r^2}, \text{ away from } q\right) \tag{20.4}$$

Equation 20.3 defines the electric field in terms of the force on the probe charge as $\vec{E} = \vec{F}_{\text{on } q'}/q'$, so for a positive charge q,

$$\vec{E} = \left(\frac{Kq}{r^2}, \text{ away from } q\right) \tag{20.5}$$

The electric field is shown in **FIGURE 20.24c**.

If q is negative, the magnitude of the force on the probe charge is the same as in Equation 20.5, but the direction is toward q, so the general expression for the field is

$$\vec{E} = \left(\frac{K|q|}{r^2}, \begin{bmatrix} \text{away from } q \text{ if } q > 0 \\ \text{toward } q \text{ if } q < 0 \end{bmatrix}\right) \tag{20.6}$$

Electric field of point charge q at a distance r from the charge

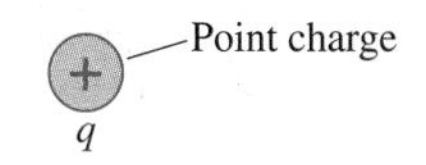

NOTE ▸ The expression for the electric field is similar to Coulomb's law. To distinguish the two, remember that Coulomb's law has the product of two charges in the numerator. It describes the force between *two* charges. The electric field has a single charge in the numerator. It is the field of a *single* charge. ◂

FIGURE 20.24 Charge q' is used to probe the electric field of point charge q.

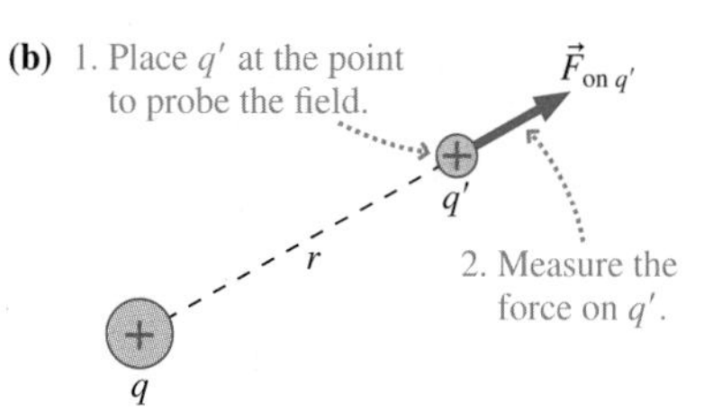

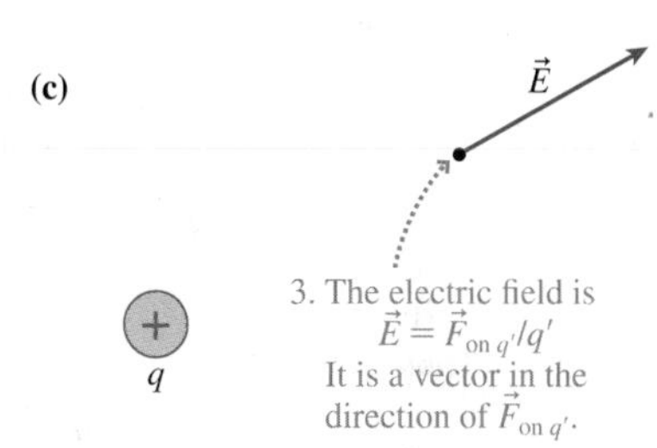

EXAMPLE 20.6 Finding the electric field of a proton

The electron in a hydrogen atom orbits the proton at a radius of 0.053 nm. What is the electric field due to the proton at the position of the electron?

SOLVE The proton's charge is $q = e$. At the distance of the electron, the magnitude of the field is

$$E = \frac{Ke}{r^2} = \frac{(9.0 \times 10^9 \text{ N} \cdot \text{m}^2/\text{C}^2)(1.60 \times 10^{-19} \text{ C})}{(5.3 \times 10^{-11} \text{ m})^2}$$

$$= 5.1 \times 10^{11} \text{ N/C}$$

Because the proton is positive, the electric field is directed away from the proton:

$$\vec{E} = (5.1 \times 10^{11} \text{ N/C, outward from the proton})$$

ASSESS This is a large field, but Table 20.2 shows that this is the correct magnitude for the field within an atom.

By drawing electric field vectors at a number of points around a positive point charge, we can construct an **electric field diagram** such as the one shown in **FIGURE 20.25a**. Notice that the field vectors all point straight away from charge q. We can draw a field diagram for a negative point charge in a similar fashion, as in **FIGURE 20.25b**. In this case, the field vectors point toward the charge, as this would be the direction of the force on a positive probe charge.

In the coming sections, as we use electric field diagrams, keep these points in mind:

1. The diagram is just a representative sample of electric field vectors. The field exists at all the other points. A well-drawn diagram gives a good indication of what the field would be like at a neighboring point.
2. The arrow indicates the direction and the strength of the electric field *at the point to which it is attached*—at the point where the *tail* of the vector is placed. The length of any vector is significant only relative to the lengths of other vectors.
3. Although we have to draw a vector across the page, from one point to another, an electric field vector does not "stretch" from one point to another. Each vector represents the electric field at *one point* in space.

FIGURE 20.25 The electric field near a point charge.

(a) The electric field diagram of a positive point charge

The field vectors point away from a positive charge.

(b) The electric field diagram of a negative point charge

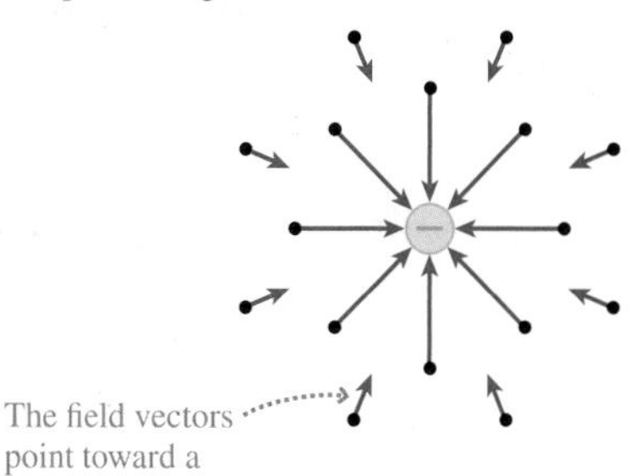

STOP TO THINK 20.4 Rank in order, from largest to smallest, the electric field strengths E_A to E_D at points A to D.

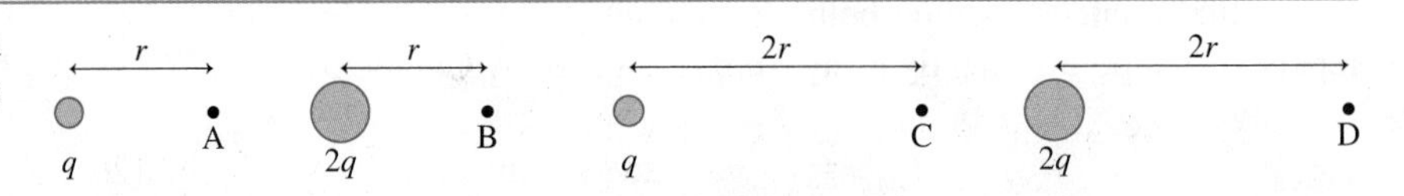

r
r
2r
2r
q
A
2q
B
q
C
2q
D

20.5 Applications of the Electric Field

Suppose we want to find the electric field due to more than one source charge. No matter what the number of source charges, the electric field at a point in space can be found by looking at the force on a probe charge. Because the net force on the probe charge is the vector sum of the forces due to all of the individual charges, **the electric field due to multiple charges is the vector sum of the electric field due to each of the charges.**

EXAMPLE 20.7 Finding the field near a dipole

A dipole consists of a positive and negative charge separated by 1.2 cm, as shown in FIGURE 20.26. What is the electric field strength along the line connecting the charges at a point 1.2 cm to the right of the positive charge?

FIGURE 20.26 Charges and distances for a dipole.

−1.5 nC 1.5 nC

1.2 cm 1.2 cm

PREPARE We define the x-axis to be along the line connecting the two charges, as in FIGURE 20.27. The dipole has no net charge, but it does have a net electric field. The point at which we calculate the field is 1.2 cm from the positive charge and 2.4 cm from the negative charge. Thus the electric field of the positive charge will be larger, as shown in Figure 20.27. The net electric field of the dipole is the vector sum of these two fields, so the electric field of the dipole at this point is in the positive x-direction.

FIGURE 20.27 Visual overview for finding the electric field.

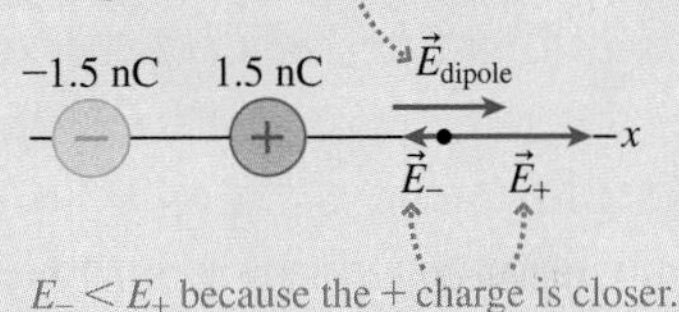

The dipole electric field at this point is in the positive x-direction.
−1.5 nC
1.5 nC
$\vec{E}_{dipole}$
x
$\vec{E}_-$
$\vec{E}_+$
$E_- < E_+$ because the + charge is closer.

SOLVE The magnitudes of the fields of the two charges are given by Equation 20.6, so the magnitude of the dipole field is

$$E_{dipole} = E_+ - E_-$$

$$= \frac{\left(9.0 \times 10^9 \dfrac{\text{N} \cdot \text{m}^2}{\text{C}^2}\right)(1.5 \times 10^{-9}\ \text{C})}{(0.012\ \text{m})^2} - \frac{\left(9.0 \times 10^9 \dfrac{\text{N} \cdot \text{m}^2}{\text{C}^2}\right)(1.5 \times 10^{-9}\ \text{C})}{(0.024\ \text{m})^2}$$

$$= 7.0 \times 10^4\ \text{N/C}$$

ASSESS Table 20.2 lists the fields due to objects charged by rubbing as typically 10^3 to 10^6 N/C, and we've already seen that charges caused by rubbing are in the range of 1–10 nC. Our answer is in this range and thus is reasonable.

FIGURE 20.28 The electric field of a dipole.

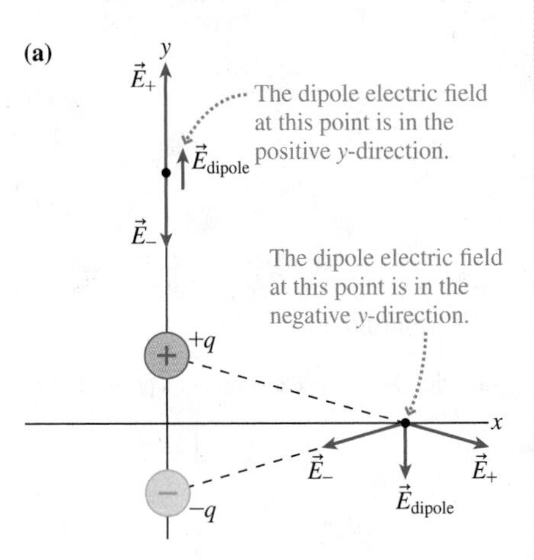

(a)
y
$\vec{E}_+$
The dipole electric field at this point is in the positive y-direction.
$\vec{E}_{dipole}$
$\vec{E}_-$
The dipole electric field at this point is in the negative y-direction.
+q
x
$\vec{E}_-$
$\vec{E}_+$
$\vec{E}_{dipole}$
−q

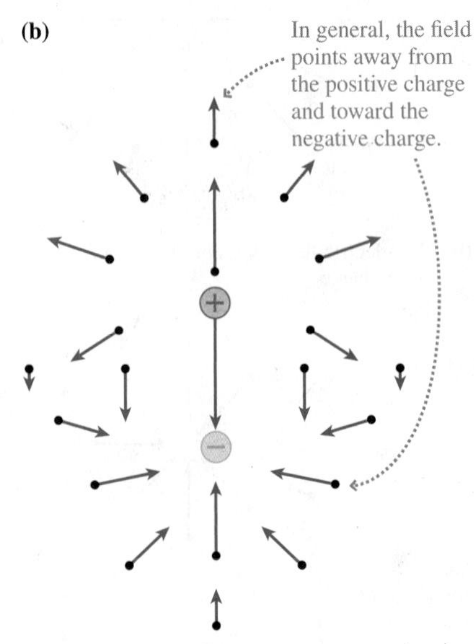

(b)
In general, the field points away from the positive charge and toward the negative charge.

The electric dipole is an important charge distribution that we will see many times, so it's worth exploring the full field diagram. FIGURE 20.28 shows a dipole oriented along the y-axis. We can determine the field at any point by a vector addition of the fields of the two charges, as shown in Figure 20.28a. If we repeat this process at many points, we end up with the field diagram of Figure 20.28b. This is more complex than the field of a single charge, but it accurately shows how two charges alter the space around them.

Uniform Electric Fields

FIGURE 20.29a shows another important practical situation, one we'll meet many times. Two conducting plates, called **electrodes,** are face-to-face with a narrow gap between them. One electrode has total charge $+Q$ and the other has total charge $-Q$. This arrangement of two electrodes, closely spaced and charged equally but oppositely, is called a **parallel-plate capacitor.** What is the electric field between the two plates? To keep things simple, we will focus on the field in the central region, far from the edges, as **FIGURE 20.29b** shows in a blown-up cross-section view.

At any point, the electric field is the vector sum of the fields from all of the positive charges and all of the negative charges on the plates. However, the field of a point charge decreases inversely with the square of its distance, so in practice only the nearby charges contribute to the field. As **FIGURE 20.30a** shows, the horizontal components of the individual fields cancel, while the vertical components add to give an electric field vector pointing from the positive plate toward the negative plate.

By mapping the electric field at many points, we find that the field inside a parallel-plate capacitor is the same—in both strength and direction—at every point. This is called a **uniform electric field.** **FIGURE 20.30b** shows that a uniform electric field is represented with parallel electric field vectors of equal length. A more detailed analysis finds that the electric field inside a parallel-plate capacitor is

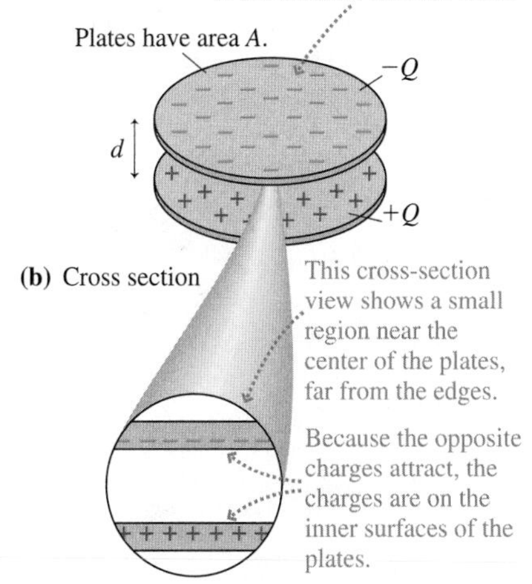
(a) Parallel-plate capacitor
The plates are wide compared to the distance between them.
Plates have area A.
$-Q$
d
$+Q$
(b) Cross section
This cross-section view shows a small region near the center of the plates, far from the edges.
Because the opposite charges attract, the charges are on the inner surfaces of the plates.

FIGURE 20.29 A parallel-plate capacitor.

$$\vec{E}_{\text{capacitor}} = \left(\frac{Q}{\epsilon_0 A}, \text{ from positive to negative}\right) \tag{20.7}$$

Electric field in a parallel-plate capacitor, plate area A and charge Q

E / Q PROPORTIONAL p. 34

Equation 20.7 introduces a new constant ϵ_0, pronounced "epsilon zero" or "epsilon naught," called the **permittivity constant.** Its value is related to the electrostatic constant as

$$\epsilon_0 = \frac{1}{4\pi K} = 8.85 \times 10^{-12}\ \text{C}^2/\text{N} \cdot \text{m}^2$$

There are a few things to note about the field in a parallel-plate capacitor:

- The field depends on the charge-to-area ratio Q/A, which is often called the *charge density.* If the charges are packed more closely, the field will be larger.
- Our analysis requires that the separation of the plates be small compared to their size. If this is true, the spacing between the plates does not affect the electric field, *and this spacing does not appear in Equation 20.7.*
- Although Figure 20.29 shows circular electrodes, the shape of the electrodes—circular or square or any other shape—is not relevant as long as the electrodes are very close together.

NOTE ▸ The charges on the plates are equal and opposite, $+Q$ and $-Q$, so the net charge is zero. The symbol Q in Equation 20.7 is the *magnitude* of the charge on each plate. ◂

FIGURE 20.30 The electric field inside a parallel-plate capacitor.

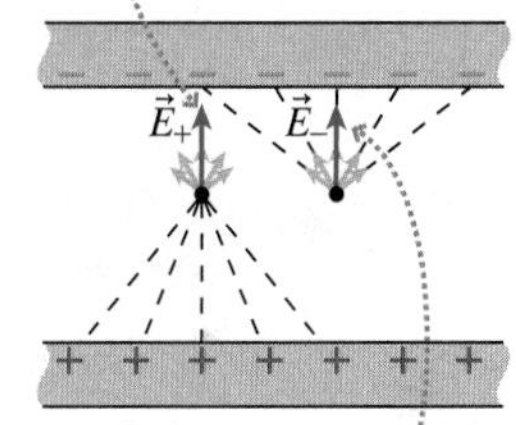
(a) The vector sum of the fields from the positive charges is directed from the positive plate to the negative . . .
$\vec{E}_+$
$\vec{E}_-$
. . . as is the vector sum of the fields from the negative charges.

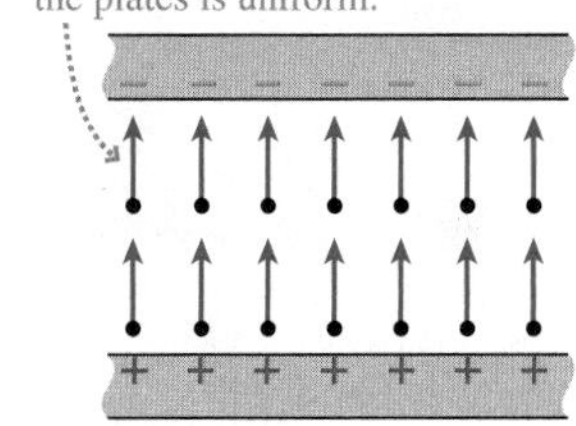
(b) The electric field between the plates is uniform.

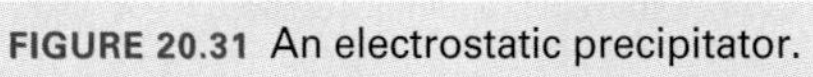

EXAMPLE 20.8 Finding the field in an air cleaner

Long highway tunnels must have air cleaners to remove dust and soot coming from passing cars and trucks. In one type, known as an *electrostatic precipitator,* air passes between two oppositely charged metal plates, as in **FIGURE 20.31**. The large electric field between the plates ionizes dust and soot particles, which then feel a force due to the field. This force causes the charged particles to move toward and stick to one or the other plate, removing them from the air. A typical unit has dimensions and charges as shown in Figure 20.31. What is the electric field between the plates?

FIGURE 20.31 An electrostatic precipitator.

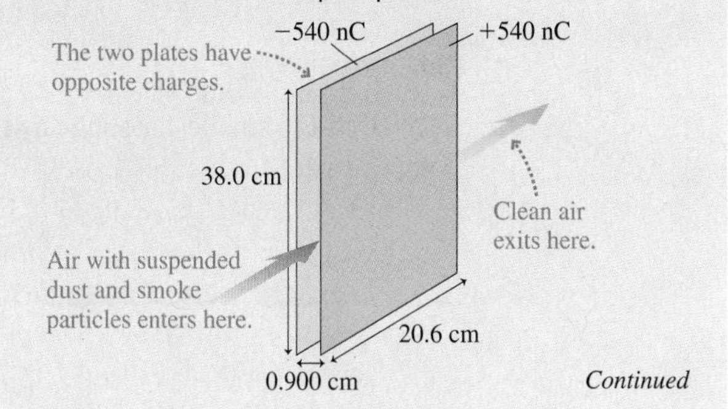
−540 nC
+540 nC
The two plates have opposite charges.
38.0 cm
Clean air exits here.
Air with suspended dust and smoke particles enters here.
20.6 cm
0.900 cm

Continued

PREPARE Because the spacing between the plates is much smaller than their size, this is a parallel-plate capacitor with a uniform electric field between the plates.

SOLVE We find the field using Equation 20.7. The direction is from the positive to the negative plate, which is to the left. The area of the plates is $A = (0.206\text{ m})(0.380\text{ m}) = 0.0783\text{ m}^2$, so the field strength between the plates is

$$E = \frac{Q}{\epsilon_0 A} = \frac{540 \times 10^{-9}\text{ C}}{(8.85 \times 10^{-12}\text{ C}^2/\text{N}\cdot\text{m}^2)(0.0783\text{ m}^2)}$$
$$= 7.79 \times 10^5\text{ N/C}$$

The question asked for the electric field, a vector, not just for the field strength. The electric field between the plates is

$$\vec{E} = (7.79 \times 10^5\text{ N/C, to the left})$$

ASSESS Table 20.2 shows that a field of 10^6 N/C will create a spark in air. The field we calculated between the plates is just a bit smaller than this, which makes sense. The field should be large, but not large enough to make a spark jump between the plates!

Electric Field Lines

We can't see the electric field, so we use pictorial tools like electric field diagrams to help us visualize the electric field in a region of space. Another way to picture the field is to draw **electric field lines.** These are imaginary lines drawn through a region of space so that

- The tangent to a field line at any point is in the direction of the electric field $\vec{E}$ at that point, and
- The field lines are closer together where the electric field strength is greater.

FIGURE 20.32a shows the relationship between electric field lines and electric field vectors in one region of space. If we know what the field vectors look like, we can extrapolate to the field lines, as in FIGURES 20.32b and 20.32c for the electric field lines near a positive charge and between the plates of a capacitor.

FIGURE 20.32 Field vectors and field lines.

(a) Relationship between field vectors and field lines

The electric field vector is tangent to the electric field line.

Field vector

Field line

The electric field is stronger where the electric field vectors are longer and where the electric field lines are closer together.

(b) Field lines of a positive point charge

The field is directed away from the positive charge, so the field lines are directed radially outward.

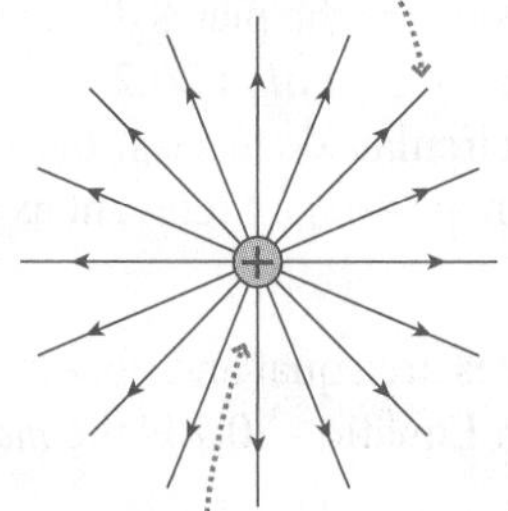

The field lines are closest together near the charge, where the field strength is greatest.

(c) Field lines in an ideal capacitor

The field vectors are directed from the positive to the negative plate, so the field lines are as well.

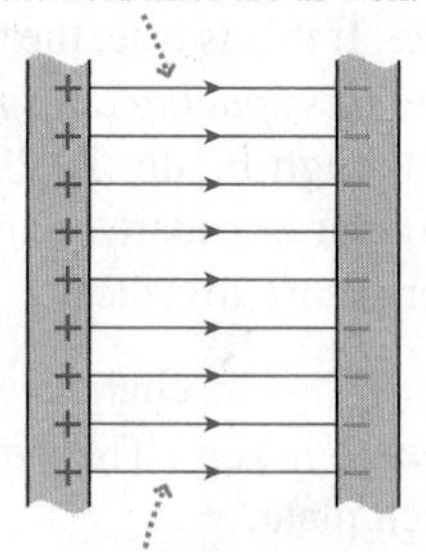

The field is constant, so the field lines are evenly spaced.

STOP TO THINK 20.5 Using what you learned in Figure 20.32, rank in order, from largest to smallest, noting any ties, the electric field strengths E_1 to E_4 at points 1 to 4.

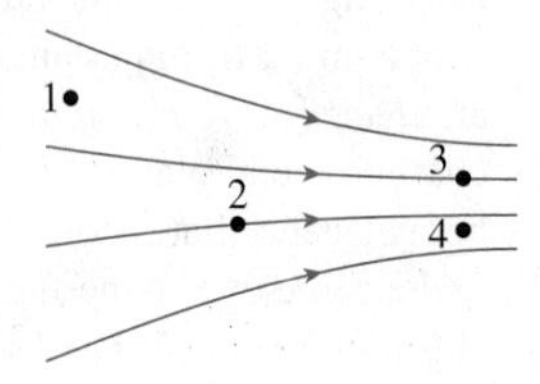

If we have an arrangement of charges, we can draw field lines as a guide to what the field looks like. As you generate a field line picture, there are two rules to keep in mind:

- Field lines cannot cross. The tangent to the field line is the electric field vector, which indicates the direction of the force on a positive charge. The force must be in a unique, well-defined direction, so two field lines cannot cross.
- The electric field is created by charges. Field lines start on a positive charge and end on a negative charge.

You can use the above information as the basis of a technique for sketching a field-line picture for an arrangement of charges. Draw field lines starting on positive charges and moving toward negative charges. Draw the lines tangent to the field vector at each point. Make the lines close together where the field is strong, far apart where the field is weak. For example, **FIGURE 20.33** pictures the electric field of a dipole using electric field lines. You should compare this to Figure 20.28b, which illustrated the field with field vectors.

FIGURE 20.33 Electric field lines for a dipole.

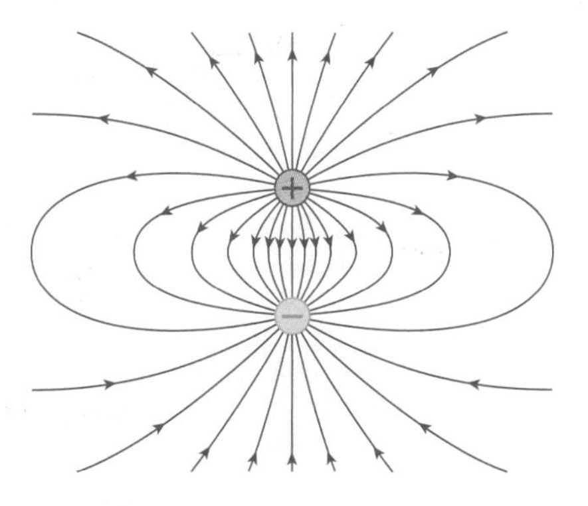

SYNTHESIS 20.1 Two key electric fields

Two important examples of the electric field are those for a point charge and for a parallel-plate capacitor.

The electric field due to a point charge

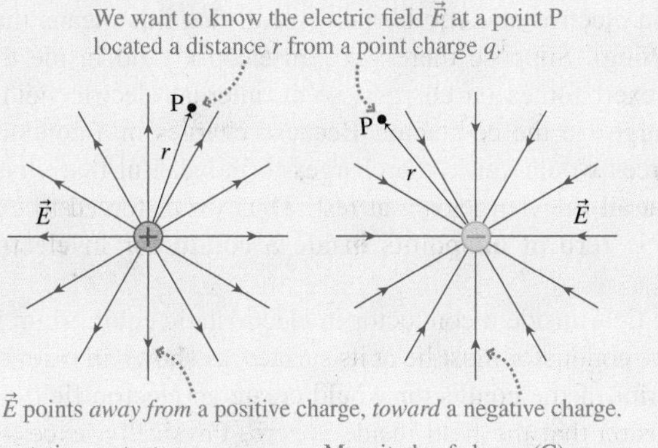

We want to know the electric field $\vec{E}$ at a point P located a distance r from a point charge q.
P
r
$\vec{E}$
$\vec{E}$ points *away from* a positive charge, *toward* a negative charge.

$$E = \frac{K|q|}{r^2}$$

Electrostatic constant: K. Magnitude of charge (C): $|q|$. Magnitude of electric field (N/C): E. Distance from charge to point P (m): r.

$$K = 9.0 \times 10^9\ \text{N}\cdot\text{m}^2/\text{C}^2$$

The electric field inside a parallel-plate capacitor

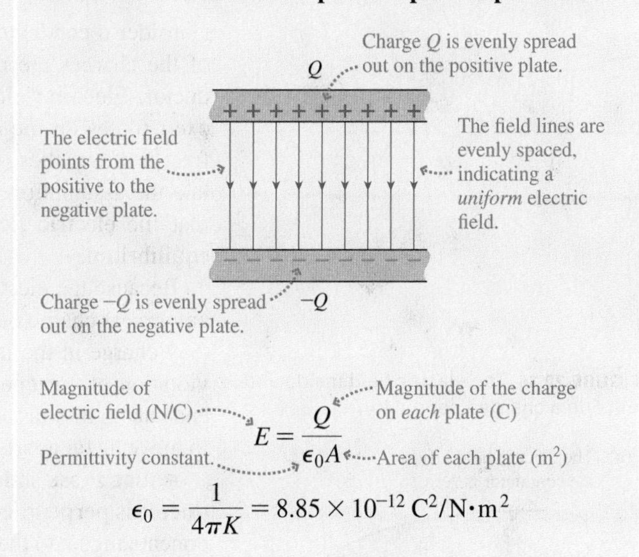

$$E = \frac{Q}{\epsilon_0 A}$$

Magnitude of electric field (N/C): E. Magnitude of the charge on *each* plate (C): Q. Permittivity constant: ϵ_0. Area of each plate (m^2): A.

$$\epsilon_0 = \frac{1}{4\pi K} = 8.85 \times 10^{-12}\ \text{C}^2/\text{N}\cdot\text{m}^2$$

The Electric Field of the Heart BIO

Nerve and muscle cells have a prominent electrical nature. As we will see in detail in Chapter 23, a cell membrane is an insulator that encloses a conducting fluid and is surrounded by conducting fluid. While resting, the membrane is *polarized* with positive charges on the outside of the cell, negative charges on the inside. When a nerve or a muscle cell is stimulated, the polarity of the membrane switches; we say that the cell *depolarizes*. Later, when the charge balance is restored, we say that the cell *repolarizes*.

All nerve and muscle cells generate an electrical signal when depolarization occurs, but the largest electrical signal in the body comes from the heart. The rhythmic beating of the heart is produced by a highly coordinated wave of depolarization that sweeps across the tissue of the heart. As **FIGURE 20.34a** on the next page shows, the surface of the heart is positive on one side of the boundary between tissue that is depolarized and tissue that is not yet depolarized, negative on the other. In other words, the

FIGURE 20.34 BIO The beating heart generates a dipole electric field.

(a) The electric dipole of the heart

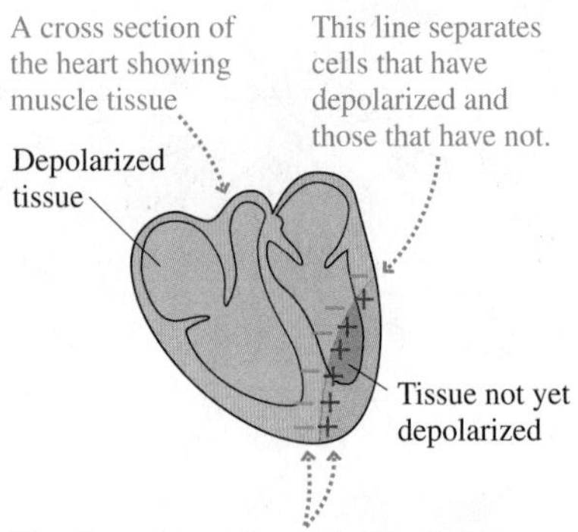
A cross section of the heart showing muscle tissue
This line separates cells that have depolarized and those that have not.
Depolarized tissue
Tissue not yet depolarized
The charge separation at the line between the two regions creates an electric dipole.

(b) The field of the heart in the body

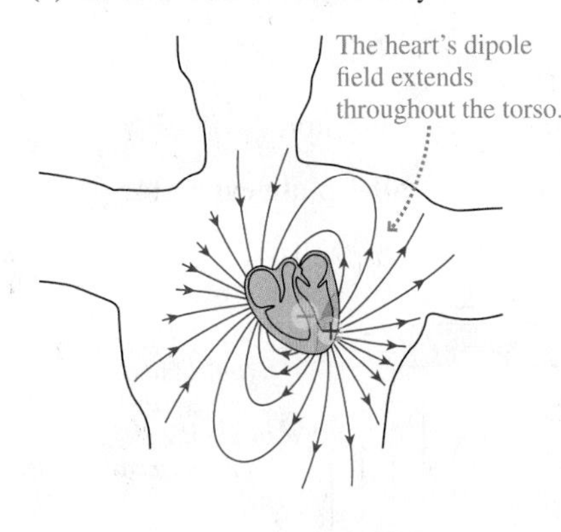
The heart's dipole field extends throughout the torso.

heart is a large electric dipole. The orientation and strength of the dipole change during each beat of the heart as the depolarization wave sweeps across it.

The electric dipole of the heart generates a dipole electric field that extends throughout the torso, as shown in FIGURE 20.34b. As we will see in Chapter 21, an *electrocardiogram* measures the changing electric field of the heart as it beats. Measurement of the heart's electric field can be used to diagnose the operation of the heart.

STOP TO THINK 20.6 Which of the following is the correct representation of the electric field created by two positive charges?

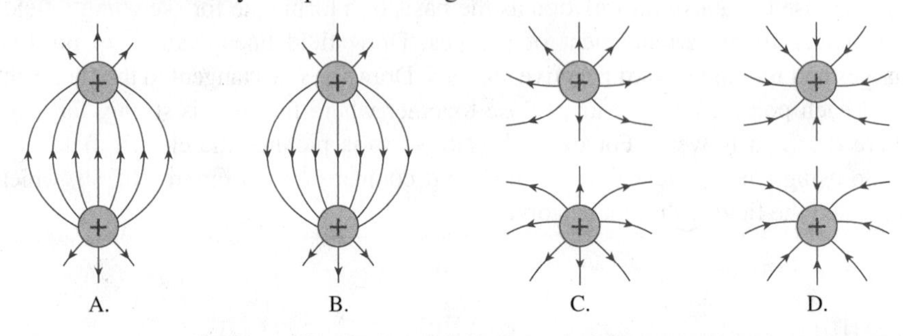
A.
B.
C.
D.

20.6 Conductors and Electric Fields

Consider a conductor in electrostatic equilibrium (recall that this means that none of the charges are moving). Suppose there were an electric field inside the conductor. Electric fields exert forces on charges, so an internal electric field would exert forces on the charges in the conductor. Because charges in a conductor are free to move, these forces would cause the charges to move. But that would violate the assumption that all the charges are at rest. Thus we're forced to conclude that **the electric field is zero at all points inside a conductor in electrostatic equilibrium.**

Because the electric field inside a conductor in electrostatic equilibrium is zero, any *excess* charge on the conductor must lie at its surface, as shown in FIGURE 20.35a. Any charge in the interior of the conductor would create an electric field there, in violation of our conclusion that the field inside is zero. Physically, excess charge ends up on the surface because the repulsive forces between like charges cause them to move as far apart as possible without leaving the conductor.

FIGURE 20.35b shows that **the electric field right at the surface of a charged conductor is perpendicular to the surface.** To see that this is so, suppose $\vec{E}$ had a component tangent to the surface. This component of $\vec{E}$ would exert a force on charges at the surface and cause them to move along the surface, thus violating the assumption that all charges are at rest. The only exterior electric field consistent with electrostatic equilibrium is one that is perpendicular to the surface.

FIGURE 20.35 The electric field inside and outside a charged conductor.

(a) The electric field inside the conductor is zero.

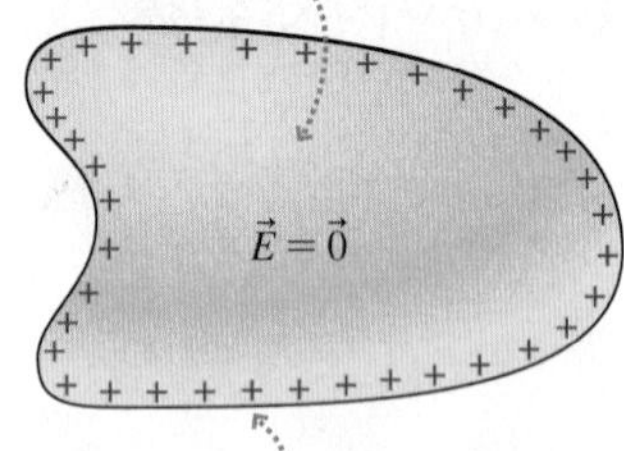
$\vec{E} = \vec{0}$
All excess charge is on the surface.

(b) The electric field at the surface is perpendicular to the surface.

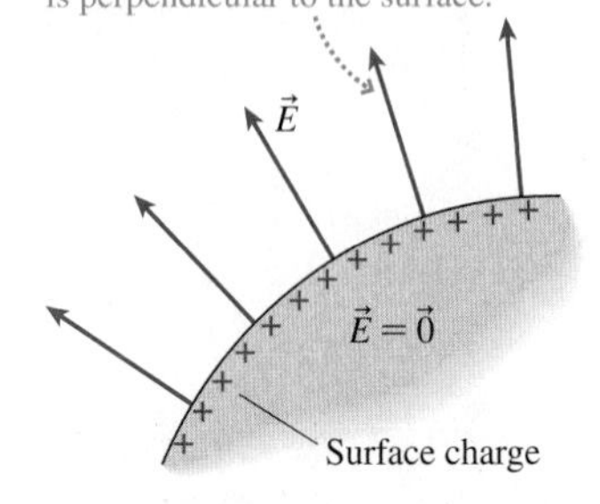
$\vec{E}$
$\vec{E} = \vec{0}$
Surface charge

CONCEPTUAL EXAMPLE 20.9 Drawing electric field lines for a charged sphere and a plate

FIGURE 20.36 shows a positively charged metal sphere above a conducting plate with a negative charge. Sketch the electric field lines.

REASON Field lines start on positive charges and end on negative charges. Thus we draw the field lines from the positive sphere to the negative plate, perpendicular to both surfaces, as shown in FIGURE 20.37. The single field line that goes upward tells us that there is a field above the sphere, but that it is weak.

FIGURE 20.36 The charged sphere and plate.

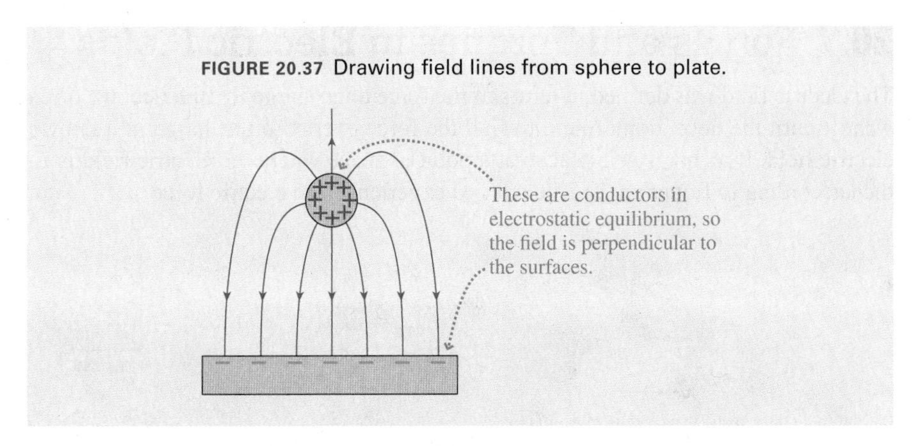

These are conductors in electrostatic equilibrium, so the field is perpendicular to the surfaces.

FIGURE 20.37 Drawing field lines from sphere to plate.

FIGURE 20.38 shows a practical use of these ideas. Here we see a charged conductor with a completely enclosed void. The excess charge on the conductor is at the surface and the electric field within the conductor is zero, so there's nothing that could create an electric field within the enclosure. We can conclude that **the electric field within a conducting enclosure is zero.**

A conducting box can be used to exclude electric fields from a region of space; this is called **screening**. Solid metal walls are ideal, but in practice wire screen or wire mesh provides sufficient screening for all but the most sensitive applications.

FIGURE 20.38 A region of space enclosed by conducting walls is screened from electric fields.

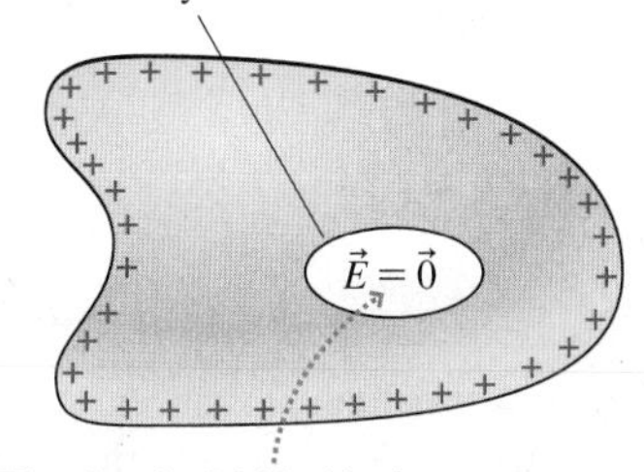

CONCEPTUAL EXAMPLE 20.10 Analyzing static protection

Computer chips and other electronic components are very sensitive to electric charges and fields. Even a small static charge or field may damage them. Such components are shipped and stored in conducting bags. How do these bags protect the components stored inside?

REASON Such a bag, when sealed, forms a conducting shell around its interior. All excess charge is on the surface of the bag, and the electric field inside is zero. A chip or component inside the bag is protected from damaging charges and fields.

Although any excess charge on a conductor will be found on the surface, it may not be uniformly distributed. **FIGURE 20.39** shows a charged conductor that is more pointed at one end than the other. It turns out that the density of charge is highest—and thus the electric field is strongest—at the pointed end.

The sharper the point, the stronger the field. The electric field near very sharp points may be strong enough to ionize the air around it. Lightning rods on buildings have such a point at the top. If charge begins to accumulate on the building, meaning a lightning strike might be imminent, a large field develops at the tip of the rod. Once the field ionizes the air, excess charge from the building can dissipate into the air, reducing the electric field and thus reducing the probability of a lightning strike. A lightning rod is intended to *prevent* a lightning strike.

FIGURE 20.39 The electric field is strongest at the pointed end.

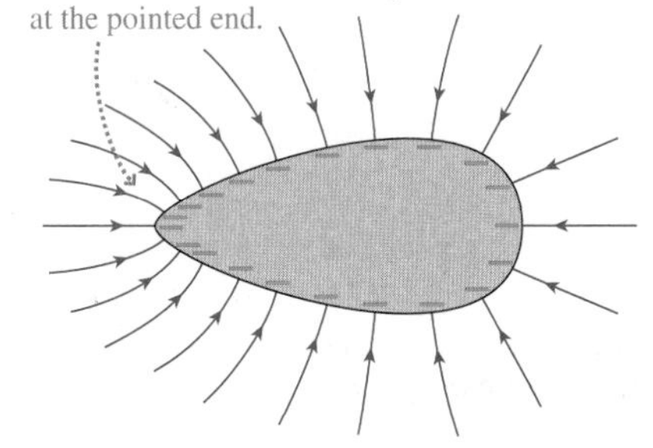

▶ **Electrolocation** BIO Many fish have stacks of specially adapted cells called *electrocytes* that develop electric charges across them. The electrocytes in the tail of this elephant nose fish form an electric dipole that produces an electric field in the water around it. The elephant nose has sensors along its body that can detect very small changes in this electric field. A nearby conductor—such as another fish—will alter this field. The elephant nose uses these changes in the field to "see" around it. These fish live in very murky water where vision is of little use, so you can see the advantage of having such an alternative form of perception.

20.7 Forces and Torques in Electric Fields

The electric field was defined in terms of the force on a charge. In practice, we often want to turn the definition around to find the force exerted on a charge in a known electric field. If a charge q is placed at a point in space where the electric field is $\vec{E}$, then according to Equation 20.3 the charge experiences an electric force

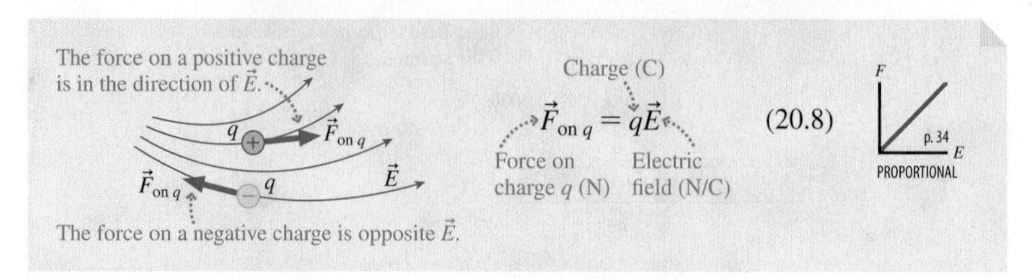
The force on a positive charge is in the direction of $\vec{E}$.
q
$\vec{F}_{\text{on } q}$
$\vec{E}$
$\vec{F}_{\text{on } q}$
q
The force on a negative charge is opposite $\vec{E}$.
Charge (C)
$\vec{F}_{\text{on } q} = q\vec{E}$ (20.8)
Force on charge q (N)
Electric field (N/C)
F
E
p. 34
PROPORTIONAL

If q is positive, the force on charge q is in the direction of $\vec{E}$. The force on a negative charge is *opposite* the direction of $\vec{E}$.

EXAMPLE 20.11 Finding the force on an electron in the atmosphere

Under normal circumstances, the earth's electric field outdoors near ground level is uniform, about 100 N/C, directed down. What is the electric force on a free electron in the atmosphere? What acceleration does this force cause?

PREPARE The electric field is uniform, as shown in the field diagram of **FIGURE 20.40**. Whatever the position of the electron, it experiences the same field. Because the electron is negative, the force on it is opposite the field—upward.

FIGURE 20.40 An electron in the earth's electric field.

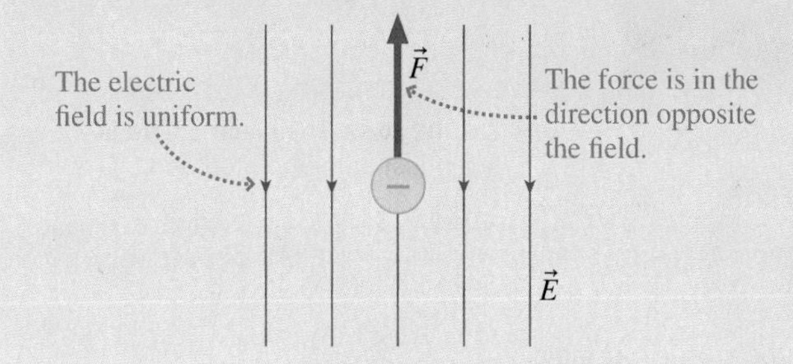
The electric field is uniform.
$\vec{F}$
The force is in the direction opposite the field.
$\vec{E}$

SOLVE The magnitude of the force is given by Equation 20.8:

$$F = eE = (1.6 \times 10^{-19}\text{ C})(100\text{ N/C}) = 1.6 \times 10^{-17}\text{ N}$$

Thus the force on the electron is

$$\vec{F} = (1.6 \times 10^{-17}\text{ N, upward})$$

The electron will accelerate upward, in the direction of the force. The magnitude of the acceleration is

$$a = \frac{F}{m} = \frac{1.6 \times 10^{-17}\text{ N}}{9.1 \times 10^{-31}\text{ kg}} = 1.8 \times 10^{13}\text{ m/s}^2$$

ASSESS This everyday field produces an extremely large acceleration on a free electron. Forces and accelerations at the atomic scale are quite different from what we are used to for macroscopic objects.

FIGURE 20.41 BIO Electrophoresis of DNA samples.

The electric field between the electrodes exerts a force on the negatively charged DNA fragments.

DNA samples begin here.

$\vec{F}$ $\vec{v}$

Different fragments have different sizes and migrate at different rates.

Gel

The photo at the start of the chapter showed the colored lines produced by gel electrophoresis of a sample of DNA. The first step of the analysis is to put the sample of DNA into solution. The DNA is then cut into fragments by enzymes. In solution, these fragments have a negative charge. Drops of solution containing the charged DNA fragments are placed in wells at one end of a container of gel. Electrodes at opposite ends of the gel create an electric field that exerts an electric force on the DNA fragments in the solution, as illustrated in **FIGURE 20.41**. The electric force makes the fragments move through the gel, but drag forces cause fragments of different sizes to migrate at different rates, with smaller fragments migrating faster than larger ones. After some time, the fragments sort themselves into distinct lines, creating a "genetic fingerprint." Two identical samples of DNA will produce the same set of fragments and thus the same pattern in the gel, but the odds are extremely small that two unrelated DNA samples would produce the same pattern.

If an electric dipole is placed in a uniform electric field, as shown in **FIGURE 20.42a**, the electric force on its negative charge is equal in magnitude but opposite in direction to the force on its positive charge. Thus an electric dipole in a uniform electric field experiences *no net force*. However, as you can see from **FIGURE 20.42b**, there is a net *torque* on the dipole that causes it to *rotate*.

FIGURE 20.42 Forces and torques on an electric dipole.

(a) Because the forces on the positive and negative charges are equal in magnitude but oppositely directed, there is no net force on the dipole.

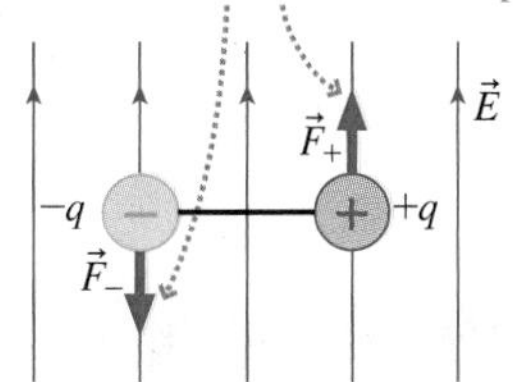

(b) However, there is a net *torque* on the dipole that causes it to *rotate*.

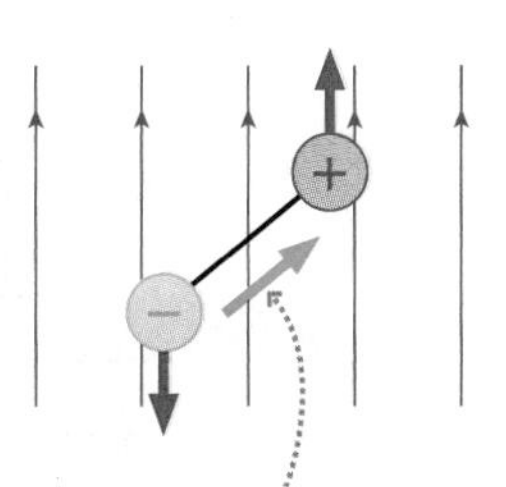

The *dipole moment* is a vector that points from the negative to the positive charge.

(c) When the dipole lines up with the field, the net torque is zero. The dipole is in static equilibrium.

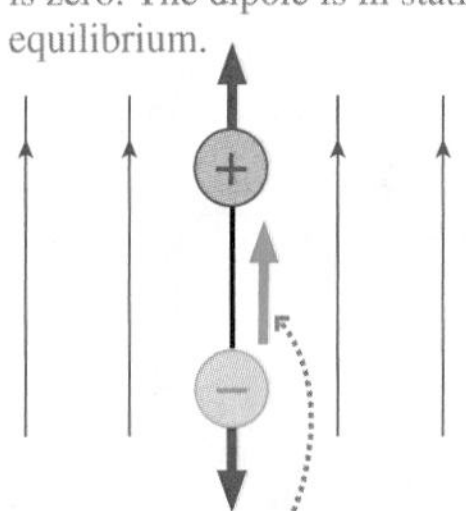

We can say that the dipole moment tries to align itself with the field.

It is useful to define the **electric dipole moment,** a vector pointing from the negative to the positive charge of a dipole. As Figures 20.42b and 20.42c show, an electric dipole in a uniform electric field experiences a torque that causes it to rotate. **The equilibrium position of a dipole in an electric field is with the electric dipole moment aligned with the field.**

Earlier, we saw a photo of grass seeds lined up with the electric field from two charged electrodes. Now we can understand why the seeds line up as they do. First, the electric field polarizes the seeds, inducing opposite charges on their ends. The seeds are induced electric dipoles, with dipole moments along the axis of each seed. Second, torques on the dipole moments cause them to line up with the electric field, revealing its structure.

STOP TO THINK 20.7 Rank in order, from largest to smallest, the forces F_A to F_E a proton would experience if placed at points A to E in this parallel-plate capacitor.

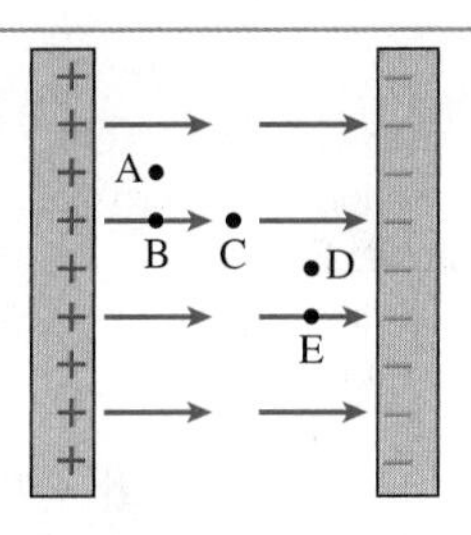

INTEGRATED EXAMPLE 20.12 A cathode-ray tube

Older televisions and computer monitors, as well as some modern electronic instruments, use a *cathode-ray tube*, or CRT, to create an image on a screen. In a CRT, electrons are accelerated by an electric field inside an electron "gun," creating a beam of electrons all moving along in a straight line at the same high speed. A second electric field then steers these electrons to a particular point on a phosphor-coated glass screen, causing the phosphor to glow brightly at that point. By rapidly varying the steering electric field and the intensity of the electron beam, the spot of electrons can be swept over the entire screen, resulting in the familiar glowing picture of a television.

FIGURE 20.43 shows a simplified model of the internal structure of a CRT. Electrons—emitted from a hot filament—start with zero speed at the negative plate of a parallel-plate capacitor. The electric field inside this capacitor accelerates the electrons toward the positive plate, where they exit the capacitor with speed v_1 through a small hole. They then coast along at this speed until they enter the steering electric field of the deflector. This field causes them to follow a curved trajectory, exiting at an angle θ with respect to their original direction.

a. The CRT designer has specified that the electrons must leave the 4.0-cm-wide electron gun with a speed of 6.0×10^7 m/s. What electric field strength is needed inside the electron-gun capacitor?

b. The steering electric field has a constant strength of 1.5×10^5 N/C over the 5.0 cm length of the deflector. By what angle θ are the electrons deflected?

Continued

FIGURE 20.43 The electron gun and electron deflector of a CRT.

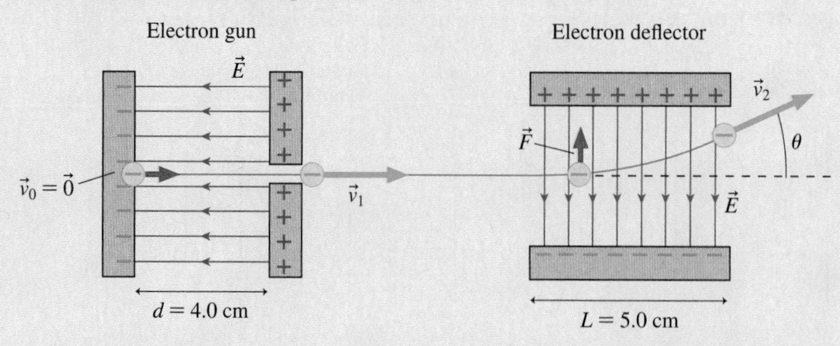

PREPARE We'll use a coordinate system in which the x-axis is horizontal and the y-axis vertical.

a. We can use constant-acceleration kinematics to find the electron's acceleration inside the electron gun. Newton's second law then gives the force on the electron, which we can relate to the electric field using Equation 20.3: $\vec{E} = \vec{F}_{\text{on } q}/q$.

b. Because the electric field is vertically down, the force on a negative electron is vertically up. An electron will accelerate vertically, but not horizontally, so the x-component of its velocity remains unchanged and equal to v_1 as it passes through the deflector. This is exactly analogous to the motion of a projectile, and the electrons follow a projectile-like parabolic trajectory. Just as with projectile motion, we'll use the horizontal motion to find the time interval, then use the time interval to find the final velocity in the y-direction. As **FIGURE 20.44** shows, the ratio of the electron's y- and x-components of velocity can be used to find θ.

FIGURE 20.44 The exit velocity of the electron.

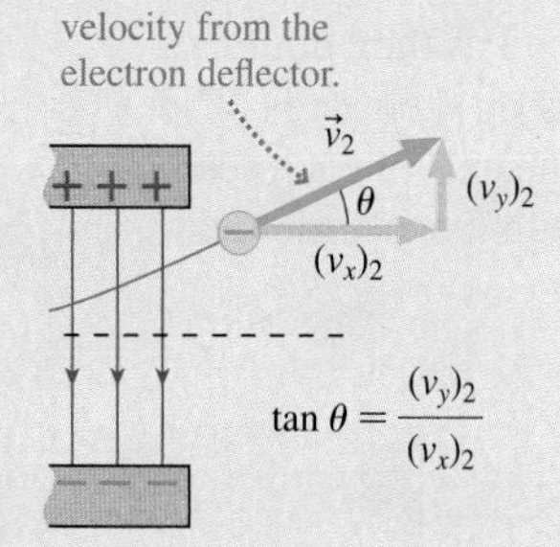

SOLVE a. One of the constant-acceleration kinematic equations from Chapter 2 was $(v_x)_1^2 = (v_x)_0^2 + 2a_x\Delta x$. Using $(v_x)_0 = 0$ and $\Delta x = d = 4.0$ cm, we find that an electron's acceleration inside the electron gun is

$$a_x = \frac{(v_x)_1^2}{2d} = \frac{(6.0\times10^7 \text{ m/s})^2}{2(0.040 \text{ m})} = 4.5\times10^{16} \text{ m/s}^2$$

Newton's second law tells us that the force causing this acceleration is

$$F_x = ma_x = (9.1\times10^{-31} \text{ kg})(4.5\times10^{16} \text{ m/s}^2)$$
$$= 4.1\times10^{-14} \text{ N}$$

Then, by Equation 20.3, the electric field is

$$E_x = \frac{F_x}{q} = \frac{4.1\times10^{-14} \text{ N}}{-1.6\times10^{-19} \text{ C}} = -2.6\times10^5 \text{ N/C}$$

This is a field in the negative x-direction, as we can see in Figure 20.43, with strength 2.6×10^5 N/C.

b. The y-component of the electron's acceleration in the deflector is

$$a_y = \frac{F_y}{m} = \frac{eE_y}{m} = \frac{(-1.6\times10^{-19} \text{ C})(-1.5\times10^5 \text{ N/C})}{9.1\times10^{-31} \text{ kg}}$$
$$= 2.6\times10^{16} \text{ m/s}^2$$

The negative electrons have an upward (positive) acceleration. This acceleration causes an electron to leave the deflector with a y-component of velocity

$$(v_y)_2 = a_y\Delta t$$

where Δt is the time the electron spends in the deflector. Because the x-component of the velocity is constant, this time is simply

$$\Delta t = \frac{L}{(v_x)_1} = \frac{0.050 \text{ m}}{6.0\times10^7 \text{ m/s}} = 8.3\times10^{-10} \text{ s}$$

Thus

$$(v_y)_2 = a_y\Delta t = (2.6\times10^{16} \text{ m/s}^2)(8.3\times10^{-10} \text{ s})$$
$$= 2.2\times10^7 \text{ m/s}$$

Referring to Figure 20.44, and using $(v_x)_2 = (v_x)_1$ because there's no horizontal acceleration, we see that

$$\tan\theta = \frac{(v_y)_2}{(v_x)_2} = \frac{2.2\times10^7 \text{ m/s}}{6.0\times10^7 \text{ m/s}} = 0.37$$

so that

$$\theta = \tan^{-1}(0.37) = 20°$$

ASSESS A strong field accelerates the electrons in the x-direction, while a weaker one accelerates them in the y-direction. Thus it is reasonable that the ratio of the y- to the x-component of velocity is significantly less than 1.

The CRT shown in Figure 20.43 deflects electrons only vertically. A real CRT has a second electron deflector, rotated 90°, to provide a horizontal deflection. The two deflectors working together can scan the electron beam over all points on the screen.

SUMMARY

Goal: To develop a basic understanding of electric phenomena in terms of charges, forces, and fields.

GENERAL PRINCIPLES

Charge

There are two kinds of charges, called **positive** and **negative.**

- Atoms consist of a nucleus containing positively charged protons surrounded by a cloud of negatively charged electrons.
- The **fundamental charge** e is the magnitude of the charge on an electron or proton: $e = 1.60 \times 10^{-19}$ C.
- Matter with equal amounts of positive and negative charge is **neutral.**
- Charge is conserved; it can't be created or destroyed.

Coulomb's Law

The forces between two charged particles q_1 and q_2 separated by distance r are

$$F_{1\,\text{on}\,2} = F_{2\,\text{on}\,1} = \frac{K|q_1||q_2|}{r^2}$$

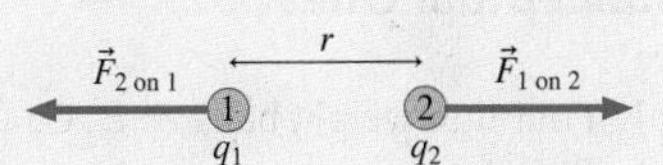

where $K = 8.99 \times 10^9\ \text{N}\cdot\text{m}^2/\text{C}^2$ is the **electrostatic constant.** These forces are an action/reaction pair directed along the line joining the particles.

- The forces are repulsive for two like charges, attractive for two opposite charges.
- The net force on a charge is the vector sum of the forces from all other charges.
- The unit of charge is the coulomb (C).

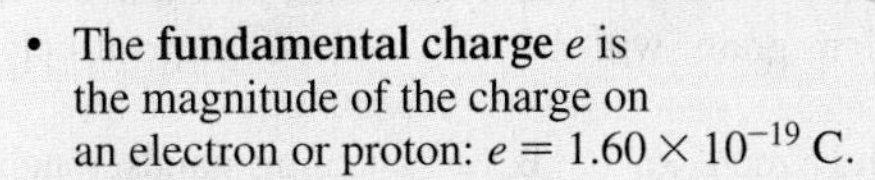

IMPORTANT CONCEPTS

The Electric Field

Charges interact with each other via the electric field $\vec{E}$.

- Charge A alters the space around it by creating an electric field.

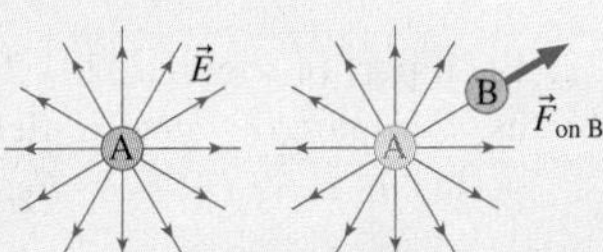

- The field is the agent that exerts a force on charge B. $\vec{F}_{\text{on B}} = q_B\vec{E}$
- An electric field is identified and measured in terms of the force on a probe charge q. The unit of the electric field is N/C. $\vec{E} = \dfrac{\vec{F}_{\text{on}\,q}}{q}$
- The electric field is a vector. The electric field from multiple charges is the vector sum of the fields from the individual charges. $\vec{E}_{\text{total}} = \vec{E}_1 + \vec{E}_2 + \cdots$

Visualizing the electric field

The electric field exists at all points in space.

- An electric field vector shows the field at only one point, the point at the tail of the vector.
- A **field diagram** shows field vectors at several points.

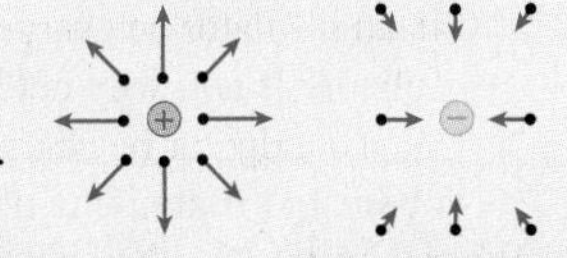

- **Electric field lines:**
  - are always parallel to the field vectors.
  - are close where the field is strong, far apart where the field is weak.
  - go from positive to negative charges.

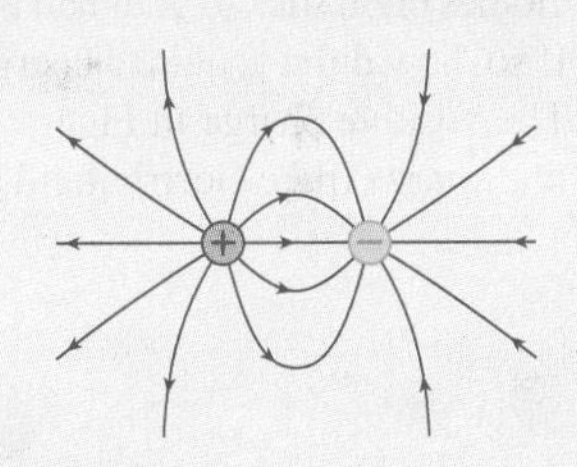

APPLICATIONS

There are two types of material, **insulators** and **conductors.**

- Charge remains fixed on an insulator.
- Charge moves easily through conductors.
- Charge is transferred by contact between objects.

A **dipole** has no net charge, but has a field because the two charges are separated.

A dipole will rotate to align with an electric field.

Electric fields: important cases

The electric field of a **point charge** is

$$\vec{E} = \left(\frac{K|q|}{r^2}, \begin{bmatrix} \text{away from } q \text{ if } q > 0 \\ \text{toward } q \text{ if } q < 0 \end{bmatrix}\right)$$

The electric field inside a **parallel-plate capacitor** is uniform:

$$\vec{E} = \left(\frac{Q}{\epsilon_0 A}, \text{from positive to negative}\right)$$

where $\epsilon_0 = 8.85 \times 10^{-12}\ \text{C}^2/\text{N}\cdot\text{m}^2$ is the **permittivity constant.**

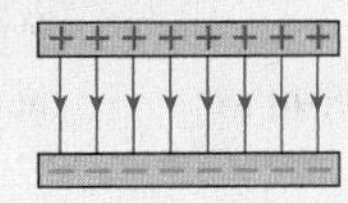

Conductors in electric fields

- The electric field inside a conductor in **electrostatic equilibrium** is zero.
- Any excess charge is on the surface.

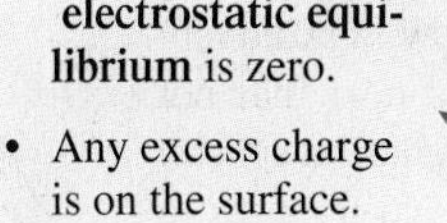

- The electric field is perpendicular to the surface.
- The density of charge and the electric field are highest near a pointed end.

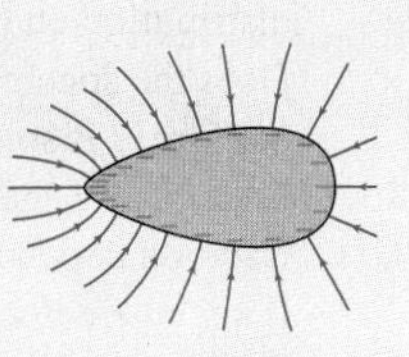

Problem difficulty is labeled as | (straightforward) to ||||| (challenging). Problems labeled INT integrate significant material from earlier chapters; BIO are of biological or medical interest.

For assigned homework and other learning materials, go to MasteringPhysics®

Scan this QR code to launch a Video Tutor Solution that will help you solve problems for this chapter.

QUESTIONS

Conceptual Questions

1. Four lightweight balls A, B, C, and D are suspended by threads. Ball A has been touched by a plastic rod that was rubbed with wool. When the balls are brought close together, without touching, the following observations are made:
   - Balls B, C, and D are attracted to ball A.
   - Balls B and D have no effect on each other.
   - Ball B is attracted to ball C.

   What are the charge states (positive, negative, or neutral) of balls A, B, C, and D? Explain.
2. Plastic and glass rods that have been charged by rubbing with wool and silk, respectively, hang by threads.
   a. An object repels the plastic rod. Can you predict what it will do to the glass rod? If so, what? If not, why not? Explain.
   b. A different object attracts the plastic rod. Can you predict what it will do to the glass rod? If so, what? If not, why not? Explain.
3. a. Can an insulator be charged? If so, how would you charge an insulator? If not, why not?
   b. Can a conductor be charged? If so, how would you charge a conductor? If not, why not?
4. When you take clothes out of the drier right after it stops, the clothes often stick to your hands and arms. Is your body charged? If so, how did it acquire a charge? If not, why does this happen?
5. The positive charge in Figure Q20.5 is $+Q$. What is the negative charge if the electric field at the dot is zero?

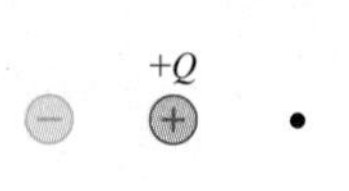

FIGURE Q20.5

FIGURE Q20.6

6. As shown in Figure Q20.6, metal sphere A has 4 units of negative charge and metal sphere B has 2 units of positive charge. The two spheres are brought into contact. What is the final charge state of each sphere? Explain.
7. Figure Q20.7 shows a positively charged rod held near, but not touching, a neutral metal sphere.
   a. Add plusses and minuses to the figure to show the charge distribution on the sphere.
   b. Does the sphere experience a net force? If so, in which direction? Explain.

   Metal

   FIGURE Q20.7
8. A plastic balloon that has been rubbed with wool will stick to a wall.
   a. Can you conclude that the wall is charged? If not, why not? If so, where does the charge come from?
   b. Draw a charge diagram showing how the balloon is held to the wall.
9. You are given two metal spheres on portable insulating stands, a glass rod, and a piece of silk. Explain how to give the spheres *exactly* equal but opposite charges.
10. BIO A honeybee acquires a positive electric charge as it flies through the air. This charge causes pollen grains to be attracted to the bee. Explain, using words and diagrams, how a neutral, conducting pollen grain will be attracted to a positively charged bee.
11. A metal rod A and a metal sphere B, on insulating stands, touch each other as shown in Figure Q20.11. They are originally neutral. A positively charged rod is brought near (but not touching) the far end of A. While the charged rod is still close, A and B are separated. The charged rod is then withdrawn. Is the sphere then positively charged, negatively charged, or neutral? Explain.

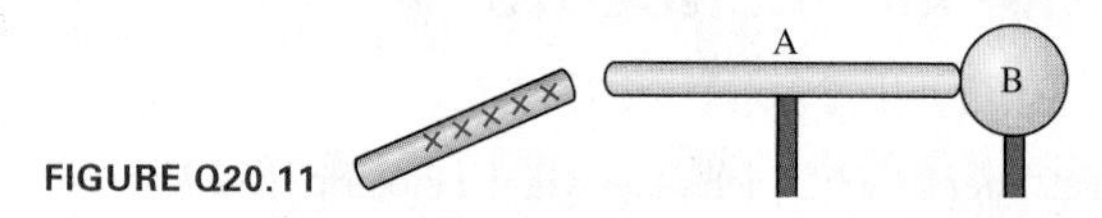

FIGURE Q20.11

12. Each part of Figure Q20.12 shows two points near two charges. Compare the electric field strengths E_1 and E_2 at these two points. Is $E_1 > E_2$, $E_1 = E_2$, or $E_1 < E_2$?

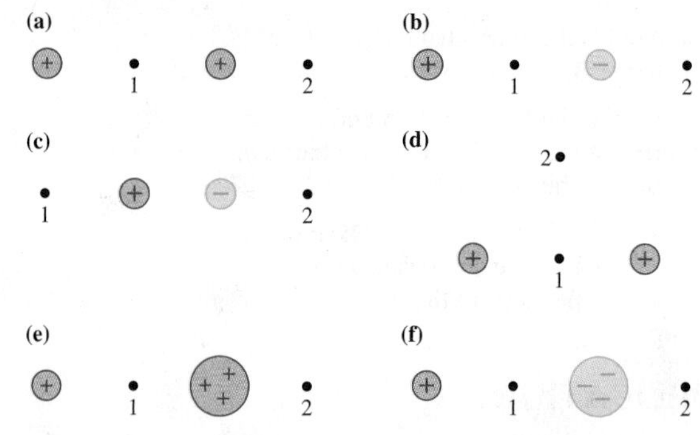

FIGURE Q20.12

13. BIO Iontophoresis is a noninvasive process that transports drugs through the skin without needles. In the photo, the red electrode is positive and the black electrode is negative. The electric field between the electrodes will drive the negatively charged molecules of an anesthetic through the skin. Should the drug be placed at the red or the black electrode? Explain.

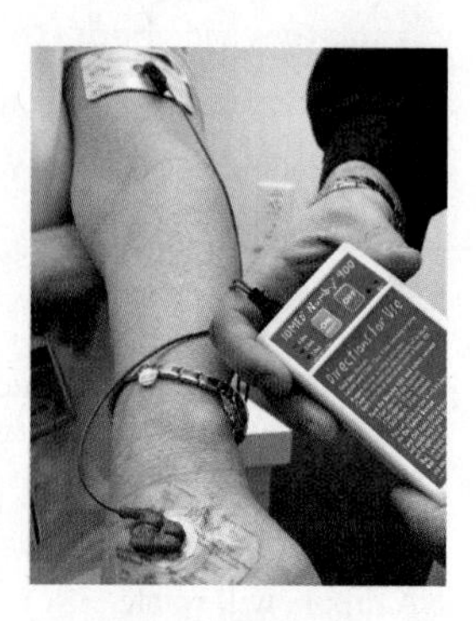

14. A 10 nC charge sits at a point in space where the magnitude of the electric field is 1200 N/C. What will the magnitude of the field be if the 10 nC charge is replaced by a 20 nC charge?

15. A hollow soda straw is uniformly charged, as shown in Figure Q20.15. What is the electric field at the center (inside) of the straw? Explain.

FIGURE Q20.15

16. A positively charged particle is in the center of a parallel-plate capacitor that has charge $\pm Q$ on its plates. Suppose the distance between the plates is doubled, with the charged particle remaining in the center. Does the force on this particle increase, decrease, or stay the same? Explain.
17. Two charged particles are separated by 10 cm. Suppose the charge on each particle is doubled. By what factor does the electric force between the particles change?
18. A small positive charge q experiences a force of magnitude F_1 when placed at point 1 in Figure Q20.18. In terms of F_1:
    a. What is the magnitude of the force on charge q at point 3?
    b. What is the magnitude of the force on a charge $3q$ at point 1?
    c. What is the magnitude of the force on a charge $2q$ at point 2?
    d. What is the magnitude of the force on a charge $-2q$ at point 2?

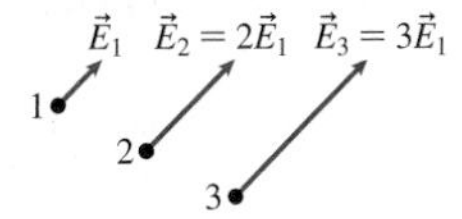

FIGURE Q20.18

19. A typical commercial airplane is struck by lightning about once per year. When this happens, the external metal skin of the airplane might be burned, but the people and equipment inside the aircraft experience no ill effects. Explain why this is so.
20. BIO Microbes such as bacteria have small positive charges when in solution. Public health agencies are exploring a new way to measure the presence of small numbers of microbes in drinking water by using electric forces to concentrate the microbes. Water is sent between the two oppositely charged electrodes of a parallel-plate capacitor. Any microbes in the water will collect on one of the electrodes.
    a. On which electrode will the microbes collect?
    b. How could the microbes be easily removed from the electrodes for analysis?
21. a. Is there a point between a 10 nC charge and a 20 nC charge at which the electric field is zero? If so, which charge is this point closer to? If not, why not?
    b. Repeat part a for the case of a 10 nC charge and a -20 nC charge.

Multiple-Choice Questions

22. | Two lightweight, electrically neutral conducting balls hang from threads. Choose the diagram in Figure Q20.22 that shows how the balls hang after:
    a. Both are touched by a negatively charged rod.
    b. Ball 1 is touched by a negatively charged rod and ball 2 is touched by a positively charged rod.
    c. Both are touched by a negatively charged rod but ball 2 picks up more charge than ball 1.
    d. Only ball 1 is touched by a negatively charged rod.
    Note that parts a through d are independent; these are not actions taken in sequence.

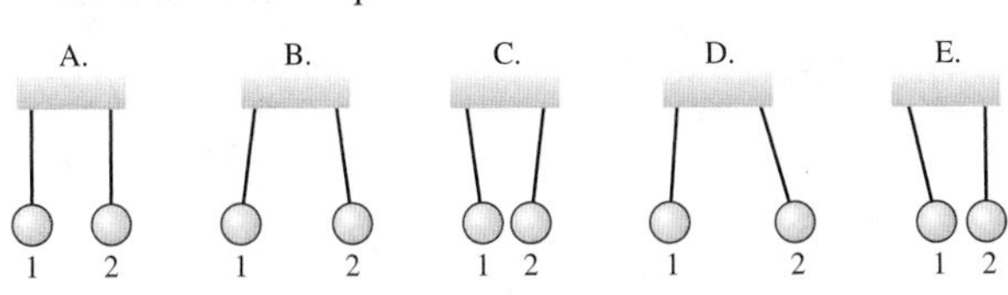

FIGURE Q20.22

23. | All the charges in Figure Q20.23 have the same magnitude. In which case does the electric field at the dot have the largest magnitude?

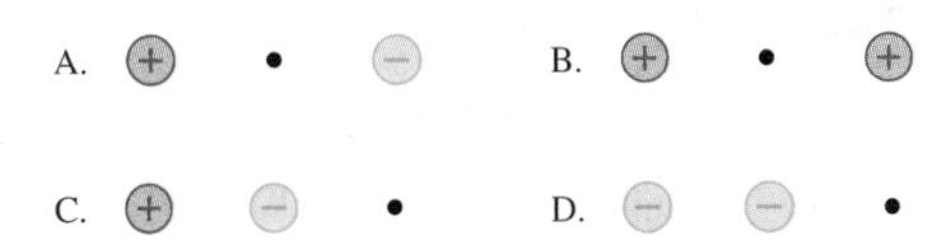

FIGURE Q20.23

24. | All the charges in Figure Q20.24 have the same magnitude. In which case does the electric field at the dot have the largest magnitude?

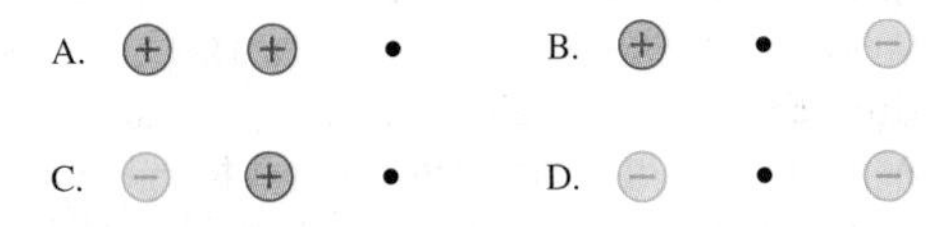

FIGURE Q20.24

25. | All the charges in Figure Q20.25 have the same magnitude. In which case does the electric field at the dot have the largest magnitude?

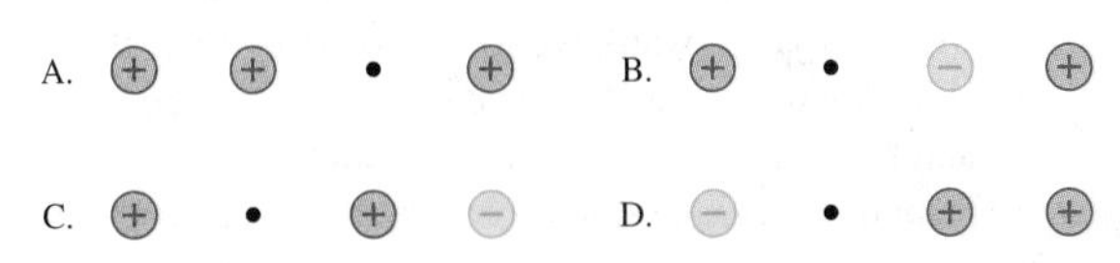

FIGURE Q20.25

26. | A glass bead charged to +3.5 nC exerts an 8.0×10^{-4} N repulsive electric force on a plastic bead 2.9 cm away. What is the charge on the plastic bead?
    A. +2.1 nC
    B. +7.4 nC
    C. +21 nC
    D. +740 nC
27. | A +7.5 nC point charge and a -2.0 nC point charge are 3.0 cm apart. What is the electric field strength at the midpoint between the two charges?
    A. 3.3×10^3 N/C
    B. 5.7×10^3 N/C
    C. 2.2×10^5 N/C
    D. 3.8×10^5 N/C
28. || Three point charges are arranged as shown in Figure Q20.28. Which arrow best represents the direction of the electric field vector at the position of the dot?

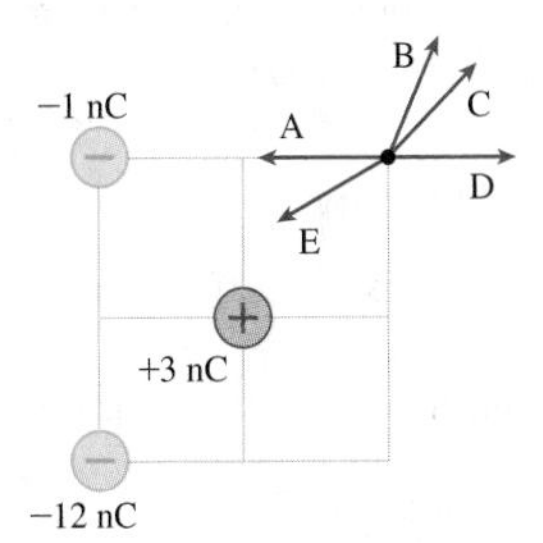

FIGURE Q20.28

29. | A positive charge is brought near to a dipole, as shown in Figure Q20.29. If the dipole is free to rotate, it
    A. Begins to rotate in a clockwise direction.
    B. Begins to rotate in a counterclockwise direction.
    C. Remains stationary.

FIGURE Q20.29

Watch video solution for problem: 20.11

PROBLEMS

Section 20.1 Charges and Forces

Section 20.2 Charges, Atoms, and Molecules

1. ||| A glass rod is charged to +5.0 nC by rubbing.
   a. Have electrons been removed from the rod or protons added? Explain.
   b. How many electrons have been removed or protons added?
2. ||| A plastic rod is charged to −20 nC by rubbing.
   a. Have electrons been added to the rod or protons removed? Explain.
   b. How many electrons have been added or protons removed?
3. ||| Suppose you have 1.0 mol of O_2 gas. How many coulombs of positive charge are contained in the atomic nuclei of this gas?
INT
4. || A plastic rod that has been charged to −15.0 nC touches a metal sphere. Afterward, the rod's charge is −10.0 nC.
   a. What kind of charged particle was transferred between the rod and the sphere, and in which direction? That is, did it move from the rod to the sphere or from the sphere to the rod?
   b. How many charged particles were transferred?
5. || A glass rod that has been charged to +12.0 nC touches a metal sphere. Afterward, the rod's charge is +8.0 nC.
   a. What kind of charged particle was transferred between the rod and the sphere, and in which direction? That is, did it move from the rod to the sphere or from the sphere to the rod?
   b. How many charged particles were transferred?
6. ||| Two identical metal spheres A and B are in contact. Both are initially neutral. 1.0×10^{12} electrons are added to sphere A, then the two spheres are separated. Afterward, what are the charge of A and the charge of B?
7. || Two identical metal spheres A and B are connected by a plastic rod. Both are initially neutral. 1.0×10^{12} electrons are added to sphere A, then the connecting rod is removed. Afterward, what are the charge of A and the charge of B?
8. || If two identical conducting spheres are in contact, any excess charge will be evenly distributed between the two. Three identical metal spheres are labeled A, B, and C. Initially, A has charge q, B has charge $-q/2$, and C is uncharged.
   a. What is the final charge on each sphere if C is touched to B, removed, and then touched to A?
   b. Starting again from the initial conditions, what is the charge on each sphere if C is touched to A, removed, and then touched to B?

Section 20.3 Coulomb's Law

9. || Two 1.0 kg masses are 1.0 m apart on a frictionless table. Each has +1.0 μC of charge.
INT
   a. What is the magnitude of the electric force on one of the masses?
   b. What is the initial acceleration of each mass if they are released and allowed to move?
10. ||| A small metal sphere has a mass of 0.15 g and a charge of −23.0 nC. It is 10.0 cm directly above an identical sphere with the same charge. This lower sphere is fixed and cannot move.
INT
    a. What is the magnitude of the force between the spheres?
    b. If the upper sphere is released, it will begin to fall. What is the magnitude of its initial acceleration?
11. ||| A small plastic sphere with a charge of −5.0 nC is near another small plastic sphere with a charge of −12 nC. If the spheres repel one another with a force of magnitude 8.2×10^{-4} N, what is the distance between the spheres?
12. ||| A small metal bead, labeled A, has a charge of 25 nC. It is touched to metal bead B, initially neutral, so that the two beads share the 25 nC charge, but not necessarily equally. When the two beads are then placed 5.0 cm apart, the force between them is 5.4×10^{-4} N. What are the charges q_A and q_B on the beads?
13. |||| A small glass bead has been charged to +20 nC. A tiny ball bearing 1.0 cm above the bead feels a 0.018 N downward electric force. What is the charge on the ball bearing?
14. | What are the magnitude and direction of the electric force on charge A in Figure P20.14?

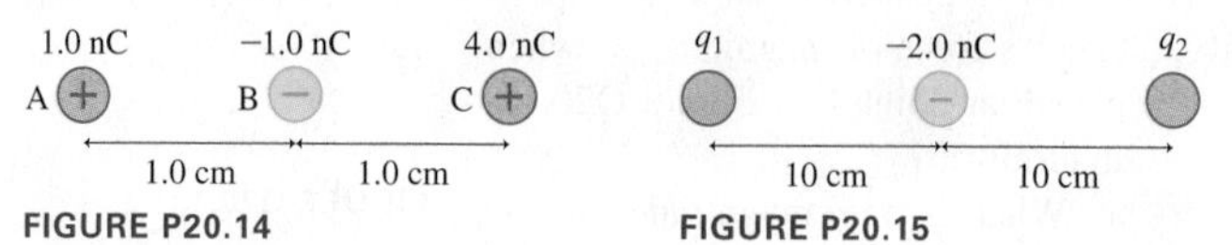

FIGURE P20.14

FIGURE P20.15

15. ||| In Figure P20.15, charge q_2 experiences no net electric force. What is q_1?
16. | Object A, which has been charged to +10 nC, is at the origin. Object B, which has been charged to −20 nC, is at $(x, y) = (0.0\text{ cm}, 2.0\text{ cm})$. What are the magnitude and direction of the electric force on each object?
17. | A small glass bead has been charged to +20 nC. What are the magnitude and direction of the acceleration of (a) a proton and (b) an electron that is 1.0 cm from the center of the bead?
INT

Section 20.4 The Concept of the Electric Field

18. || What magnitude charge creates a 1.0 N/C electric field at a point 1.0 m away?
19. || What are the strength and direction of the electric field 2.0 cm from a small glass bead that has been charged to +6.0 nC?
20. | A 30 nC charge experiences a 0.035 N electric force. What is the magnitude of electric field at the position of this charge?
21. | What are the strength and direction of the electric field 1.0 mm from (a) a proton and (b) an electron?
22. | A +10 nC charge is located at the origin.
    a. What are the strengths of the electric fields at the positions $(x, y) = (5.0\text{ cm}, 0.0\text{ cm})$, $(-5.0\text{ cm}, 5.0\text{ cm})$, and $(-5.0\text{ cm}, -5.0\text{ cm})$?
    b. Draw a field diagram showing the electric field vectors at these points.
23. | A −10 nC charge is located at the origin.
    a. What are the strengths of the electric fields at the positions $(x, y) = (0.0\text{ cm}, 5.0\text{ cm})$, $(-5.0\text{ cm}, -5.0\text{ cm})$, and $(-5.0\text{ cm}, 5.0\text{ cm})$?
    b. Draw a field diagram showing the electric field vectors at these points.
24. ||| What are the strength and direction of the electric field at the position indicated by the dot in Figure P20.24? Specify the direction as an angle above or below horizontal.

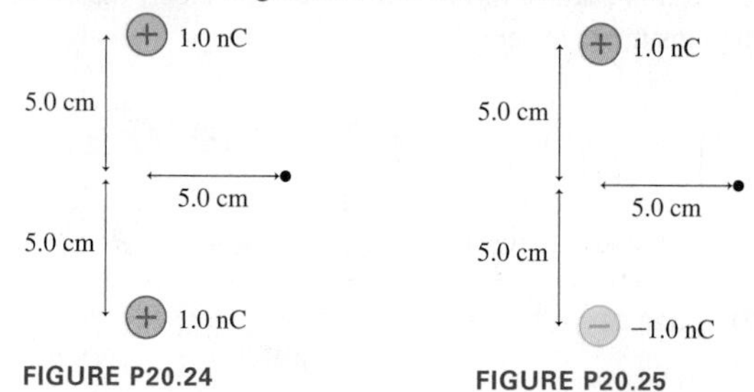

FIGURE P20.24

FIGURE P20.25

25. || What are the strength and direction of the electric field at the position indicated by the dot in Figure P20.25? Specify the direction as an angle above or below horizontal.

Section 20.5 Applications of the Electric Field

26. | What are the strength and direction of an electric field that INT will balance the weight of (a) a proton and (b) an electron?
27. || A 0.10 g plastic bead is charged by the addition of 1.0×10^{10} INT excess electrons. What electric field $\vec{E}$ (strength and direction) will cause the bead to hang suspended in the air?
28. || A parallel-plate capacitor is constructed of two square plates, size $L \times L$, separated by distance d. The plates are given charge $\pm Q$. What is the ratio E_f/E_i of the final electric field strength E_f to the initial electric field strength E_i if:
    a. Q is doubled?
    b. L is doubled?
    c. d is doubled?
29. || A parallel-plate capacitor is formed from two 4.0 cm × 4.0 cm electrodes spaced 2.0 mm apart. The electric field strength inside the capacitor is 1.0×10^6 N/C. What is the charge (in nC) on each electrode?
30. | Two identical closely spaced circular disks form a parallel-plate capacitor. Transferring 1.5×10^9 electrons from one disk to the other causes the electric field strength between them to be 1.0×10^5 N/C. What are the diameters of the disks?

Section 20.6 Conductors and Electric Fields

31. || A parallel-plate capacitor is constructed of two horizontal 12.0-cm-diameter circular plates. A 1.0 g plastic bead, with a charge of -6.0 nC, is suspended between the two plates by the force of the electric field between them.
    a. Which plate, the upper or the lower, is positively charged?
    b. What is the charge on the positive plate?
32. ||| Storm clouds may build up large negative charges near their bottom edges. The earth is a good conductor, so the charge on the cloud attracts an equal and opposite charge on the earth under the cloud. The electric field strength near the earth depends on the shape of the earth's surface, as we can explain with a simple model. The top metal plate in Figure P20.32 has uniformly distributed negative charge. The bottom metal plate, which has a high point, has an equal and opposite charge that is free to move.
    a. Sketch the two plates and the region between them, showing the distribution of positive charge on the bottom plate.
    b. Complete your diagram by sketching electric field lines between the two plates. Be sure to note the direction of the field. Where is the field strongest?
    c. Explain why it is more dangerous to be on top of a hill or mountain during a lightning storm than on level ground.

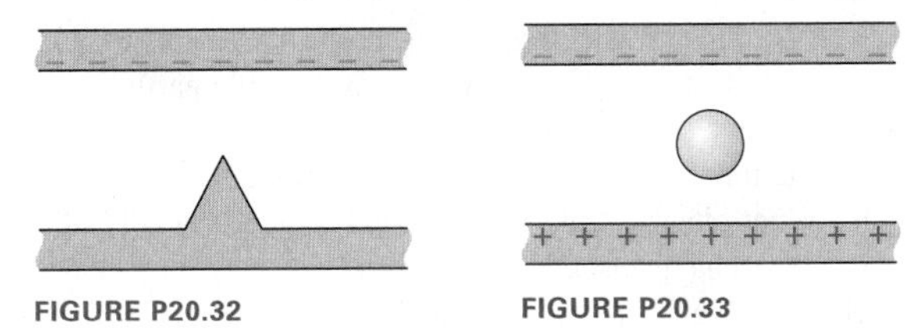

FIGURE P20.32 FIGURE P20.33

33. ||| A neutral conducting sphere is between two parallel charged plates, as shown in Figure P20.33. Sketch the electric field lines in the region between the plates. Be sure to include the effect of the conducting sphere.

Section 20.7 Forces and Torques in Electric Fields

34. || One kind of e-book display consists of millions of very small INT spheres that float in a thin fluid layer between two conducting, transparent plates. Each sphere is black on one side and white on the other, and possesses an electric dipole moment directed from the white side to the black. When an electric field is applied between the plates, the spheres rotate so that their dipole moment lines up with the field. Depending on the field's direction, either the black or the white sides of the spheres can be made visible.

    The dipole moment of the spheres can be modeled as two opposite charges of magnitude 3.5×10^{-15} C, separated by a distance of 100 μm. What is the maximum possible torque on a sphere if the electric field between the transparent plates is 4.0×10^5 N/C?
35. || A protein molecule in an electrophoresis gel has a negative BIO charge. The exact charge depends on the pH of the solution, but 30 excess electrons is typical. What is the magnitude of the electric force on a protein with this charge in a 1500 N/C electric field?
36. || Large electric fields in cell membranes cause ions to move BIO through the cell wall, as we will explore in Chapter 23. The field strength in a typical membrane is 1.0×10^7 N/C. What is the magnitude of the force on a calcium ion with charge $+e$?
37. ||| Molecules of carbon monoxide are permanent electric INT dipoles due to unequal sharing of electrons between the carbon and oxygen atoms. Figure P20.37 shows the distance and charges. Suppose a carbon monoxide molecule with a horizontal axis is in a vertical electric field of strength 15,000 N/C.
    a. What is the magnitude of the net force on the molecule?
    b. What is the magnitude of the torque on the molecule?

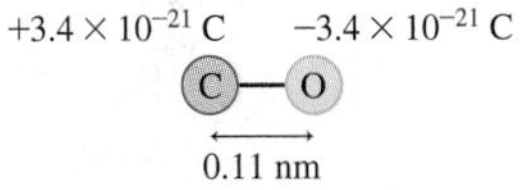

$+3.4 \times 10^{-21}$ C -3.4×10^{-21} C
C O
0.11 nm

FIGURE P20.37

General Problems

38. |||| A 2.0-mm-diameter copper ball is charged to +50 nC. What INT fraction of its electrons have been removed? The density of copper is 8900 kg/m^3.
39. | Pennies today are copper-covered zinc, but older pennies are INT 3.1 g of solid copper. What are the total positive charge and total negative charge in a solid copper penny that is electrically neutral?
40. || Two protons are 2.0 fm apart. (1 fm = 1 femtometer = INT 1×10^{-15} m.)
    a. What is the magnitude of the electric force on one proton due to the other proton?
    b. What is the magnitude of the gravitational force on one proton due to the other proton?
    c. What is the ratio of the electric force to the gravitational force?
41. |||| The nucleus of a ^{125}Xe atom (an isotope of the element xenon INT with mass 125 u) is 6.0 fm in diameter. It has 54 protons and charge $q = +54e$. (1 fm = 1 femtometer = 1×10^{-15} m.)
    a. What is the electric force on a proton 2.0 fm from the surface of the nucleus?
    b. What is the proton's acceleration?

    Hint: Treat the spherical nucleus as a point charge.

Watch video solutions for problems: 20.45  20.57 

42. ||| Two equally charged, 1.00 g spheres are placed with 2.00 cm
INT between their centers. When released, each begins to accelerate at 225 m/s^2. What is the magnitude of the charge on each sphere?

43. || Objects A and B are both positively charged. Both have a mass of 100 g, but A has twice the charge of B. When A and B are placed with 10 cm between their centers, B experiences an electric force of 0.45 N.
   a. How large is the force on A?
   b. What are the charges q_A and q_B?

44. |||| An electric dipole is formed from ± 1.0 nC point charges spaced 2.0 mm apart. The dipole is centered at the origin, oriented along the y-axis. What is the electric field strength at the points (a) $(x, y) = (10 \text{ mm}, 0 \text{ mm})$ and (b) $(x, y) = (0 \text{ mm}, 10 \text{ mm})$?

45. ||| What are the strength and direction of the electric field at the position indicated by the dot in Figure P20.45? Specify the direction as an angle above or below horizontal.

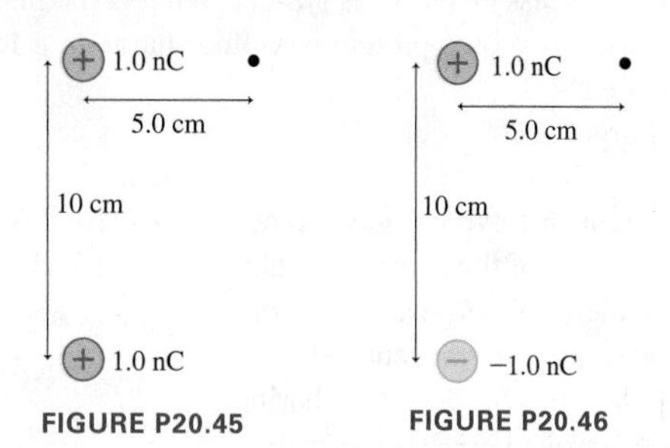

FIGURE P20.45 FIGURE P20.46

46. |||| What are the strength and direction of the electric field at the position indicated by the dot in Figure P20.46? Specify the direction as an angle above or below horizontal.

47. || What is the force on the 1.0 nC charge in Figure P20.47? Give your answer as a magnitude and a direction.

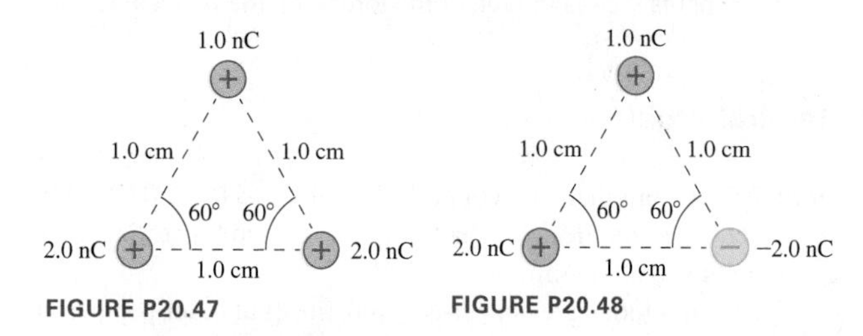

FIGURE P20.47 FIGURE P20.48

48. || What is the force on the 1.0 nC charge in Figure P20.48? Give your answer as a magnitude and a direction.

49. | What is the magnitude of the force on the 1.0 nC charge in the middle of Figure P20.49 due to the four other charges?

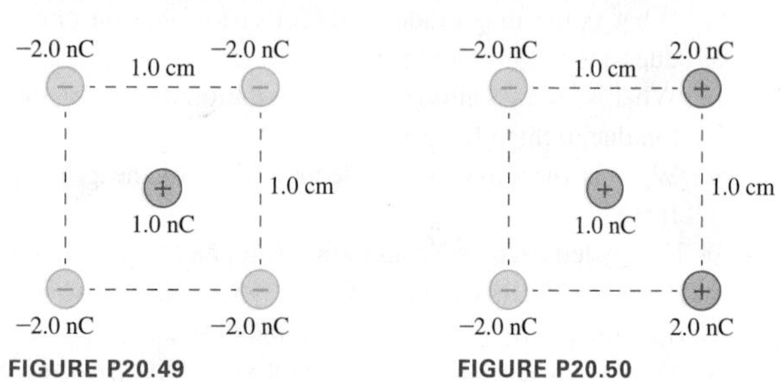

FIGURE P20.49 FIGURE P20.50

50. ||| What are the magnitude and direction of the force on the 1.0 nC charge in the middle of Figure P20.50 due to the four other charges?

51. || What are the magnitude and direction of the force on the 1.0 nC charge at the bottom of Figure P20.51?

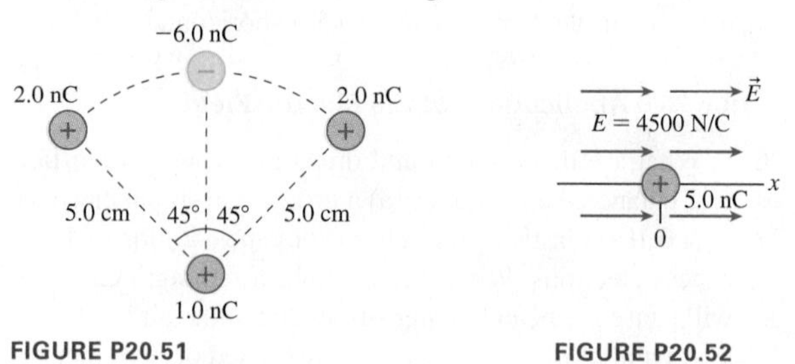

FIGURE P20.51 FIGURE P20.52

52. ||| As shown in Figure P20.52, a 5.0 nC charge sits at $x = 0$ in a uniform 4500 N/C electric field directed to the right. At what point along the x-axis would (a) a proton and (b) an electron experience no net force?

53. || The net force on the 1.0 nC charge in Figure P20.53 is zero. What is q?

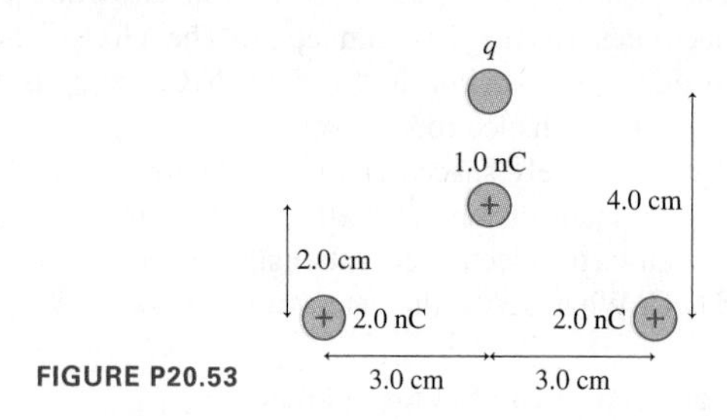

FIGURE P20.53

54. || Two particles have positive charges q and Q. A third charged particle is placed halfway between them. What must this particle's charge be so that the net force on charge Q is zero?

55. |||| Figure P20.55 shows four charges at the corners of a square of side L. What magnitude and sign of charge Q will make the force on charge q zero?

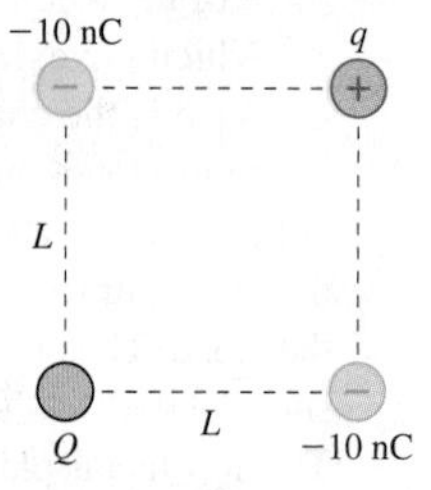

FIGURE P20.55

56. ||| Model a pollen grain as a sphere
INT of carbon 0.10 mm in diameter. If two such grains, spaced 1.0 m apart, were somehow stripped of all their electrons, leaving only the protons, what would be the force between them? Compare this force to the weight of an aircraft carrier, which is about 9×10^8 N. The density of carbon is 2300 kg/m^3.

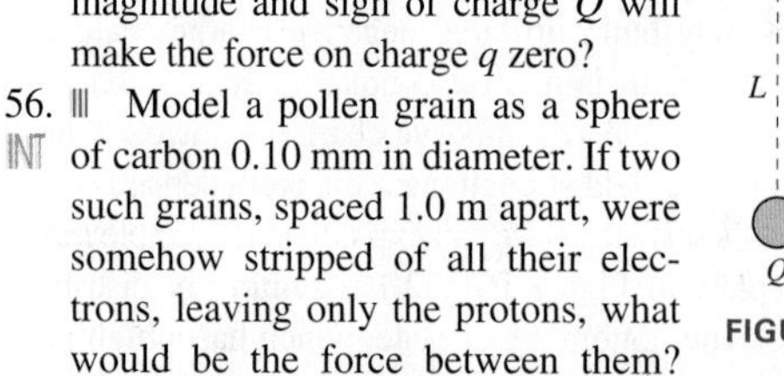

57. ||| In a simple model of the hydrogen atom, the electron
INT moves in a circular orbit of radius 0.053 nm around a stationary proton. How many revolutions per second does the electron make?
Hint: What must be true for a force that causes circular motion?

58. ||| A 0.10 g honeybee acquires a charge of +23 pC while flying.
BIO
   a. The electric field near the surface of the earth is typically 100 N/C, directed downward. What is the ratio of the electric force on the bee to the bee's weight?
   b. What electric field strength and direction would allow the bee to hang suspended in the air?

59. |||| A +10 nC charge is located at $(x, y) = (0 \text{ cm}, 10 \text{ cm})$ and a −5.0 nC charge is located at $(x, y) = (5.0 \text{ cm}, 0 \text{ cm})$. Where would a −10 nC charge need to be located in order that the electric field at the origin be zero?

60. ||| Two 2.0-cm-diameter disks face each other, 1.0 mm apart. INT They are charged to ± 10 nC.
   a. What is the electric field strength between the disks?
   b. A proton is shot from the negative disk toward the positive disk. What launch speed must the proton have to just barely reach the positive disk?
61. |||| The electron gun in a television tube uses a uniform electric INT field to accelerate electrons from rest to 5.0×10^7 m/s in a distance of 1.2 cm. What is the electric field strength?
62. |||| A 0.020 g plastic bead hangs from a INT lightweight thread. Another bead is fixed in position beneath the point where the thread is tied. If both beads have charge q, the moveable bead swings out to the position shown in Figure P20.62. What is q?

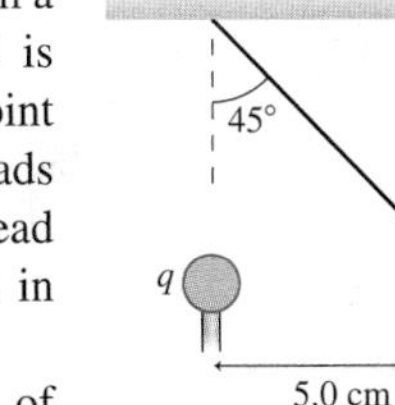

FIGURE P20.62

63. ||| A 4.0 mg bead with a charge of INT 2.5 nC rests on a table. A second bead, with a charge of -5.6 nC, is directly above the first bead and is slowly lowered toward it. What is the closest the centers of the two beads can be brought together before the lower bead is lifted off the table?
64. ||||| Two 3.0 g spheres on 1.0-m-long threads repel each other INT after being equally charged, as shown in Figure P20.64. What is the charge q?

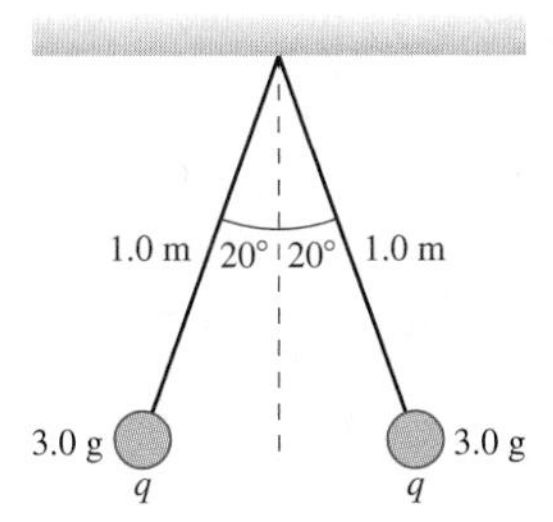

FIGURE P20.64

65. ||||| An electric field $\vec{E} = (100{,}000$ N/C, right) causes the 5.0 g INT ball in Figure P20.65 to hang at a 20° angle. What is the charge on the ball?

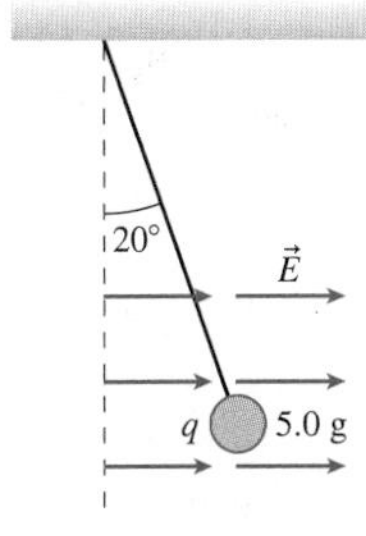

FIGURE P20.65

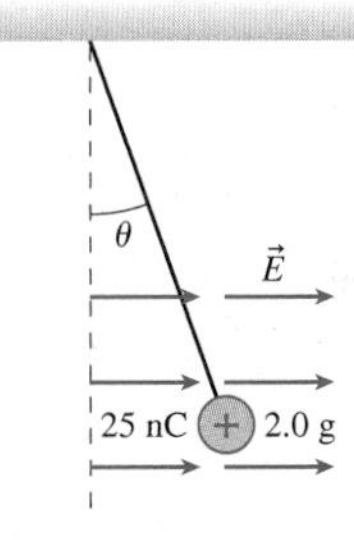

FIGURE P20.66

66. |||| An electric field $\vec{E} = (200{,}000$ N/C, right) causes the 2.0 g INT ball in Figure P20.66 to hang at an angle. What is θ?
67. || A small charged bead has a mass of 1.0 g. It is held in a INT uniform electric field $\vec{E} = (200{,}000$ N/C, up). When the bead is released, it accelerates upward with an acceleration of 20 m/s^2. What is the charge on the bead?
68. || A bead with a mass of 0.050 g and a charge of 15 nC is free INT to slide on a vertical rod. At the base of the rod is a fixed 10 nC charge. In equilibrium, at what height above the fixed charge does the bead rest?
69. || A small bead with a positive charge q is free to slide on a INT horizontal wire of length 4.0 cm. At the left end of the wire is a fixed charge q, and at the right end is a fixed charge $4q$. How far from the left end of the wire does the bead come to rest?
70. ||| A parallel-plate capacitor consists of two plates, each with an INT area of 28 cm^2, separated by 3.0 mm. The charge on the capacitor is 8.3 nC. A proton is released from rest next to the positive plate. How long does it take for the proton to reach the negative plate?

MCAT-Style Passage Problems

Flow Cytometry BIO

Flow cytometry, illustrated in Figure P20.71, is a technique used to sort cells by type. The cells are placed in a conducting saline solution which is then forced from a nozzle. The stream breaks up into small droplets, each containing one cell. A metal collar surrounds the stream right at the point where the droplets separate from the stream. Charging the collar polarizes the conducting liquid, causing the droplets to become charged as they break off from the stream. A laser beam probes the solution just upstream from the charging collar, looking for the presence of certain types of cells. All droplets containing one particular type of cell are given the same charge by the charging collar. Droplets with other desired types of cells receive a different charge, and droplets with no desired cell receive no charge. The charged droplets then pass between two parallel charged electrodes where they receive a horizontal force that directs them into different collection tubes, depending on their charge.

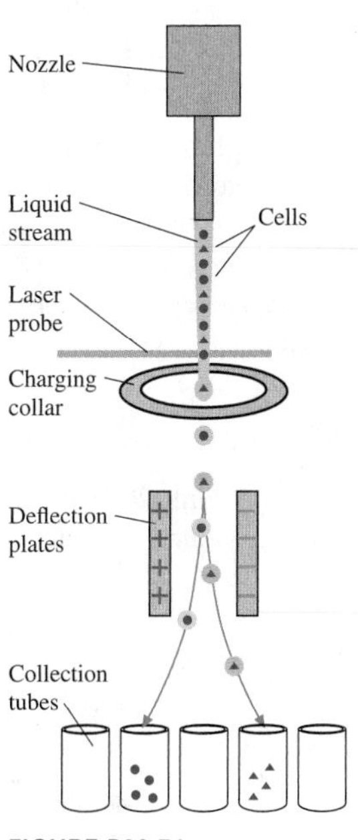

FIGURE P20.71

71. | If the charging collar has a positive charge, the net charge on a droplet separating from the stream will be
   A. Positive.
   B. Negative.
   C. Neutral.
   D. The charge will depend on the type of cell.
72. | Which of the following describes the charges on the droplets that end up in the five tubes, moving from left to right?
   A. $+2q, +q, 0, -q, -2q$
   B. $+q, +2q, 0, -2q, -q$
   C. $-q, -2q, 0, +2q, +q$
   D. $-2q, -q, 0, +q, +2q$
73. | Because the droplets are conductors, a droplet's positive and negative charges will separate while the droplet is in the region between the deflection plates. Suppose a neutral droplet passes between the plates. The droplet's dipole moment will point
   A. Up. B. Down. C. Left. D. Right.

74. I Another way to sort the droplets would be to give each droplet the same charge, then vary the electric field between the deflection plates. For the apparatus as sketched, this technique will not work because
   A. Several droplets are between the plates at one time, and they would all feel the same force.
   B. The cells in the solution have net charges that would affect the droplet charge.
   C. A droplet with a net charge would always experience a net force between the plates.
   D. The droplets would all repel each other, and this force would dominate the deflecting force.

STOP TO THINK ANSWERS

Chapter Preview Stop to Think: A. The x-component of the tension is directed to the left, so it is negative. This rules out answers B, D, and E. Also, the x-component of the tension is related to the cosine of the 30° angle, which, according to the question, is greater than the sine of 30°. Thus answer A, where the magnitude of the x-component is greater than that of the y-component, must be correct.

Stop to Think 20.1: A. The electroscope is originally given a positive charge. The charge spreads out, and the leaves repel each other. When a rod with a negative charge is brought near, some of the positive charge is attracted to the top of the electroscope, away from the leaves. There is less charge on the leaves, and so they move closer together.

Stop to Think 20.2: $q_E(+3e) > q_A(+1e) > q_D(0) > q_B(-1e) > q_C(-2e)$.

Stop to Think 20.3: B. The two forces are an action/reaction pair, opposite in direction but *equal* in magnitude.

Stop to Think 20.4: $E_B > E_A > E_D > E_C$. The field is proportional to the charge, and inversely proportional to the square of the distance.

Stop to Think 20.5: $E_3 = E_4 > E_2 > E_1$. The electric field is strongest where the electric field lines are closest together. At points 3 and 4 the spacing is the same, so the electric field strengths at these two points are equal.

Stop to Think 20.6: C. Electric field lines *start* on positive charges. Very near to each of the positive charges, the field lines should look like the field lines of a single positive charge.

Stop to Think 20.7: $F_A = F_B = F_C = F_D = F_E$. The field inside a capacitor is the same at all points. Because the field is uniform, the force on the proton will be the same at all points. The electric field exists at all points whether or not a vector is shown at that point.

21 Electric Potential

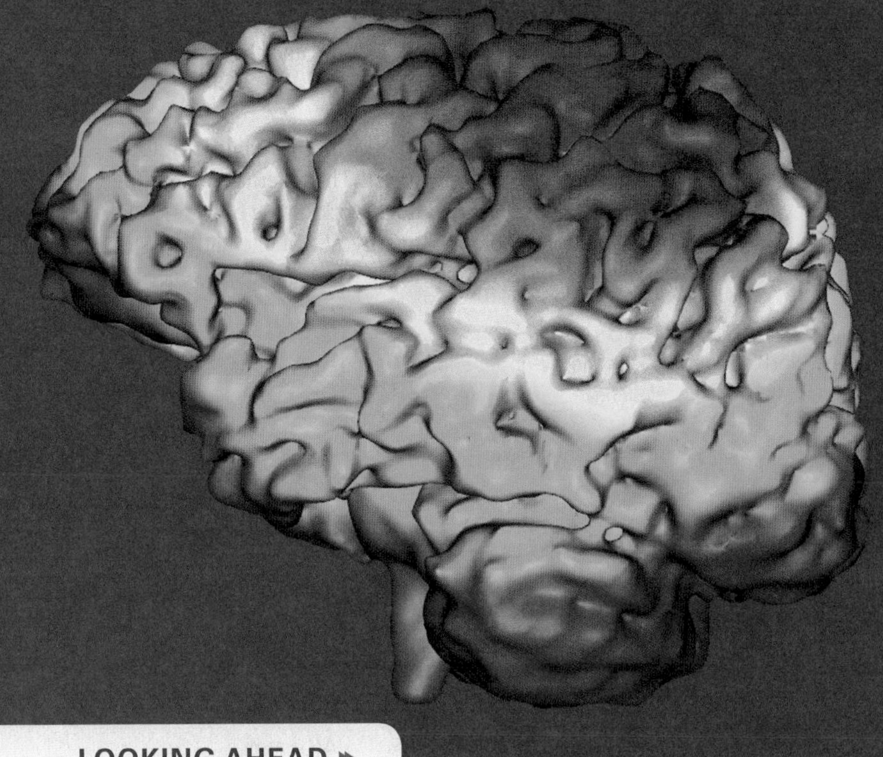

The colors on this patient's brain are a map of the electric potential on the brain's surface after the patient was given a particular sensory stimulus. How does visualizing the electric potential help us understand the electric properties of the brain?

LOOKING AHEAD ▸▸

Goal: To calculate and use the electric potential and electric potential energy.

Electric Potential

The *voltage* of a battery is the difference in **electric potential** between its two terminals.

You'll learn how an electric potential is created when positive and negative charges are separated.

Capacitors

The capacitors on this circuit board store charge and **electric potential energy.**

You'll learn how the energy stored in a capacitor depends on its charge.

Potential and Field

There is an intimate connection between the electric potential and the electric field.

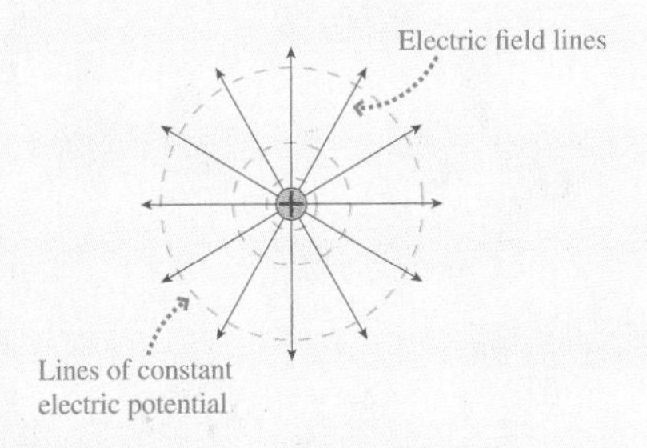

You'll learn how to move back and forth between field and potential representations.

LOOKING BACK ◂◂

Work and Potential Energy

In Section 10.4 you learned that it is possible to store *potential energy* in a system of interacting objects. In this chapter, we'll learn about a new form of potential energy, electric potential energy.

This roller coaster is pulled to the top of the first hill by a chain. The tension in the chain does work on the coaster, increasing its gravitational potential energy.

STOP TO THINK

You lift a book at a constant speed. Which statement is true about the work W done by your hand and the change in the book's gravitational potential energy ΔU_g?

A. $W > \Delta U_g > 0$
B. $W < \Delta U_g < 0$
C. $W = \Delta U_g > 0$
D. $W = \Delta U_g < 0$

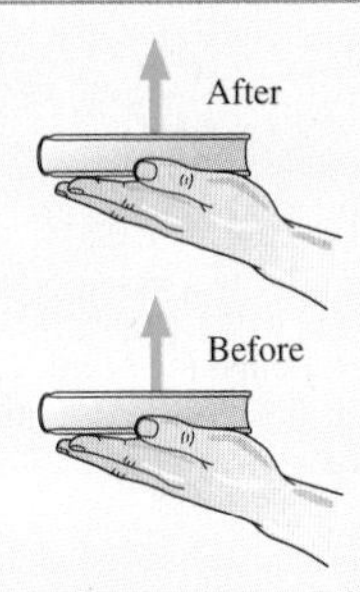

21.1 Electric Potential Energy and Electric Potential

Conservation of energy was a powerful tool for understanding the motion of mechanical systems. In ◀◀ SECTION 10.2 you learned that a system's energy can be changed by doing *work* on it; you will recall that work is done, and thus a system's energy is changed, when an external force acts on the system as the system undergoes a *displacement*.

To remind ourselves of how conservation of energy works for mechanical systems, FIGURE 21.1a shows a hand lifting a book at a constant speed. The external force of the hand pushes on the book as the book rises, so the hand does work on the book, increasing its energy—in this case, its gravitational potential energy U_g. Mathematically, we can write this process as $\Delta U_g = W$: The work W *changes* the book's potential energy.

FIGURE 21.1 The work done increases the potential energy for three systems.

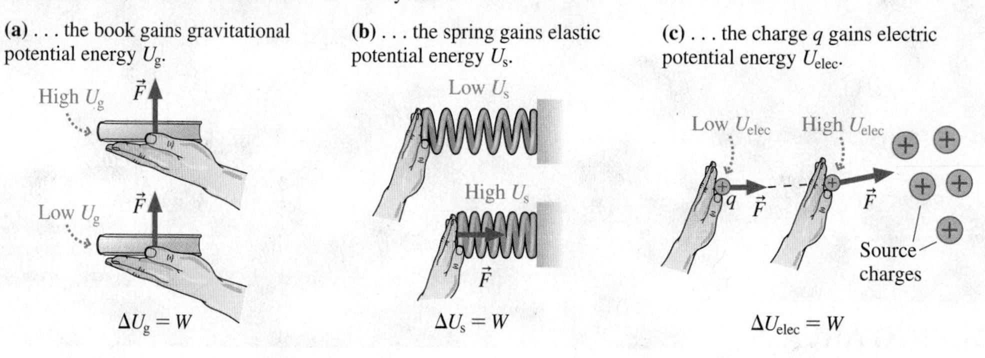
As the force of the hand does work W on each system . . .
(a) . . . the book gains gravitational potential energy U_g.
High U_g
$\vec{F}$
Low U_g
$\vec{F}$
$\Delta U_g = W$
(b) . . . the spring gains elastic potential energy U_s.
Low U_s
High U_s
$\vec{F}$
$\Delta U_s = W$
(c) . . . the charge q gains electric potential energy U_{elec}.
Low U_{elec}
High U_{elec}
q
$\vec{F}$
$\vec{F}$
Source charges
$\Delta U_{elec} = W$

A similar process occurs when a spring is compressed, as in FIGURE 21.1b. Here, the hand pushes on the end of the spring as it compresses, doing work on the spring and increasing its elastic potential energy U_s, a process we can write as $\Delta U_s = W$.

Now let's apply these same ideas to the system of charged particles in FIGURE 21.1c. Here, several charges have been identified as *source charges* that are fixed and don't move. Because charge q is repelled by the source charges, the hand has to *push* on charge q in order to move it closer to the source charges. This means that the hand does work on q as it moves it, transferring energy into the system of charges. Just as in the book and spring systems, this energy is in the form of potential energy—in this case, **electric potential energy** U_{elec}. Writing this process as $\Delta U_{elec} = W$, we see that **we can determine the electric potential energy of a charge when it's at a particular position by computing how much work it took to move the charge to that position.**

Electric Potential

We introduced the concept of the *electric field* in Chapter 20. In the field model, the electric field is the agent by which charges exert a long-range force on another charge q. FIGURE 21.2 reminds you that the source charges first alter the space around them by creating an electric field $\vec{E}$ at every point in space. It is then this electric field—not the source charges themselves—that exerts a force $\vec{F}_{elec} = q\vec{E}$ on charge q. An important idea was that the electric field of the source charges is present throughout space whether or not charge q is present to experience it. In other words, the electric field tells us what the force on the charge *would be* if the charge were placed there.

FIGURE 21.2 What is the electric potential energy near some source charges?

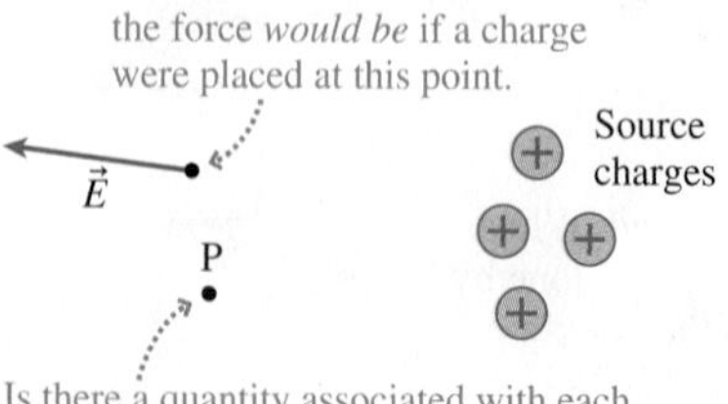
The electric field tells us what the force *would be* if a charge were placed at this point.
$\vec{E}$
Source charges
P
Is there a quantity associated with each point around the source charges that would tell us the electric potential energy of a charge placed at that point?

Can we apply similar reasoning to electric potential energy? Consider again the source charges in Figure 21.2. If we place a charge q at point P near the source charges, charge q will have electric potential energy. If we place a different charge q' at P, charge q' will have a different electric potential energy. Is there a quantity associated with P that would tell us what the electric potential energy of q, q', or any other charge *would have* at point P, without actually having to place the charge there?

To find out if this is possible, we'll have to understand in a bit more detail how to find the electric potential energy of a charge q. Suppose, to be specific, we take $q = 10$ nC in FIGURE 21.3a and, for convenience, we let $(U_{elec})_A = 0$ J when the charge is at point A.

As we've already seen, to find charge q's electric potential energy at any other point, such as point B or C, we need to find the amount of work it takes to move the charge from A to B, or A to C. In FIGURE 21.3b it takes the hand 4 μJ of work to move the charge from A to B; thus its electric potential energy at B is $(U_{elec})_B = 4\ \mu\text{J}$. (Recall that the SI unit of energy is the joule, J.) Similarly, q's electric potential energy at point C is $(U_{elec})_C = 6\ \mu\text{J}$ because it took 6 μJ of work to move it to point C.

What if we were to repeat this experiment with a different charge—say, $q = 20$ nC? According to Coulomb's law, the electric force on this charge due to the source charges will be twice that on the 10 nC charge. Consequently, the hand will have to push with twice as much force and thus do twice as much work in moving this charged particle from A to B. As a result, the 20 nC particle has $(U_{elec})_B = 8\ \mu\text{J}$ at B. A 5 nC particle would have $(U_{elec})_B = 2\ \mu\text{J}$ at B because the hand would have to work only half as hard to move it to B as it did to move the 10 nC particle to B. In general, **a charged particle's potential energy is proportional to its charge.**

When two quantities are proportional to each other, their *ratio* is constant. We can see this directly for the electric potential energy of a charged particle at point B by calculating the ratio

$$\frac{U_{(elec)B}}{q} = \frac{2\ \mu\text{J}}{5\ \text{nC}} = \frac{4\ \mu\text{J}}{10\ \text{nC}} = \frac{8\ \mu\text{J}}{20\ \text{nC}} = 400\ \frac{\text{J}}{\text{C}}$$

U/q for $q = 5$ nC; U/q for $q = 10$ nC; U/q for $q = 20$ nC; All three ratios are the *same*.

Thus we can write a simple expression for the electric potential energy that *any* charge q *would have* if placed at point B:

$$(U_{elec})_B = \left(400\ \frac{\text{J}}{\text{C}}\right)q$$

This number is associated with point B. This part depends on the charge we place at B.

The number 400 J/C, which is associated with point B, tells us the *potential* for creating potential energy (there's a mouthful!) if a charge q is placed at point B. Thus this value is called the **electric potential,** and it is given the symbol V. At B, then, the electric potential is 400 J/C, as shown in FIGURE 21.3c. By similar reasoning, the electric potential at point C is 600 J/C.

This idea can be generalized. Any source charges create an electric potential at every point in the space around them. At a point where the potential is V, the electric potential energy of a charged particle q is

$$U_{elec} = qV \quad (21.1)$$

Relationship between electric potential and electric potential energy

U_{elec} — V — p. 34 PROPORTIONAL

Notice the similarity to $\vec{F}_{elec} = q\vec{E}$. The electric potential, like the electric field, is created by the source charges and is present at all points in space. The electric potential is there whether or not charge q is present to experience it. While the electric field tells us how the source charges would exert a *force* on q, the electric potential tells us

FIGURE 21.3 Finding the electric potential energy and the electric potential.

(a) The electric potential energy of a 10 nC charge at A is zero. What is its potential energy at point B or C?

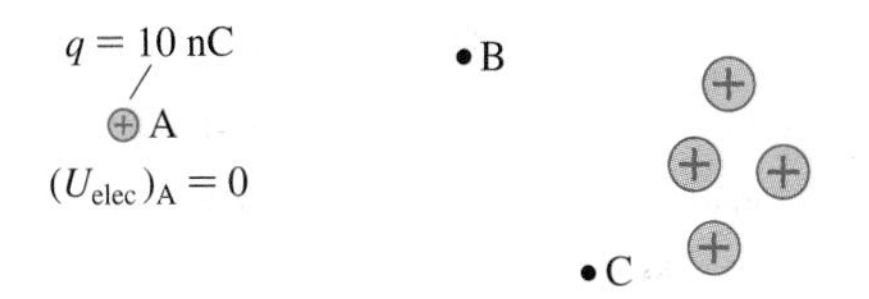

(b) The charge's electric potential energy at any point is equal to the amount of work done in moving it there from point A.

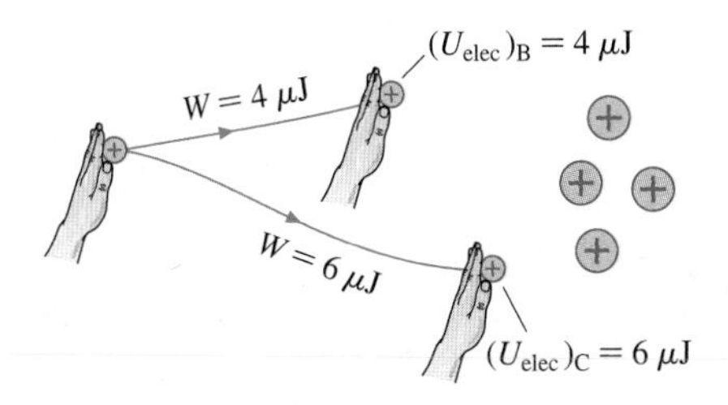

(c) The electric potential is created by the source charges. It exists at *every* point in space, not only at A, B, and C.

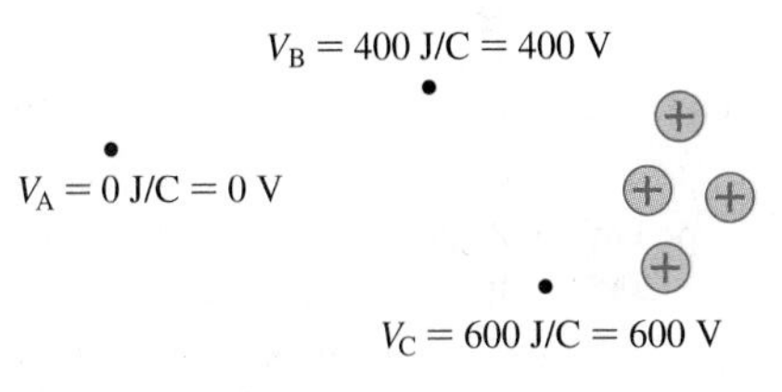

Class Video

TABLE 21.1 Typical electric potentials

Source of potential	Approximate potential
Brain activity at scalp (EEG)	10–100 μV
Cells in human body	100 mV
Battery	1–10 V
Household electricity	100 V
Static electricity	10 kV
Transmission lines	500 kV

how the source charges would provide q with *potential energy*. Although we used the work done on a positive charge to justify Equation 21.1, it is also valid if q is negative.

NOTE ▶ For potential energy, we found that we could choose a particular configuration of the system to have $U = 0$, the zero of potential energy. The same idea holds for electric potential. We can choose any point in space, wherever is convenient, to be $V = 0$. It will turn out that only *changes* in the electric potential are important, so the choice of a point to be $V = 0$ has no physical consequences. ◀

The unit of electric potential is the joule per coulomb, called the **volt** V:

$$1 \text{ volt} = 1 \text{ V} = 1 \text{ J/C}$$

Microvolts (μV), millivolts (mV), and kilovolts (kV) are commonly used units. Table 21.1 lists some typical electric potentials. We can now recognize that the electric potential in Figure 21.2—a potential due to the source charges—is 0 V at A and 400 V at B. This is shown in Figure 21.3c.

NOTE ▶ The symbol V is widely used to represent the *volume* of an object, and now we're introducing the same symbol to mean *potential.* To make matters more confusing, V is the abbreviation for *volts.* In printed text, V for potential is italicized while V for volts is not, but you can't make such a distinction in handwritten work. This is not a pleasant state of affairs, but these are the commonly accepted symbols. You must be especially alert to the *context* in which a symbol is used. ◀

EXAMPLE 21.1 Finding the change in a charge's electric potential energy

A 15 nC charged particle moves from point A, where the electric potential is 300 V, to point B, where the electric potential is −200 V. By how much does the electric potential change? By how much does the particle's electric potential energy change? How would your answers differ if the particle's charge were −15 nC?

PREPARE The change in the electric potential ΔV is the potential at the final point B minus the potential at the initial point A. From Equation 21.1, we can find the change in the electric potential energy by noting that $\Delta U_{\text{elec}} = (U_{\text{elec}})_\text{B} - (U_{\text{elec}})_\text{A} = q(V_\text{B} - V_\text{A}) = q\Delta V$.

SOLVE We have

$$\Delta V = V_\text{B} - V_\text{A} = (-200 \text{ V}) - (300 \text{ V}) = -500 \text{ V}$$

This change is *independent* of the charge q because the electric potential is created by source charges.

The change in the particle's electric potential energy is

$$\Delta U_{\text{elec}} = q\Delta V = (15 \times 10^{-9} \text{ C})(-500 \text{ V}) = -7.5 \ \mu\text{J}$$

A −15 nC charge would have ΔU_{elec} + 7.5 μJ because q changes sign while ΔV remains unchanged.

ASSESS Because the electric potential at B is lower than that at A, the positive (+15 nC) charge will lose electric potential energy, while the negative (−15 nC) charge will gain energy.

STOP TO THINK 21.1 A positively charged particle moves from point 1 to point 2. As it does, its electric potential energy

A. Increases.
B. Decreases.
C. Stays the same.

21.2 Sources of Electric Potential

Class Video

How is an electric potential created in the first place? Consider the uncharged capacitor shown in **FIGURE 21.4a**. There's no force on charge q, so no work is required to move it from A to B. Consequently, charge q's electric potential energy remains *unchanged* as it is moved from A to B, so that $(U_{\text{elec}})_\text{B} = (U_{\text{elec}})_\text{A}$. Then, because $U_{\text{elec}} = qV$, it must be the case that $V_\text{B} = V_\text{A}$. We say that the **potential difference** $\Delta V = V_\text{B} - V_\text{A}$ is *zero*.

FIGURE 21.4 Potential differences are created by charge separation.

(a) The force on charge q is zero. No work is needed to move it from A to B, so there is no potential difference between A and B.

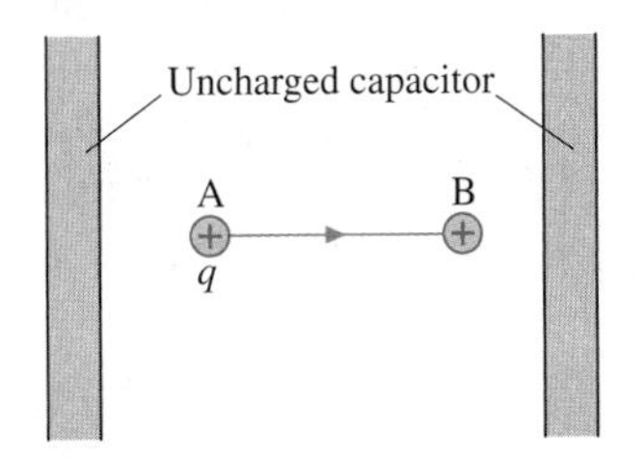

(b) The capacitor still has no net charge, but charge has been *separated* to give the plates charges $+Q$ and $-Q$.

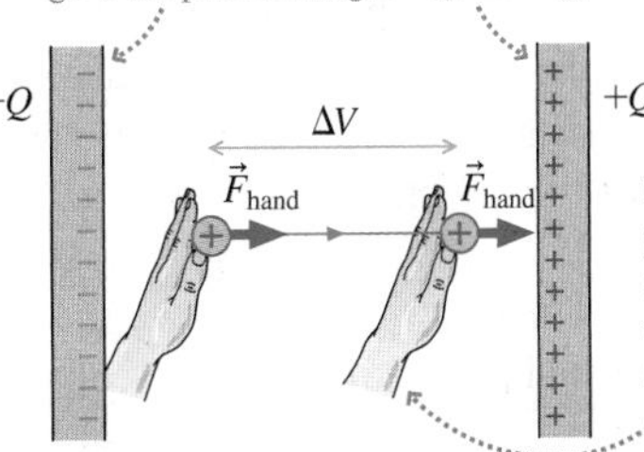

Now, because q is repelled from the positive plate and attracted to the negative plate, the hand must do work on q to push it from A to B, so there must be an *electric potential difference* ΔV between A and B.

Now consider what happens if electrons are transferred from the right side of the capacitor to the left, giving the left electrode charge $-Q$ and the right electrode charge $+Q$. The capacitor still has no net charge, but the charge has been *separated*. These separated charges exert a force $\vec{F}_{elec}$ on q, so, as FIGURE 21.4b shows, the hand must now do work on q to move it from A to B, increasing its electric potential energy so that $(U_{elec})_B > (U_{elec})_A$. This means that the potential difference ΔV between A and B is no longer zero. What we've shown here is quite general: **A potential difference is created by *separating* positive charge from negative charge.**

Perhaps the most straightforward way to create a separation of charge is by the frictional transfer of charge discussed in ◀◀ SECTION 20.1. As you shuffle your feet across a carpet, the friction between your feet and the carpet transfers charge to your body, causing a potential difference between your body and, say, a nearby doorknob. The potential difference between you and a doorknob can be many tens of thousands of volts—enough to create a spark as the excess charge on your body moves from higher to lower potential.

Lightning is the result of large potential differences built up by charge separation within clouds.

Lightning is the result of a charge separation that occurs in clouds. As small ice particles in the clouds collide, they become charged by frictional rubbing. The details are still not well understood, but heavier particles, which fall to the bottom of the cloud, gain a negative charge, while lighter particles, which are lifted to the cloud's top, become positive. Thus the top of the cloud becomes positively charged and the bottom negatively charged. This natural charge separation creates a huge potential difference—as much as 100 million volts—between the top and bottom of the cloud. The negative charge in the bottom of the cloud causes positive charge to accumulate in the ground below. A lightning strike occurs when the potential difference between the cloud and the ground becomes too large for the air to sustain.

Charge separation, and hence potential differences, can also be created by chemical processes. A common and important means of creating a fixed potential difference is the **battery.** We'll study batteries in Chapter 22, but all batteries use chemical reactions to create an internal charge separation. This separation proceeds until a characteristic potential difference—about 1.5 V for a standard alkaline battery—appears between the two terminals of the battery. Different kinds of batteries maintain different potential differences between their terminals.

NOTE ▶ The potential difference between two points is often called the **voltage.** Thus we say that a battery's voltage is 1.5 V or 12 V, and we speak of the potential difference between a battery's terminals as the voltage "across" the battery. ◀

Chemical means of producing potential differences are also crucial in biological systems. For example, there's a potential difference of about 70 mV between the inside and outside of a cell, with the inside of the cell more negative than the outside.

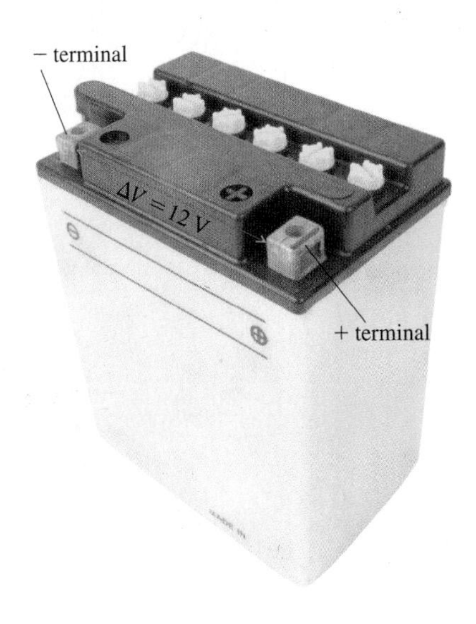

▶A car battery maintains a fixed potential difference of 12 V between its + and − terminals.

FIGURE 21.5 BIO The membrane potential of a cell is due to a charge separation.

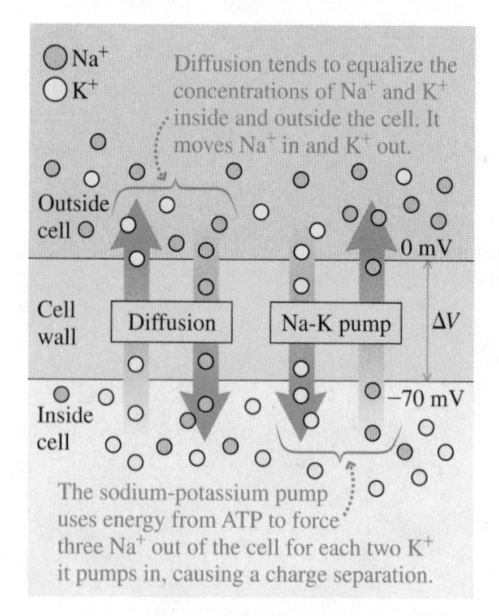

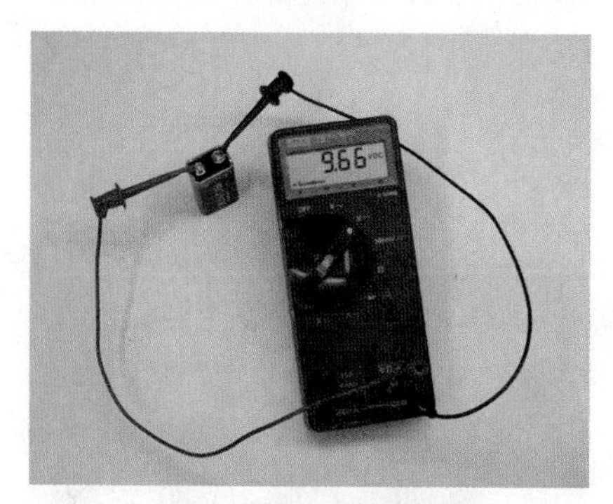

A voltmeter always uses two probes to measure a potential difference. Here, we see that the potential difference of a fresh 9 V battery is closer to 9.7 V.

As illustrated in FIGURE 21.5, this *membrane potential* is caused by an imbalance of potassium (K^+) and sodium (Na^+) ions. The molar concentration of K^+ is higher inside the cell than outside, while the molar concentration of Na^+ is higher outside than inside. To keep the charge separated in the face of diffusion, which tends to equalize the ion concentrations, a *sodium-potassium exchange pump* continuously pumps sodium out of the cell and potassium into the cell. During one pumping cycle, three Na^+ are pushed out of the cell but only two K^+ are pushed in, giving a net transfer of one positive charge out of the cell. This continuous pumping leads to the charge separation that causes the membrane potential. We'll have a careful look at the electrical properties of nerve cells in Chapter 23.

Measuring Electric Potential

Measurements of the electric potential play an important role in a broad range of applications. An electrocardiogram measures the potential difference between several locations on the body to diagnose possible heart problems; temperature is often measured using a *thermocouple,* a device that develops a potential difference proportional to temperature; your digital camera can sense when its battery is low by measuring the potential difference between its terminals.

Note that in all these applications, it's the potential *difference* between two points that's measured. The actual value of the potential at a given point depends on where we choose V to be zero, but the difference in potential between two points is independent of this choice. Because of this, a **voltmeter,** the basic instrument for measuring potential differences, always has *two* inputs. Probes are connected from these inputs to the two points between which the potential difference is to be measured. We'll learn more in Chapter 23 about how voltmeters work.

As small as cells are, the membrane potential difference between the inside and outside of a cell can be measured by a (very small) probe connected to a voltmeter. FIGURE 21.6 is a micrograph of a nerve cell whose membrane potential is being measured. A very small glass pipette, filled with conductive fluid, is actually inserted through the cell's membrane. This pipette is one of the probes. The second probe need not be so small; it is simply immersed in the conducting fluid that surrounds the cell and can be quite far from the cell.

FIGURE 21.6 BIO Measuring the membrane potential.

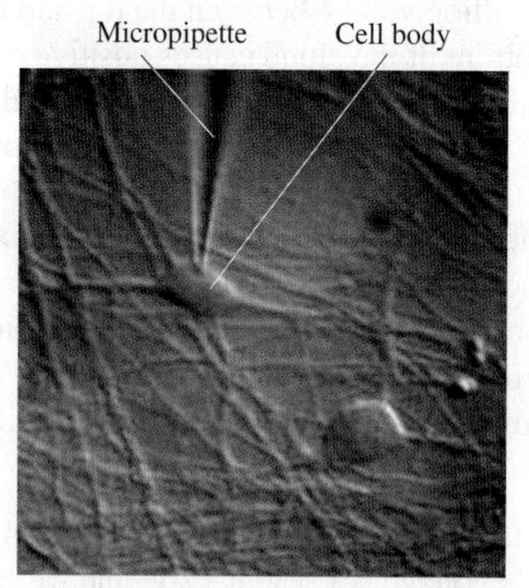

21.3 Electric Potential and Conservation of Energy

The potential energy of a charged particle is determined by the electric potential: $U_{elec} = qV$. Although potential and potential energy are related and have similar names, they are not the same thing. Table 21.2 will help you distinguish between electric potential and electric potential energy.

As a roller coaster car moves from the top of the track to the bottom, its gravitational potential energy decreases. Because energy is conserved, this decrease in potential energy is accompanied by an *increase* in the car's kinetic energy K—the car speeds up. Similarly, when a particle with positive charge q moves from a region of high electric potential to a region of low electric potential, its electric potential energy decreases and its kinetic energy increases. The conservation of energy equation for a charged particle is

$$K_f + (U_{elec})_f = K_i + (U_{elec})_i$$

which we can write in terms of the electric potential V as

$$K_f + qV_f = K_i + qV_i \tag{21.2}$$

Conservation of energy for a charged particle moving in an electric potential V

where, as usual, the subscripts i and f stand for the initial and final situations.

FIGURE 21.7 shows two positive charges moving through a region of changing electric potential. This potential has been created by source charges that aren't shown; our concern is only with the *effect* of this potential on the moving charge. Notice that we've used the before-and-after visual overview introduced earlier when we studied conservation of momentum and energy.

TABLE 21.2 Distinguishing electric potential and potential energy

The *electric potential* is created by the source charges. The electric potential is present whether or not a charged particle is there to experience it. Potential is measured in J/C, or V.

The *electric potential energy* is the interaction energy of a charged particle with the source charges. Potential energy is measured in J.

FIGURE 21.7 A charged particle speeds up or slows down as it moves through a potential difference.

(a) A positive charge speeds up ($\Delta K > 0$) as it moves from higher to lower potential ($\Delta V < 0$). Electric potential energy is transformed into kinetic energy.

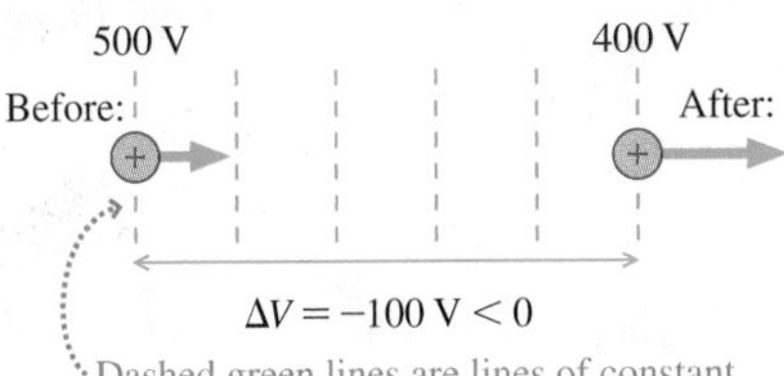

Dashed green lines are lines of constant electric potential. The potential is 500 V at all points on this line.

(b) A positive charge slows down ($\Delta K < 0$) as it moves from lower to higher potential ($\Delta V > 0$). Kinetic energy is transformed into electric potential energy.

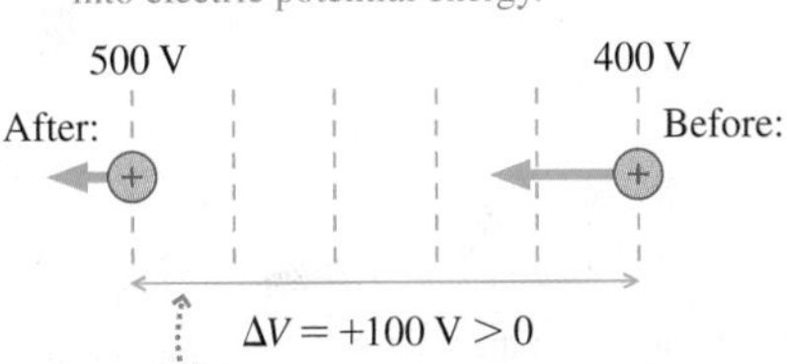

A double-headed green arrow is used to represent a potential difference.

STOP TO THINK 21.2 At a point where the electric potential is 400 V, a positive charge is moving to the left. As the charge continues to move, it slows down, stops, and then begins to move back to the right. Using what you learned in Figure 21.7, its speed as it passes its initial position is

A. Greater than its initial speed.
B. Less than its initial speed.
C. Equal to its initial speed.
D. There is not enough information to tell.

We can understand the motion of the charges if we rewrite Equation 21.2 as $K_f - K_i = -q(V_f - V_i)$, or

$$\Delta K = -q\Delta V \tag{21.3}$$

For the charge in Figure 21.7a, the change in potential—that is, the potential difference—as it moves from left to right is

$$\Delta V = V_f - V_i = 400\text{ V} - 500\text{ V} = -100\text{ V}$$

which is negative. Equation 21.3 then shows that ΔK is *positive,* indicating that the particle *speeds up* as it moves from higher to lower potential. Conversely, for the charge in Figure 21.7b the potential difference is $+100$ V; Equation 21.3 then shows that ΔK is negative, so the particle *slows down* in moving from lower to higher potential.

NOTE ▸ The situation is reversed for a negative charge. If $q < 0$, Equation 21.3 requires K to increase as V increases. A negative charge speeds up if it moves into a region of higher potential. ◂

PROBLEM-SOLVING STRATEGY 21.1 **Conservation of energy in charge interactions**

We use the principle of conservation of energy for electric charges in exactly the same way as we did for mechanical systems. The only difference is that we now need to consider electric potential energy.

PREPARE Draw a before-and-after visual overview. Define symbols that will be used in the problem, list known values, and identify what you're trying to find.

SOLVE The mathematical representation is based on the law of conservation of mechanical energy:

$$K_f + qV_f = K_i + qV_i$$

- Find the electric potential at both the initial and final positions. You may need to calculate it from a known expression for the potential, such as that of a point charge.
- K_i and K_f are the total kinetic energies of all moving particles.
- Some problems may need additional conservation laws, such as conservation of charge or conservation of momentum.

ASSESS Check that your result has the correct units, is reasonable, and answers the question.

Exercise 18 

EXAMPLE 21.2 A speeding proton

A proton moves through an electric potential created by a number of source charges. Its speed is 2.5×10^5 m/s at a point where the potential is 1500 V. What will be the proton's speed a short time later when it reaches a point where the potential is -500 V?

PREPARE The mass of a proton is $m = 1.67 \times 10^{-27}$ kg. The positively charged proton moves from a region of higher potential to one of lower potential, so the change in potential is negative and the proton loses electric potential energy. Conservation of energy then requires its kinetic energy to increase. We can use Equation 21.2 to find the proton's final speed.

SOLVE Conservation of energy gives

$$K_f + qV_f = K_i + qV_i$$

or

$$\tfrac{1}{2}mv_f^2 + qV_f = \tfrac{1}{2}mv_i^2 + qV_i$$

which we can write as

$$\tfrac{1}{2}mv_f^2 = \tfrac{1}{2}mv_i^2 + (qV_i - qV_f)$$

We can solve for v_f^2 by multiplying both sides by $2/m$ to get

$$v_f^2 = v_i^2 + \frac{2}{m}(qV_i - qV_f) = v_i^2 + \frac{2q}{m}(V_i - V_f)$$

or

$$v_f^2 = (2.5 \times 10^5 \text{ m/s})^2 + \frac{2(1.6 \times 10^{-19}\text{ C})}{1.67 \times 10^{-27}\text{ kg}}[1500\text{ V} - (-500\text{ V})]$$

$$= 4.46 \times 10^{11}\ (\text{m/s})^2$$

Solving for the final speed gives

$$v_f = 6.7 \times 10^5 \text{ m/s}$$

ASSESS This problem is very similar to the situation in Figure 21.7a. A positively charged particle speeds up as it moves from higher to lower potential, analogous to a particle speeding up as it slides down a hill from higher gravitational potential energy to lower gravitational potential energy.

So far we've considered only the transformation of electric potential energy into kinetic energy as a charged particle moves from higher to lower electric potential. That is, we've studied the energy transformation $\Delta K = -q\Delta V$. But it's worth noting that electric potential energy can also be transformed into other kinds of energy; this is the basis of many applications of electricity. In **FIGURE 21.8**, for example, charges move in the wires from the high-potential terminal of the battery, through the lightbulb, and back to the low-potential terminal. In the bulb, their electric potential energy is transformed into thermal energy E_{th}, making the bulb hot enough to glow brightly. This energy transformation is $\Delta E_{th} = -q\Delta V$. Or, as charges move from the high- to the low-potential terminals of an elevator motor, their electric potential energy is transformed into gravitational potential energy as the elevator and its passengers are lifted, so in this case $\Delta U_g = -q\Delta V$.

We'll have much more to say about these and other transformations of electric energy in chapters to come!

FIGURE 21.8 A transformation of electric potential energy into thermal energy.

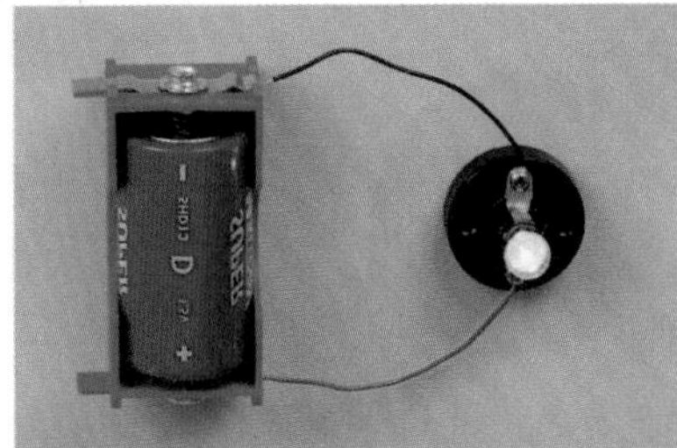

The Electron Volt

The joule is a unit of appropriate size in mechanics and thermodynamics, where we deal with macroscopic objects, but it will be very useful to have an energy unit appropriate to atomic and nuclear events.

Suppose an electron accelerates through a potential difference $\Delta V = 1$ V. The electron might be accelerating from 0 V to 1 V, or from 1000 V to 1001 V. Regardless of the actual voltages, an electron, being negative, *speeds up* when it moves toward a higher potential, so that, according to Equation 21.3, the 1 V potential difference causes the electron, with $q = -e$, to gain kinetic energy

$$\Delta K = -q\Delta V = e\Delta V = (1.60 \times 10^{-19}\text{ C})(1\text{ V}) = 1.60 \times 10^{-19}\text{ J}$$

Let us define a new unit of energy, called the **electron volt,** as

$$1\text{ electron volt} = 1\text{ eV} = 1.60 \times 10^{-19}\text{ J}$$

With this definition, the kinetic energy gained by the electron in our example is

$$\Delta K = 1\text{ eV}$$

In other words, **1 electron volt is the kinetic energy gained by an electron (or proton) if it accelerates through a potential difference of 1 volt.**

NOTE ► The abbreviation eV uses a lowercase e but an uppercase V. Units of keV (10^3 eV), MeV (10^6 eV), and GeV (10^9 eV) are common. ◄

The electron volt can be a troublesome unit. One difficulty is its unusual name, which looks less like a unit than, say, "meter" or "second." A more significant difficulty is that the name suggests a relationship to volts. But *volts* are units of electric potential, whereas this new unit is a unit of energy! It is crucial to distinguish between the *potential* V, measured in volts, and an *energy* that can be measured either in joules or in electron volts. You can now use electron volts anywhere that you would previously have used joules. Doing so is no different from converting back and forth between units of centimeters and inches.

NOTE ► The joule remains the SI unit of energy. It will be useful to express energies in eV, but you *must* convert this energy to joules before doing most calculations. ◄

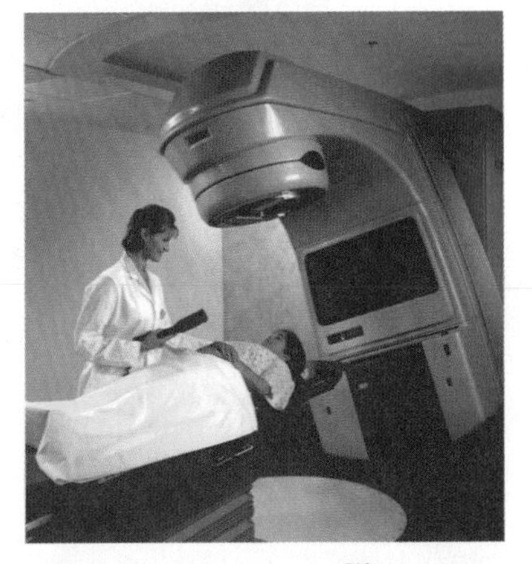

Cancer-fighting electrons BIO Tightly focused beams of x rays can be used in radiation therapy for cancer patients. The x rays are generated by directing a high-energy beam of electrons at a metal target. The electrons gain their energy by being accelerated in a *linear accelerator* through a potential difference of 20 MV, so their final kinetic energy is 20 MeV.

Class Video

EXAMPLE 21.3 The speed of a proton

Atomic particles are often characterized by their kinetic energy in MeV. What is the speed of an 8.7 MeV proton?

SOLVE The kinetic energy of this particle is 8.7×10^6 eV. First, we convert the energy to joules:

$$K = 8.7 \times 10^6\text{ eV} \times \frac{1.60 \times 10^{-19}\text{ J}}{1.0\text{ eV}} = 1.39 \times 10^{-12}\text{ J}$$

Now we can find the speed from

$$K = \frac{1}{2}mv^2$$

which gives

$$v = \sqrt{\frac{2K}{m}} = \sqrt{\frac{2(1.39 \times 10^{-12}\text{ J})}{1.67 \times 10^{-27}\text{ kg}}} = 4.1 \times 10^7\text{ m/s}$$

Because the proton's charge and the electron's charge have the same magnitude, a general rule is that a proton or electron that accelerates (decelerates) through a potential difference of V volts gains (loses) V eV of kinetic energy. In Example 21.2, with a 2000 V potential difference, the proton gained 2000 eV of kinetic energy. In Example 21.3, the proton had to accelerate through a 8.7×10^6 V = 8.7 MV potential difference to acquire 8.7 MeV of kinetic energy.

STOP TO THINK 21.3 A proton is released from rest at point Q, where the potential is 0 V. Afterward, the proton

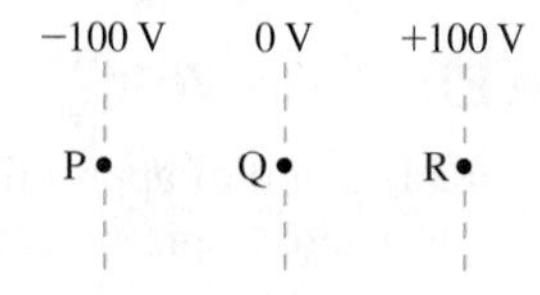

A. Remains at rest at Q.
B. Moves toward P with a steady speed.
C. Moves toward P with an increasing speed.
D. Moves toward R with a steady speed.
E. Moves toward R with an increasing speed.

21.4 Calculating the Electric Potential

Now that we understand what the electric potential is, let's go ahead and calculate the electric potential for some important cases. We'll do so using Equation 21.1, the relationship between the potential energy of a charge q at a point in space and the electric potential at that point. Rewriting Equation 21.1 slightly, we have

$$V = \frac{U_{\text{elec}}}{q} \tag{21.4}$$

Our prescription for finding the potential at a certain point in space, then, is to first calculate the electric potential *energy* of a charge q placed at that point. Then we can use Equation 21.4 to find the electric potential.

FIGURE 21.9 Finding the potential of a parallel-plate capacitor.

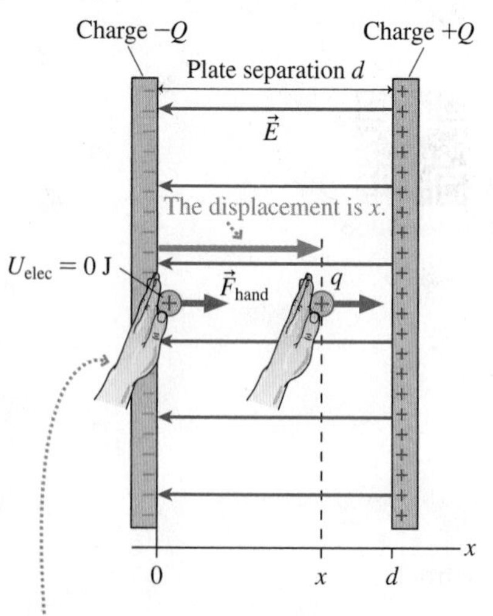

The Electric Potential Inside a Parallel-Plate Capacitor

In Chapter 20 we learned that a *uniform* electric field can be created by placing equal but opposite charges on two parallel conducting plates—the **parallel-plate capacitor.** Thus finding the electric potential inside a parallel-plate capacitor is equivalent to finding the potential for the very important case of a uniform electric field.

FIGURE 21.9 shows a cross-section view of a charged parallel-plate capacitor with separation d between the plates. The charges $\pm Q$ on the plates are the source charges that create both the electric field and the electric potential in the space between the plates. As we found in ◀◀ SECTION 20.5, the electric field is $\vec{E} = (Q/\epsilon_0 A$, from positive to negative). We'll choose a coordinate system with $x = 0$ at the negative plate and $x = d$ at the positive plate.

We're free to choose the point of zero potential energy anywhere that's convenient, so let $U_{\text{elec}} = 0$ when a mobile charge q is at the negative plate. The charge's potential energy at any other position x is then the amount of work an external force must do to move the charge at steady speed from the negative plate to that position. We'll represent the external force by a hand, although that's not really how charges get moved around.

The electric field in Figure 21.9 points to the left, so the force $\vec{F}_{elec} = q\vec{E}$ of the field on the charge is also to the left. To move the charge to the right at constant speed ($\vec{F}_{net} = \vec{0}$), the external force $\vec{F}_{hand}$ must push to the right with a force of the same magnitude: $F_{hand} = qE$. This force does work on the system as the charge is moved, changing the potential energy of the system.

Because the external force is constant and is parallel to the displacement, the work to move the charge to position x is

$$W = \text{force} \times \text{displacement} = F_{hand}x = qEx$$

Consequently, the electric potential energy when charge q is at position x is

$$U_{elec} = W = qEx$$

As the final step, we can use Equation 21.4 to find that the electric potential of the parallel-plate capacitor at position x, measured from the negative plate, is

$$V = \frac{U_{elec}}{q} = Ex = \frac{Q}{\epsilon_0 A}x \qquad (21.5)$$

where, in the last step, we wrote the electric field strength in terms of the amount of charge on the capacitor plates.

A first point to notice is that the electric potential increases linearly from the negative plate at $x = 0$, where $V = V_- = 0$, to the positive plate at $x = d$, where $V = V_+ = Ed$. Let's define the potential difference ΔV_C between the two capacitor plates to be

$$\Delta V_C = V_+ - V_- = Ed \qquad (21.6)$$

In many cases, the capacitor voltage is fixed at some value ΔV_C by connecting its plates to a battery with a known voltage. In this case, the electric field strength inside the capacitor is determined from Equation 21.6 as

$$E = \frac{\Delta V_C}{d} \qquad (21.7)$$

This means that we can establish an electric field of known strength by applying a voltage across a capacitor whose plate spacing is known.

Equation 21.7 implies that the units of electric field are volts per meter, or V/m. We have been using electric field units of newtons per coulomb. It's straightforward to show that these units are in fact equivalent. Because volts per meter are the electric field units used by scientists and engineers in practice, we will now adopt them as our standard electric field units.

Returning to the electric potential, we can substitute Equation 21.7 for E into Equation 21.5 for V. In terms of the capacitor voltage ΔV_C, the electric potential at position x inside the capacitor is

$$V = \frac{x}{d}\Delta V_C \qquad (21.8)$$

You can see that the potential increases linearly from $V = 0$ at $x = 0$ (the negative plate) to $V = \Delta V_C$ at $x = d$ (the positive plate).

Let's explore the electric potential inside the capacitor by looking at several different, but related, ways that the potential can be represented. In the table on the next page, a battery has established a 1.5 V potential difference across a parallel-plate capacitor with a 3 mm plate spacing.

Graphical representations of the electric potential inside a capacitor

A graph of potential versus x. You can see the potential increasing from 0 V at the negative plate to 1.5 V at the positive plate.

A three-dimensional view showing **equipotential surfaces.** These are mathematical surfaces, not physical surfaces, that have the same value of V at every point. The equipotential surfaces of a capacitor are planes parallel to the capacitor plates. The capacitor plates are also equipotential surfaces.

A two-dimensional **equipotential map.** The green dashed lines represent slices through the equipotential surfaces, so V has the same value everywhere along such a line. We call these lines of constant potential **equipotential lines** or simply **equipotentials.**

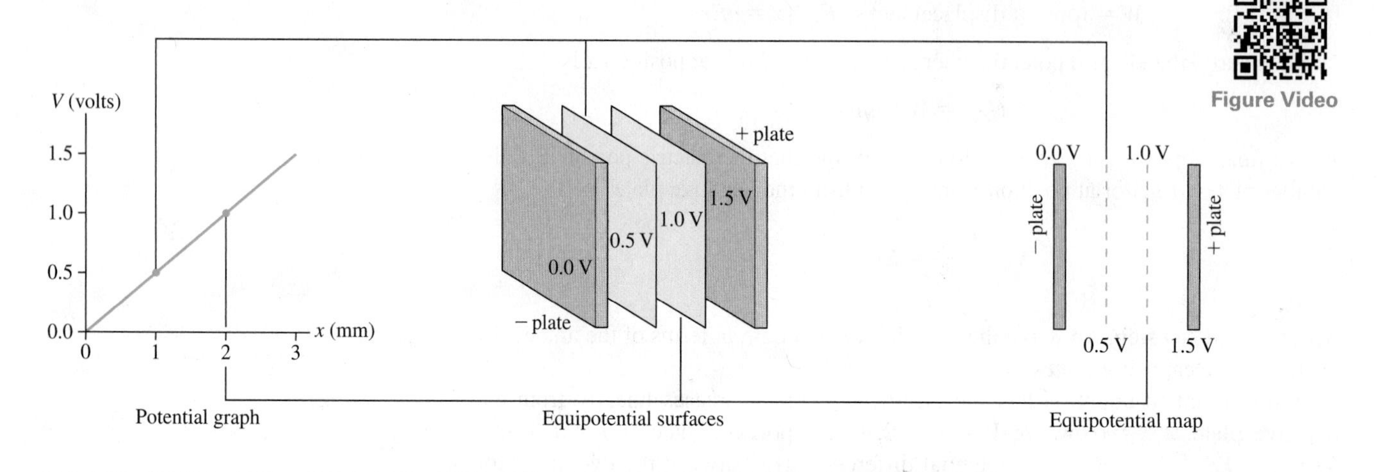

NOTE ► Equipotential lines are just the intersections of equipotential surfaces with the two-dimensional plane of the paper, and so are really just another way of representing equipotential surfaces. Because of this, we'll often use the terms "equipotential lines," "equipotential surfaces," and "equipotentials" interchangeably. ◄

EXAMPLE 21.4 A proton in a capacitor

A parallel-plate capacitor is constructed of two disks spaced 2.00 mm apart. It is charged to a potential difference of 500 V. A proton is shot through a small hole in the negative plate with a speed of 2.0×10^5 m/s. What is the farthest distance from the negative plate that the proton reaches?

PREPARE Energy is conserved. The proton's potential energy inside the capacitor can be found from the capacitor's electric potential. FIGURE 21.10 is a before-and-after visual overview of the proton in the capacitor.

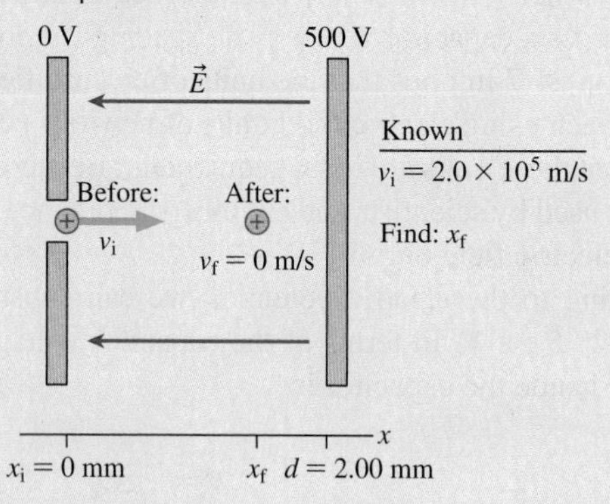

FIGURE 21.10 A before-and-after visual overview of a proton moving in a capacitor.

SOLVE The proton starts at the negative plate, where $x_i = 0$ mm. Let the final point, where $v_f = 0$ m/s, be at x_f. The potential inside the capacitor is given by $V = \Delta V_C x/d$ with $d = 0.0020$ m and $\Delta V_C = 500$ V. Conservation of energy requires $K_f + eV_f = K_i + eV_i$. This is

$$0 + e\,\Delta V_C \frac{x_f}{d} = \frac{1}{2}mv_i^2 + 0$$

where we used $V_i = 0$ V at the negative plate ($x_i = 0$) and $K_f = 0$ at the final point. The solution for the final point is

$$x_f = \frac{mdv_i^2}{2e\,\Delta V_C} = 0.84 \text{ mm}$$

The proton travels 0.84 mm, less than halfway across, before stopping and being turned back.

ASSESS We were able to use the electric potential inside the capacitor to determine the proton's potential energy.

Synthesis 21.1 brings together what we've learned about the electric field and electric potential of a parallel-plate capacitor.

SYNTHESIS 21.1 The parallel-plate capacitor: Potential and electric field

There are several related expressions for the electric potential and field inside a parallel-plate capacitor.

The variables used in describing the parallel-plate capacitor

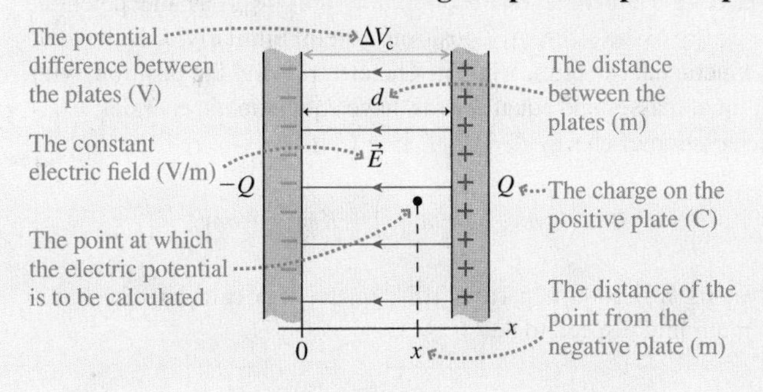

Three equivalent expressions for the electric potential are related by two expressions for the electric field inside the capacitor:

Potential V $\quad V = E x \quad V = \dfrac{Q}{\epsilon_0 A} x \quad V = \dfrac{x}{d} \Delta V_C$

Field E $\quad E = \dfrac{Q}{\epsilon_0 A} \quad E = \dfrac{\Delta V_C}{d}$

The Potential of a Point Charge

The simplest possible source charge is a single fixed point charge q. To find the electric potential, we'll again start by first finding the electric potential energy when a second charge, which we'll call q', is distance r from charge q. As usual, we'll do this by calculating the work needed to bring q' from a point where $U_{elec} = 0$ to distance r from q. We're free to choose $U_{elec} = 0$ at any point that's convenient. Because the influence of a point charge goes to zero infinitely far from the charge, it is natural to choose $U_{elec} = 0$ (and hence $V = 0$) at a point that is infinitely distant from q.

We can't use the simple expression $W = Fd$ to find the work done in moving q'; this expression is valid for only a *constant* force F and, as we know from Coulomb's law, the force on q' gets larger and larger as it approaches q. To do this calculation properly requires the methods of calculus. However, we can understand *qualitatively* how the potential energy depends on the distance r between the two charges.

FIGURE 21.11 shows q' at two different distances r from the fixed charge q. When q' is relatively far from q, the electric force on q' is small. Not much external force is needed to push q' closer to q by a small displacement d, so the work done on q' is small and the *change* ΔU_{elec} in the electric potential energy is small as well. This implies, as Figure 21.11 shows, that the graph of U_{elec} versus r is fairly flat when q' is far from q.

On the other hand, the force on q' is quite large when it gets near q, so the work required to move it through the same small displacement is much greater than before. The change in U_{elec} is large, so the graph of U_{elec} versus r is steeper when q' is near q.

The general shape of the graph of U_{elec} must be as shown in Figure 21.11. When q' is far from q, the potential energy is small. As q' approaches q, the potential energy increases rapidly. An exact calculation finds the potential energy of two point charges to be

FIGURE 21.11 The electric potential energy of two point charges.

Near q the force is large. To move q' by displacement d takes a lot of work, so ΔU_{elec} is large.

Far from q the force is small. To move q' by d takes little work, so ΔU_{elec} is small.

$$U_{elec} = K\frac{qq'}{r} = \frac{1}{4\pi\epsilon_0}\frac{qq'}{r} \qquad (21.9)$$

Electric potential energy of two charges q and q' separated by distance r

INVERSE p. 108

NOTE ▶ This expression is very similar to Coulomb's law. The difference is that the electric potential energy depends on the *inverse* of r—that is, on $1/r$—instead of the inverse-square dependence of Coulomb's law. Make sure you remember which is which! ◀

Figure 21.11 was a graph of the potential energy for two like charges, where qq' is positive. But Equation 21.9 is equally valid for *opposite* charges. In this case, the potential energy of the charges is *negative.* As **FIGURE 21.12** shows, the potential energy of the two charges *decreases* as they get closer together. A particle speeds up as its potential energy decreases ($U \rightarrow K$), so charge q' accelerates toward the fixed charge q.

FIGURE 21.12 Potential-energy diagram for two opposite charges.

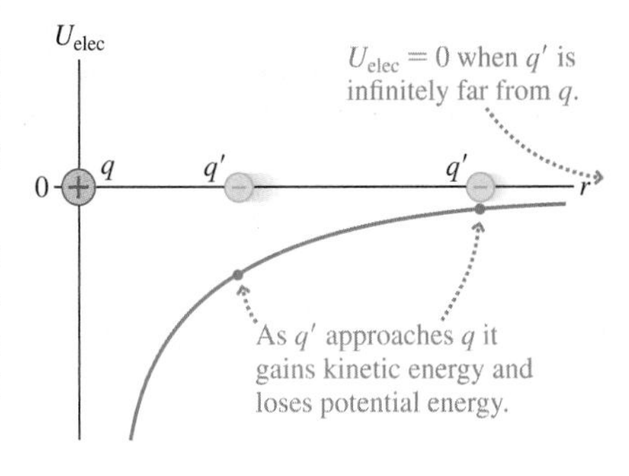

EXAMPLE 21.5 Finding the escape velocity

An interaction between two elementary particles causes an electron and a positron (a positively charged electron) to be shot out back-to-back with equal speeds. What minimum speed must each particle have when they are 100 fm apart in order to end up far from each other? (Recall that 1 fm = 10^{-15} m.)

PREPARE Energy is conserved. The particles end up "far from each other," which we interpret as sufficiently far to make $(U_{elec})_f \approx 0$ J. FIGURE 21.13 shows the before-and-after visual overview. The minimum speed to escape is the speed that allows the particles to reach $r_f = \infty$ with $v_f = 0$.

FIGURE 21.13 The before-and-after visual overview of an electron and a positron flying apart.

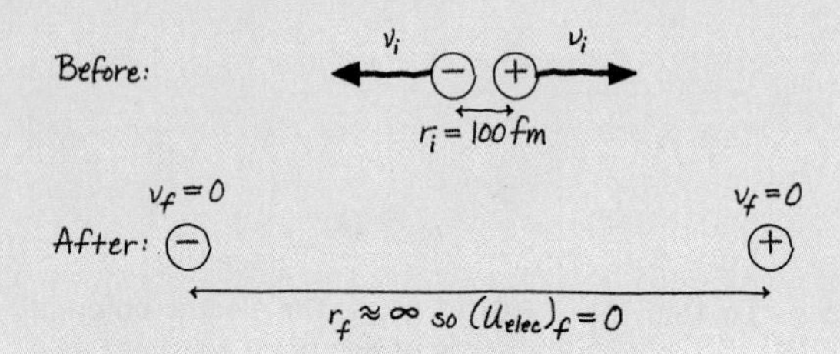

SOLVE Here it is essential to interpret U_{elec} as the potential energy of the electron + positron system. Similarly, K is the *total* kinetic energy of the system. The electron and the positron, with equal masses and equal speeds, have equal kinetic energies. Conservation of energy $K_f + U_f = K_i + U_i$ is

$$0 + 0 = \left(\frac{1}{2}mv_i^2 + \frac{1}{2}mv_i^2\right) + K\frac{q_e q_p}{r_i} = mv_i^2 - \frac{Ke^2}{r_i}$$

Using $r_i = 100 \text{ fm} = 1.0 \times 10^{-13}$ m, we can calculate the minimum initial speed to be

$$v_i = \sqrt{\frac{Ke^2}{mr_i}} = 5.0 \times 10^7 \text{ m/s}$$

ASSESS v_i is a little more than 10% the speed of light, just about the limit of what a "classical" calculation can predict. We would need to use the theory of relativity if v_i were much larger.

Equation 21.9 gives the potential energy of a charge q' when it is a distance r from a point charge q. We know that the electric *potential* is related to the potential energy by $V = U_{elec}/q'$. Thus the electric potential of charge q is

$$V = K\frac{q}{r} = \frac{1}{4\pi\epsilon_0}\frac{q}{r} \tag{21.10}$$

Electric potential at distance r from a point charge q

V p. 108 r INVERSE

Notice that only the *source* charge q appears in this expression. This is in line with our picture that a source charge *creates* the electric potential around it.

EXAMPLE 21.6 Calculating the potential of a point charge

What is the electric potential 1.0 cm from a 1.0 nC charge? What is the potential difference between a point 1.0 cm away and a second point 3.0 cm away?

PREPARE We can use Equation 21.10 to find the potential at the two distances from the charge.

SOLVE The potential at $r = 1.0$ cm is

$$V_{1\,cm} = K\frac{q}{r} = (9.0 \times 10^9 \text{ N} \cdot \text{m}^2/\text{C}^2)\left(\frac{1.0 \times 10^{-9} \text{ C}}{0.010 \text{ m}}\right) = 900 \text{ V}$$

We can similarly calculate $V_{3\,cm} = 300$ V. Thus the potential difference between these two points is $\Delta V = V_{1\,cm} - V_{3\,cm} = 600$ V.

ASSESS 1 nC is typical of the electrostatic charge produced by rubbing, and you can see that such a charge creates a fairly large potential nearby. Why aren't we shocked and injured when working with the "high voltages" of such charges? As we'll learn in Chapter 26, the sensation of being shocked is a result of current, not potential. Some high-potential sources simply do not have the ability to generate much current.

FIGURE 21.14 shows three graphical representations of the electric potential of a point charge. These match the three representations of the electric potential inside a capacitor,

► FIGURE 21.14 Three graphical representations of the electric potential of a point charge.

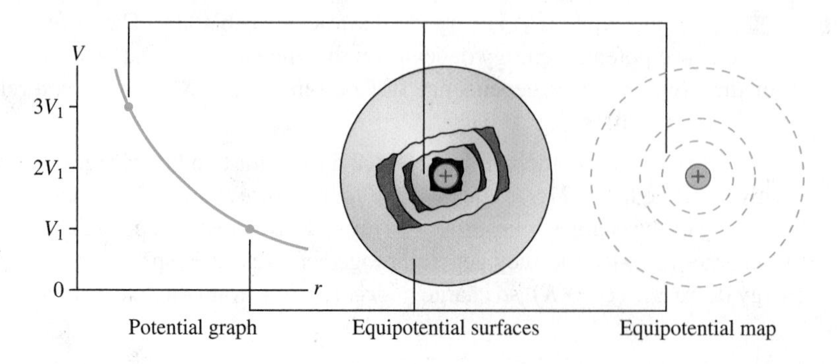

and a comparison of the two is worthwhile. This figure assumes that q is positive; you may want to think about how the representations would change if q were negative.

The Electric Potential of a Charged Sphere

Equation 21.10 gives the electric potential of a point charge. It can be shown that the electric potential outside a charged sphere is the *same* as that of a point charge; that is,

$$V = K\frac{Q}{r} = \frac{1}{4\pi\epsilon_0}\frac{Q}{r} \tag{21.11}$$

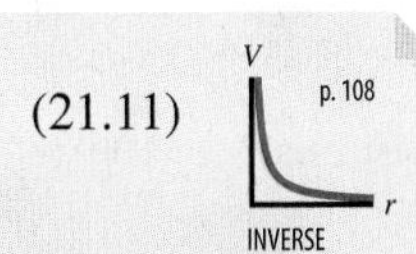

Electric potential at a distance $r > R$ from the center of a sphere of radius R and with charge Q

We can cast this result in a more useful form. It is common to charge a metal object, such as a sphere, "to" a certain potential; for instance, this can be done by connecting the sphere to, say, a 30 volt battery. This potential, which we will call V_0, is the potential right on the surface of the sphere. We can see from Equation 21.11 that

$$V_0 = V(\text{at } r = R) = \frac{Q}{4\pi\epsilon_0 R} \tag{21.12}$$

Consequently, a sphere of radius R that is charged to potential V_0 has total charge

$$Q = 4\pi\epsilon_0 R V_0 \tag{21.13}$$

If we substitute this expression for Q into Equation 21.11, we can write the potential outside a sphere that is charged to potential V_0 as

$$V = \frac{R}{r}V_0 \tag{21.14}$$

Equation 21.14 tells us that the potential of a sphere is V_0 on the surface and decreases inversely with the distance from the center. Thus the potential at $r = 3R$ is $\frac{1}{3}V_0$.

EXAMPLE 21.7 Releasing a proton from a charged sphere

A proton is released from rest at the surface of a 1.0-cm-diameter sphere that has been charged to +1000 V.

a. What is the charge of the sphere?
b. What is the proton's speed when it is 1.0 cm from the sphere?
c. When the proton is 1.0 cm from the sphere, what is its kinetic energy in eV?

PREPARE Energy is conserved. The potential outside the charged sphere is the same as the potential of a point charge at the center. **FIGURE 21.15** is a before-and-after visual overview.

FIGURE 21.15 A before-and-after visual overview of a sphere and a proton.

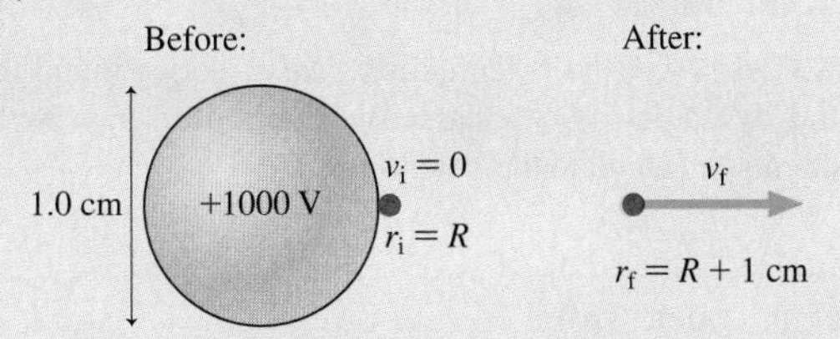

SOLVE

a. We can use the sphere's potential in Equation 21.13 to find that the charge of the sphere is

$$Q = 4\pi\epsilon_0 R V_0 = 0.56 \times 10^{-9}\text{ C} = 0.56\text{ nC}$$

b. A sphere charged to $V_0 = +1000$ V is positively charged. The proton will be repelled by this charge and move away from the sphere. The conservation of energy equation $K_f + eV_f = K_i + eV_i$, with Equation 21.14 for the potential of a sphere, is

$$\frac{1}{2}mv_f^2 + \frac{eR}{r_f}V_0 = \frac{1}{2}mv_i^2 + \frac{eR}{r_i}V_0$$

The proton starts from the surface of the sphere, $r_i = R$, with $v_i = 0$. When the proton is 1.0 cm from the *surface* of the sphere, it has $r_f = 1.0\text{ cm} + R = 1.5$ cm. Using these values, we can solve for v_f:

$$v_f = \sqrt{\frac{2eV_0}{m}\left(1 - \frac{R}{r_f}\right)} = 3.57 \times 10^5\text{ m/s}$$

c. The kinetic energy is

$$K_f = \tfrac{1}{2}mv_f^2 = \tfrac{1}{2}(1.67 \times 10^{-27}\text{ kg})(3.57 \times 10^5\text{ m/s})^2$$
$$= 1.07 \times 10^{-16}\text{ J}$$

Converting to electron volts, we find

$$1.07 \times 10^{-16}\text{ J} \times \frac{1\text{ eV}}{1.6 \times 10^{-19}\text{ J}} = 670\text{ eV}$$

ASSESS A proton at the surface of the sphere, where $V = 1000$ V, would, by definition, have 1000 eV of electric potential energy. It is reasonable that after it's moved some distance away, it has transformed 670 eV of this to kinetic energy.

Synthesis 21.2 highlights the similarity of the potential due to a point charge and the potential outside a charged sphere.

SYNTHESIS 21.2 Potential of a point charge and a charged sphere

The electric potential V on or outside a charged sphere is the same as for a point charge: It depends only on the distance and the charge on the sphere.

Point charge q

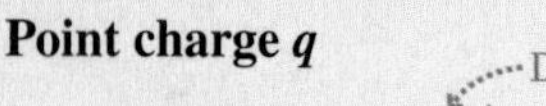

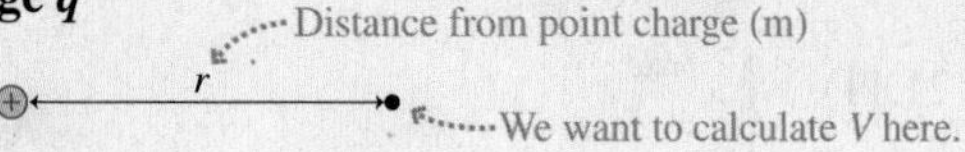

$$V = K\frac{q}{r} = \frac{1}{4\pi\epsilon_0}\frac{q}{r}$$

Electrostatic constant; Permittivity constant

Sphere with total charge Q

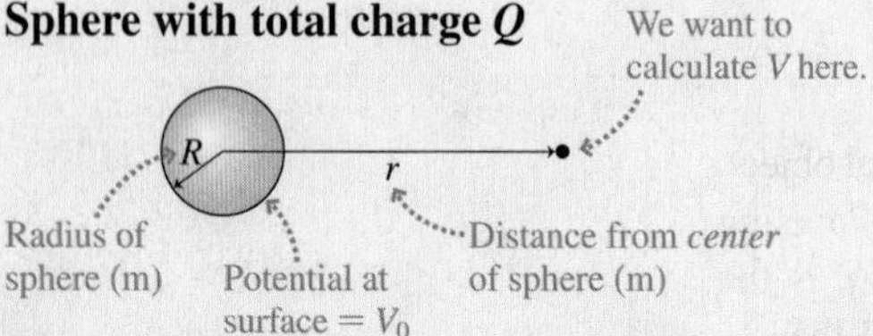

When the total charge Q on the sphere is known:

$$V = K\frac{Q}{r} = \frac{1}{4\pi\epsilon_0}\frac{Q}{r}$$

When the potential V_0 at the surface of the sphere is known:

$$V = \frac{R}{r}V_0$$

These expressions are valid only *on* or *outside* the sphere, so $r \geq R$.

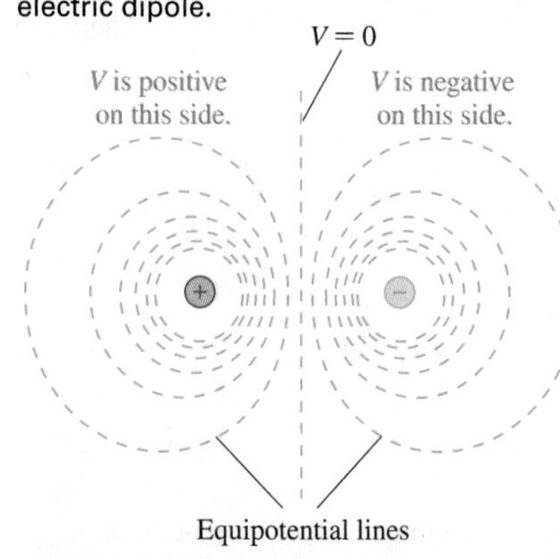

V = 0
V is positive on this side.
V is negative on this side.
Equipotential lines

FIGURE 21.16 The electric potential of an electric dipole.

The Electric Potential of Many Charges

Suppose there are many source charges $q_1, q_2, \ldots$. The electric potential V at a point in space is the *sum* of the potentials due to each charge:

$$V = \sum_i K\frac{q_i}{r_i} = \sum_i \frac{1}{4\pi\epsilon_0}\frac{q_i}{r_i} \tag{21.15}$$

where r_i is the distance from charge q_i to the point in space where the potential is being calculated. Unlike the electric field of multiple charges, which required vector addition, the electric potential is a simple scalar sum. This makes finding the potential of many charges considerably easier than finding the corresponding electric field. As an example, the equipotential map in FIGURE 21.16 shows that the potential of an electric dipole is the sum of the potentials of the positive and negative charges. We'll see later that the electric potential of the heart has the form of an electric dipole.

EXAMPLE 21.8 Finding the potential of two charges

What is the electric potential at the point indicated in FIGURE 21.17?

FIGURE 21.17 Finding the potential of two charges.

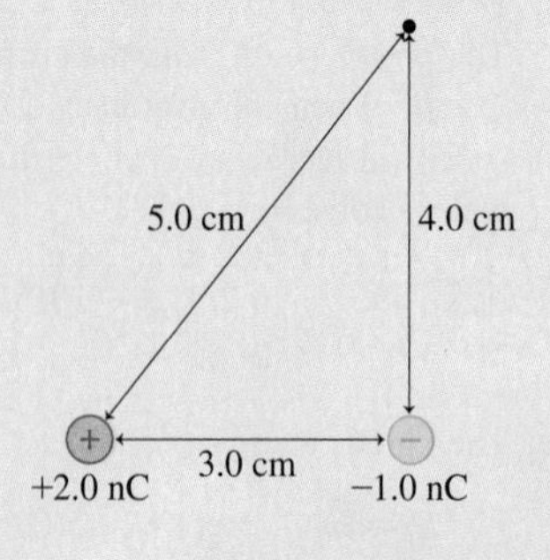

PREPARE The potential is the sum of the potentials due to each charge.

SOLVE The potential at the indicated point is

$$V = \frac{Kq_1}{r_1} + \frac{Kq_2}{r_2}$$

$$= (9.0 \times 10^9\ \text{N}\cdot\text{m}^2/\text{C}^2)\left(\frac{2.0 \times 10^{-9}\ \text{C}}{0.050\ \text{m}} + \frac{-1.0 \times 10^{-9}\ \text{C}}{0.040\ \text{m}}\right)$$

$$= 140\ \text{V}$$

ASSESS As noted, the potential is a *scalar*, so we found the net potential by adding two scalars. We don't need any angles or components to calculate the potential.

STOP TO THINK 21.4 Rank in order, from largest to smallest, the potential differences ΔV_{12}, ΔV_{13}, and ΔV_{23} between points 1 and 2, points 1 and 3, and points 2 and 3.

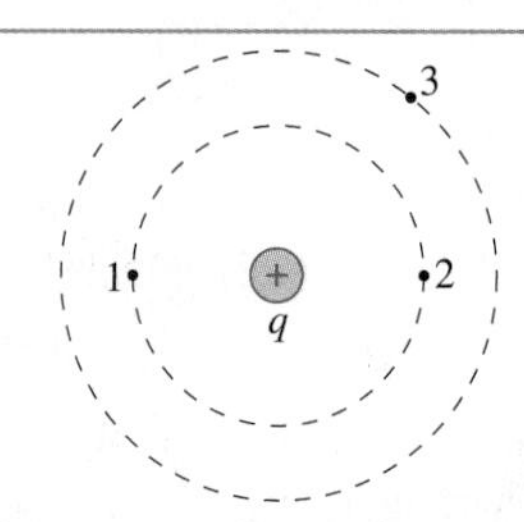

21.5 Connecting Potential and Field

In Chapter 20, we learned how source charges create an electric field around them; in this chapter, we found that source charges also create an electric potential everywhere in the space around them. But these two concepts of field and potential are clearly linked. For instance, we can calculate potential differences by considering the work done on a charge as it is pushed against the electric force due to the field. In this section, we'll build on this idea and will find that **the electric potential and electric field are not two distinct entities but, instead, two different perspectives or two different mathematical representations of how source charges alter the space around them.**

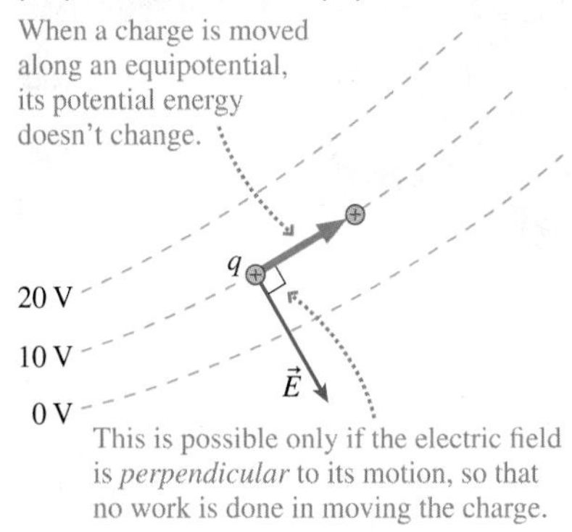

When a charge is moved along an equipotential, its potential energy doesn't change.
q
20 V
10 V
0 V
$\vec{E}$
This is possible only if the electric field is *perpendicular* to its motion, so that no work is done in moving the charge.

FIGURE 21.18 The electric field is always perpendicular to an equipotential surface.

To make the connection between potential and field, FIGURE 21.18 shows an equipotential map of the electric potential due to some source charges (which aren't shown here). Suppose a charge q moves a short distance along one of the equipotential surfaces. Because it moves along an equipotential, its potential and hence its potential energy are the *same* at the beginning and end of its displacement. This means that no work is done in moving the charge. As Figure 21.18 shows, the only way that no work can be done is if the electric field is *perpendicular* to the equipotential. (You should recall, from ◀◀ SECTION 10.2, that no work is done by a force perpendicular to a particle's displacement.) This, then, is our first discovery about the connection between field and potential: **The electric field at a point is perpendicular to the equipotential surface at that point.**

In Figure 21.18, there are actually two directions that are perpendicular to the equipotential surface: the one shown by $\vec{E}$ in the figure, and the other pointing opposite $\vec{E}$. Which direction is correct?

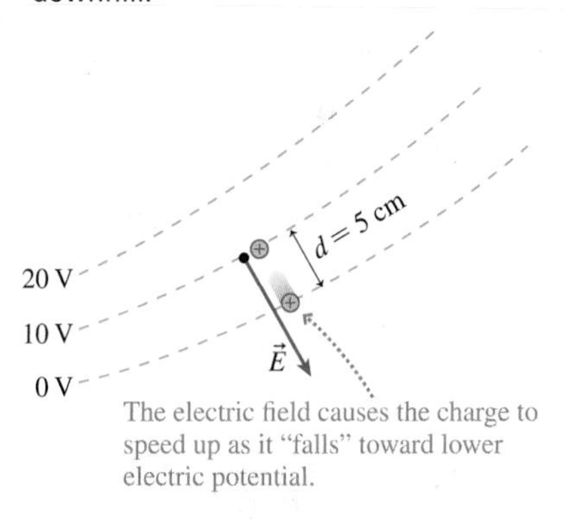

FIGURE 21.19 The electric field points "downhill."

FIGURE 21.19 shows a positive charge released from rest starting at the 10 V equipotential. You learned in Section 21.3 that a positive charge speeds up as it moves from higher to lower potential, so this charge will speed up as it moves toward the lower 0 V equipotential. Because it is the electric field $\vec{E}$ in Figure 21.19 that pushes on the charge, causing it to speed up, $\vec{E}$ must point as shown, from higher potential (10 V) to lower potential (0 V). This is our second discovery about the connection between field and potential: **The electric field points in the direction of** decreasing **potential.**

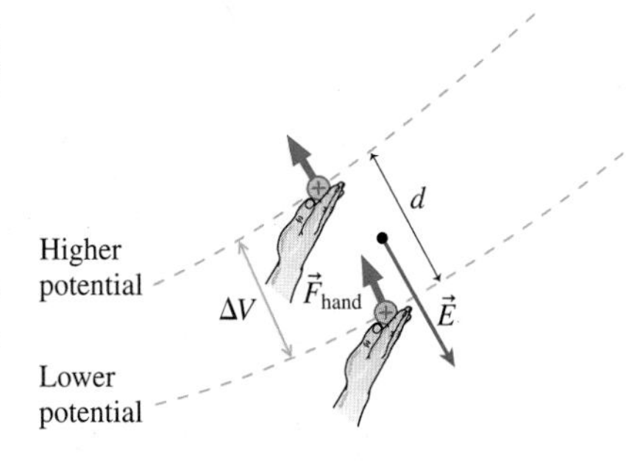

FIGURE 21.20 Finding the strength of the electric field.

Finally, we can find an expression for the *magnitude* of $\vec{E}$ by considering the work required to move the charge, at constant speed, through a displacement that is directed *opposite* $\vec{E}$, as shown in FIGURE 21.20. The displacement is small enough that the electric field in this region can be considered as nearly constant. Conservation of energy requires that

$$W = \Delta U_{elec} = q\Delta V \tag{21.16}$$

Because the charge moves at a constant speed, the magnitude of the force of the hand $\vec{F}_{hand}$ is exactly equal to that of the electric force $q\vec{E}$. Thus, the work done on the charge in moving it through displacement d is

$$W = F_{hand}d = qEd$$

Comparing this result with Equation 21.16 shows that the strength of the electric field is

$$E = \frac{\Delta V}{d} \tag{21.17}$$

Electric field strength in terms of the potential difference ΔV between two equipotential surfaces a distance d apart

Ampullae of Lorenzini

◂ **Electroreception in sharks** BIO Sharks have a "sixth sense" that aids them in detecting prey: They are able to detect the weak electric potentials created by other fish. A shark has an array of sensor cells, the *ampullae of Lorenzini,* distributed around its snout. These cells are highly sensitive to potential differences and can measure voltages as small as a few nV! As we saw in Chapter 20, there are electric potentials associated with every beat of the heart; muscle contractions also create potential differences. The shark can detect these small potential differences, even using them to sense animals buried in the sand.

For example, in Figure 21.19 the magnitude of the potential difference is $\Delta V = (10\text{ V}) - (0\text{ V}) = 10\text{ V}$, while $d = 5\text{ cm} = 0.05\text{ m}$. Thus the electric field strength between these two equipotential surfaces is

$$E = \frac{\Delta V}{d} = \frac{10\text{ V}}{0.05\text{ m}} = 200\text{ V/m}$$

NOTE ▸ This expression for the electric field strength is similar to the result $E = \Delta V_C/d$ for the electric field strength inside a parallel-plate capacitor. A capacitor has a uniform field, the same at all points, so we can calculate the field using the full potential difference ΔV_C and the full spacing d. In contrast, Equation 21.17 applies only *locally,* at a point where the spacing between two nearby equipotential lines is d. ◂

FIGURE 21.21 summarizes what we've learned about the connection between potential and field.

FIGURE 21.21 The geometry of potential and field.

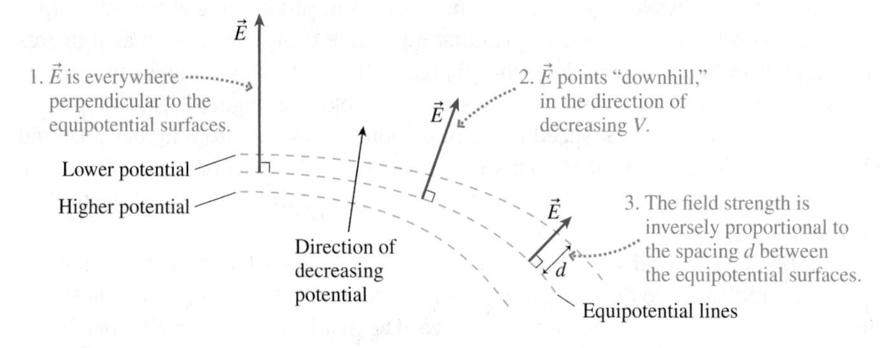
$\vec{E}$
1. $\vec{E}$ is everywhere perpendicular to the equipotential surfaces.
2. $\vec{E}$ points "downhill," in the direction of decreasing V.
Lower potential
Higher potential
Direction of decreasing potential
3. The field strength is inversely proportional to the spacing d between the equipotential surfaces.
d
Equipotential lines

FIGURE 21.22 shows three important arrangements of charges that we've studied in this chapter and Chapter 20. Both electric field lines and equipotentials are shown, and you can see how the connections between field and potential summarized in Figure 21.21 apply in each case.

FIGURE 21.22 Electric field lines and equipotentials for three important cases.

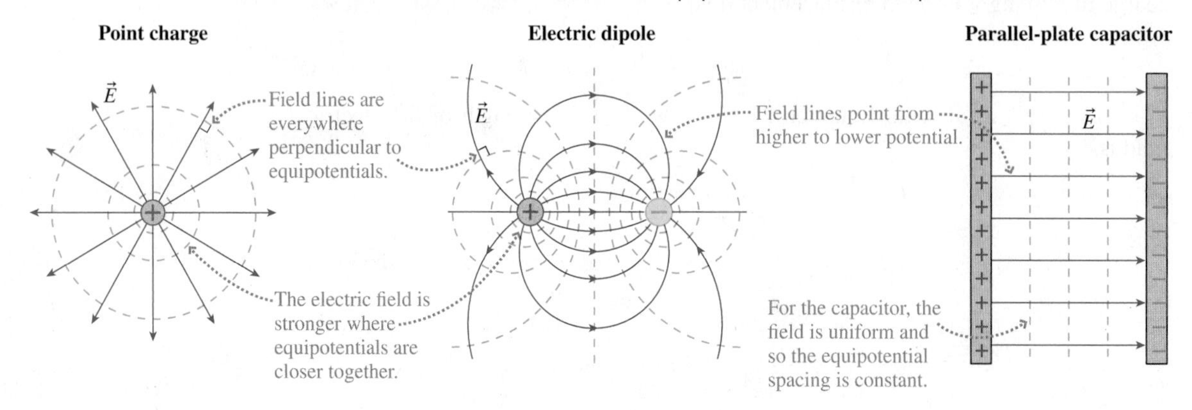
Point charge
Electric dipole
Parallel-plate capacitor
$\vec{E}$
Field lines are everywhere perpendicular to equipotentials.
The electric field is stronger where equipotentials are closer together.
Field lines point from higher to lower potential.
For the capacitor, the field is uniform and so the equipotential spacing is constant.

EXAMPLE 21.9 Finding the electric field from equipotential lines

In **FIGURE 21.23** a 1 cm × 1 cm grid is superimposed on an equipotential map of the potential. Estimate the strength and direction of the electric field at points 1, 2, and 3. Show your results graphically by drawing the electric field vectors on the equipotential map.

FIGURE 21.23 Equipotential lines.

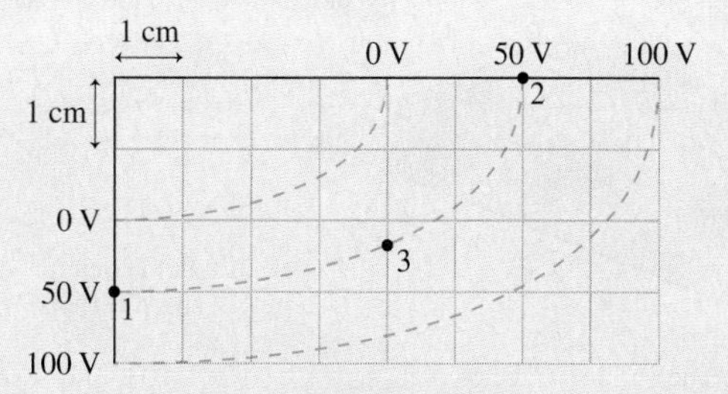

PREPARE The electric field is perpendicular to the equipotential lines, points "downhill," and depends on the spacing between the equipotential lines. The potential is highest on the bottom and the right.

SOLVE Some distant but unseen source charges have created an electric field and potential. We do not need to see the source charges to relate the field to the potential. Because $E = \Delta V/d$, the electric field is stronger where the equipotential lines are closer together and weaker where they are farther apart.

FIGURE 21.24 shows how measurements of d from the grid are combined with values of ΔV to determine $\vec{E}$. Point 3 requires an estimate of the spacing between the 0 V and the 100 V lines. Notice that we're using the 0 V and 100 V equipotential lines to determine $\vec{E}$ at a point on the 50 V equipotential.

FIGURE 21.24 The electric field at points 1, 2, and 3.

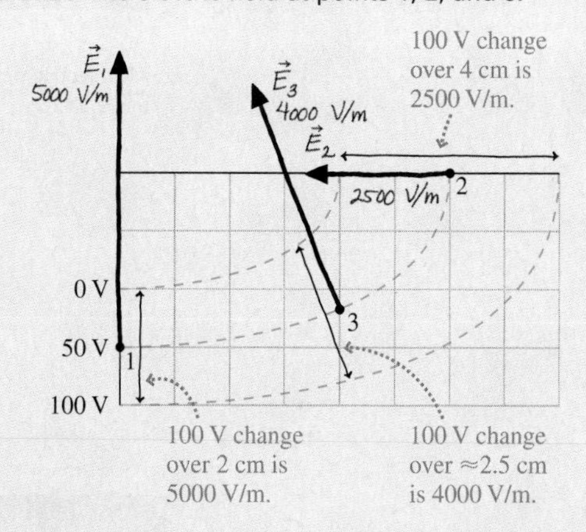

ASSESS The *directions* of $\vec{E}$ are found by drawing downhill vectors perpendicular to the equipotentials. The distances between the equipotential lines are needed to determine the field strengths.

A Conductor in Electrostatic Equilibrium

In ◀◀SECTION 20.6, you learned four important properties about conductors in electrostatic equilibrium:

1. Any excess charge is on the surface.
2. The electric field inside is zero.
3. The exterior electric field is perpendicular to the surface.
4. The field strength is largest at sharp corners.

Now we can add a fifth important property:

5. The entire conductor is at the same potential, and thus the surface is an equipotential surface.

To see why this is so, **FIGURE 21.25** shows two points inside a conductor connected by a line that remains entirely inside the conductor. We can find the potential difference $\Delta V = V_2 - V_1$ between these points by using an external force to push a charge along the line from 1 to 2 and calculating the work done. But because $\vec{E} = \vec{0}$, there is no force on the charge. The work is zero, and so $\Delta V = 0$. In other words, **any two points inside a conductor in electrostatic equilibrium are at the same potential.**

When a conductor is in electrostatic equilibrium, the *entire conductor* is at the same potential. If we charge a metal electrode, the entire electrode is at a single potential. The facts that the surface is an equipotential surface and that the exterior electric field is perpendicular to the surface can now be seen as a special case of our conclusion from the preceding section that electric fields are always perpendicular to equipotentials.

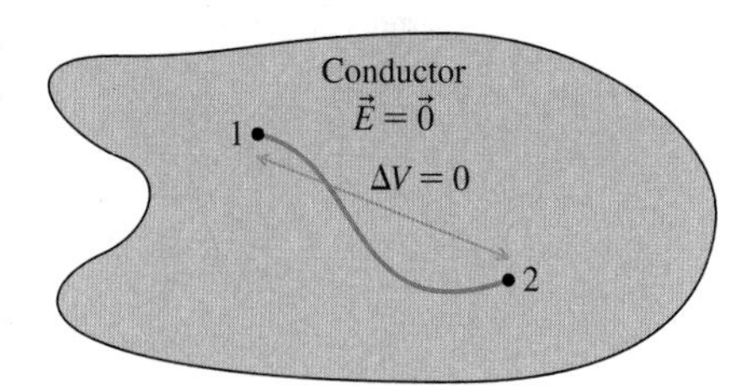

▶ **FIGURE 21.25** All points inside a conductor in electrostatic equilibrium are at the same potential.

Video Tutor Demo

FIGURE 21.26 summarizes what we know about conductors in electrostatic equilibrium.

FIGURE 21.26 Electrical properties of a conductor in electrostatic equilibrium.

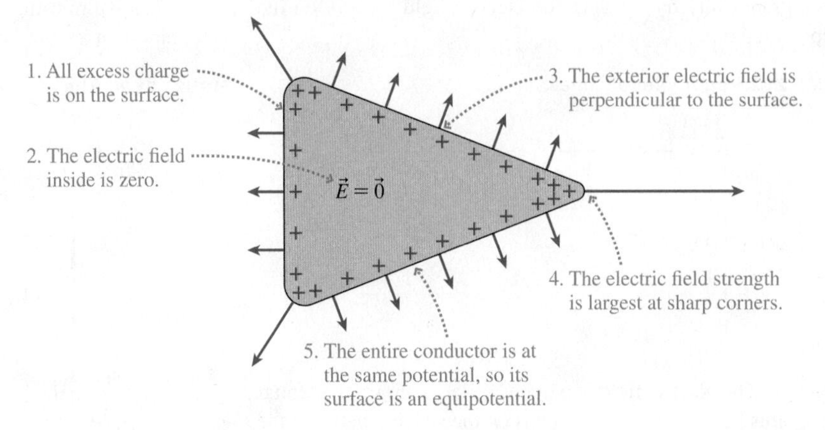

STOP TO THINK 21.5 Which set of equipotential surfaces matches this electric field?

$\vec{E}$

A. 0 V … 50 V
B. 0 V … 50 V
C. 0 V … 50 V
D. 50 V … 0 V
E. 50 V … 0 V
F. 50 V … 0 V

FIGURE 21.27 A contracting heart is an electric dipole.

(a)

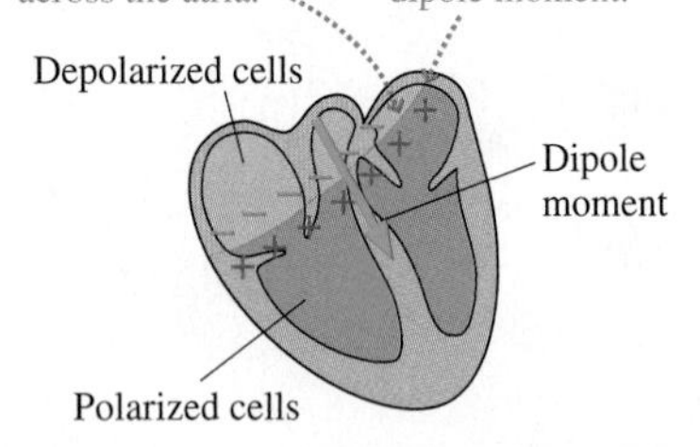

The boundary between polarized and depolarized cells sweeps rapidly across the atria.
At the boundary there is a charge separation. This creates an electric dipole and an associated dipole moment.
Depolarized cells
Dipole moment
Polarized cells

(b)

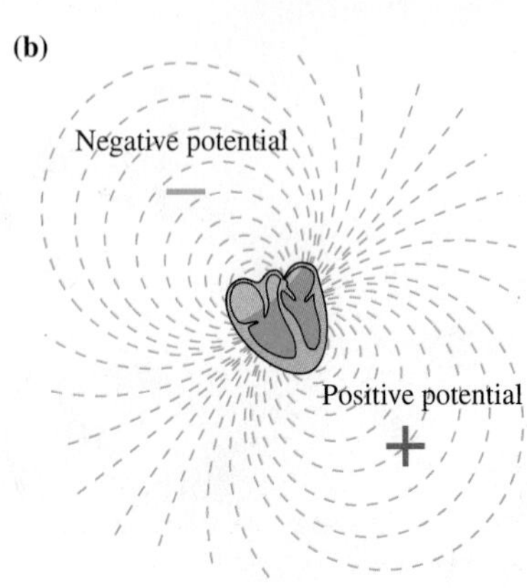

Negative potential
−
Positive potential
+

21.6 The Electrocardiogram BIO

As we saw in ◀◀ SECTION 20.5, the electrical activity of cardiac muscle cells makes the beating heart an electric dipole. A resting nerve or muscle cell is *polarized,* meaning that the outside is positive and the inside negative. Figure 21.5 showed this situation. Initially, all muscle cells in the heart are polarized. When triggered by an electrical impulse from the heart's sino-auricular node in the right atrium, heart cells begin to *depolarize,* moving ions through the cell wall until the outside becomes negative. This causes the muscle to contract. The depolarization of one cell triggers depolarization in an adjacent cell, causing a "wave" of depolarization to spread across the tissues of the heart.

At any instant during this process, a boundary divides the negative charges of depolarized cells from the positive charges of cells that have not depolarized. As **FIGURE 21.27a** shows, this separation of charges creates an electric dipole and produces a dipole electric field and potential. **FIGURE 21.27b** shows the equipotential surfaces of the heart's dipole at one instant of time. These equipotential surfaces match those shown earlier in Figure 21.16.

NOTE ▶ The convention is to draw diagrams of the heart as if you were facing the person whose heart is being drawn. The left side of the heart is thus on the right side of the diagram. ◀

A measurement of the electric potential of the heart is an invaluable diagnostic tool. Recall, however, that only potential *differences* are meaningful, so we need to measure the potential difference between two points on the torso. In practice, as **FIGURE 21.28** shows, the potential difference is measured between several pairs of *electrodes* (often called *leads*). A chart of the potential differences is known as an **electrocardiogram,** abbreviated either ECG or, from its European origin, EKG. A common method of performing an EKG uses 12 leads and records 12 pairs of potential differences.

FIGURE 21.29 shows a simplified model of electrocardiogram measurement using only two electrodes, one on each arm. As the wave of depolarization moves across the heart muscle during each heart beat, the dipole moment vector of the heart changes its magnitude and direction. As Figure 21.29 shows, both of these affect the potential difference between the electrodes, so each point on the EKG graph corresponds to a particular magnitude and orientation of the dipole moment.

FIGURE 21.28 Measuring an EKG.

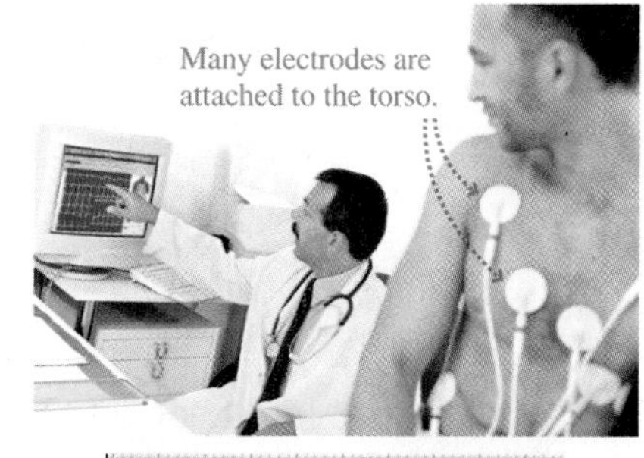
Many electrodes are attached to the torso.

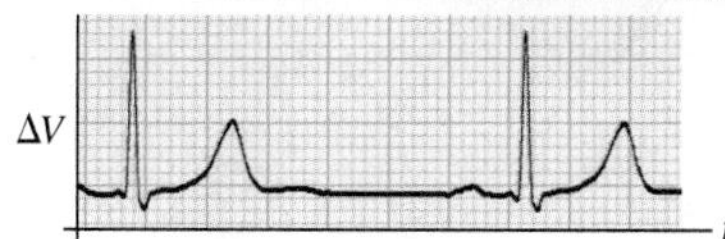
ΔV
t

Records of the potential differences between various pairs of electrodes allow the doctor to analyze the heart's condition.

FIGURE 21.29 The potential difference between the electrodes changes as the heart beats.

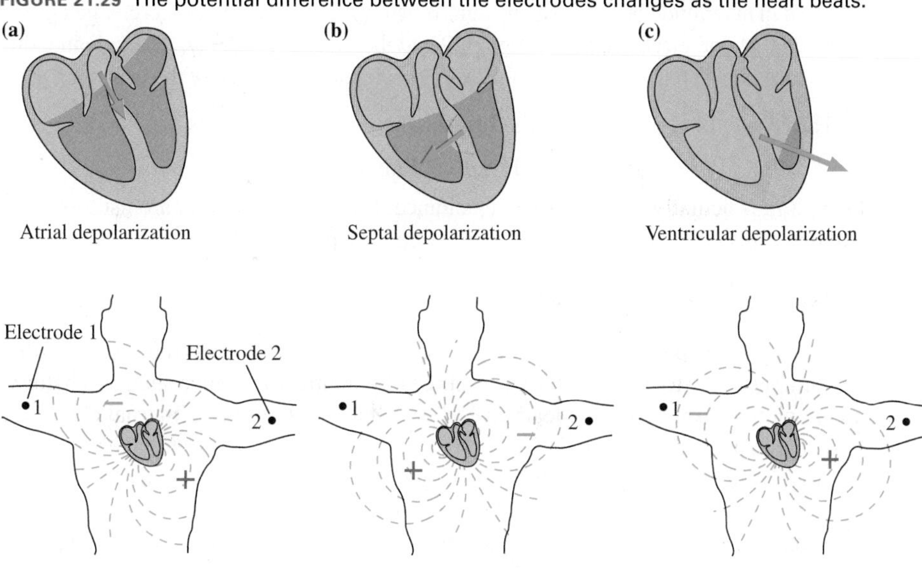
(a)
(b)
(c)
Atrial depolarization
Septal depolarization
Ventricular depolarization
Electrode 1
Electrode 2
1
2
V_2 is positive, V_1 negative.
V_2 is negative, V_1 positive.
V_2 is positive, V_1 negative.

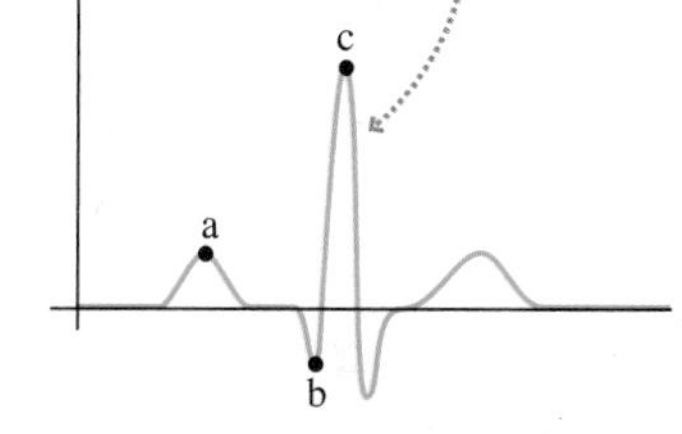
(d)
The record of the potential difference between the two electrodes is the electrocardiogram.
$\Delta V = V_2 - V_1$
a
b
c
t
The potential differences at a, b, and c correspond to those measured in the three stages shown to the left.

21.7 Capacitance and Capacitors

In Section 21.2 we found that potential differences are caused by the separation of charge. One common method of creating a charge separation, shown in **FIGURE 21.30** on the next page, is to move charge Q from one initially uncharged conductor to a second initially uncharged conductor. This results in charge $+Q$ on one conductor and $-Q$ on the other. Two conductors with equal but opposite charge form a **capacitor.** The two conductors that make up a capacitor are its *electrodes* or *plates.* We've already looked at some of the properties of parallel-plate capacitors; now we want to study capacitors that might have any shape. Capacitors can be used to store charge, making them invaluable in all kinds of electronic circuits.

As Figure 21.30 shows, the electric field strength E and the potential difference ΔV_C increase as the charge on each electrode increases. If we double the amount of charge on each electrode, the work required to move a charge from one electrode to the other doubles. This implies a doubling of the potential difference between the electrodes. Thus **the potential difference between the electrodes is directly proportional to their charge.**

Capacitors are important elements in electric circuits. They come in a wide variety of sizes and shapes.

FIGURE 21.30 Charging a capacitor.

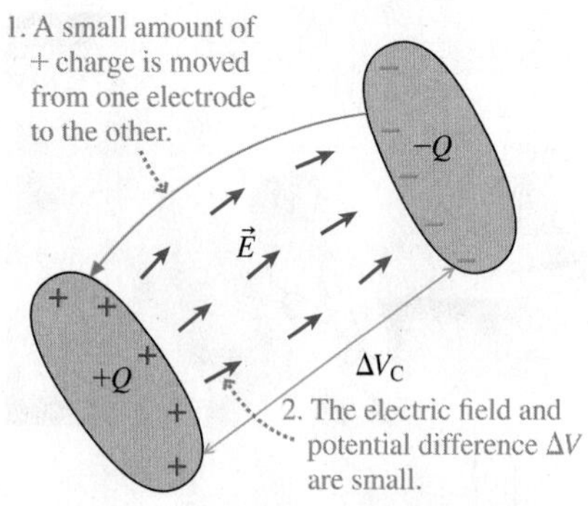

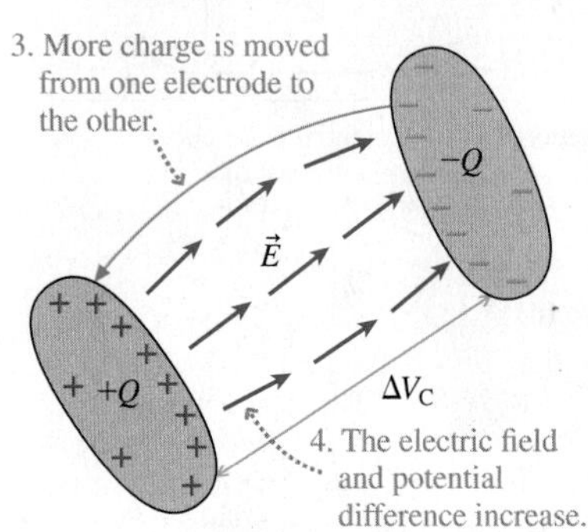

Stated another way, **the charge of a capacitor is directly proportional to the potential difference between its electrodes.** As we'll see, this is actually the most common way that a capacitor is used: A source of potential difference, such as a battery, is connected between the electrodes, causing a charge proportional to the potential difference to be moved from one electrode to the other. We can write the relationship between charge and potential difference as

$$Q = C\,\Delta V_C \tag{21.18}$$

Charge on a capacitor with potential difference ΔV_C

Q ΔV_C p. 34 PROPORTIONAL

The constant of proportionality C between Q and ΔV_C is called the **capacitance** of the capacitor. Capacitance depends on the shape, size, and spacing of the two electrodes. A capacitor with a large capacitance holds more charge for a given potential difference than one with a small capacitance.

NOTE ▶ We will consider only situations where the charges on the electrodes are equal in magnitude but opposite in sign. When we say "A capacitor has charge Q," we mean that one electrode has charge $+Q$ and the other charge $-Q$. The potential difference between the electrodes is called the potential difference *of* the capacitor. ◀

The SI unit of capacitance is the **farad.** One farad is defined as

$$1 \text{ farad} = 1 \text{ F} = 1 \text{ coulomb/volt} = 1 \text{ C/V}$$

One farad is actually a very large capacitance. Practical capacitors are usually measured in units of microfarads (μF) or picofarads (1 pF $= 10^{-12}$ F).

Charging a Capacitor

To "charge" a capacitor, we need to move charge from one electrode to the other. The simplest way to do this is to use a source of potential difference such as a battery, as shown in FIGURE 21.31. We learned earlier that a battery uses its internal chemistry to maintain a fixed potential difference between its terminals. If we connect a capacitor to a battery, charge flows from the negative electrode of the capacitor, through the battery, and onto the positive electrode. This flow of charge continues until the potential difference between the capacitor's electrodes is the same as the fixed potential difference of the battery. If the battery is then removed, the capacitor remains charged with a potential equal to that of the battery that charged it because there's no conducting path for charge on the positive electrode to move back to the negative electrode. Thus **a capacitor can be used to store charge.**

FIGURE 21.31 Charging a capacitor using a battery.

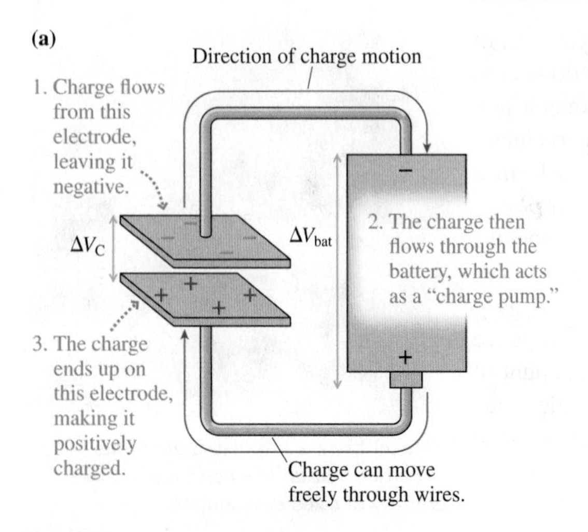

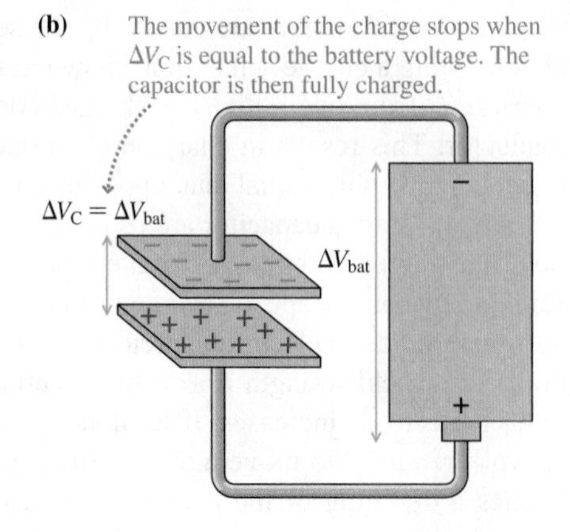

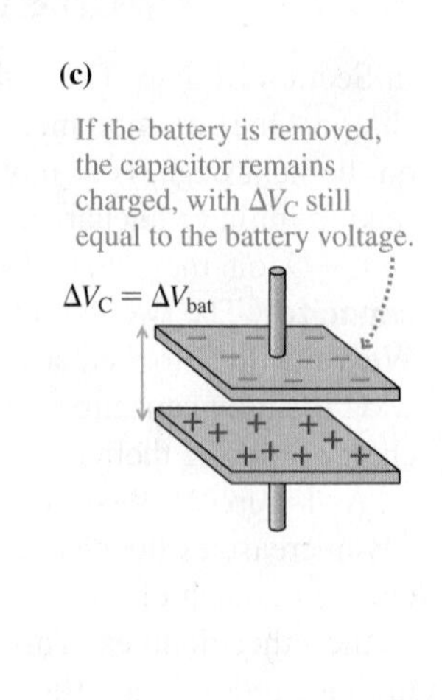

(c)
If the battery is removed, the capacitor remains charged, with ΔV_C still equal to the battery voltage.
$\Delta V_C = \Delta V_{bat}$

EXAMPLE 21.10 Charging a capacitor

A 1.3 μF capacitor is connected to a 1.5 V battery. What is the charge on the capacitor?

PREPARE Charge flows through the battery from one capacitor electrode to the other until the potential difference ΔV_C between the electrodes equals that of the battery, or 1.5 V.

SOLVE The charge on the capacitor is given by Equation 21.18:

$$Q = C\,\Delta V_C = (1.3 \times 10^{-6}\ \text{F})(1.5\ \text{V}) = 2.0 \times 10^{-6}\ \text{C}$$

ASSESS This is the charge on the positive electrode; the other electrode has a charge of -2.0×10^{-6} C.

The Parallel-Plate Capacitor

As we've seen, the *parallel-plate capacitor* is important because it creates a uniform electric field between its flat electrodes. In ◀◀ SECTION 20.5, we found that the electric field of a parallel-plate capacitor is

$$\vec{E} = \left(\frac{Q}{\epsilon_0 A}, \text{ from positive to negative}\right)$$

where A is the surface area of the electrodes and Q is the charge on the capacitor. We can use this result to find the capacitance of a parallel-plate capacitor.

Earlier in this chapter, we found that the electric field strength of a parallel-plate capacitor is related to the potential difference ΔV_C and the plate spacing d by

$$E = \frac{\Delta V_C}{d}$$

Combining these two results, we see that

$$\frac{Q}{\epsilon_0 A} = \frac{\Delta V_C}{d}$$

or, equivalently,

$$Q = \frac{\epsilon_0 A}{d}\Delta V_C \tag{21.19}$$

If we compare Equation 21.19 to Equation 21.18, the definition of capacitance, we see that the capacitance of the parallel-plate capacitor is

$$C = \frac{\epsilon_0 A}{d} \tag{21.20}$$

Capacitance of a parallel-plate capacitor with plate area A and separation d

PROPORTIONAL p. 34 INVERSE p. 108

NOTE ▶ From Equation 21.20 you can see that the units of ϵ_0 can be written as F/m. These units are useful when working with capacitors. ◀

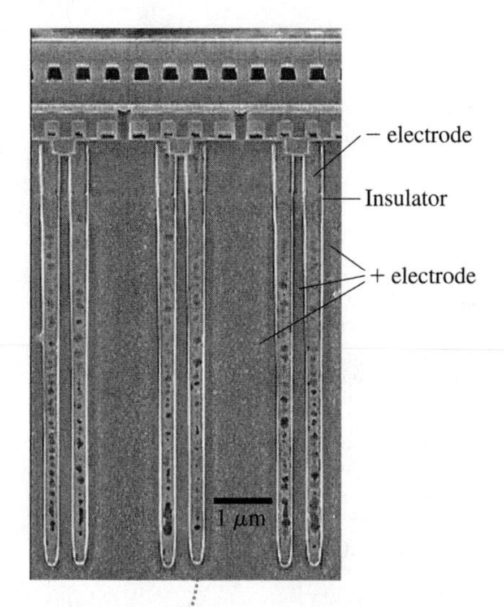

Each long structure is one capacitor.

A capacity for memory Your computer's random-access memory, or RAM, uses tiny capacitors to store the digital ones and zeroes that make up your data. A charged capacitor represents a one and an uncharged capacitor a zero. For a billion or more capacitors to fit on a single chip they must be very small. The micrograph is a cross section through the silicon wafer that makes up the memory chip. Each capacitor consists of a very long electrode separated by a thin insulating layer from the common electrode shared by all capacitors. Each capacitor's capacitance is only about 30×10^{-15} F!

EXAMPLE 21.11 Charging a parallel-plate capacitor

The spacing between the plates of a 1.0 μF parallel-plate capacitor is 0.070 mm.

a. What is the surface area of the plates?
b. How much charge is on the plates if this capacitor is attached to a 1.5 V battery?

SOLVE a. From the definition of capacitance,

$$A = \frac{dC}{\epsilon_0} = \frac{(0.070 \times 10^{-3}\ \text{m})(1.0 \times 10^{-6}\ \text{F})}{8.85 \times 10^{-12}\ \text{F/m}} = 7.9\ \text{m}^2$$

b. The charge is $Q = C\,\Delta V_C = (1.0 \times 10^{-6}\ \text{F})(1.5\ \text{V}) = 1.5 \times 10^{-6}\ \text{C} = 1.5\ \mu\text{C}$.

ASSESS The surface area needed to construct a 1.0 μF capacitor (a fairly typical value) is enormous and hardly practical. We'll see in the next section that real capacitors can be reduced to a more manageable size by placing an insulator between the capacitor plates.

FIGURE 21.32 An insulator in an electric field becomes polarized.

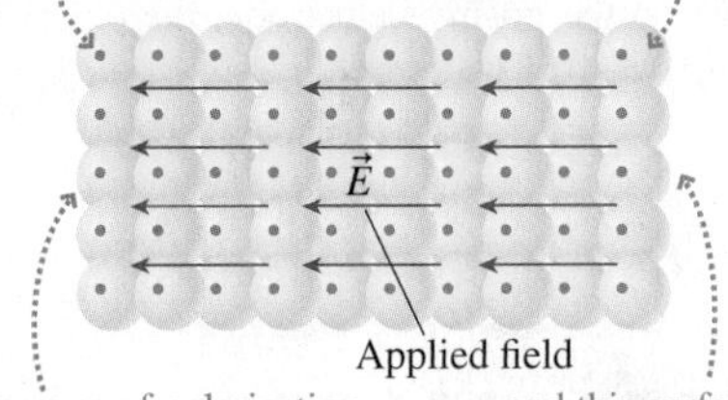
The positive nuclei shift very slightly in the direction of the field.
The negative electron cloud shifts very slightly in the direction opposite the field.
$\vec{E}$
Applied field
Because of polarization, this surface has an excess of positive charge . . .
. . . and this surface has an excess of negative charge.

FIGURE 21.33 The electric field inside a dielectric.

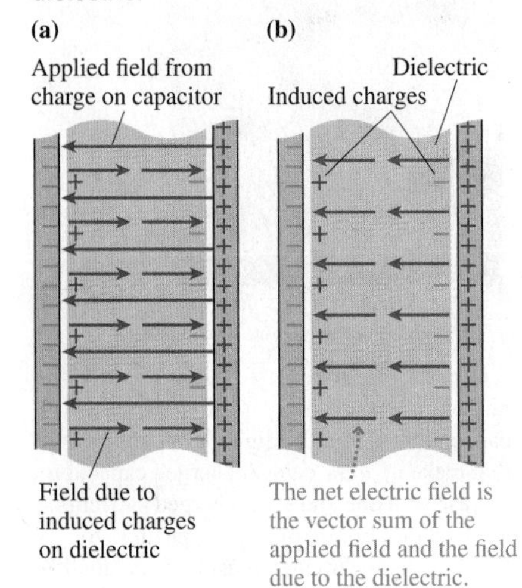
(a)
(b)
Applied field from charge on capacitor
Dielectric
Induced charges
Field due to induced charges on dielectric
The net electric field is the vector sum of the applied field and the field due to the dielectric.

Dielectrics and Capacitors

An insulator consists of vast numbers of atoms. When an insulator is placed in an electric field, each of its atoms polarizes. Recall, from ◀◀ SECTION 20.2, that *polarization* occurs when an atom's negative electron cloud and positive nucleus shift very slightly in opposite directions in response to an applied electric field. The net effect of all these tiny atomic polarizations is shown in FIGURE 21.32: An *induced* positive charge builds up on one surface of the insulator, and an induced negative charge on the other surface.

Notice that this distribution of charge—two equal but opposite layers—is identical to that of a parallel-plate capacitor, so this induced charge will create a uniform electric field, but one that is directed *opposite* to the applied electric field.

An insulator placed between the plates of a capacitor is called a **dielectric.** FIGURE 21.33a shows a dielectric in the uniform field of a parallel-plate capacitor with charge Q. The capacitor's electric field polarizes the dielectric; the polarized dielectric creates a field of its own directed opposite to the capacitor's field. As FIGURE 21.33b shows, these two fields add to give a net field in the same direction as the applied field, but *smaller*. Thus the electric field between the capacitor plates is smaller than it was without the dielectric.

When the dielectric is inserted, the capacitor's charge Q doesn't change because its plates are isolated from each other. But the electric field between the plates decreases, which implies that the potential difference $\Delta V_C = Ed$ between the plates decreases as well. This implies that the capacitance $C = Q/\Delta V_C$ *increases.* **The presence of a dielectric results in an increased capacitance.**

Because some atoms are more easily polarized than others, the factor by which the capacitance is increased depends on the dielectric material. This factor is called the **dielectric constant** of the material, and it is given the symbol κ (Greek letter kappa). If the capacitance without a dielectric present is C_0, then the capacitance with a dielectric present is

$$C = \kappa C_0 \tag{21.21}$$

Capacitance of a parallel-plate capacitor with a dielectric of dielectric constant κ

TABLE 21.3 Dielectric constants of some materials at 20°C

Material	Dielectric constant κ
Vacuum	1 (exactly)
Air	1.00054*
Teflon	2.0
Paper	3.0
Pyrex glass	4.8
Cell membrane	9.0
Ethanol	24
Water	80
Strontium titanate	300

*Use 1.00 in all calculations.

Table 21.3 lists the dielectric constants for a number of common substances. Note the high dielectric constant of water, which is of great importance in regulating the chemistry of biological processes. As we saw in ◀◀ SECTION 20.2, a molecule of water is a *permanent* electric dipole because the oxygen atom has a slight negative charge while the two hydrogen atoms have a slight positive charge. Because the charge in a water molecule is *already* separated, the molecules' dipoles easily turn to line up with an applied electric field, leading to water's very high dielectric constant.

EXAMPLE 21.12 Finding the dielectric constant

A parallel-plate capacitor is charged using a 100 V battery; then the battery is removed. If a dielectric slab is slid between the plates, filling the space inside, the capacitor voltage drops to 30 V. What is the dielectric constant of the dielectric?

PREPARE The capacitor voltage remains $(\Delta V_C)_1 = 100$ V when it is disconnected from the battery. Placing the dielectric between the plates reduces the voltage to $(\Delta V_C)_2 = 30$ V. Because the plates are not connected when the dielectric is inserted, the charge on the plates remains constant.

SOLVE Because the plates are not connected, the charge on the capacitor is constant, so we have

$$Q_1 = C_1(\Delta V_C)_1 = Q_2 = C_2(\Delta V_C)_2$$

Inserting the dielectric increases the capacitance by a factor of κ, so that $C_2 = \kappa C_1$. Thus $C_1(\Delta V_C)_1 = \kappa C_1(\Delta V_C)_2$ or, canceling C_1, $(\Delta V_C)_1 = \kappa(\Delta V_C)_2$. The dielectric constant is then

$$\kappa = \frac{(\Delta V_C)_1}{(\Delta V_C)_2} = \frac{100\ \text{V}}{30\ \text{V}} = 3.3$$

ASSESS The dielectric constant is greater than 1, as must be the case.

In Example 21.11, we found that impractically large electrodes would be needed to create a parallel-plate capacitor with $C = 1\ \mu\text{F}$. Practical capacitors take advantage of dielectrics with $\kappa > 100$. As shown in **FIGURE 21.34**, a typical capacitor is a "sandwich" of two modest-sized pieces of aluminum foil separated by a very thin dielectric that increases the capacitance by a large factor. This sandwich is then folded, rolled up, and sealed in a small plastic cylinder. The two wires extending from a capacitor connect to the two electrodes and allow the capacitor to be charged. Even though the electrodes are no longer planes, the capacitance is reasonably well predicted from the parallel-plate-capacitor equation if A is the area of the foils before being folded and rolled.

FIGURE 21.34 A capacitor disassembled to show its internal rolled-up structure.

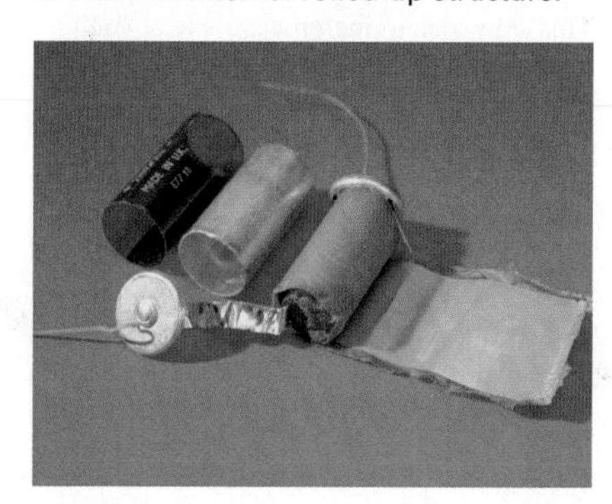

STOP TO THINK 21.6 If the potential difference across a capacitor is doubled, its capacitance

A. Doubles. B. Halves. C. Remains the same.

21.8 Energy and Capacitors

We've seen that a practical way of charging a capacitor is to attach its plates to a battery. Charge then flows from one plate, through the battery, and onto the other plate, leaving one plate with charge $-Q$ and the other with charge $+Q$. The battery must do *work* to transfer the charge; this work increases the electric potential energy of the charge on the capacitor. **A charged capacitor stores energy as electric potential energy.**

To find out how much energy is stored in a charged capacitor, recall that a charge q moved through a potential difference ΔV gains potential energy $U = q\Delta V$. When a capacitor is charged, total charge Q is moved from the negative plate to the positive plate. At first, when the capacitor is uncharged, the potential difference is $\Delta V = 0$. The last little bit of charge to be moved, as the capacitor reaches full charge, has to move through a potential difference $\Delta V \approx \Delta V_C$. *On average,* the potential difference across a capacitor while it's being charged is $\Delta V_{\text{average}} = \frac{1}{2}\Delta V_C$. Thus it seems plausible (and can be proved in a more advanced treatment) that the potential energy U_C stored in a charged capacitor is

$$U_C = Q\,\Delta V_{\text{average}} = \frac{1}{2}Q\,\Delta V_C$$

Taking a picture in a flash When you take a flash picture, the flash is fired using electric potential energy stored in a capacitor. Batteries are unable to deliver the required energy rapidly enough, but capacitors can discharge all their energy in only microseconds. A battery is used to slowly charge up the capacitor, which then rapidly discharges through the flashlamp. This slow recharging process is why you must wait some time between taking flash pictures.

We can use $Q = C\Delta V_C$ to write this in two different ways:

$$U_C = \frac{1}{2}\frac{Q^2}{C} = \frac{1}{2}C(\Delta V_C)^2 \tag{21.22}$$

Electric potential energy of a capacitor with charge Q and potential difference ΔV_C

U_C p.44 ΔV_C QUADRATIC

The potential energy stored in a capacitor depends on the *square* of the potential difference across it. This result is reminiscent of the potential energy $U_s = \frac{1}{2}k(\Delta x)^2$ stored in a spring, and a charged capacitor really is analogous to a stretched spring. A stretched spring holds energy until we release it; then that potential energy is transformed into kinetic energy. Likewise, a charged capacitor holds energy until we discharge it.

EXAMPLE 21.13 Energy in a camera flash

How much energy is stored in a 220 μF camera-flash capacitor that has been charged to 330 V? What is the average power delivered to the flash lamp if this capacitor is discharged in 1.0 ms?

SOLVE The energy stored in the capacitor is

$$U_C = \frac{1}{2}C(\Delta V_C)^2 = \frac{1}{2}(220 \times 10^{-6}\ \text{F})(330\ \text{V})^2 = 12\ \text{J}$$

If this energy is released in 1.0 ms, the average power is

$$P = \frac{\Delta E}{\Delta t} = \frac{12\ \text{J}}{1.0 \times 10^{-3}\ \text{s}} = 12{,}000\ \text{W}$$

ASSESS The stored energy is equivalent to raising a 1 kg mass by 1.2 m. This is a rather large amount of energy; imagine the damage a 1 kg object could do after falling 1.2 m. When this energy is released very quickly, as is possible in an electronic circuit, the power is very high.

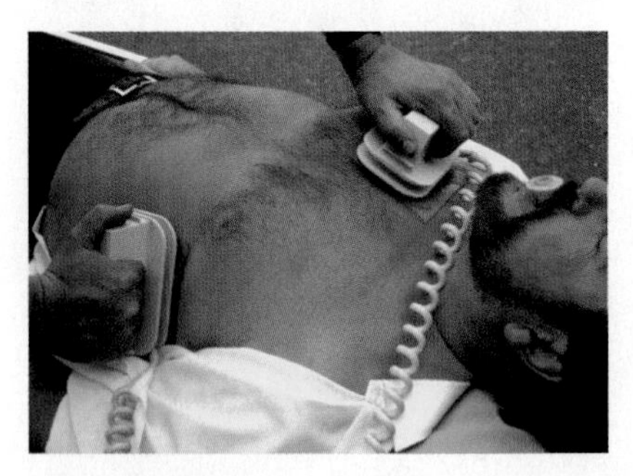

BIO A defibrillator, which can restore a heart-beat, discharges a capacitor through the patient's chest.

The usefulness of a capacitor stems from the fact that it can be charged very slowly, over many seconds, and then can release the energy very quickly. A mechanical analogy would be using a crank to slowly stretch the spring of a catapult, then quickly releasing the energy to launch a massive rock.

An important medical application of the ability of capacitors to rapidly deliver energy is the *defibrillator.* A heart attack or a serious injury can cause the heart to enter a state known as *fibrillation* in which the heart muscles twitch randomly and cannot pump blood. A strong electric shock through the chest can sometimes stop the fibrillation and allow a normal heart rhythm to be restored. A defibrillator has a large capacitor that can store up to 360 J of energy. This energy is released in about 2 ms through two "paddles" pressed against the patient's chest. It takes several seconds to charge the capacitor, which is why, on television medical shows, you hear an emergency room doctor or nurse shout, "Charging!"

The Energy in the Electric Field

We can "see" the potential energy of a stretched spring in the tension of the coils. If a charged capacitor is analogous to a stretched spring, where is the stored energy? It's in the electric field!

FIGURE 21.35 on the next page shows a parallel-plate capacitor, filled with a dielectric with dielectric constant κ, in which the plates have area A and are separated by distance d. Its capacitance is then κ times that of a parallel-plate capacitor without a dielectric, or $C = \kappa\epsilon_0 A/d$. The potential difference across the capacitor is related to the electric field inside the capacitor by $\Delta V_C = Ed$. Substituting into Equation 21.22, we find that the energy stored in the capacitor is

$$U_C = \frac{1}{2}C(\Delta V_C)^2 = \frac{1}{2}\frac{\kappa\epsilon_0 A}{d}(Ed)^2 = \frac{1}{2}\kappa\epsilon_0(Ad)E^2 \qquad (21.23)$$

The quantity Ad is the volume *inside* the capacitor, the region in which the capacitor's electric field exists. (Recall that an ideal capacitor has $\vec{E} = \vec{0}$ everywhere except between the plates.) Although we talk about "the energy stored in the capacitor," Equation 21.23 suggests that, strictly speaking, **the energy is stored in the capacitor's electric field.**

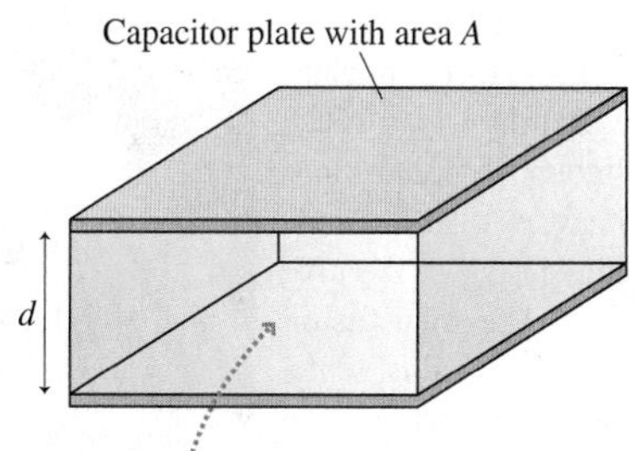

FIGURE 21.35 A capacitor's energy is stored in the electric field.

Because Ad is the volume in which the energy is stored, we can define an **energy density** u_E of the electric field:

$$u_E = \frac{\text{energy stored}}{\text{volume in which it is stored}} = \frac{U_C}{Ad} = \frac{1}{2}\kappa\epsilon_0 E^2 \qquad (21.24)$$

The energy density has units J/m^3. We've derived Equation 21.24 for a parallel-plate capacitor, but it turns out to be the correct expression for any electric field.

From this perspective, charging a capacitor stores energy in the capacitor's electric field as the field grows in strength. Later, when the capacitor is discharged, the energy is released as the field collapses.

We first introduced the electric field as a way to visualize how a long-range force operates. But if the field can store energy, the field must be real, not merely a pictorial device. We'll explore this idea further in Chapter 25, where we'll find that the energy transported by a light wave—the very real energy of warm sunshine—is the energy of electric and magnetic fields.

EXAMPLE 21.14 Finding the energy density for a defibrillator

A defibrillator unit contains a 150 μF capacitor that is charged to 2000 V. The capacitor plates are separated by a 0.010-mm-thick dielectric with $\kappa = 300$.

a. What is the total area of the capacitor plates?
b. What is the energy density stored in the electric field when the capacitor is charged?

PREPARE Assume the capacitor can be modeled as a parallel-plate capacitor with a dielectric.

SOLVE a. The surface area of the electrodes is

$$A = \frac{dC}{\kappa\epsilon_0} = \frac{(1.0 \times 10^{-5}\ \text{m})(150 \times 10^{-6}\ \text{F})}{(300)(8.85 \times 10^{-12}\ \text{F/m})} = 0.56\ \text{m}^2$$

b. The electric field strength is

$$E = \frac{\Delta V_C}{d} = \frac{2000\ \text{V}}{1.0 \times 10^{-5}\ \text{m}} = 2.0 \times 10^8\ \text{V/m}$$

Consequently, the energy density in the electric field is

$$u_E = \frac{1}{2}\kappa\epsilon_0 E^2$$
$$= \frac{1}{2}(300)(8.85 \times 10^{-12}\ \text{F/m})(2.0 \times 10^8\ \text{V/m})^2$$
$$= 5.3 \times 10^7\ \text{J/m}^3$$

ASSESS For comparison, the energy density of gasoline is about $3 \times 10^9\ \text{J/m}^3$, about 60 times higher than this capacitor. Capacitors store less energy than some other devices, but they can deliver this energy *very* rapidly.

STOP TO THINK 21.7 The plates of a parallel-plate capacitor are connected to a battery. If the distance between the plates is halved, the energy of the capacitor

A. Increases by a factor of 4.
B. Doubles.
C. Remains the same.
D. Is halved.
E. Decreases by a factor of 4.

INTEGRATED EXAMPLE 21.15 Proton fusion in the sun

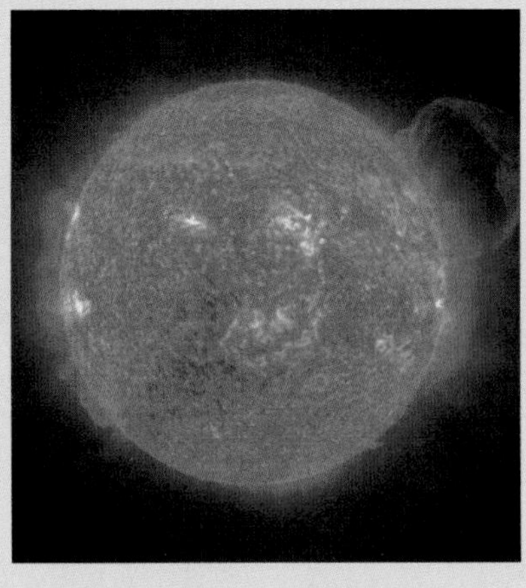

The sun's energy comes from nuclear reactions that fuse lighter nuclei into heavier ones, releasing energy in the process. The solar fusion process begins when two protons (the nuclei of hydrogen atoms) merge to produce a *deuterium* nucleus. Deuterium is the "heavy" isotope of hydrogen, with a nucleus consisting of a proton *and* a neutron. To become deuterium, one of the protons that fused has to turn into a neutron. The nuclear-physics process by which this occurs—and which releases the energy—will be studied in Chapter 30. Our interest for now lies not with the nuclear physics but with the conditions that allow fusion to occur.

Before two protons can fuse, they must come into contact. However, the energy required to bring two protons into contact is considerable because the electric potential energy of the two protons increases rapidly as they approach each other. Fusion occurs in the core of the sun because the ultra-high temperature there gives the protons the kinetic energy they need to come together.

a. A proton can be modeled as a charged sphere of diameter $d_p = 1.6 \times 10^{-15}$ m with total charge e. When two protons are in contact, what is the electric potential of one proton at the center of the other?

b. Two protons are approaching each other head-on, each with the same speed v_0. What value of v_0 is required for the protons to just come into contact with each other?

c. What does the temperature of the sun's core need to be so that the rms speed v_{rms} of protons is equal to v_0?

PREPARE Energy is conserved, so Problem-Solving Strategy 21.1 is the basis of our solution. **FIGURE 21.36** shows a before-and-after visual overview. Both protons are initially moving with speeds $v_i = v_0$, so both contribute to the initial kinetic energy. We will assume that they start out so far apart that $U_i \approx 0$. To "just touch" means that they've instantaneously come to rest ($K_f = 0$) at the point where the distance between their centers is equal to the diameter of a proton. We can use the potential of a charged sphere and the energy-conservation equation to find the speed v_0 required to achieve contact. Then we can use the results of Chapter 12 to find the temperature at which v_0 is the rms speed of the protons.

FIGURE 21.36 Visual overview of two protons coming into contact.

SOLVE a. The electric potential at distance r from a charged sphere was found to be $V = KQ/r$. When the protons are in contact, the distance between their centers is $r_f = d_p = 1.6 \times 10^{-15}$ m. Thus the potential of one proton, with $Q = e$, at the center of the other is

$$V = \frac{Ke}{r_f} = \frac{Ke}{d_p} = \frac{(9.0 \times 10^9 \text{ N} \cdot \text{m}^2/\text{C}^2)(1.6 \times 10^{-19} \text{ C})}{1.6 \times 10^{-15} \text{ m}}$$
$$= 9.0 \times 10^5 \text{ V}$$

b. The conservation of energy equation $K_f + qV_f = K_i + qV_i$ is

$$(0 + 0) + eV_f = \left(\frac{1}{2}mv_0^2 + \frac{1}{2}mv_0^2\right) + 0$$

where, as noted above, both protons contribute to the initial kinetic energy, both end up at rest as the protons touch, and they started far enough apart that the initial potential energy (and potential) is effectively zero. When the protons meet, their potential energy is the charge of one proton (e) multiplied by the electric potential of the other—namely, the potential found in part a: $V_f = 9.0 \times 10^5$ V. Solving the energy equation for v_0, we get

$$v_0 = \sqrt{\frac{eV_f}{m}} = \sqrt{\frac{(1.6 \times 10^{-19} \text{ C})(9.0 \times 10^5 \text{ V})}{1.67 \times 10^{-27} \text{ kg}}}$$
$$= 9.29 \times 10^6 \text{ m/s}$$

c. In Section 12.2 we found that the temperature of a gas is related to the average kinetic energy of the particles and thus to the rms speed of the particles by the equation

$$T = \frac{mv_{rms}^2}{3k_B}$$

It may seem strange to think of protons as a gas, but in the center of the sun, where all the atoms are ionized into nuclei and electrons, the protons are zooming around and do, indeed, act like a gas. For v_{rms} of the protons to be equal to v_0 that we calculated in part b, the temperature would have to be

$$T = \frac{mv_0^2}{3k_B} = \frac{(1.67 \times 10^{-27} \text{ kg})(9.29 \times 10^6 \text{ m/s})^2}{3(1.38 \times 10^{-23} \text{ J/K})} = 3.5 \times 10^9 \text{ K}$$

ASSESS An extraordinarily high temperature—over 3 billion kelvin—is required to give an average solar proton a speed of 9.29×10^6 m/s. In fact, the core temperature of the sun is "only" about 14 million kelvin, a factor of ≈ 200 less than we calculated. Protons can fuse at this lower temperature both because there are always a few protons moving much faster than average and because protons can reach each other even if their speeds are too low by the quantum-mechanical process of *tunneling*, which you'll learn about in Chapter 28. Still, because of the core's relatively "low" temperature, most protons bounce around in the sun for several billion years before fusing!

SUMMARY

Goal: To calculate and use the electric potential and electric potential energy.

GENERAL PRINCIPLES

Electric Potential and Potential Energy

The electric potential V is created by charges and exists at every point surrounding those charges.

These charges create the electric potential.

$V = -100$ V $V = 300$ V $V = 200$ V

When a charge q is brought near these charges, it acquires an electric potential energy

$$U_{elec} = qV$$

at a point where the other charges have created an electric potential V.

Energy is conserved for a charged particle in an electric potential:

$$K_f + qV_f = K_i + qV_i$$

or

$$\Delta K = -q\Delta V$$

Sources of Potential

Potential differences ΔV are created by a *separation of charge.* Two important sources of potential difference are

- A *battery,* which uses chemical means to separate charge and produce a potential difference.
- The opposite charges on the plates of a *capacitor,* which create a potential difference between the plates.

The electric potential of a point charge q is $V = K\dfrac{q}{r}$

Connecting potential and field

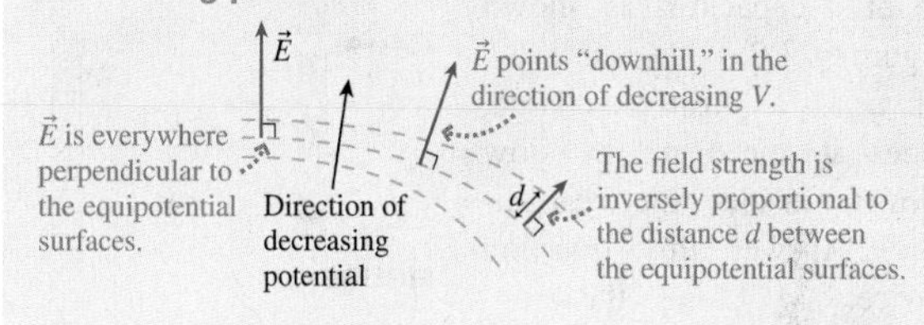

$\vec{E}$
$\vec{E}$ points "downhill," in the direction of decreasing V.
$\vec{E}$ is everywhere perpendicular to the equipotential surfaces.
Direction of decreasing potential
d
The field strength is inversely proportional to the distance d between the equipotential surfaces.

IMPORTANT CONCEPTS

For a **conductor in electrostatic equilibrium**

- Any excess charge is on the surface.
- The electric field inside is zero.
- The exterior electric field is perpendicular to the surface.
- The field strength is largest at sharp corners.
- The entire conductor is at the same potential and so the surface is an equipotential.

$\vec{E}$ $\vec{E} = \vec{0}$ The surface is an equipotential.

Graphical representations of the potential

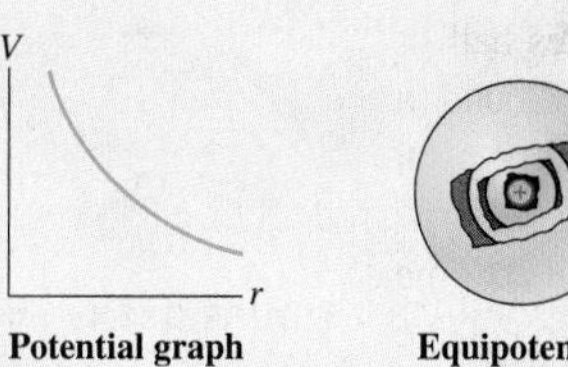

V
r
Potential graph
Equipotential surfaces

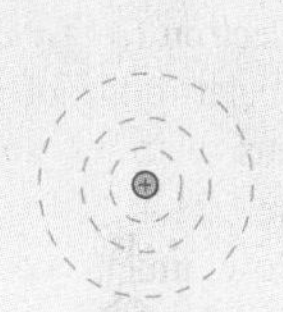

Equipotential map

APPLICATIONS

Capacitors and dielectrics

The charge $\pm Q$ on two conductors and the potential difference ΔV_C between them are proportional:

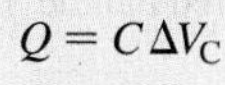

$$Q = C\Delta V_C$$

$+Q$ $-Q$ ΔV_C

where C is the **capacitance** of the two conductors.

A **parallel-plate capacitor** with plates of area A and separation d has a capacitance

$$C = \epsilon_0 A/d$$

When a **dielectric** is inserted between the plates of a capacitor, its capacitance is increased by a factor κ, the **dielectric constant** of the material.

The **energy stored in a capacitor** is $U_C = \frac{1}{2}C(\Delta V_C)^2$.

This energy is stored in the electric field, which has energy density

$$u_E = \tfrac{1}{2}\kappa\epsilon_0 E^2$$

Parallel-plate capacitor

For a capacitor charged to ΔV_C the potential at distance x from the negative plate is

$$V = \frac{x}{d}\Delta V_C$$

The electric field inside is

$$E = \Delta V_C/d$$

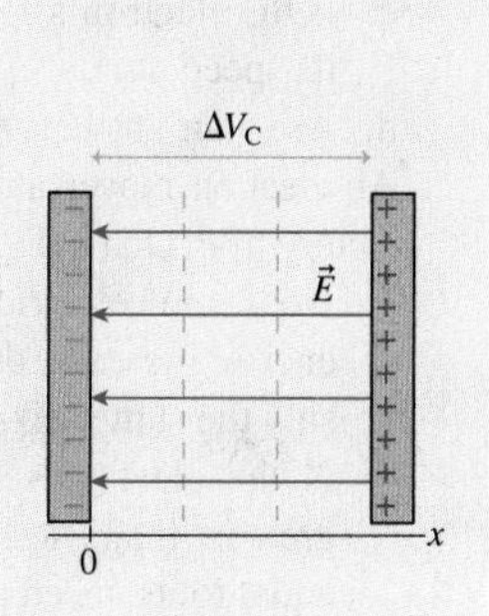

ΔV_C
$\vec{E}$
0
x

Units

- Electric potential: 1 V = 1 J/C
- Electric field: 1 V/m = 1 N/C
- Energy: 1 electron volt = 1 eV = 1.60×10^{-19} J is the kinetic energy gained by an electron upon accelerating through a potential difference of 1 V.

Problem difficulty is labeled as | (straightforward) to ||||| (challenging). Problems labeled INT integrate significant material from earlier chapters; BIO are of biological or medical interest.

For assigned homework and other learning materials, go to MasteringPhysics®

Scan this QR code to launch a Video Tutor Solution that will help you solve problems for this chapter.

QUESTONS

Conceptual Questions

1. By moving a 10 nC charge from point A to point B, you determine that the electric potential at B is 150 V. What would be the potential at B if a 20 nC charge were moved from A to B?
2. Charge q is fired through a small hole in the positive plate of a capacitor, as shown in Figure Q21.2.
   a. If q is a positive charge, does it speed up or slow down inside the capacitor? Answer this question twice: (i) Using the concept of force. (ii) Using the concept of energy.
   b. Repeat part a if q is a negative charge.

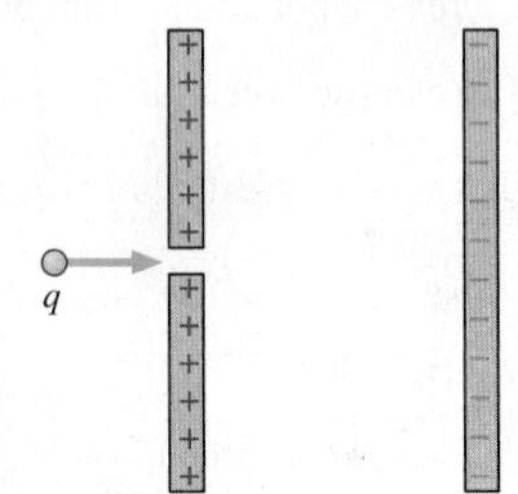

FIGURE Q21.2

3. *Why* is the potential energy of two opposite charges a negative number? (Note: Saying that the formula gives a negative number is not an explanation.)
4. An electron ($q = -e$) completes half of a circular orbit of radius r around a nucleus with $Q = +3e$, as shown in Figure Q21.4.
   a. How much work is done on the electron as it moves from i to f? Give either a numerical value or an expression from which you could calculate the value if you knew the radius. Justify your answer.
   b. By how much does the electric potential energy change as the electron moves from i to f?
   c. Is the electron's speed at f greater than, less than, or equal to its speed at i?
   d. Are your answers to parts a and c consistent with each other?

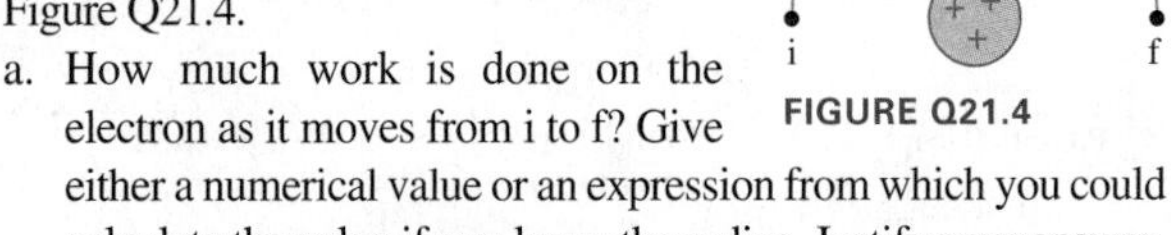

FIGURE Q21.4

5. An electron moves along the trajectory from i to f in Figure Q21.5.
   a. Does the electric potential energy increase, decrease, or stay the same? Explain.
   b. Is the electron's speed at f greater than, less than, or equal to its speed at i? Explain.

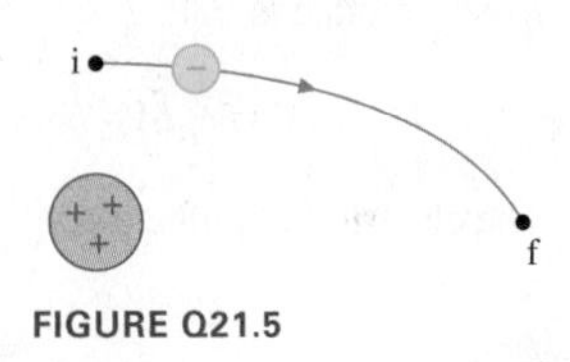

FIGURE Q21.5

6. The graph in Figure Q21.6 shows the electric potential along the x-axis. Draw a graph of the potential energy of a 0.10 C charged particle in this region of space, providing a numerical scale on the energy axis.

FIGURE Q21.6

7. As shown in Figure Q21.7, two protons are launched with the *same* speed from point 1 inside a parallel-plate capacitor. One proton moves along the path from 1 to 2, the other from 1 to 3. Points 2 and 3 are the *same* distance from the positive plate.
   a. Is $\Delta U_{1\to 2}$, the change in potential energy along the path $1 \to 2$, larger than, smaller than, or equal to $\Delta U_{1\to 3}$? Explain.
   b. Is the proton's speed v_2 at point 2 larger than, smaller than, or equal to the proton's speed v_3 at point 3? Explain.

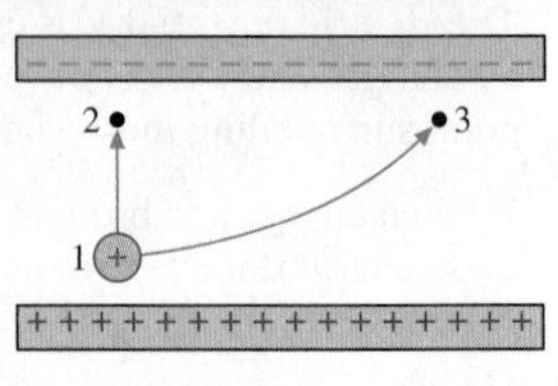

FIGURE Q21.7

8. Each part of Figure Q21.8 shows one or more point charges. The charges have equal magnitudes. If a positive charge is moved from position i to position f, does the electric potential energy increase, decrease, or stay the same? Explain.

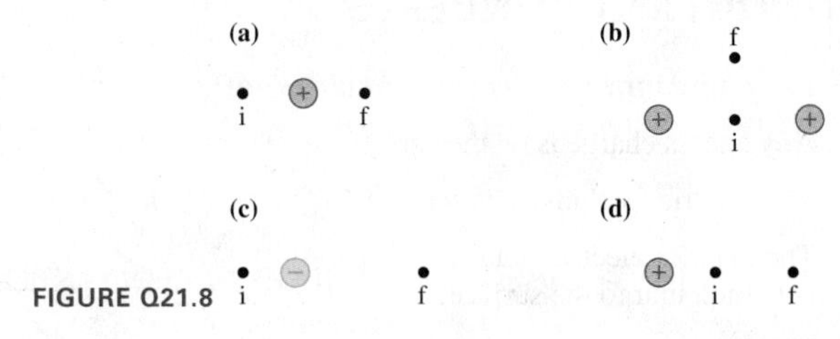

FIGURE Q21.8

9. Figure Q21.9 shows two points inside a capacitor. Let $V = 0$ V at the negative plate.
   a. What is the ratio V_2/V_1 of the electric potential at these two points? Explain.
   b. What is the ratio E_2/E_1 of the electric field strength at these two points? Explain.

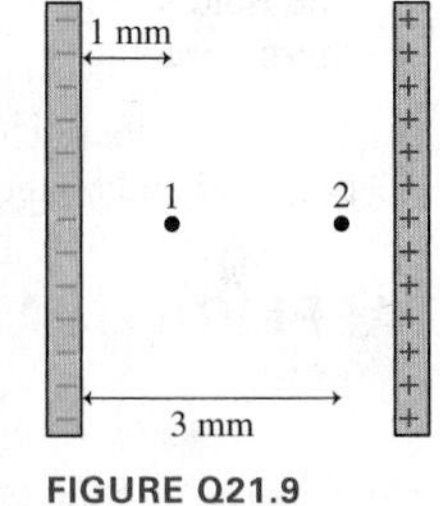

FIGURE Q21.9

10. A capacitor with plates separated by distance d is charged to a potential difference ΔV_C. All wires and batteries are disconnected, then the two plates are pulled apart (with insulated handles) to a new separation of distance $2d$.
    a. Does the capacitor charge Q change as the separation increases? If so, by what factor? If not, why not?
    b. Does the electric field strength E change as the separation increases? If so, by what factor? If not, why not?
    c. Does the potential difference ΔV_C change as the separation increases? If so, by what factor? If not, why not?
11. Rank in order, from most positive to most negative, the electric potentials V_1 to V_5 at points 1 to 5 in Figure Q21.11. Explain.

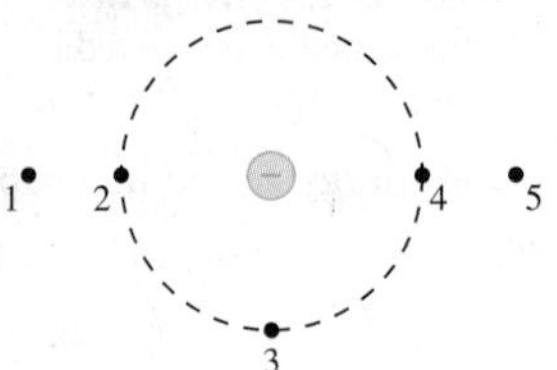

FIGURE Q21.11

12. Figure Q21.12 shows two points near a positive point charge.
   a. What is the ratio V_1/V_2 of the electric potentials at these two points? Explain.
   b. What is the ratio E_1/E_2 of the electric field strengths at these two points? Explain.

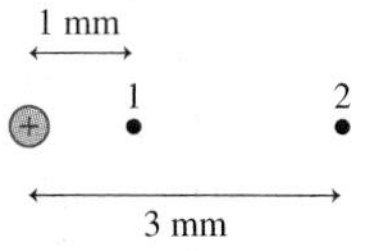

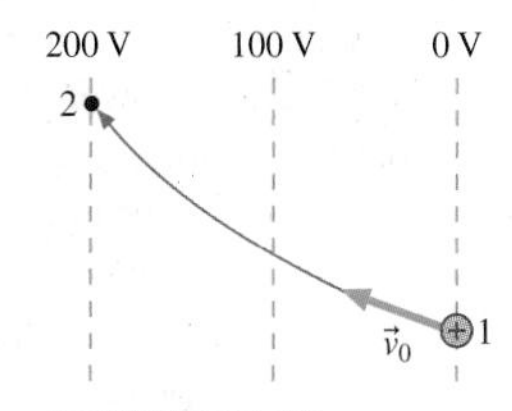

FIGURE Q21.12

13. Each part of Figure Q21.13 shows three points in the vicinity of two point charges. The charges have equal magnitudes. Rank in order, from largest to smallest, the potentials V_1, V_2, and V_3.

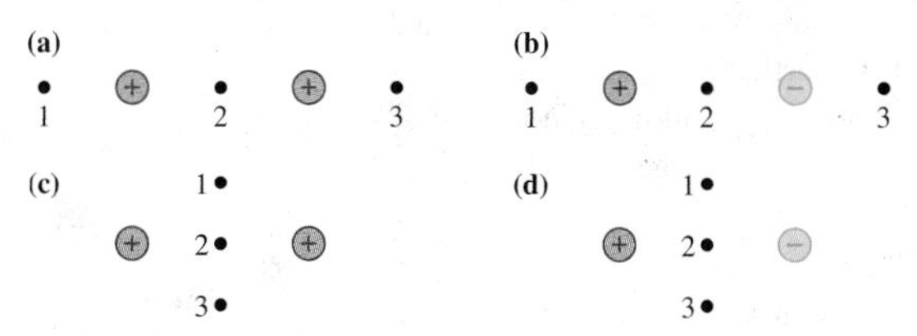

FIGURE Q21.13

14. a. Suppose that $\vec{E} = \vec{0}$ throughout some region of space. Can you conclude that $V = 0$ V in this region? Explain.
   b. Suppose that $V = 0$ V throughout some region of space. Can you conclude that $\vec{E} = \vec{0}$ in this region? Explain.
15. Rank in order, from largest to smallest, the electric field strengths E_1, E_2, E_3, and E_4 at the four labeled points in Figure Q21.15. Explain.

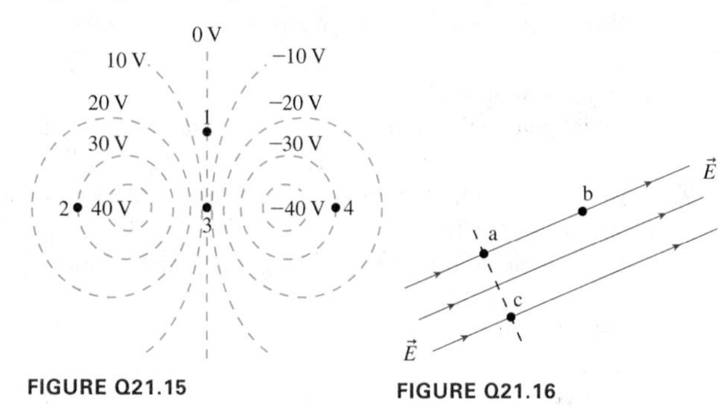

FIGURE Q21.15

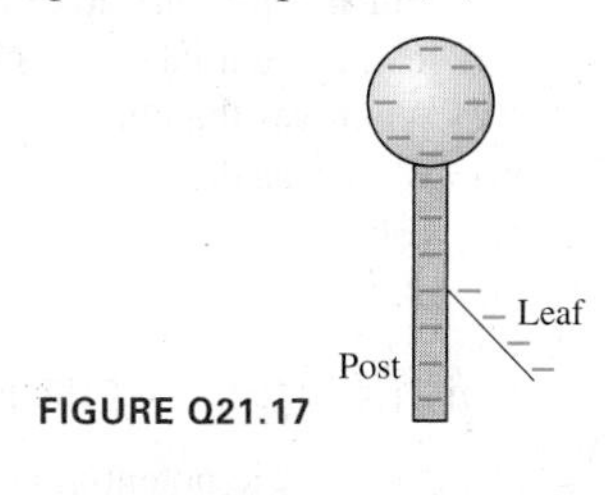

FIGURE Q21.16

16. Figure Q21.16 shows an electric field diagram. Rank in order, from highest to lowest, the electric potentials at points a, b, and c.
17. Figure Q21.17 shows a negatively charged electroscope. The gold leaf stands away from the rigid metal post. Is the electric potential of the leaf higher than, lower than, or equal to the potential of the post? Explain.

FIGURE Q21.17

18. Rank in order, from largest to smallest, the energies $(U_C)_1$ to $(U_C)_4$ stored in each of the capacitors in Figure Q21.18. Explain.

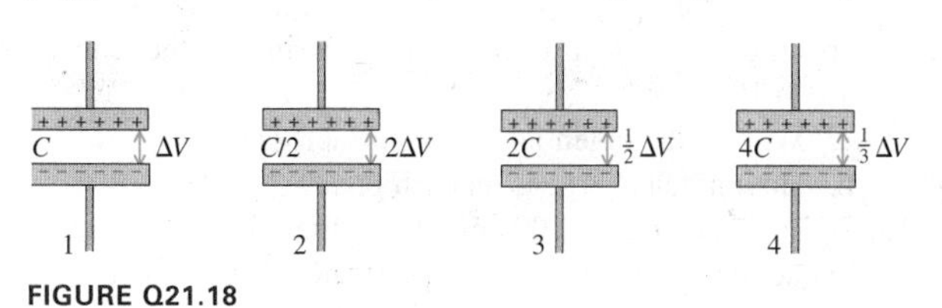

FIGURE Q21.18

19. A parallel-plate capacitor with plate separation d is connected to a battery that has potential difference ΔV_{bat}. Without breaking any of the connections, insulating handles are used to increase the plate separation to $2d$.
   a. Does the potential difference ΔV_C change as the separation increases? If so, by what factor? If not, why not?
   b. Does the capacitance change? If so, by what factor? If not, why not?
   c. Does the capacitor charge Q change? If so, by what factor? If not, why not?
20. A proton is launched from point 1 in Figure Q21.20 with an initial velocity of 3.9×10^5 m/s. By how much has its kinetic energy changed, in eV, by the time it passes through point 2?

FIGURE Q21.20

Multiple-Choice Questions

21. | A 1.0 nC positive point charge is located at point A in Figure Q21.21. The electric potential at point B is
   A. 9.0 V
   B. $9.0 \sin 30°$ V
   C. $9.0 \cos 30°$ V
   D. $9.0 \tan 30°$ V

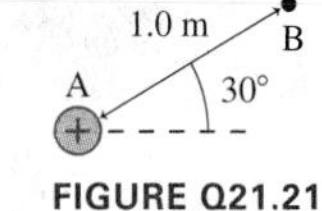

FIGURE Q21.21

22. || A 100 V battery is connected across the plates of a parallel-plate capacitor. If a sheet of Teflon is slid between the plates, without disconnecting the battery, the electric field between the plates
   A. Increases.
   B. Decreases.
   C. Remains the same.
23. | The electric potential is 300 V at $x = 0$ cm, is -100 V at $x = 5$ cm, and varies linearly with x. If a positive charge is released from rest at $x = 2.5$ cm, and is subject only to electric forces, the charge will
   A. Move to the right.
   B. Move to the left.
   C. Stay at $x = 2.5$ cm.
   D. Not enough information to tell.

Questions 24 through 28 refer to Figure Q21.24, which shows equipotential lines in a region of space. The equipotential lines are spaced by the same difference in potential, and several of the potentials are given.

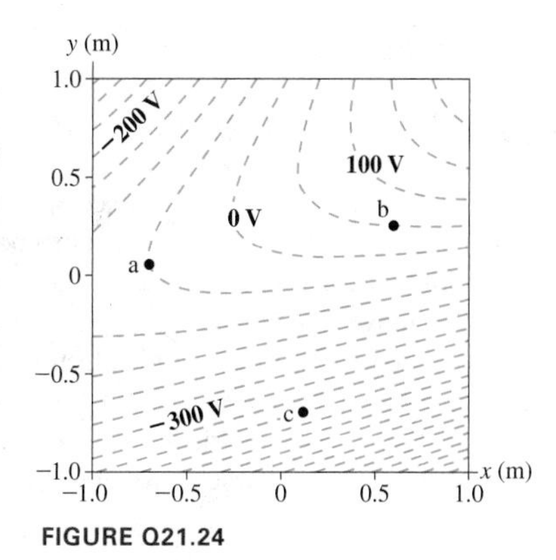

FIGURE Q21.24

24. | What is the potential at point c?
   A. −400 V
   B. −350 V
   C. −100 V
   D. 350 V
   E. 400 V
25. | At which point, a, b, or c, is the magnitude of the electric field the greatest?

Watch video solution for problem: 21.3

26. | What is the approximate magnitude of the electric field at point c?
   A. 100 V/m B. 300 V/m C. 800 V/m
   D. 1500 V/m E. 3000 V/m
27. | The direction of the electric field at point b is closest to which direction?
   A. Right B. Up C. Left D. Down
28. || A +10 nC charge is moved from point c to point a. How much work is required in order to do this?
   A. 3.5×10^{-6} J B. 4.0×10^{-6} J C. 3.5×10^{-3} J
   D. 4.0×10^{-3} J E. 3.5 J
29. | A bug zapper consists of two metal plates connected to a high-voltage power supply. The voltage between the plates is set to give an electric field slightly less than 1×10^6 V/m. When a bug flies between the two plates, it increases the field enough to initiate a spark that incinerates the bug. If a bug zapper has a 4000 V power supply, what is the approximate separation between the plates?
   A. 0.05 cm B. 0.5 cm C. 5 cm D. 50 cm
30. | An atom of helium and one of argon are singly ionized—one electron is removed from each. The two ions are then accelerated from rest by the electric field between two plates with a potential difference of 150 V. After accelerating from one plate to the other,
   A. The helium ion has more kinetic energy.
   B. The argon ion has more kinetic energy.
   C. Both ions have the same kinetic energy.
   D. There is not enough information to say which ion has more kinetic energy.
31. || The dipole moment of the heart is shown
BIO at a particular instant in Figure Q21.31. Which of the following potential differences will have the largest positive value?
   A. $V_1 - V_2$
   B. $V_1 - V_3$
   C. $V_2 - V_1$
   D. $V_3 - V_1$

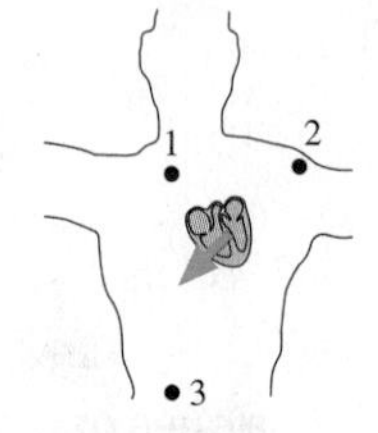

FIGURE Q21.31

PROBLEMS

Section 21.1 Electric Potential Energy and Electric Potential

Section 21.2 Sources of Electric Potential

1. ||| Moving a charge from point A, where the potential is 300 V, to point B, where the potential is 150 V, takes 4.5×10^{-4} J of work. What is the value of the charge?
2. |||| The graph in Figure P21.2 shows the electric potential energy as a function of separation for two point charges. If one charge is +0.44 nC, what is the other charge?

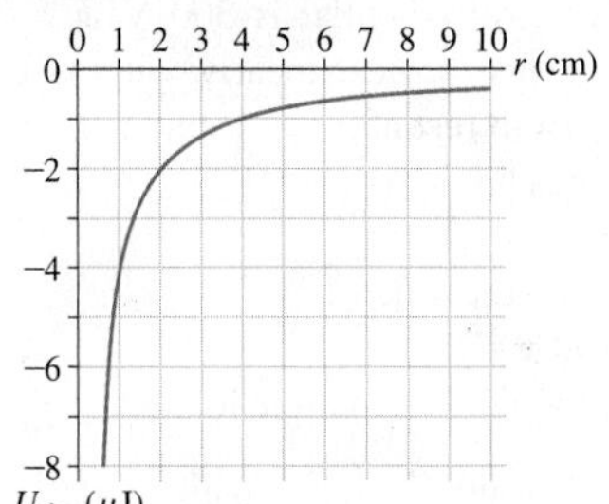

FIGURE P21.2

3. ||| It takes 3.0 μJ of work to move a 15 nC charge from point A to B. It takes −5.0 μJ of work to move the charge from C to B. What is the potential difference $V_C - V_A$?
4. | A 20 nC charge is moved from a point where $V = 150$ V to a point where $V = -50$ V. How much work is done by the force that moves the charge?
5. | At one point in space, the electric potential energy of a 15 nC charge is 45 μJ.
   a. What is the electric potential at this point?
   b. If a 25 nC charge were placed at this point, what would its electric potential energy be?

Section 21.3 Electric Potential and Conservation of Energy

6. | An electron has been accelerated from rest through a potential difference of 1000 V.
   a. What is its kinetic energy, in electron volts?
   b. What is its kinetic energy, in joules?
   c. What is its speed?
7. | A proton has been accelerated from rest through a potential difference of −1000 V.
   a. What is its kinetic energy, in electron volts?
   b. What is its kinetic energy, in joules?
   c. What is its speed?
8. ||||| What potential difference is needed to accelerate a He^+ ion (charge $+e$, mass 4 u) from rest to a speed of 1.0×10^6 m/s?
9. || An electron with an initial speed of 500,000 m/s is brought to rest by an electric field.
   a. Did the electron move into a region of higher potential or lower potential?
   b. What was the potential difference that stopped the electron?
   c. What was the initial kinetic energy of the electron, in electron volts?
10. ||| A proton with an initial speed of 800,000 m/s is brought to rest by an electric field.
   a. Did the proton move into a region of higher potential or lower potential?
   b. What was the potential difference that stopped the proton?
   c. What was the initial kinetic energy of the proton, in electron volts?

Section 21.4 Calculating the Electric Potential

11. || The electric potential at a point that is halfway between two identical charged particles is 300 V. What is the potential at a point that is 25% of the way from one particle to the other?
12. || A 2.0 cm × 2.0 cm parallel-plate capacitor has a 2.0 mm spacing. The electric field strength inside the capacitor is 1.0×10^5 V/m.
   a. What is the potential difference across the capacitor?
   b. How much charge is on each plate?
13. ||| Two 2.00 cm × 2.00 cm plates that form a parallel-plate capacitor are charged to ±0.708 nC. What are the electric field strength inside and the potential difference across the capacitor if the spacing between the plates is (a) 1.00 mm and (b) 2.00 mm?

14. | a. In Figure P21.14, which capacitor plate, left or right, is the positive plate?
    b. What is the electric field strength inside the capacitor?
    c. What is the potential energy of a proton at the midpoint of the capacitor?

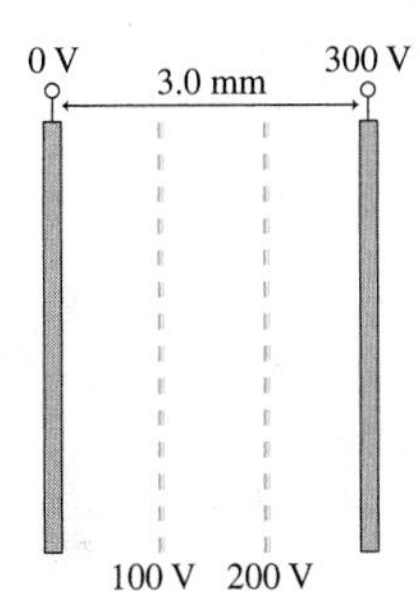

FIGURE P21.14

15. | A +25 nC charge is at the origin. How much farther from the charge is the 2000 V equipotential surface than the 3000 V surface?
16. || a. What is the electric potential at points A, B, and C in Figure P21.16?
    b. What is the potential energy of an electron at each of these points?
    c. What are the potential differences ΔV_{AB} and ΔV_{BC}?

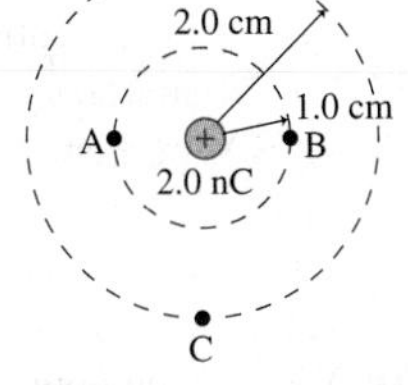

FIGURE P21.16

17. | A 1.0-cm-diameter sphere is charged to a potential of 3400 V. How much charge is on the sphere?
18. || What is the electric potential at the point indicated with the dot in Figure P21.18?

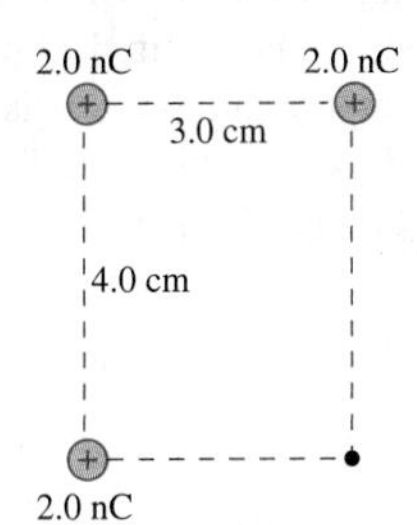

FIGURE P21.18

19. || a. What is the potential difference between the terminals of an ordinary AA or AAA battery? (If you're not sure, find one and look at the label.)
    b. An AA battery is connected to a parallel-plate capacitor having 4.0-cm-diameter plates spaced 2 mm apart. How much charge does the battery move from one plate to the other?

Section 21.5 Connecting Potential and Field

20. || a. In Figure P21.20, which point, A or B, has a higher electric potential?
    b. What is the potential difference between A and B?

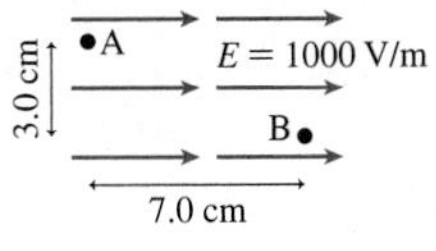

FIGURE P21.20

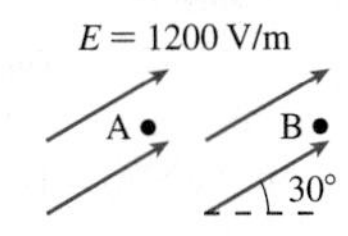

FIGURE P21.21

21. ||| In Figure P21.21, the electric potential at point A is −300 V. What is the potential at point B, which is 5.0 cm to the right of A?
22. || What is the potential difference between $x_i = 10$ cm and $x_f = 30$ cm in the uniform electric field $E_x = 1000$ V/m?
23. | What are the magnitude and direction of the electric field at the dot in Figure P21.23?

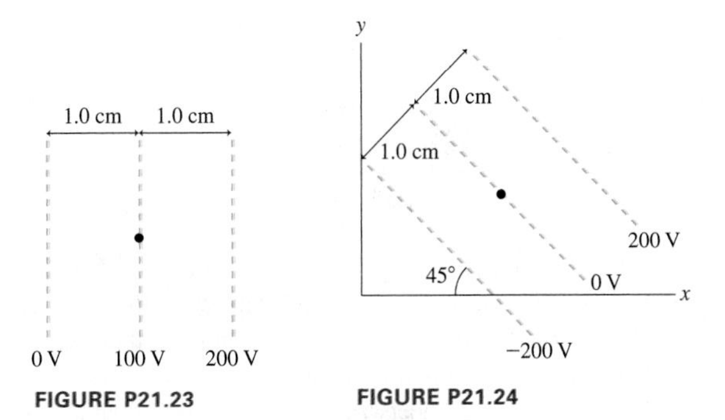

FIGURE P21.23

FIGURE P21.24

24. | What are the magnitude and direction of the electric field at the dot in Figure P21.24?

Section 21.6 The Electrocardiogram

25. | BIO One standard location for a pair of electrodes during an EKG is shown in Figure P21.25. The potential difference $\Delta V_{31} = V_3 - V_1$ is recorded. For each of the three instants a, b, and c during the heart's cycle shown in Figure 21.29, will ΔV_{31} be positive or negative? Explain.

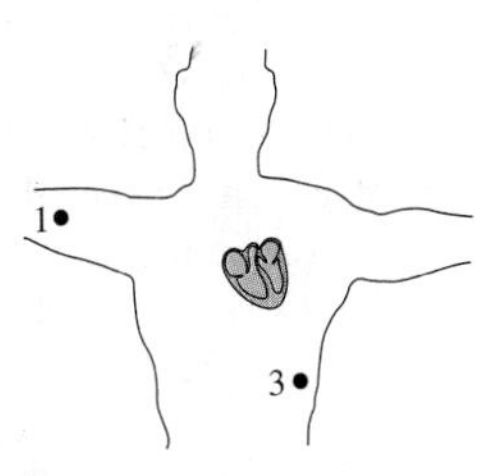

FIGURE P21.25

26. | BIO Three electrodes, 1–3, are attached to a patient as shown in Figure P21.26. During ventricular depolarization (see Figure 21.29), across which pair of electrodes is the magnitude of the potential difference likely to be the smallest? Explain.

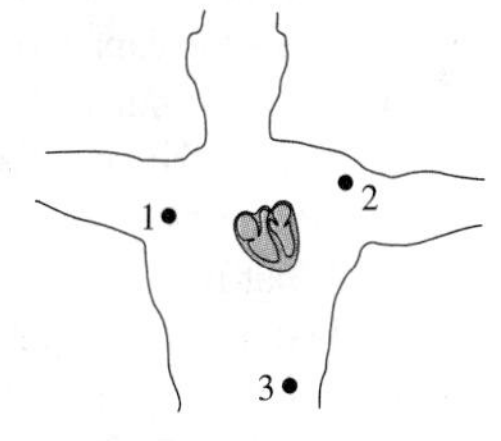

FIGURE P21.26

Section 21.7 Capacitance and Capacitors

27. ||| Two 2.0 cm × 2.0 cm square aluminum electrodes, spaced 0.50 mm apart, are connected to a 100 V battery.
    a. What is the capacitance?
    b. What is the charge on the positive electrode?
28. ||| An uncharged capacitor is connected to the terminals of a 3.0 V battery, and 6.0 μC flows to the positive plate. The 3.0 V battery is then disconnected and replaced with a 5.0 V battery, with the positive and negative terminals connected in the same manner as before. How much additional charge flows to the positive plate?
29. ||| You need to construct a 100 pF capacitor for a science project. You plan to cut two $L \times L$ metal squares and place spacers between them. The thinnest spacers you have are 0.20 mm thick. What is the proper value of L?
30. | A switch that connects a battery to a 10 μF capacitor is closed. Several seconds later you find that the capacitor plates are charged to ± 30 μC. What is the battery voltage?

31. | What is the voltage of a battery that will charge a 2.0 μF capacitor to $\pm 48\ \mu$C?

32. | Two electrodes connected to a 9.0 V battery are charged to ± 45 nC. What is the capacitance of the electrodes?

33. | Initially, the switch in Figure P21.33 is open and the capacitor is uncharged. How much charge flows through the switch after the switch is closed?

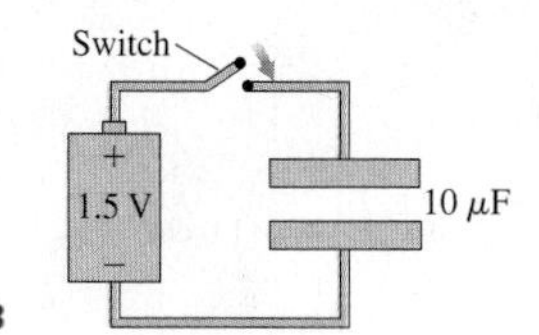

FIGURE P21.33

34. || A 1.2 nF parallel-plate capacitor has an air gap between its plates. Its capacitance increases by 3.0 nF when the gap is filled by a dielectric. What is the dielectric constant of that dielectric?

35. ||| A science-fair radio uses a homemade capacitor made of two 35 cm $\times$ 35 cm sheets of aluminum foil separated by a 0.25-mm-thick sheet of paper. What is its capacitance?

36. ||| A 25 pF parallel-plate capacitor with an air gap between the plates is connected to a 100 V battery. A Teflon slab is then inserted between the plates and completely fills the gap. What is the change in the charge on the positive plate when the Teflon is inserted?

37. || Two 2.0-cm-diameter electrodes with a 0.10-mm-thick sheet of Teflon between them are attached to a 9.0 V battery. Without disconnecting the battery, the Teflon is removed. What are the charge, potential difference, and electric field (a) before and (b) after the Teflon is removed?

38. ||| A parallel-plate capacitor is connected to a battery and stores 4.4 nC of charge. Then, while the battery remains connected, a sheet of Teflon is inserted between the plates.
    a. Does the capacitor's charge increase or decrease?
    b. By how much does the charge change?

39. ||| A parallel-plate capacitor is charged by a 12.0 V battery, then the battery is removed.
    a. What is the potential difference between the plates after the battery is disconnected?
    b. What is the potential difference between the plates after a sheet of Teflon is inserted between them?

Section 21.8 Energy and Capacitors

40. ||| To what potential should you charge a 1.0 μF capacitor to store 1.0 J of energy?

41. || A pair of 10 μF capacitors in a high-power laser are charged to 1.7 kV.
    a. What charge is stored in each capacitor?
    b. How much energy is stored in each capacitor?

42. | Capacitor 2 has half the capacitance and twice the potential difference as capacitor 1. What is the ratio $(U_C)_1/(U_C)_2$?

43. |||| Two uncharged metal spheres, spaced 15.0 cm apart, have a capacitance of 24.0 pF. How much work would it take to move 12.0 nC of charge from one sphere to the other?

44. |||| 50 pJ of energy is stored in a 2.0 cm $\times$ 2.0 cm $\times$ 2.0 cm region of uniform electric field. What is the electric field strength?

General Problems

45. || A 2.0-cm-diameter parallel-plate capacitor with a spacing of 0.50 mm is charged to 200 V. What are (a) the total energy stored in the electric field and (b) the energy density?

46. ||| What is the change in electric potential energy of a 3.0 nC point charge when it is moved from point A to point B in Figure P21.46?

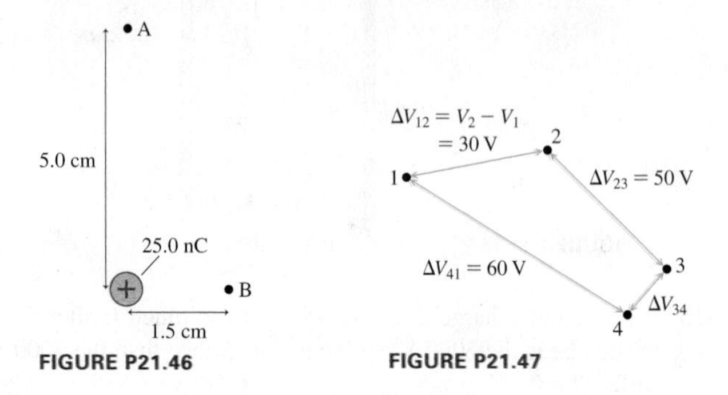

FIGURE P21.46 FIGURE P21.47

47. || What is the potential difference ΔV_{34} in Figure P21.47?

48. || INT A -50 nC charged particle is in a uniform electric field $\vec{E} = (10\text{ V/m, east})$. An external force moves the particle 1.0 m north, then 5.0 m east, then 2.0 m south, and finally 3.0 m west. The particle begins and ends its motion with zero velocity.
    a. How much work is done on it by the external force?
    b. What is the potential difference between the particle's final and initial positions?

49. || INT At a distance r from a point charge, the electric potential is 3000 V and the magnitude of the electric field is 2.0×10^5 V/m.
    a. What is the distance r?
    b. What are the electric potential and the magnitude of the electric field at distance $r/2$ from the charge?

50. || The 4000 V equipotential surface is 10.0 cm farther from a positively charged particle than the 5000 V equipotential surface. What is the charge on the particle?

51. || What is the electric potential energy of the electron in Figure P21.51? The protons are fixed and can't move.

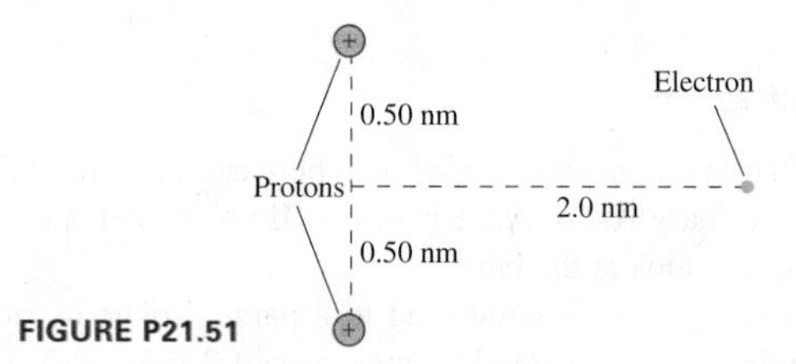

FIGURE P21.51

52. |||| Two point charges 2.0 cm apart have an electric potential energy $-180\ \mu$J. The total charge is 30 nC. What are the two charges?

53. || INT Two positive point charges are 5.0 cm apart. If the electric potential energy is 72 μJ, what is the magnitude of the force between the two charges?

54. ||||| A +3.0 nC charge is at $x = 0$ cm and a -1.0 nC charge is at $x = 4$ cm. At what point or points on the x-axis is the electric potential zero?

55. ||||| A -3.0 nC charge is on the x-axis at $x = -9$ cm and a +4.0 nC charge is on the x-axis at $x = 16$ cm. At what point or points on the y-axis is the electric potential zero?

Watch video solution for problem: 21.71 

56. II A -2.0 nC charge and a $+2.0$ nC charge are located on the INT x-axis at $x = -1.0$ cm and $x = +1.0$ cm, respectively.
    a. At what position or positions on the x-axis is the electric field zero?
    b. At what position or positions on the x-axis is the electric potential zero?
    c. Draw graphs of the electric field strength and the electric potential along the x-axis.
57. III A -10.0 nC point charge and a $+20.0$ nC point charge are INT 15.0 cm apart on the x-axis.
    a. What is the electric potential at the point on the x-axis where the electric field is zero?
    b. What are the magnitude and direction of the electric field at the point on the x-axis, between the charges, where the electric potential is zero?
58. IIIII A 2.0-mm-diameter glass bead is positively charged. The potential difference between a point 2.0 mm from the bead and a point 4.0 mm from the bead is 500 V. What is the charge on the bead?
59. I In a semiclassical model of the hydrogen atom, the electron orbits the proton at a distance of 0.053 nm.
    a. What is the electric potential of the proton at the position of the electron?
    b. What is the electron's potential energy?
60. I What is the electric potential at the point indicated with the dot in Figure P21.60?

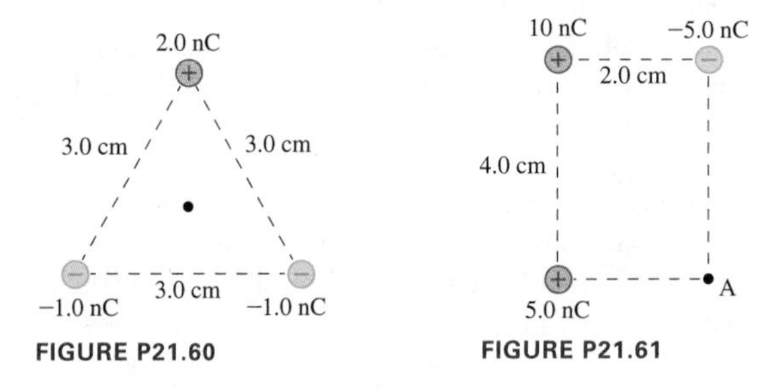

FIGURE P21.60 FIGURE P21.61

61. I a. What is the electric potential at point A in Figure P21.61?
    b. What is the potential energy of a proton at point A?
62. III A proton's speed as it passes point A is 50,000 m/s. It follows the trajectory shown in Figure P21.62. What is the proton's speed at point B?

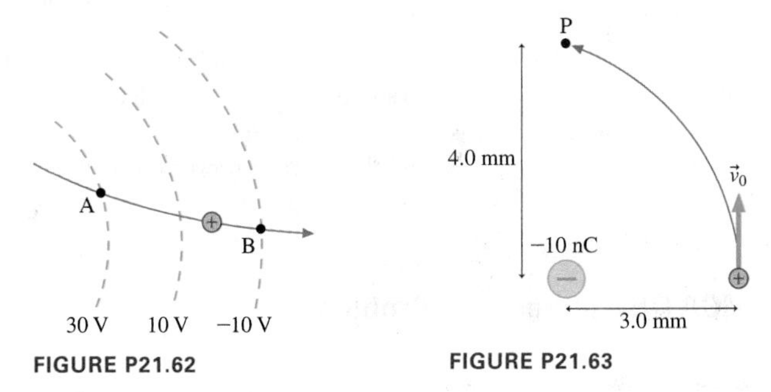

FIGURE P21.62 FIGURE P21.63

63. III A proton follows the path shown in Figure P21.63. Its initial speed is $v_0 = 1.9 \times 10^6$ m/s. What is the proton's speed as it passes through point P?
64. II Electric outlets have a voltage of approximately 120 V between the two parallel slots. Estimate the electric field strength between these two slots.
65. I Estimate the magnitude of the electric field in a cell membrane with a thickness of 8 nm. BIO
66. II A Na^+ ion moves from inside a cell, where the electric potential is -70 mV, to outside the cell, where the potential is 0 V. What is the change in the ion's electric potential energy as it moves from inside to outside the cell? Does its energy increase or decrease? BIO
67. III Suppose that a molecular ion with charge $-10e$ is embedded within the 5.0-nm-thick cell membrane of a cell with membrane potential -70 mV. What is the electric force on the molecule? BIO
68. III The electric field strength is 50,000 V/m inside a parallel-plate capacitor with a 2.0 mm spacing. A proton is released from rest at the positive plate. What is the proton's speed when it reaches the negative plate?
69. III A parallel-plate capacitor is charged to 5000 V. A proton is fired into the center of the capacitor at a speed of 3.0×10^5 m/s, as shown in Figure P21.69. The proton is deflected while inside the capacitor, and the plates are long enough that the proton will hit one of them before emerging from the far side of the capacitor. What is the impact speed of the proton?

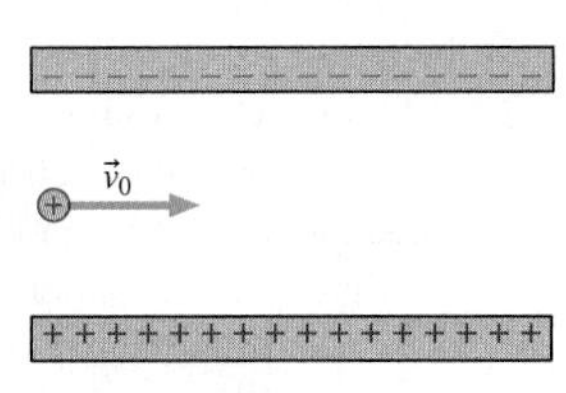

FIGURE P21.69

70. III A proton is released from rest at the positive plate of a parallel-plate capacitor. It crosses the capacitor and reaches the negative plate with a speed of 50,000 m/s. What will be the proton's final speed if the experiment is repeated with double the amount of charge on each capacitor plate?
71. II The electric field strength is 20,000 V/m inside a parallel-plate capacitor with a 1.0 mm spacing. An electron is released from rest at the negative plate. What is the electron's speed when it reaches the positive plate?
72. II In the early 1900s, Robert Millikan used small charged droplets of oil, suspended in an electric field, to make the first quantitative measurements of the electron's charge. A 0.70-μm-diameter droplet of oil, having a charge of $+e$, is suspended in midair between two horizontal plates of a parallel-plate capacitor. The upward electric force on the droplet is exactly balanced by the downward force of gravity. The oil has a density of 860 kg/m^3, and the capacitor plates are 5.0 mm apart. What must the potential difference between the plates be to hold the droplet in equilibrium? INT
73. IIII Two 2.0-cm-diameter disks spaced 2.0 mm apart form a parallel-plate capacitor. The electric field between the disks is 5.0×10^5 V/m.
    a. What is the voltage across the capacitor?
    b. How much charge is on each disk?
    c. An electron is launched from the negative plate. It strikes the positive plate at a speed of 2.0×10^7 m/s. What was the electron's speed as it left the negative plate?
74. II In *proton-beam therapy,* a high-energy beam of protons is fired at a tumor. The protons come to rest in the tumor, depositing their kinetic energy and breaking apart the tumor's DNA, thus killing its cells. For one patient, it is desired that 0.10 J of proton energy be deposited in a tumor. To create the proton beam, the protons are accelerated from rest through a 10 MV potential difference. What is the total charge of the protons that must be fired at the tumor to deposit the required energy? BIO

75. |||| A 2.5-mm-diameter sphere is charged to −4.5 nC. An electron fired directly at the sphere from far away comes to within 0.30 mm of the surface of the target before being reflected.
   a. What was the electron's initial speed?
   b. At what distance from the surface of the sphere is the electron's speed half of its initial value?
   c. What is the acceleration of the electron at its turning point?
76. || A proton is fired from far away toward the nucleus of an iron atom. Iron is element number 26, and the diameter of the nucleus is 9.0 fm. (1 fm $= 10^{-15}$ m.) What initial speed does the proton need to just reach the surface of the nucleus? Assume the nucleus remains at rest.
77. || Two 10.0-cm-diameter electrodes 0.50 cm apart form a parallel-plate capacitor. The electrodes are attached by metal wires to the terminals of a 15 V battery. After a long time, the capacitor is disconnected from the battery but is not discharged. What are the charge on each electrode, the electric field strength inside the capacitor, and the potential difference between the electrodes
   a. Right after the battery is disconnected?
   b. After insulating handles are used to pull the electrodes away from each other until they are 1.0 cm apart?
78. || Two 10.0-cm-diameter electrodes 0.50 cm apart form a parallel-plate capacitor. The electrodes are attached by metal wires to the terminals of a 15 V battery. What are the charge on each electrode, the electric field strength inside the capacitor, and the potential difference between the electrodes
   a. While the capacitor is attached to the battery?
   b. After insulating handles are used to pull the electrodes away from each other until they are 1.0 cm apart? The electrodes remain connected to the battery during this process.
79. ||| Determine the magnitude and direction of the electric field at points 1 and 2 in Figure P21.79.

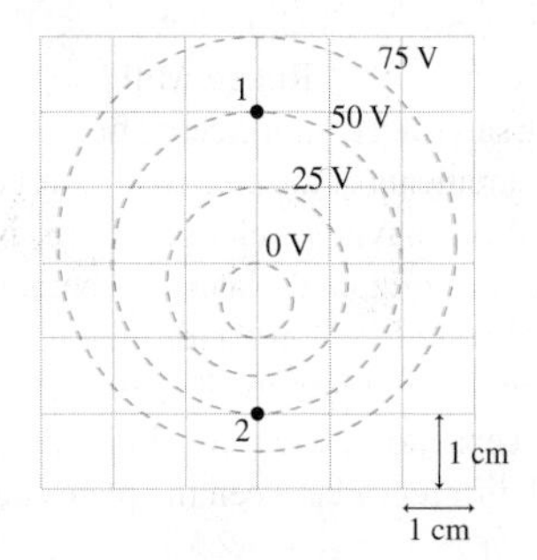

FIGURE P21.79

80. | Figure P21.80 shows a series of equipotential curves.
   a. Is the electric field strength at point A larger than, smaller than, or equal to the field strength at point B? Explain.
   b. Is the electric field strength at point C larger than, smaller than, or equal to the field strength at point D? Explain.
   c. Determine the electric field $\vec{E}$ at point D. Express your answer as a magnitude and direction.

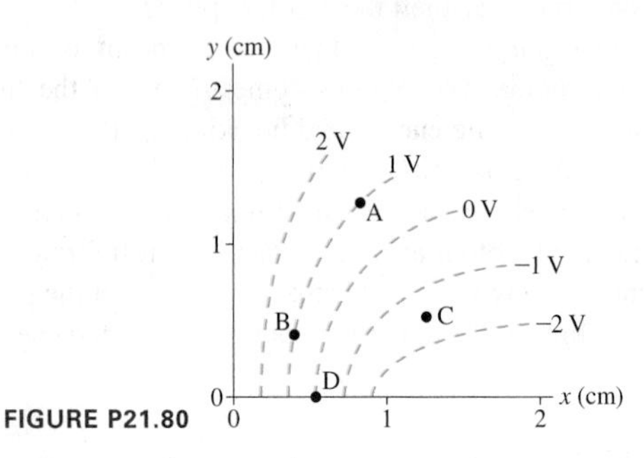

FIGURE P21.80

81. || Figure P21.81 shows the electric potential on a grid whose squares are 5.0 cm on a side.
   a. Reproduce this figure on your paper, then draw the 50 V, 75 V, and 100 V equipotential surfaces.
   b. Estimate the electric field (strength and direction) at points A, B, C, and D.
   c. Draw the electric field vectors at points A, B, C, and D on your diagram.

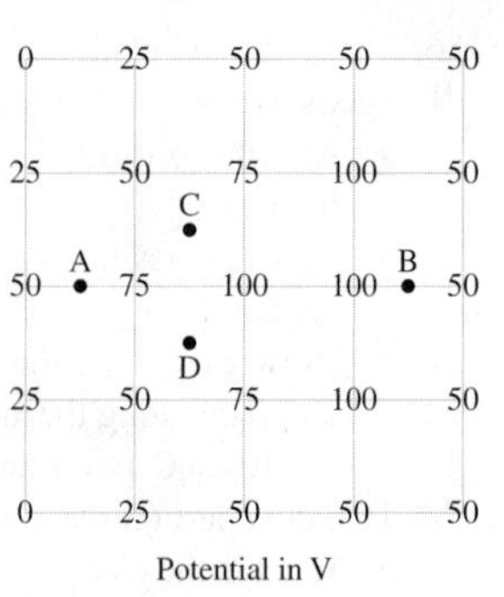

FIGURE P21.81

82. ||| A capacitor consists of two 6.0-cm-diameter circular plates separated by 1.0 mm. The plates are charged to 150 V, then the battery is removed. INT
   a. How much energy is stored in the capacitor?
   b. How much work must be done to pull the plates apart to where the distance between them is 2.0 mm?
83. |||| The dielectric in a capacitor serves two purposes. It increases the capacitance, compared to an otherwise identical capacitor with an air gap, and it increases the maximum potential difference the capacitor can support. If the electric field in a material is sufficiently strong, the material will suddenly become able to conduct, creating a spark. The critical field strength, at which breakdown occurs, is 3.0 MV/m for air, but 60 MV/m for Teflon.
   a. A parallel-plate capacitor consists of two square plates, 15 cm on a side, spaced 0.50 mm apart with only air between them. What is the maximum energy that can be stored by the capacitor?
   b. What is the maximum energy that can be stored if the plates are separated by a 0.50-mm-thick Teflon sheet?
84. | The highest magnetic fields in the world are generated when large arrays, or "banks," of capacitors are discharged through the copper coils of an electromagnet. At the National High Magnetic Field Laboratory, the total capacitance of the capacitor bank is 32 mF. These capacitors can be charged to 16 kV. INT
   a. What is the energy stored in the capacitor bank when it is fully charged?
   b. When discharged, the entire energy from this bank flows through the magnet coil in 10 ms. What is the average power delivered to the coils during this time?
85. ||||| The flash unit in a camera uses a special circuit to "step up" the 3.0 V from the batteries to 300 V, which charges a capacitor. The capacitor is then discharged through a flashlamp. The discharge takes 10 μs, and the average power dissipated in the flashlamp is 10^5 W. What is the capacitance of the capacitor? INT

MCAT-Style Passage Problems

A Lightning Strike

Storm clouds build up large negative charges, as described in the chapter. The charges dwell in *charge centers,* regions of concentrated charge. Suppose a cloud has −25 C in a 1.0-km-diameter spherical charge center located 10 km above the ground, as sketched in Figure P21.86. The negative charge center attracts a similar amount of positive charge that is spread on the ground below the cloud.

The charge center and the ground function as a charged capacitor, with a potential difference of approximately 4×10^8 V. The large electric field between these two "electrodes" may ionize the air, leading to a conducting path between the cloud and the ground. Charges will flow along this conducting path, causing a discharge of the capacitor—a lightning strike.

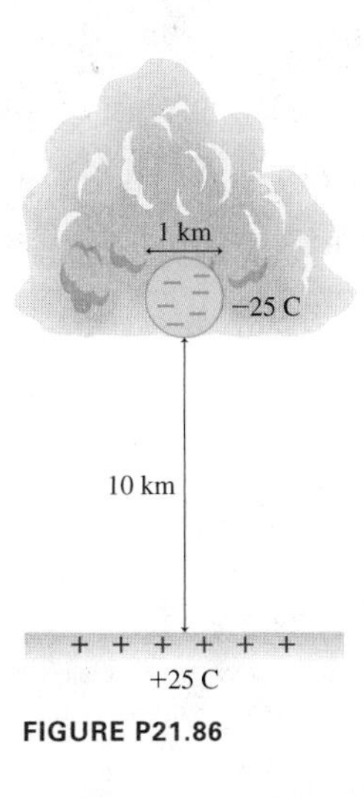

FIGURE P21.86

86. | What is the approximate magnitude of the electric field between the charge center and the ground?
A. 4×10^4 V/m
B. 4×10^5 V/m
C. 4×10^6 V/m
D. 4×10^7 V/m

87. | Which of the curves sketched in Figure P21.87 best approximates the shape of an equipotential drawn halfway between the charge center and the ground?

A. B. C. D.

FIGURE P21.87

88. | What is the approximate capacitance of the charge center + ground system?
A. 6×10^{-8} F B. 2×10^7 F
C. 4×10^6 F D. 8×10^6 F

89. | If 12.5 C of charge is transferred from the cloud to the ground in a lightning strike, what fraction of the stored energy is dissipated?
A. 12% B. 25% C. 50% D. 75%

90. | If the cloud transfers all of its charge to the ground via several rapid lightning flashes lasting a total of 1 s, what is the average power?
A. 1 GW B. 2 GW C. 5 GW D. 10 GW

STOP TO THINK ANSWERS

Chapter Preview Stop to Think: C. In general, conservation of energy dictates that the work done on a system increases its energy by the amount of work done. Because the only energy that changes for the book is its gravitational potential energy, we can write this as $W = \Delta U_g$, indicating that the two terms are equal. Furthermore, the force of your hand on the book points in the same direction as the book's displacement, so the work done by the hand is positive. Thus $W = \Delta U_g > 0$.

Stop to Think 21.1: B. If the charge were moved from 1 to 2 at a constant speed by a hand, the force exerted by the hand would need to be to the left, to oppose the rightward-directed electric force on the charge due to the source charges. Because the force due to the hand would be opposite the displacement, the hand would do *negative* work on the charge, decreasing its electric potential energy.

Stop to Think 21.2: C. If the particle starts and ends at the same point, where $V = 400$ V, then $\Delta V = 0$ and so ΔK must be zero as well.

Stop to Think 21.3: C. The proton gains speed by losing potential energy. It loses potential energy by moving in the direction of decreasing electric potential.

Stop to Think 21.4: $\Delta V_{13} = \Delta V_{23} > \Delta V_{12}$. The potential depends on only the *distance* from the charge, not the direction. $\Delta V_{12} = 0$ because these points are at the same distance.

Stop to Think 21.5: C. $\vec{E}$ points "downhill," so V must decrease from right to left. E is larger on the left than on the right, so the equipotential lines must be closer together on the left.

Stop to Think 21.6: C. Capacitance is a property of the shape and position of the electrodes. It does not depend on the potential difference or charge.

Stop to Think 21.7: B. The energy is $\frac{1}{2}C(\Delta V_C)^2$. ΔV_C is constant, but C doubles when the distance is halved.

22 Current and Resistance

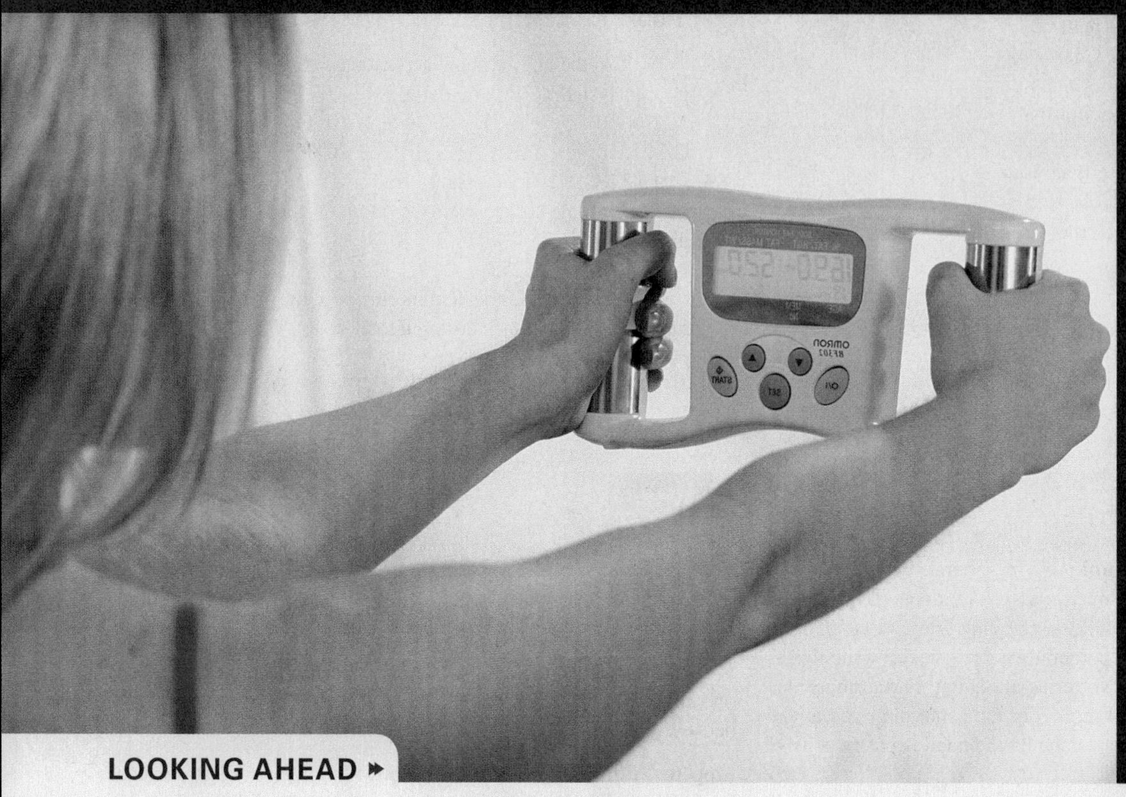

This woman is measuring her percentage body fat by gripping a device that passes a small electric current through her body. How does a measurement of the current reveal such details of the body's structure?

LOOKING AHEAD ▸▸

Goal: To learn how and why charge moves through a conductor as what we call a current.

Current

The motion of charge through a conductor, like the wires connecting these lights, is called a **current.**

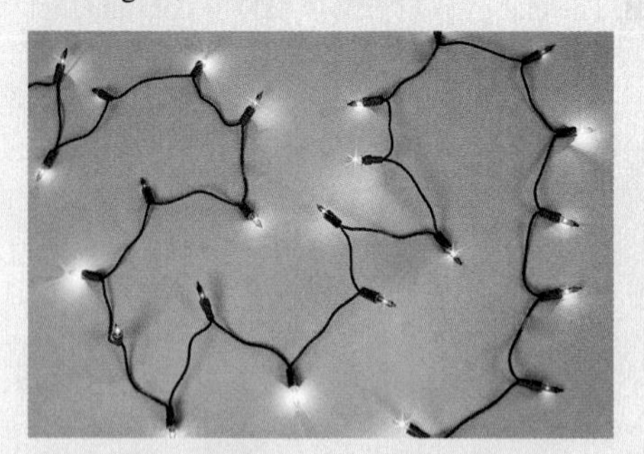

You'll learn why **conservation of current** ensures that each bulb is equally bright.

Resistance and Ohm's Law

Ohm's law relates the current in this bulb to the battery's voltage and the bulb's **resistance** to the flow of charge.

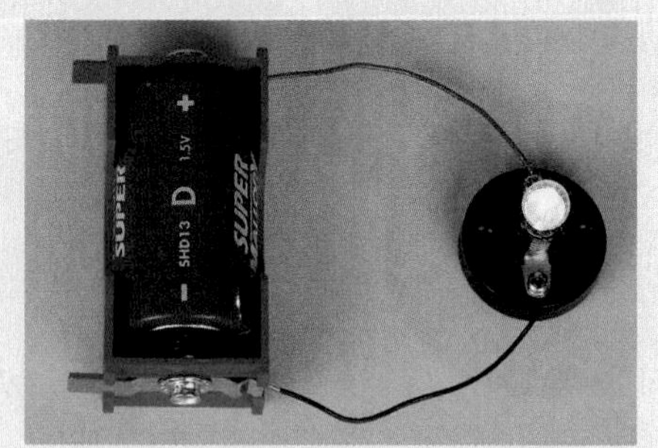

You'll learn how to relate the resistance of a wire to its size and composition.

Electric Power

A hair dryer converts electric energy to thermal energy, leading to a blast of hot air.

You'll learn the relationship of electric power to current, resistance, and voltage.

LOOKING BACK ◂◂

Electric Potential and Electric Field

In Section 21.5, you learned the connection between the electric potential and the electric field. In this chapter, you'll use this connection to understand why charges move in conductors.

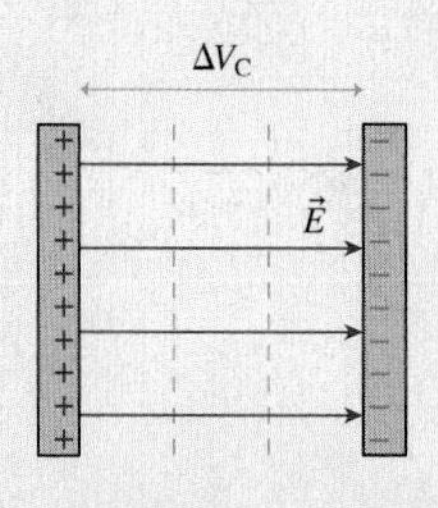

The electric field always points "downhill," from higher to lower potential, as in this parallel-plate capacitor.

STOP TO THINK

An electron is released from rest at the dot. Afterward, the electron

A. Starts moving to the right.
B. Starts moving to the left.
C. Remains at rest.

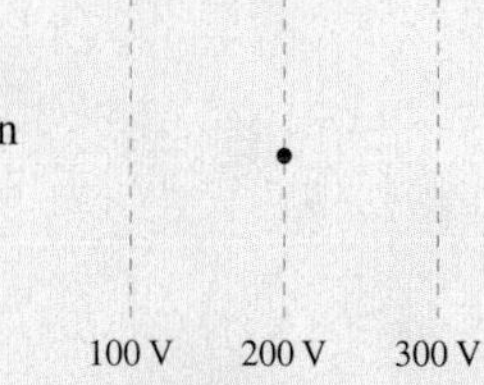

22.1 A Model of Current

Let's start our exploration of current with a very simple experiment. **FIGURE 22.1a** shows a charged parallel-plate capacitor. If we connect the two capacitor plates to each other with a metal wire, as shown in **FIGURE 22.1b**, the plates quickly become neutral. We say that the capacitor has been *discharged.*

The wire is a conductor, a material through which charge easily moves. Apparently the excess charge on one capacitor plate is able to move through the wire to the other plate, neutralizing both plates. The motion of charges through a material is called a *current*, so the capacitor is discharged by a current in the connecting wire. Later in this chapter we will develop a quantitative expression for current, but for now the simple idea of current as charges in motion will suffice.

If we observe the capacitor discharge, we see other effects. As **FIGURE 22.2** shows, the connecting wire gets warmer. If the wire is very thin in places, such as the thin filament in a lightbulb, the wire gets hot enough to glow. The current-carrying wire also deflects a compass needle. We will explore the connection between currents and magnetism in Chapter 24. For now, we will use "makes the wire warmer" and "deflects a compass needle" as *indicators* that a current is present in a wire. We can use the brightness of a lightbulb to tell us the magnitude of the current; **more current means a brighter bulb.**

FIGURE 22.1 A capacitor is discharged by a metal wire.

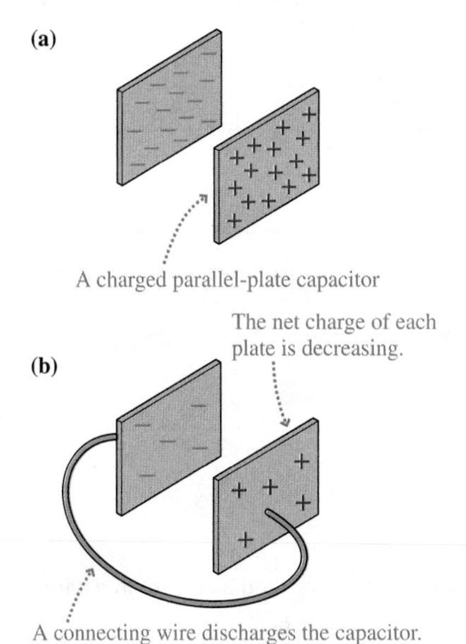
(a)
A charged parallel-plate capacitor
(b)
The net charge of each plate is decreasing.
A connecting wire discharges the capacitor.

FIGURE 22.2 Properties of a current.

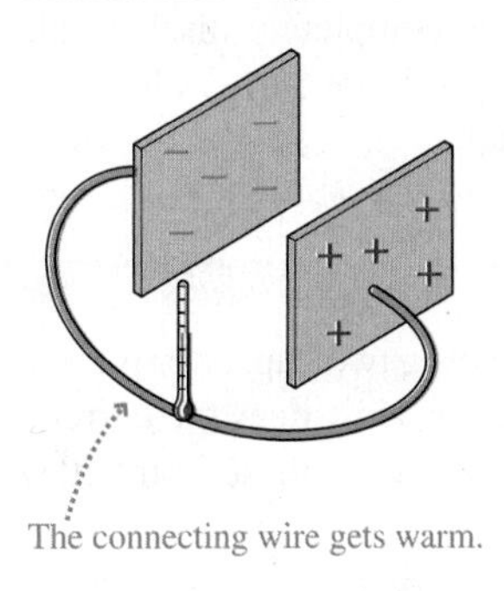
The connecting wire gets warm.

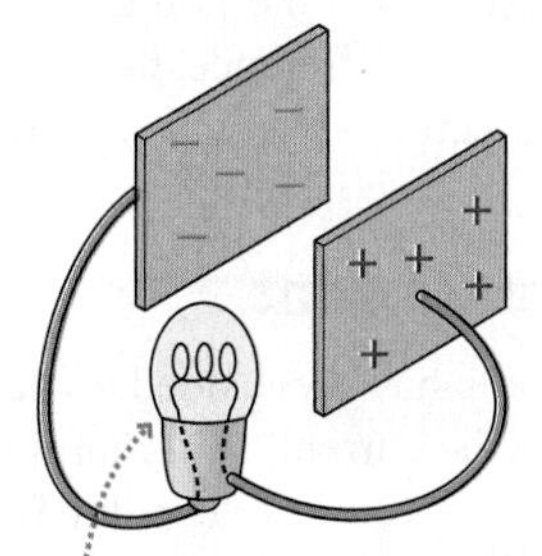
A lightbulb glows. The lightbulb filament is part of the connecting wire.

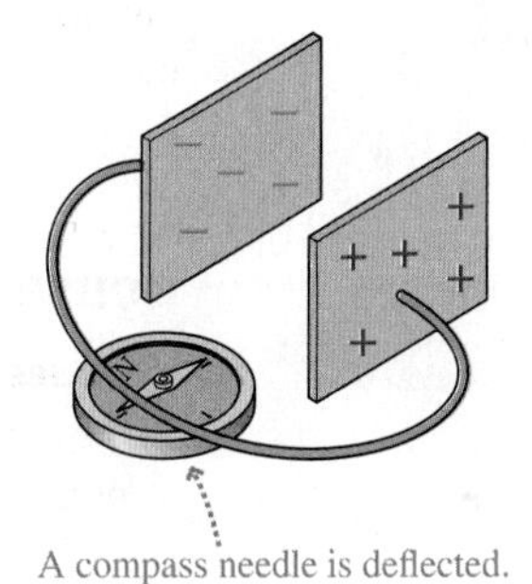
A compass needle is deflected.

Figure Video

Charge Carriers

The charges that move in a current are called the *charge carriers.* In a metal, the charge carriers are electrons. As **FIGURE 22.3** shows, it is the motion of the *conduction electrons,* which are free to move around, that forms a current—a flow of charge—in the metal. An *insulator* does not have such free charges and cannot carry a current. Although electrons are the charge carriers in metals, other materials may have different charge carriers. For instance, in ionic solutions such as seawater, blood, and intercellular fluids, the charge carriers are ions, both positive and negative.

FIGURE 22.3 Conduction electrons in a metal.

Ions (the metal atoms minus the conduction electrons) occupy fixed positions.

The conduction electrons (generally a few per atom) are bound to the solid as a whole, not to any particular atom. They are free to move around.

The metal as a whole is electrically neutral.

Creating a Current

Suppose you want to slide a book across the table to your friend. You give it a quick push to start it moving, but it begins slowing down because of friction as soon as you take your hand off of it. The book's kinetic energy is transformed into thermal energy, leaving the book and the table slightly warmer. The only way to keep the book moving at a *constant* speed is to continue pushing it.

Something similar happens in a conductor. As we saw in ◀◀ **SECTION 20.7**, we can use an electric field to push on the electrons in a conductor. Suppose we take a piece of metal and apply an electric field, as in **FIGURE 22.4** on the next page. The

FIGURE 22.4 The motion of an electron in a conductor.

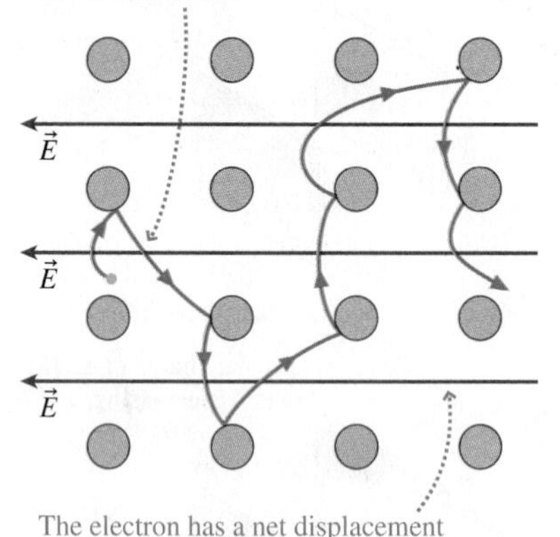
The collisions "reset" the motion of the electron. It then accelerates until the next collision.
$\vec{E}$
$\vec{E}$
$\vec{E}$
The electron has a net displacement opposite the electric field.

field exerts a force on the electrons, and they begin to accelerate. But the electrons aren't moving in a vacuum. Collisions between the electrons and the atoms of the metal slow them down, transforming the electrons' kinetic energy into the thermal energy of the metal, making the metal warmer. (Recall that "makes the wire warmer" is one of our indicators of a current.) The motion of the electrons will cease *unless you continue pushing*. To keep the electrons moving, we must maintain an electric field. In a constant field, an electron's average motion will be opposite the field. We call this motion the electron's *drift velocity*. If the field goes to zero, so does the drift velocity.

How can you have an electric field inside a conductor? The important conclusion of ◀◀ SECTION 20.6 was that $\vec{E} = \vec{0}$ inside a conductor in electrostatic equilibrium. But a conductor with electrons moving through it is *not* in electrostatic equilibrium. The charges are in motion, so the field need not be zero.

FIGURE 22.5 shows how a wire connected between the plates of a charged capacitor causes it to discharge. (We've "unfolded" the plates to make the flow of charge easier to see.) The initial separation of charges creates a potential difference between the two plates. We saw in ◀◀ SECTION 21.5 that whenever there's a potential difference, an electric field points from higher potential toward lower potential. Connecting a wire between the plates establishes an electric field in the wire, and this electric field causes electrons to flow from the negative plate (which has an excess of electrons) toward the positive plate. **The potential difference creates the electric field that drives the current in the wire.**

FIGURE 22.5 Creating a current in a wire.

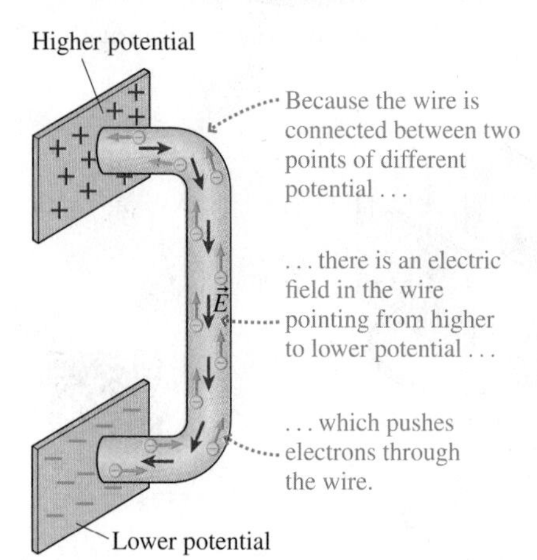
Higher potential
Because the wire is connected between two points of different potential . . .
. . . there is an electric field in the wire pointing from higher to lower potential . . .
$\vec{E}$
. . . which pushes electrons through the wire.
Lower potential

As the current continues, and the charges flow, the plates discharge and the potential difference decreases. At some point, the plates will be completely discharged, meaning no more potential difference, no more field—and no more current. Finally, $\vec{E} = \vec{0}$ inside the conducting wire, and we have equilibrium.

Conservation of Current

In FIGURE 22.6 a lightbulb has been added to the wire connecting two capacitor plates. The bulb glows while the current is discharging the capacitor. How does the current at point A compare to the current at point B? Are the currents at these points the same? Or is one larger than the other?

FIGURE 22.6 How does the current at A compare to the current at B?

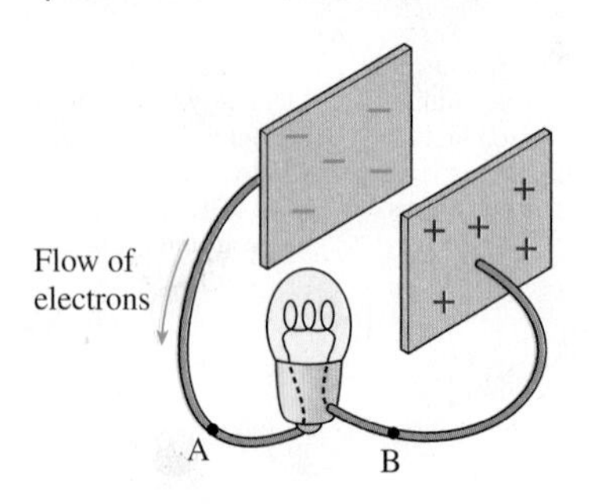
Flow of electrons
A
B

You might have predicted that the current at B is less than the current at A because the bulb, in order to glow, must use up some of the current. It's easy to test this prediction; for instance, we could compare the currents at A and B by comparing how far two compass needles at these positions are deflected. Any such test gives the same result: The current at point B is *exactly equal* to the current at point A. **The current leaving a lightbulb is exactly the same as the current entering the lightbulb.**

This is an important observation, one that demands an explanation. After all, "something" makes the bulb glow, so why don't we observe a decrease in the current? Electrons are charged particles. The lightbulb can't destroy electrons without violating both the law of conservation of mass and the law of conservation of charge. Thus the *number* of electrons is not changed by the lightbulb. Further, the lightbulb can't store electrons. Were it to do so, the bulb would become increasingly negative until its repulsive force stopped the flow of new electrons and the bulb would go out. This doesn't happen. Every electron entering the lightbulb must be matched by an electron leaving the bulb, and thus the current at B is the same as at A.

Let's consider an analogy with water flowing through a pipe. Suppose we put a turbine in the middle of the pipe so that the flow of the water turns the turbine, as in FIGURE 22.7. Water flows *through* the turbine. It is not consumed by the turbine, and the number of gallons of water per minute leaving the pipe is exactly the same as the number entering. Nonetheless, the water must do work to turn the turbine, so there is an energy change as the water passes through.

Similarly, the lightbulb doesn't "use up" current, but, like the turbine, it *does* use energy. The energy is dissipated by atomic-level friction as the electrons move through the wire, making the wire hotter until, in the case of the lightbulb filament, it glows.

There are many other issues we'll need to examine, but we can draw a first important conclusion:

Law of conservation of current The current is the same at all points in a current-carrying wire.

FIGURE 22.7 Water in a pipe turns a turbine.

v
v
The amount of water leaving the turbine equals the amount entering; the number of electrons leaving the bulb equals the number entering.

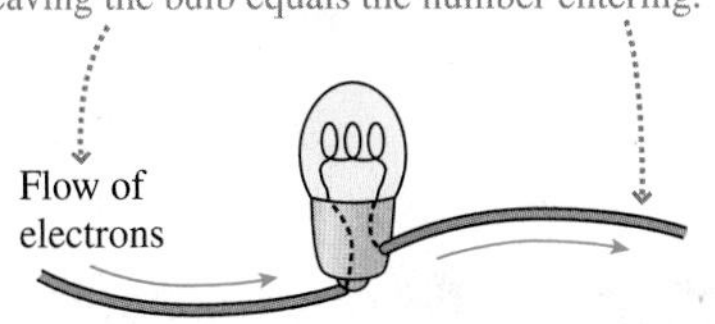

Flow of electrons

CONCEPTUAL EXAMPLE 22.1 Which bulb is brighter?

The discharge of a capacitor lights two identical bulbs, as shown in **FIGURE 22.8**. Compare the brightness of the two bulbs.

FIGURE 22.8 Two bulbs lit by the current discharging a capacitor.

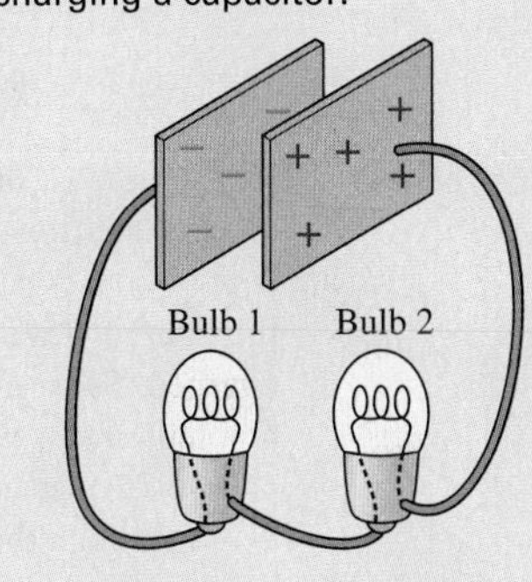

Bulb 1
Bulb 2

REASON Current is conserved, so any current that goes through bulb 1 must go through bulb 2 as well—the currents in the two bulbs are equal. We've noted that the brightness of a bulb is proportional to the current it carries. Identical bulbs carrying equal currents must have the same brightness.

ASSESS This result makes sense in terms of what we've seen about the conservation of current. No charge is "used up" by either bulb.

22.2 Defining and Describing Current

As we've seen, the current in a metal consists of electrons moving from low electric potential to high potential, in a direction opposite the electric field. But, as **FIGURE 22.9** shows, a capacitor discharges in exactly the same way whether we consider the current to be negative charges moving opposite the field or positive charges moving in the same direction as the field. In general, for *any* circuit, the current is in the same direction independent of which of these two perspectives we choose. We thus adopt the convention that **current is the flow of positive charge.** All our calculations will be correct, and all our circuits will work perfectly well, with this positive-charge-current convention.

FIGURE 22.9 Negative and positive charges moving in opposite directions give the same current.

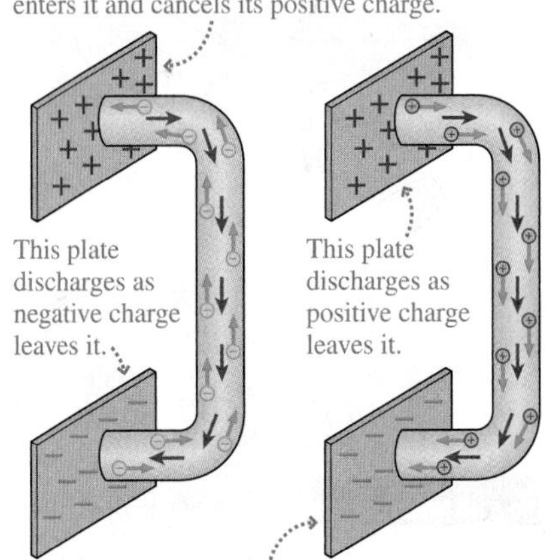

This plate discharges as negative charge enters it and cancels its positive charge.
This plate discharges as negative charge leaves it.
This plate discharges as positive charge leaves it.
This plate discharges as positive charge enters it and cancels its negative charge.

Definition of Current

Because the coulomb is the SI unit of charge, and because currents are charges in motion, we define current as the *rate,* in coulombs per second, at which charge moves through a wire. **FIGURE 22.10** shows a wire in which the electric field is $\vec{E}$. This electric field causes charges to move through the wire. Because we are considering current as the motion of positive charges, the motion is in the direction of the field.

The flow rate of water in a pipe measures the amount of water passing a cross-section area of the pipe per second. We use a similar convention for current. As illustrated in Figure 22.10, we can measure the amount of charge Δq that passes through a cross section of the wire in a time interval Δt. We then define the current in the wire as

$$I = \frac{\Delta q}{\Delta t} \tag{22.1}$$

Definition of current

FIGURE 22.10 The current I.

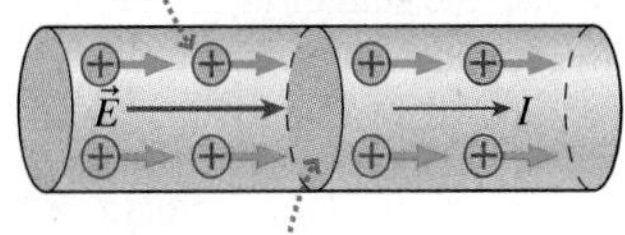

The current I is due to the motion of charges in the electric field.
$\vec{E}$
I
We imagine an area across the wire through which the charges move. In a time Δt, charge Δq moves through this area.

The current direction in a wire is from higher potential to lower potential or, equivalently, in the direction of the electric field $\vec{E}$. The SI unit for current is based on the units for charge and time according to Equation 22.1. Current is measured in coulombs per second, which we define as the **ampere,** with the abbreviation A:

$$1 \text{ ampere} = 1 \text{ A} = 1 \text{ coulomb per second} = 1 \text{ C/s}$$

The *amp* is an informal shortening of ampere. Household currents are typically ≈ 1 A. For example, the current through a 100 watt lightbulb is 0.83 A. The smaller currents in electronic devices are typically measured in milliamps (1 mA $= 10^{-3}$ A) or microamps (1 μA $= 10^{-6}$ A).

For a *steady current,* which will be our primary focus, the total amount of charge delivered by current I during the time interval Δt is

$$q = I\,\Delta t \tag{22.2}$$

EXAMPLE 22.2 **Charge flow in a lightbulb**

A 100 W lightbulb carries a current of 0.83 A. How much charge flows through the bulb in 1 minute?

SOLVE According to Equation 22.2, the total charge passing through the bulb in 1 min = 60 s is

$$q = I\,\Delta t = (0.83 \text{ A})(60 \text{ s}) = 50 \text{ C}$$

ASSESS The current corresponds to a flow of a bit less than 1 C per second, so our calculation seems reasonable, but the result is still somewhat surprising. That's a lot of charge! The enormous charge that flows through the bulb is a good check on the concept of conservation of current. If even a minuscule fraction of the charge stayed in the bulb, the bulb would become highly charged. For comparison, a Van de Graaff generator develops a potential of several hundred thousand volts due to an excess charge of just a few μC, a ten-millionth of the charge that flows through the bulb in 1 minute. Lightbulbs do not develop a noticeable charge, so the current into and out of the bulb must be exactly the same.

FIGURE 22.11 Kirchhoff's junction law.

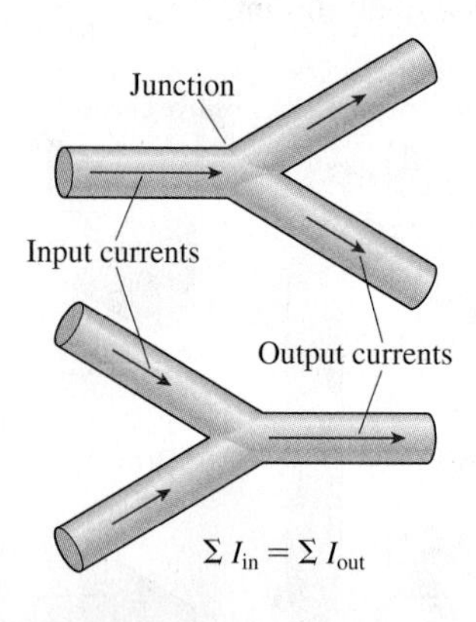

Conservation of Current at a Junction

FIGURE 22.11 shows a wire splitting into two and two wires merging into one. A point where a wire branches is called a **junction.** The presence of a junction doesn't change the fact that current is conserved. We cannot create or destroy charges in the wire, and neither can we store them in the junction. The rate at which electrons flow into one *or many* wires must be exactly balanced by the rate at which they flow out of others. For a *junction,* the law of conservation of charge requires that

$$\Sigma I_{in} = \Sigma I_{out} \tag{22.3}$$

where, as usual, the Σ symbol means "the sum of."

This basic conservation statement—that the sum of the currents into a junction equals the sum of the currents leaving—is called **Kirchhoff's junction law.** The junction law isn't a new law of physics; it is a consequence of the conservation of charge.

EXAMPLE 22.3 **Currents in a junction**

Four wires have currents as noted in **FIGURE 22.12**. What are the direction and the magnitude of the current in the fifth wire?

PREPARE This is a conservation of current problem. We compute the sum of the currents coming into the junction and the sum of the currents going out of the junction, and then compare these two sums. The unknown current is whatever is required to make the currents into and out of the junction "balance."

FIGURE 22.12 The junction of five wires.

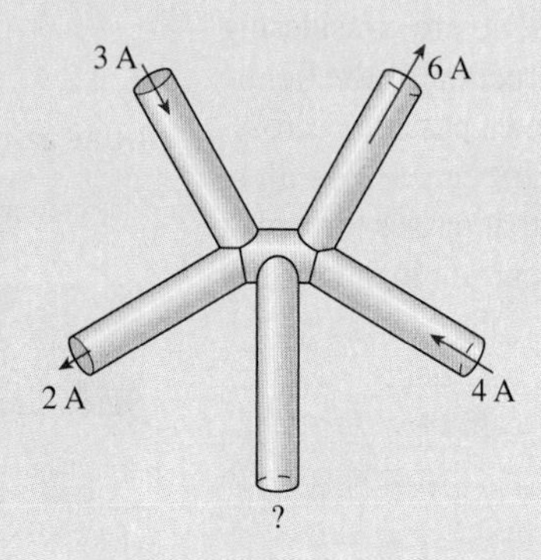

SOLVE Two of the wires have currents into the junction:

$$\Sigma I_{in} = 3 \text{ A} + 4 \text{ A} = 7 \text{ A}$$

Two of the wires have currents out of the junction:

$$\Sigma I_{out} = 6 \text{ A} + 2 \text{ A} = 8 \text{ A}$$

To conserve current, the fifth wire must carry a current of 1 A into the junction.

ASSESS If the unknown current is 1 A into the junction, a total of 8 A flows in—exactly what is needed to balance the current going out.

STOP TO THINK 22.1 The discharge of a capacitor lights three bulbs. Comparing the current in bulbs 1 and 2, we can say that

A. The current in bulb 1 is greater than the current in bulb 2.
B. The current in bulb 1 is less than the current in bulb 2.
C. The current in bulb 1 is equal to the current in bulb 2.

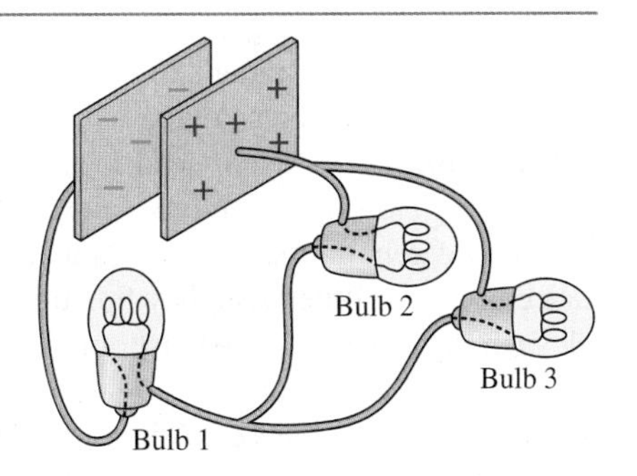

22.3 Batteries and emf

There are practical devices, such as a camera flash, that use the charge on a capacitor to create a current. But a camera flash gives a single, bright flash of light; the capacitor discharges and the current ceases. If you want a light to illuminate your way along a dark path, you need a *continuous* source of light like a flashlight. Continuous light requires the current to be continuous as well.

FIGURE 22.13 shows a wire connecting the two terminals of a battery, much like the wire that connected the capacitor plates in Figure 22.1. Just like that wire, the wire connecting the battery terminals gets warm, deflects a compass needle, and makes a lightbulb inserted into it glow brightly. These indicators tell us that charges flow through the wire from one terminal to the other. The current in the wire is the same whether it is supplied by a capacitor or a battery. Everything you've learned so far about current applies equally well to the current supplied by a battery, with one important difference—the duration of the current.

The wire connecting the battery terminals *continues* to deflect the compass needle and *continues* to light the lightbulb. The capacitor quickly runs out of excess charge, but the battery can keep the charges in motion.

How does a battery produce this sustained motion of charge? A real battery involves a series of chemical reactions, but FIGURE 22.14 shows a simple model of a battery that illustrates the motion of charges. The inner workings of a battery act like a *charge escalator* between the two terminals. Charges are removed from the negative terminal and "lifted" to the positive terminal. It is the charge escalator that sustains the current in the wire by providing a continuously renewed supply of charges at the positive terminal.

Once a charge reaches the positive terminal, it is able to flow downhill through the wire as a current until it reaches the negative terminal. The charge escalator then lifts the charge back to the positive terminal where it can start the loop all over again. This flow of charge in a continuous loop is what we call a **complete circuit.**

The charge escalator in the battery must be powered by some external source of energy. It is lifting the electrons "uphill" against an electric field. A battery consists of chemicals, called *electrolytes,* sandwiched between two electrodes made of different materials. The energy to move charges comes from chemical reactions between the electrolytes and the electrodes. These chemical reactions separate charge by moving positive ions to one electrode and negative ions to the other. In other words, chemical reactions, rather than a mechanical conveyor belt, transport charge from one electrode to the other.

As a battery creates a current in a circuit, the reactions that run the charge escalator deplete chemicals in the battery. A dead battery is one in which the supply of chemicals, and thus the supply of chemical energy, has been exhausted. You can "recharge" some types of batteries by forcing a current into the positive terminal, reversing the chemical reactions that move the charges, thus replenishing the chemicals and storing energy as chemical energy.

FIGURE 22.13 There is a current in a wire connecting the terminals of a battery.

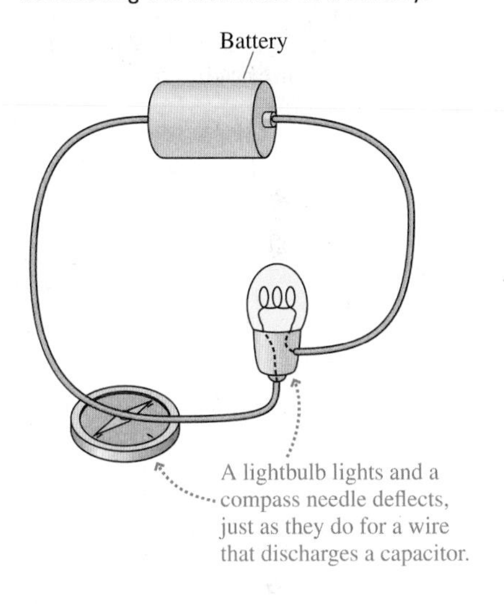

Battery
A lightbulb lights and a compass needle deflects, just as they do for a wire that discharges a capacitor.

FIGURE 22.14 The charge escalator model of a battery.

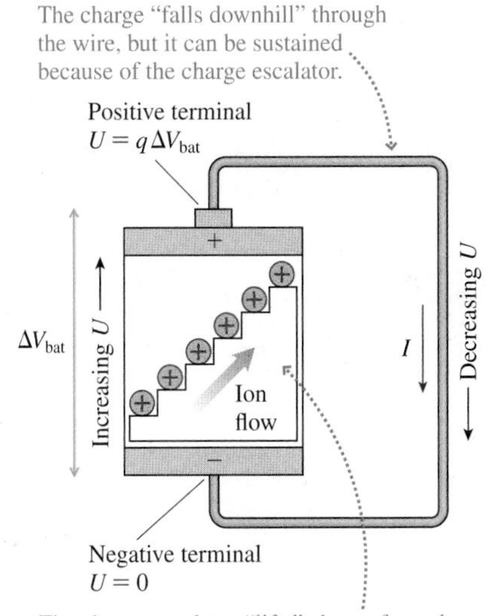

A shocking predator? BIO The torpedo ray captures and eats fish by paralyzing them with electricity. As we will see in Chapter 23, cells in the body use chemical energy to separate charge, just as in a battery. Special cells in the body of the ray called *electrocytes* produce an emf of a bit more than 0.10 V for a short time when stimulated. Such a small emf will not produce a large effect, but the torpedo ray has organs that contain clusters of hundreds of these electrocytes connected in a row. The total emf can be 50 V or more, enough to immobilize nearby prey.

By separating charge, the charge escalator establishes the potential difference ΔV_{bat} between the terminals of the battery shown in Figure 22.14. The potential difference established by a device, such as a battery, that can actively separate charge is called its **emf.** It is pronounced as the sequence of three letters "e-m-f." The symbol for emf is $\mathcal{E}$, a script E, and its units are volts.

Electric generators, photocells, and power supplies use different means to separate charge, but otherwise they function much like a battery. The common feature of all such devices is that they use some source of energy to separate charge and, thus, to create a potential difference—they create an emf. In contrast, a capacitor *stores* separated charges, but a capacitor has no means to do the separation. Hence a charged capacitor has a potential difference but not an emf.

The *rating* of a battery, such as 1.5 V, is the battery's emf. It is determined by the specific chemical reactions employed by the battery. An alkaline battery has an emf of 1.5 V; a rechargeable nickel-metal hydride (NiMH) battery has an emf of 1.2 V. Larger emfs are created by using several smaller "cells" in a row, much like going from the first to the fourth floor of a building by taking three separate escalators.

A battery with no current in it has a potential difference equal to its emf. With a current, the battery's potential difference is slightly less than the emf. We'll overlook this small difference and assume $\Delta V_{\text{bat}} = \mathcal{E}$.

CONCEPTUAL EXAMPLE 22.4 Potential difference for batteries in series

Three batteries are connected one after the other as shown in **FIGURE 22.15**; we say they are connected in *series.* What's the total potential difference?

FIGURE 22.15 Three batteries in series.

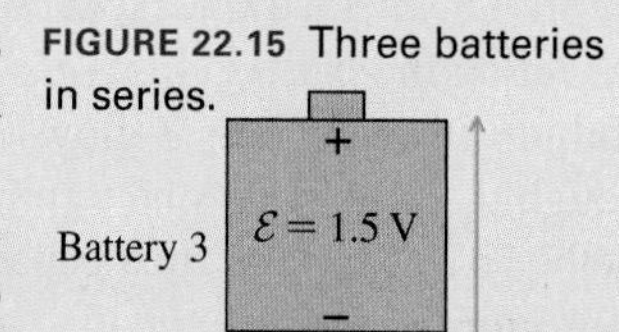

REASON We can think of this as three charge escalators, one after the other. Each one lifts charges to a higher potential. Because each battery raises the potential by 1.5 V, the total potential difference of the three batteries in series is 4.5 V.

ASSESS Common AA and AAA batteries are 1.5 V batteries. Many consumer electronics, such as digital cameras, use two or four of these batteries. Wires inside the device connect the batteries in series to produce a total 3.0 V or 6.0 V potential difference.

STOP TO THINK 22.2 A battery produces a current in a wire. As the current continues, which of the following quantities (perhaps more than one) decreases?

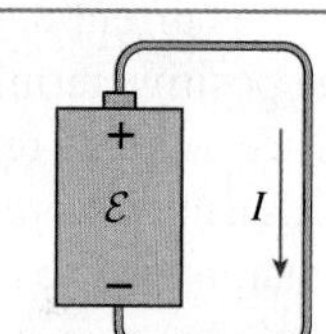

A. The positive charge in the battery
B. The emf of the battery
C. The chemical energy in the battery

22.4 Connecting Potential and Current

An important conclusion of the charge escalator model is that **a battery is a source of potential difference.** As **FIGURE 22.16** shows, when charges flow through a wire that connects the battery terminals, this current is a *consequence* of the battery's potential difference.

Because the ends of the wire are connected to the terminals of the battery, the potential difference between the two ends of the wire is equal to the potential difference between the battery terminals:

$$\Delta V_{\text{wire}} = \Delta V_{\text{bat}} \tag{22.4}$$

This potential difference causes a current in the direction of decreasing potential.

FIGURE 22.16 The electric field and the current inside the wire.

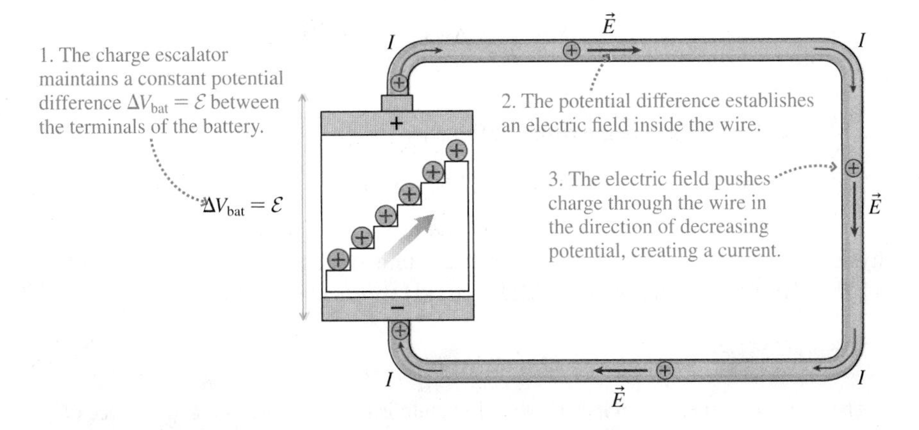
1. The charge escalator maintains a constant potential difference $\Delta V_{bat} = \mathcal{E}$ between the terminals of the battery.
$\Delta V_{bat} = \mathcal{E}$
2. The potential difference establishes an electric field inside the wire.
3. The electric field pushes charge through the wire in the direction of decreasing potential, creating a current.
$\vec{E}$
I

STOP TO THINK 22.3 In Figure 22.16 the wire is changed to a new one, leading to an increased current in the wire. Compared to the original wire, the potential difference across the ends of this new wire is

A. Larger. B. Smaller. C. The same.

Resistance

Figure 22.16 explained the *direction* of the current in a wire connected between the terminals of a battery. **FIGURE 22.17** shows a series of experiments to determine what factors affect the current's *magnitude*. The experiments show that there are two factors that determine the current: the potential difference and the properties of the wire.

Figure 22.17a shows that adding a second battery in series increases the current, as you would expect. A larger potential difference creates a larger electric field that pushes charges through the wire faster. Careful measurements would show that the current I is proportional to ΔV_{wire}.

Figures 22.17b and 22.17c illustrate the two properties of the wire that affect the current: the dimensions and the material of which the wire is made. Figure 22.17b shows that increasing the length of the wire decreases the current, while increasing the thickness of the wire increases the current. This seems reasonable because it should be harder to push charges through a long wire than a short one, and an electric field should be able to push more charges through a fat wire than a skinny one. Figure 22.17c shows that wires of different materials will carry different currents—some materials are better conductors than others.

For any particular wire, we can define a quantity called the **resistance** that is a measure of how hard it is to push charges through the wire. We use the symbol R for resistance. A large resistance implies that it is hard to move the charges through the wire; in a wire with small resistance, the charges move much more easily. The current in the wire depends on the potential difference ΔV_{wire} between the ends of the wire and the wire's resistance R:

$$I = \frac{\Delta V_{wire}}{R} \quad (22.5)$$

Establishing a potential difference ΔV_{wire} between the ends of a wire of resistance R creates an electric field that, in turn, causes a current $I = \Delta V_{wire}/R$ in the wire. As we would expect, the smaller the resistance, the larger the current.

FIGURE 22.17 Factors affecting the current in a wire.

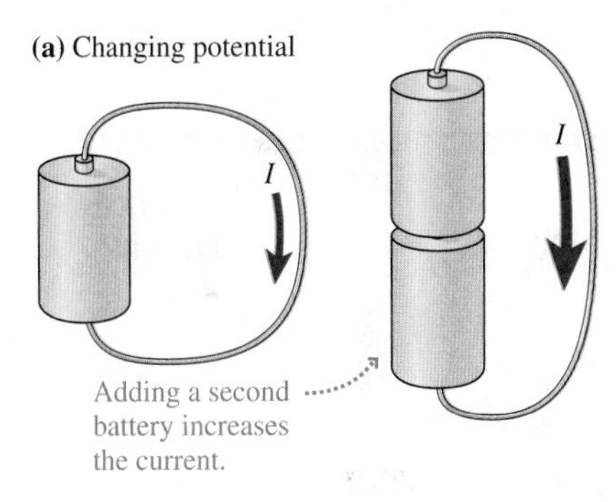
(a) Changing potential
I
Adding a second battery increases the current.

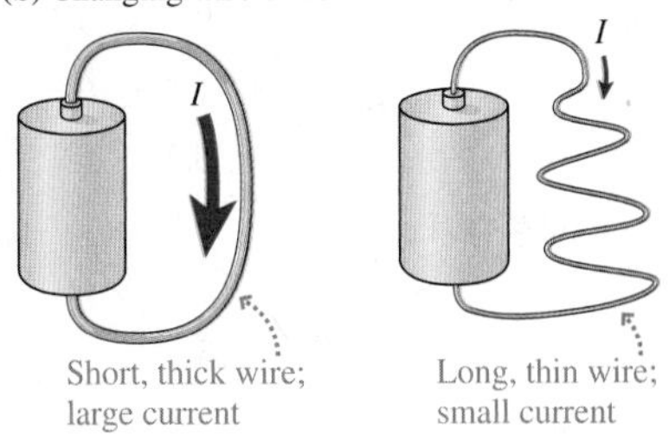
(b) Changing wire dimensions
I
Short, thick wire; large current
Long, thin wire; small current

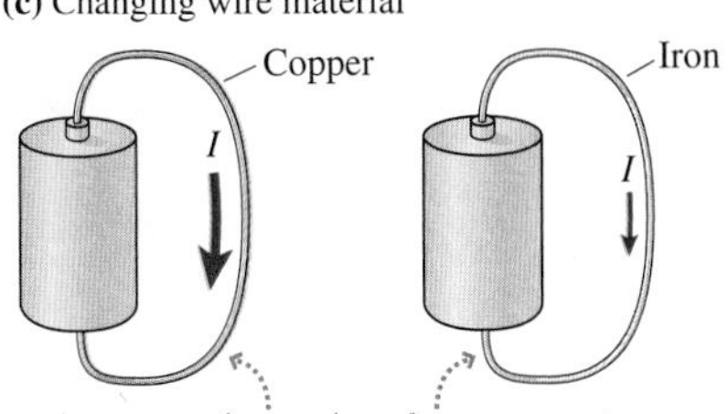
(c) Changing wire material
Copper
Iron
I
A copper wire carries a larger current than an iron wire of the same dimensions.

We can think of Equation 22.5 as the definition of resistance. If a potential difference ΔV_{wire} causes current I in a wire, the wire's resistance is

$$R = \frac{\Delta V_{wire}}{I} \tag{22.6}$$

The SI unit of resistance is the **ohm,** defined as

$$1 \text{ ohm} = 1\ \Omega = 1\ \text{V/A}$$

where Ω is an uppercase Greek omega. The unit takes its name from the German physicist Georg Ohm. The ohm is the basic unit of resistance, although kilohms ($1\ \text{k}\Omega = 10^3\ \Omega$) and megohms ($1\ \text{M}\Omega = 10^6\ \Omega$) are widely used.

EXAMPLE 22.5 Resistance of a lightbulb

The glowing element in an incandescent lightbulb is the *filament,* a long, thin piece of tungsten wire that is heated by the electric current through it. When connected to the 120 V of an electric outlet, a 60 W bulb carries a current of 0.50 A. What is the resistance of the filament in the lamp?

SOLVE We can use Equation 22.6 to compute the resistance:

$$R = \frac{\Delta V_{wire}}{I} = \frac{120\ \text{V}}{0.50\ \text{A}} = 240\ \Omega$$

ASSESS As we will see below, the resistance of the filament varies with temperature. This value holds for the lightbulb only when the bulb is glowing and the filament is hot.

Resistivity

TABLE 22.1 Resistivities of materials

Material	Resistivity ($\Omega \cdot \text{m}$)
Copper	1.7×10^{-8}
Aluminum	2.7×10^{-8}
Tungsten (20°C)	5.6×10^{-8}
Tungsten (1500°C)	5.0×10^{-7}
Iron	9.7×10^{-8}
Nichrome	1.5×10^{-6}
Seawater	0.22
Blood (average)	1.6
Muscle	13
Fat	25
Pure water	2.4×10^{5}
Cell membrane	3.6×10^{7}

Figure 22.17c showed that the resistance of a wire depends on what it is made of. We define a quantity called **resistivity,** for which we use the symbol ρ (lowercase Greek rho), to characterize the electrical properties of materials. Materials that are good conductors have low resistivities; materials that are poor conductors (and thus that are good insulators) have high resistivities. The resistivity ρ has units of $\Omega \cdot \text{m}$. The resistivities of some common materials are listed in Table 22.1.

Metals are generally good conductors (and so have very low resistivities), but metals such as copper are much better conductors than metals such as nichrome, an alloy of nickel and chromium that is used to make heating wires. Water is a poor conductor, but the dissolved salts in seawater produce ions that can carry charge, so seawater is a good conductor, with a resistivity one million times less than that of pure water. Glass is an excellent insulator with a resistivity in excess of $10^{14}\ \Omega \cdot \text{m}$, 10^{22} times that of copper.

The resistivity of a material depends on the temperature, as you can see from the two values for tungsten listed in Table 22.1. As the temperature increases, so do the thermal vibrations of the atoms. This makes them "bigger targets" for the moving electrons, causing collisions to be more frequent. Thus the resistivity of a metal increases with increasing temperature.

The resistance of a wire depends both on the resistivity of its material and on the dimensions of the wire. A wire made of a material of resistivity ρ, with length L and cross-section area A, has resistance

$$R = \frac{\rho L}{A} \tag{22.7}$$

Resistance of a wire in terms of resistivity and dimensions

R L p. 34 PROPORTIONAL

R A p. 108 INVERSE

Resistance is a property of a *specific* wire or conductor because it depends on the conductor's length and diameter as well as on the resistivity of the material from which it is made.

Class Video

NOTE ► It is important to distinguish between resistivity and resistance. *Resistivity* is a property of the *material,* not any particular piece of it. All copper wires (at the same temperature) have the same resistivity. *Resistance* characterizes a specific piece of the conductor having a specific geometry. A short, thick copper wire has a smaller resistance than a long, thin copper wire. The relationship between resistivity and resistance is analogous to that between density and mass. ◄

EXAMPLE 22.6 The length of a lightbulb filament

We calculated in Example 22.5 that a 60 W lightbulb has a resistance of 240 Ω. At the operating temperature of the tungsten filament, the resistivity is approximately $5.0 \times 10^{-7}\ \Omega \cdot \text{m}$. If the wire used to make the filament is 0.040 mm in diameter (a typical value), how long must the filament be?

PREPARE The resistance of a wire depends on its length, its cross-section area, and the material of which it is made.

SOLVE The cross-section area of the wire is $A = \pi r^2 = \pi(2.0 \times 10^{-5}\ \text{m})^2 = 1.26 \times 10^{-9}\ \text{m}^2$. Rearranging Equation 22.7 shows us that the filament must be of length

$$L = \frac{AR}{\rho} = \frac{(1.26 \times 10^{-9}\ \text{m}^2)(240\ \Omega)}{5.0 \times 10^{-7}\ \Omega \cdot \text{m}} = 0.60\ \text{m}$$

ASSESS This is quite long—nearly 2 feet. This result may seem surprising, but some reflection shows that it makes sense. The resistivity of tungsten is low, so the filament must be quite thin and long.

Coils of coils A close view of a lightbulb's filament shows that it is made of very thin wire that is coiled and then coiled again. The double-coil structure is necessary to fit the great length of the filament into the small space of the bulb's globe.

EXAMPLE 22.7 Making a heater

An amateur astronomer uses a heater to warm her telescope eyepiece so moisture does not collect on it. The heater is a 20-cm-long, 0.50-mm-diameter nichrome wire that wraps around the eyepiece. When the wire is connected to a 1.5 V battery, what is the current in the wire?

PREPARE The current in the wire depends on the emf of the battery and the resistance of the wire. The resistance of the wire depends on the resistivity of nichrome, given in Table 22.1, and the dimensions of the wire. Converted to meters, the relevant dimensions of the wire are $L = 0.20$ m and $r = 2.5 \times 10^{-4}$ m.

SOLVE The wire's resistance is

$$R = \frac{\rho L}{A} = \frac{\rho L}{\pi r^2} = \frac{(1.5 \times 10^{-6}\ \Omega \cdot \text{m})(0.20\ \text{m})}{\pi(2.5 \times 10^{-4}\ \text{m})^2} = 1.53\ \Omega$$

The wire is connected to the battery, so $\Delta V_{\text{wire}} = \Delta V_{\text{bat}} = 1.5$ V. The current in the wire is

$$I = \frac{\Delta V_{\text{wire}}}{R} = \frac{1.5\ \text{V}}{1.53\ \Omega} = 0.98\ \text{A}$$

ASSESS The emf of the battery is small, but so is the resistance of the wire, so this is a reasonable current, enough to warm the wire and the eyepiece.

Electrical Measurements of Physical Properties

Measuring resistance is quite straightforward. Because resistance depends sensitively on the properties of materials, a measurement of resistance can be a simple but effective probe of other quantities of interest. For example, the resistivity of water is strongly dependent on dissolved substances in the water, so it is easy to make a quick test of water purity by making a measurement of resistivity.

CONCEPTUAL EXAMPLE 22.8 Testing drinking water

A house gets its drinking water from a well that has an intermittent problem with salinity. Before the water is pumped into the house, it passes between two electrodes in the circuit shown in **FIGURE 22.18**. The current passing through the water is measured with a meter. Which corresponds to increased salinity—an increased current or a decreased current?

REASON Increased salinity causes the water's resistivity to decrease. This decrease causes a decrease in resistance between the electrodes. Current is inversely proportional to resistance, so this leads to an increase in current.

ASSESS Increasing salinity means more ions in solution and thus more charge carriers, so an increase in current is expected. Electrical systems similar to this can therefore provide a quick check of water purity.

FIGURE 22.18 A water-testing circuit.

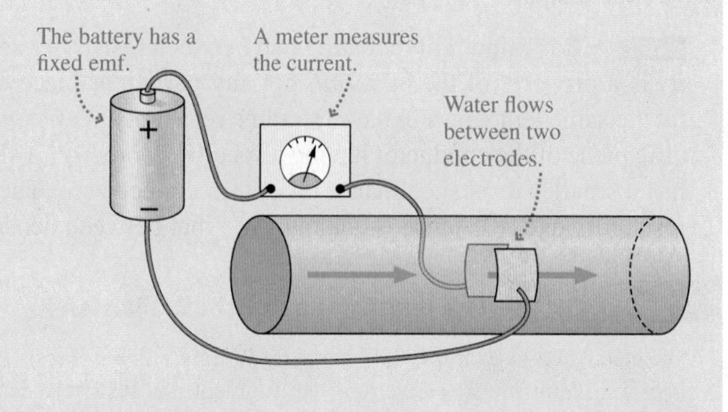
The battery has a fixed emf.
A meter measures the current.
Water flows between two electrodes.
+
−

Different tissues in the body have different resistivities, as we see in Table 22.1. For example, fat has a higher resistivity than muscle. Consequently, a routine test to estimate the percentage of fat in a person's body is based on a measurement of the body's resistance, as illustrated in the photo at the start of the chapter. A higher resistance of the body means a higher proportion of fat.

FIGURE 22.19 BIO An electrical impedance map showing the cross section of a healthy patient's torso.

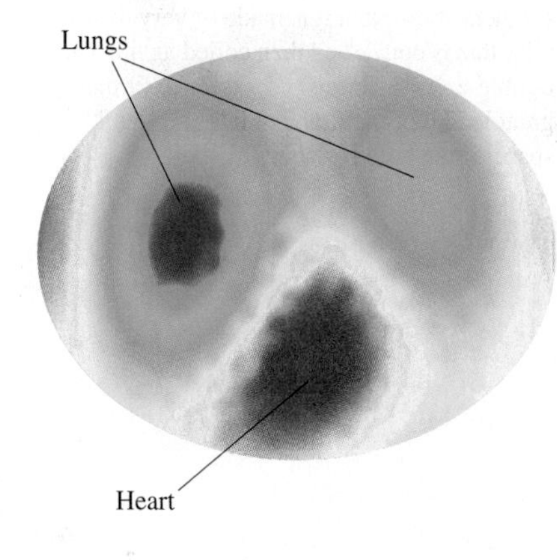
Lungs
Heart

More careful measurements of resistance can provide more detailed diagnostic information. Passing a small, safe current between pairs of electrodes on opposite sides of a person's torso permits a measurement of the resistance of the intervening tissue, a technique known as *electrical impedance tomography*. (Impedance is similar to resistance, but it also applies to AC circuits, which we will explore in Chapter 26.) **FIGURE 22.19** shows an image of a patient's torso generated from measurements of resistance between many pairs of electrodes. The image shows the change in resistance between two subsequent measurements; decreasing resistance shows in red, increasing resistance in blue. This image was created during the resting phase of the heart, when blood was leaving the lungs and entering the heart. Blood is a better conductor than the tissues of the heart and lungs, so the motion of blood decreased the resistance of the heart and increased that of the lungs. This patient was healthy, but in a patient with circulatory problems any deviation from normal blood flow would lead to abnormal patterns of resistance that would be revealed in such an image.

STOP TO THINK 22.4 A wire connected between the terminals of a battery carries a current. The wire is removed and stretched, decreasing its cross-section area and increasing its length. When the wire is reconnected to the battery, the new current is

A. Larger than the original current.
B. The same as the original current.
C. Smaller than the original current.

22.5 Ohm's Law and Resistor Circuits

The relationship between the potential difference across a conductor and the current passing through it that we saw in the preceding section was first deduced by Georg Ohm and is known as **Ohm's law:**

Class Video

$$I = \frac{\Delta V}{R} \qquad (22.8)$$

Ohm's law for a conductor of resistance R

PROPORTIONAL p. 34 INVERSE p. 108

If we know that a wire of resistance R carries a current I, we can compute the potential difference between the ends of the wire as $\Delta V = IR$.

Despite its name, Ohm's law is *not* a law of nature. It is limited to those materials whose resistance R remains constant—or very nearly so—during use. Materials to which Ohm's law applies are called **ohmic.** FIGURE 22.20 shows that the current through an ohmic material is directly proportional to the potential difference; doubling the potential difference results in a doubling of the current. This is a linear relationship, and the resistance R can be determined from the slope of the graph.

Other materials and devices are **nonohmic,** meaning that the current through the device is *not* directly proportional to the potential difference. Two important examples of nonohmic devices are batteries, where $\Delta V = \mathcal{E}$ is determined by chemical reactions, independent of I, and capacitors, where, as you'll learn in Chapter 26, the relationship between I and ΔV is very different from that of a resistor.

FIGURE 22.20 Current-versus-potential-difference for a conductor with resistance R.

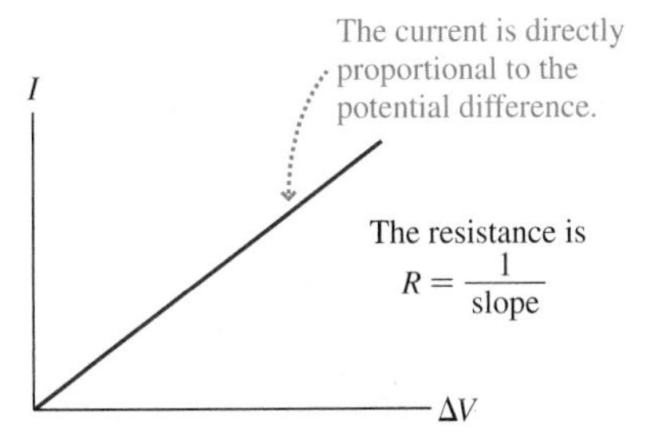

I
The current is directly proportional to the potential difference.
The resistance is $R = \frac{1}{\text{slope}}$
ΔV

NOTE ▶ Ohm's law applies only to resistive devices like wires or lightbulb filaments. It does *not* apply to nonohmic devices such as batteries or capacitors. ◀

Resistors

The word "resistance" may have negative connotations—who needs something that slows charges and robs energy? In some cases resistance *is* undesirable. But in many other cases, circuit elements are designed to have a certain resistance for very practical reasons. We call these circuit elements **resistors.** There are a few basic types that will be very important as we start to look at electric circuits in detail.

Examples of resistors

Heating elements

As charges move through a resistive wire, their electric energy is transformed into thermal energy, heating the wire. Wires in a toaster, a stove burner, or the rear window defroster of a car are practical examples of this electric heating.

R113
R118
R112
R111
R117
R116
Resistors

Circuit elements

Inside many electronic devices is a circuit board with many small cylinders. These cylinders are resistors that help control currents and voltages in the circuit. The colored bands on the resistors indicate their resistance values.

Light-sensitive resistor

Sensor elements

A resistor whose resistance changes in response to changing circumstances can be used as a sensor. The resistance of this night-light sensor changes when daylight strikes it. A circuit detects this change and turns off the light during the day.

CONCEPTUAL EXAMPLE 22.9 The changing current in a toaster

When you press the lever on a toaster, a switch connects the heating wires to 120 V. The wires are initially cool, but the current in the wires raises the temperature until they are hot enough to glow. As the wire heats up, how does the current in the toaster change?

REASON As the wire heats up, its resistivity increases, as noted above, so the resistance of the wires increases. Because the potential difference stays the same, an increasing resistance causes the current to decrease. The current through a toaster is largest when the toaster is first turned on.

ASSESS This result makes sense. As the wire's temperature increases, the current decreases. This makes the system stable. If, instead, the current increased as the temperature increased, higher temperature could lead to more current, leading to even higher temperatures, and the toaster could overheat.

Analyzing a Simple Circuit

FIGURE 22.21a shows the anatomy of a lightbulb. The important point is that a lightbulb, like a wire, has two "ends" and current passes *through* the bulb. Connections to the filament in the bulb are made at the tip and along the side of the metal cylinder. It is often useful to think of a lightbulb as a resistor that happens to give off light when a current is present. Now, let's look at a circuit using a battery, a lightbulb, and wires to make connections, as in FIGURE 22.21b. This is the basic circuit in a flashlight.

FIGURE 22.21 The basic circuit of a battery and a bulb.

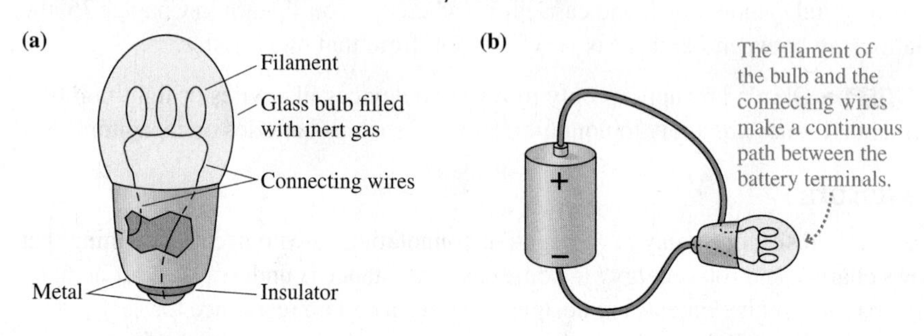
(a)
Filament
Glass bulb filled with inert gas
Connecting wires
Metal
Insulator
(b)
The filament of the bulb and the connecting wires make a continuous path between the battery terminals.
+
−

FIGURE 22.22 The potential along a wire-resistor-wire combination.

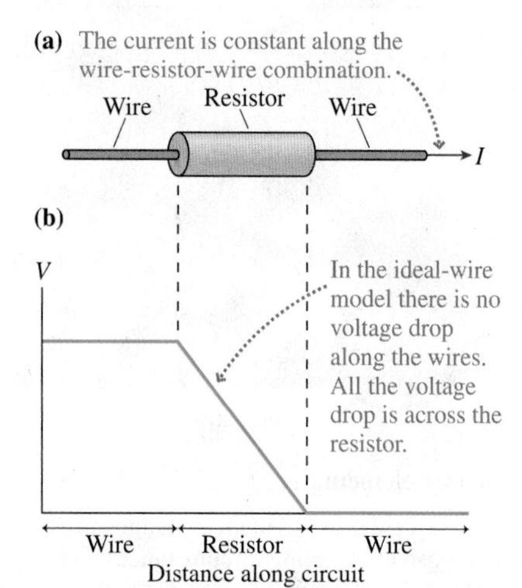
(a) The current is constant along the wire-resistor-wire combination.
Wire
Resistor
Wire
I
(b)
V
In the ideal-wire model there is no voltage drop along the wires. All the voltage drop is across the resistor.
Wire
Resistor
Wire
Distance along circuit

A typical flashlight bulb has a resistance of $\approx 3\ \Omega$, while a wire that one would use to connect such a bulb to a battery has a resistance of $\approx 0.01\ \Omega$. The resistance of the wires is so much less than that of the bulb that we can, with very little error, assume that the resistance of the wires is *zero*. In this model of an **ideal wire,** with $R_{\text{wire}} = 0\ \Omega$, the potential difference ΔV between the ends of a connecting wire is zero, *even if there is a current in it.*

We know that, physically, the potential difference can't be zero. There must be an electric field in the wire for the charges to move, so it must have a potential difference. But in practice this potential difference is so small that we can assume it to be zero.

FIGURE 22.22 shows how the ideal-wire model is used in the analysis of circuits. The resistor in Figure 22.22a is connected at each end to a wire, and current I flows through all three. The current requires a potential difference $\Delta V_{\text{resist}} = IR_{\text{resist}}$ across the resistor, but there's no potential difference ($\Delta V_{\text{wire}} = 0$) for the ideal wires. Figure 22.22b shows this idea graphically by displaying the potential along the wire-resistor-wire combination. Current moves in the direction of decreasing potential, so there is a large *voltage drop*—a decrease in potential—across the resistor as we go from left to right, the direction of the current. The segments of the graph corresponding to the wires are horizontal because there's no voltage change along an ideal wire.

FIGURE 22.23 Electric field inside a resistor.

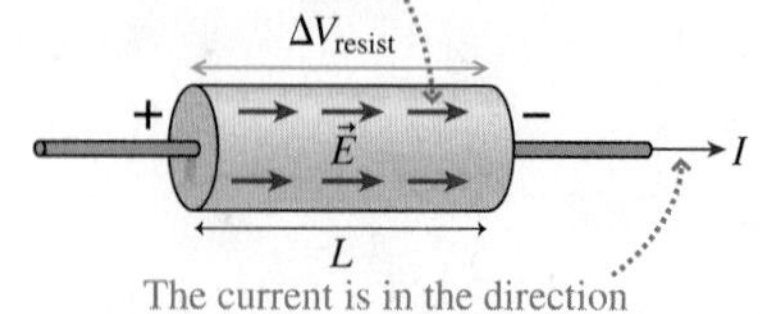
The electric field inside the resistor is uniform and points from high to low potential.
ΔV_{resist}
+
−
$\vec{E}$
I
L
The current is in the direction of decreasing potential.

The linear variation in the potential across the resistor is similar to the linear variation in potential between the plates of a parallel-plate capacitor. In a capacitor, this linear variation in potential corresponds to a uniform electric field; the same will be true here. As we see in FIGURE 22.23, the electric field in a resistor carrying a current in a circuit is uniform; the strength of the electric field is

$$E = \frac{\Delta V}{L}$$

in analogy to Equation 21.6 for a parallel-plate capacitor. A larger potential difference corresponds to a larger field, as we would expect.

EXAMPLE 22.10 Analyzing a single-resistor circuit

A 15 Ω resistor is connected to the terminals of a 1.5 V battery.

a. Draw a graph showing the potential as a function of distance traveled through the circuit, starting from $V = 0$ V at the negative terminal of the battery.
b. What is the current in the circuit?

PREPARE To help us visualize the change in potential as charges move through the circuit, we begin with the sketch of the circuit in **FIGURE 22.24**. The zero point of potential is noted. We have drawn our sketch so that "up" corresponds to higher potential, which will help us make sense of the circuit. Charges are raised to higher potential in the battery, then travel "downhill" from the positive terminal through the resistor and back to the negative terminal. We assume ideal wires.

FIGURE 22.24 A single-resistor circuit.

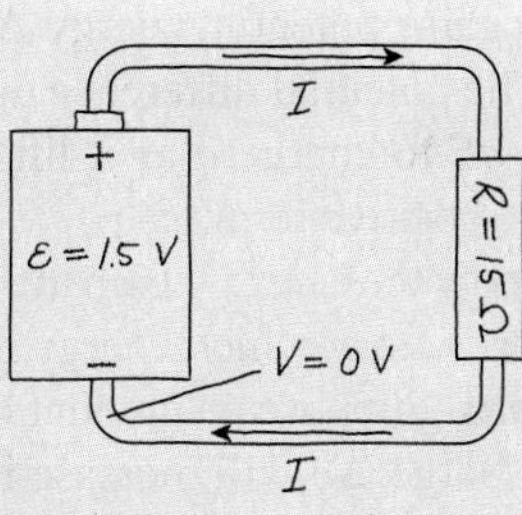

SOLVE a. **FIGURE 22.25** is a graphical representation of the potential in the circuit. The distance s is measured from the battery's negative terminal, where $V = 0$ V. As we move around the circuit to the starting point, the potential must return to its original value. Because the wires are ideal, there is no change in potential along the wires. This means that the potential difference across the resistor must be equal to the potential difference across the battery: $\Delta V_R = \mathcal{E} = 1.5$ V.

FIGURE 22.25 Potential-versus-position graph.

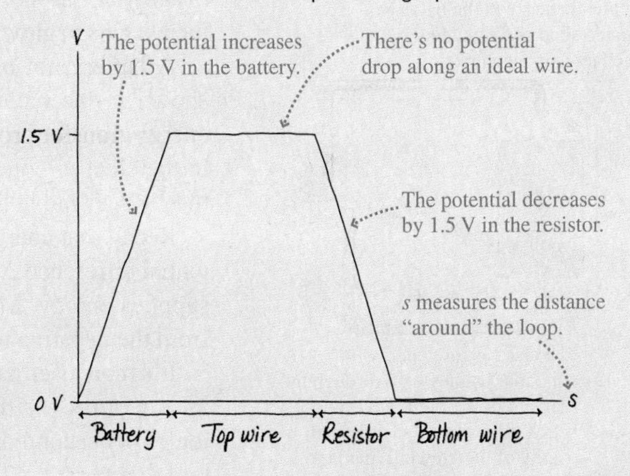

b. Now that we know the potential difference of the resistor, we can compute the current in the resistor by using Ohm's law:

$$I = \frac{\Delta V_R}{R} = \frac{1.5\text{ V}}{15\ \Omega} = 0.10\text{ A}$$

Because current is conserved, this is the current at any point in the circuit. In other words, the battery's charge escalator lifts charge at the rate 0.10 C/s, and charge flows through the wires and the resistor at the rate 0.10 C/s.

ASSESS This is a reasonable value of the current in a battery-powered circuit.

As we noted, there are many devices whose resistance varies as a function of a physical variable that we might like to measure, such as light intensity, temperature, or sound intensity. As the following example shows, we can use resistance measurements to monitor a physical variable.

EXAMPLE 22.11 Using a thermistor

A thermistor is a device whose resistance varies with temperature in a well-defined way. A certain thermistor has a resistance of 2.8 kΩ at 20°C and 0.39 kΩ at 70°C. This thermistor is used in a water bath in a lab to monitor the temperature. The thermistor is connected in a circuit with a 1.5 V battery, and the current measured. What is the change in current in the circuit as the temperature rises from 20°C to 70°C?

SOLVE We can use Ohm's law to find the current in each case:

$$I(20°\text{C}) = \frac{\Delta V}{R} = \frac{1.5\text{ V}}{2.8 \times 10^3\ \Omega} = 0.54\text{ mA}$$

$$I(70°\text{C}) = \frac{\Delta V}{R} = \frac{1.5\text{ V}}{0.39 \times 10^3\ \Omega} = 3.8\text{ mA}$$

The change in current is thus 3.3 mA.

ASSESS A modest change in temperature leads to a large change in current, which is reasonable—this is a device intended to provide a sensitive indication of a temperature change.

STOP TO THINK 22.5 Two identical batteries are connected in series in a circuit with a single resistor. $V = 0$ V at the negative terminal of the lower battery. Rank in order, from highest to lowest, the potentials V_A to V_E at the labeled points, noting any ties. Assume the wires are ideal.

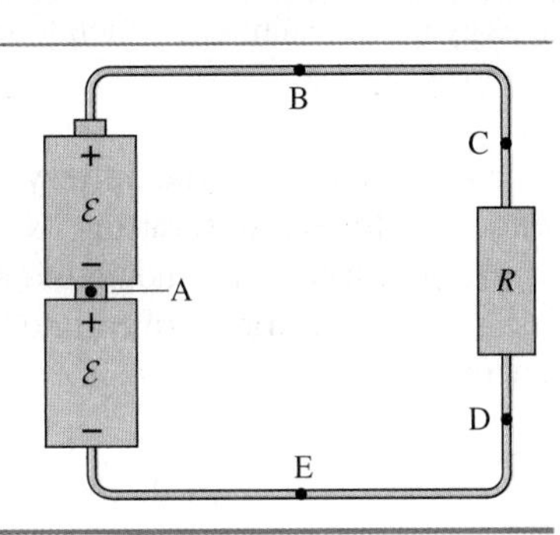

FIGURE 22.26 Energy transformations in a circuit with a battery and a lightbulb.

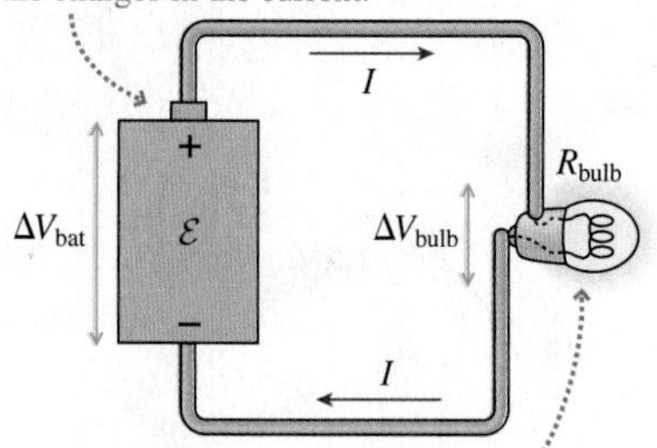

22.6 Energy and Power

When you flip the switch on a flashlight, a battery is connected to a lightbulb, which then begins to glow. The bulb is radiating energy. Where does this energy come from?

A battery not only supplies a potential difference but also supplies energy, as shown in the battery and bulb circuit of FIGURE 22.26. The charge escalator is an energy-transfer process, transferring the chemical energy E_{chem} stored in the battery to the electric potential energy U of the charges. That energy is then dissipated as the charges move through the lightbulb, keeping the filament warm and glowing.

Recall that charge q gains potential energy $\Delta U = q\Delta V$ as it moves through a potential difference ΔV. The potential difference of a battery is $\Delta V_{bat} = \mathcal{E}$, so the battery supplies energy $\Delta U = q\mathcal{E}$ to charge q as it lifts the charge up the charge escalator from the negative to the positive terminal.

It's more useful to know the *rate* at which the battery supplies energy. You learned in ◀◀ SECTION 10.8 that the rate at which energy is transformed is *power,* measured in joules per second or *watts.* Suppose an amount of charge Δq moves through the battery in a time Δt. The charge Δq will increase its potential energy by $\Delta U = (\Delta q)\mathcal{E}$. The *rate* at which energy is transferred from the battery to the moving charges is

$$P_{bat} = \text{rate of energy transfer} = \frac{\Delta U}{\Delta t} = \frac{\Delta q}{\Delta t}\mathcal{E} \tag{22.9}$$

But $\Delta q/\Delta t$, the rate at which charge moves through the battery, is the current I. Hence the power supplied by a battery or any source of emf is

$$P_{emf} = I\mathcal{E} \tag{22.10}$$

Power delivered by a source of emf

$I\mathcal{E}$ has units of J/s, or W.

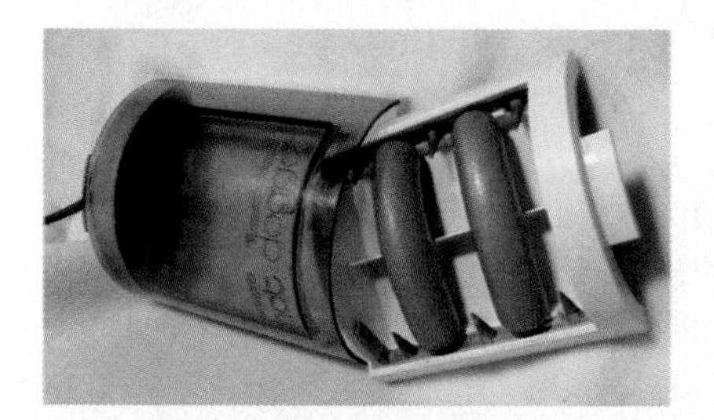

Hot dog resistors Before microwave ovens were common, there were devices that used a decidedly lower-tech approach to cook hot dogs. Prongs connected the hot dog to the 120 V of household electricity, making it the resistor in a circuit. The current through the hot dog dissipated energy as thermal energy, cooking the hot dog in about 2 minutes.

EXAMPLE 22.12 Power delivered by a car battery

A car battery has $\mathcal{E} = 12$ V. When the car's starter motor is running, the battery current is 320 A. What power does the battery supply?

SOLVE The power is the product of the emf of the battery and the current:

$$P_{bat} = I\mathcal{E} = (320\text{ A})(12\text{ V}) = 3.8\text{ kW}$$

ASSESS This is a lot of power (about 5 hp), but this amount makes sense because turning over a car's engine is hard work. Car batteries are designed to reliably provide such intense bursts of power for starting the engine.

FIGURE 22.27 The power from the battery is dissipated in the resistor.

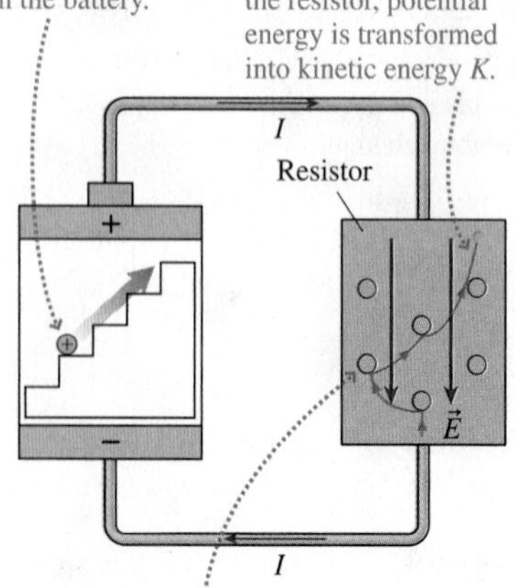

1. Charges gain potential energy U in the battery.
2. As charges accelerate in the electric field in the resistor, potential energy is transformed into kinetic energy K.
I
Resistor
+
−
$\vec{E}$
I
3. Collisions with atoms in the resistor transform the kinetic energy of the charges into thermal energy E_{th} of the resistor.

Suppose we consider a circuit consisting of a battery and a single resistor. $P_{bat} = I\mathcal{E}$ is the energy transferred per second from the battery's store of chemicals to the moving charges that make up the current. FIGURE 22.27 shows the entire sequence of energy transformations, which looks like

$$E_{chem} \rightarrow U \rightarrow K \rightarrow E_{th}$$

The net result is that **the battery's chemical energy is transferred to the thermal energy of the resistor,** raising its temperature.

In the resistor, the amount of charge Δq loses potential energy $\Delta U = (\Delta q)(\Delta V_R)$ as this energy is transformed into kinetic energy and then into the resistor's thermal energy. Thus the rate at which energy is transferred from the current to the resistor is

$$P_R = \frac{\Delta U}{\Delta t} = \frac{\Delta q}{\Delta t}\Delta V_R = I\,\Delta V_R \tag{22.11}$$

We say that this power—so many joules per second—is *dissipated* by the resistor as charge flows through it.

Our analysis of the single-resistor circuit in Example 22.10 found that $\Delta V_R = \mathcal{E}$. That is, the potential difference across the resistor is exactly the emf supplied by the battery. Because the current is the same in the battery and the resistor, a comparison of Equations 22.10 and 22.11 shows that

$$P_R = P_{bat} \qquad (22.12)$$

The power dissipated in the resistor is exactly equal to the power supplied by the battery. The *rate* at which the battery supplies energy is exactly equal to the *rate* at which the resistor dissipates energy. This is, of course, exactly what we would have expected from energy conservation.

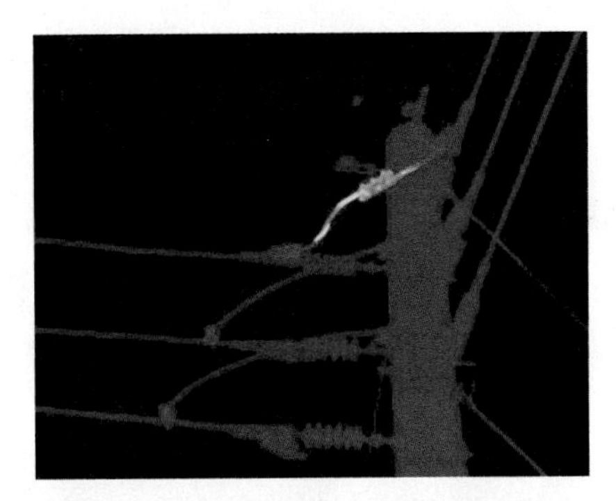

Hot wire This thermal camera image of power lines shows that the lines are warm—as we'd expect, given the large currents that they carry. Spots where corrosion has thinned the wires get especially warm, making such images helpful for monitoring the condition of power lines.

Most household appliances, such as a 100 W lightbulb or a 1500 W hair dryer, have a power rating. These appliances are intended for use at a standard household voltage of 120 V, and their rating is the power they will dissipate if operated with a potential difference of 120 V. Their power consumption will differ from the rating if they are operated at any other potential difference—for instance, if you use a lightbulb with a dimmer switch.

EXAMPLE 22.13 **Finding the current in a lightbulb**

How much current is "drawn" by a 75 W lightbulb connected to a 120 V outlet?

PREPARE We can model the lightbulb as a resistor.

SOLVE Because the lightbulb is operating as intended, it will dissipate 75 W of power. We can rearrange Equation 22.11 to find

$$I = \frac{P_R}{\Delta V_R} = \frac{75\text{ W}}{120\text{ V}} = 0.63\text{ A}$$

ASSESS We've said that we expect currents on the order of 1 A for lightbulbs and other household items, so our result seems reasonable.

A resistor obeys Ohm's law: $I = \Delta V_R/R$. This gives us two alternative ways of writing the power dissipated by a resistor. We can either substitute IR for ΔV_R or substitute $\Delta V_R/R$ for I. Thus

$$P_R = I\Delta V_R = I^2R = \frac{(\Delta V_R)^2}{R} \qquad (22.13)$$

Power dissipated by resistance R with current I and potential difference ΔV_R

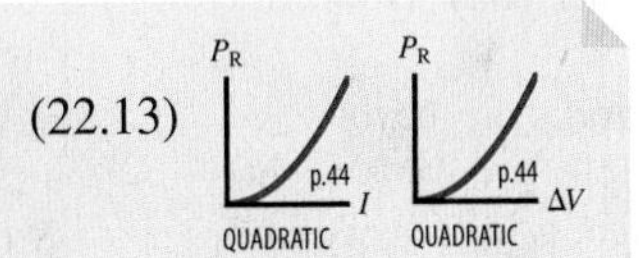

Class Video

Video Tutor Demo

It is worth writing the different forms of this equation to illustrate that the power varies as the square of both the current and the potential difference.

EXAMPLE 22.14 **Finding the power of a dim bulb**

How much power is dissipated by a 60 W (120 V) lightbulb when operated, using a dimmer switch, at 100 V?

PREPARE The 60 W rating is for operation at 120 V. We will assume that the resistance doesn't change if the bulb is run at a lower power—not quite right, but a reasonable approximation for this case in which the voltage is only slightly different from the rated value. We can compute the resistance for this case and then compute the power with the dimmer switch.

SOLVE The lightbulb dissipates 60 W at $\Delta V_R = 120$ V. Thus the filament's resistance is

$$R = \frac{(\Delta V_R)^2}{P_R} = \frac{(120\text{ V})^2}{60\text{ W}} = 240\ \Omega$$

The power dissipation when operated at $\Delta V_R = 100$ V is

$$P_R = \frac{(\Delta V_R)^2}{R} = \frac{(100\text{ V})^2}{240\ \Omega} = 42\text{ W}$$

ASSESS Reducing the voltage by 17% leads to a 30% reduction of the power. This makes sense; the power is proportional to the square of the voltage, so we expect a proportionally larger change in power.

EXAMPLE 22.15 **Determining the voltage of a stereo**

Most stereo speakers are designed to have a resistance of 8.0 Ω. If an 8.0 Ω speaker is connected to a stereo amplifier with a rating of 100 W, what is the maximum possible potential difference the amplifier can apply to the speakers?

PREPARE The rating of an amplifier is the *maximum* power it can deliver. Most of the time it delivers far less, but the maximum might be needed for brief, intense sounds. The maximum potential difference will occur when the amplifier is providing the maximum power, so we will make our computation with this figure. We can model the speaker as a resistor.

SOLVE The maximum potential difference occurs when the power is a maximum. At the maximum power of 100 W,

$$P_R = 100\ \text{W} = \frac{(\Delta V_R)^2}{R} = \frac{(\Delta V_R)^2}{8.0\ \Omega}$$

$$\Delta V_R = \sqrt{(8.0\ \Omega)(100\ \text{W})} = 28\ \text{V}$$

This is the maximum potential difference the amplifier might provide.

ASSESS As a check on our result, we note that the resistance of the speaker is less than that of a lightbulb, so a smaller potential difference can provide 100 W of power.

STOP TO THINK 22.6 Rank in order, from largest to smallest, the powers P_A to P_D dissipated in resistors A to D.

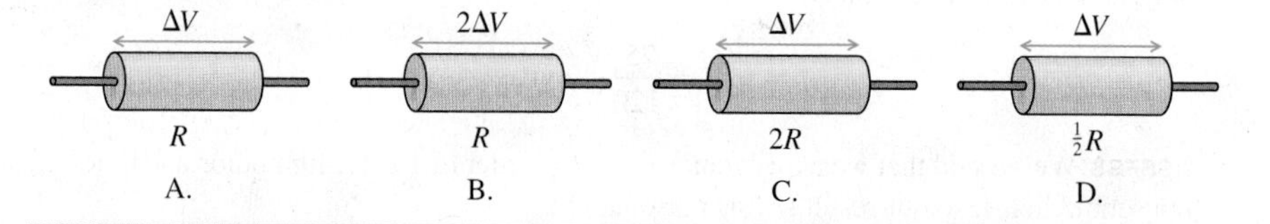

INTEGRATED EXAMPLE 22.16 **Electrical measurements of body composition** BIO

The woman in the photo at the start of the chapter is gripping a device that passes a small current through her body. How does this permit a determination of body fat?

The exact details of how the device works are beyond the scope of this chapter, but the basic principle is quite straightforward: The device applies a small potential difference and measures the resulting current. Comparing multiple measurements allows the device to determine the resistance of one part of the body, the upper arm. The resistance of the upper arm depends sensitively on the percentage of body fat in the upper arm, and the percentage of body fat in the upper arm is a good predictor of the percentage of fat in the body overall. Let's make a simple model of the upper arm to show how the resistance of the upper arm varies with percentage body fat.

The model of a person's upper arm in **FIGURE 22.28** ignores the nonconductive elements (such as the skin and the mineralized portion of the bone) and groups the conductive elements into two distinct sections—muscle and fat—that form two parallel segments. The resistivity of each tissue type is shown. This simple model isn't a good description of the actual structure of the arm, but it predicts the electrical character quite well.

FIGURE 22.28 A simple model of the resistance of the upper arm.

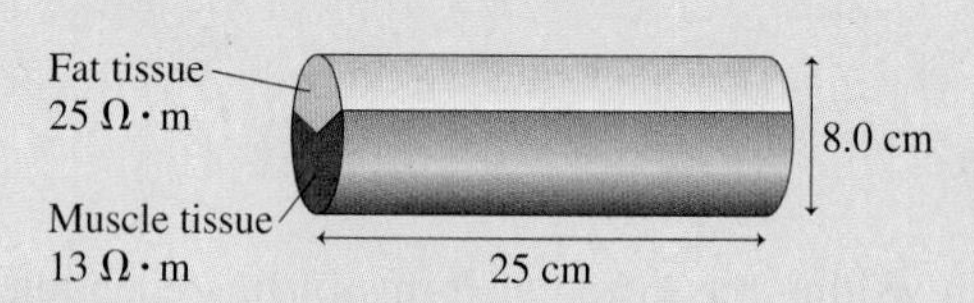

a. An experimental subject's upper arm, with the dimensions shown in the figure, is 40% fat and 60% muscle. A potential difference of 0.60 V is applied between the elbow and the shoulder. What current is measured?

b. A 0.60 V potential difference applied to the upper arm of a second subject with an arm of similar dimensions gives a current of 0.87 mA. What are the percentages of muscle and fat in this person's upper arm?

PREPARE **FIGURE 22.29** shows how we can model the upper arm as two resistors that are connected together at the ends. We use the ideal-wire model in which there's no "loss" of potential along the wires. Consequently, the potential difference across each of the two segments is the full 0.60 V of the battery. The current "splits" at the junction between the two resistors, and conservation of current tells us that

$$I_{\text{total}} = I_{\text{muscle}} + I_{\text{fat}}$$

The resistance of each segment depends on its resistivity (given in Figure 22.28) and on its length and cross-section area. The cross-section area of the whole arm is $A = \pi r^2 = \pi(0.040\text{ m})^2 = 0.00503\text{ m}^2$; the area of each segment is this number multiplied by the appropriate fraction.

SOLVE a. An object's resistance is related to its geometry and the resistivity of the material by $R = \rho L/A$. Thus the resistances of the muscle (60% of the area) and fat (40% of the area) segments are

FIGURE 22.29 Current through the tissues of the upper arm.

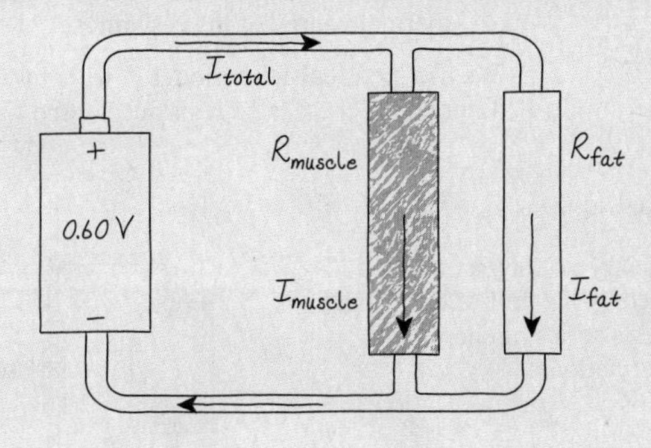

$$R_{\text{muscle}} = \frac{\rho_{\text{muscle}} L}{A_{\text{muscle}}} = \frac{(13\ \Omega \cdot \text{m})(0.25\text{ m})}{(0.60)(0.00503\text{ m}^2)} = 1080\ \Omega$$

$$R_{\text{fat}} = \frac{\rho_{\text{fat}} L}{A_{\text{fat}}} = \frac{(25\ \Omega \cdot \text{m})(0.25\text{ m})}{(0.40)(0.00503\text{ m}^2)} = 3110\ \Omega$$

The potential difference across each segment is 0.60 V. We can then use Ohm's law, $I = \Delta V/R$, to find that the current in each segment is

$$I_{\text{muscle}} = \frac{0.60\text{ V}}{1080\ \Omega} = 0.556\text{ mA}$$

$$I_{\text{fat}} = \frac{0.60\text{ V}}{3110\ \Omega} = 0.193\text{ mA}$$

The conservation of current equation then gives the total current as the sum of these two values:

$$I_{\text{total}} = 0.556\text{ mA} + 0.193\text{ mA} = 0.75\text{ mA}$$

b. If we know the current, we can determine the amount of muscle and fat. Let the fraction of muscle tissue be x; the fraction of fat tissue is then $(1 - x)$. We repeat the steps of the above calculation with these expressions in place:

$$R_{\text{muscle}} = \frac{\rho_{\text{muscle}} L}{A} = \frac{(13\ \Omega \cdot \text{m})(0.25\text{ m})}{(x)(0.00503\text{ m}^2)} = \frac{646\ \Omega}{x}$$

$$R_{\text{fat}} = \frac{\rho_{\text{fat}} L}{A} = \frac{(25\ \Omega \cdot \text{m})(0.25\text{ m})}{(1 - x)(0.00503\text{ m}^2)} = \frac{1240\ \Omega}{1 - x}$$

In terms of these values, the current in each segment is:

$$I_{\text{muscle}} = \frac{0.60\text{ V}}{646\ \Omega} x = 0.929(x)\text{ mA}$$

$$I_{\text{fat}} = \frac{0.60\text{ V}}{1240\ \Omega}(1 - x) = 0.484(1 - x)\text{ mA}$$

The sum of these currents is the total current:

$$I_{\text{total}} = 0.87\text{ mA} = 0.929(x)\text{ mA} + 0.484(1 - x)\text{ mA}$$

Rearranging the terms on the right side gives

$$0.87\text{ mA} = (0.484 + 0.445x)\text{ mA}$$

Finally, we can solve for x:

$$x = 0.87$$

This person therefore has 87% muscle and 13% fatty tissue in the upper arm.

ASSESS A good check on our work is that the total current we find in part a is small—important for safety—and reasonably close to the value given in part b; the arms are the same size, and the variation in body fat between individuals isn't all that large, so we expect the numbers to be similar. The current given in part b is greater than we found in part a. Muscle has a lower resistance than fat, so the subject of part b must have a higher percentage of muscle—exactly what we found.

SUMMARY

Goal: To learn how and why charge moves through a conductor as what we call a current.

GENERAL PRINCIPLES

Batteries, Resistors, and Current

The **current** is defined to be the motion of positive charges

$$I = \frac{\Delta q}{\Delta t}$$

The battery transforms its chemical energy into electric potential energy of the charges passing through it, raising their electric potential. The potential difference of the battery is its **emf** $\mathcal{E}$.

A **battery** is a source of potential difference. Chemical processes in the battery separate charges. We use a **charge escalator** model to show the lifting of charges to higher potential.

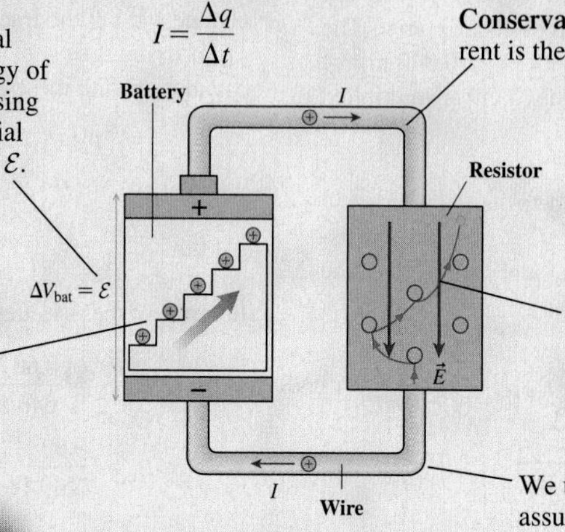

Conservation of current dictates that the current is the same at all points in the circuit.

The battery creates an electric field in the circuit that causes charges to move. Positive charges move in the direction of the electric field, which is the direction of decreasing potential.

The actual charge carriers are electrons. Their random collisions with atoms impede the flow of charge and are the source of **resistance.** The collisions increase the thermal energy of the **resistor.**

We use the **ideal-wire model** in which we assume that there is no resistance in the wires.

IMPORTANT CONCEPTS

Resistance, resistivity, and Ohm's law

The **resistivity** ρ is a property of a material, a measure of how good a conductor the material is.

- Good conductors have low resistivity.
- Poor conductors have high resistivity.

The **resistance** is a property of a particular wire or conductor. The resistance of a wire depends on its resistivity and dimensions.

$$R = \frac{\rho L}{A}$$

Cross-section area A — Length L

Ohm's law describes the relationship between potential difference and current in a resistor:

$$I = \frac{\Delta V}{R}$$

ΔV — R — I

Energy and power

The energy used by a circuit is supplied by the emf of the battery through a series of energy transformations:

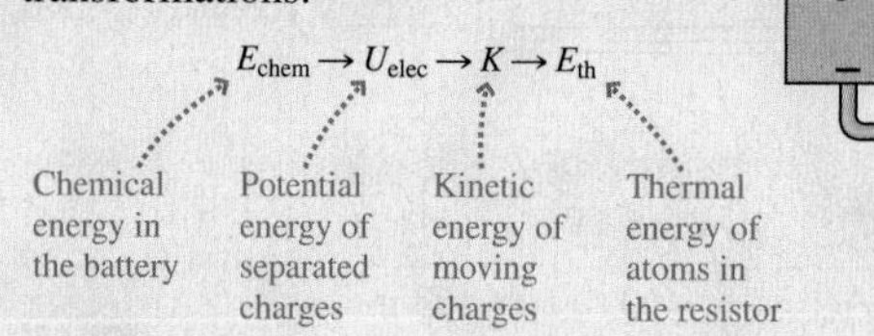

I, +, $\mathcal{E}$, −, R, ΔV, I

The battery *supplies* power at the rate

$$P_{emf} = I\mathcal{E}$$

The resistor *dissipates* power at the rate

$$P_R = I\Delta V_R = I^2 R = \frac{(\Delta V_R)^2}{R}$$

APPLICATIONS

Conducting materials

When a potential difference is applied to a wire, if the relationship between potential difference and current is linear, the material is **ohmic.**

The resistance is

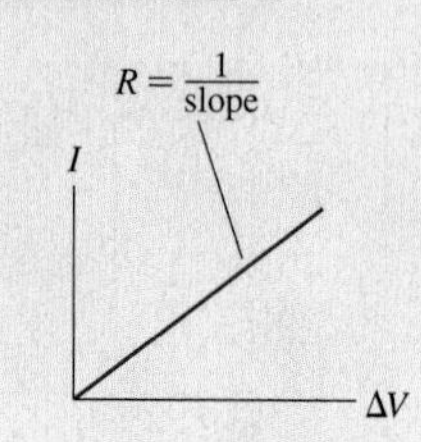

Resistors are made of ohmic materials and have a well-defined value of resistance:

$$R = \frac{\Delta V}{I}$$

Batteries in series

Batteries connected one after the other are in *series*. The total potential difference is the sum of the potential differences of each battery.

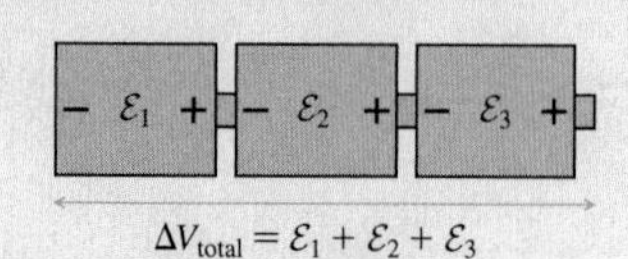

Problem difficulty is labeled as I (straightforward) to IIIII (challenging). Problems labeled INT integrate significant material from earlier chapters; BIO are of biological or medical interest.

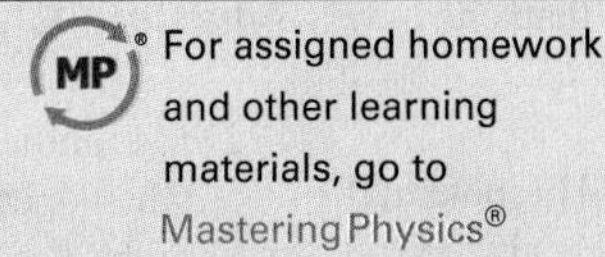

Scan this QR code to launch a Video Tutor Solution that will help you solve problems for this chapter.

QUESTIONS

Conceptual Questions

1. Two wires connect a lightbulb to a battery, completing a circuit and causing the bulb to glow. Do the simple observations and measurements that you can make on this circuit prove that something is *flowing* through the wires? If so, state the observations and/or measurements that are relevant and the steps by which you can then infer that something must be flowing. If not, can you offer an alternative hypothesis about why the bulb glows that is at least plausible and that could be tested?
2. Two wires connect a lightbulb to a battery, completing a circuit and causing the bulb to glow. Are the simple observations and measurements you can make on this circuit able to distinguish a current composed of positive charge carriers from a current composed of negative charge carriers? If so, describe how you can tell which it is. If not, why not?
3. What *causes* electrons to move through a wire as a current?
4. A lightbulb is connected to a battery by two copper wires of equal lengths but different thicknesses. A thick wire connects one side of the lightbulb to the positive terminal of the battery and a thin wire connects the other side of the bulb to the negative terminal.
   a. Which wire carries a greater current? Or is the current the same in both? Explain.
   b. If the two wires are switched, will the bulb get brighter, dimmer, or stay the same? Explain.
5. All wires in Figure Q22.5 are made of the same material and have the same diameter. Rank in order, from largest to smallest, the currents I_1 to I_4. Explain.

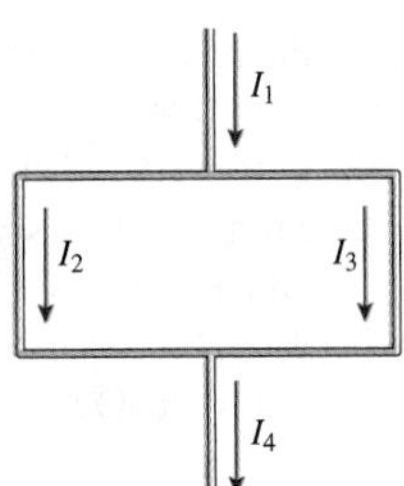

FIGURE Q22.5

6. A wire carries a 4 A current. What is the current in a second wire that delivers twice as much charge in half the time?
7. An electrolyte contains equal numbers of positive and negative ions in solution. When an electric field is applied, the positive ions move in the direction of the field and the negative ions move opposite the field. If the negative ions move faster than the positive ions, is the net current in the direction of the field, is it opposite the field, or do the two ion currents cancel to give no net current? Explain.
8. BIO Cells in the nervous system have a potential difference of 70 mV across the cell membrane separating the interior of the cell from the extracellular fluid. This potential difference is maintained by ion pumps that move charged ions across the membrane. Is this an emf?
9. a. Which direction—clockwise or counterclockwise—does an electron travel through the wire in Figure Q22.9? Explain.
   b. Does an electron's electric potential energy increase, decrease, or stay the same as it moves through the wire? Explain.
   c. If you answered "decrease" in part b, where does the energy go? If you answered "increase" in part b, where does the energy come from?

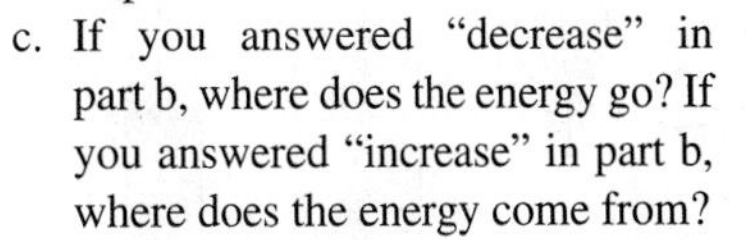

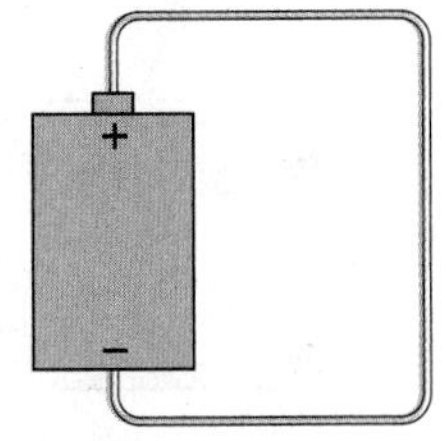

FIGURE Q22.9

   d. Which way—up or down—does an electron move through the *battery?* Explain.
   e. Does an electron's electric potential energy increase, decrease, or stay the same as it moves through the battery? Explain.
   f. If you answered "decrease" in part e, where does the energy go? If you answered "increase" in part e, where does the energy come from?
10. If you change the temperature of a segment of metal wire, the dimensions change and the resistivity changes. How does each of these changes affect the resistance of the wire?
11. The wires in Figure Q22.11 are all made of the same material; the length and radius of each wire are noted. Rank in order, from largest to smallest, the resistances R_1 to R_5 of these wires. Explain.

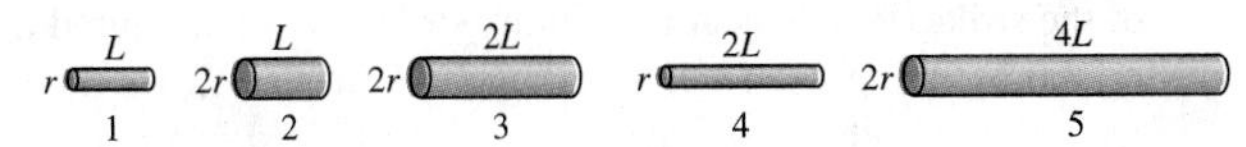

FIGURE Q22.11

12. The two circuits in Figure Q22.12 use identical batteries and wires made of the same material and of equal diameters. Rank in order, from largest to smallest, the currents I_1, I_2, I_3, and I_4 at points 1 to 4.

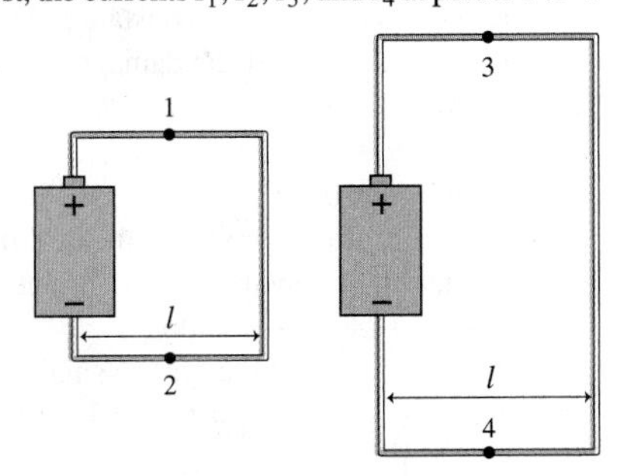

FIGURE Q22.12

13. The two circuits in Figure Q22.13 use identical batteries and wires made of the same material and of equal diameters. Rank in order, from largest to smallest, the currents I_1 to I_7 at points 1 to 7. Explain.

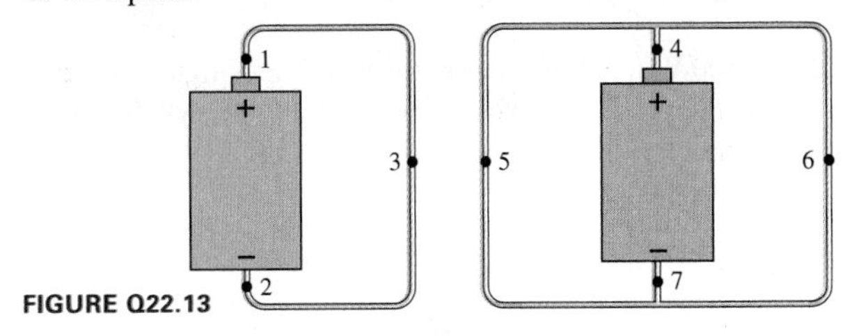

FIGURE Q22.13

14. Which, if any, of these statements are true? (More than one may be true.) Explain your choice or choices.
    a. A battery supplies energy to a circuit.
    b. A battery is a source of potential difference. The potential difference between the terminals of the battery is always the same.
    c. A battery is a source of current. The current leaving the battery is always the same.
15. Rank in order, from largest to smallest, the currents I_1 to I_4 through the four resistors in Figure Q22.15. Explain.

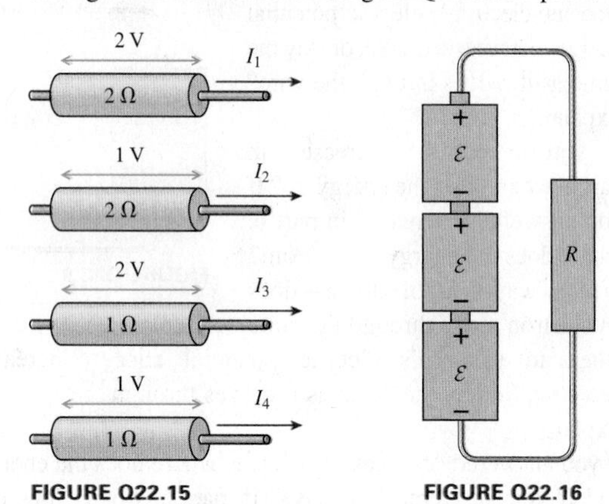

FIGURE Q22.15

FIGURE Q22.16

16. The circuit in Figure Q22.16 has three batteries of emf $\mathcal{E}$ in series. Assuming the wires are ideal, sketch a graph of the potential as a function of distance traveled around the circuit, starting from $V = 0$ V at the negative terminal of the bottom battery. Note all important points on your graph.
17. When lightning strikes the ground, it generates a large electric field along the surface of the ground directed toward the point of the strike. People near a lightning strike are often injured not by the lightning itself but by a large current that flows up one leg and down the other due to this electric field. To minimize this possibility, you are advised to stand with your feet close together if you are trapped outside during a lightning storm. Explain why this is beneficial.
    Hint: The current path through your body, up one leg and down the other, has a certain resistance. The larger the current along this path, the greater the damage.
18. One way to find out if a wire has corroded is to measure its resistance. Explain why the resistance of a wire increases if it becomes corroded.
19. Over time, atoms "boil off" the hot filament in an incandescent bulb and the filament becomes thinner. How does this affect the brightness of the lightbulb?
20. Rank in order, from largest to smallest, the powers P_1 to P_4 dissipated by the four resistors in Figure Q22.20.

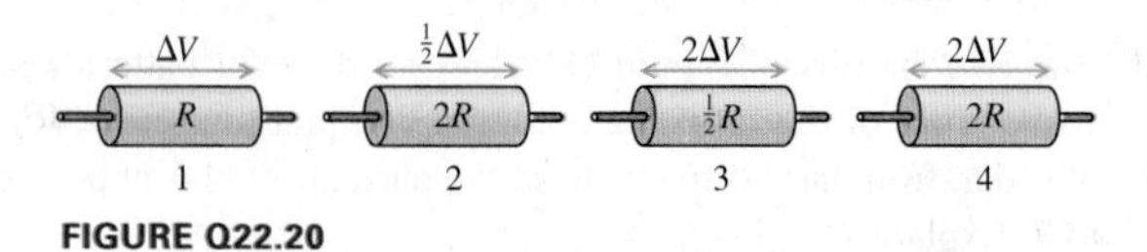

FIGURE Q22.20

21. A 100 W lightbulb is brighter than a 60 W lightbulb when both operate at the same voltage of 120 V. If, instead, they were both operated at the same current of 0.5 A, which would be brighter? Explain.

Multiple-Choice Questions

22. | Lightbulbs are typically rated by their power dissipation when operated at a given voltage. Which of the following lightbulbs has the largest current through it when operated at the voltage for which it's rated?
    A. 0.8 W, 1.5 V B. 6 W, 3 V
    C. 4 W, 4.5 V D. 8 W, 6 V
23. || Lightbulbs are typically rated by their power dissipation when operated at a given voltage. Which of the following lightbulbs has the largest resistance when operated at the voltage for which it's rated?
    A. 0.8 W, 1.5 V B. 6 W, 3 V
    C. 4 W, 4.5 V D. 8 W, 6 V
24. | A copper wire is stretched so that its length increases and its diameter decreases. As a result,
    A. The wire's resistance decreases, but its resistivity stays the same.
    B. The wire's resistivity decreases, but its resistance stays the same.
    C. The wire's resistance increases, but its resistivity stays the same.
    D. The wire's resistivity increases, but its resistance stays the same.
25. | The potential difference across a length of wire is increased. Which of the following does *not* increase as well?
    A. The electric field in the wire
    B. The power dissipated in the wire
    C. The resistance of the wire
    D. The current in the wire
26. ||| A stereo amplifier creates a 5.0 V potential difference across a speaker. To double the power output of the speaker, the amplifier's potential difference must be increased to
    A. 7.1 V B. 10 V C. 14 V D. 25 V
27. | A resistor connected to a 3.0 V battery dissipates 1.0 W. If the battery is replaced by a 6.0 V battery, the power dissipated by the resistor will be
    A. 1.0 W B. 2.0 W C. 3.0 W D. 4.0 W
28. | If a 1.5 V battery stores 5.0 kJ of energy (a reasonable value for an inexpensive C cell), for how many minutes could it sustain a current of 1.2 A?
    A. 2.7 B. 6.9 C. 9.0 D. 46
29. | Figure Q22.29 shows a side view of a wire of varying circular cross section. Rank in order the currents flowing in the three sections.
    A. $I_1 > I_2 > I_3$
    B. $I_2 > I_3 > I_1$
    C. $I_1 = I_2 = I_3$
    D. $I_1 > I_3 > I_2$

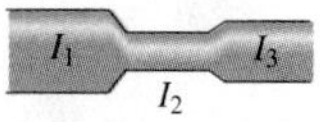

I_1 I_2 I_3

FIGURE Q22.29

30. ||| A person gains weight by adding fat—and therefore adding BIO girth—to his body and his limbs, with the amount of muscle remaining constant. How will this affect the electrical resistance of his limbs?
    A. The resistance will increase.
    B. The resistance will stay the same.
    C. The resistance will decrease.

PROBLEMS

Section 22.1 A Model of Current

Section 22.2 Defining and Describing Current

1. ||| The current in an electric hair dryer is 10 A. How much charge and how many electrons flow through the hair dryer in 5.0 min?
2. || 2.0×10^{13} electrons flow through a transistor in 1.0 ms. What is the current through the transistor?
3. | Three wires meet at a junction. Wire 1 has a current of 0.40 A into the junction. The current of wire 2 is 0.65 A out of the junction. (a) How many electrons per second move past a point in wire 3? (b) In which direction do the electrons move—into or out of the junction?
4. | When a nerve cell depolarizes, charge is transferred across BIO the cell membrane, changing the potential difference. For a typical nerve cell, 9.0 pC of charge flows in a time of 0.50 ms. What is the average current?
5. | A wire carries a 15 μA current. How many electrons pass a given point on the wire in 1.0 s?
6. || In a typical lightning strike, 2.5 C flows from cloud to ground in 0.20 ms. What is the current during the strike?
7. || A capacitor is charged to 6.0×10^{-4} C, then discharged by connecting a wire between the two plates. 40 μs after the discharge begins, the capacitor still holds 13% of its original charge. What was the average current during the first 40 μs of the discharge?
8. || In an ionic solution, 5.0×10^{15} positive ions with charge $+2e$ pass to the right each second while 6.0×10^{15} negative ions with charge $-e$ pass to the left. What are the magnitude and direction of current in the solution?
9. || The starter motor of a car engine draws a current of 150 A from the battery. The copper wire to the motor is 5.0 mm in diameter and 1.2 m long. The starter motor runs for 0.80 s until the car engine starts. How much charge passes through the starter motor?
10. | A car battery is rated at 90 A · h, meaning that it can supply a 90 A current for 1 h before being completely discharged. If you leave your headlights on until the battery is completely dead, how much charge leaves the positive terminal of the battery?
11. || What are the values of currents I_B and I_C in Figure P22.11? The directions of the currents are as noted.

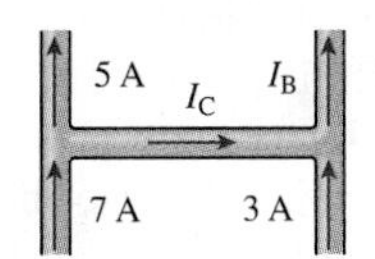

FIGURE P22.11

12. | The currents through several segments of a wire object are shown in Figure P22.12. What are the magnitudes and directions of the currents I_B and I_C in segments B and C?

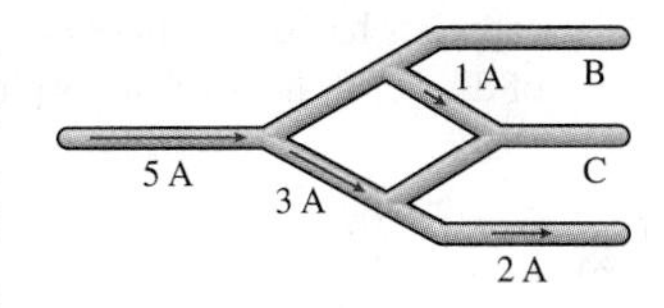

FIGURE P22.12

Section 22.3 Batteries and emf

13. | How much electric potential energy does 1.0 μC of charge gain as it moves from the negative terminal to the positive terminal of a 1.5 V battery?
14. | What is the emf of a battery that increases the electric potential energy of 0.050 C of charge by 0.60 J as it moves it from the negative to the positive terminal?
15. || A 9.0 V battery supplies a 2.5 mA current to a circuit for 5.0 h.
    a. How much charge has been transferred from the negative to the positive terminal?
    b. How much electric potential energy has been gained by the charges that passed through the battery?
16. | An individual hydrogen-oxygen fuel cell has an output of 0.75 V. How many cells must be connected in series to drive a 24.0 V motor?
17. | An electric catfish can generate BIO a significant potential difference using stacks of special cells called *electrocytes*. Each electrocyte develops a potential difference of 110 mV. How many cells must be connected in series to give the 350 V a large catfish can produce?

Section 22.4 Connecting Potential and Current

18. | A wire with resistance R is connected to the terminals of a 6.0 V battery. What is the potential difference ΔV_{ends} between the ends of the wire and the current I through it if the wire has the following resistances? (a) 1.0 Ω (b) 2.0 Ω (c) 3.0 Ω.
19. | Wires 1 and 2 are made of the same metal. Wire 2 has twice the length and twice the diameter of wire 1. What are the ratios (a) ρ_2/ρ_1 of the resistivities and (b) R_2/R_1 of the resistances of the two wires?
20. ||| A wire has a resistance of 0.010 Ω. What will the wire's resistance be if it is stretched to twice its original length without changing the volume of the wire?
21. ||| Resistivity measurements on the leaves of corn plants are a BIO good way to assess stress and overall health. The leaf of a corn plant has a resistance of 2.0 MΩ measured between two electrodes placed 20 cm apart along the leaf. The leaf has a width of 2.5 cm and is 0.20 mm thick. What is the resistivity of the leaf tissue? Is this greater than or less than the resistivity of muscle tissue in the human body?
22. || What is the resistance of
    a. A 1.0-m-long copper wire that is 0.50 mm in diameter?
    b. A 10-cm-long piece of iron with a 1.0 mm × 1.0 mm square cross section?
23. ||| A motorcyclist is making an electric vest that, when connected to the motorcycle's 12 V battery, will warm her on cold rides. She is using 0.25-mm-diameter copper wire, and she wants a current of 4.0 A in the wire. What length wire must she use?
24. || The femoral artery is the large artery that carries blood to the leg. BIO A person's femoral artery has an inner diameter of 1.0 cm. What is the resistance of a 20-cm-long column of blood in this artery?
25. || A 3.0 V potential difference is applied between the ends of a 0.80-mm-diameter, 50-cm-long nichrome wire. What is the current in the wire?
26. ||| A 1.0-mm-diameter, 20-cm-long copper wire carries a 3.0 A current. What is the potential difference between the ends of the wire?

27. IIII The relatively high resistivity of dry skin, about BIO $1 \times 10^6\ \Omega \cdot \text{m}$, can safely limit the flow of current into deeper tissues of the body. Suppose an electrical worker places his palm on an instrument whose metal case is accidentally connected to a high voltage. The skin of the palm is about 1.5 mm thick. Estimate the area of skin on the worker's palm that would contact a flat panel, then calculate the approximate resistance of the skin of the palm.
28. II The aluminum wire in a high-voltage transmission line is 2.7 cm in diameter. It is designed to carry a current of 1100 A. What is the voltage drop across a 37-km-long segment of this power line?

Section 22.5 Ohm's Law and Resistor Circuits

29. I Figure P22.29 shows the current-versus-potential-difference graph for a resistor.
    a. What is the resistance of this resistor?
    b. Suppose the length of the resistor is doubled while keeping its cross section the same. (This requires doubling the amount of material the resistor is made of.) Copy the figure and add to it the current-versus-potential-difference graph for the longer resistor.

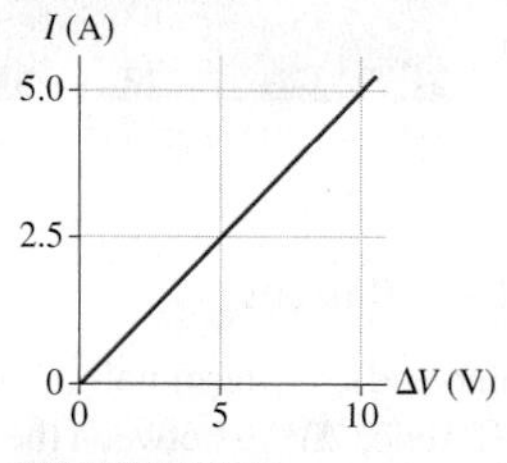

FIGURE P22.29

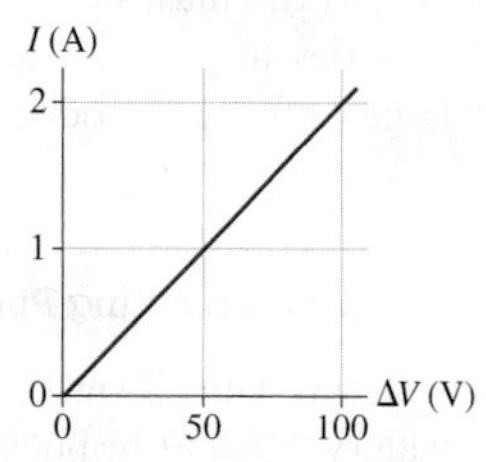

FIGURE P22.30

30. I Figure P22.30 is a current-versus-potential-difference graph for a cylinder. What is the cylinder's resistance?
31. I In Example 22.6 the length of a 60 W, 240 Ω lightbulb filament was calculated to be 60 cm.
    a. If the potential difference across the filament is 120 V, what is the strength of the electric field inside the filament?
    b. Suppose the length of the bulb's filament were doubled without changing its diameter or the potential difference across it. What would the electric field strength be in this case?
    c. Remembering that the current in the filament is proportional to the electric field, what is the current in the filament following the doubling of its length?
    d. What is the resistance of the filament following the doubling of its length?
32. I The electric field inside a 30-cm-long copper wire is 0.010 V/m. What is the potential difference between the ends of the wire?
33. II A copper wire is 1.0 mm in diameter and carries a current of 20 A. What is the electric field strength inside this wire?
34. III Two identical lightbulbs are connected in series to a single 9.0 V battery.
    a. Sketch the circuit.
    b. Sketch a graph showing the potential as a function of distance through the circuit, starting with $V = 0\ \text{V}$ at the negative terminal of the battery.

Section 22.6 Energy and Power

35. I A 1.5 V battery moves 2000 C of charge around a circuit. By how much does the chemical energy of the battery decrease?
36. I a. What is the resistance of a 1500 W (120 V) hair dryer?
    b. What is the current in the hair dryer when it is used?
37. II Every second, a battery increases the electric potential energy of the 1.2 C of charge passing through it by 7.2 J. What is the battery's emf?
38. II A 70 W electric blanket runs at 18 V.
    a. What is the resistance of the wire in the blanket?
    b. How much current does the wire carry?
39. I A 60-cm-long heating wire is connected to a 120 V outlet. If the wire dissipates 45 W, what are (a) the current in and (b) the resistance of the wire?
40. III An electric eel develops a potential difference of 450 V, BIO driving a current of 0.80 A for a 1.0 ms pulse. For this pulse, find (a) the power, (b) the total energy, and (c) the total charge that flows.
41. IIIII The total charge a household battery can supply is given in units of $\text{mA} \cdot \text{h}$. For example, a 9.0 V alkaline battery is rated $450\ \text{mA} \cdot \text{h}$, meaning that such a battery could supply a 1 mA current for 450 h, a 2 mA current for 225 h, etc. How much energy, in joules, is this battery capable of supplying?

General Problems

42. I A 3.0 V battery powers a flashlight bulb that has a resistance of 6.0 Ω. How much charge moves through the battery in 10 min?
43. I A heating element in a toaster dissipates 900 W when run at 120 V. How much charge passes through the heating element in 1 minute?
44. IIII Older freezers developed a coating of ice inside that had INT to be melted periodically; an electric heater could speed this defrosting process. Suppose you're melting ice from your freezer using a heating wire that carries a current of 5.0 A when connected to 120 V.
    a. What is the resistance of the wire?
    b. How long will it take the heater to melt 720 g of accumulated ice at −10°C? Assume that all of the heat goes into warming and melting the ice, and that the melt water runs out and doesn't warm further.
45. IIII For a science experiment you need to electroplate a 100-nm- INT thick zinc coating onto both sides of a very thin, 2.0 cm × 2.0 cm copper sheet. You know that the charge carriers in the ionic solution are divalent (charge $2e$) zinc ions. The density of zinc is $7140\ \text{kg/m}^3$. If the electroplating apparatus operates at 1.0 mA, how long will it take the zinc to reach the desired thickness?
46. III The hot dog cooker described in the chapter heats hot dogs INT by connecting them to 120 V household electricity. A typical hot dog has a mass of 60 g and a resistance of 150 Ω. How long will it take for the cooker to raise the temperature of the hot dog from 20°C to 80°C? The specific heat of a hot dog is approximately $2500\ \text{J/kg} \cdot \text{K}$.
47. IIII Air isn't a perfect electric insulator, but it has a very high resistiv- INT ity. Dry air has a resistivity of approximately $3 \times 10^{13}\ \Omega \cdot \text{m}$. A capacitor has square plates 10 cm on a side separated by 1.2 mm of dry air. If the capacitor is charged to 250 V, what fraction of the charge will flow across the air gap in 1 minute? Make the approximation that the potential difference doesn't change as the charge flows.
48. II The biochemistry that takes place inside cells depends on BIO various elements, such as sodium, potassium, and calcium, that are dissolved in water as ions. These ions enter cells through narrow pores in the cell membrane known as *ion channels*. Each ion channel, which is formed from a specialized protein molecule, is selective for one type of ion. Measurements with microelectrodes have shown that a 0.30-nm-diameter potassium ion (K^+) channel carries a current of 1.8 pA. How many potassium ions pass through if the ion channel opens for 1.0 ms?

49. || High-resolution measurements have shown that an ion channel (see Problem 48) is a 0.30-nm-diameter cylinder with length of 5.0 nm. The intracellular fluid filling the ion channel has resistivity $0.60\ \Omega \cdot \text{m}$. What is the resistance of the ion channel?
BIO

50. | When an ion channel opens in a cell wall (see Problem 48), monovalent (charge e) ions flow through the channel at a rate of 1.0×10^7 ions/s.
BIO
   a. What is the current through the channel?
   b. The potential difference across the ion channel is 70 mV. What is the power dissipation in the channel?

51. ||| The total charge a battery can supply is rated in mA · h, the product of the current (in mA) and the time (in h) that the battery can provide this current. A battery rated at 1000 mA · h can supply a current of 1000 mA for 1.0 h, 500 mA current for 2.0 h, and so on. A typical AA rechargeable battery has a voltage of 1.2 V and a rating of 1800 mA · h. For how long could this battery drive current through a long, thin wire of resistance 22 Ω?

52. || A 1.5 V D-cell battery is rated at 15,000 mA · h (see Problem 51). For how long could such a battery power a flashlight bulb rated at 1.5 W?

53. || The heating element of a simple heater consists of a 2.0-m-long, 0.60-mm-diameter nichrome wire. When plugged into a 120 V outlet, the heater draws 8.0 A of current when hot.
   a. What is the wire's resistance when it is hot?
   b. Use your answer to part a to calculate the resistivity of nichrome in this situation. Why is it not the same as the value of ρ given for nichrome in Table 22.1?

54. ||| Variations in the resistivity of blood can give valuable clues to changes in the blood's viscosity and other properties. The resistivity is measured by applying a small potential difference and measuring the current. Suppose a medical device attaches electrodes into a 1.5-mm-diameter vein at two points 5.0 cm apart. What is the blood resistivity if a 9.0 V potential difference causes a 230 μA current through the blood in the vein?
BIO

55. |||| A 40 W (120 V) lightbulb has a tungsten filament of diameter 0.040 mm. The filament's operating temperature is 1500°C.
   a. How long is the filament?
   b. What is the resistance of the filament at 20°C?

56. || Wires aren't really ideal. The voltage drop across a current-carrying wire can be significant unless the resistance of the wire is quite low. Suppose a 50 ft extension cord is being used to provide power to an electric lawn mower. The cord carries a 10 A current. The copper wire in a typical extension cord has a 1.3 mm diameter. What is the voltage drop across a 50 ft length of wire at this current?

57. ||| When the starter motor on a car is engaged, there is a 300 A current in the wires between the battery and the motor. Suppose the wires are made of copper and have a total length of 1.0 m. What minimum diameter can the wires have if the voltage drop along the wires is to be less than 0.50 V?

58. ||| The electron beam inside a television picture tube is 0.40 mm in diameter and carries a current of 50 μA. This electron beam impinges on the inside of the picture tube screen.
INT
   a. How many electrons strike the screen each second?
   b. The electrons move with a velocity of 4.0×10^7 m/s. What electric field strength is needed to accelerate electrons from rest to this velocity in a distance of 5.0 mm?
   c. Each electron transfers its kinetic energy to the picture tube screen upon impact. What is the *power* delivered to the screen by the electron beam?

   Hint: What potential difference produced the field that accelerated electrons? This is an emf.

59. | The two segments of the wire in Figure P22.59 have equal diameters and equal lengths but different resistivities ρ_1 and ρ_2. Current I passes through this wire. If the resistivities have the ratio $\rho_2/\rho_1 = 2$, what is the ratio $\Delta V_1/\Delta V_2$ of the potential differences across the two segments of the wire?

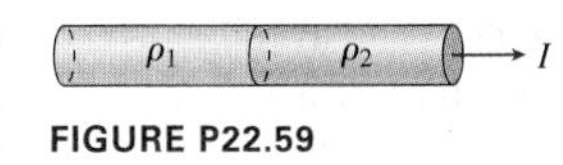

FIGURE P22.59

60. || A long wire used as a heating element carries a current of 0.80 A. It dissipates 6.0 W for every meter of length. What is the electric field strength inside this wire?

61. || A wire is 2.3 m long and has a diameter of 0.38 mm. When connected to a 1.2 V battery, there is a current of 0.61 A. What material is the wire likely made of?

62. | The filament of a 100 W (120 V) lightbulb is a tungsten wire 0.035 mm in diameter. At the filament's operating temperature, the resistivity is $5.0 \times 10^{-7}\ \Omega \cdot \text{m}$. How long is the filament?

63. ||| You've made the finals of the Science Olympics! As one of your tasks, you're given 1.0 g of copper and asked to make a wire, using all the metal, with a resistance of 1.0 Ω. Copper has a density of 8900 kg/m^3. What length and diameter will you choose for your wire?
INT

64. ||| If resistors 1 and 2 are connected to identical batteries, resistor 1 dissipates 3.0 times more power than resistor 2. What is the ratio P_1/P_2 of their power dissipations if the same current passes through each resistor?

65. || An immersion heater used to boil water for a single cup of tea plugs into a 120 V outlet and is rated at 300 W.
INT
   a. What is the resistance of the heater?
   b. Suppose your super-size, super-insulated tea mug contains 400 g of water at a temperature of 18°C. How long will this heater take to bring the water to a boil? You can ignore the energy needed to raise the temperature of the mug and the heater itself.

66. |||| The graph in Figure P22.66 shows the current through a 1.0 Ω resistor as a function of time.
   a. How much charge flowed through the resistor during the 10 s interval shown?
   b. What was the total energy dissipated by the resistor during this time?

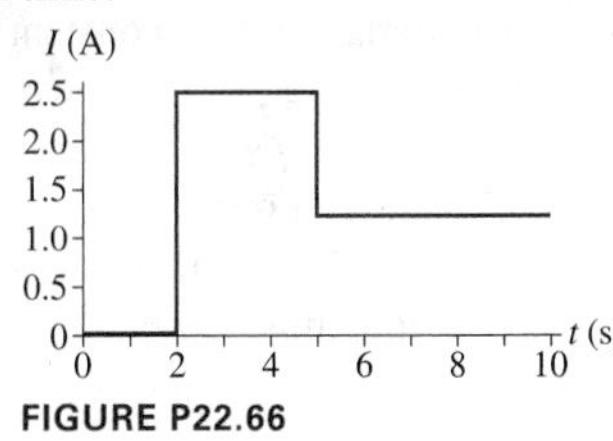

FIGURE P22.66

67. ||| It's possible to estimate the percentage of fat in the body by measuring the resistance of the upper leg rather than the upper arm; the calculation is similar. A person's leg measures 40 cm between the knee and the hip, with an average leg diameter (ignoring bone and other poorly conducting tissue) of 12 cm. A potential difference of 0.75 V causes a current of 1.6 mA. What are the fractions of (a) muscle and (b) fat in the leg?
BIO

68. | If you touch the two terminals of a power supply with your two fingertips on opposite hands, the potential difference will produce a current through your torso. The maximum safe current is approximately 5 mA.
BIO
   a. If your hands are completely dry, the resistance of your body from fingertip to fingertip is approximately 500 kΩ. If you accidentally touch both terminals of your 120 V household electricity supply with dry fingers, will you receive a dangerous shock?
   b. If your hands are moist, your resistance drops to approximately 1 kΩ. If you accidentally touch both terminals of your 120 V household supply with moist fingers, will you receive a dangerous shock?

69. | The average resistivity of the human body (apart from surface resistance of the skin) is about 5.0 $\Omega \cdot m$. The conducting path between the right and left hands can be approximated as a cylinder 1.6 m long and 0.10 m in diameter. The skin resistance can be made negligible by soaking the hands in salt water.
BIO
a. What is the resistance between the hands if the skin resistance is negligible?
b. If skin resistance is negligible, what potential difference between the hands is needed for a lethal shock current of 100 mA? Your result shows that even small potential differences can produce dangerous currents when skin is damp.

MCAT-Style Passage Problems

Lightbulb Failure

You've probably observed that the most common time for an incandescent lightbulb to fail is the moment when it is turned on. Let's look at the properties of the bulb's filament to see why this happens.

The current in the tungsten filament of a lightbulb heats the filament until it glows. The filament is so hot that some of the atoms on its surface fly off and end up sticking on a cooler part of the bulb. Thus the filament gets progressively thinner as the bulb ages. There will certainly be one spot on the filament that is a bit thinner than elsewhere. This thin segment will have a higher resistance than the surrounding filament. More power will be dissipated at this spot, so it won't only be a thin spot, it also will be a hot spot.

Now, let's look at the resistance of the filament. The graph in Figure P22.70 shows data for the current in a lightbulb as a function of the potential difference across it. The graph is not linear, so the filament is not an ohmic material with a constant resistance. However, we can define the resistance at any particular potential difference ΔV to be $R = \Delta V/I$. This ratio, and hence the resistance, increases with ΔV and thus with temperature.

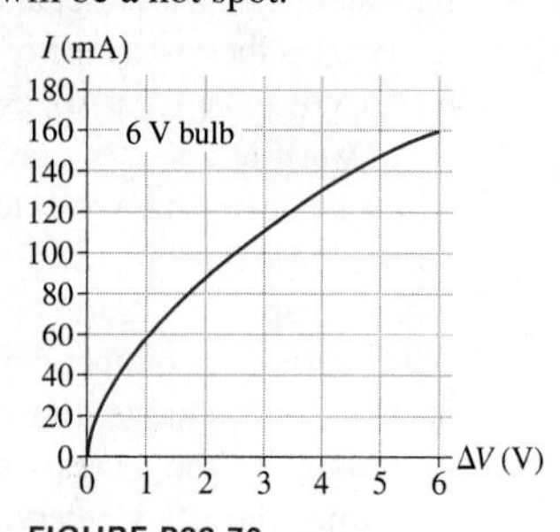

FIGURE P22.70

When the bulb is turned on, the filament is cold and its resistance is much lower than during normal, high-temperature operation. The low resistance causes a surge of higher-than-normal current lasting a fraction of a second until the filament heats up. Because power dissipation is I^2R, the power dissipated during this first fraction of a second is much larger than the bulb's rated power. This current surge concentrates the power dissipation at the high-resistance thin spot, perhaps melting it and breaking the filament.

70. | For the bulb in Figure P22.70, what is the approximate resistance of the bulb at a potential difference of 6.0 V?
A. 7.0 Ω
B. 17 Ω
C. 27 Ω
D. 37 Ω

71. | As the bulb ages, the resistance of the filament
A. Increases.
B. Decreases.
C. Stays the same.

72. | Which of the curves in Figure P22.72 best represents the expected variation in current as a function of time in the short time interval immediately after the bulb is turned on?

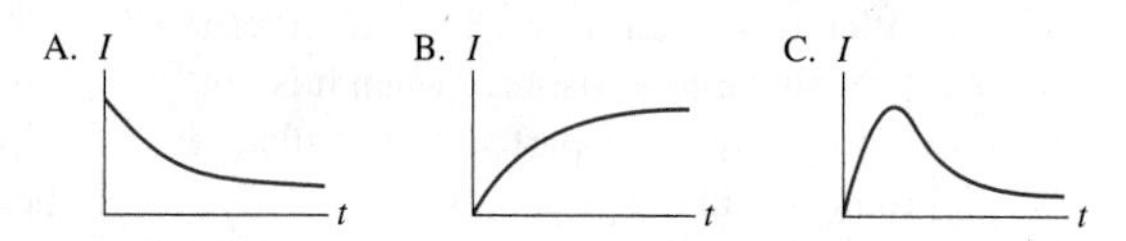

FIGURE P22.72

73. | There are devices to put in a light socket that control the current through a lightbulb, thereby increasing its lifetime. Which of the following strategies would increase the lifetime of a bulb without making it dimmer?
A. Reducing the average current through the bulb
B. Limiting the maximum current through the bulb
C. Increasing the average current through the bulb
D. Limiting the minimum current through the bulb

STOP TO THINK ANSWERS

Chapter Preview Stop to Think: A. The electric field points from high to low potential—here, from right to left. The electron, being negatively charged, feels a force opposite the field, or to the right. It will start moving in that direction.

Stop to Think 22.1: A. From Kirchhoff's junction law the current through bulb 1 is the sum of the currents through bulbs 2 and 3. Bulb 1 carries a larger current than bulb 2, so it will be brighter.

Stop to Think 22.2: C. Charge flows out of one terminal of the battery but back into the other; the amount of charge in the battery does not change. The emf is determined by the chemical reactions in the battery and is constant. But the chemical energy in the battery steadily decreases as the battery converts it to the potential energy of charges.

Stop to Think 22.3: C. Because they are connected end-to-end, the potential difference across the wire is equal to that across the battery. But the battery supplies a *constant* potential difference, its emf, independent of the current it provides.

Stop to Think 22.4: C. Stretching the wire decreases the area and increases the length. Both of these changes increase the resistance of the wire. When the wire is reconnected to the battery, the resistance is greater but the potential difference is the same as in the original case, so the current will be smaller.

Stop to Think 22.5: $V_B = V_C > V_A > V_D = V_E$. There's no potential difference along ideal wires, so $V_B = V_C$ and $V_D = V_E$. Potential increases in going from the $-$ to the $+$ terminal of a battery, so $V_A > V_E$ and $V_B > V_A$. These imply $V_C > V_D$, which was expected because potential decreases as current passes through a resistor.

Stop to Think 22.6: $P_B > P_D > P_A > P_C$. The power dissipated by a resistor is $P_R = (\Delta V_R)^2/R$. Increasing R decreases P_R; increasing ΔV_R increases P_R. But changing the potential has a larger effect because P_R depends on the square of ΔV_R.

23 Circuits

The electric eel isn't really an eel; it's a fish. But it is electric, producing pulses of up to 600 V that it uses to stun prey. How does the fish produce such a large potential difference?

LOOKING AHEAD ▸

Goal: To understand the fundamental physical principles that govern electric circuits.

Analyzing Circuits

Practical circuits consist of many elements—resistors, batteries, capacitors—connected together.

You'll learn how to analyze complex circuits by breaking them into simpler pieces.

Series and Parallel Circuits

There are two basic ways to connect resistors together and capacitors together: **series circuits** and **parallel circuits.**

You'll learn why holiday lights are wired in series but headlights are in parallel.

Electricity in the Body

Your nervous system works by transmitting electrical signals along *axons*, the long nerve fibers shown here.

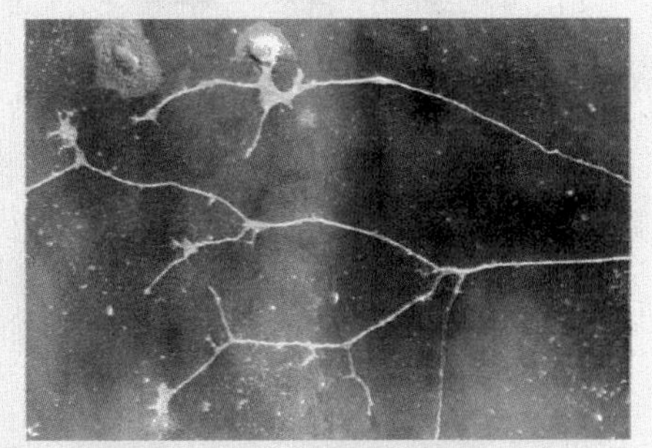

You'll learn how to understand nerve impulses in terms of the resistance, capacitance, and electric potential of individual nerve cells.

LOOKING BACK ◂

Ohm's Law

In Section 22.5 you learned Ohm's law, the relationship between the current through a resistor and the potential difference across it.

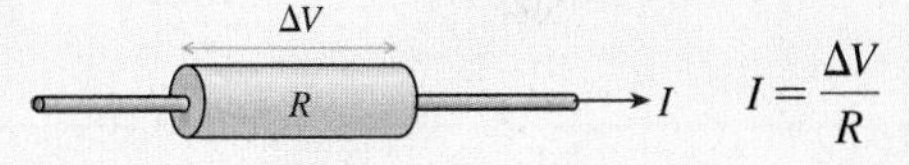

$$I = \frac{\Delta V}{R}$$

In this chapter, you'll use Ohm's law when analyzing more complex circuits consisting of multiple resistors and batteries.

STOP TO THINK

Rank in order, from smallest to largest, the resistances R_1 to R_4 of the four resistors.

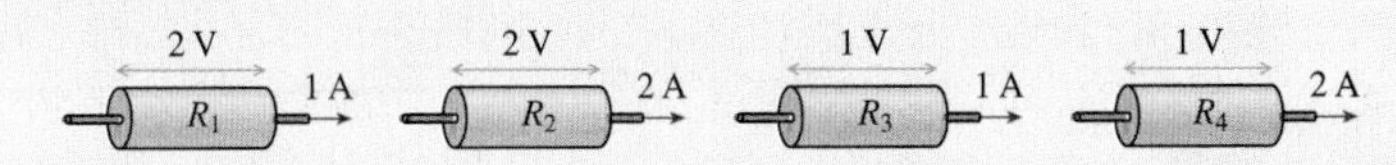

23.1 Circuit Elements and Diagrams

In Chapter 22 we analyzed a very simple circuit, a resistor connected to a battery. In this chapter, we will explore more complex circuits involving more and different elements. As was the case with other topics in this textbook, we will learn a good deal by making appropriate drawings. To do so, we need a system for representing circuits symbolically in a manner that highlights their essential features.

FIGURE 23.1 An electric circuit.

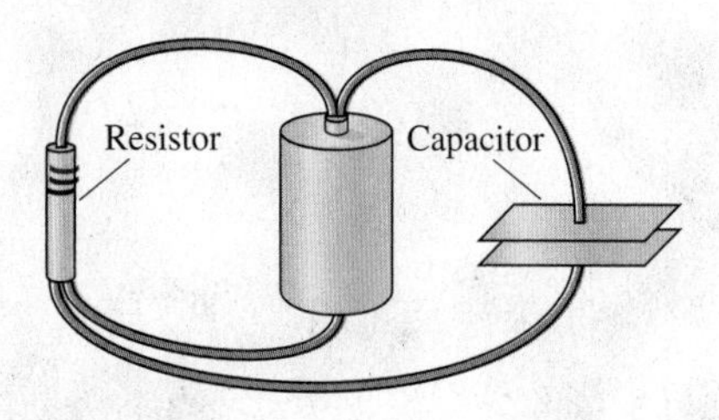

FIGURE 23.1 shows an electric circuit in which a resistor and a capacitor are connected by wires to a battery. To understand the operation of this circuit, we do not need to know whether the wires are bent or straight, or whether the battery is to the right or to the left of the resistor. The literal picture of Figure 23.1 provides many irrelevant details. It is customary when describing or analyzing circuits to use a more abstract picture called a **circuit diagram.** A circuit diagram is a *logical* picture of what is connected to what. The actual circuit, once it is built, may *look* quite different from the circuit diagram, but it will have the same logic and connections.

In a circuit diagram we replace pictures of the circuit elements with symbols. **FIGURE 23.2** shows the basic symbols that we will need.

FIGURE 23.2 A library of basic symbols used for electric circuit drawings.

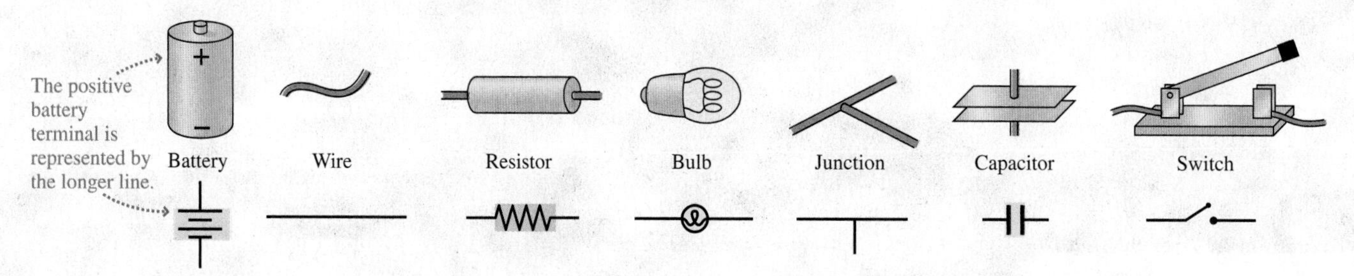

FIGURE 23.3 A circuit diagram for the circuit of Figure 23.1.

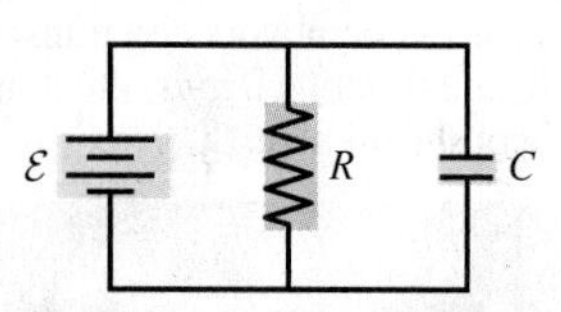

FIGURE 23.3 is a circuit diagram of the circuit shown in Figure 23.1. Notice how circuit elements are labeled. The battery's emf $\mathcal{E}$ is shown beside the battery, and the resistance R of the resistor and capacitance C of the capacitor are written beside them. We would use numerical values for $\mathcal{E}$, R, and C if we knew them. The wires, which in practice may bend and curve, are shown as straight-line connections between the circuit elements. The positive potential of the battery is at the top of the diagram; in general, we try to put higher potentials toward the top. You should get into the habit of drawing your own circuit diagrams in a similar fashion.

STOP TO THINK 23.1 Which of these diagrams represent the same circuit?

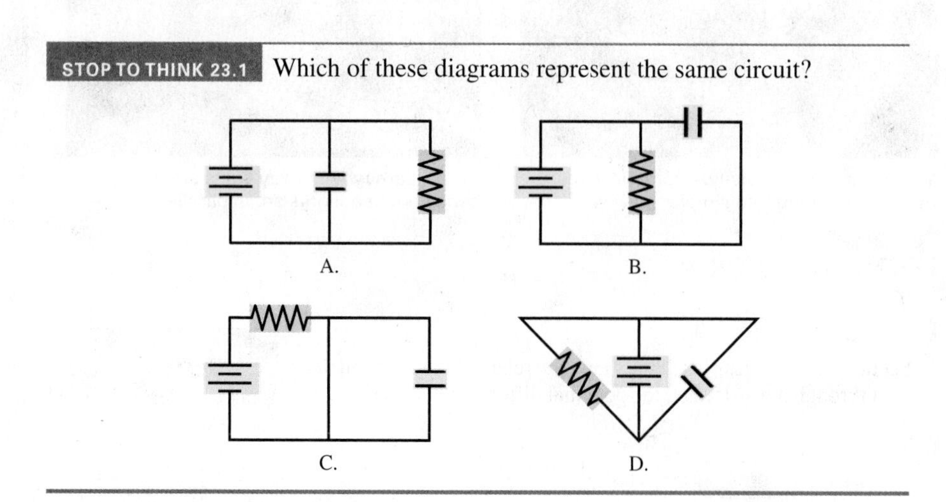

23.2 Kirchhoff's Laws

Once we have a diagram for a circuit, we can analyze it. Our tools and techniques for analyzing circuits will be based on the physical principles of potential differences and currents.

You learned in ◀◀ SECTION 22.2 that, as a result of charge and current conservation, the total current into a junction must equal the total current leaving the junction, as in FIGURE 23.4. This result was called *Kirchhoff's junction law,* which we wrote as

FIGURE 23.4 Kirchhoff's junction law.

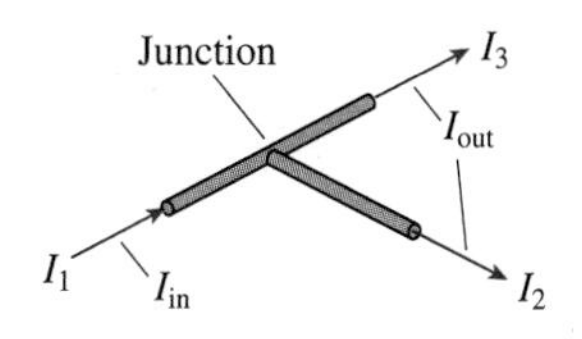

Junction law: $I_1 = I_2 + I_3$

$$\sum I_{in} = \sum I_{out} \tag{23.1}$$

Kirchhoff's junction law

Kirchhoff's junction law isn't a new law of nature. It's an application of a law we already know: the conservation of charge. We can also apply the law of conservation of energy to circuits. When we learned about gravitational potential energy in Chapter 10, we saw that the gravitational potential energy of an object depends only on its position, not on the path it took to get to that position. The same is true of electric potential energy, as you learned in Chapter 21 and as we discussed in ◀◀ SECTION 22.5. If a charged particle moves around a closed loop and returns to its starting point, there is no net change in its electric potential energy: $\Delta U_{elec} = 0$. Because $V = U_{elec}/q$, **the net change in the electric potential around any loop or closed path must be zero** as well.

FIGURE 23.5a shows a circuit consisting of a battery and two resistors. If we start at point a in the lower left corner, at the negative terminal of the battery, and plot the potential around the loop, we get the graph shown in the figure. The potential increases as we move "uphill" through the battery, then decreases in two "downhill" steps, one for each resistor. Ultimately, the potential ends up where it started, as it must. This is a general principle that we can apply to any circuit, as shown in FIGURE 23.5b. If we add all of the potential differences around the loop formed by the circuit, the sum must be zero. This result is known as **Kirchhoff's loop law:**

FIGURE 23.5 Kirchhoff's loop law.

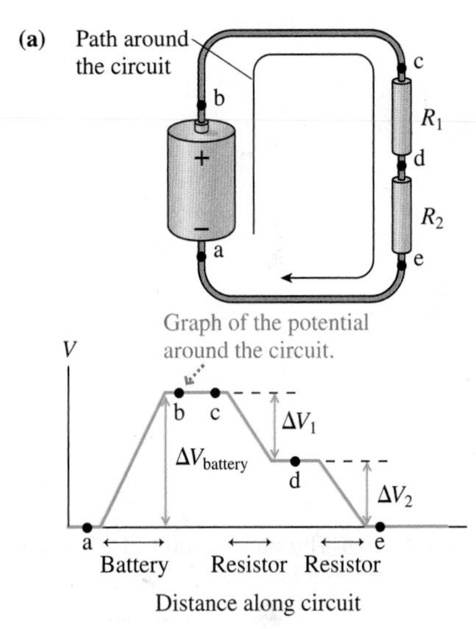

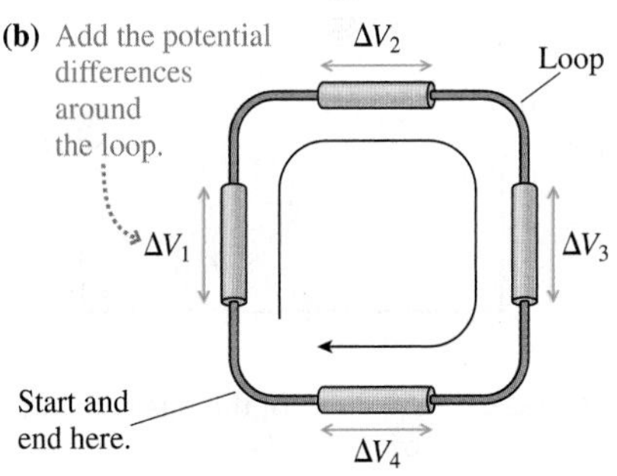

Loop law: $\Delta V_1 + \Delta V_2 + \Delta V_3 + \Delta V_4 = 0$

$$\Delta V_{loop} = \sum_i \Delta V_i = 0 \tag{23.2}$$

Kirchhoff's loop law

In Equation 23.2, ΔV_i is the potential difference of the ith component in the loop.

Kirchhoff's loop law can be true only if at least one of the potential differences ΔV_i is negative. To apply the loop law, we need to explicitly identify which potential differences are positive and which are negative.

TACTICS BOX 23.1 Using Kirchhoff's loop law

1. **Draw a circuit diagram.** Label all known and unknown quantities.
2. **Assign a direction to the current.** Draw and label a current arrow I to show your choice. Choose the direction of the current based on how the batteries or sources of emf "want" the current to go. If you choose the current direction opposite the actual direction, the final value for the current that you calculate will have the correct magnitude but will be negative, letting you know that the direction is opposite the direction you chose.

Continued

❸ **"Travel" around the loop.** Start at any point in the circuit, then go all the way around the loop in the direction you assigned to the current in step 2. As you go through each circuit element, ΔV is interpreted to mean $\Delta V = V_{\text{downstream}} - V_{\text{upstream}}$.

- For a battery with current in the negative-to-positive direction:

$$\Delta V_{\text{bat}} = +\mathcal{E}$$

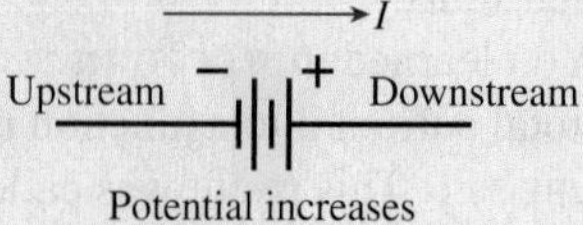

I
Upstream
−
+
Downstream
Potential increases

- For a battery with current in the positive-to-negative direction:

$$\Delta V_{\text{bat}} = -\mathcal{E}$$

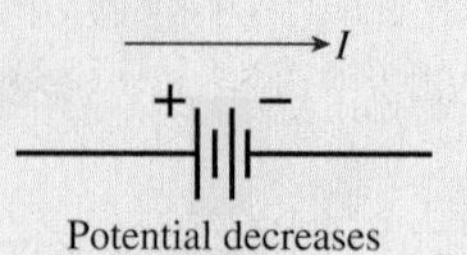

I
+
−
Potential decreases

- For a resistor: $\Delta V_{\text{R}} = -IR$

I
\+ −
Potential decreases

❹ **Apply the loop law:** $\sum \Delta V_i = 0$

Exercises 8, 9 

ΔV_{bat} can be positive or negative for a battery, but ΔV_{R} for a resistor is always negative because the potential in a resistor *decreases* along the direction of the current—charge flows "downhill," as we saw in ◀◀ SECTION 22.4. Because the potential across a resistor always decreases, we often speak of the *voltage drop* across the resistor.

NOTE ▶ The equation for ΔV_{R} in Tactics Box 23.1 seems to be the opposite of Ohm's law, but Ohm's law was concerned with only the *magnitude* of the potential difference. Kirchhoff's law requires us to recognize that the electric potential inside a resistor *decreases* in the direction of the current. ◀

FIGURE 23.6 The basic circuit of a resistor connected to a battery.

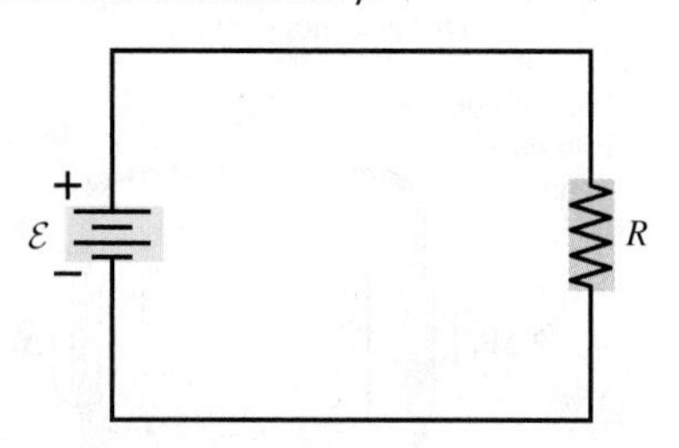

+
ℰ
−
R

The most basic electric circuit is a single resistor connected to the two terminals of a battery, as in **FIGURE 23.6**. We considered this circuit in ◀◀ SECTION 22.5, but let's now apply Kirchhoff's laws to its analysis.

This circuit of Figure 23.6 has no junctions, so the current is the same in all parts of the circuit. Kirchhoff's junction law is not needed. Kirchhoff's loop law is the tool we need to analyze this circuit, and **FIGURE 23.7** shows the first three steps of Tactics Box 23.1. Notice that we're assuming the ideal-wire model in which there are no potential differences along the connecting wire. The fourth step is to apply Kirchhoff's loop law, $\sum \Delta V_i = 0$:

$$\Delta V_{\text{loop}} = \sum_i \Delta V_i = \Delta V_{\text{bat}} + \Delta V_{\text{R}} = 0 \tag{23.3}$$

FIGURE 23.7 Analysis of the basic circuit using Kirchhoff's loop law.

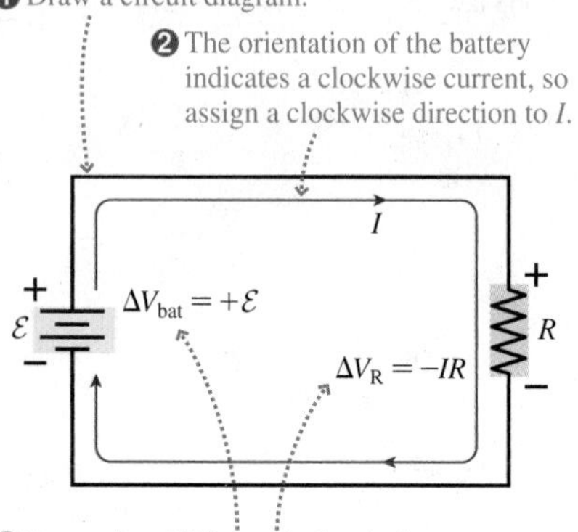

❶ Draw a circuit diagram.
❷ The orientation of the battery indicates a clockwise current, so assign a clockwise direction to I.
I
+
ℰ
−
ΔV_bat = +ℰ
ΔV_R = −IR
+
R
−
❸ Determine ΔV for each circuit element.

Let's look at each of the two terms in Equation 23.3:

1. The potential *increases* as we travel through the battery on our clockwise journey around the loop, as we see in the conventions in Tactics Box 23.1. We enter the negative terminal and, farther downstream, exit the positive terminal after having gained potential $\mathcal{E}$. Thus

$$\Delta V_{\text{bat}} = +\mathcal{E}$$

2. The *magnitude* of the potential difference across the resistor is $\Delta V = IR$, but Ohm's law does not tell us whether this should be positive or negative—and the difference is crucial. The potential of a resistor *decreases* in the direction of the current, which we've indicated with the + and − signs in Figure 23.7. Thus

$$\Delta V_{\text{R}} = -IR$$

With this information about ΔV_{bat} and ΔV_R, the loop equation becomes

$$\mathcal{E} - IR = 0 \tag{23.4}$$

We can solve the loop equation to find that the current in the circuit is

$$I = \frac{\mathcal{E}}{R} \tag{23.5}$$

This is exactly the result we saw in Chapter 22. Notice again that the current in the circuit depends on the size of the resistance. The emf of a battery is a fixed quantity; the current that the battery delivers depends jointly on the emf and the resistance.

EXAMPLE 23.1 Analyzing a circuit with two batteries

What is the current in the circuit of **FIGURE 23.8**? What is the potential difference across each resistor?

FIGURE 23.8 The circuit with two batteries.

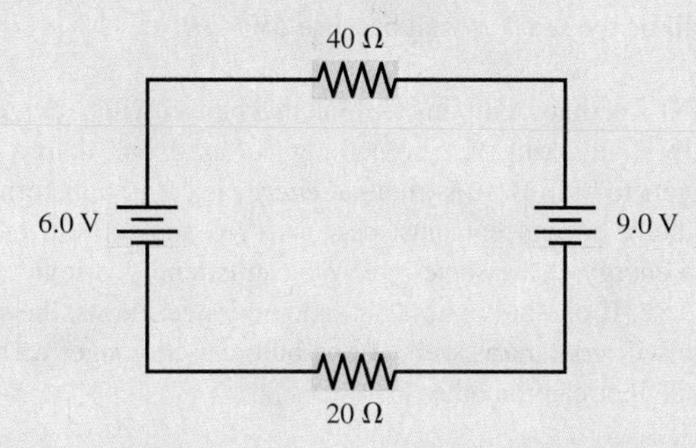

PREPARE We will solve this circuit using Kirchhoff's loop law, as outlined in Tactics Box 23.1. But how do we deal with *two* batteries? What happens when charge flows "backward" through a battery, from positive to negative? Consider the charge escalator analogy. Left to itself, a charge escalator lifts charge from lower to higher potential. But it *is* possible to run down an up escalator, as many of you have probably done. If two escalators are placed "head to head," whichever is "stronger" will, indeed, force the charge to run down the up escalator of the other battery. The current in a battery *can* be from positive to negative if driven in that direction by a larger emf from a second battery. Indeed, this is how batteries are "recharged." In this circuit, the current goes in the direction that the larger emf—the 9.0 V battery—"wants" it to go. We have redrawn the circuit in **FIGURE 23.9**, showing the direction of the current and the direction of the potential difference for each circuit element.

FIGURE 23.9 Analyzing the circuit.

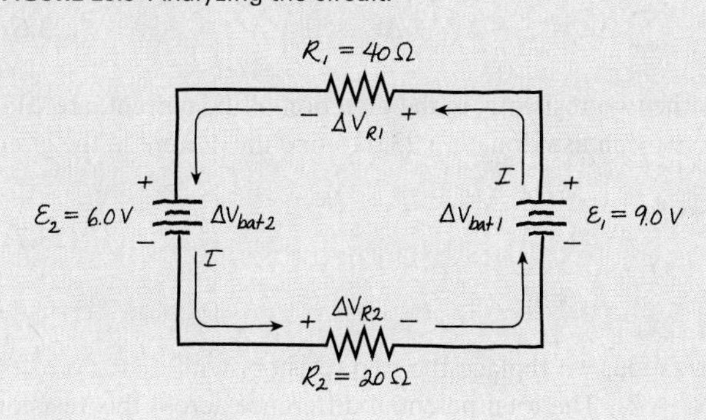

SOLVE Kirchhoff's loop law requires us to add the potential differences as we travel around the circuit in the direction of the current. Let's do this starting at the negative terminal of the 9.0 V battery:

$$\sum_i \Delta V_i = +9.0\text{ V} - I(40\ \Omega) - 6.0\text{ V} - I(20\ \Omega) = 0$$

The 6.0 V battery has $\Delta V_{bat} = -\mathcal{E}$, in accord with Tactics Box 23.1, because the potential decreases as we travel through this battery in the positive-to-negative direction. We can solve this equation for the current:

$$I = \frac{3.0\text{ V}}{60\ \Omega} = 0.050\text{ A} = 50\text{ mA}$$

Now that the current is known, we can use Ohm's law, $\Delta V = IR$, to find the magnitude of the potential difference across each resistor. For the 40 Ω resistor,

$$\Delta V_1 = (0.050\text{ A})(40\ \Omega) = 2.0\text{ V}$$

and for the 20 Ω resistor,

$$\Delta V_2 = (0.050\text{ A})(20\ \Omega) = 1.0\text{ V}$$

ASSESS The Assess step will be very important in circuit problems. There are generally other ways that you can analyze a circuit to check your work. In this case, you can do a final application of the loop law. If we start at the lower right-hand corner of the circuit and travel counterclockwise around the loop, the potential increases by 9.0 V in the first battery, then decreases by 2.0 V in the first resistor, decreases by 6.0 V in the second battery, and decreases by 1.0 V in the second resistor. The total decrease is 9.0 V, so the charge returns to its starting potential, a good check on our calculations.

STOP TO THINK 23.2 What is the potential difference across resistor R?

A. −3.0 V B. −4.0 V
C. −5.0 V D. −6.0 V
E. −10 V

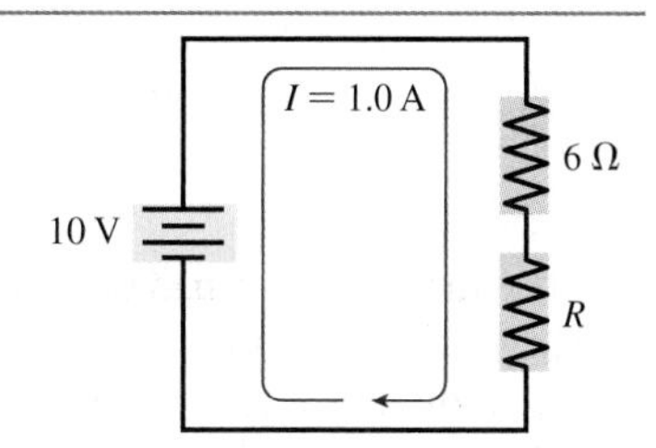

23.3 Series and Parallel Circuits

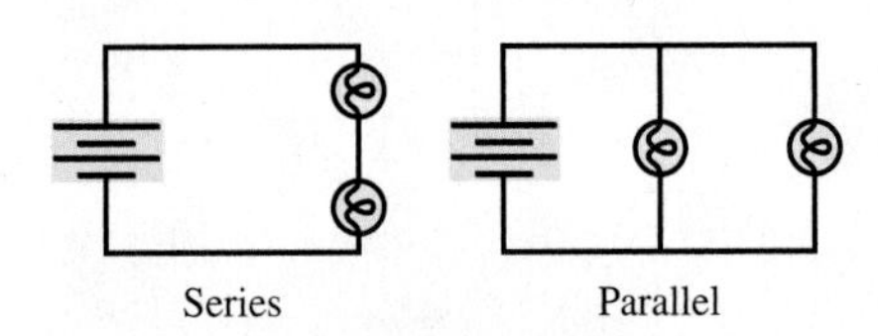

FIGURE 23.10 Series and parallel circuits.

Example 23.1 involved a circuit with multiple elements—two batteries and two resistors. As we introduce more circuit elements, we have possibilities for different types of connections. Suppose you use a single battery to light two lightbulbs. There are two possible ways that you can connect the circuit, as shown in FIGURE 23.10. These *series* and *parallel* circuits have very different properties. We will consider these two cases in turn.

We say two bulbs are connected in **series** if they are connected directly to each other with no junction in between. All series circuits share certain characteristics.

CONCEPTUAL EXAMPLE 23.2 **Brightness of bulbs in series**

FIGURE 23.11 shows two identical lightbulbs connected in series. Which bulb is brighter: A or B? Or are they equally bright?

FIGURE 23.11 Two bulbs in series.

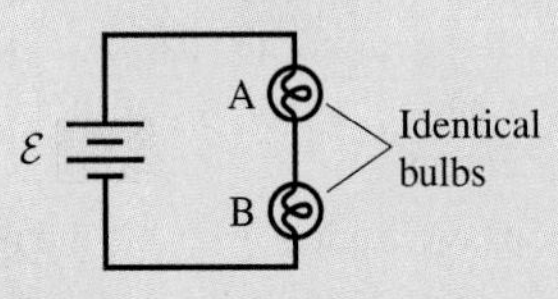

REASON Current is conserved, and there are no junctions in the circuit. Thus, as FIGURE 23.12 shows, the current is the same at all points.

FIGURE 23.12 The current in the series circuit.

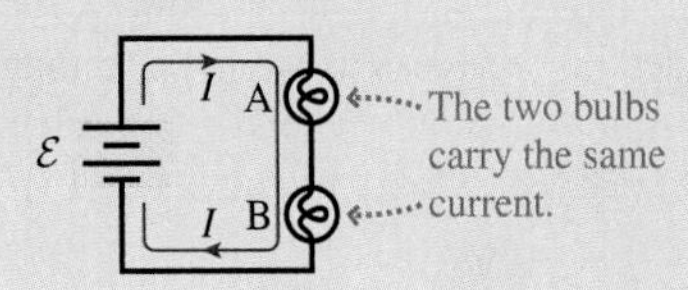

We learned in ◀◀ SECTION 22.6 that the power dissipated by a resistor is $P = I^2R$. If the two bulbs are identical (i.e., the same resistance) and have the same current through them, the power dissipated by each bulb is the same. This means that the brightness of the bulbs must be the same. The voltage across each of the bulbs will be the same as well because $\Delta V = IR$.

ASSESS It's perhaps tempting to think that bulb A will be brighter than bulb B, thinking that something is "used up" before the current gets to bulb B. It is true that *energy* is being transformed in each bulb, but current must be conserved and so both bulbs dissipate energy at the same rate. We can extend this logic to a special case: If one bulb burns out, and no longer lights, the second bulb will go dark as well. If one bulb can no longer carry a current, neither can the other.

Series Resistors

FIGURE 23.13 Replacing two series resistors with an equivalent resistor.

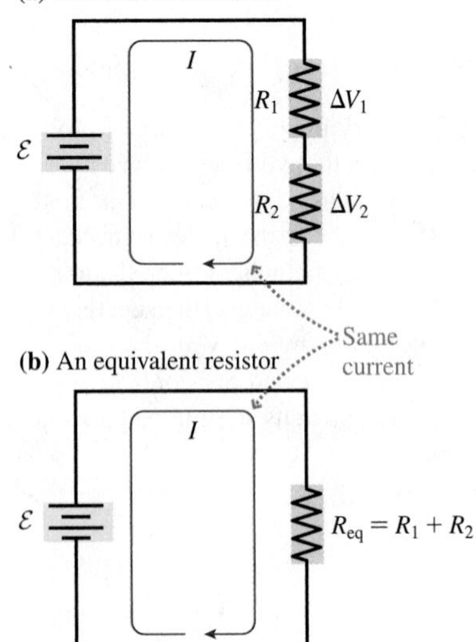

FIGURE 23.13a shows two resistors in series connected to a battery. Because there are no junctions, the current I must be the same in both resistors.

We can use Kirchhoff's loop law to look at the potential differences. Starting at the battery's negative terminal and following the current clockwise around the circuit, we find

$$\sum_i \Delta V_i = \mathcal{E} + \Delta V_1 + \Delta V_2 = 0 \qquad (23.6)$$

The voltage drops across the two resistors, in the direction of the current, are $\Delta V_1 = -IR_1$ and $\Delta V_2 = -IR_2$, so we can use Equation 23.6 to find the current in the circuit:

$$\mathcal{E} = -\Delta V_1 - \Delta V_2 = IR_1 + IR_2$$
$$I = \frac{\mathcal{E}}{R_1 + R_2} \qquad (23.7)$$

Suppose, as in FIGURE 23.13b, we replace the two resistors with a single resistor having the value $R_{eq} = R_1 + R_2$. The total potential difference across this resistor is still $\mathcal{E}$ because the potential difference is established by the battery. Further, the current in this single-resistor circuit is

$$I = \frac{\mathcal{E}}{R_{eq}} = \frac{\mathcal{E}}{R_1 + R_2}$$

which is the same as it had been in the two-resistor circuit. In other words, this single resistor is *equivalent* to the two series resistors in the sense that the circuit's current and potential difference are the same in both cases. Nothing anywhere else in the circuit would differ if we took out resistors R_1 and R_2 and replaced them with resistor R_{eq}.

We can extend this analysis to a case with more resistors. If we have N resistors in series, their **equivalent resistance** is the sum of the N individual resistances:

$$R_{\text{eq}} = R_1 + R_2 + \cdots + R_N \qquad (23.8)$$

Equivalent resistance of N series resistors

The current and the power output of the battery will be unchanged if the N series resistors are replaced by the single resistor R_{eq}.

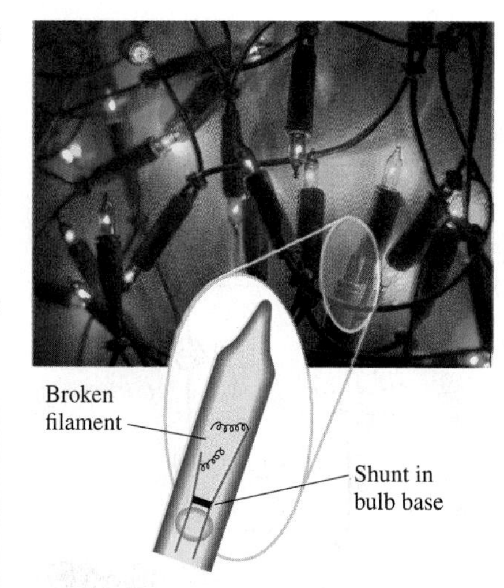

A seasonal series circuit puzzle
Christmas-tree minilights are connected in series. This is easy to verify: When you remove one bulb from a string of lights, the circuit is not complete, and the entire string of lights goes out. But when one bulb *burns* out, meaning its filament has broken, the string of lights stays lit. How is this possible? The secret is a *shunt* in the base of the bulb. Initially, the shunt is a good insulator. But if the filament breaks, the shunt is activated, and its resistance drops. The shunt can now carry the current, so the other bulbs will stay lit.

EXAMPLE 23.3 Potential difference of Christmas-tree minilights

A string of Christmas-tree minilights consists of 50 bulbs wired in series. What is the potential difference across each bulb when the string is plugged into a 120 V outlet?

FIGURE 23.14 50 bulbs connected in series.

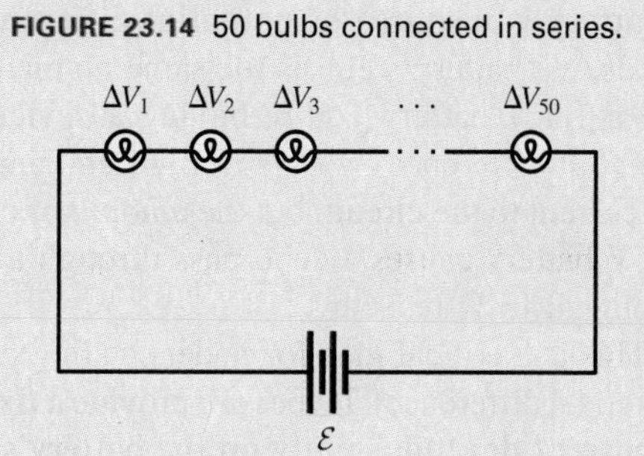

PREPARE FIGURE 23.14 shows the minilight circuit, which has 50 bulbs in series. The current in each of the bulbs is the same because they are in series.

SOLVE Applying Kirchhoff's loop law around the circuit, we find

$$\mathcal{E} = \Delta V_1 + \Delta V_2 + \cdots + \Delta V_{50}$$

The bulbs are all identical and, because the current in the bulbs is the same, all of the bulbs have the same potential difference. The potential difference across a single bulb is thus

$$\Delta V_1 = \frac{\mathcal{E}}{50} = \frac{120\text{ V}}{50} = 2.4\text{ V}$$

ASSESS This result seems reasonable. The potential difference is "shared" by the bulbs in the circuit. Since the potential difference is shared among 50 bulbs, the potential difference across each bulb will be quite small.

Minilights are wired in series because the bulbs can be inexpensive low-voltage bulbs. But there is a drawback that is true of all series circuits: If one bulb is removed, there is no longer a complete circuit, and there will be no current. Indeed, if you remove a bulb from a string of minilights, the entire string will go dark.

Class Video

EXAMPLE 23.4 Analyzing a series resistor circuit

What is the current in the circuit of **FIGURE 23.15**?

PREPARE The three resistors are in series, so we can replace them with a single equivalent resistor as shown in **FIGURE 23.16**.

SOLVE The equivalent resistance is calculated using Equation 23.8:

$$R_{\text{eq}} = 25\ \Omega + 31\ \Omega + 19\ \Omega = 75\ \Omega$$

The current in the equivalent circuit of Figure 23.16 is

$$I = \frac{\mathcal{E}}{R_{\text{eq}}} = \frac{9.0\text{ V}}{75\ \Omega} = 0.12\text{ A}$$

This is also the current in the original circuit.

ASSESS The current in the circuit is the same whether there are three resistors or a single equivalent resistor. The equivalent resistance is the sum of the individual resistance values, and so it is always greater than any of the individual values. This is a good check on your work.

FIGURE 23.15 A series resistor circuit.

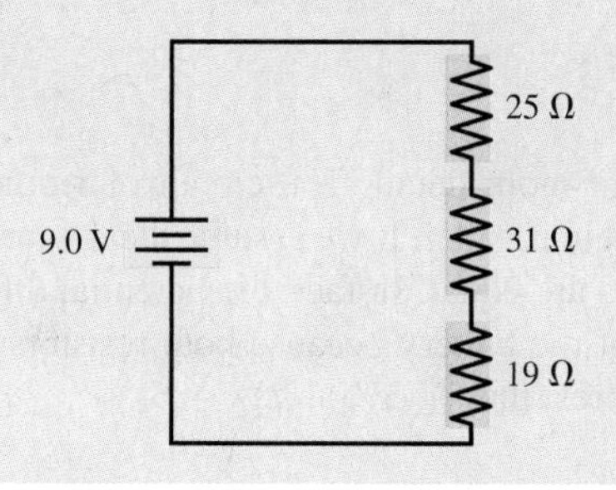

FIGURE 23.16 Analyzing a circuit with series resistors.

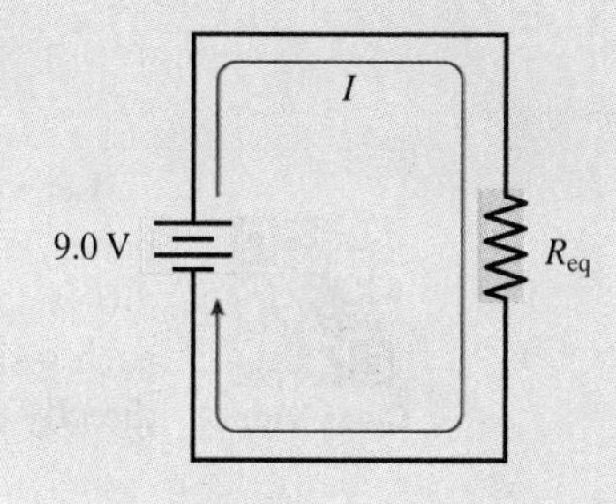

FIGURE 23.17 How does the brightness of bulb B compare to that of bulb A?

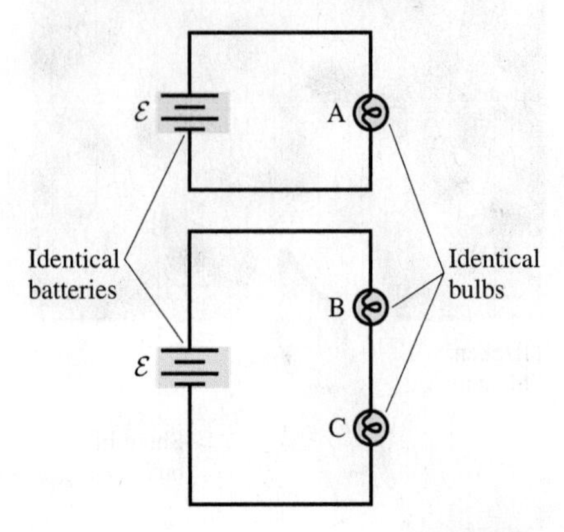
$\mathcal{E}$
A
Identical batteries
Identical bulbs
B
$\mathcal{E}$
C

Figure Video

Let's use our knowledge of series circuits to look at another lightbulb puzzle. **FIGURE 23.17** shows two different circuits, one with one battery and one lightbulb and a second with one battery and two lightbulbs. All of the batteries and bulbs are identical. You now know that B and C, which are connected in series, are equally bright, but how does the brightness of B compare to that of A?

Suppose the resistance of each identical lightbulb is R. In the first circuit, the battery drives current $I_A = \mathcal{E}/R$ through bulb A. In the second circuit, bulbs B and C are in series, with an equivalent resistance $R_{eq} = R_B + R_C = 2R$, but the battery has the same emf $\mathcal{E}$. Thus the current through bulbs B and C is $I_{B+C} = \mathcal{E}/R_{eq} = \mathcal{E}/2R = \frac{1}{2}I_A$. Bulb B has only half the current of bulb A, so B is dimmer.

Many people predict that A and B should be equally bright. It's the same battery, so shouldn't it provide the same current to both circuits? No—recall that **a battery is a source of potential difference, *not* a source of current.** In other words, the battery's emf is the same no matter how the battery is used. When you buy a 1.5 V battery you're buying a device that provides a specified amount of potential difference, not a specified amount of current. The battery does provide the current to the circuit, but the *amount* of current depends on the resistance. Your 1.5 V battery causes 1 A to pass through a 1.5 Ω resistor but only 0.1 A to pass through a 15 Ω resistor.

This is a critical idea for understanding circuits. A battery provides a fixed emf (potential difference). It does *not* provide a fixed and unvarying current. **The amount of current depends jointly on the battery's emf *and* the resistance of the circuit attached to the battery.**

Parallel Resistors

In the next example, we consider the second way of connecting two bulbs in a circuit. The two bulbs in Figure 23.18 are connected at *both* ends. We say that they are connected in **parallel.**

CONCEPTUAL EXAMPLE 23.5 **Brightness of bulbs in parallel**

Which lightbulb in the circuit of **FIGURE 23.18** is brighter: A or B? Or are they equally bright?

FIGURE 23.18 Two bulbs in parallel.

REASON Both ends of the two lightbulbs are connected together by wires. Because there's no potential difference along ideal wires, the potential at the top of bulb A must be the same as the potential at the top of bulb B. Similarly, the potentials at the bottoms of the bulbs must be the same. This means that the potential *difference* ΔV across the two bulbs must be the same, as we see in **FIGURE 23.19**. Because the bulbs are identical (i.e., equal resistances), the currents $I = \Delta V/R$ through the two bulbs are equal and thus the bulbs are equally bright.

FIGURE 23.19 The potential differences of the bulbs.

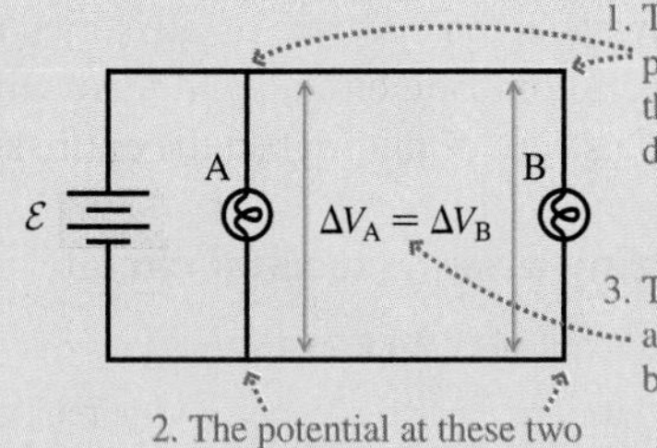
1. The potential at these two points is the same because there is no potential difference across the wire.
A
B
$\mathcal{E}$
$\Delta V_A = \Delta V_B$
3. The potential differences across the two bulbs must be equal.
2. The potential at these two points is the same as well.

ASSESS One might think that A would be brighter than B because current takes the "shortest route." But current is determined by potential difference, and two bulbs connected in parallel have the same potential difference.

Class Video

Let's look at parallel circuits in more detail. The circuit of **FIGURE 23.20a** has a battery and two resistors connected in parallel. If we assume ideal wires, the potential differences across the two resistors are equal. In fact, the potential difference across each resistor is equal to the emf of the battery because both resistors are connected directly to the battery with ideal wires; that is, $\Delta V_1 = \Delta V_2 = \mathcal{E}$.

Now we apply Kirchhoff's junction law. The current I_{bat} from the battery splits into currents I_1 and I_2 at the top junction noted in **FIGURE 23.20b**. According to the junction law,

$$I_{bat} = I_1 + I_2 \tag{23.9}$$

We can apply Ohm's law to each resistor to find that the battery current is

$$I_{bat} = \frac{\Delta V_1}{R_1} + \frac{\Delta V_2}{R_2} = \frac{\mathcal{E}}{R_1} + \frac{\mathcal{E}}{R_2} = \mathcal{E}\left(\frac{1}{R_1} + \frac{1}{R_2}\right) \tag{23.10}$$

FIGURE 23.20 Replacing two parallel resistors with an equivalent resistor.

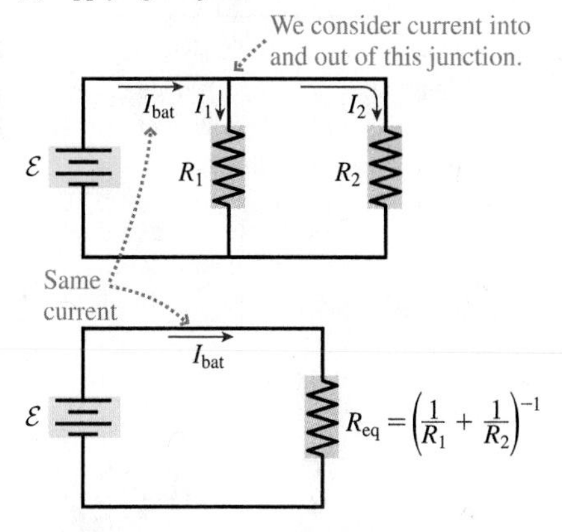

Can we replace a group of parallel resistors with a single equivalent resistor as we did for series resistors? To be equivalent, the potential difference across the equivalent resistor must be $\Delta V = \mathcal{E}$, the same as for the two resistors it replaces. Further, so that the battery can't know there's been any change, the current through the equivalent resistor must be $I = I_{bat}$. A resistor with this current and potential difference must have resistance

$$R_{eq} = \frac{\Delta V}{I} = \frac{\mathcal{E}}{I_{bat}} = \left(\frac{1}{R_1} + \frac{1}{R_2}\right)^{-1} \tag{23.11}$$

where we used Equation 23.10 for I_{bat}. This is the *equivalent resistance,* so a single resistor R_{eq} acts exactly the same as the two resistors R_1 and R_2 as shown in **FIGURE 23.20c.**

We can extend this analysis to the case of N resistors in parallel. For this circuit, the equivalent resistance is the inverse of the sum of the inverses of the N individual resistances:

$$R_{eq} = \left(\frac{1}{R_1} + \frac{1}{R_2} + \cdots + \frac{1}{R_N}\right)^{-1} \tag{23.12}$$

Equivalent resistance of N parallel resistors

The current and the power output of the battery will be unchanged if the N parallel resistors are replaced by the single resistor R_{eq}.

NOTE ► When you use Equation 23.12, don't forget to take the inverse of the sum that you compute. ◄

In Figure 23.20 each of the resistors "sees" the full potential difference of the battery. If one resistor were removed, the conditions of the second resistor would not change. This is an important property of parallel circuits.

► **Parallel circuits for safety** You have certainly seen cars with only one headlight lit. This tells us that automobile headlights are connected in parallel: The currents in the two bulbs are independent, so the loss of one bulb doesn't affect the other. The parallel wiring is very important so that the failure of one headlight will not leave the car without illumination.

Now, let's look at another lightbulb puzzle. **FIGURE 23.21** shows two different circuits: one with one battery and one lightbulb and a second with one battery and two lightbulbs. As before, the batteries and the bulbs are identical. You know that B and C, which are connected in parallel, are equally bright, but how does the brightness of B compare to that of A?

FIGURE 23.21 How does the brightness of bulb B compare to that of bulb A?

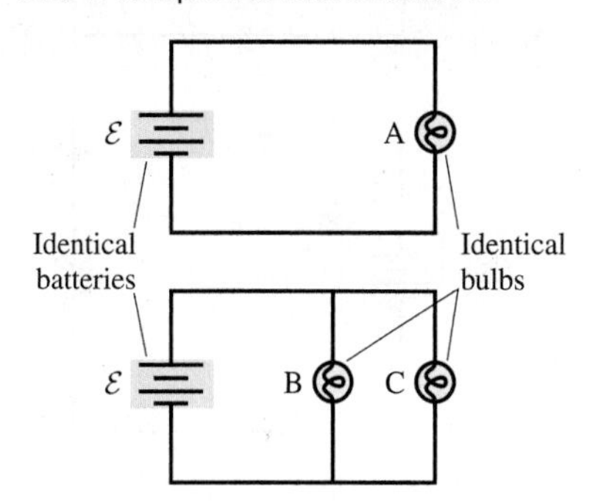

Each of the bulbs A, B, and C is connected to the same potential difference, that of the battery, so they each have the *same* brightness. Though all of the bulbs have the same brightness, there is a difference between the circuits. In the second circuit, the battery must power two lightbulbs, and so it must provide twice as much current. Recall that the battery is a source of fixed potential difference; the current depends on the circuit that is connected to the battery. Adding a second lightbulb doesn't change the potential difference, but it does increase the current from the battery.

EXAMPLE 23.6 Current in a parallel resistor circuit

The three resistors of FIGURE 23.22 are connected to a 12 V battery. What current is provided by the battery?

FIGURE 23.22 A parallel resistor circuit.

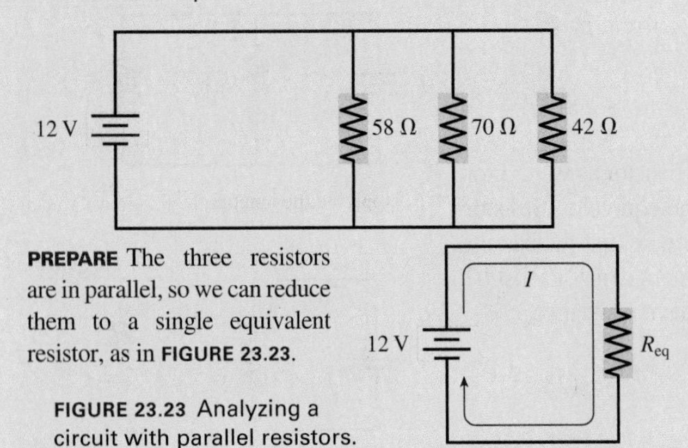

PREPARE The three resistors are in parallel, so we can reduce them to a single equivalent resistor, as in FIGURE 23.23.

FIGURE 23.23 Analyzing a circuit with parallel resistors.

SOLVE We can use Equation 23.12 to calculate the equivalent resistance:

$$R_{eq} = \left(\frac{1}{58\ \Omega} + \frac{1}{70\ \Omega} + \frac{1}{42\ \Omega}\right)^{-1} = 18.1\ \Omega$$

Once we know the equivalent resistance, we can use Ohm's law to calculate the current leaving the battery:

$$I = \frac{\mathcal{E}}{R_{eq}} = \frac{12\ \text{V}}{18.1\ \Omega} = 0.66\ \text{A}$$

Because the battery can't tell the difference between the original three resistors and this single equivalent resistor, the battery in Figure 23.22 provides a current of 0.66 A to the circuit.

ASSESS As we'll see, the equivalent resistance of a group of parallel resistors is less than the resistance of any of the resistors in the group. 18 Ω is less than any of the individual values, a good check on our work.

The value of the total resistance in this example may seem surprising. The equivalent of a parallel combination of 58 Ω, 70 Ω, and 42 Ω is 18 Ω. Shouldn't more resistors imply more resistance? The answer is yes for resistors in series, but not for resistors in parallel. Even though a resistor is an obstacle to the flow of charge, parallel resistors provide more pathways for charge to get through. Consequently, **the equivalent of several resistors in parallel is always *less* than any single resistor in the group.** As an analogy, think about driving in heavy traffic. If there is an alternate route or an extra lane for cars to travel, more cars will be able to "flow."

Video Tutor Demo

STOP TO THINK 23.3 Rank in order, from brightest to dimmest, the identical bulbs A to D.

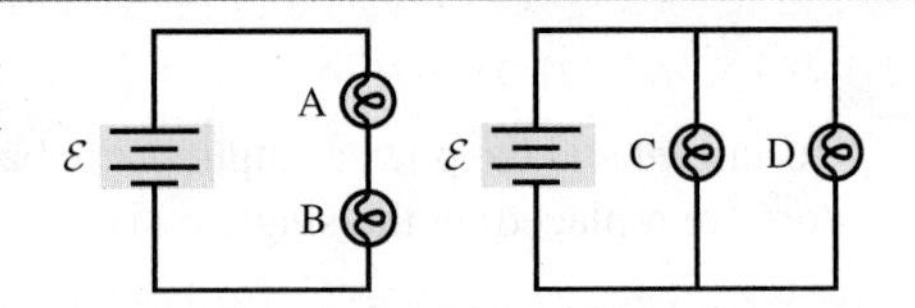

23.4 Measuring Voltage and Current

When you use a meter to measure the voltage or the current in a circuit, how do you connect the meter? The connection depends on the quantity you wish to measure.

A device that measures the current in a circuit element is called an **ammeter.** Because charge flows *through* circuit elements, an ammeter must be placed *in series* with the circuit element whose current is to be measured.

FIGURE 23.24 An ammeter measures the current in a circuit.

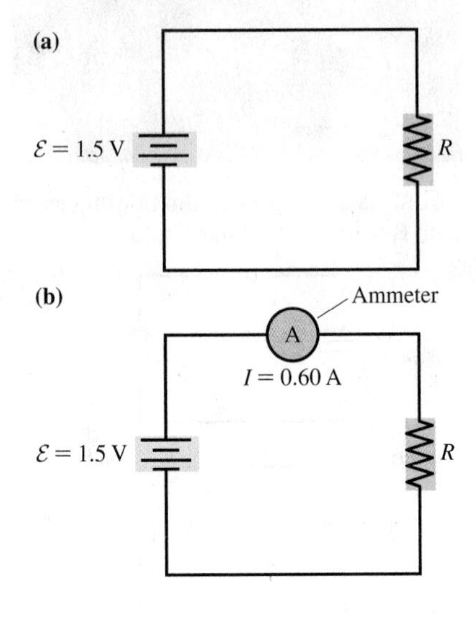

FIGURE 23.24a shows a simple one-resistor circuit with a fixed emf $\mathcal{E} = 1.5$ V and an unknown resistance R. To determine the resistance, we must know the current in the circuit, which we measure using an ammeter. We insert the ammeter in the circuit as shown in FIGURE 23.24b. We have to *break the connection* between the battery and the resistor in order to insert the ammeter. The resistor and the ammeter now have the same current because they are in series, so the reading of the ammeter is the current through the resistor.

Because the ammeter is in series with resistor R, the total resistance seen by the battery is $R_{eq} = R + R_{meter}$. In order to *measure* the current without *changing* the current, the ammeter's resistance must be much less than R. Thus **the resistance of an ideal ammeter is zero.** Real ammeters come quite close to this ideal.

The ammeter in Figure 23.24b reads 0.60 A, meaning that the current in the ammeter—and in the resistor—is $I = 0.60$ A. If the ammeter is ideal, which we will assume, then there is no potential difference across the ammeter ($\Delta V = IR = 0$ if

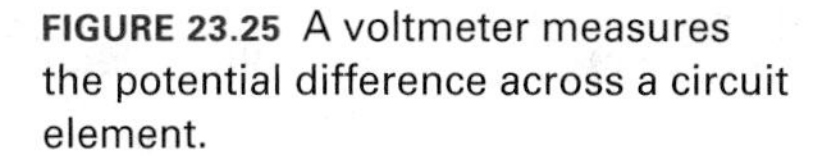

$R = 0\ \Omega$) and thus the potential difference across the resistor is $\Delta V = \mathcal{E}$. The resistance can then be calculated as

$$R = \frac{\mathcal{E}}{I} = \frac{1.5\text{ V}}{0.60\text{ A}} = 2.5\ \Omega$$

As we saw in ◀◀SECTION 21.2, we can use a **voltmeter** to measure potential differences in a circuit. Because a potential difference is measured *across* a circuit element, from one side to the other, a voltmeter is placed in *parallel* with the circuit element whose potential difference is to be measured. We want to *measure* the voltage without *changing* the voltage—without affecting the circuit. Because the voltmeter is in parallel with the resistor, the voltmeter's resistance must be very large so that it draws very little current. **An ideal voltmeter has infinite resistance.** Real voltmeters come quite close to this ideal.

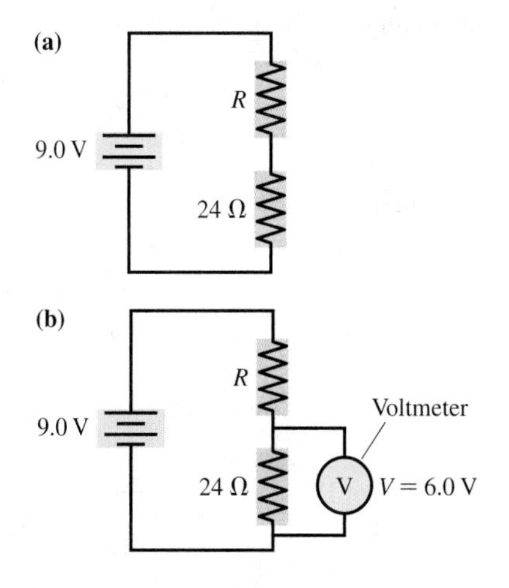

FIGURE 23.25 A voltmeter measures the potential difference across a circuit element.

FIGURE 23.25a shows a simple circuit in which a 24 Ω resistor is connected in series with an unknown resistance, with the pair of resistors connected to a 9.0 V battery. To determine the unknown resistance, we first characterize the circuit by measuring the potential difference across the known resistor with a voltmeter as shown in **FIGURE 23.25b**. The voltmeter is connected in parallel with the resistor; using a voltmeter does *not* require that we break the connections. The resistor and the voltmeter have the same potential difference because they are in parallel, so the reading of the voltmeter is the voltage across the resistor.

The voltmeter in Figure 23.25b tells us that the potential difference across the 24 Ω resistor is 6.0 V, so the current through the resistor is

$$I = \frac{\Delta V}{R} = \frac{6.0\text{ V}}{24\ \Omega} = 0.25\text{ A} \tag{23.13}$$

The two resistors are in series, so this is also the current in unknown resistor R. We can use Kirchhoff's loop law and the voltmeter reading to find the potential difference across the unknown resistor:

$$\sum_i \Delta V_i = 9.0\text{ V} + \Delta V_R - 6.0\text{ V} = 0 \tag{23.14}$$

from which we find $\Delta V_R = -3.0$ V. We can now use $\Delta V_R = -IR$ to calculate

$$R = \frac{-\Delta V_R}{I} = -\frac{(-3.0\text{ V})}{0.25\text{ A}} = 12\ \Omega \tag{23.15}$$

STOP TO THINK 23.4 Which is the right way to connect the meters to measure the potential difference across and the current through the resistor?

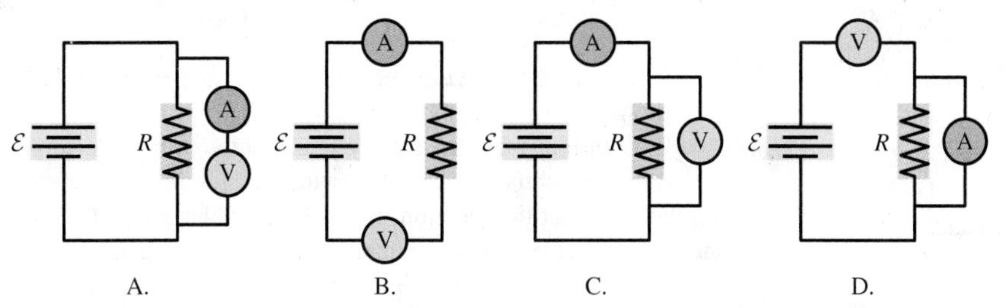

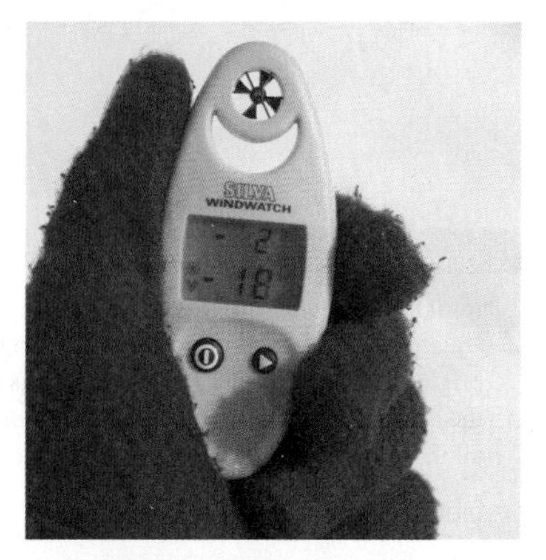

A circuit for all seasons This device displays wind speed and temperature, but these are computed from basic measurements of voltage and current. The wind turns a propeller attached to a generator; a rapid spin means a high voltage. A circuit in the device contains a *thermistor,* whose resistance varies with temperature; low temperatures mean high resistance and thus a small current.

23.5 More Complex Circuits

In this section, we will consider circuits that involve both series and parallel resistors. Combinations of resistors can often be reduced to a single equivalent resistance through a step-by-step application of the series and parallel rules.

EXAMPLE 23.7 Combining resistors

What is the equivalent resistance of the group of resistors shown in **FIGURE 23.26**?

FIGURE 23.26 A resistor circuit.

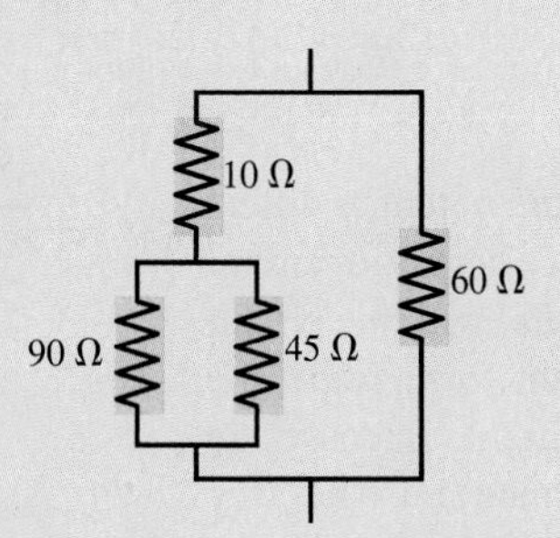

PREPARE We can analyze this circuit by reducing combinations of series and parallel resistors. We will do this in a series of steps, redrawing the circuit after each step.

SOLVE The process of simplifying the circuit is shown in **FIGURE 23.27**. Note that the 10 Ω and 60 Ω resistors are *not* in parallel. They are connected at their top ends but not at their bottom ends. Resistors must be connected at *both* ends to be in parallel. Similarly, the 10 Ω and 45 Ω resistors are *not* in series because of the junction between them.

ASSESS The last step in the process is to reduce a combination of parallel resistors. The resistance of parallel resistors is always less than the smallest of the individual resistance values, so our final result must be less than 40 Ω. This is a good check on the result.

FIGURE 23.27 A combination of resistors is reduced to a single equivalent resistor.

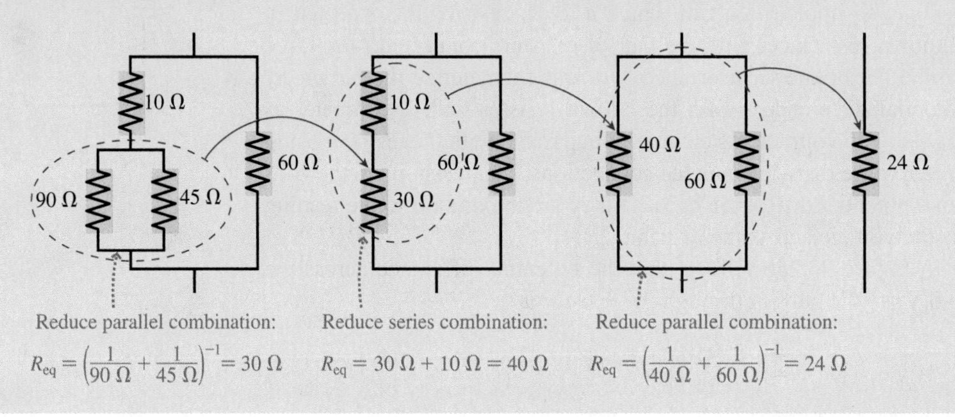

Two special cases (worth remembering for reducing circuits) are the equivalent resistances of two identical resistors $R_1 = R_2 = R$ in series and in parallel:

$$\text{Two identical resistors in series:} \quad R_{eq} = 2R$$

$$\text{Two identical resistors in parallel:} \quad R_{eq} = \frac{R}{2}$$

EXAMPLE 23.8 How does the brightness change?

Initially the switch in **FIGURE 23.28** is open. Bulbs A and B are equally bright, and bulb C is not glowing. What happens to the brightness of A and B when the switch is closed? And how does the brightness of C then compare to that of A and B? Assume that all bulbs are identical.

FIGURE 23.28 A lightbulb circuit.

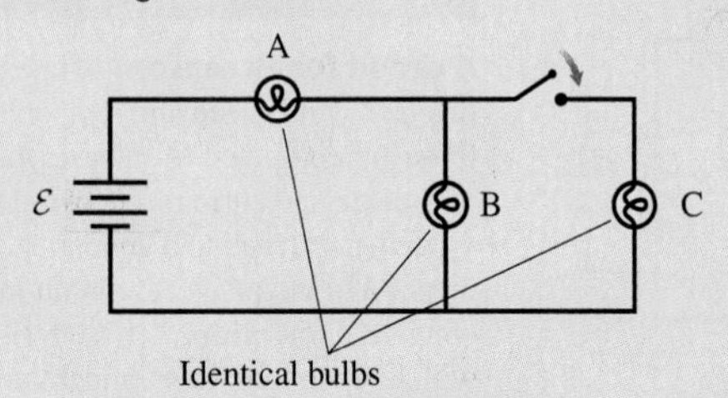

SOLVE Suppose the resistance of each bulb is R. Initially, before the switch is closed, bulbs A and B are in series; bulb C is not part of the circuit. A and B are identical resistors in series, so their equivalent resistance is $2R$ and the current from the battery is

$$I_{before} = \frac{\mathcal{E}}{R_{eq}} = \frac{\mathcal{E}}{2R} = \frac{1}{2}\frac{\mathcal{E}}{R}$$

This is the initial current in bulbs A and B, so they are equally bright.

Closing the switch places bulbs B and C in parallel with each other. The equivalent resistance of the two identical resistors in parallel is $R_{B+C} = R/2$. This equivalent resistance of B and C is in series with bulb A; hence the total resistance of the circuit is $R_{eq} = R + \frac{1}{2}R = \frac{3}{2}R$, and the current leaving the battery is

$$I_{after} = \frac{\mathcal{E}}{R_{eq}} = \frac{\mathcal{E}}{3R/2} = \frac{2}{3}\frac{\mathcal{E}}{R} > I_{before}$$

Closing the switch *decreases* the total circuit resistance and thus *increases* the current leaving the battery.

All the current from the battery passes through bulb A, so A *increases* in brightness when the switch is closed. The current I_{after} then splits at the junction. Bulbs B and C have equal resistance, so the current divides equally. The current in B is $\frac{1}{3}(\mathcal{E}/R)$, which is *less* than I_{before}. Thus B *decreases* in brightness when the switch is closed. With the switch closed, bulbs B and C are in parallel, so bulb C has the same brightness as bulb B.

ASSESS Our final results make sense. Initially, bulbs A and B are in series, and all of the current that goes through bulb A goes through bulb B. But when we add bulb C, the current has another option—it can go through bulb C. This will increase the total current, and all that current must go through bulb A, so we expect a brighter bulb A. But now the current through bulb A can go through bulbs B and C. The current splits, so we'd expect that bulb B will be dimmer than before.

Analyzing Complex Circuits

We can use the information in this chapter to analyze more complex but more realistic circuits. This will give us a chance to bring together the many ideas of this chapter and to see how they are used in practice. The techniques that we use for this analysis are quite general.

This x-ray image of a cell phone shows the complex circuitry inside. Though there are thousands of components, the analysis of such a circuit starts with the same basic rules we are studying in this chapter.

PROBLEM-SOLVING STRATEGY 23.1 **Resistor circuits** 

We can analyze any resistor circuit by sequentially reducing parallel and series resistor combinations to their equivalent resistors until only the battery and a single equivalent resistor are left.

PREPARE Draw a circuit diagram. Label all known and unknown quantities.

SOLVE Base your mathematical analysis on Kirchhoff's laws and on the rules for series and parallel resistors:

- Step by step, reduce the circuit to the smallest possible number of equivalent resistors.
- Determine the current through and potential difference across the equivalent resistors.
- Rebuild the circuit, using the facts that the current is the same through all resistors in series and the potential difference is the same across all parallel resistors.

ASSESS Use two important checks as you rebuild the circuit.

- Verify that the sum of the potential differences across series resistors matches ΔV for the equivalent resistor.
- Verify that the sum of the currents through parallel resistors matches I for the equivalent resistor.

Exercise 23 

EXAMPLE 23.9 **Analyzing a complex circuit**

Find the current through and the potential difference across each of the four resistors in the circuit shown in **FIGURE 23.29**.

FIGURE 23.29 A multiple-resistor circuit.

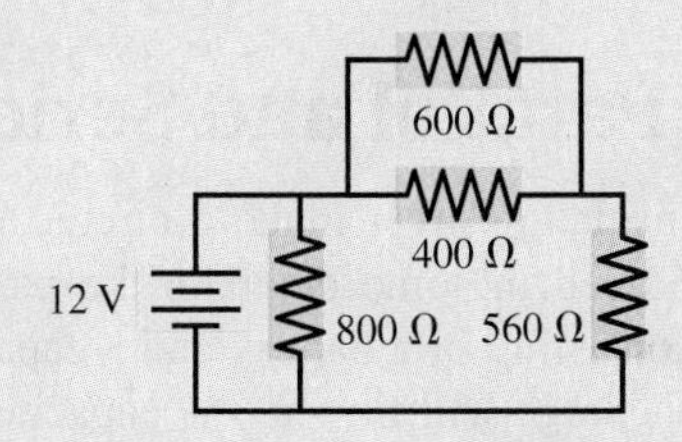

PREPARE FIGURE 23.30 on the next page shows the circuit diagram. We'll keep redrawing the diagram as we analyze the circuit.

SOLVE First, we break down the circuit, step-by-step, into one with a single resistor. Figure 23.30a does this in three steps, using the rules for series and parallel resistors. The final battery-and-resistor circuit is one that is easy to analyze. The potential difference across the 400 Ω equivalent resistor is $\Delta V_{400} = \Delta V_{bat} = \mathcal{E} = 12$ V. The current is

$$I = \frac{\mathcal{E}}{R} = \frac{12\text{ V}}{400\ \Omega} = 0.030\text{ A} = 30\text{ mA}$$

Second, we rebuild the circuit, step-by-step, finding the currents and potential differences at each step. Figure 23.30b repeats the steps of Figure 23.30a exactly, but in reverse order. The 400 Ω resistor came from two 800 Ω resistors in parallel. Because $\Delta V_{400} = 12$ V, it must be true that each $\Delta V_{800} = 12$ V. The current through each 800 Ω is then $I = \Delta V/R = 15$ mA. A check on our work is to note that 15 mA + 15 mA = 30 mA.

The right 800 Ω resistor was formed by combining 240 Ω and 560 Ω in series. Because $I_{800} = 15$ mA, it must be true that $I_{240} = I_{560} = 15$ mA. The potential difference across each is $\Delta V = IR$, so $\Delta V_{240} = 3.6$ V and $\Delta V_{560} = 8.4$ V. Here the check on our work is to note that 3.6 V + 8.4 V = 12 V = ΔV_{800}, so the potential differences add as they should.

Finally, the 240 Ω resistor came from 600 Ω and 400 Ω in parallel, so they each have the same 3.6 V potential difference as their 240 Ω equivalent. The currents are $I_{600} = 6.0$ mA and $I_{400} = 9.0$ mA. Note that 6.0 mA + 9.0 mA = 15 mA, which is a third check on our work. We now know all currents and potential differences.

ASSESS We *checked our work* at each step of the rebuilding process by verifying that currents summed properly at junctions and that potential differences summed properly along a series of resistances. This "check as you go" procedure is extremely important. It provides you, the problem solver, with a built-in error finder that will immediately inform you if a mistake has been made.

Continued

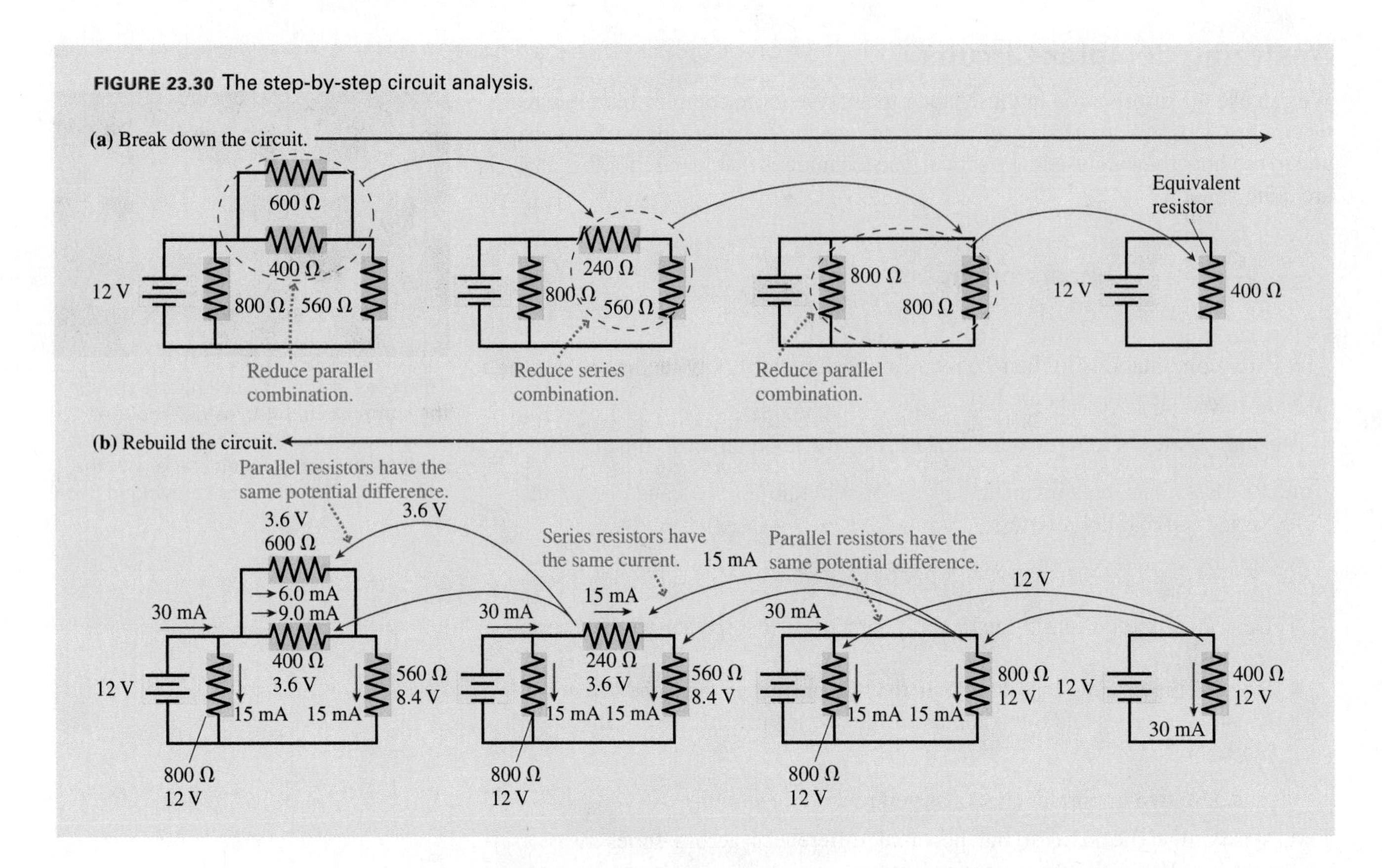

FIGURE 23.30 The step-by-step circuit analysis.

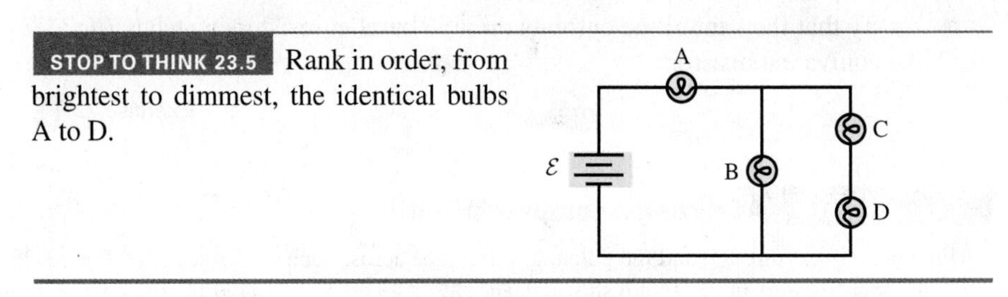

STOP TO THINK 23.5 Rank in order, from brightest to dimmest, the identical bulbs A to D.

FIGURE 23.31 Simple capacitor circuit.

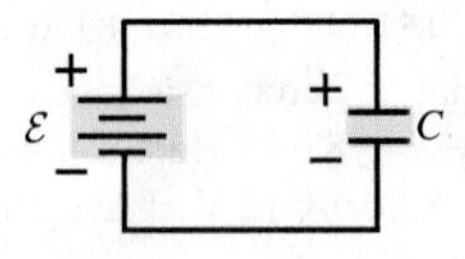

FIGURE 23.32 Parallel and series capacitors.

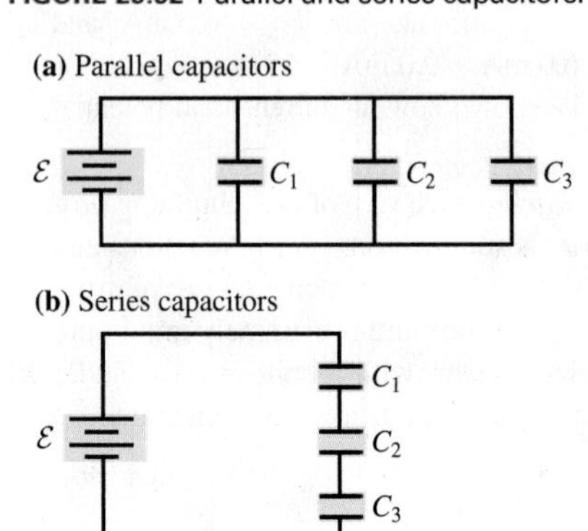

23.6 Capacitors in Parallel and Series

Two conductors separated by an insulating layer make a circuit element called a *capacitor,* a device we have considered in some detail in the past few chapters. **FIGURE 23.31** shows a basic circuit consisting of a battery and a capacitor. When we connect the capacitor to the battery, charge will flow to the capacitor, increasing its potential difference until $\Delta V_C = \mathcal{E}$. Once the capacitor is fully charged, there will be no further current. We saw in ◀◀SECTION 21.7 that the magnitude of the charge on each plate of the capacitor at this point will be $Q = C\Delta V_C = C\mathcal{E}$.

In resistor circuits, we often combine multiple resistors; we can do the same with capacitors. **FIGURE 23.32** illustrates two basic combinations: parallel capacitors and series capacitors.

NOTE ▸ The terms "parallel capacitors" and "parallel-plate capacitor" do not describe the same thing. The former term describes how two or more capacitors are connected to each other; the latter describes how a particular capacitor is constructed. ◂

Parallel or series capacitors can be represented by a single **equivalent capacitance,** though the rules for the combinations are different from those for resistors. Let's start our analysis with the two parallel capacitors C_1 and C_2 of **FIGURE 23.33a**.

The charge on C_1 is $Q_1 = C_1 \Delta V_C$ and the charge on C_2 is $Q_2 = C_2 \Delta V_C$. Note that both capacitors have the *same* potential difference ΔV_C across them because they are connected in parallel. The total charge Q on the two capacitors is then

$$Q = Q_1 + Q_2 = C_1 \Delta V_C + C_2 \Delta V_C = (C_1 + C_2)\Delta V_C$$

We can replace the two capacitors by a single equivalent capacitance C_{eq}, as shown in FIGURE 23.33b. The equivalent capacitance is the sum of the individual capacitance values:

$$C_{eq} = \frac{Q}{\Delta V_C} = \frac{(C_1 + C_2)\Delta V_C}{\Delta V_C} = C_1 + C_2 \qquad (23.16)$$

We can easily extend this analysis to more than two capacitors. If N capacitors are in parallel, their equivalent capacitance is the sum of the individual capacitances:

$$C_{eq} = C_1 + C_2 + C_3 + \cdots + C_N \qquad (23.17)$$

Equivalent capacitance of N parallel capacitors

Neither the battery nor any other part of a circuit can tell if the parallel capacitors are replaced by a single capacitor having capacitance C_{eq}.

NOTE ▶ Adding another capacitor in parallel adds more capacitance. The formula for *parallel* capacitors is thus similar to the formula for *series* resistors. ◀

FIGURE 23.33 Replacing two parallel capacitors with an equivalent capacitor.

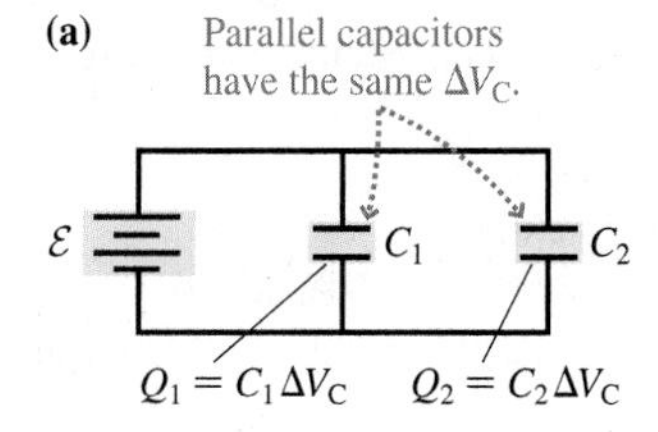

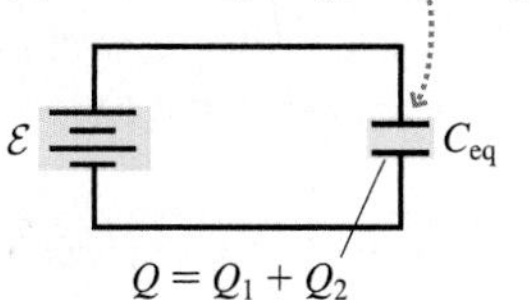

Now let's look at two capacitors connected in series. As we'll see, in this case the capacitors have the same charge, but each sees less than the full voltage of the battery.

Take a look at the circuit consisting of two series capacitors in FIGURE 23.34a. The center section, consisting of the bottom plate of C_1, the top plate of C_2, and the connecting wire, is electrically isolated. The battery cannot remove charge from or add charge to this section. If it starts out with no net charge, it must end up with no net charge. As a consequence, the two capacitors in series have equal charges $\pm Q$. The battery transfers Q from the bottom of C_2 to the top of C_1. This transfer polarizes the center section, but it still has $Q_{net} = 0$.

The potential differences across the two capacitors are $\Delta V_1 = Q/C_1$ and $\Delta V_2 = Q/C_2$. The total potential difference across both capacitors is $\Delta V_C = \Delta V_1 + \Delta V_2$. Suppose, as in FIGURE 23.34b, we replaced the two capacitors with a single capacitor having charge Q and potential difference $\Delta V_C = \Delta V_1 + \Delta V_2$. This capacitor is equivalent to the original two because the battery has to establish the same potential difference and move the same amount of charge in either case.

The inverse of the capacitance of this equivalent capacitor is

$$\frac{1}{C_{eq}} = \frac{\Delta V_C}{Q} = \frac{\Delta V_1 + \Delta V_2}{Q} = \frac{\Delta V_1}{Q} + \frac{\Delta V_2}{Q} = \frac{1}{C_1} + \frac{1}{C_2} \qquad (23.18)$$

FIGURE 23.34 Replacing two series capacitors with an equivalent capacitor.

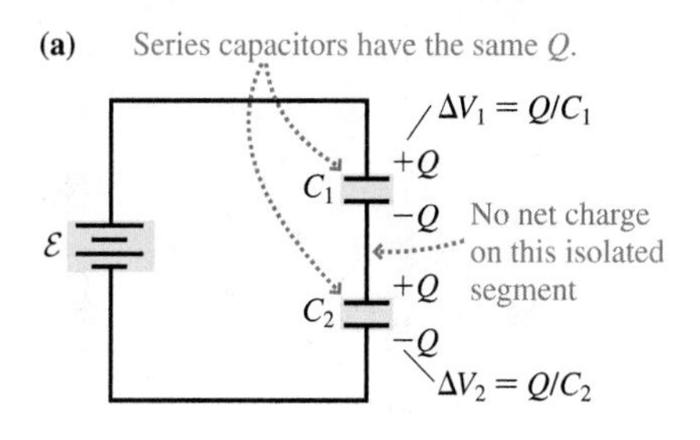

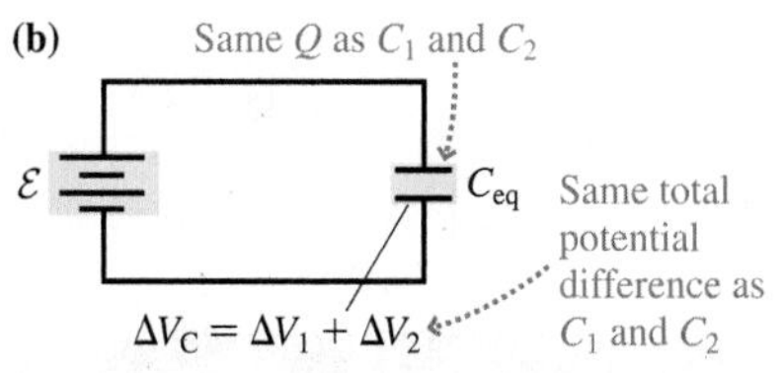

This analysis hinges on the fact that **series capacitors each have the same charge Q.**

We can easily extend this analysis to more than two capacitors. If N capacitors are in series, their equivalent capacitance is the inverse of the sum of the inverses of the individual capacitances:

$$C_{eq} = \left(\frac{1}{C_1} + \frac{1}{C_2} + \frac{1}{C_3} + \cdots + \frac{1}{C_N}\right)^{-1} \qquad (23.19)$$

Equivalent capacitance of N series capacitors

For series capacitors, the equivalent capacitance is less than that of the individual capacitors.

NOTE ▶ The total charge on the capacitors is the charge on each individual capacitor, but each capacitor sees only a fraction of the voltage. Adding capacitors in *series* reduces the total capacitance, just like adding resistors in *parallel*. ◀

EXAMPLE 23.10 Analyzing a capacitor circuit

a. Find the equivalent capacitance of the combination of capacitors in the circuit of FIGURE 23.35.

b. What charge flows through the battery as the capacitors are being charged?

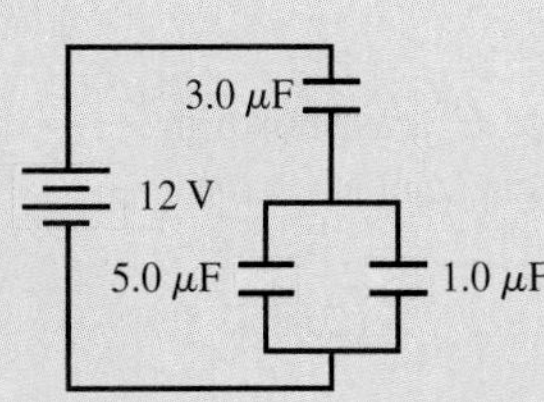

FIGURE 23.35 A capacitor circuit.

PREPARE We can use the relationships for parallel and series capacitors to reduce the capacitors to a single equivalent capacitance, much as we did for resistor circuits. We can then compute the charge through the battery using this value of capacitance.

SOLVE

a. FIGURE 23.36 shows how we find the equivalent capacitance by reducing parallel and series combinations.

b. The battery sees a capacitance of 2.0 μF. To establish a potential difference of 12 V, the charge that must flow is

$$Q = C_{eq}\Delta V_C = (2.0 \times 10^{-6}\ \text{F})(12\ \text{V}) = 2.4 \times 10^{-5}\ \text{C}$$

ASSESS We solve capacitor circuit problems in a manner very similar to what we followed for resistor circuits.

FIGURE 23.36 Analyzing a capacitor circuit.

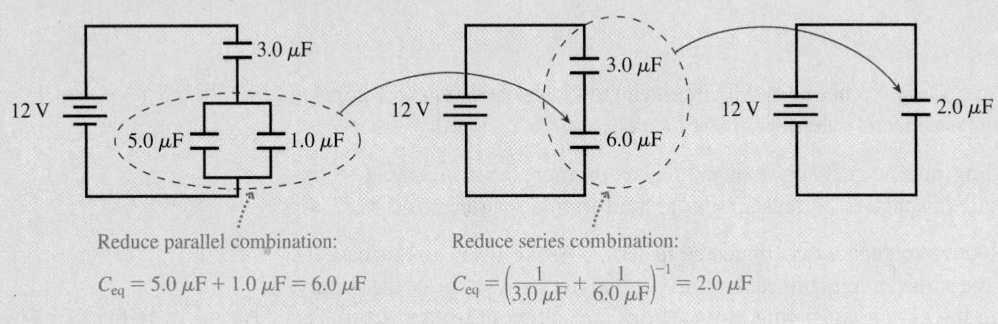

STOP TO THINK 23.6 Rank in order, from largest to smallest, the equivalent capacitance $(C_{eq})_A$ to $(C_{eq})_C$ of circuits A to C.

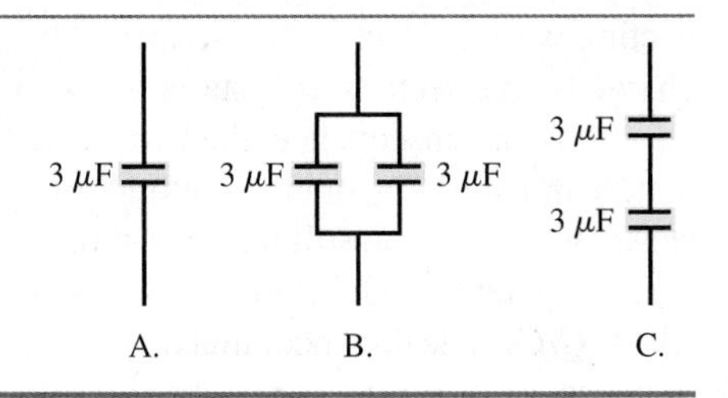

23.7 *RC* Circuits

The resistor circuits we have seen have a steady current. If we add a capacitor to a resistor circuit, we can make a circuit in which the current varies with time. Circuits containing resistors and capacitors are known as ***RC* circuits.** They are very common in electronic equipment. A simple example of an *RC* circuit is the flashing bike light in the photograph. As we will see, the values of the resistance and capacitance in an *RC* circuit determine the *time* it takes the capacitor to charge or discharge. In the case of the bike light, this time determines the time between flashes. A large capacitance causes a slow cycle of on-off-on; a smaller capacitance means a more rapid flicker.

The rear flasher on a bike blinks on and off. The timing is controlled by an *RC* circuit.

FIGURE 23.37a shows an *RC* circuit consisting of a charged capacitor, an open switch, and a resistor. The capacitor has initial charge Q_0 and potential difference $(\Delta V_C)_0 = Q_0/C$. There is no current, so the potential difference across the resistor is zero. Then, at $t = 0$, the switch closes and the capacitor begins to discharge through the resistor.

FIGURE 23.37b shows the circuit *immediately* after the switch closes. The capacitor voltage is still $(\Delta V_C)_0$ because the capacitor hasn't yet had time to lose any charge, but now there's a current I_0 in the circuit that's starting to discharge the capacitor. Applying Kirchhoff's loop law, going around the loop clockwise, we find

$$\sum_i \Delta V_i = \Delta V_C + \Delta V_R = (\Delta V_C)_0 - I_0 R = 0$$

FIGURE 23.37 Discharging an *RC* circuit.

(a) Before the switch closes

The switch will close at $t = 0$.

Charge Q_0
$(\Delta V_C)_0 = Q_0/C$

(b) Immediately after the switch closes

The charge separation on the capacitor produces a potential difference, which causes a current.

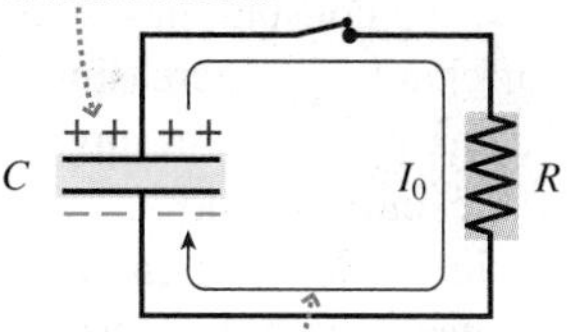

Current is the flow of charge, so the current discharges the capacitor.

(c) At a later time

The current has reduced the charge on the capacitor. This reduces the potential difference.

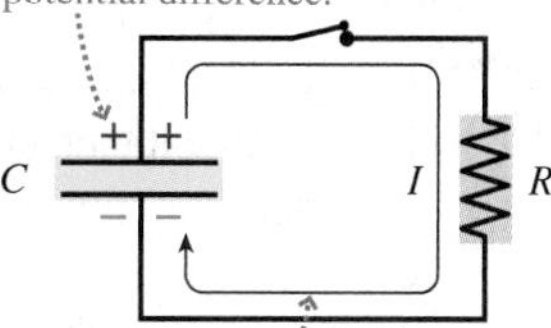

The reduced potential difference leads to a reduced current.

Thus the *initial* current—the initial rate at which the capacitor begins to discharge—is

$$I_0 = \frac{(\Delta V_C)_0}{R} \tag{23.20}$$

As time goes by, the current continues and the charge on the capacitor decreases. **FIGURE 23.37c** shows the circuit some time after the switch is closed; as we can see, both the charge on the capacitor (and thus the potential difference) and the current in the circuit have decreased. When the capacitor voltage has decreased to ΔV_C, the current has decreased to

$$I = \frac{\Delta V_C}{R} \tag{23.21}$$

The current discharges the capacitor, which causes ΔV_C to decrease. The capacitor voltage ΔV_C drives the current, so the current I decreases as well. The current I and the voltage ΔV_C both decrease until the capacitor is fully discharged and the current is zero.

If we use a voltmeter and an ammeter to measure the capacitor voltage and the current in the circuit of Figure 23.37 as a function of time, we find the variation shown in the graphs of **FIGURE 23.38**. At $t = 0$, when the switch closes, the potential difference across the capacitor is $(\Delta V_C)_0$ and the current suddenly jumps to I_0. The current and the capacitor voltage then "decay" to zero, but *not* linearly.

FIGURE 23.38 Current and capacitor voltage in an *RC* discharge circuit.

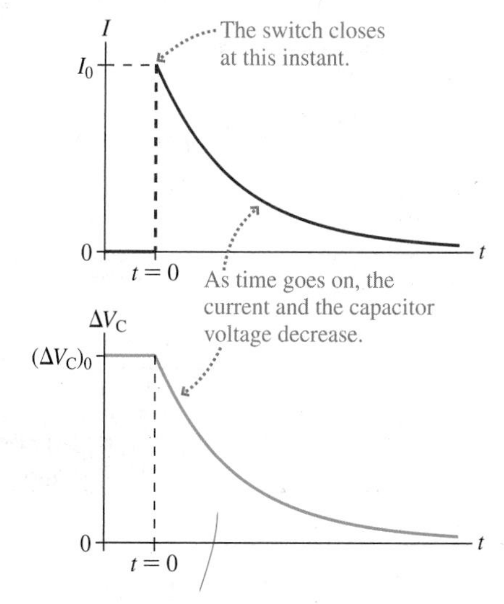

The graphs in Figure 23.38 have the same shape as the graph for the decay of the amplitude of a damped simple harmonic oscillator we saw in Chapter 14. Both the voltage and the current are *exponential decays* given by the equations

$$I = I_0 e^{-t/RC}$$
$$\Delta V_C = (\Delta V_C)_0 e^{-t/RC} \tag{23.22}$$

Current and voltage during a capacitor discharge

p. 456 EXPONENTIAL

In ◀◀ **SECTION 14.6**, we saw that we could characterize exponential decay by a **time constant** τ. The time constant is really a *characteristic time* for the circuit. A long time constant implies a slow decay; a short time constant, a rapid decay. The time constant for the decay of current and voltage in an *RC* circuit is

$$\tau = RC \tag{23.23}$$

If you work with the units, you can show that the product of ohms and farads is seconds, so the quantity *RC* really is a time. In terms of this time constant, the current and voltage equations are

$$I = I_0 e^{-t/\tau}$$
$$\Delta V_C = (\Delta V_C)_0 e^{-t/\tau} \tag{23.24}$$

Video Tutor Demo

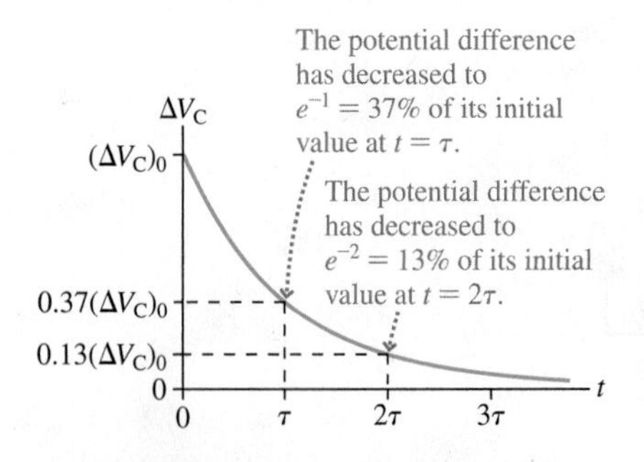

FIGURE 23.39 The meaning of the time constant in an *RC* circuit.

The current and voltage in the circuit do not drop to zero after one time constant; that's not what the time constant means. Instead, each increase in time by one time constant causes the voltage and current to decrease by a factor of $e^{-1} = 0.37$, as we see in **FIGURE 23.39**.

We can understand why the time constant has the form $\tau = RC$. A large value of resistance opposes the flow of charge, so increasing R increases the decay time. A larger capacitance stores more charge, so increasing C also increases the decay time.

After one time constant, the current and voltage in a capacitor circuit have decreased to 37% of their initial values. When is the capacitor fully discharged? There is no exact time that we can specify because ΔV_C approaches zero gradually. But after 5τ the voltage and current have decayed to less than 1% of their initial values. For most purposes, we can say that the capacitor is discharged at this time.

EXAMPLE 23.11 **Finding the current in an *RC* circuit**

The switch in the circuit of **FIGURE 23.40** has been in position a for a long time, so the capacitor is fully charged. The switch is changed to position b at $t = 0$. What is the current in the circuit immediately after the switch is closed? What is the current in the circuit 25 μs later?

FIGURE 23.40 The *RC* circuit.

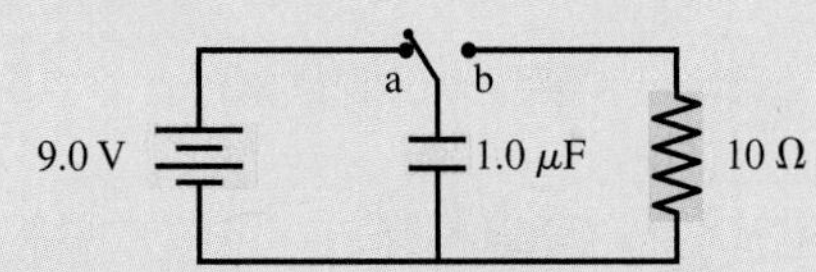

SOLVE The capacitor is connected across the battery terminals, so initially it is charged to $(\Delta V_C)_0 = 9.0$ V. When the switch is closed, the initial current is given by Equation 23.20:

$$I_0 = \frac{(\Delta V_C)_0}{R} = \frac{9.0 \text{ V}}{10 \ \Omega} = 0.90 \text{ A}$$

As charge flows, the capacitor discharges. The time constant for the decay is given by Equation 23.23:

$$\tau = (10 \ \Omega)(1.0 \times 10^{-6} \text{ F}) = 1.0 \times 10^{-5} \text{ s} = 10 \ \mu\text{s}$$

The current in the circuit as a function of time is given by Equation 23.22. 25 μs after the switch is closed, the current is

$$I = I_0 e^{-t/\tau} = (0.90 \text{ A})e^{-(25 \mu\text{s})/(10 \mu\text{s})} = 0.074 \text{ A}$$

ASSESS This result makes sense. 25 μs after the switch has closed is 2.5 time constants, so we expect the current to decrease to a small fraction of the initial current. Notice that we left times in units of μs; this is one of the rare cases where we needn't convert to SI units. Because the exponent is $-t/\tau$, which involves a ratio of two times, we need only be certain that both t and τ are in the same units.

FIGURE 23.41 A circuit for charging a capacitor.

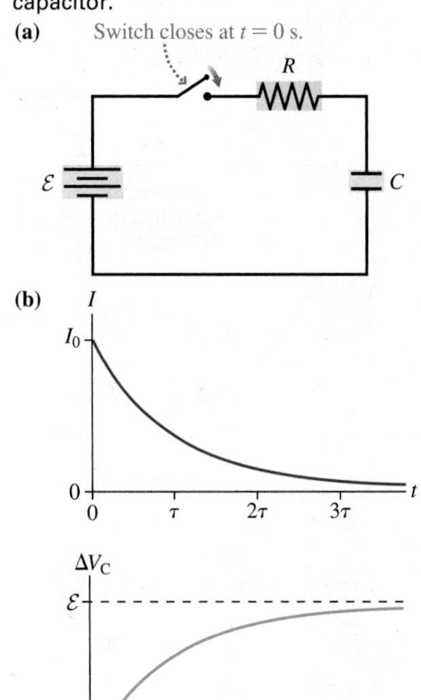

Charging a Capacitor

FIGURE 23.41a shows a circuit that charges a capacitor. After the switch is closed, the potential difference of the battery causes a current in the circuit, and the capacitor begins to charge. As the capacitor charges, it develops a potential difference that opposes the current, so the current decreases. As the current decreases, so does the rate of charging of the capacitor. The capacitor charges until $\Delta V_C = \mathcal{E}$, when the charging current ceases.

If we measure the current in the circuit and the potential difference across the capacitor as a function of time, we find that they vary according to the graphs in **FIGURE 23.41b**. The characteristic time for this charging circuit is the same as for the discharge, the time constant $\tau = RC$.

When the switch is first closed, the potential difference across the uncharged capacitor is zero, so the initial current is

$$I_0 = \frac{\mathcal{E}}{R}$$

The equations that describe the capacitor voltage and the current as a function of time are

$$I = I_0 e^{-t/RC}$$
$$\Delta V_C = \mathcal{E}(1 - e^{-t/RC}) \qquad (23.25)$$

Current and voltage while charging a capacitor

The time constant τ in an RC circuit can be used to control the behavior of a circuit. For example, a bike flasher uses an RC circuit that alternately charges and discharges, over and over, as a switch opens and closes. A separate circuit turns the light on when the capacitor voltage exceeds some threshold voltage and turns the light off when the capacitor voltage goes below this threshold. The time constant of the RC circuit determines how long the capacitor voltage stays above the threshold and thus sets the length of the flashes. More complex RC circuits provide timing in computers and other digital electronics. As we will see in the next section, we can also use RC circuits to model the transmission of nerve impulses, and the time constant will be a key factor in determining the speed at which signals can be propagated in the nervous system.

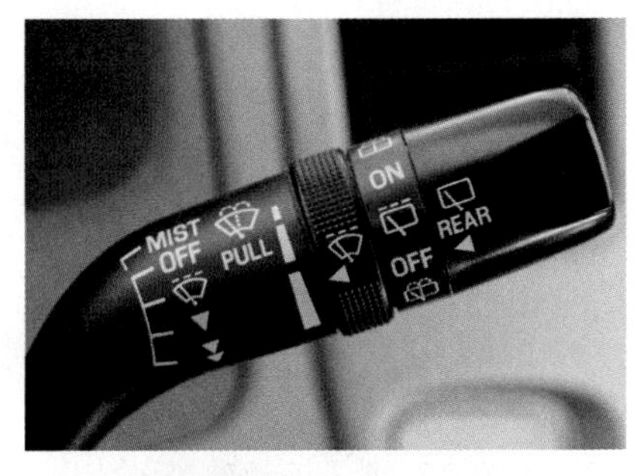
MIST
OFF
PULL
ON
REAR
OFF

A rainy-day *RC* circuit When you adjust the dial to control the delay of the intermittent windshield wipers in your car, you are adjusting a variable resistor in an RC circuit that triggers the wipers. Increasing the resistance increases the time constant and thus produces a longer delay between swipes of the blades. A light mist calls for a long time constant and thus a large resistance.

STOP TO THINK 23.7 The time constant for the discharge of this capacitor is

A. 5 s
B. 4 s
C. 2 s
D. 1 s

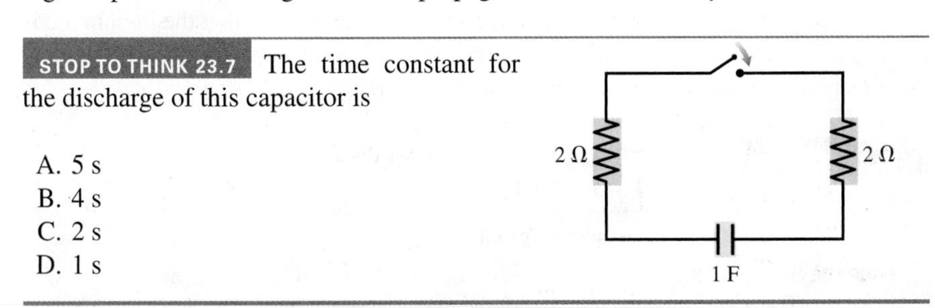
2 Ω
2 Ω
1 F

23.8 Electricity in the Nervous System BIO

In the late 1700s, the Italian scientist Galvani discovered that animal tissue has an electrical nature. He found that a frog's leg would twitch when stimulated with electricity, even when no longer attached to the frog. Further investigations by Galvani and others revealed that electrical signals can animate muscle cells, and that a small potential applied to the *axon* of a nerve cell can produce a signal that propagates down its length.

Our goal in this section will be to understand the nature of electrical signals in the nervous system. When your brain orders your hand to move, how does the signal get from your brain to your hand? Answering this question will use our knowledge of fields, potential, resistance, capacitance, and circuits, all of the knowledge and techniques that we have learned so far in Part VI.

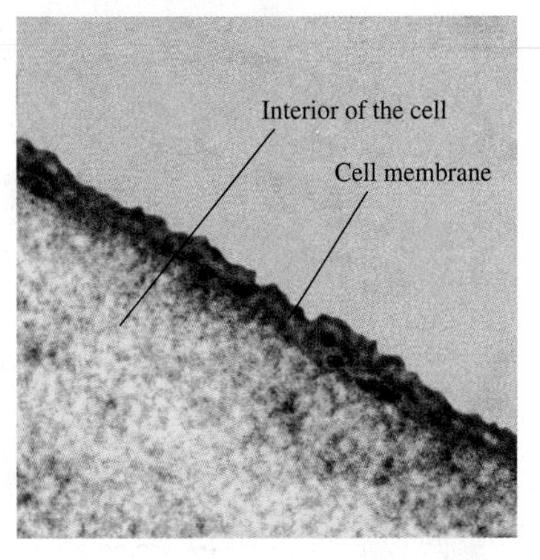
Interior of the cell
Cell membrane

A close-up view of the cell membrane, the insulating layer that divides the interior of a cell from the conducting fluid outside.

The Electrical Nature of Nerve Cells

We start our analysis with a very simple *model* of a nerve cell that allows us to describe its electrical properties. The model begins with a *cell membrane,* an insulating layer of lipids approximately 7 nm thick that separates regions of conducting fluid inside and outside the cell.

As we saw in ◀◀ SECTION 21.2, the cell membrane is not a passive structure. It has channels and pumps that transport ions between the inside and the outside of the cell. Ions, rather than electrons, are the charge carriers of the cell. In our simple model we will consider the transport of only two positive ions, sodium (Na^+) and potassium (K^+), though other ions are also important to cell function. Ions can slowly diffuse through the cell membrane. In addition, sodium and potassium ions are transported via the following structures:

- *Sodium-potassium exchange pumps.* These pump Na^+ ions out of the cell and K^+ ions in. In the cell's resting state, the concentration of sodium ions outside the cell is about ten times the concentration on the inside. Potassium ions are more concentrated on the inside.
- *Sodium and potassium channels.* These channels in the cell membrane are usually closed. When they are open, ions move in the direction of lower concentration. Thus Na^+ ions flow into the cell and K^+ ions flow out.

Our simple model, illustrated in **FIGURE 23.42**, ignores many of the features of real cells, but it allows us to accurately describe the reaction of nerve cells to a stimulus and the conduction of electrical signals.

FIGURE 23.42 A simple model of a nerve cell.

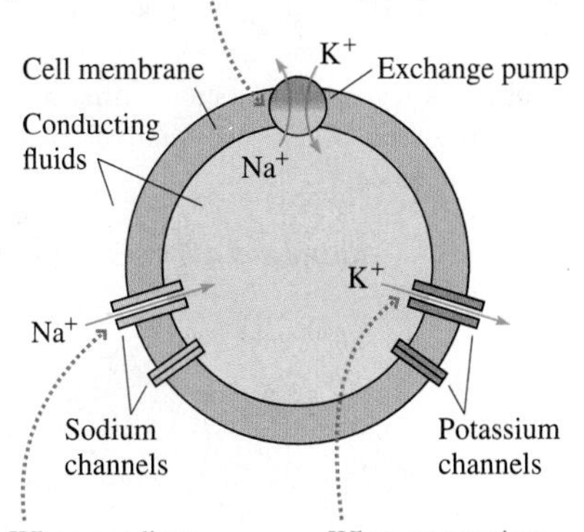
The pump moves sodium out of the cell and potassium in, so the sodium concentration is higher outside the cell, the potassium concentration is higher inside.
Cell membrane
K+
Exchange pump
Conducting fluids
Na+
K+
Na+
Sodium channels
Potassium channels
When a sodium channel is open, the higher sodium concentration outside the cell causes ions to flow into the cell.
When a potassium channel is open, the higher potassium concentration inside the cell causes ions to flow out of the cell.

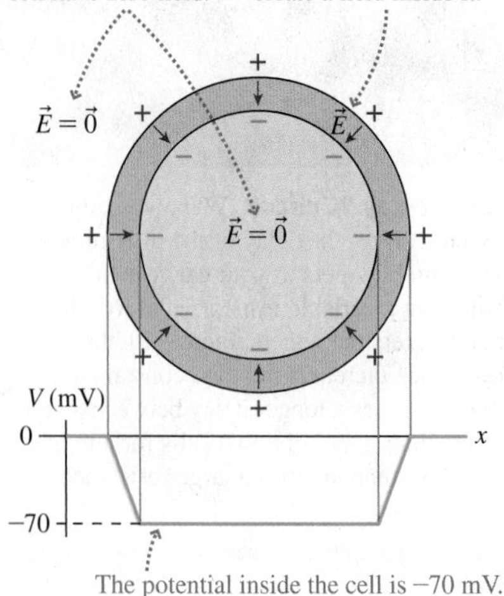
The conducting fluids inside and outside the cell have zero field.
Charges on the inside and outside surfaces of the insulating membrane create a field inside it.
$\vec{E} = \vec{0}$
$\vec{E}$
$\vec{E} = \vec{0}$
V (mV)
0
x
−70
The potential inside the cell is −70 mV.

FIGURE 23.43 The resting potential of a nerve cell.

The ion exchange pumps act much like the charge escalator of a battery, using chemical energy to separate charge by transporting ions. The transport and subsequent diffusion of charged ions lead to a separation in charge across the cell membrane. Consequently, **a living cell generates an emf.** This emf takes energy to create and maintain. The ion pumps that produce the emf of neural cells account for 25–40% of the energy usage of the brain.

The charge separation produces an electric field inside the cell membrane and results in a potential difference between the inside and the outside of the cell, as shown in **FIGURE 23.43**. The potential inside a nerve cell is typically 70 mV less than that outside the cell. This is called the cell's *resting potential.* Because this potential difference is produced by a charge separation across the membrane, we say that the membrane is *polarized.* Because the potential difference is entirely across the membrane, we may call this potential difference the *membrane potential.*

EXAMPLE 23.12 Electric field in a cell membrane BIO

The thickness of a typical nerve cell membrane is 7.0 nm. What is the electric field inside the membrane of a resting nerve cell?

PREPARE The potential difference across the membrane of a resting nerve cell is −70 mV. The inner and outer surfaces of the membrane are equipotentials. We learned in Chapter 21 that the electric field is perpendicular to the equipotentials and is related to the potential difference by $E = \Delta V/d$.

SOLVE The magnitude of the potential difference between the inside and the outside of the cell is 70 mV. The field strength is thus

$$E = \frac{\Delta V}{d} = \frac{70 \times 10^{-3}\text{ V}}{7.0 \times 10^{-9}\text{ m}} = 1.0 \times 10^{7}\text{ V/m}$$

The field points from positive to negative, so the electric field is

$$\vec{E} = (1.0 \times 10^{7}\text{ V/m, inward})$$

ASSESS This is a very large electric field; in air it would be large enough to cause a spark! But we expect the fields to be large to explain the cell's strong electrical character.

EXAMPLE 23.13 Finding the resistance of a cell membrane BIO

Charges can move across the cell membrane, so it is not a perfect insulator; the cell membrane will have a certain resistance. The resistivity of the cell membrane was given in Chapter 22 as $3.6 \times 10^{7}\ \Omega \cdot \text{m}$. What is the resistance of the 7.0-nm-thick membrane of a spherical cell with diameter 0.050 mm?

PREPARE The membrane potential will cause charges to move *through* the membrane. As we learned in ◀◀ SECTION 22.4, an object's resistance depends on its resistivity, length, and cross-section area. What this means for a cell membrane is noted in **FIGURE 23.44**.

SOLVE The area of the membrane is the surface area of a sphere, $4\pi r^2$. We can calculate the resistance using the equation for the resistance of a conductor of length L and cross-section area A from Chapter 22:

$$R_{\text{membrane}} = \frac{\rho L}{A} = \frac{(3.6 \times 10^{7}\ \Omega \cdot \text{m})(7.0 \times 10^{-9}\text{ m})}{4\pi(2.5 \times 10^{-5}\text{ m})^2}$$
$$= 3.2 \times 10^{7}\ \Omega = 32\text{ M}\Omega$$

ASSESS The resistance is quite high; the membrane is a good insulator, as we noted.

FIGURE 23.44 The cell membrane can be modeled as a resistor.

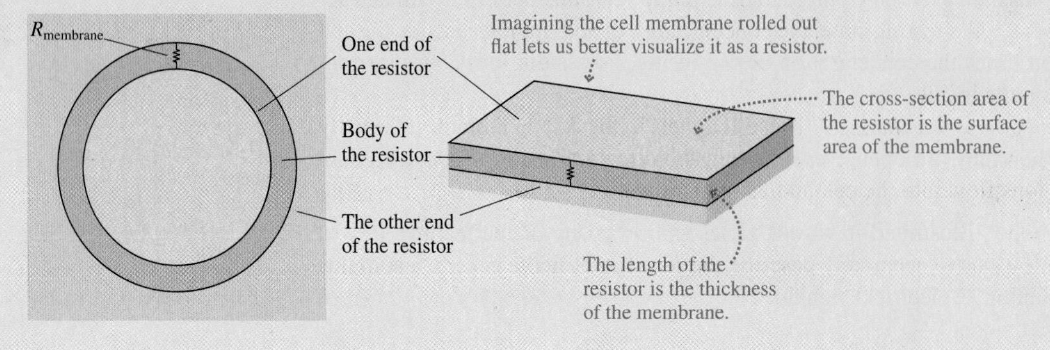
R_{membrane}
One end of the resistor
Imagining the cell membrane rolled out flat lets us better visualize it as a resistor.
Body of the resistor
The cross-section area of the resistor is the surface area of the membrane.
The other end of the resistor
The length of the resistor is the thickness of the membrane.

We can associate a resistance with the cell membrane, but we can associate other electrical quantities as well. The fluids inside and outside of the membrane are good conductors; they are separated by the membrane, which is not. Charges therefore accumulate on the inside and outside surfaces of the membrane. A cell thus looks like two charged conductors separated by an insulator—a capacitor.

EXAMPLE 23.14 **Finding the capacitance of a cell membrane** BIO

What is the capacitance of the membrane of the spherical cell specified in Example 23.13? The dielectric constant of a cell membrane is approximately 9.0.

PREPARE If we imagine opening up a cell membrane and flattening it out, we would get something that looks like a parallel-plate capacitor with the plates separated by a dielectric as illustrated in **FIGURE 23.45**. The relevant dimensions are the same as those in Example 23.13.

SOLVE The capacitance of the membrane is that of a parallel-plate capacitor filled with a dielectric, so its capacitance is κ times that of a parallel-plate capacitor without a dielectric, or $C_{membrane} = \kappa\epsilon_0 A/d$. Inserting the dimensions from Example 23.13, we find

$$C_{membrane} = \frac{\kappa\epsilon_0 A}{d} = \frac{9.0(8.85 \times 10^{-12}\ \mathrm{C^2/N \cdot m^2})4\pi(2.5 \times 10^{-5}\ \mathrm{m})^2}{7.0 \times 10^{-9}\ \mathrm{m}}$$
$$= 8.9 \times 10^{-11}\ \mathrm{F}$$

ASSESS Though the cell is small, the cell membrane has a reasonably large capacitance of ≈ 90 pF. This makes sense because the membrane is quite thin.

FIGURE 23.45 The cell membrane can also be modeled as a capacitor.

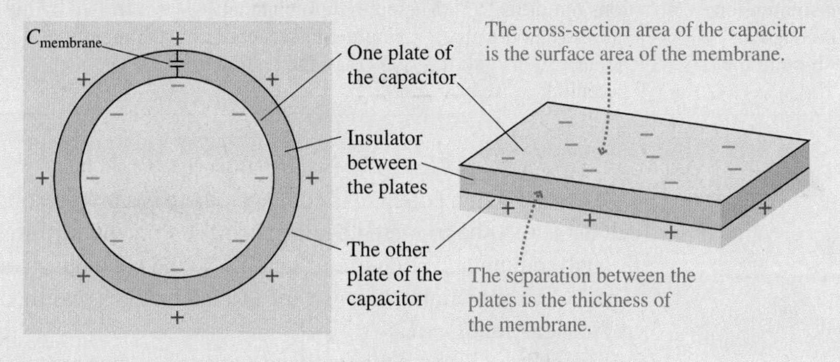

Because the cell membrane has both resistance and capacitance, it can be modeled as an *RC* circuit, as shown in **FIGURE 23.46**. The membrane, like any *RC* circuit, has a time constant. The previous examples calculated the resistance and capacitance of the 7.0-nm-thick membrane of a 0.050-mm-diameter cell. We can use these numbers to compute the membrane's time constant:

$$\tau = RC = (3.2 \times 10^7\ \Omega)(8.9 \times 10^{-11}\ \mathrm{F}) = 2.8 \times 10^{-3}\ \mathrm{s} \approx 3\ \mathrm{ms}$$

Indeed, if we raise the membrane potential of a real nerve cell by 10 mV (large enough to easily measure but not enough to trigger a response in the cell), the potential will decay back to its resting value with a time constant of a few ms.

But the real action happens when some stimulus *is* large enough to trigger a response in the cell. In this case, ion channels open and the potential changes in much less time than the cell's time constant, as we will see next.

FIGURE 23.46 The cell membrane can be modeled as an *RC* circuit.

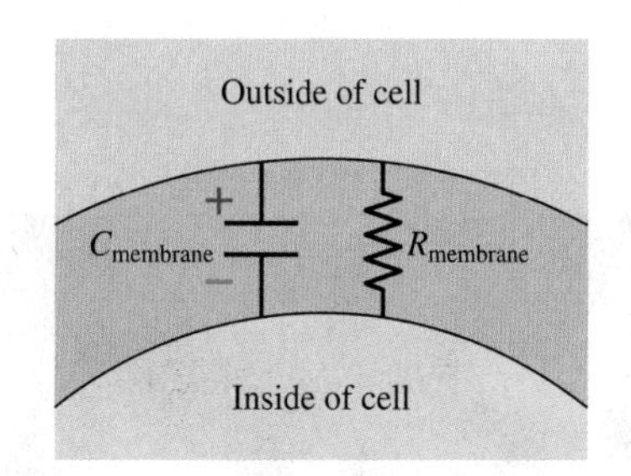

The Action Potential

Suppose a nerve cell is sitting quietly at its resting potential. The membrane potential is approximately -70 mV. However, this potential can change drastically in response to a *stimulus*. *Neurons*—nerve cells—can be stimulated by neurotransmitter chemicals released at synapse junctions. A neuron can also be electrically stimulated by a changing potential, which is why Galvani saw the frog's leg jump. Whatever the stimulus, the result is a rapid change called an *action potential;* this is the "firing" of a nerve cell. There are three phases in the action potential, as outlined on the next page.

The action potential

Depolarization	Repolarization	Reestablishing resting potential
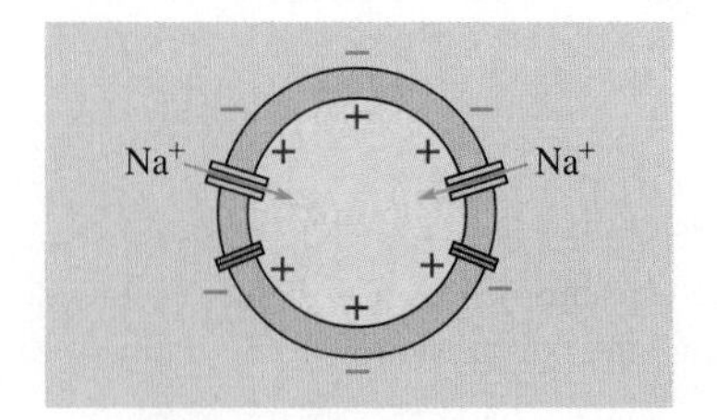	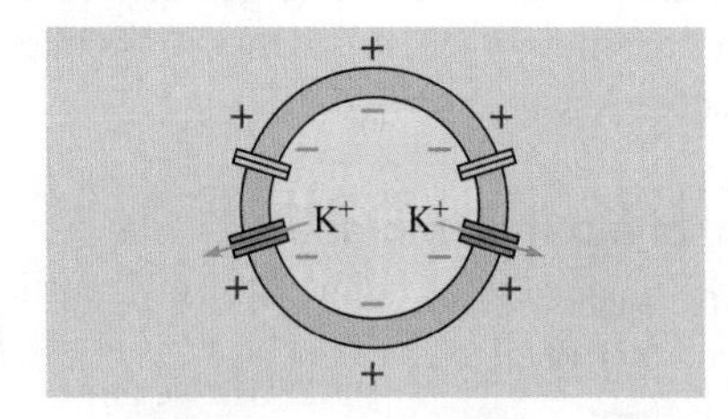	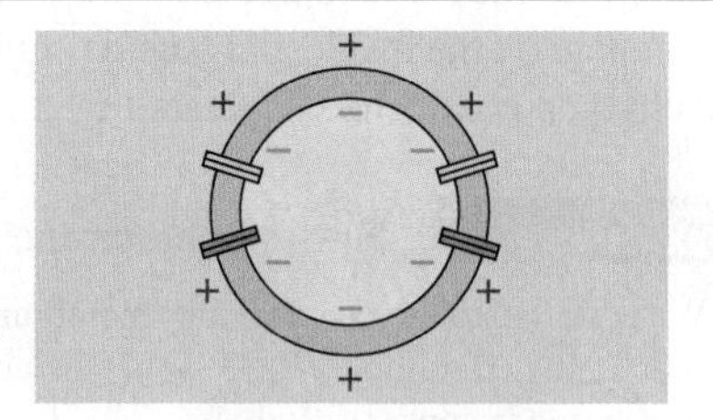
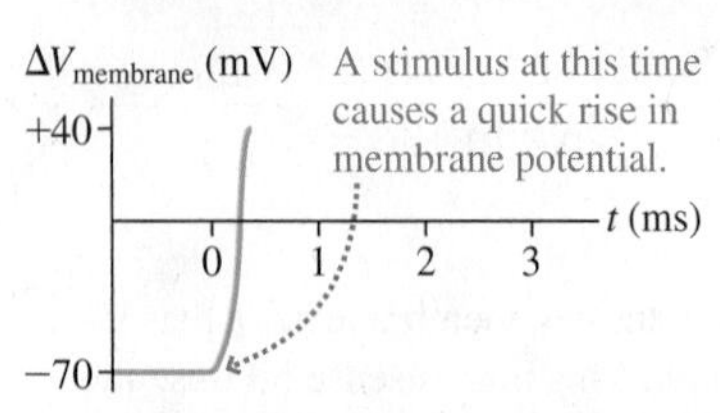	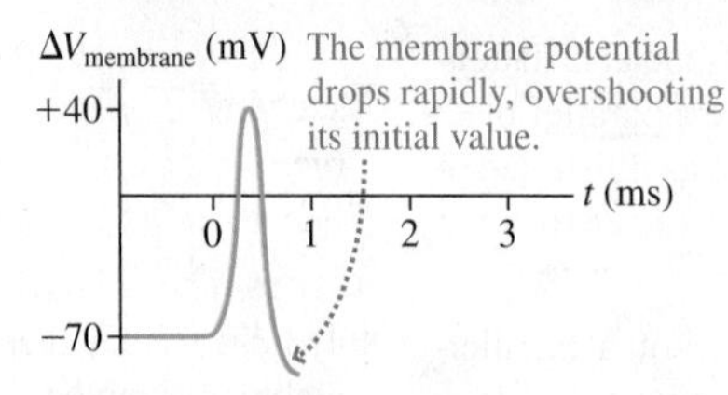	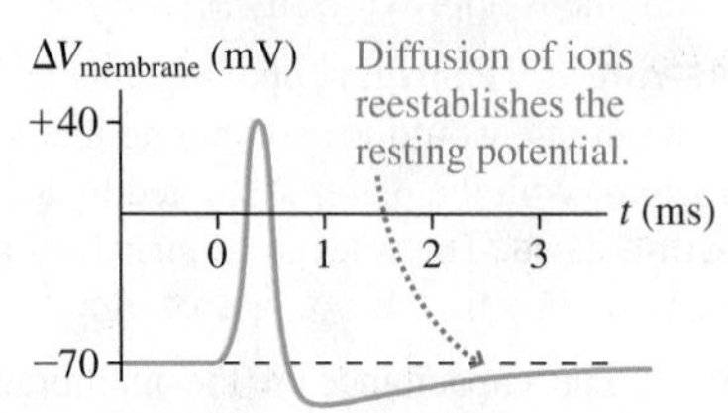
A cell *depolarizes* when a stimulus causes the opening of the sodium channels. The concentration of sodium ions is much higher outside the cell, so positive sodium ions flow into the cell, rapidly raising its potential to 40 mV, at which point the sodium channels close.	The cell *repolarizes* as the potassium channels open. The higher potassium concentration inside the cell drives these ions out of the cell. The potassium channels close when the membrane potential reaches about −80 mV, slightly *less* than the resting potential.	The reestablishment of the resting potential after the sodium and potassium channels close is a relatively slow process controlled by the motion of ions across the membrane.

After the action potential is complete, there is a brief resting period, after which the cell is ready to be triggered again. The action potential is driven by ionic conduction through sodium and potassium channels, so the potential changes are quite rapid. The time for the potential to rise and then to fall is much less than the 3 ms time constant of the membrane.

The above discussion concerned nerve cells, but muscle cells undergo a similar cycle of depolarization and repolarization. The resulting potential changes are responsible for the signal that is measured by an electrocardiogram, which we learned about in ◀◀SECTION 21.6. The potential differences in the human body are small because the changes in potential are small. But some fish have electric organs in which the action potentials of thousands of specially adapted cells are added in series, leading to very large potential differences—hundreds of volts in the case of the electric eel.

◀ **Touchless typing** Different thought processes lead to different patterns of action potentials among the many neurons of the brain. The electrical activity of the cells and the motion of ions through the conducting fluid surrounding them lead to measurable differences in potential between points on the scalp. You can't use these potential differences to "read someone's mind," but it is possible to program a computer to recognize patterns and perform actions when they are detected. This man is using his thoughts—and the resulting pattern of electric potentials—to select and enter letters.

EXAMPLE 23.15 Counting ions through a channel BIO

Investigators can measure the ion flow through a single ion channel with the *patch clamp* technique, as illustrated in FIGURE 23.47. A micropipette, a glass tube $\approx 1\ \mu m$ in diameter, makes a seal on a patch of cell membrane that includes one sodium channel. This tube is filled with a conducting saltwater solution, and a very sensitive ammeter measures the current as sodium ions flow into the cell. A sodium channel passes an average current of 4.0 pA during the 0.40 ms that the channel is open during an action potential. How many sodium ions pass through the channel?

FIGURE 23.47 Measuring the current in a single sodium channel.

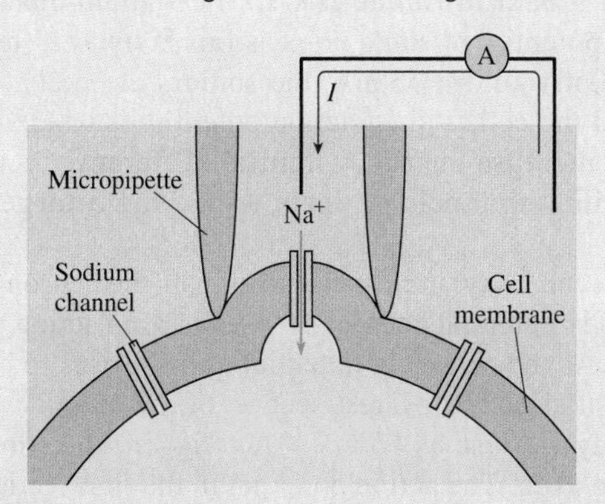
A
I
Micropipette
Na+
Sodium channel
Cell membrane

PREPARE Current is the rate of flow of charge. Each ion has charge $q = e$.

SOLVE In ◀◀ SECTION 22.2, we saw that the charge delivered by a steady current in time Δt is $Q = I\Delta t$. The amount of charge flowing through the channel in $\Delta t = 4.0 \times 10^{-4}$ s is

$$Q = I\Delta t = (4.0 \times 10^{-12}\text{ A})(4.0 \times 10^{-4}\text{ s}) = 1.6 \times 10^{-15}\text{ C}$$

This charge is due to N ions, each with $q = e$, so the number of ions is

$$N = \frac{Q}{e} = \frac{1.6 \times 10^{-15}\text{ C}}{1.6 \times 10^{-19}\text{ C}} = 10{,}000$$

ASSESS The number of ions flowing through one channel is not large, but a cell has a great many channels. The patch clamp technique and other similar procedures have allowed investigators to elucidate the details of the response of the cell membrane to a stimulus.

The Propagation of Nerve Impulses

Let's return to the question posed at the start of the section: How is a signal transmitted from the brain to a muscle in the hand? The primary cells of the nervous system responsible for signal transmission are known as *neurons*. The transmission of a signal to a muscle is the function of a *motor neuron,* whose structure is sketched in **FIGURE 23.48**. The transmission of signals takes place along the *axon* of the neuron, a long fiber—up to 1 m in length—that connects the cell body to a muscle fiber. This particular neuron has a myelin sheath around the axon, though not all neurons do.

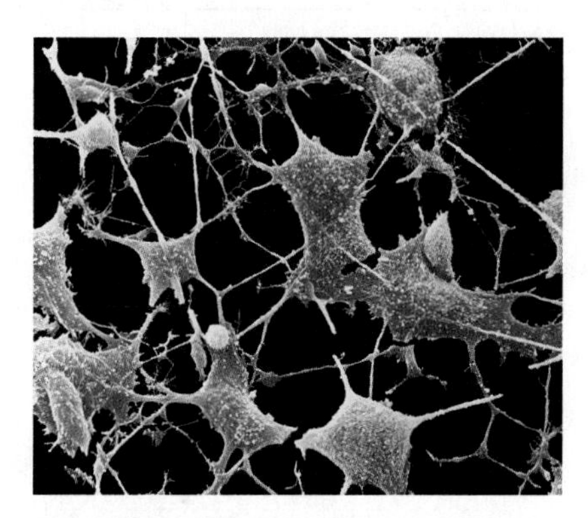

The axons connecting these nerve cells are clearly visible in the micrograph.

How is a signal transmitted along an axon? The axon is long enough that different points on its membrane may have different potentials. When one point on the axon's membrane is stimulated, the membrane will depolarize at this point. The resulting action potential may trigger depolarization in adjacent parts of the membrane. Stimulating the axon's membrane at one point can trigger a *wave* of action potential—a nerve impulse—that travels along the axon. When this signal reaches a muscle cell, the muscle cell depolarizes and produces a mechanical response.

FIGURE 23.48 A motor neuron.

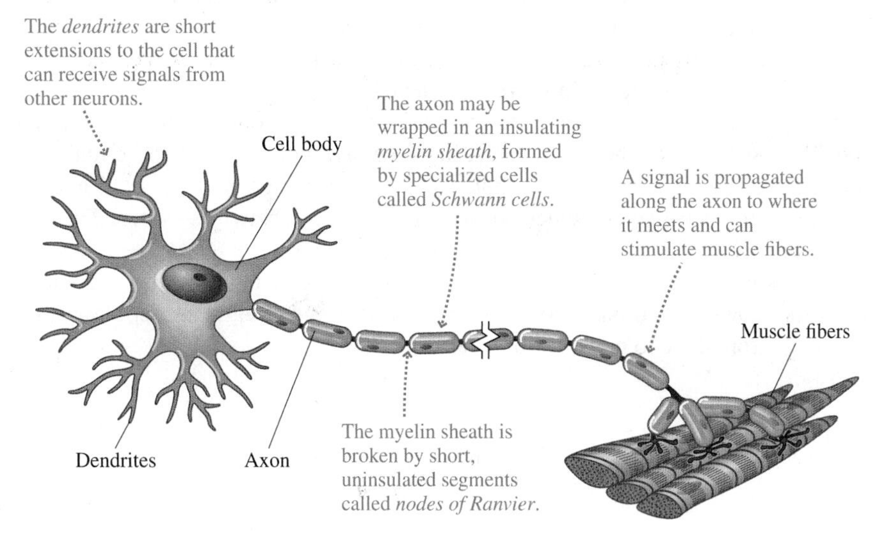
The *dendrites* are short extensions to the cell that can receive signals from other neurons.
Cell body
The axon may be wrapped in an insulating *myelin sheath*, formed by specialized cells called *Schwann cells*.
A signal is propagated along the axon to where it meets and can stimulate muscle fibers.
Muscle fibers
Dendrites
Axon
The myelin sheath is broken by short, uninsulated segments called *nodes of Ranvier*.

FIGURE 23.49 Propagation of a nerve impulse.

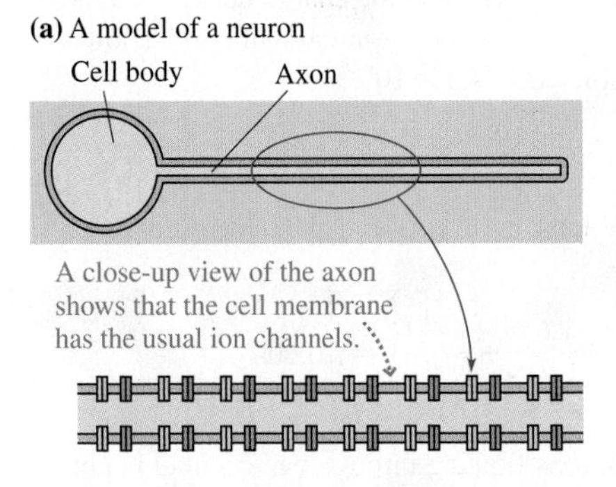
(a) A model of a neuron
Cell body
Axon
A close-up view of the axon shows that the cell membrane has the usual ion channels.

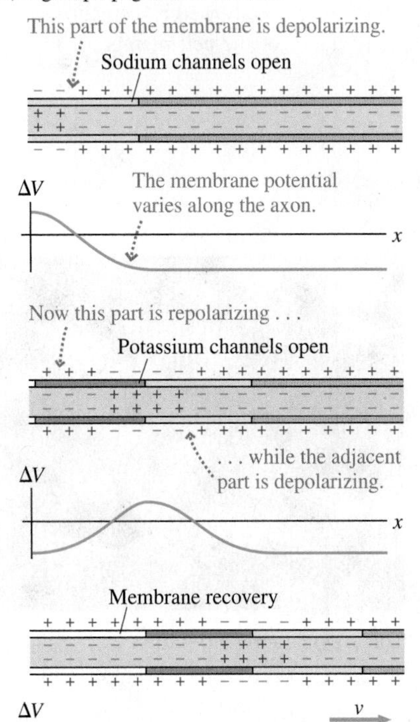
(b) Signal propagation in the axon
This part of the membrane is depolarizing.
Sodium channels open
ΔV
The membrane potential varies along the axon.
x
Now this part is repolarizing . . .
Potassium channels open
. . . while the adjacent part is depolarizing.
ΔV
x
Membrane recovery
ΔV
v
x
A changing potential at one point triggers the membrane to the right, leading to a wave of action potential that moves along the axon.

Let's look at this process in more detail. We will start with a simple model of an axon with no myelin sheath in FIGURE 23.49a. The sodium channels are normally closed, but if the potential at some point is raised by ≈ 15 mV, from the resting potential of -70 mV to ≈ -55 mV, the sodium channels suddenly open, sodium ions rush into the cell, and an action potential is triggered. This is the key idea: **A small increase in the potential difference across the membrane causes the sodium channels to open, triggering a large action-potential response.**

This process begins at the cell body, in response to signals the neuron receives at its dendrites. If the cell body potential goes up by ≈ 15 mV, an action potential is initiated in the cell body. As the cell body potential quickly rises to a peak of $+40$ mV, it causes the potential on the nearest section of the axon—where the axon attaches to the cell body—to rise by 15 mV. This triggers an action potential in this first section of the axon. The action potential in the first section of the axon triggers an action potential in the next section of the axon, which triggers an action potential in the next section, and so on down the axon until reaching the end.

As FIGURE 23.49b shows, this process causes a wave of action potential to propagate down the axon. The signal moves relatively slowly. At each point on the membrane, channels must open and ions must diffuse through, which takes time. On a typical axon with no myelin sheath, the action potential propagates at a speed of about 1 m/s. If all nerve signals traveled at this speed, a signal telling your hand to move would take about 1 s to travel from your brain to your hand. Clearly, at least some neurons in the nervous system must transmit signals at a higher speed than this!

One way to make the signals travel more quickly is to increase an axon's diameter. The giant axon in the squid triggers a rapid escape response when the squid is threatened. This axon may have a diameter of 1 mm, a thousand times that of a typical axon, providing for the necessary rapid signal transmission. But your nervous system consists of 300 billion neurons, and they can't all be 1 mm in diameter—there simply isn't enough space in your body. In your nervous system, higher neuron signal speed is achieved in a totally different manner.

Increasing Speed by Insulation

The axons of motor neurons and most other neurons in your body can transmit signals at very high speeds because they are insulated with a myelin sheath. Look back at the structure of a motor neuron in Figure 23.48. Schwann cells wrap the axon with myelin, insulating it electrically and chemically, with breaks at the nodes of Ranvier. The ion channels are concentrated in these nodes because this is the only place where the extracellular fluid is in contact with the cell membrane. In an insulated axon, a signal propagates by jumping from one node to the next. This process is called *saltatory conduction,* from the Latin *saltare,* "to leap."

FIGURE 23.50a shows an electrical model for saltatory conduction in which we model one segment of the axon, between successive nodes, as an RC circuit. As shown in FIGURE 23.50b, when the potential at a node increases by ≈ 15 mV an action potential is triggered, which we model as closing a battery's switch. The emf of this battery charges the capacitance C through resistance R. As the capacitor charges, its potential soon reaches 15 mV, triggering the action potential in the next node. As shown in FIGURE 23.51, this process continues, with the depolarization "jumping" from node to node down the axon.

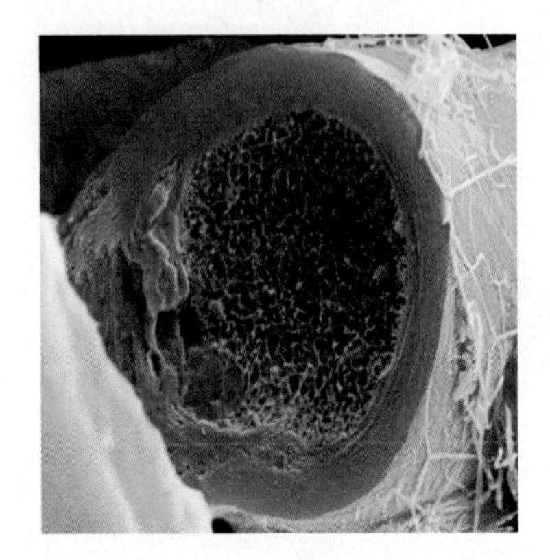

◀ The interior of this axon is insulated from the surrounding intercellular fluids by a thick myelin sheath, clearly visible in the cross-section view.

FIGURE 23.50 A circuit model of nerve-impulse propagation along myelinated axons.

(a) An electrical model of a myelinated axon

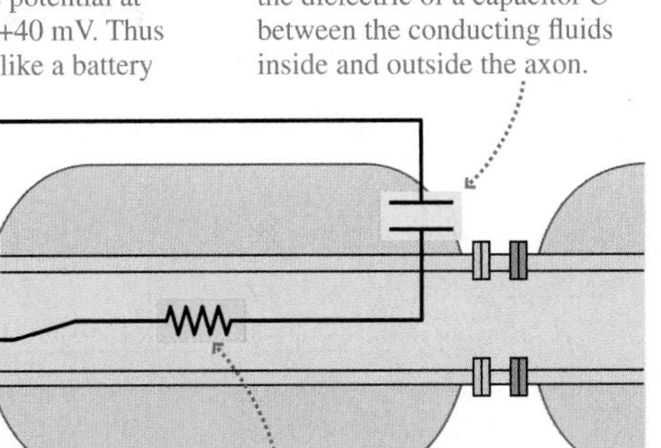

(b) Signal propagation in the myelinated axon

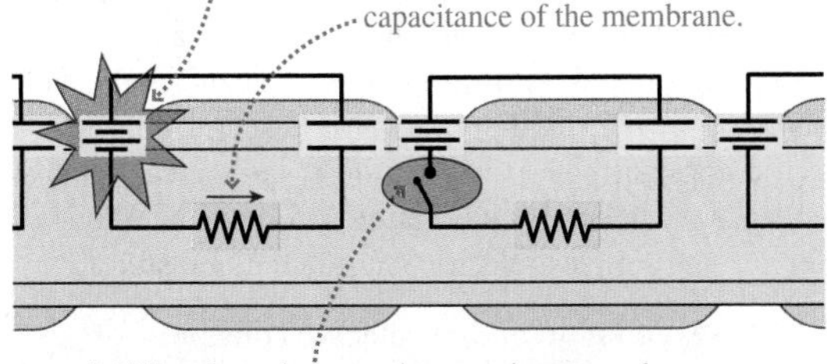

1. An action potential is triggered at this node; we close the switch.
2. Once the switch is closed, the action potential emf drives a current down the axon and charges the capacitance of the membrane.
3. When the voltage on the capacitor exceeds a threshold, it triggers an action potential at this node—the next switch is closed.

STOP TO THINK 23.8 In the axon model of Figure 23.50, if the thickness of the myelin sheath were increased, the propagation speed of nerve impulses would

A. Increase. B. Decrease. C. Remain the same.

How rapidly does a pulse move down a myelinated axon? The critical time for propagation is the time constant $\tau = RC$ for charging the capacitance of the segments of the axon.

The resistance of an axon between one node and the next is $\approx 25\ \text{M}\Omega$. The myelin insulation increases the separation between the inner conducting fluid and the outer conducting fluid. Because the capacitance of a capacitor depends inversely on the electrode spacing d, the myelin reduces the capacitance of the membrane from the ≈ 90 pF we calculated earlier to ≈ 1.6 pF per segment. With these values, the time constant for charging the capacitor in one segment is

$$\tau = R_{\text{axon}} C_{\text{membrane}} = (25 \times 10^6\ \Omega)(1.6 \times 10^{-12}\ \text{F}) = 40\ \mu\text{s}$$

We've modeled the axon as a series of such segments, and the time constant is a good estimate of how much time it takes for a signal to jump from one node to the next. Because the nodes of Ranvier are spaced about 1 mm apart, the speed at which the nerve impulse travels down the axon is approximately

$$v = \frac{L_{\text{node}}}{\tau} = \frac{1.0 \times 10^{-3}\ \text{m}}{40 \times 10^{-6}\ \text{s}} = 25\ \text{m/s}$$

Although our model of nerve-impulse propagation is very simple, this predicted speed is just about right for saltatory conduction of signals in myelinated axons. This speed is 25 times faster than that in unmyelinated axons; at this speed, your brain can send a signal to your hand in $\approx \frac{1}{25}$ s.

Your electrical nature might not be as apparent as that of the electric eel, but the operation of your nervous system is inherently electrical. When you decide to move your hand, the signal from your brain travels to your hand in a process that is governed by the electrical nature of the cells in your body.

FIGURE 23.51 Nerve propagation along a myelinated axon.

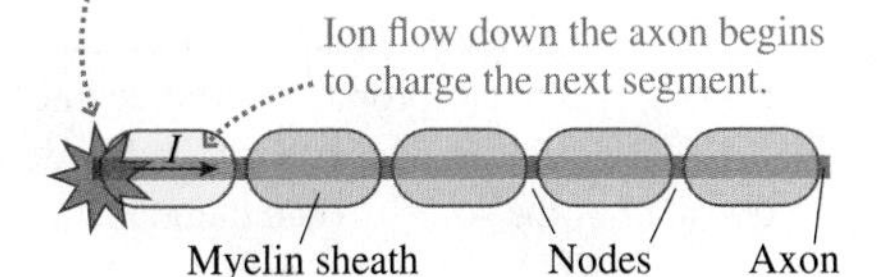

The ion channels at this node are triggered, generating an action potential.
Ion flow down the axon begins to charge the next segment.
I
Myelin sheath
Nodes
Axon

Once the potential reaches a threshold value, an action potential is triggered at the next node.

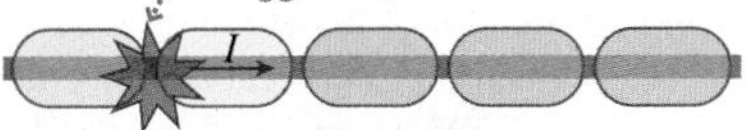

The process continues, with the signal triggering each node in sequence . . .

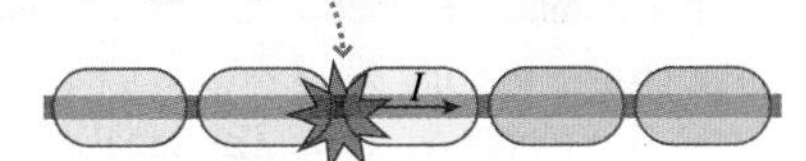

. . . so the signal moves rapidly along the axon from node to node.

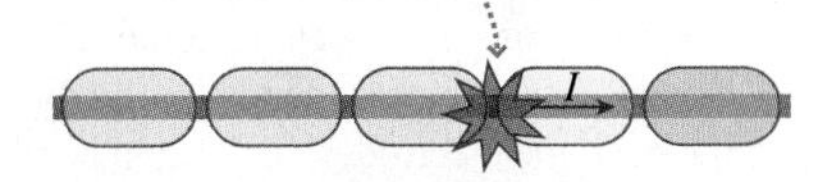

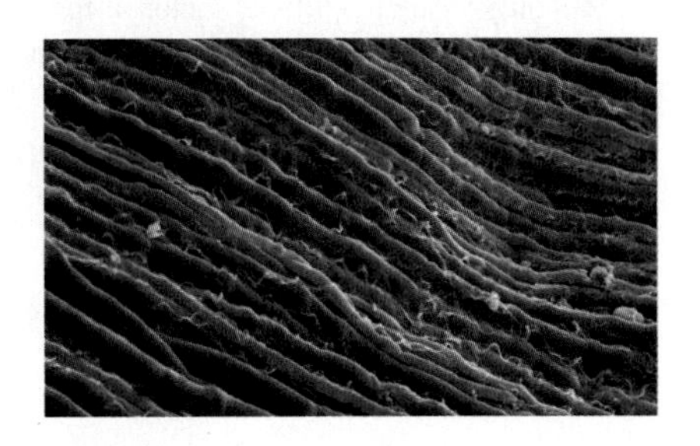

▶ Myelinated axons in the spinal cord carry electrical signals between the brain and body. You can see the nodes of Ranvier on some of the axons.

INTEGRATED EXAMPLE 23.16 Soil moisture measurement

The moisture content of soil is given in terms of its *volumetric fraction,* the ratio of the volume of water in the soil to the volume of the soil itself. Making this measurement directly is quite time-consuming, so soil scientists are eager to find other means to reliably measure soil moisture. Water has a very large dielectric constant, so the dielectric constant of soil increases as its moisture content increases. **FIGURE 23.52** shows data for the dielectric constant of soil versus the volumetric fraction; the increase in dielectric constant with soil moisture is quite clear. This strong dependence of dielectric constant on soil moisture allows a very sensitive—and simple—electrical test of soil moisture.

FIGURE 23.52 Variation of the dielectric constant with soil moisture.

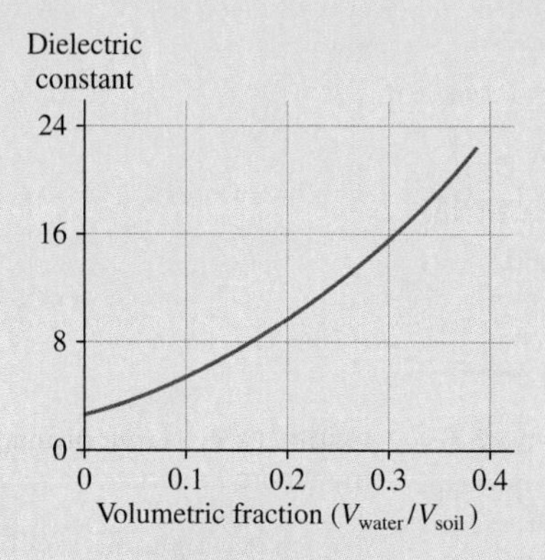

A soil moisture meter has a probe with two separated electrodes. When the probe is inserted into the soil, the electrodes form a capacitor whose capacitance depends on the dielectric constant of the soil between them. A circuit charges the capacitor probe to 3.0 V, then discharges it through a resistor. The decay time depends on the capacitance—and thus the soil moisture—so a measurement of the time for the capacitor to discharge allows a determination of the amount of moisture in the soil.

In air, the probe's capacitance takes 15 μs to discharge from 3.0 V to 1.0 V. In one particular test, when the probe was inserted into the ground, this discharge required 150 μs. What was the approximate volumetric fraction of water for the soil in this test?

PREPARE FIGURE 23.53 is a sketch of the measurement circuit of the soil moisture meter. The capacitor is first charged, then connected across a resistor to form an *RC* circuit. The decay of the capacitor voltage is governed by the time constant for the circuit. The time constant depends on the resistance and on the electrode capacitance, which depends on the dielectric constant of the soil between the electrodes.

The capacitance of the probe in air is that of a parallel-plate capacitor: $C_{air} = \epsilon_0 A/d$. The capacitance of the probe in soil differs only by the additional factor of the dielectric constant of the medium between the plates—in this case, the soil:

$$C_{soil} = \kappa_{soil}\left(\frac{\epsilon_0 A}{d}\right) = \kappa_{soil} C_{air}$$

FIGURE 23.53 The measurement circuit of the soil moisture meter.

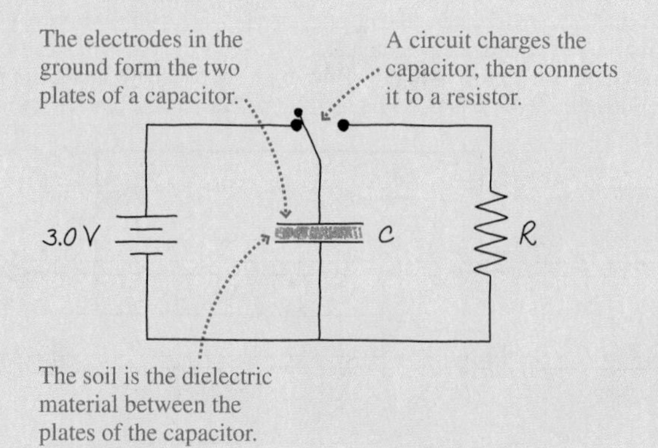

The ratio of the capacitance values gives the dielectric constant:

$$\kappa_{soil} = \frac{C_{soil}}{C_{air}}$$

Once we know the dielectric constant, we can determine the volumetric fraction of water from the graph.

SOLVE We are given the times for the decay in air and in soil, but not the capacitance or the resistance of the probe. That's not a problem, though; we don't actually need the capacitance, only the *ratio* of the capacitances in air and in soil. Equation 23.22 gives the voltage decay of an *RC* circuit: $\Delta V_C = (\Delta V_C)_0 e^{-t/RC}$. In air, the decay is

$$1.0 \text{ V} = (3.0 \text{ V})e^{-(15\times10^{-6} \text{ s})/RC_{air}}$$

In soil, the decay is

$$1.0 \text{ V} = (3.0 \text{ V})e^{-(150\times10^{-6} \text{ s})/RC_{soil}}$$

Because the starting and ending points for the decay are the same, the exponents of the two expressions must be equal:

$$\frac{15 \times 10^{-6} \text{ s}}{RC_{air}} = \frac{150 \times 10^{-6} \text{ s}}{RC_{soil}}$$

We can solve this for the ratio of the capacitances in soil and air:

$$\frac{C_{soil}}{C_{air}} = \frac{150 \times 10^{-6} \text{ s}}{15 \times 10^{-6} \text{ s}} = 10$$

We saw above that this ratio is the dielectric constant, so $\kappa_{soil} = 10$. We then use the graph of Figure 23.52 to determine that this dielectric constant corresponds to a volumetric water fraction of approximately 0.20.

ASSESS The decay times in air and in soil differ by a factor of 10, so the capacitance in the soil is much larger than that in air. This implies a large dielectric constant, meaning that there is a lot of water in the soil. A volumetric fraction of 0.20 means that 20% of the soil's volume is water (that is, 1.0 cm^3 of soil contains 0.20 cm^3 of water)—which is quite a bit, so our result seems reasonable.

SUMMARY

Goal: To understand the fundamental physical principles that govern electric circuits.

GENERAL PRINCIPLES

Kirchhoff's loop law

For a closed loop:

- Assign a direction to the current.
- Add potential differences around the loop:

$$\sum_i \Delta V_i = 0$$

Kirchhoff's junction law

For a junction:

$$\sum I_{in} = \sum I_{out}$$

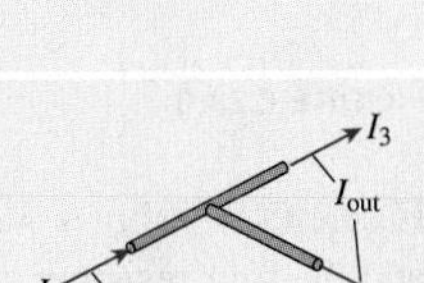
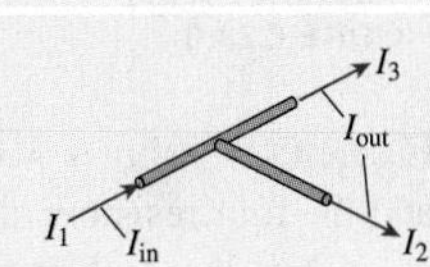

Analyzing Circuits

PREPARE Draw a circuit diagram.

SOLVE *Break the circuit down:*

- Reduce the circuit to the smallest possible number of equivalent resistors.
- Find the current and potential difference.

Rebuild the circuit:

- Find current and potential difference for each resistor.

ASSESS Verify that

- The sum of the potential differences across series resistors matches that for the equivalent resistor.
- The sum of the currents through parallel resistors matches that for the equivalent resistor.

IMPORTANT CONCEPTS

Series elements

A series connection has no junction. The current in each element is the same.

Resistors in series can be reduced to an equivalent resistance:

$$R_{eq} = R_1 + R_2 + R_3 + \cdots$$

Capacitors in series can be reduced to an equivalent capacitance:

$$C_{eq} = \left(\frac{1}{C_1} + \frac{1}{C_2} + \frac{1}{C_3} + \cdots\right)^{-1}$$

Parallel elements

Elements connected in parallel are connected by wires at both ends. The potential difference across each element is the same.

Resistors in parallel can be reduced to an equivalent resistance:

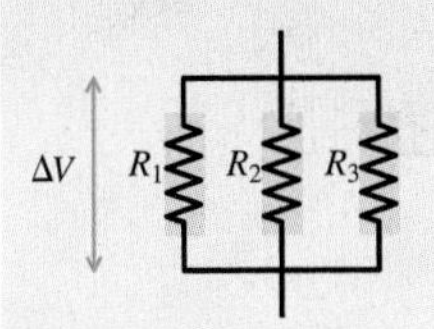

$$R_{eq} = \left(\frac{1}{R_1} + \frac{1}{R_2} + \frac{1}{R_3} + \cdots\right)^{-1}$$

Capacitors in parallel can be reduced to an equivalent capacitance:

$$C_{eq} = C_1 + C_2 + C_3 + \cdots$$

APPLICATIONS

RC circuits

The discharge of a capacitor through a resistor is an exponential decay:

$$\Delta V_C = (\Delta V_C)_0 e^{-t/RC}$$

The **time constant** for the decay is

$$\tau = RC$$

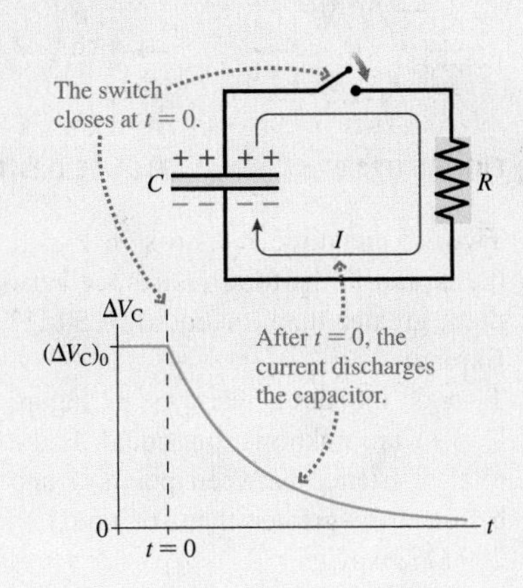

Electricity in the nervous system

Cells in the nervous system maintain a negative potential inside the cell membrane. When triggered, the membrane depolarizes and generates an *action potential.*

An action potential travels as a wave along the axon of a neuron. More rapid saltatory conduction can be achieved by insulating the axon with myelin, causing the action potential to jump from node to node.

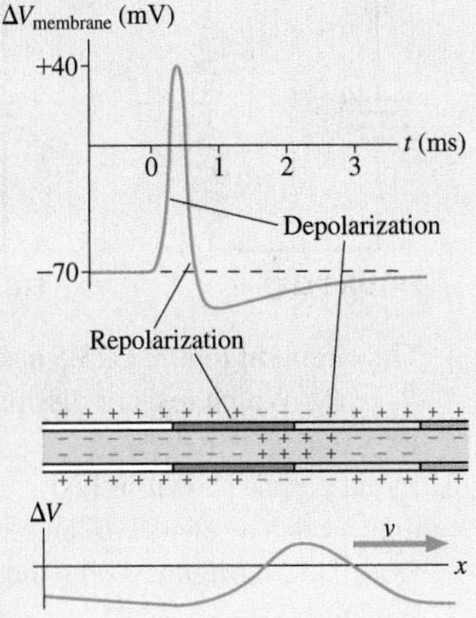

Problem difficulty is labeled as | (straightforward) to ||||| (challenging). Problems labeled INT integrate significant material from earlier chapters; BIO are of biological or medical interest.

For assigned homework and other learning materials, go to MasteringPhysics®

Scan this QR code to launch a **Video Tutor Solution** that will help you solve problems for this chapter.

QUESTIONS

Conceptual Questions

1. The tip of a flashlight bulb is touching the top of a 3 V battery as shown in Figure Q23.1. Does the bulb light? Why or why not?

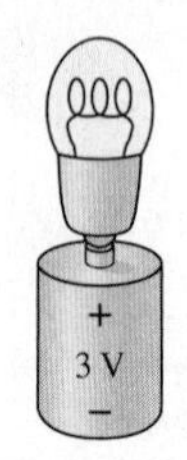

FIGURE Q23.1

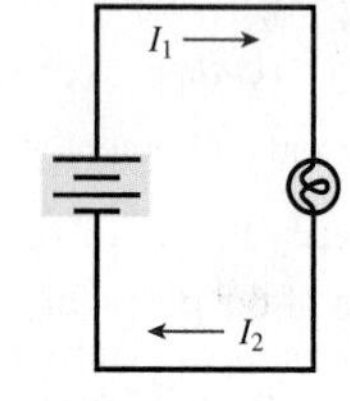

FIGURE Q23.2

2. A flashlight bulb is connected to a battery and is glowing; the circuit is shown in Figure Q23.2. Is current I_2 greater than, less than, or equal to current I_1? Explain.
3. Current I_{in} flows into three resistors connected together one after the other as shown in Figure Q23.3. The accompanying graph shows the value of the potential as a function of position.
   a. Is I_{out} greater than, less than, or equal to I_{in}? Explain.
   b. Rank in order, from largest to smallest, the three resistances R_1, R_2, and R_3. Explain.

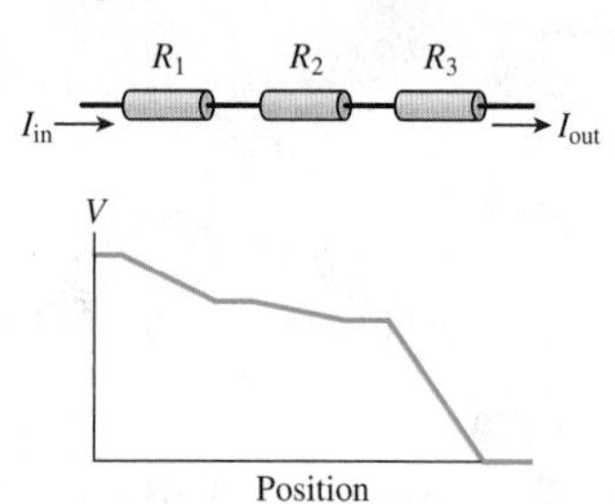

FIGURE Q23.3

4. The circuit in Figure Q23.4 has two resistors, with $R_1 > R_2$. Which resistor dissipates the larger amount of power? Explain.

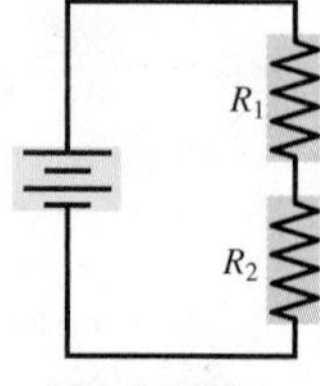

FIGURE Q23.4

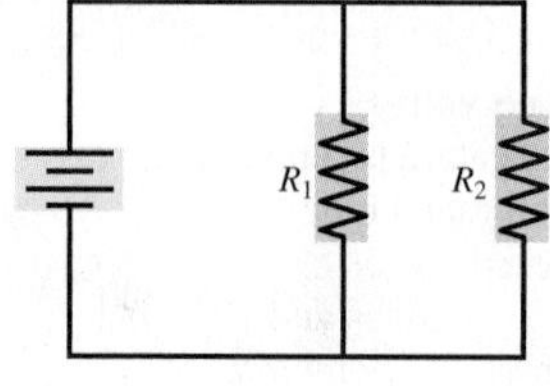

FIGURE Q23.5

5. The circuit in Figure Q23.5 has a battery and two resistors, with $R_1 > R_2$. Which resistor dissipates the larger amount of power? Explain.
6. In the circuit shown in Figure Q23.6, bulbs A and B are glowing. Then the switch is closed. What happens to each bulb? Does it get brighter, stay the same, get dimmer, or go out? Explain.

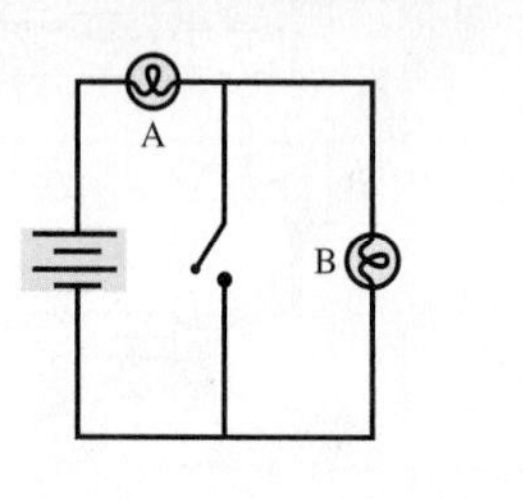

FIGURE Q23.6

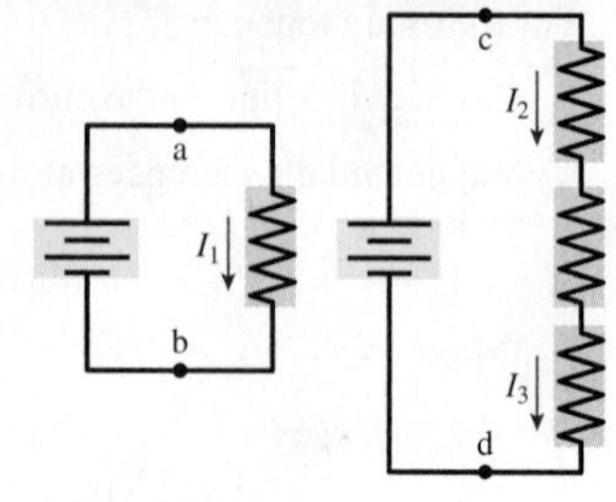

FIGURE Q23.7

7. Figure Q23.7 shows two circuits. The two batteries are identical and the four resistors all have exactly the same resistance.
   a. Is ΔV_{ab} larger than, smaller than, or equal to ΔV_{cd}? Explain.
   b. Rank in order, from largest to smallest, the currents I_1, I_2, and I_3. Explain.
8. Figure Q23.8 shows two circuits. The two batteries are identical and the four resistors all have exactly the same resistance.
   a. Compare ΔV_{ab}, ΔV_{cd}, and ΔV_{ef}. Are they all the same? If not, rank them in order from largest to smallest. Explain.
   b. Rank in order, from largest to smallest, the five currents I_1 to I_5. Explain.

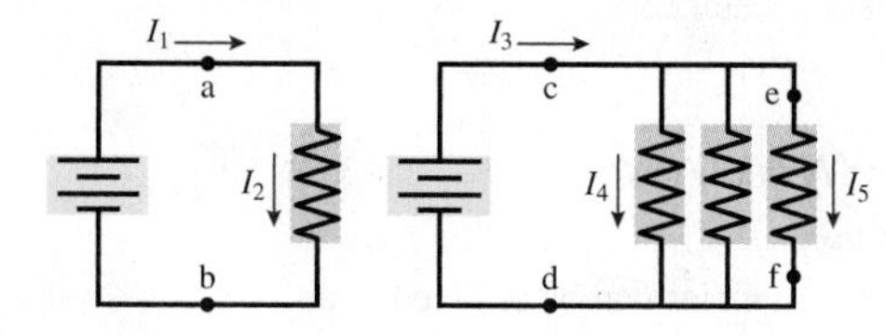

FIGURE Q23.8

9. a. In Figure Q23.9, what fraction of current I goes through the 3 Ω resistor?
   b. If the 9 Ω resistor is replaced with a larger resistor, will the fraction of current going through the 3 Ω resistor increase, decrease, or stay the same?

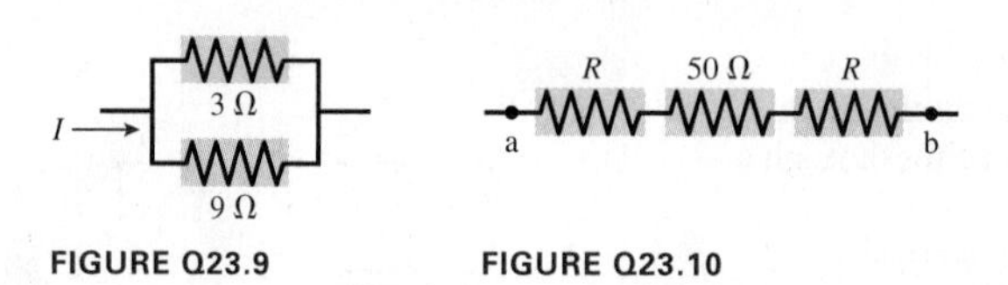

FIGURE Q23.9 FIGURE Q23.10

10. Two of the three resistors in Figure Q23.10 are unknown but equal. Is the total resistance between points a and b less than, greater than, or equal to 50 Ω? Explain.
11. Two of the three resistors in Figure Q23.11 are unknown but equal. Is the total resistance between points a and b less than, greater than, or equal to 200 Ω? Explain.

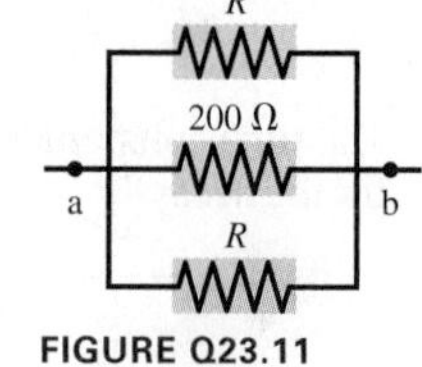

FIGURE Q23.11

12. Rank in order, from largest to smallest, the currents I_1, I_2, and I_3 in the circuit diagram in Figure Q23.12.

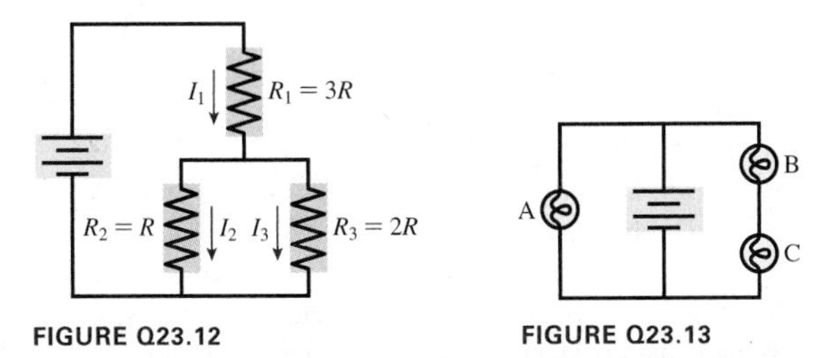

FIGURE Q23.12 FIGURE Q23.13

13. The three bulbs in Figure Q23.13 are identical. Rank the bulbs from brightest to dimmest. Explain.
14. The four bulbs in Figure Q23.14 are identical. Rank the bulbs from brightest to dimmest. Explain.

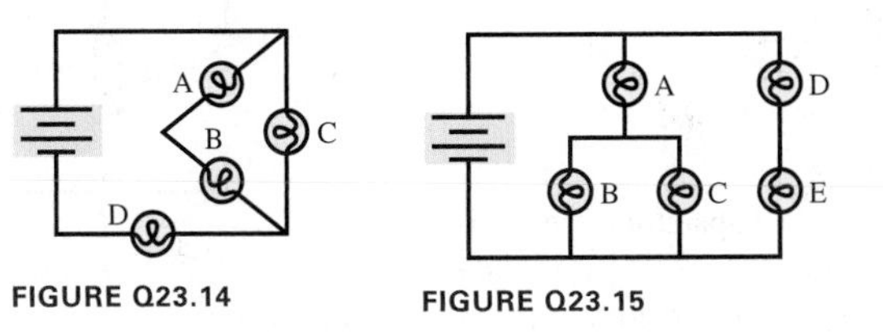

FIGURE Q23.14 FIGURE Q23.15

15. Figure Q23.15 shows five identical bulbs connected to a battery. All the bulbs are glowing. Rank the bulbs from brightest to dimmest. Explain.
16. a. The three bulbs in Figure Q23.16 are identical. Rank the bulbs from brightest to dimmest. Explain.
    b. Suppose a wire is connected between points 1 and 2. What happens to each bulb? Does it get brighter, stay the same, get dimmer, or go out? Explain.

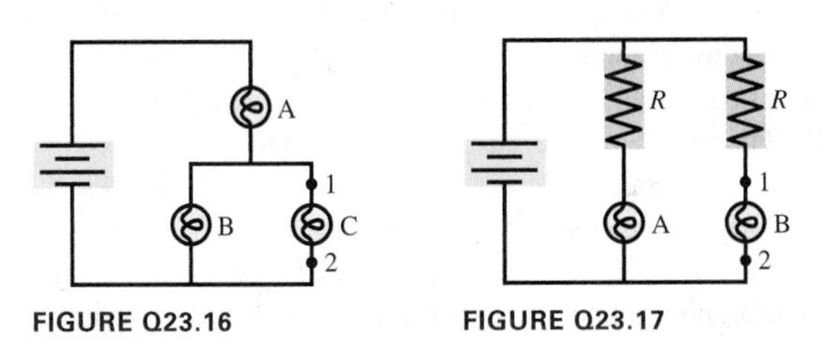

FIGURE Q23.16 FIGURE Q23.17

17. Initially, bulbs A and B in Figure Q23.17 are both glowing. Bulb B is then removed from its socket. Does removing bulb B cause the potential difference ΔV_{12} between points 1 and 2 to increase, decrease, stay the same, or become zero? Explain.
18. a. Consider the points a and b in Figure Q23.18. Is the potential difference ΔV_{ab} between points a and b zero? If so, why? If not, which point is more positive?
    b. If a wire is connected between points a and b, does it carry a current? If so, in which direction—to the right or to the left? Explain.

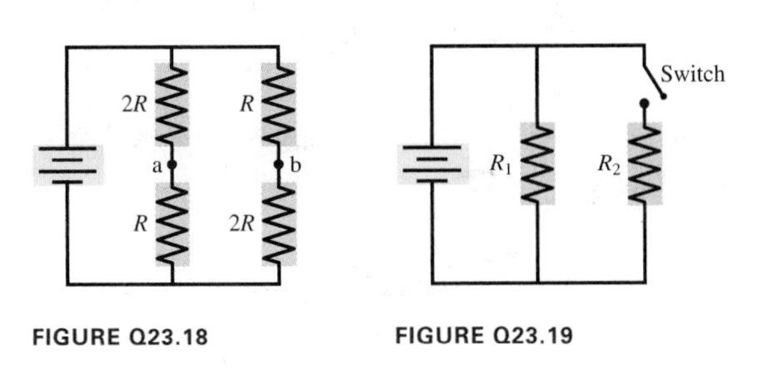

FIGURE Q23.18 FIGURE Q23.19

19. When the switch in Figure Q23.19 is closed,
    a. Does the current through the battery increase, decrease, or stay the same? Explain.
    b. Does the current through R_1 increase, decrease, or stay the same? Explain.
20. A voltmeter is (incorrectly) inserted into a circuit as shown in Figure Q23.20.
    a. What is the current in the circuit?
    b. What does the voltmeter read?
    c. How would you change the circuit to correctly connect the voltmeter to measure the potential difference across the resistor?

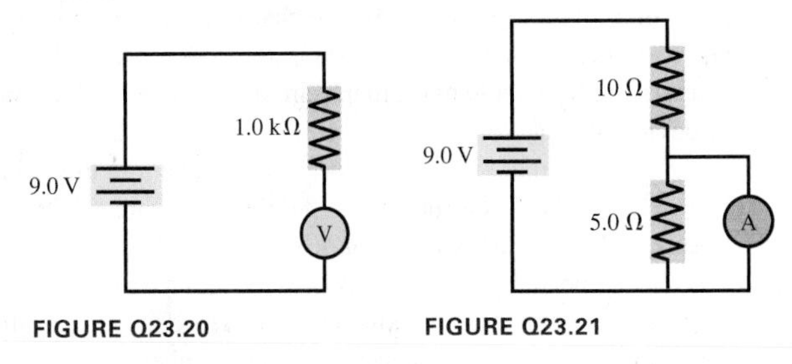

FIGURE Q23.20 FIGURE Q23.21

21. An ammeter is (incorrectly) inserted into a circuit as shown in Figure Q23.21.
    a. What is the current through the 5.0 Ω resistor?
    b. How would you change the circuit to correctly connect the ammeter to measure the current through the 5.0 Ω resistor?
22. Rank in order, from largest to smallest, the equivalent capacitances $(C_{eq})_1$ to $(C_{eq})_4$ of the four groups of capacitors shown in Figure Q23.22.

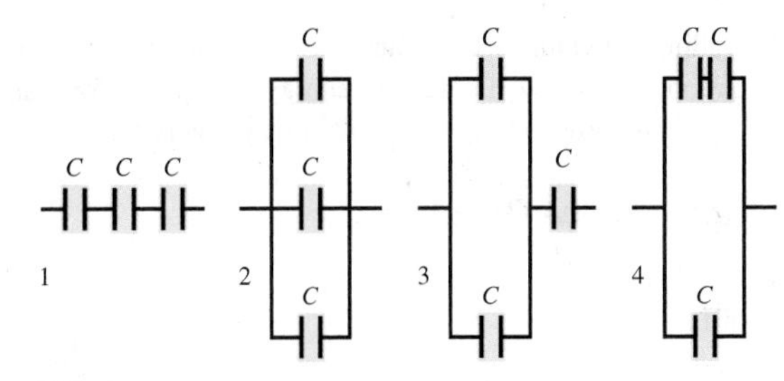

FIGURE Q23.22

23. Figure Q23.23 shows a circuit consisting of a battery, a switch, two identical lightbulbs, and a capacitor that is initially uncharged.
    a. *Immediately* after the switch is closed, are either or both bulbs glowing? Explain.
    b. If both bulbs are glowing, which is brighter? Or are they equally bright? Explain.
    c. For any bulb (A or B or both) that lights up immediately after the switch is closed, does its brightness increase with time, decrease with time, or remain unchanged? Explain.

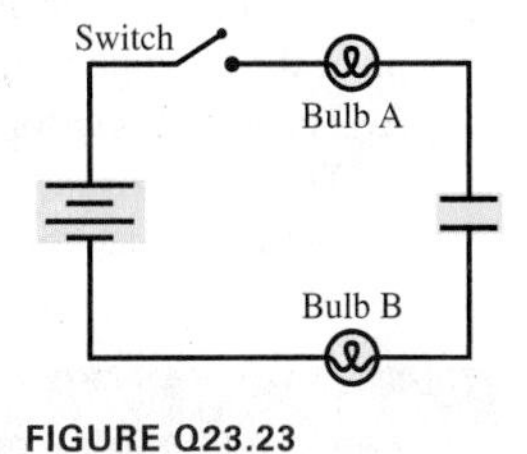

FIGURE Q23.23

24. Figure Q23.24 shows the voltage as a function of time across a capacitor as it is discharged (separately) through three different resistors. Rank in order, from largest to smallest, the values of the resistances R_1 to R_3.

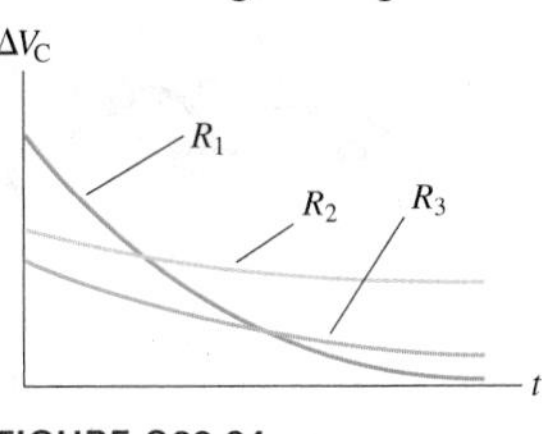

FIGURE Q23.24

25. A charged capacitor could be connected to two identical resistors in either of the two ways shown in Figure Q23.25. Which configuration will discharge the capacitor in the shortest time once the switch is closed? Explain.

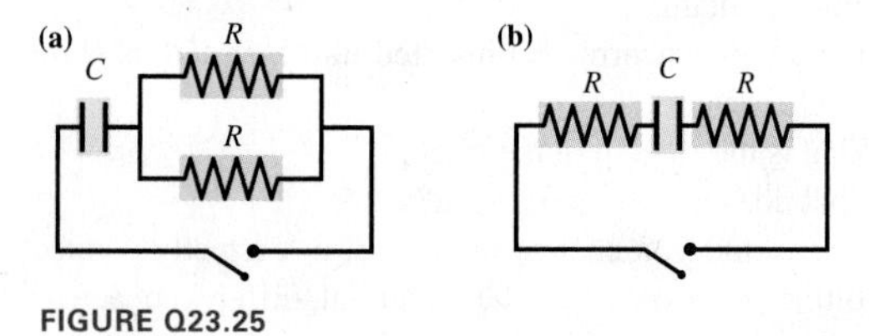

FIGURE Q23.25

26. A flashing light is controlled by the charging and discharging of an *RC* circuit. If the light is flashing too rapidly, describe two changes that you could make to the circuit to reduce the flash rate.

27. A device to make an electrical measurement of skin moisture has electrodes that form two plates of a capacitor; the skin is the dielectric between the plates. Adding moisture to the skin means adding water, which has a large dielectric constant. If a circuit repeatedly charges and discharges the capacitor to determine the capacitance, how will an increase in skin moisture affect the charging and discharging time? Explain. BIO

28. Consider the model of nerve conduction in myelinated axons presented in the chapter. Suppose the distance between the nodes of Ranvier was halved for a particular axon. BIO
    a. How would this affect the resistance and the capacitance of one segment of the axon?
    b. How would this affect the time constant for the charging of one segment?
    c. How would this affect the signal propagation speed for the axon?

29. Adding a myelin sheath to an axon results in faster signal propagation. It also means that less energy is required for a signal to propagate down the axon. Explain why this is so. BIO

Multiple-Choice Questions

30. | What is the current in the circuit of Figure Q23.30?
    A. 1.0 A B. 1.7 A
    C. 2.5 A D. 4.2 A

31. | Which resistor in Figure Q23.30 dissipates the most power?
    A. The 4.0 Ω resistor.
    B. The 6.0 Ω resistor.
    C. Both dissipate the same power.

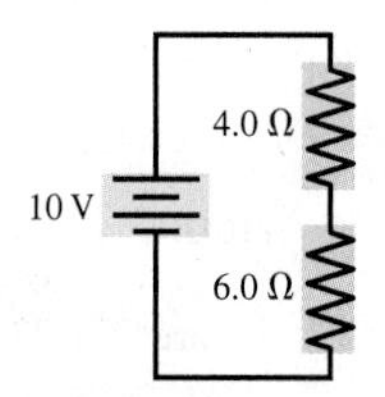

FIGURE Q23.30

32. || Normally, household lightbulbs are connected in parallel to a power supply. Suppose a 40 W and a 60 W lightbulb are, instead, connected in series, as shown in Figure Q23.32. Which bulb is brighter?
    A. The 60 W bulb.
    B. The 40 W bulb.
    C. The bulbs are equally bright.

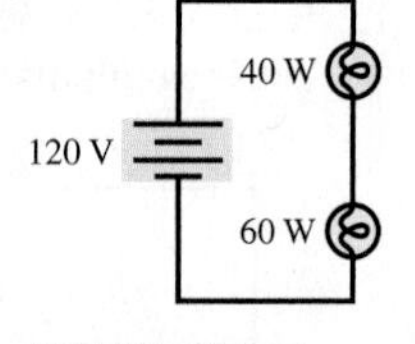

FIGURE Q23.32

33. ||| A metal wire of resistance R is cut into two pieces of equal length. The two pieces are connected together side by side. What is the resistance of the two connected wires?
    A. $R/4$ B. $R/2$ C. R
    D. $2R$ E. $4R$

34. | What is the value of resistor R in Figure Q23.34?
    A. 4.0 Ω
    B. 12 Ω
    C. 36 Ω
    D. 72 Ω
    E. 96 Ω

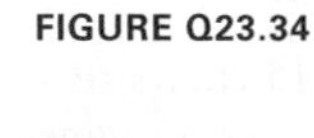

FIGURE Q23.34

35. | Two capacitors are connected in series. They are then reconnected to be in parallel. The capacitance of the parallel combination
    A. Is less than that of the series combination.
    B. Is more than that of the series combination.
    C. Is the same as that of the series combination.
    D. Could be more or less than that of the series combination depending on the values of the capacitances.

36. | If a cell's membrane thickness doubles but the cell stays the same size, how do the resistance and the capacitance of the cell membrane change? BIO
    A. The resistance and the capacitance increase.
    B. The resistance increases, the capacitance decreases.
    C. The resistance decreases, the capacitance increases.
    D. The resistance and the capacitance decrease.

37. ||| If a cell's diameter is reduced by 50% without changing the membrane thickness, how do the resistance and capacitance of the cell membrane change? BIO
    A. The resistance and the capacitance increase.
    B. The resistance increases, the capacitance decreases.
    C. The resistance decreases, the capacitance increases.
    D. The resistance and the capacitance decrease.

PROBLEMS

Section 23.1 Circuit Elements and Diagrams

1. || Draw a circuit diagram for the circuit of Figure P23.1.

2. || Draw a circuit diagram for the circuit of Figure P23.2.

3. || Draw a circuit diagram for the circuit of Figure P23.3.

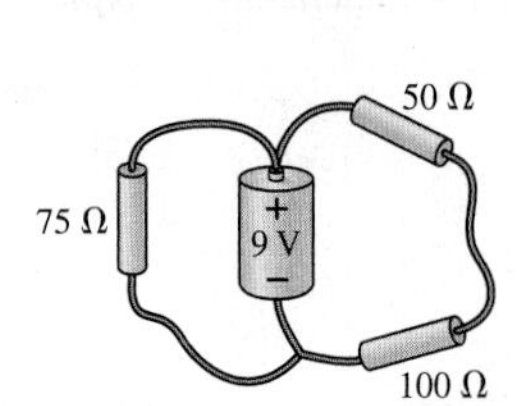

FIGURE P23.1

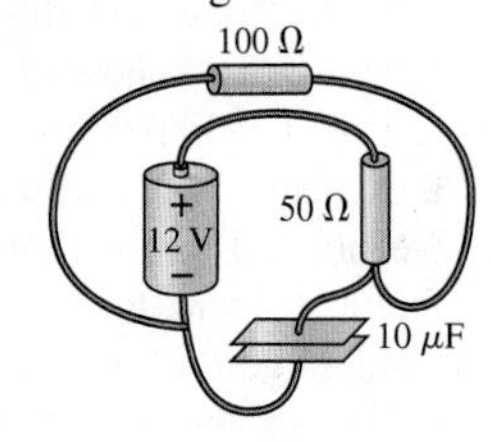

FIGURE P23.2

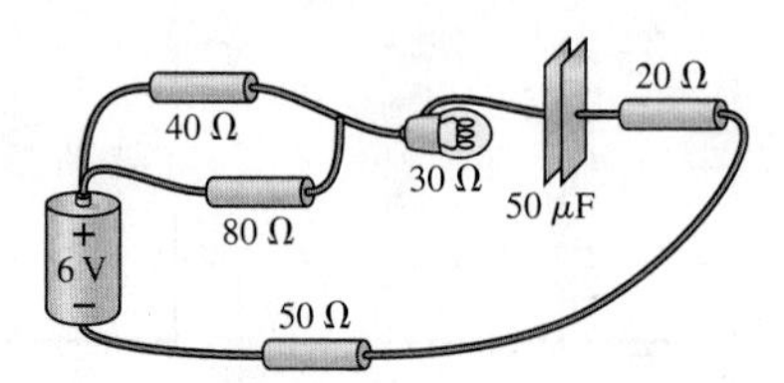

FIGURE P23.3

Section 23.2 Kirchhoff's Laws

4. || In Figure P23.4, what is the current in the wire above the junction? Does charge flow toward or away from the junction?

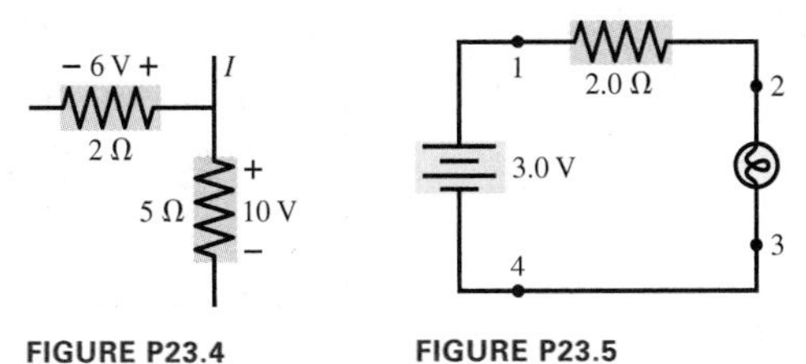

FIGURE P23.4

FIGURE P23.5

5. || The lightbulb in the circuit diagram of Figure P23.5 has a resistance of 1.0 Ω. Consider the potential difference between pairs of points in the figure.
   a. What are the magnitudes of ΔV_{12}, ΔV_{23}, and ΔV_{34}?
   b. What are the magnitudes if the bulb is removed?
6. | a. What are the magnitude and direction of the current in the 30 Ω resistor in Figure P23.6?
   b. Draw a graph of the potential as a function of the distance traveled through the circuit, traveling clockwise from $V = 0$ V at the lower left corner. See Figure P23.9 for an example of such a graph.

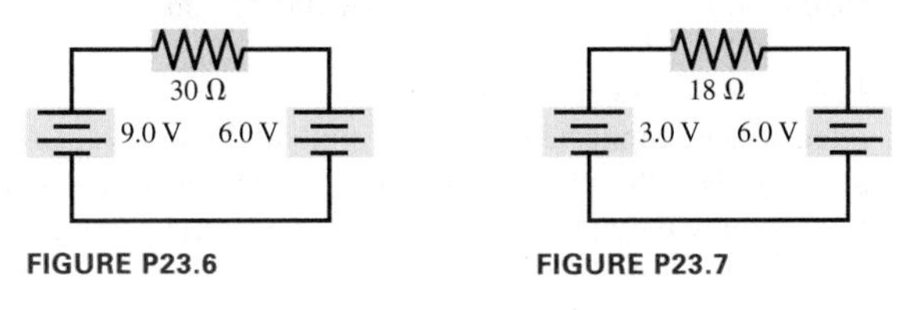

FIGURE P23.6

FIGURE P23.7

7. || a. What are the magnitude and direction of the current in the 18 Ω resistor in Figure P23.7?
   b. Draw a graph of the potential as a function of the distance traveled through the circuit, traveling clockwise from $V = 0$ V at the lower left corner. See Figure P23.9 for an example of such a graph.
8. | a. What is the potential difference across each resistor in Figure P23.8?
   b. Draw a graph of the potential as a function of the distance traveled through the circuit, traveling clockwise from $V = 0$ V at the lower left corner. See Figure P23.9 for an example of such a graph.

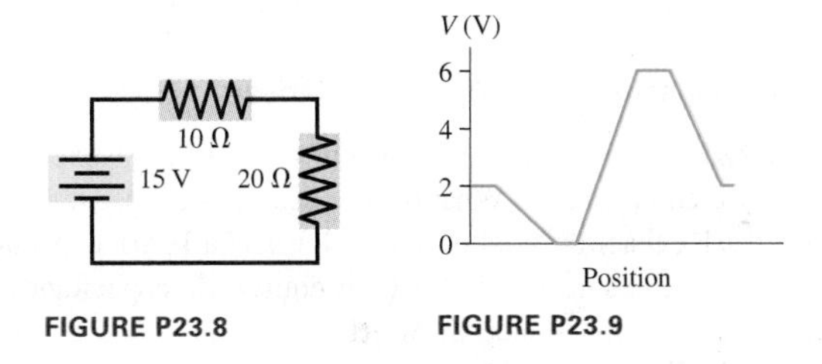

FIGURE P23.8

FIGURE P23.9

9. | The current in a circuit with only one battery is 2.0 A. Figure P23.9 shows how the potential changes when going around the circuit in the clockwise direction, starting from the lower left corner. Draw the circuit diagram.

Section 23.3 Series and Parallel Circuits

10. | What is the equivalent resistance of each group of resistors shown in Figure P23.10?

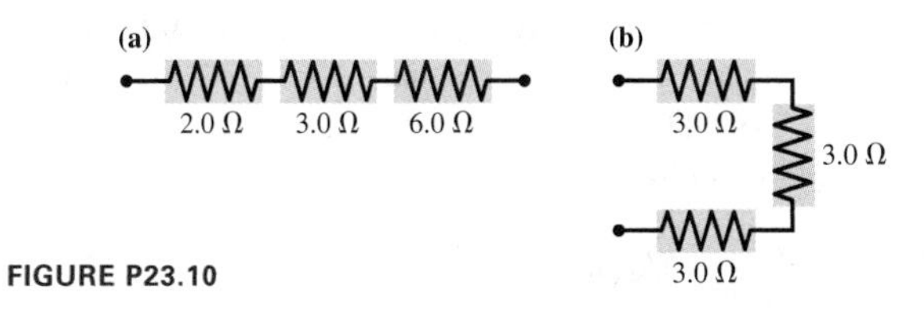

FIGURE P23.10

11. | What is the equivalent resistance of each group of resistors shown in Figure P23.11?

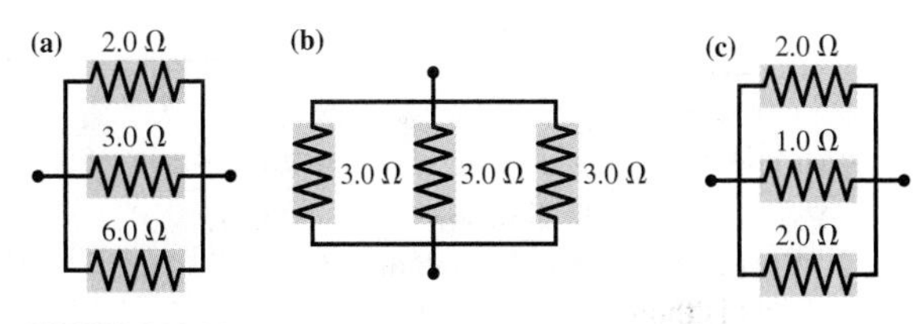

FIGURE P23.11

12. | Three resistors in parallel have an equivalent resistance of 5.0 Ω. Two of the resistors have resistances of 10 Ω and 30 Ω. What is the resistance of the third resistor?
13. | Three identical resistors have an equivalent resistance of 30 Ω when connected in parallel. What is their equivalent resistance when connected in series?
14. | You have a collection of 1.0 kΩ resistors. How can you connect four of them to produce an equivalent resistance of 0.25 kΩ?
15. | You have a collection of six 1.0 kΩ resistors. What is the smallest resistance you can make by combining them?
16. || You have six 1.0 kΩ resistors. How can you connect them to produce a total equivalent resistance of 1.5 kΩ?

Section 23.4 Measuring Voltage and Current

Section 23.5 More Complex Circuits

17. || What is the equivalent resistance between points a and b in Figure P23.17?

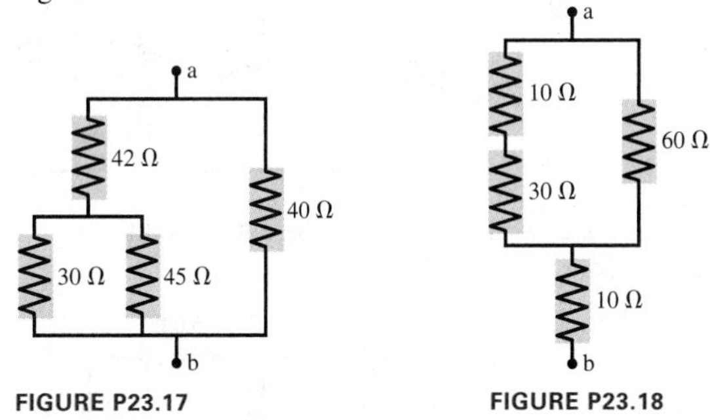

FIGURE P23.17

FIGURE P23.18

18. | What is the equivalent resistance between points a and b in Figure P23.18?
19. || The currents in two resistors in a circuit are shown in Figure P23.19. What is the value of resistor R?

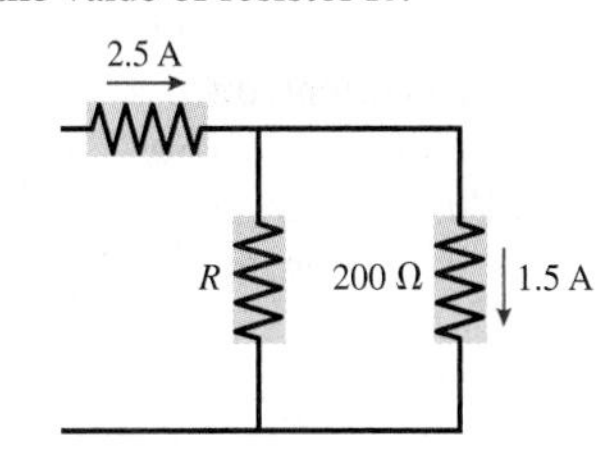

FIGURE P23.19

Watch video solution for problem: 23.25

20. || Two batteries supply current to the circuit in Figure P23.20. The figure shows the potential difference across two of the resistors and the value of the third resistor. What current is supplied by the batteries?

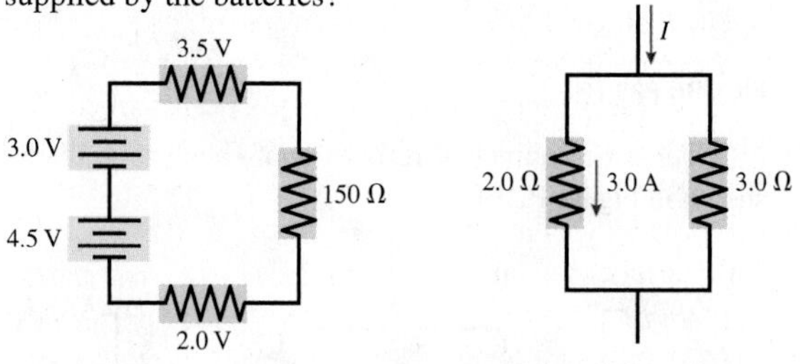

FIGURE P23.20

FIGURE P23.21

21. | Part of a circuit is shown in Figure P23.21.
 a. What is the current through the 3.0 Ω resistor?
 b. What is the value of the current I?

22. | What is the value of resistor R in Figure P23.22?

$\Delta V = 5.0$ V, 10 Ω, 15 Ω, R, $I = 100$ mA

FIGURE P23.22

23. || What are the resistances R and the emf of the battery in Figure P23.23?

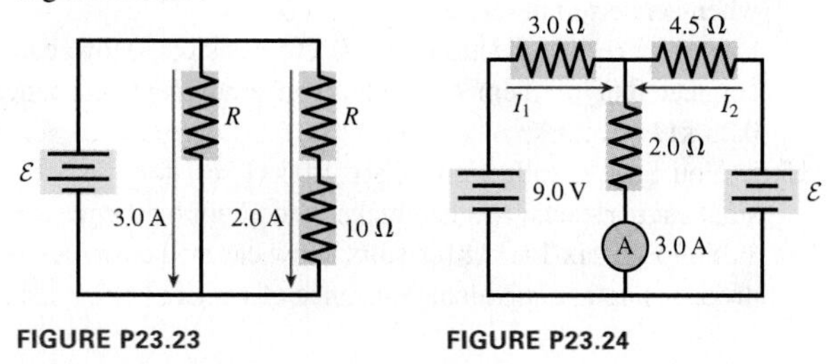

FIGURE P23.23

FIGURE P23.24

24. || The ammeter in Figure P23.24 reads 3.0 A. Find I_1, I_2, and $\mathcal{E}$.

25. ||| Find the current through and the potential difference across each resistor in Figure P23.25.

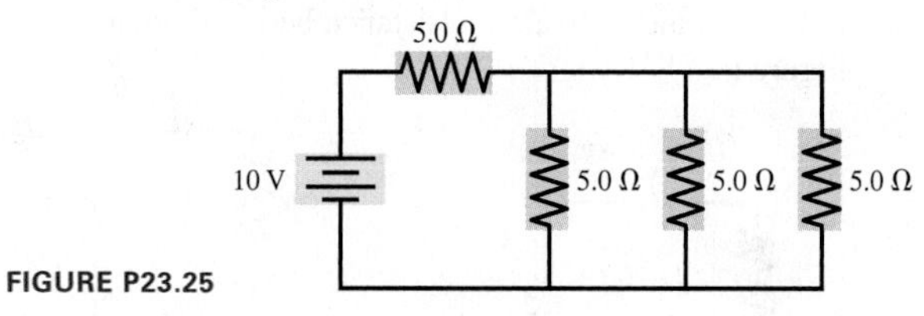

FIGURE P23.25

26. ||| Find the current through and the potential difference across each resistor in Figure P23.26.

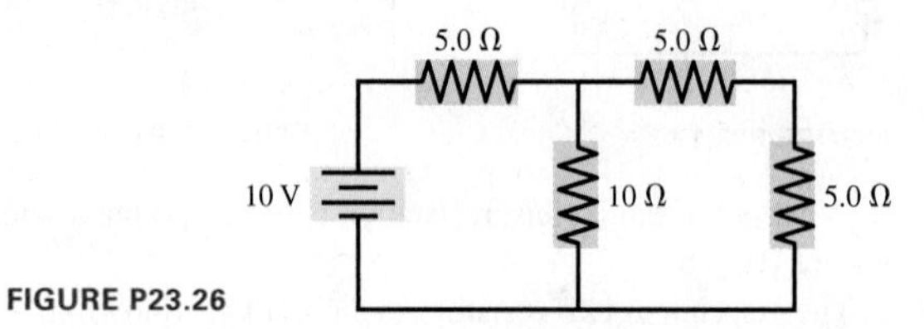

FIGURE P23.26

27. || For the circuit shown in Figure P23.27, find the current through and the potential difference across each resistor. Place your results in a table for ease of reading.

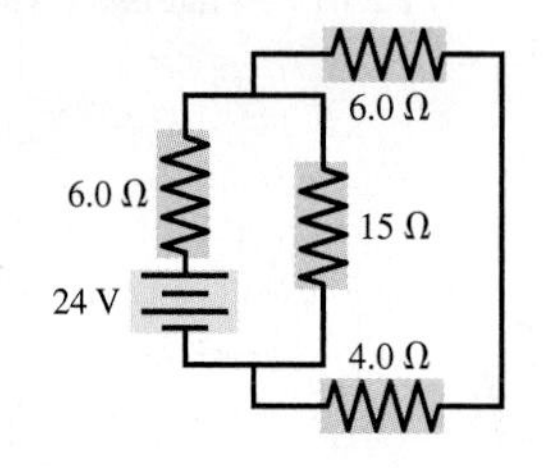

FIGURE P23.27

28. |||| Consider the potential differences between pairs of points in Figure P23.28. What are the magnitudes of the potential differences ΔV_{14}, ΔV_{24}, and ΔV_{34}?

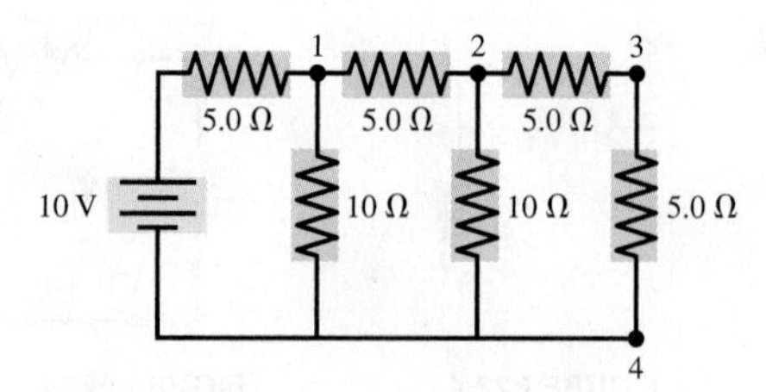

FIGURE P23.28

29. || For the circuit shown in Figure P23.29, find the current through and the potential difference across each resistor. Place your results in a table for ease of reading.

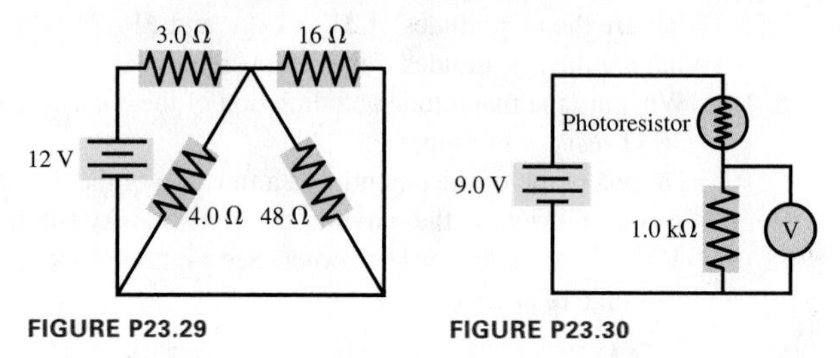

FIGURE P23.29

FIGURE P23.30

30. || A photoresistor, whose resistance decreases with light intensity, is connected in the circuit of Figure P23.30. On a sunny day, the photoresistor has a resistance of 0.56 kΩ. On a cloudy day, the resistance rises to 4.0 kΩ. At night, the resistance is 20 kΩ.
 a. What does the voltmeter read for each of these conditions?
 b. Does the voltmeter reading increase or decrease as the light intensity increases?

31. ||| The two unknown resistors in Figure P23.31 have the same resistance R. When the switch is closed, the current through the battery increases by 50%. What is R?

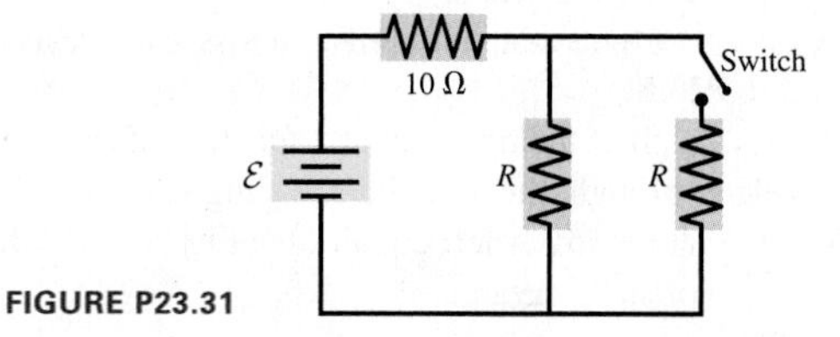

FIGURE P23.31

Section 23.6 Capacitors in Parallel and Series

32. | A 6.0 μF capacitor, a 10 μF capacitor, and a 16 μF capacitor are connected in parallel. What is their equivalent capacitance?

33. | A 6.0 μF capacitor, a 10 μF capacitor, and a 16 μF capacitor are connected in series. What is their equivalent capacitance?

34. | You need a capacitance of 50 μF, but you don't happen to have a 50 μF capacitor. You do have a 30 μF capacitor. What additional capacitor do you need to produce a total capacitance of 50 μF? Should you join the two capacitors in parallel or in series?

35. | You need a capacitance of 50 μF, but you don't happen to have a 50 μF capacitor. You do have a 75 μF capacitor. What additional capacitor do you need to produce a total capacitance of 50 μF? Should you join the two capacitors in parallel or in series?

36. || What is the equivalent capacitance of the three capacitors in Figure P23.36?

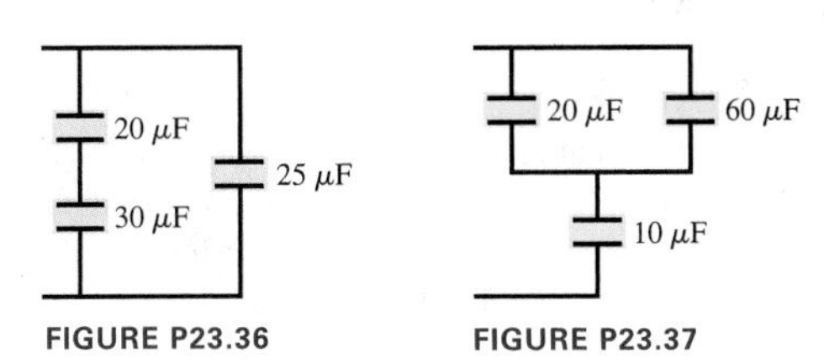

FIGURE P23.36

FIGURE P23.37

37. | What is the equivalent capacitance of the three capacitors in Figure P23.37?

38. ||| For the circuit of Figure P23.38,

a. What is the equivalent capacitance?

b. How much charge flows through the battery as the capacitors are being charged?

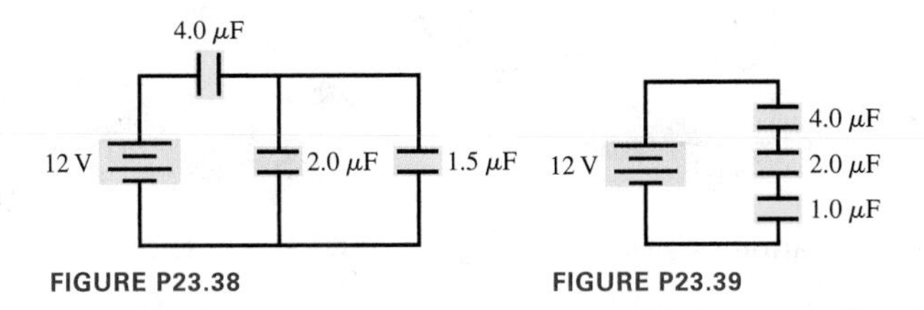

FIGURE P23.38

FIGURE P23.39

39. ||| For the circuit of Figure P23.39,

a. What is the equivalent capacitance?

b. What is the charge of each of the capacitors?

Section 23.7 *RC* Circuits

40. || What is the time constant for the discharge of the capacitor in Figure P23.40?

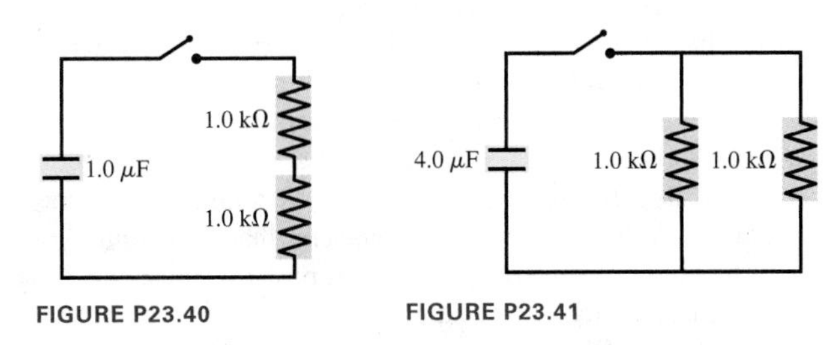

FIGURE P23.40

FIGURE P23.41

41. || What is the time constant for the discharge of the capacitor in Figure P23.41?

42. || After how many time constants has the voltage across a discharging capacitor decayed to 0.10% of its initial value?

43. |||| A 10 μF capacitor initially charged to 20 μC is discharged through a 1.0 kΩ resistor. How long does it take to reduce the capacitor's charge to 10 μC?

44. | A capacitor charging circuit consists of a battery, an uncharged 20 μF capacitor, and a 4.0 kΩ resistor. At $t = 0$ s, the switch is closed; 0.15 s later, the current is 0.46 mA. What is the battery's emf?

45. || The switch in Figure P23.45 has been in position a for a long time. It is changed to position b at $t = 0$ s. What are the charge Q on the capacitor and the current I through the resistor (a) immediately after the switch is closed? (b) At $t = 50$ μs? (c) At $t = 200$ μs?

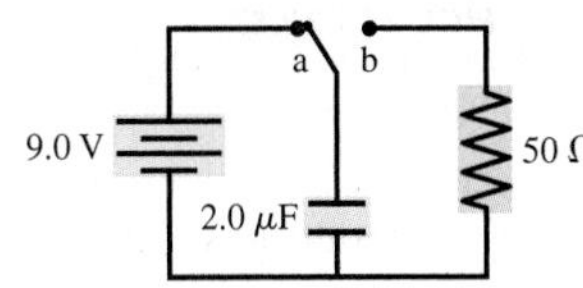

FIGURE P23.45

Section 23.8 Electricity in the Nervous System

46. | A 9.0-nm-thick cell membrane undergoes an action potential BIO that follows the curve in the table on page 748. What is the strength of the electric field inside the membrane just before the action potential and at the peak of the depolarization?

47. ||| A cell membrane has a resistance and a capacitance and thus a BIO characteristic time constant. What is the time constant of a 9.0-nm-thick membrane surrounding a 0.040-mm-diameter spherical cell?

48. | Changing the thickness of the myelin sheath surrounding an BIO axon changes its capacitance and thus the conduction speed. A myelinated nerve fiber has a conduction speed of 55 m/s. If the spacing between nodes is 1.0 mm and the resistance of segments between nodes is 25 MΩ, what is the capacitance of each segment?

49. ||| A particular myelinated axon has nodes spaced 0.80 mm BIO apart. The resistance between nodes is 20 MΩ; the capacitance of each insulated segment is 1.2 pF. What is the conduction speed of a nerve impulse along this axon?

50. | To measure signal propagation in a nerve in the arm, the BIO nerve is triggered near the armpit. The peak of the action potential is measured at the elbow and then, 4.0 ms later, 24 cm away from the elbow at the wrist.

a. What is the speed of propagation along this nerve?

b. A determination of the speed made by measuring the time between the application of a stimulus at the armpit and the peak of an action potential at the elbow or the wrist would be inaccurate. Explain the problem with this approach, and why the noted technique is preferable.

51. || A myelinated axon conducts nerve impulses at a speed of BIO 40 m/s. What is the signal speed if the thickness of the myelin sheath is halved but no other changes are made to the axon?

General Problems

52. || How much power is dissipated by INT each resistor in Figure P23.52?

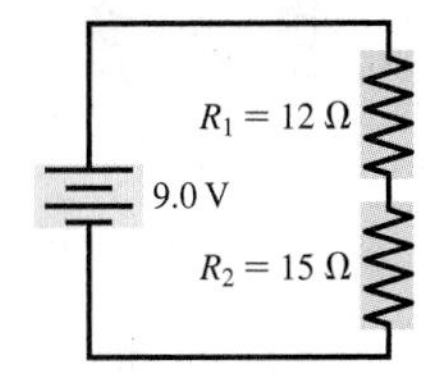

FIGURE P23.52

53. |||| Two 75 W (120 V) lightbulbs are wired in series, then the INT combination is connected to a 120 V supply. How much power is dissipated by each bulb?

54. |||| The corroded contacts in a lightbulb socket have 5.0 Ω total INT resistance. How much actual power is dissipated by a 100 W (120V) lightbulb screwed into this socket?

55. |||| A real battery is not just an emf. We can INT model a real 1.5 V battery as a 1.5 V emf in series with a resistor known as the "internal resistance," as shown in Figure P23.55. A typical battery has 1.0 Ω internal resistance due to imperfections that limit current through the battery. When there's no current through the battery, and thus no voltage drop across the internal resistance, the potential difference between its terminals is 1.5 V, the value of the emf. Suppose the terminals of this battery are connected to a 2.0 Ω resistor.

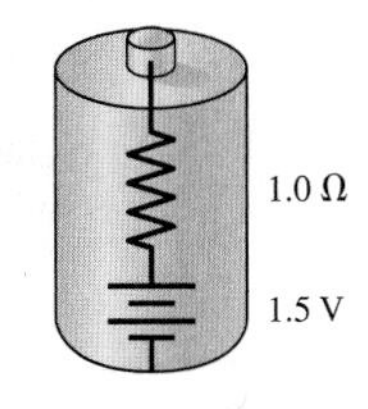

FIGURE P23.55

a. What is the potential difference between the terminals of the battery?

b. What fraction of the battery's power is dissipated by the internal resistance?

Watch video solution for problem: 23.59

56. |||| For the real battery shown in Figure P23.55, calculate the power dissipated by a resistor R connected to the battery when (a) $R = 0.25\ \Omega$, (b) $R = 0.50\ \Omega$, (c) $R = 1.0\ \Omega$, (d) $R = 2.0\ \Omega$, and (e) $R = 4.0\ \Omega$. (Your results should suggest that maximum power dissipation is achieved when the external resistance R equals the internal resistance. This is true in general.)

57. |||| Batteries are recharged by connecting them to a power INT supply (i.e., another battery) of greater emf in such a way that the current flows *into* the positive terminal of the battery being recharged, as was shown in Example 23.1. This reverse current through the battery replenishes its chemicals. The current is kept fairly low so as not to overheat the battery being recharged by dissipating energy in its internal resistance.
 a. Suppose the real battery of Figure P23.55 is rechargeable. What emf power supply should be used for a 0.75 A recharging current?
 b. If this power supply charges the battery for 10 minutes, how much energy goes into the battery? How much is dissipated as thermal energy in the internal resistance?

58. ||| When two resistors are connected in parallel across a battery of unknown voltage, one resistor carries a current of 3.2 A while the second carries a current of 1.8 A. What current will be supplied by the same battery if these two resistors are connected to it in series?

59. || The 10 Ω resistor in Figure P23.59 is dissipating 40 W INT of power. How much power are the other two resistors dissipating?

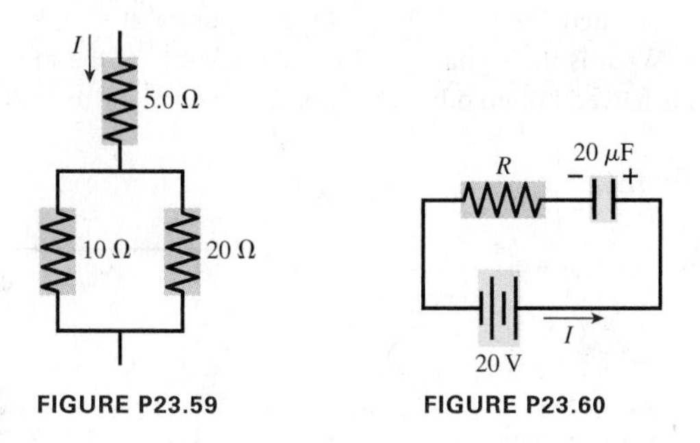

FIGURE P23.59   FIGURE P23.60

60. |||| At this instant, the current in the circuit of Figure P23.60 is INT 20 mA in the direction shown and the capacitor charge is 200 μC. What is the resistance R?

61. | What is the equivalent resistance between points a and b in Figure P23.61?

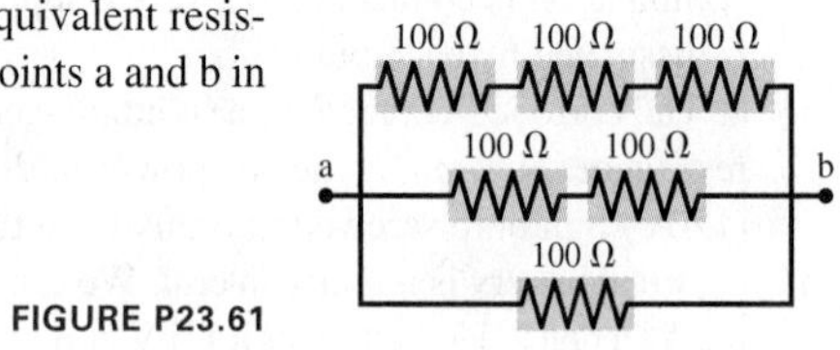

FIGURE P23.61

62. ||| What is the current through the battery in Figure P23.62 when the switch is (a) open and (b) closed?

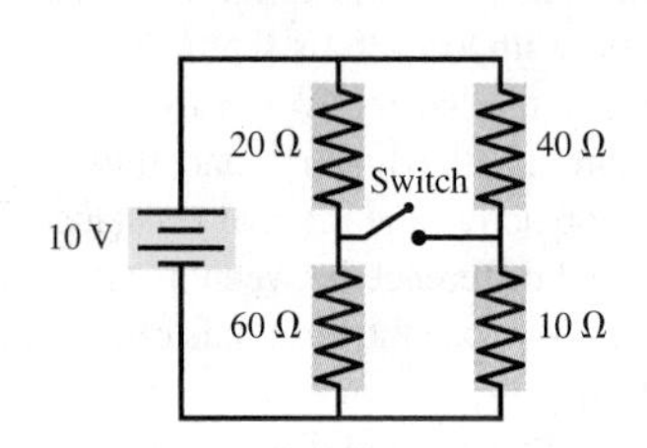

FIGURE P23.62

63. || What is the ratio $P_{parallel}/P_{series}$ of the total power dissipated INT by two identical resistors connected in parallel to a battery to the total power when they are connected in series to the same battery?

64. ||| You have a device that needs a voltage reference of 3.0 V, but you have only a 9.0 V battery. Fortunately, you also have several 10 kΩ resistors. Show how you can use the resistors and the battery to make a circuit that provides a potential difference of 3.0 V.

65. | There is a current of 0.25 A in the circuit of Figure P23.65.
INT a. What is the direction of the current? Explain.
 b. What is the value of the resistance R?
 c. What is the power dissipated by R?
 d. Make a graph of potential versus position, starting from $V = 0$ V in the lower left corner and proceeding clockwise. See Figure P23.9 for an example.

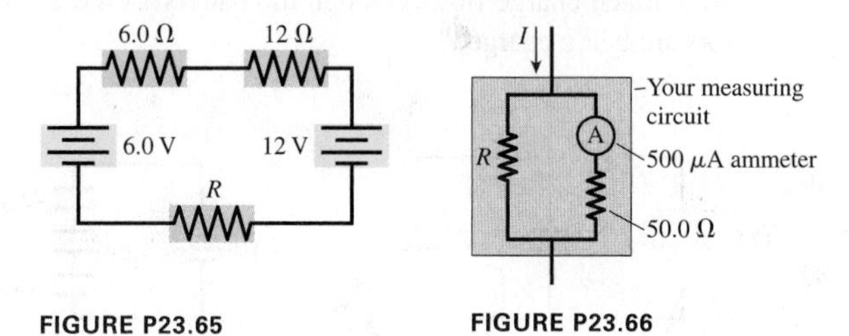

FIGURE P23.65   FIGURE P23.66

66. || A circuit you're building needs an ammeter that goes from 0 mA to a full-scale reading of 50.0 mA. Unfortunately, the only ammeter in the storeroom goes from 0 μA to a full-scale reading of only 500 μA. Fortunately, you can make this ammeter work by putting it in a measuring circuit, as shown in Figure P23.66. This lets a certain fraction of the current pass through the meter; knowing this value, you can deduce the total current. Assume that the ammeter is ideal.
 a. What value of R must you use so that the meter will go to full scale when the current I is 50.0 mA?
 Hint: When $I = 50.0$ mA, the ammeter should be reading its maximum value.
 b. What is the equivalent resistance of your measuring circuit?

67. || A circuit you're building needs a voltmeter that goes from 0 V to a full-scale reading of 5.0 V. Unfortunately, the only meter in the storeroom is an *ammeter* that goes from 0 μA to a full-scale reading of 500 μA. It is possible to use this meter to measure voltages by putting it in a measuring circuit as shown in Figure P23.67. What value of R must you use so that the meter will go to full scale when the potential difference ΔV is 5.0 V? Assume that the ammeter is ideal.

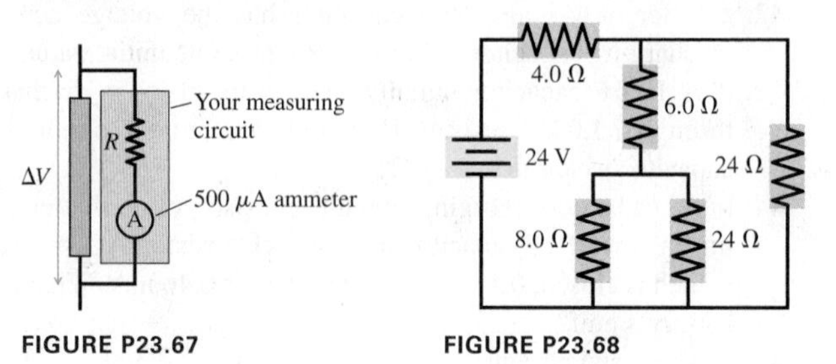

FIGURE P23.67   FIGURE P23.68

68. || For the circuit shown in Figure P23.68, find the current through and the potential difference across each resistor. Place your results in a table for ease of reading.

69. ‖ You have three 12 μF capacitors. Draw diagrams showing how you could arrange all three so that their equivalent capacitance is (a) 4.0 μF, (b) 8.0 μF, (c) 18 μF, and (d) 36 μF.

70. ‖ Initially, the switch in Figure P23.70 is in position a and capacitors C_2 and C_3 are uncharged. Then the switch is flipped to position b. Afterward, what are the charge on and the potential difference across each capacitor?

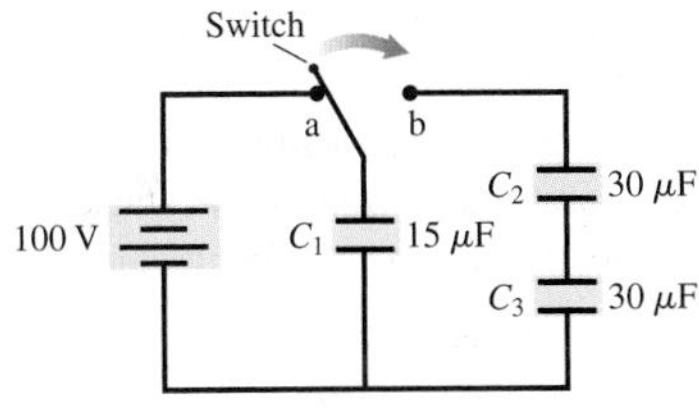

FIGURE P23.70

71. ‖ The capacitor in an RC circuit with a time constant of 15 ms INT is charged to 10 V. The capacitor begins to discharge at $t = 0$ s.
   a. At what time will the charge on the capacitor be reduced to half its initial value?
   b. At what time will the energy stored in the capacitor be reduced to half its initial value?

72. ||| The capacitor in Figure P23.72 is initially uncharged and the switch, in position c, is not connected to either side of the circuit. The switch is now flipped to position a for 10 ms, then to position b for 10 ms, and then brought back to position c. What is the final potential difference across the capacitor?

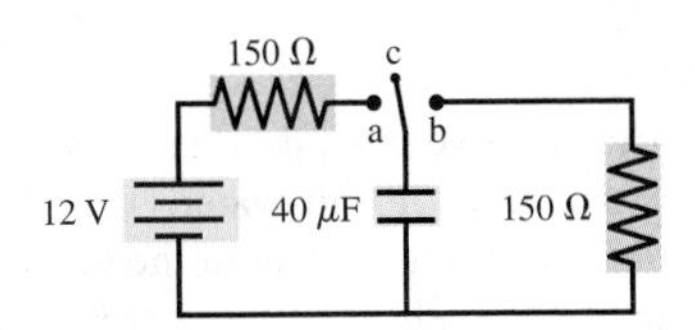

FIGURE P23.72

73. ‖ What value resistor will discharge a 1.0 μF capacitor to 10% of its initial charge in 2.0 ms?

74. |||| The charging circuit for the flash system of a camera uses a 100 μF capacitor that is charged from a 250 V power supply. What is the most resistance that can be in series with the capacitor if the capacitor is to charge to at least 87% of its final voltage in no more than 8.0 s?

75. ‖ A capacitor is discharged through a 100 Ω resistor. The discharge current decreases to 25% of its initial value in 2.5 ms. What is the value of the capacitor?

76. ||| A 50 μF capacitor that had been charged to 30 V is discharged through a resistor. Figure P23.76 shows the capacitor voltage as a function of time. What is the value of the resistance?

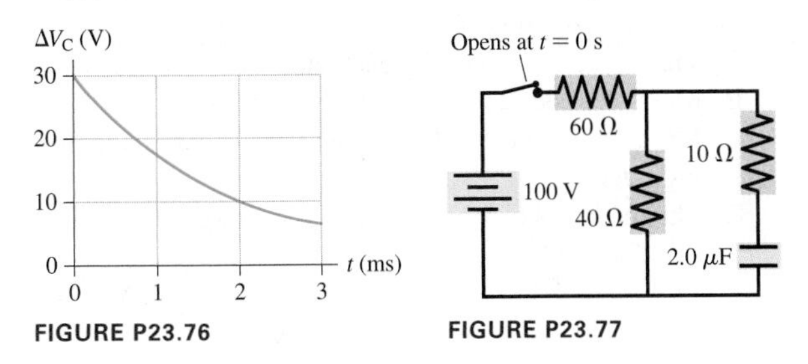

FIGURE P23.76

FIGURE P23.77

77. |||| The switch in Figure P23.77 has been closed for a very long time.
   a. What is the charge on the capacitor?
   b. The switch is opened at $t = 0$ s. At what time has the charge on the capacitor decreased to 10% of its initial value?

78. ||||| Intermittent windshield wipers use a variable resistor in an RC circuit to set the delay between successive passes of the wipers. A typical circuit is shown in Figure P23.78. When the switch closes, the capacitor (initially uncharged) begins to charge and the potential at point b begins to increase. A sensor measures the potential difference between points a and b, triggering a pass of the wipers when $V_b = V_a$. (Another part of the circuit, not shown, discharges the capacitor at this time so that the cycle can start again.)
   a. What value of the variable resistor will give 12 seconds from the start of a cycle to a pass of the wipers?
   b. To decrease the time, should the variable resistance be increased or decreased?

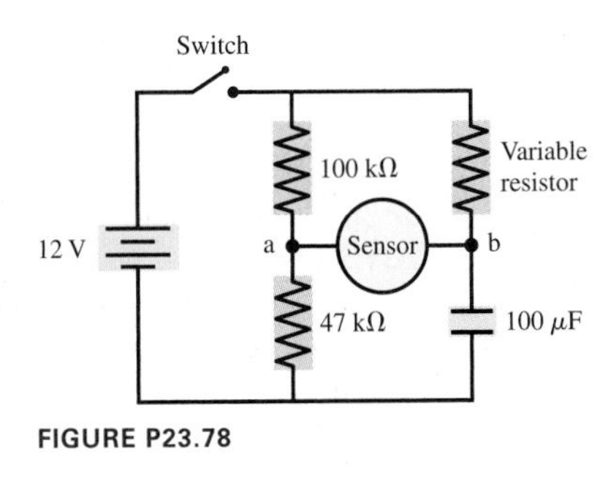

FIGURE P23.78

79. ||| In Example 23.14 we estimated the capacitance of the cell BIO membrane to be 89 pF, and in Example 23.15 we found that approximately 10,000 Na^+ ions flow through an ion channel when it opens. Based on this information and what you learned in this chapter about the action potential, estimate the total number of sodium ion channels in the membrane of a nerve cell.

80. |||| The giant axon of a squid is 0.5 mm in diameter, 10 cm BIO INT long, and not myelinated. Unmyelinated cell membranes behave as capacitors with 1 μF of capacitance per square centimeter of membrane area. When the axon is charged to the −70 mV resting potential, what is the energy stored in this capacitance?

81. ‖ A cell has a 7.0-nm-thick membrane with a total membrane BIO area of 6.0×10^{-9} m^2.
   a. We can model the cell as a capacitor, as we have seen. What is the magnitude of the charge on each "plate" when the membrane is at its resting potential of −70 mV?
   b. How many sodium ions does this charge correspond to?

MCAT-Style Passage Problems

The Defibrillator BIO

A defibrillator is designed to pass a large current through a patient's torso in order to stop dangerous heart rhythms. Its key part is a capacitor that is charged to a high voltage. The patient's torso plays the role of a resistor in an RC circuit. When a switch is closed, the capacitor discharges through the patient's torso. A jolt from a defibrillator is intended to be intense and rapid; the maximum current is very large, so the capacitor discharges quickly. This rapid pulse depolarizes the heart, stopping all electrical activity. This allows the heart's internal nerve circuitry to reestablish a healthy rhythm.

A typical defibrillator has a 32 μF capacitor charged to 5000 V. The electrodes connected to the patient are coated with a conducting gel that reduces the resistance of the skin to where the effective resistance of the patient's torso is 100 Ω.

82. | Which pair of graphs in Figure P23.82 best represents the capacitor voltage and the current through the torso as a function of time after the switch is closed?

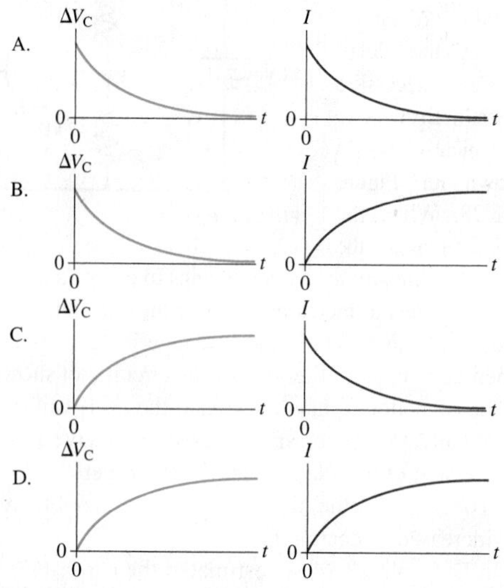

FIGURE P23.82

83. | For the values noted in the passage above, what is the time constant for the discharge of the capacitor?
A. 3.2 μs B. 160 μs C. 3.2 ms D. 160 ms

84. | If a patient receives a series of jolts, the resistance of the torso may increase. How does such a change affect the initial current and the time constant of subsequent jolts?
A. The initial current and the time constant both increase.
B. The initial current decreases, the time constant increases.
C. The initial current increases, the time constant decreases.
D. The initial current and the time constant both decrease.

85. | In some cases, the defibrillator may be charged to a lower voltage. How will this affect the time constant of the discharge?
A. The time constant will increase.
B. The time constant will not change.
C. The time constant will decrease.

Electric Fish BIO INT

The voltage produced by a single nerve or muscle cell is quite small, but there are many species of fish that use multiple action potentials in series to produce significant voltages. The electric organs in these fish are composed of specialized disk-shaped cells called *electrocytes.* The cell at rest has the usual potential difference between the inside and the outside, but the net potential difference *across* the cell is zero. An electrocyte is connected to nerve fibers that initially trigger a depolarization in one side of the cell but not the other. For the very short time of this depolarization, there is a net potential difference across the cell, as shown in Figure P23.86. Stacks of these cells connected in series can produce a large total voltage. Each stack can produce a small current; for more total current, more stacks are needed, connected in parallel.

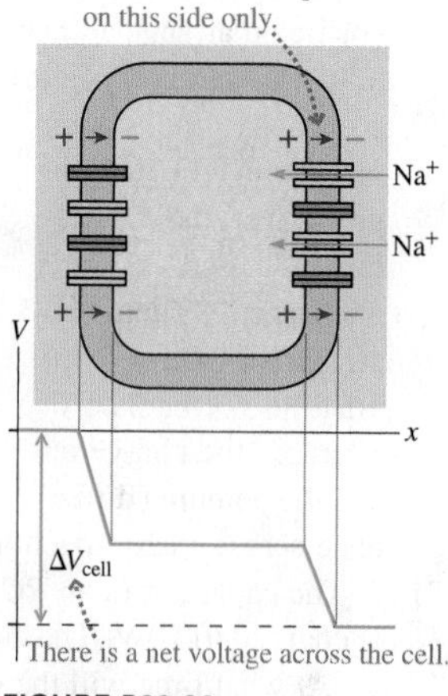

FIGURE P23.86

86. | In an electric eel, each electrocyte can develop a voltage of 150 mV for a short time. For a total voltage of 450 V, how many electrocytes must be connected in series?
A. 300
B. 450
C. 1500
D. 3000

87. | An electric eel produces a pulse of current of 0.80 A at a voltage of 500 V. For the short time of the pulse, what is the instantaneous power?
A. 400 W
B. 500 W
C. 625 W
D. 800 W

88. | Electric eels live in fresh water. The torpedo ray is an electric fish that lives in salt water. The electrocytes in the ray are grouped differently than in the eel; each stack of electrocytes has fewer cells, but there are more stacks in parallel. Which of the following best explains the ray's electrocyte arrangement?
A. The lower resistivity of salt water requires more current but lower voltage.
B. The lower resistivity of salt water requires more voltage but lower current.
C. The higher resistivity of salt water requires more current but lower voltage.
D. The higher resistivity of salt water requires more voltage but lower current.

89. | The electric catfish is another electric fish that produces a voltage pulse by means of stacks of electrocytes. As the fish grows in length, the magnitude of the voltage pulse the fish produces grows as well. The best explanation for this change is that, as the fish grows,
A. The voltage produced by each electrocyte increases.
B. More electrocytes are added to each stack.
C. More stacks of electrocytes are added in parallel to the existing stacks.
D. The thickness of the electrocytes increases.

STOP TO THINK ANSWERS

Chapter Preview Stop to Think: $R_4 < R_3 = R_2 < R_1$. Ohm's law can be used to solve for R: $R = \Delta V/I$. Thus we have $R_1 = (2\text{ V})/(1\text{ A}) = 2\ \Omega$, $R_2 = (2\text{ V})/(2\text{ A}) = 1\ \Omega$, $R_3 = (1\text{ V})/(1\text{ A}) = 1\ \Omega$, and $R_4 = (1\text{ V})/(2\text{ A}) = \frac{1}{2}\ \Omega$.

Stop to Think 23.1: A, B, and D. These three are the same circuit because the logic of the connections is the same. In each case, there is a junction that connects one side of each circuit element and a second junction that connects the other side. In C, the functioning of the circuit is changed by the extra wire connecting the two sides of the capacitor.

Stop to Think 23.2: B. The potential difference in crossing the battery is its emf, or 10 V. The potential drop across the 6 Ω resistor is $\Delta V = -IR = -6.0$ V. To make the sum of the potential differences zero, as Kirchhoff's loop law requires, the potential difference across resistor R must be -4.0 V.

Stop to Think 23.3: C = D > A = B. The two bulbs in series are of equal brightness, as are the two bulbs in parallel. But the two bulbs in series have a larger resistance than a single bulb, so there will be less current through the bulbs in series than the bulbs in parallel.

Stop to Think 23.4: C. The voltmeter must be connected in parallel with the resistor, and the ammeter in series.

Stop to Think 23.5: A > B > C = D. All the current from the battery goes through A, so it is brightest. The current divides at the junction, but not equally. Because B is in parallel with C + D, but has half the resistance of the two bulbs together, twice as much current travels through B as through C + D. So B is dimmer than A but brighter than C and D. C and D are equally bright because of conservation of current.

Stop to Think 23.6: $(C_{eq})_B > (C_{eq})_A > (C_{eq})_C$. Two capacitors in parallel have a larger capacitance than either alone; two capacitors in series have a smaller capacitance than either alone.

Stop to Think 23.7: B. The two 2 Ω resistors are in series and equivalent to a 4 Ω resistor. Thus $\tau = RC = 4$ s.

Stop to Think 23.8: A. A thicker sheath would space the "plates" of the capacitor—the inner and outer surfaces of the sheath—farther apart, thereby reducing the capacitance C. The RC time constant τ would thus decrease, increasing the speed $v = L_{node}/\tau$.

24 Magnetic Fields and Forces

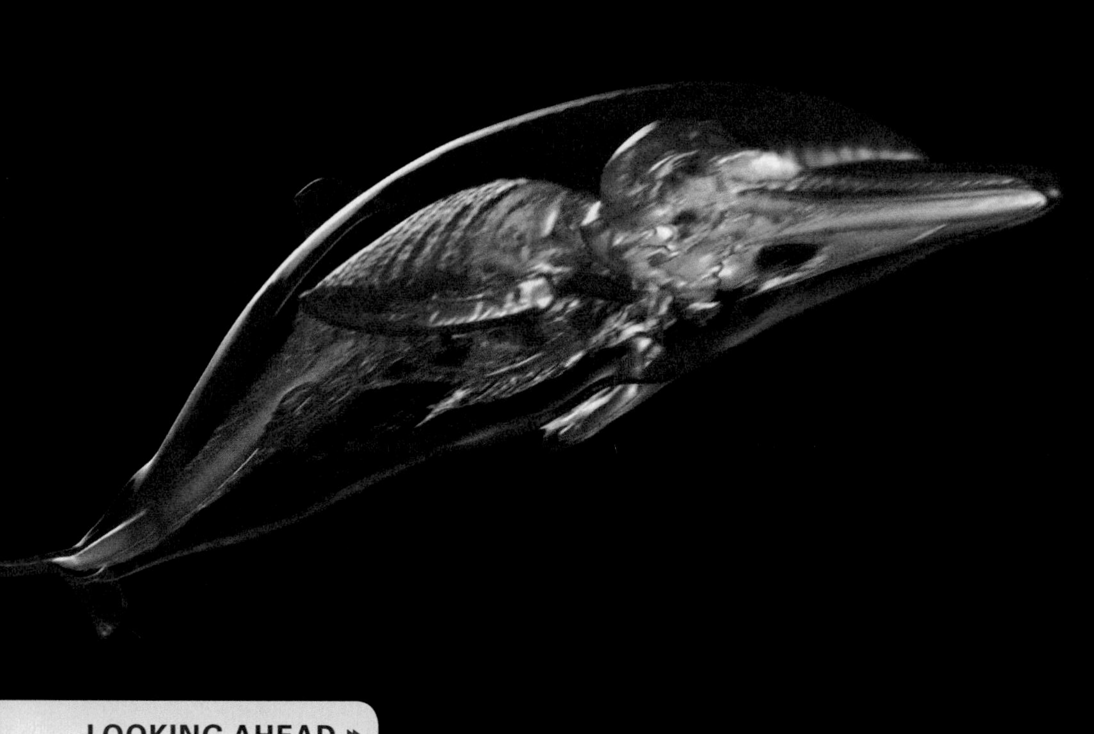

This detailed image of the skeletal system of a dolphin wasn't made with x rays; it was made with magnetism. How is this done?

LOOKING AHEAD ▸▸

Goal: To learn about magnetic fields and how magnetic fields exert forces on currents and moving charges.

Magnetic Fields

A compass is a magnetic dipole. It will rotate to line up with a magnetic field.

You'll learn how to use compasses and other tools to map magnetic fields.

Sources of the Field

Magnets produce a magnetic field; so do current-carrying wires, loops, and coils.

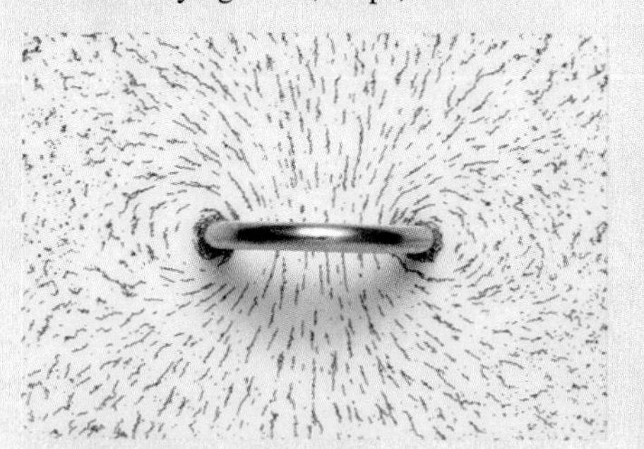

You'll learn to describe the magnetic fields created by currents. These iron filings show the magnetic-field shape for this current-carrying wire.

Effects of the Field

Magnetic fields exert forces on moving charged particles and electric currents.

You'll see how the motion of charged particles in the earth's magnetic field gives rise to the aurora.

LOOKING BACK ◂◂

Electric Fields

In Chapter 20, we described electric interactions between charged objects in terms of the field model.

You learned how to draw and interpret the electric field of a dipole. In this chapter, you'll see how a magnetic dipole creates a magnetic field with a similar structure.

STOP TO THINK

An electric dipole in a uniform electric field experiences no net force, but it does experience a net torque. The rotation of this dipole will be

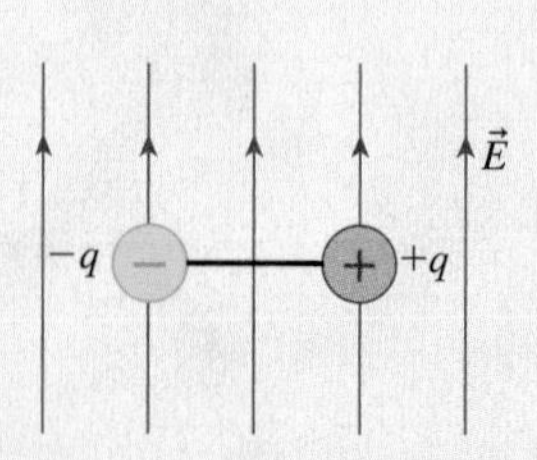

A. Clockwise.
B. Counterclockwise.

24.1 Magnetism

We began our investigation of electricity in Chapter 20 by looking at the results of simple experiments with charged rods. We'll do the same with magnetism.

Exploring magnetism

Experiment 1

If a bar magnet is taped to a piece of cork and allowed to float in a dish of water, it turns to align itself in an approximate north-south direction. The end of a magnet that points north is the **north pole.** The other end is the **south pole.**

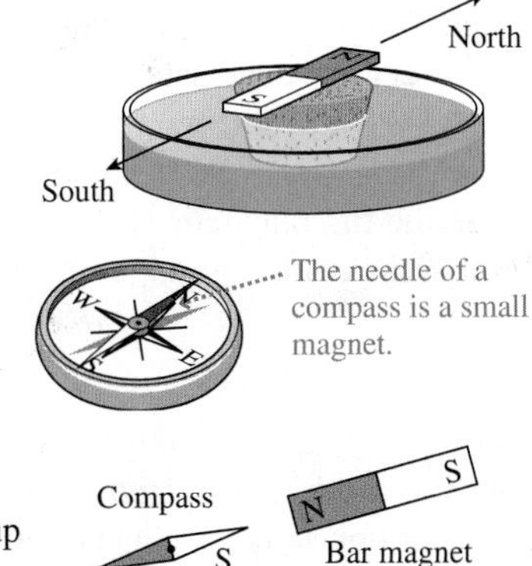

A magnet that is free to pivot like this is called a **compass.** A compass will pivot to line up with a nearby magnet.

Experiment 2

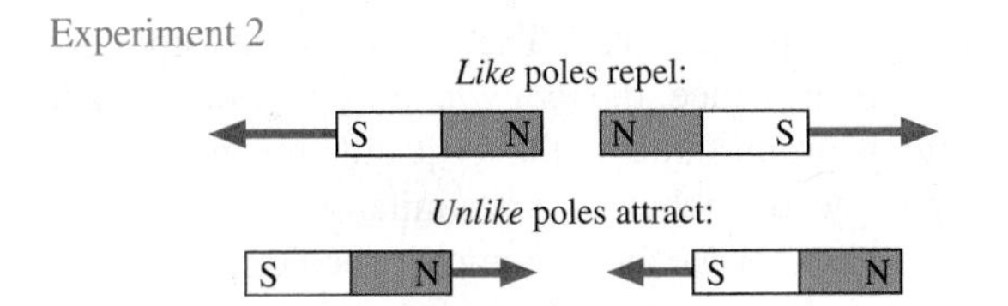

If the north pole of one magnet is brought near the north pole of another magnet, they repel each other. Two south poles also repel each other, but the north pole of one magnet exerts an attractive force on the south pole of another magnet.

Experiment 3

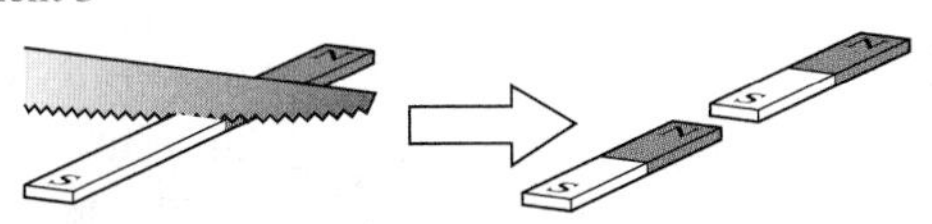

Cutting a bar magnet in half produces two weaker but still complete magnets, each with a north pole and a south pole.

Experiment 4

Magnets can pick up some objects, such as paper clips, but not all. If an object is attracted to one pole of a magnet, it is also attracted to the other pole. Most materials, including copper, aluminum, glass, and plastic, experience no force from a magnet.

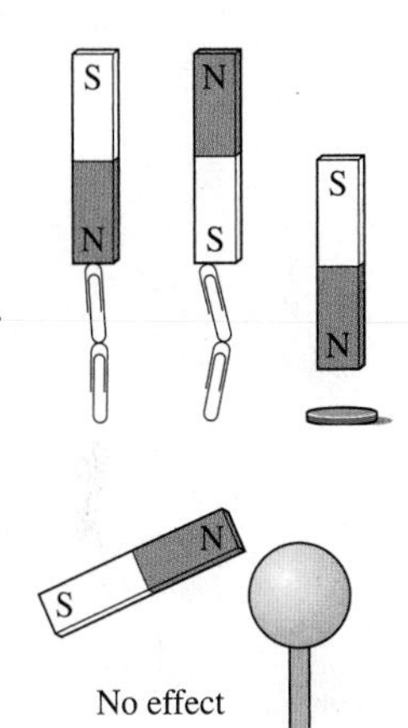

Experiment 5

When a magnet is brought near an electroscope, the leaves of the electroscope remain undeflected. If a charged rod is brought near a magnet, there is a small polarization force like the ones we studied in Chapter 21, as there would be on any metal bar, but there is no other effect.

No effect

What do these experiments tell us?

- Experiment 5 reveals that magnetism is not the same as electricity. **Magnetic poles and electric charges share some similar behavior, but they are not the same.**
- Experiment 2 shows that magnetism is a long-range force. Magnets need not touch each other to exert a force on each other.
- Experiments 1 and 3 show that magnets have two types of poles, called north and south poles, and thus are **magnetic dipoles.** Cutting a magnet in half yields two weaker but still complete magnets, each with a north pole and a south pole. The basic unit of magnetism is thus a magnetic dipole.
- Experiments 1 and 2 show how the poles of a bar magnet can be identified by using it as a compass. Other magnets can be identified by testing them against a bar magnet. A pole that repels a known south pole and attracts a known north pole must be a south magnetic pole.
- Experiment 4 reveals that only certain materials, called **magnetic materials,** are attracted to a magnet. The most common magnetic material is iron. Magnetic materials are attracted to both poles of a magnet.

STOP TO THINK 24.1 Does the compass needle rotate?

A. Yes, clockwise.
B. Yes, counterclockwise.
C. No, not at all.

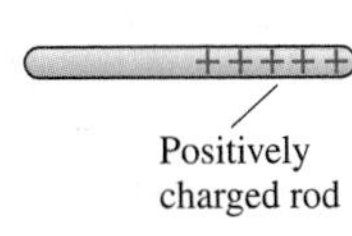

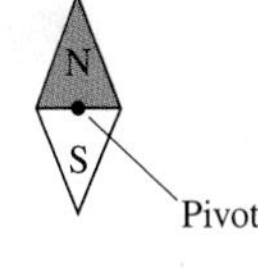

24.2 The Magnetic Field

When we studied the *electric* force between two charges in ◀◀ SECTION 20.4, we developed a new way to think about forces between charges—the *field model.* In this viewpoint, the space around a charge is not empty: The charge alters the space around it by creating an *electric field.* A second charge brought into this electric field then feels a force due to the *field.*

The concept of a field can also be used to describe the force that turns a compass to line up with a magnet: **Every magnet sets up a *magnetic* field in the space around it**. If another magnet—such as a compass needle—is then brought into this field, the second magnet will feel the effects of the *field* of the first magnet. In this section, we'll see how to define the magnetic field, and then we'll study what the magnetic field looks like for some common shapes and arrangements of magnets.

Measuring the Magnetic Field

What does the direction a compass needle points tell us about the magnetic field at the position of the compass? Recall how an *electric* dipole behaves when placed in an electric field, as shown in FIGURE 24.1a. In Chapter 20 we learned that an electric dipole experiences a *torque* when placed in an electric field, a torque that tends to align the axis of the dipole with the field. This means that the *direction* of the electric field is the same as the direction of the dipole's axis. The torque on the dipole is greater when the electric field is stronger; hence, the *magnitude* of the field, which we also call the *strength* of the field, is proportional to the torque on the dipole.

The magnetic dipole of a compass needle behaves very similarly when it is in a magnetic field. The magnetic field exerts a torque on the compass needle, causing the needle to point in the field direction, as shown in FIGURE 24.1b.

FIGURE 24.1 Dipoles in electric and magnetic fields.

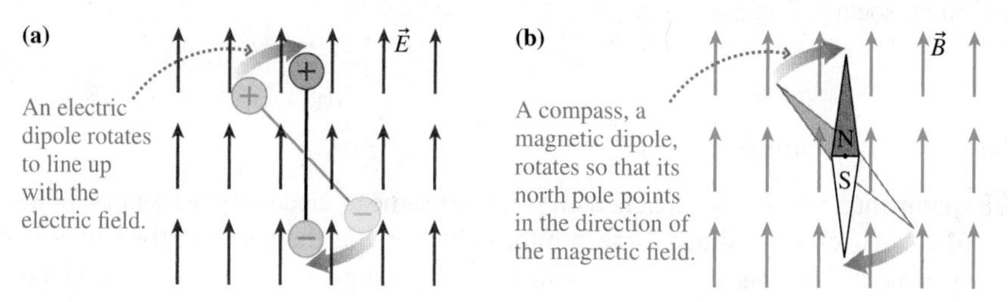

Because the magnetic field has both a direction and a magnitude, we represent it using a *vector*. We will use the symbol $\vec{B}$ to represent the magnetic field and B to represent the magnitude or strength of the field. FIGURE 24.2 shows how to use a compass to determine the magnitude and direction of the magnetic field. The direction of the magnetic field is the direction that the north pole of a compass needle points; the strength of the magnetic field is proportional to the torque felt by the compass needle as it turns to line up with the field direction.

FIGURE 24.2 Determining the direction and strength of a magnetic field.

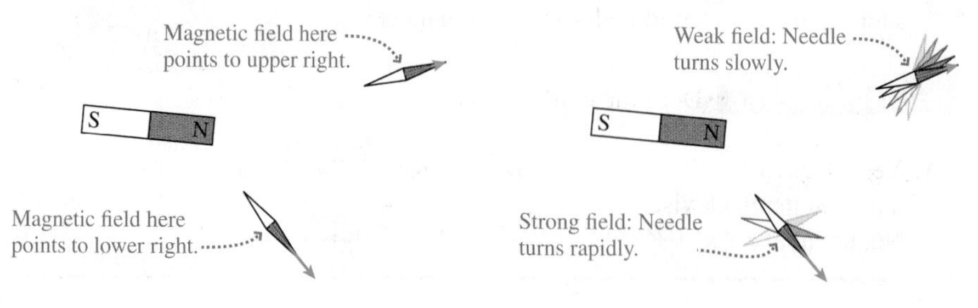

We can produce a "picture" of the magnetic field by using *iron filings*—very small elongated grains of iron. If there are enough grains, iron filings can give a very detailed representation of the magnetic field, as shown in FIGURE 24.3. The compasses that we use to determine field direction show us that **the magnetic field of a magnet points *away* from the north pole and *toward* the south pole.**

FIGURE 24.3 Revealing the field of a bar magnet using iron filings.

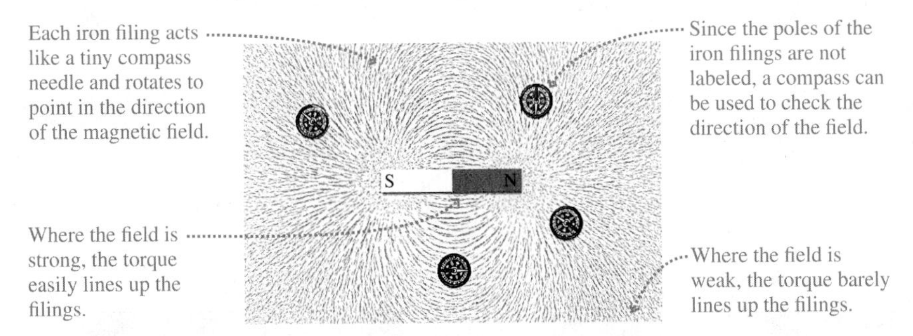

Each iron filing acts like a tiny compass needle and rotates to point in the direction of the magnetic field.
Since the poles of the iron filings are not labeled, a compass can be used to check the direction of the field.
S
N
Where the field is strong, the torque easily lines up the filings.
Where the field is weak, the torque barely lines up the filings.

Magnetic Field Vectors and Field Lines

We can draw the field of a magnet such as the one shown in Figure 24.3 in either of two ways. When we want to represent the magnetic field at one particular point, the **magnetic field vector** representation is especially useful. But if we want an overall representation of the field, **magnetic field lines** are often simpler to use. These two representations are similar to the *electric field* vectors and lines used in Chapter 20, and we'll use similar rules to draw them.

As shown in FIGURE 24.4, we can imagine placing a number of compasses near the magnet to measure the direction and magnitude of the magnetic field. To represent the field at the location of one of the compasses, we then draw a vector with its *tail* at that location. Figure 24.4 shows how to choose the direction and magnitude of this vector. Although we've drawn magnetic field vectors at only a few points around the magnet, it's important to remember that the magnetic field exists at *every* point around the magnet.

FIGURE 24.4 Mapping out the field of a bar magnet using compasses.

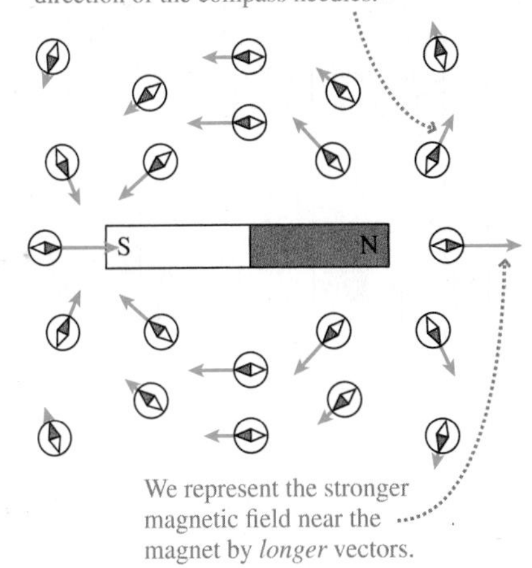

The magnetic field vectors point in the direction of the compass needles.
S
N
We represent the stronger magnetic field near the magnet by *longer* vectors.

We can also represent the magnetic field using magnetic field lines. The rules for drawing these lines are similar to those for drawing the electric field lines of Chapter 20. Electric field lines begin on positive charges and end on negative charges; magnetic field lines go from a north magnetic pole to a south magnetic pole. The direction and the spacing of the field lines show the direction and the strength of the field, as illustrated in FIGURE 24.5.

FIGURE 24.5 Drawing the magnetic field lines of a bar magnet.

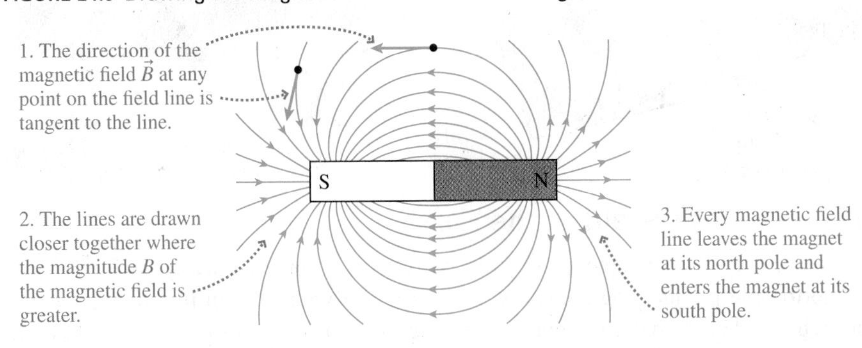

Now that we know how to think about magnetic fields, let's look at magnetic fields from magnets of different arrangements. We'll use the iron filing method to show the lines from real magnets, along with a drawing of the field lines.

An atlas of magnetic fields produced by magnets

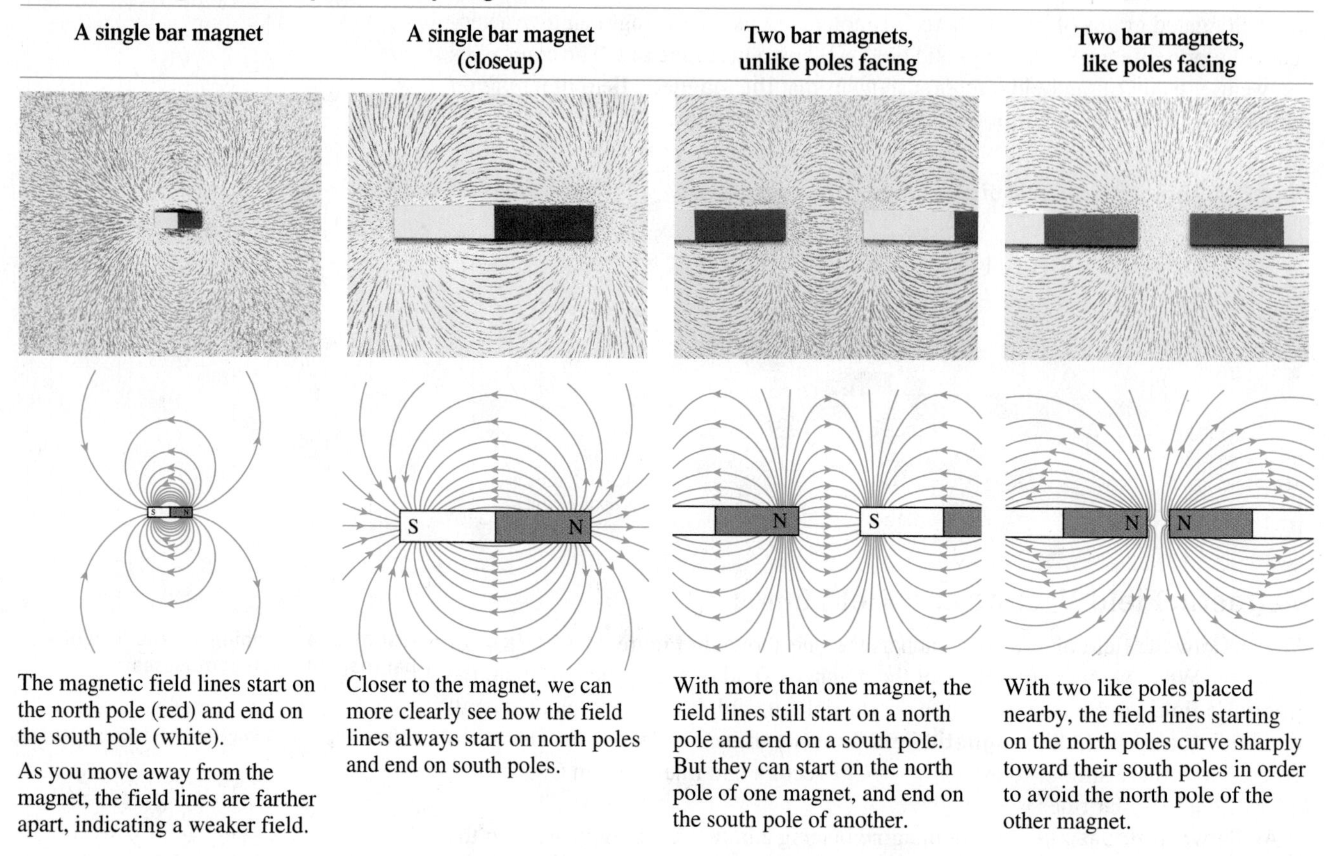
A single bar magnet
A single bar magnet (closeup)
Two bar magnets, unlike poles facing
Two bar magnets, like poles facing
S
N
S
N
N
S
N
N

The magnetic field lines start on the north pole (red) and end on the south pole (white).

As you move away from the magnet, the field lines are farther apart, indicating a weaker field.

Closer to the magnet, we can more clearly see how the field lines always start on north poles and end on south poles.

With more than one magnet, the field lines still start on a north pole and end on a south pole. But they can start on the north pole of one magnet, and end on the south pole of another.

With two like poles placed nearby, the field lines starting on the north poles curve sharply toward their south poles in order to avoid the north pole of the other magnet.

You are familiar with bar magnets that have a north pole on one end and a south pole on the other, but magnets can have more than one pair of north-south poles, and the poles need not be at the ends of the magnet. Flexible refrigerator magnets have an unusual arrangement of long, striped poles, as shown in **FIGURE 24.6**. Most of the field exits the plain side of the magnet, so this side sticks better to your refrigerator than the label side does.

FIGURE 24.6 The magnetic field of a refrigerator magnet.

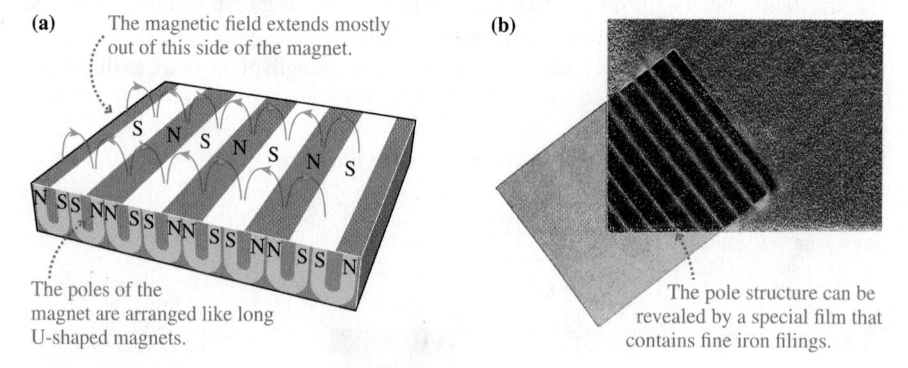
(a)
The magnetic field extends mostly out of this side of the magnet.
S N S N S N S
N S S N N S S N N S S N N S S N
The poles of the magnet are arranged like long U-shaped magnets.
(b)
The pole structure can be revealed by a special film that contains fine iron filings.

The Magnetic Field of the Earth

As we've seen, a bar magnet that is free to pivot—a compass—always swings so that its north pole points geographically north. But we've also seen that if a magnet is brought near a compass, the compass swings so that its north pole faces the south pole of the magnet. These observations can be reconciled if the earth itself is a large magnet, as shown in **FIGURE 24.7a**, where **the south pole of the earth's magnet is located near—but not exactly coincident with—the north geographic pole of the earth**. The north pole of a compass needle placed at the equator will point toward the south pole of the earth's magnet—that is, to the north.

The earth's magnetic field has components both parallel to the ground (horizontal) and perpendicular to the ground (vertical). An ordinary north-pointing compass responds to only the horizontal component of the field, but a compass free to pivot vertically will tilt downward as well. **FIGURE 24.7b** shows that, near the equator, the earth's magnetic field is nearly parallel to the ground. The angle from the horizontal, called the **dip angle,** is quite small. Near the poles, the field is more nearly vertical; the field points down more than it points north. Sea turtles seem to use the dip angle of the earth's field to determine their latitude.

FIGURE 24.7 The earth's magnetic field.

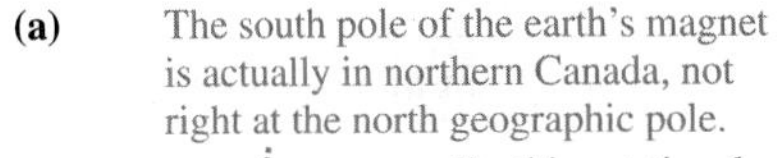

(a)
The south pole of the earth's magnet is actually in northern Canada, not right at the north geographic pole.

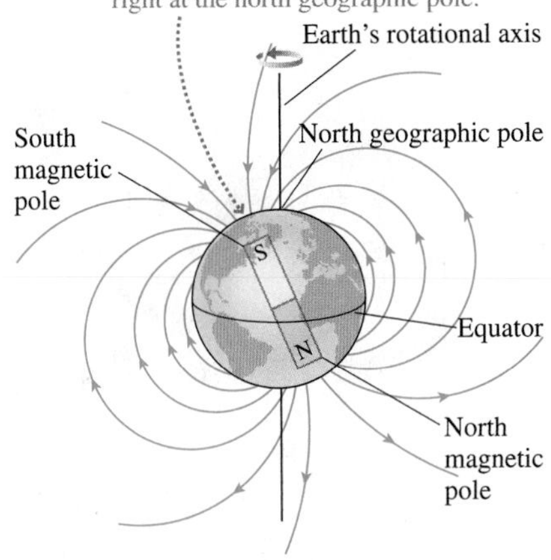

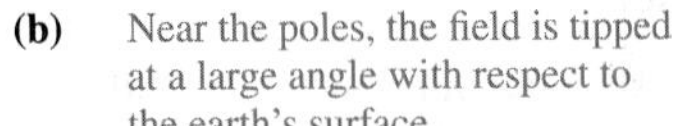

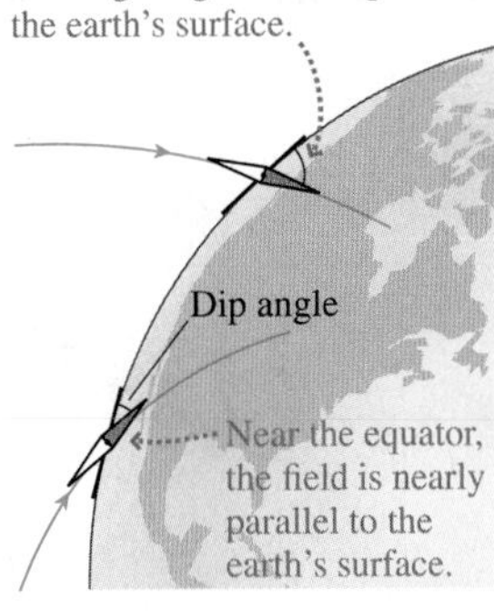

There are other animals that use the earth's magnetic field to navigate. You might imagine that animals would use an internal compass to determine orientation, to know which way is north, but the most unequivocal example of a creature that navigates using the earth's field is one that uses the field to tell up from down, not north from south. Magnetotactic bacteria have strongly magnetized bits of iron in their bodies, as you can see in **FIGURE 24.8**. Their bodies rotate to line up parallel to the earth's field, just like a compass. These bacteria prefer the low-oxygen zone near the bottom of bodies of water. If the water is disturbed and they are displaced upward, they travel in the direction of the earth's field, following the vertical component of the field downward.

FIGURE 24.8 BIO The internal compass of a magnetotactic bacterium.

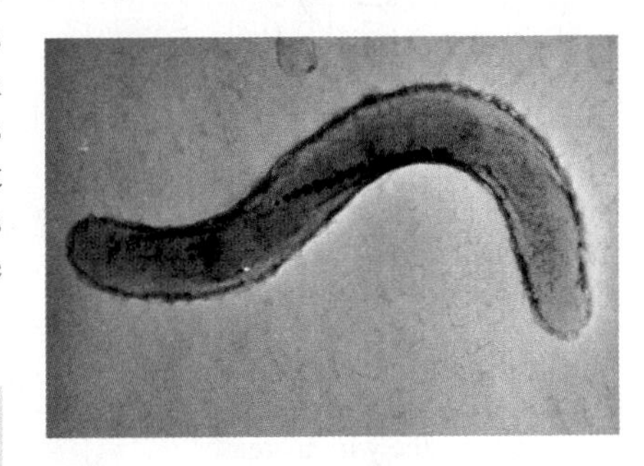

CONCEPTUAL EXAMPLE 24.1 **Balancing a compass**

Compasses made for use in northern latitudes are weighted so that the south pole of their needle is slightly heavier than the north pole. Explain why this is done.

REASON Figure 24.7b shows that, at northern latitudes, the magnetic field of the earth has a large vertical component. A compass needle that pivots to line up with the field has its north pole pointing north, but the north pole also tips down to follow the field. To keep the compass balanced, there must be an extra force on the south end of the compass. A small weight on the south pole provides a force that keeps the needle balanced.

ASSESS This strategy makes sense. Keeping the needle horizontal when the field is not horizontal requires some extra force.

STOP TO THINK 24.2 A compass is placed next to a bar magnet as shown. Which figure shows the correct alignment of the compass?

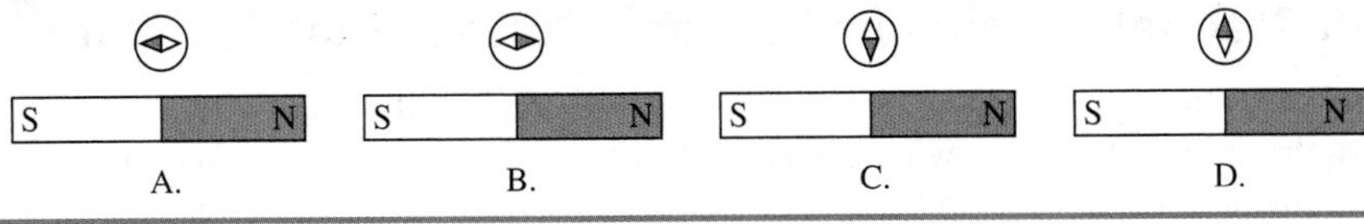

24.3 Electric Currents Also Create Magnetic Fields

You expect a compass to react to the presence of a bar magnet. But a compass will also deflect if you place the compass near a wire and pass a current through the wire. When the current stops, the compass goes back to its original orientation. This shows us that an *electric* current produces a *magnetic* field. The shape of the field lines depends on the shape of the current-carrying wire. The following table shows the most important configurations we'll study.

An atlas of magnetic fields produced by currents

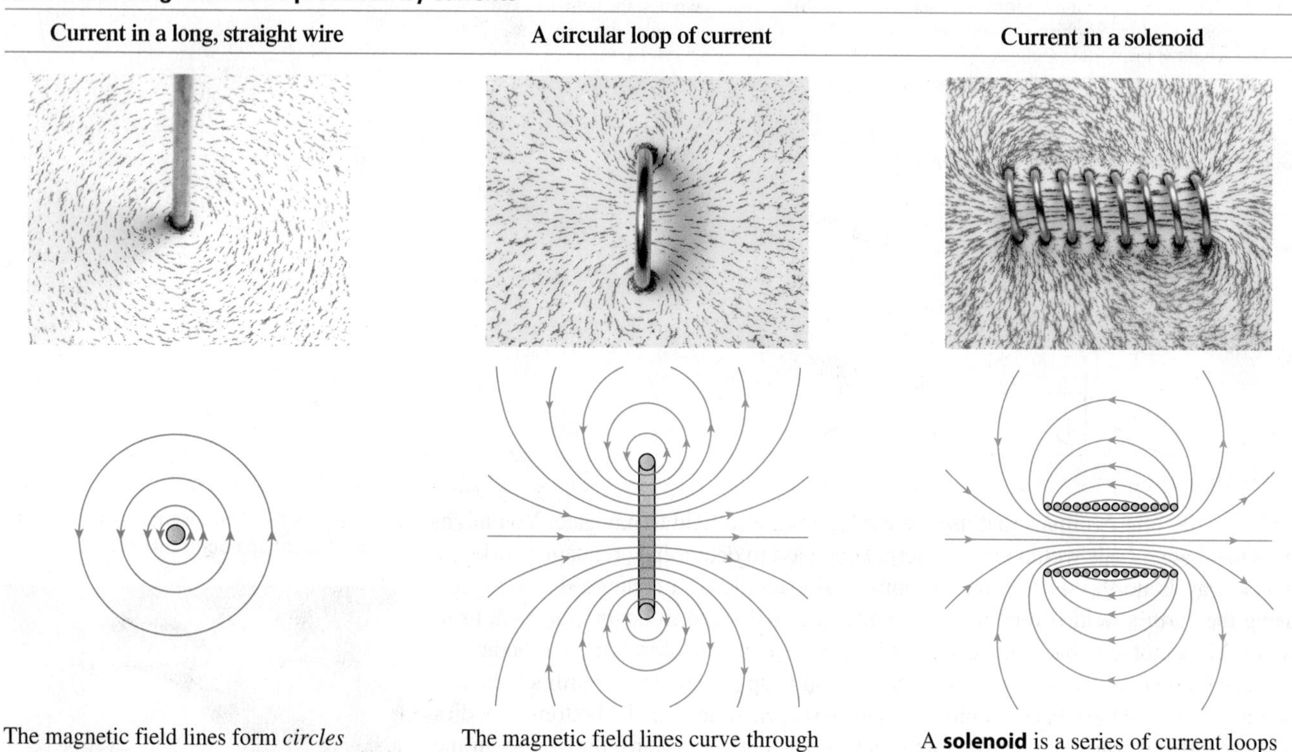

Current in a long, straight wire	A circular loop of current	Current in a solenoid
The magnetic field lines form *circles* around the wire. The iron filings are less affected by the field as the distance from the wire increases, indicating that the field is getting weaker as the distance from the wire increases.	The magnetic field lines curve through the center of the loop, around the outside, and back through the loop's center, forming complete closed curves. The field lines far from the loop look like the field lines far from a bar magnet.	A **solenoid** is a series of current loops placed along a common axis. The field outside is very weak compared to the field inside. Inside the solenoid, the magnetic field lines are reasonably evenly spaced; the field inside is nearly uniform.

Earlier we noted that the field lines of magnets start and end on magnetic poles. However, the field lines due to currents have no start or end: They form complete closed curves. If we consider the field lines continuing *inside* a magnet, however, we find that these lines also form complete closed curves, as shown in **FIGURE 24.9**.

FIGURE 24.9 Field lines form closed curves for magnets, too.

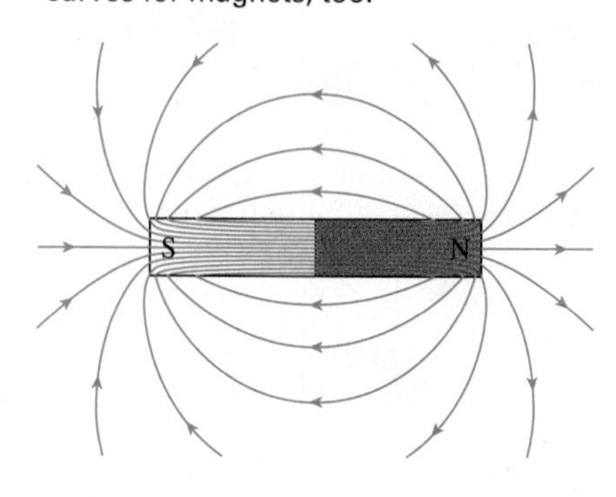

In this and the next section, we'll explore in some detail the magnetic fields created by currents; later, we'll look again at fields due to magnets. Ordinary magnets are often called **permanent magnets** to distinguish their unchanging magnetism from that caused by currents that can be switched on and off. We look at magnetism in this order because magnetism from currents is easier to understand, but keep in mind that currents and magnets are both equally important sources of magnetic fields.

The Magnetic Field of a Straight, Current-Carrying Wire

From the atlas picture above, we see that the iron filings line up in *circles* around a straight, current-carrying wire. As **FIGURE 24.10** shows, we also can use our basic instrument, the compass, to determine the direction of the magnetic field.

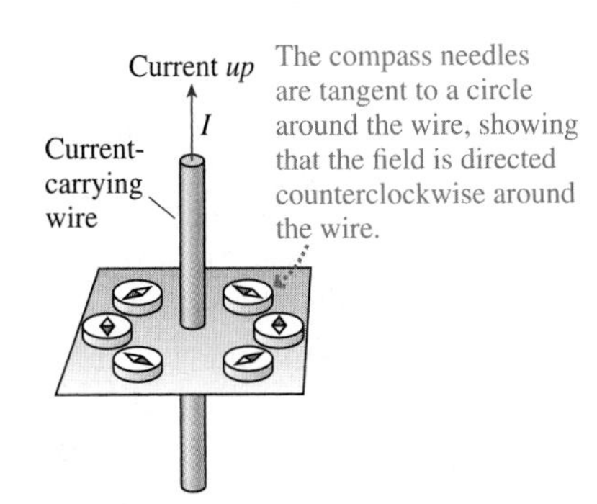
Current *up*
I
Current-carrying wire
The compass needles are tangent to a circle around the wire, showing that the field is directed counterclockwise around the wire.

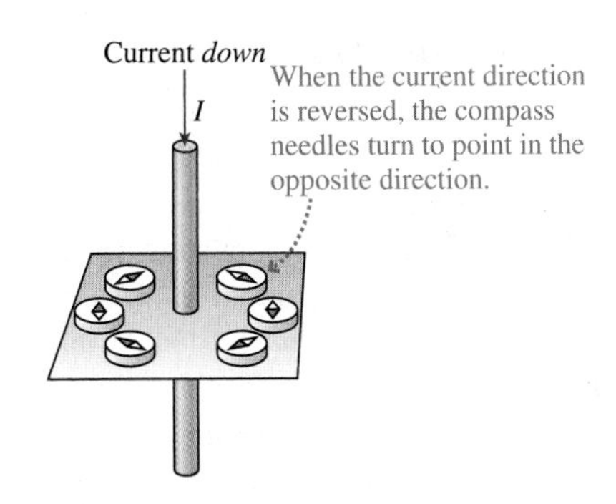
Current *down*
I
When the current direction is reversed, the compass needles turn to point in the opposite direction.

◄ **FIGURE 24.10** How compasses respond to a current-carrying wire.

To help remember in which direction compasses will point, we use the *right-hand rule* shown in Tactics Box 24.1. We'll use this same rule later to find the direction of the magnetic field due to several other shapes of current-carrying wire, so we'll call this rule the **right-hand rule for fields.**

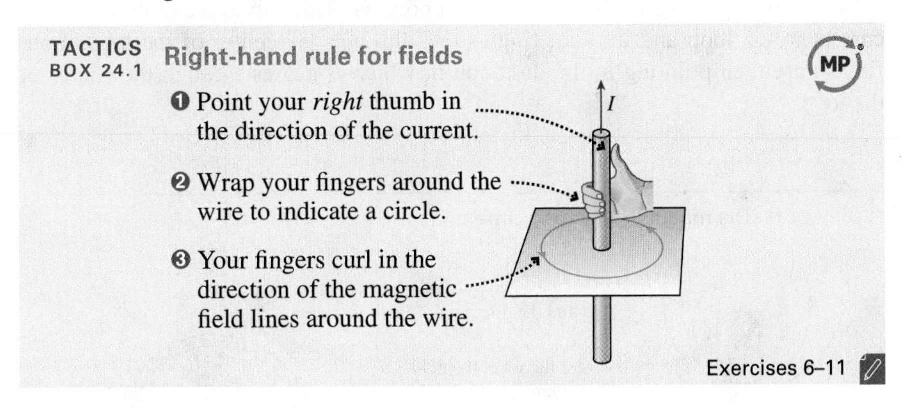

TACTICS BOX 24.1 Right-hand rule for fields

❶ Point your *right* thumb in the direction of the current.

❷ Wrap your fingers around the wire to indicate a circle.

❸ Your fingers curl in the direction of the magnetic field lines around the wire.

Exercises 6–11

Magnetism often requires a three-dimensional perspective of the sort shown in Tactics Box 24.1. But since two-dimensional figures are easier to draw, we will make as much use of them as we can. Consequently, we will often need to indicate field vectors or currents that are perpendicular to the page. **FIGURE 24.11a** shows the notation we will use. **FIGURE 24.11b** demonstrates this notation by showing the compasses around a current that is directed into the page. To use the right-hand rule with this drawing, point your right thumb into the page. Your fingers will curl clockwise, giving the direction in which the north poles of the compass needles point.

FIGURE 24.11 The notation for vectors and currents that are perpendicular to the page.

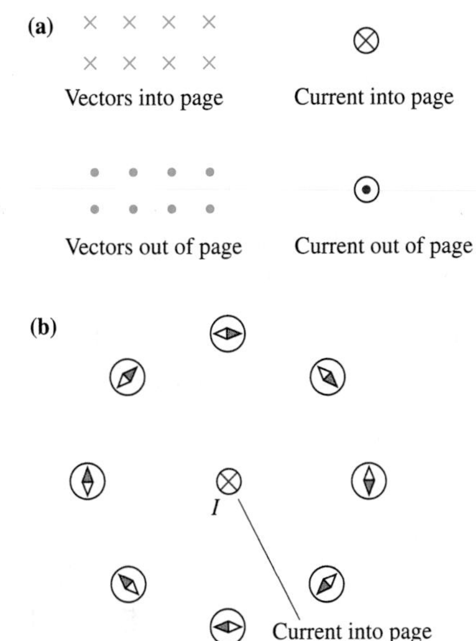
(a)
Vectors into page
Current into page
Vectors out of page
Current out of page
(b)
I
Current into page

CONCEPTUAL EXAMPLE 24.2 Drawing the magnetic field of a current-carrying wire

Sketch the magnetic field of a long, current-carrying wire, with the current going into the paper. Draw both magnetic field line and magnetic field vector representations.

REASON From the iron filing picture in the atlas, we have seen that the field lines form circles around the wire, and the magnetic field becomes weaker as the distance from the wire is increased. **FIGURE 24.12** shows how we construct both field line and field vector representations of such a field.

FIGURE 24.12 Drawing the magnetic field of a long, straight, current-carrying wire.

❶ ⊗ means current goes *into* the page: point your right thumb in this direction.

❷ Your fingers curl clockwise . . .

❸ . . . so the magnetic field lines are clockwise circles around the wire.

Magnetic field vectors are longer where the field is stronger.

Magnetic field vectors are tangent to the field lines.

Field lines are closer together where the field is stronger.

$\vec{B}$

ASSESS Figure 24.12 illustrates the key features of the field. The direction of the field vectors and field lines matches what we saw in Figure 24.11, and the field strength drops off with distance, as we learned in the atlas figure. We don't expect you to draw such a figure, but it's worth looking at the full 3-D picture of the field in **FIGURE 24.13**. This conveys the idea that the field lines exist in every plane along the length of the wire.

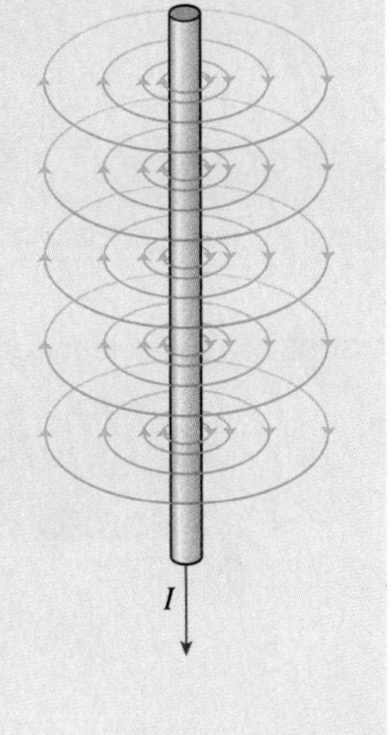
I

FIGURE 24.13 Field lines exist everywhere along the wire.

The Magnetic Field of a Current Loop

We can extend our understanding of the field from a long, straight, current-carrying wire to the fields due to other shapes of current-carrying wires. Let's start with the simple circular current-carrying loop, shown in three views in **FIGURE 24.14.** To see what the field due to a current loop looks like, we can imagine bending a straight wire into a loop, as shown in **FIGURE 24.15.** As we do so, the field lines near the wire will remain similar to what they looked like when the wire was still straight: circles going around the wire. Farther from the wires the field lines are no longer circles, but they still curve through the center of the loop, back around the outside, and then return through the center.

If we reverse the direction of the current in the loop, all the field lines reverse direction as well. Because a current loop is essentially a straight wire bent into a circle, the same right-hand rule of Tactics Box 24.1, used to find the field direction for a long, straight wire, can also be used to find the field direction for a current loop. As shown in Figure 24.15, you again point your thumb in the direction of the current in the loop and let your fingers curl through the center of the loop. Your fingers are then pointing in the direction in which $\vec{B}$ passes through the *center* of the loop.

FIGURE 24.14 Three views of a current loop.

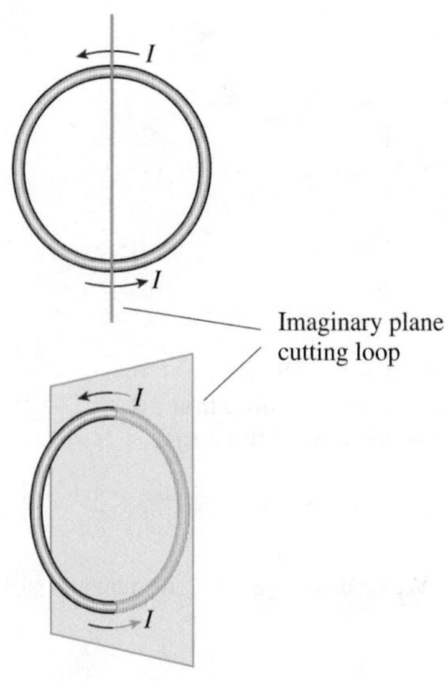
I
I
Imaginary plane
cutting loop
I
I

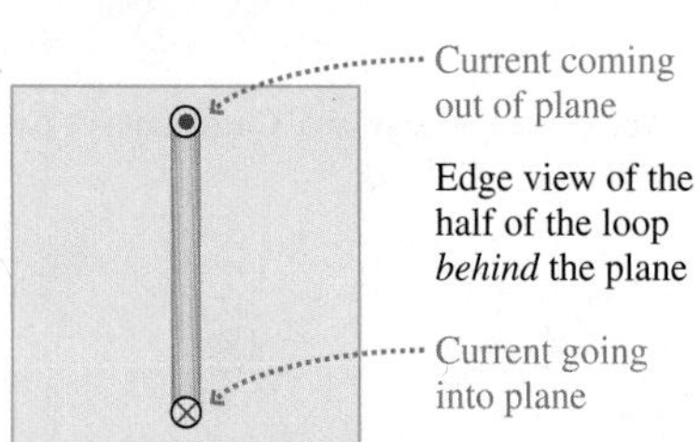
Current coming
out of plane
Edge view of the
half of the loop
behind the plane
Current going
into plane

Figure Video

FIGURE 24.15 The magnetic field of a current loop.

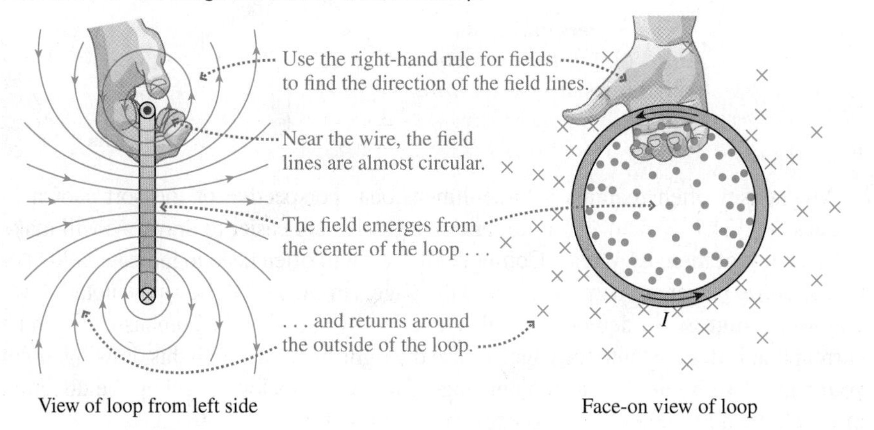
Use the right-hand rule for fields
to find the direction of the field lines.
Near the wire, the field
lines are almost circular.
The field emerges from
the center of the loop . . .
. . . and returns around
the outside of the loop.
I
View of loop from left side
Face-on view of loop

STOP TO THINK 24.3 The figure shows an edge view of a current loop. Using what you've learned in Figure 24.15, which arrow gives the direction of the field in the center of the loop?

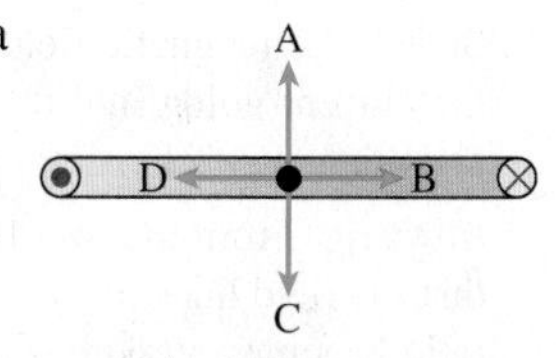
A
D
B
C

The Magnetic Field of a Solenoid

There are many applications of magnetism, such as the MRI system used to make the image at the beginning of this chapter, for which we would like to generate a **uniform magnetic field,** a field that has the same magnitude and the same direction at every point within some region of space. As we've seen, a reasonably uniform magnetic field can be generated with a solenoid. A solenoid, as shown in **FIGURE 24.16,** is a long coil of wire with the same current I passing through each loop in the coil. Solenoids may have hundreds or thousands of loops, often called *turns,* sometimes wrapped in several layers.

FIGURE 24.16 A solenoid.

I
I

The iron filing picture in the atlas on page 782 shows us that **the field within the solenoid is strong, mainly parallel to the axis, and reasonably uniform, whereas the field outside the solenoid is very weak.** FIGURE 24.17 reviews these points and shows why the field inside is much stronger than the field outside. The field direction inside the solenoid can be determined by using the right-hand rule for any of the loops that form it.

FIGURE 24.17 The field of a solenoid.

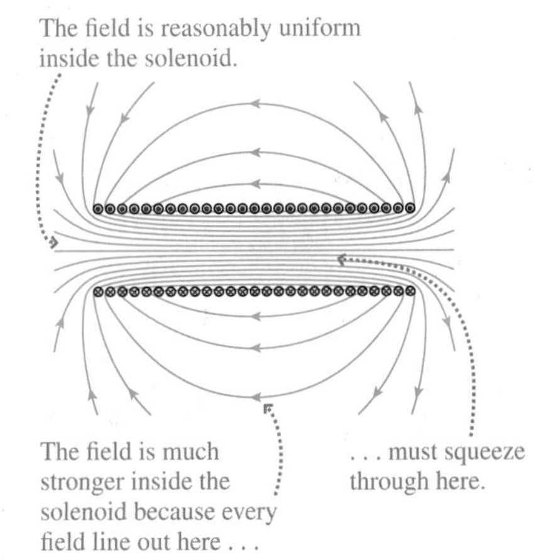
The field is reasonably uniform inside the solenoid.
The field is much stronger inside the solenoid because every field line out here . . .
. . . must squeeze through here.

STOP TO THINK 24.4 A compass is placed above a long wire. When a large current is turned on in the direction shown, in which direction will the compass point?

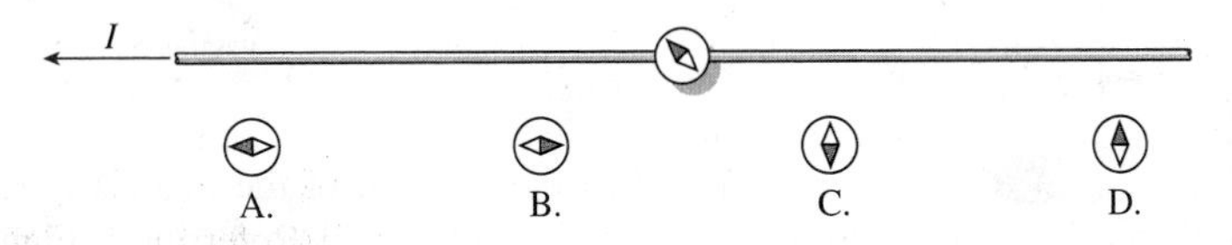
I
A.
B.
C.
D.

24.4 Calculating the Magnetic Field Due to a Current

We've now seen the appearance of the magnetic field for several shapes of wires. Next, we'll learn how to calculate magnetic fields. In SI units, magnetic field strengths are measured in **tesla,** abbreviated as T. One tesla is quite a large field, so most of the field strengths you will work with are much less than 1 T, as you can see from the field strengths listed in Table 24.1, so we often express field strengths in μT (10^{-6} T) or mT (10^{-3} T). The earth's magnetic field strength varies from place to place, but 50 μT is a good average value that you can use for solving problems.

Now, let's look at the expression for the field around a long, straight, current-carrying wire. We've seen already that the magnetic field lines form circles around the wire and that the field gets weaker as the distance from the wire increases. Not surprisingly, the magnitude of the field also depends on the *current* through the wire, increasing in proportion to the current. Putting these experimental observations together, we find that the magnitude of the magnetic field (in T) a distance r from the wire carrying current I is given by

$$B = \frac{\mu_0 I}{2\pi r} \tag{24.1}$$

Magnetic field due to a long, straight, current-carrying wire

B, r — INVERSE p. 108

TABLE 24.1 Typical magnetic field strengths

Field source and location	Field strength (T)
10 cm from a wire with 1 A current	2×10^{-6}
Surface of the earth	5×10^{-5}
1 cm from a wire with 10 A current	2×10^{-4}
Refrigerator magnet	5×10^{-3}
100-turn coil, 1 cm diameter, with 1 A current	1×10^{-2}
Surface of the sun, in a sunspot	1×10^{-1}
Near a rare-earth magnet	1
MRI solenoid	1

Equation 24.1 is only exact for an infinitely long wire, but it is quite accurate as well if the length of the wire is much greater than the distance r from the wire.

The constant μ_0, which relates the strength of the magnetic field to the currents that produce it, is called the **permeability constant.** Its role in magnetic field expressions is similar to the role of the permittivity constant in electric field expressions. Its value is

$$\mu_0 = 1.26 \times 10^{-6}\,\text{T} \cdot \text{m/A}$$

According to Equation 24.1, the magnetic field strength near a current-carrying wire decreases with distance; the strength of the field is inversely proportional to the distance from the wire. FIGURE 24.18 on the next page illustrates the meaning of this relationship.

FIGURE 24.18 The strength of the magnetic field due to a long, current-carrying wire vs. distance from the wire.

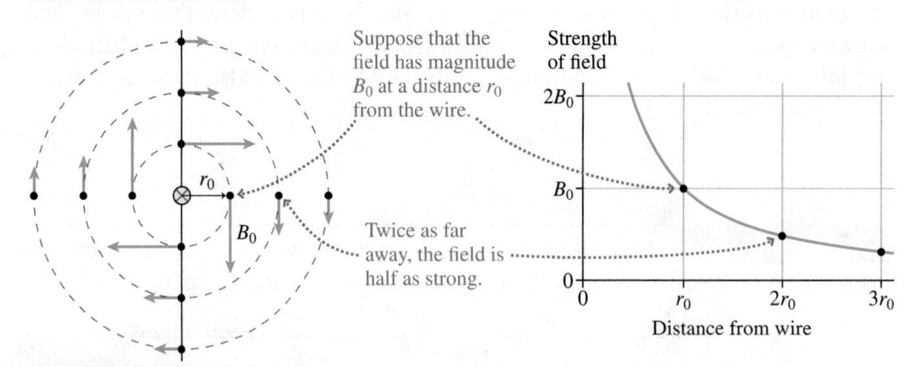

STOP TO THINK 24.5 Table 24.1 shows that the strength of the magnetic field 1 cm from a wire carrying a 10 A current is 4 times the earth's field. Referring to Figure 24.18 for guidance, at what distance from the wire will the field equal that of the earth?

A. 2 cm B. 4 cm C. 8 cm D. 16 cm

FIGURE 24.19 Adding fields due to more than one source.

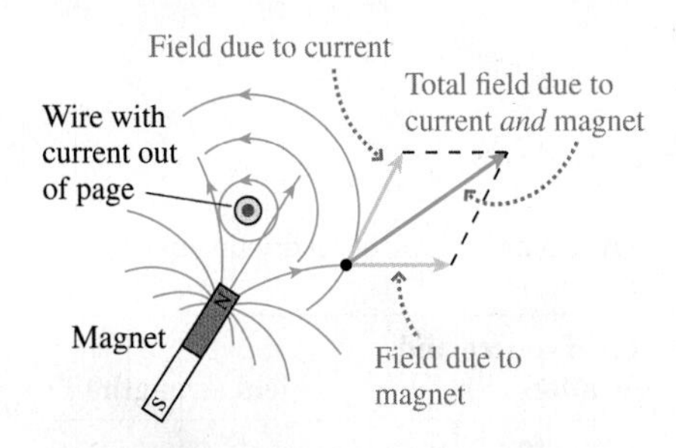

Magnetic Fields from More Than One Source

When two or more sources of magnetic fields are brought near each other, how do we find the total magnetic field at any particular point in space? For electric fields, we used the principle of superposition: The total electric field at any point is the *vector* sum of the individual fields at that point. The same principle holds for magnetic fields as well. FIGURE 24.19 illustrates the principle of superposition applied to *magnetic* fields.

EXAMPLE 24.3 Finding the magnetic field of two parallel wires

Two long, straight wires lie parallel to each other, as shown in FIGURE 24.20. They each carry a current of 5.0 A, but in opposite directions. What is the magnetic field at point P?

FIGURE 24.20 Two parallel, current-carrying wires.

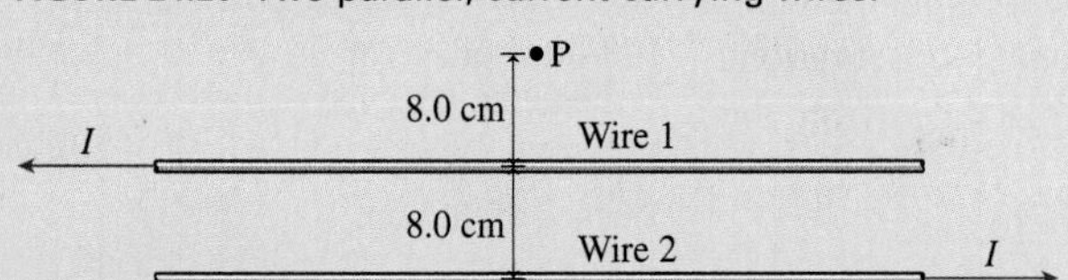

PREPARE Redraw the wires as seen from their right ends, as in FIGURE 24.21, and add a coordinate system. In this view, the right-hand rule for fields tells us that the magnetic field at P from wire 1 points to the right and that from wire 2 to the left. Because wire 2 is twice as far from P as wire 1, we've drawn its field $\vec{B}_2$ half as long as the field $\vec{B}_1$ from wire 1.

FIGURE 24.21 View of the wires from their right ends.

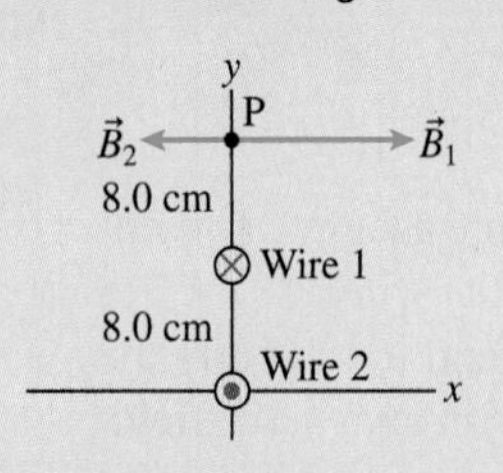

SOLVE From Figure 24.21, we can see that the total field $\vec{B}$—the vector sum of the two fields $\vec{B}_1$ and $\vec{B}_2$—points to the right. To find the magnitude of $\vec{B}$, we'll need the magnitudes of $\vec{B}_1$ and $\vec{B}_2$.

We can use Equation 24.1 to find these magnitudes. From Figure 24.21, r for wire 1 is 8.0 cm (or 0.080 m), while r for wire 2 is 16 cm.

We then have

$$B_1 = \frac{\mu_0 I}{2\pi r} = \frac{(1.26 \times 10^{-6}\ \text{T}\cdot\text{m/A})\,(5.0\ \text{A})}{2\pi\,(0.080\ \text{m})} = 1.253 \times 10^{-5}\ \text{T}$$

and

$$B_2 = \frac{\mu_0 I}{2\pi r} = \frac{(1.26 \times 10^{-6}\ \text{T}\cdot\text{m/A})\,(5.0\ \text{A})}{2\pi\,(0.16\ \text{m})} = 0.627 \times 10^{-5}\ \text{T}$$

Because $\vec{B}_2$ points to the left, its x-component is negative. Thus, the x-component of the total field $\vec{B}$ is

$$B_x = (B_1)_x + (B_2)_x$$
$$= 1.253 \times 10^{-5}\ \text{T} - 0.627 \times 10^{-5}\ \text{T} = 0.63 \times 10^{-5}\ \text{T}$$
$$= 6.3\ \mu\text{T}$$

In terms of the original view of the problem in Figure 24.20, we can write

$$\vec{B} = (6.3\ \mu\text{T, into the page})$$

ASSESS B_x is positive, which tells us that the total field points to the right, in the same direction as the field due to wire 1. This makes sense because P is closer to wire 1 than to wire 2. Table 24.1 gives the field 10 cm from a wire carrying 1 A as 2 μT. The field in this problem should be about the same size. The distances are about the same; the currents are larger, but there are two opposing fields. So, the magnitude of our answer makes sense.

Current Loops

The magnetic field due to a current loop is more complex than that of a straight wire, as we can see from **FIGURE 24.22**, but there is a simple expression for the field at the *center* of the loop. Because the loop can be thought of as a wire bent into a circle, the expression for the strength of the field is similar to that of a wire. The magnitude of the field at the center is

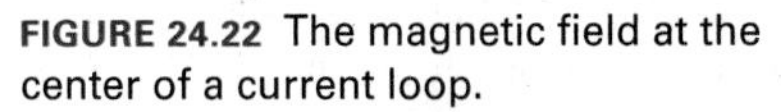

FIGURE 24.22 The magnetic field at the center of a current loop.

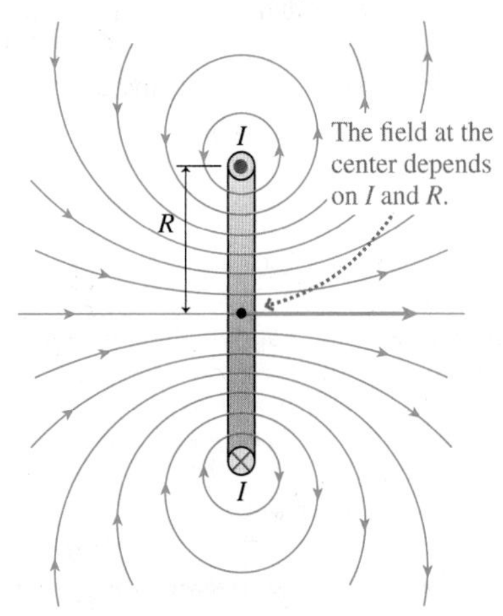

$$B = \frac{\mu_0 I}{2R} \quad (24.2)$$

Magnetic field at the center of a current loop of radius R

where I is the current in the loop.

If N loops of wire carrying the same current I are all tightly wound into a single flat coil, then the magnitude of the field at the center is just N times bigger (since we're superimposing N individual current loops):

$$B = \frac{\mu_0 NI}{2R} \quad (24.3)$$

Magnetic field at the center of a thin coil with N turns

▶ **The magnetocardiogram** BIO When the heart muscle contracts, action potentials create a dipole electric field that can be measured to create an electrocardiogram. These action potentials also cause charges to circulate around the heart, creating a current loop. This current loop creates a small, but measurable, magnetic field. A record of the heart's magnetic field, a *magnetocardiogram,* can provide useful information about the heart in cases where an electrocardiogram is not possible. This image shows the magnetic field of a fetal heartbeat measured at the surface of the mother's abdomen. Here, blue represents a field pointing into the body and red a field pointing out of the body. This is just the field expected from a current loop whose plane lies along the black line between the two colored areas.

EXAMPLE 24.4 **Canceling the earth's field**

Green turtles are thought to navigate by using the dip angle of the earth's magnetic field. To test this hypothesis, green turtle hatchlings were placed in a 72-cm-diameter tank with a 60-turn coil of wire wrapped around the outside. A current in the coil created a magnetic field at the center of the tank that exactly canceled the vertical component of the earth's 50 μT field. At the location of the test, the earth's field was directed 60° below the horizontal. What was the current in the coil?

PREPARE **FIGURE 24.23** shows the earth's field passing downward through the coil. To cancel the vertical component of this field, the current in the coil must generate an upward field of equal magnitude. We can use the right-hand rule (see Figure 24.15) to find that the current must circulate around the coil as shown. Viewed from above, the current will be counterclockwise.

FIGURE 24.23 The coil field needed to cancel the vertical component of the earth's field.

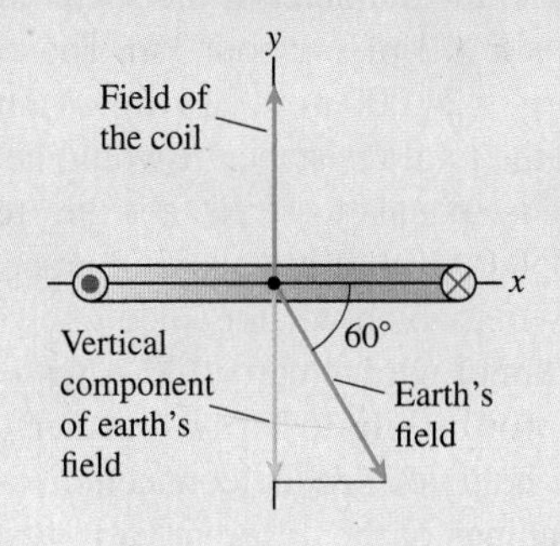

SOLVE The vertical component of the earth's field is

$$(B_{\text{earth}})_y = -(50 \times 10^{-6}\ \text{T})\sin(60°) = -4.33 \times 10^{-5}\ \text{T}$$

The field of the coil, given by Equation 24.3, must have the same magnitude at the center. The $2R$ in the equation is just the diameter of the coil, 72 cm or 0.72 m. Thus

$$B_{\text{coil}} = \frac{\mu_0 NI}{2R} = 4.33 \times 10^{-5}\ \text{T}$$

$$I = \frac{(4.33 \times 10^{-5}\ \text{T})(2R)}{(\mu_0 N)}$$

$$= \frac{(4.33 \times 10^{-5}\ \text{T})(0.72\ \text{m})}{(1.26 \times 10^{-6}\ \text{T}\cdot\text{m/A})(60)} = 0.41\ \text{A}$$

As noted, this current is counterclockwise as viewed from above.

Continued

ASSESS Equation 24.3 shows that the field in the center of a coil is proportional to the number of turns in the coil, proportional to the current, and inversely proportional to the radius of the coil. Table 24.1 gives 0.01 T for the field in the center of a 100-turn coil that is 1 cm in diameter and carries a current of 1.0 A. The coil in this problem is nearly 100 times larger, carries about half as much current, and has about half as many turns. We'd therefore predict a field that is less than the Table 24.1 value by a factor of a few hundred, which it is. A rough estimate of the answer agrees with our result, so it seems reasonable.

Solenoids

FIGURE 24.24 The magnetic field inside a solenoid.

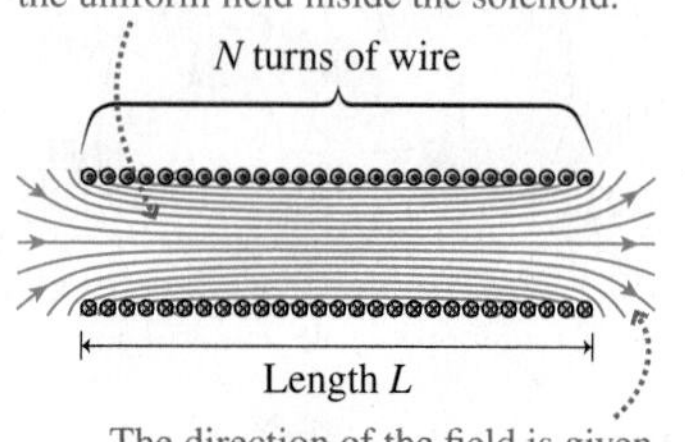
Equation 24.4 gives the magnitude of the uniform field inside the solenoid.
N turns of wire
Length L
The direction of the field is given by the right-hand rule for fields.

As we've seen, the field inside a solenoid is fairly uniform, while the field outside is quite small; the greater a solenoid's length in comparison to its diameter, the better these statements hold. Measurements that need a uniform magnetic field are often conducted inside a solenoid, which can be built quite large. The cylinder that surrounds a patient undergoing magnetic resonance imaging (MRI), such as the one shown at the bottom of the page, contains a large solenoid. Consider a solenoid of length L with N turns of wire, as in **FIGURE 24.24**. We expect that the more turns we can pack into a solenoid of a given length—that is, the greater the ratio N/L—the stronger the field inside will be. We further expect that the strength of the field will be proportional to the current I in the turns. Somewhat surprisingly, the field inside a solenoid does *not* depend on its radius. For this reason, the radius R doesn't appear in the equation for the field inside a solenoid:

$$B = \frac{\mu_0 NI}{L} \tag{24.4}$$

Magnetic field inside a solenoid of length L with N turns

EXAMPLE 24.5 Generating an MRI magnetic field BIO

A typical MRI solenoid has a length of about 1 m and a diameter of about 1 m. A typical field inside such a solenoid is about 1 T. How many turns of wire must the solenoid have to produce this field if the largest current the wire can carry is 100 A?

PREPARE This solenoid is not very long compared to its diameter, so using Equation 24.4 will give only an approximate result. This is acceptable, since we have only rough estimates of the field B and the length L.

Equation 24.4 gives the magnetic field B of a solenoid in terms of the current I, the number of turns N, and the length L. Here, however, we want to find the number of turns in terms of the other variables. We'll need $B = 1$ T, $I = 100$ A, and $L = 1$ m.

SOLVE We can solve Equation 24.4 for N to get

$$N = \frac{LB}{\mu_0 I} = \frac{(1\text{ m})(1\text{ T})}{(1.26 \times 10^{-6}\text{ T}\cdot\text{m/A})(100\text{ A})} = 8000\text{ turns}$$

to one significant figure.

ASSESS The number of turns required is quite large, but the field is quite large, so this makes sense.

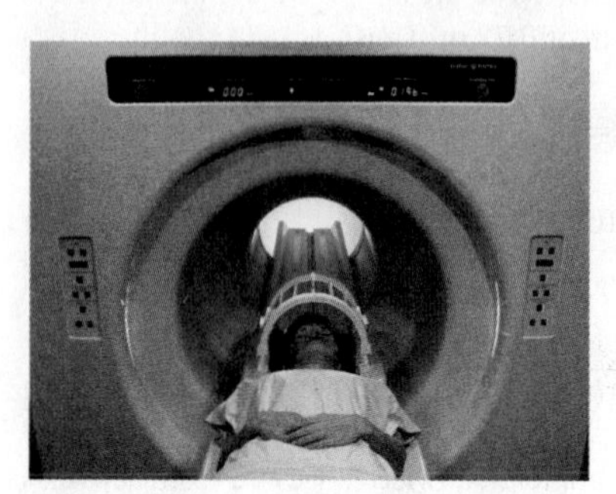

A patient's head in the solenoid of an MRI scanner.

In the above example, the diameter of the solenoid was about 1 m. The length of wire in each turn is thus $\pi \times 1$ m, or about 3 m. The total length of wire is then about 8000 turns $\times$ 3 m/turn = 24,000 m = 24 km $\approx$ 15 miles! If this magnet used ordinary copper wire, the total resistance R would be about 35 Ω. The power dissipated by a resistor is equal to I^2R, so the total power would be about $(100\text{ A})^2(35\ \Omega) = 350{,}000$ W, a huge and impractical value. MRI magnets must use *superconducting* wire, which when cooled to near absolute zero has *zero* resistance. This allows 24 km of wire to carry 100 A with no power dissipation at all.

At this point, it's worthwhile to bring together the relationships for magnetic fields from currents to help you see the connections between the different situations and remember the meanings of the different variables:

SYNTHESIS 24.1 Fields from currents

An electric current produces a magnetic field. The geometry of the field depends on the geometry of the conductor.

Current in a **long, straight wire** creates a magnetic field that encircles the wire.

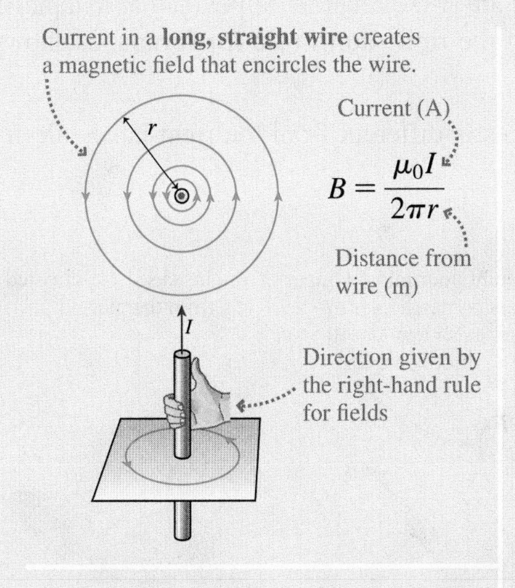

Current (A)

$$B = \frac{\mu_0 I}{2\pi r}$$

Distance from wire (m)

Direction given by the right-hand rule for fields

Permittivity constant

$$\mu_0 = 1.26 \times 10^{-6}\ \mathrm{T \cdot m/A}$$

A **coil** is a stack of loops with the length small compared to the radius. We treat this as a loop with multiple turns of wire.

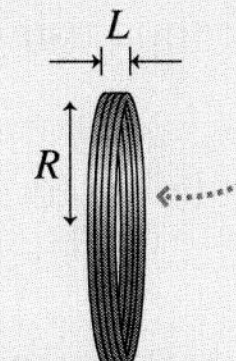

For current in a coil or a loop, field lines emerge from the center of the loop and wrap around it in a direction given by the right-hand rule for fields.

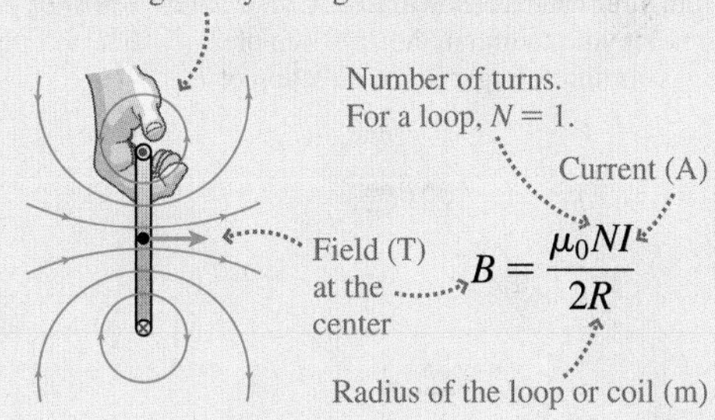

Number of turns. For a loop, $N = 1$.

Current (A)

Field (T) at the center

$$B = \frac{\mu_0 N I}{2R}$$

Radius of the loop or coil (m)

A **solenoid** is a stack of loops with the length much greater than the radius.

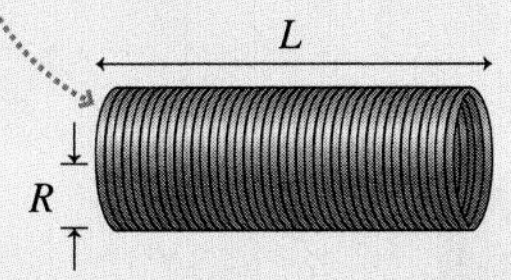

The field (T) inside a solenoid is uniform.

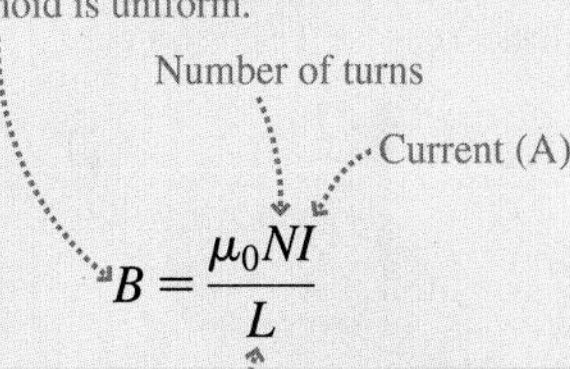

Number of turns

Current (A)

$$B = \frac{\mu_0 N I}{L}$$

Length of the solenoid (m)

STOP TO THINK 24.6 An investigator has made a solenoid by wrapping 100 turns of wire on a tube that is 10 cm long and 2 cm in diameter. The power supply is providing as much current as it can, but a stronger field is needed, so the solenoid must be rewrapped. Which of the following will result in a stronger field?

A. Wrapping 100 turns of wire on a tube that is 20 cm long and 2 cm in diameter.
B. Wrapping 100 turns of wire on a tube that is 10 cm long and 1 cm in diameter.
C. Wrapping 100 turns of wire on a tube that is 5 cm long and 2 cm in diameter.

24.5 Magnetic Fields Exert Forces on Moving Charges

It's time to switch our attention from what magnetic fields look like and how they are created to what they actually *do*. We've already seen that magnetic fields exert forces and torques on magnets, such as the torque that causes a compass needle to line up with the field. Now we'll see that magnetic fields also exert forces on moving charged particles and, later on, on electric currents in wires. The following series of experiments illustrates the nature of the magnetic force on a moving charged particle.

The force on a charged particle moving in a magnetic field

$\vec{F} = \vec{0}$

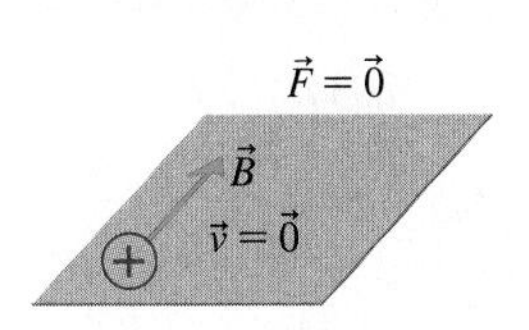

There is no magnetic force on a charged particle at rest.

$\vec{F} = \vec{0}$

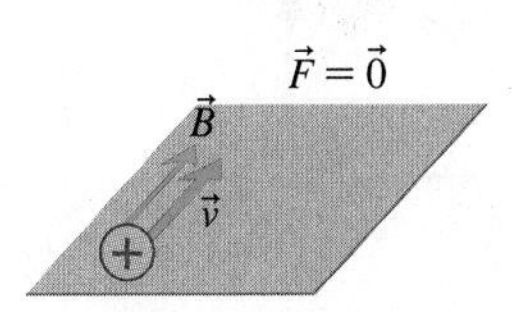

There is no magnetic force on a charged particle moving *parallel* to a magnetic field.

Force perpendicular to plane

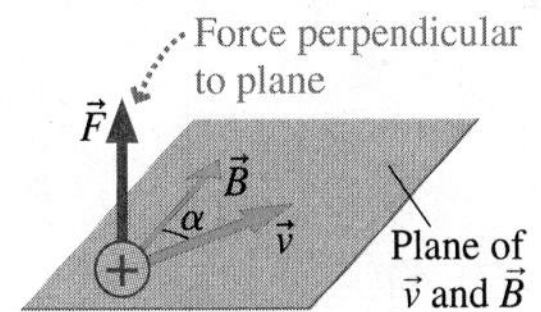

Plane of $\vec{v}$ and $\vec{B}$

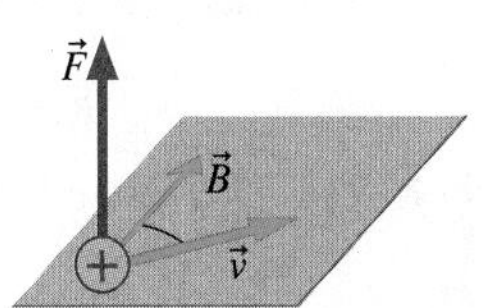

Force greatest at $\alpha = 90°$

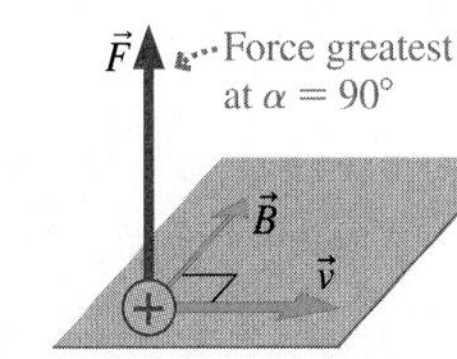

As the angle α between the velocity and the magnetic field increases, the magnetic force also increases. The force is greatest when the angle is 90°. The magnetic force is always perpendicular to the plane containing $\vec{v}$ and $\vec{B}$.

As the experiments show, the magnetic force is quite different from the electric force. First, there is no magnetic force if the charged particle is at rest or if it's moving parallel to the magnetic field. Second, the force is always *perpendicular* to the plane containing $\vec{v}$ and $\vec{B}$. Because there are two directions perpendicular to a plane, we determine the correct direction using the **right-hand rule for forces,** as shown in **FIGURE 24.25**.

NOTE ► The right-hand rule for forces is different from the right-hand rule for fields. ◄

FIGURE 24.25 The right-hand rule for forces.

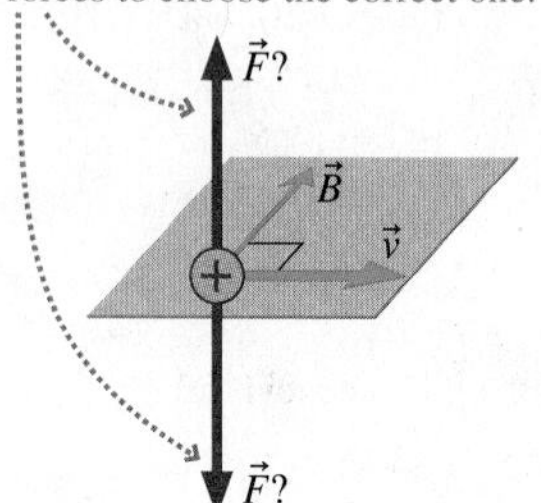

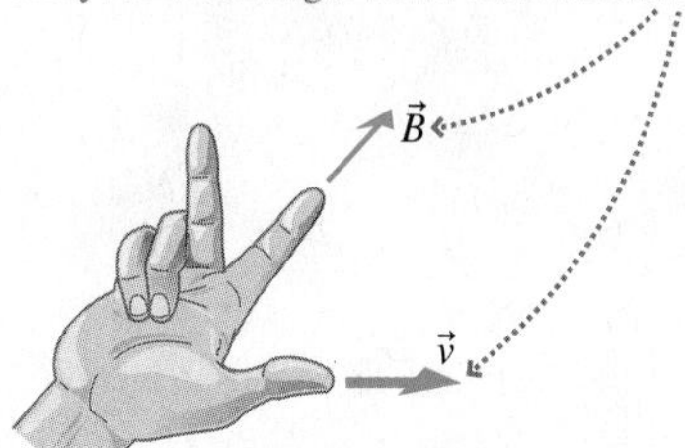

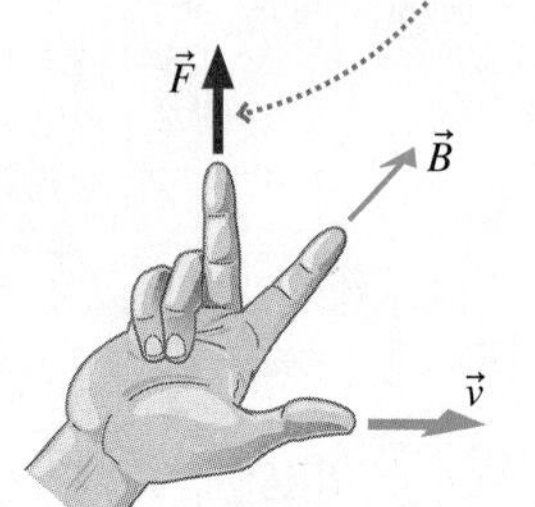

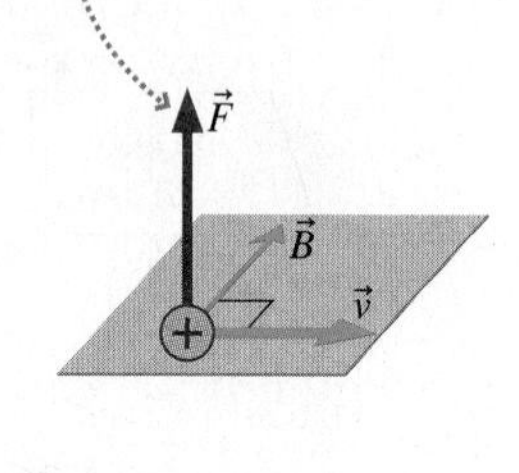

NOTE ► The right-hand rule for forces gives the direction of the force on a *positive* charge. For a negative charge, the force is in the opposite direction. ◄

We can organize all of the experimental information about the magnetic force on a moving charged particle into a single equation. The magnitude and direction of the force $\vec{F}$ on a charged particle moving in a magnetic field are given by

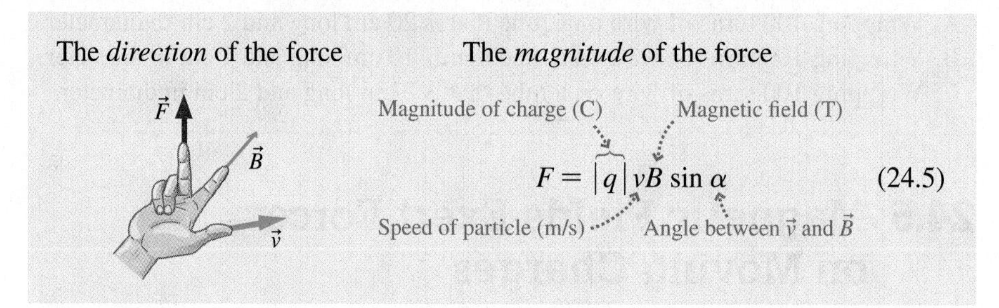

$$F = |q|vB \sin \alpha \tag{24.5}$$

Video Tutor Demo

The velocity and the magnetic field are perpendicular in many practical situations. In this case α is 90°, and the magnitude of the magnetic field simplifies to

$$F = |q|vB \tag{24.6}$$

with the direction of the field again determined by the right-hand rule for forces. The following Tactics Box summarizes and shows how to use the above information.

TACTICS BOX 24.2 **Determining the magnetic force on a moving charged particle**

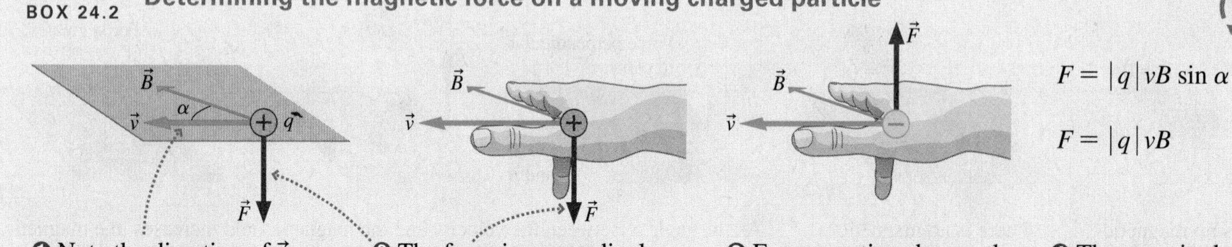

$$F = |q|vB \sin \alpha$$

$$F = |q|vB$$

❶ Note the direction of $\vec{v}$ and $\vec{B}$, and find the angle α between them.

❷ The force is perpendicular to the plane containing $\vec{v}$ and $\vec{B}$. The direction of $\vec{F}$ is given by the right-hand rule.

❸ For a negative charge, the force is in the direction opposite that predicted by the right-hand rule.

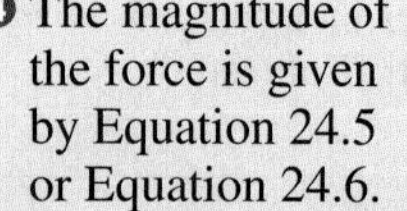

❹ The magnitude of the force is given by Equation 24.5 or Equation 24.6.

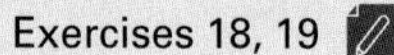

Exercises 18, 19

CONCEPTUAL EXAMPLE 24.6 Determining the force on a moving electron

An electron is moving to the right in a magnetic field that points upward, as in **FIGURE 24.26**. What is the direction of the magnetic force?

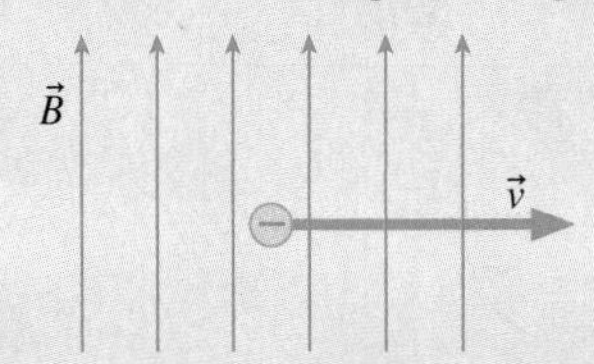

FIGURE 24.26 An electron moving in a magnetic field.

REASON FIGURE 24.27 shows how the right-hand rule for forces is applied to this situation:

- Point your right thumb in the direction of the electron's velocity and your index finger in the direction of the magnetic field.
- Bend your middle finger to be perpendicular to your index finger. Your middle finger, which now points out of the page, is the direction of the force on a positive charge. But the electron is negative, so the force on the electron is *into* the page.

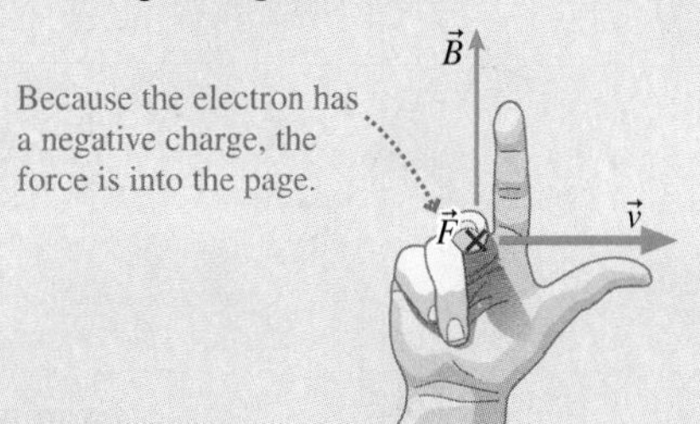

FIGURE 24.27 Using the right-hand rule.

ASSESS The force is perpendicular to both the velocity and the magnetic field, as it must be. The force on an electron is into the page; the force on a proton would be out of the page.

CONCEPTUAL EXAMPLE 24.7 Determining the force on a charged particle moving near a current-carrying wire

A proton is moving to the right above a horizontal wire that carries a current to the right. What is the direction of the magnetic force on the proton?

REASON The current in the wire creates a magnetic field; this magnetic field exerts a force on the moving proton. We follow three steps to solve the problem:

1. Sketch the situation, as shown in **FIGURE 24.28a**.
2. Determine the direction of the field at the position of the proton due to the current in the wire (**FIGURE 24.28b**).
3. Determine the direction of the force that this field exerts on the proton (**FIGURE 24.28c**). In this case, the force is down.

ASSESS As it should be, the magnetic force is perpendicular to both $\vec{v}$ and $\vec{B}$. This example also gives insight into how two parallel, current-carrying wires exert forces on each other, a topic we'll study in a later section. The proton moving to the right, which is essentially a small current in the same direction as the current in the wire, is *attracted* toward the wire.

FIGURE 24.28 Determining the direction of the force.

(a)

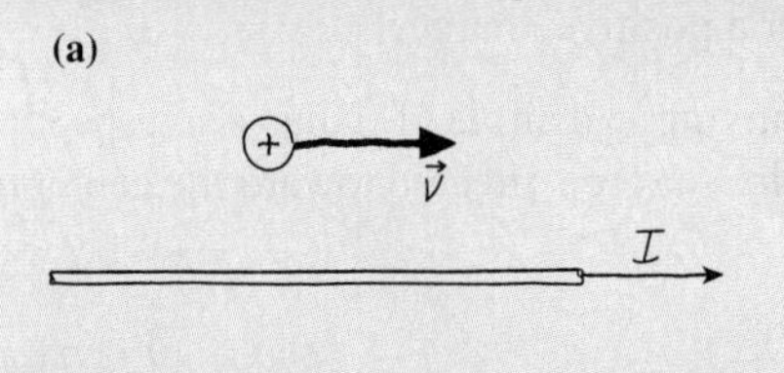

1. The proton moves above the wire. The proton velocity and the current in the wire are shown.

(b)

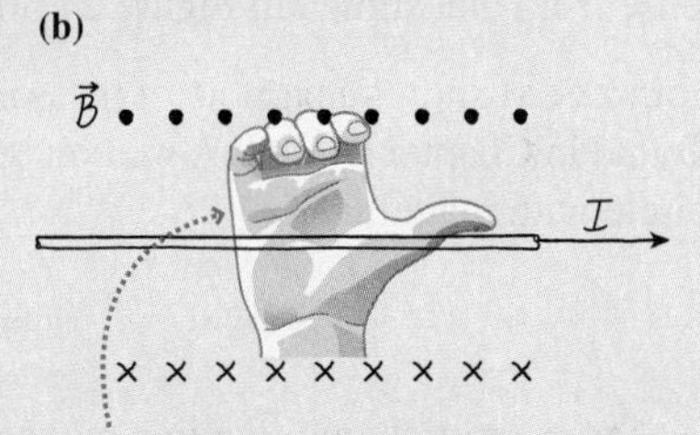

2. The right-hand rule for fields shows that, at the position of the proton, above the wire, the field points out of the page.

(c)

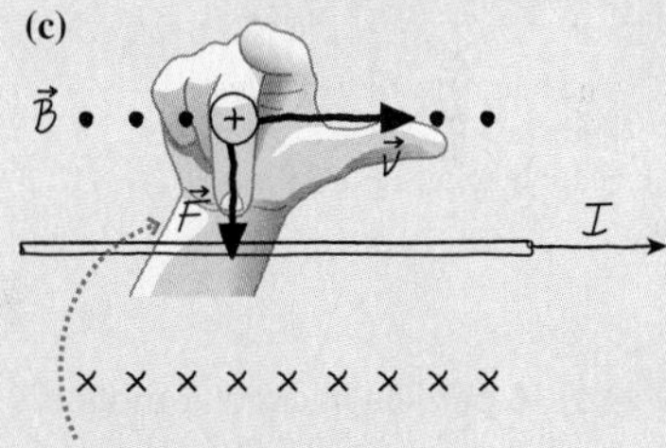

3. The right-hand rule for forces shows that the field of the wire exerts a force on the moving proton that points down, toward the wire.

EXAMPLE 24.8 Force on a charged particle in the earth's field

The sun emits streams of charged particles (in what is called the *solar wind*) that move toward the earth at very high speeds. A proton is moving toward the equator of the earth at a speed of 500 km/s. At this point, the earth's field is 5.0×10^{-5} T directed parallel to the earth's surface. What are the direction and the magnitude of the force on the proton?

PREPARE Our first step is, as usual, to draw a picture. As we saw in Figure 24.7a, the field lines of the earth go from the earth's south pole to the earth's north pole. **FIGURE 24.29** shows the proton entering the field, which is directed north.

We also need to convert the proton's velocity to m/s:

$$500 \text{ km/s} = 5.0 \times 10^2 \text{ km/s} \times \frac{1 \times 10^3 \text{ m}}{1 \text{ km}} = 5.0 \times 10^5 \text{ m/s}$$

SOLVE We use the steps of Tactics Box 24.2 to determine the force.

❶ $\vec{v}$ and $\vec{B}$ are perpendicular, so $\alpha = 90°$.

❷ The right-hand rule for forces tells us that the force will be into the page in Figure 24.29. That is, the force is toward the east.

❹ We compute the magnitude of the force using Equation 24.6:

$$F = |q|vB = (1.6 \times 10^{-19}\text{ C})(5.0 \times 10^5 \text{ m/s})(5.0 \times 10^{-5}\text{ T})$$
$$= 4.0 \times 10^{-18}\text{ N}$$

ASSESS This is a small force, but the proton has an extremely small mass of 1.67×10^{-27} kg. Consequently, this force produces a very large acceleration: approximately 200 million times the acceleration due to gravity!

Continued

FIGURE 24.29 A proton in the field of the earth.

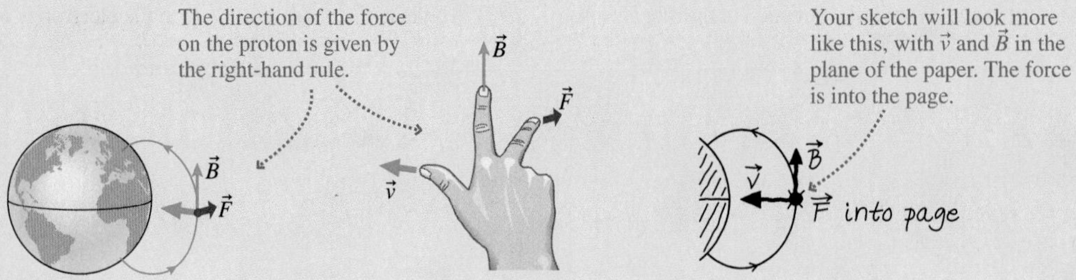

The direction of the force on the proton is given by the right-hand rule.

Your sketch will look more like this, with $\vec{v}$ and $\vec{B}$ in the plane of the paper. The force is into the page.

$\vec{F}$ into page

Example 24.8 looked at the force on a proton from the sun as it reaches the magnetic field of the earth. How does this force affect the *motion* of the proton? Let's consider the general question of how charged particles move in magnetic fields.

Paths of Charged Particles in Magnetic Fields

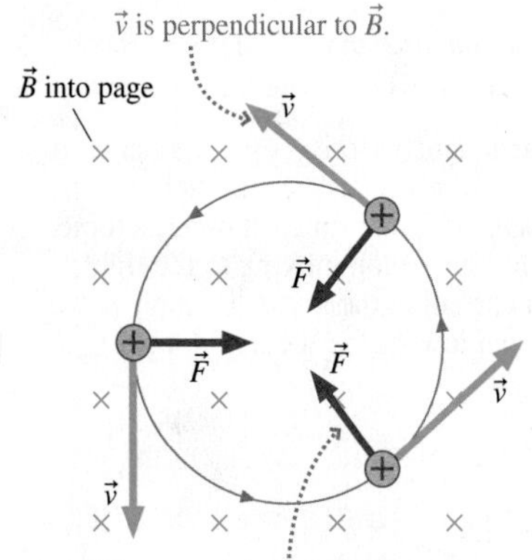

FIGURE 24.30 A charged particle moving perpendicular to a uniform magnetic field.

We know that the magnetic force on a moving charged particle is always perpendicular to its velocity. Suppose a positively charged particle is moving perpendicular to a uniform magnetic field $\vec{B}$, as shown in FIGURE 24.30. In ◀◀ SECTION 6.3, we looked at the motion of objects subject to a force that was always perpendicular to the velocity. The result was *circular motion at a constant speed.* For a ball moving in a circle at the end of a string, the tension force is always perpendicular to $\vec{v}$. For a satellite moving in a circular orbit, the gravitational force is always perpendicular to $\vec{v}$. Now, for a charged particle moving in a magnetic field, the magnetic force is always perpendicular to $\vec{v}$ and so causes the particle to move in a circle, as Figure 24.30 shows. Thus, **a particle moving perpendicular to a uniform magnetic field undergoes uniform circular motion at constant speed.**

NOTE ▶ The direction of the force on a negative charge is opposite to that on a positive charge, so a particle with a negative charge will orbit in the opposite sense from that shown in Figure 24.30 for a positive charge. ◀

FIGURE 24.31 shows a particle of mass m moving in a circle of radius r at a speed v. We found in Chapter 6 that this motion requires a force directed toward the center of the circle with magnitude

$$F = \frac{mv^2}{r} \tag{24.7}$$

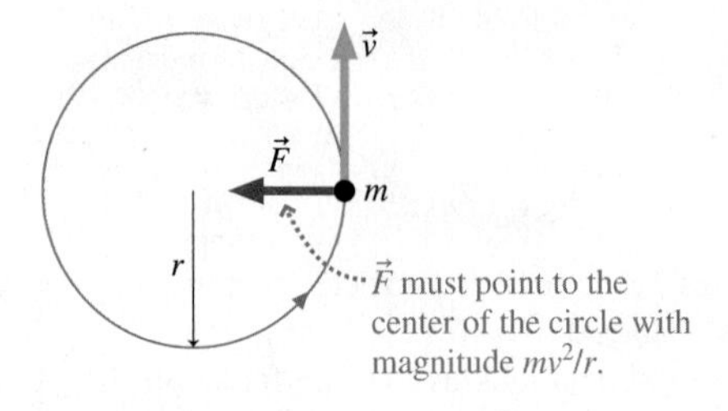

FIGURE 24.31 A particle in circular motion.

For a charged particle moving in a magnetic field, this force is provided by the magnetic force. In Figure 24.30 we assumed that the velocity was perpendicular to the magnetic field, so the magnitude of the force on the charged particle due to the magnetic field is given by Equation 24.6. This is the force that produces the circular motion, so we can equate it to the force in Equation 24.7:

$$F = |q|vB = \frac{mv^2}{r}$$

Solving for r, we find that the radius of the circular orbit for a charged particle moving in a magnetic field is given by

$$r = \frac{mv}{|q|B} \tag{24.8}$$

Particle mass (kg); Particle velocity (m/s); Particle charge (C); Magnetic field strength (T)

A particle moving *perpendicular* to a magnetic field moves in a circle. In the table at the start of this section, we saw that a particle moving *parallel* to a magnetic field experiences no magnetic force, and so continues in a straight line. A more general situation in which a charged particle's velocity $\vec{v}$ is neither parallel to nor perpendicular to the field $\vec{B}$ is shown in **FIGURE 24.32**. The net result is a circular motion due to the perpendicular component of the velocity coupled with a constant velocity parallel to the field: The charged particle spirals around the magnetic field lines in a helical trajectory.

High-energy particles stream out from the sun in the solar wind. Some of the charged particles of the solar wind become trapped in the earth's magnetic field. As **FIGURE 24.33** shows, the particles spiral in helical trajectories along the earth's magnetic field lines. Some of these particles enter the atmosphere near the north and south poles, ionizing gas and creating the ghostly glow of the **aurora.**

FIGURE 24.32 A charged particle in a magnetic field follows a helical trajectory.

(a) The velocity can be broken into components parallel and perpendicular to the field. The parallel component will continue without change.

(b) A top view shows that the perpendicular component will change, leading to circular motion.

(c) The net result is a helical path that spirals around the field lines.

FIGURE 24.33 Charged particles in the earth's magnetic field create the aurora.

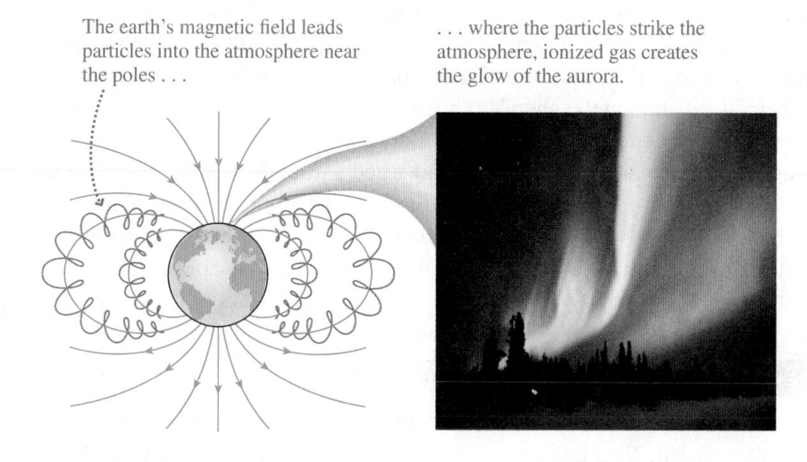

The earth's magnetic field leads particles into the atmosphere near the poles . . .
. . . where the particles strike the atmosphere, ionized gas creates the glow of the aurora.

EXAMPLE 24.9 Force on a charged particle in the earth's field, revisited

In Example 24.8, we considered a proton in the solar wind moving toward the equator of the earth, where the earth's field is 5.0×10^{-5} T, at a speed of 500 km/s (5.0×10^5 m/s). We now know that the proton will move in a circular orbit around the earth's field lines. What are the radius and the period of the orbit?

PREPARE We begin with a sketch of the situation, noting the proton's orbit, as shown in **FIGURE 24.34**.

FIGURE 24.34 A proton orbits the earth's field lines.

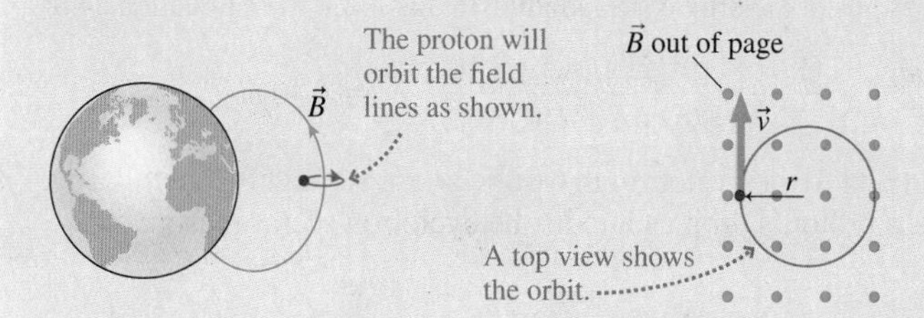

SOLVE Before we use any numbers, we will do some work with symbols. The radius r of the orbit of the proton is given by Equation 24.8. The period T for one orbit is just the distance of one orbit (the circumference $2\pi r$) divided by the speed:

$$T = \frac{2\pi r}{v}$$

We substitute Equation 24.8 for the radius of the orbit of a charged particle moving in a magnetic field to get

$$T = \frac{2\pi}{v} r = \frac{2\pi}{\not{v}}\left(\frac{m\not{v}}{qB}\right) = \frac{2\pi m}{qB}$$

The speed cancels, and doesn't appear in the final expression. All protons in the earth's field orbit with the same period, regardless of their speed. A higher speed just means a larger circle, completed in the same time. Using values for mass, charge, and field, we compute the radius and the period of the orbit:

$$r = \frac{(1.67 \times 10^{-27}\ \text{kg})(5.0 \times 10^5\ \text{m/s})}{(1.60 \times 10^{-19}\ \text{C})(5.0 \times 10^{-5}\ \text{T})} = 100\ \text{m}$$

$$T = \frac{2\pi(1.67 \times 10^{-27}\ \text{kg})}{(1.60 \times 10^{-19}\ \text{C})(5.0 \times 10^{-5}\ \text{T})} = 0.0013\ \text{s}$$

ASSESS We can do a quick check on our math. We've found the radius of the orbit and the period, so we can compute the speed:

$$v = \frac{2\pi r}{T} = \frac{2\pi(100\ \text{m})}{0.0013\ \text{s}} = 5 \times 10^5\ \text{m/s}$$

This is the speed that we were given in the problem statement, which is a good check on our work.

FIGURE 24.35 The cyclotron.

(a) Interior view of a cyclotron

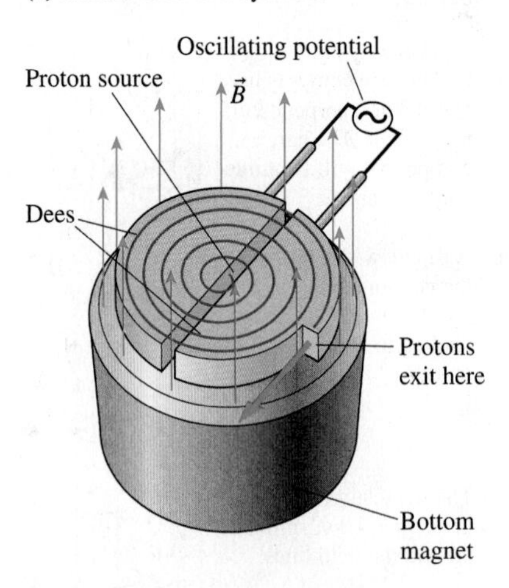

(b) A cyclotron used for the production of medical isotopes

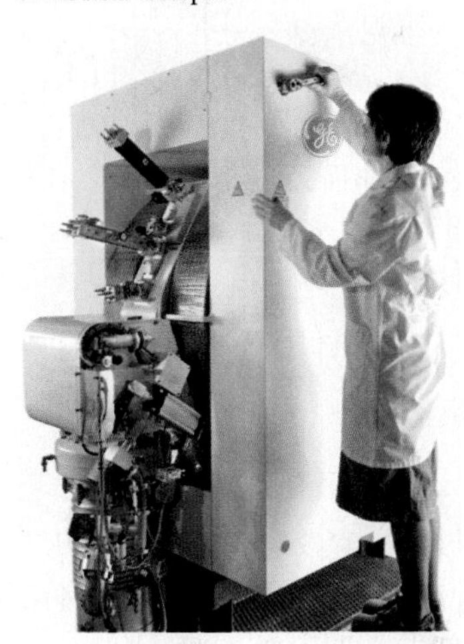

The Cyclotron

The medical imaging technique of *positron-emission tomography* (*PET*), which you'll learn about in Chapter 30, is used to make images of the internal *biological activity* of the body. PET requires the fluorine isotope ^{18}F, which, because of its short half-life of only 110 minutes, must be created at or near the medical facility in which it's required. This is done using a **cyclotron** to fire energetic protons at ^{18}O atoms in water, replacing a neutron with a proton to create ^{18}F. A cyclotron, shown in FIGURE 24.35a, consists of an evacuated chamber within a large, uniform magnetic field. Inside the chamber are two hollow conductors called "dees," separated by a small gap. Protons are injected into the magnetic field from a source near the center of the cyclotron, and they begin to move in a circular orbit in the magnetic field.

The key to accelerating the protons lies in the oscillating electric potential applied between the dees. This potential creates a strong electric field in the gap between the dees, which accelerates the protons passing through the gap. Soon, these protons, traveling in a circular path in the magnetic field, reach the opposite gap. By this time, the potential has changed sign, so the electric field again accelerates the protons, increasing their speed and kinetic energy. This process repeats each time the protons cross a gap, so that their energy increases continuously.

According to Equation 24.8, the radius of a charged particle's orbit is proportional to its speed, so as the protons speed up, the radius of their orbit increases. When the protons reach the outer edge of the magnet, where their speed and energy are the highest, they exit the cyclotron as a continuous high-energy beam that can be directed to an ^{18}O target.

EXAMPLE 24.10 A medical cyclotron BIO

It takes a proton with a kinetic energy of 11 MeV to efficiently change ^{18}O nuclei into ^{18}F. If the magnetic field inside the cyclotron is 1.2 T, what is the radius of the protons' orbit just before they exit the cyclotron?

PREPARE Equation 24.8 relates the radius of a charged particle's orbit to its speed. We can find the speed of the protons from their kinetic energy; to do so, we'll first need to convert their energy from MeV to J.

SOLVE An 11 MeV proton's kinetic energy in J is

$$K = (11 \times 10^6 \text{ eV}) \times \frac{1.60 \times 10^{-19}\text{ J}}{1 \text{ eV}} = 1.76 \times 10^{-12}\text{ J}$$

The kinetic energy of a particle is $K = (1/2)mv^2$, so the proton's speed is

$$v = \sqrt{\frac{2K}{m}} = \sqrt{\frac{2(1.76 \times 10^{-12}\text{ J})}{1.67 \times 10^{-27}\text{ kg}}} = 4.59 \times 10^7 \text{ m/s}$$

We can then find the radius of the orbit corresponding to this speed from Equation 24.8:

$$r = \frac{mv}{|q|B} = \frac{(1.67 \times 10^{-27}\text{ kg})(4.59 \times 10^7\text{ m/s})}{(1.60 \times 10^{-19}\text{ C})(1.2\text{ T})} = 0.40\text{ m}$$

ASSESS The woman next to the cyclotron in FIGURE 24.35b is roughly 2 m tall, so a 0.80 m diameter for a proton's orbit as it exits the cyclotron seems reasonable.

Electromagnetic Flowmeters

Class Video

Blood contains many kinds of ions, such as Na^+ and Cl^-. When blood flows through a vessel, these ions move with the blood. An applied magnetic field will produce a force on these moving charges. We can use this principle to make a completely noninvasive device for measuring the blood flow in an artery: an *electromagnetic flowmeter*.

A flowmeter probe clamped to an artery has two active elements: magnets that apply a strong field across the artery and electrodes that contact the artery on opposite sides, as shown in FIGURE 24.36. The blood flowing in an artery carries a mix of positive and negative ions. Because these ions are in motion, the magnetic field exerts a force on them that produces a measurable voltage. We know from Equation 24.5 that the faster the blood's ions are moving, the greater the forces separating the positive and negative ions. The greater the forces, the greater the degree of separation and the higher the voltage. The measured voltage is therefore directly proportional to the velocity of the blood.

Go with the flow Many scientists and resource managers rely on accurate measurements of stream flows. The easiest way to get a quick measurement of the speed of a river or creek is to use an electromagnetic flowmeter similar to the one used for measuring flow in blood vessels. Water flows between the poles of a strong magnet. Two electrodes measure the resulting potential difference, which is proportional to the flow speed.

FIGURE 24.36 The operation of an electromagnetic flowmeter.

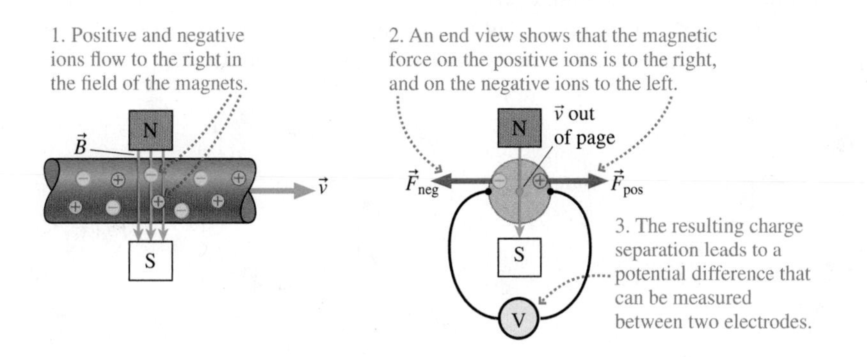

STOP TO THINK 24.7 These charged particles are traveling in circular orbits with velocities and field directions as noted. Which particles have a negative charge?

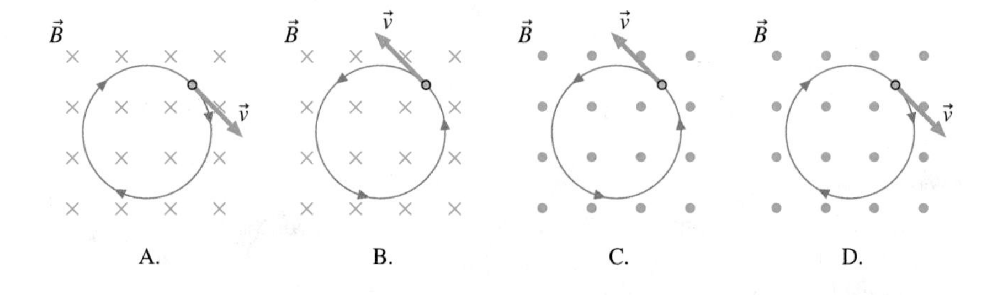

24.6 Magnetic Fields Exert Forces on Currents

We have seen that a magnetic field exerts a force on a moving charge. Because the current in a wire consists of charges moving through the wire, we expect that a magnetic field will exert a force on a current-carrying wire as well. This force is responsible for the operation of loudspeakers, electric motors, and many other devices.

The Form of the Magnetic Force on a Current

In the table at the start of Section 24.5, we saw that a magnetic field exerts no force on a charged particle moving parallel to a magnetic field. If a current-carrying wire is *parallel* to a magnetic field, we also find that the force on it is zero.

FIGURE 24.37 Magnetic force on a current-carrying wire.

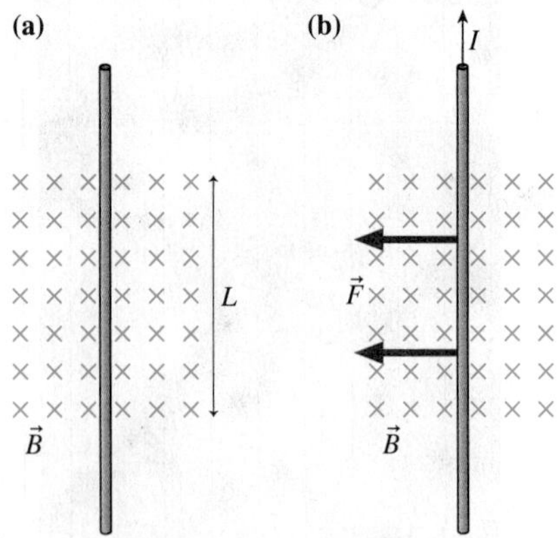

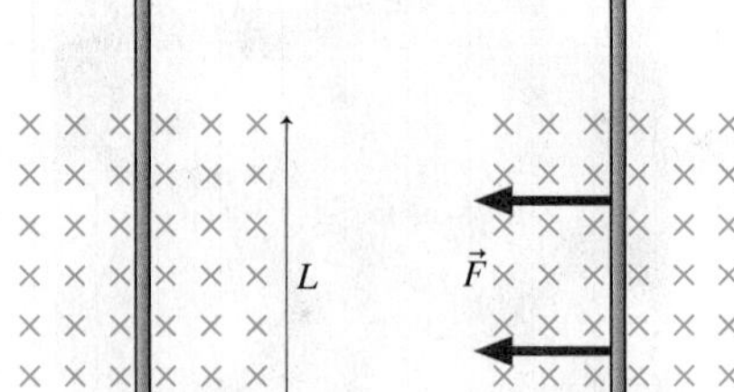

A wire is perpendicular to an externally created magnetic field.

If the wire carries a current, the magnetic field will exert a force on the wire.

(c)

$\vec{B}$

$\vec{F}$

I

The right-hand rule for forces applies. Your thumb should point in the direction of the current.

However, there is a force on a current-carrying wire that is *perpendicular* to a magnetic field, as shown in **FIGURE 24.37**.

NOTE ► The magnetic field is an external field, created by a permanent magnet or by other currents; it is *not* the field of the current I in the wire. ◄

The direction of the force on the current is found by considering the force on each charge in the current. We model current as the flow of positive charge, so **the right-hand rule for forces applies to currents in the same way it does for moving charges.** With your fingers aligned as usual, point your right thumb in the direction of the current (the direction of the motion of positive charges) and your index finger in the direction of $\vec{B}$. Your middle finger is then pointing in the direction of the force $\vec{F}$ on the wire, as in **FIGURE 24.37c**. Consequently, the entire length of wire within the magnetic field experiences a force perpendicular to both the current direction and the field direction, as shown in **FIGURE 24.37b**.

If the length of the wire L, the current I, or the magnetic field B is increased, then the magnitude of the force on the wire will also increase. We can show that the magnetic force on a current-carrying wire is given by

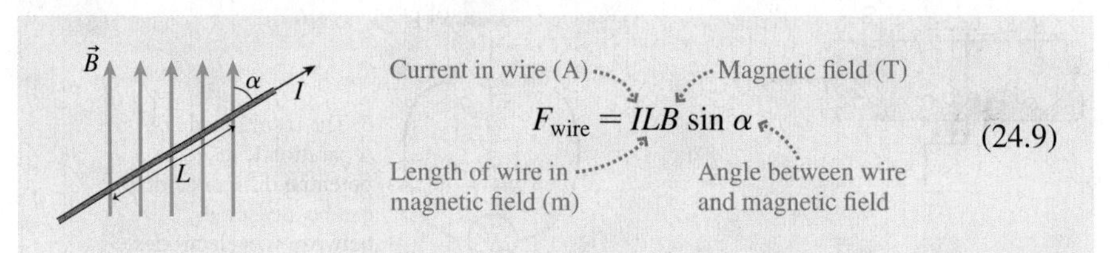

$$F_{wire} = ILB \sin \alpha \qquad (24.9)$$

In many practical cases the wire will be perpendicular to the field, so that $\alpha = 90°$. In this case,

$$F_{wire} = ILB \qquad (24.10)$$

Video Tutor Demo

If we rewrite Equation 24.10 as $B = F_{wire}/IL$, we can see that the unit for magnetic field, the tesla, can be defined in terms of other units:

$$1\ \text{T} = 1\ \frac{\text{N}}{\text{A} \cdot \text{m}}$$

EXAMPLE 24.11 Magnetic force on a power line

A DC power line near the equator runs east-west. At this location, the earth's magnetic field is parallel to the ground, points north, and has magnitude 50 μT. A 400 m length of the heavy cable that spans the distance between two towers has a mass of 1000 kg. What direction and magnitude of current would be necessary to offset the force of gravity and "levitate" the wire? (The power line will actually carry a current that is much less than this; 850 A is a typical value.)

PREPARE First, we sketch a top view of the situation, as in **FIGURE 24.38**. The magnetic force on the wire must be opposite that of gravity. An application of the right-hand rule for forces shows that a current to the east will result in an upward force—out of the page.

FIGURE 24.38 Top view of a power line near the equator.

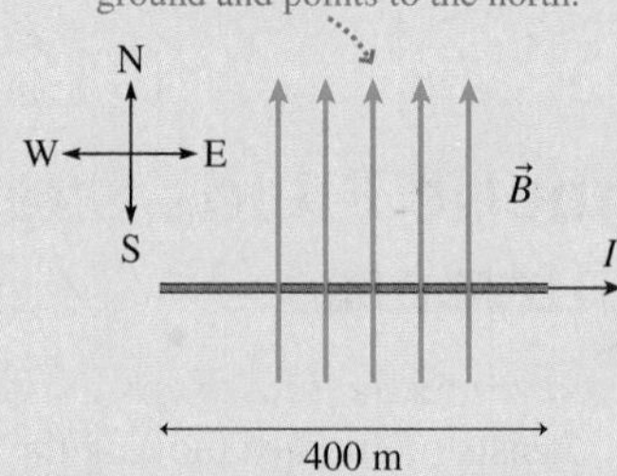

SOLVE The magnetic field is perpendicular to the current, so the magnitude of the magnetic force is given by Equation 24.10. To levitate the wire, this force must be opposite to the weight force but equal in magnitude, so we can write

$$mg = ILB$$

where m and L are the mass and length of the wire and B is the magnitude of the earth's field. Solving for the current, we find

$$I = \frac{mg}{LB} = \frac{(1000\ \text{kg})(9.8\ \text{m/s}^2)}{(400\ \text{m})(50 \times 10^{-6}\ \text{T})} = 4.9 \times 10^5\ \text{A}$$

directed to the east.

ASSESS The current is much larger than a typical current, as we expected.

Forces Between Currents

Because a current produces a magnetic field, and a magnetic field exerts a force on a current, it follows that two current-carrying wires will exert forces on each other. To see how, **FIGURE 24.39a** shows a wire carrying current I_1. From the right-hand rule for fields, we know that this current creates a magnetic field $\vec{B}_1$ pointing out of the page to the left of the wire and into the page to its right. If, as in **FIGURE 24.39b**, a second wire with current I_2 is placed next to the first wire, the current in the second wire will feel a magnetic force due to $\vec{B}_1$. Using the right-hand rule for forces, we can see that when I_2 is in the same direction as I_1 the force on the second wire is to the left; that is, the second wire is *attracted* to the first wire.

We can repeat this process to find the force on the first wire due to the magnetic field of the current I_2 in the second wire; you should be able to show that this force points to the right, so the first wire is also attracted to the second wire. Note that the forces on the two wires form a Newton's third law action/reaction pair. Thus, not only are they directed oppositely, as we've just seen, but they must have the *same* magnitude.

The attractive forces between two wires when the currents are in the same direction are shown in **FIGURE 24.40a**. By a similar argument, we can show that the forces between two wires are *repulsive* when the currents are in opposite directions, as shown in **FIGURE 24.40b**.

FIGURE 24.39 How one current exerts a force on another current.

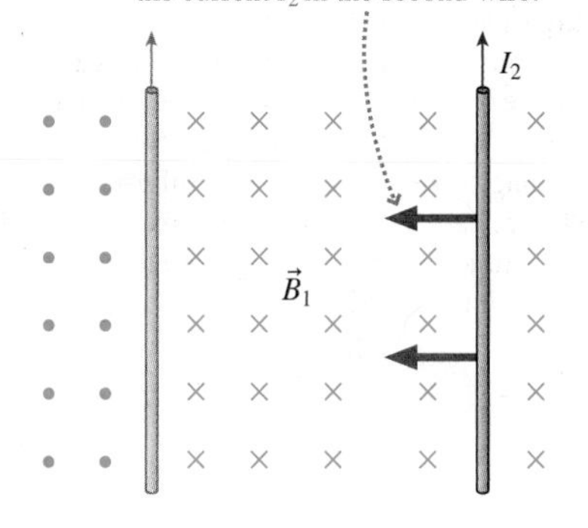

FIGURE 24.40 Forces between currents.

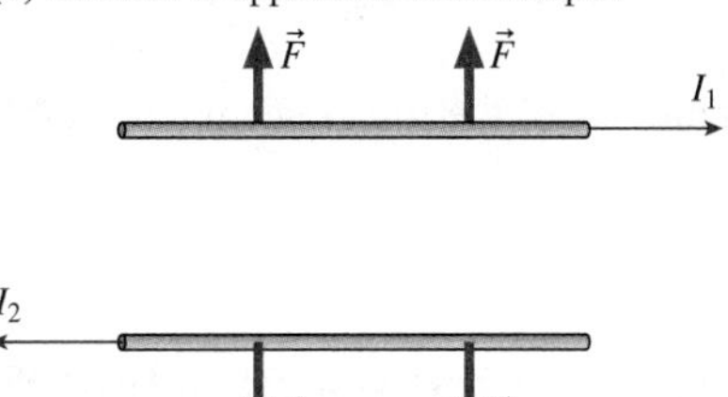

The following example shows how to calculate the *magnitude* of the force between two current-carrying wires.

EXAMPLE 24.12 Finding the force between wires in jumper cables

You may have used a set of jumper cables connected to a running vehicle to start a car with a dead battery. Jumper cables are a matched pair of wires, red and black, joined together along their length. Suppose we have a set of jumper cables in which the two wires are separated by 1.2 cm along their 3.7 m (12 ft) length. While starting a car, the wires each carry a current of 150 A, in opposite directions. What is the force between the two wires?

PREPARE Our first step is to sketch the situation, noting distances and currents, as shown in **FIGURE 24.41**. Let's find the force on the red wire; from the discussion above, the force on the black wire has the same magnitude but is in the opposite direction. The force on the red wire is found using a two-step process. First, we find the magnetic field due to the current in the black wire at the position of the red wire. Then, we find the force on the current in the red wire due to this magnetic field.

FIGURE 24.41 Jumper cables carrying opposite currents.

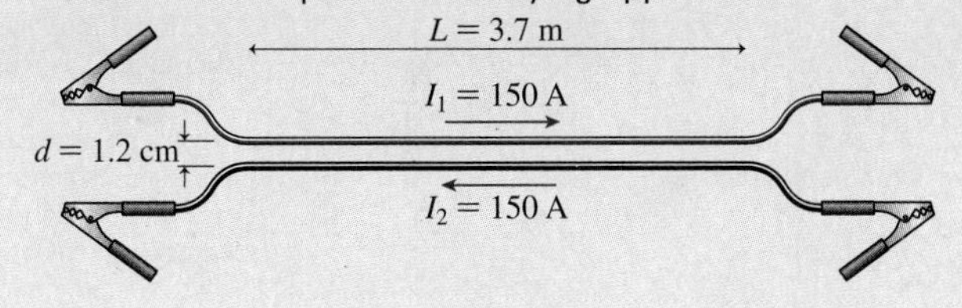

SOLVE The magnetic field at the position of the red wire, due to the current in the black wire, is

$$B = \frac{\mu_0 I}{2\pi d} = \frac{(1.26 \times 10^{-6}\,\text{T}\cdot\text{m/A})(150\,\text{A})}{2\pi(0.012\,\text{m})} = 2.51 \times 10^{-3}\,\text{T}$$

According to the right-hand rule for fields, this magnetic field is directed into the page. The magnitude of the force on the red wire is then

$$F_{\text{wire}} = ILB = (150\,\text{A})(3.7\,\text{m})(2.51 \times 10^{-3}\,\text{T}) = 1.4\,\text{N}$$

The direction of the force can be found using the right-hand rule for forces. The magnetic field at the position of the red wire is into the page, while the current is to the right. This means that the force on the red wire is in the plane of the page, directed *away* from the black wire. Thus the force between the two wires is repulsive, as we expect when their currents are directed oppositely.

ASSESS These wires are long, close together, and carry very large currents. But the force between them is quite small—much less than the weight of the wires. In practice, the forces between currents are not an important consideration unless there are many coils of wire, leading to a large total force. This is the case in an MRI solenoid.

Crushed by currents The forces between currents are normally quite small, but the enormous currents in a lightning strike can produce remarkable forces. This hollow rod carried tens of thousands of amps following a lightning strike. The currents in all parts of the rod were parallel, and the size of the current led to attractive forces strong enough to actually crush the rod.

PROBLEM-SOLVING STRATEGY 24.1 **Magnetic force problems**

Magnetic fields exert forces on both moving charges and current-carrying wires, but in both cases the force can be found using the same strategy.

PREPARE There are two key factors to identify in magnetic force problems:

- The source of the magnetic field.
- The charges or currents that feel a force due to this magnetic field.

SOLVE First, determine the magnitude and direction of the magnetic field at the position of the charges or currents that are of interest.

- In some problems the magnetic field is given.
- If the field is due to a current, use the right-hand rule for fields to determine the field direction and the appropriate equation to determine its magnitude.

Next, determine the force this field produces. Working with the charges or currents you identified previously,

- Use the right-hand rule for forces to determine the direction of the force on any moving charge or current.
- Use the appropriate equation to determine the magnitude of the force on any moving charge or current.

ASSESS Are the forces you determine perpendicular to velocities of moving charges and to currents? Are the forces perpendicular to the fields? Do the magnitudes of the forces seem reasonable?

Exercise 34 

Forces Between Current Loops

We will now consider the forces between two current loops. Doing so will allow us to begin to make connections with some of the basic phenomena of magnetism that we saw earlier in the chapter.

We've seen that there is an attractive force between two parallel wires that have their currents in the same direction. If these two wires are bent into loops, as in FIGURE 24.42, then the force between the two loops will also be attractive. The forces will be repulsive if the currents are in opposite directions.

Early in the chapter, we examined the fields from various permanent magnets and current arrangements. FIGURE 24.43a reminds us that a bar magnet is a magnetic dipole, with a north and a south pole. Field lines come out of its north pole, loop back around, and go into the south pole. FIGURE 24.43b shows that the field of a current loop is very similar to that of a bar magnet. This leads us to the conclusion that **a current loop, like a bar magnet, is a magnetic dipole,** with a north and a south pole, as indicated in Figure 24.43b. We can use this conclusion to understand the forces between current loops. In FIGURE 24.44, we show how the poles of each current loop can be represented by a magnet. This model helps us understand the forces between current loops; in the next section, we'll use this model to understand torques on current loops as well.

FIGURE 24.42 Forces between parallel current loops.

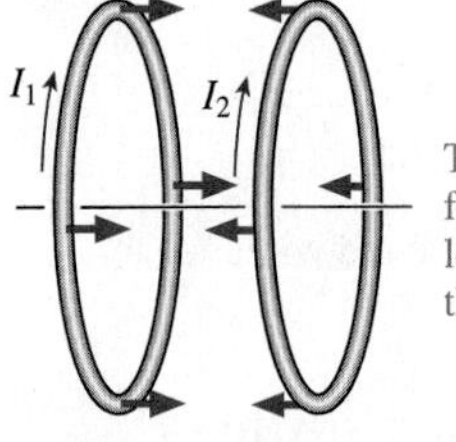

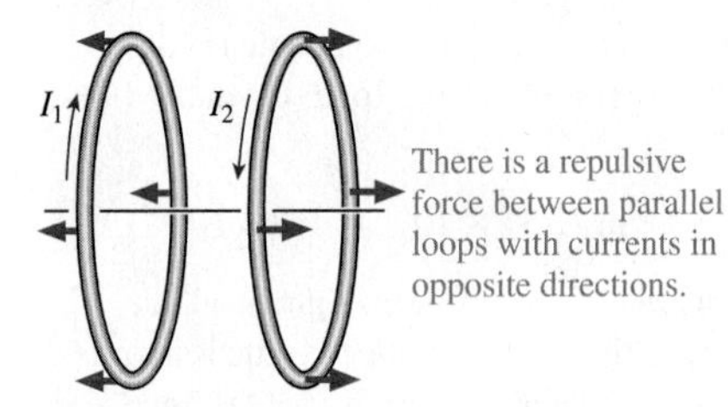

FIGURE 24.43 We can picture the loop as a small bar magnet.

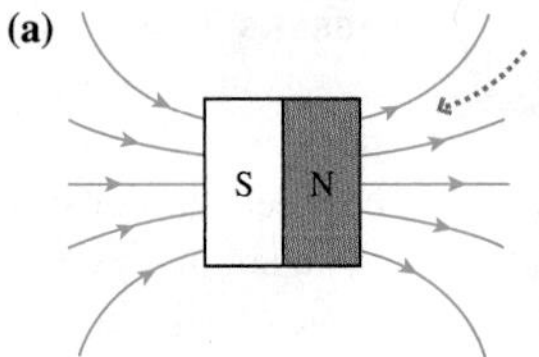

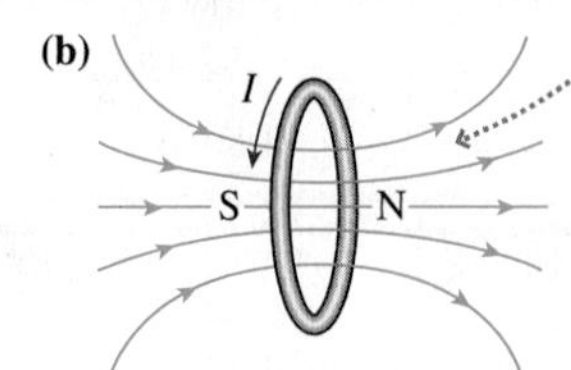

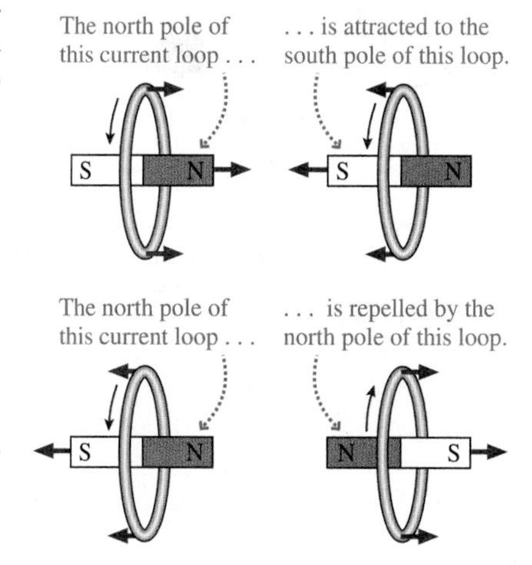
Because currents loops have north and south poles, we can picture a current loop as a small bar magnet.
The north pole of this current loop . . .
. . . is attracted to the south pole of this loop.
S
N
S
N
The north pole of this current loop . . .
. . . is repelled by the north pole of this loop.
S
N
N
S

▶ **FIGURE 24.44** Forces between current loops can be understood in terms of their magnetic poles.

STOP TO THINK 24.8 Four wires carry currents in the directions shown. A uniform magnetic field is directed into the paper as shown. Which wire experiences a force to the left?

A. B. C. D.

24.7 Magnetic Fields Exert Torques on Dipoles

One of our first observations at the beginning of the chapter was that a compass needle is a small magnet. The fact that it pivots to line up with an external magnetic field means that it experiences a *torque*. In this section, we'll use what we've learned about magnetic forces to understand the torque on a magnetic dipole. We will consider only the case of a current loop (which you'll recall is a magnetic dipole), but the results will be equally applicable to permanent magnets and magnetic dipoles such as compass needles.

A Current Loop in a Uniform Field

FIGURE 24.45 shows a current loop—a magnetic dipole—in a uniform magnetic field $\vec{B}$. The current in each of the four sides of the loop experiences a magnetic force due to $\vec{B}$. Because the field is uniform, the forces on opposite sides of the loop are of equal magnitude. The direction of each force is determined by the right-hand rule for forces. The forces $\vec{F}_{front}$ and $\vec{F}_{back}$ produce no net force or torque. The forces $\vec{F}_{top}$ and $\vec{F}_{bottom}$ also give no net force, but they will rotate the loop by exerting a torque on it.

Although we've shown a current loop, the conclusion is true for any magnetic dipole: **In a uniform field, a dipole experiences a torque but no net force.** For example, a compass needle is not attracted to the earth's poles; it merely feels a torque that lines it up with the earth's field.

We can calculate the torque by looking at a side view of the current loop of Figure 24.45. This is shown in **FIGURE 24.46**, where you can see that the forces F_{top} and F_{bottom} act to rotate the loop clockwise. The angle θ between the axis of the loop and the field will be important. We've drawn a vector from the center of the loop that we'll use to define this angle.

In ◀◀ **SECTION 7.2**, we calculated the torque due to a force F as $rF \sin\theta$, where r is the distance from the pivot to the point where the force acts, and θ is the angle between the force vector and a line drawn from the pivot to that point. In Figure 24.46, the net torque is then

$$\tau = \tau_{top} + \tau_{bottom} = (\tfrac{1}{2}L)F_{top}\sin\theta + (\tfrac{1}{2}L)F_{bottom}\sin\theta$$
$$= (\tfrac{1}{2}L)(ILB)\sin\theta + (\tfrac{1}{2}L)(ILB)\sin\theta$$
$$= (IL^2)B\sin\theta$$

L^2 is the area A of the square loop. Using this, we can generalize the result to any loop of area A:

$$\tau = (IA)B\sin\theta \qquad (24.11)$$

FIGURE 24.45 A loop in a uniform magnetic field experiences a torque.

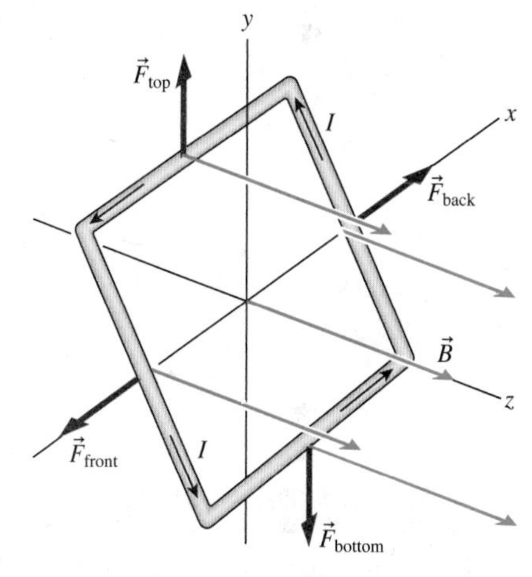
y
$\vec{F}_{top}$
I
x
$\vec{F}_{back}$
$\vec{B}$
z
$\vec{F}_{front}$
I
$\vec{F}_{bottom}$

FIGURE 24.46 Calculating the torque on a current loop in a uniform magnetic field.

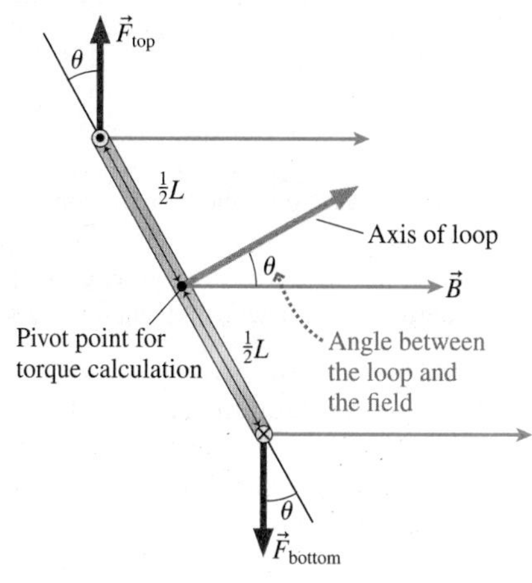
$\vec{F}_{top}$
θ
½L
Axis of loop
θ
$\vec{B}$
Pivot point for torque calculation
½L
Angle between the loop and the field
θ
$\vec{F}_{bottom}$

FIGURE 24.47 The dipole moment vector.

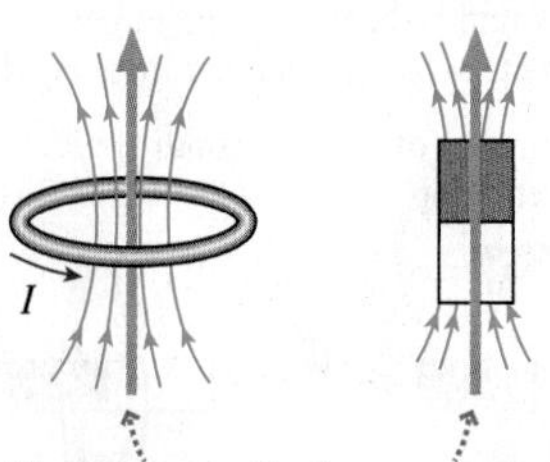

The magnetic dipole moment is represented as a vector that points in the direction of the dipole's field. A longer vector means a stronger field.

There are two things to note about this torque:

1. **The torque depends on properties of the current loop:** its area (A) and the current (I). The quantity IA, known as the **magnetic dipole moment** (or simply *magnetic moment*), is a measure of how much torque a dipole will feel in a magnetic field. A permanent magnet can be assigned a magnetic dipole moment as well. We'll represent the magnetic moment as an arrow pointing in the direction of the dipole's magnetic field, as in FIGURE 24.47. This figure shows the magnetic dipole moment for a current loop and a bar magnet. You can now see that the vector defining the axis of the loop in Figure 24.46 is simply the magnetic dipole moment.
2. **The torque depends on the angle between the magnetic dipole moment and the magnetic field.** The torque is maximum when θ is 90°, when the magnetic moment is perpendicular to the field. The torque is zero when θ is 0° (or 180°), when the magnetic moment is parallel to the field. As we see in FIGURE 24.48, a magnetic dipole free to rotate in a field will do so until θ is zero, at which point it will be stable. **A magnetic dipole will rotate to line up with a magnetic field** just as an electric dipole will rotate to line up with an electric field.

FIGURE 24.48 Torque on a dipole in an externally created magnetic field.

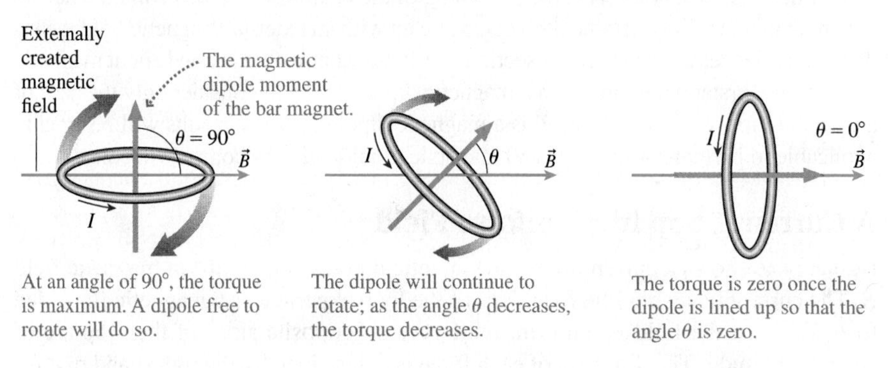

Externally created magnetic field
The magnetic dipole moment of the bar magnet.
$\theta = 90°$
$\vec{B}$
I
θ
$\theta = 0°$
At an angle of 90°, the torque is maximum. A dipole free to rotate will do so.
The dipole will continue to rotate; as the angle θ decreases, the torque decreases.
The torque is zero once the dipole is lined up so that the angle θ is zero.

A compass needle, which is a dipole, also rotates until its north pole is in the direction of the magnetic field, as we noted at the very start of the chapter.

CONCEPTUAL EXAMPLE 24.13 Does the loop rotate?

Two nearby current loops are oriented as shown in FIGURE 24.49. Loop 1 is fixed in place; loop 2 is free to rotate. Will it do so?

FIGURE 24.49 Will loop 2 rotate?

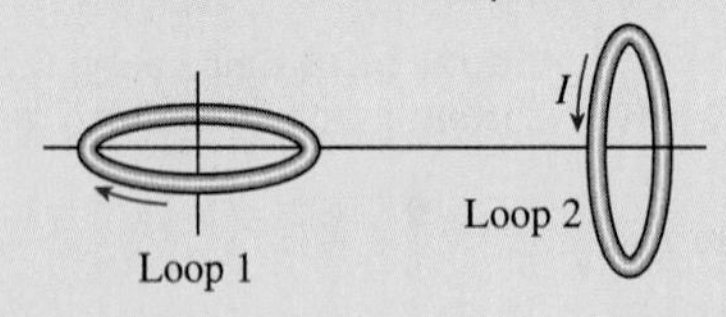

REASON The current in loop 1 generates a magnetic field. As FIGURE 24.50 shows, the field of loop 1 is upward as it passes loop 2. Because the field is perpendicular to the magnetic moment of loop 2, the field exerts a torque on loop 2 and causes loop 2 to rotate until its magnetic moment lines up with the field from loop 1.

ASSESS These two loops align so that their magnetic moments point in opposite directions. We can think of this in terms of their poles: When aligned this way, the north pole of loop 1 is closest to the south pole of loop 2. This makes sense because these opposite poles attract each other.

FIGURE 24.50 How the field of loop 1 affects loop 2.

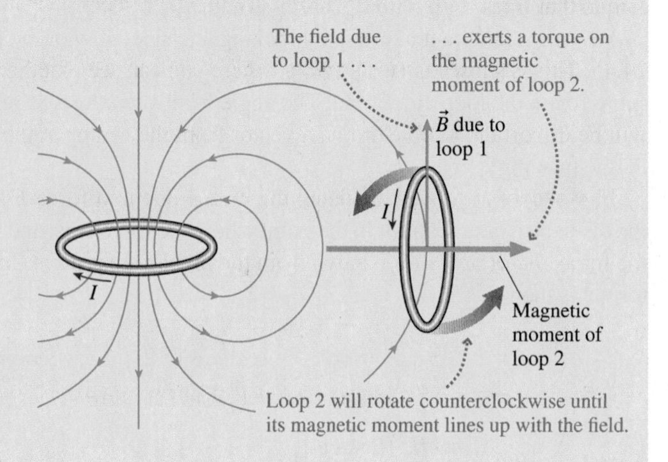

For any dipole in a field, there are actually two angles for which the torque is zero, $\theta = 0°$ and $\theta = 180°$, but there is a difference between these two cases. The $\theta = 0°$ case is *stable:* Once the dipole is in this configuration, it will stay there. The $\theta = 180°$ case is *unstable.* There is no torque if the alignment is perfect, but the slightest rotation will result in a torque that rotates the dipole until it reaches $\theta = 0°$.

We can make a gravitational analogy with this situation in **FIGURE 24.51**. For an upside-down pendulum, there will be no torque if the mass is directly above the pivot point. But, as we saw in ◀◀ SECTION 7.4, this is a position of unstable equilibrium. If displaced even slightly, the mass will rotate until it is below the pivot point. This is the point of lowest potential energy.

We can see, by analogy with the upside-down pendulum, that the unstable alignment of a magnetic dipole has a higher energy. Given a chance, the magnet will rotate "downhill" to the position of lower energy and stable equilibrium. This difference in energy is the key to understanding how the magnetic properties of atoms can be used to image tissues in the body in MRI.

FIGURE 24.51 Going from unstable to stable equilibrium.

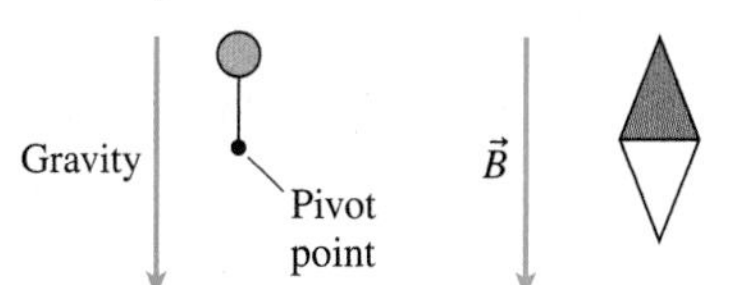
Gravity
Pivot point
$\vec{B}$

The pendulum balancing upside down and the magnet aligned opposite the field are in unstable equilibrium. A small nudge . . .

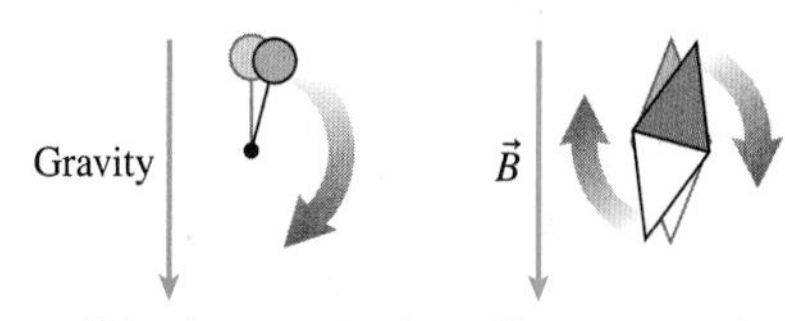
Gravity
$\vec{B}$

. . . will lead to a torque that will cause a rotation that will continue until . . .

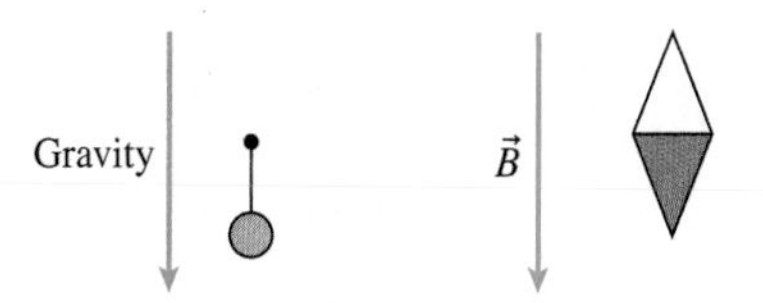
Gravity
$\vec{B}$

. . . the condition of stable equilibrium is reached.

Magnetic Resonance Imaging (MRI)

Magnetic resonance imaging is a modern diagnostic tool that provides detailed images of tissues and structures in the body with no radiation exposure. The key to this imaging technique is the magnetic nature of atoms. **The nuclei of individual atoms have magnetic moments and behave like magnetic dipoles.** Atoms of different elements have different magnetic moments; therefore, a magnetic field exerts different torques on different kinds of atoms.

A person receiving an MRI scan is placed in a large solenoid that has a strong magnetic field along its length. Think about one hydrogen atom in this person's body. The nucleus of the hydrogen atom is a single proton. The proton has a magnetic moment, but the proton is a bit different from a simple bar magnet: It is subject to the rules of quantum mechanics, which we will learn about in Chapter 28. A bar magnet can have any angle with the field, but the proton can only line up either *with the field* (the low-energy state) or *opposed to the field* (the high-energy state), as we see in **FIGURE 24.52**.

The energy difference ΔE between these two orientations of the proton depends on two key parameters: the magnetic moment of the proton and the strength of the magnetic field.

A tissue sample will have many hydrogen atoms with many protons. Some of the protons will be in the high-energy state and some in the low-energy state. A second magnetic field, called a *probe field,* can be applied to "flip" the dipoles from the low-energy state to the high-energy state. The probe field must be precisely tuned to the energy difference ΔE for this to occur. The probe field is selected to correspond to the magnetic moment of a particular nucleus, in this case hydrogen. If the tuning is correct, dipoles will change state and a signal is measured. A strong signal means that many atoms of this kind are present.

How is this measurement turned into an image? The magnetic field strengths of the solenoid and the probe field are varied so that the correct tuning occurs at only one point in the patient's body. The position of the point of correct tuning is swept across a "slice" of tissue. Combining atoms into molecules slightly changes the energy levels. Different tissues in the body have different concentrations of atoms in different chemical states, so the strength of the signal at a point will vary depending on the nature of the tissue at that point. As the point of correct tuning is moved, the intensity of the signal is measured at each point; a record of the intensity versus position gives an image of the structure of the interior of the body, as in **FIGURE 24.53**.

FIGURE 24.52 The energy difference between the two possible orientations of a proton's magnetic moment during MRI.

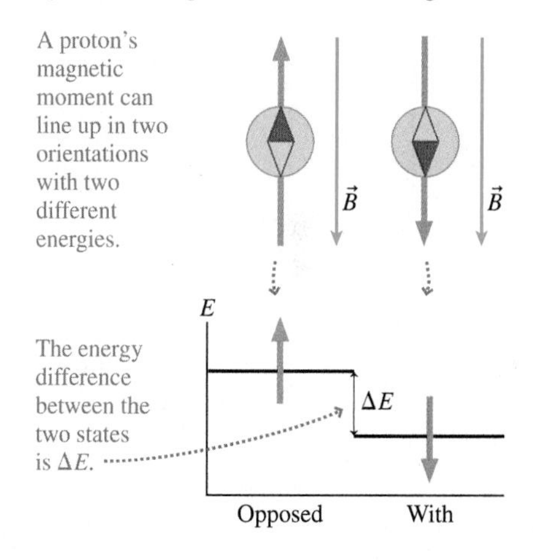
A proton's magnetic moment can line up in two orientations with two different energies.
$\vec{B}$
$\vec{B}$
E
The energy difference between the two states is ΔE.
ΔE
Opposed
With

FIGURE 24.53 BIO Cross-sectional image from an MRI scan.

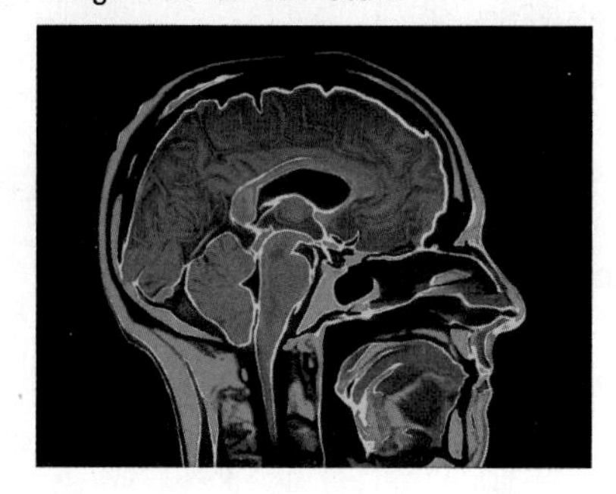

Electric Motors

The torque on a current loop in a magnetic field is the basis for how an electric motor works. The *armature* of a motor is a loop of wire wound on an axle that is free to rotate. This loop is in a strong magnetic field. A current in the loop causes it to feel a torque due to this field, as **FIGURE 24.54** shows. The loop will rotate to align itself with the external field. If the current were steady, the armature would simply rotate until its magnetic moment was in its stable position. To keep the motor turning, a device called a *commutator* reverses the current direction in the loop every 180°. As the loop reaches its stable configuration, the direction of current in the loop switches, putting the loop back into the unstable configuration, so it will keep rotating to line up in the other direction. This process continues: Each time the loop nears a stable point, the current switches. The loop will keep rotating as long as the current continues.

FIGURE 24.54 The operation of a simple motor depends on the torque on a current loop in a magnetic field.

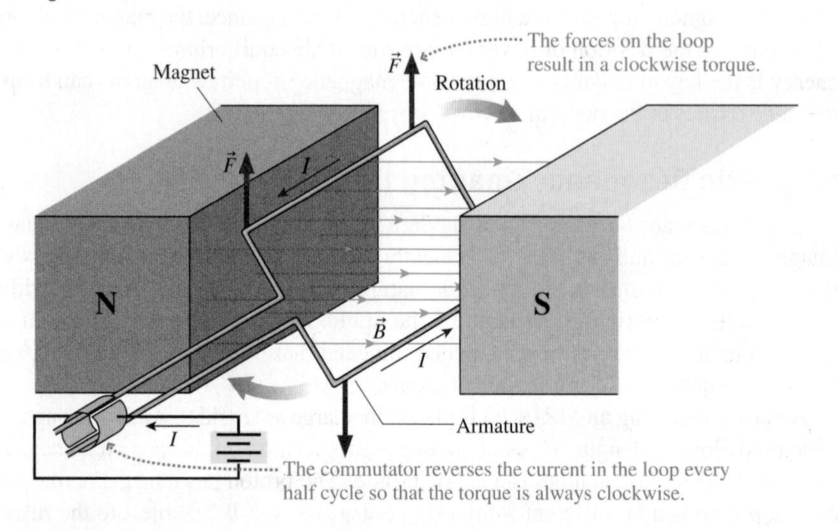

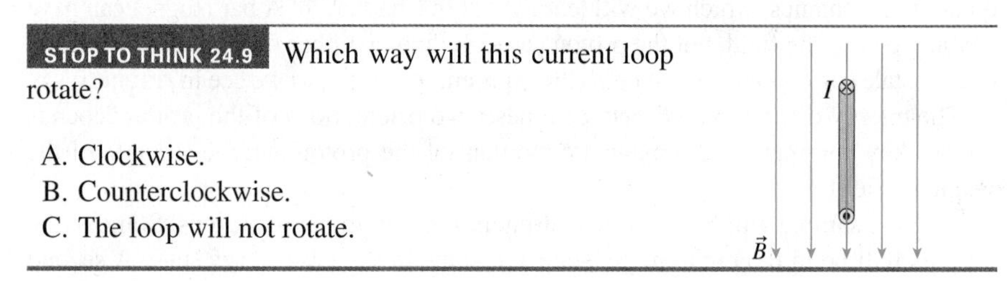

STOP TO THINK 24.9 Which way will this current loop rotate?

A. Clockwise.
B. Counterclockwise.
C. The loop will not rotate.

24.8 Magnets and Magnetic Materials

We started the chapter by looking at permanent magnets. We know that permanent magnets produce a magnetic field, but what is the source of this field? There are no electric currents in these magnets. Why can you make a magnet out of certain materials but not others? Why does a magnet stick to the refrigerator? The goal of this section is to answer these questions by developing an atomic-level view of the magnetic properties of matter.

Ferromagnetism

Iron, nickel, and cobalt are elements that have very strong magnetic behavior: A chunk of iron (or steel, which is mostly iron) will stick to a magnet, and the chunk can be magnetized so that it is itself a magnet. Other metals—such as aluminum and copper—do not exhibit this property. We call materials that are strongly attracted to magnets and that can be magnetized **ferromagnetic** (from the Latin for iron, *ferrum*).

FIGURE 24.55 Magnetic moment of the electron.

The arrow represents the inherent magnetic moment of the electron.

The key to understanding magnetism at the atomic level is that electrons, just like protons and nuclei, have an *inherent magnetic moment*, as we see in **FIGURE 24.55**. Magnetism, at an atomic level, is due to the inherent magnetic moment of electrons.

If the magnetic moments of all the electrons in an atom pointed in the same direction, the atom would have a very strong magnetic moment. But this doesn't happen. In atoms with many electrons, the electrons usually occur in pairs with magnetic moments in opposite directions, as we'll see in Chapter 29. Only the electrons that are unpaired are able to give the atom a net magnetic moment.

Even so, atoms with magnetic moments don't necessarily form a solid with magnetic properties. For most elements whose atoms have magnetic moments, the magnetic moments of the atoms are randomly arranged when the atoms join together to form a solid. As **FIGURE 24.56** shows, this random arrangement produces a solid whose net magnetic moment is very close to zero. Ferromagnetic materials have atoms with net magnetic moments that tend to line up and reinforce each other as in **FIGURE 24.57**. This alignment of moments occurs in only a few elements and alloys, and this is why a small piece of iron has such a strong overall magnetic moment. Such a piece has a north and a south magnetic pole, generates a magnetic field, and aligns parallel to an external magnetic field. In other words, it is a magnet—and a very strong one at that.

In a large sample of iron, the magnetic moments will be lined up in local regions called **domains,** each looking like the ordered situation of Figure 24.57, but there will be no long-range ordering. Inside a domain, the atomic magnetic moments will be aligned, but the magnetic moments of individual domains will be randomly oriented. The individual domains are quite small—on the order of 0.1 mm or so—so a piece of iron the size of a nail has thousands of domains. The random orientation of the magnetic moments of the domains, as shown in **FIGURE 24.58**, means that there is no overall magnetic moment. So, how can you magnetize a nail?

FIGURE 24.56 The random magnetic moments of the atoms in a typical solid.

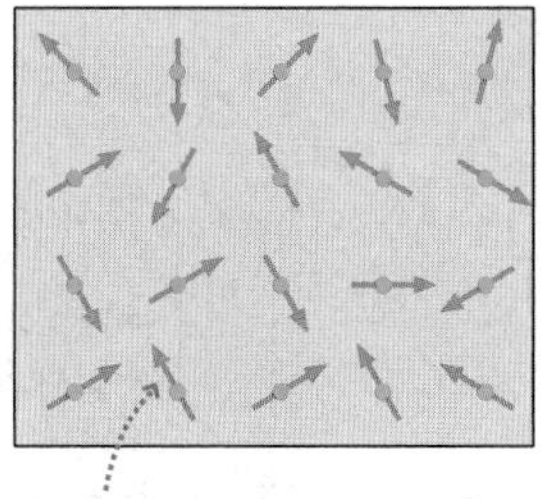

FIGURE 24.57 In a ferromagnetic solid, the atomic magnetic moments align.

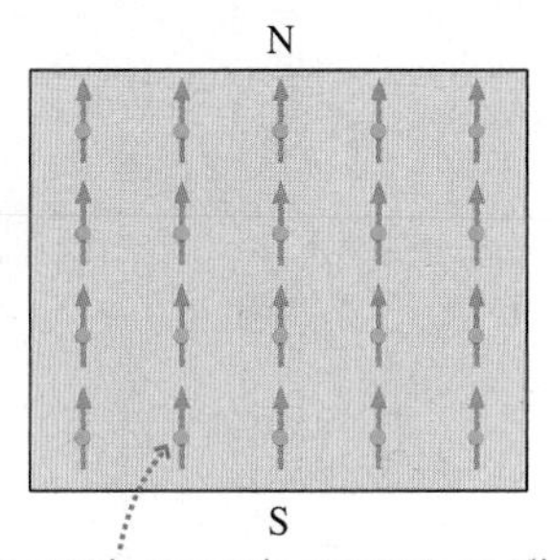

FIGURE 24.58 Magnetic domains in a ferromagnetic material.

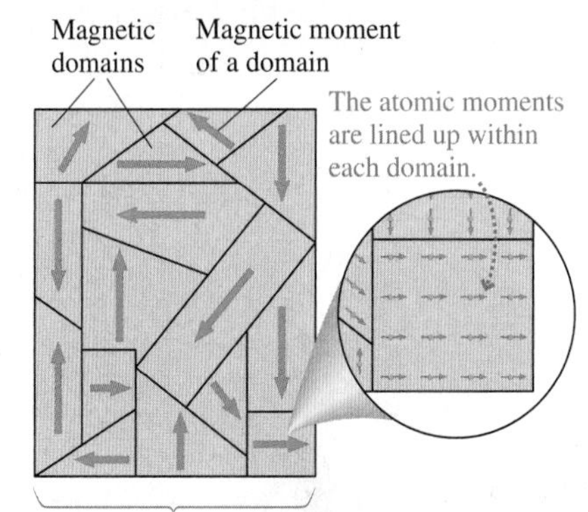

Magnetic domains
Magnetic moment of a domain
The atomic moments are lined up within each domain.
The magnetic moments of the domains tend to cancel one another. The sample as a whole possesses no net magnetic moment.

Induced Magnetic Moments

When you bring a magnet near a piece of iron, as in **FIGURE 24.59** on the next page, the magnetic field of the magnet penetrates the iron and creates torques on the atomic magnetic moments. The atoms will stay organized in domains, but domains aligned with the external field become larger at the expense of domains opposed to the field. After this shift in domain boundaries, the magnetic moments of the domains no longer cancel out. The iron will have a net magnetic moment that is aligned with the external field. The iron will have developed an *induced magnetic moment*.

NOTE ► Inducing a magnetic moment with a magnetic field is analogous to inducing an electric dipole with an electric field, which we saw in Chapter 21. ◄

Looking at the pole structure of the induced magnetic moment in the iron, we can see that the iron will now be attracted to the magnet. The fact that a magnet attracts and picks up ferromagnetic objects was one of the basic observations about magnetism with which we started the chapter. Now we have an explanation of how this works, based on three facts:

1. Electrons are microscopic magnets due to their inherent magnetic moment.
2. In a ferromagnetic material, these atomic magnetic moments are aligned. Regions of aligned moments form magnetic domains.
3. The individual domains shift in response to an external magnetic field to produce an induced magnetic moment for the entire object. This induced magnetic moment will be attracted by a magnet that produced the orientation.

When a piece of iron is near a magnet, the iron becomes a magnet as well. But when the applied field is taken away, the domain structure will (generally) return to where it began: The induced magnetic moment will disappear. In the presence of a *very* strong field, however, a piece of iron can undergo more significant changes to its domain structure, and some domains may permanently change orientation. When the field is removed, the iron may retain some of this magnetic character: The iron will have become permanently magnetized. But pure iron is a rather poor permanent magnetic material; it is very easy to disrupt the ordering of the domains that has

FIGURE 24.59 Inducing a magnetic moment in a piece of iron.

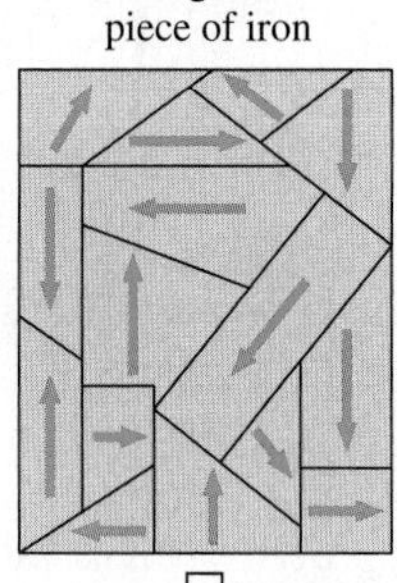
Unmagnetized piece of iron
1. Initially, the magnetic moments of the domains cancel each other; there is no net magnetic moment.

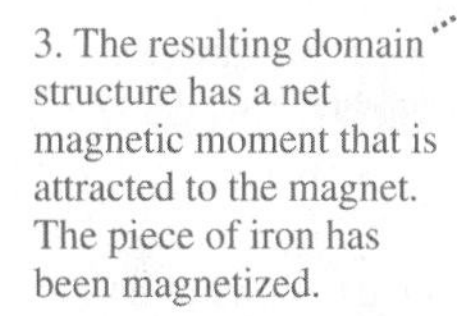
N
S
2. The magnetic field from a magnet causes favorably oriented domains (shown in green) to grow at the expense of other domains.
3. The resulting domain structure has a net magnetic moment that is attracted to the magnet. The piece of iron has been magnetized.
N
S

created the magnetic moment of the iron. For instance, if you heat (or even just drop!) a piece of magnetized iron, the resulting random atomic motions tend to destroy the alignment of the domains, destroying the magnetic character in the process.

Alloys of ferromagnetic materials often possess more robust magnetic characters. Alloys of iron and other ferromagnetic elements with rare-earth elements can make permanent magnets of incredible strength.

CONCEPTUAL EXAMPLE 24.14 Sticking things to the refrigerator

Everyone has used a magnet to stick papers to the fridge. Why does the magnet stick to the fridge through a layer of paper?

REASON When you bring a magnet near the steel door of the refrigerator, the magnetic field of the magnet induces a magnetic moment in the steel. The direction of this induced moment will be such that it is attracted to the magnet—thus the magnet will stick to the fridge. Magnetic fields go through nonmagnetic materials such as paper, so the magnet will hold the paper to the fridge.

ASSESS This result helps makes sense of another observation you've no doubt made: You can't stick a thick stack of papers to the fridge. This is because the magnetic field of the magnet decreases rapidly with distance. Because the field is weaker, the induced magnetic moment is smaller.

Induced magnetic moments are used to store information on computer hard disk drives. As shown in **FIGURE 24.60**, a hard drive consists of a rapidly rotating disk with a thin magnetic coating on its surface. Information for computers is stored as digital zeros and ones, and on the disk these are stored as tiny magnetic domains, each less than 100 nm long! The direction of the domains can be changed by the write head—a tiny switchable magnet that skims over the surface of the disk. The magnetic field of the write head changes the orientation of the domains, encoding information on the disk. The information can then be retrieved by the read head—a small probe that is sensitive to the magnetic fields of the tiny domains.

FIGURE 24.60 Computer hard disks store information using magnetic fields.

Cross section of the magnetic coating on a hard disk
S N S N N S N S
"0" "1" "0"
Zeros are stored as one longer magnet.
Ones are stored as two short magnets.
This arm rapidly moves the read and write heads to the required position over the disk.

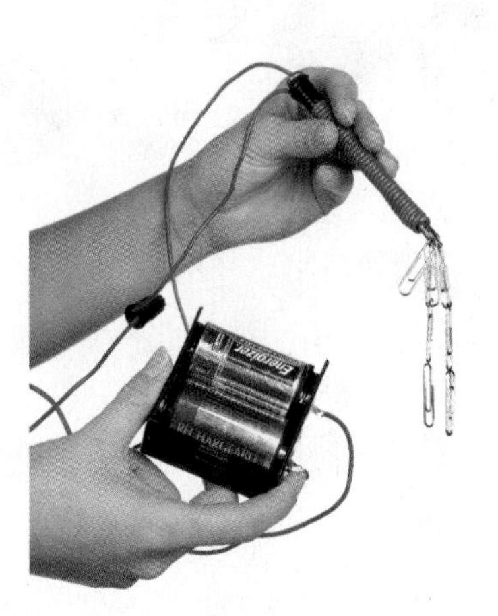

An electric current in the wire produces a magnetic field that magnetizes the nail around which the wire is wound.

Electromagnets

The magnetic domains in a ferromagnetic material have a strong tendency to line up with an applied magnetic field. This means that it is possible to use a piece of iron or other ferromagnetic material to increase the strength of the field from a current-carrying wire. For example, suppose a solenoid is wound around a piece of iron. When current is passed through the wire, the solenoid's magnetic field lines up the domains in the iron, thus magnetizing it. The resulting **electromagnet** may produce a field that is hundreds of times stronger than the field due to the solenoid itself.

In the past few sections, we've begun to see examples of the deep connection between electricity and magnetism. This connection was one of the most important scientific discoveries of the 1800s and is something we will explore in more detail in the next chapter.

Class Video

Class Video

STOP TO THINK 24.10 A chain of paper clips is hung from a permanent magnet. Which diagram shows the correct induced pole structure of the paper clips?

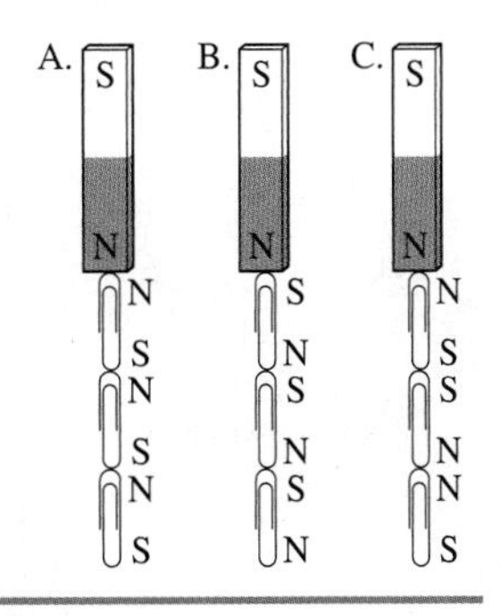

INTEGRATED EXAMPLE 24.15 Making music with magnetism

A loudspeaker creates sound by pushing air back and forth with a paper cone that is driven by a magnetic force on a wire coil at the base of the cone. **FIGURE 24.61** shows the details. The bottom of the cone is wrapped with several turns of fine wire. This coil of wire sits in the gap between the poles of a circular magnet, the black disk in the photo. The magnetic field exerts a force on a current in the wire, pushing the cone and thus pushing the air.

FIGURE 24.61 The arrangement of the coil and magnet poles in a loudspeaker.

(a)

(b)

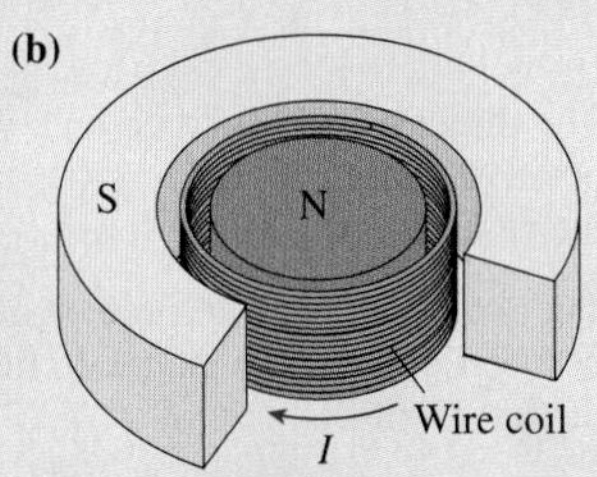

There is a 0.18 T field in the gap between the poles. The coil of wire that sits in this gap has a diameter of 5.0 cm, contains 20 turns of wire, and has a resistance of 8.0 Ω. The speaker is connected to an amplifier whose instantaneous output voltage of 6.0 V creates a clockwise current in the coil as seen from above. What is the magnetic force on the coil at this instant?

PREPARE The current in the coil experiences a force due to the magnetic field between the poles. Let's start with a sketch of the field to determine the direction of this force. Magnetic field lines go from the north pole to the south pole of a magnet, so the field lines for the loudspeaker magnet appear as in **FIGURE 24.62**. The field is at all points perpendicular to the current, and the right-hand rule shows us that, for a clockwise current, the force at each point of the wire is out of the page.

FIGURE 24.62 The magnetic field in the gap and the current in the coil.

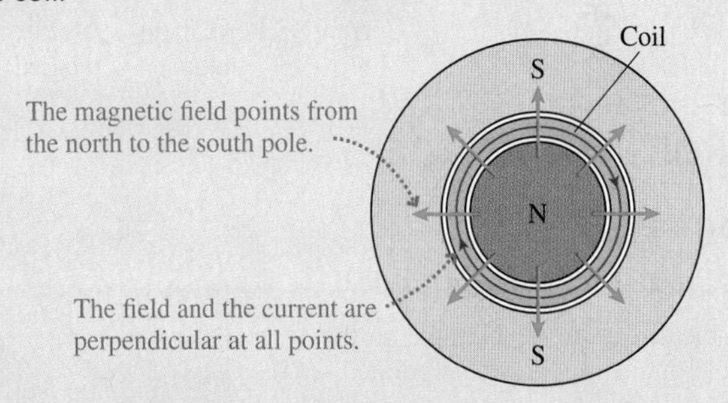

SOLVE The current in the wire is produced by the amplifier. The current is related to the potential difference and the resistance of the wire by Ohm's law:

$$I = \frac{\Delta V}{R} = \frac{6.0\ \text{V}}{8.0\ \Omega} = 0.75\ \text{A}$$

Because the current is perpendicular to the field, we can use Equation 24.10 to determine the force on this current. We know the field and the current, but we need to know the length of the wire in the field region. The coil has diameter 5.0 cm and thus circumference $\pi(0.050\ \text{m})$. The coil has 20 turns, so the total length of the wire in the field is

$$L = 20\pi(0.050\ \text{m}) = 3.1\ \text{m}$$

The magnitude of the force is then given by Equation 24.10 as

$$F = ILB = (0.75\ \text{A})(3.1\ \text{m})(0.18\ \text{T}) = 0.42\ \text{N}$$

This force is directed out of the page, as already noted.

ASSESS The force is small, but this is reasonable. A loudspeaker cone is quite light, so only a small force is needed for a large acceleration. The force for a clockwise current is out of the page, but when the current switches direction to counterclockwise, the force will switch directions as well. A current that alternates direction will cause the cone to oscillate in and out—just what is needed for making music.

SUMMARY

Goal: To learn about magnetic fields and how magnetic fields exert forces on currents and moving charges.

GENERAL PRINCIPLES

Sources of Magnetism

Magnetic fields can be created by either:

- Electric currents or
- Permanent magnets

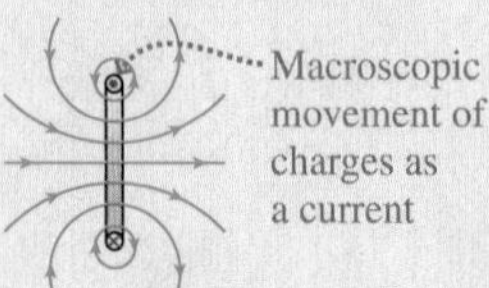

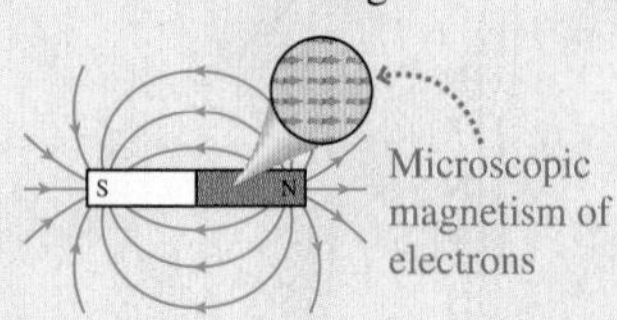

The most basic unit of magnetism is the **magnetic dipole,** which consists of a north and a south pole.

Three basic kinds of dipoles are:

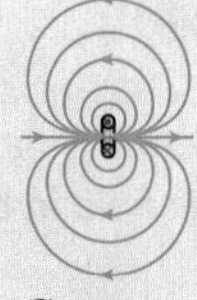
Current loop

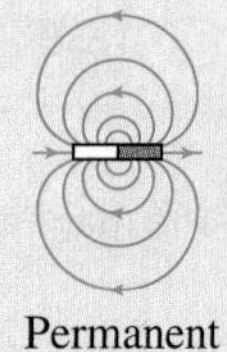
Permanent magnet

Atomic magnet

Consequences of Magnetism

Magnetic fields exert long-range forces on magnetic materials and on moving charges or currents.

- Unlike poles of magnets attract each other; like poles repel each other.
- A magnetic field exerts a force on a moving charged particle.

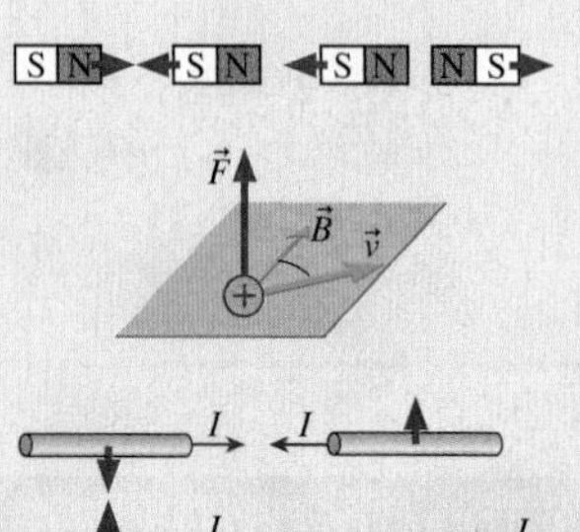

- Parallel wires with currents in the same direction attract each other; when the currents are in opposite directions, the wires repel each other.

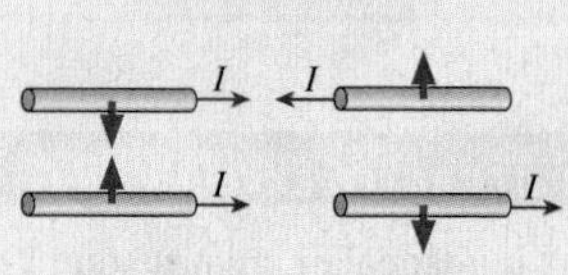

Magnetic fields exert torques on magnetic dipoles, aligning their axes with the field.

IMPORTANT CONCEPTS

Magnetic Fields

The **direction of the magnetic field**

- is the direction in which the north pole of a compass needle points.
- due to a current can be found from the **right-hand rule for fields.**

The **strength of the magnetic field** is

- proportional to the torque on a compass needle when turned slightly from the field direction.
- measured in tesla (T).

Magnetic Forces and Torques

The magnitude of the magnetic force on a *moving* charge depends on its charge q, its speed v, and the angle α between the velocity and the field:

$$F = |q|vB\sin\alpha$$

The direction of this force on a positive charge is given by the **right-hand rule for forces.**

The magnitude of the force on a *current-carrying wire* perpendicular to the magnetic field depends on the current and the length of the wire: $F = ILB$.

The torque on a *current loop* in a magnetic field depends on the current, the loop's area, and how the loop is oriented in the field: $\tau = (IA)B\sin\theta$.

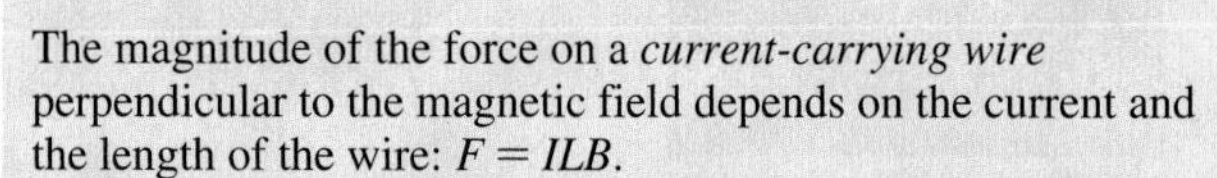

APPLICATIONS

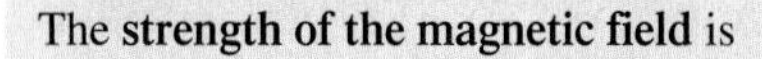

Fields due to common currents

Long, straight wire $B = \dfrac{\mu_0 I}{2\pi r}$

Current loop $B = \dfrac{\mu_0 I}{2R}$

Solenoid $B = \dfrac{\mu_0 NI}{L}$

Charged-particle motion

There is no force if $\vec{v}$ is parallel to $\vec{B}$.

If $\vec{v}$ is perpendicular to $\vec{B}$, the particle undergoes uniform circular motion with radius $r = mv/|q|B$.

Stability of magnetic dipoles

A magnetic dipole is stable (in a lower energy state) when aligned with the external magnetic field. It is unstable (in a higher energy state) when aligned opposite to the field.

The probe field of an MRI scanner measures the flipping of magnetic dipoles between these two orientations.

Problem difficulty is labeled as | (straightforward) to ||||| (challenging). Problems labeled INT integrate significant material from earlier chapters; BIO are of biological or medical interest.

For assigned homework and other learning materials, go to MasteringPhysics®

Scan this QR code to launch a Video Tutor Solution that will help you solve problems for this chapter.

QUESTIONS

Conceptual Questions

1. In Figure Q24.1, suppose the magnet on the right is fixed in place and the magnet on the left is free to pivot about its center. Will the magnet on the left start to rotate? If so, will it initially rotate clockwise or counterclockwise?

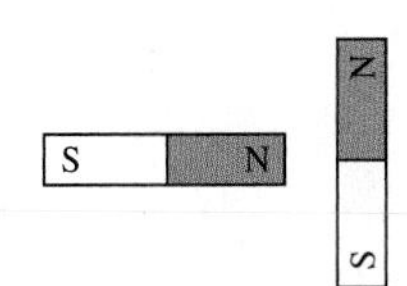

FIGURE Q24.1

2. You have a bar magnet whose poles are not marked. How can you find which pole is north and which is south by using only a piece of string?
3. When you are in the southern hemisphere, does a compass point north or south?
4. If you were standing directly at the earth's north magnetic pole, in what direction would a compass point if it were free to swivel in any direction? Explain.
5. BIO If you took a sample of magnetotactic bacteria from the northern hemisphere to the southern hemisphere, would you expect them to survive? Explain.
6. BIO Green turtles use the earth's magnetic field to navigate. They seem to use the field to tell them their latitude—how far north or south of the equator they are. Explain how knowing the direction of the earth's field could give this information.
7. What is the current direction in the wire of Figure Q24.7? Explain.

FIGURE Q24.7 FIGURE Q24.8

8. What is the current direction in the wire of Figure Q24.8?
9. Since the wires in the walls of your house carry current, you might expect that you could use a compass to detect the positions of the wires. In fact, a compass will experience no deflection when brought near a current-carrying wire because the current is AC (meaning "alternating current"—the current switches direction 120 times each second). Explain why a compass doesn't react to an AC current.
10. As shown in Figure Q24.10, a uniform magnetic field points upward, in the plane of the paper. A long wire perpendicular to the paper initially carries no current. When a current is turned on in the wire in the direction shown, the magnetic field at point 1 is found to be zero. Draw the magnetic field vector at point 2 when the current is on.

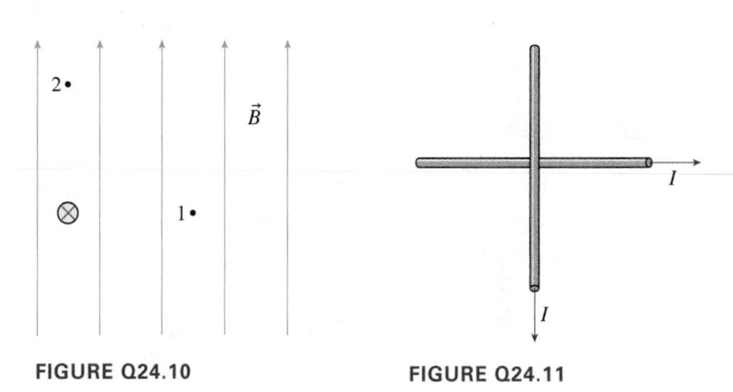

FIGURE Q24.10 FIGURE Q24.11

11. Two long wires carry currents in the directions shown in Figure Q24.11. One wire is 10 cm above the other. In which direction is the magnetic field at a point halfway between them?
12. An electron is moving in a circular orbit in a uniform magnetic field. Is the kinetic energy of the electron changing? Explain.
13. Figure Q24.13 shows a solenoid as seen in cross section. Compasses are placed at points 1, 2, and 3. In which direction will each compass point when there is a large current in the direction shown? Explain.

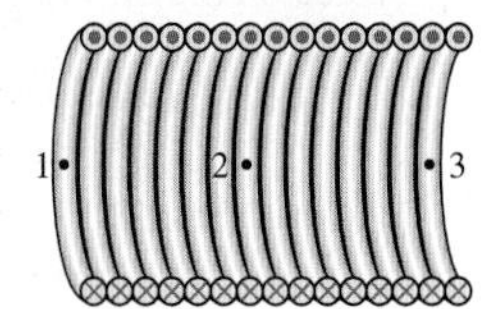

FIGURE Q24.13

14. One long solenoid is placed inside another solenoid. Both solenoids have the same length and the same number of turns of wire, but the outer solenoid has twice the diameter of the inner solenoid. Each solenoid carries the same current, but the two currents are in opposite directions, as shown in Figure Q24.14. What is the magnetic field at the center of the inner solenoid? Explain.

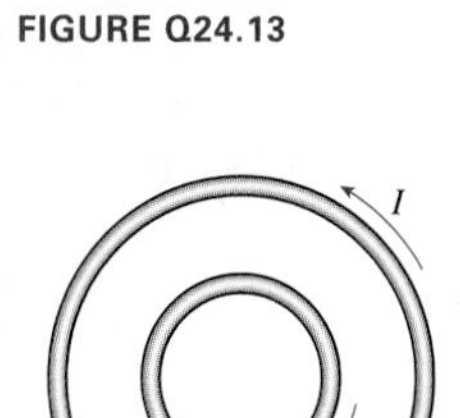

FIGURE Q24.14

15. What is the *initial* direction of deflection for the charged particles entering the magnetic fields shown in Figure Q24.15?

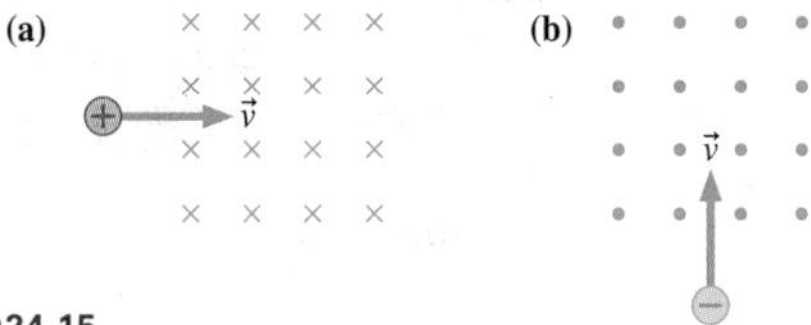

FIGURE Q24.15

16. Describe the force on the charged particles after they enter the magnetic fields shown in Figure Q24.16.

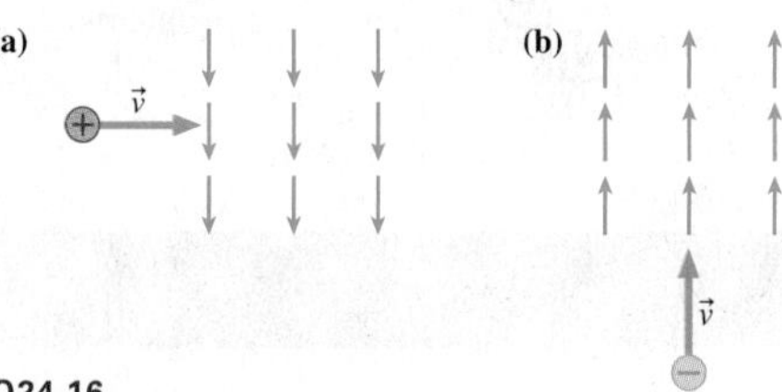

FIGURE Q24.16

17. Determine the magnetic field direction that causes the charged particles shown in Figure Q24.17 to experience the indicated magnetic forces.

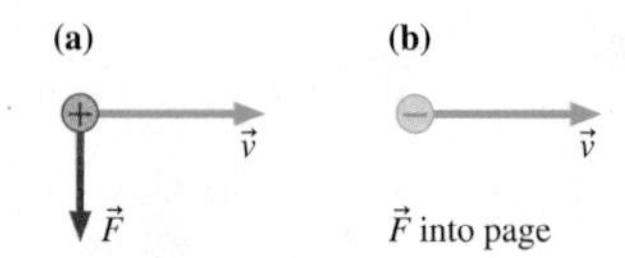

FIGURE Q24.17

18. Determine the magnetic field direction that causes the charged particles shown in Figure Q24.18 to experience the indicated magnetic forces.

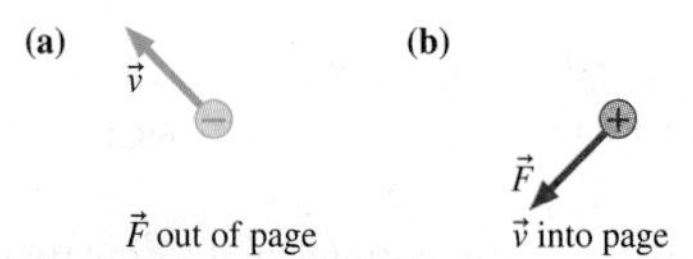

FIGURE Q24.18

19. An electron is moving near a long, current-carrying wire, as shown in Figure Q24.19. What is the direction of the magnetic force on the electron?

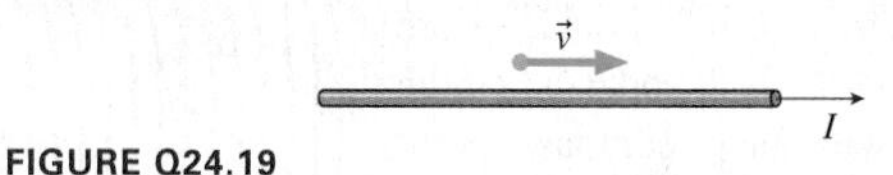

FIGURE Q24.19

20. Two positive charges are moving in a uniform magnetic field with velocities as shown in Figure Q24.20. The magnetic force on each charge is also shown. In which direction does the magnetic field point?

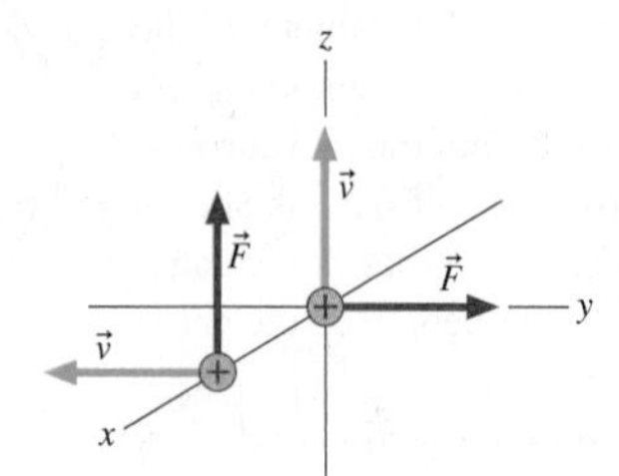

FIGURE Q24.20

21. An electron is moving in a circular orbit in the earth's magnetic field directly above the north magnetic pole. Viewed from above, is the rotation clockwise or counterclockwise?

22. An electron and a proton are moving in circular orbits in the earth's magnetic field, high above the earth's atmosphere. The two particles move at the same speed. Which particle takes more time to complete one orbit? Explain.

23. A proton moves in a region of uniform magnetic field, as shown in Figure Q24.23. The velocity at one instant is shown. Will the subsequent motion be a clockwise or counterclockwise orbit?

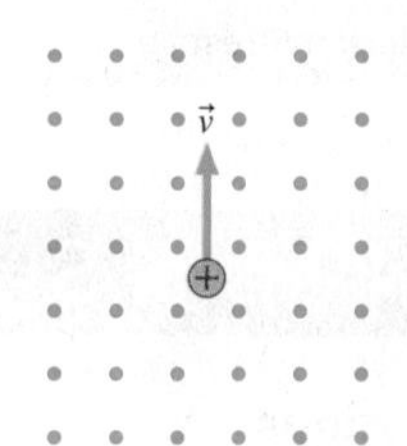

FIGURE Q24.23

24. A proton is moving near a long, current-carrying wire. When the proton is at the point shown in Figure Q24.24, in which direction is the force on it?

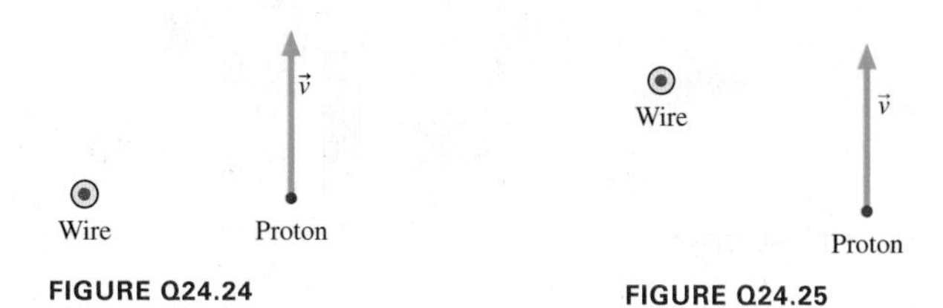

FIGURE Q24.24

FIGURE Q24.25

25. A proton is moving near a long, current-carrying wire. When the proton is at the point shown in Figure Q24.25, in which direction is the force on it?

26. A long wire and a square loop lie in the plane of the paper. Both carry a current in the direction shown in Figure Q24.26. In which direction is the net force on the loop? Explain.

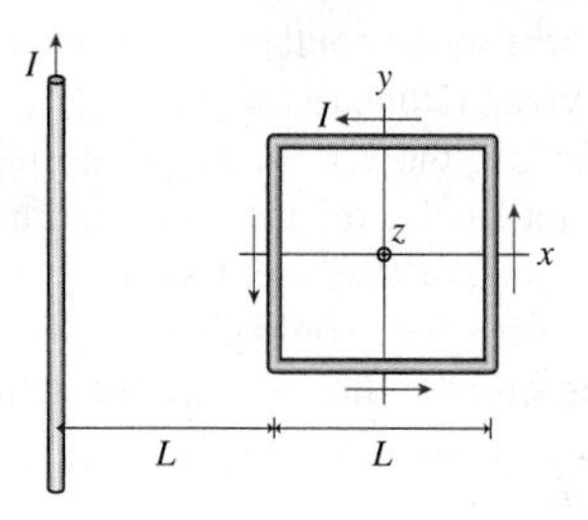

FIGURE Q24.26

27. A solenoid carries a current that produces a field inside it. A wire carrying a current lies inside the solenoid, at the center, carrying a current along the solenoid's axis. Is there a force on this wire due to the field of the solenoid? Explain.

28. Archaeologists can use instruments that measure small variations in magnetic field to locate buried walls made of fired brick, as shown in Figure Q24.28. When fired, the magnetic moments in the clay become randomly aligned; as the clay cools, the magnetic moments line up with the earth's field and retain this alignment even if the bricks are subsequently moved. Explain how this leads to a measurable magnetic field variation over a buried wall.

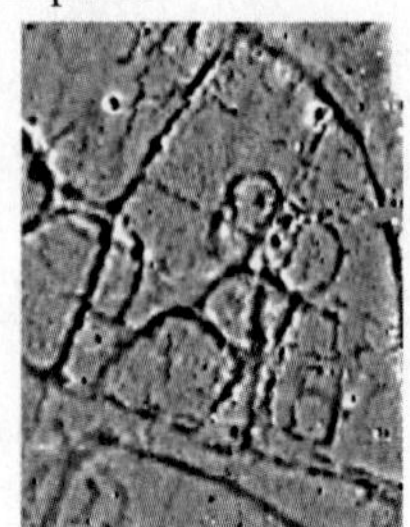

FIGURE Q24.28

Watch video solution for problem: 24.7

Multiple-Choice Questions

29. ‖ An unmagnetized metal sphere hangs by a thread. When the north pole of a bar magnet is brought near, the sphere is strongly attracted to the magnet, as shown in Figure Q24.29. Then the magnet is reversed and its south pole is brought near the sphere. How does the sphere respond?

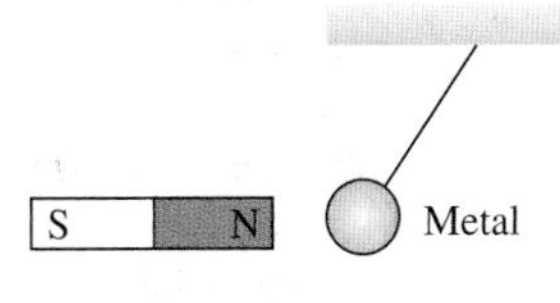

FIGURE Q24.29

   A. It is strongly attracted to the magnet.
   B. It is weakly attracted to the magnet.
   C. It does not respond.
   D. It is weakly repelled by the magnet.
   E. It is strongly repelled by the magnet.

30. ‖ If a compass is placed above a current-carrying wire, as in Figure Q24.30, the needle will line up with the field of the wire. Which of the views shows the correct orientation of the needle for the noted current direction?

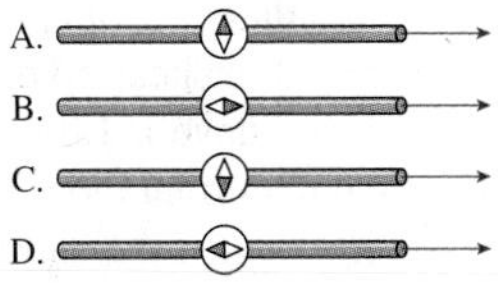

FIGURE Q24.30

31. | Two wires carry equal and opposite currents, as shown in Figure Q24.31. At a point directly between the two wires, the field is

FIGURE Q24.31

   A. Directed up, toward the top of the page.
   B. Directed down, toward the bottom of the page.
   C. Directed to the left.
   D. Directed to the right.
   E. Zero.

32. | Figure Q24.32 shows four particles moving to the right as they enter a region of uniform magnetic field, directed into the paper as noted. All particles move at the same speed and have the same charge. Which particle has the largest mass?

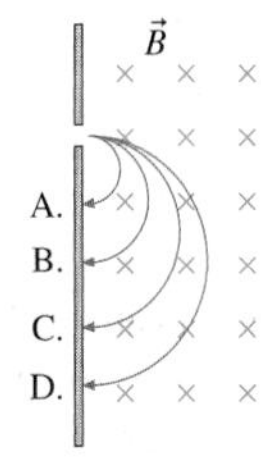

FIGURE Q24.32

33. | Four particles of identical charge and mass enter a region of uniform magnetic field and follow the trajectories shown in Figure Q24.33. Which particle has the highest velocity?

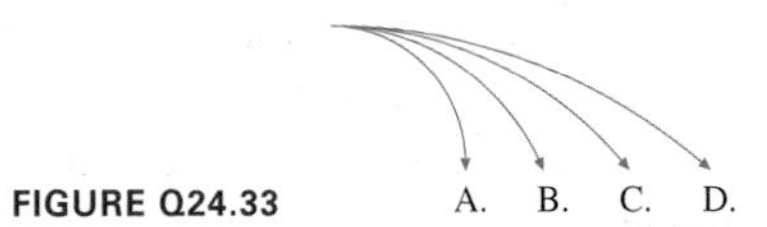

FIGURE Q24.33

34. | If all of the particles shown in Figure Q24.33 are electrons, what is the direction of the magnetic field that produced the indicated deflection?
   A. Up (toward the top of the page).
   B. Down (toward the bottom of the page).
   C. Out of the plane of the paper.
   D. Into the plane of the paper.

35. | If two compasses are brought near enough to each other, the magnetic fields of the compasses themselves will be larger than the field of the earth, and the needles will line up with each other. Which of the arrangements of two compasses shown in Figure Q24.35 is a possible stable arrangement?

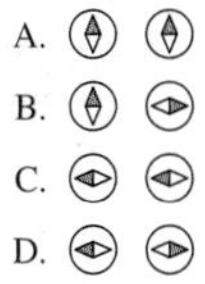

FIGURE Q24.35

PROBLEMS

Section 24.1 Magnetism

Section 24.2 The Magnetic Field

Section 24.3 Electric Currents Also Create Magnetic Fields

Section 24.4 Calculating the Magnetic Field Due to a Current

1. | Table 24.1 notes that the magnetic field 10 cm from a wire carrying a 1 A current is 2 μT. What is the field 1 cm from the wire?
2. | Use the data from Table 24.1 to determine at what approximate distance from a wire carrying a current of 10 A the magnetic field strength would equal that of a refrigerator magnet.
3. ‖ The magnetic field at the center of a 1.0-cm-diameter loop is 2.5 mT.
   a. What is the current in the loop?
   b. A long, straight wire carries the same current you found in part a. At what distance from the wire is the magnetic field 2.5 mT?
4. ‖ For a particular scientific experiment, it is important to be completely isolated from any magnetic field, including the earth's field. The earth's field is approximately 50 μT, but at any particular location it may be a bit more or less than this. A 1.00-m-diameter current loop with 200 turns of wire is adjusted to carry a current of 0.215 A; at this current, the coil's field at the center is exactly equal to the earth's field in magnitude but opposite in direction, so that the total field at the center of the coil is zero. What is the strength of the earth's magnetic field at this location?
5. ‖ Two wires carry equal currents in opposite directions, as in Figure P24.5. The field is 2.0 mT at a point below the lower wire. What are the strength and direction of the field at point 1 and at point 2, as shown in the figure?

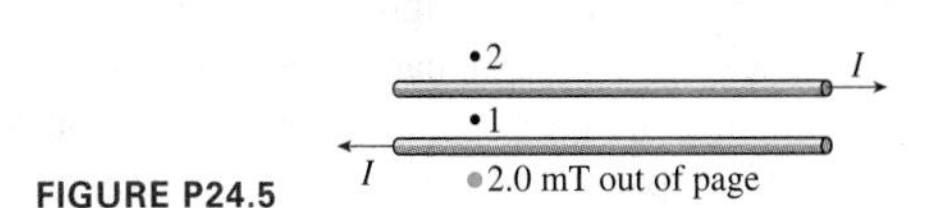

FIGURE P24.5

6. | What are the magnetic field strength and direction at points 1, 2, and 3 in Figure P24.6?

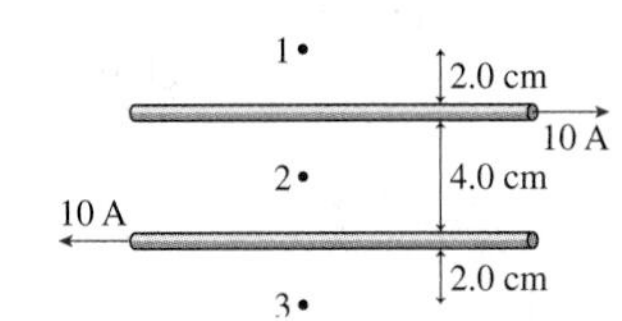

FIGURE P24.6

7. ‖ An investigator places a sample 1.0 cm from a wire carrying a large current; the strength of the magnetic field has a particular value at this point. Later, she must move the sample to a 5.0 cm distance, but she would like to keep the field the same. By what factor must she increase the current?

Watch video solution for problem: 24.23

8. | Although the evidence is weak, there has been concern in BIO recent years over possible health effects from the magnetic fields generated by transmission lines. A typical high-voltage transmission line is 20 m off the ground and carries a current of 200 A. Estimate the magnetic field strength on the ground underneath such a line. What percentage of the earth's magnetic field does this represent?

9. | Some consumer groups urge pregnant women not to use elec- BIO tric blankets, in case there is a health risk from the magnetic fields from the approximately 1 A current in the heater wires.
   a. Estimate, stating any assumptions you make, the magnetic field strength a fetus might experience. What percentage of the earth's magnetic field is this?
   b. It is becoming standard practice to make electric blankets with minimal external magnetic field. Each wire is paired with another wire that carries current in the opposite direction. How does this reduce the external magnetic field?

10. ||| A long wire carrying a 5.0 A current perpendicular to the xy-plane intersects the x-axis at $x = -2.0$ cm. A second, parallel wire carrying a 3.0 A current intersects the x-axis at $x = +2.0$ cm. At what point or points on the x-axis is the magnetic field zero if (a) the two currents are in the same direction and (b) the two currents are in opposite directions?

11. || The element niobium, which is a metal, is a superconductor (i.e., no electrical resistance) at temperatures below 9 K. However, the superconductivity is destroyed if the magnetic field at the surface of the wire of the metal reaches or exceeds 0.10 T. What is the maximum current in a straight, 3.0-mm-diameter superconducting niobium wire?
Hint: You can assume that all the current flows in the center of the wire.

12. | The small currents in axons corresponding to nerve impulses BIO produce measurable magnetic fields. A typical axon carries a peak current of 0.040 μA. What is the strength of the field at a distance of 1.0 mm?

13. || A solenoid used to produce magnetic fields for research purposes is 2.0 m long, with an inner radius of 30 cm and 1000 turns of wire. When running, the solenoid produces a field of 1.0 T in the center. Given this, how large a current does it carry?

14. | Two concentric current loops lie in the same plane. The smaller loop has a radius of 3.0 cm and a current of 12 A. The bigger loop has a current of 20 A. The magnetic field at the center of the loops is found to be zero. What is the radius of the bigger loop?

15. | The magnetic field of the brain has been measured to be BIO approximately 3.0×10^{-12} T. Although the currents that cause this field are quite complicated, we can get a rough estimate of their size by modeling them as a single circular current loop 16 cm (the width of a typical head) in diameter. What current is needed to produce such a field at the center of the loop?

16. || A researcher would like to perform an experiment in zero magnetic field, which means that the field of the earth must be canceled. Suppose the experiment is done inside a solenoid of diameter 1.0 m, length 4.0 m, with a total of 5000 turns of wire. The solenoid is oriented to produce a field that opposes and exactly cancels the 52 μT local value of the earth's field. What current is needed in the solenoid's wire?

17. ||| What is the magnetic field at the center of the loop in Figure P24.17?

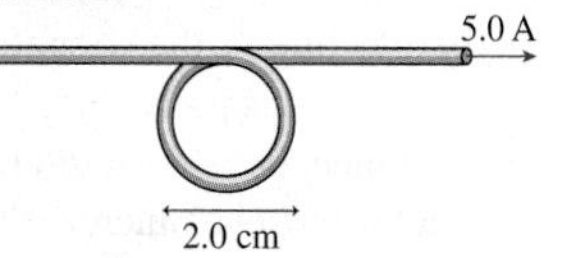

FIGURE P24.17

18. | Experimental tests have shown that hammerhead sharks BIO can detect magnetic fields. In one such test, 100 turns of wire were wrapped around a 7.0-m-diameter cylindrical shark tank. A magnetic field was created inside the tank when this coil of wire carried a current of 1.5 A. Sharks trained by getting a food reward when the field was present would later unambiguously respond when the field was turned on.
   a. What was the magnetic field strength in the center of the tank due to the current in the coil?
   b. Is the strength of the coil's field at the center of the tank larger or smaller than that of the earth?

19. | We have seen that the heart produces a magnetic field that BIO can be used to diagnose problems with the heart. The magnetic field of the heart is a dipole field produced by a loop current in the outer layers of the heart. Suppose that the field at the center of the heart is 90 pT (a pT is 10^{-12} T) and that the heart has a diameter of approximately 12 cm. What current circulates around the heart to produce this field?

20. |||| You have a 1.0-m-long copper wire. You want to make an N-turn current loop that generates a 1.0 mT magnetic field at the center when the current is 1.0 A. You must use the entire wire. What will be the diameter of your coil?

21. ||||| In the Bohr model of the hydrogen atom, the electron moves in a circular orbit of radius 5.3×10^{-11} m with speed 2.2×10^{6} m/s. According to this model, what is the magnetic field at the center of a hydrogen atom due to the motion of the electron?
Hint: Determine the *average* current of the orbiting electron.

Section 24.5 Magnetic Fields Exert Forces on Moving Charges

22. | A proton moves with a speed of 1.0×10^{7} m/s in the directions shown in Figure P24.22. A 0.50 T magnetic field points in the positive x-direction. For each, what is the magnetic force on the proton? Give your answers as a magnitude and a direction.

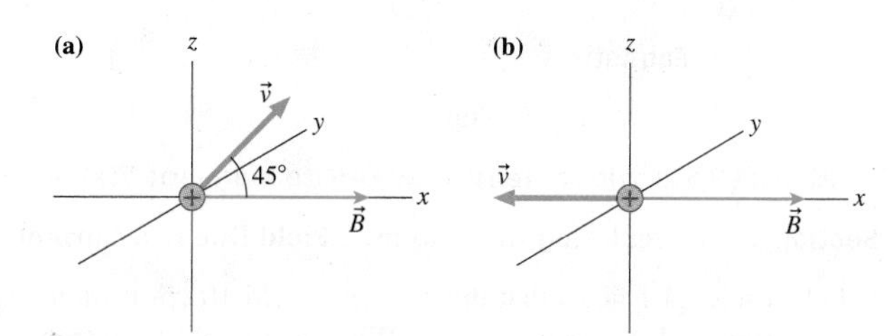

FIGURE P24.22

23. || An electron moves with a speed of 1.0×10^{7} m/s in the directions shown in Figure P24.23. A 0.50 T magnetic field points in the positive x-direction. For each, what is the magnetic force on the electron? Give your answers as a magnitude and a direction.

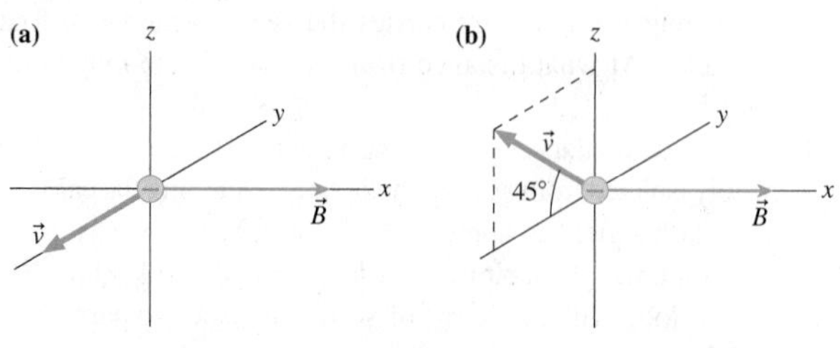

FIGURE P24.23

24. | An electromagnetic flowmeter applies a magnetic field of 0.20 T to blood flowing through a coronary artery at a speed of 15 cm/s. What force is felt by a chlorine ion with a single negative charge?
BIO

25. | The aurora is caused when electrons and protons, moving in the earth's magnetic field of $\approx 5.0 \times 10^{-5}$ T, collide with molecules of the atmosphere and cause them to glow. What is the radius of the circular orbit for
    a. An electron with speed 1.0×10^6 m/s?
    b. A proton with speed 5.0×10^4 m/s?

26. ||| Problem 24.25 describes two particles that orbit the earth's magnetic field lines. What is the *frequency* of the circular orbit for
    a. An electron with speed 1.0×10^6 m/s?
    b. A proton with speed 5.0×10^4 m/s?

27. ||| A muon (a charged particle with a mass between that of an electron and a proton that is found in cosmic rays) follows a circular orbit in earth's 50 μT field at a frequency of 6.8 kHz. At what frequency would a muon orbit in Jupiter's more intense 430 μT field?

28. ||| Charged particles orbit magnetic field lines in the space above Mars just as they do in the space above earth, but Mars's magnetic field is much weaker so the period of an orbit is much longer. One spacecraft observed a 1.0 ms period for electrons orbiting in Mars's magnetic field. What is the orbital period of a proton?

29. |||| The microwaves in a microwave oven are produced in a special tube called a *magnetron.* The electrons orbit in a magnetic field at a frequency of 2.4 GHz, and as they do so they emit 2.4 GHz electromagnetic waves. What is the strength of the magnetic field?

30. ||| A cyclotron is used to produce a beam of high-energy deuterons that then collide with a target to produce radioactive isotopes for a medical procedure. Deuterons are nuclei of deuterium, an isotope of hydrogen, consisting of one neutron and one proton, with total mass 3.34×10^{-27} kg. The deuterons exit the cyclotron with a kinetic energy of 5.00 MeV.
BIO INT
    a. What is the speed of the deuterons when they exit?
    b. If the magnetic field inside the cyclotron is 1.25 T, what is the diameter of the deuterons' largest orbit, just before they exit?
    c. If the beam current is 400 μA, how many deuterons strike the target each second?

31. ||| A medical cyclotron used in the production of medical isotopes accelerates protons to 6.5 MeV. The magnetic field in the cyclotron is 1.2 T.
BIO
    a. What is the diameter of the largest orbit, just before the protons exit the cyclotron?
    b. A proton exits the cyclotron 1.0 ms after starting its spiral trajectory in the center of the cyclotron. How many orbits does the proton complete during this 1.0 ms?

32. ||| Early black-and-white television sets used an electron beam to draw a picture on the screen. The electrons in the beam were accelerated by a voltage of 3.0 kV; the beam was then steered to different points on the screen by coils of wire that produced a magnetic field of up to 0.65 T.
INT
    a. What is the speed of electrons in the beam?
    b. What acceleration do they experience due to the magnetic field, assuming that it is perpendicular to their path? What is this acceleration in units of g?
    c. If the electrons were to complete a full circular orbit, what would be the radius?
    d. A magnetic field can be used to redirect the beam, but the electrons are brought to high speed by an electric field. Why can't we use a magnetic field for this task?

Section 24.6 Magnetic Fields Exert Forces on Currents

33. | What magnetic field strength and direction will levitate the 2.0 g wire in Figure P24.33?

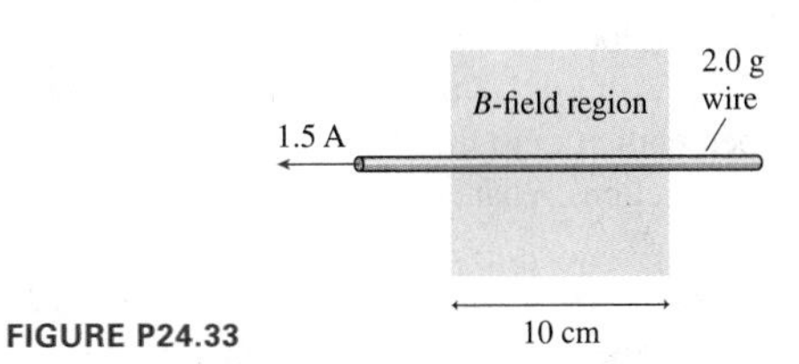

FIGURE P24.33

34. | What is the net force (magnitude and direction) on each wire in Figure P24.34?

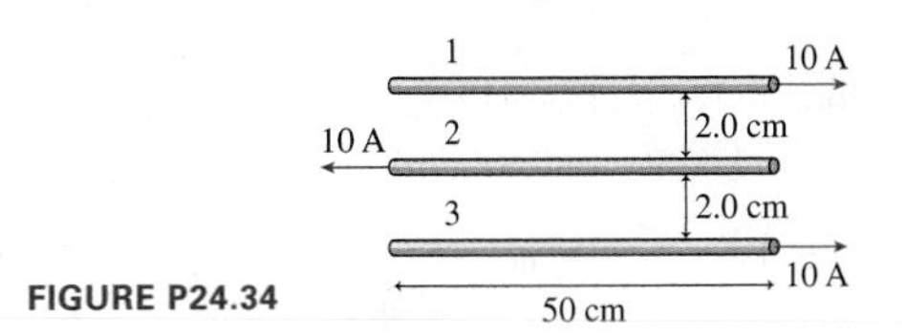

FIGURE P24.34

35. | The unit of current, the ampere, is defined in terms of the force between currents. If two 1.0-meter-long sections of very long wires a distance 1.0 m apart each carry a current of 1.0 A, what is the force between them? (If the force between two actual wires has this value, the current is defined to be exactly 1 A.)

36. | A uniform 2.5 T magnetic field points to the right. A 3.0-m-long wire, carrying 15 A, is placed at an angle of 30° to the field, as shown in Figure P24.36. What is the force (magnitude and direction) on the wire?

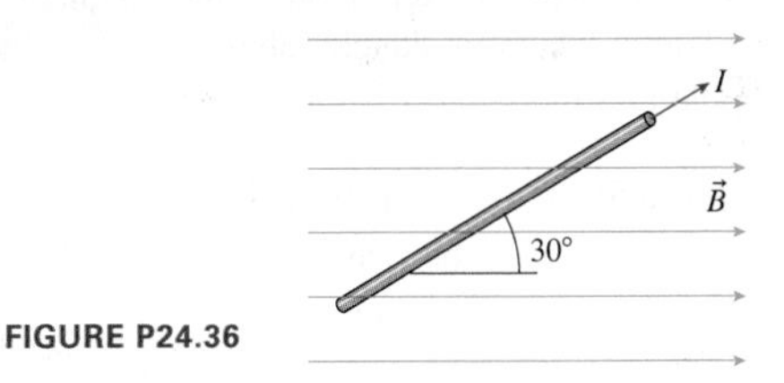

FIGURE P24.36

37. || Magnetic information on hard drives is accessed by a read head that must move rapidly back and forth across the disk. The force to move the head is generally created with a *voice coil actuator,* a flat coil of fine wire that moves between the poles of a strong magnet, as in Figure P24.37. Assume that the coil is a square 1.0 cm on a side made of 200 turns of fine wire with total resistance 1.5 Ω. The field between the poles of the magnet is 0.30 T; assume that the field does not extend beyond the edge of the magnet. The coil and the mount that it rides on have a total mass of 12 g.
INT
    a. If a voltage of 5.0 V is applied to the coil, what is the current?
    b. If the current is clockwise viewed from above, what are the magnitude and direction of the net force on the coil?
    c. What is the magnitude of the acceleration of the coil?

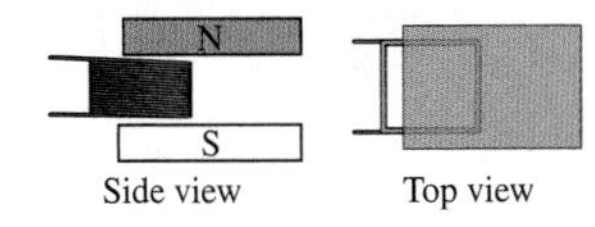

FIGURE P24.37

Watch video solutions for problems: 24.41 24.47

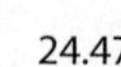

Section 24.7 Magnetic Fields Exert Torques on Dipoles

38. || A current loop in a motor has an area of 0.85 cm^2. It carries a 240 mA current in a uniform field of 0.62 T. What is the magnitude of the maximum torque on the current loop?

39. || A square current loop 5.0 cm on each side carries a 500 mA current. The loop is in a 1.2 T uniform magnetic field. The axis of the loop, perpendicular to the plane of the loop, is 30° away from the field direction. What is the magnitude of the torque on the current loop?

40. || People have proposed driving motors with the earth's magnetic field. This is possible in principle, but the small field means that unrealistically large currents are needed to produce noticeable torques. Suppose a 20-cm-diameter loop of wire is oriented for maximum torque in the earth's field. What current would it need to carry in order to experience a very modest 1.0×10^{-3} N · m torque?

41. || a. What is the magnitude of the torque on the circular current loop in Figure P24.41?
    b. What is the loop's stable equilibrium position?

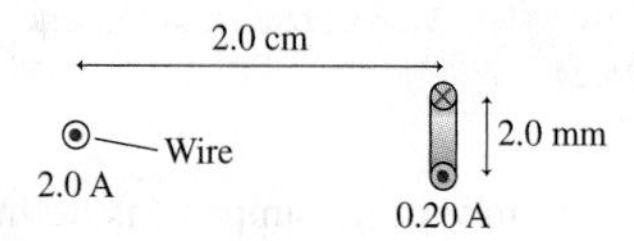

FIGURE P24.41

Section 24.8 Magnets and Magnetic Materials

42. | All ferromagnetic materials have a *Curie temperature,* a temperature above which they will cease to be magnetic. Explain in some detail why you might expect this to be so.

43. | A solenoid is near a piece of iron, as shown in Figure P24.43. When a current is present in the solenoid, a magnetic field is created. This magnetic field will magnetize the iron, and there will be a net force between the solenoid and the iron.
    a. Make a sketch showing the direction of the magnetic field from the solenoid. On your sketch, label the induced north magnetic pole and the induced south magnetic pole in the iron.
    b. Will the force on the iron be attractive or repulsive?
    c. Suppose this force moves the iron. Which way will the iron move?

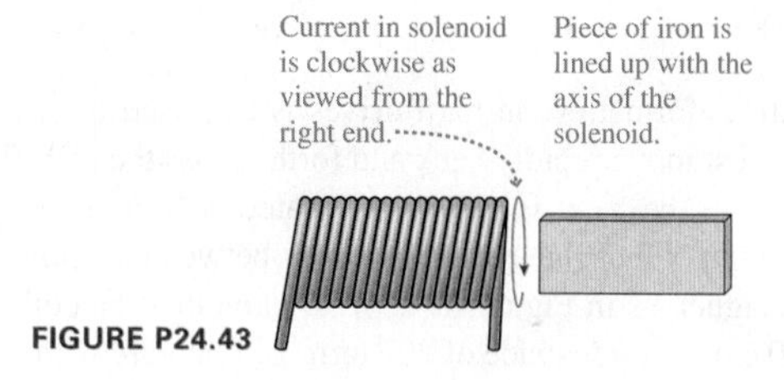

FIGURE P24.43

General Problems

44. | The right edge of the circuit in Figure P24.44 extends into INT a 50 mT uniform magnetic field. What are the magnitude and direction of the net force on the circuit?

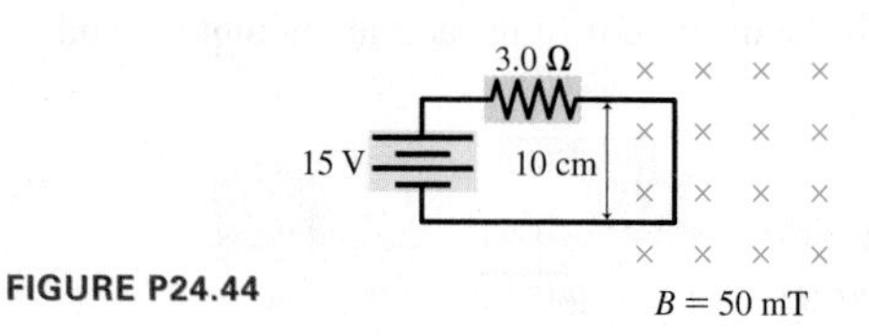

FIGURE P24.44

45. || The two 10-cm-long parallel wires in Figure P24.45 are sepa- INT rated by 5.0 mm. For what value of the resistor R will the force between the two wires be 5.4×10^{-5} N?

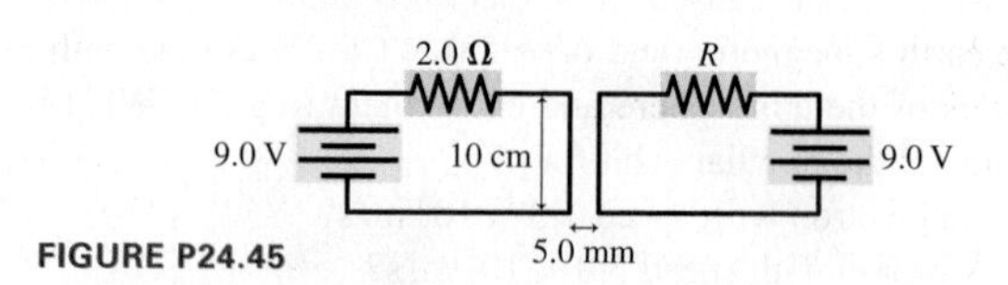

FIGURE P24.45

46. || An electron travels with speed 1.0×10^7 m/s between the INT two parallel charged plates shown in Figure P24.46. The plates are separated by 1.0 cm and are charged by a 200 V battery. What magnetic field strength and direction will allow the electron to pass between the plates without being deflected?

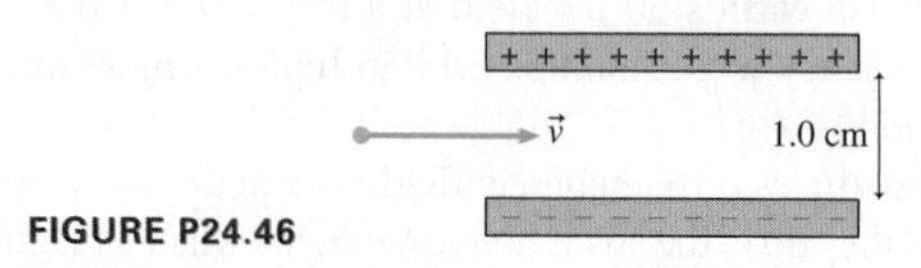

FIGURE P24.46

47. || The two springs in Figure P24.47 each have a spring constant INT of 10 N/m. They are stretched by 1.0 cm when a current passes through the wire. How big is the current?

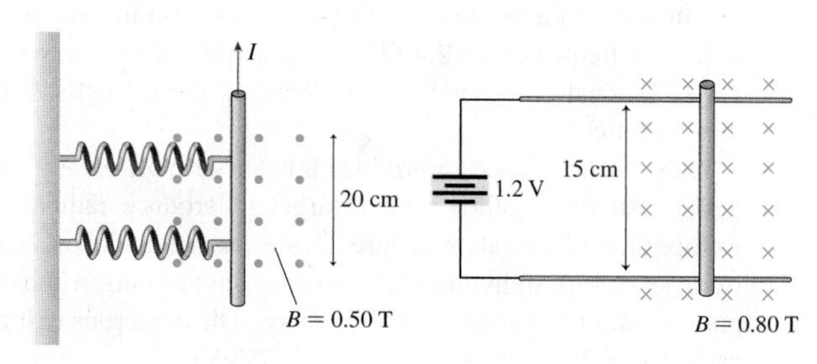

FIGURE P24.47 FIGURE P24.48

48. || A device called a *railgun* uses the magnetic force on currents INT to launch projectiles at very high speeds. An idealized model of a railgun is illustrated in Figure P24.48. A 1.2 V power supply is connected to two conducting rails. A segment of copper wire, in a region of uniform magnetic field, slides freely on the rails. The wire has a 0.85 mΩ resistance and a mass of 5.0 g. Ignore the resistance of the rails. When the power supply is switched on,
    a. What is the current?
    b. What are the magnitude and direction of the force on the wire?
    c. What will be the wire's speed after it has slid a distance of 6.0 cm?

49. ||| Irrigation channels that require regular flow monitoring are often equipped with electromagnetic flowmeters in which the magnetic field is produced by horizontal coils embedded in the bottom of the channel. A particular coil has 100 turns and a diameter of 6.0 m. When it's time for a measurement, a 5.0 A current is turned on. The large diameter of the coil means that the field in the water flowing directly above the center of the coil is approximately equal to the field in the center of the coil.
    a. What is the magnitude of the field at the center of the coil?
    b. If the field is directed downward and the water is flowing east, what is the direction of the force on a positive ion in the water above the center of the coil?
    c. If the water is flowing above the center of the coil at 1.5 m/s, what is the magnitude of the force on an ion with a charge $+e$?

50. | Typical blood velocities in the coronary arteries range from BIO INT 10 to 30 cm/s. An electromagnetic flowmeter applies a magnetic field of 0.25 T to a coronary artery with a blood velocity of 15 cm/s. As we saw in Figure 24.36, this field exerts a force on ions in the blood, which will separate. The ions will separate until they make an electric field that exactly balances the magnetic force. This electric field produces a voltage that can be measured.
    a. What force is felt by a singly ionized (positive) sodium ion?
    b. Charges in the blood will separate until they produce an electric field that cancels this magnetic force. What will be the resulting electric field?
    c. What voltage will this electric field produce across an artery with a diameter of 3.0 mm?
51. | A power line consists of two wires, each carrying a current of 400 A in the same direction. The lines are perpendicular to the earth's magnetic field and are separated by a distance of 5.0 m. Which is larger: the force of the earth's magnetic field on each wire or the magnetic force between the wires?
52. || Consider the long rectangular loop in Figure P24.52. If you pass a 1 A current through the loop, there is a repulsive force between the two long sides of the loop. If the current is increased to 3 A, by what factor does the magnitude of the force increase?

FIGURE P24.52

53. ||| Bats are capable of navigating using the earth's field—a plus BIO for an animal that may fly great distances from its roost at night. If, while sleeping during the day, bats are exposed to a field of a similar magnitude but different direction than the earth's field, they are more likely to lose their way during their next lengthy night flight. Suppose you are a researcher doing such an experiment in a location where the earth's field is 50 μT at a 60° angle below horizontal. You make a 50-cm-diameter, 100-turn coil around a roosting box; the sleeping bats are at the center of the coil. You wish to pass a current through the coil to produce a field that, when combined with the earth's field, creates a net field with the same strength and dip angle (60° below horizontal) as the earth's field but with a horizontal component that points south rather than north. What are the proper orientation of the coil and the necessary current?
54. ||| At the equator, the earth's field is essentially horizontal; near BIO the north pole, it is nearly vertical. In between, the angle varies. As you move farther north, the dip angle, the angle of the earth's field below horizontal, steadily increases. Green turtles seem to use this dip angle to determine their latitude. Suppose you are a researcher wanting to test this idea. You have gathered green turtle hatchlings from a beach where the magnetic field strength is 50 μT and the dip angle is 56°. You then put the turtles in a 1.2-m-diameter circular tank and monitor the direction in which they swim as you vary the magnetic field in the tank. You change the field by passing a current through a 100-turn horizontal coil wrapped around the tank. This creates a field that adds to that of the earth. What current should you pass through the coil, and in what direction, to produce a net field in the center of the tank that has a dip angle of 62°?
55. || Internal components of cathode-ray-tube televisions and computer monitors can become magnetized; the resulting magnetic field can deflect the electron beam and distort the colors on the screen. Demagnetization can be accomplished with a coil of wire whose current switches direction rapidly and gradually decreases in amplitude. Explain what effect this will have on the magnetic moments of the magnetic materials in the device, and how this might eliminate any magnetic ordering.
56. ||| A 1.0-m-long, 1.0-mm-diameter copper wire carries a current INT of 50.0 A to the east. Suppose we create a magnetic field that produces an upward force on the wire exactly equal in magnitude to the wire's weight, causing the wire to "levitate." What are the field's direction and magnitude?
57. || An insulated copper wire is wrapped around an iron nail. The resulting coil of wire consists of 240 turns of wire that cover 1.8 cm of the nail, as shown in Figure P24.57. A current of 0.60 A passes through the wire. If the ferromagnetic properties of the nail increase the field by a factor of 100, what is the magnetic field strength inside the nail?

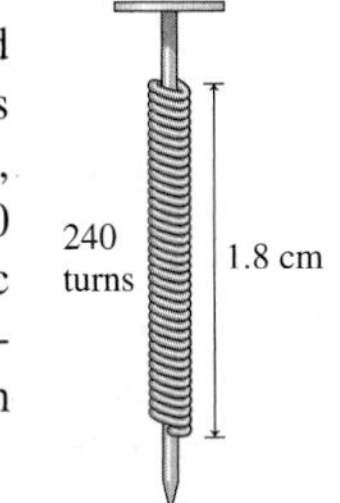

FIGURE P24.57

58. ||| A long, straight wire with a linear mass density of 50 g/m INT is suspended by threads, as shown in Figure P24.58. There is a uniform magnetic field pointing vertically downward. A 10 A current in the wire experiences a horizontal magnetic force that deflects it to an equilibrium angle of 10°. What is the strength of the magnetic field $\vec{B}$?

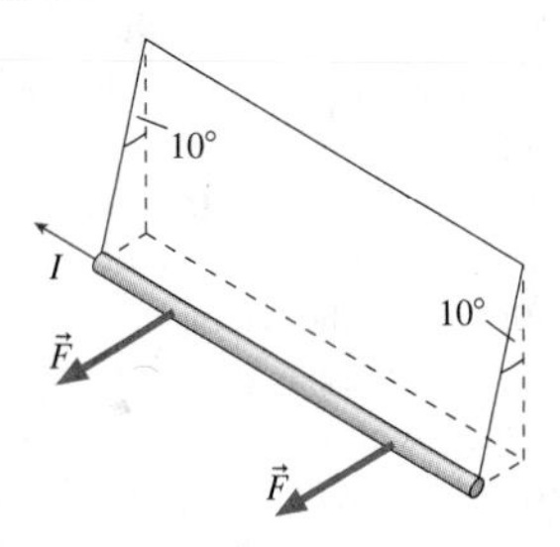

FIGURE P24.58

MCAT-Style Passage Problems

The Velocity Selector INT

In experiments where all the charged particles in a beam are required to have the same velocity (for example, when entering a mass spectrometer), scientists use a *velocity selector.* A velocity selector has a region of uniform electric and magnetic fields that are perpendicular to each other and perpendicular to the motion of the charged particles. Both the electric and magnetic fields exert a force on the charged particles. If a particle has precisely the right velocity, the two forces exactly cancel and the particle is not deflected. Equating the forces due to the electric field and the magnetic field gives the following equation:

$$qE = qvB$$

Solving for the velocity, we get:

$$v = \frac{E}{B}$$

A particle moving at this velocity will pass through the region of uniform fields with no deflection, as shown in Figure P24.59. For higher or lower velocities than this, the particles will feel a net force and will be deflected. A slit at the end of the region allows only the particles with the correct velocity to pass.

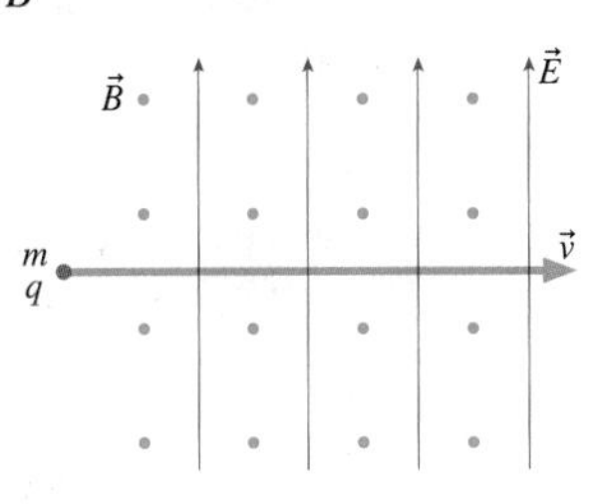

FIGURE P24.59

59. | Assuming the particle in Figure P24.59 is positively charged, what are the directions of the forces due to the electric field and to the magnetic field?
    A. The force due to the electric field is directed up (toward the top of the page); the force due to the magnetic field is directed down (toward the bottom of the page).
    B. The force due to the electric field is directed down (toward the bottom of the page); the force due to the magnetic field is directed up (toward the top of the page).
    C. The force due to the electric field is directed out of the plane of the paper; the force due to the magnetic field is directed into the plane of the paper.
    D. The force due to the electric field is directed into the plane of the paper; the force due to the magnetic field is directed out of the plane of the paper.
60. | How does the kinetic energy of the particle in Figure P24.59 change as it traverses the velocity selector?
    A. The kinetic energy increases.
    B. The kinetic energy does not change.
    C. The kinetic energy decreases.
61. | Suppose a particle with twice the velocity of the particle in Figure P24.59 enters the velocity selector. The path of this particle will curve
    A. Upward (toward the top of the page).
    B. Downward (toward the bottom of the page).
    C. Out of the plane of the paper.
    D. Into the plane of the paper.
62. | Next, a particle with the same mass and velocity as the particle in Figure P24.59 enters the velocity selector. This particle has a charge of $2q$—twice the charge of the particle in Figure P24.59. In this case, we can say that
    A. The force of the electric field on the particle is greater than the force of the magnetic field.
    B. The force of the magnetic field on the particle is greater than the force of the electric field.
    C. The forces of the electric and magnetic fields on the particle are still equal.

Ocean Potentials INT

The ocean is salty because it contains many dissolved ions. As these charged particles move with the water in strong ocean currents, they feel a force from the earth's magnetic field. Positive and negative charges are separated until an electric field develops that balances this magnetic force. This field produces measurable potential differences that can be monitored by ocean researchers.

The Gulf Stream moves northward off the east coast of the United States at a speed of up to 3.5 m/s. Assume that the current flows at this maximum speed and that the earth's field is 50 μT tipped 60° below horizontal.

63. | What is the direction of the magnetic force on a singly ionized negative chlorine ion moving in this ocean current?
    A. East B. West
    C. Up D. Down
64. | What is the magnitude of the force on this ion?
    A. 2.8×10^{-23} N B. 2.4×10^{-23} N
    C. 1.6×10^{-23} N D. 1.4×10^{-23} N
65. | What magnitude electric field is necessary to exactly balance this magnetic force?
    A. 1.8×10^{-4} N/C B. 1.5×10^{-4} N/C
    C. 1.0×10^{-4} N/C D. 0.9×10^{-4} N/C
66. | The electric field produces a potential difference. If you place one electrode 10 m below the surface of the water, you will measure the greatest potential difference if you place the second electrode
    A. At the surface.
    B. At a depth of 20 m.
    C. At the same depth 10 m to the north.
    D. At the same depth 10 m to the east.

The Mass Spectrometer

If you have a sample of unknown composition, a first step at analysis might be a determination of the masses of the atoms and molecules in the sample. A *mass spectrometer* to make such an analysis can take various forms, but for many years the best technique was to determine the masses of ionized atoms and molecules in a sample by observing their circular paths in a uniform magnetic field, as illustrated in Figure P24.67. A sample to be analyzed is vaporized, then singly ionized. The ions are accelerated through an electric field, and ions of a known speed selected. These ions travel into a region of uniform magnetic field, where they follow circular paths. An exit slit allows ions that have followed a particular path to be counted by a detector, producing a record of the masses of the particles in the sample.

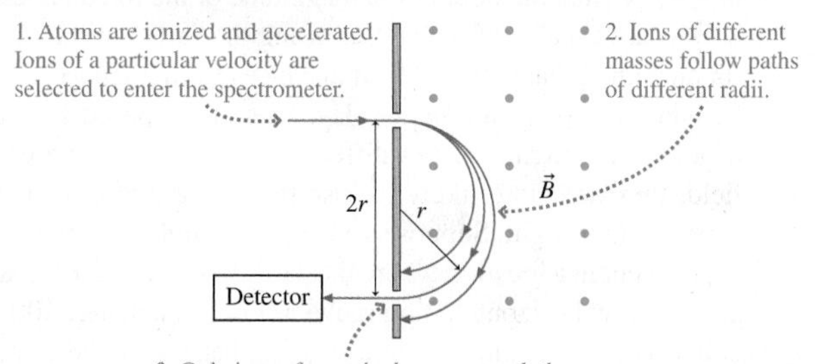
1. Atoms are ionized and accelerated. Ions of a particular velocity are selected to enter the spectrometer.
2. Ions of different masses follow paths of different radii.
$\vec{B}$
2r
r
Detector
3. Only ions of a particular mass reach the exit slit and continue to the detector.

FIGURE P24.67

67. | In the spectrometer shown in Figure P24.67, do the ions have positive or negative charge?
    A. Positive B. Negative
68. || The moving ions can be thought of as a current loop, and it will produce its own magnetic field. The direction of this field at the center of the particles' circular orbit is
    A. In the same direction as the spectrometer's magnetic field.
    B. Opposite the direction of the spectrometer's magnetic field.
69. || Why is it important that the ions have a known speed?
    A. The radius of the orbit depends on the mass, the charge, and the speed. If the charge and the speed are the same, the orbit depends on only the mass.
    B. The orbit must be circular, and this is the case for only a certain range of speeds.
    C. If the ions are moving too fast, the magnetic field will not be able to bend their path to the detector.
    D. The ions are all accelerated by the same electric field, and so will all have the same speed anyway.
70. || A mass spectrometer similar to the one in Figure P24.67 is designed to analyze biological samples. Molecules in the sample are singly ionized, then they enter a 0.80 T uniform magnetic field at a speed of 2.3×10^5 m/s. If a molecule has a mass 85 times the mass of the proton, what will be the approximate distance between the points where the ion enters and exits the magnetic field?
    A. 25 cm B. 50 cm
    C. 75 cm D. 100 cm

STOP TO THINK ANSWERS

Chapter Preview Stop to Think: B. The positive charge feels a force in the direction of the field, the negative charge feels a force opposite the field. There is no net force, but there will be a torque that rotates the dipole counterclockwise.

Stop to Think 24.1: C. The compass needle will not rotate, since there is no force between the stationary charges on the rod and the magnetic poles of the compass needle.

Stop to Think 24.2: A. The compass needle will rotate to line up with the field of the magnet, which goes from the north to the south pole.

Stop to Think 24.3: A. Using the right-hand rule for currents on any part of the loop shows that the field at the center of the loop is directed toward the top of the page.

Stop to Think 24.4: D. The compass needle will rotate to line up with the field circling the wire. The right-hand rule for fields shows this to be toward the top of the paper in the figure.

Stop to Think 24.5: B. The strength of the field is inversely proportional to the distance from the wire. If the distance from the wire is increased by a factor of 4, the field strength decreases by a factor of 4.

Stop to Think 24.6: C. If you keep the number of turns of wire the same and decrease the overall length, the field strength will increase. The field doesn't depend on the diameter.

Stop to Think 24.7: A, C. The force to produce these circular orbits is directed toward the center of the circle. Using the right-hand rule for forces, we see that this will be true for the situations in A and C if the particles are negatively charged.

Stop to Think 24.8: C. The right-hand rule for forces gives the direction of the force. With the field into the paper, the force is to the left if the current is toward the top of the paper.

Stop to Think 24.9: B. Looking at the forces on the top and the bottom of the loop, we can see that the loop will rotate counterclockwise. Alternatively, we can look at the dipole structure of the loop: With a north pole on the left and a south pole on the right, the loop will rotate counterclockwise.

Stop to Think 24.10: B. All of the induced dipoles will be aligned with the field of the bar magnet.

25 EM Induction and EM Waves

The photo of the flower on the left shows how it appears to our eyes, in visible light. But there's more to the story! The false-color view of the flower on the right shows its appearance in the ultraviolet, beyond the range of human vision, revealing pigments we can't see. Whose eyes are these pigments intended for?

LOOKING AHEAD ▸▸

Goal: To understand the nature of electromagnetic induction and electromagnetic waves.

Magnetism and Electricity

The turning windmill blades rotate a wire coil in a *magnetic* field, producing an *electric* current.

You'll continue to explore the deep connections between magnetism and electricity.

Induction

A physician programs a pacemaker by using a rapidly changing magnetic field to **induce** a voltage in the implanted device.

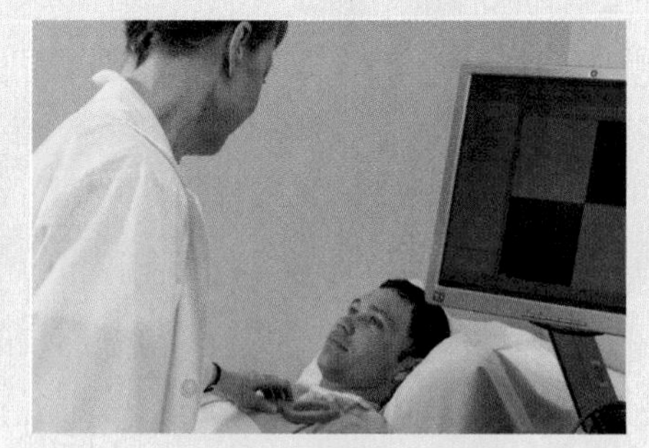

You'll learn to analyze **electromagnetic induction** qualitatively and quantitatively.

Electromagnetic Waves

This antenna detects **electromagnetic waves**, waves of electric and magnetic fields.

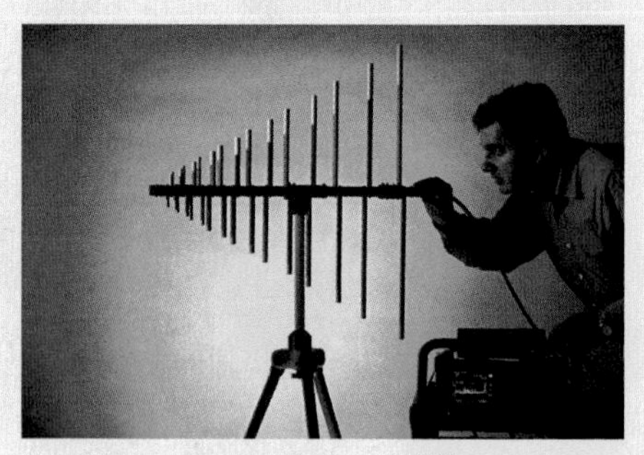

You'll learn the properties of different electromagnetic waves, from radio waves to light waves.

LOOKING BACK ◂◂

Traveling Waves

In Chapter 15 you learned the properties of traveling waves. For sinusoidal waves, the wave speed is the product of the wave's frequency and wavelength.

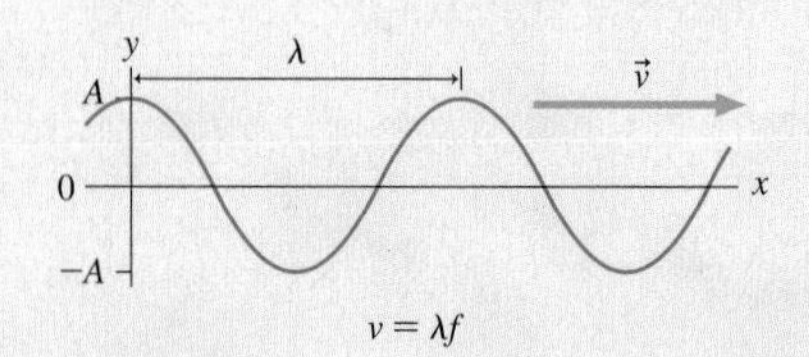

In this chapter, you'll see how the properties of traveling waves are used to describe electromagnetic waves.

STOP TO THINK

A microwave oven uses 2.4 GHz electromagnetic waves. A cell phone uses electromagnetic waves at a slightly lower 1.9 GHz frequency. What can you say about the wavelengths of the two?

A. The waves from the oven have a longer wavelength.
B. The waves from the phone have a longer wavelength.
C. The waves from the oven and the phone have the same wavelength.

25.1 Induced Currents

In Chapter 24, we learned that a current can create a magnetic field. As soon as this discovery was widely known, investigators began considering a related question: Can a magnetic field create a current?

One of the early investigators was Michael Faraday, who was experimenting with two coils of wire wrapped around an iron ring, as shown in **FIGURE 25.1**, when he made a remarkable discovery. He had hoped that the magnetic field generated by a current in the coil on the left would create a magnetic field in the iron, and that the magnetic field in the iron might then somehow produce a current in the circuit on the right.

This technique failed to generate a steady current, but Faraday noticed that the needle of the current meter jumped ever so slightly at the instant when he closed the switch in the circuit on the left. After the switch was closed, the needle immediately returned to zero. Faraday's observation suggested to him that a current was generated only if the magnetic field was *changing* as it passed through the coil. Faraday set out to test this hypothesis through a series of experiments.

FIGURE 25.1 Faraday's discovery of electromagnetic induction.

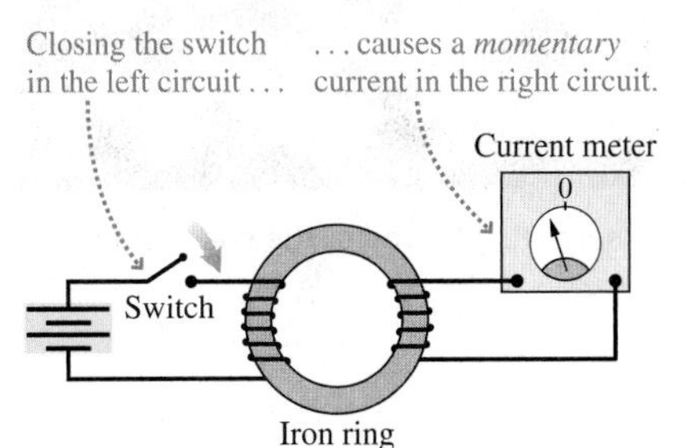

Faraday investigates electromagnetic induction

Faraday placed one coil directly above the other, without the iron ring. There was no current in the lower circuit while the switch was in the closed position, but a momentary current appeared whenever the switch was opened or closed.

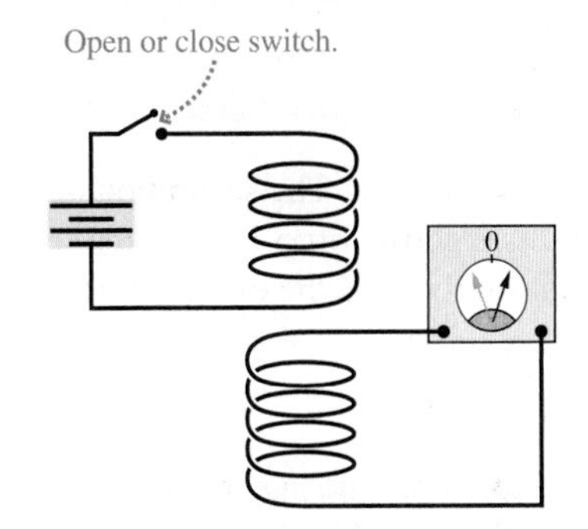

Opening or closing the switch creates a momentary current.

Faraday pushed a bar magnet into a coil of wire. This action caused a momentary deflection of the needle in the current meter, although *holding* the magnet inside the coil had no effect. A quick withdrawal of the magnet deflected the needle in the other direction.

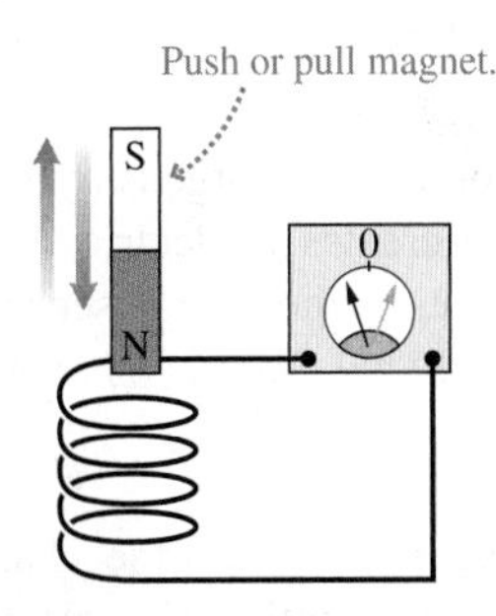

Pushing the magnet into the coil or pulling it out creates a momentary current.

Must the magnet move? Faraday created a momentary current by rapidly pulling a coil of wire out of a magnetic field, although there was no current if the coil was stationary in the magnetic field. Pushing the coil *into* the magnet caused the needle to deflect in the opposite direction.

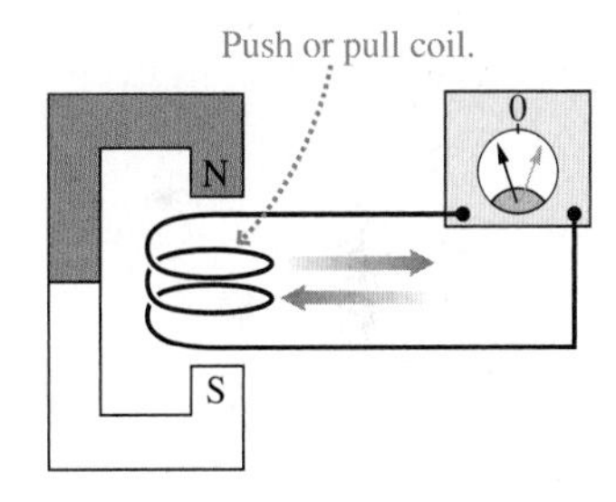

Pushing the coil into the magnet or pulling it out creates a momentary current.

All of these experiments served to bolster Faraday's hypothesis: **Faraday found that there is a current in a coil of wire if and only if the magnetic field passing through the coil is *changing*.** It makes no difference what causes the magnetic field to change: current stopping or starting in a nearby circuit, moving a magnet through the coil, or moving the coil into and out of a magnet. The effect is the same in all cases. There is no current if the field through the coil is not changing, so it's not the magnetic field itself that is responsible for the current but, instead, it is the *changing of the magnetic field.*

The current in a circuit due to a changing magnetic field is called an **induced current.** Opening the switch or moving the magnet *induces* a current in a nearby circuit. An induced current is not caused by a battery; it is a completely new way to generate a current. The creation of an electric current by a changing magnetic field is our first example of **electromagnetic induction.**

A satellite tethered to the space shuttle.

25.2 Motional emf

In 1996, astronauts on the space shuttle deployed a satellite at the end of a 20-km-long conducting tether. A potential difference of up to 3500 V developed between the shuttle and the satellite as the wire between the two swept through the earth's magnetic field. Why would the motion of a wire in a magnetic field produce such a large voltage? Let's explore the mechanism behind this *motional emf.*

To begin, consider a conductor of length l that moves with velocity $\vec{v}$ through a uniform magnetic field $\vec{B}$, as shown in **FIGURE 25.2**. The charge carriers inside the conductor—assumed to be positive, as in our definition of current—also move with velocity $\vec{v}$, so they each experience a magnetic force. For simplicity, we will assume that $\vec{v}$ is perpendicular to $\vec{B}$, in which case the magnitude of the force is $F_B = qvB$. This force causes the charge carriers to move. For the geometry of Figure 25.2, the right-hand rule tells us that the positive charges move toward the top of the moving conductor, leaving an excess of negative charge at the bottom.

FIGURE 25.2 The magnetic force on the charge carriers in a moving conductor creates an electric field inside the conductor.

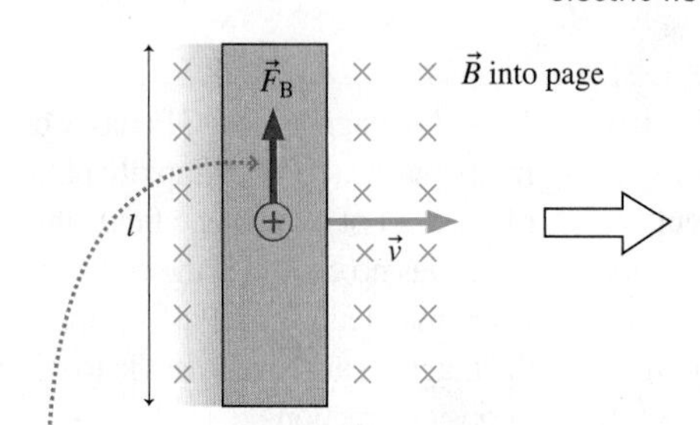

Charge carriers in the conductor experience a force of magnitude $F_B = qvB$. Positive charges are free to move and drift upward.

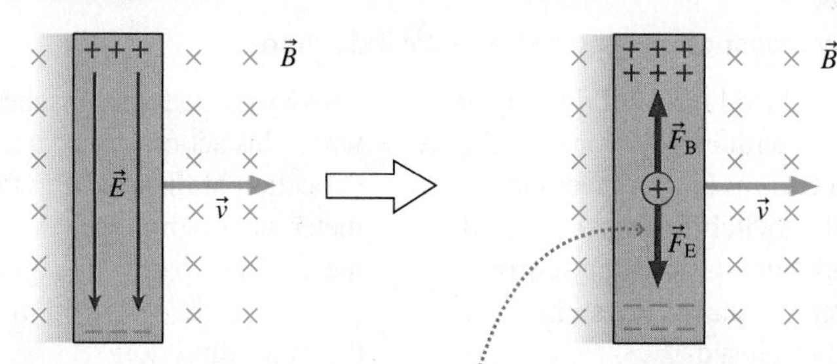

The resulting charge separation creates an electric field in the conductor. $\vec{E}$ increases as more charge flows.

The charge flow continues until the electric and magnetic forces balance. For a positive charge carrier, the upward magnetic force $\vec{F}_B$ is equal to the downward electric force $\vec{F}_E$.

This motion of the charge carriers cannot continue forever. The separation of the charge carriers creates an electric field. The resulting electric force *opposes* the separation of charge, so the charge separation continues only until the electric force has grown to exactly balance the magnetic force:

$$F_E = qE = F_B = qvB$$

When this balance occurs, the charge carriers experience no net force and thus undergo no further motion. The electric field strength at equilibrium is

$$E = vB \tag{25.1}$$

Thus, **the magnetic force on the charge carriers in a moving conductor creates an electric field $E = vB$ inside the conductor.**

The electric field, in turn, creates an electric potential difference between the two ends of the moving conductor. We found in ◀◀ SECTION 21.5 that the potential difference between two points separated by distance l parallel to an electric field E is $\Delta V = El$. Thus the motion of the wire through a magnetic field *induces* a potential difference

$$\Delta V = vlB \tag{25.2}$$

between the ends of the conductor. The potential difference depends on the strength of the magnetic field and on the wire's speed through the field. This is similar to the action of the electromagnetic flowmeter that we saw in the preceding chapter.

There's an important analogy between this potential difference and the potential difference of a battery. **FIGURE 25.3a** reminds you that a battery uses a nonelectric force—which we called the charge escalator—to separate positive and negative charges. We refer to a battery, where the charges are separated by chemical reactions, as a source of *chemical emf*. The moving conductor of **FIGURE 25.3b** develops a potential difference because of the work done to separate the charges. The emf of the conductor is due to its motion, rather than to chemical reactions inside, so we can define the **motional emf** of a conductor of length l moving with velocity $\vec{v}$ perpendicular to a magnetic field $\vec{B}$ to be

$$\mathcal{E} = vlB \tag{25.3}$$

emf due to motion of conductor (V); Speed of conductor perpendicular to magnetic field (m/s); Length of conductor (m); Magnetic field strength (T)

FIGURE 25.3 Two different ways to generate an emf.

(a) Chemical reactions separate the charges and cause a potential difference between the ends. This is a chemical emf.

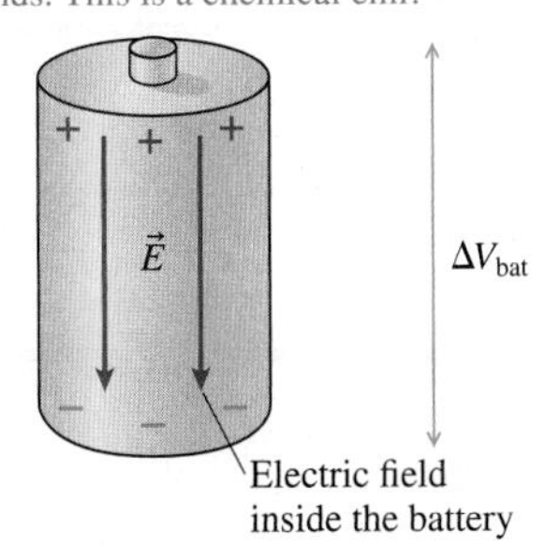

$\vec{E}$
ΔV_{bat}
Electric field inside the battery

(b) Magnetic forces separate the charges and cause a potential difference between the ends. This is a motional emf.

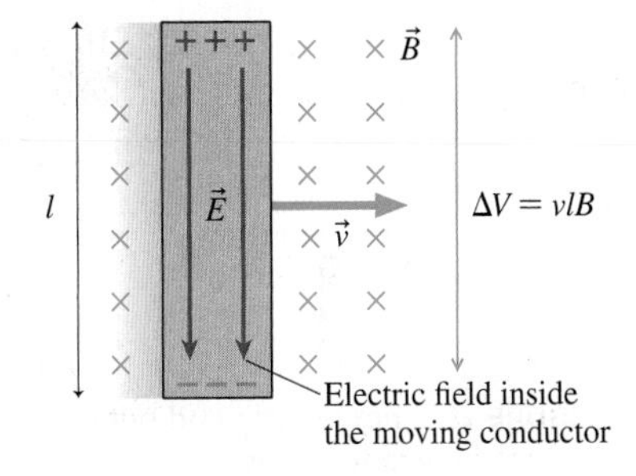

$\vec{B}$
l
$\vec{E}$
$\vec{v}$
$\Delta V = vlB$
Electric field inside the moving conductor

EXAMPLE 25.1 Finding the motional emf for an airplane

A Boeing 747 aircraft with a wingspan of 65 m is cruising at 260 m/s over northern Canada, where the magnetic field of the earth (magnitude 5.0×10^{-5} T) is directed straight down. What is the potential difference between the tips of the wings?

PREPARE The wing is a conductor moving through a magnetic field, so there will be a motional emf. We can visualize a top view of this situation exactly as in Figure 25.3b, with the wing as the moving conductor.

SOLVE The magnetic field is perpendicular to the velocity, so we can compute the potential difference using Equation 25.3:

$$\Delta V = vlB = (260 \text{ m/s})(65 \text{ m})(5.0 \times 10^{-5} \text{ T}) = 0.85 \text{ V}$$

ASSESS The earth's magnetic field is small, so the motional emf will be small as well unless the speed and the length are quite large. The tethered satellite generated a much higher voltage due to its much greater speed and the great length of the tether, the moving conductor.

▶**A head for magnetism?** BIO Hammerhead sharks seem to navigate using the earth's magnetic field. It's likely that they detect the *magnetic* field using their keen *electric* sense, detecting the earth's magnetic field by sensing the motional emf as they move through the water. The width of their oddly shaped heads is an asset because the magnitude of the potential difference ($\mathcal{E}$) is proportional to the length of the moving conductor (l).

Induced Current in a Circuit

The moving conductor of Figure 25.3 had an emf, but it couldn't sustain a current because the charges had nowhere to go. We can change this by including the moving conductor in a circuit.

FIGURE 25.4 shows a length of wire with resistance R sliding with speed v along a fixed U-shaped conducting rail. The wire and the rail together form a closed conducting loop—a circuit.

Suppose a magnetic field $\vec{B}$ is perpendicular to the plane of the circuit. Charges in the moving wire will be pushed to the ends of the wire by the magnetic force, just as they were in Figure 25.3, but now the charges can continue to flow around the circuit. The moving wire acts like the battery in a circuit.

The current in the circuit is an induced current, due to magnetic forces on moving charges. In this example, the induced current is counterclockwise. The total resistance of the circuit is just the resistance R of the moving wire, so the induced current is given by Ohm's law:

$$I = \frac{\mathcal{E}}{R} = \frac{vlB}{R} \tag{25.4}$$

FIGURE 25.4 A current is induced in the circuit as the wire moves through a magnetic field.

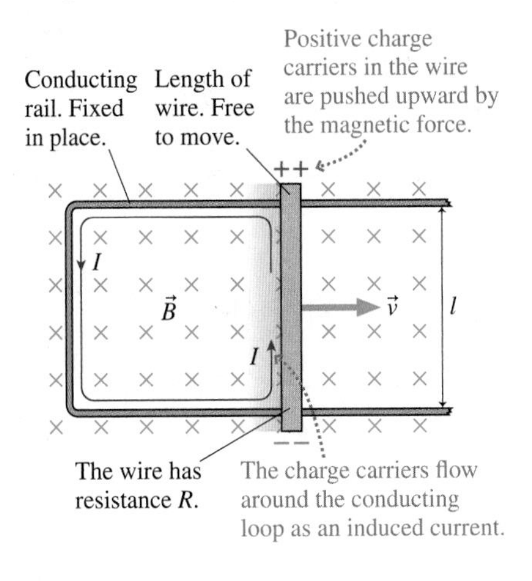

Conducting rail. Fixed in place.
Length of wire. Free to move.
Positive charge carriers in the wire are pushed upward by the magnetic force.
I
$\vec{B}$
I
$\vec{v}$
l
The wire has resistance R.
The charge carriers flow around the conducting loop as an induced current.

FIGURE 25.5 A pulling force is needed to move the wire to the right.

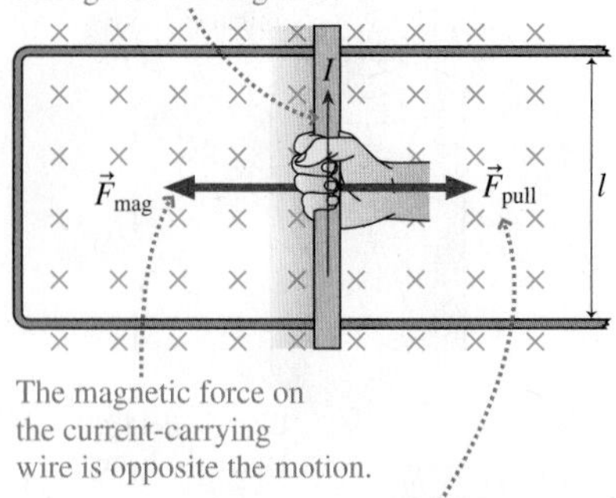
The induced current flows through the moving wire.
I
$\vec{F}_{mag}$
$\vec{F}_{pull}$
l
The magnetic force on the current-carrying wire is opposite the motion.
A pulling force to the right must balance the magnetic force to keep the wire moving at constant speed.

Video Tutor Demo

We've assumed that the wire is moving along the rail at constant speed. But we must apply a continuous pulling force $\vec{F}_{pull}$ to make this happen; FIGURE 25.5 shows why. The moving wire, which now carries induced current I, is in a magnetic field. You learned in Chapter 24 that a magnetic field exerts a force on a current-carrying wire. According to the right-hand rule, the magnetic force $\vec{F}_{mag}$ on the moving wire points to the left. This "magnetic drag" will cause the wire to slow down and stop *unless* we exert an equal but opposite pulling force $\vec{F}_{pull}$ to keep the wire moving.

NOTE ▶ Think about this carefully. As the wire moves to the right, the magnetic force $\vec{F}_B$ pushes the charge carriers *parallel* to the wire. Their motion, as they continue around the circuit, is the induced current I. Now, because we have a current, a second magnetic force $\vec{F}_{mag}$ enters the picture. This force on the current is *perpendicular* to the wire and acts to slow the wire's motion. ◀

The magnitude of the magnetic force on a current-carrying wire was found in Chapter 24 to be $F_{mag} = IlB$. Using that result, along with Equation 25.4 for the induced current, we find that the force required to pull the wire with a constant speed v is

$$F_{pull} = F_{mag} = IlB = \left(\frac{vlB}{R}\right)lB = \frac{vl^2B^2}{R} \tag{25.5}$$

Energy Considerations

FIGURE 25.6 Power into and out of an induced-current circuit.

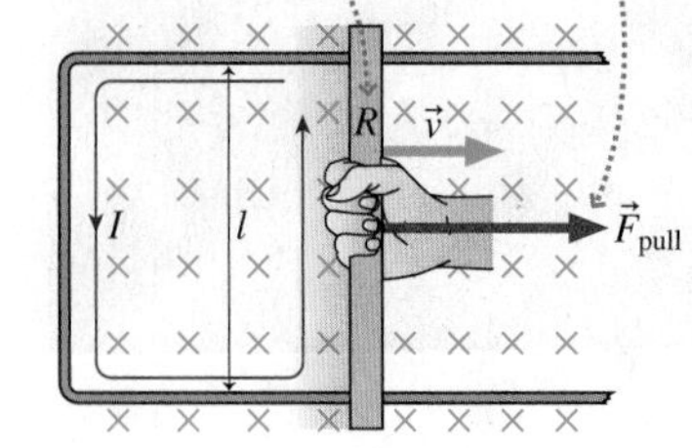
Because there is a current, power is dissipated in the resistance of the rail.
Pulling to the right takes work. This is a power input to the system.
R
$\vec{v}$
I
l
$\vec{F}_{pull}$

FIGURE 25.6 is another look at the wire moving on a conducting rail. Because a force is needed to pull the wire through the magnetic field at a constant speed, we must do work to keep the wire moving. You learned in Chapter 10 that the power exerted by a force pushing or pulling an object with velocity v is $P = Fv$, so the power provided to the circuit by the force pulling on the wire is

$$P_{input} = F_{pull}v = \frac{v^2l^2B^2}{R} \tag{25.6}$$

This is the rate at which energy is added to the circuit by the pulling force.

But the circuit dissipates energy in the resistance of the circuit. You learned in Chapter 22 that the power dissipated by current I as it passes through resistance R is $P = I^2R$. Equation 25.4 for the induced current I gives us the power dissipated by the circuit of Figure 25.6:

$$P_{dissipated} = I^2R = \frac{v^2l^2B^2}{R} \tag{25.7}$$

Equations 25.6 and 25.7 have identical results. This makes sense: The rate at which work is done on the circuit is exactly balanced by the rate at which energy is dissipated. The fact that our final result is consistent with energy conservation is a good check on our work.

Generators

A device that converts mechanical energy to electric energy is called a **generator.** The example of Figure 25.6 is a simple generator, but it is not very practical. Rather than move a straight wire, it's more practical to rotate a coil of wire, as in FIGURE 25.7. As the coil rotates, the left edge always moves upward through the magnetic field while the right edge always moves downward. The motion of the wires through the magnetic field induces a current to flow as noted in the figure. The induced current is removed from the rotating loop by *brushes* that press up against rotating *slip rings.* The circuit is completed as shown in the figure.

FIGURE 25.7 A generator using a rotating loop of wire.

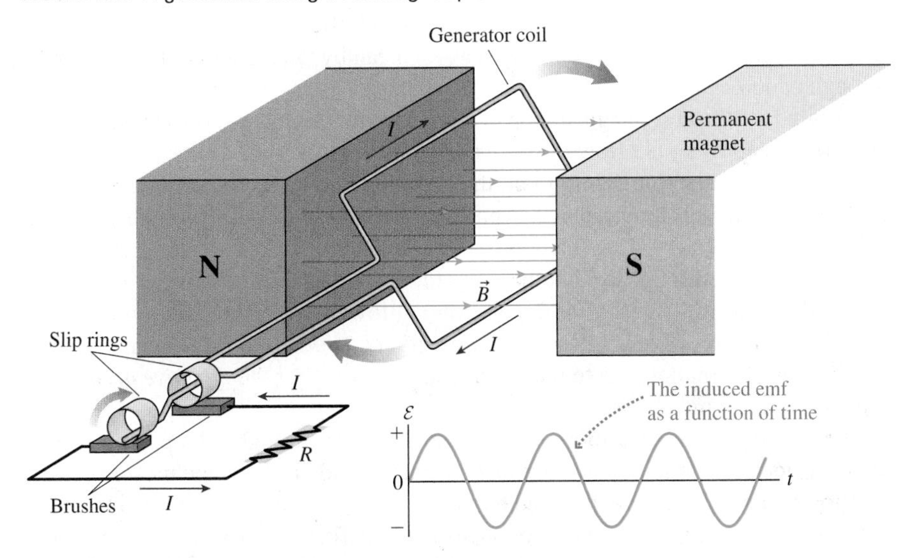

As the coil in the generator of Figure 25.7 rotates, the sense of the emf changes, giving a sinusoidal variation of emf as a function of time. Electricity is produced using generators of this sort; the electricity in your house has a varying voltage. The alternating sign of the voltage produces an *alternating current,* so we call such electricity AC. We'll have more to say about this type of electricity in Chapter 26.

▶ **No work, no light** Turning the crank on a generator flashlight rotates a coil of wire in the magnetic field of a permanent magnet. With the switch off, there is no current and no drag force; it's easy to turn the crank. Closing the switch allows an induced current to flow through the coil, so the bulb lights. But the current in the wire experiences a drag force in the magnetic field, so you must do work to keep the crank turning. This is the source of the output power of the circuit, the light of the bulb.

STOP TO THINK 25.1 A square conductor moves through a uniform magnetic field directed out of the page. Which of the figures shows the correct charge distribution on the conductor?

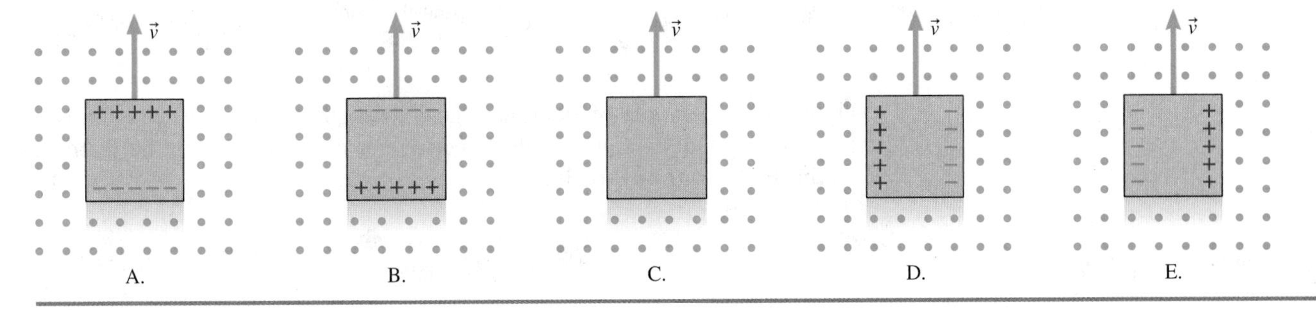

25.3 Magnetic Flux

We've begun our exploration of electromagnetic induction by analyzing a circuit in which one wire moves through a magnetic field. You might be wondering what this has to do with Faraday's discovery. Faraday found that a current is induced when the amount of magnetic field passing through a coil or a loop of wire changes. But that's exactly what happens as the slide wire moves down the rail in Figure 25.4! As the circuit expands, more magnetic field passes through the larger loop. It's time to define more clearly what we mean by "the amount of field passing through a loop."

Imagine holding a rectangular loop of wire in front of a fan, as shown in **FIGURE 25.8a**. The arrows represent the flow of the air. If you want to get the most air through the loop, you know that you should hold the loop perpendicular to the direction of the flow. If you tip the loop from this position, less air will pass through the loop. **FIGURE 25.8b** is a side view that makes this reduction clear—fewer arrows pass through the tipped loop. Yet another way to visualize this situation is **FIGURE 25.8c**, which shows a front view with the air coming toward you; the dots represent the front of the arrows. From this point of view it's clear why the flow is smaller: The loop presents a smaller area to the moving air. We say that the *effective area* of the loop has been reduced.

FIGURE 25.8 The amount of air flowing through a loop depends on the angle.

(a) A fan blows air through a loop

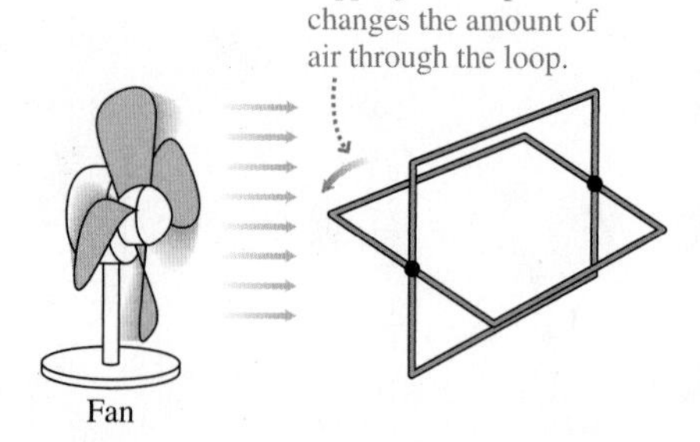

(b) Side view of the air through the loop

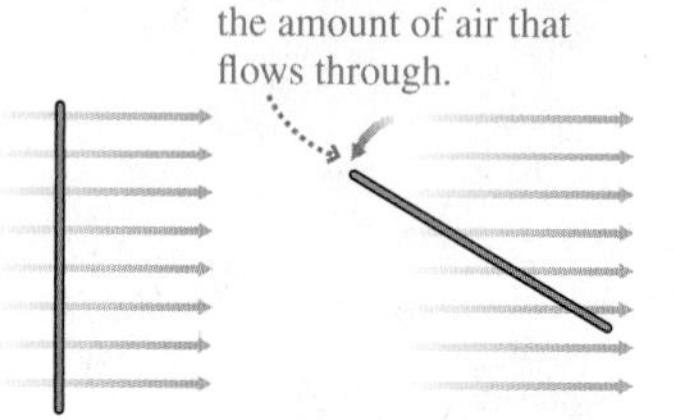

Tipping the loop reduces the amount of air that flows through.

(c) Front view of the air through the loop

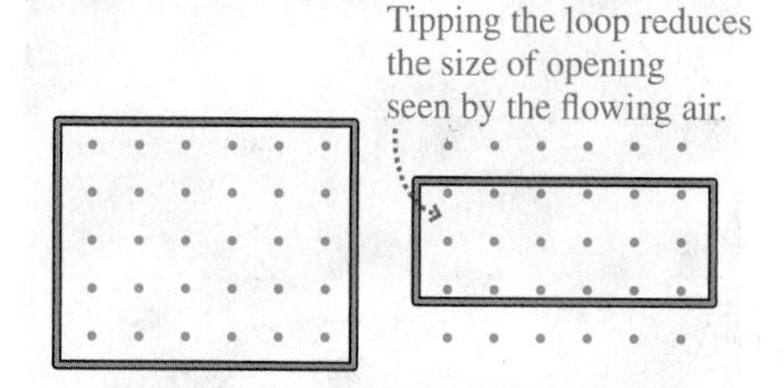

Tipping the loop reduces the size of opening seen by the flowing air.

We can apply this idea to a magnetic field passing through a loop. **FIGURE 25.9a** shows a side view of a loop in a uniform magnetic field. To have the most field vectors going through the loop, we need to turn the loop to be perpendicular to the magnetic field vectors, just as we did for airflow. We define the *axis* of the loop to be a line through the center of the loop that is perpendicular to the plane of the loop. We see that the largest number of field vectors go through the loop when its axis is lined up with the field. Tipping the loop by an angle θ reduces the number of vectors passing through the loop, just as for the air from the fan.

FIGURE 25.9b is a front view of the loop with the dimensions noted. When the loop is tipped by an angle θ, fewer field vectors pass through the loop because the effective area is smaller. We can define the effective area as

$$A_{\text{eff}} = ab\cos\theta = A\cos\theta \tag{25.8}$$

FIGURE 25.9 The amount of magnetic field passing through a loop depends on the angle.

(a) Loop seen from the side

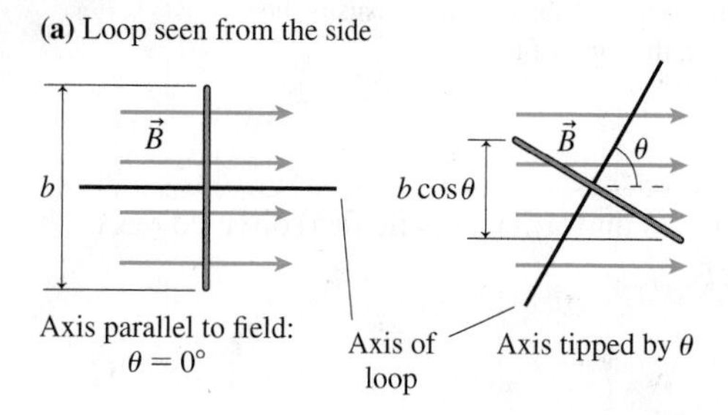

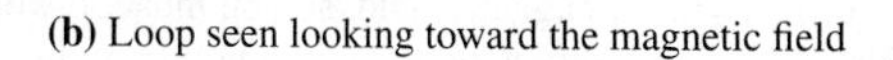

(b) Loop seen looking toward the magnetic field

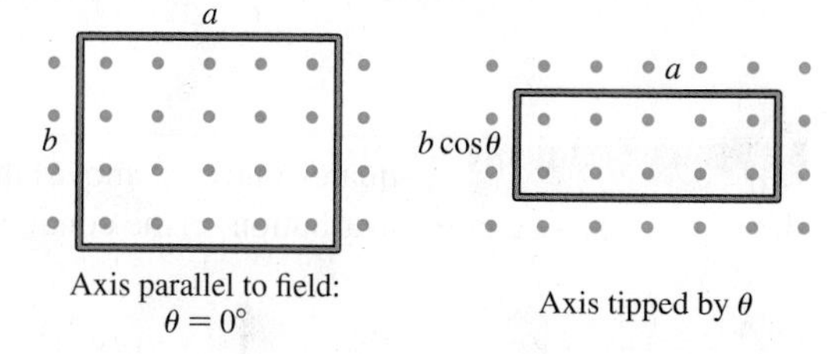

Ultimately, the amount of field that "goes through" the loop depends on two things: the strength of the field and the effective area of the loop. With this in mind, let's define the **magnetic flux** Φ as

$$\Phi = A_{\text{eff}}B = AB\cos\theta \tag{25.9}$$

Magnetic flux through area A at angle θ to field B

FIGURE 25.10 Definition of magnetic flux.

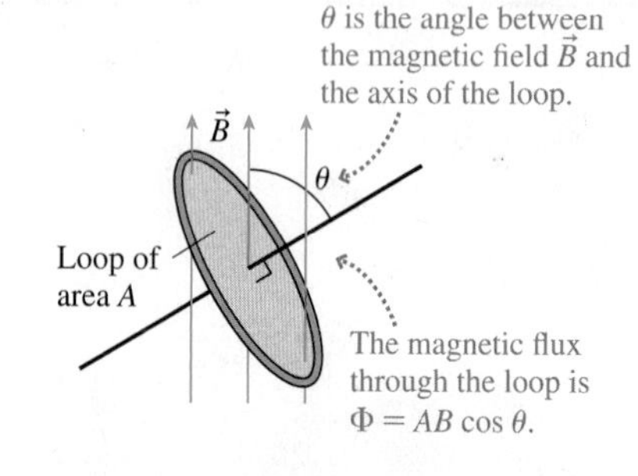

The magnetic flux measures the amount of magnetic field passing through a loop of area A if the loop is tilted at angle θ from the field. The SI unit of magnetic flux is the **weber.** From Equation 25.9 you can see that

$$1 \text{ weber} = 1 \text{ Wb} = 1 \text{ T} \cdot \text{m}^2$$

The relationship of Equation 25.9 is illustrated in **FIGURE 25.10**.

EXAMPLE 25.2 **Finding the flux of the earth's field through a vertical loop**

At a particular location, the earth's magnetic field is 50 μT tipped at an angle of 60° below horizontal. A 10-cm-diameter circular loop of wire sits flat on a table. What is the magnetic flux through the loop?

PREPARE **FIGURE 25.11** shows the loop and the field of the earth. The field is tipped by 60°, so the angle of the field with respect to the axis of the loop is $\theta = 30°$. The radius of the loop is 5.0 cm, so the area of the loop is $A = \pi r^2 = \pi(0.050\text{ m})^2 = 0.0079\text{ m}^2$.

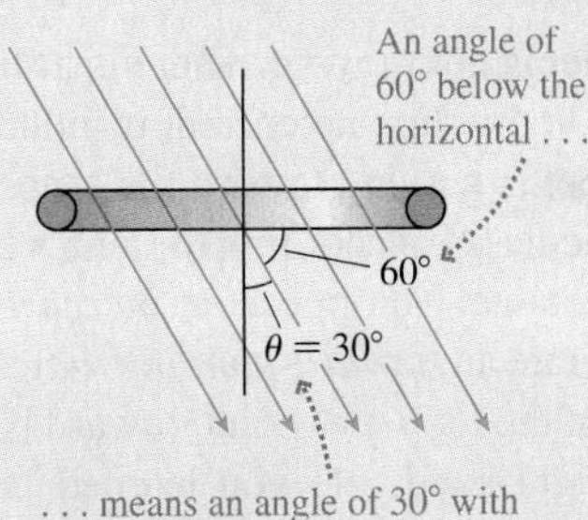
An angle of 60° below the horizontal . . .
60°
$\theta = 30°$
. . . means an angle of 30° with respect to the axis of the loop.

FIGURE 25.11 Finding the flux of the earth's field through a loop.

SOLVE The flux through the loop is given by Equation 25.9, with the angle and area as above:

$$\Phi = AB\cos\theta = (0.0079\text{ m}^2)(50 \times 10^{-6}\text{ T})\cos 30°$$
$$= 3.4 \times 10^{-7}\text{ Wb}$$

ASSESS It's a small loop and a small field, so a very small flux seems reasonable.

Lenz's Law

Some of the induction experiments from earlier in the chapter could be explained in terms of motional emf, but others had no motion. What they all have in common, though, is that one way or another the magnetic flux through the coil or loop *changes*. We can summarize all of the discoveries as follows: **Current is induced in a loop of wire when the magnetic flux through the loop changes.**

For example, a momentary current is induced in the loop of **FIGURE 25.12** as the bar magnet is pushed toward the loop because the flux through the loop increases. Pulling the magnet away from the loop, which decreases the flux, causes the current meter to deflect in the opposite direction. How can we predict the *direction* of the current in the loop?

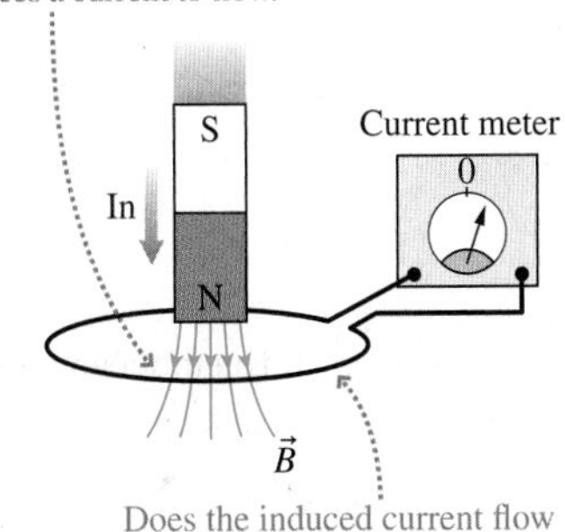
Pushing a bar magnet toward a loop increases the flux through the loop and induces a current to flow.
S
N
In
Current meter
0
$\vec{B}$
Does the induced current flow clockwise or counterclockwise?

FIGURE 25.12 Pushing a bar magnet toward the loop induces a current in the loop.

The German physicist Heinrich Lenz began to study electromagnetic induction after learning of Faraday's discovery. Lenz developed a rule for determining the direction of the induced current. We now call his rule **Lenz's law,** and it can be stated as follows:

> **Lenz's law** There is an induced current in a closed, conducting loop if and only if the magnetic flux through the loop is changing. The direction of the induced current is such that the induced magnetic field opposes the *change* in the flux.

Lenz's law is rather subtle, and it takes some practice to see how to apply it.

NOTE ► One difficulty with Lenz's law is the term "flux," from a Latin root meaning "flow." In everyday language, the word "flux" may imply that something is changing. Think of the phrase "The situation is in flux." In physics, "flux" simply means "passes through." A steady magnetic field through a loop creates a steady, *un*changing magnetic flux. ◄

Lenz's law tells us to look for situations where the flux is *changing*. We'll see three ways in which this comes about:

1. The magnetic field through the loop changes (increases or decreases).
2. The loop changes in area or angle.
3. The loop moves into or out of a magnetic field.

We can understand Lenz's law this way: If the flux through a loop changes, a current is induced in a loop. That current generates *its own* magnetic field $\vec{B}_{\text{induced}}$. **It is this induced field that opposes the flux change.** Let's look at an example to clarify what we mean by this statement.

FIGURE 25.13 shows a magnet above a coil of wire. The field of the magnet creates a downward flux through the loop. The three parts of the figure show how the induced current in the loop develops in response to the motion of the magnet and thus the changing flux in the coil.

- In part a, the magnet is not moving. There is a flux through the loop, but because there is no *change* in the flux, no current is induced in the loop.
- In part b, the magnet is moving toward the loop, so the downward magnetic flux through the loop increases. According to Lenz's law, the loop will generate a field that opposes this change. To oppose an *increase in the downward flux*, the loop itself needs to generate an *upward-pointing magnetic field*. The induced magnetic field at the center of the loop will point upward if the current is counterclockwise, according to the right-hand rule you learned in Chapter 24. Thus pushing the north end of a bar magnet toward the loop induces a counterclockwise current around the loop. This induced current ceases as soon as the magnet stops moving.
- Now suppose the bar magnet is pulled back away from the loop, as in part c. There is a downward magnetic flux through the loop, but the flux decreases as the magnet moves away. According to Lenz's law, the induced magnetic field of the loop will oppose this decrease. To oppose a *decrease in the downward flux*, the loop itself needs to generate a *downward-pointing magnetic field*. The induced current is clockwise, opposite the induced current of part b.

The magnetic field of the bar magnet is pointing downward in each part of the figure, but the induced magnetic field can be zero, directed upward, or directed downward depending on the motion of the magnet. It is not the *flux* due to the magnet that the induced current opposes, but the *change in the flux*. This is a subtle but critical distinction. When the field of the magnet points downward and is increasing, the induced current *opposes the increase* by generating an upward field. When the field of the magnet points downward but is decreasing, the induced current *opposes the decrease* by generating a downward field.

FIGURE 25.13 The induced current depends on the motion of the magnet.

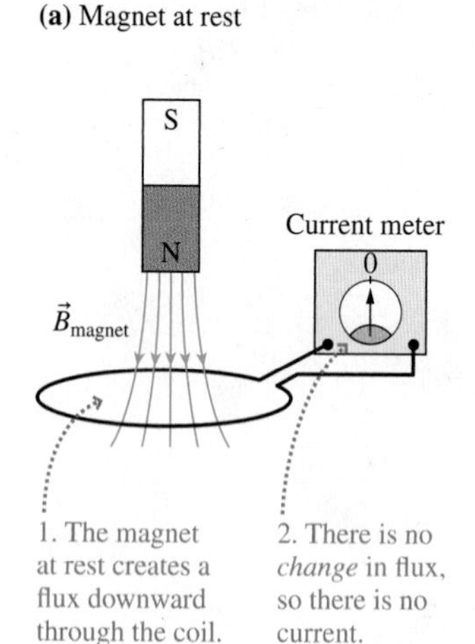
(a) Magnet at rest
S
N
Current meter
0
$\vec{B}_{\text{magnet}}$
1. The magnet at rest creates a flux downward through the coil.
2. There is no *change* in flux, so there is no current.

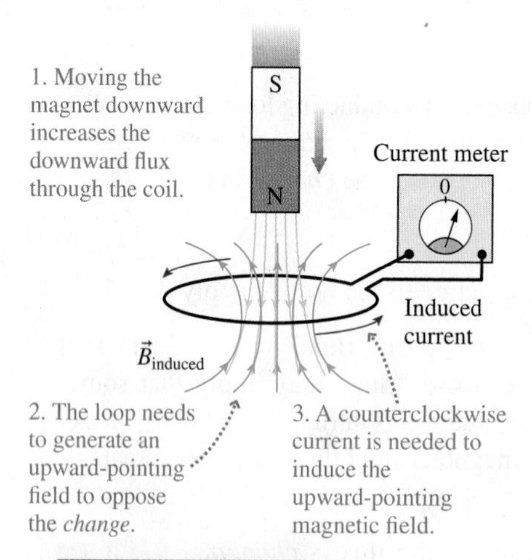
(b) Magnet moving down
1. Moving the magnet downward increases the downward flux through the coil.
S
N
Current meter
0
Induced current
$\vec{B}_{\text{induced}}$
2. The loop needs to generate an upward-pointing field to oppose the *change*.
3. A counterclockwise current is needed to induce the upward-pointing magnetic field.

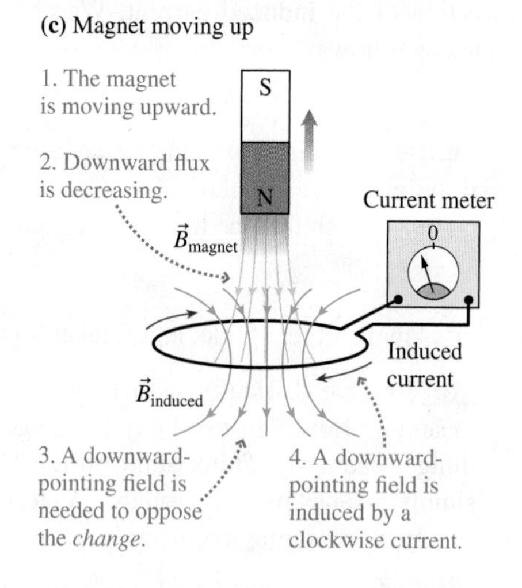
(c) Magnet moving up
1. The magnet is moving upward.
S
2. Downward flux is decreasing.
N
Current meter
0
$\vec{B}_{\text{magnet}}$
Induced current
$\vec{B}_{\text{induced}}$
3. A downward-pointing field is needed to oppose the *change*.
4. A downward-pointing field is induced by a clockwise current.

STOP TO THINK 25.2 A magnet is below a coil, as shown at right. Referring to Figure 25.13 for guidance, what is the direction of the induced current in the coil if

- The magnet is moving toward the coil?
- The magnet is held stationary near the coil?
- The magnet is moving away from the coil?

A. Clockwise. B. Counterclockwise.
C. There is no induced current.

Current meter
0
N
S

The approach we took above for using Lenz's law to determine the direction of an induced current is a general one; it's worthwhile to spell out the steps.

TACTICS BOX 25.1 Using Lenz's law

1. Determine the direction of the applied magnetic field. The field must pass through the loop.
2. Determine how the flux is changing. Is it increasing, decreasing, or staying the same?
3. Determine the direction of an induced magnetic field that will oppose the *change* in the flux:
   - Increasing flux: The induced magnetic field points opposite the applied magnetic field.
   - Decreasing flux: The induced magnetic field points in the same direction as the applied magnetic field.
   - Steady flux: There is no induced magnetic field.
4. Determine the direction of the induced current. Use the right-hand rule to determine the current direction in the loop that generates the induced magnetic field you found in step 3.

Exercises 9–11 

EXAMPLE 25.3 Applying Lenz's law 1

The switch in the top circuit of **FIGURE 25.14** has been closed for a long time. What happens in the lower loop when the switch is opened?

PREPARE The current in the upper loop creates a magnetic field. This magnetic field produces a flux through the lower loop. When you open a switch, the current doesn't immediately drop to zero; it falls off over a short time. As the current changes in the upper loop, the flux in the lower loop changes.

SOLVE **FIGURE 25.15** shows the four steps of using Lenz's law to find the current in the lower loop. Opening the switch induces a counterclockwise current in the lower loop. This is a momentary current, lasting only until the magnetic field of the upper loop drops to zero.

ASSESS The induced current is in the same direction as the original current. This makes sense, because the induced current is opposing the change, a decrease in the current.

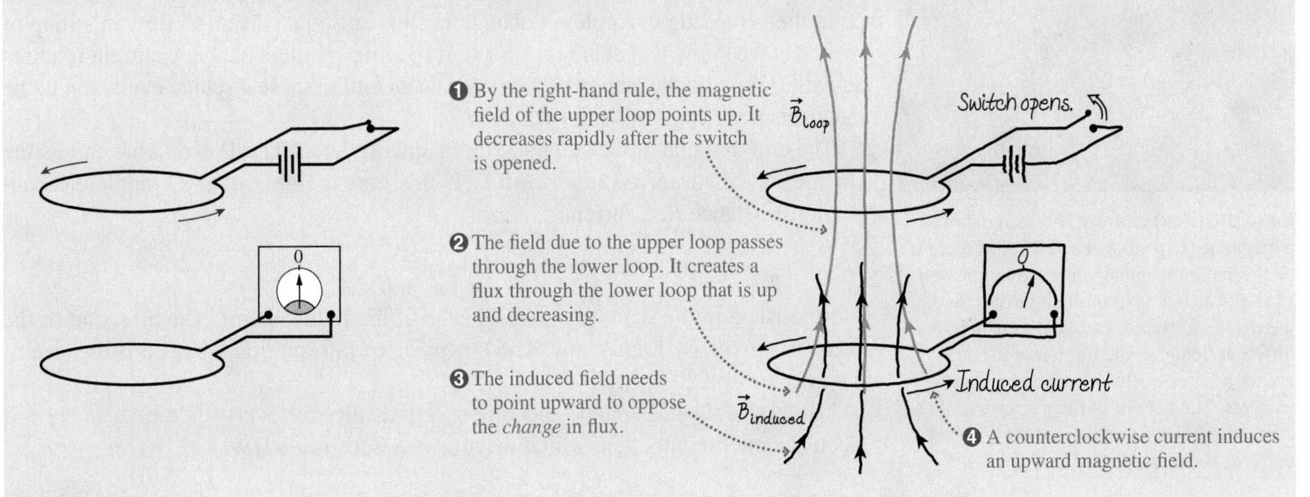

FIGURE 25.14 Circuits for Example 25.3. **FIGURE 25.15** Finding the induced current.

EXAMPLE 25.4 Applying Lenz's law 2

A loop is moved toward a current-carrying wire as shown in **FIGURE 25.16**. As the wire is moving, is there a clockwise current around the loop, a counterclockwise current, or no current?

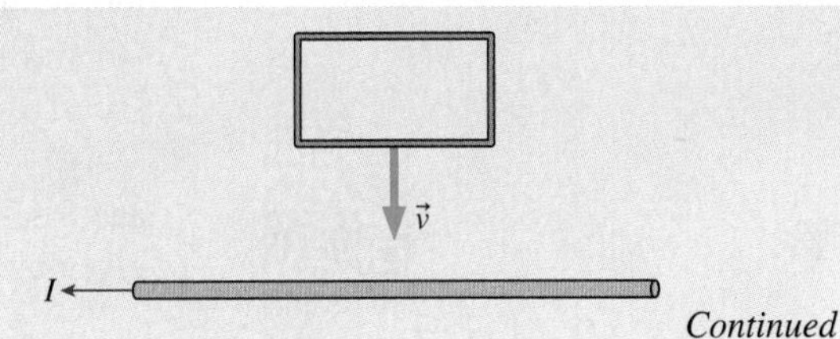

FIGURE 25.16 The moving loop.

Continued

PREPARE FIGURE 25.17 shows that the magnetic field above the wire points into the page. We learned in Chapter 24 that the magnetic field of a straight, current-carrying wire is proportional to $1/r$, where r is the distance away from the wire, so the field is stronger closer to the wire.

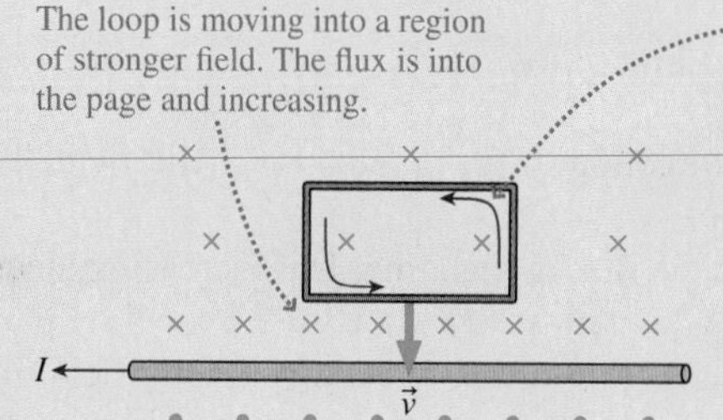

The loop is moving into a region of stronger field. The flux is into the page and increasing.
The induced current must create a magnetic field out of the page to oppose the change, so the right-hand rule tells us that the induced current is counterclockwise.
I
v

FIGURE 25.17 The motion of the loop changes the flux through the loop and induces a current.

SOLVE As the loop moves toward the wire, the flux through the loop increases. To oppose the *change* in the flux—the increase into the page—the magnetic field of the induced current must point out of the page. Thus, according to the right-hand rule, a counterclockwise current is induced, as shown in Figure 25.17.

ASSESS The loop moves into a region of stronger field. To oppose the increasing flux, the induced field should be opposite the existing field, so our answer makes sense.

STOP TO THINK 25.3 As a coil moves to the right at constant speed, it passes over the north pole of a magnet and then moves beyond it. Which graph best represents the current in the loop for the time of the motion? A counterclockwise current as viewed from above the loop is a positive current, clockwise is a negative current.

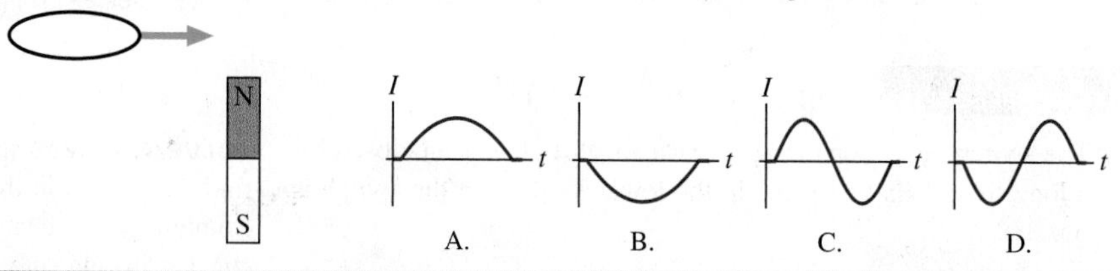

25.4 Faraday's Law

Keep the (flux) change A credit card has a magnetic strip on the back that has regions of alternating magnetization. "Swiping" the card moves this strip through the circuit of the reader. Different parts of the strip have different fields, so the motion of the strip causes a series of flux changes that induce currents. The pattern of magnetization of the card determines the pattern of the induced currents—and so the card is "read."

A current is induced when the magnetic flux through a conducting loop changes. Lenz's law allows us to find the direction of the induced current. To put electromagnetic induction to practical use, we also need to know the *size* of the induced current.

In the preceding examples, a change in flux caused a current to flow in a loop of wire. But we know that charges don't start moving spontaneously. A current requires an emf to provide the energy. There *must* be an emf in these circuits, even though the mechanism for this emf is not yet clear.

The emf associated with a changing magnetic flux, regardless of what causes the change, is called an **induced emf** $\mathcal{E}$. If this emf is induced in a complete circuit having resistance R, a current

$$I_{\text{induced}} = \frac{\mathcal{E}}{R} \tag{25.10}$$

is established in the wire as a *consequence* of the induced emf. The direction of the current is given by Lenz's law. The last piece of information we need is the size of the induced emf $\mathcal{E}$.

The research of Faraday and others led to the discovery of the basic law of electromagnetic induction, which we now call **Faraday's law.**

Faraday's law An emf $\mathcal{E}$ is induced in a conducting loop if the magnetic flux through the loop changes. If the flux changes by $\Delta\Phi$ during time interval Δt, the magnitude of the emf is

$$\mathcal{E} = \left|\frac{\Delta\Phi}{\Delta t}\right| \tag{25.11}$$

and the direction of the emf is such as to drive an induced current in the direction given by Lenz's law.

Class Video

In other words, the magnitude of the induced emf is the *rate of change* of the magnetic flux through the loop.

A coil of wire consisting of N turns in a changing magnetic field acts like N batteries in series, so the induced emf of the entire coil is

$$\mathcal{E}_{\text{coil}} = N\left|\frac{\Delta\Phi_{\text{per coil}}}{\Delta t}\right| \tag{25.12}$$

PROBLEM-SOLVING STRATEGY 25.1 **Electromagnetic induction**

Faraday's law allows us to find the *magnitude* of induced emfs and currents; Lenz's law allows us to determine the *direction.*

PREPARE Make simplifying assumptions about wires and magnetic fields. Draw a picture or a circuit diagram. Use Lenz's law to determine the direction of the induced current.

SOLVE The mathematical representation is based on Faraday's law

$$\mathcal{E} = \left|\frac{\Delta\Phi}{\Delta t}\right|$$

For an N-turn coil, multiply by N. The size of the induced current is $I = \mathcal{E}/R$.

ASSESS Check that your result has the correct units, is reasonable, and answers the question.

Exercise 16

Class Video

Class Video

Let's return to the situation of Figure 25.4, where a wire moves through a magnetic field by sliding on a U-shaped conducting rail. We looked at this problem as an example of motional emf; now, let's look at it using Faraday's law.

EXAMPLE 25.5 Finding the emf using Faraday's law

FIGURE 25.18 shows a wire of resistance R sliding on a U-shaped conducting rail. Assume that the conducting rail is an ideal wire. Use Faraday's law and the steps of Problem-Solving Strategy 25.1 to derive an expression for the current in the wire.

FIGURE 25.18 A wire sliding on a rail.

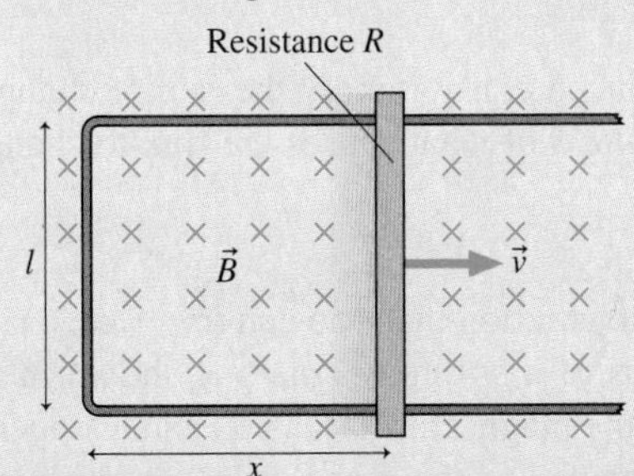

PREPARE FIGURE 25.19 shows the current loop formed by the wire and the rail. Even though the magnetic field is constant, the flux is changing because the loop is increasing in area. The flux is into the loop and increasing. According to Lenz's law, the induced current must be counterclockwise so as to oppose the change, because the induced magnetic field must be out of the loop.

FIGURE 25.19 Induced current in the sliding wire.

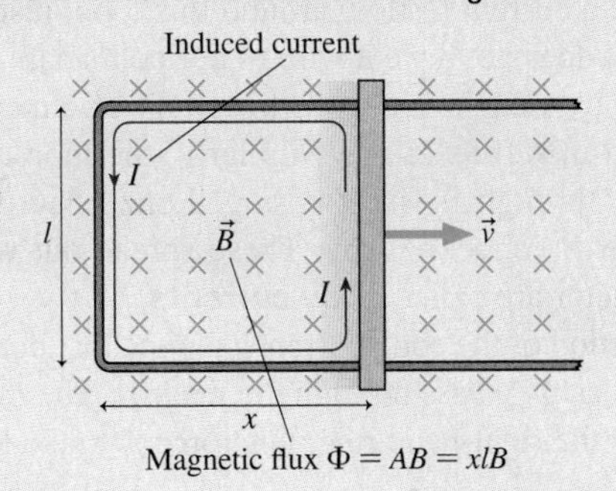

SOLVE The magnetic field $\vec{B}$ is perpendicular to the plane of the loop, so $\theta = 0°$ and the magnetic flux is $\Phi = AB$, where A is the area of the loop. If the sliding wire is distance x from the end, as in Figure 25.19, the area of the loop is $A = xl$ and the flux at that instant of time is

$$\Phi = AB = xlB$$

The flux through the loop increases as the wire moves and x increases. This flux change induces an emf, according to Faraday's law, so we write

$$\mathcal{E} = \left|\frac{\Delta\Phi}{\Delta t}\right| = \left|\frac{\Delta(AB)}{\Delta t}\right| = \left|\frac{\Delta(xlB)}{\Delta t}\right|$$

The only quantity in the final ratio that is changing is the position x, so we can write

$$\mathcal{E} = \left|\frac{\Delta(xlB)}{\Delta t}\right| = lB\left|\frac{\Delta x}{\Delta t}\right|$$

But $|\Delta x/\Delta t|$ is the wire's speed v, so the induced emf is

$$\mathcal{E} = vlB$$

The wire and the loop have a total resistance R; thus the magnitude of the induced current is

$$I = \frac{\mathcal{E}}{R} = \frac{vlB}{R}$$

ASSESS This is exactly the same result we found in Section 25.2, where we analyzed this situation by considering the force on moving charge carriers. This is a good check on our work and a nice connection between the ideas of motional emf and Faraday's law.

EXAMPLE 25.6 Finding the induced current in a circular loop BIO

A patient having an MRI scan has neglected to remove a copper bracelet. The bracelet is 6.0 cm in diameter and has a resistance of 0.010 Ω. The magnetic field in the MRI solenoid is directed along the person's body from head to foot; her bracelet is perpendicular to $\vec{B}$. As a scan is taken, the magnetic field in the solenoid decreases from 1.00 T to 0.40 T in 1.2 s. What are the magnitude and direction of the current induced in the bracelet?

PREPARE We follow the steps in Problem-Solving Strategy 25.1, beginning with a sketch of the situation. **FIGURE 25.20** shows the bracelet and the applied field looking down along the patient's body. The field is directed down through the loop; as the applied field decreases, the flux into the loop decreases. To oppose the decreasing flux, as required by Lenz's law, the field from the induced current must be in the direction of the applied field. Thus, from the right-hand rule, the induced current in the bracelet must be clockwise.

FIGURE 25.20 A circular conducting loop in a decreasing magnetic field.

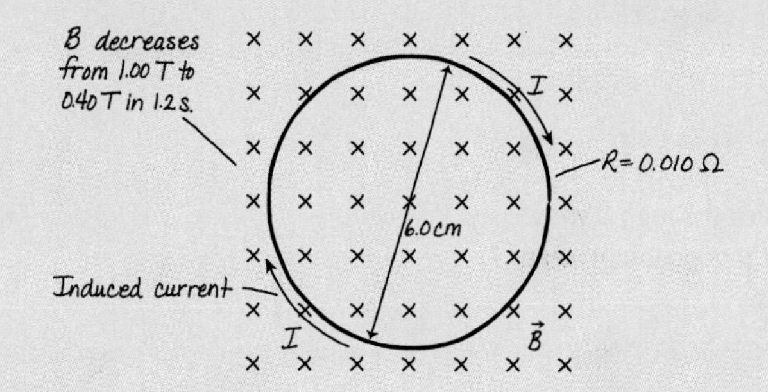

SOLVE The magnetic field is perpendicular to the plane of the loop; hence $\theta = 0°$ and the magnetic flux is $\Phi = AB = \pi r^2 B$. The area of the loop doesn't change with time, but B does, so $\Delta\Phi = \Delta(AB) = A\Delta B = \pi r^2\,\Delta B$. The change in the magnetic field during $\Delta t = 1.2$ s is $\Delta B = 0.40\text{ T} - 1.00\text{ T} = -0.60\text{ T}$. According to Faraday's law, the magnitude of the induced emf is

$$\mathcal{E} = \left|\frac{\Delta\Phi}{\Delta t}\right| = \pi r^2\left|\frac{\Delta B}{\Delta t}\right| = \pi(0.030\text{ m})^2\left|\frac{-0.60\text{ T}}{1.2\text{ s}}\right|$$

$$= \pi(0.030\text{ m})^2(0.50\text{ T/s}) = 0.0014\text{ V}$$

The current induced by this emf is

$$I = \frac{\mathcal{E}}{R} = \frac{0.0014\text{ V}}{0.010\ \Omega} = 0.14\text{ A}$$

The decreasing magnetic field causes a 0.14 A clockwise current during the 1.2 s that the field is decreasing.

ASSESS The emf is quite small, but, because the resistance of the metal bracelet is also very small, the current is respectable. Electromagnetic induction produces currents large enough for practical applications, so this result seems plausible. The induced current could easily distort the readings of the MRI machine, and a larger current could cause enough heating to be potentially dangerous, so operators are careful to have patients remove all metal before an MRI scan.

FIGURE 25.21 Eddy currents.

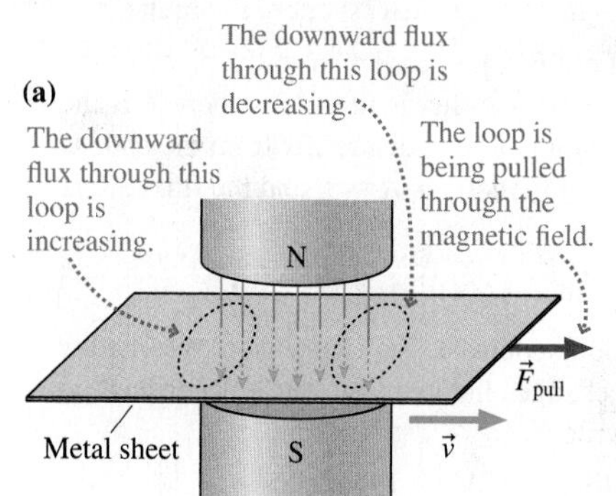

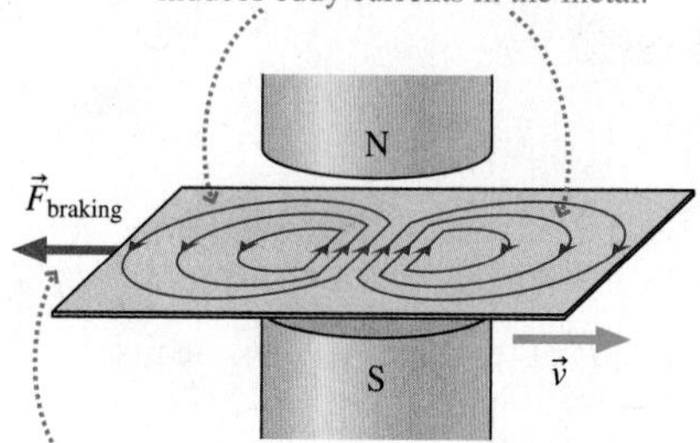

As these two examples show, there are two fundamentally different ways to change the magnetic flux through a conducting loop:

1. The loop can move or expand or rotate, creating a motional emf.
2. The magnetic field can change.

Faraday's law tells us that the induced emf is simply the rate of change of the magnetic flux through the loop, *regardless* of what causes the flux to change.

Eddy Currents

Here is a remarkable physics demonstration that you can try: Take a sheet of copper and place it between the pole tips of a strong magnet, as shown in **FIGURE 25.21a**. Now, pull the copper sheet out of the magnet as fast as you can. Copper is not a magnetic material and thus is not attracted to the magnet, but, surprisingly, it takes a significant effort to pull the metal through the magnetic field.

Let's analyze this situation to discover the origin of the force. Figure 25.21a shows two "loops" lying entirely inside the metal sheet. The loop on the right is leaving the magnetic field, and the flux through it is decreasing. According to Faraday's law, the flux change will induce a current to flow around this loop, just as in a loop of wire, even though this current does not have a wire to define its path. As a consequence, a clockwise—as given by Lenz's law— "whirlpool" of current begins to circulate in the metal, as shown in **FIGURE 25.21b**. Similarly, the loop on the left is entering the field, and the flux through it is increasing. Lenz's law requires this whirlpool of current to circulate the opposite way. These spread-out whirlpools of induced current in a solid conductor are called **eddy currents.**

Figure 25.21b shows the direction of the eddy currents. Notice that both whirlpools are moving in the same direction as they pass through the magnet. The magnet's field exerts a force on this current. By the right-hand rule, this force is to the left, opposite the direction of the pull, and thus it acts as a *braking* force. Because of the braking force, **an external force is required to pull a metal through a magnetic field.** If the pulling force ceases, the magnetic braking force quickly causes the metal to decelerate until it stops. No matter which way the metal is moved, the magnetic forces on the eddy currents act to oppose the motion of the metal. *Magnetic braking* uses the braking force associated with eddy currents to slow trains and transit-system vehicles.

Eddy currents can also be induced by changing fields; this has practical applications as well. In a technique known as *transcranial magnetic stimulation* (TMS), a large oscillating magnetic field is applied to the head via a current-carrying coil. **FIGURE 25.22** illustrates how this field produces small eddy currents that stimulate neurons in the tissue of the brain. This produces a short-term inhibitory effect on the neurons in the stimulated region that can produce long-term clinical effects. And by inhibiting the action of specific regions of the brain, researchers can determine the importance of these regions to certain perceptions or tasks.

FIGURE 25.22 BIO Transcranial magnetic stimulation.

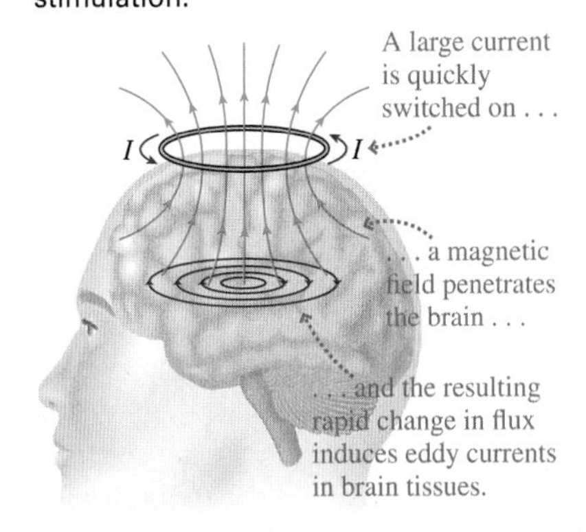

STOP TO THINK 25.4 A loop of wire rests in a region of uniform magnetic field. The magnitude of the field is increasing, thus inducing a current in the loop. Which of the following changes to the situation would make the induced current larger? Choose all that apply.

A. Increase the rate at which the field is increasing.
B. Replace the loop with one of the same resistance but larger diameter.
C. Replace the loop with one of the same resistance but smaller diameter.
D. Orient the loop parallel to the magnetic field.
E. Replace the loop with one of lower resistance.

25.5 Electromagnetic Waves

We will start this section with the puzzle shown in **FIGURE 25.23**. A long, tightly wound solenoid of radius r_1 passes through the center of a conducting loop having a larger radius r_2. The solenoid carries a current and generates a magnetic field. What happens to the loop if the solenoid current changes?

You learned in Chapter 24 that the magnetic field is strong inside a long solenoid but essentially zero outside. Even so, changing the field inside the solenoid causes the flux through the loop to change, so our theory predicts an induced current in the loop. But the loop is completely outside the solenoid, where the magnetic field is zero. How can the charge carriers in the conducting loop possibly "know" that the magnetic field inside the solenoid is changing?

FIGURE 25.23 A changing current in the solenoid induces a current in the loop.

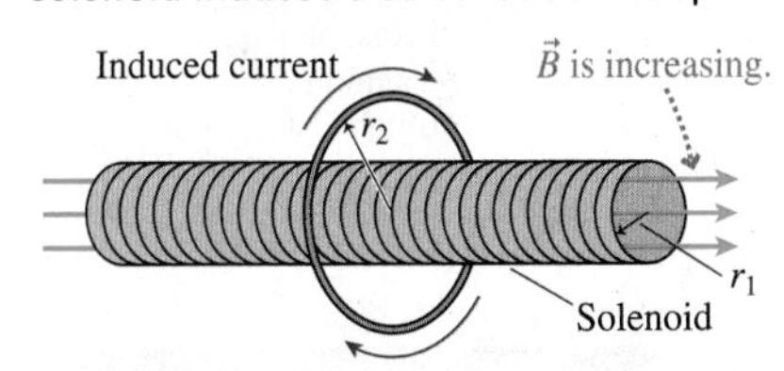

Induced Fields

In order to answer this question, we will first consider another related question: When a changing flux through a loop induces a current, what actually *causes* the current? What *force* pushes the charges around the loop against the resistive forces of the metal? When we considered currents in Chapter 22, it was an *electric field* that moved charges through a conductor. Somehow, changing a magnetic field must create an electric field.

In fact, a changing magnetic field *does* cause what we call an **induced electric field.** **FIGURE 25.24a** shows a conducting loop in an increasing magnetic field. According to Lenz's law, there is an induced current in the counterclockwise direction. Something has to act on the charge carriers to make them move, so we can infer that the current is produced by an induced electric field tangent to the loop at all points.

But the induced electric field exists whether there is a conducting loop or not. The space in which the magnetic field is changing is filled with the pinwheel pattern of induced electric fields shown in **FIGURE 25.24b**.

FIGURE 25.24 An induced electric field creates a current in the loop.

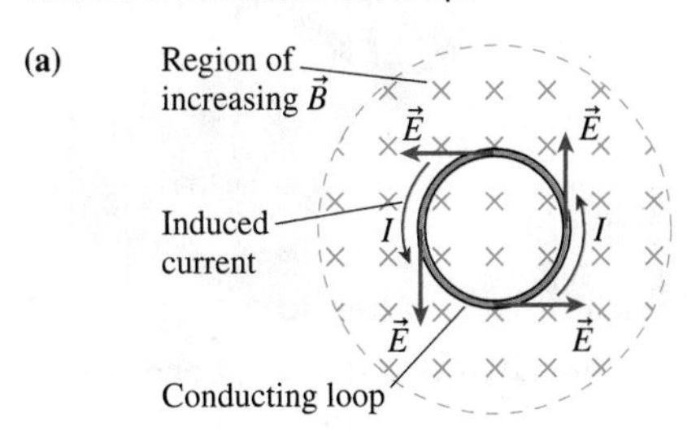

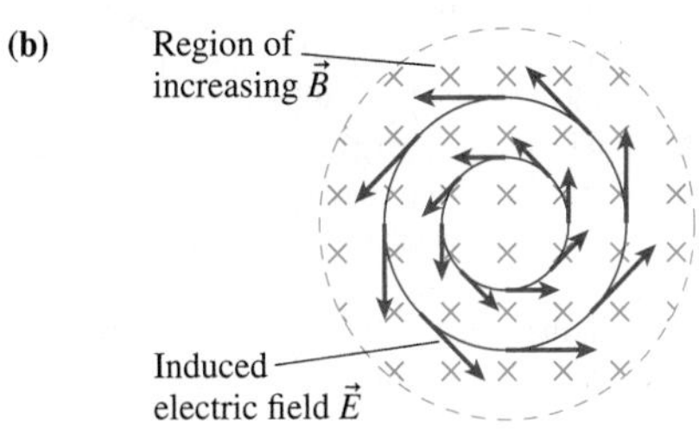

A changing magnetic field induces an electric field. What about a changing *electric* field? Early investigators looking for connections between electricity and magnetism wondered about this and hypothesized that **a changing electric field creates an induced magnetic field.**

This hypothesis leads to a surprising conclusion: If a changing magnetic field can induce an electric field in the absence of any charges, and if a changing electric field can induce a magnetic field in the absence of any currents, then it should be possible to establish self-sustaining electric and magnetic fields *independent of any charges or currents.* A changing electric field $\vec{E}$ creates a magnetic field $\vec{B}$, which then changes in just the right way to recreate the electric field, which then changes in just the right way to again recreate the magnetic field, with the fields continuously recreated through electromagnetic induction. In fact, electric and magnetic fields *can* sustain themselves, free from any charges or currents, if they take the form of an **electromagnetic wave.**

Properties of Electromagnetic Waves

In order to sustain itself and travel through space, an electromagnetic wave must have a very specific geometry, shown in FIGURE 25.25. You can see that **an electromagnetic wave is a transverse wave.** $\vec{E}$ and $\vec{B}$ are perpendicular to each other as well as perpendicular to the direction of travel.

FIGURE 25.25 A sinusoidal electromagnetic wave.

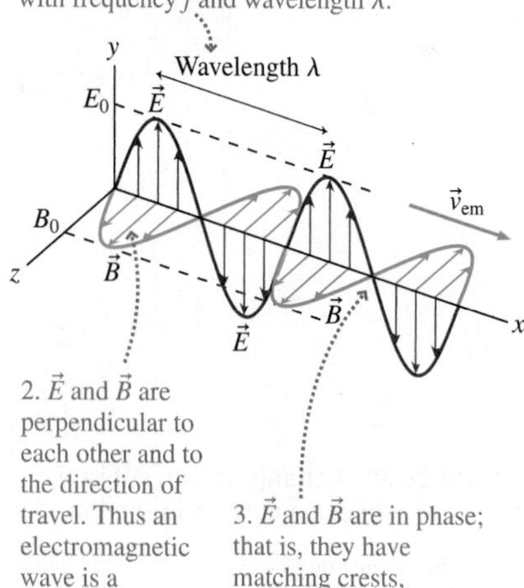

A mathematical analysis shows that such a wave travels with speed

$$v_{em} = \frac{1}{\sqrt{\epsilon_0 \mu_0}} \tag{25.13}$$

where ϵ_0 and μ_0 are the permittivity and permeability constants from our expressions for electric and magnetic fields. If you insert the values we've seen for these constants, you find $v_{em} = 3.00 \times 10^8\ \text{m/s}$. This is a value you have seen before—it is the speed of light!

In a vacuum, all electromagnetic waves must travel at the same speed, a speed that we call the speed of light and for which we use the symbol c. James Clerk Maxwell, the first to make this analysis, made a bold leap and concluded that **light is an electromagnetic wave.** We studied the wave properties of light in Part V, but at that time we didn't discuss just what is "waving." Now we know—light is a wave of electric and magnetic fields.

The amplitudes of the electric and magnetic fields vary in tandem; at every point on the wave, the electric and magnetic field strengths are related by

$$\frac{E}{B} = c \tag{25.14}$$

Figure 25.25 shows the values of the electric and magnetic fields at points along a single line, the x-axis.

The grandfather of the cell phone Not long after Maxwell determined the properties of electromagnetic waves, experimenters used these waves to carry signals across a laboratory. By the late 1800s, Guglielmo Marconi and other investigators had developed the technology for reliably producing and detecting these electromagnetic waves. Marconi was able to send signals with no wires at the speed of light to others at great distances away—something we have come to take for granted.

NOTE ▸ An $\vec{E}$ vector pointing in the y-direction says that at that point on the x-axis, where the vector's tail is, the electric field points in the y-direction and has a certain strength. Nothing is "reaching" to a point in space above the x-axis. ◂

Electromagnetic waves are oscillations of the electric and magnetic fields, but they are still waves, and all the general principles we have learned about waves apply. In ◂◂ SECTION 15.3, we learned that we could characterize sinusoidal waves by their speed, wavelength, and frequency, with these variables related by the fundamental relationship $v = \lambda f$. All electromagnetic waves move at the speed of light, $v_{em} = c$, so for sinusoidal electromagnetic waves this relationship becomes

$$c = \lambda f \tag{25.15}$$

The spectrum of electromagnetic waves ranges from waves of long wavelength and (relatively) low frequency (radio waves and microwaves) to waves of short wavelength and high frequency (visible light, ultraviolet, and x rays). We'll have more to say about the electromagnetic spectrum later in the chapter.

Suppose the electromagnetic wave of Figure 25.25 is a plane wave traveling in the x-direction. Recall from Chapter 15 that the displacement of a plane wave is the same at *all points* in any plane perpendicular to the direction of motion. In this case, the fields are the same in any yz-plane. If you were standing on the x-axis as the wave moves toward you, the electric and magnetic fields would vary as in the series of pictures of **FIGURE 25.26**. The $\vec{E}$ and $\vec{B}$ fields at each point in the yz-plane oscillate in time, but they are always synchronized with all the other points in the plane. As the plane wave passed you, you would see a uniform oscillation of the $\vec{E}$ and $\vec{B}$ fields of the wave.

FIGURE 25.26 The fields of an electromagnetic plane wave moving toward you, shown every one-eighth period for half a cycle.

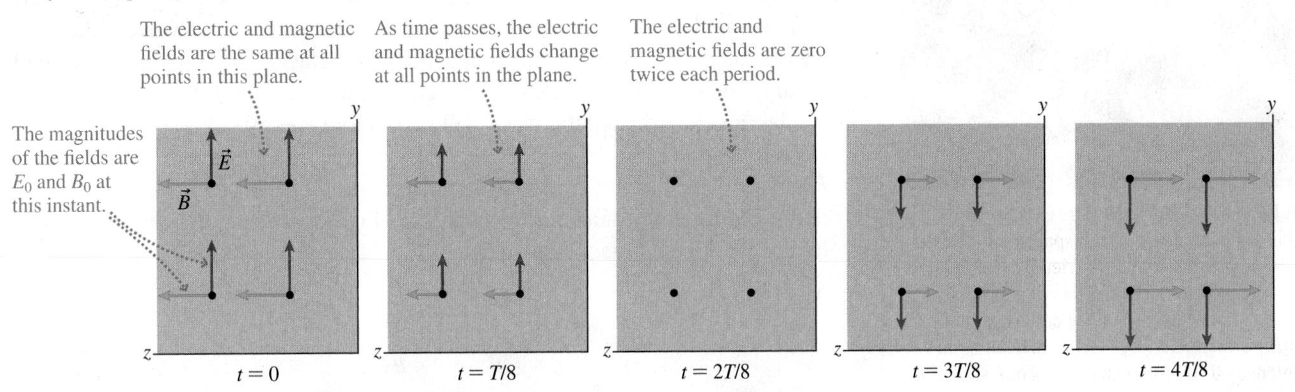

We can adapt our equation for traveling waves from Chapter 15 to electromagnetic waves. If a plane electromagnetic wave moves in the x-direction with the electric field along the y-axis, then the magnetic field is along the z-axis. The equations for the electric and magnetic fields of a wave with wavelength λ and period T are

$$E_y = E_0 \sin\left(2\pi\left(\frac{x}{\lambda} - \frac{t}{T}\right)\right) \qquad B_z = B_0 \sin\left(2\pi\left(\frac{x}{\lambda} - \frac{t}{T}\right)\right) \tag{25.16}$$

E_0 and B_0 are the amplitudes of the oscillating fields. The amplitudes of the fields must match the relationship between the fields given in Equation 25.14, so they must be related like this:

$$\frac{E_0}{B_0} = c \tag{25.17}$$

Relationship between field amplitudes for an electromagnetic wave

Since the value of c is so large, the electric field amplitude in N/C is much larger than the magnetic field amplitude in T.

Polarization

As we've seen, the electric field vectors of an electromagnetic wave lie in a plane perpendicular to the direction of propagation. The plane containing the electric field vectors is called the **plane of polarization.** **FIGURE 25.27a** shows a wave traveling along the x-axis; the plane of polarization is the xy-plane. If the wave were moving toward you, it would appear as in the first diagram in **FIGURE 25.27b**. This particular wave is *vertically polarized* ($\vec{E}$ oscillating along the y-axis). The second diagram of Figure 25.27b shows a wave that is *horizontally polarized* ($\vec{E}$ oscillating along the z-axis). The plane of polarization needn't be horizontal or vertical; it can have any orientation, as shown in the third diagram.

NOTE ▶ This use of the term "polarization" is completely independent of the idea of *charge polarization* that you learned about in Chapter 20. ◀

Each atom in the sun's hot atmosphere emits light independently of all the other atoms, as does each tiny piece of metal in the filament of an incandescent lightbulb. An

FIGURE 25.27 The polarization of an electromagnetic wave.

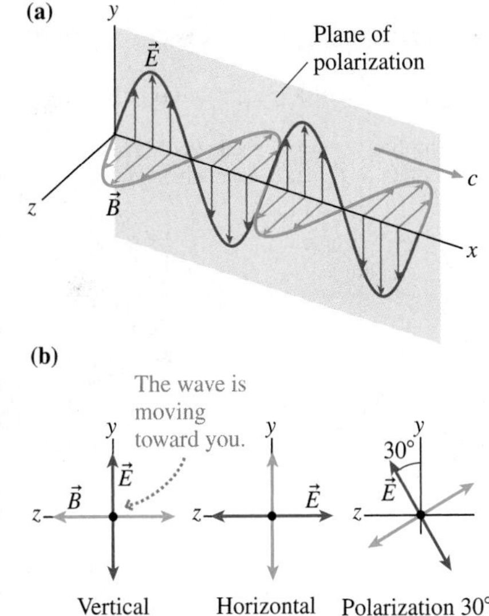

Video Tutor Demo

electromagnetic wave that you see or measure is a superposition of waves from each of these tiny emitters. Although the wave from each individual emitter is polarized, it is polarized in a random direction with respect to the waves from all its neighbors. The resulting wave, a superposition of waves with electric fields in all possible directions, is *unpolarized*, as is the radiation from most sources of electromagnetic radiation.

Energy of Electromagnetic Waves

A field-fired furnace The energy from the sun is carried through space by electromagnetic waves—it is carried by electric and magnetic fields. The mirrors of this solar furnace in southern France concentrate the electromagnetic waves from the sun to an intensity 1000 times that of normal sunlight, allowing researchers to test the properties of materials at up to 3800°C.

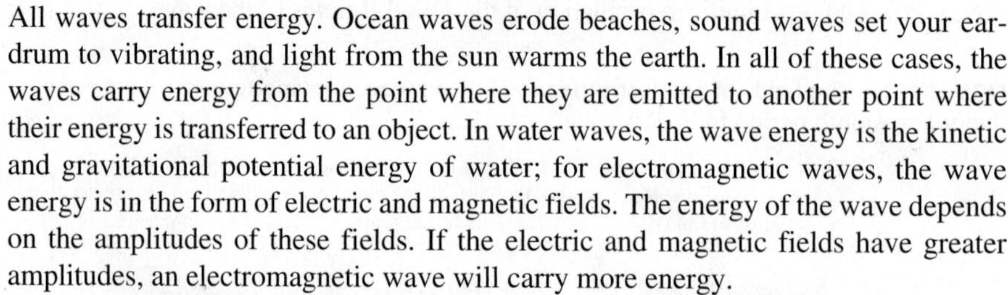

All waves transfer energy. Ocean waves erode beaches, sound waves set your eardrum to vibrating, and light from the sun warms the earth. In all of these cases, the waves carry energy from the point where they are emitted to another point where their energy is transferred to an object. In water waves, the wave energy is the kinetic and gravitational potential energy of water; for electromagnetic waves, the wave energy is in the form of electric and magnetic fields. The energy of the wave depends on the amplitudes of these fields. If the electric and magnetic fields have greater amplitudes, an electromagnetic wave will carry more energy.

In ◀◀ SECTION 15.5, we defined the *intensity* of a wave (measured in W/m^2) to be $I = P/A$, where P is the power (energy transferred per second) of a wave that impinges on area A. The intensity of an electromagnetic wave depends on the amplitudes of the oscillating electric and magnetic fields:

$$I = \frac{P}{A} = \frac{1}{2}c\epsilon_0 E_0^2 = \frac{1}{2}\frac{c}{\mu_0}B_0^2 \tag{25.18}$$

Intensity of an electromagnetic wave with field amplitudes E_0 and B_0

The intensity of a plane wave, such as that of a laser beam, does not change with distance. As we saw, the intensity of a spherical wave, spreading out from a point, must decrease with the square of the distance to conserve energy. If a source with power P_{source} emits waves *uniformly* in all directions, the wave intensity at distance r is

$$I = \frac{P_{\text{source}}}{4\pi r^2} \tag{25.19}$$

The intensities of the electromagnetic waves from antennas, cell phones, and other "point sources" are reasonably well described by Equation 25.19.

It's worthwhile to pull together all of the details that we've seen for electromagnetic waves. Seeing the "big picture" will help us as we start to solve electromagnetic wave problems.

SYNTHESIS 25.1 Electromagnetic waves

An electromagnetic wave is a transverse wave of oscillating electric and magnetic fields.

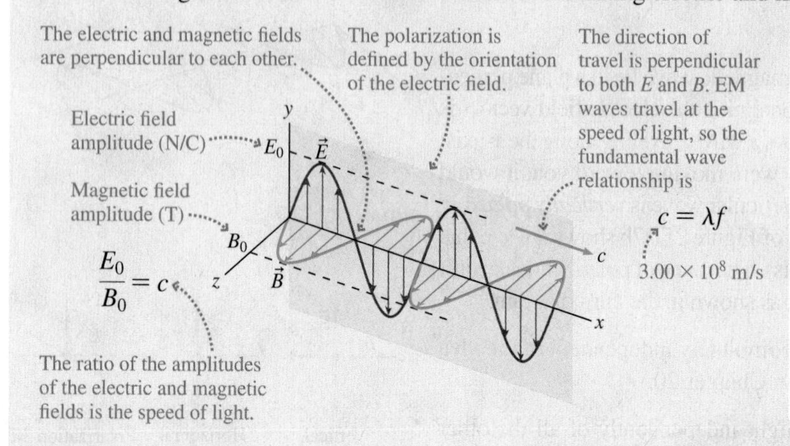

The intensity of the EM wave depends on the field amplitudes:

Intensity (W/m^2)

$$I = \frac{1}{2}c\epsilon_0 E_0^2 = \frac{1}{2}\frac{c}{\mu_0}B_0^2$$

Permittivity constant, $8.85 \times 10^{-12}\ C^2/N \cdot m^2$

Permeability constant, $1.26 \times 10^{-6}\ T \cdot m/A$

The intensity of an EM point source decreases with distance from the source:

$$I = \frac{P_{\text{source}}}{4\pi r^2}$$

Source power (W)

Distance from the source (m)

Video Tutor Demo

EXAMPLE 25.7 Electric and magnetic fields of a cell phone

A digital cell phone emits 0.60 W of 1.9 GHz radio waves. What are the amplitudes of the electric and magnetic fields at a distance of 10 cm?

PREPARE We can solve this problem using details from Synthesis 25.1. We can approximate the cell phone as a point source, so we can use the second intensity equation to find the intensity at 10 cm. Once we know the intensity, we can use the first intensity equation to compute the field amplitudes.

SOLVE The intensity at a distance of 10 cm is

$$I = \frac{P_{\text{source}}}{4\pi r^2} = \frac{0.60\ \text{W}}{4\pi(0.10\ \text{m})^2} = 4.8\ \text{W/m}^2$$

We can rearrange the first intensity equation to solve for the amplitude of the electric field:

$$E_0 = \sqrt{\frac{2I}{c\epsilon_0}} = \sqrt{\frac{2(4.8\ \text{W/m}^2)}{(3.0 \times 10^8\ \text{m/s})(8.85 \times 10^{-12}\ \text{C}^2/\text{N}\cdot\text{m}^2)}} = 60\ \text{V/m}$$

We can then use the relationship between field amplitudes to find the amplitude of the magnetic field:

$$B_0 = \frac{E_0}{c} = 2.0 \times 10^{-7}\ \text{T}$$

ASSESS The electric field amplitude is reasonably small. For comparison, the typical electric field due to atmospheric electricity is 100 V/m; the field near a charged Van de Graaff generator can be 1000 times larger than this. The scale of the result thus seems reasonable; we know that the electric fields near a cell phone's antenna aren't large enough to produce significant effects. The magnetic field is smaller yet, only 1/250th of the earth's field, which, as you know, is quite weak. This makes sense as well; you haven't noticed magnetic effects while making a phone call!

Polarizers and Changing Polarization

The fields of an electromagnetic wave will interact with matter. The interaction depends on—and can affect—the polarization of an electromagnetic wave.

If we have an unpolarized source of light, we can transform it into polarized light by sending it through a *polarizing filter*. A typical polarizing filter is a plastic sheet containing long organic molecules called polymers, as shown in **FIGURE 25.28**. The molecules are aligned to form a grid, like the metal bars in a barbecue grill, then treated so they conduct electrons along their length.

FIGURE 25.28 A polarizing filter.

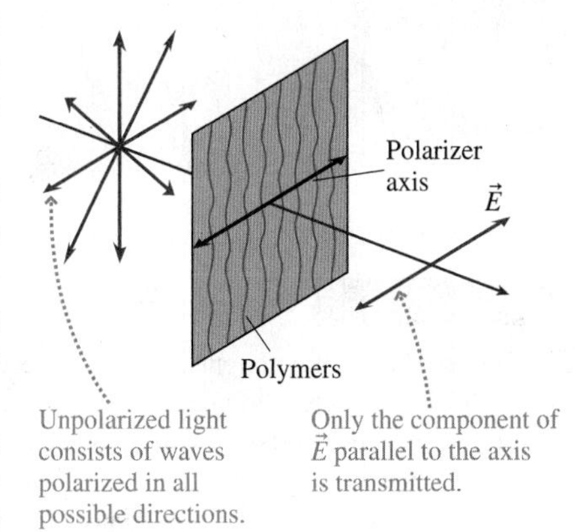
Polarizer axis
$\vec{E}$
Polymers
Unpolarized light consists of waves polarized in all possible directions.
Only the component of $\vec{E}$ parallel to the axis is transmitted.

As a light wave travels through a polarizing filter, the component of the electric field oscillating parallel to the polymer grid drives the electrons up and down the molecules. The electrons absorb energy from the light wave, so the parallel component of $\vec{E}$ is absorbed in the filter. Thus the light wave emerging from a polarizing filter is polarized perpendicular to the polymer grid. We call the direction of the transmitted polarization the axis of the polarizer.

Suppose a *polarized* light wave with electric field amplitude E_{incident} approaches a polarizing filter with a vertical axis (that is, the filter transmits only vertically polarized light). What is the intensity of the light that passes through the filter? **FIGURE 25.29** on the next page shows that the oscillating electric field of the polarized light can be decomposed into horizontal and vertical components. The vertical component will pass; the horizontal component will be blocked. As we see in Figure 25.29, the magnitude of the electric field of the light transmitted by the filter is

$$E_{\text{transmitted}} = E_{\text{incident}} \cos\theta \qquad (25.20)$$

FIGURE 25.29 Components of the electric field parallel and perpendicular to a polarizer's axis.

y The incident light is polarized at angle θ with respect to the polarizer's axis.

Polarizer axis

$\vec{E}$

$E_0 \cos\theta$ θ

$E_0 \sin\theta$

θ

x

Only the component of $\vec{E}$ in the direction of the axis is transmitted.

Because the intensity depends on the square of the electric field amplitude, the transmitted intensity is related to the incident intensity by what is known as **Malus's law:**

$$I_{\text{transmitted}} = I_{\text{incident}}(\cos\theta)^2 \quad (25.21)$$

Malus's law for transmission of polarized light by a polarizing filter

FIGURE 25.30a shows how Malus's law can be demonstrated with two polarizing filters. The first, called the *polarizer,* is used to produce polarized light of intensity I_0. The second, called the *analyzer,* is rotated by angle θ relative to the polarizer. As the photographs of FIGURE 25.30b show, the transmission of the analyzer is (ideally) 100% when $\theta = 0°$ and steadily decreases to zero when $\theta = 90°$. Two polarizing filters with perpendicular axes, called *crossed polarizers,* block all the light.

FIGURE 25.30 The intensity of the transmitted light depends on the angle between the polarizing filters.

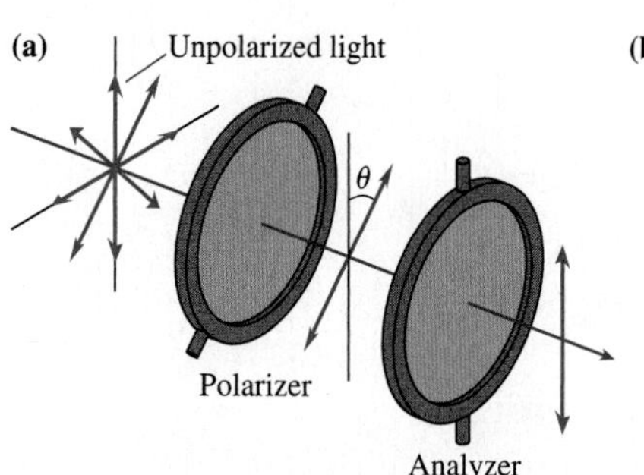

(b)

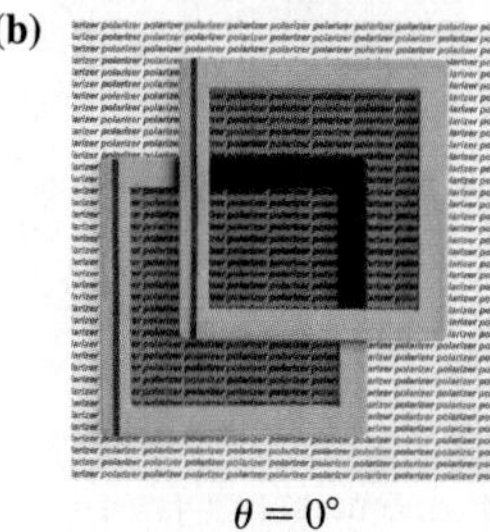

$\theta = 0°$

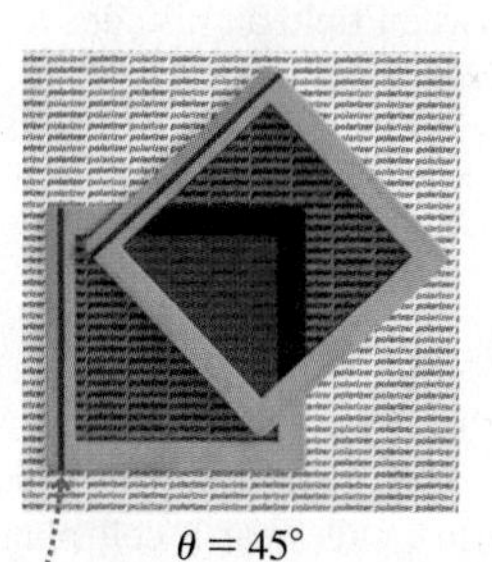

$\theta = 45°$

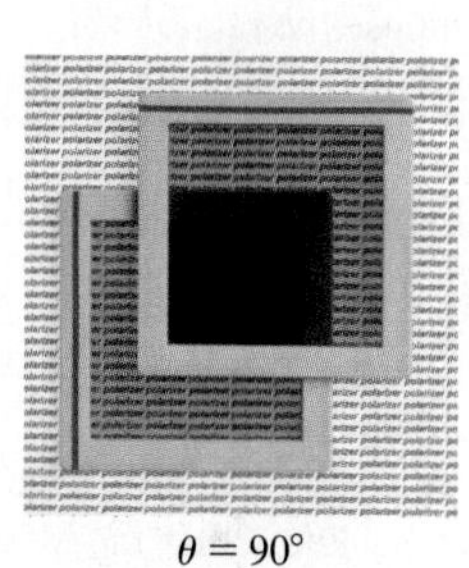

$\theta = 90°$

The red lines show the axes of the polarizers.

Figure Video

FIGURE 25.31 Polarized light micrograph of a thin section of molar teeth.

Suppose you place an object between two crossed polarizers. Normally, no light would make it through the analyzer, and the object would appear black. But if the object is able to *change* the polarization of the light, some of the light emerging from the object will be able to pass through the analyzer. This can be a valuable analytical technique. FIGURE 25.31 shows a micrograph of a very thin section of molar teeth as it appears when viewed between crossed polarizers. Different minerals and different materials in the teeth affect the polarization of the light in different ways, giving an image that clearly highlights the different tissues in the teeth.

When light passes through a glucose solution, the plane of polarization is rotated by an amount that is proportional to the concentration of glucose. Higher concentrations of glucose mean larger angles of rotation. In principle, this could be used as a test of blood glucose—a test that doesn't involve poking a fingertip.

EXAMPLE 25.8 Using polarization to test for blood glucose BIO

Doctors have long sought a noninvasive technique for patients, such as diabetics, who need to monitor their blood glucose levels several times a day. One possibility is to measure the change in the polarization of light transmitted through the clear aqueous humor in the front of the eye, where glucose levels are known to be almost the same as those in the blood. At the expected glucose levels, the change in polarization is quite small, but measurable, as this example—using realistic numbers—shows.

A laser is passed through a polarizer and emerges with intensity 55 W/m^2. In a first measurement, the polarized laser light is passed through an analyzer, which is adjusted to an angle of 90°, giving zero transmitted intensity. Then, in a second measurement, the polarized laser light is passed through the aqueous humor in a patient's eye before reaching the analyzer, which is still set at 90°; in this case, the rotation of the plane of polarization by the glucose allows some light to pass, giving a transmitted intensity of 8.7×10^{-7} W/m^2. What is the angle of rotation of the polarization by the glucose?

PREPARE We can use Malus's law (Equation 25.21) to relate the intensity transmitted through the analyzer to the intensity incident on the analyzer. For the initial measurement, $\theta = 90°$ exactly because the analyzer is adjusted so that no light passes. In the second test, some light is transmitted because the glucose rotates the polarization so that the angle is smaller than 90°. The difference between the two angles is the rotation of the plane of polarization.

SOLVE We can adjust Equation 25.21 to find the rotation angle for the second test. The incident intensity is the 55 W/m^2 of the polarized laser light and the transmitted intensity is 8.7×10^{-7} W/m^2, so we find

$$\cos\theta = \sqrt{\frac{I_{\text{transmitted}}}{I_{\text{incident}}}} = \sqrt{\frac{8.7 \times 10^{-7}\ \text{W/m}^2}{55\ \text{W/m}^2}} = 1.26 \times 10^{-4}$$

The polarization angle is thus

$$\theta = \cos^{-1}(1.26 \times 10^{-4}) = 89.9928°$$

We've kept extra significant figures because we are interested in the small difference between this angle and 90°. The angle 90° − 89.9928° = 0.0072° is the angle by which the glucose has rotated the plane of polarization.

ASSESS The plane of polarization rotates by a very small angle, but we expect this from the problem statement. The rotation, though small, can be readily measured with modern optical equipment.

In polarizing sunglasses, the polarization axis is vertical (when the glasses are in the normal orientation) so that the glasses transmit only vertically polarized light. *Glare*—the reflection of the sun and the skylight from lakes and other horizontal surfaces—has a strong horizontal polarization. This light is almost completely blocked, so the sunglasses "cut glare" without affecting the main scene.

▶ **Making a beeline** BIO At an angle of 90° to the sun, skylight is strongly polarized. In the left photo, this polarization causes different transmission through polarizers with polarization axes given by the red arrows. Opposite the sun in the sky, the skylight is unpolarized, causing equal transmission for all of the polarizers in the right photo. In the photo of the sky at 90° to the sun, the arrows in the dark polarizers define a line that points toward the sun; you can tell where the sun is though it does not appear in the photo. Honeybees and other insects—unlike you—have eyes that are sensitive to polarization. Bees can reliably navigate in dense forest cover; as long as they can see a small patch of blue sky, they can determine the position of the sun.

STOP TO THINK 25.5 Unpolarized light of equal intensity is incident on four pairs of polarizing filters. Rank in order, from largest to smallest, the intensities I_A to I_D transmitted through the second polarizer of each pair.

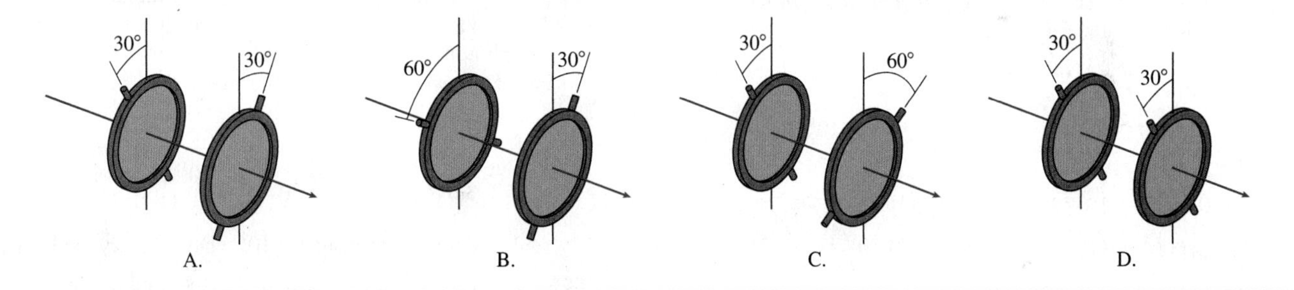

25.6 The Photon Model of Electromagnetic Waves

FIGURE 25.32 on the next page shows three photographs made with a camera in which the film has been replaced by a special high-sensitivity detector. A correct exposure, at the bottom, shows a perfectly normal photograph of a woman. But with very faint illumination (top), the picture is *not* just a dim version of the properly exposed photo. Instead, it is a collection of dots. A few points on the detector have registered the presence of light, but most have not. As the illumination increases, the density of these dots increases until the dots form a full picture.

This is not what we might expect. If light is a wave, reducing its intensity should cause the picture to grow dimmer and dimmer until it disappears, but the entire

FIGURE 25.32 Photographs made with an increasing level of light intensity.

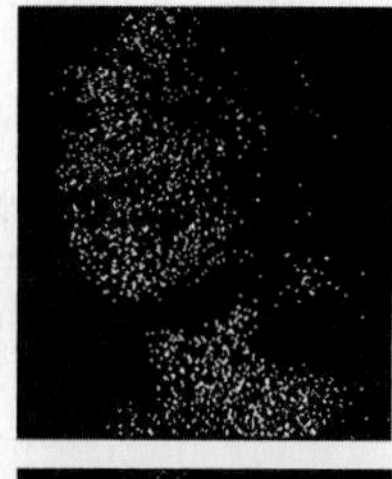

At low light levels, we see individual points, as if from particles.

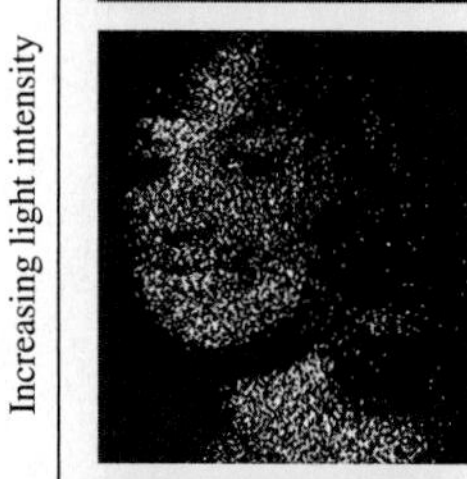

Increasing light intensity

At higher light levels, we don't see the particle-like behavior.

picture would remain present. Instead, the top photograph in Figure 25.32 looks as if someone randomly threw "pieces" of light at the detector, causing full exposure at some points but no exposure at others.

If we did not know that light is a wave, we would interpret the results of this experiment as evidence that light is a stream of some type of particle-like object. If these particles arrive frequently enough, they overwhelm the detector and it senses a steady "river" instead of the individual particles in the stream. Only at very low intensities do we become aware of the individual particles.

As we will see in Chapter 28, many experiments convincingly lead to the surprising result that **electromagnetic waves have a particle-like nature.** These particle-like components of electromagnetic waves are called **photons.**

The **photon model** of electromagnetic waves consists of three basic postulates:

1. Electromagnetic waves consist of discrete, massless units called photons. A photon travels in vacuum at the speed of light, 3.00×10^8 m/s.
2. Each photon has energy

$$E_{\text{photon}} = hf \qquad (25.22)$$

   where f is the frequency of the wave and h is a *universal constant* called **Planck's constant.** The value of Planck's constant is

$$h = 6.63 \times 10^{-34}\ \text{J} \cdot \text{s}$$

   In other words, the electromagnetic waves come in discrete "chunks" of energy hf. The higher the frequency, the more energetic the chunks.
3. The superposition of a sufficiently large number of photons has the characteristics of a continuous electromagnetic wave.

EXAMPLE 25.9 Finding the energy of a photon of visible light

550 nm is the approximate average wavelength of visible light.

a. What is the energy of a photon with a wavelength of 550 nm?

b. A 40 W incandescent lightbulb emits about 1 J of visible light energy every second. Estimate the number of visible light photons emitted per second.

SOLVE a. The frequency of the photon is

$$f = \frac{c}{\lambda} = \frac{3.00 \times 10^8\ \text{m/s}}{550 \times 10^{-9}\ \text{m}} = 5.4 \times 10^{14}\ \text{Hz}$$

Equation 25.22 gives us the energy of this photon:

$$E_{\text{photon}} = hf = (6.63 \times 10^{-34}\ \text{J} \cdot \text{s})(5.4 \times 10^{14}\ \text{Hz}) = 3.6 \times 10^{-19}\ \text{J}$$

This is an extremely small energy! In fact, photon energies are so small that they are usually measured in electron volts (eV) rather than joules. Recall that 1 eV $= 1.60 \times 10^{-19}$ J. With this, we find that the photon energy is

$$E_{\text{photon}} = 3.6 \times 10^{-19}\ \text{J} \times \frac{1\ \text{eV}}{1.60 \times 10^{-19}\ \text{J}} = 2.3\ \text{eV}$$

b. The photons emitted by a lightbulb span a range of energies, because the light spans a range of wavelengths, but the *average* photon energy corresponds to a wavelength near 550 nm. Thus we can estimate the number of photons in 1 J of light as

$$N \approx \frac{1\ \text{J}}{3.6 \times 10^{-19}\ \text{J/photon}} \approx 3 \times 10^{18}\ \text{photons}$$

A typical lightbulb emits about 3×10^{18} photons every second.

ASSESS The number of photons emitted per second is staggeringly large. It's not surprising that in our everyday life we sense only the river and not the individual particles within the flow.

TABLE 25.1 Energies of some atomic and molecular processes

Process	Energy
Breaking a hydrogen bond between two water molecules	0.24 eV
Energy released in metabolizing one molecule of ATP	0.32 eV
Breaking the bond between atoms in a water molecule	4.7 eV
Ionizing a hydrogen atom	13.6 eV

As we saw, a single photon of light at a wavelength of 550 nm has an energy of 2.3 eV. It is worthwhile to see just what 2.3 eV "buys" in interactions with atoms and molecules. Table 25.1 shows some energies required for typical atomic and molecular processes. These values show that 2.3 eV is a significant amount of energy on an atomic scale. It is certainly enough to cause a molecular transformation (as it does in the sensory system of your eye), and photons with a bit more energy (higher frequency and thus shorter wavelength) can break a covalent bond. The photon model of light will be essential as we explore the interaction of electromagnetic waves with matter.

STOP TO THINK 25.6 Two FM radio stations emit radio waves at frequencies of 90.5 MHz and 107.9 MHz. Each station emits the same total power. If you think of the radio waves as photons, which station emits the larger number of photons per second?

A. The 90.5 MHz station.
B. The 107.9 MHz station.
C. Both stations emit the same number of photons per second.

25.7 The Electromagnetic Spectrum

We have now seen two very different ways to look at electromagnetic waves: as oscillating waves of the electric and magnetic fields, and as particle-like units of the electromagnetic field called photons. This dual nature of electromagnetic waves is something we will discuss at length in Chapter 28. For now, we will note that each view is appropriate in certain circumstances. For example, we speak of radio *waves* but of x *rays.* The "ray" terminology tells us that x rays are generally better described as photons than as waves.

FIGURE 25.33 shows the *electromagnetic spectrum* with photon energy (in eV) and wavelength (in m) scales. As you can see, electromagnetic waves span an extraordinarily wide range of wavelengths and energies. Radio waves have wavelengths of many meters but very low photon energies—only a few billionths of an eV. Because the photon energies are so small, radio waves are well described by Maxwell's theory of electromagnetic waves. At the other end of the spectrum, x rays and gamma rays have very short wavelengths and very high photon energies—large enough to ionize atoms and break molecular bonds. Consequently, x rays and gamma rays, although they do have wave-like characteristics, are best described as photons. Visible light is in the middle. As we will see in Chapter 28, we must consider *both* views to fully understand the nature of visible light.

FIGURE 25.33 The electromagnetic spectrum.

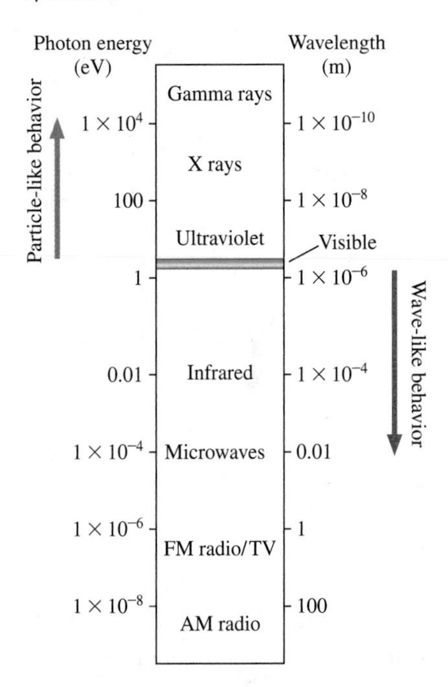

Radio Waves and Microwaves

An electromagnetic wave is self-sustaining, independent of charges or currents. However, charges and currents are needed at the *source* of an electromagnetic wave. Radio waves and microwaves are generally produced by the motion of charged particles in an antenna.

FIGURE 25.34 reminds you what the electric field of an electric dipole looks like. If the dipole is vertical, the electric field $\vec{E}$ at points along the horizontal axis in the figure is also vertical. Reversing the dipole, by switching the charges, reverses $\vec{E}$. If the charges were to *oscillate* back and forth, switching position at frequency f, then $\vec{E}$ would oscillate in a vertical plane. The changing $\vec{E}$ would then create an induced magnetic field $\vec{B}$, which could then create an $\vec{E}$, which could then create a $\vec{B}$, . . . , and a vertically polarized electromagnetic wave at frequency f would radiate out into space.

This is exactly what an **antenna** does. **FIGURE 25.35** on the next page shows two metal wires attached to the terminals of an oscillating voltage source. The figure shows an instant when the top wire is negative and the bottom is positive, but these will reverse in half a cycle. The wire is basically an oscillating dipole, and it creates an oscillating electric field. The oscillating $\vec{E}$ induces an oscillating $\vec{B}$, and they take off as an electromagnetic wave at speed $v_{em} = c$. The wave does need oscillating charges as a *wave source,* but once created it is self-sustaining and independent of the source.

Radio waves are *detected* by antennas as well. The electric field of a vertically polarized radio wave drives a current up and down a vertical conductor, producing a potential difference that can be amplified. For best reception, the antenna length should be about $\frac{1}{4}$ of a wavelength. A typical cell phone works at 1.9 GHz, with wavelength $\lambda = c/f = 16$ cm. Thus a cell phone antenna should be about 4 cm long, or about $1\frac{1}{2}$ inches; it is generally hidden inside the phone itself.

FIGURE 25.34 The electric field of an oscillating dipole.

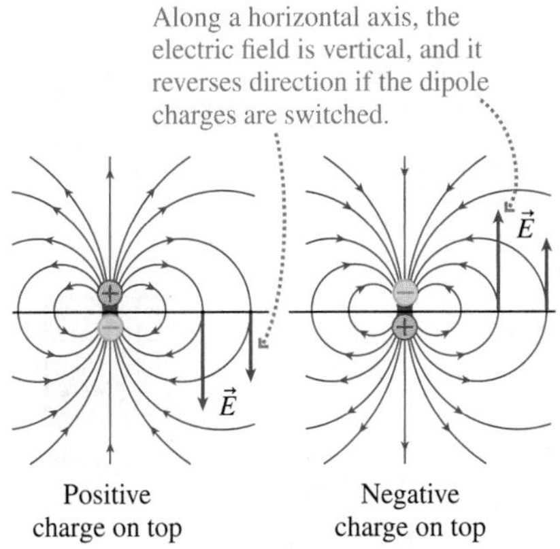

Along a horizontal axis, the electric field is vertical, and it reverses direction if the dipole charges are switched.

$\vec{E}$ $\vec{E}$

Positive charge on top

Negative charge on top

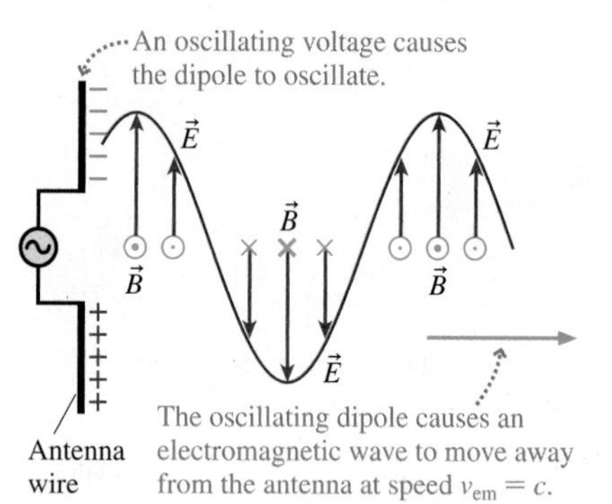

FIGURE 25.35 An antenna generates a self-sustaining electromagnetic wave.

AM radio has a lower frequency and thus a longer wavelength—typically 300 m. Having an antenna that is $\frac{1}{4}$ of a wavelength—75 m long!—is simply not practical. Instead, the antenna in an AM radio consists of a coil of wire wrapped around a core of magnetic material. This antenna detects the *magnetic* field of the radio wave. The changing flux of the wave's magnetic field induces an emf in the coil that is detected and amplified by the receiver.

CONCEPTUAL EXAMPLE 25.10 **Orienting a coil antenna**

A vertically polarized AM radio wave is traveling to the right. How should you orient a coil antenna to detect the oscillating magnetic field component of the wave?

REASON You want the oscillating magnetic field of the wave to produce the maximum possible induced emf in the coil, which requires the maximum changing flux. The flux is maximum when the coil is perpendicular to the magnetic field of the electromagnetic wave, as in **FIGURE 25.36**. Thus the plane of the coil should match the wave's plane of polarization.

FIGURE 25.36 A coil antenna.

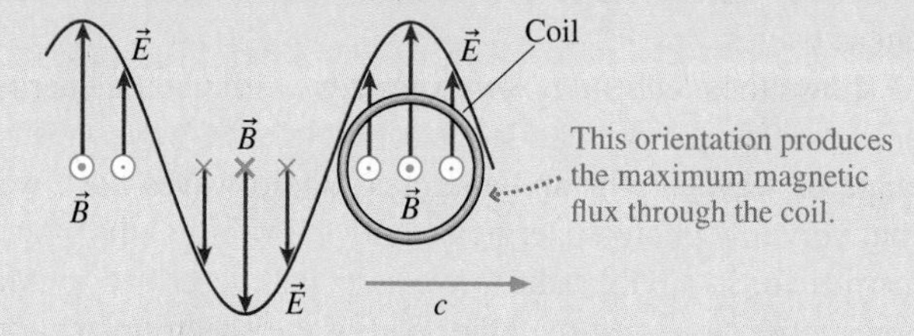

Coil
This orientation produces the maximum magnetic flux through the coil.
c

ASSESS Coil antennas are highly directional. If you turn an AM radio—and thus the antenna—in certain directions, you will no longer have the correct orientation of the magnetic field and the coil, and reception will be poor.

FIGURE 25.37 A radio wave interacts with matter.

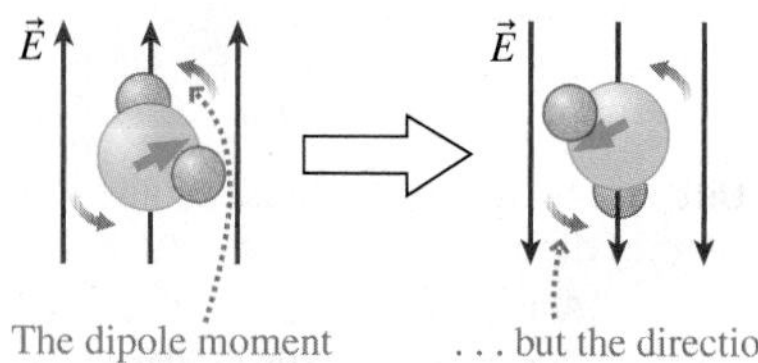

In materials with no free charges, the electric fields of radio waves and microwaves can still interact with matter by exerting a torque on molecules, such as water, that have a permanent electric dipole moment, as shown in **FIGURE 25.37**. The molecules acquire kinetic energy from the wave; then their collisions with other molecules transform that energy into thermal energy, increasing the temperature.

This is how a microwave oven heats food. Water molecules, with their large dipole moment, rotate in response to the electric field of the microwaves, then transfer this energy to the food via molecular collisions.

Infrared, Visible Light, and Ultraviolet

Radio waves can be produced by oscillating charges in an antenna. At the higher frequencies of infrared, visible light, and ultraviolet, the "antennas" are individual atoms. This portion of the electromagnetic spectrum is *atomic radiation.*

Nearly all the atomic radiation in our environment is *thermal radiation* due to the thermal motion of the atoms in an object. As we saw in Chapter 12, thermal radiation—a form of heat transfer—is described by Stefan's law: If heat energy Q is radiated in a time interval Δt by an object with surface area A and absolute temperature T, the *rate* of heat transfer $Q/\Delta t$ (joules per second) is

$$\frac{Q}{\Delta t} = e\sigma AT^4 \tag{25.23}$$

The constant e in this equation is the object's emissivity, a measure of its effectiveness at emitting electromagnetic waves, and σ is the Stefan-Boltzmann constant, $\sigma = 5.67 \times 10^{-8}\ \mathrm{W/(m^2 \cdot K^4)}$.

In Chapter 12 we considered the amount of energy radiated and its dependence on temperature. The filament of an incandescent bulb glows simply because it is hot. If you increase the current through a lightbulb filament, the filament temperature increases and so does the total energy emitted by the bulb, in accordance with Stefan's law. The three pictures in **FIGURE 25.38** show a glowing lightbulb with the filament at successively higher temperatures. We can clearly see an increase in brightness in the sequence of three photographs.

But it's not just the brightness that varies. The *color* of the emitted radiation changes as well. At low temperatures, the light from the bulb is quite red. (A dim bulb doesn't look this red to your eye because your brain, knowing that the light "should" be white, compensates. But the camera doesn't lie.) Looking at the change in color as the temperature of the bulb rises in Figure 25.38, we see that **the spectrum of thermal radiation changes with temperature.**

FIGURE 25.38 The brightness of the bulb varies with the temperature of the filament.

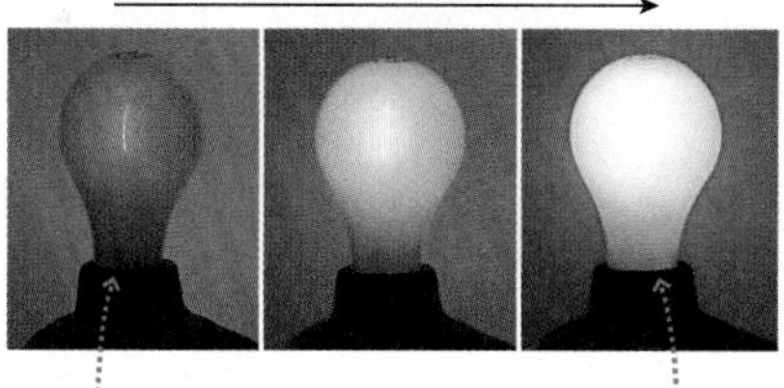

Increasing filament temperature
At lower filament temperatures, the bulb is dim and the light is noticeably reddish.
When the filament is hotter, the bulb is brighter and the light is whiter.

If we measure the intensity of thermal radiation as a function of wavelength for an object at three temperatures, 3500 K, 4500 K, and 5500 K, the data appear as in **FIGURE 25.39**. Notice two important features:

- Increasing the temperature increases the intensity at all wavelengths. **Making the object hotter causes it to emit more radiation across the entire spectrum.**
- Increasing the temperature causes the peak intensity to shift to a shorter wavelength. **The higher the temperature, the shorter the wavelength of the peak of the spectrum.**

FIGURE 25.39 A thermal emission spectrum depends on the temperature.

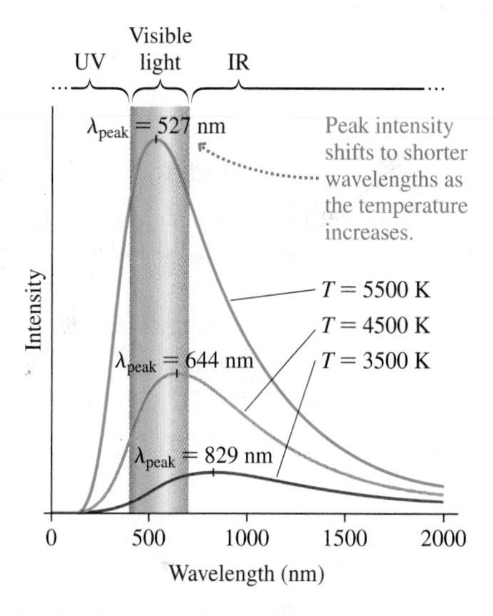

It is this variation of the peak wavelength that causes the change in color of the glowing filament in Figure 25.38. The temperature dependence of the peak wavelength of thermal radiation is known as *Wien's law,* which appears as follows:

$$\lambda_{\text{peak}}\text{ (in nm)} = \frac{2.9 \times 10^6\text{ nm}\cdot\text{K}}{T\text{ (in K)}} \qquad (25.24)$$

Wien's law for the peak wavelength of a thermal emission spectrum

λ_{peak} p. 108 T INVERSE

EXAMPLE 25.11 Finding peak wavelengths

What are the wavelengths of peak intensity and the corresponding spectral regions for radiating objects at (a) normal human body temperature of 37°C, (b) the temperature of the filament in an incandescent lamp, 1500°C, and (c) the temperature of the surface of the sun, 5800 K?

PREPARE All of the objects emit thermal radiation, so the peak wavelengths are given by Equation 25.24.

SOLVE First, we convert temperatures to kelvin. The temperature of the human body is $T = 37 + 273 = 310$ K, and the filament temperature is $T = 1500 + 273 = 1773$ K. Equation 25.24 then gives the wavelengths of peak intensity as

a. $\lambda_{\text{peak}}(\text{body}) = \dfrac{2.9 \times 10^6\text{ nm}\cdot\text{K}}{310\text{ K}} = 9.4 \times 10^3\text{ nm} = 9.4\ \mu\text{m}$

b. $\lambda_{\text{peak}}(\text{filament}) = \dfrac{2.9 \times 10^6\text{ nm}\cdot\text{K}}{1773\text{ K}} = 1600\text{ nm}$

c. $\lambda_{\text{peak}}(\text{sun}) = \dfrac{2.9 \times 10^6\text{ nm}\cdot\text{K}}{5800\text{ K}} = 500\text{ nm}$

ASSESS The peak of the emission curve at body temperature is far into the infrared region of the spectrum, well below the range of sensitivity of human vision. You don't see someone "glow," although people do indeed emit significant energy in the form of electromagnetic waves, as we saw in Chapter 12. The sun's emission peaks right in the middle of the visible spectrum, which seems reasonable. Interestingly, most of the energy radiated by an incandescent bulb is *not* visible light. The tail of the emission curve extends into the visible region, but the peak of the emission curve—and most of the emitted energy—is in the infrared region of the spectrum. A 100 W bulb emits only a few watts of visible light.

► **It's the pits . . .** BIO Certain snakes—including rattlesnakes and other pit vipers—can hunt in total darkness. The viper in the left photo has pits in front of its eyes. These are a second set of vision organs; they have sensitive tissue at the bottom that allows them to detect the thermal radiation emitted by warm-blooded prey, such as the thermal radiation emitted by the mouse in the thermal image on the right. These snakes need no light to "see" you. You emit a "glow" they can detect.

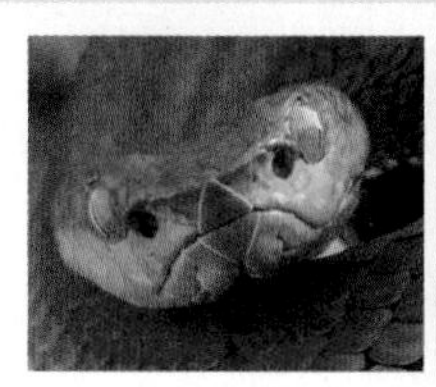

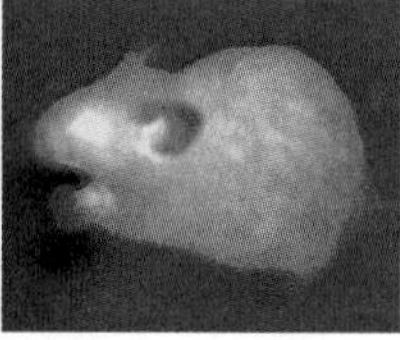

Infrared radiation, with its relatively long wavelength and low photon energy, produces effects in tissue similar to those of microwaves—heating—but the penetration is much less than for microwaves. Infrared is absorbed mostly by the top layer of your skin and simply warms you up, as you know from sitting in the sun or under a heat lamp. The wave picture is generally most appropriate for infrared.

In contrast, ultraviolet photons have enough energy to interact with molecules in entirely different ways, ionizing molecules and breaking molecular bonds. The cells in skin are altered by ultraviolet radiation, causing sun tanning and sun burning. DNA molecules can be permanently damaged by ultraviolet radiation. There is a sharp threshold for such damage at 290 nm (corresponding to 4.3 eV photon energy). At longer wavelengths, damage to cells is slight; at shorter wavelengths, it is extensive. The interactions of ultraviolet radiation with matter are best understood from the photon perspective, with the absorption of each photon being associated with a particular molecular event.

Visible light is at a transition point in the electromagnetic spectrum. Your studies of wave optics in Chapter 17 showed you that light has a wave nature. At the same time, the energy of photons of visible light is large enough to cause molecular transitions—which is how your eye detects light. When we work with visible light, we will often move back and forth between the wave and photon models.

EXAMPLE 25.12 Finding the photon energy for ultraviolet light

Ultraviolet radiation with a wavelength of 254 nm is used in germicidal lamps. What is the photon energy in eV for such a lamp?

SOLVE The photon energy is $E = hf$:

$$E = hf = \frac{hc}{\lambda} = \frac{(6.63 \times 10^{-34}\ \text{J}\cdot\text{s})(3.00 \times 10^{8}\ \text{m/s})}{254 \times 10^{-9}\ \text{m}}$$
$$= 7.83 \times 10^{-19}\ \text{J}$$

In eV, this is

$$E = 7.83 \times 10^{-19}\ \text{J} \times \frac{1\ \text{eV}}{1.60 \times 10^{-19}\ \text{J}} = 4.89\ \text{eV}$$

ASSESS Table 25.1 shows that this energy is sufficient to break the bonds in a water molecule. It will be enough energy to break other bonds as well, leading to damage on a cellular level.

Color Vision

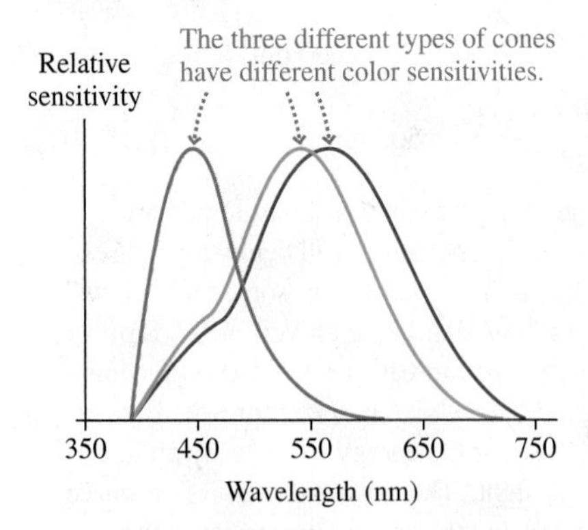
The three different types of cones have different color sensitivities.
Relative sensitivity
350
450
550
650
750
Wavelength (nm)

FIGURE 25.40 The sensitivity of different cones in the human eye.

The cones, the color-sensitive cells in the retina of the eye, each contain one of three slightly different forms of a light-sensitive photopigment. A single photon of light can trigger a reaction in a photopigment molecule, which ultimately leads to a signal being produced by a cell in the retina. The energy of the photon must be matched to the energy of a molecular transition for absorption of the photon energy to take place. Each photopigment has a range of photon energies to which it is sensitive. Our color vision is a result of the differential response of three types of cones containing three slightly different pigments, shown in FIGURE 25.40.

Humans have three types of cone in the eye, mice have two, and chickens four—giving a chicken keener color vision than a human. The three color photopigments that bees possess give them excellent color vision, but a bee's color sense is different from a human's. The peak sensitivities of a bee's photopigments are in the yellow, blue, and ultraviolet regions of the spectrum. A bee can't see the red of a rose, but it is quite sensitive to ultraviolet wavelengths well beyond the range of human vision. The flower in the right-hand photo at the start of the chapter looks pretty to us, but its coloration is really intended for other eyes. The ring of ultraviolet-absorbing pigments near the center of the flower, which is invisible to humans, helps bees zero in on the pollen.

X Rays and Gamma Rays

At the highest energies of the electromagnetic spectrum we find x rays and gamma rays. There is no sharp dividing line between these two regions of the spectrum; the difference is the source of radiation. High-energy photons emitted by electrons are called x rays. If the source is a nuclear process, we call them gamma rays.

We will look at the emission of x rays in atomic processes and gamma rays in nuclear processes in Part VII. For now, we will focus on the "artificial" production of x rays in an x-ray tube, such as the one shown in **FIGURE 25.41**. Electrons are emitted from a cathode and accelerated to a kinetic energy of several thousand eV by the electric field between two electrodes connected to a high-voltage power supply. The electrons make a sudden stop when they hit a metal target electrode. The rapid deceleration of an electron can cause the emission of a single photon with a significant fraction of the electron's kinetic energy. These photons, with energies well in excess of 1000 eV, are x rays. The x rays pass through a window in the tube and then may be used to produce an image or to treat a disease.

FIGURE 25.41 A simple x-ray tube.

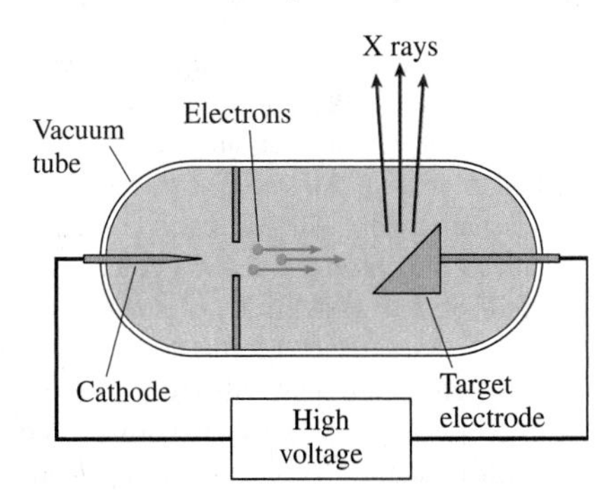

EXAMPLE 25.13 Determining x-ray energies

An x-ray tube used for medical work has an accelerating voltage of 30 kV. What is the maximum energy of an x-ray photon that can be produced in this tube? What is the wavelength of this x ray?

SOLVE An electron accelerated through a potential difference of 30 kV acquires a kinetic energy of 30 keV. When this electron hits the metal target and stops, energy may be converted to an x ray. The maximum energy that could be converted is 30 keV, so this is the maximum possible energy of an x-ray photon from the tube. In joules, this energy is

$$E = 30 \times 10^3 \text{ eV} \times \frac{1.60 \times 10^{-19}\text{ J}}{1 \text{ eV}} = 4.8 \times 10^{-15}\text{ J}$$

For electromagnetic waves, $c = \lambda f$, so we can calculate

$$\lambda = \frac{c}{f} = \frac{c}{E/h} = \frac{hc}{E} = \frac{(6.63 \times 10^{-34}\text{ J}\cdot\text{s})(3.00 \times 10^8\text{ m/s})}{4.8 \times 10^{-15}\text{ J}}$$
$$= 4.1 \times 10^{-11}\text{ m} = 0.041\text{ nm}$$

ASSESS This is a very short wavelength, comparable to the spacing between atoms in a solid.

X rays and gamma rays (and the short-wavelength part of the ultraviolet spectrum) are **ionizing radiation;** the individual photons have sufficient energy to ionize atoms. When such radiation strikes tissue, the resulting ionization can produce cellular damage. When people speak of "radiation" they often mean "ionizing radiation."

At several points in this chapter we have hinted at places where a full understanding of the phenomena requires some new physics. We have used the photon model of electromagnetic waves, and we have mentioned that nuclear processes can give rise to gamma rays. There are other questions that we did not raise, such as why the electromagnetic spectrum of a hot object has the shape that it does. These puzzles began to arise in the late 1800s and early 1900s, and it soon became clear that the physics of Newton and Maxwell was not sufficient to fully describe the nature of matter and energy. Some new rules, some new models, were needed. We will return to these puzzles as we begin to explore the exciting notions of quantum physics in Part VII.

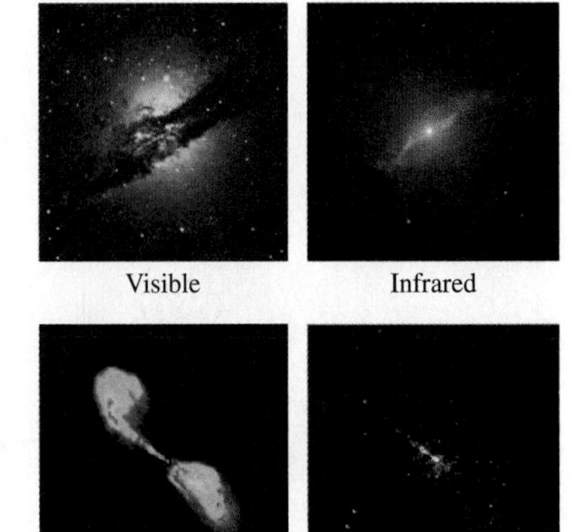

► **Seeing the universe in a different light** These four images of the Centaurus A galaxy have the same magnification and orientation, but they are records of different types of electromagnetic waves. (All but the visible-light image are false-color images.) The visible-light image shows a dark dust lane cutting across the galaxy. In the infrared, this dust lane glows quite brightly—telling us that the dust particles are hot. The radio and x-ray images show jets of matter streaming out of the galaxy's center, hinting at the presence of a massive black hole. Views of the cosmos beyond the visible range are important tools of modern astronomy.

STOP TO THINK 25.7 A group of four stars, all the same size, have the four different surface temperatures given below. Which of these stars emits the most red light?

A. 3000 K B. 4000 K C. 5000 K D. 6000 K

INTEGRATED EXAMPLE 25.14 Space circuits

The very upper part of the atmosphere, where the space shuttle orbits, is called the *ionosphere*. The few atoms and molecules that remain at this altitude are mostly ionized by intense ultraviolet radiation from the sun. The thin gas of the ionosphere thus consists largely of positive ions and negative electrons, so it can carry an electric current. This was crucial to the operation of the tethered satellite system that was tested on the space shuttle in the 1990s. As described in the chapter, a probe was deployed on a conducting wire that tethered the probe at a great distance from the craft. As the space shuttle orbited, the wire moved through the earth's magnetic field, creating a potential difference between the ends of the wire. Charge flowing through the ionosphere back to the shuttle created a complete circuit.

In the final test of the tethered satellite system, a potential difference of 3500 V was generated across 20 km of cable as the shuttle orbited at 7800 m/s.

a. To produce the noted potential difference, what was the component of the magnetic field perpendicular to the wire?
b. The 3500 V potential created a current of 480 mA in the ionosphere. What was the total resistance of the circuit thus formed?
c. How much power was dissipated in this circuit?
d. What is the drag force on the wire due to its motion in the earth's field?
e. The ionization of the upper atmosphere is due to solar radiation at wavelengths of 95 nm and shorter. In what part of the spectrum is this radiation? What is the lowest-energy photon, in eV, that contributes to the ionization?

PREPARE The motion of the wire connecting the tethered satellite to the space shuttle leads to a motional emf that drives a current through this wire and back through the ionosphere. **FIGURE 25.42** shows how we can model this process as an electric circuit. We know the voltage and the current in this circuit, so we can find the resistance and the power. Because the wire carries a current, the earth's field will exert a force on it, which is the drag force we are asked to find.

FIGURE 25.42 The tethered satellite circuit.

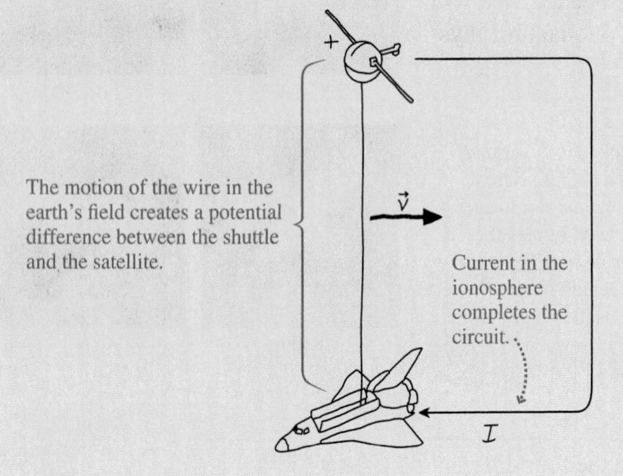
+
The motion of the wire in the earth's field creates a potential difference between the shuttle and the satellite.
$\vec{v}$
Current in the ionosphere completes the circuit.
I

SOLVE a. We know the magnitude of the velocity and the length of the wire, so we can use the equation for the motional emf, Equation 25.3, to find the magnitude of the component of the field perpendicular to the wire:

$$B = \frac{\mathcal{E}}{vl} = \frac{3500\text{ V}}{(7800\text{ m/s})(20 \times 10^3\text{ m})} = 2.24 \times 10^{-5}\text{ T}$$

To two significant figures, this is $B = 22\ \mu\text{T}$.

b. The 3500 V potential difference produced a current of 480 mA. From Ohm's law, the resistance of the circuit was thus

$$R = \frac{\Delta V}{I} = \frac{3500\text{ V}}{480 \times 10^{-3}\text{ A}} = 7300\ \Omega$$

c. We know the voltage and the current, so we can compute the power:

$$P = I\,\Delta V = (0.48\text{ A})(3500\text{ V}) = 1700\text{ W}$$

d. The component of the magnetic field perpendicular to the current-carrying wire exerts a drag force on the wire. We learned in Chapter 24 that the force on a current-carrying wire is $F = IlB$. Thus

$$F = IlB = (0.48\text{ A})(20 \times 10^3\text{ m})(2.24 \times 10^{-5}\text{ T}) = 0.22\text{ N}$$

e. Radiation with wavelengths of 95 nm and shorter is in the ultraviolet region of the spectrum. The lowest-energy photon in this region has the lowest frequency and thus the longest wavelength—namely, 95 nm. The frequency is

$$f = \frac{c}{\lambda} = \frac{3.0 \times 10^8\text{ m/s}}{95 \times 10^{-9}\text{ m}} = 3.2 \times 10^{15}\text{ Hz}$$

The photon energy is then given by Equation 25.22:

$$E_{\text{photon}} = hf = (6.63 \times 10^{-34}\text{ J}\cdot\text{s})(3.2 \times 10^{15}\text{ Hz})$$
$$= 2.1 \times 10^{-18}\text{ J}$$

Converting to eV, we find

$$E_{\text{photon}} = 2.1 \times 10^{-18}\text{ J} \times \frac{1\text{ eV}}{1.6 \times 10^{-19}\text{ J}} = 13\text{ eV}$$

ASSESS We have many good chances to check our work to verify that it makes sense. First, the field component that we calculate is about half the value we typically use for the earth's field, which seems reasonable—we'd be suspicious if the field we calculated was more than the earth's field.

The product of the drag force and the speed is the power dissipated by the drag force:

$$P = Fv = (0.22\text{ N})(7800\text{ m/s}) = 1700\text{ W}$$

This is exactly what we found for the electric power dissipated in the circuit, a good check on our work. The two values must be equal, as they are.

A final check on our work is the value we calculate for the photon energy. Table 25.1 shows that it takes about 13 eV to ionize a hydrogen atom. Photons with wavelengths shorter than 95 nm are able to ionize hydrogen atoms, so it seems likely they would also ionize the nitrogen and oxygen molecules of the upper atmosphere.

SUMMARY

Goal: To understand the nature of electromagnetic induction and electromagnetic waves.

GENERAL PRINCIPLES

Electromagnetic Induction

The **magnetic flux** measures the amount of magnetic field passing through a surface:

$$\Phi = AB\cos\theta$$

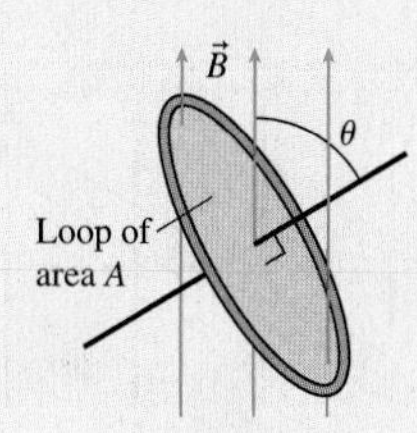

Lenz's law specifies that there is an induced current in a closed conducting loop if the magnetic flux through the loop is changing. The direction of the induced current is such that the induced magnetic field opposes the *change* in flux.

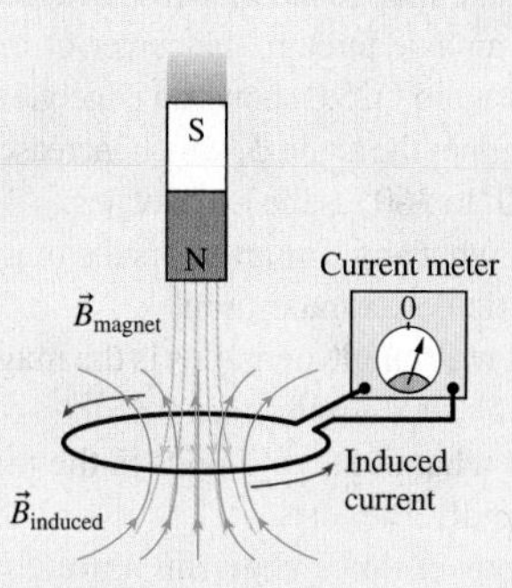

Faraday's law specifies the magnitude of the induced emf in a closed loop:

$$\mathcal{E} = \left|\frac{\Delta\Phi}{\Delta t}\right|$$

Multiply by N for an N-turn coil.

The size of the induced current is

$$I = \frac{\mathcal{E}}{R}$$

Electromagnetic Waves

An electromagnetic wave is a self-sustaining oscillation of electric and magnetic fields.

- The wave is a transverse wave with $\vec{E}$, $\vec{B}$, and $\vec{v}$ mutually perpendicular.
- The wave propagates with speed
$$v_{em} = c = \frac{1}{\sqrt{\epsilon_0\mu_0}} = 3.00 \times 10^8 \text{ m/s}$$
- The wavelength, frequency, and speed are related by
$$c = \lambda f$$
- The amplitudes of the fields are related by
$$\frac{E_0}{B_0} = c$$

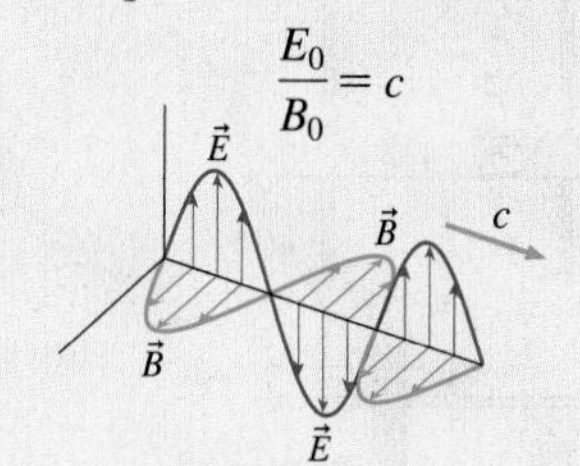

IMPORTANT CONCEPTS

Motional emf

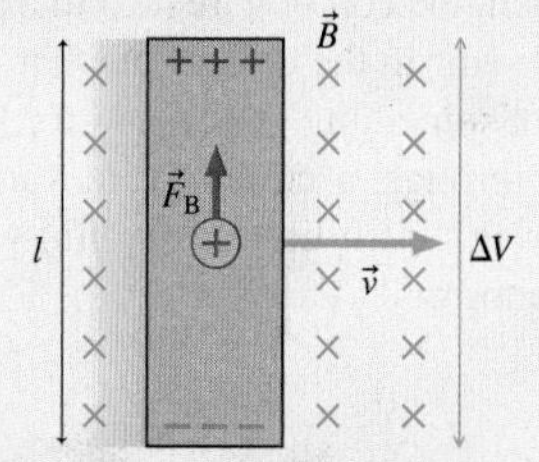

The motion of a conductor through a magnetic field produces a force on the charges. The separation of charges leads to an emf:

$$\mathcal{E} = vlB$$

The photon model

Electromagnetic waves appear to be made of discrete units called photons. The energy of a photon of frequency f is

$$E = hf$$

This photon view becomes increasingly important as the photon energy increases.

The electromagnetic spectrum

Electromagnetic waves come in a wide range of wavelengths and photon energies.

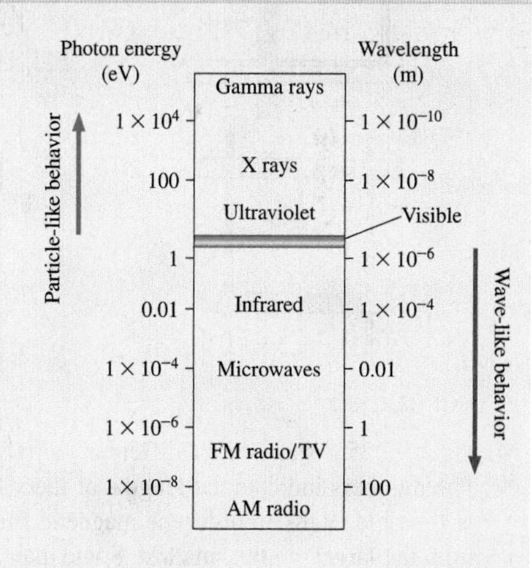

APPLICATIONS

A changing flux in a solid conductor creates **eddy currents.**

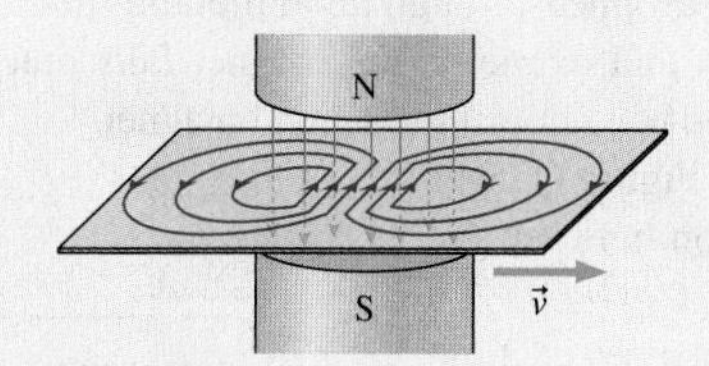

The plane of the electric field of an electromagnetic wave defines its **polarization.** The intensity of polarized light transmitted through a polarizing filter is given by **Malus's law:**

$$I = I_0\cos^2\theta$$

where θ is the angle between the electric field and the polarizer axis.

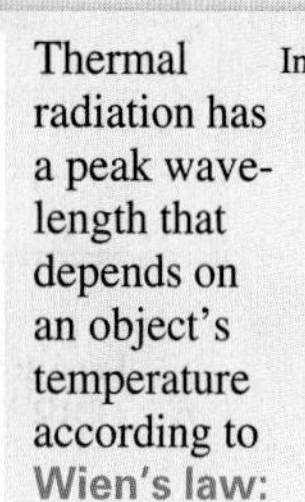

Thermal radiation has a peak wavelength that depends on an object's temperature according to **Wien's law:**

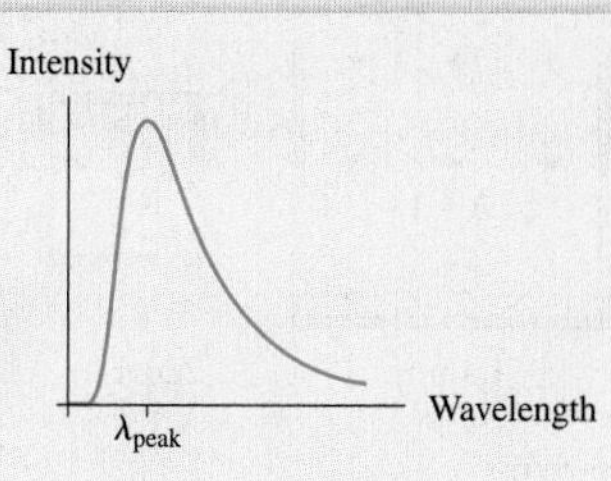

$$\lambda_{peak}\text{(in nm)} = \frac{2.9 \times 10^6 \text{ nm}\cdot\text{K}}{T}$$

Problem difficulty is labeled as | (straightforward) to ||||| (challenging). Problems labeled INT integrate significant material from earlier chapters; BIO are of biological or medical interest.

For assigned homework and other learning materials, go to MasteringPhysics®

Scan this QR code to launch a Video Tutor Solution that will help you solve problems for this chapter.

QUESTIONS

Conceptual Questions

1. The world's strongest magnet can produce a steady field of 45 T. If a circular wire loop of radius 10 cm were held in this magnetic field, what current would be induced in the loop?
2. The rapid vibration accompanying the swimming motions of mayflies has been measured by gluing a small magnet to a swimming mayfly and recording the emf in a small coil of wire placed nearby. Explain how this technique works. BIO
3. Parts a through f of Figure Q25.3 show one or more metal wires sliding on fixed metal rails in a magnetic field. For each, determine if the induced current is clockwise, counterclockwise, or zero.

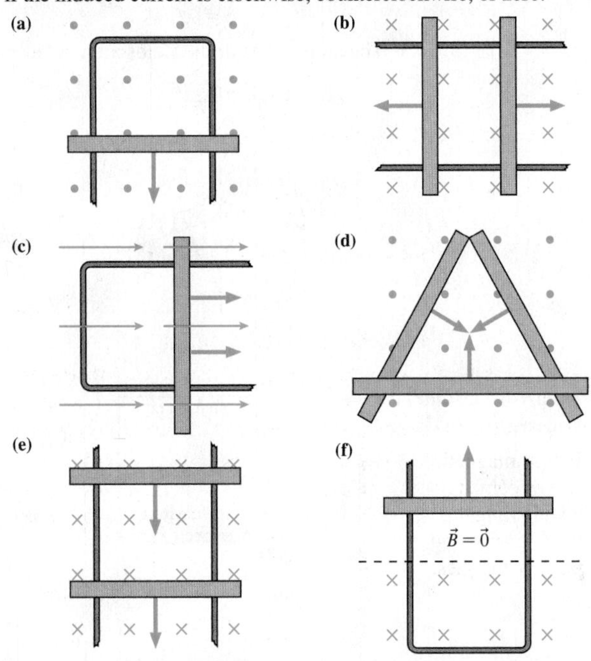

FIGURE Q25.3

4. Figure Q25.4 shows four different loops in a magnetic field. The numbers indicate the lengths of the sides and the strength of the field. Rank in order the magnetic fluxes Φ_1 through Φ_4, from the largest to the smallest. Some may be equal. Explain.

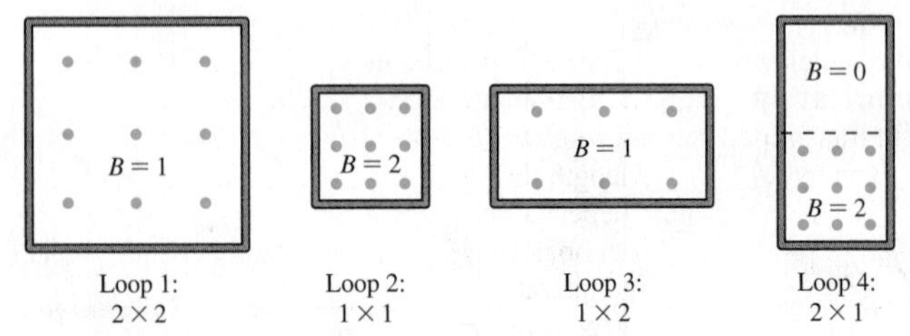

FIGURE Q25.4

5. Figure Q25.5 shows four different circular loops that are perpendicular to the page. The radius of loops 3 and 4 is twice that of loops 1 and 2. The magnetic field is the same for each. Rank in order the magnetic fluxes Φ_1 through Φ_4, from the largest to the smallest. Some may be equal. Explain.

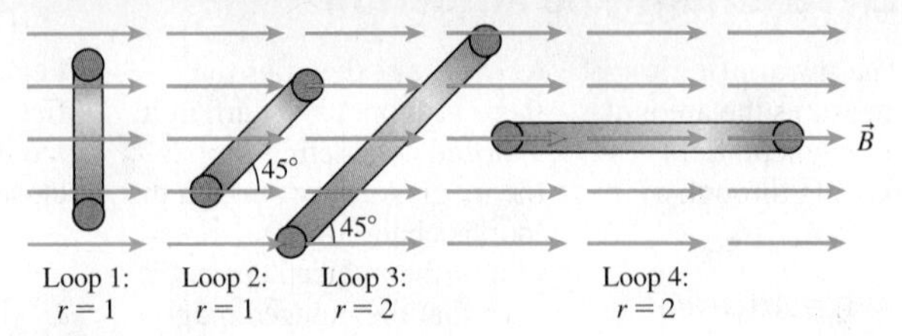

FIGURE Q25.5

6. A circular loop rotates at constant speed about an axle through the center of the loop. Figure Q25.6 shows an edge view and defines the angle ϕ, which increases from 0° to 360° as the loop rotates.
   a. At what angle or angles is the magnetic flux a maximum?
   b. At what angle or angles is the magnetic flux a minimum?
   c. At what angle or angles is the magnetic flux *changing* most rapidly?

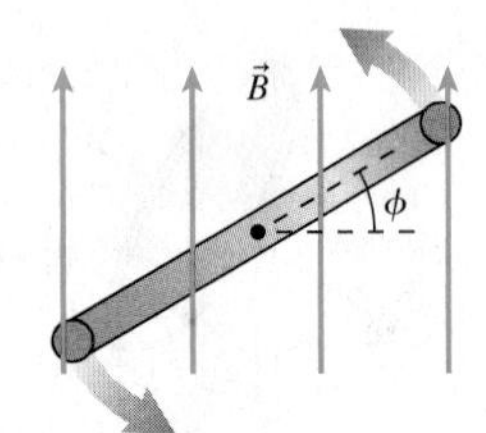

FIGURE Q25.6

7. The power lines that run through your neighborhood carry *alternating currents* that reverse direction 120 times per second. As the current changes, so does the magnetic field around a line. Suppose you wanted to put a loop of wire up near the power line to extract power by "tapping" the magnetic field. Sketch a picture of how you would orient the coil of wire next to a power line to develop the maximum emf in the coil. (Note that this is dangerous and illegal, and not something you should try.)
8. The magnetic flux passing through a coil of wire varies as shown in Figure Q25.8. During which time interval(s) will an induced current be present in the coil? Explain.

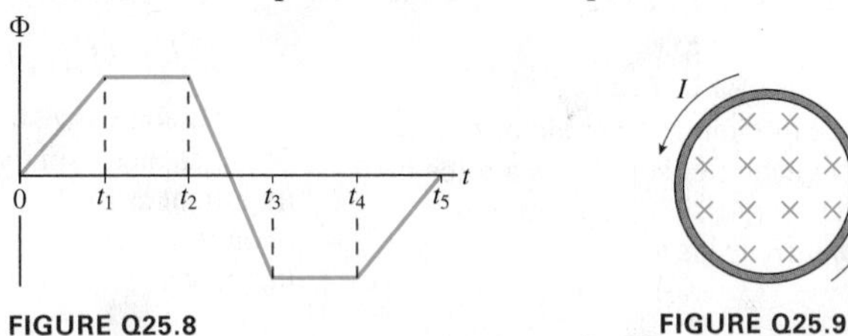

FIGURE Q25.8

FIGURE Q25.9

9. There is a counterclockwise induced current in the conducting loop shown in Figure Q25.9. Is the magnetic field inside the loop increasing in strength, decreasing in strength, or steady?
10. A magnet dropped through a clear plastic tube accelerates as expected in free fall. If dropped through an aluminum tube of exactly the same length and diameter, the magnet falls much more slowly. Explain the behavior of the second magnet.
11. The conducting loop in Figure Q25.11 is moving into the region between the magnetic poles shown.
    a. Is the induced current (viewed from above) clockwise or counterclockwise?
    b. Is there an attractive magnetic force that tends to pull the loop in, like a magnet pulls on a paper clip? Or do you need to push the loop in against a repulsive force?

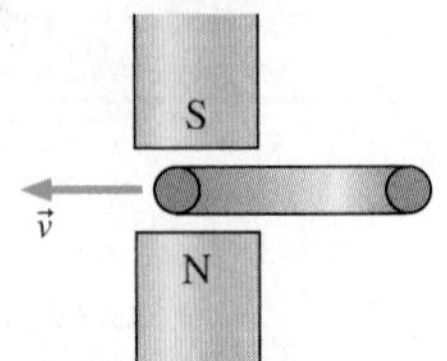

FIGURE Q25.11

12. Figure Q25.12 shows two concentric, conducting loops. We will define a counterclockwise current (viewed from above) to be positive, a clockwise current to be negative. The graph shows the current in the outer loop as a function of time. Sketch a graph that shows the induced current in the inner loop. Explain.

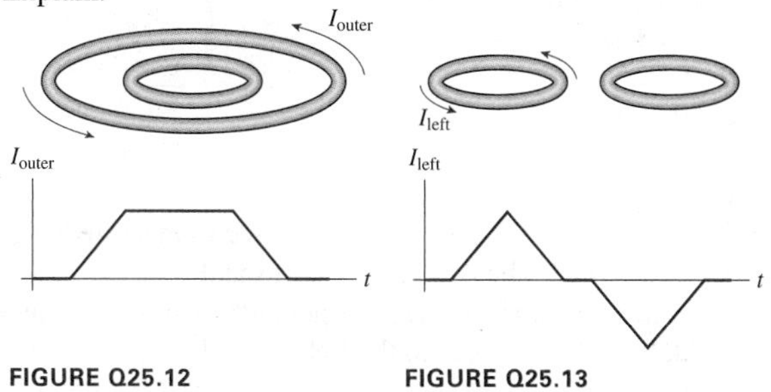

FIGURE Q25.12 **FIGURE Q25.13**

13. Figure Q25.13 shows conducting loops next to each other. We will define a counterclockwise current (viewed from above) to be positive, a clockwise current to be negative. The graph shows the current in the left loop as a function of time. Sketch a graph that shows the induced current in the right loop. Explain.
14. Two loops of wire are stacked vertically, one above the other, as shown in Figure Q25.14. Does the upper loop have a clockwise current, a counterclockwise current, or no current at the following times? Explain your reasoning.

FIGURE Q25.14

    a. Before the switch is closed
    b. Immediately after the switch is closed
    c. Long after the switch is closed
    d. Immediately after the switch is reopened
15. A loop of wire is horizontal. A bar magnet is pushed toward the loop from below, along the axis of the loop, as shown in Figure Q25.15.
    a. In what direction is the current in the loop? Explain.
    b. Is there a magnetic force on the loop? If so, in which direction? Explain.

    Hint: Recall that a current loop is a magnetic dipole.
    c. Is there a magnetic force on the magnet? If so, in which direction?

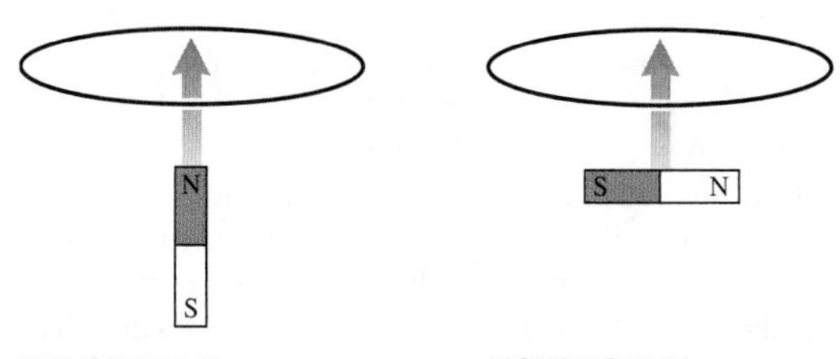

FIGURE Q25.15 **FIGURE Q25.16**

16. A bar magnet is pushed toward a loop of wire, as shown in Figure Q25.16. Is there a current in the loop? If so, in which direction? If not, why not?
17. A conducting loop around a region of strong magnetic field contains two light bulbs, as shown in Figure Q25.17. The wires connecting the bulbs are ideal. The magnetic field is increasing rapidly.
    a. Do the bulbs glow? Why or why not?
    b. If they glow, which bulb is brighter? Or are they equally bright? Explain.

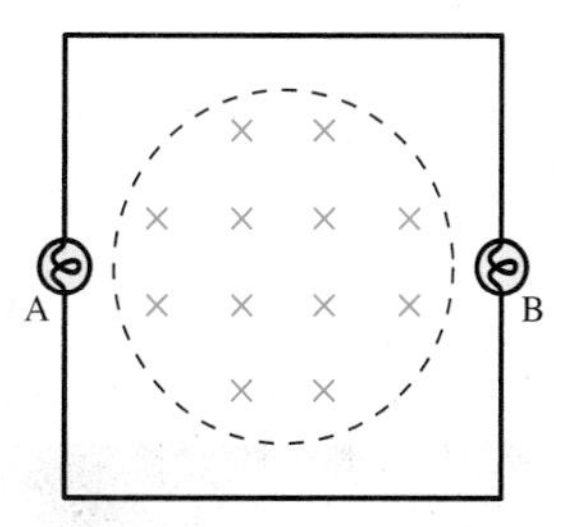

FIGURE Q25.17

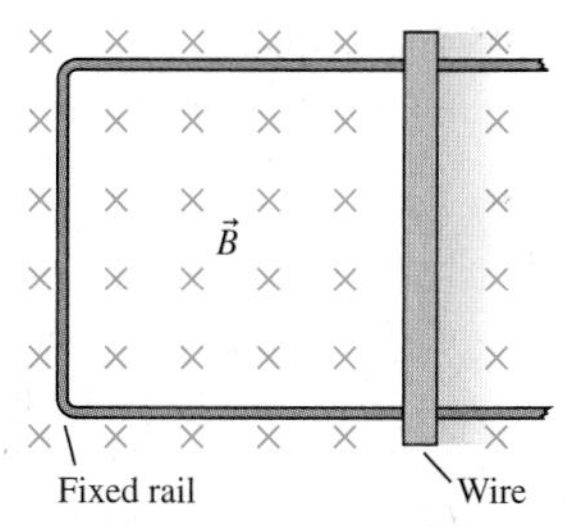

FIGURE Q25.18

18. A metal wire is resting on a U-shaped conducting rail, as shown in Figure Q25.18. The rail is fixed in position, but the wire is free to move.
    a. If the magnetic field is increasing in strength, which way does the wire move?
    b. If the magnetic field is decreasing in strength, which way does the wire move?
19. Although sunlight is unpolarized, the light that reflects from smooth surfaces may be partially polarized in the direction parallel to the plane of the reflecting surface. How should the axis of the polarizers in sunglasses be oriented—vertically or horizontally—to reduce the glare from a horizontal surface such as a road or a lake?
20. Old-fashioned roof-mounted television antennas were designed to pick up signals across a broad frequency range. Explain why these antennas had metal bars of many different lengths.

21. An AM radio detects the oscillating magnetic field of the radio wave with an antenna consisting of a coil of wire wrapped around a ferrite bar, as shown in Figure Q25.21. Ferrite is a magnetic material that "amplifies" the magnetic field of the wave. The radio antenna broadcasts waves with the electric field vertical, the magnetic field horizontal. If a radio antenna is located due north of you, how must the ferrite bar be oriented for best reception?

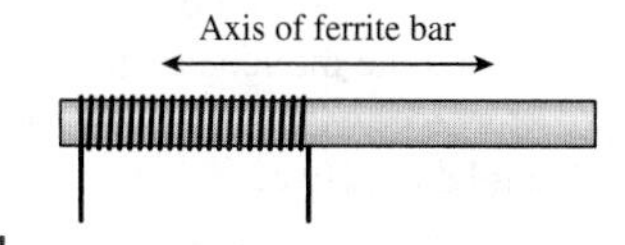

FIGURE Q25.21

22. Three laser beams have wavelengths $\lambda_1 = 400$ nm, $\lambda_2 = 600$ nm, and $\lambda_3 = 800$ nm. The power of each laser beam is 1 W.
    a. Rank in order, from largest to smallest, the photon energies E_1, E_2, and E_3 in these three laser beams. Explain.
    b. Rank in order, from largest to smallest, the number of photons per second N_1, N_2, and N_3 delivered by the three laser beams. Explain.
23. The intensity of a beam of light is increased but the light's frequency is unchanged. As a result, which of the following (perhaps more than one) are true? Explain.
    A. The photons travel faster.
    B. Each photon has more energy.
    C. The photons are larger.
    D. There are more photons per second.

24. The frequency of a beam of light is increased but the light's intensity is unchanged. As a result, which of the following (perhaps more than one) are true? Explain.
    A. The photons travel faster.
    B. Each photon has more energy.
    C. There are fewer photons per second.
    D. There are more photons per second.
25. Arc welding uses electric current to make an extremely hot electric arc that can melt metal. The arc emits ultraviolet light that can cause sunburn and eye damage if a welder is not wearing protective gear. Why does the arc give off ultraviolet light?

Multiple-Choice Questions

26. | A circular loop of wire has an area of 0.30 m^2. It is tilted by 45° with respect to a uniform 0.40 T magnetic field. What is the magnetic flux through the loop?
    A. 0.085 T · m^2
    B. 0.12 T · m^2
    C. 0.38 T · m^2
    D. 0.75 T · m^2
    E. 1.3 T · m^2
27. | In Figure Q25.27, a square loop is rotating in the plane of the page around an axis through its center. A uniform magnetic field is directed into the page. What is the direction of the induced current in the loop?
    A. Clockwise.
    B. Counterclockwise.
    C. There is no induced current.

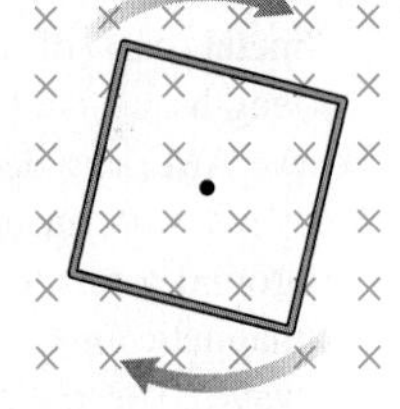

FIGURE Q25.27

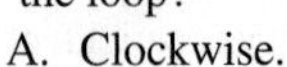

28. | A diamond-shaped loop of wire is pulled at a constant velocity through a region where the magnetic field is directed into the paper in the left half and is zero in the right half, as shown in Figure Q25.28. As the loop moves from left to right, which graph best represents the induced current in the loop as a function of time? Let a clockwise current be positive and a counterclockwise current be negative.

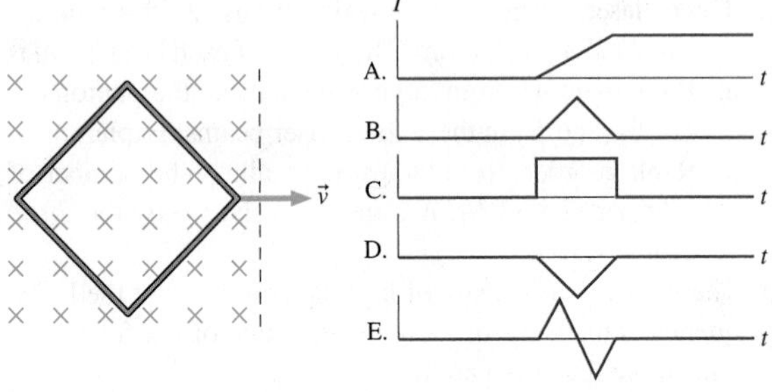

FIGURE Q25.28

29. || Figure Q25.29 shows a triangular loop of wire in a uniform magnetic field. If the field strength changes from 0.30 to 0.10 T in 50 ms, what is the induced emf in the loop?
    A. 0.08 V
    B. 0.12 V
    C. 0.16 V
    D. 0.24 V
    E. 0.36 V

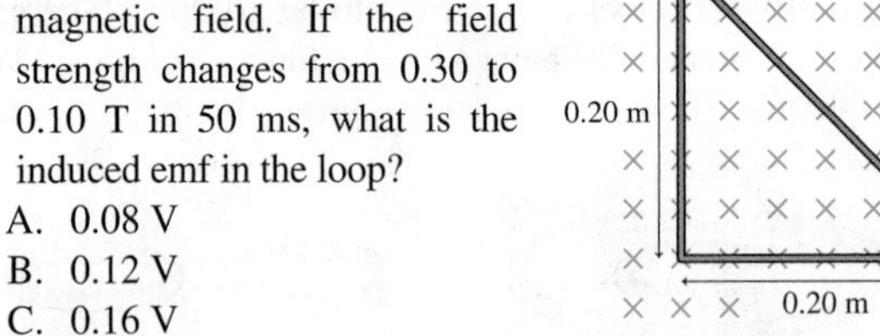

FIGURE Q25.29

30. || A device called a *flip coil* can be used to measure the earth's magnetic field. The coil has 100 turns and an area of 0.010 m^2. It is oriented with its plane perpendicular to the earth's magnetic field, then flipped 180° so the field goes through the coil in the opposite direction. The earth's magnetic field is 0.050 mT, and the coil flips over in 0.50 s. What is the average emf induced in the coil during the flip?
    A. 0.050 mV B. 0.10 mV
    C. 0.20 mV D. 1.0 mV
31. | The electromagnetic waves that carry FM radio range in frequency from 87.9 MHz to 107.9 MHz. What is the range of wavelengths of these radio waves?
    A. 500–750 nm B. 0.87–91.08 m
    C. 2.78–3.41 m D. 278–341 m
    E. 234–410 km
32. || The beam from a laser is focused with a lens, reducing the area of the beam by a factor of 2. By what factor does the amplitude of the electric field increase?
    A. The amplitude does not change.
    B. The amplitude increases by a factor of $\sqrt{2}$.
    C. The amplitude increases by a factor of 2.
    D. The amplitude increases by a factor of $2\sqrt{2}$.
    E. The amplitude increases by a factor of 4.
33. | A spacecraft in orbit around the moon measures its altitude by reflecting a pulsed 10 MHz radio signal from the surface. If the spacecraft is 10 km high, what is the time between the emission of the pulse and the detection of the echo?
    A. 33 ns B. 67 ns
    C. 33 μs D. 67 μs
34. | A 6.0 mW vertically polarized laser beam passes through a polarizing filter whose axis is 75° from vertical. What is the laser-beam power after passing through the filter?
    A. 0.40 mW B. 1.0 mW
    C. 1.6 mW D. 5.6 mW
35. | Communication with submerged submarines via radio waves is difficult because seawater is conductive and absorbs electromagnetic waves. Penetration into the ocean is greater at longer wavelengths, so the United States has radio installations that transmit at 76 Hz for submarine communications. What is the approximate wavelength of those extremely low-frequency waves?
    A. 500 km B. 1000 km
    C. 2000 km D. 4000 km
36. || How many photons are emitted during 5.0 s of operation of a red laser pointer? The device outputs 2.8 mW at a 635 nm wavelength.
    A. 4.5×10^{10} B. 4.5×10^{11}
    C. 4.5×10^{15} D. 4.5×10^{16}

PROBLEMS

Section 25.1 Induced Currents

Section 25.2 Motional emf

1. | A potential difference of 0.050 V is developed across the 10-cm-long wire in Figure P25.1 as it moves through a magnetic field at 5.0 m/s. The magnetic field is perpendicular to the axis of the wire. What are the direction and strength of the magnetic field?

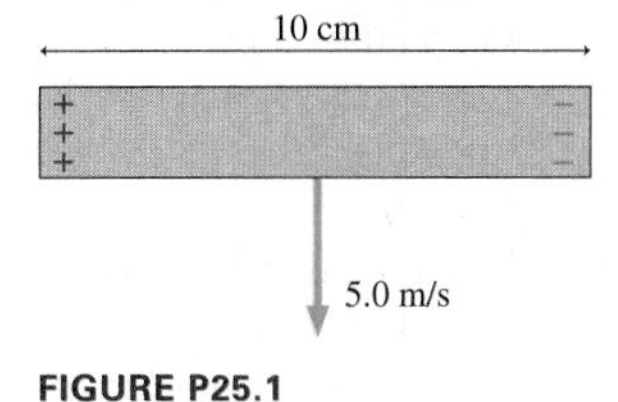

FIGURE P25.1

2. || A scalloped hammerhead shark swims at a steady speed of 1.5 m/s with its 85-cm-wide head perpendicular to the earth's 50 μT magnetic field. What is the magnitude of the emf induced between the two sides of the shark's head? BIO
3. || A 10-cm-long wire is pulled along a U-shaped conducting rail in a perpendicular magnetic field. The total resistance of the wire and rail is 0.20 Ω. Pulling the wire with a force of 1.0 N causes 4.0 W of power to be dissipated in the circuit.
   a. What is the speed of the wire when pulled with a force of 1.0 N?
   b. What is the strength of the magnetic field?
4. | Figure P25.4 shows a 15-cm-long metal rod pulled along two frictionless, conducting rails at a constant speed of 3.5 m/s. The rails have negligible resistance, but the rod has a resistance of 0.65 Ω.
   a. What is the current induced in the rod?
   b. What force is required to keep the rod moving at a constant speed?

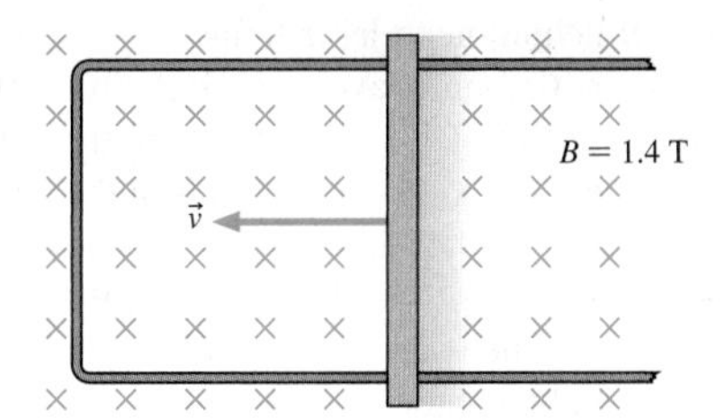

FIGURE P25.4

5. ||| A 50 g horizontal metal bar, 12 cm long, is free to slide up and down between two tall, vertical metal rods that are 12 cm apart. A 0.060 T magnetic field is directed perpendicular to the plane of the rods. The bar is raised to near the top of the rods, and a 1.0 Ω resistor is connected across the two rods at the top. Then the bar is dropped. What is the terminal speed at which the bar falls? Assume the bar remains horizontal and in contact with the rods at all times.
6. || In the rainy season, the Amazon flows fast and runs deep. In one location, the river is 23 m deep and moves at a speed of 4.0 m/s toward the east. The earth's 50 μT magnetic field is parallel to the ground and directed northward. If the bottom of the river is at 0 V, what is the potential (magnitude and sign) at the surface?
7. || A delivery truck with 2.8-m-high aluminum sides is driving west at 75 km/h in a region where the earth's magnetic field is $\vec{B} = (5.0 \times 10^{-5}$ T, north).
   a. What is the potential difference between the top and the bottom of the truck's side panels?
   b. Will the tops of the panels be positive or negative relative to the bottoms?
8. ||| Michael Faraday unsuccessfully attempted to measure the potential difference across the River Thames due to the flow of water (which is a good conductor due to dissolved salts) in the earth's field. At Waterloo Bridge, where Faraday made his attempt, the river is 200 m wide and flows eastward at 2.0 m/s. The earth's magnetic field here has a magnitude of 50 μT and is tipped 65° from the horizontal, so it has both horizontal (north) and vertical (down) components. Is the north or south riverbank at a higher potential? What is the magnitude of the potential difference?

Section 25.3 Magnetic Flux

9. | Figure P25.9 is an edge-on view of a 10-cm-diameter circular loop rotating in a uniform 0.050 T magnetic field. What is the magnetic flux through the loop when θ is 0°, 30°, 60°, and 90°?

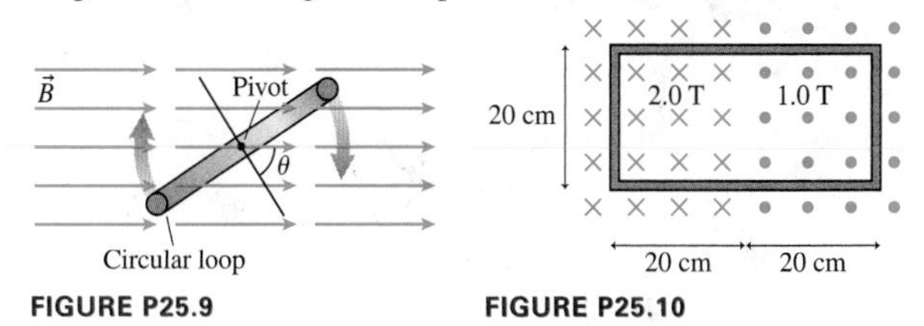

FIGURE P25.9 FIGURE P25.10

10. || What is the magnetic flux through the loop shown in Figure P25.10?
11. ||| The 2.0-cm-diameter solenoid in Figure P25.11 passes through the center of a 6.0-cm-diameter loop. The magnetic field inside the solenoid is 0.20 T. What is the magnetic flux through the loop (a) when it is perpendicular to the solenoid and (b) when it is tilted at a 60° angle?

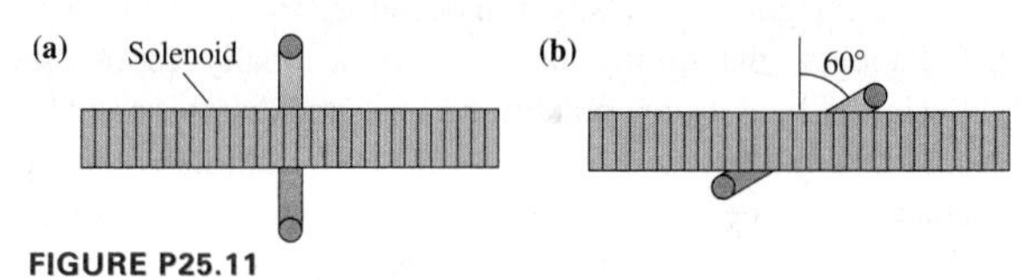

FIGURE P25.11

12. || At a typical location in the United States, the earth's magnetic field has a magnitude of 5.0×10^{-5} T and is at a 65° angle from the horizontal. What is the flux through the 22 cm $\times$ 28 cm front cover of your textbook if it is flat on your desk?
13. | The metal equilateral triangle in Figure P25.13, 20 cm on each side, is halfway into a 0.10 T magnetic field.
    a. What is the magnetic flux through the triangle?
    b. If the magnetic field strength decreases, what is the direction of the induced current in the triangle?

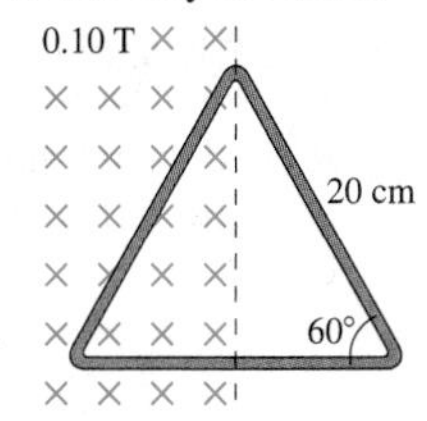

FIGURE P25.13

Section 25.4 Faraday's Law

14. || A magnet and a coil are oriented as shown in Figure P25.14. The magnet is moved rapidly into the coil, held stationary in the coil for a short time, and then rapidly pulled back out of the coil. Sketch a graph showing the reading of the ammeter as a function of time. The ammeter registers a positive value when current goes into the "+" terminal.

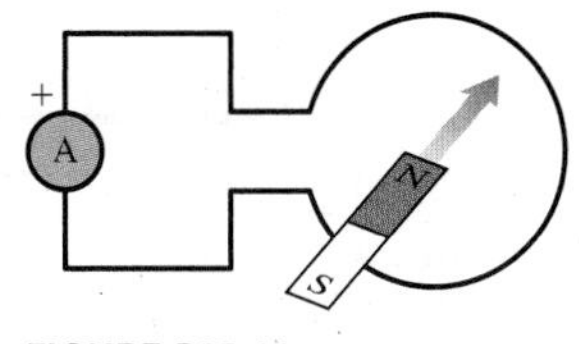

FIGURE P25.14

Watch video solution for problem: 25.19

15. ||| A 1000-turn coil of wire 2.0 cm in diameter is in a magnetic field that drops from 0.10 T to 0 T in 10 ms. The axis of the coil is parallel to the field. What is the emf of the coil?
16. | Figure P25.16 shows a 100-turn coil of wire of radius 12 cm in a 0.15 T magnetic field. The coil is rotated 90° in 0.30 s, ending up parallel to the field. What is the average emf induced in the coil as it rotates?

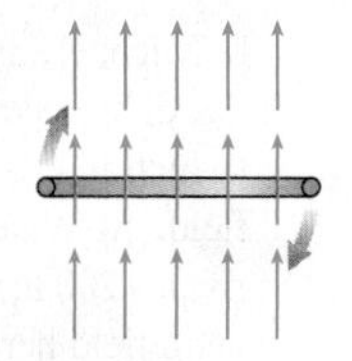
FIGURE P25.16

17. | Figure P25.17 shows a 10-cm-diameter loop in three different magnetic fields. The loop's resistance is 0.10 Ω. For each case, determine the induced emf, the induced current, and the direction of the current.

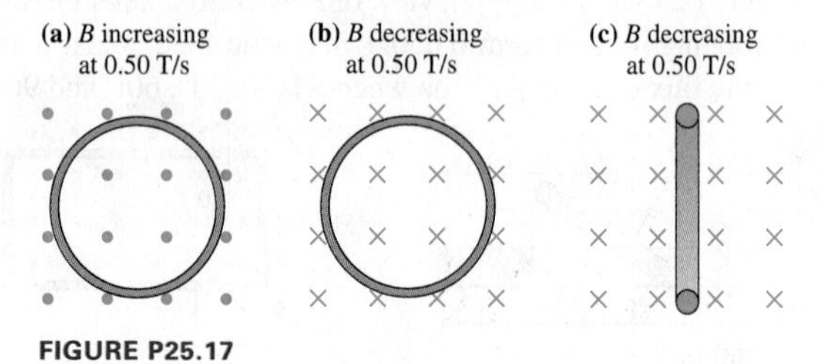

FIGURE P25.17

18. || The plane of a loop of wire is perpendicular to a magnetic field. Rank, from greatest to least, the magnitudes of the loop's induced emf for the following situations:
    A. The magnetic field strength increases from 0 to 1 T in 6 s.
    B. The magnetic field strength increases from 1 T to 4 T in 2 s.
    C. The magnetic field strength remains at 4 T for 1 min.
    D. The magnetic field strength decreases from 4 T to 3 T in 4 s.
    E. The magnetic field strength decreases from 3 T to 0 T in 1 s.
19. || Patients undergoing an MRI occasionally report seeing BIO flashes of light. Some practitioners assume that this results from electric stimulation of the eye by the emf induced by the rapidly changing fields of an MRI solenoid. We can do a quick calculation to see if this is a reasonable assumption. The human eyeball has a diameter of approximately 25 mm. Rapid changes in current in an MRI solenoid can produce rapid changes in field, with $\Delta B/\Delta t$ as large as 50 T/s. What emf would this induce in a loop circling the eyeball? How does this compare to the 15 mV necessary to trigger an action potential?
20. || The loop in Figure P25.20 has an induced current as shown. The loop has a resistance of 0.10 Ω. Is the magnetic field strength increasing or decreasing? What is the rate of change of the field, $\Delta B/\Delta t$?

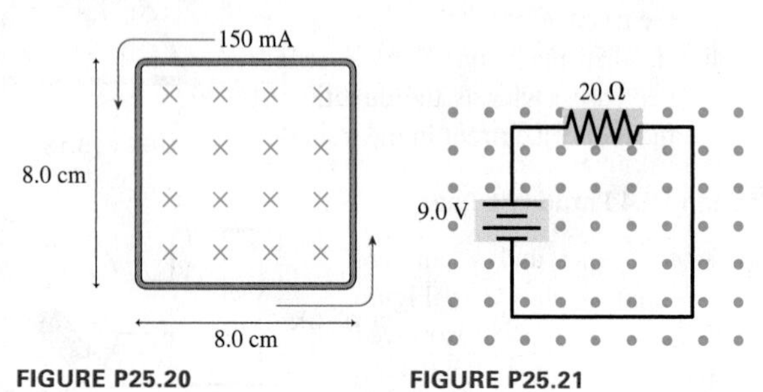

FIGURE P25.20 FIGURE P25.21

21. || The circuit of Figure P25.21 is a square 20 cm on a side. The magnetic field increases steadily from 0 T to 0.50 T in 10 ms. What is the current in the resistor during this time?
22. ||| A 5.0-cm-diameter loop of wire has resistance 1.2 Ω. A nearby solenoid generates a uniform magnetic field along the axis of the loop that varies with time as shown in Figure P25.22. Graph the magnitude of the current in the loop over the same time interval.

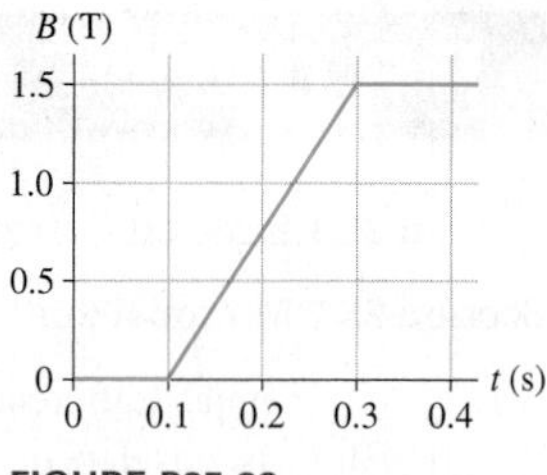

FIGURE P25.22

23. || A 20-cm-circumference loop of wire has a resistance of 0.12 Ω. The loop is placed between the poles of an electromagnet, and a field of 0.55 T is switched on in a time of 15 ms. What is the induced current in the loop?
24. || The magnetic field at the earth's surface can vary in response to solar activity. During one intense solar storm, the vertical component of the magnetic field changed by 2.8 μT per minute, causing voltage spikes in large loops of the power grid that knocked out power in parts of Canada. What emf is induced in a square 100 km on a side by this rate of change of field?

Section 25.5 Electromagnetic Waves

25. | What is the electric field amplitude of an electromagnetic wave whose magnetic field amplitude is 2.0 mT?
26. | What is the magnetic field amplitude of an electromagnetic wave whose electric field amplitude is 10 V/m?
27. || A microwave oven operates at 2.4 GHz with an intensity inside the oven of 2500 W/m^2. What are the amplitudes of the oscillating electric and magnetic fields?
28. || The maximum allowed leakage of microwave radiation from a microwave oven is 5.0 mW/cm^2. If microwave radiation outside an oven has the maximum value, what is the amplitude of the oscillating electric field?
29. ||| A typical helium-neon laser found in supermarket checkout scanners emits 633-nm-wavelength light in a 1.0-mm-diameter beam with a power of 1.0 mW. What are the amplitudes of the oscillating electric and magnetic fields in the laser beam?
30. | The magnetic field of an electromagnetic wave in a vacuum is $B_z = (3.0\ \mu\text{T})\sin((1.0 \times 10^7)x - 2\pi ft)$, where x is in m and t is in s. What are the wave's (a) wavelength, (b) frequency, and (c) electric field amplitude?
31. || At what distance from a 10 mW point source of electromagnetic waves is the electric field amplitude 0.010 V/m?
32. || A radio receiver can detect signals with electric field amplitudes as small as 300 μV/m. What is the intensity of the smallest detectable signal?
33. ||| A radio antenna broadcasts a 1.0 MHz radio wave with 25 kW of power. Assume that the radiation is emitted uniformly in all directions.
    a. What is the wave's intensity 30 km from the antenna?
    b. What is the electric field amplitude at this distance?
34. || A 200 MW laser pulse is focused with a lens to a diameter of INT 2.0 μm. What is the laser beam's electric field amplitude at the focal point?
35. | The intensity of a polarized electromagnetic wave is 10 W/m^2. What will be the intensity after passing through a polarizing filter whose axis makes the following angles with the plane of polarization? (a) $\theta = 0°$, (b) $\theta = 30°$, (c) $\theta = 45°$, (d) $\theta = 60°$, (e) $\theta = 90°$.
36. || A low-power college radio station broadcasts 10 W of electromagnetic waves. At what distance from the antenna is the electric field amplitude 2.0×10^{-3} V/m, the lower limit at which good reception is possible?

Watch video solution for problem: 25.49 

37. || Only 25% of the intensity of a polarized light wave passes through a polarizing filter. What is the angle between the electric field and the axis of the filter?
38. || A 200 mW horizontally polarized laser beam passes through a polarizing filter whose axis is 25° from vertical. What is the power of the laser beam as it emerges from the filter?
39. || The polarization of a helium-neon laser can change with time. The light from a 1.5 mW laser is initially horizontally polarized; as the laser warms up, the light changes to be vertically polarized. Suppose the laser beam passes through a polarizer whose axis is 30° from horizontal. By what percent does the light intensity transmitted through the polarizer decrease as the laser warms up?

Section 25.6 The Photon Model of Electromagnetic Waves

40. | What is the energy (in eV) of a photon of visible light that has a wavelength of 500 nm?
41. | What is the energy (in eV) of an x-ray photon that has a wavelength of 1.0 nm?
42. | What is the wavelength of a photon whose energy is twice that of a photon with a 600 nm wavelength?
43. || One recent study has shown that x rays with a wavelength of BIO 0.0050 nm can produce mutations in human cells.
    a. Calculate the energy in eV of a photon of radiation with this wavelength.
    b. Assuming that the bond energy holding together a water molecule is typical, use Table 25.1 to estimate how many molecular bonds could be broken with this energy.
44. | Rod cells in the retina of the eye detect light using a pho- BIO topigment called rhodopsin. 1.8 eV is the lowest photon energy that can trigger a response in rhodopsin. What is the maximum wavelength of electromagnetic radiation that can cause a transition? In what part of the spectrum is this?
45. || The thermal emission of the human body has maximum BIO intensity at a wavelength of approximately 9.5 μm. What photon energy corresponds to this wavelength?
46. | Extremely high-energy photons of 2.0×10^{13} eV are occasionally observed from distant astronomical sources. What are the corresponding wavelength and frequency?
47. || The intensity of electromagnetic radiation from the sun reaching the earth's upper atmosphere is 1.37 kW/m^2. Assuming an average wavelength of 680 nm for this radiation, find the number of photons per second that strike a 1.00 m^2 solar panel directly facing the sun on an orbiting satellite.
48. || A 193-nm-wavelength UV laser for eye surgery emits a BIO 0.500 mJ pulse. How many photons does the light pulse contain?
49. ||| The human eye can barely detect a star whose intensity at the BIO earth's surface is 1.6×10^{-11} W/m^2. If the dark-adapted eye has a pupil diameter of 7.0 mm, how many photons per second enter the eye from the star? Assume the starlight has a wavelength of 550 nm.
50. || A particular species of copepod, a small marine crustacean, BIO emits a flash of light consisting of 1.4×10^{10} photons at a wavelength of 490 nm. The flash lasts for 2.4 s. What is the power of the flash?

Section 25.7 The Electromagnetic Spectrum

51. || The spectrum of a glowing filament has its peak at a wavelength of 1200 nm. What is the temperature of the filament, in °C?
52. | Our sun's 5800 K surface temperature gives a peak wavelength in the middle of the visible spectrum. What is the minimum surface temperature for a star whose emission peaks at some wavelength less than 400 nm—that is, in the ultraviolet?
53. ||| While using a dimmer switch to investigate a new type of incandescent light bulb, you notice that the light changes both its spectral characteristics and its brightness as the voltage is increased.
    a. If the wavelength of maximum intensity decreases from 1800 nm to 1600 nm as the bulb's voltage is increased, by how many °C does the filament temperature increase?
    b. By what factor does the total radiation from the filament increase due to this temperature change?
54. || The star Sirius is much hotter than the sun, with a peak wavelength of 290 nm compared to the sun's 500 nm. It is also larger, with a diameter 1.7 times that of the sun. By what factor does the energy emitted by Sirius exceed that of the sun?
55. | The photon energies used in different types of medical x-ray BIO imaging vary widely, depending upon the application. Single dental x rays use photons with energies of about 25 keV. The photon energy used for x-ray microtomography, a process that allows repeated imaging in single planes at varying depths within the sample, is 2.5 times greater. What are the wavelengths of the x rays used for these two purposes?
56. | A python can detect thermal radiation with intensity greater BIO than 0.60 W/m^2. A typical human body has a surface area of 1.8 m^2, a surface temperature of 30°C, and an emissivity $e = 0.97$ at infrared wavelengths. What is the maximum distance from which a python can detect your presence? You can model the human body as a point source of radiation.
57. | If astronomers look toward any point in outer space, they see radiation that matches the emission spectrum of an object at 2.7 K, a remnant of the Big Bang. What is the peak wavelength of this radiation? What part of the electromagnetic spectrum is it in?

General Problems

58. ||| A 100-turn, 2.0-cm-diameter coil is at rest in a horizontal plane. A uniform magnetic field 60° away from vertical increases from 0.50 T to 1.50 T in 0.60 s. What is the induced emf in the coil?
59. |||| A 25-turn, 10.0-cm-diameter coil is oriented in a vertical plane with its axis aligned east-west. A magnetic field pointing to the northeast decreases from 0.80 T to 0.20 T in 2.0 s. What is the emf induced in the coil?
60. |||| People immersed in strong unchanging magnetic fields occa- BIO sionally report sensing a metallic taste. Some investigators suspect that motion in the constant field could produce a changing flux and a resulting emf that could stimulate nerves in the tongue. We can make a simple model to see if this is reasonable by imagining a somewhat extreme case. Suppose a patient having an MRI is immersed in a 3.0 T field along the axis of his body. He then quickly tips his head to the side, toward his right shoulder, tipping his head by 30° in the rather short time of 0.15 s. Estimate the area of the tongue; then calculate the emf that could be induced in a loop around the outside of the tongue by this motion of the head. How does this emf compare to the approximately 15 mV necessary to trigger an action potential? Does it seem reasonable to suppose that an induced emf is responsible for the noted effect?

Watch video solutions for problems: 25.65 25.67

61. |||| Currents induced by rapid field changes in an MRI solenoid can, in some cases, heat tissues in the body, but under normal circumstances the heating is small. We can do a quick estimate to show this. Consider the "loop" of muscle tissue shown in Figure P25.61. This might be muscle circling the bone of your arm or leg. Muscle tissue is not a great conductor, but current will pass through muscle and so we can consider this a conducting loop with a rather high resistance. Suppose the magnetic field along the axis of the loop drops from 1.6 T to 0 T in 0.30 s, as it might in an MRI solenoid. BIO INT
   a. How much energy is dissipated in the loop?
   b. By how much will the temperature of the tissue increase?

   Assume that muscle tissue has resistivity 13 $\Omega \cdot$ m, density 1.1×10^3 kg/m^3, and specific heat 3600 J/kg $\cdot$ K.

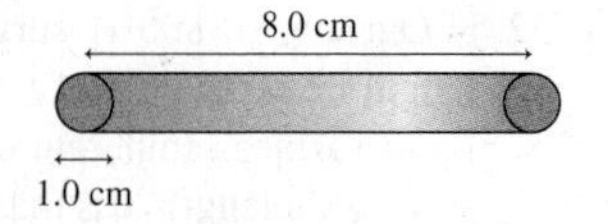

FIGURE P25.61

62. || The loop in Figure P25.62 is being pushed into the 0.20 T magnetic field at 50 m/s. The resistance of the loop is 0.10 Ω. What are the direction and magnitude of the current in the loop?

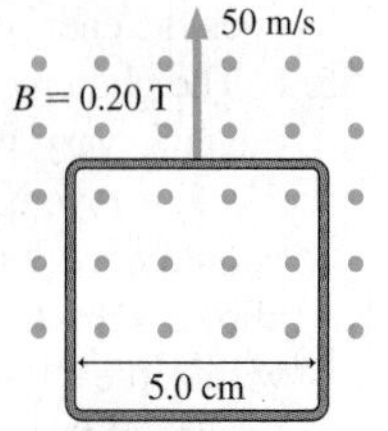

FIGURE P25.62

63. ||| A 20-cm-long, zero-resistance wire is pulled outward, on zero-resistance rails, at a steady speed of 10 m/s in a 0.10 T magnetic field. (See Figure P25.63.) On the opposite side, a 1.0 Ω carbon resistor completes the circuit by connecting the two rails. The mass of the resistor is 50 mg. INT
   a. What is the induced current in the circuit?
   b. How much force is needed to pull the wire at this speed?
   c. How much does the temperature of the carbon increase if the wire is pulled for 10 s? The specific heat of carbon is 710 J/kg $\cdot$ K. Neglect thermal energy transfer out of the resistor.

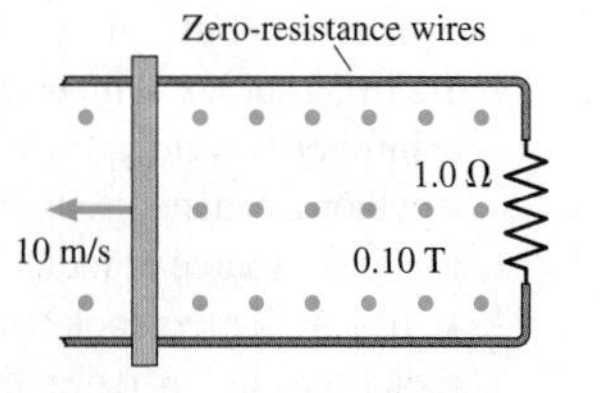

FIGURE P25.63

64. |||| A TMS (transcranial magnetic stimulation) device creates very rapidly changing magnetic fields. The field near a typical pulsed-field machine rises from 0 T to 2.5 T in 200 μs. Suppose a technician holds his hand near the device so that the axis of his 2.0-cm-diameter wedding band is parallel to the field. BIO INT
   a. What emf is induced in the ring as the field changes?
   b. If the band is made of a gold alloy with resistivity 6.2×10^{-8} $\Omega \cdot$ m and has a cross-section area of 4.0 mm^2, what is the induced current?

65. || The 10-cm-wide, zero-resistance wire shown in Figure P25.65 is pushed toward the 2.0 Ω resistor at a steady speed of 0.50 m/s. The magnetic field strength is 0.50 T.
   a. What is the magnitude of the pushing force?
   b. How much power does the pushing force supply to the wire?
   c. What are the direction and magnitude of the induced current?
   d. How much power is dissipated in the resistor?

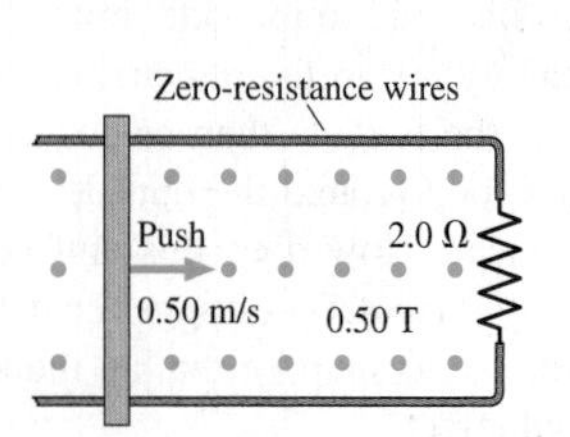

FIGURE P25.65

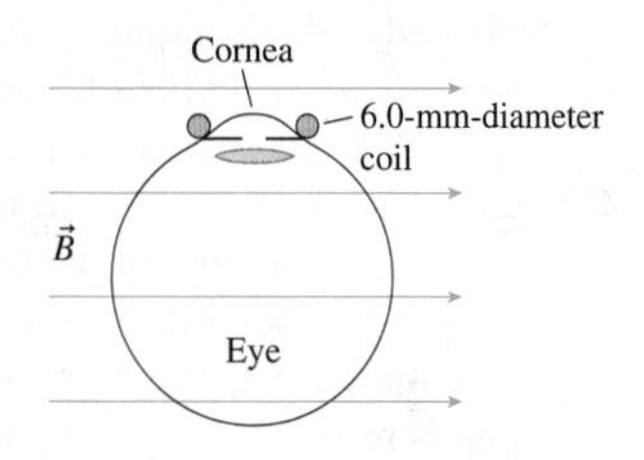

FIGURE Q25.66

66. |||| Experiments to study vision often need to track the movements of a subject's eye. One way of doing so is to have the subject sit in a magnetic field while wearing special contact lenses that have a coil of very fine wire circling the edge. A current is induced in the coil each time the subject rotates his eye. Consider an experiment in which a 20-turn, 6.0-mm-diameter coil of wire circles the subject's cornea while a 1.0 T magnetic field is directed as shown in Figure P25.66. The subject begins by looking straight ahead. What emf is induced in the coil if the subject shifts his gaze by 5.0° in 0.20 s? BIO

67. || A LASIK vision correction system uses a laser that emits 10-ns-long pulses of light, each with 2.5 mJ of energy. The laser is focused to a 0.85-mm-diameter circle. (a) What is the average power of each laser pulse? (b) What is the electric field strength of the laser light at the focus point? BIO INT

68. | When the Voyager 2 spacecraft passed Neptune in 1989, it was 4.5×10^9 km from the earth. Its radio transmitter, with which it sent back data and images, broadcast with a mere 21 W of power. Assuming that the transmitter broadcast equally in all directions,
   a. What signal intensity was received on the earth?
   b. What electric field amplitude was detected?

   (The received signal was slightly stronger than your result because the spacecraft used a directional antenna.)

69. || A new cordless phone emits 4.0 mW at 5.8 GHz. The manufacturer claims that the phone has a range of 100 feet. If we assume that the wave spreads out evenly with no obstructions, what is the electric field strength at the base unit 100 feet from the phone?

70. || In reading the instruction manual that came with your garage-door opener, you see that the transmitter unit in your car produces a 250 mW signal and that the receiver unit is supposed to respond to a radio wave of the correct frequency if the electric field amplitude exceeds 0.10 V/m. You wonder if this is really true. To find out, you put fresh batteries in the transmitter and start walking away from your garage while opening and closing the door. Your garage door finally fails to respond when you're 42 m away. Are the manufacturer's claims true?

71. ||| Unpolarized light passes through a vertical polarizing filter, emerging with an intensity I_0. The light then passes through a horizontal filter, which blocks all of the light; the intensity transmitted through the pair of filters is zero. Suppose a third polarizer with axis 45° from vertical is inserted between the first two. What is the transmitted intensity now?

72. ||| A light-emitting diode (LED) connected to a 3.0 V power supply emits 440 nm blue light. The current in the LED is 10 mA, and the LED is 60% efficient at converting electric power input into light power output. How many photons per second does the LED emit? INT

73. | A 1000 kHz AM radio station broadcasts with a power of 20 kW. How many photons does the transmitting antenna emit each second?

74. |||| The human body has a surface area of approximately 1.8 m^2, a surface temperature of approximately 30°C, and a typical emissivity at infrared wavelengths of $e = 0.97$. If we make the approximation that all photons are emitted at the wavelength of peak intensity, how many photons per second does the body emit? BIO

MCAT-Style Passage Problems

Electromagnetic Wave Penetration BIO

Radio waves and microwaves are used in therapy to provide "deep heating" of tissue because the waves penetrate beneath the surface of the body and deposit energy. We define the *penetration depth* as the depth at which the wave intensity has decreased to 37% of its value at the surface. The penetration depth is 15 cm for 27 MHz radio waves. For radio frequencies such as this, the penetration depth is proportional to $\sqrt{\lambda}$, the square root of the wavelength.

75. | What is the wavelength of 27 MHz radio waves?
 A. 11 m B. 9.0 m C. 0.011 m D. 0.009 m
76. | If the frequency of the radio waves is increased, the depth of penetration
 A. Increases. B. Does not change. C. Decreases.
77. | For 27 MHz radio waves, the wave intensity has been reduced by a factor of 3 at a depth of approximately 15 cm. At this point in the tissue, the electric field amplitude has decreased by a factor of
 A. 9 B. $3\sqrt{3}$ C. 3 D. $\sqrt{3}$

The Metal Detector

Metal detectors use induced currents to sense the presence of any metal—not just magnetic materials such as iron. A metal detector, shown in Figure P25.78, consists of two coils: a transmitter coil and a receiver coil. A high-frequency oscillating current in the transmitter coil generates an oscillating magnetic field along the axis and a changing flux through the receiver coil. Consequently, there is an oscillating induced current in the receiver coil.

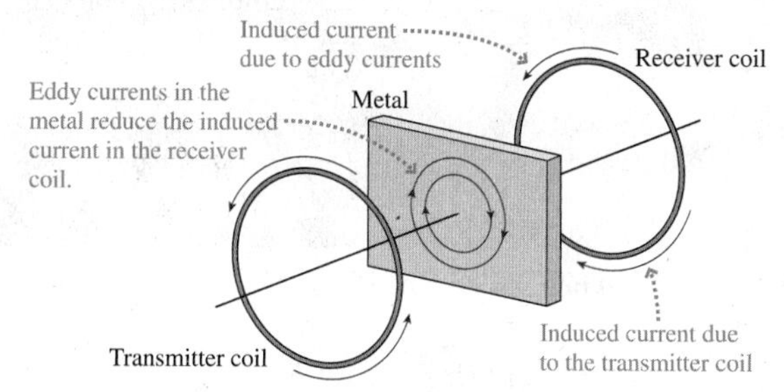

FIGURE P25.78

If a piece of metal is placed between the transmitter and the receiver, the oscillating magnetic field in the metal induces eddy currents in a plane parallel to the transmitter and receiver coils. The receiver coil then responds to the superposition of the transmitter's magnetic field and the magnetic field of the eddy currents. Because the eddy currents attempt to prevent the flux from changing, in accordance with Lenz's law, the net field at the receiver decreases when a piece of metal is inserted between the coils. Electronic circuits detect the current decrease in the receiver coil and set off an alarm.

78. | The metal detector will not detect insulators because
 A. Insulators block magnetic fields.
 B. No eddy current can be produced in an insulator.
 C. No emf can be produced in an insulator.
 D. An insulator will increase the field at the receiver.
79. | A metal detector can detect the presence of metal screws used to repair a broken bone inside the body. This tells us that
 A. The screws are made of magnetic materials.
 B. The tissues of the body are conducting.
 C. The magnetic fields of the device can penetrate the tissues of the body.
 D. The screws must be perfectly aligned with the axis of the device.
80. | Which of the following changes would *not* produce a larger eddy current in the metal?
 A. Increasing the frequency of the oscillating current in the transmitter coil
 B. Increasing the magnitude of the oscillating current in the transmitter coil
 C. Increasing the resistivity of the metal
 D. Decreasing the distance between the metal and the transmitter

STOP TO THINK ANSWERS

Chapter Preview Stop to Think: B. We can rewrite the relationship given in the Looking Back box as $\lambda = c/f$. Lower frequencies correspond to longer wavelengths, so the 1.9 GHz cell phone waves have a longer wavelength than the 2.4 GHz oven waves.

Stop to Think 25.1: E. According to the right-hand rule, the magnetic force on a positive charge carrier is to the right.

Stop To Think 25.2: A, C, B. The magnetic field is directed upward out of the north pole of the magnet, so:

- As the magnet moves toward the coil, the upward flux through the loop increases. Opposing this change requires an induced downward field, meaning a clockwise current, choice A.
- The magnet isn't moving, so there is no change in the flux and no induced current, choice C.
- As the magnet moves away from the coil, the upward flux through the loop decreases. Opposing this change requires an induced upward field, meaning a counterclockwise current, choice B.

Stop to Think 25.3: D. The field of the bar magnet emerges from the north pole and points upward. As the coil moves toward the pole, the flux through it is upward and increasing. To oppose the increase, the induced field must point downward. This requires a clockwise (negative) current. As the coil moves away from the pole, the upward flux is decreasing. To oppose the decrease, the induced field must point upward. This requires a counterclockwise (positive) current.

Stop to Think 25.4: A, B, E. The induced emf is equal to the rate of change of the flux; the current is proportional to this emf and inversely proportional to the resistance. Increasing the rate of change of the field will increase the rate of change of the flux. If the loop is larger, the flux will be greater, and therefore the rate of change of the flux will also be greater. And a given emf will induce a larger current in the loop if the resistance is decreased.

Stop to Think 25.5: $I_D > I_A > I_B = I_C$. The intensity depends upon $\cos^2\theta$, where θ is the angle *between* the axes of the two filters. The filters in D have $\theta = 0°$, so all light is transmitted. The two filters in both B and C are crossed ($\theta = 90°$) and transmit no light at all.

Stop to Think 25.6: A. The photon energy is proportional to the frequency. The photons of the 90.5 MHz station each have lower energy, so more photons must be emitted per second.

Stop to Think 25.7: D. A hotter object emits more radiation across the *entire* spectrum than a cooler object. The 6000 K star has its maximum intensity in the blue region of the spectrum, but it still emits more red radiation than the somewhat cooler stars.

26 AC Electricity

Transmission lines carry alternating current at voltages that can exceed 500,000 V. Why are such high voltages used? And how can birds perch safely on these high-voltage wires?

LOOKING AHEAD ▸▸

Goal: To understand and apply basic principles of AC electricity, electrical safety, and household electricity.

Household Electricity

Electricity in your home is **alternating current**, abbreviated as AC. The current reverses direction 120 times per second.

AC allows the use of transformers, like the ones shown, to convert voltages for electronic devices. You'll learn the details of how this is done.

Electrical Safety

Nerve and muscle cells respond to electric signals, so exposure can be dangerous. Insulating gloves protect this lineman.

You'll learn a model for electrical conduction in the body that allows you to determine the risks from different electrical exposures.

AC Circuits

Tuning a radio means adjusting the resonant frequency of an **oscillation circuit** to match a station's broadcast frequency.

You'll learn how these oscillation circuits are used in radios, televisions, cell phones, and other devices.

LOOKING BACK ◂◂

DC Circuits

In Chapter 23, you learned to calculate the voltage and current in DC circuits with resistors.

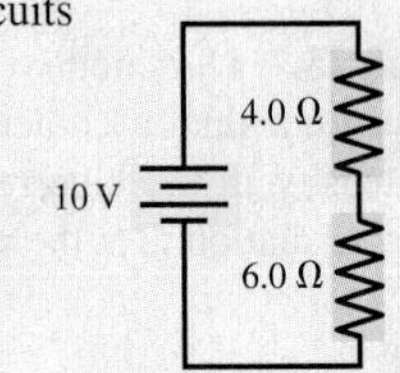

In this chapter, you'll extend the analysis to AC circuits that also include capacitors and **inductors**.

STOP TO THINK

Circuit 1 and circuit 2 at right each has a combination of resistors connected between points a and b. Which circuit has the greater equivalent resistance between points a and b?

A. Circuit 1
B. Circuit 2
C. Both circuits have the same resistance.

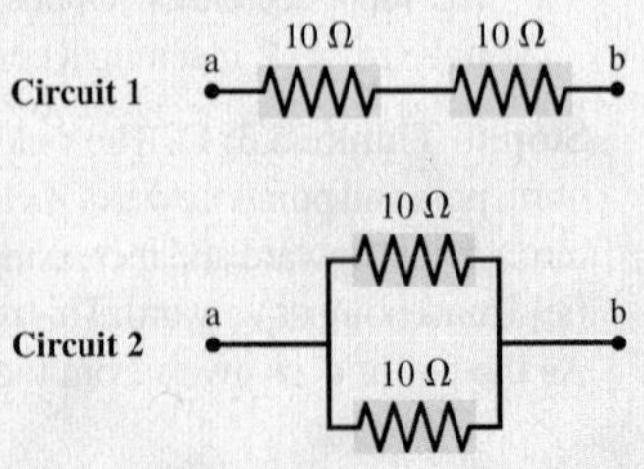

26.1 Alternating Current

A battery creates a constant emf. In a battery-powered flashlight, the bulb carries a constant current and glows with a steady light. The electricity distributed to homes in your neighborhood is different. The picture on the right is a long-exposure photo of a string of LED minilights swung through the air. Each bulb appears as a series of dashes because each bulb in the string flashes on and off 60 times each second. This isn't a special property of the bulbs, but of the electricity that runs them. Household electricity does not have a constant emf; it has a sinusoidal variation that causes the light output of the bulbs to vary. The bulbs light when the emf is positive but not when the emf is negative. The resulting flicker is too rapid to notice under normal circumstances.

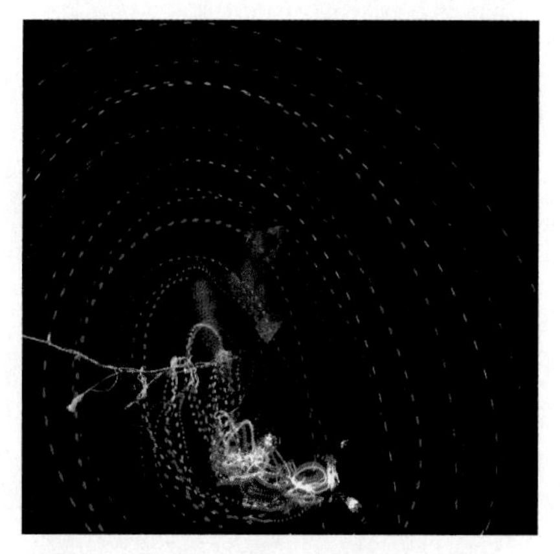

LED minilights flash on and off 60 times a second.

In ◀◀SECTION 25.2 we saw that an electrical generator—whether powered by steam, water, or wind—works by rotating a coil of wire in a magnetic field. The steady rotation of the coil causes the emf and the induced current in the coil to oscillate sinusoidally, alternately positive and then negative. This oscillation forces the charges to flow first in one direction and then, a half cycle later, in the other—an **alternating current,** abbreviated as AC. (If the emf is constant and the current is always in the same direction, we call the electricity *direct current,* abbreviated as DC.) The electricity from power outlets in your house is *AC electricity,* with an emf oscillating at a frequency of 60 Hz. Audio, radio, television, computer, and telecommunication equipment also make extensive use of AC circuits, with frequencies ranging from approximately 10^2 Hz in audio circuits to approximately 10^9 Hz in cell phones.

The instantaneous emf of an AC voltage source, shown graphically in **FIGURE 26.1**, can be written as

$$\mathcal{E} = \mathcal{E}_0 \cos(2\pi f t) = \mathcal{E}_0 \cos\left(\frac{2\pi t}{T}\right) \qquad (26.1)$$

emf of an AC voltage source

SINUSOIDAL p. 445

where $\mathcal{E}_0$ is the peak or maximum emf (recall that the units of emf are volts), T is the period of oscillation (in s), and $f = 1/T$ is the oscillation frequency (in cycles per second, or Hz).

FIGURE 26.1 The emf of an AC voltage source.

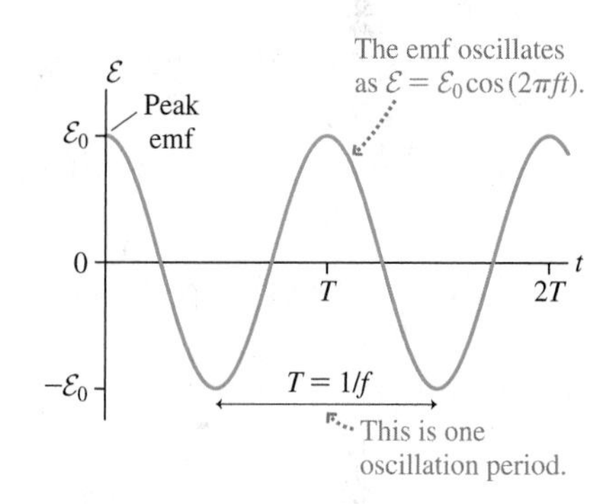

Resistor Circuits

In Chapter 23 you learned to analyze a circuit in terms of the current I and potential difference ΔV. Now, because the current and voltage are oscillating, we will use a lowercase i to represent the *instantaneous* current through a circuit element, the value of the current at a particular instant of time. Similarly, we will use a lowercase v for the circuit element's instantaneous voltage.

FIGURE 26.2 shows the instantaneous current i_R through a resistor R. The potential difference across the resistor, which we call the *resistor voltage* v_R, is given by Ohm's law:

$$v_R = i_R R \qquad (26.2)$$

FIGURE 26.2 The instantaneous current through a resistor.

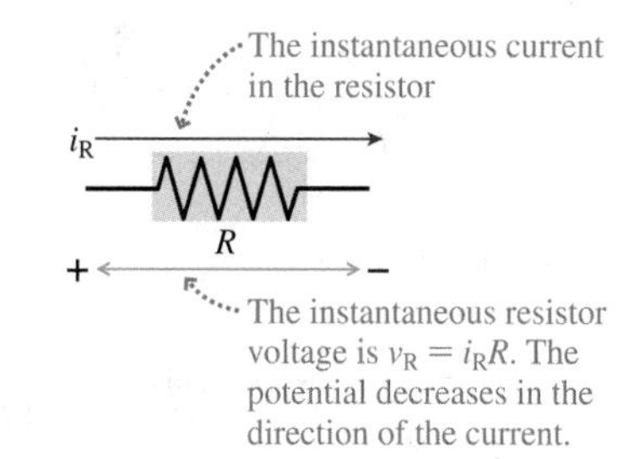

FIGURE 26.3 shows a resistor R connected across an AC emf $\mathcal{E}$. The circuit symbol for an AC generator is -(~)-. We can analyze this circuit in exactly the same way we analyzed a DC resistor circuit. Kirchhoff's loop law says that the sum of all the potential differences around a closed path is zero, so we can write:

$$\sum \Delta V = \Delta V_{source} + \Delta V_R = \mathcal{E} - v_R = 0 \qquad (26.3)$$

The minus sign appears, just as it did in the equation for a DC circuit, because the potential *decreases* when we travel through a resistor in the direction of the current. Thus we find from the loop law that $v_R = \mathcal{E} = \mathcal{E}_0 \cos(2\pi f t)$. This isn't surprising because the resistor is connected directly across the terminals of the emf.

FIGURE 26.3 An AC resistor circuit.

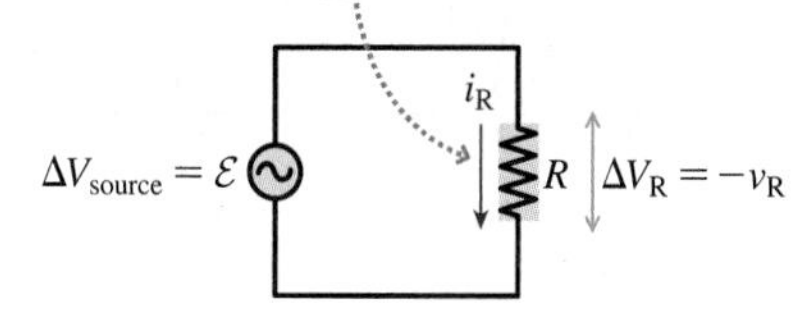

Because the resistor voltage is a sinusoidal voltage at frequency f, it is useful to write

$$v_R = V_R \cos(2\pi ft) \tag{26.4}$$

In this equation V_R is the peak or maximum voltage, the amplitude of the sinusoidally varying voltage. You can see that $V_R = \mathcal{E}_0$ in the single-resistor circuit of Figure 26.3. Thus the current through the resistor is

$$i_R = \frac{v_R}{R} = \frac{V_R \cos(2\pi ft)}{R} = I_R \cos(2\pi ft) \tag{26.5}$$

where $I_R = V_R / R$ is the peak current.

NOTE ▶ It is important to understand the distinction between instantaneous and peak quantities. The instantaneous current i_R, for example, is a quantity that is changing with time according to Equation 26.5. The peak current I_R is the maximum value that the instantaneous current reaches. The instantaneous current oscillates between $+I_R$ and $-I_R$. ◀

FIGURE 26.4 Graph of the current through and voltage across a resistor.

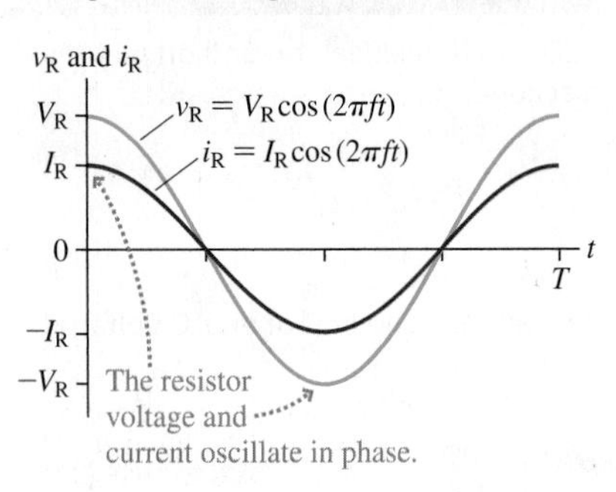

The resistor's instantaneous current and voltage both oscillate as $\cos(2\pi ft)$. **FIGURE 26.4** shows the voltage and the current simultaneously on the same graph. The fact that the peak current I_R is drawn as being less than V_R has no significance. Current and voltage are measured in different units, so in a graph like this you can't compare the value of one to the value of the other. Showing the two different quantities on a single graph—a tactic that can be misleading if you're not careful—simply illustrates that they oscillate *in phase:* **The current is at its maximum value when the voltage is at its maximum, and the current is at its minimum value when the voltage is at its minimum.**

AC Power in Resistors

In Chapter 23 you learned that the power dissipated by a resistor is $P = I\Delta V_R = I^2 R$. In an AC circuit, the resistor current i_R and voltage v_R are constantly changing, as we saw in Figure 26.4, so the instantaneous power loss $p = i_R v_R = i_R^2 R$ (note the lowercase p) is constantly changing as well. We can use Equations 26.4 and 26.5 to write this instantaneous power as

$$p = i_R^2 R = [I_R \cos(2\pi ft)]^2 R = I_R^2 R[\cos(2\pi ft)]^2 \tag{26.6}$$

FIGURE 26.5 The instantaneous power loss in a resistor.

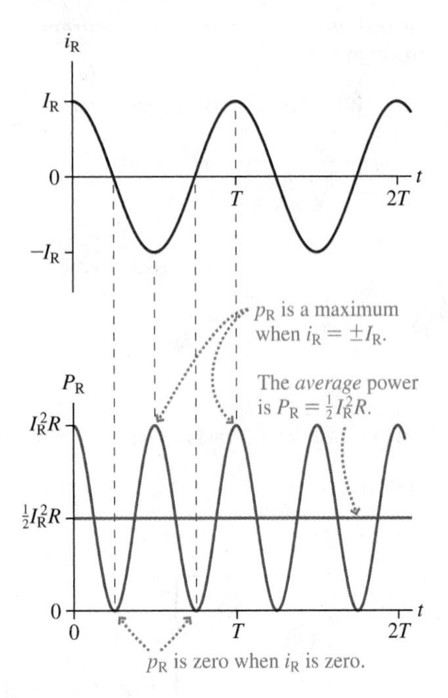

FIGURE 26.5 shows the instantaneous power graphically. You can see that, because the cosine is squared, the power oscillates *twice* during every cycle of the emf: The energy dissipation peaks both when $i_R = I_R$ and when $i_R = -I_R$. The energy dissipation doesn't depend on the current's direction through the resistor.

The current in an incandescent lightbulb reverses direction 120 times per second (twice per cycle), so the power reaches a maximum 120 times a second. But the hot filament of the bulb glows steadily, so it makes more sense to pay attention to the *average power* than the instantaneous power. Figure 26.5 shows that the **average power** P_R is related to the peak power as follows:

$$P_R = \frac{1}{2} I_R^2 R \tag{26.7}$$

We could do a similar analysis for voltage to show that the power is $P_R = \frac{1}{2} V_R{}^2 / R$. Recall that, in a DC circuit, the power is $P_R = I^2 R = V^2 / R$. We can write more useful expressions for power that mirror the DC expressions by defining a **root-mean-square current** and a **root-mean-square voltage** as follows. (We use the subscript "rms" to signify root-mean-square current and voltage.)

$$I_{rms} = \frac{I_R}{\sqrt{2}} \qquad V_{rms} = \frac{V_R}{\sqrt{2}} \tag{26.8}$$

Root-mean-square current and root-mean-square voltage

An rms value is a single number that serves as the measure of the size of a varying quantity. We saw this idea in Chapter 12, where we introduced the rms speed of molecules in a gas. If we rewrite the expressions for the average power dissipated by a resistor in an AC circuit in terms of the rms expressions, we obtain

$$P_R = (I_{rms})^2 R = \frac{(V_{rms})^2}{R} = I_{rms} V_{rms} \tag{26.9}$$

Average power loss in a resistor

The "120 V" on this lightbulb is its operating rms voltage. The "100 W" is its average power dissipation at this voltage.

These expressions are just the same as the DC expressions, with I_{rms} replacing I and V_{rms} replacing V. **As long as you work with rms voltages and currents, all the expressions you learned for DC power carry over to AC power.** The average power loss for a resistor in an AC circuit with $I_{rms} = 1$ A is the same as in a DC circuit with $I = 1$ A. In what follows, we'll use rms values as a general rule; if you are given a single voltage for an AC power supply, you can assume that this is the rms voltage. For instance, we've noted that household lamps and appliances in the United States operate at the 120 V present at wall outlets. This voltage is the rms value $\mathcal{E}_{rms}$; the peak voltage is higher by a factor of $\sqrt{2}$, so $\mathcal{E}_0 = 170$ V.

EXAMPLE 26.1 The resistance and current of a toaster

The hot wire in a toaster dissipates 580 W when plugged into a 120 V outlet.

a. What is the wire's resistance?
b. What are the rms and peak currents through the wire?

PREPARE We've seen that the 120 V outlet voltage is an rms value. The filament has resistance R. It dissipates 580 W when there's an rms voltage of 120 V across it. We can solve Equation 26.9 for R and then use Equations 26.9 and 26.8 to find the rms current and the peak current.

SOLVE

a. We rearrange Equation 26.9 to find the resistance from the rms voltage and the average power:

$$R = \frac{(V_{rms})^2}{P_R} = \frac{(120\text{ V})^2}{580\text{ W}} = 25\ \Omega$$

b. A second rearrangement of Equation 26.9 allows us to find the current in terms of the power and the resistance, both of which are known:

$$I_{rms} = \sqrt{\frac{P_R}{R}} = \sqrt{\frac{580\text{ W}}{25\Omega}} = 4.8\text{ A}$$

From Equation 26.8, the peak current is

$$I_R = \sqrt{2}\, I_{rms} = \sqrt{2}\,(4.8\text{ A}) = 6.8\text{ A}$$

ASSESS We can do a quick check on our work by calculating the power for the rated voltage and computed current:

$$P_R = I_{rms} V_{rms} = (4.8\text{ A})(120\text{ V}) = 580\text{ W}$$

This agrees with the value given in the problem statement, giving us confidence in our solution.

STOP TO THINK 26.1 An AC current with a peak value of 1.0 A passes through bulb A. A DC current of 1.0 A passes through an identical bulb B. Which bulb is brighter?

A. Bulb A B. Bulb B
C. Both bulbs are equally bright.

26.2 AC Electricity and Transformers

Your cell phone runs at about 3.5 V. To charge it up from a wall outlet at $V_{rms} = 120$ V requires the use of a **transformer,** a device that takes an *alternating* voltage as an input and produces either a higher or lower voltage as its output. As we'll see, the operation of a transformer is based on the emf produced by changing magnetic fields, so the input must be AC electricity.

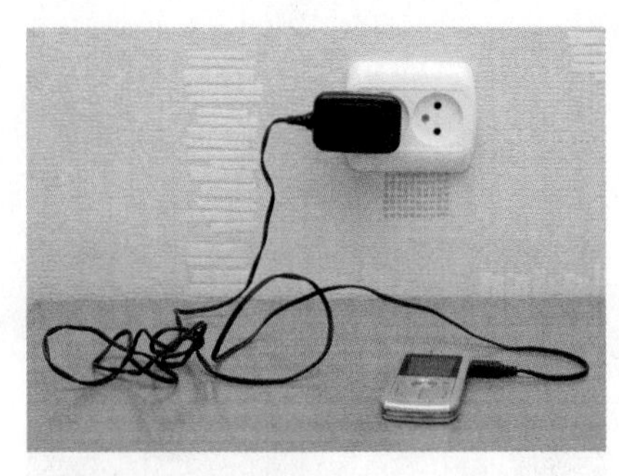

Your cell phone charger incorporates a transformer that provides the necessary voltage reduction.

Transformer Operation

FIGURE 26.6a shows a simplified version of a transformer, consisting of two coils of wire wrapped on a single iron core. The left coil is called the **primary coil** (or simply the *primary*). It has N_1 turns of wire connected to an AC voltage of amplitude V_1. This AC voltage creates an alternating current. The current in the coil creates a magnetic field that magnetizes the iron of the core to produce a much stronger net field—and a large flux through the primary coil. The field lines tend to follow the iron core, as shown in the figure, so nearly all of the flux from the left primary coil also goes through the right coil of wire, which has N_2 turns and is called the **secondary coil** (or the *secondary*). FIGURE 26.6b shows the symbol we'll use to represent a transformer in a circuit, with N_1 and N_2 specifying the numbers of turns of wire in the primary and secondary coils. The current in the primary coil is an alternating current, so it creates an oscillating magnetic field in the iron core; the changing magnetic field means that there is a changing flux in the secondary coil. This changing flux induces an emf, an AC voltage of amplitude V_2, in the secondary coil. To complete the picture, we connect this emf to a resistor R, which we call the *load*. Current in this resistor will dissipate power.

FIGURE 26.6 A transformer and its circuit symbol.

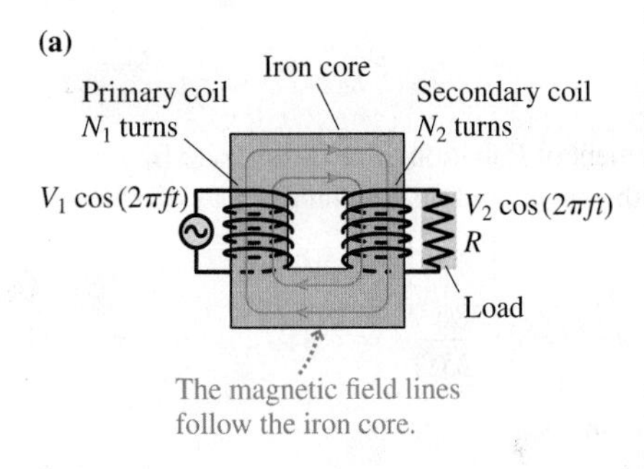

(b) We will use this symbol to represent a transformer in a circuit.

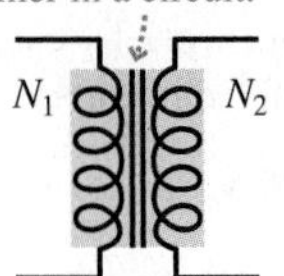

The point of a transformer is to change the voltage, so we need to define how the voltage V_2 at the secondary is related to the voltage V_1 at the primary. Suppose at some instant of time the instantaneous current through the primary is i_1. This current creates a magnetic flux Φ through the primary coil. This flux is changing, because i_1 is changing, and the change induces an emf across the coil. According to Faraday's law (Equation 25.12), the instantaneous voltage v_1 across the N_1 turns of the primary coil is

$$v_1 = \mathcal{E}_1 = N_1 \frac{\Delta \Phi}{\Delta t} \tag{26.10}$$

In an **ideal transformer,** all of the flux is "guided" by the iron core through the secondary coil. Consequently, the rate at which the flux changes through the secondary coil is also $\Delta\Phi/\Delta t$. This changing flux induces an emf across the secondary coil given by

$$v_2 = \mathcal{E}_2 = N_2 \frac{\Delta \Phi}{\Delta t} \tag{26.11}$$

Because $\Delta\Phi/\Delta t$ is the same in Equations 26.10 and 26.11, we can write

$$\frac{v_1}{v_2} = \frac{N_1}{N_2} \tag{26.12}$$

Equation 26.12 gives the ratio of the instantaneous voltages. This equation also applies at the instant at which these voltages are at their peak values V_1 and V_2, and the rms voltages are related to these peak values by a factor of $1/\sqrt{2}$. We can thus relate the peak and rms voltages of the primary and secondary coils:

$$V_2 = \frac{N_2}{N_1} V_1 \quad \text{and} \quad (V_2)_{\text{rms}} = \frac{N_2}{N_1} (V_1)_{\text{rms}} \tag{26.13}$$

Transformer voltages for primary and secondary coils with N_1 and N_2 turns

Depending on the ratio N_2/N_1, the voltage V_2 across the load can be transformed to a higher or a lower voltage than V_1. A *step-up transformer,* with $N_2 > N_1$, increases the voltage, while a *step-down transformer,* with $N_2 < N_1$, lowers the voltage.

Class Video

Equation 26.13 relates the voltage at a transformer's secondary to the voltage at its primary. We can also relate the *currents* in the secondary and primary by considering energy conservation. If a transformer is connected to a load that draws an rms current $(I_2)_{rms}$ from the secondary, the average power supplied to the load by the transformer, given by Equation 26.9, is

$$P_2 = (V_2)_{rms}\,(I_2)_{rms}$$

The primary coil draws a current $(I_1)_{rms}$ from the voltage source to which it's connected. This source provides power

$$P_1 = (V_1)_{rms}\,(I_1)_{rms}$$

to the transformer.

We'll assume that our ideal transformer has no loss of electric energy, so $P_1 = P_2$, or $(V_1)_{rms}(I_1)_{rms} = (V_2)_{rms}(I_2)_{rms}$. We can solve this for the current in the secondary:

$$(I_2)_{rms} = \frac{(V_1)_{rms}}{(V_2)_{rms}}(I_1)_{rms} = \frac{N_1}{N_2}(I_1)_{rms}$$

A similar expression holds for the peak current as well. We can therefore write expressions for the current in the coils of a transformer that mirror the expressions for the voltage in Equation 26.13:

$$I_2 = \frac{N_1}{N_2}I_1 \quad \text{and} \quad (I_2)_{rms} = \frac{N_1}{N_2}(I_1)_{rms} \tag{26.14}$$

Transformer currents for primary and secondary coils with N_1 and N_2 turns

Comparing Equations 26.13 and 26.14, you can see that **a step-up transformer raises voltage but lowers current; a step-down transformer lowers voltage but raises current.** This must be the case in order to conserve energy.

NOTE ► We've made some assumptions about the ideal transformer in completing this derivation. Real transformers come quite close to this ideal, so you can use the above equations for computations on real transformers. ◄

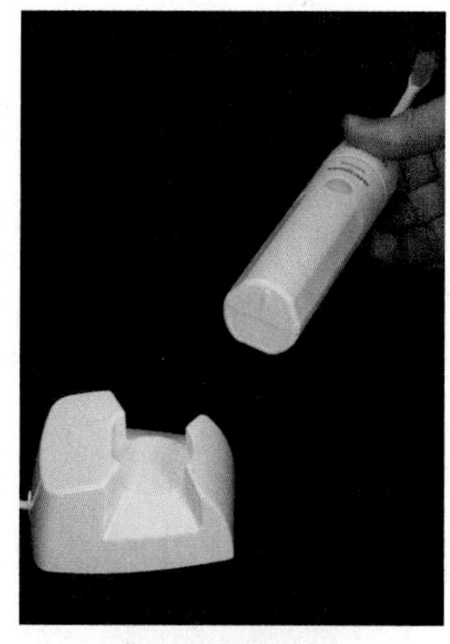

Getting a charge There is no direct electrical contact between the primary and secondary coils in a transformer; the energy is carried from one coil to the other by the magnetic field. This makes it possible to charge devices that are completely sealed, with no external electrical contacts, such as the electric toothbrush shown. A primary coil in the base creates an alternating magnetic field that induces an alternating emf in a secondary coil in the brush's handle. This emf, after conversion to DC, is used to charge the toothbrush's battery.

Class Video

EXAMPLE 26.2 Analyzing a step-down transformer

A book light has a 1.4 W, 4.8 V bulb that is powered by a transformer connected to a 120 V electric outlet. The secondary coil of the transformer has 20 turns of wire. How many turns does the primary coil have? What is the current in the primary coil?

PREPARE FIGURE 26.7 is a sketch of the circuit: this is the basic transformer circuit of Figure 26.6a, with the bulb as the load. All of the circuit quantities given in the problem statement are included in the sketch.

FIGURE 26.7 A transformer provides power to a book light.

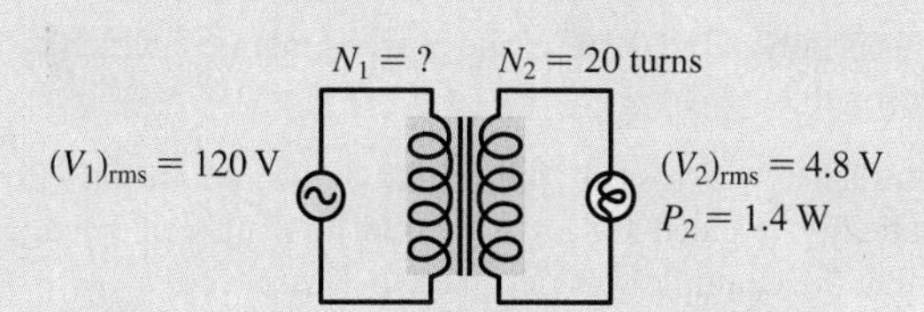

We know the voltages of the primary and the secondary, so we can compute the turns in the primary coil using Equation 26.13. We know the voltage and the power of the bulb, so we can find the current in the bulb. This is the current in the secondary, which we can use in Equation 26.14 to find the current in the primary, the current provided by the outlet.

SOLVE The bulb is rated at 4.8 V; this is the rms voltage at the secondary, so $(V_2)_{rms} = 4.8$ V. The power outlet has the usual $(V_1)_{rms} = 120$ V, so we can rearrange Equation 26.13 to find

$$N_1 = N_2\frac{(V_1)_{rms}}{(V_2)_{rms}} = (20 \text{ turns})\left(\frac{120 \text{ V}}{4.8 \text{ V}}\right) = 500 \text{ turns}$$

The bulb connected to the secondary dissipates 1.4 W at 4.8 V; this is an rms voltage, so the rms current in the secondary is

$$(I_2)_{rms} = \frac{P_2}{(V_2)_{rms}} = \frac{1.4 \text{ W}}{4.8 \text{ V}} = 0.29 \text{ A}$$

We can then rearrange Equation 26.14 to find the rms current in the primary:

$$(I_1)_{rms} = (I_2)_{rms}\frac{N_2}{N_1} = (0.29 \text{ A})\frac{20}{500} = 0.012 \text{ A}$$

ASSESS We can check our results by looking at the power supplied by the wall outlet. This is $P_1 = (120 \text{ V})(0.012 \text{ A}) = 1.4$ W, the same as the power dissipated by the bulb, as must be the case because we've assumed the transformer is ideal.

Power Transmission

Long-distance electrical transmission lines run at very high voltages—up to 1,000,000 volts! FIGURE 26.8 outlines the steps in transmitting electricity from a power plant to a city; the voltages shown are typical values. A generator in the power plant produces electricity at 25 kV, which is stepped up to 500 kV for transmission, then stepped down in two stages to give you 120 V at the outlet. There are three stages of transformers involved. Why go to all this trouble to transmit electric power at such high voltages? In a word: efficiency, as we show in the next example.

FIGURE 26.8 Electricity transmission.

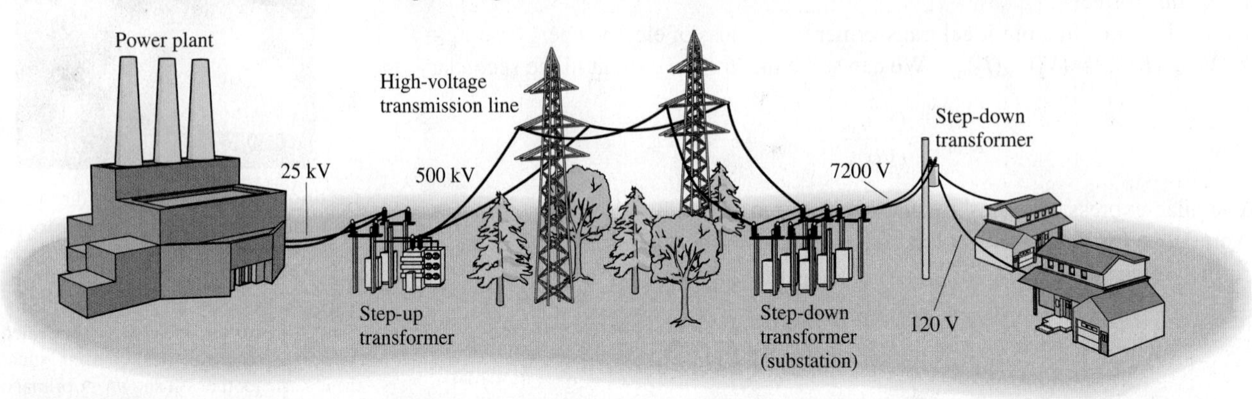

EXAMPLE 26.3 Practical power transmission

To provide power to a small city, a power plant generates 40 MW of AC electricity. The power plant is 50 km from the city (a typical distance), and the 100 km of wire used in the transmission line (to the city and back) has a resistance of 7.0 Ω.

a. To provide 40 MW of power at the generator voltage of 25,000 V, what current is required?
b. What is the power dissipated in the resistance of the transmission line for this current?
c. To provide 40 MW of power at 500,000 V, what current is required?
d. What is the power dissipated in the resistance of the transmission line for this higher voltage?

PREPARE We can treat the city—and the wires that transmit power to it—as a load. All of the voltages are rms values and the power is an average power, so we can find the current to provide a given power and the power dissipated using the relationships in Equation 26.9.

SOLVE

a. To provide 40 MW at the generator voltage of 25,000 V, the current is

$$I_{rms} = \frac{40 \times 10^6 \text{ W}}{25 \times 10^3 \text{ V}} = 1600 \text{ A}$$

b. Passing this current through the transmission lines will result in power dissipation in the 7.0 Ω resistance of the wires. We don't know the voltage drop across the wires, but we do know the current and resistance, so we can compute

$$P_{\text{dissipated in wires}} = (I_{rms})^2 R = (1600 \text{ A})^2 (7.0 \ \Omega) = 18 \text{ MW}$$

This is nearly half the power generated, clearly an unacceptable loss.

c. Increasing the transmission voltage to 500 kV reduces the necessary current:

$$I_{rms} = \frac{40 \times 10^6 \text{ W}}{500 \times 10^3 \text{ V}} = 80 \text{ A}$$

This is a remarkably small current to supply a city. If you use several high-power appliances at one time, you could easily use this much current in your house. But the necessary current for the city can be so small because the voltage is so large.

d. The relatively small current means that the power dissipated in the resistance of the wires will be small as well:

$$P_{\text{dissipated in wires}} = (I_{rms})^2 R = (80 \text{ A})^2 (7.0 \ \Omega) = 0.045 \text{ MW}$$

This is only about 0.1% of the power generated, which is quite reasonable.

ASSESS The final result—the power dissipated in the wires is dramatically reduced for an increased transmission voltage—is just what the example was designed to illustrate.

Transmitting electricity at high voltages means that the current in the transmission lines can be decreased, which results in manageable power losses. High-voltage transmission requires transformers, as shown in Figure 26.8. Transformers increase the voltage from the generator to the higher transmission voltage and then, once the transmission lines reach their destination, decrease the voltage to a safe value for household use. Transformers require AC electricity because they rely on changing flux for their operation. This is the main reason we use AC power even though (as you'll see in the next section) it is slightly more dangerous than DC.

STOP TO THINK 26.2 Each of the transformers shown here has its primary connected to a 120 V AC power supply and its secondary connected to a 1000 Ω load resistor. Rank the transformers, from highest to lowest, by the rms voltage they create across the resistor.

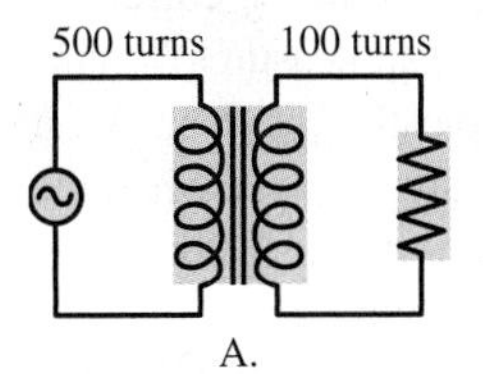

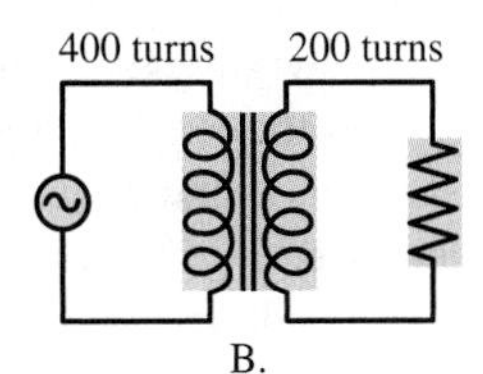

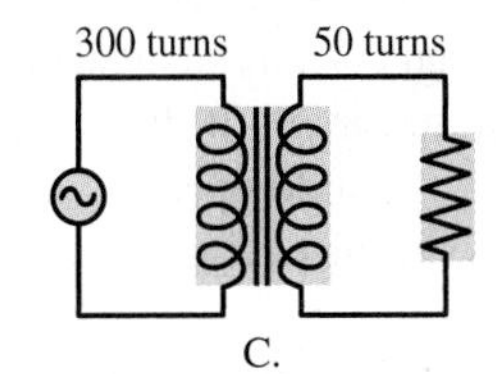

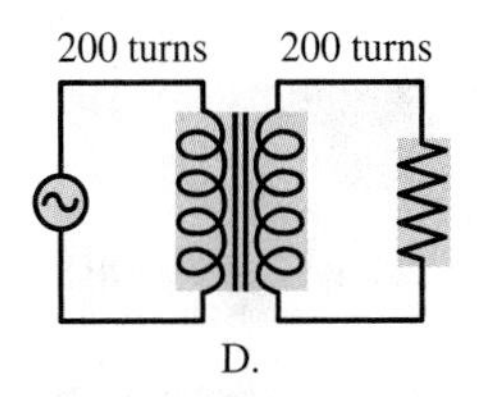

26.3 Household Electricity

The electricity in your home can be understood using the techniques of circuit analysis we've developed, but we need to add one more concept. So far we've dealt with potential *differences*. Although we are free to choose the zero point of potential anywhere that is convenient, our analysis of circuits has not suggested any need to establish a zero point. Potential differences are all we have needed.

Difficulties can begin to arise, however, if you want to connect two different circuits together. Perhaps you would like to connect your CD player to your amplifier or connect your computer monitor to the computer itself. Incompatibilities can arise unless all the circuits to be connected have a common reference point for the potential. This is the reason for having an electric *ground*.

Getting Grounded

You learned previously that the earth itself is a conductor. Suppose we have two circuits. If we connect one point of each circuit to the earth by an ideal wire, then both circuits have a common reference point. A circuit connected to the earth in this way is said to be **grounded.** In practice, we also agree to call the potential of the earth $V_{earth} = 0$ V. **FIGURE 26.9** shows a circuit with a ground connection. Under normal circumstances, the ground connection does not carry any current because it is not part of a complete circuit. In this case, it does not alter the behavior of the circuit.

FIGURE 26.9 A grounded circuit.

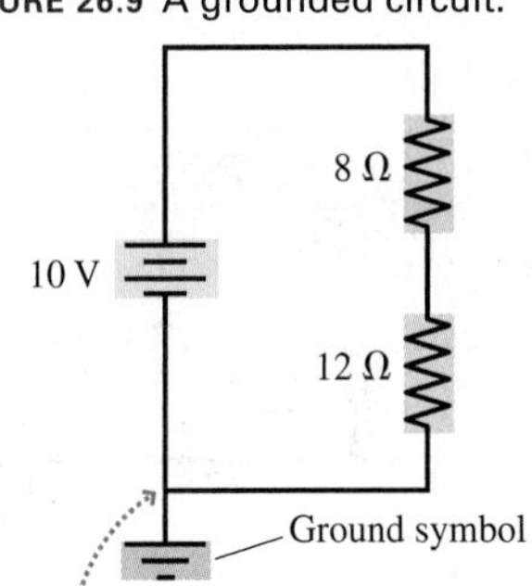

Electric Outlets Are Grounded Parallel Circuits

FIGURE 26.10 shows a circuit diagram for the outlets in your house. The 120 V electric supply is provided by the power company. It is transmitted to outlets throughout your house by wires in the walls. One terminal of the electric supply is grounded; we call this the **neutral** side. The other side is at a varying potential; we call this the **hot** side. Each electric outlet has two slots, one connected to the hot side and one connected to the neutral side. When you insert a plug into an electric outlet, the prongs of the plug connect to the two terminals of the electric supply. The device you've plugged in completes a circuit between the two terminals, the potential difference across the device leads to a current, and the device turns on.

The multiple outlets in a room or area of your house are connected in parallel so that each works when the others are not being used. Because the outlets on a single circuit are in parallel, when you plug in another device, the total current in the circuit increases.

FIGURE 26.10 Multiple outlets on one circuit.

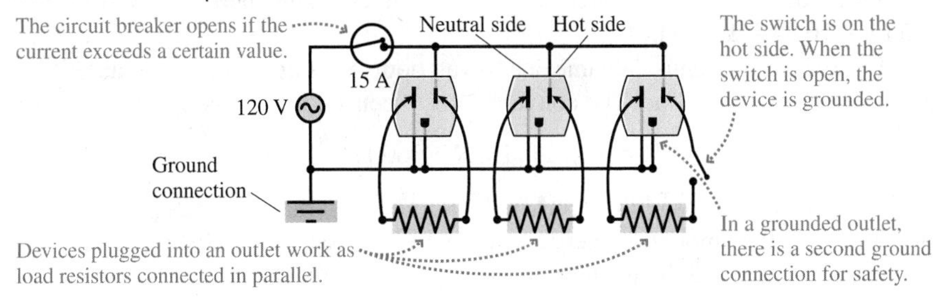

The circuit breaker opens if the current exceeds a certain value.
Neutral side
Hot side
15 A
120 V
Ground connection
The switch is on the hot side. When the switch is open, the device is grounded.
In a grounded outlet, there is a second ground connection for safety.
Devices plugged into an outlet work as load resistors connected in parallel.

Figure Video

This can create a problem; although the wires in your walls are good conductors, they aren't ideal. The wires have a small resistance and heat up when carrying a current. If there is too much current, the wires could get hot enough to cause a fire.

The circuits in your house are protected with circuit breakers to limit the current in each circuit. A circuit breaker consists of a switch and an ammeter that measures the current in the circuit. If the ammeter measures too much current (typically $I_{rms} \geq 15$ A), it sends a signal to open the switch to disconnect the circuit. You have probably had the experience of having a circuit breaker "trip" if you have too many things plugged in. Some things need to be turned off, unplugged, or moved to a different circuit.

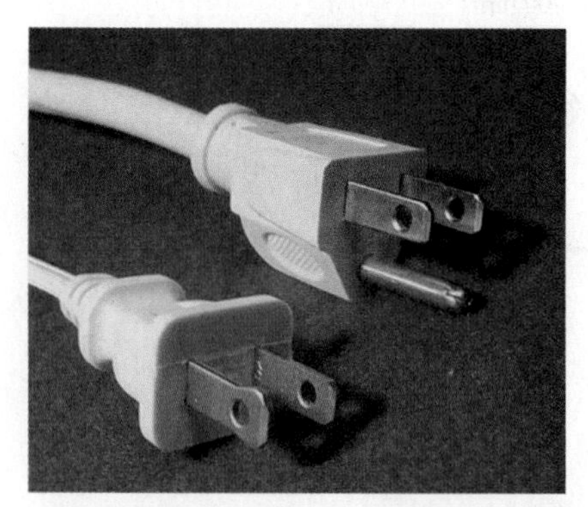

A two-prong "polarized" plug has one large prong and one small one. When plugged into a standard outlet, the large prong is grounded. A three-prong plug has a round pin that makes a second ground connection.

Grounding of household circuits provides an important reference potential, but the main reason for grounding is safety. The two slots in a standard outlet are different sizes; the neutral slot is a bit larger. Most electric devices are fitted with plugs that can be inserted into an outlet in only one orientation. A lamp will almost certainly have this sort of plug. When you turn the lamp off, the switch disconnects the hot wire, not the neutral wire. The lamp is then grounded, and thus safe, when it is switched off.

The round hole in a standard electric outlet is a second ground connection. If a device has a metal case, the case will likely be connected to the ground. If a wire comes loose inside the device and contacts the metal case, a person touching the case could get a shock. But if the case is grounded, its potential is always 0 V and it is always safe to touch. In addition, a hot wire touching the grounded case would be a short circuit, causing a sudden very large current that would trip the circuit breaker, disconnecting the hot wire and preventing any danger.

EXAMPLE 26.4 Will the circuit breaker open?

A circuit in a student's room has a 15 A circuit breaker. One evening, she plugs in a computer (240 W), a lamp (with two 60 W bulbs), and a space heater (1200 W). Will this be enough to trip the circuit breaker?

PREPARE We start by sketching the circuit, as in **FIGURE 26.11**. Because the three devices are in the same circuit, they are connected in parallel. We can model each of them as a resistor.

FIGURE 26.11 The circuit with the circuit breaker.

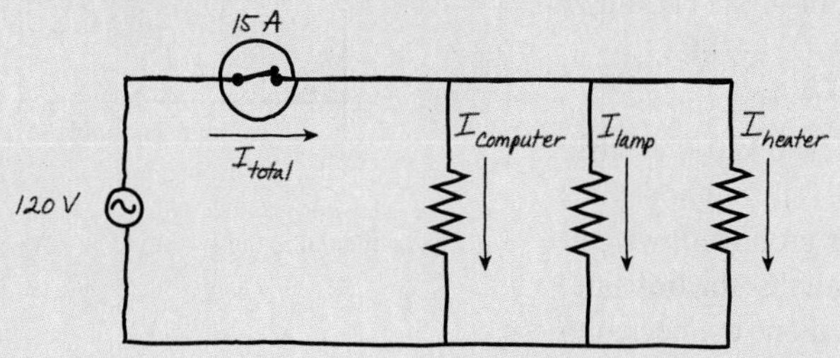

SOLVE The current in the circuit is the sum of the currents in the individual devices:

$$(I_{total})_{rms} = (I_{computer})_{rms} + (I_{lamp})_{rms} + (I_{heater})_{rms}$$

Equation 26.9 gives the power as the rms current times the rms voltage, so the current in each device is the power divided by the rms voltage:

$$(I_{total})_{rms} = \frac{240\text{ W}}{120\text{ V}} + \frac{120\text{ W}}{120\text{ V}} + \frac{1200\text{ W}}{120\text{ V}} = 13\text{ A}$$

This is almost but not quite enough to trip the circuit breaker.

ASSESS Generally all of the outlets in one room (and perhaps the lights as well) are on the same circuit. You have quite possibly used electric devices with this much total power in one room without problems, so this result seems reasonable.

Kilowatt Hours

The product of watts and seconds is joules, the SI unit of energy. However, your local electric company prefers to use a different unit, called *kilowatt hours,* to measure the energy you use each month.

A device in your home that consumes P kW of electricity for Δt hours has used $P\Delta t$ kilowatt hours of energy, abbreviated kWh. For example, suppose you run a 1500 W electric water heater for 10 hours. The energy used in kWh is (1.5 kW)(10 h) = 15 kWh.

Your home's electric meter records the kilowatt hours of electricity you use.

Despite the rather unusual name, a kilowatt hour is a unit of energy because it is a power multiplied by a time. The conversion between kWh and J is

$$1.00\text{ kWh} = (1.00 \times 10^3\text{ W})(3600\text{ s}) = 3.60 \times 10^6\text{ J}$$

Your monthly electric bill specifies the number of kilowatt hours you used last month. This is the amount of energy that the electric company delivered to you that you transformed into light and thermal energy inside your home.

EXAMPLE 26.5 **Computing the cost of electric energy**

A typical electric space heater draws an rms current of 12.5 A on its highest setting. If electricity costs 12¢ per kilowatt hour (an approximate national average), how much does it cost to run the heater for 2 hours?

SOLVE The power dissipated by the heater is

$$P = V_{\text{rms}} I_{\text{rms}} = (120\text{ V})(12.5\text{ A}) = 1500\text{ W} = 1.5\text{ kW}$$

In 2 hours, the energy used is $(1.5\text{ kW})(2.0\text{ h}) = 3.0\text{ kWh}$. At 12¢ per kWh, the cost is 36¢.

STOP TO THINK 26.3 Rank the following four devices by the energy they use for the time period given, from highest to lowest:

A. Electric blanket (60 W) used for 8 h
B. Flat-screen television (120 W) for 3 h
C. Whole-house fan (250 W) for 8 h
D. Air conditioner (1000 W) for 3 h

26.4 Biological Effects and Electrical Safety BIO

You can handle an ordinary 9 V battery without the slightest danger, but the 120 volts from an electric outlet can lead to a nasty shock. Yet the girl in **FIGURE 26.12** is safely touching a Van de Graaff generator at a potential of 400,000 V. What makes electricity either safe or dangerous?

FIGURE 26.12 High voltages are not necessarily dangerous.

The relative safety of electric sources isn't governed by the voltage but by the current. Current—the flow of charges through the body—is what produces physiological effects and damage because it mimics nerve impulses and causes muscles to involuntarily contract.

Higher voltages are generally more dangerous than lower voltages because they tend to produce larger currents, but the amount of current also depends on resistance and on the ability of the voltage source to deliver current. The Van de Graaff generator in Figure 26.12 is at a high potential with respect to the ground, but the girl is standing on an insulating platform. The high resistance of the platform means that very little current is passing through her to the ground; she won't feel a thing. Even if she touches a grounded object, the current will be modest—the total charge on the generator is quite small, so a dangerous current simply isn't possible. You can get a much worse shock from a 120 V household circuit because it is capable of providing a much larger current.

TABLE 26.1 Physiological effects of currents passing through the body

Physiological effect	AC current (rms) (mA)	DC current (mA)
Threshold of sensation	1	3
Paralysis of respiratory muscles	15	60
Heart fibrillation, likely fatal	> 100	> 500

Table 26.1 lists approximate values of current that produce different physiological effects. Currents through the chest cavity are particularly dangerous because they can interfere with respiration and the proper rhythm of the heart. It requires less AC current than DC current to produce physiological effects. An AC current larger than 100 mA can induce fibrillation of the heart, in which it beats in a rapid, chaotic, uncontrolled fashion.

To calculate likely currents through the body, we model the body as several connected resistors, as shown in **FIGURE 26.13**. (These are average values; there is significant individual variation.) Because the interior of the body has a high salt-water content, its resistance is fairly low. But current must pass through the skin before getting inside the body, and the skin generally has a fairly high resistance. If you touch a wire with the dry skin of a finger, the skin's resistance might be greater than 1 MΩ. Moist skin and larger contact areas can reduce the skin's resistance to less than 10 kΩ.

FIGURE 26.13 Resistance model of the body.

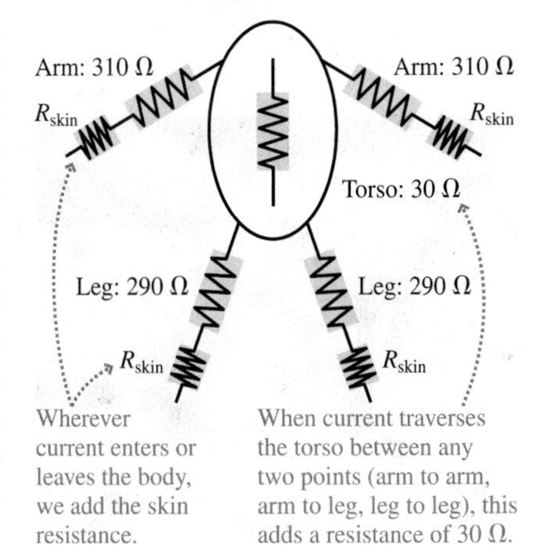

Arm: 310 Ω
Arm: 310 Ω
R_{skin}
R_{skin}
Torso: 30 Ω
Leg: 290 Ω
Leg: 290 Ω
R_{skin}
R_{skin}
Wherever current enters or leaves the body, we add the skin resistance.
When current traverses the torso between any two points (arm to arm, arm to leg, leg to leg), this adds a resistance of 30 Ω.

EXAMPLE 26.6 Is the worker in danger?

A worker in a plant grabs a bare wire that he does not know is connected to a 480 V AC supply. His other hand is holding a grounded metal railing. The skin resistance of each of his hands, in full contact with a conductor, is 2200 Ω. He will receive a shock. Is it large enough to be dangerous?

PREPARE We can draw a circuit model for this situation as in **FIGURE 26.14a**; the worker's body completes a circuit between two points at a potential difference of 480 V. The current will depend on this potential difference and the resistance of his body, including the resistance of the skin.

SOLVE Following the model of Figure 26.13, the current path goes through the skin of one hand, up one arm, across the torso, down the other arm, and through the skin of the other hand, as in **FIGURE 26.14b**. The equivalent resistance of the series combination is 5050 Ω, so the AC current through his body is

$$I = \frac{\Delta V}{R_{eq}} = \frac{480\ \text{V}}{5050\,\Omega} = 95\ \text{mA}$$

FIGURE 26.14 A circuit model for the worker.

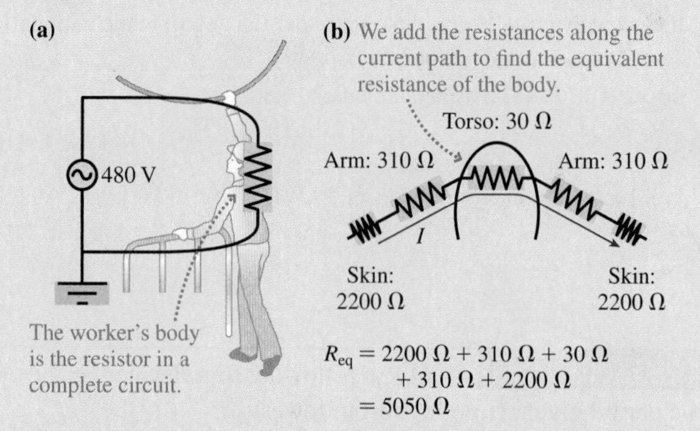
(a)
480 V
The worker's body is the resistor in a complete circuit.
(b) We add the resistances along the current path to find the equivalent resistance of the body.
Torso: 30 Ω
Arm: 310 Ω
Arm: 310 Ω
I
Skin: 2200 Ω
Skin: 2200 Ω
R_{eq} = 2200 Ω + 310 Ω + 30 Ω + 310 Ω + 2200 Ω = 5050 Ω

From Table 26.1 we see that this is a very dangerous, possibly fatal, current.

ASSESS The voltage is high and the resistance relatively low, so it's no surprise to find a dangerous level of current.

FIGURE 26.15 Wearing electrically insulating boots increases resistance.

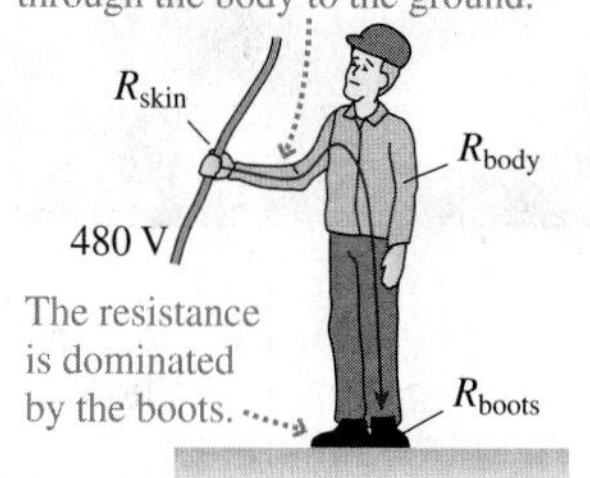
There is a current path from the wire through the body to the ground.
R_{skin}
R_{body}
480 V
The resistance is dominated by the boots.
R_{boots}

Because of the danger of electric shock, workers who might accidentally contact electric lines wear protective clothing. The resistance of the protective clothing is much greater than that of the body or the skin. Boots, in particular, are made with soles that have very high electrical resistance, because workers are usually standing on the ground. Suppose a worker wearing protective boots touches a live wire at 480 V, as shown in **FIGURE 26.15**. There is a 480 V potential difference between his hand and the ground, so there will be a current through his body. It will travel through his skin, then his body, then his boots. The total resistance is the sum of these series resistances. For typical boots, the electrical resistance of the soles is at least 10 MΩ; this is so much greater than the resistance of the skin and the resistance of the body that the total resistance can be approximated as being equal to the resistance of the soles of the boots. If we assume a resistance of 10 MΩ for the current path, then the current that passes through the worker's body is only

$$I = \frac{\Delta V}{R_{eq}} \approx \frac{\Delta V}{R_{boots}} = \frac{480\ \text{V}}{10\ \text{M}\Omega} = 48\ \mu\text{A}$$

In this case, the worker will be fine. Because the current is much less than the threshold for sensation, he won't even feel a shock!

Class Video

Now, think about the birds sitting on the wire at the start of the chapter. How are they able to perch on the high-voltage wire? Think about the examples we've seen; we always consider the *potential difference* between two parts of the body. The wire is at an elevated potential with respect to the ground, so each foot is at a high potential, but there is only a very small potential *difference* between the feet due to the potential decrease along the wire. For the high-voltage power line considered earlier in the chapter, the potential difference between the two points along the wire where a bird's feet touch is less than 1 mV, not something the bird will notice! If a bird touches a wire and a grounded pole, or two neighboring wires, however, the result can be very different. This is a problem for birds with large wingspans that can establish a connection between widely separated conductors at very different voltages.

◀ **The lightning crouch** Most lightning injuries come from the resulting *ground current*, not the lightning itself. If you are standing near a point where lightning strikes the ground, there will be potential difference $\Delta V = Ed$ between your feet, which can cause a dangerous current up one leg and down the other. If you are caught outdoors in a lightning storm, you can minimize the potential difference ΔV by minimizing d, the distance between your feet. Experts advise you to assume the lightning crouch, shown in the photo, with your feet close together and your heels off the ground.

GFI Circuits

If you are standing in good electrical contact with the ground, you are grounded. If you then accidentally touch a hot wire with your hand, a dangerous current could pass through you to ground. In kitchens and bathrooms, where grounding on damp floors is a good possibility, building codes require *ground fault interrupter* outlets, abbreviated GFI. A GFI outlet, as shown in **FIGURE 26.16**, has two buttons, one red and one black, for testing and resetting the circuitry. Some devices, such as hair dryers, are generally constructed with a GFI in the power cord.

GFI outlets have a built-in sensing circuit that compares the currents in the hot and neutral wires of the outlet. In normal operation, all the current coming in through the hot wire passes through the device and then back out through the neutral wire, so the currents in the hot and neutral wires should always be equal. If the current in the hot wire does *not* equal the current in the neutral wire, some current from the hot wire is finding an alternative path to ground—perhaps through a person. This is a *ground fault,* and the GFI disconnects the circuit. GFIs are set to trip at current differences of about 5 mA—large enough to feel, but not large enough to be dangerous.

FIGURE 26.16 A GFI outlet.

STOP TO THINK 26.4 Suppose a current enters the body at one point and exits at another. Using the model of resistance of the body presented in Figure 26.13 rank the following current paths in terms of resistance, from highest to lowest. Assume that the skin resistance is zero.

A. Right hand to left hand
B. Right hand to right foot
C. Right hand to left foot
D. Right foot to left foot

26.5 Capacitor Circuits

In ◀◀ SECTION 23.7 we analyzed the one-time charging or discharging of a capacitor in an RC circuit. In this chapter we will look at capacitors in circuits with an AC source of emf that repeatedly charges and discharges the capacitor.

FIGURE 26.17a shows a current i_C charging a capacitor with capacitance C. The instantaneous capacitor voltage is $v_C = q/C$, where $\pm q$ is the charge on the two capacitor plates at this instant of time. **FIGURE 26.17b**, where capacitance C is connected across an AC source of emf $\mathcal{E}$, is the most basic capacitor circuit. The capacitor is in parallel with the source, so the capacitor voltage equals the emf: $v_C = \mathcal{E} = \mathcal{E}_0 \cos(2\pi f t)$. It is useful to write

$$v_C = V_C \cos(2\pi f t) \tag{26.15}$$

where V_C is the peak or maximum voltage across the capacitor. $V_C = \mathcal{E}_0$ in this single-capacitor circuit.

Charge flows to and from the capacitor plates but not through the gap between the plates. But the charges $\pm q$ on the opposite plates are always of equal magnitude, so the currents into and out of the capacitor must be equal. We can find the current i_C by considering how the charge q on the capacitor varies with time. In **FIGURE 26.18a** on the next page we have plotted the oscillating voltage v_C across the capacitor. Because the charge and the capacitor voltage are directly proportional, with $q = Cv_C$, the graph of q, shown in **FIGURE 26.18b**, looks like the graph of v_C. We say that the charge is *in phase* with the voltage.

FIGURE 26.17 An AC capacitor circuit.

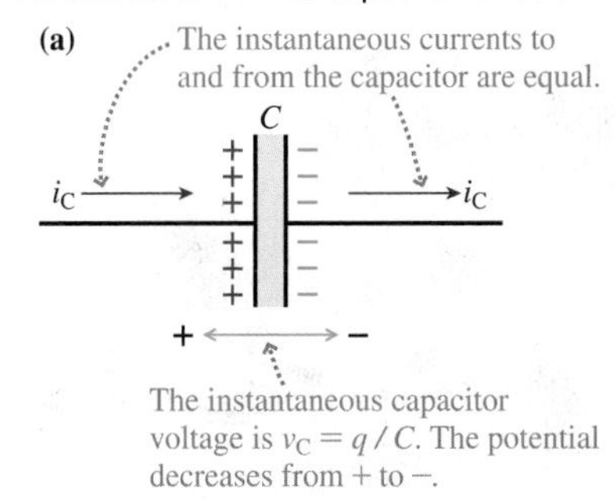

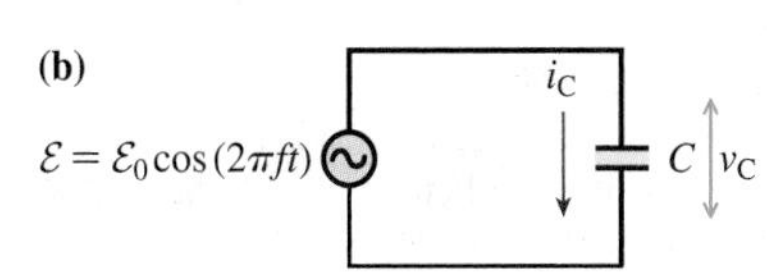

FIGURE 26.18 Voltage, charge, and current graphs for a capacitor in an AC circuit.

(a)

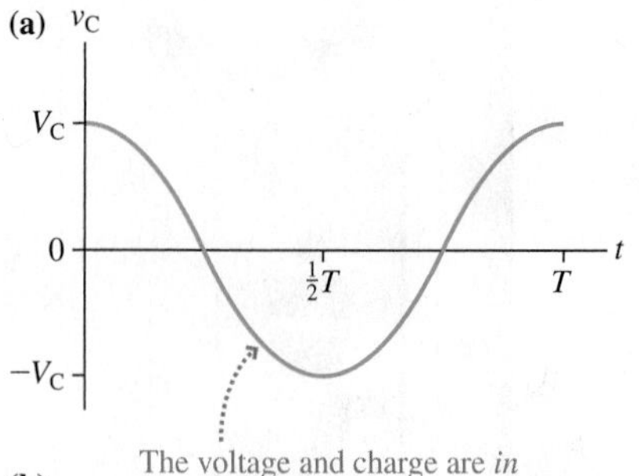

(b)

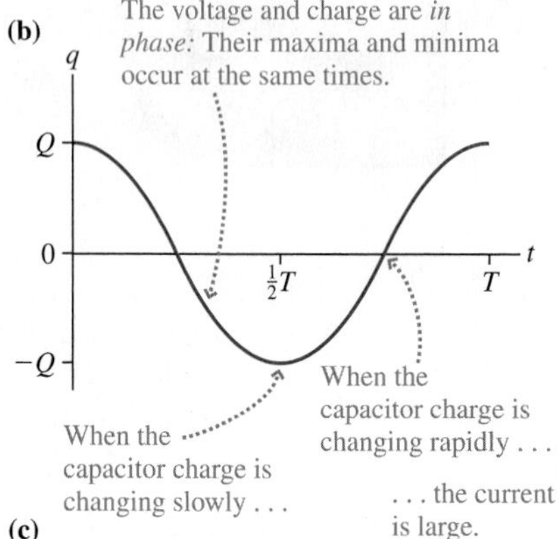

(c)

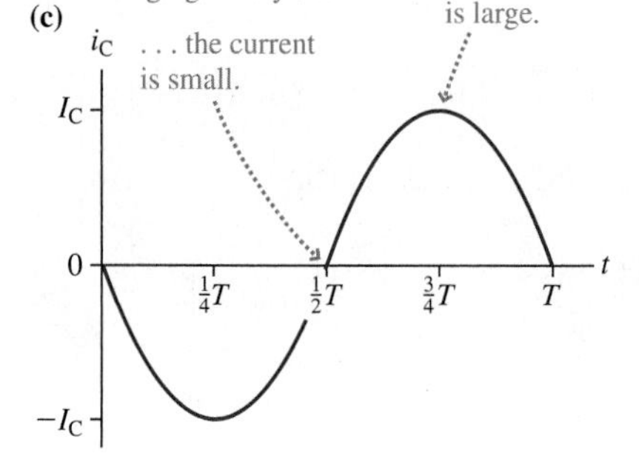

Detecting digits Under the surface of a trackpad or a touchscreen is an array of tiny capacitors. When you touch the pad or screen, your finger's high dielectric constant (it's largely water) changes the capacitance of nearby capacitors, altering the current in the capacitor circuits and thereby revealing the location of your finger. The eraser end of a pencil, with its small dielectric constant, won't work—try it!

In Chapter 22, we defined current to be $\Delta q/\Delta t$. The capacitor current, i_C, will thus be related to the charge on the capacitor by

$$i_C = \frac{\Delta q}{\Delta t}$$

The main factor determining i_C is Δq, the *change* of charge during a time interval Δt, not the amount of charge q. As the figure shows, when the current is large, the charge on the capacitor is changing rapidly, which makes sense. But these times of large current and rapid change occur when the voltage on the capacitor is small. The maximum current doesn't occur at the same time as the maximum voltage.

A capacitor's voltage and current are *not* in phase, as they were for a resistor. You can see from **FIGURE 26.18c** that the current peaks at $\frac{3}{4}$ T, one-quarter period *before* the voltage peaks. We say that **the AC current through a capacitor *leads* the capacitor voltage.**

Capacitive Reactance

The phase of the current is not the whole story. We would like to know its peak value, which we can also determine from $i_C = \Delta q/\Delta t$. Because $q = Cv_C$, this can also be written as

$$i_C = C\frac{\Delta v_C}{\Delta t} \tag{26.16}$$

Let's now reason by analogy. In ◀◀ SECTION 14.3 the position of a simple harmonic oscillator was given by $x = A\cos(2\pi ft)$. Further, the oscillator's velocity $v = \Delta x/\Delta t$ was found to be $v = -v_{max}\sin(2\pi ft)$, with a maximum or peak value $v_{max} = 2\pi fA$. Here we have an oscillating voltage $v_C = V_C\cos(2\pi ft)$, analogous to position, and we need to find the quantity $\Delta v_C/\Delta t$, analogous to velocity. Thus $\Delta v_C/\Delta t$ must have a maximum or peak value $(\Delta v_C/\Delta t)_{max} = 2\pi fV_C$.

The peak capacitor current I_C occurs when $\Delta v_C/\Delta t$ is maximum, so the peak current is

$$I_C = C(2\pi fV_C) = (2\pi fC)V_C \tag{26.17}$$

For a resistor, the peak current and voltage are related through Ohm's law:

$$I_R = \frac{V_R}{R}$$

We can write Equation 26.17 in a form similar to Ohm's law if we define the **capacitive reactance** X_C to be

$$X_C = \frac{1}{2\pi fC} \tag{26.18}$$

X_C p. 108 f INVERSE

With this definition of capacitive reactance, Equation 26.17 becomes

$$I_C = \frac{V_C}{X_C} \quad \text{or} \quad V_C = I_C X_C \tag{26.19}$$

Peak current through or voltage across a capacitor

The units of reactance, like those of resistance, are ohms.

NOTE ▶ Reactance relates the *peak* voltage V_C and current I_C. It does *not* relate the *instantaneous* capacitor voltage and current because they are out of phase; that is, $v_C \neq i_C X_C$. ◀

A capacitor's reactance X_C depends inversely on the frequency. The reactance becomes very large at low frequencies (i.e., the capacitor is a large impediment to current). The reactance decreases as the frequency increases until, at very high frequencies, $X_C \approx 0$ and the capacitor begins to act like an ideal wire.

EXAMPLE 26.7 Finding the capacitive reactance

What is the capacitive reactance of a 0.100 μF capacitor at a 100 Hz audio frequency and at a 100 MHz FM-radio frequency?

SOLVE At 100 Hz,

$$X_C(\text{at } 100\text{ Hz}) = \frac{1}{2\pi fC} = \frac{1}{2\pi(100\text{ Hz})(1.00 \times 10^{-7}\text{ F})} = 15{,}900\ \Omega$$

Increasing the frequency by a factor of 10^6 decreases X_C by a factor of 10^6, giving

$$X_C(\text{at } 100\text{ MHz}) = 0.0159\ \Omega$$

ASSESS A capacitor with a substantial reactance at audio frequencies has virtually no reactance at FM radio frequencies.

EXAMPLE 26.8 Finding a capacitor's current

A 10 μF capacitor is connected to a 1000 Hz oscillator with a peak emf of 5.0 V. What is the peak current through the capacitor?

PREPARE The circuit diagram is as in Figure 26.17b. This is a simple one-capacitor circuit.

SOLVE The capacitive reactance at $f = 1000$ Hz is

$$X_C = \frac{1}{2\pi fC} = \frac{1}{2\pi(1000\text{ Hz})(10 \times 10^{-6}\text{ F})} = 16\ \Omega$$

The peak voltage across the capacitor is $V_C = \mathcal{E}_0 = 5.0$ V; hence the peak current is

$$I_C = \frac{V_C}{X_C} = \frac{5.0\text{ V}}{16\ \Omega} = 0.31\text{ A}$$

ASSESS Using reactance and Equation 26.19 is just like using resistance and Ohm's law, but don't forget that it applies to only the *peak* current and voltage, not the instantaneous values. Further, reactance, unlike resistance, depends on the frequency of the signal.

STOP TO THINK 26.5 A capacitor is attached to an AC voltage source. Which change will result in a doubling of the current?

A. Halving the voltage and doubling the frequency
B. Doubling the frequency
C. Halving the frequency
D. Doubling the voltage and halving the frequency

26.6 Inductors and Inductor Circuits

FIGURE 26.19 shows a length of wire formed into a coil, making a solenoid. In Chapter 24 you learned that current in a solenoid creates a magnetic field inside the solenoid. If this current is *increasing,* as shown in Figure 26.19a, then the magnetic field—and thus the flux—inside the coil increases as well. According to Faraday's law, this changing flux causes an emf—a potential difference—to develop across the coil. The *direction* of the emf can be inferred from Lenz's law: Its direction will *oppose* the increase in the flux; that is, it will oppose the increase in the current. The emf will have the opposite sign if the current through the coil is decreasing, as shown in Figure 26.19b.

Coils of this kind, called **inductors,** are widely used in AC circuits. The circuit symbol for an inductor is —ꝏꝏ—. There are two primary things to remember about an inductor. First, **an inductor develops a potential difference across it if the current through it is *changing*.** Second, because the direction of this potential difference opposes the change in the current, **an inductor resists changes in the current through it.**

FIGURE 26.19 A changing current through a solenoid induces an emf across the solenoid.

(a)

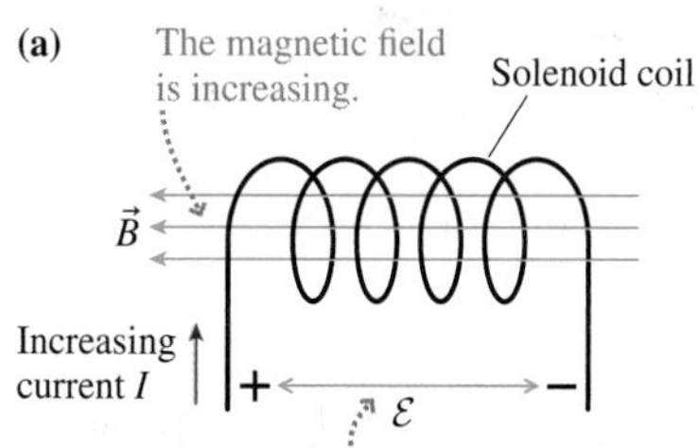

The increasing flux through the loop causes an emf to develop. By Lenz's law, the sign of the emf is such as to oppose further increases in I.

(b)

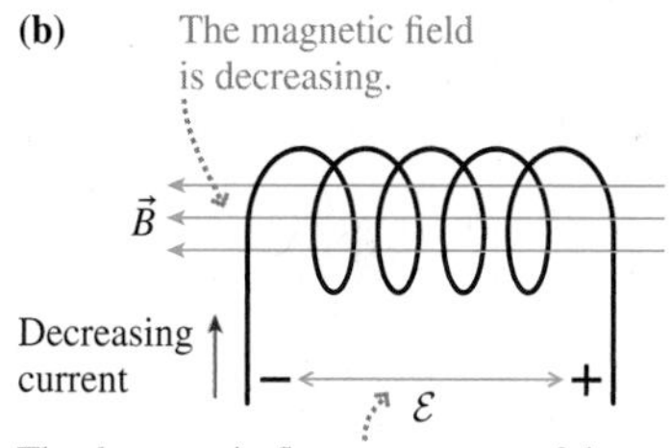

The decrease in flux causes an emf that opposes further decreases in I.

Inductance

Class Video

By Faraday's law, the voltage developed across an inductor is proportional to the rate at which the flux through the coil changes. And, because the flux is proportional to the coil's current, the instantaneous inductor voltage v_L must be proportional to $\Delta i_L/\Delta t$, the rate at which the current through the inductor changes. Thus we can write

$$v_L = L\frac{\Delta i_L}{\Delta t} \tag{26.20}$$

The constant of proportionality L is called the **inductance** of the inductor. A coil with many turns has a higher inductance than a similarly sized coil with fewer turns. Inductors often have an iron core inside their windings to increase their inductance. The magnetic field from the current magnetizes the iron core, which greatly increases the overall field through the windings. This gives a larger change in flux through the windings and hence a larger induced emf. Equation 26.20 shows that this implies a larger value of L.

From Equation 26.20 we see that inductance has units of V · s/A. It's convenient to define an SI unit of inductance called the **henry,** in honor of Joseph Henry, an early investigator of magnetism. We have

$$1 \text{ henry} = 1 \text{ H} = 1 \text{ V} \cdot \text{s/A}$$

Practical inductances are usually in the range of millihenries (mH) or microhenries (μH).

◄ **Is anybody up there?** Detectors beneath the pavement can sense the presence of cars waiting at intersections and on stretches of road before traffic lights. A slot is cut in the pavement and a coil of wire is sealed into the slot; this coil of wire is the inductor in a detection circuit. The presence of the steel in a car above the coil greatly increases the coil's inductance, signaling that a car is present.

Inductor Circuits

FIGURE 26.20 An AC inductor circuit.

(a) The instantaneous current through the inductor

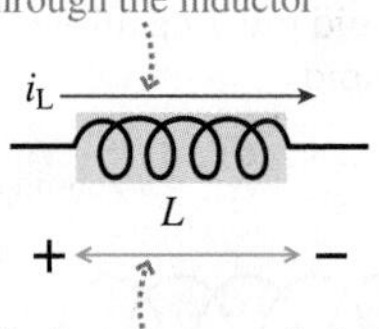

The instantaneous inductor voltage is $v_L = L(\Delta i_L/\Delta t)$.

(b)

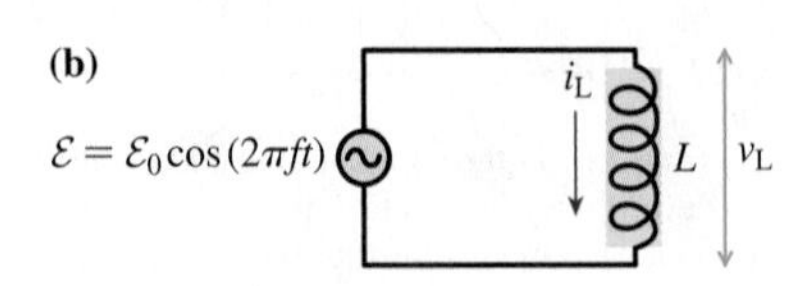

FIGURE 26.20a shows the instantaneous current i_L through an inductor. If the current is changing, the instantaneous inductor voltage is given by Equation 26.20.

FIGURE 26.20b, where inductance L is connected across an AC source of emf $\mathcal{E}$, is the simplest inductor circuit. The inductor is in parallel with the source, so the inductor voltage equals the emf: $v_L = \mathcal{E} = \mathcal{E}_0 \cos(2\pi ft)$. We can write

$$v_L = V_L \cos(2\pi ft) \tag{26.21}$$

where V_L is the peak or maximum voltage across the inductor. You can see that $V_L = \mathcal{E}_0$ in this single-inductor circuit.

We can find the inductor current i_L by considering again Equation 26.20, which tells us that the inductor voltage is high when the current is changing rapidly (i.e., $\Delta i_L / \Delta t$ is large) and is low when the current is changing slowly (i.e., $\Delta i_L / \Delta t$ is small). From this, we can graphically find the current, as shown in FIGURE 26.21.

Just as for a capacitor, the current and voltage are not in phase. There is again a phase difference of one-quarter cycle, but for an inductor the current peaks one-quarter period *after* the voltage peaks. **The AC current through an inductor *lags* the inductor voltage.**

Inductive Reactance

To find the relationship between the peak values of the inductor's current and voltage, we can use Equation 26.20 and, as we did for the capacitor, the analogy with simple harmonic motion. For the oscillating capacitor voltage, with peak value $(v_C)_{max} = V_C$, the maximum value of $\Delta v_C / \Delta t$ was $(\Delta v_C / \Delta t)_{max} = 2\pi f V_C$. If an oscillating inductor current has peak value $(i_L)_{max} = I_L$, then, by exactly the same reasoning, the maximum value of $\Delta i_L / \Delta t$ is $(\Delta i_L / \Delta t)_{max} = 2\pi f I_L$. If we use this result in Equation 26.20, we see that the maximum or peak value of the inductor voltage is

$$V_L = L(2\pi f I_L) = (2\pi f L) I_L \tag{26.22}$$

We can write Equation 26.22 in a form reminiscent of Ohm's law:

$$I_L = \frac{V_L}{X_L} \quad \text{or} \quad V_L = I_L X_L \tag{26.23}$$

Peak current through or voltage across an inductor

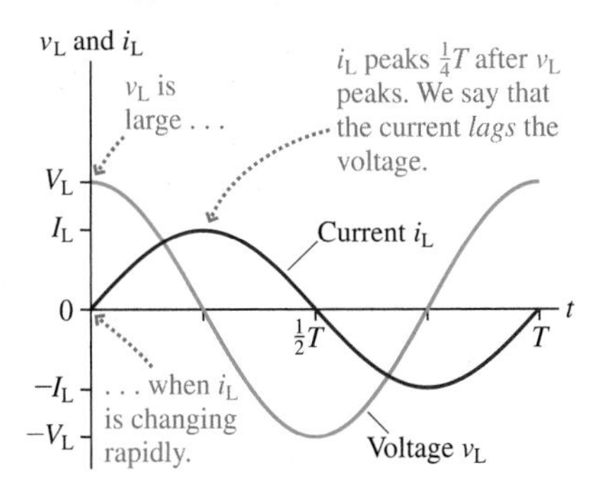

FIGURE 26.21 Voltage and current graphs for an inductor in an AC circuit.

where the **inductive reactance,** analogous to the capacitive reactance, is defined as

$$X_L = 2\pi f L \tag{26.24}$$

X_L vs. f PROPORTIONAL p. 34

The inductive reactance increases linearly as the frequency increases. This makes sense. Faraday's law tells us that the induced voltage across a coil increases as the rate of change of B increases, and B is directly proportional to the inductor current.

▶ **Clean power** The digital circuits inside computers generate high-frequency AC signals that can "leak" through the power supply and propagate through a household's electricity supply. To help prevent this, there are one or more inductors—in this case, a copper coil with an iron core—on the board inside a computer that connect it to your household electricity. At high frequencies, the inductive reactance is high enough to significantly reduce the unwanted transmissions.

Now that we've seen how resistors, capacitors, and inductors behave in circuits, it's worthwhile to collect the details to compare the similarities and differences.

SYNTHESIS 26.1 AC circuit elements

The graphs of current and voltage are always sinusoidal, but the phase is different for the three circuit elements.

Resistor The current and voltage are *in phase:* they peak or are zero at the same time.

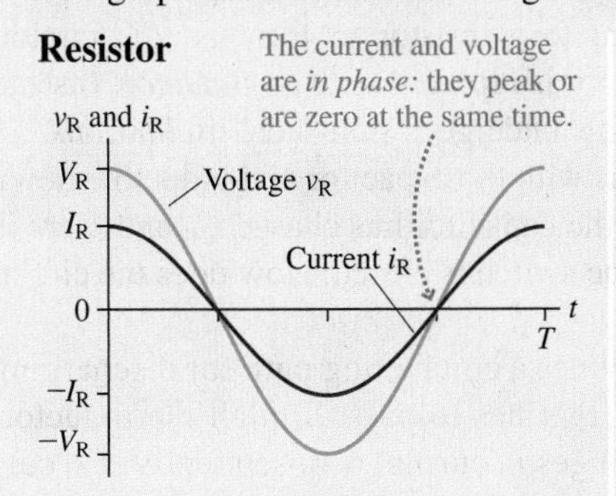

Capacitor The current *leads* the voltage: the current peaks before the voltage does.

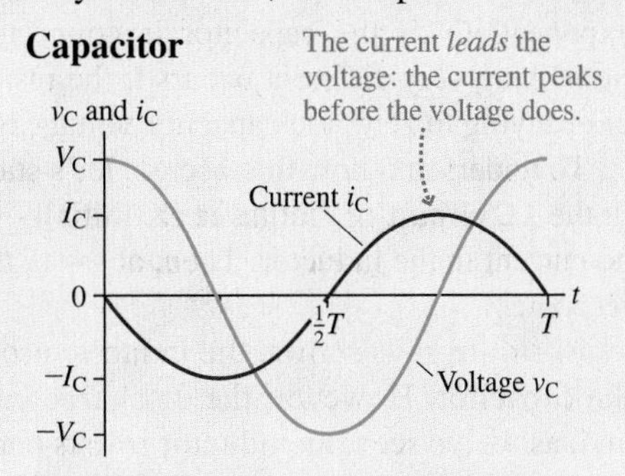

Inductor The current *lags* the voltage: the current peaks after the voltage does.

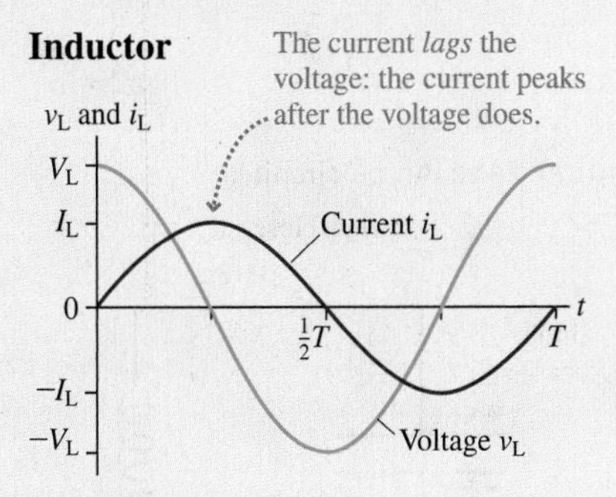

The *peak* values of current and voltage are related by the resistance or the reactance:

$V_R = I_R R$ ← Resistance (Ω)

$V_C = I_C X_C$ ← Capacitive reactance (Ω)

$V_L = I_L X_L$ ← Inductive reactance (Ω)

Resistance is constant, but capacitive and inductive reactance vary with frequency:

Does not depend on frequency → $R = \text{constant}$

Inversely proportional to frequency → $X_C = \dfrac{1}{2\pi f C}$ (Capacitance (F); Frequency (Hz))

Proportional to frequency → $X_L = 2\pi f L$ (Inductance (H); Frequency (Hz))

EXAMPLE 26.9 Finding the current and voltage of a radio's inductor

A 0.25 μH inductor is used in an FM radio circuit that oscillates at 100 MHz. The current through the inductor reaches a peak value of 2.0 mA at $t = 5.0$ ns. What is the peak inductor voltage, and when, closest to $t = 5.0$ ns, does it occur?

PREPARE The inductor is in a circuit that is driven at a certain frequency, just like that in Figure 26.20b. Synthesis 26.1 shows that the inductor current lags the voltage; the voltage reaches its peak value one-quarter period before the current reaches its peak.

SOLVE The inductive reactance at $f = 100\text{ MHz} = 1.0 \times 10^8$ Hz is

$$X_L = 2\pi fL = 2\pi(1.0 \times 10^8\text{ Hz})(0.25 \times 10^{-6}\text{ H}) = 157\ \Omega$$

Thus the peak voltage is $V_L = I_L X_L = (0.0020\text{ A})(157\ \Omega) = 0.31$ V. The voltage peak occurs one-quarter period before the current peaks, and we know that the current peaks at $t = 5.0$ ns. The period of a 100 MHz oscillation is 10 ns, so the voltage peaks at

$$t = 5.0\text{ ns} - \frac{10\text{ ns}}{4} = 2.5\text{ ns}$$

STOP TO THINK 26.6 An inductor is attached to an AC voltage source. Which change will result in a halving of the current?

A. Halving the voltage and doubling the frequency
B. Doubling the frequency
C. Halving the frequency
D. Doubling the voltage and halving the frequency

26.7 Oscillation Circuits

All of the radio stations in your city are broadcasting all the time, but you can tune a radio to pick up one station and no other. This is done using an *oscillation circuit,* a circuit that is designed to have a particular frequency at which it "wants" to oscillate. Tuning your radio means adjusting the frequency of the oscillation circuit to equal that of the station you want to listen to.

LC Circuits

You learned in Chapter 23 that the voltage across a charged capacitor decays exponentially if the capacitor is connected to a resistor to form an *RC* circuit. Something very different occurs if the resistor is replaced with an *inductor.* Instead of decaying to zero, the capacitor voltage now undergoes sinusoidal *oscillations.*

FIGURE 26.22 An *LC* circuit.

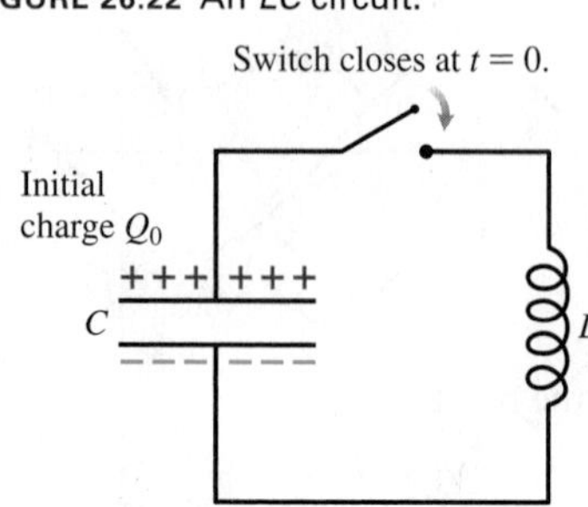

To understand how this occurs, let's start with the capacitor and inductor shown in the ***LC* circuit** of FIGURE 26.22. Initially, the capacitor has charge Q_0 and there is no current in the inductor. Then, at $t = 0$, the switch is closed. How does the circuit respond?

As FIGURE 26.23 shows, the inductor provides a conducting path for discharging the capacitor. However, the discharge current has to pass through the inductor, and, as we've seen, an inductor resists changes in current. Consequently, the current doesn't stop when the capacitor charge reaches zero. A block attached to a stretched spring is a useful mechanical analogy. The capacitor starts with a charge, like starting with the block pulled to the side and the spring stretched. Closing the switch to discharge the capacitor is like releasing the block. But the block doesn't stop when it reaches the origin—it keeps it going until the spring is fully compressed. Likewise, the current continues until it has recharged the capacitor with the opposite polarization. This process repeats over and over, charging the capacitor first one way, then the other. The charge and current *oscillate.*

FIGURE 26.23 The capacitor charge oscillates much like a block attached to a spring.

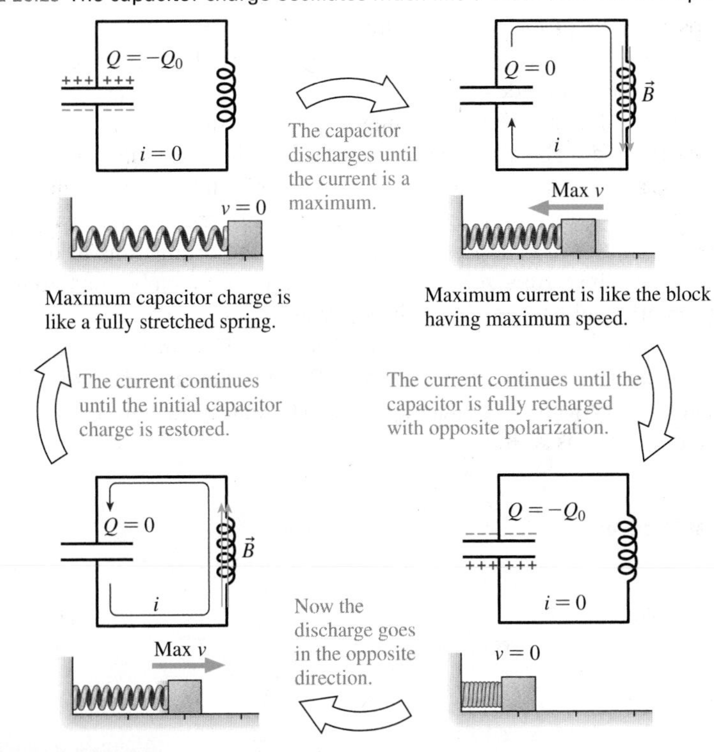

Q = −Q0
i = 0
v = 0
The capacitor discharges until the current is a maximum.
Q = 0
i
B
Max v
Maximum capacitor charge is like a fully stretched spring.
Maximum current is like the block having maximum speed.
The current continues until the initial capacitor charge is restored.
The current continues until the capacitor is fully recharged with opposite polarization.
Q = 0
B
i
Max v
Now the discharge goes in the opposite direction.
Q = −Q0
i = 0
v = 0

STOP TO THINK 26.7 Consider the oscillation circuit shown in Figure 26.23. During the portion of the cycle when the capacitor's top plate is positive and the amount of charge is increasing, the current through the inductor is

A. Clockwise and increasing.
B. Clockwise and decreasing.
C. Zero.
D. Counterclockwise and increasing.
E. Counterclockwise and decreasing.

Recall that the oscillation frequency of a mass m on a spring with spring constant k is

$$f = \frac{1}{2\pi}\sqrt{\frac{k}{m}}$$

The oscillation frequency thus depends only on the two basic parameters k and m of the system and not on the amplitude of the oscillation. Similarly, the frequency of an LC oscillator is determined solely by the values of its inductance and capacitance. We would expect larger values of L and C to cause an oscillation with a lower frequency, because a larger inductance means the current changes more slowly and a larger capacitance takes longer to discharge. A detailed analysis shows that the frequency has a form reminiscent of that for a mass and spring:

$$f = \frac{1}{2\pi\sqrt{LC}} \qquad (26.25)$$

Frequency of an LC oscillator

EXAMPLE 26.10 The frequency of an *LC* oscillator

An LC circuit consists of a 10 μH inductor and a 500 pF capacitor. What is the oscillator's frequency?

SOLVE From Equation 26.25 we have

$$f = \frac{1}{2\pi\sqrt{LC}} = \frac{1}{2\pi\sqrt{(10 \times 10^{-6}\ \text{H})(500 \times 10^{-12}\ \text{F})}}$$

which gives $f = 2.3 \times 10^6$ Hz $= 2.3$ MHz.

ASSESS The frequencies of LC oscillators are generally *much* higher than those of mechanical oscillators. Frequencies in the MHz or even GHz (10^9 Hz) range are typical.

FIGURE 26.24 An *RLC* circuit.

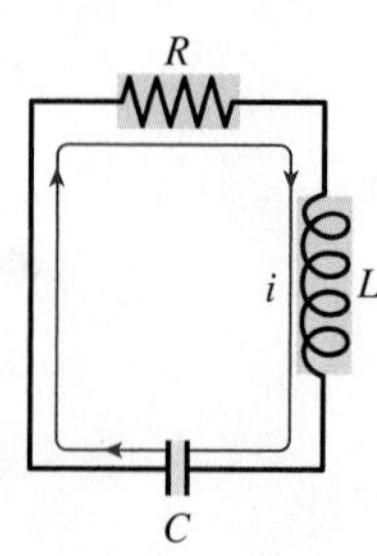

RLC Circuits

Once the oscillations of the ideal *LC* circuit of Figure 26.23 are started, they will continue forever. All real circuits, however, have some resistance, which dissipates energy, causing the amplitudes of the voltages and currents to decay.

We can model this situation by adding a series resistor R to our *LC* circuit, as shown in FIGURE 26.24. The circuit still oscillates, but the peak values of the voltage and current decrease with time as the current through the resistor transforms the electric and magnetic energy of the circuit into thermal energy. Because the power loss is proportional to R, a larger resistance causes the oscillations to decay more rapidly, as shown in FIGURE 26.25. This behavior is completely analogous to the damped harmonic oscillator discussed in ◀◀ SECTION 14.6.

FIGURE 26.25 An *RLC* circuit exhibits damped oscillations.

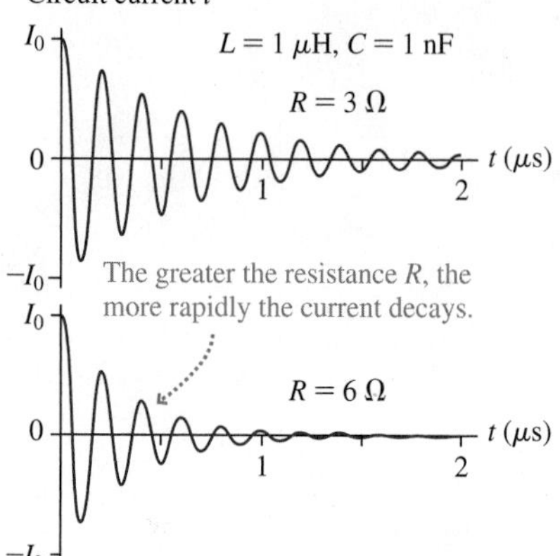

To keep the circuit oscillating, we need to add an AC source to the circuit. This makes the **driven *RLC* circuit** of FIGURE 26.26. Because the reactances of the capacitor and inductor vary with the frequency of the AC source, the current in this circuit varies with frequency as well.

FIGURE 26.26 A driven *RLC* circuit.

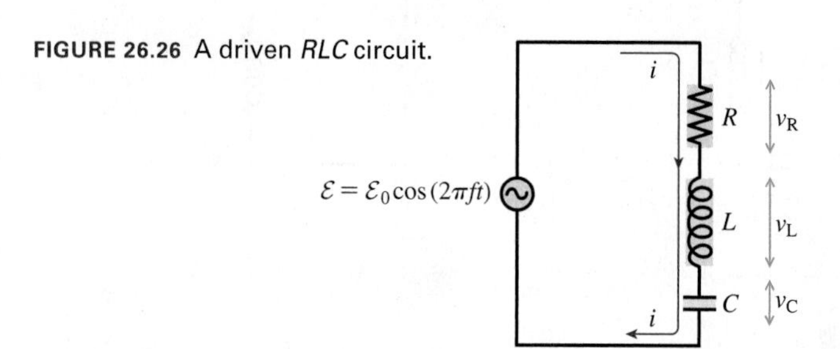

Recall that the reactance of a capacitor or inductor plays the same role for the peak quantities I and V as does the resistance of a resistor. When the reactance is large, the current through the capacitor or inductor is small. Because this is a series circuit, with the same current throughout, any circuit element with a large resistance or reactance can block the current.

If the AC source frequency f is very small, the capacitor's reactance $X_C = 1/(2\pi f C)$ is extremely large. If the source frequency f becomes very large, the inductor's reactance $X_L = 2\pi f L$ becomes extremely large. This has two consequences. First, the current in the driven *RLC* circuit will approach zero at very low and very high frequencies. Second, there must be some intermediate frequency, where neither X_C nor X_L is too large, at which the circuit current I is a maximum. The frequency f_0 at which the current is at its maximum value is called the **resonance frequency.**

Nuclear magnetic resonance, or *NMR,* is an important analytic technique in chemistry and biology. In a large magnetic field, the magnetic moments of atomic nuclei rotate at tens to hundreds of MHz. This motion generates a changing magnetic field that induces an AC emf in a coil placed around the sample. Each kind of nucleus rotates at a characteristic frequency. Adding a capacitor to the coil creates an *RLC* resonance circuit that responds strongly to only one frequency—and hence to one kind of nucleus.

Resonance occurs at the frequency at which the capacitive reactance equals the inductive reactance. If the reactances are equal, the capacitor voltage and the inductor voltage are the same. But these two voltages are out of phase with the current, with the current *leading* the capacitor voltage and *lagging* the inductor voltage. At resonance, the capacitor and inductor voltages have equal magnitudes but are exactly out of phase with each other. When we add all the voltages around the loop in Kirchhoff's loop law, the capacitor and inductor voltages cancel and the current is then limited only by the resistance R. Thus the condition for resonance is $X_C = X_L$, or

$$\frac{1}{2\pi f_0 C} = 2\pi f_0 L$$

which gives

$$f_0 = \frac{1}{2\pi\sqrt{LC}} \qquad (26.26)$$

This is the same frequency at which the circuit would oscillate if it had no resistor. At this frequency, maximum current is determined by the magnitude of the emf and the resistance:

$$I_{max} = \frac{\mathcal{E}_0}{R} \tag{26.27}$$

The driven *RLC* circuit is directly analogous to the driven, damped oscillator that you studied in ◀◀ **SECTION 14.7**. A mechanical oscillator exhibits resonance by having a large-amplitude response when the driving frequency matches the system's natural frequency. Equation 26.26 is the natural frequency of the driven *RLC* circuit, the frequency at which the current would like to oscillate. The circuit has a large current response when the oscillating emf matches this frequency.

FIGURE 26.27 shows the peak current I of a driven *RLC* circuit as the emf frequency f is varied. Notice how the current increases until reaching a maximum at frequency f_0, then decreases. This is the hallmark of resonance.

As R decreases, causing the damping to decrease, the maximum current becomes larger and the curve in Figure 26.27 becomes narrower. You saw exactly the same behavior for a driven mechanical oscillator. The emf frequency must be very close to f_0 in order for a lightly damped system to respond, but the response at resonance is very large—exactly what is needed for a tuning circuit.

It is possible to derive an expression for the current graphs shown in Figure 26.27. The peak current in the *RLC* circuit is

$$I = \frac{\mathcal{E}_0}{\sqrt{R^2 + (X_L - X_C)^2}} = \frac{\mathcal{E}_0}{\sqrt{R^2 + (2\pi f L - 1/2\pi f C)^2}} \tag{26.28}$$

Peak current in an *RLC* circuit

Note that the denominator is smallest, and hence the current is largest, when $X_L = X_C$; that is, at resonance. The three peak voltages, if you need them, are then found from $V_R = IR$, $V_L = IX_L$, and $V_C = IX_C$.

FIGURE 26.27 A graph of the current I versus emf frequency for a series *RLC* circuit.

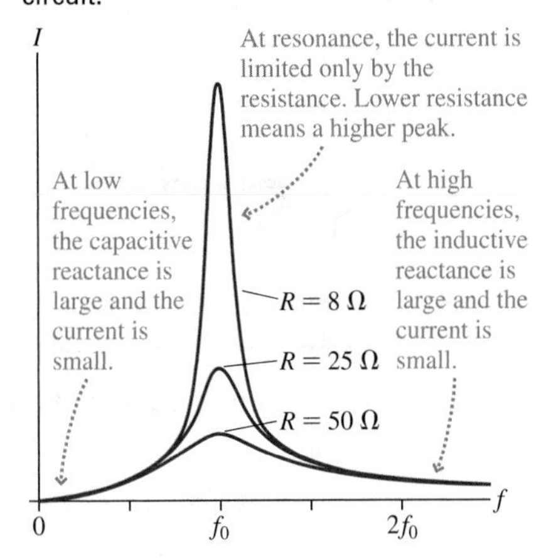

EXAMPLE 26.11 Designing a radio receiver

An AM radio antenna picks up a 1000 kHz signal with a peak voltage of 5.0 mV. The tuning circuit consists of a 60 μH inductor in series with a variable capacitor. The inductor coil has a resistance of 0.25 Ω, and the resistance of the rest of the circuit is negligible.

a. To what value should the capacitor be tuned to listen to this radio station?
b. What is the peak current through the circuit at resonance?
c. A stronger station at 1050 kHz produces a 10 mV antenna signal. What is the current at this frequency when the radio is tuned to 1000 MHz?

PREPARE The inductor's 0.25 Ω resistance can be modeled as a resistance in series with the inductance; hence we have a series *RLC* circuit. The antenna signal at $f = 1000$ kHz $= 10^6$ Hz is the emf. The circuit looks like that in Figure 26.26.

SOLVE

a. The capacitor needs to be tuned so that the resonant frequency of the circuit is $f_0 = 1000$ kHz. Because $f_0 = 1/2\pi\sqrt{LC}$, the appropriate capacitance is

$$C = \frac{1}{L(2\pi f_0)^2} = \frac{1}{(60 \times 10^{-6}\ \text{H})(2\pi \times 10^6\ \text{Hz})^2} = 4.22 \times 10^{-10}\ \text{F} = 420\ \text{pF}$$

b. $X_L = X_C$ at resonance, so the maximum current is

$$I_{max} = \frac{\mathcal{E}_0}{R} = \frac{5.0 \times 10^{-3}\ \text{V}}{0.25\ \Omega} = 0.020\ \text{A} = 20\ \text{mA}$$

c. The 1050 kHz signal is "off resonance," so the reactances X_L and X_C are not equal: $X_L = 2\pi f L = 396\ \Omega$ and $X_C = 1/2\pi f C = 359\ \Omega$ at $f = 1050$ kHz. The peak voltage of this signal is $\mathcal{E}_0 = 10$ mV. With these values, Equation 26.28 for the peak current is

$$I = \frac{\mathcal{E}_0}{\sqrt{R^2 + (X_L - X_C)^2}} = 0.27\ \text{mA}$$

ASSESS These are realistic values for the input stage of an AM radio. You can see that the signal from the 1050 kHz station is strongly suppressed when the radio is tuned to 1000 kHz. The resonant circuit has a large response to the selected station, but not nearby stations.

INTEGRATED EXAMPLE 26.12 The ground fault interrupter

As we've seen, a GFI disconnects a household circuit when the currents in the hot and neutral wires are unequal. Let's look inside a GFI outlet to see how this is done.

When you plug something in, the load completes the circuit between the hot and neutral wires. The current varies in the AC circuit thus produced, but at any instant the currents in the hot and neutral wires are equal and opposite, as **FIGURE 26.28** shows.

If someone accidentally touches a hot wire—a potentially dangerous situation—some of the current in the hot wire is diverted to a different path. The currents in the hot and neutral wires are no longer equal. This difference is sensed by the GFI, and the outlet is switched off.

FIGURE 26.28 Currents in the wires in an electric outlet.

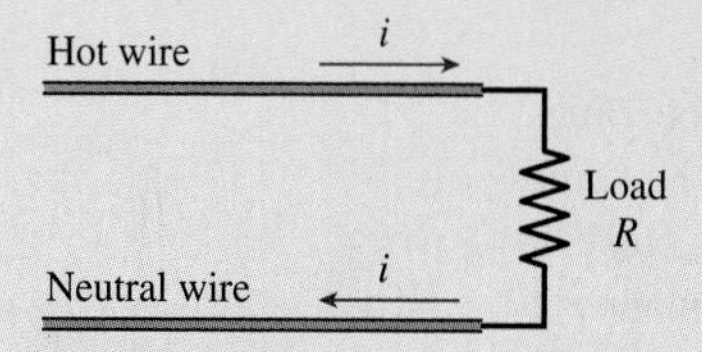

This determination is made inside the GFI outlet as shown in **FIGURE 26.29**. The hot and neutral wires thread through an iron ring. A coil of fine wire wraps around the ring; this coil is connected to a circuit that detects any current in the coil.

FIGURE 26.29 The working elements of a GFI outlet.

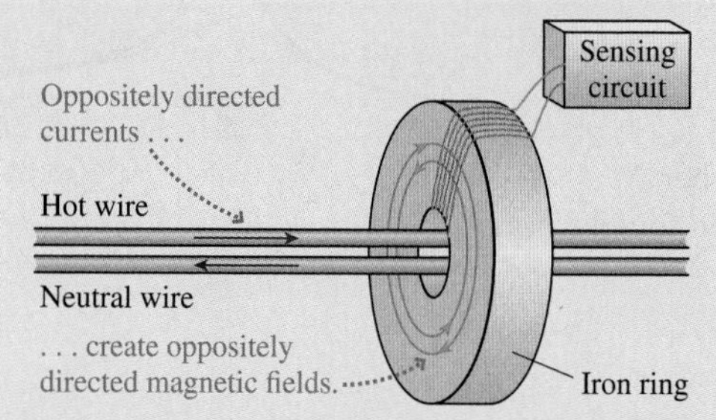

If the currents in the hot and neutral wires are equal and opposite, the magnetic fields of the two wires are equal and opposite. There is no net magnetic field in the iron ring, and thus no flux through the sensing coil. But if the two currents aren't equal, there is a net field. The current is AC, so this field isn't constant; it is changing. This field magnetizes the iron ring, increasing the overall field strength and producing a significant (and changing) flux through the sensing coil. According to Faraday's law, this changing flux induces a current in the coil. The sensing circuit detects this current and opens a small circuit breaker to turn the outlet off within a few milliseconds, in time to prevent any injury.

a. A GFI will break the circuit if the difference in current between the hot and neutral wires is 5.0 mA. Suppose that, at some instant, there is an excess current of 5.0 mA to the right in the hot wire of Figure 26.29. What are the direction and the magnitude of the magnetic field in the iron ring due to the current in the wire? The ring has an average diameter of 0.80 cm.

b. Consider a worst-case scenario: A person immersed in the bathtub reaches out of the tub and accidentally touches a hot wire with a wet, soapy hand that has minimal skin resistance. What would be the peak current through the person's body? Would this be dangerous? Would a GFI disconnect the circuit?

c. If you connect a capacitor to an outlet, the current and voltage will be out of phase, as we've seen. Would this trigger a GFI?

PREPARE Part a is about the magnetic fields of the wires. The currents in the hot and neutral wires create a field around the wires, as we saw in Chapter 24. If there is an excess current in the hot wire to the right, the right-hand rule tells us that the field will be clockwise through the iron ring in the view of Figure 26.29. We can use Equation 24.1 to find the magnitude of the magnetic field. Part b is an electrical safety question; we will need to find the resistance of the current path through the body to find the current.

SOLVE a. The iron ring has an average diameter of 0.80 cm, so we need to find the field at a distance of 0.40 cm from the wire. Equal currents produce no net field, so we need only consider the field due to the excess current in the hot wire. We learned in Chapter 24 that the field of a long, straight, current-carrying wire is $B = \mu_0 I/2\pi r$. Thus

$$B = \frac{\mu_0 I}{2\pi r} = \frac{(4\pi \times 10^{-7}\ \text{T}\cdot\text{m/A})(5.0 \times 10^{-3}\ \text{A})}{2\pi(0.0040\ \text{m})}$$
$$= 2.5 \times 10^{-7}\ \text{T} = 0.25\ \mu\text{T}$$

We noted above that the direction of this field is clockwise.

b. The person in the tub is sitting in conducting water that is well grounded. If this person touches a hot wire, there is a current path from the hot wire to ground, as **FIGURE 26.30a** shows. Resistance values for elements of the body are as noted in Figure 26.13. Because of the negligible skin resistance of a wet, soapy hand, the equivalent resistance of the body is the sum of the resistance of the arm and the torso: $R_{eq} = 340\ \Omega$.

FIGURE 26.30 The resistance of the path through the body and the resulting circuit.

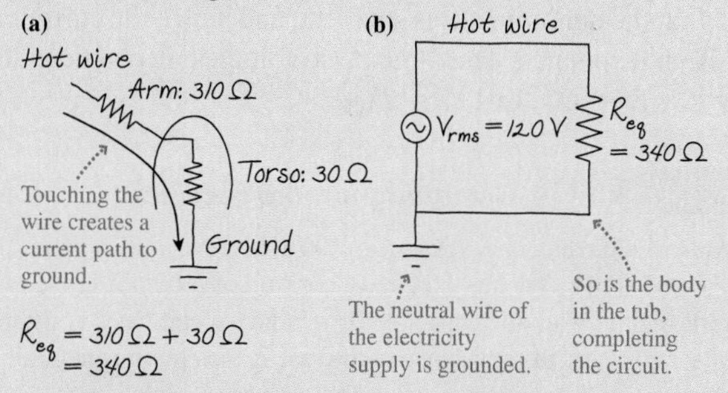

Given that the neutral wire of the electricity supply is also grounded, there is a complete circuit through the person, as shown in **FIGURE 26.30b**. The peak current in this circuit, occurring at the peak of the sinusoidal household AC voltage ($\mathcal{E}_0 = 170$ V), is

$$I_R = \frac{\mathcal{E}_0}{R_{eq}} = \frac{170\ \text{V}}{340\ \Omega} = 0.50\ \text{A}$$

Table 26.1 shows that this is a very dangerous level of current, likely to be fatal. Fortunately, it's also well above the threshold for detection by the GFI, which will quickly disconnect the circuit.

c. A difference in phase between voltage and current won't affect the GFI. The GFI detects a difference in current between two wires. As we've seen, the currents into and out of a capacitor are always equal, with no difference between them, so the presence of a capacitor in the circuit will not affect the operation of the GFI.

ASSESS The worst-case scenario is one that you could imagine happening if a person in a bathtub touches a faulty radio or picks up a faulty hair dryer. It's a situation you would expect to be dangerous, in which we'd expect a GFI to come into play.

SUMMARY

Goal: To understand and apply basic principles of AC electricity, electrical safety, and household electricity.

IMPORTANT CONCEPTS

AC circuits are driven by an emf that oscillates with frequency f.

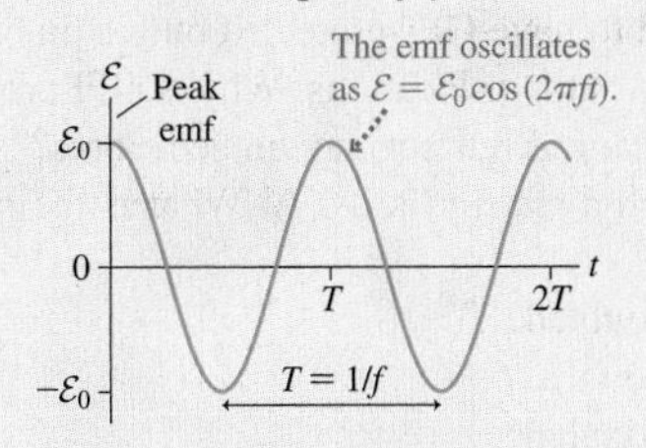

- Peak values of voltages and currents are denoted by capital letters: I, V.
- Instantaneous values of voltages and currents are denoted by lowercase letters: i, v.

Circuit elements used in AC circuits

	Resistor	Capacitor	Inductor
Symbol			
Reactance	Resistance R is constant	$X_C = 1/(2\pi f C)$	$X_L = 2\pi f L$
I and V	$V_R = I_R R$	$V_C = I_C X_C$	$V_L = I_L X_L$
Graph	v_R and i_R; i_R is in phase with v.	v_C and i_C; i_C leads v_C	v_L and i_L; i_L lags v_L

Power in AC resistor circuits

The average power dissipated by a resistor is

$$P_R = (I_{rms})^2 R = \frac{(V_{rms})^2}{R} = I_{rms} V_{rms}$$

where $I_{rms} = I_R/\sqrt{2}$ and $V_{rms} = V_R/\sqrt{2}$ are the root-mean-square (rms) voltage and current.

Electrical safety and biological effects

Currents passing through the body can produce dangerous effects. Currents larger than 15 mA (AC) and 60 mA (DC) are potentially fatal.

The body can be modeled as a network of resistors. If the body forms a circuit between two different voltages, the current is given by Ohm's law: $I = \Delta V/R$.

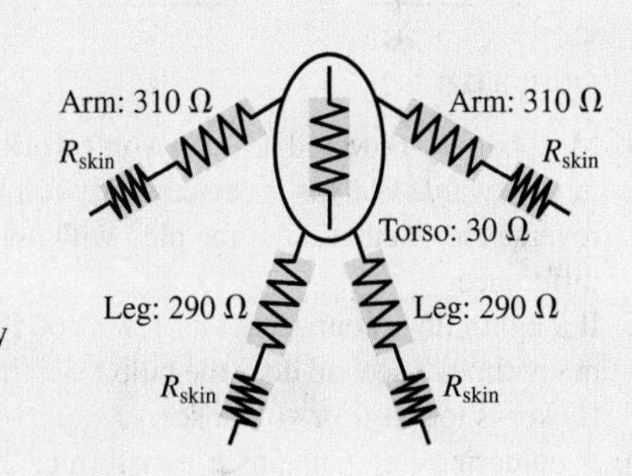

APPLICATIONS

Transformers are used to increase or decrease an AC voltage. The rms voltage at the secondary is related to the rms voltage at the primary by

$$(V_2)_{rms} = \frac{N_2}{N_1}(V_1)_{rms}$$

and the currents are related by

$$(I_2)_{rms} = \frac{N_1}{N_2}(I_1)_{rms}$$

Primary coil, N_1 turns

Secondary coil, N_2 turns

LC and *RLC* circuits

In an LC circuit, the current and voltages oscillate with frequency

$$f = \frac{1}{2\pi\sqrt{LC}}$$

In the RLC circuit, the oscillations decay as energy is dissipated in the resistor.

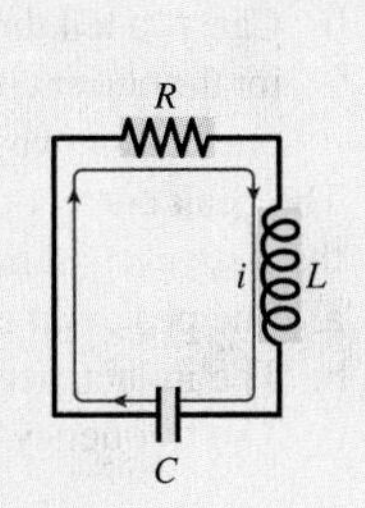

The driven *RLC* circuit

If an AC source of amplitude $\mathcal{E}_0$ is placed in an RLC circuit, then the voltages and currents oscillate continuously.

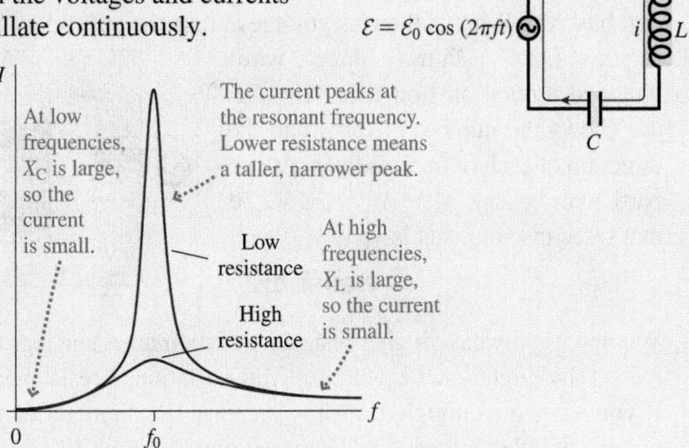

- The **resonance frequency** is $f_0 = \dfrac{1}{2\pi\sqrt{LC}}$.
- The maximum value of the current is $I_{max} = \mathcal{E}_0/R$.
- The peak current at any frequency f is given by

$$I = \frac{\mathcal{E}_0}{\sqrt{R^2 + (X_L - X_C)^2}}$$

Problem difficulty is labeled as | (straightforward) to ||||| (challenging). Problems labeled INT integrate significant material from earlier chapters; BIO are of biological or medical interest.

For assigned homework and other learning materials, go to MasteringPhysics®

Scan this QR code to launch a Video Tutor Solution that will help you solve problems for this chapter.

QUESTIONS

Conceptual Questions

1. Identical resistors are connected to separate 12 V AC sources. One source operates at 60 Hz, the other at 120 Hz. In which circuit, if either, does the resistor dissipate the greater average power?
2. Consider the three circuits in Figure Q26.2. Rank in order, from largest to smallest, the average total powers P_1 to P_3 dissipated by all the resistors in each circuit. Explain.

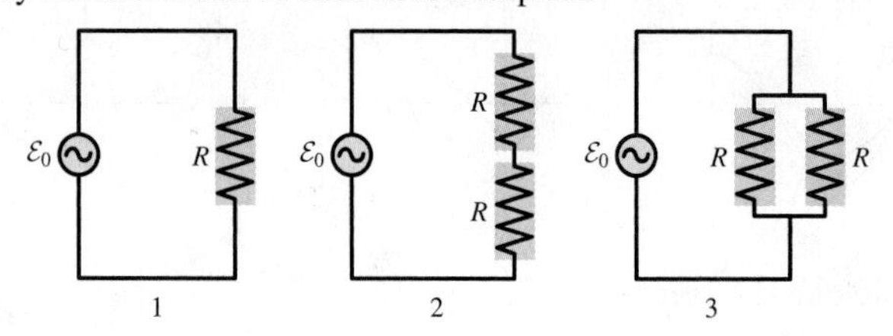

FIGURE Q26.2

3. Most battery-powered devices won't work if you put the battery in backward. But for a device that you plug in, you can often reverse the orientation of the plug with no problem. Explain the difference.
4. If a lightbulb is connected to a 120 V, 60 Hz electric outlet, how many times a second does the bulb reach peak brightness? (This flicker is too fast for you to see.)
5. A soldering gun contains a transformer that lowers the 120 V from an outlet to a few volts. It's possible to have a current of 150 A in the tip of the soldering iron even though the outlet has a circuit breaker that permits no more than 15 A. Explain how this is possible.
6. A 12 V DC power supply is connected to the primary coil of a transformer. The primary coil has 100 turns and the secondary coil has 200. What is the rms voltage across the secondary?
7. Figure Q26.7 shows three wires wrapped around an iron core. The figure shows the number of turns and the direction of each of the windings. At one particular instant, $V_A - V_B = 20$ V. At that same instant, what is $V_C - V_D$?

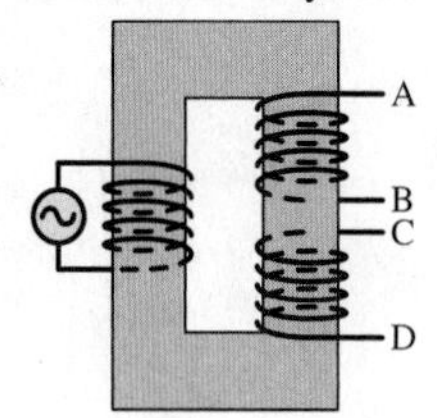

FIGURE Q26.7

8. Women usually have higher resistance of their arms and legs than men. Why might you expect to see this variation in resistance? BIO
9. If you work out enough to visibly increase the diameter of your biceps, will this increase or decrease your susceptibility to electric shock? Explain. BIO
10. A circuit breaker won't keep you from getting a shock; a GFI will. Explain why this is so. BIO
11. New homes are required to have GFI-protected outlets in bathrooms, kitchens, and any outdoor locations. Why is GFI protection required in these locations but not, say, in bedrooms? BIO
12. The peak current through a resistor is 2.0 A. What is the peak current if
    a. The resistance R is doubled?
    b. The peak emf $\mathcal{E}_0$ is doubled?
    c. The frequency f is doubled?
13. The peak current through a capacitor is 2.0 A. What is the peak current if
    a. The peak emf $\mathcal{E}_0$ is doubled?
    b. The capacitance C is doubled?
    c. The frequency f is doubled?
14. Consider the four circuits in Figure Q26.14. Rank in order, from largest to smallest, the capacitive reactances $(X_C)_1$ to $(X_C)_4$. Explain.

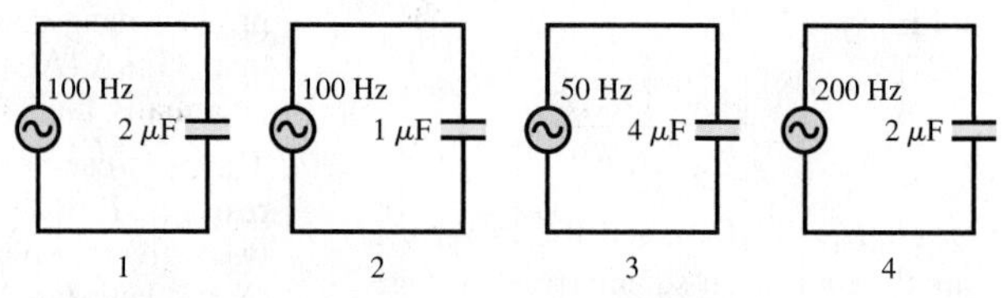

FIGURE Q26.14

15. An inductor is plugged into a 120 V, 60 Hz wall outlet in the U.S. Would the peak current be larger, smaller, or unchanged if this inductor were plugged into a wall outlet in a country where the voltage is 120 V at 50 Hz? Explain.
16. An inductive loop buried in a roadway detects the presence of cars above it, as described in the chapter. If the loop is connected to an AC supply, will the current increase or decrease when a car drives above the loop?
17. Figure Q26.17 shows two inductors and the potential difference across them at time $t = 0$ s.
    a. Can you tell which of these inductors has the larger current flowing through it at $t = 0$ s? If so, which one? If not, why not?
    b. Can you tell through which inductor the current is changing more rapidly at $t = 0$ s? If so, which one? If not, why not?

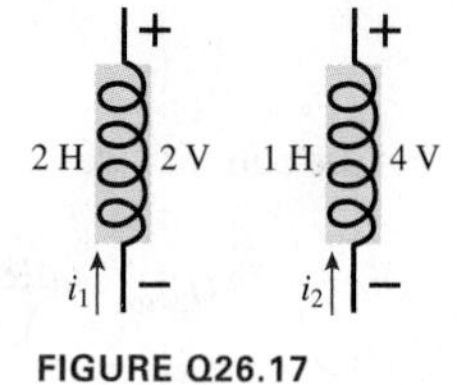

FIGURE Q26.17

18. The peak current passing through an inductor is 2.0 A. What is the peak current if
    a. The peak emf $\mathcal{E}_0$ is doubled?
    b. The inductance L is doubled?
    c. The frequency f is doubled?

19. Consider the four circuits in Figure Q26.19. Rank in order, from largest to smallest, the inductive reactances $(X_L)_1$ to $(X_L)_4$. Explain.

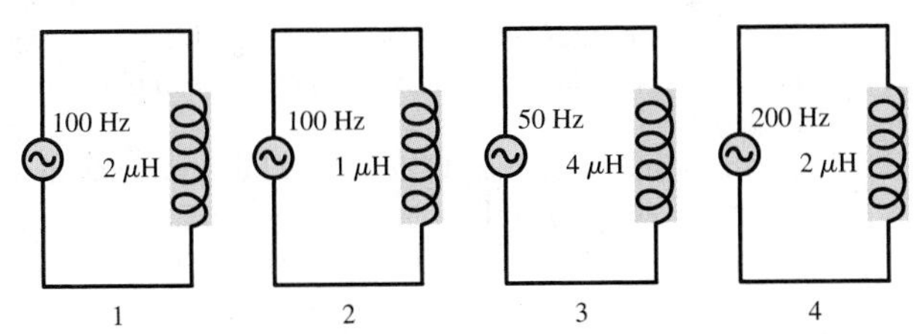

FIGURE Q26.19

20. The tuning circuit in a radio uses an *RLC* circuit. Will adjusting the resistance have any effect on the resonance frequency?
21. The resonance frequency of a driven *RLC* circuit is 1000 Hz. What is the resonance frequency if
    a. The resistance *R* is doubled?
    b. The inductance *L* is doubled?
    c. The capacitance *C* is doubled?
    d. The peak emf $\mathcal{E}_0$ is doubled?
    e. The emf frequency *f* is doubled?
22. Consider the four circuits in Figure Q26.22. They all have the same resonance frequency f_0 and are driven by the same emf. Rank in order, from largest to smallest, the maximum currents $(I_{max})_1$ to $(I_{max})_4$. Explain.

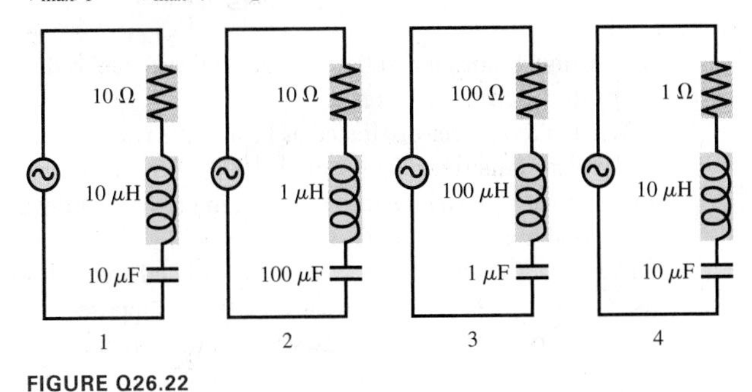

FIGURE Q26.22

Multiple-Choice Questions

23. | A transformer has 1000 turns in the primary coil and 100 turns in the secondary coil. If the primary coil is connected to a 120 V outlet and draws 0.050 A, what are the voltage and current of the secondary coil?
    A. 1200 V, 0.0050 A
    B. 1200 V, 0.50 A
    C. 12 V, 0.0050 A
    D. 12 V, 0.50 A
24. | An inductor is connected to an AC generator. As the generator's frequency is increased, the current in the inductor
    A. Increases.
    B. Decreases.
    C. Does not change.
25. | A capacitor is connected to an AC generator. As the generator's frequency is increased, the current in the capacitor
    A. Increases.
    B. Decreases.
    C. Does not change.
26. | An AC source is connected to a series combination of a lightbulb and a variable inductor. If the inductance is increased, the bulb's brightness
    A. Increases.
    B. Decreases.
    C. Does not change.
27. || An AC source is connected to a series combination of a lightbulb and a variable capacitor. If the capacitance is increased, the bulb's brightness
    A. Increases.
    B. Decreases.
    C. Does not change.
28. | The circuit shown in Figure Q26.28 has a resonance frequency of 15 kHz. What is the value of *L*?
    A. 1.6 μH
    B. 2.4 μH
    C. 5.2 μH
    D. 18 μH
    E. 59 μH

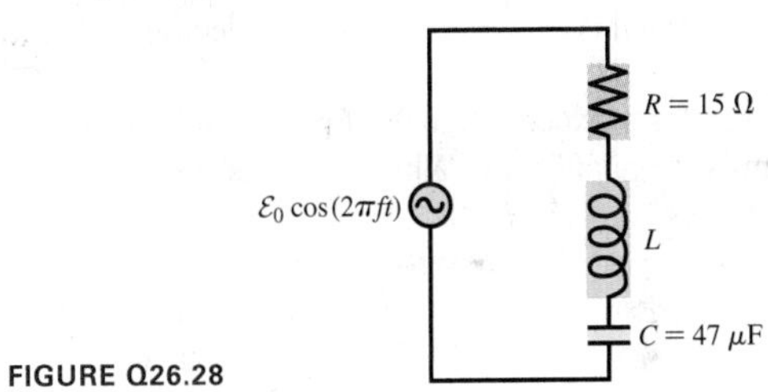

FIGURE Q26.28

29. ||| At resonance, a driven *RLC* circuit has $V_C = 5.0$ V, $V_R = 8.0$ V, and $V_L = 5.0$ V. What is the peak voltage across the entire circuit?
    A. 18 V B. 10 V C. 8.0 V D. 5.0 V E. 3.0 V
30. ||| A driven *RLC* circuit has $V_C = 5.0$ V, $V_R = 7.0$ V, and $V_L = 9.0$ V. The driving frequency is
    A. Higher than the resonance frequency.
    B. Equal to the resonance frequency.
    C. Lower than the resonance frequency.

Watch video solutions for problems: 26.5 26.11 26.19 

PROBLEMS

Section 26.1 Alternating Current

1\. | A 200 Ω resistor is connected to an AC source with $\mathcal{E}_0 = 10$ V. What is the peak current through the resistor if the emf frequency is (a) 100 Hz? (b) 100 kHz?

2\. | Figure P26.2 shows voltage and current graphs for a resistor.
a. What is the value of the resistance R?
b. What is the emf frequency f?

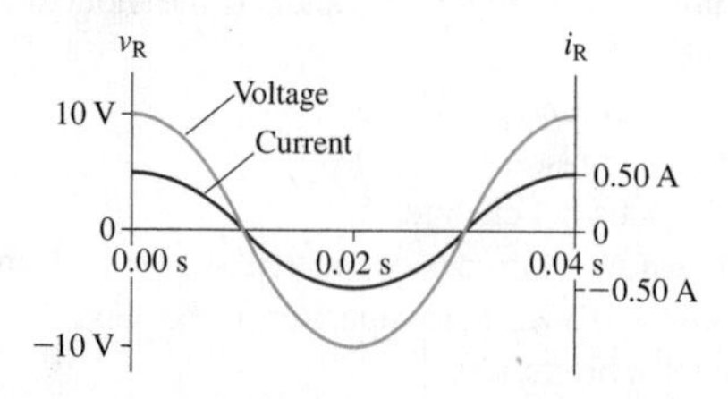

FIGURE P26.2

3\. || A resistor dissipates 2.00 W when the rms voltage of the emf is 10.0 V. At what rms voltage will the resistor dissipate 10.0 W?

4\. | The heating element of a hair dryer dissipates 1500 W when connected to a 120 V outlet. What is the resistance?

5\. || A toaster oven is rated at 1600 W for operation at 120 V, 60 Hz.
a. What is the resistance of the oven heater element?
b. What is the peak current through it?
c. What is the peak power dissipated by the oven?

6\. | A small electric space heater uses a wire that has a resistance of 24 Ω. When it is connected to a 120 V electric outlet, what is the power output?

7\. || A generator produces 40 MW of power and sends it to town at an rms voltage of 75 kV. What is the rms current in the transmission lines?

8\. | Soles of boots that are designed to protect workers from electric shock are rated to pass a maximum rms current of 1.0 mA when connected across an 18,000 V AC source. What is the minimum allowed resistance of the sole?
BIO

Section 26.2 AC Electricity and Transformers

9\. | The primary coil of a transformer is connected to a 120 V wall outlet. The secondary coil is connected to a lamp that dissipates 60 W. What is the rms current in the primary coil?

10\. || A soldering iron uses an electric current in a wire to heat the tip. A transformer with 100 turns on the secondary coil provides 50 W at an rms voltage of 24 V.
a. What is the resistance of the wire in the iron?
b. How many turns are in the primary coil?
c. What is the current in the primary coil?

11\. || A power pack charging a cell phone battery has an output of 0.40 A at 5.2 V (both rms). What is the rms current at the 120 V wall outlet where the power pack is plugged in?

12\. || A neon sign transformer has a 450 W AC output with an rms voltage of 15 kV when connected to a normal household outlet. There are 500 turns of wire in the primary coil.
a. How many turns of wire does the secondary coil have?
b. When the transformer is running at full power, what is the current in the secondary coil? The current in the primary coil?

13\. || The "power cube" transformer for a portable CD player has an output of 4.5 V and 600 mA (both rms) when plugged into a 120 V outlet.
a. If the primary coil has 400 turns of wire, how many turns are on the secondary coil?
b. What is the peak current in the primary coil?

14\. | A science hobbyist has purchased a surplus transformer that converts 7.2 kV from neighborhood distribution lines into 120 V for homes. He connects the transformer "backward," connecting 120 V power to the secondary coil and connecting a load to the primary coil. What is the voltage across the load?

15\. | A generator produces 250 kW of electric power at 7.2 kV. The current is transmitted to a remote village through wires with a total resistance of 15 Ω.
a. What is the power loss due to resistance in the wires?
b. What is the power loss if the voltage is increased to 30 kV?

Section 26.3 Household Electricity

16\. || In an old house, the wires leading to a 120 V outlet have a total resistance of 0.45 Ω. When you plug in a hair dryer, it draws a 12 A current.
a. How much does the outlet voltage decrease due to the voltage drop across the wires?
b. What is the power dissipated as heat in the wires?

17\. || A typical American family uses 1000 kWh of electricity a month. What is the average rms current in the 120 V power line to a typical house?

18\. || The wiring in the wall of your house to and from an outlet has a total resistance of typically 0.10 Ω. Suppose a device plugged into a 120 V outlet draws 10.0 A of current.
INT
a. What is the voltage drop along the wire?
b. How much power is dissipated in the wire?
c. What is the voltage drop across the device?

19\. | The following appliances are connected to a single 120 V, 15 A circuit in a kitchen: a 330 W blender, a 1000 W coffeepot, a 150 W coffee grinder, and a 750 W microwave oven. If these are all turned on at the same time, will they trip the circuit breaker?
INT

20\. | Your refrigerator uses 220 W when the compressor is running, and the compressor runs 30% of the time. How many kWh will your refrigerator use over the course of a year?

21\. || A 60 W (120 V) night light is turned on for an average of 12 h a day year round. What is the annual cost of electricity at a billing rate of \$0.10/kWh?

22\. || Suppose you leave a 110 W television and two 100 W lightbulbs on in your house to scare off burglars while you go out dancing. If the cost of electric energy in your town is \$0.12/kWh and you stay out for 4.0 h, how much does this robbery-prevention measure cost?

23\. ||| The manufacturer of an electric table saw claims that it has a 3.0 horsepower motor. 1 horsepower is approximately 750 W. It is designed to be used on a normal 120 V outlet with a 15 A circuit breaker. Is this claim reasonable? Explain.
INT

Section 26.4 Biological Effects and Electrical Safety

24. ‖ John is changing a lightbulb in a lamp. It's a warm summer evening, and the resistance of his damp skin is only 4000 Ω. While one hand is holding the grounded metal frame of the lamp, the other hand accidentally touches the hot electrode in the base of the socket. What is the current through his torso? BIO

25. ‖ In some countries AC outlets near bathtubs are restricted to a maximum of 25 V to minimize the chance of dangerous shocks while bathing. A man is in the tub; the lower end of his torso is well grounded, and the skin resistance of his wet, soapy hands is negligible. He reaches out and accidentally touches a live electric wire. What voltage on the wire would produce a dangerous 100 mA current? BIO

26. ‖ If you touch the terminal of a battery, the small area of contact means that the skin resistance will be relatively large; 50 kΩ is a reasonable value. What current will pass through your body if you touch the two terminals of a 9.0 V battery with your two hands? Will you feel it? Will it be dangerous? BIO

27. ‖ A person standing barefoot on the ground 20 m from the point of a lightning strike experiences an instantaneous potential difference of 300 V between his feet. If we assume a skin resistance of 1.0 kΩ, how much current goes up one leg and back down the other? BIO

28. ‖ Electrodes used to make electrical measurements of the body (such as those used when recording an electrocardiogram) use a conductive paste to reduce the skin resistance to very low values. Great care must be exercised when a patient is wearing such electrodes. If a patient is wearing one such electrode on each wrist, what is the minimum AC voltage that could provide a potentially fatal shock? BIO

29. ‖ A fisherman has netted a torpedo ray. As he picks it up, this electric fish creates a short-duration 50 V potential difference between his hands. His hands are wet with salt water, and so his skin resistance is a very low 100 Ω. What current passes through his body? Will he feel this DC pulse? BIO

Problems 30 and 31 concern a high-voltage transmission line. Such lines are made of bare wire; they are not insulated. Assume that the wire is 100 km long, has a resistance of 7.0 Ω, and carries 200 A.

30. ||| A bird is perched on the wire with its feet 2.0 cm apart. What is the potential difference between its feet? INT BIO

31. ||| Would it be possible for a person to safely hang from this wire? Assume that the hands are 15 cm apart, and assume a skin resistance of 2200 Ω. BIO INT

Section 26.5 Capacitor Circuits

32. ‖ A 0.30 μF capacitor is connected across an AC generator that produces a peak voltage of 10.0 V. What is the peak current through the capacitor if the emf frequency is (a) 100 Hz? (b) 100 kHz?

33. ‖ A 20 μF capacitor is connected across an AC generator that produces a peak voltage of 6.0 V. The peak current is 0.20 A. What is the oscillation frequency in Hz?

34. | The peak current through a capacitor is 10.0 mA. What is the current if
    a. The emf frequency is doubled?
    b. The emf peak voltage is doubled (at the original frequency)?
    c. The frequency is halved and, at the same time, the emf is doubled?

35. ‖ A 20 nF capacitor is connected across an AC generator that produces a peak voltage of 5.0 V.
    a. At what frequency f is the peak current 50 mA?
    b. What is the instantaneous value of the emf at the instant when $i_C = I_C$?

36. ||| A capacitor is connected to a 15 kHz oscillator that produces an rms voltage of 6.0 V. The peak current is 65 mA. What is the value of the capacitance C?

37. | The peak current through a capacitor is 8.0 mA when connected to an AC source with a peak voltage of 1.0 V. What is the capacitive reactance of the capacitor?

Section 26.6 Inductors and Inductor Circuits

38. ||| What is the potential difference across a 10 mH inductor if the current through the inductor drops from 150 mA to 50 mA in 10 μs?

39. ‖ A 20 mH inductor is connected across an AC generator that produces a peak voltage of 10.0 V. What is the peak current through the inductor if the emf frequency is (a) 100 Hz? (b) 100 kHz?

40. | The peak current through an inductor is 10.0 mA. What is the current if
    a. The emf frequency is doubled?
    b. The emf peak voltage is doubled (at the original frequency)?
    c. The frequency is halved and, at the same time, the emf is doubled?

41. | A 500 μH inductor is connected across an AC generator that produces a peak voltage of 5.0 V.
    a. At what frequency f is the peak current 50 mA?
    b. What is the instantaneous value of the emf at the instant when $i_L = I_L$?

42. ||| An inductor is connected to a 15 kHz oscillator that produces an rms voltage of 6.0 V. The peak current is 65 mA. What is the value of the inductance L?

43. | The peak current through an inductor is 12.5 mA when connected to an AC source with a peak voltage of 1.0 V. What is the inductive reactance of the inductor?

Section 26.7 Oscillation Circuits

44. ‖ A 2.0 mH inductor is connected in parallel with a variable capacitor. The capacitor can be varied from 100 pF to 200 pF. What is the range of oscillation frequencies for this circuit?

45. ‖ An FM radio station broadcasts at a frequency of 100 MHz. What inductance should be paired with a 10 pF capacitor to build a receiver circuit for this station?

46. ‖ The inductor in the *RLC* tuning circuit of an AM radio has a value of 350 mH. What should be the value of the variable capacitor in the circuit to tune the radio to 740 kHz?

47. ‖ At what frequency f do a 1.0 μF capacitor and a 1.0 μH inductor have the same reactance? What is the value of the reactance at this frequency?

48. ‖ What capacitor in series with a 100 Ω resistor and a 20 mH inductor will give a resonance frequency of 1000 Hz?

49. ‖ What inductor in series with a 100 Ω resistor and a 2.5 μF capacitor will give a resonance frequency of 1000 Hz?

50. | A series *RLC* circuit has a 200 kHz resonance frequency. What is the resonance frequency if the capacitor value is doubled and the inductor value is halved?

Watch video solution for 26.59 problem:

51. ||| An RLC circuit with a 10 μF capacitor is connected to a variable-frequency power supply with an rms output voltage of 6.0 V. The rms current in the circuit as a function of the driving frequency appears as in Figure P26.51. What are the values of the resistor and the inductor?

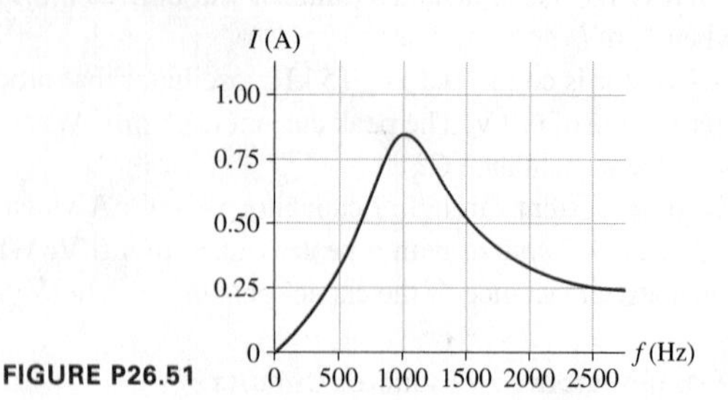

FIGURE P26.51

52. |||| A series RLC circuit consists of a 280 Ω resistor, a 25 μH inductor, and an 18 μF capacitor. What is the rms current if the emf is supplied by a standard 120 V, 60 Hz wall outlet?

General Problems

53. || Electric outlets in England are 230 V. Alice brings her electric kettle from England, where it draws 13 A, and wants to use it in the United States. She uses a step-up transformer to increase the 120 V outlet voltage to 230 V, then plugs her kettle into the secondary. What is the current in the primary in the few seconds before the 15 A circuit breaker trips?
54. ||| The voltage-to-current ratio in the primary coil of a transformer can be thought of as the transformer's effective resistance. A step-down transformer converts 120 V at the primary to 24 V at the secondary, which is connected to a load of resistance 8.0 Ω. What is the effective resistance of this transformer when connected to this load?
    Hint: Resistance is defined as the ratio of two circuit quantities.
55. |||| A 15-km-long, 230 kV aluminum transmission line delivers INT 34 MW to a city. If we assume a solid cylindrical cable, what minimum diameter is needed if the voltage decrease along this run is to be no more than 1.0% of the transmission voltage? The resistivity of aluminum is $2.7 \times 10^{-8}\ \Omega \cdot \text{m}$.
56. || The voltage across a 60 μF capacitor is described by the equation $v_C = (18\text{ V})\cos(200t)$, where t is in seconds.
    a. What is the voltage across the capacitor at $t = 0.010$ s?
    b. What is the capacitive reactance?
    c. What is the peak current?
57. || The voltage across a 75 μH inductor is described by the equation $v_L = (25\text{ V})\cos(60t)$, where t is in seconds.
    a. What is the voltage across the inductor at $t = 0.10$ s?
    b. What is the inductive reactance?
    c. What is the peak current?
58. ||| An electronics hobbyist is building a radio set to receive the AM band, with frequencies from 520 kHz to 1700 kHz. The antenna, which also serves as the inductor in an LC circuit, has an inductance of 230 μH. She needs to add a variable capacitor whose capacitance she can adjust to tune the radio. What is the minimum capacitance the capacitor must have? The maximum value?
59. || For the circuit of Figure P26.59,
    a. What is the resonance frequency?
    b. At resonance, what is the peak current through the circuit?

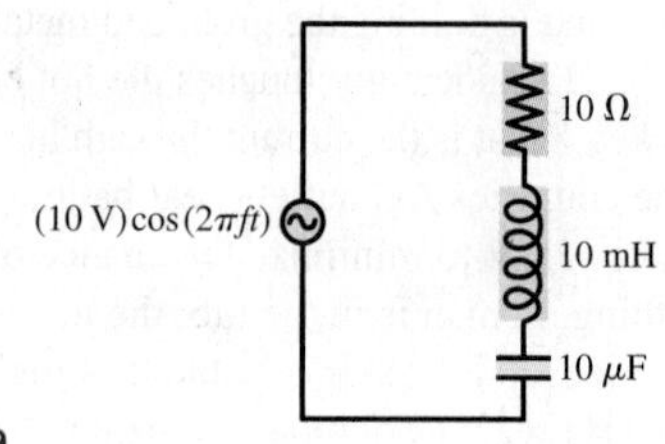

FIGURE P26.59

60. | For the circuit of Figure P26.60,
    a. What is the resonance frequency?
    b. At resonance, what is the peak current through the circuit?

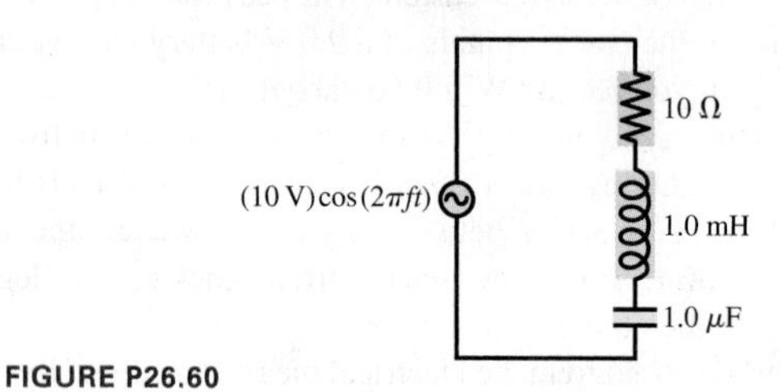

FIGURE P26.60

61. || An RLC circuit consists of a 48 Ω resistor, a 200 μF capacitor, and an inductor. The rms current is 2.5 A when the circuit is connected to a 120 V, 60 Hz outlet. What is the inductance?

MCAT-Style Passage Problems

Cell Membrane Resistance BIO

The capacitance of biological membranes is about 1.0 μF per cm^2 of membrane area, so investigators can determine the surface area of a cell membrane by using intracellular electrodes to measure the membrane's capacitive reactance. An investigator applies a 1.0 μA peak current at 40 kHz to a cell and measures the peak out-of-phase voltage—that is, the component of the voltage due to the capacitive reactance of the cell membrane—to be 0.16 V.

62. || If the frequency is doubled to 80 kHz and the current is kept the same, what will be the peak out-of-phase voltage?
    A. 0.32 V B. 0.16 V
    C. 0.080 V D. 0.040 V
63. || What is the approximate capacitance of the cell membrane?
    A. 20×10^{-11} F B. 10×10^{-11} F
    C. 5.0×10^{-11} F D. 2.5×10^{-11} F
64. || If the capacitance of a cell membrane is measured to be 6.0×10^{-11} F, what is the area?
    A. 6.0×10^{-13} m^2 B. 6.0×10^{-11} m^2
    C. 6.0×10^{-9} m^2 D. 6.0×10^{-7} m^2
65. || If the investigator applies a 1.0 μA peak current at 40 kHz to a cell with twice the membrane area of the cell noted in the passage, what will be the peak out-of-phase voltage?
    A. 0.32 V B. 0.16 V
    C. 0.080 V D. 0.040 V

Halogen Bulbs

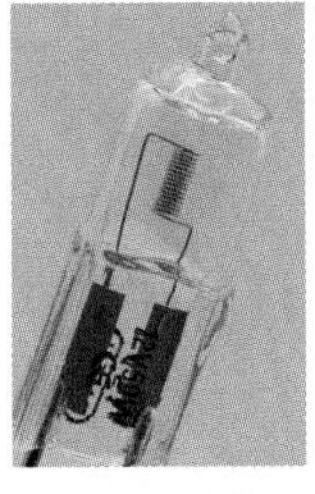

Halogen bulbs have some differences from standard incandescent lightbulbs. They are generally smaller, the filament runs at a higher temperature, and they have a quartz (rather than glass) envelope. They may also operate at lower voltage. Consider a 12 V, 50 W halogen bulb for use in a desk lamp. The lamp plugs into a 120 V, 60 Hz outlet, and it has a transformer in its base.

66. | The 12 V rating of the bulb refers to the rms voltage. What is the peak voltage across the bulb?
    A. 8.5 V
    B. 12 V
    C. 17 V
    D. 24 V

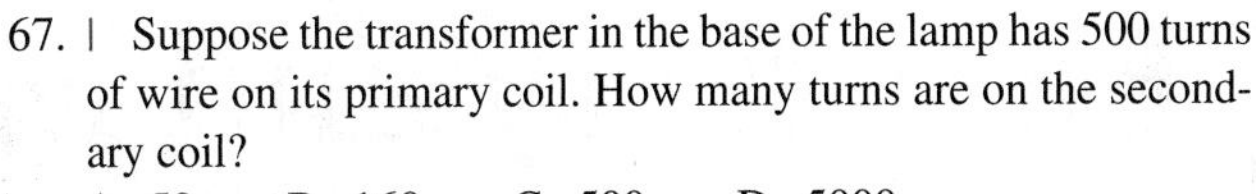

67. | Suppose the transformer in the base of the lamp has 500 turns of wire on its primary coil. How many turns are on the secondary coil?
    A. 50 B. 160 C. 500 D. 5000
68. | How much current is drawn by the lamp at the outlet? That is, what is the rms current in the primary?
    A. 0.42 A
    B. 1.3 A
    C. 4.2 A
    D. 13 A
69. | What will be the voltage across the bulb if the lamp's power cord is accidentally plugged into a 240 V, 60 Hz outlet?
    A. 2 V
    B. 24 V
    C. 36 V
    D. 48 V

STOP TO THINK ANSWERS

Chapter Preview Stop to Think: A. When two resistors are connected in series, the equivalent resistance is the sum of the two resistances and is greater than either of the individual resistances. When two resistors are connected in parallel, the equivalent resistance is less than either of the individual resistances. The series combination thus has the greater resistance.

Stop to Think 26.1: B. The power in the AC circuit is proportional to the square of the rms current, or to $(I_{rms})^2 = (I_R/\sqrt{2})^2 = \frac{1}{2}{I_R}^2 = \frac{1}{2}(1\ \text{A})^2$. The power in the DC circuit is proportional to the square of the DC current, or to ${I_R}^2 = (1\ \text{A})^2$. Thus the power in the DC circuit is twice that in the AC circuit, so bulb B is brighter.

Stop to Think 26.2: D > B > A > C. The rms voltage on the load resistor is the voltage on the secondary, given by Equation 26.13. The primary voltage is the same for each transformer, so the secondary voltage is determined by the ratio of the turns on the secondary to the turns on the primary, N_2/N_1. A higher ratio means a higher voltage.

Stop to Think 26.3: D > C > A > B. The energy the devices use, in kWh, is the power in kW multiplied by the time used in hours: A. 0.48 kWh, B. 0.36 kWh, C. 2.0 kWh, D. 3.0 kWh.

Stop to Think 26.4: A > B = C > D. Any way that current goes through the body, it goes through the torso; this contributes 30 Ω to the current path in every case. Figure 26.13 shows the resistance of each arm to be 310 Ω and that of each leg to be slightly lower, 290 Ω. The highest resistance is thus a path through two arms, the lowest is a path through two legs, with a path through one arm and one leg in the middle.

Stop to Think 26.5: B. The current is $I_C = V_C/X_C = 2\pi f C V_C$. Thus I_C is proportional to both f and V_C. Doubling the frequency while keeping V_C constant will double the current.

Stop to Think 26.6: B. The current is $I_L = V_L/X_L = V_L/2\pi f L$. Thus I_L is *inversely* proportional to f. Doubling the frequency while keeping V_L constant will halve the current.

Stop to Think 26.7: E. The charge on the capacitor is increasing, so the current must be counterclockwise. As the capacitor charges up, the capacitor voltage opposes the current, so the current is decreasing. This corresponds to the time between the two figures on the left side of Figure 26.23.

Electricity and Magnetism

Mass and charge are the two most fundamental properties of matter. The first four parts of this text were about properties and interactions of masses. Part VI has been a study of charge—what charge is and how charges interact.

Electric and magnetic fields were introduced to enable us to understand the long-range forces of electricity and magnetism. One charge—the source charge—alters the space around it by creating an electric field and, if the charge is moving, a magnetic field. Other charges experience forces exerted *by the fields.* Electric and magnetic fields are the agents by which charges interact.

In addition to the electric field, we often describe electric interactions in terms of the electric potential. This is a particularly fruitful concept for dealing with electric circuits in which charges flow through wires, resistors, etc.

Electric and magnetic fields are *real* and can exist independently of charges. The clearest evidence for their independent existence is electromagnetic waves—the quintessential electromagnetic phenomenon. All electromagnetic waves, including light, are similar in structure, but they span a wide range of wavelengths and frequencies. In Part VI, we got our first hints that electromagnetic waves might have a particle-like nature, a concept we will explore further in Part VII.

Part VI has introduced many new phenomena, concepts, and laws. The knowledge structure below draws together the major ideas. These ideas build on what we've learned in the first five parts of this text. All the pieces are now in place to support the development of the ideas of modern physics in Part VII.

KNOWLEDGE STRUCTURE VI Electricity and Magnetism

BASIC GOALS How do charged particles interact? How do electric circuits work? What are the properties and characteristics of electric and magnetic fields?

GENERAL PRINCIPLES

Forces between charges: **Coulomb's law**	$F_{1\text{ on }2} = F_{2\text{ on }1} = \dfrac{K\lvert q_1\rvert\lvert q_2\rvert}{r^2}$	The force is along the line connecting the charges. For like charges, the force is repulsive; for opposite charges, attractive.
Electric force on a charge:	$\vec{F} = q\vec{E}$	The force is in the direction of the field for a positive charge; opposite the field for a negative charge.
Magnetic force on a moving charge:	$F = \lvert q\rvert vB\sin\alpha$	The force is perpendicular to the velocity and the field, with direction as specified by the right-hand rule for forces.
Induced emf: **Faraday's law**	$\mathcal{E} = \left\lvert \dfrac{\Delta\Phi}{\Delta t} \right\rvert$	The induced current $I = \mathcal{E}/R$ is such that the induced magnetic field opposes the change in the magnetic flux. This is **Lenz's law.**

Electric and magnetic fields

Charges and changing magnetic fields create electric fields.

- Electric fields exert forces on charges and torques on dipoles.
- The electric field is perpendicular to equipotential surfaces and points in the direction of decreasing potential.
- The electric field causes charges to move in conductors but not insulators.

Currents and permanent magnets create magnetic fields.

- Magnetic fields exert forces on currents (and moving charged particles) and torques on magnetic dipoles.
- A compass needle or other magnetic dipole will line up with a magnetic field.

Electric potential

The interaction of charged particles can also be described in terms of an electric potential V.

- Only potential differences ΔV are important.
- If the potential of a particle of charge q changes by ΔV, its potential energy changes by $\Delta U = q\Delta V$.
- Where two equipotential surfaces with potential difference ΔV are separated by distance d, the electric field strength is $E = \Delta V/d$.

Current and circuits

Potential differences ΔV drive current in circuits. Though electrons are the charge carriers in metals, the **current** I is defined to be the motion of positive charges.

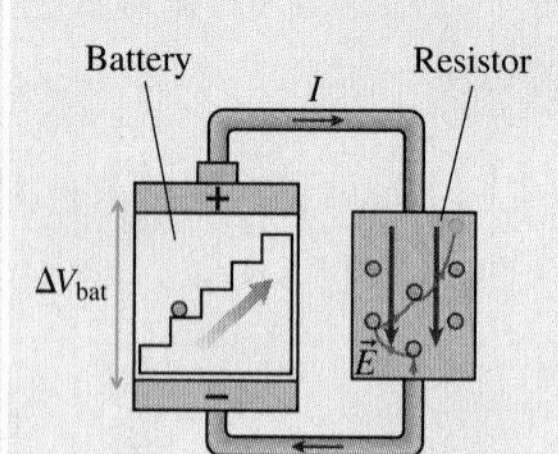

- Circuits obey Kirchhoff's loop law (conservation of energy) and Kirchhoff's junction law (conservation of charge).
- The current through a resistor is $I = \Delta V_R/R$. This is **Ohm's law.**

Electromagnetic waves

An electromagnetic wave is a self-sustaining oscillation of electric and magnetic fields.

- $\vec{E}$ and $\vec{B}$ are perpendicular to each other and to the direction of travel.
- All electromagnetic waves travel at the same speed, c.
- The **electromagnetic spectrum** is the spread of wavelengths and frequencies of electromagnetic waves, from radio waves through visible light to gamma rays.

The Greenhouse Effect and Global Warming

Electromagnetic waves are real, and we depend on them for our very existence; energy carried by electromagnetic waves from the sun provides the basis for all life on earth. Because of the sun's high surface temperature, it emits most of its thermal radiation in the visible portion of the electromagnetic spectrum. As the figure below shows, the earth's atmosphere is transparent to the visible and near-infrared radiation, so most of this energy travels through the atmosphere and warms the earth's surface.

Although seasons come and go, *on average* the earth's climate is very steady. To maintain this stability, the earth must radiate thermal energy—electromagnetic waves—back into space at exactly the same average rate that it receives energy from the sun. Because the earth is much cooler than the sun, its thermal radiation is long-wavelength infrared radiation that we cannot see. A straightforward calculation using Stefan's law finds that the average temperature of the earth should be $-18°C$, or 0°F, for the incoming and outgoing radiation to be in balance.

This result is clearly not correct; at this temperature, the entire earth would be covered in snow and ice. The measured global average temperature is actually a balmier 15°C, or 59°F. The straightforward calculation fails because it neglects to consider the earth's atmosphere. At visible wavelengths, as the figure shows, the atmosphere has a wide "window" of transparency, but this is not true at the infrared wavelengths of the earth's thermal radiation. The atmosphere lets in the visible radiation from the sun, but the outgoing thermal radiation from the earth sees a much smaller "window." Most of this radiation is absorbed in the atmosphere.

The radiation from the sun is mostly at visible wavelengths. The atmosphere is mostly transparent to the incoming radiation.
Outgoing radiation from the much cooler earth is mostly at infrared wavelengths. The atmosphere is mostly opaque to the outgoing radiation.
Atmospheric transmission:
More than 50%
Less than 50%
Relative intensity
Solar radiation
Terrestrial radiation
100 nm
1000 nm
10 μm
100 μm
Wavelength

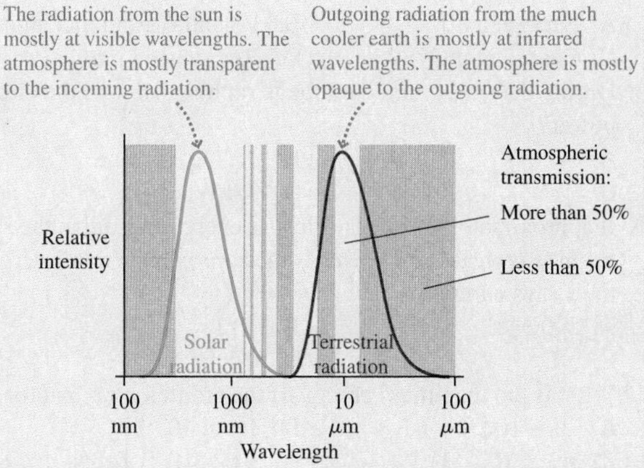

Thermal radiation curves for the sun and the earth. The shaded bands show regions for which the atmosphere is transparent (no shading) or opaque (shaded) to electromagnetic radiation.

Because it's easier for visible radiant energy to get in than for infrared to get out, the earth is warmer than it would be without the atmosphere. The additional warming of the earth's surface because of the atmosphere is called the **greenhouse effect.** The greenhouse effect is a natural part of the earth's physics; it has nothing to do with human activities, although it's doubtful any advanced life forms would have evolved without it.

The atmospheric gases most responsible for the greenhouse effect are carbon dioxide and water vapor, both strong absorbers of infrared radiation. These **greenhouse gases** are of concern today because humans, through the burning of fossil fuels (oil, coal, and natural gas), are rapidly increasing the amount of carbon dioxide in the atmosphere. Preserved air samples show that carbon dioxide made up 0.027% of the atmosphere before the industrial revolution. In the last 150 years, human activities have increased the amount of carbon dioxide by nearly 50%, to about 0.040%. By 2050, the carbon dioxide concentration will likely increase to 0.054%, double the pre-industrial value, unless the use of fossil fuels is substantially reduced.

Carbon dioxide is a powerful absorber of infrared radiation. And good absorbers are also good emitters. The carbon dioxide in the atmosphere radiates energy back to the surface of the earth, warming it. Increasing the concentration of carbon dioxide in the atmosphere means more radiation; this increases the average surface temperature of the earth. The net result is **global warming.**

There is strong evidence that the earth has warmed nearly 1°C in the last 100 years because of increased greenhouse gases. What happens next? Climate scientists, using sophisticated models of the earth's atmosphere and oceans, calculate that a doubling of the carbon dioxide concentration will likely increase the earth's average temperature by an additional 2°C ($\approx$3°F) to 6°C ($\approx$9°F) There is some uncertainty in these calculations; the earth is a large and complex system. Perhaps the earth will get cloudier as the temperature increases, moderating the increase. Or perhaps the arctic ice cap will melt, making the earth less reflective and leading to an even more dramatic temperature increase.

But the basic physics that leads to the greenhouse effect, and to global warming, is quite straightforward. Carbon dioxide in the atmosphere keeps the earth warm; more carbon dioxide will make it warmer. How much warmer? That's an important question, one that many scientists around the world are attempting to answer with ongoing research. But large or small, change *is* coming. Global warming is one of the most serious challenges facing scientists, engineers, and all citizens in the 21st century.

The following questions are related to the passage "The Greenhouse Effect and Global Warming" on the previous page.

1. The intensity of sunlight at the top of the earth's atmosphere is approximately 1400 W/m^2. Mars is about 1.5 times as far from the sun as the earth. What is the approximate intensity of sunlight at the top of Mars's atmosphere?
   A. 930 W/m^2
   B. 620 W/m^2
   C. 410 W/m^2
   D. 280 W/m^2
2. Averaged over day, night, seasons, and weather conditions, a square meter of the earth's surface receives an average of 240 W of radiant energy from the sun. The average power radiated back to space is
   A. Less than 240 W.
   B. More than 240 W.
   C. Approximately 240 W.
3. The thermal radiation from the earth's surface peaks at a wavelength of approximately 10 μm. If the surface of the earth warms, this peak will
   A. Shift to a longer wavelength.
   B. Stay the same.
   C. Shift to a shorter wavelength.
4. The thermal radiation from the earth's surface peaks at a wavelength of approximately 10 μm. What is the energy of a photon at this wavelength?
   A. 2.4 eV B. 1.2 eV
   C. 0.24 eV D. 0.12 eV
5. Electromagnetic waves in certain wavelength ranges interact with water molecules because the molecules have a large electric dipole moment. The electric field of the wave
   A. Exerts a net force on the water molecules.
   B. Exerts a net torque on the water molecules.
   C. Exerts a net force and a net torque on the water molecules.

The following passages and associated questions are based on the material of Part VI.

Taking an X Ray

X rays are a very penetrating form of electromagnetic radiation. X rays pass through the soft tissue of the body but are largely stopped by bones and other more dense tissues. This makes x rays very useful for medical and dental purposes, as you know.

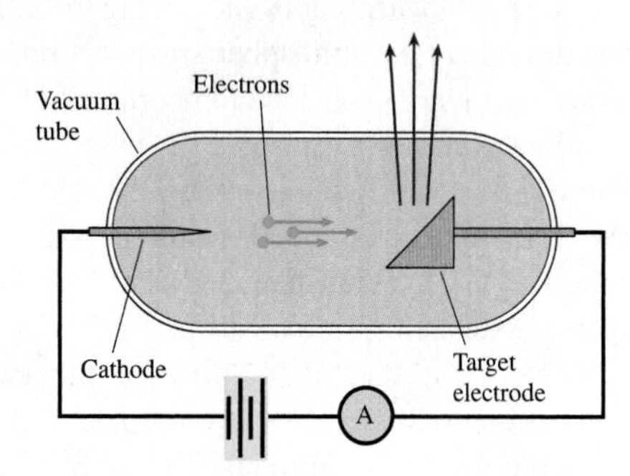

FIGURE VI.1

A schematic view of an x-ray tube and a driver circuit is given in Figure VI.1. A filament warms the cathode, freeing electrons. These electrons are accelerated by the electric field established by a high-voltage power supply connected between the cathode and a metal target. The electrons accelerate in the direction of the target. The rapid deceleration when they strike the target generates x rays. Each electron will emit one or more x rays as it comes to rest.

An x-ray image is essentially a shadow; x rays darken the film where they pass, but the film stays unexposed, and thus light, where bones or dense tissues block x rays. An x-ray technician adjusts the quality of an image by adjusting the energy and the intensity of the x-ray beam. This is done by adjusting two parameters: the accelerating voltage and the current through the tube. The accelerating voltage determines the energy of the x-ray photons, which can't be greater than the energy of the electrons. The current through the tube determines the number of electrons per second and thus the number of photons emitted. In clinical practice, the exposure is characterized by two values: "kVp" and "mAs." kVp is the peak voltage in kV. The value mAs is the product of the current (in mA) and the time (in s) to give a reading in mA · s. This is a measure of the total number of electrons that hit the target and thus the number of x rays emitted.

Typical values for a dental x ray are a kVp of 70 (meaning a peak voltage of 70 kV) and mAs of 7.5 (which comes from a current of 10 mA for 0.75 s, for a total of 7.5 mAs). Assume these values in all of the problems that follow.

6. In Figure VI.1, what is the direction of the electric field in the region between the cathode and the target electrode?
   A. To the left
   B. To the right
   C. Toward the top of the page
   D. Toward the bottom of the page
7. If the distance between the cathode and the target electrode is approximately 1.0 cm, what will be the maximum acceleration of the free electrons? Assume that the electric field is uniform.
   A. 1.2×10^{18} m/s^2 B. 1.2×10^{16} m/s^2
   C. 1.2×10^{15} m/s^2 D. 1.2×10^{12} m/s^2
8. What, physically, does the product of a current (in mA) and a time (in s) represent?
   A. Energy in mJ B. Potential difference in mV
   C. Charge in mC D. Resistance in mΩ
9. During the 0.75 s that the tube is running, what is the electric power?
   A. 7.0 kW B. 700 W
   C. 70 W D. 7.0 W
10. If approximately 1% of the electric energy ends up in the x-ray beam (a typical value), what is the approximate total energy of the x rays emitted?
    A. 500 J B. 50 J
    C. 5 J D. 0.5 J
11. What is the maximum energy of the emitted x-ray photons?
    A. 70×10^3 J B. 1.1×10^{-11} J
    C. 1.1×10^{-14} J D. 1.6×10^{-18} J

Electric Cars

In recent years, practical hybrid cars have hit the road—cars in which the gasoline engine runs a generator that charges batteries that run an electric motor. These cars offer increased efficiency, but significantly greater efficiency could be provided by a purely electric car run by batteries that you charge by plugging into an electric outlet in your house.

But there's a practical problem with such vehicles: the time necessary to recharge the batteries. If you refuel your car with gas at the pump, you add 130 MJ of energy per gallon. If you add 20 gallons, you add a total of 2.6 GJ in about 5 minutes. That's a lot of energy in a short time; the electric system of your house simply can't provide power at this rate.

There's another snag as well. Suppose there were electric filling stations that could provide very high currents to recharge your electric car. Conventional batteries can't recharge very quickly; it would still take longer for a recharge than to refill with gas.

One possible solution is to use capacitors instead of batteries to store energy. Capacitors can be charged much more quickly, and as an added benefit, they can provide energy at a much greater rate—allowing for peppier acceleration. Today's capacitors can't store enough energy to be practical, but future generations will.

12. A typical home's electric system can provide 100 A at a voltage of 220 V. If you had a charger that ran at this full power, approximately how long would it take to charge a battery with the equivalent of the energy in one gallon of gas?
    A. 100 min B. 50 min
    C. 20 min D. 5 min
13. The Tesla Roadster, a production electric car, has a 375 V battery system that can provide a power of 200 kW. At this peak power, what is the current supplied by the batteries?
    A. 75 kA B. 1900 A
    C. 530 A D. 75 A
14. To charge the batteries in a Tesla Roadster, a transformer is used to step up the voltage of the household supply. If you step a 220 V, 100 A system up to 400 V, what is the maximum current you can draw at this voltage?
    A. 180 A B. 100 A
    C. 55 A D. 45 A
15. One design challenge for a capacitor-powered electric car is that the voltage would change with time as the capacitors discharged. If the capacitors in a car were discharged to half their initial voltage, what fraction of energy would still be left?
    A. 75% B. 67%
    C. 50% D. 25%

Wireless Power Transmission

Your laptop has wireless communications connectivity, and you might even have a wireless keyboard or mouse. But there's one wire you haven't been able to get rid of yet—the power cord.

Researchers are working on ways to circumvent the need for a direct electrical connection for power, and they are experiencing some success. Recently, investigators were able to use current flowing through a primary coil to power a 60 W lightbulb connected to a secondary coil 2.0 m away, with approximately 15% efficiency. The coils were large and the efficiency low, but it's a start.

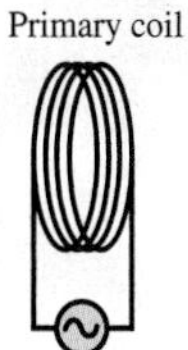

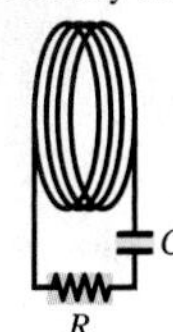

FIGURE VI.2

The wireless power transfer system is outlined in Figure VI.2. An AC supply generates a current through the primary coil, creating a varying magnetic field. This field induces a current in the secondary coil, which is connected to a resistance (the lightbulb) and a capacitor that sets the resonance frequency of the secondary circuit to match the frequency of the primary circuit.

16. At a particular moment, the current in the primary coil is clockwise, as viewed from the secondary coil. At the center of the secondary coil, the field from the primary coil is
    A. To the right.
    B. To the left.
    C. Zero.
17. At a particular moment, the magnetic field from the primary coil points to the right and is increasing in strength. The field due to the induced current in the secondary coil is
    A. To the right.
    B. To the left.
    C. Zero.
18. The power supply drives the primary coil at 9.9 MHz. If this frequency is doubled, how must the capacitor in the secondary circuit be changed?
    A. Increase by a factor of 2
    B. Increase by a factor of $\sqrt{2}$
    C. Decrease by a factor of 2
    D. Decrease by a factor of 4
19. What are the rms and peak currents for a 60 W bulb? (The rms voltage is the usual 120 V.)
    A. 0.71 A, 0.71 A B. 0.71 A, 0.50 A
    C. 0.50 A, 0.71 A D. 0.50 A, 0.50 A

Additional Integrated Problems

20. A 20 Ω resistor is connected across a 120 V source. The resistor is then lowered into an insulated beaker, containing 1.0 L of water at 20° C, for 60 s. What is the final temperature of the water?
21. As shown in Figure VI.3, a square loop of wire, with a mass of 200 g, is free to pivot about a horizontal axis through one of its sides. A 0.50 T horizontal magnetic field is directed as shown. What current I in the loop, and in what direction, is needed to hold the loop steady in a horizontal plane?

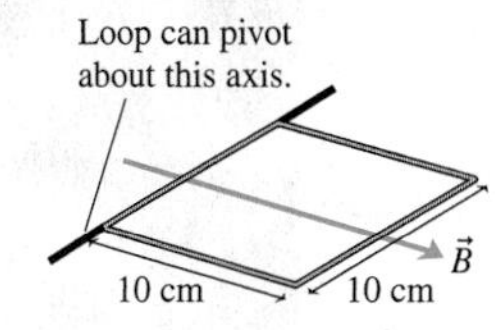

FIGURE VI.3

PART VII

Modern Physics

The eerie glow of this comb jelly is due to *bioluminescence.* Energy released in chemical reactions in special cells is turned directly into a cool blue light. How is this light different from the light from a hot, incandescent filament?

New Ways of Looking at the World

Newton's mechanics and Maxwell's electromagnetism are remarkable theories that explain a wide range of physical phenomena, as we have seen in the past 26 chapters—but our story doesn't stop there. In the early 20th century, a series of discoveries profoundly altered our understanding of the universe at the most fundamental level, forcing scientists to reconsider the very nature of space and time and to develop new models of light and matter.

Relativity

The idea of measuring distance with a meter stick and time with a clock or stopwatch seems self-evident. But Albert Einstein, as a young, unknown scientist, realized that Maxwell's theory of electromagnetism could be consistent only if an additional rather odd assumption was made: that the speed of light is the same for all observers, no matter how they might be moving with respect to each other or to the source of the light. This assumption changes the way that we think about space and time. When we study Einstein's theory of *relativity,* you will see how different observers can disagree about lengths and time intervals. We need to go beyond stopwatches and meter sticks. Time can pass at different rates for different observers; time is, as you will see, relative. Our exploration will end with the most famous equation in physics, Einstein's $E = mc^2$. Matter can be converted to energy, and energy to matter.

Quantum Physics

We've seen that light, a wave, sometimes acts like a particle, a photon. We'll now find that particles such as electrons or atoms sometimes behave like waves. All of the characteristics of waves, such as diffraction and interference, will also apply to particles. This odd notion—that there's no clear distinction between particles and waves—is the core of a new model of light and matter called *quantum physics.* The wave nature of particles will lead to the *quantization* of energy. A particle confined in a box—or an electron in an atom—can have only certain energies. This idea will be a fruitful one for us, allowing us to understand the spectra of gases and phenomena such as bioluminescence.

Atoms, Nuclei, and Particles

We have frequently used the atomic model, explaining the properties of matter by considering the behavior of the atoms that comprise it. But when you get right down to it, what *is* an atom? And what's inside an atom? As you know, an atom has a tiny core called a *nucleus.* We'll look at what goes on inside the nucleus. One remarkable discovery will be that the nuclei of certain atoms spontaneously decay, turning the atom from one element to another. This phenomenon of *radioactivity* will give us a window into the nature of atoms and the particles that comprise them.

Once we know that the nucleus of an atom is composed of protons and neutrons, there's a natural next question to ask: What's inside a proton or neutron? There is an answer to this question, an answer we will learn in the final chapter of this textbook.

27 Relativity

For a wildlife conservation study, this turtle has been fitted with a collar containing a global positioning system (GPS) receiver, allowing the turtle to be tracked over thousands of miles with an accuracy of ±15 m. How does the theory of relativity affect the accuracy of a GPS receiver?

LOOKING AHEAD ➧

Goal: To understand how Einstein's theory of relativity changes our concepts of time and space.

Simultaneity

The lightning strikes are simultaneous to you, but to someone who is moving relative to you they occur at different times.

You'll learn how to compute the order in which two events occur according to observers moving relative to each other.

Time and Space

Time itself runs faster on the surface of the earth than on GPS satellites that are in rapid motion relative to the earth.

You'll learn how moving clocks run slower and moving objects are shorter than when they are at rest.

Mass and Energy

The sun's energy comes from converting 4 billion kilograms of matter into energy every second.

You'll learn how Einstein's famous equation $E = mc^2$ shows that mass and energy are essentially equivalent.

LOOKING BACK ➧

Relative Motion

In Section 3.5 you learned how to find the velocity of a ball relative to Ana given its velocity relative to Carlos.

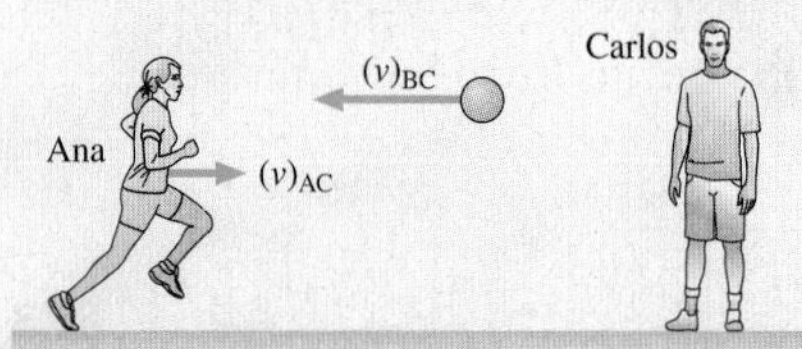

In this chapter, we'll see how our commonsense ideas about relative motion break down when one or more of the velocities approach the speed of light.

STOP TO THINK

The car is moving at 10 m/s relative to Bill. How fast does Amy see the car as moving?

A. 5 m/s
B. 10 m/s
C. 15 m/s
D. 20 m/s

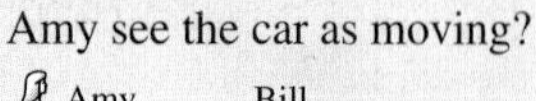

27.1 Relativity: What's It All About?

Albert Einstein (1879–1955) was one of the most influential thinkers in history.

What do you think of when you hear the phrase "theory of relativity"? A white-haired Einstein? $E = mc^2$? Black holes? Time travel? There is, without doubt, a certain mystique associated with relativity, an aura of the strange and exotic. The good news is that understanding the ideas of relativity is well within your grasp. Einstein's *special theory of relativity,* the portion of relativity we'll study, is not mathematically difficult at all. The challenge is conceptual because relativity questions deeply held assumptions about the nature of space and time. In fact, that's what relativity is all about—space and time.

In Newtonian mechanics, space and time are absolute quantities; the length of a meter stick and the time between ticks of a clock are the same to any observer, whether moving or not. But relativity challenges these commonsense notions. As we'll see, ground-based observers measure the length of a fast-moving rocket to be *shorter,* and a clock on the rocket to run *slower,* compared to when the rocket is at rest.

Because relativity challenges our basic notions about space and time, to make progress we will have to exercise the utmost care with regard to logic and precision. We will need to state very precisely just how it is that we know things about the physical world and then ruthlessly follow the logical consequences. The challenge is to stay on this path and not to let our prior assumptions—assumptions that are deeply ingrained in all of us—lead us astray.

What's Special About Special Relativity?

Einstein's first paper on relativity, in 1905, dealt exclusively with inertial reference frames, reference frames that move relative to each other with constant velocity. Ten years later, Einstein published a more encompassing theory of relativity that considers accelerated motion and its connection to gravity. The second theory, because it's more general in scope, is called *general relativity.* General relativity is the theory that describes black holes, curved spacetime, and the evolution of the universe. It is a fascinating theory but, alas, very mathematical and outside the scope of this textbook.

Motion at constant velocity is a "special case" of motion—namely, motion for which the acceleration is zero. Hence Einstein's first theory of relativity has come to be known as *special relativity.* It is special in the sense of being a restricted, special case of his more general theory, not special in the everyday sense of meaning distinctive or exceptional. Special relativity, with its conclusions about time dilation and length contraction, is what we will study.

27.2 Galilean Relativity

You may have had the experience of sitting on a train and looking up to see another train moving slowly past. It can be hard to tell if the other train is moving past your stationary train, or if you're moving in the opposite direction past a stationary train. Only the *relative* velocity between the trains has meaning.

Galileo was the first to understand how the laws of physics depended on the relative motion between different observers. A firm grasp of *Galilean relativity* is necessary if we are to appreciate and understand what is new in Einstein's theory. Thus we begin with the ideas of relativity that are embodied in Newtonian mechanics.

Reference Frames

Suppose you're passing me as we both drive in the same direction along a freeway. My car's speedometer reads 55 mph while your speedometer shows 60 mph. Is 60 mph your "true" speed? That is certainly your speed relative to someone standing beside the road, but your speed relative to me is only 5 mph. Your speed is 120 mph relative to a driver approaching from the other direction at 60 mph.

An object does not have a "true" speed or velocity. The very definition of velocity, $v = \Delta x/\Delta t$, assumes the existence of a coordinate system in which, during some time interval Δt, the displacement Δx is measured. The best we can manage is to specify an object's velocity relative to, or with respect to, the coordinate system in which it is measured.

Let's define a **reference frame** to be a coordinate system in which experimenters equipped with meter sticks, stopwatches, and any other needed equipment make position and time measurements on moving objects. Three ideas are implicit in our definition of a reference frame:

- A reference frame extends infinitely far in all directions.
- The experimenters are at rest in the reference frame.
- The number of experimenters and the quality of their equipment are sufficient to measure positions and velocities to any level of accuracy needed.

FIGURE 27.1 The standard reference frames S and S′.

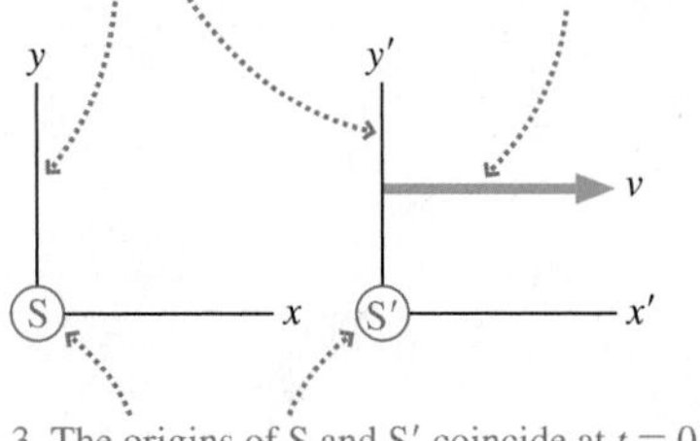
1. The axes of S and S′ have the same orientation.
2. Frame S′ moves with velocity v relative to frame S. The relative motion is parallel to the x- and x′-axes.
y
y′
v
S
x
S′
x′
3. The origins of S and S′ coincide at t = 0. This is our definition of t = 0.

The first two points are especially important. It is often convenient to say "the laboratory reference frame" or "the reference frame of the rocket." These are shorthand expressions for "a reference frame, infinite in all directions, in which the laboratory (or the rocket) and a set of experimenters happen to be at rest."

NOTE ► A reference frame is not the same thing as a "point of view." That is, each experimenter does not have his or her own private reference frame. **All experimenters at rest relative to each other share the same reference frame.** ◄

FIGURE 27.1 shows how we represent two reference frames, S and S′, that are in relative motion. The coordinate axes in S are x, y, z and those in S′ are x', y', z'. Reference frame S′ moves with velocity v relative to S or, equivalently, S moves with velocity $-v$ relative to S′. There's no implication that either reference frame is "at rest." Notice that the zero of time, when experimenters start their stopwatches, is the instant when the origins of S and S′ coincide.

Inertial Reference Frames

Certain reference frames are especially simple. FIGURE 27.2a shows a student cruising at constant velocity in an airplane. If he places a ball on the floor, it stays there. There are no horizontal forces, and the ball remains at rest relative to the airplane. That is, $\vec{a} = \vec{0}$ in the airplane's coordinate system when $\vec{F}_{net} = \vec{0}$, so Newton's first law is satisfied. Similarly, if the student drops the ball, it falls straight down—relative to the student—with an acceleration of magnitude g, satisfying Newton's second law.

FIGURE 27.2 Two reference frames.

(a)

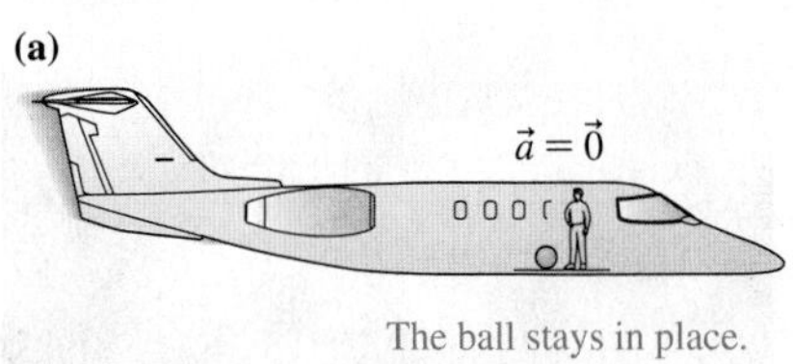
$\vec{a} = \vec{0}$
The ball stays in place.

A ball with no horizontal forces stays at rest in an airplane cruising at constant velocity. The airplane is an inertial reference frame.

(b)

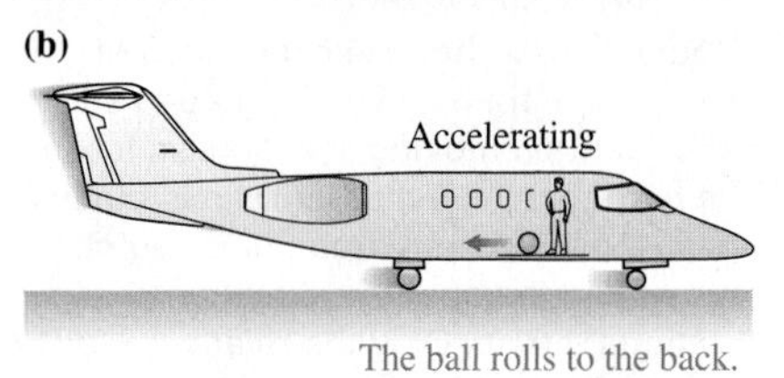
Accelerating
The ball rolls to the back.

The ball rolls to the back of the plane during takeoff. An accelerating plane is not an inertial reference frame.

We define an **inertial reference frame** as one in which Newton's first law is valid. That is, an inertial reference frame is one in which an isolated particle, one on which there are no forces, either remains at rest or moves in a straight line at constant speed, as measured by experimenters at rest in that frame.

Not all reference frames are inertial. The student in FIGURE 27.2b conducts the same experiment during takeoff. He carefully places the ball on the floor just as the airplane starts to accelerate down the runway. You can imagine what happens. The ball rolls to the back of the plane as the passengers are being pressed back into their seats. If the student measures the ball's motion using a meter stick attached to the plane, he will find that the ball accelerates *in the plane's reference frame.* Yet he would be unable to identify any force on the ball that would act to accelerate it toward the back of the plane. This violates Newton's first law, so the plane is *not* an inertial reference frame during takeoff. **In general, accelerating reference frames are not inertial reference frames.**

NOTE ► An inertial reference frame is an idealization. A true inertial reference frame would need to be floating in deep space, far from any gravitational influence. In practice, an earthbound laboratory is a good approximation of an inertial reference frame because the accelerations associated with the earth's rotation and motion around the sun are too small to influence most experiments. ◄

These ideas are in accord with your everyday experience. If you're in a jet flying smoothly at 600 mph—an inertial reference frame—Newton's laws are valid: You can toss and catch a ball, or pour a cup of coffee, exactly as you would on the ground. But if the plane were diving, or shaking from turbulence, simple "experiments" like these would fail. A ball thrown straight up would land far from your hand, and the stream of coffee would bend and turn on its way to missing the cup. These apparently simple observations can be stated as the *Galilean principle of relativity:*

This flight attendant pours wine on a smoothly flying airplane moving at 600 mph just as easily as she does at the terminal. These ideas were first discussed by Galileo in 1632 in the context of pouring water while on a moving ship.

Galilean principle of relativity Newton's laws of motion are valid in all inertial reference frames.

In our study of relativity, we will restrict our attention to inertial reference frames. This implies that the relative velocity v between two reference frames is constant. Any reference frame that moves at constant velocity with respect to an inertial reference frame is itself an inertial reference frame. Conversely, a reference frame that accelerates with respect to an inertial reference frame is not an inertial reference frame. Although special relativity can be used for accelerating reference frames, we will confine ourselves here to the simple case of inertial reference frames moving with respect to each other at constant velocity.

STOP TO THINK 27.1 Which of these is an inertial reference frame (or a very good approximation)?

A. Your bedroom
B. A car rolling down a steep hill
C. A train coasting along a level track
D. A rocket being launched
E. A roller coaster going over the top of a hill
F. A sky diver falling at terminal speed

The Galilean Velocity Transformation

Special relativity is largely concerned with how physical quantities such as position and time are measured by experimenters in different reference frames. Let's begin by studying how, within Galilean relativity, the *velocity* of an object is measured in different reference frames. We have already touched on these ideas back in Section 3.5.

Figure Video

Suppose Sue is standing beside a highway as Jim drives by at 50 mph, as shown in **FIGURE 27.3**. Let S be Sue's reference frame—a reference frame attached to the ground—and let S′ be the reference frame moving with Jim, attached to his car. We see that the velocity of reference frame S′ relative to S is $v = 50$ mph.

Now suppose a motorcyclist blasts down the highway, traveling in the same direction as Jim. Sue measures the motorcycle's velocity to be $u = 75$ mph. What is the cycle's velocity u' measured relative to Jim? We can answer this on the basis of common sense. If you're driving at 50 mph, and someone passes you going 75 mph, then his speed *relative to you* is 25 mph. This is the *difference* between his speed relative to the ground and your speed relative to the ground. Thus Jim measures the motorcycle's velocity to be $u' = 25$ mph.

FIGURE 27.3 A motorcycle's velocity as seen by Sue and by Jim.

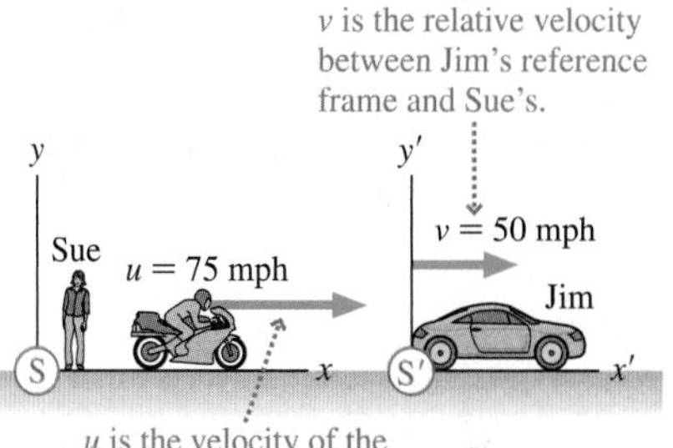
v is the relative velocity between Jim's reference frame and Sue's.
y
y′
Sue
u = 75 mph
v = 50 mph
Jim
S
x
S′
x′
u is the velocity of the motorcycle as measured in Sue's frame.

NOTE ▶ In this chapter, we will use v to represent the velocity of one reference frame relative to another. We will use u and u' to represent the velocities of objects with respect to reference frames S and S′. In addition, we will assume that all motion is parallel to the x-axis. ◀

We can state this idea as a general rule. An object's velocity measured in frame S is related to its velocity measured in frame S′ by

Velocity of object as measured in frame S

$$u' = u - v \quad \text{and} \quad u = u' + v \qquad (27.1)$$

The relative velocity between the two frames

Velocity of object as measured in frame S′

Equations 27.1 are the **Galilean velocity transformations.** If you know the velocity of a particle as measured by the experimenters in one inertial reference frame, you can use Equations 27.1 to find the velocity that would be measured by experimenters in any other inertial reference frame.

EXAMPLE 27.1 Finding the speed of sound

An airplane is flying at speed 200 m/s with respect to the ground. Sound wave 1 is approaching the plane from the front, while sound wave 2 is catching up from behind. Both waves travel at 340 m/s relative to the ground. What is the velocity of each wave relative to the plane?

PREPARE Assume that the earth (frame S) and the airplane (frame S′) are inertial reference frames. Frame S′, in which the airplane is at rest, moves with velocity $v = 200$ m/s relative to frame S. **FIGURE 27.4** shows the airplane and the sound waves.

FIGURE 27.4 Experimenters in the plane measure different speeds for the sound waves than do experimenters on the ground.

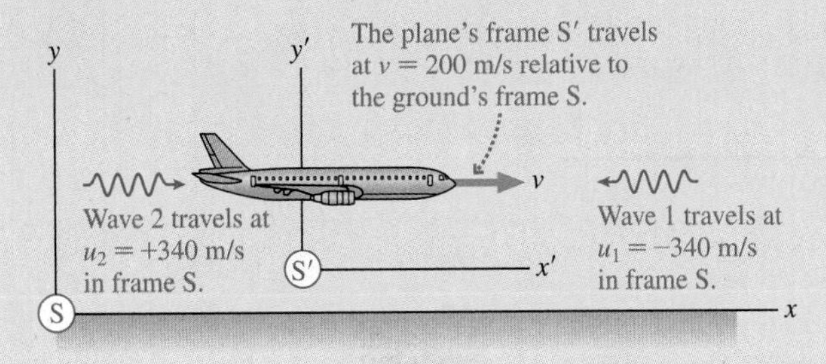

SOLVE The speed of a mechanical wave, such as a sound wave or a wave on a string, is its speed *relative to its medium.* Thus the *speed of sound* is the speed of a sound wave through a reference frame in which the air is at rest. This is reference frame S, where wave 1 travels with velocity $u_1 = -340$ m/s and wave 2 travels with velocity $u_2 = +340$ m/s. Notice that the Galilean transformations use *velocities,* with appropriate signs, not just speeds.

The airplane travels to the right with reference frame S′ at velocity v. We can use the Galilean transformations of velocity to find the velocities of the two sound waves in frame S′:

$$u_1' = u_1 - v = -340 \text{ m/s} - 200 \text{ m/s} = -540 \text{ m/s}$$
$$u_2' = u_2 - v = 340 \text{ m/s} - 200 \text{ m/s} = 140 \text{ m/s}$$

Thus wave 1 approaches the plane with a *speed* of 540 m/s, while wave 2 approaches with a speed of 140 m/s.

ASSESS This isn't surprising. If you're driving at 50 mph, a car coming the other way at 55 mph is approaching you at 105 mph. A car coming up behind you at 55 mph seems to be gaining on you at the rate of only 5 mph. Wave speeds behave the same. Notice that a mechanical wave would appear to be stationary to a person moving at the wave speed. To a surfer, the crest of the ocean wave remains at rest under his or her feet.

STOP TO THINK 27.2 Ocean waves are approaching the beach at 10 m/s. A boat heading out to sea travels at 6 m/s. How fast are the waves moving in the boat's reference frame?

A. 16 m/s B. 10 m/s C. 6 m/s D. 4 m/s

27.3 Einstein's Principle of Relativity

The 19th century was an era of optics and electromagnetism. Thomas Young demonstrated in 1801 that light is a wave, and by midcentury scientists had devised techniques for measuring the speed of light. Faraday discovered electromagnetic induction in 1831, setting in motion a series of events leading to Maxwell's conclusion, in 1864, that light is an electromagnetic wave.

If light is a wave, what is the medium in which it travels? This was perhaps the most important scientific question in the second half of the 19th century. The medium in which light waves were assumed to travel was called the *ether.* Experiments to measure the speed of light were assumed to be measuring its speed through the ether. But just what is the ether? What are its properties? Can we collect a jar full of ether to study? Despite the significance of these questions, experimental efforts to detect the ether or measure its properties kept coming up empty-handed.

Maxwell's theory of electromagnetism didn't help the situation. The crowning success of Maxwell's theory was his prediction that light waves travel with speed

$$c = \frac{1}{\sqrt{\epsilon_0 \mu_0}} = 3.00 \times 10^8 \text{ m/s}$$

This is a very specific prediction with no wiggle room. The difficulty with such a specific prediction was the implication that Maxwell's laws of electromagnetism are valid *only* in the reference frame of the ether. After all, as **FIGURE 27.5** shows, the light speed should certainly be faster or slower than c in a reference frame moving through the ether, just as the sound speed is different to someone moving through the air.

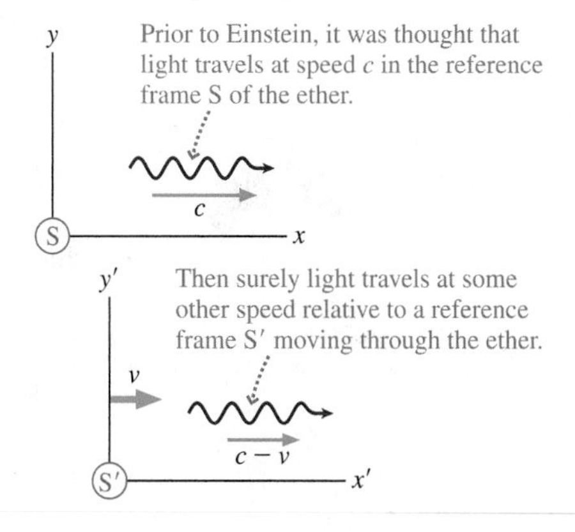

Prior to Einstein, it was thought that light travels at speed c in the reference frame S of the ether.
Then surely light travels at some other speed relative to a reference frame S′ moving through the ether.
y
x
c
S
y'
x'
v
$c - v$
S′

FIGURE 27.5 It seems as if the speed of light should differ from c in a reference frame moving through the ether.

As the 19th century closed, it appeared that Maxwell's theory did not obey the classical principle of relativity. There was just one reference frame, the reference frame of the ether, in which the laws of electromagnetism seemed to be true. And to make matters worse, the fact that no one had been able to detect the ether meant that no one could identify the one reference frame in which Maxwell's equations "worked."

It was in this muddled state of affairs that a young Albert Einstein made his mark on the world. Even as a teenager, Einstein had wondered how a light wave would look to someone "surfing" the wave, traveling alongside the wave at the wave speed. You can do that with a water wave or a sound wave, but light waves seemed to present a logical difficulty. An electromagnetic wave sustains itself by virtue of the fact that a changing magnetic field induces an electric field and a changing electric field induces a magnetic field. But to someone moving with the wave, *the fields would not change.* How could there be an electromagnetic wave under these circumstances?

Several years of thinking about the connection between electromagnetism and reference frames led Einstein to the conclusion that *all* the laws of physics, not just the laws of mechanics, should obey the principle of relativity. In other words, the principle of relativity is a fundamental statement about the nature of the physical universe. The Galilean principle of relativity stated only that Newton's laws hold in any inertial reference frame. Einstein was able to state a much more general principle:

In certain rivers, tides send waves upriver that can be surfed for miles. From the reference frame of these surfers, the waves are standing still. If you could move along with a light wave, would the electric and magnetic fields appear motionless?

Principle of relativity All the laws of physics are the same in all inertial reference frames.

All of the results of Einstein's theory of relativity flow from this one simple statement.

The Constancy of the Speed of Light

If Maxwell's equations of electromagnetism are laws of physics, and there's every reason to think they are, then, according to the principle of relativity, Maxwell's equations must be true in *every* inertial reference frame. On the surface this seems to be an innocuous statement, equivalent to saying that the law of conservation of momentum is true in every inertial reference frame. But follow the logic:

1. Maxwell's equations are true in all inertial reference frames.
2. Maxwell's equations predict that electromagnetic waves, including light, travel at speed $c = 3.00 \times 10^8$ m/s.
3. Therefore, **light travels at speed c in all inertial reference frames.**

FIGURE 27.6 Light travels at speed c in all inertial reference frames, regardless of how the reference frames are moving with respect to the light source.

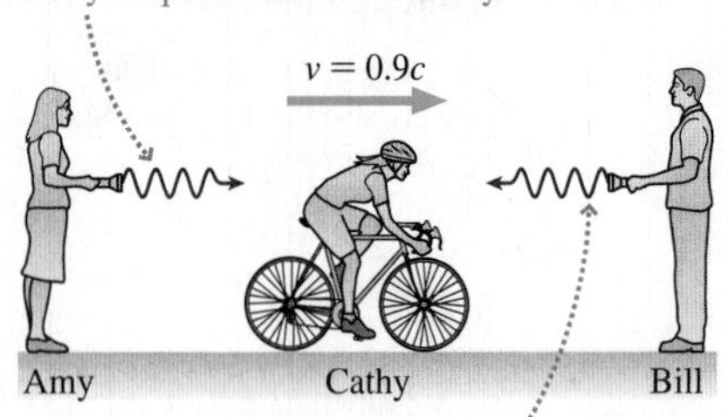
This light wave leaves Amy at speed c relative to Amy. It approaches Cathy at speed c relative to Cathy.
$v = 0.9c$
Amy
Cathy
Bill
This light wave leaves Bill at speed c relative to Bill. It approaches Cathy at speed c relative to Cathy.

FIGURE 27.6 shows the implications of this conclusion. *All* experimenters, regardless of how they move with respect to each other, find that *all* light waves, regardless of the source, travel in their reference frame with the *same* speed c. If Cathy's velocity toward Bill and away from Amy is $v = 0.9c$, Cathy finds, by making measurements in her reference frame, that the light from Bill approaches her at speed c, not at $c + v = 1.9c$. And the light from Amy, which left Amy at speed c, catches up from behind at speed *c relative to Cathy,* not the $c - v = 0.1c$ you would have expected.

Although this prediction goes against all shreds of common sense, the experimental evidence for it is strong. Laboratory experiments are difficult because even the highest laboratory speed is insignificant in comparison to c. In the 1930s, however, the physicists R. J. Kennedy and E. M. Thorndike realized that they could use the earth itself as a laboratory. The earth's speed as it circles the sun is about 30,000 m/s. The velocity of the earth in January differs by 60,000 m/s from its velocity in July, when the earth is moving in the opposite direction. Kennedy and Thorndike were able to use a very sensitive and stable interferometer to show that the numerical values of the speed of light in January and July differ by less than 2 m/s, well within experimental error.

More recent experiments have used unstable elementary particles, called π mesons, that decay into high-energy photons, or particles of light. The π mesons, created in a particle accelerator, move through the laboratory at 99.975% the speed of light, or $v = 0.99975c$, as they emit photons at speed c in the π meson's reference frame. As **FIGURE 27.7** shows, you would expect the photons to travel through the laboratory with speed $c + v = 1.99975c$. Instead, the measured speed of the photons in the laboratory was, within experimental error, 3.00×10^8 m/s.

FIGURE 27.7 Experiments find that the photons travel through the laboratory with speed c, not the speed $1.99975c$ that you might expect.

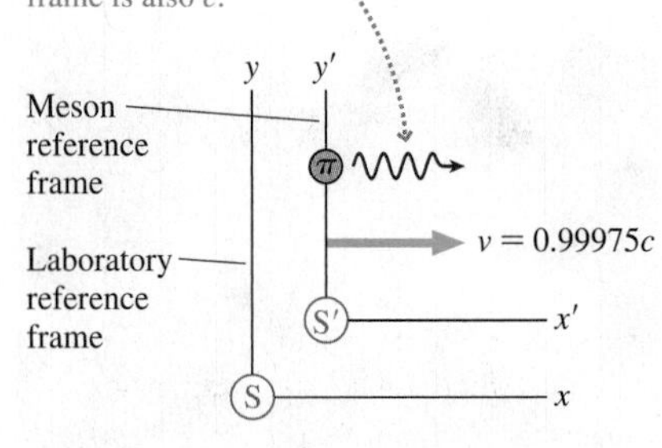
A photon is emitted at speed c relative to the π meson. Measurements find that the photon's speed in the laboratory reference frame is also c.
Meson reference frame
Laboratory reference frame
$v = 0.99975c$
S
S'

In summary, *every* experiment designed to compare the speed of light in different reference frames has found that light travels at 3.00×10^8 m/s in every inertial reference frame, regardless of how the reference frames are moving with respect to each other.

How Can This Be?

You're in good company if you find this impossible to believe. Suppose I shot a ball forward at 50 m/s while driving past you at 30 m/s. You would certainly see the ball traveling at 80 m/s relative to you and the ground. What we're saying with regard to light is equivalent to saying that the ball travels at 50 m/s relative to my car and *at the same time* travels at 50 m/s relative to the ground, even though the car is moving across the ground at 30 m/s. It seems logically impossible.

You might think that this is merely a matter of semantics. If we can just get our definitions and use of words straight, then the mystery and confusion will disappear. Or perhaps the difficulty is a confusion between what we "see" versus what "really happens." In other words, a better analysis, one that focuses on what really happens, would find that light "really" travels at different speeds in different reference frames.

Alas, what "really happens" is that light travels at 3.00×10^8 m/s in every inertial reference frame, regardless of how the reference frames are moving with respect to each other. It's not a trick. There remains only one way to escape the logical contradictions.

FIGURE 27.8 shows how the speed of a light ray would be measured by Laura, in stationary frame S, and by Dan and Eric, in frame S′ which is moving to the right at

FIGURE 27.8 Measuring the speed of light in two different reference frames.

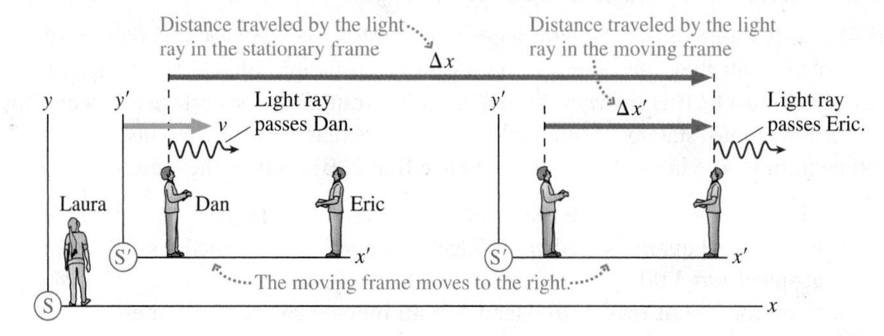
Distance traveled by the light ray in the stationary frame
Δx
Distance traveled by the light ray in the moving frame
$\Delta x'$
Light ray passes Dan.
Light ray passes Eric.
Laura
Dan
Eric
The moving frame moves to the right.

relative velocity v. As the ray moves from Dan to Eric, they measure it as having traveled a distance $\Delta x'$, the distance between them in their frame. To Laura, however, the ray will have traveled the longer distance Δx, simply because Eric is moving to the right so, as seen by Laura, the ray has to travel farther before it reaches him.

Recall that the definition of velocity is $\Delta x/\Delta t$, the ratio of the distance traveled to the time interval in which the travel occurs. Laura measures a *different* value for the distance traveled by the light ray than do Dan and Eric, but according to the principle of relativity, both sets of observers must measure the *same* value for the speed of light c. The only way this is possible is that the time Δt that, according to Laura, it takes light to make the trip is *not the same* as the time $\Delta t'$ that Dan and Eric measure.

We've assumed, since the beginning of this text, that time is simply time. It flows along like a river, and all experimenters in all reference frames simply use it. But this is evidently not the case. Something must be wrong with assumptions that we've made about the nature of time. The principle of relativity has painted us into a corner, and our only way out is to reexamine our understanding of time.

27.4 Events and Measurements

To question some of our most basic assumptions about space and time requires extreme care. We need to be certain that no assumptions slip into our analysis unnoticed. Our goal is to describe the motion of a particle in a clear and precise way, making the barest minimum of assumptions.

Events

The fundamental element of relativity is called an **event.** An event is a physical activity that takes place at a definite point in space and at a definite instant of time. A firecracker exploding is an event. A collision between two particles is an event. A light wave hitting a detector is an event.

Events can be observed and measured by experimenters in different reference frames. An exploding firecracker is as clear to you as you drive by in your car as it is to me standing on the street corner. We can quantify where and when an event occurs with four numbers: the coordinates (x, y, z) and the instant of time t. These four numbers, illustrated in FIGURE 27.9, are called the **spacetime coordinates** of the event.

FIGURE 27.9 The location and time of an event are described by its spacetime coordinates.

An event has spacetime coordinates (x, y, z, t) in frame S and different spacetime coordinates (x', y', z', t') in frame S$'$.

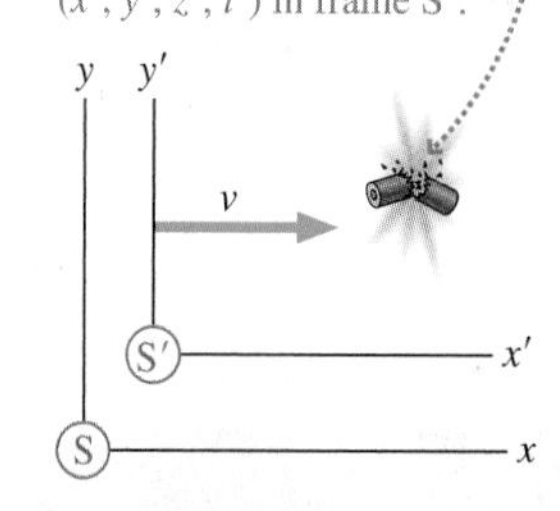

The spatial coordinates of an event measured in reference frames S and S$'$ may differ. But it now appears that the instant of time recorded in S and S$'$ may also differ. Thus the spacetime coordinates of an event measured by experimenters in frame S are (x, y, z, t), and the spacetime coordinates of the *same event* measured by experimenters in frame S$'$ are (x', y', z', t').

Measurements

Events are what "really happen," but how do we learn about an event? That is, how do the experimenters in a reference frame determine the spacetime coordinates of an event? This is a problem of *measurement.*

We defined a reference frame to be a coordinate system in which experimenters can make position and time measurements. That's a good start, but now we need to be more precise as to *how* the measurements are made. Imagine that a reference frame is filled with a cubic lattice of meter sticks, as shown in FIGURE 27.10. At every intersection is a clock, and all the clocks in a reference frame are *synchronized.* We'll return in a moment to consider how to synchronize the clocks, but assume for the moment it can be done.

FIGURE 27.10 The spacetime coordinates of an event are measured by a lattice of meter sticks and clocks.

The spacetime coordinates of this event are measured by the nearest meter stick intersection and the nearest clock.

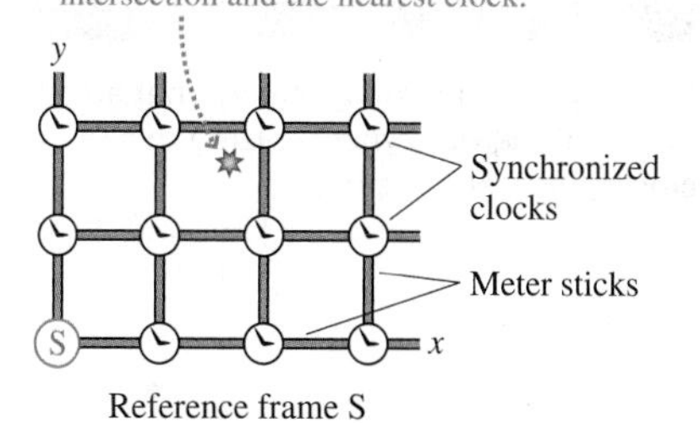

Reference frame S$'$ has its own meter sticks and its own clocks.

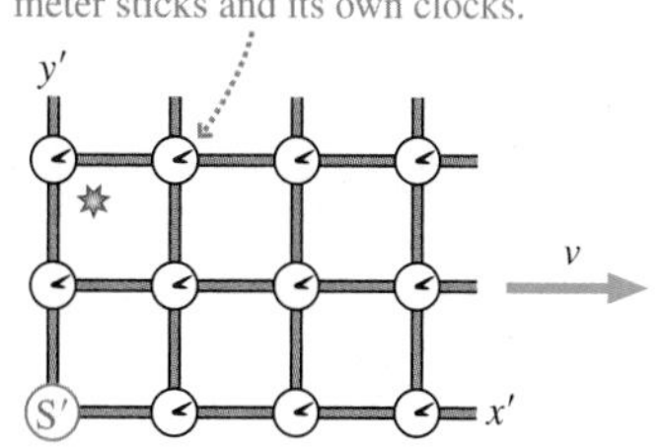

Now, with our meter sticks and clocks in place, we can use a two-part measurement scheme:

- The (x, y, z) coordinates of an event are determined by the intersection of meter sticks closest to the event.
- The event's time t is the time displayed on the clock nearest the event.

Several important issues need to be noted:

1. The clocks and meter sticks in each reference frame are imaginary, so they have no difficulty passing through each other.
2. Measurements of position and time made in one reference frame must use only the clocks and meter sticks in that reference frame.
3. There's nothing special about the sticks being 1 m long and the clocks 1 m apart. The lattice spacing can be altered to achieve whatever level of measurement accuracy is desired.
4. We'll assume that the experimenters in each reference frame have assistants sitting beside every clock to record the position and time of nearby events.
5. Perhaps most important, t is the time at which the event *actually happens,* not the time at which an experimenter sees the event or at which information about the event reaches an experimenter.
6. All experimenters in one reference frame agree on the spacetime coordinates of an event. In other words, **an event has a unique set of spacetime coordinates in each reference frame.**

STOP TO THINK 27.3 A carpenter is working on a house two blocks away. You notice a slight delay between seeing the carpenter's hammer hit the nail and hearing the blow. At what time does the event "hammer hits nail" occur?

A. At the instant you hear the blow
B. At the instant you see the hammer hit
C. Very slightly before you see the hammer hit
D. Very slightly after you see the hammer hit

Clock Synchronization

It's easy to synchronize clocks that are all in one place, but synchronizing distant clocks takes some care.

It's important that all the clocks in a reference frame be **synchronized,** meaning that all clocks in the reference frame have the same reading at any one instant of time. We would not be able to use a sequence of events to track the motion of a particle if the clocks differed in their readings. Thus we need a method of synchronization. One idea that comes to mind is to designate the clock at the origin as the *master clock.* We could then carry this clock around to every clock in the lattice, adjust that clock to match the master clock, and finally return the master clock to the origin.

This would be a perfectly good method of clock synchronization in Newtonian mechanics, where time flows along smoothly, the same for everyone. But we've been driven to reexamine the nature of time by the possibility that time is different in reference frames moving relative to each other. Because the master clock would *move,* we cannot assume that the master clock keeps time in the same way as the stationary clocks.

We need a synchronization method that does not require moving the clocks. Fortunately, such a method is easy to devise. Each clock is resting at the intersection of meter sticks, so by looking at the meter sticks, the assistant knows, or can calculate, exactly how far each clock is from the origin. Once the distance is known, the assistant can calculate exactly how long a light wave will take to travel from the origin to each clock. For example, light will take 1.00 μs to travel to a clock 300 m from the origin.

NOTE ► It's handy for many relativity problems to know that the speed of light is $c = 300$ m/μs. ◄

To synchronize the clocks, the assistants begin by setting each clock to display the light travel time from the origin, but they don't start the clocks. Next, as

FIGURE 27.11 shows, a light flashes at the origin and, simultaneously, the clock at the origin starts running from $t = 0$ s. The light wave spreads out in all directions at speed c. A photodetector on each clock recognizes the arrival of the light wave and, without delay, starts the clock. The clock had been preset with the light travel time, so each clock as it starts reads exactly the same as the clock at the origin. Thus all the clocks will be synchronized after the light wave has passed by.

FIGURE 27.11 Synchronizing the clocks.

1. This clock is preset to 1.00 μs, the time it takes light to travel 300 m.

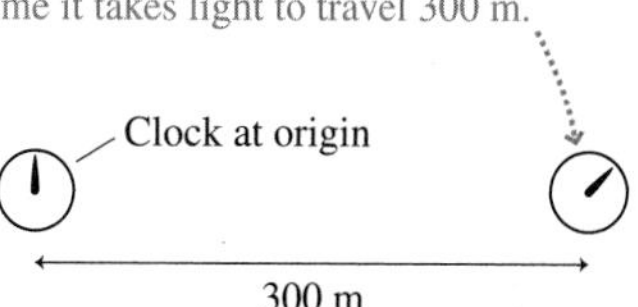

2. At $t = 0$ s, a light flashes at the origin and the origin clock starts running. A very short time later, seen here, a light wave has begun to move outward.

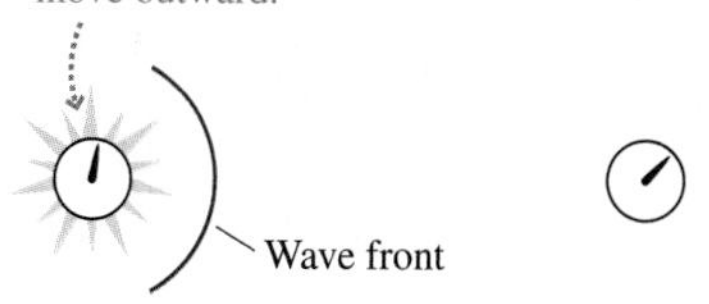

3. The clock starts when the light wave reaches it. It is now synchronized with the origin clock.

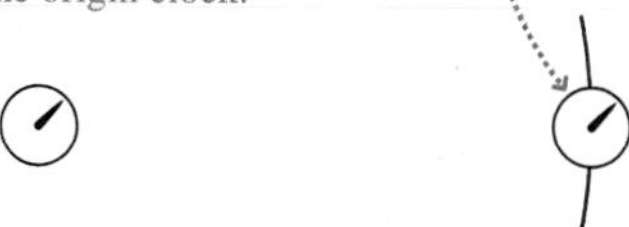

Events and Observations

We noted above that t is the time the event *actually happens.* This is an important point, one that bears further discussion. Light waves take time to travel. Messages, whether they're transmitted by light pulses, telephone, or courier on horseback, take time to be delivered. An experimenter *observes* an event, such as an exploding firecracker, only *at a later time* when light waves reach his or her eyes. But our interest is in the event itself, not the experimenter's observation of the event. The time at which the experimenter sees the event or receives information about the event is not when the event actually occurred.

Suppose at $t = 0$ s a firecracker explodes at $x = 300$ m. The flash of light from the firecracker will reach an experimenter at the origin at $t_1 = 1.0\ \mu$s. The sound of the explosion will reach the experimenter at the origin at $t_2 = 0.88$ s. Neither of these is the time t_{event} of the explosion, although the experimenter can work backward from these times, using known wave speeds, to determine t_{event}. In this example, the spacetime coordinates of the event—the explosion—are (300 m, 0 m, 0 m, 0 s).

EXAMPLE 27.2 Finding the time of an event

Experimenter A in reference frame S stands at the origin looking in the positive x-direction. Experimenter B stands at $x = 900$ m looking in the negative x-direction. A firecracker explodes somewhere between them. Experimenter B sees the light flash at $t = 3.00\ \mu$s. Experimenter A sees the light flash at $t = 4.00\ \mu$s. What are the spacetime coordinates of the explosion?

PREPARE Experimenters A and B are in the same reference frame and have synchronized clocks. FIGURE 27.12 shows the two experimenters and the explosion at unknown position x.

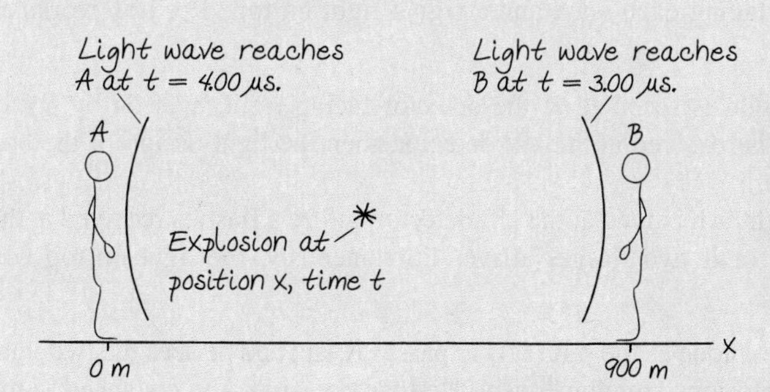

SOLVE The two experimenters observe light flashes at two different instants, but there's only one event. Light travels at 300 m/μs, so the additional 1.00 μs needed for the light to reach experimenter A implies that distance $(x - 0\text{ m})$ from x to A is 300 m longer than distance $(900\text{ m} - x)$ from B to x; that is,

$$(x - 0\text{ m}) = (900\text{ m} - x) + 300\text{ m}$$

This is easily solved to give $x = 600$ m as the position coordinate of the explosion. The light takes 1.00 μs to travel 300 m to experimenter B and 2.00 μs to travel 600 m to experimenter A. The light is received at 3.00 μs and 4.00 μs, respectively; hence it was emitted by the explosion at $t = 2.00\ \mu$s. The spacetime coordinates of the explosion are (600 m, 0 m, 0 m, 2.00 μs).

ASSESS Although the experimenters *see* the explosion at different times, they agree that the explosion actually *happened* at $t = 2.00\ \mu$s.

◀ **FIGURE 27.12** The light wave reaches the experimenters at different times. Neither of these is the time at which the event actually happened.

Simultaneity

Two events 1 and 2 that take place at different positions x_1 and x_2 but at the *same time* $t_1 = t_2$, as measured in some reference frame, are said to be **simultaneous** in that reference frame. Simultaneity is determined by when the events actually happen, not when they are seen or observed. In general, simultaneous events are not *seen* at the same time because of the difference in light travel times from the events to an experimenter.

EXAMPLE 27.3 Are the explosions simultaneous?

An experimenter in reference frame S stands at the origin looking in the positive x-direction. At $t = 3.0\ \mu s$ she sees firecracker 1 explode at $x = 600$ m. A short time later, at $t = 5.0\ \mu s$, she sees firecracker 2 explode at $x = 1200$ m. Are the two explosions simultaneous? If not, which firecracker exploded first?

PREPARE Light from both explosions travels toward the experimenter at 300 m/μs.

SOLVE The experimenter *sees* two different explosions, but perceptions of the events are not the events themselves. When did the explosions *actually* occur? Using the fact that light travels at 300 m/μs, it's easy to see that firecracker 1 exploded at $t_1 = 1.0\ \mu s$ and firecracker 2 also exploded at $t_2 = 1.0\ \mu s$. The events *are* simultaneous.

STOP TO THINK 27.4 A tree and a pole are 3000 m apart. Each is suddenly hit by a bolt of lightning. Mark, who is standing at rest midway between the two, sees the two lightning bolts at the same instant of time. Nancy is at rest under the tree. Define event 1 to be "lightning strikes tree" and event 2 to be "lightning strikes pole." For Nancy, does event 1 occur before, after, or at the same time as event 2?

27.5 The Relativity of Simultaneity

We've now established a means for measuring the time of an event in a reference frame, so let's begin to investigate the nature of time. The following "thought experiment" is very similar to one suggested by Einstein.

FIGURE 27.13 A railroad car traveling to the right with velocity v.

The firecrackers will make burn marks on the ground at the positions where they explode.

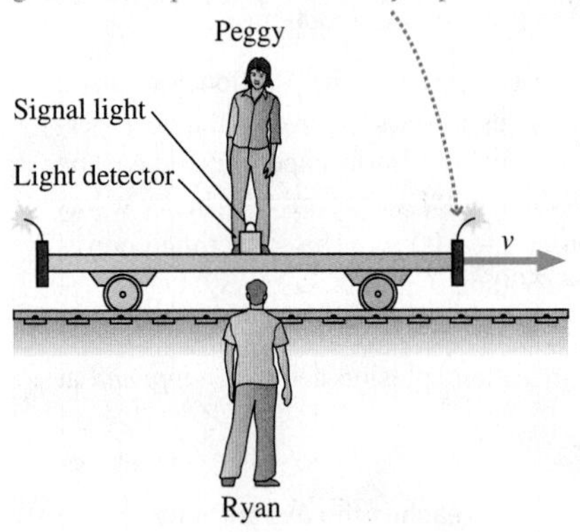

FIGURE 27.13 shows a long railroad car traveling to the right with a velocity v that may be an appreciable fraction of the speed of light. A firecracker is tied to each end of the car, just above the ground. Each firecracker is powerful enough that, when it explodes, it will make a burn mark on the ground at the position of the explosion.

Ryan is standing on the ground, watching the railroad car go by. Peggy is standing in the exact center of the car with a special box at her feet. This box has two light detectors, one facing each way, and a signal light on top. The box works as follows:

1. If a flash of light is received at the detector facing right, as seen by Ryan, before a flash is received at the left detector, then the light on top of the box will turn green.
2. If a flash of light is received at the left detector before a flash is received at the right detector, or if two flashes arrive simultaneously, the light on top will turn red.

The firecrackers explode as the railroad car passes Ryan, and he sees the two light flashes from the explosions simultaneously. He then measures the distances to the two burn marks and finds that he was standing exactly halfway between the marks. Because light travels equal distances in equal times, Ryan concludes that the two explosions were simultaneous in his reference frame, the reference frame of the ground. Further, because he was midway between the two ends of the car, he was directly opposite Peggy when the explosions occurred.

FIGURE 27.14a shows the sequence of events in Ryan's reference frame. Light travels at speed c in all inertial reference frames, so, although the firecrackers were moving, the light waves are spheres centered on the burn marks. Ryan determines that the light wave coming from the right reaches Peggy and the box before the light wave coming from the left. Thus, according to Ryan, the signal light on top of the box turns green.

FIGURE 27.14 Exploding firecrackers seen in two different reference frames.

(a) The events in Ryan's frame

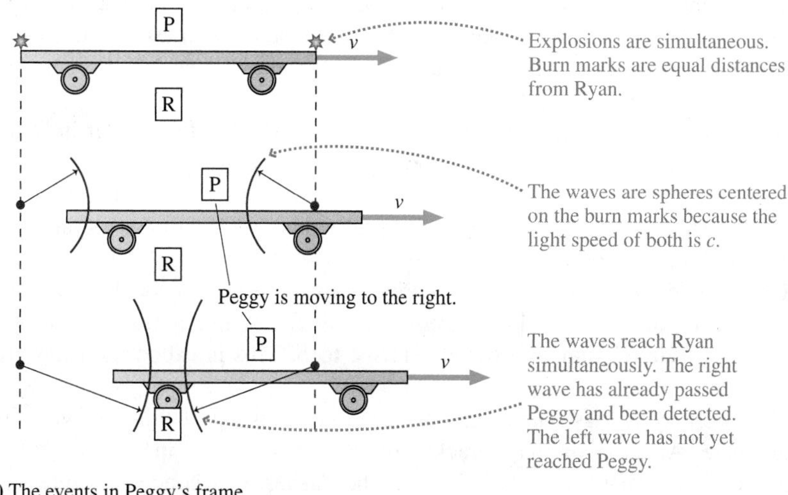
P
R
v
Explosions are simultaneous. Burn marks are equal distances from Ryan.
The waves are spheres centered on the burn marks because the light speed of both is c.
Peggy is moving to the right.
The waves reach Ryan simultaneously. The right wave has already passed Peggy and been detected. The left wave has not yet reached Peggy.

(b) The events in Peggy's frame

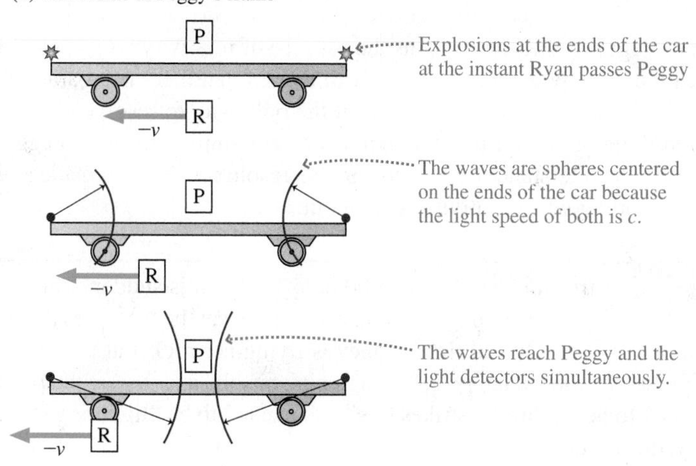
P
R
−v
Explosions at the ends of the car at the instant Ryan passes Peggy
The waves are spheres centered on the ends of the car because the light speed of both is c.
The waves reach Peggy and the light detectors simultaneously.

How do things look in Peggy's reference frame, a reference frame moving to the right at velocity v relative to the ground? As **FIGURE 27.14b** shows, Peggy sees Ryan moving to the left with speed v. Light travels at speed c in all inertial reference frames, so the light waves are spheres centered on the ends of the car. If the explosions are simultaneous, as Ryan has determined, the two light waves reach her and the box simultaneously. Thus, according to Peggy, the signal light on top of the box turns red!

Now the light on top must be either green or red. *It can't be both!* Later, after the railroad car has stopped, Ryan and Peggy can place the box in front of them. It has either a red light or a green light. Ryan can't see one color while Peggy sees the other. Hence we have a paradox. It's impossible for Peggy and Ryan both to be right. But who is wrong, and why?

What do we know with absolute certainty?

1. Ryan detected the flashes simultaneously.
2. Ryan was halfway between the firecrackers when they exploded.
3. The light from the two explosions traveled toward Ryan at equal speeds.

The conclusion that the explosions were simultaneous in Ryan's reference frame is unassailable. The light is green.

Peggy, however, made an assumption. It's a perfectly ordinary assumption, one that seems sufficiently obvious that you probably didn't notice, but an assumption nonetheless. Peggy assumed that the explosions were simultaneous.

Didn't Ryan find them to be simultaneous? Indeed, he did. Suppose we call Ryan's reference frame S, the explosion on the right event R, and the explosion on the left event L. Ryan found that $t_R = t_L$. But Peggy has to use a different set of clocks, the clocks in her reference frame S′, to measure the times t'_R and t'_L at which the explosions occurred. The fact that $t_R = t_L$ in frame S does *not* allow us to conclude that $t'_R = t'_L$ in frame S′.

In fact, the right firecracker must explode *before* the left firecracker in frame S′. Figure 27.14b, with its assumption about simultaneity, was incorrect. **FIGURE 27.15** shows the situation in Peggy's reference frame with the right firecracker exploding first. Now the wave from the right reaches Peggy and the box first, as Ryan had concluded, and the light on top turns green.

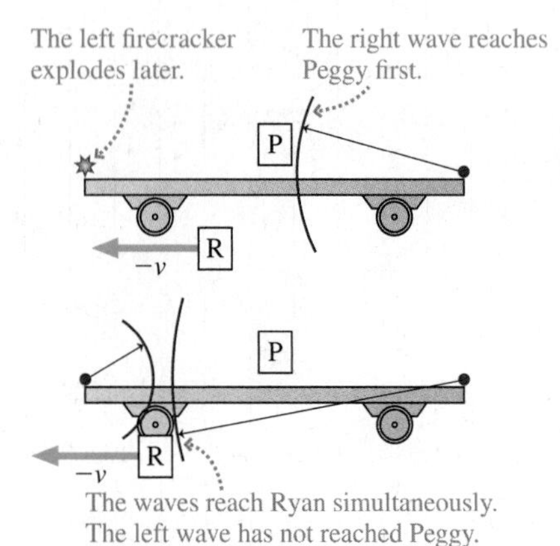

The right firecracker explodes first.
P
R
−v
The left firecracker explodes later.
The right wave reaches Peggy first.
P
R
−v
P
R
−v
The waves reach Ryan simultaneously. The left wave has not reached Peggy.

FIGURE 27.15 The real sequence of events in Peggy's reference frame.

One of the most disconcerting conclusions of relativity is that **two events occurring simultaneously in reference frame S are *not* simultaneous in any reference frame S′ that is moving relative to S.** This is called the **relativity of simultaneity.**

The two firecrackers *really* explode at the same instant of time in Ryan's reference frame. And the right firecracker *really* explodes first in Peggy's reference frame. It's not a matter of when they see the flashes. Our conclusion refers to the times at which the explosions actually occur.

The paradox of Peggy and Ryan contains the essence of relativity, and it's worth careful thought. First, review the logic until you're certain that there *is* a paradox, a logical impossibility. Then convince yourself that the only way to resolve the paradox is to abandon the assumption that the explosions are simultaneous in Peggy's reference frame. If you understand the paradox and its resolution, you've made a big step toward understanding what relativity is all about.

STOP TO THINK 27.5 A tree and a pole are 3000 m apart. Each is suddenly hit by a bolt of lightning. Mark, who is standing at rest midway between the two, sees the two lightning bolts at the same instant of time. Nancy is flying her rocket at $v = 0.5c$ in the direction from the tree toward the pole. The lightning hits the tree just as she passes by it. Define event 1 to be "lightning strikes tree" and event 2 to be "lightning strikes pole." For Nancy, does event 1 occur before, after, or at the same time as event 2?

27.6 Time Dilation

The principle of relativity has driven us to the logical conclusion that the time at which an event occurs may not be the same for two reference frames moving relative to each other. Our analysis thus far has been mostly qualitative. It's time to start developing some quantitative tools that will allow us to compare measurements in one reference frame to measurements in another reference frame.

FIGURE 27.16 The ticking of a light clock can be measured by experimenters in two different reference frames.

(a) A light clock

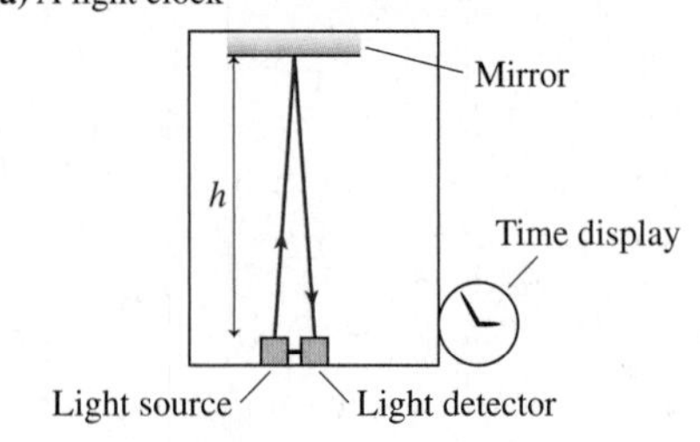

(b) The clock is at rest in frame S′.

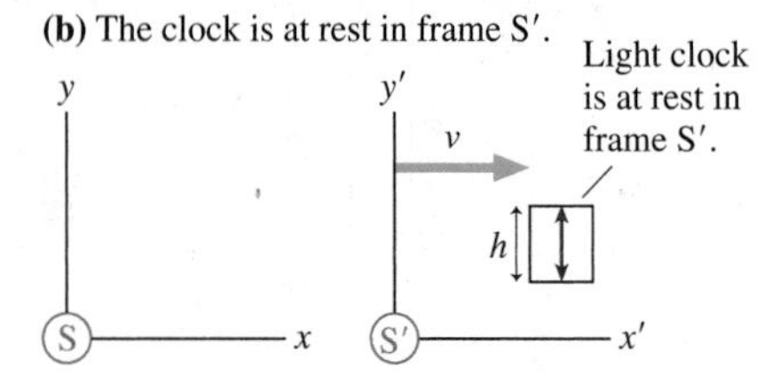

FIGURE 27.16a shows a special clock called a **light clock.** The light clock is a box of height h with a light source at the bottom and a mirror at the top. The light source emits a very short pulse of light that travels to the mirror and reflects back to a light detector beside the source. The clock advances one "tick" each time the detector receives a light pulse, and it immediately, with no delay, causes the light source to emit the next light pulse.

Our goal is to compare two measurements of the interval between two ticks of the clock: one taken by an experimenter standing next to the clock and the other by an experimenter moving with respect to the clock. To be specific, **FIGURE 27.16b** shows the clock at rest in reference frame S′. We call this the **rest frame** of the clock. Reference frame S′ moves to the right with velocity v relative to reference frame S.

Relativity requires us to measure *events,* so let's define event 1 to be the emission of a light pulse and event 2 to be the detection of that light pulse. Experimenters in both reference frames are able to measure where and when these events occur *in their frame.* In frame S, the time interval $\Delta t = t_2 - t_1$ is one tick of the clock. Similarly, one tick in frame S′ is $\Delta t' = t_2' - t_1'$.

It's simple to calculate the tick interval $\Delta t'$ observed in frame S′, the rest frame of the clock, because the light simply goes straight up and back down. The total distance traveled by the light is $2h$, so the time of one tick is

$$\Delta t' = \frac{2h}{c} \tag{27.2}$$

FIGURE 27.17 shows the light clock as seen in frame S. As seen in S, the clock is moving to the right at speed v. Thus the mirror has moved a distance $\frac{1}{2}v(\Delta t)$ during the time $\frac{1}{2}(\Delta t)$ in which the light pulse moves from the source to the mirror. To move from the source to the mirror, as seen from frame S the light must move along the *diagonal path* shown. That is, the light must travel *farther* from source to mirror than it did in the rest frame of the clock.

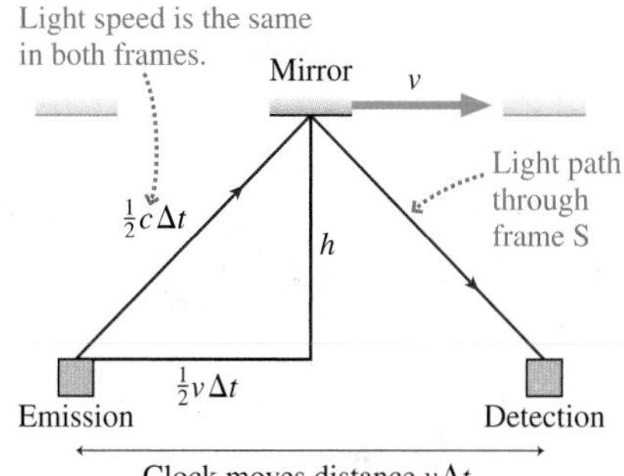

FIGURE 27.17 A light clock analysis in which the speed of light is the same in all reference frames.

The length of this diagonal is easy to calculate because, according to special relativity, the speed of light is equal to c in all inertial frames. Thus the diagonal length is simply

$$\text{distance} = \text{speed} \times \text{time} = c\left(\frac{1}{2}\Delta t\right) = \frac{1}{2}c\,\Delta t$$

We can then apply the Pythagorean theorem to the right triangle in Figure 27.17 to find that

$$h^2 + \left(\frac{1}{2}v\,\Delta t\right)^2 = \left(\frac{1}{2}c\,\Delta t\right)^2 \tag{27.3}$$

We can solve for Δt by first rewriting Equation 27.3 as

$$h^2 = \left(\frac{1}{2}c\,\Delta t\right)^2 - \left(\frac{1}{2}v\,\Delta t\right)^2 = \left[\left(\frac{1}{2}c\right)^2 - \left(\frac{1}{2}v\right)^2\right]\Delta t^2 = \left(\frac{1}{2}\right)^2(c^2 - v^2)\,\Delta t^2$$

so that

$$\Delta t^2 = \frac{h^2}{\left(\frac{1}{2}\right)^2(c^2 - v^2)} = \frac{(2h)^2}{c^2 - v^2} = \frac{(2h/c)^2}{1 - v^2/c^2}$$

from which we find

$$\Delta t = \frac{2h/c}{\sqrt{1 - v^2/c^2}} = \frac{\Delta t'}{\sqrt{1 - v^2/c^2}} \tag{27.4}$$

where we have used Equation 27.2 to write $2h/c$ as $\Delta t'$. The time interval between two ticks in frame S is *not* the same as in frame S′.

It's useful to define $\beta = v/c$, the speed as a fraction of the speed of light. For example, a reference frame moving with $v = 2.4 \times 10^8$ m/s has $\beta = 0.80$. In terms of β, Equation 27.4 is

$$\Delta t = \frac{\Delta t'}{\sqrt{1 - \beta^2}} \tag{27.5}$$

If reference frame S′ is at rest relative to frame S, then $\beta = 0$ and $\Delta t = \Delta t'$. In other words, experimenters in both reference frames measure time to be the same. But two experimenters moving relative to each other will measure *different* time intervals between the same two events. We're unaware of these differences in our everyday lives because our typical speeds are so small compared to c. But the differences are easily measured in a laboratory, and they do affect the precise timekeeping needed to make accurate location measurements with a GPS receiver.

Proper Time

Frame S′ has one important distinction. It is the *one and only* inertial reference frame in which the clock is at rest. Consequently, it is the one and only inertial reference frame in which the times of both events—the emission of the light and the detection of the light—are measured at the *same* position. You can see that the light pulse in Figure 27.16a, the rest frame of the clock, starts and ends at the same position, while in Figure 27.17 the emission and detection take place at *different* positions in frame S.

The time interval between two events that occur at the *same position* is called the **proper time** $\Delta\tau$. Only one inertial reference frame measures the proper time, and it can do so with a single clock that is present at both events. Experimenters in an inertial reference frame moving with speed $v = \beta c$ relative to the proper-time frame must use two clocks to measure the time interval: one at the position of the first event, the other at the position of the second event. We can rewrite Equation 27.5, where the time interval $\Delta t'$ is the proper time $\Delta\tau$, to find that the time interval in any other reference frame is

$$\Delta t = \frac{\Delta\tau}{\sqrt{1-\beta^2}} \geq \Delta\tau \qquad (27.6)$$

Time dilation in terms of proper time $\Delta\tau$ (where $\beta = v/c$)

Because $\beta = v/c$, and v is always less than c, β is always less than 1. This means that the factor $1/\sqrt{1-\beta^2}$ appearing in Equation 27.6 is always *equal to* (when $v = 0$) *or greater than 1.* Thus $\Delta t \geq \Delta\tau$. Recalling that Δt is the time between clock ticks in a frame such as S in which the clock is moving, while the proper time $\Delta\tau$ is the time between ticks in a frame at which the clock is at rest, we can interpret Equation 27.6 as saying that **the time interval between two ticks is the shortest in the reference frame in which the clock is at rest.** The time interval between two ticks is longer when it is measured in any reference frame in which the clock is moving. Because a longer tick interval implies a clock that runs more slowly, we can also say that a **moving clock runs slowly compared to an identical clock at rest.** This "stretching out" of the time interval implied by Equation 27.6 is called **time dilation.** FIGURE 27.18 illustrates the sense in which a moving clock runs slow.

FIGURE 27.18 A moving clock runs slow relative to a stationary one.

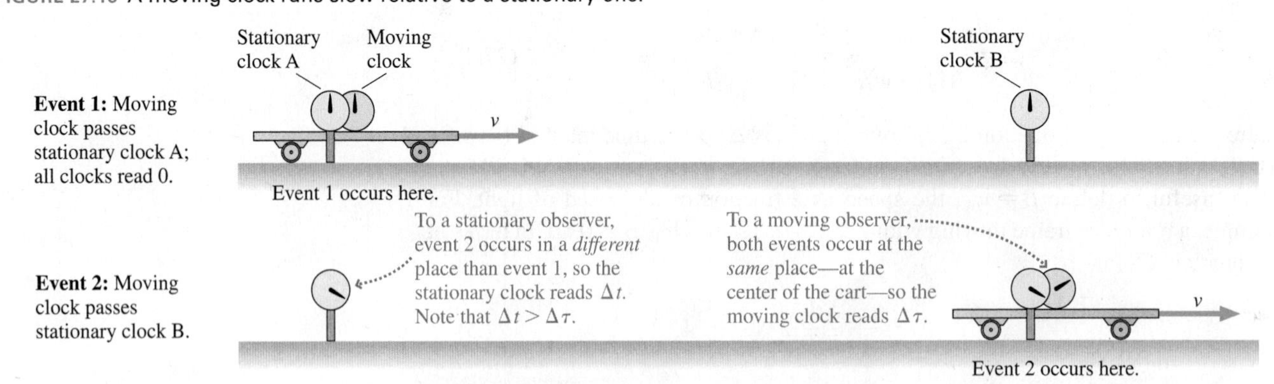
Stationary clock A
Moving clock
Stationary clock B
Event 1: Moving clock passes stationary clock A; all clocks read 0.
v
Event 1 occurs here.
To a stationary observer, event 2 occurs in a different place than event 1, so the stationary clock reads Δt. Note that Δt > Δτ.
To a moving observer, both events occur at the same place—at the center of the cart—so the moving clock reads Δτ.
Event 2: Moving clock passes stationary clock B.
v
Event 2 occurs here.

STOP TO THINK 27.6 A firecracker explodes on the ground in front of Martin just as Lisa passes by at $0.5c$. 30 μs later, according to Martin, a second firecracker explodes in front of him. Using what you've learned from Figure 27.18, the time Lisa measures between these two events is

A. 30 μs B. Greater than 30 μs C. Less than 30 μs

Equation 27.6 was derived using a light clock because the operation of a light clock is clear and easy to analyze. But the conclusion is really about time itself. Any clock, regardless of how it operates, behaves the same. For example, suppose you and a light clock are traveling in a very fast spaceship. The light clock happens to tick at the same rate as your heart beats—say, 60 times a minute, or once a second. Because the light clock is at rest in your frame, it measures the proper time between two successive beats of your heart; that is, $\Delta\tau = 1$ s. But to an experimenter stationed on the ground, watching you pass by at an enormous speed, the time interval $\Delta t = \Delta\tau/\sqrt{1-\beta^2}$ between two ticks of the clock—and hence two beats of your heart—would be *longer*. If, for instance, $\Delta t = 2$ s, the ground-based experimenter would conclude that your heart is beating only 30 times per minute. To the experimenter, *all* processes on your spaceship, including all your biological processes, would appear to run slowly.

EXAMPLE 27.4 **Journey time from the sun to Saturn**

Saturn is 1.43×10^{12} m from the sun. A rocket travels along a line from the sun to Saturn at a constant speed of exactly $0.9c$ relative to the solar system. How long does the journey take as measured by an experimenter on earth? As measured by an astronaut on the rocket?

PREPARE Let the solar system be in reference frame S and the rocket be in reference frame S′ that travels with velocity $v = 0.9c$ relative to S. Relativity problems must be stated in terms of *events*. Let event 1 be "the rocket and the sun coincide" (the experimenter on earth says that the rocket passes the sun; the astronaut on the rocket says that the sun passes the rocket) and event 2 be "the rocket and Saturn coincide."

FIGURE 27.19 shows the two events as seen from the two reference frames. Notice that the two events occur at the *same position* in S′, the position of the rocket.

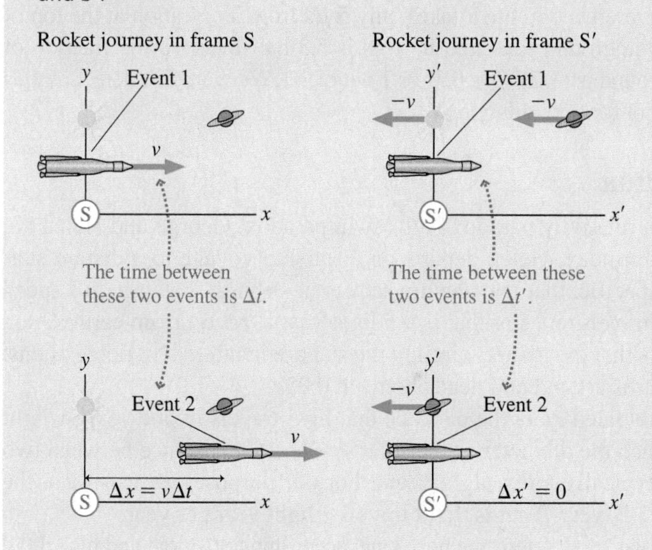

FIGURE 27.19 Visual overview of the trip as seen in frames S and S′.

SOLVE The time interval measured in the solar system reference frame, which includes the earth, is simply

$$\Delta t = \frac{\Delta x}{v} = \frac{1.43 \times 10^{12}\ \text{m}}{0.9 \times (3.00 \times 10^8\ \text{m/s})} = 5300\ \text{s}$$

Relativity hasn't abandoned the basic definition $v = \Delta x/\Delta t$, although we do have to be sure that Δx and Δt are measured in just one reference frame and refer to the same two events.

How are things in the rocket's reference frame? The two events occur at the *same position* in S′. Thus the time measured by the astronauts is the *proper time* $\Delta\tau$ between the two events. We can then use Equation 27.6 with $\beta = 0.9$ to find

$$\Delta\tau = \sqrt{1-\beta^2}\,\Delta t = \sqrt{1-0.9^2}\,(5300\ \text{s}) = 2310\ \text{s}$$

ASSESS The time interval measured between these two events by the astronauts is less than half the time interval measured by experimenters on earth. The difference has nothing to do with when earthbound astronomers *see* the rocket pass the sun and Saturn. Δt is the time interval from when the rocket actually passes the sun, as measured by a clock at the sun, until it actually passes Saturn, as measured by a synchronized clock at Saturn. The interval between *seeing* the events from earth, which would have to allow for light travel times, would be something other than 5300 s. Δt and $\Delta\tau$ are different because *time is different* in two reference frames moving relative to each other.

STOP TO THINK 27.7 Molly flies her rocket past Nick at constant velocity v. Molly and Nick both measure the time it takes the rocket, from nose to tail, to pass Nick. Which of the following is true?

A. Both Molly and Nick measure the same amount of time.
B. Molly measures a shorter time interval than Nick.
C. Nick measures a shorter time interval than Molly.

The global positioning system (GPS) If you've ever used a GPS receiver, you know it can pinpoint your location anywhere in the world. The system uses a set of orbiting satellites whose positions are very accurately known. The satellites orbit at a speed of about 14,000 km/h, enough to make the moving satellite's clocks run slow by about 7 μs a day. This may not seem like much, but it would introduce an error of 2000 m into your position! To function properly, the clocks are carefully corrected for effects due to relativity (including effects due to general relativity).

Experimental Evidence

Is there any evidence for the crazy idea that clocks moving relative to each other tell time differently? Indeed, there's plenty. An experiment in 1971 sent an atomic clock around the world on a jet plane while an identical clock remained in the laboratory. This was a difficult experiment because the traveling clock's speed was so small compared to c, but measuring the small differences between the time intervals was just barely within the capabilities of atomic clocks. It was also a more complex experiment than we've analyzed because the clock accelerated as it moved around a circle. The scientists found that, upon its return, the eastbound clock, traveling faster than the laboratory on a rotating earth, was 60 ns behind the stay-at-home clock, which was exactly as predicted by relativity.

Very detailed studies have been done on unstable particles called *muons* that are created at the top of the atmosphere, at a height of about 60 km, when high-energy cosmic rays collide with air molecules. It is well known, from laboratory studies, that stationary muons decay with a *half-life* of 1.5 μs. That is, half the muons decay within 1.5 μs, half of those remaining decay in the next 1.5 μs, and so on. The decays can be used as a clock.

The muons travel down through the atmosphere at very nearly the speed of light. The time needed to reach the ground, assuming $v \approx c$, is $\Delta t \approx (60{,}000 \text{ m})/(3 \times 10^8 \text{ m/s}) = 200\ \mu\text{s}$. This is 133 half-lives, so the fraction of muons reaching the ground should be $\approx (1/2)^{133} = 10^{-40}$. That is, only 1 out of every 10^{40} muons should reach the ground. In fact, experiments find that about 1 in 10 muons reach the ground, an experimental result that differs by a factor of 10^{39} from our prediction!

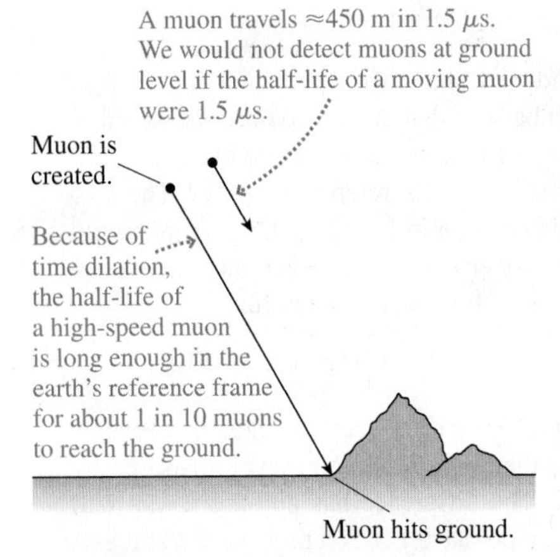
A muon travels ≈450 m in 1.5 μs. We would not detect muons at ground level if the half-life of a moving muon were 1.5 μs.
Muon is created.
Because of time dilation, the half-life of a high-speed muon is long enough in the earth's reference frame for about 1 in 10 muons to reach the ground.
Muon hits ground.

FIGURE 27.20 We wouldn't detect muons at the ground if not for time dilation.

The discrepancy is due to time dilation. In **FIGURE 27.20**, the two events "muon is created" and "muon hits ground" take place at two different places in the earth's reference frame. However, these two events occur at the *same position* in the muon's reference frame. (The muon is like the rocket in Example 27.4.) Thus the muon's internal clock measures the proper time. The time-dilated interval $\Delta t = 200\ \mu\text{s}$ in the earth's reference frame corresponds to a time of only $\Delta t' \approx 5\ \mu\text{s}$ in the muon's reference frame. That is, in the muon's reference frame it takes only 5 μs from its creation at the top of the atmosphere until the ground runs into it. This is 3.3 half-lives, so the fraction of muons reaching the ground is $(1/2)^{3.3} = 0.1$, or 1 out of 10. We wouldn't detect muons at the ground at all if not for time dilation.

The Twin Paradox

The most well-known relativity paradox is the twin paradox. George and Helen are twins. On their 25th birthday, Helen departs on a starship voyage to a distant star. Let's imagine, to be specific, that her starship accelerates almost instantly to a speed of $0.95c$ and that she travels to a star that is 9.5 light years (9.5 ly) from earth. Upon arriving, she discovers that the planets circling the star are inhabited by fierce aliens, so she immediately turns around and heads home at $0.95c$.

Alpha Centauri (arrow) is one of the closest stars to the sun, at a distance of 4.3 ly. If you traveled there and back at $0.99c$, your earthbound friends would be 8.6 years older, while you would have aged only 1.2 years.

A **light year,** abbreviated ly, is the distance that light travels in one year. A light year is vastly larger than the diameter of the solar system. The distance between two neighboring stars is typically a few light years. For our purpose, we can write the speed of light as $c = 1$ ly/year. That is, light travels 1 light year per year.

This value for c allows us to determine how long, according to George and his fellow earthlings, it takes Helen to travel out and back. Her total distance is 19 ly and, due to her rapid acceleration and rapid turn around, she travels essentially the entire distance at speed $v = 0.95c = 0.95$ ly/year. Thus the time she's away, as measured by George, is

$$\Delta t_{\text{G}} = \frac{19 \text{ ly}}{0.95 \text{ ly/year}} = 20 \text{ years} \tag{27.7}$$

George will be 45 years old when his sister Helen returns with tales of adventure.

While she's away, George takes a physics class and studies Einstein's theory of relativity. He realizes that time dilation will make Helen's clocks run more slowly than his clocks, which are at rest relative to him. Her heart—a clock—will beat fewer times and the minute hand on her watch will go around fewer times. In other words, she's aging more slowly than he is. Although she is his twin, she will be younger than he is when she returns.

Calculating Helen's age is not hard. We simply have to identify Helen's clock, because it's always with Helen as she travels, as the clock that measures proper time $\Delta\tau$. From Equation 27.6,

$$\Delta t_H = \Delta\tau = \sqrt{1-\beta^2}\,\Delta t_G = \sqrt{1-0.95^2}\,(20 \text{ years}) = 6.25 \text{ years} \qquad (27.8)$$

George will have just celebrated his 45th birthday as he welcomes home his 31-year-and-3-month-old twin sister.

This may be unsettling, because it violates our commonsense notion of time, but it's not a paradox. There's no logical inconsistency in this outcome. So why is it called "the twin paradox"? Read on.

Helen, knowing that she had quite of bit of time to kill on her journey, brought along several physics books to read. As she learns about relativity, she begins to think about George and her friends back on earth. Relative to her, they are all moving away at $0.95c$. Later they'll come rushing toward her at $0.95c$. Time dilation will cause their clocks to run more slowly than her clocks, which are at rest relative to her. In other words, as **FIGURE 27.21** shows, Helen concludes that people on earth are aging more slowly than she is. Alas, she will be much older than they when she returns.

FIGURE 27.21 The twin paradox.

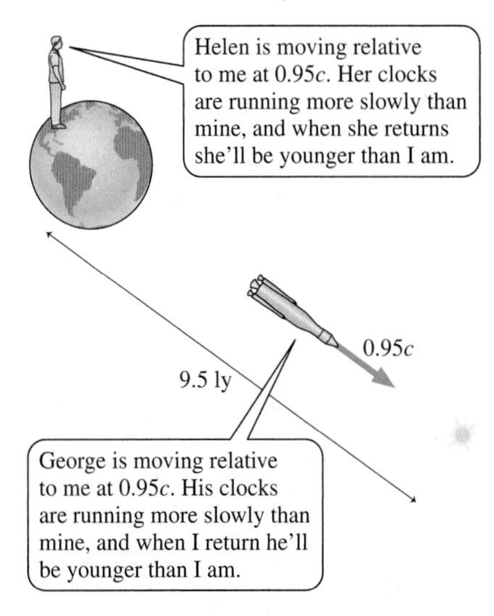

Helen is moving relative to me at 0.95c. Her clocks are running more slowly than mine, and when she returns she'll be younger than I am.
9.5 ly
0.95c
George is moving relative to me at 0.95c. His clocks are running more slowly than mine, and when I return he'll be younger than I am.

Finally, the big day arrives. Helen lands back on earth and steps out of the starship. George is expecting Helen to be younger than he is. Helen is expecting George to be younger than she is.

Here's the paradox! It's logically impossible for each to be younger than the other at the time when they are reunited. Where, then, is the flaw in our reasoning? It seems to be a symmetrical situation—Helen moves relative to George and George moves relative to Helen—but symmetrical reasoning has led to a conundrum.

But are the situations really symmetrical? George goes about his business day after day without noticing anything unusual. Helen, on the other hand, experiences three distinct periods during which the starship engines fire, she's crushed into her seat, and free dust particles that had been floating inside the starship are no longer, in the starship's reference frame, at rest or traveling in a straight line at constant speed. In other words, George spends the entire time in an inertial reference frame, *but Helen does not.* The situation is *not* symmetrical.

The principle of relativity applies *only* to inertial reference frames. Our discussion of time dilation was for inertial reference frames. Thus George's analysis and calculations are correct. Helen's analysis and calculations are *not* correct because she was trying to apply an inertial reference frame result to a noninertial reference frame.

Helen is younger than George when she returns. This is strange, but not a paradox. It is a consequence of the fact that time flows differently in two reference frames moving relative to each other.

27.7 Length Contraction

We've seen that relativity requires us to rethink our idea of time. Now let's turn our attention to the concepts of space and distance. Consider again Peggy on her train car, which is reference frame S′, moving past Ryan, who is at rest in frame S, at relative speed v. Ryan wants to measure the length L of Peggy's car as it moves past him. As shown in **FIGURE 27.22a**, he can do so by measuring the time Δt that it takes the car to pass the fixed cone; he then calculates that $L = v\,\Delta t$.

FIGURE 27.22b shows the situation in Peggy's reference frame S′, where the car is at rest. Peggy wants to measure the length L' of her car; we'll soon see that L' need not be the same as L. Peggy can measure L' by finding the time $\Delta t'$ that the cone, moving at speed v, takes to move from one end of the car to the other. In this way, she finds that $L' = v\,\Delta t'$.

FIGURE 27.22 The length of a train car as measured by Ryan and by Peggy.

(a) In Ryan's frame S, Peggy moves to the right at speed v.

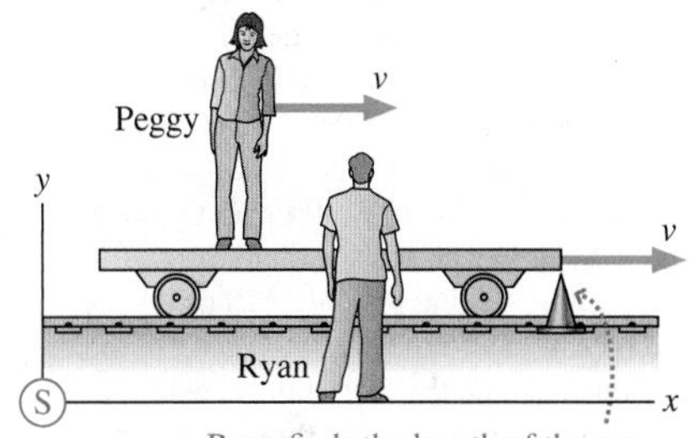

Ryan finds the length of the car by measuring the time Δt it takes to pass the cone. Then $L = v\,\Delta t$.

(b) In Peggy's frame S′, Ryan moves to the left at speed v.

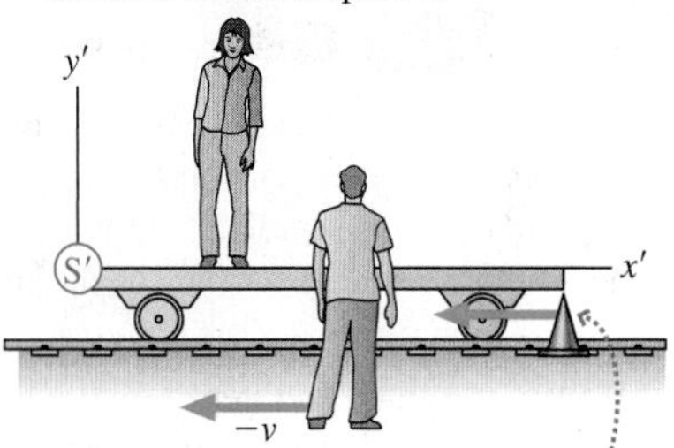

Peggy finds the length of the car by measuring the time $\Delta t'$ it takes the cone, moving at speed v, to pass her car. Then $L' = v\,\Delta t'$.

Speed v is the relative speed between S and S′ and is the same for both Ryan and Peggy. From Ryan's and Peggy's measurements of the car's length we can then write

$$v = \frac{L}{\Delta t} = \frac{L'}{\Delta t'} \tag{27.9}$$

The time interval Δt measured in Ryan's frame S is the proper time $\Delta\tau$ because the two events that define the time intervals—the front end and back end of the car passing the cone—occur at the same position (the cone) in Ryan's frame. We can use the time-dilation result, Equation 27.6, to relate $\Delta\tau$ measured by Ryan to $\Delta t'$ measured by Peggy. Equation 27.9 then becomes

$$\frac{L}{\Delta\tau} = \frac{L'}{\Delta t'} = \frac{L'}{\Delta\tau/\sqrt{1-\beta^2}} \tag{27.10}$$

The $\Delta\tau$ cancels, so the car's length L in Ryan's frame is related to its length L' in Peggy's frame by

$$L = \sqrt{1-\beta^2}\,L' \tag{27.11}$$

Surprisingly, we find that **the length of the car in Ryan's frame is different from its length in Peggy's frame.**

Peggy's frame S′, in which the car's length is L', has one important distinction. It is the *one and only* inertial reference frame in which the car is at rest. Experimenters in frame S′ can take all the time they need to measure L' because the car isn't moving. The length of an object measured in the reference frame in which the object is at rest is called the **proper length** ℓ. When you measure the length of an everyday object, such as a curtain rod or tabletop, it is usually at rest in your reference frame, so everyday length measurements are of proper length.

We can use the proper length ℓ to write Equation 27.11 as

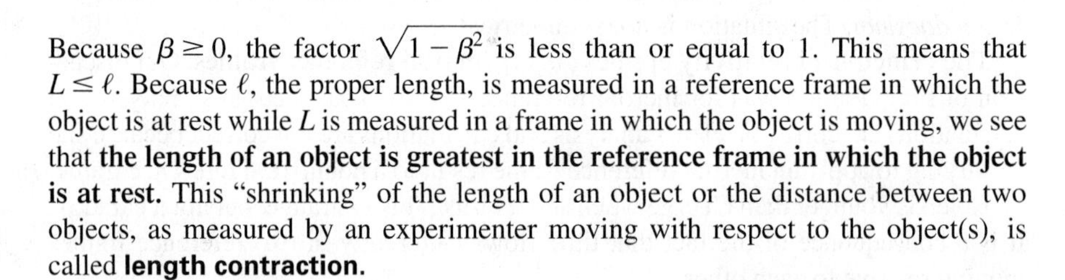

$$L = \sqrt{1-\beta^2}\,\ell \le \ell \tag{27.12}$$

Length contraction in terms of proper length ℓ

Two perspectives on a relativistic trip The Stanford Linear Accelerator is a 3.2-km-long electron accelerator that accelerates electrons to a speed of $0.99999999995c$. From our perspective, the electrons take a time $\Delta t = (3200\text{ m})/c = 11\ \mu\text{s}$ to make the trip. However, we see their "clocks" run slowly, ticking off only 110 ps during the trip. What do the electrons see? In their reference frame the end of the accelerator is coming toward them at $0.99999999995c$, but its length is contracted to only 3.3 cm. Thus it arrives in a time $(3.3\text{ cm})/c = 110$ ps. The same result, but from a different perspective.

Because $\beta \ge 0$, the factor $\sqrt{1-\beta^2}$ is less than or equal to 1. This means that $L \le \ell$. Because ℓ, the proper length, is measured in a reference frame in which the object is at rest while L is measured in a frame in which the object is moving, we see that **the length of an object is greatest in the reference frame in which the object is at rest.** This "shrinking" of the length of an object or the distance between two objects, as measured by an experimenter moving with respect to the object(s), is called **length contraction.**

NOTE ► A moving object's length is contracted only in the direction in which it's moving (its length along the x-axis in Figure 27.22). The object's length in the y- and z-directions doesn't change. ◄

EXAMPLE 27.5 Length contraction of a ladder

Dan holds a 5.0-m-long ladder parallel to the ground. He then gets up to a good sprint, eventually reaching 98% of the speed of light. How long is the ladder according to Dan, once he is running, and according to Carmen, who is standing on the ground as Dan goes by?

PREPARE Let reference frame S′ be attached to Dan. The ladder is at rest in this reference frame, so Dan measures the proper length of the ladder: $\ell = 5.0$ m. Dan's frame S′ moves relative to Carmen's frame S with velocity $v = 0.98c$.

SOLVE We can find the length of the ladder in Carmen's frame from Equation 27.12. We have

$$L = \sqrt{1-\beta^2}\,\ell = \sqrt{1-0.98^2}\,(5.0\text{ m}) = 1.0\text{ m}$$

ASSESS The length of the moving ladder as measured by Carmen is only one-fifth its length as measured by Dan. These lengths are different because *space is different* in two reference frames moving relative to each other.

The conclusion that space is different in reference frames moving relative to each other is a direct consequence of the fact that time is different. Experimenters in both reference frames agree on the relative velocity v, leading to Equation 27.9: $v = L/\Delta t = L'/\Delta t'$. Because of time dilation, Ryan (who measures proper time) finds that $\Delta t < \Delta t'$. Thus L *has* to be less than L'. That is the only way Ryan and Peggy can reconcile their measurements.

The Binomial Approximation

A useful mathematical tool is the **binomial approximation.** Suppose we need to evaluate the quantity $(1 + x)^n$. If x is much less than 1, it turns out that an excellent approximation is

$$(1 + x)^n \approx 1 + nx \qquad \text{if} \qquad x \ll 1 \qquad (27.13)$$

You can try this on your calculator. Suppose you need to calculate 1.01^2. Comparing this expression with Equation 27.13, we see that $x = 0.01$ and $n = 2$. Equation 27.13 then tells us that

$$1.01^2 \approx 1 + 2 \times 0.01 = 1.02$$

The exact result, using a calculator, is 1.0201. The approximate answer is good to about 99.99%! The smaller the value of x, the better the approximation.

The binomial approximation is very useful when we need to calculate a relativistic expression for a speed much less than c, so that $v \ll c$. Because $\beta = v/c$, a reference frame moving with $v^2/c^2 \ll 1$ has $\beta^2 \ll 1$. In these cases, we can write

$$\sqrt{1 - \beta^2} = (1 - v^2/c^2)^{1/2} \approx 1 - \frac{1}{2}\frac{v^2}{c^2}$$
$$\frac{1}{\sqrt{1 - \beta^2}} = (1 - v^2/c^2)^{-1/2} \approx 1 + \frac{1}{2}\frac{v^2}{c^2} \qquad (27.14)$$

The following example illustrates the use of the binomial approximation.

EXAMPLE 27.6 The shrinking school bus

An 8.0-m-long school bus drives past at 30 m/s. By how much is its length contracted?

PREPARE The school bus is at rest in an inertial reference frame S′ moving at velocity $v = 30$ m/s relative to the ground frame S. The given length, 8.0 m, is the proper length ℓ in frame S′.

SOLVE In frame S, the school bus is length-contracted to

$$L = \sqrt{1 - \beta^2}\,\ell$$

The bus's speed v is much less than c, so we can use the binomial approximation to write

$$L \approx \left(1 - \frac{1}{2}\frac{v^2}{c^2}\right)\ell = \ell - \frac{1}{2}\frac{v^2}{c^2}\ell$$

The *amount* of the length contraction is

$$\ell - L = \frac{1}{2}\frac{v^2}{c^2}\ell = \frac{1}{2}\left(\frac{30 \text{ m/s}}{3.0 \times 10^8 \text{ m/s}}\right)^2 (8.0 \text{ m})$$
$$= 4.0 \times 10^{-14} \text{ m} = 40 \text{ fm}$$

where 1 fm = 1 femtometer = 10^{-15} m.

ASSESS The amount the bus "shrinks" is only slightly larger than the diameter of the nucleus of an atom. It's no wonder that we're not aware of length contraction in our everyday lives. If you had tried to calculate this number exactly, your calculator would have shown $\ell - L = 0$. The difficulty is that the difference between ℓ and L shows up only in the 14th decimal place. A scientific calculator determines numbers to 10 or 12 decimal places, but that isn't sufficient to show the difference. The binomial approximation provides an invaluable tool for finding the very tiny difference between two numbers that are nearly identical.

STOP TO THINK 27.8 Peggy, standing on a moving railroad car, passes Ryan at velocity v. Peggy and Ryan both measure the length of the car, from one end to the other. The length Peggy measures is ______ the length Ryan measures.

A. Longer than B. The same as C. Shorter than

27.8 Velocities of Objects in Special Relativity

In Section 27.2 we discussed Galilean relativity, which is applicable to objects that are moving at speeds much less than the speed of light. We found that if the velocity of an object is u in reference frame S, then its velocity measured in frame S′, moving at velocity v relative to frame S, is $u' = u - v$.

But we soon learned that this expression is invalid for objects moving at an appreciable fraction of the speed of light. In particular, light itself moves at speed c as measured by *all* observers, independent of their relative velocities. The Galilean transformation of velocity needs to be modified for objects moving at relativistic speeds.

Although a proof is beyond the scope of this text, Einstein's relativity includes a velocity-addition expression valid for *any* velocities. An object's velocity measured in frame S is related to its velocity measured in frame S′ by the Lorentz velocity transformation:

Velocity of object as measured in frame S

$$u' = \frac{u - v}{1 - uv/c^2} \quad \text{or} \quad u = \frac{u' + v}{1 + u'v/c^2} \tag{27.15}$$

The relative velocity between the two frames

Velocity of object as measured in frame S′

These equations were discovered by Dutch physicist H. A. Lorentz a few years before Einstein published his theory of relativity, but Lorentz didn't completely understand their implications for space and time. Notice that the denominator is ≈ 1 if either u or v is much less than c. In other words, these equations agree with the Galilean velocity transformation when velocities are nonrelativistic (i.e., $\ll c$), but they differ as velocities approach the speed of light.

EXAMPLE 27.7 A speeding bullet

A rocket flies past the earth at precisely $0.9c$. As it goes by, the rocket fires a bullet in the forward direction at precisely $0.95c$ with respect to the rocket. What is the bullet's speed with respect to the earth?

PREPARE The rocket and the earth are inertial reference frames. Let the earth be frame S and the rocket be frame S′. The velocity of frame S′ relative to frame S is $v = 0.9c$. The bullet's velocity in frame S′ is $u' = 0.95c$.

SOLVE We can use the Lorentz velocity transformation to find

$$u = \frac{u' + v}{1 + u'v/c^2} = \frac{0.95c + 0.90c}{1 + (0.95c)(0.90c)/c^2} = 0.997c$$

The bullet's speed with respect to the earth is 99.7% of the speed of light.

NOTE ▸ Many relativistic calculations are much easier when velocities are specified as a fraction of c. ◂

ASSESS The Galilean transformation of velocity would give $u = 1.85c$. Now, despite the very high speed of the rocket and of the bullet with respect to the rocket, the bullet's speed with respect to the earth remains less than c. This is yet more evidence that objects cannot travel faster than the speed of light.

Suppose the rocket in Example 27.7 fired a laser beam in the forward direction as it traveled past the earth at velocity v. The laser beam would travel away from the rocket at speed $u' = c$ in the rocket's reference frame S′. What is the laser beam's speed in the earth's frame S? According to the Lorentz velocity transformation, it must be

$$u = \frac{u' + v}{1 + u'v/c^2} = \frac{c + v}{1 + cv/c^2} = \frac{c + v}{1 + v/c} = \frac{c + v}{(c + v)/c} = c$$

Light travels at speed c in both frame S and frame S$'$. This important consequence of the principle of relativity is "built into" the Lorentz velocity transformation.

STOP TO THINK 27.9 Sam flies past earth at $0.75c$. As he goes by, he fires a bullet forward at $0.75c$. Suzy, on the earth, measures the bullet's speed to be

A. $1.5c$ B. c C. Between $0.75c$ and c
D. $0.75c$ E. Less than $0.75c$

27.9 Relativistic Momentum

In Newtonian mechanics, the total momentum of a system is a conserved quantity. Further, the law of conservation of momentum, $P_f = P_i$, is true in all inertial reference frames *if* the particle velocities in different reference frames are related by the Galilean velocity transformation.

In relativity, however, we know that particle velocities in different frames are related by the Lorentz velocity transformations, which differ dramatically from the Galilean transformations as particle speeds approach c. If we use the Lorentz transformations, it's not difficult to show that a particle's Newtonian momentum $p = mu$ is *not* conserved in a frame moving relative to a frame in which momentum *is* conserved.

Yet momentum conservation is such a central and important feature of mechanics that it seems likely to hold in relativity as well. Indeed, a relativistic analysis of particle collisions shows that momentum conservation does hold, provided that we redefine the momentum of a particle as

$$p = \frac{mu}{\sqrt{1 - u^2/c^2}} \tag{27.16}$$

In this photograph, the track (shown in red) of a high-energy proton enters from the lower right. The proton then collides with other protons, sending them in all directions, where further collisions occur. Momentum is conserved, but at these high speeds the relativistic expression for momentum must be used.

You can see that Equation 27.16 reduces to the classical expression $p = mu$ when the particle's speed $u \ll c$.

To simplify our notation, let's define the quantity

$$\gamma = \frac{1}{\sqrt{1 - u^2/c^2}} \tag{27.17}$$

With this definition of γ, the momentum of a particle is

$$p = \gamma mu \tag{27.18}$$

Relativistic momentum for a particle with mass m and speed u

EXAMPLE 27.8 Momentum of a subatomic particle

Electrons in a particle accelerator reach a speed of $0.999c$ relative to the laboratory. One collision of an electron with a target produces a muon that moves forward with a speed of $0.950c$ relative to the laboratory. The muon mass is 1.90×10^{-28} kg. What is the muon's momentum in the laboratory frame and in the frame of the electron beam?

PREPARE Let the laboratory be reference frame S. The reference frame S$'$ of the electron beam (i.e., a reference frame in which the electrons are at rest) moves in the direction of the electrons at $v = 0.999c$. The muon velocity in frame S is $u = 0.95c$.

SOLVE γ for the muon in the laboratory reference frame is

$$\gamma = \frac{1}{\sqrt{1 - u^2/c^2}} = \frac{1}{\sqrt{1 - 0.95^2}} = 3.203$$

Continued

Thus the muon's momentum in the laboratory is

$$\begin{aligned} p &= \gamma m u \\ &= (3.203)(1.90 \times 10^{-28}\ \text{kg})(0.95 \times 3.00 \times 10^8\ \text{m/s}) \\ &= 1.73 \times 10^{-19}\ \text{kg}\cdot\text{m/s} \end{aligned}$$

The momentum is a factor of 3.2 larger than the Newtonian momentum mu. To find the momentum in the electron-beam reference frame, we must first use the velocity transformation equation to find the muon's velocity in frame S′:

$$u' = \frac{u - v}{1 - uv/c^2} = \frac{0.95c - 0.999c}{1 - (0.95c)(0.999c)/c^2} = -0.9617$$

In the laboratory frame, the faster electrons are overtaking the slower muon. Hence the muon's velocity in the electron-beam frame is negative. γ' for the muon in frame S′ is

$$\gamma' = \frac{1}{\sqrt{1 - u'^2/c^2}} = \frac{1}{\sqrt{1 - 0.9617^2}} = 3.648$$

The muon's momentum in the electron-beam reference frame is

$$\begin{aligned} p' &= \gamma' m u' \\ &= (3.648)(1.90 \times 10^{-28}\ \text{kg})(-0.9617 \times 3.00 \times 10^8\ \text{m/s}) \\ &= -2.00 \times 10^{-19}\ \text{kg}\cdot\text{m/s} \end{aligned}$$

ASSESS From the laboratory perspective, the muon moves only slightly slower than the electron beam. But it turns out that the muon moves faster with respect to the electrons, although in the opposite direction, than it does with respect to the laboratory.

The Cosmic Speed Limit

FIGURE 27.23a is a graph of momentum versus velocity. For a Newtonian particle, with $p = mu$, the momentum is directly proportional to the velocity. The relativistic expression for momentum agrees with the Newtonian value if $u \ll c$, but p approaches ∞ as $u \to c$.

The implications of this graph become clear when we relate momentum to force. Consider a particle subjected to a constant force, such as a rocket that never runs out of fuel. From the impulse-momentum theorem we have $\Delta p = F\Delta t$, or $p = mu = Ft$ if the rocket starts from rest at $t = 0$. If Newtonian physics were correct, a particle would go faster and faster as its velocity $u = p/m = (F/m)t$ increased without limit. But the relativistic result, shown in **FIGURE 27.23b**, is that the particle's velocity approaches the speed of light ($u \to c$) as p approaches ∞. Relativity gives a very different outcome than Newtonian mechanics.

FIGURE 27.23 The speed of a particle cannot reach the speed of light.

(a)

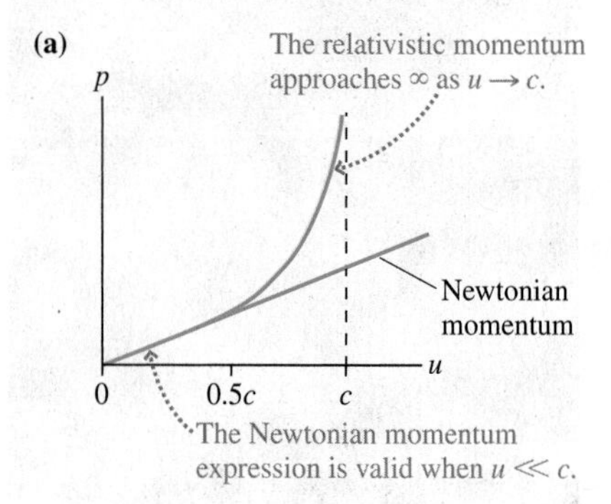
p
The relativistic momentum approaches ∞ as u → c.
Newtonian momentum
u
0
0.5c
c
The Newtonian momentum expression is valid when u ≪ c.

(b)

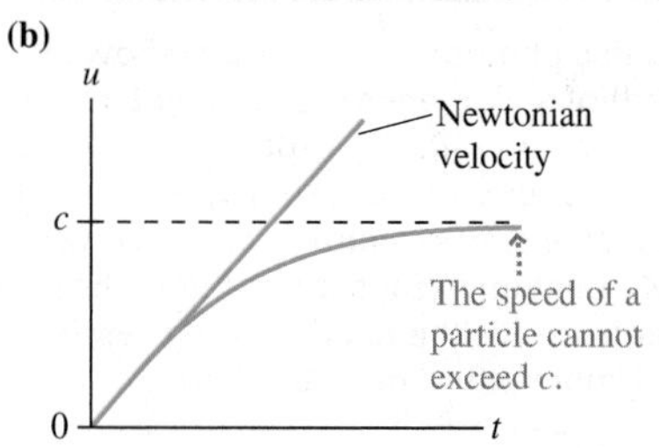
u
Newtonian velocity
c
The speed of a particle cannot exceed c.
0
t

The speed c is a "cosmic speed limit" for material particles. A force cannot accelerate a particle to a speed higher than c because the particle's momentum becomes infinitely large as the speed approaches c. The amount of effort required for each additional increment of velocity becomes larger and larger until no amount of effort can raise the velocity any higher.

Actually, at a more fundamental level, c is a speed limit for *any* kind of **causal influence.** If you throw a rock and break a window, your throw is the *cause* of the breaking window and the rock is the causal influence. A causal influence can be any kind of particle, wave, or information that travels from A to B and allows A to be the cause of B.

For two unrelated events—a firecracker explodes in Tokyo and a balloon bursts in Paris—the relativity of simultaneity tells us that in one reference frame the firecracker may explode before the balloon bursts, but in some other reference frame the balloon may burst first.

However, for two causally related events—A causes B—it would be nonsense for an experimenter in any reference frame to find that B occurs before A. No experimenter in any reference frame, no matter how it is moving, will find that you are born before your mother is born.

But according to relativity, a causal influence traveling faster than light could result in B causing A, a logical absurdity. Thus **no causal influence of any kind—a particle, wave, or other influence—can travel faster than *c*.**

The existence of a cosmic speed limit is one of the most interesting consequences of the theory of relativity. "Hyperdrive," in which a spaceship suddenly leaps to faster-than-light velocities, is simply incompatible with the theory of relativity. Rapid travel to the stars must remain in the realm of science fiction.

STOP TO THINK 27.10 Lauren is flying in her spaceship at $0.3c$. If she doubles her speed to $0.6c$, her momentum

A. Doubles. B. More than doubles. C. Less than doubles.

27.10 Relativistic Energy

Energy is our final topic in this chapter on relativity. Space, time, velocity, and momentum are changed by relativity, so it seems inevitable that we'll need a new view of energy. Indeed, one of the most profound results of relativity, and perhaps the one with the most far-reaching consequences, was Einstein's discovery of the fundamental relationship between energy and mass.

Consider an object of mass m moving at speed u. Einstein found that the **total energy** of such an object is

$$E = \frac{mc^2}{\sqrt{1 - u^2/c^2}} = \gamma mc^2 \tag{27.19}$$

Total energy of an object of mass m moving at speed u

where $\gamma = 1/\sqrt{1 - u^2/c^2}$ was defined in Equation 27.17.

To understand this expression, let's start by examining its behavior for objects traveling at speeds much less than the speed of light. In this case we can use the binomial approximation you learned in Section 27.7 to write

$$\gamma = \frac{1}{\sqrt{1 - u^2/c^2}} \approx 1 + \frac{1}{2}\frac{u^2}{c^2}$$

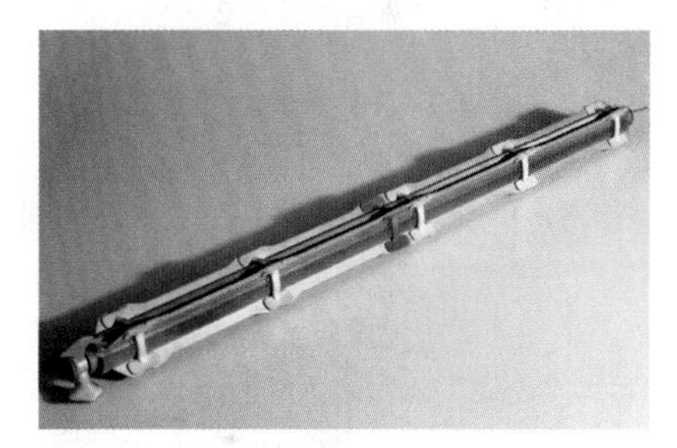

This fuel rod for a nuclear power reactor contains about 5 kg of uranium. Its usable energy content, which comes from the conversion of a small fraction of the uranium's mass to energy, is equivalent to that of about 10 million kg of coal.

For low speeds u, then, the object's total energy is

$$E \approx mc^2 + \frac{1}{2}mu^2 \tag{27.20}$$

The second term in this expression is the familiar Newtonian kinetic energy $K = \frac{1}{2}mu^2$ you studied in ◀◀ SECTION 10.3, written here in terms of velocity u rather than v. But there is an additional term in the total energy, the **rest energy** given by

$$E_0 = mc^2 \tag{27.21}$$

When a particle is at rest, with $u = 0$, it still has energy E_0. Indeed, because c is so large, the rest energy can be enormous. Equation 27.21 is, of course, Einstein's famous $E = mc^2$, perhaps the most famous equation in all of physics. It tells us that there is a fundamental equivalence between mass and energy, an idea we'll explore later in this section.

Class Video

EXAMPLE 27.9 The rest energy of an apple

What is the rest energy of a 200 g apple?

SOLVE From Equation 27.21 we have

$$E_0 = mc^2 = (0.20\text{ kg})(3.0 \times 10^8\text{ m/s})^2 = 1.8 \times 10^{16}\text{ J}$$

ASSESS This is an enormous energy, enough to power a medium-sized city for about a year.

Equation 27.20 suggests that the total energy of an object is the sum of a rest energy, which is a new idea, and the familiar kinetic energy. But Equation 27.20 is valid only when the object's speed is low compared to c. For higher speeds, we need to use the full energy expression, Equation 27.19. We can use Equation 27.19 to find a relativistic expression for the kinetic energy K by subtracting the rest energy E_0 from the total energy. Doing so gives

$$K = E - E_0 = \gamma mc^2 - mc^2 = (\gamma - 1)mc^2 = (\gamma - 1)E_0 \tag{27.22}$$

Thus we can write the total energy of an object of mass m as

$$E = \underbrace{mc^2}_{\text{Rest energy } E_0} + \underbrace{(\gamma - 1)mc^2}_{\text{Kinetic energy } K} \tag{27.23}$$

EXAMPLE 27.10 Comparing energies of a ball and an electron

Calculate the rest energy and the kinetic energy of (a) a 100 g ball moving with a speed of 100 m/s and (b) an electron with a speed of 0.999c.

PREPARE The ball, with $u \ll c$, is a classical particle. We don't need to use the relativistic expression for its kinetic energy. The electron is highly relativistic.

SOLVE

a. For the ball, with $m = 0.100$ kg,

$$E_0 = mc^2 = 9.00 \times 10^{15}\text{ J}$$

$$K = \frac{1}{2}mu^2 = 500\text{ J}$$

b. For the electron, we start by calculating

$$\gamma = \frac{1}{\sqrt{1 - u^2/c^2}} = 22.4$$

Then, using $m_e = 9.11 \times 10^{-31}$ kg,

$$E_0 = mc^2 = 8.20 \times 10^{-14}\text{ J}$$

$$K = (\gamma - 1)E_0 = 175 \times 10^{-14}\text{ J}$$

ASSESS The ball's kinetic energy is a typical kinetic energy. Its rest energy, by contrast, is a staggeringly large number. For a relativistic electron, on the other hand, the kinetic energy is more important than the rest energy.

STOP TO THINK 27.11 An electron moves through the lab at a speed such that $\gamma = 1.5$. The electron's kinetic energy is

A. Greater than its rest energy.
B. Equal to its rest energy.
C. Less than its rest energy.

The Equivalence of Mass and Energy

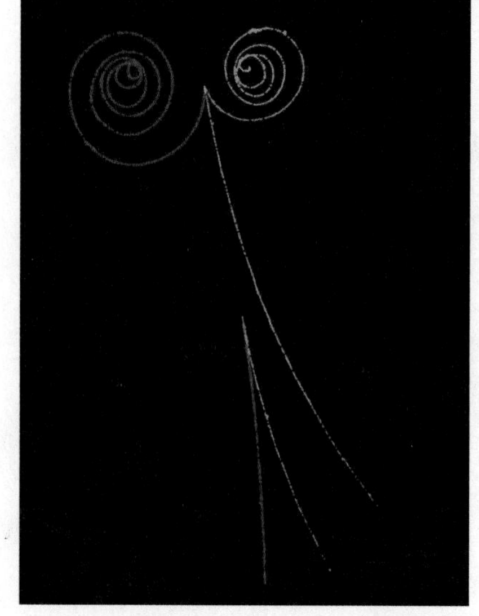

The tracks of elementary particles in a bubble chamber show the creation of two electron-positron pairs. The negative electron and positive positron curve in opposite directions in the magnetic field.

Now we're ready to explore the significance of Einstein's famous equation $E = mc^2$. **FIGURE 27.24** shows an experiment that has been done countless times in the last 50 years at particle accelerators around the world. An electron that has been accelerated to $u \approx c$ is aimed at a target material. When a high-energy electron collides with an atom in the target, it can easily knock one of the electrons out of the atom. Thus we would expect to see two electrons leaving the target: the incident electron and the ejected electron. Instead, *four* particles emerge from the target: three electrons and a positron. A *positron,* or positive electron, is the antimatter version of an electron, identical to an electron in all respects other than having charge $q = +e$. In particular, a positron has the same mass m_e as an electron.

FIGURE 27.24 An inelastic collision between electrons can create an electron-positron pair.

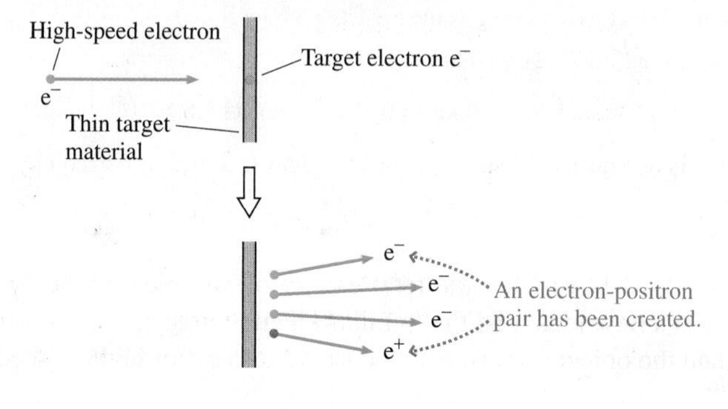

In chemical-reaction notation, the collision is

$$e^-(\text{fast}) + e^-(\text{at rest}) \rightarrow e^- + e^- + e^- + e^+$$

An electron and a positron have been created, apparently out of nothing. Mass $2m_e$ before the collision has become mass $4m_e$ after the collision. (Notice that charge has been conserved in this collision.)

Although the mass has increased, it wasn't created "out of nothing." If you measured the energies before and after the collision, you would find that the kinetic energy before the collision was *greater* than the kinetic energy after. In fact, the decrease in kinetic energy is exactly equal to the rest energy of the two particles that have been created: $\Delta K = 2m_e c^2$. The new particles have been created *out of energy!*

Not only can particles be created from energy, particles can return to energy. **FIGURE 27.25** shows an electron colliding with a positron, its antimatter partner. When a particle and its antiparticle meet, they *annihilate* each other. The mass disappears, and the energy equivalent of the mass is transformed into two high-energy photons. Photons have no mass and represent pure energy. Positron-electron annihilation is also the basis of the medical procedure known as a positron-emission tomography, or PET scans. We'll study this important diagnostic tool in detail in Chapter 30.

FIGURE 27.25 The annihilation of an electron-positron pair.

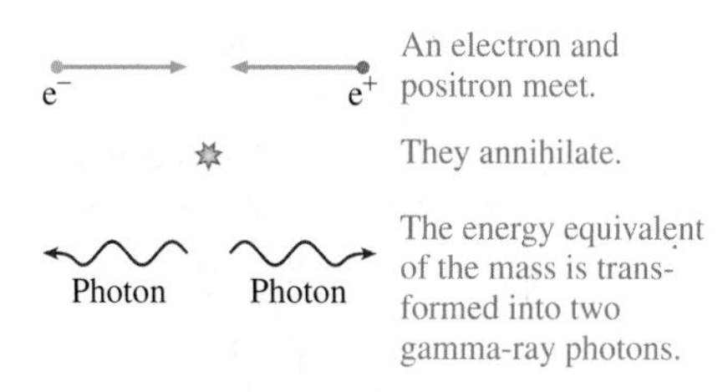

Conservation of Energy

The creation and annihilation of particles with mass, processes strictly forbidden in Newtonian mechanics, are vivid proof that neither mass nor the Newtonian definition of energy is conserved. Even so, the *total* energy—the kinetic energy *and* the energy equivalent of mass—remains a conserved quantity.

Law of conservation of total energy The energy $E = \sum E_i$ of an isolated system is conserved, where $E_i = \gamma_i m_i c^2$ is the total energy of particle i.

TABLE 27.1 Mass before and after fission of ^{235}U

Initial nucleus	Initial mass (u)	Final nucleus	Final mass (u)
^{235}U	235.0439	^{144}Ba	143.9229
n	1.0087	^{89}Kr	88.9176
		3n	3.0260
Total	236.0526		235.8665

Mass and energy are not the same thing, but, as the last few examples have shown, they are *equivalent* in the sense that mass can be transformed into energy and energy can be transformed into mass as long as the total energy is conserved.

Probably the most well-known application of the conservation of total energy is nuclear fission. The uranium isotope ^{236}U, containing 236 protons and neutrons, does not exist in nature. It can be created when a ^{235}U nucleus absorbs a neutron, increasing its atomic mass from 235 to 236. The ^{236}U nucleus quickly fragments into two smaller nuclei and several extra neutrons, a process known as **nuclear fission**. The nucleus can fragment in several ways, but one is

$$n + {}^{235}\text{U} \rightarrow {}^{236}\text{U} \rightarrow {}^{144}\text{Ba} + {}^{89}\text{Kr} + 3n \tag{27.24}$$

Ba and Kr are the atomic symbols for barium and krypton; n is the symbol for a neutron.

This reaction seems like an ordinary chemical reaction—until you check the masses. The masses of atomic isotopes are known with great precision from many decades of measurement in instruments called mass spectrometers. As shown in Table 27.1, if you add up the masses on both sides, you find that the mass of the products is 0.186 u less than the mass of the initial neutron and ^{235}U, where, you will recall, $1\text{ u} = 1.66 \times 10^{-27}$ kg is the atomic mass unit. Converting to kilograms gives us the mass loss of 3.09×10^{-28} kg.

Mass has been lost, but the energy equivalent of the mass has not. As **FIGURE 27.26** shows, the mass has been converted to kinetic energy, causing the two product nuclei and three neutrons to be ejected at very high speeds. The kinetic energy is easily calculated: $\Delta K = m_{\text{lost}} c^2 = 2.8 \times 10^{-11}$ J.

This is a very tiny amount of energy, but it is the energy released from *one* fission. The number of nuclei in a macroscopic sample of uranium is on the order of N_A, Avogadro's number. Hence the energy available if *all* the nuclei fission is enormous.

FIGURE 27.26 In nuclear fission, the energy equivalent of lost mass is converted into kinetic energy.

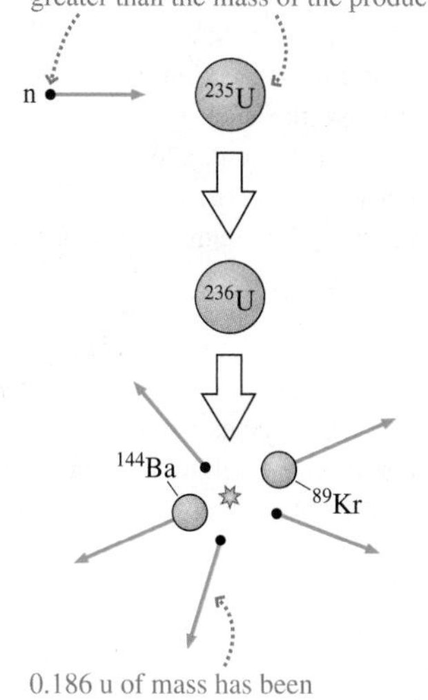

This energy, of course, is the basis for both nuclear power reactors and nuclear weapons.

We started this chapter with an expectation that relativity would challenge our basic notions of space and time. We end by finding that relativity changes our understanding of mass and energy. Most remarkable of all is that each and every one of these new ideas flows from one simple statement: The laws of physics are the same in all inertial reference frames.

INTEGRATED EXAMPLE 27.11 The global positioning system

The turtle in the photo that opens this chapter is being tracked using the global positioning system (GPS), a system of 24 satellites in circular orbits high above the earth. Each satellite, with an orbital speed of 3900 m/s, carries an atomic clock whose time is accurate to ± 1 ns per day. Every 30 s, each satellite sends out a radio signal giving its precise location in space and the exact time the signal was sent.

The turtle's GPS receiver records the time at which the signal is received. Because the signal specifies the time at which it was sent, the receiver can easily calculate how long it took the signal to reach the turtle. Then, because radio waves are electromagnetic waves traveling at the speed of light, the precise distance to the satellite can be calculated.

The signal from one satellite actually locates the turtle's position only along a large sphere centered on the satellite. To pinpoint the location exactly requires the signals from four or more satellites. In this problem, we'll ignore these complicating effects.

a. As the satellite passes overhead, suppose two lights flash: one at the front of the satellite (the side toward which the satellite is moving) and one at the rear (the side opposite the direction of motion). In the reference frame of the satellite, the two flashes are simultaneous. Are they simultaneous to an experimenter on the earth? If not, which flash occurs first?
b. In one day, how much time does the clock running on the satellite gain or lose compared to an identical clock on earth?
c. If the clock error in part b were not properly taken into account, by how much would the turtle's position on earth be in error after one day?

PREPARE Consider an astronaut standing on the satellite halfway between the lights. Because the two flashes are simultaneous in the reference frame of the satellite, the light from these flashes will reach the astronaut at the same time. An experimenter on the earth will also see these flashes reaching the astronaut at the same time, but will not agree that the flashes occurred simultaneously. We can use these observations to decide which flash occurred first to an earthbound experimenter.

In part b, because of time dilation the clock on the moving satellite runs slow compared to one on the earth, so this clock will *lose* time.

SOLVE a. **FIGURE 27.27** shows the satellite (here represented by a rod) as seen by an experimenter on earth. When the light waves from the two flashes meet at the satellite's center, the one from the rear of the satellite will have traveled farther than the one from the front. This is because the flash coming from the rear needs to "catch up" with the center of the moving satellite. Since both waves travel at speed c, but the wave from the rear travels farther, the rear light must have flashed *earlier* according to an earthbound experimenter.

FIGURE 27.27 The satellite as seen by an observer on earth.

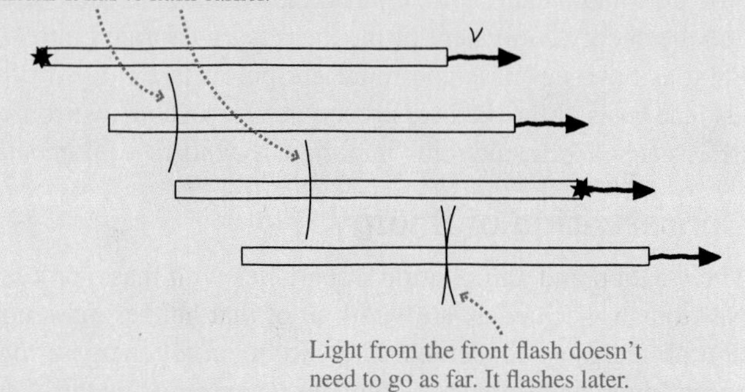

Because the satellite is moving, light from the rear flash — which is playing catch-up — has to travel farther to reach the satellite's center. This means it has to flash earlier.
v
Light from the front flash doesn't need to go as far. It flashes later.

b. The clock on the satellite measures proper time $\Delta\tau$ because this clock is at rest in the reference frame of the satellite. We want to know how much time the satellite's clock measures during an interval $\Delta t = 24\text{ h} = 86{,}400\text{ s}$ measured on the earth. We can rearrange Equation 27.6 as

$$\Delta\tau = \Delta t\sqrt{1-\beta^2}$$

Even for a fast-moving satellite, $\beta = v/c$ is so small that the term $\sqrt{1-\beta^2}$ will be exactly 1 on most calculators. We must therefore use the binomial expansion, Equation 27.14, to write

$$\Delta\tau = \Delta t\sqrt{1-\beta^2} \approx \Delta t\left(1 - \frac{1}{2}\frac{v^2}{c^2}\right) = \Delta t - \frac{1}{2}\frac{v^2}{c^2}\Delta t$$

The *difference* between the satellite and earth clocks is

$$\Delta t - \Delta\tau = \frac{1}{2}\frac{v^2}{c^2}\Delta t = \frac{1}{2}\left(\frac{3900\text{ m/s}}{3.0\times 10^8\text{ m/s}}\right)^2(86{,}400\text{ s}) = 7.3\ \mu\text{s}$$

Because this result is positive, $\Delta\tau$ is less than Δt: The moving clock *loses* 7.3 μs per day compared to a clock on earth.

c. The radio signal from the satellite travels at speed c. In 7.3 μs, the signal travels a distance

$$\Delta x = c\,\Delta t = (3.0\times 10^8\text{ m/s})(7.3\times 10^{-6}\text{ s}) = 2200\text{ m}$$

If the satellite clocks were not corrected for this relativistic effect, all GPS receivers on earth would miscalculate their positions by 2.2 km after just one day.

ASSESS Even for a very fast-moving object like a satellite, the corrections due to relativity are small. But these small corrections can have large effects on high-precision measurements. Interestingly, the relativistic correction is implemented *before* the satellites are launched by setting the clocks to run slightly fast, by a factor of 1.000000000447. Once in orbit, the clocks slow down to match an earthbound clock exactly.

SUMMARY

Goal: To understand how Einstein's theory of relativity changes our concepts of space and time.

GENERAL PRINCIPLES

Principle of Relativity

All the laws of physics are the same in all inertial reference frames.

- The speed of light c is the same in all inertial reference frames.
- No particle or causal influence can travel at a speed greater than c.

IMPORTANT CONCEPTS

Time

Time measurements depend on the motion of the experimenter relative to the events.

Proper time $\Delta\tau$ is the time interval between two events measured in a reference frame in which the events occur at the same position. The time interval Δt between the same two events, in a frame moving with relative velocity v, is

$$\Delta t = \Delta\tau/\sqrt{1-\beta^2} \geq \Delta\tau$$

where $\beta = v/c$. This is called **time dilation.**

Space

Spatial measurements depend on the motion of the experimenter relative to the events.

Proper length ℓ is the length of an object measured in a reference frame in which the object is at rest. The length L in a frame in which the object moves with velocity v is

$$L = \sqrt{1-\beta^2}\,\ell \leq \ell$$

This is called **length contraction.**

Momentum

The law of conservation of momentum is valid in all inertial reference frames if the momentum of a particle with velocity u is $p = \gamma m u$, where

$$\gamma = 1/\sqrt{1-u^2/c^2}$$

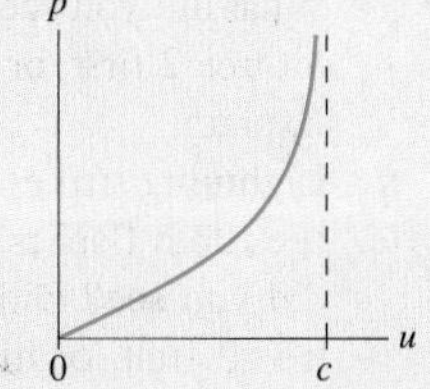

The momentum approaches ∞ as $u \rightarrow c$.

Energy

The **total energy** of a particle is $E = \gamma mc^2$. This can be written as

$$E = \underbrace{mc^2}_{\text{Rest energy } E_0} + \underbrace{(\gamma - 1)mc^2}_{\text{Kinetic energy } K}$$

K approaches ∞ as $u \rightarrow c$.

The total energy of an isolated system is conserved.

Simultaneity

Events that are simultaneous in reference frame S are not simultaneous in frame S′ moving relative to S.

Mass-energy equivalence

Mass m can be transformed into energy $E = mc^2$.

Energy can be transformed into mass $m = E/c^2$.

APPLICATIONS

An **event** happens at a specific place in space and time. Spacetime coordinates are (x, t) in frame S and (x', t') in frame S′.

A **reference frame** is a coordinate system with meter sticks and clocks for measuring events. Experimenters at rest relative to each other share the same reference frame.

If an object has velocity u in frame S and u' in frame S′, the two velocities are related by the **Lorentz velocity transformation:**

$$u' = \frac{u - v}{1 - uv/c^2} \qquad u = \frac{u' + v}{1 + u'v/c^2}$$

where v is the relative velocity between the two frames.

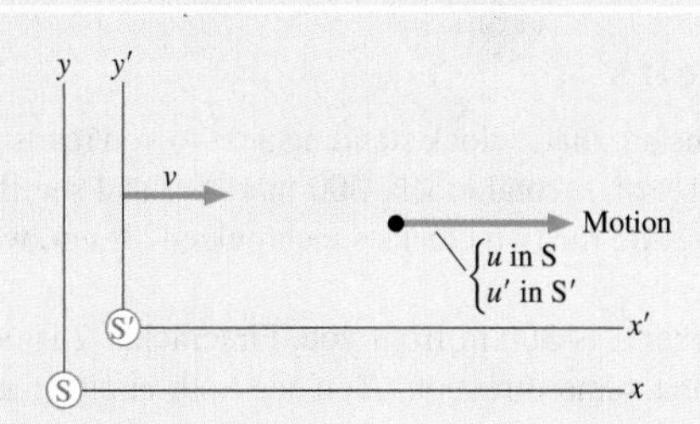

Problem difficulty is labeled as | (straightforward) to ||||| (challenging). Problems labeled INT integrate significant material from earlier chapters; BIO are of biological or medical interest.

For assigned homework and other learning materials, go to MasteringPhysics®

Scan this QR code to launch a Video Tutor Solution that will help you solve problems for this chapter.

QUESTIONS

Conceptual Questions

1. You are in an airplane cruising smoothly at 600 mph. What experiment, if any, could you do that would demonstrate that you are moving, while those on the ground are at rest?
2. Frame S′ moves relative to frame S as shown in Figure Q27.2.
   a. A ball is at rest in frame S′. What are the speed and direction of the ball in frame S?
   b. A ball is at rest in frame S. What are the speed and direction of the ball in frame S′?

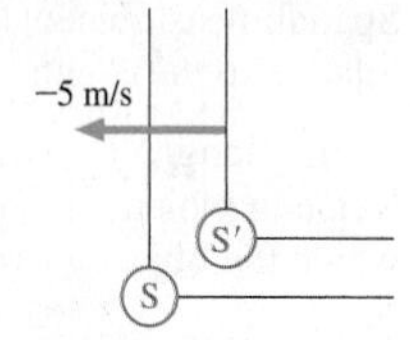

FIGURE Q27.2

3. a. Two balls move as shown in Figure Q27.3. What are the speed and direction of each ball in a reference frame that moves with ball 1?
   b. What are the speed and direction of each ball in a reference frame that moves with ball 2?

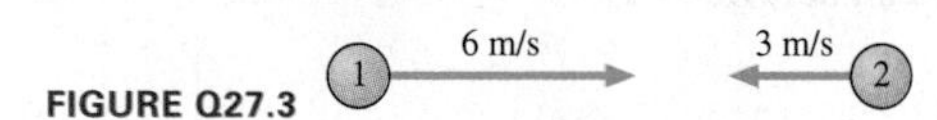

FIGURE Q27.3

4. A lighthouse beacon alerts ships to the danger of a rocky coastline.
   a. According to the lighthouse keeper, with what speed does the light leave the lighthouse?
   b. A boat is approaching the coastline at speed $0.5c$. According to the captain, with what speed is the light from the beacon approaching her boat?
5. As a rocket passes the earth at $0.75c$, it fires a laser perpendicular to its direction of travel as shown in Figure Q27.5.
   a. What is the speed of the laser beam relative to the rocket?
   b. What is the speed of the laser beam relative to the earth?

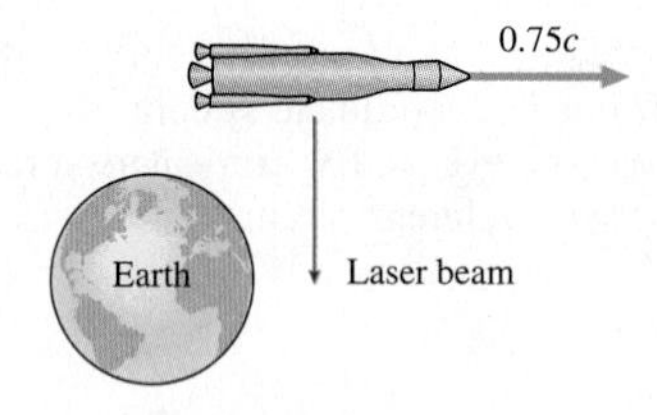

FIGURE Q27.5

6. At the instant that a clock standing next to you reads $t = 1.0\ \mu s$, you look at a second clock, 300 m away, and see that it reads $t = 0\ \mu s$. Are the two clocks synchronized? If not, which one is ahead?
7. Firecracker 1 is 300 m from you. Firecracker 2 is 600 m from you in the same direction. You see both explode at the same time. Define event 1 to be "firecracker 1 explodes" and event 2 to be "firecracker 2 explodes." Does event 1 occur before, after, or at the same time as event 2? Explain.
8. Firecrackers 1 and 2 are 600 m apart. You are standing exactly halfway between them. Your lab partner is 300 m on the other side of firecracker 1. You see two flashes of light, from the two explosions, at exactly the same instant of time. Define event 1 to be "firecracker 1 explodes" and event 2 to be "firecracker 2 explodes." According to your lab partner, based on measurements he or she makes, does event 1 occur before, after, or at the same time as event 2? Explain.
9. Your clocks and calendars are synchronized with the clocks and calendars in a star system exactly 10 ly from earth that is at rest relative to the earth. You receive a TV transmission from the star system that shows a date and time display. The date it shows is June 17, 2050. When you glance over at your own wall calendar, what date does it show?
10. Two trees are 600 m apart. You are standing exactly halfway between them and your lab partner is at the base of tree 1. Lightning strikes both trees.
    a. Your lab partner, based on measurements he makes, determines that the two lightning strikes were simultaneous. What did you see? Did you see the lightning hit tree 1 first, hit tree 2 first, or hit them both at the same instant of time? Explain.
    b. Lightning strikes the trees again. This time your lab partner sees both flashes of light at the same instant of time. What did you see? Did you see the lightning hit tree 1 first, hit tree 2 first, or hit them both at the same instant of time? Explain.
    c. In the scenario of part b, were the lightning strikes simultaneous? Explain.
11. Figure Q27.11 shows Peggy standing at the center of her railroad car as it passes Ryan on the ground. Firecrackers attached to the ends of the car explode. A short time later, the flashes from the two explosions arrive at Peggy at the same time.
    a. Were the explosions simultaneous in Peggy's reference frame? If not, which exploded first? Explain.
    b. Were the explosions simultaneous in Ryan's reference frame. If not, which exploded first? Explain.

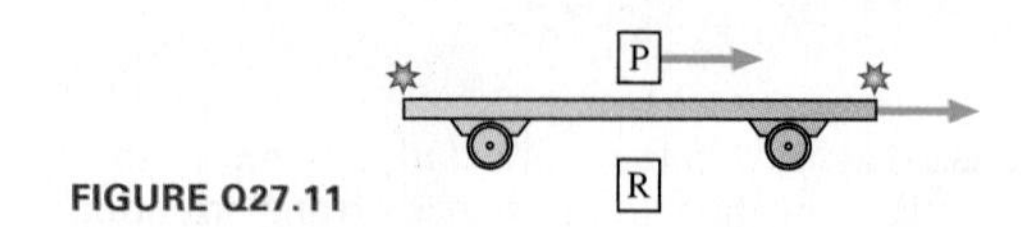

FIGURE Q27.11

12. In Figure Q27.12, clocks C_1 and C_2 in frame S are synchronized. Clock C′ moves at speed v relative to frame S. Clocks C′ and C_1 read exactly the same as C′ goes past. As C′ passes C_2, is the time shown on C′ earlier, later, or the same as the time shown on C_2? Explain.

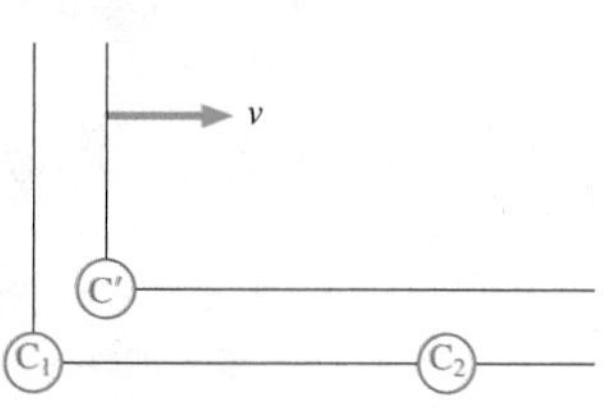

FIGURE Q27.12

13. A meter stick passes you at a speed of $0.5c$. Explain clearly how you would measure the length of this fast-moving object.
14. You're passing a car on the highway. You want to know how much time is required to completely pass the car, from no overlap between the cars to no overlap between the cars. Call your car A, the car you are passing B.
    a. Specify two events that can be given spacetime coordinates. In describing the events, refer to cars A and B and to their front bumpers and rear bumpers.
    b. In either reference frame, is there *one* clock that is present at both events?
    c. Who, if anyone, measures the proper time between the events?
15. Your friend flies from Los Angeles to New York. He determines the distance using the tried-and-true $d = vt$. You and your assistants on the ground also measure the distance using meter sticks and surveying equipment.
    a. Who, if anyone, measures the proper length?
    b. Who, if anyone, measures the shorter distance?
16. A 100-m-long train is heading for an 80-m-long tunnel. If the train moves sufficiently fast, is it possible, according to experimenters on the ground, for the entire train to be inside the tunnel at one instant of time? Explain.
17. Dan picks up a 15-m-long pole and begins running very fast, holding the pole horizontally and pointing in the direction he's running. He heads toward a barn that is 12 m long and has open doors at each end. Dan runs so fast that, to Farmer Brown standing by his barn, the ladder is only 5 m long. As soon as the pole is completely inside the barn, Farmer Brown closes both doors so that Dan and the pole are inside with both doors shut. Then, just before Dan reaches the far door, Farmer Brown opens both doors and Dan emerges, still moving at high speed. According to Dan, however, the barn is contracted to only 4 m and the pole has its full 15 m length. Farmer Brown sees the pole completely inside the barn with both doors closed. What does Dan see happening?
18. The rocket speeds shown in Figure Q27.18 are relative to the earth. Is the speed of A relative to B greater than, less than, or equal to $0.8c$?

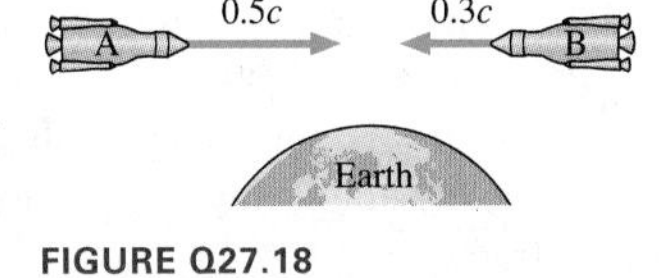

FIGURE Q27.18

19. Can a particle of mass m have total energy less than mc^2? Explain.
20. In your chemistry classes, you have probably learned that, in a chemical reaction, the mass of the products is equal to the mass of the reactants. That is, the mass of the substances produced in a chemical reaction is equal to the mass of the substances consumed in the reaction. Is this absolutely true, or is there actually a small difference? Explain.

Multiple-Choice Questions

21. | Lee and Leigh are twins. At their first birthday party, Lee is placed on a spaceship that travels away from the earth and back at a steady 0.866c. The spaceship eventually returns, landing in the swimming pool at Leigh's eleventh birthday party. When Lee emerges from the ship, it is discovered that
    A. He is still only 1 year old. B. He is 6 years old.
    C. He is also 11 years old. D. He is 21 years old.
22. || A space cowboy wants to eject from his spacecraft 100,000 km after passing a space buoy, as seen by spectators at rest with respect to the buoy. To do this, the cowboy sets a timer on his craft that will start as he passes the buoy. He plans to cruise by the buoy at $0.300c$. How much time should he allow between passing the buoy and ejecting?
    A. 1.01 s B. 1.06 s C. 1.11 s
    D. 1.33 s E. 1.58 s
23. | A lamp flashes at the origin. Then, 1 μs later, a lamp flashes at a point 150 m from the origin. How fast, as a fraction of c, would an astronaut need to travel so that he measures the proper time between these two events?
    A. $0.35c$ B. $0.50c$ C. $0.62c$ D. $0.87c$
24. | Energy in the sun is produced by the fusion of four protons into a helium nucleus. The process involves several steps, but the net reaction is simply $4p \rightarrow {}^4He$ + energy. Given this, you can say that
    A. One helium atom has more mass than four hydrogen atoms.
    B. One helium atom has less mass than four hydrogen atoms.
    C. One helium atom has the same mass as four hydrogen atoms.
25. || A particle moving at speed $0.40c$ has momentum p_0. The speed of the particle is increased to $0.80c$. Its momentum is now
    A. Less than $2p_0$ B. Exactly $2p_0$ C. Greater than $2p_0$
26. | A particle moving at speed $0.40c$ has kinetic energy K_0. The speed of the particle is increased to $0.80c$. The kinetic energy is now
    A. Less than $4K_0$ B. Exactly $4K_0$ C. Greater than $4K_0$

PROBLEMS

Section 27.2 Galilean Relativity

1. ||| A sprinter crosses the finish line of a race. The roar of the crowd in front approaches her at a speed of 355 m/s. The roar from the crowd behind her approaches at 335 m/s. What are the speed of sound and the speed of the sprinter?
2. | A baseball pitcher can throw a ball with a speed of 40 m/s. He is in the back of a pickup truck that is driving away from you. He throws the ball in your direction, and it floats toward you at a lazy 10 m/s. What is the speed of the truck?
3. | A boy on a skateboard coasts along at 5 m/s. He has a ball that he can throw at a speed of 10 m/s. What is the ball's speed relative to the ground if he throws the ball (a) forward or (b) backward?
4. || A boat takes 3.0 hours to travel 30 km down a river, then 5.0 hours to return. How fast is the river flowing?
5. || When the moving sidewalk at the airport is broken, as it often seems to be, it takes you 50 s to walk from your gate to baggage claim. When it is working and you stand on the moving sidewalk the entire way, without walking, it takes 75 s to travel the same distance. How long will it take you to travel from the gate to baggage claim if you walk while riding on the moving sidewalk?
6. || An assembly line has a staple gun that rolls to the left at 1.0 m/s while parts to be stapled roll past it to the right at 3.0 m/s. The staple gun fires 10 staples per second. How far apart are the staples in the finished part?

Section 27.3 Einstein's Principle of Relativity

7. || An out-of-control alien spacecraft is diving into a star at a speed of 1.0×10^8 m/s. At what speed, relative to the spacecraft, is the starlight approaching?
8. || A starship blasts past the earth at 2.0×10^8 m/s. Just after passing the earth, the starship fires a laser beam out its back. With what speed does the laser beam approach the earth?
9. || You are flying at $0.99c$ with respect to Kara. At the exact instant you pass Kara, she fires a very short laser pulse in the same direction you're heading.
   a. After 1.0 s has elapsed on Kara's watch, what does Kara say the distance is between you and the laser pulse?
   b. After 1.0 s has elapsed on your watch, what do you say the distance is between you and the laser pulse?

Section 27.4 Events and Measurements

Section 27.5 The Relativity of Simultaneity

10. || You are standing at the point $x = 300$ m, $y = 400$ m in a reference frame with synchronized clocks. What time does your clock show at the instant you see the clock at the origin showing 3.00 μs?
11. || Bjorn is standing at $x = 600$ m. Firecracker 1 explodes at the origin and firecracker 2 explodes at $x = 900$ m. The flashes from both explosions reach Bjorn's eye at $t = 3.0$ μs. At what time did each firecracker explode?
12. |||| Bianca is standing at $x = 600$ m. Firecracker 1, at the origin, and firecracker 2, at $x = 900$ m, explode simultaneously. The flash from firecracker 1 reaches Bianca's eye at $t = 3.0$ μs. At what time does she see the flash from firecracker 2?
13. || You are standing at $x = 9.0$ km. Lightning bolt 1 strikes at $x = 0$ km and lightning bolt 2 strikes at $x = 12.0$ km. Both flashes reach your eye at the same time. Your assistant is standing at $x = 3.0$ km. Does your assistant see the flashes at the same time? If not, which does she see first and what is the time difference between the two?

14. || A light flashes at position $x = 0$ m. One microsecond later, a light flashes at position $x = 1000$ m. In a second reference frame, moving along the x-axis at speed v, the two flashes are simultaneous. Is this second frame moving to the right or to the left relative to the original frame?
15. ||| Jose is looking to the east. Lightning bolt 1 strikes a tree 300 m from him. Lightning bolt 2 strikes a barn 900 m from him in the same direction. Jose sees the tree strike 1.0 μs before he sees the barn strike. According to Jose, were the lightning strikes simultaneous? If not, which occurred first and what was the time difference between the two?
16. || Your 1000-m-long starship has warning lights at each end that, to you, flash simultaneously every minute. You are moving directly away from the planet Zerkon at $0.70c$. To a Zerkonian, do the lights flash simultaneously? If not, which flashes first—the light at the front of your ship or the trailing one?
17. || There is a lightbulb exactly halfway between the front and rear of a long hallway in your spaceship. Your ship is traveling at $0.5c$ relative to the earth. The bulb is suddenly turned on. Let event 1 be when the light from the bulb first strikes the front end of the hall, and event 2 be when it strikes the rear end. Which event occurs first, or are they simultaneous, in the frame of (a) the spaceship and (b) a person standing on the earth?

Section 27.6 Time Dilation

18. ||| A cosmic ray travels 60 km through the earth's atmosphere in 400 μs, as measured by experimenters on the ground. How long does the journey take according to the cosmic ray?
19. || At what speed relative to a laboratory does a clock tick at half the rate of an identical clock at rest in the laboratory? Give your answer as a fraction of c.
20. || An astronaut travels to a star system 4.5 ly away at a speed of $0.90c$. Assume that the time needed to accelerate and decelerate is negligible.
   a. How long does the journey take according to Mission Control on earth?
   b. How long does the journey take according to the astronaut?
   c. How much time elapses between the launch and the arrival of the first radio message from the astronaut saying that she has arrived?
21. ||| A starship voyages to a distant planet 10 ly away. The explorers stay 1 yr, return at the same speed, and arrive back on earth 26 yr after they left. Assume that the time needed to accelerate and decelerate is negligible.
   a. What is the speed of the starship?
   b. How much time has elapsed on the astronauts' chronometers?

Section 27.7 Length Contraction

22. | At what speed, as a fraction of c, will a moving rod have a length 60% that of an identical rod at rest?
23. | Jill claims that her new rocket is 100 m long. As she flies past your house, you measure the rocket's length and find that it is only 80 m. Should Jill be cited for exceeding the $0.5c$ speed limit?
24. || A muon travels 60 km through the atmosphere at a speed of $0.9997c$. According to the muon, how thick is the atmosphere?
25. || The Stanford Linear Accelerator (SLAC) accelerates electrons to $v = 0.99999997c$ in a 3.2-km-long tube. If they travel the length of the tube at full speed (they don't, because they are accelerating), how long is the tube in the electrons' reference frame?
26. | Our Milky Way galaxy is 100,000 ly in diameter. A spaceship crossing the galaxy measures the galaxy's diameter to be a mere 1.0 ly.
   a. What is the speed of the spaceship relative to the galaxy?
   b. How long is the crossing time as measured in the galaxy's reference frame?
27. | The X-15 rocket-powered plane holds the record for the fastest speed ever attained by a manned aircraft, at 2020 m/s. At this speed, by how much is the 15.5-m-long aircraft length contracted?
   Hint: Use the binomial approximation.

Section 27.8 Velocities of Objects in Special Relativity

28. |||| A rocket cruising past earth at $0.800c$ shoots a bullet out the back door, opposite the rocket's motion, at $0.900c$ relative to the rocket. What is the bullet's speed relative to the earth?
29. ||| A base on Planet X fires a missile toward an oncoming space fighter. The missile's speed according to the base is $0.85c$. The space fighter measures the missile's speed as $0.96c$. How fast is the space fighter traveling relative to Planet X?

Watch video solutions for problems: 27.37  27.41   27.45 

30. ||| A solar flare blowing out from the sun at $0.90c$ is overtaking a rocket as it flies away from the sun at $0.80c$. According to the crew on board, with what speed is the flare gaining on the rocket?

Section 27.9 Relativistic Momentum

31. | A proton is accelerated to $0.999c$.
    a. What is the proton's momentum?
    b. By what factor does the proton's momentum exceed its Newtonian momentum?
32. ||| A 1.0 g particle has momentum 400,000 kg $\cdot$ m/s. What is the particle's speed?
33. | At what speed is a particle's momentum twice its Newtonian value?
34. || What is the speed of a particle whose momentum is mc?

Section 27.10 Relativistic Energy

35. || What are the kinetic energy, the rest energy, and the total energy of a 1.0 g particle with a speed of $0.80c$?
36. | A quarter-pound hamburger with all the fixings has a mass of 200 g. The food energy of the hamburger (480 food calories) is 2 MJ.
    a. What is the energy equivalent of the mass of the hamburger?
    b. By what factor does the energy equivalent exceed the food energy?
37. || How fast must an electron move so that its total energy is 10% more than its rest mass energy?
38. | At what speed is a particle's kinetic energy twice its rest energy?

General Problems

39. ||| A firecracker explodes at $x = 0$ m, $t = 0\ \mu$s. A second explodes at $x = 300$ m, $t = 2.0\ \mu$s. What is the proper time between these events?
40. ||| You're standing on an asteroid when you see your best friend rocketing by in her new spaceship. As she goes by, you notice that the front and rear of her ship coincide exactly with the 400-m-diameter of another nearby asteroid that is stationary with respect to you. However, you happen to know that your friend's spaceship measured 500 m long in the showroom. What is your friend's speed relative to you?
41. || A subatomic particle moves through the laboratory at $0.90c$. Laboratory experimenters measure its lifetime, from creation to annihilation, to be 2.3 ps (1 ps = 1 picosecond = 10^{-12} s). According to the particle, how long did it live?
42. || You and Maria each own identical spaceships. As you fly past Maria, you measure her ship to be 90 m long and your own ship to be 100 m long.
    a. How long does Maria measure your ship to be?
    b. How fast is Maria moving relative to you?
43. ||| A very fast-moving train car passes you, moving to the right at $0.50c$. You measure its length to be 12 m. Your friend David flies past you to the right at a speed relative to you of $0.80c$. How long does David measure the train car to be?
44. || Two events in reference frame S occur 10 μs apart at the same point in space. Frame S′ travels at speed $v = 0.90c$ relative to frame S.
    a. What is the time interval between the events in reference frame S′?
    b. What is the distance between the events in frame S′?
45. ||| A spaceship heads directly toward an asteroid at a relative speed of $0.75c$. When it is 3.0×10^8 m from the asteroid, as measured in the asteroid's frame, the rocket fires a laser pulse at the asteroid. According to the rocket captain, what is the time interval between when the pulse hits the asteroid and when the rocket passes the asteroid, barely missing it?
46. ||| Because of the earth's rotation, a person living on top of a mountain moves at a faster speed than someone at sea level. The mountain dweller's clocks thus run slowly compared to those at sea level. If the average life span of a hermit is 80 years, on average how much longer would a hermit dwelling on the top of a 3000-m-high mountain live compared to a sea-level hermit?
    INT
47. ||||| Two spaceships approach each other at speeds of $0.80c$ and $0.90c$ relative to the sun. When they are 5.0×10^9 m apart, as measured in the sun's reference frame, they each set their clocks to zero. By how much do their clocks differ at the instant they pass each other?
    Hint: First find how long it takes the spaceships to pass in the sun's reference frame.
48. ||| You fly 5000 km across the United States on an airliner at 250 m/s. You return two days later at the same speed.
    a. Have you aged more or less than your friends at home?
    b. By how much?
    Hint: Use the binomial approximation.
49. || A cube has a density of 2000 kg/m^3 while at rest in the laboratory. What is the cube's density as measured by an experimenter in the laboratory as the cube moves through the laboratory at 90% of the speed of light in a direction perpendicular to one of its faces?
    INT
50. |||| A railroad car that is 20 m long when at rest passes Rachael, who is standing on the ground. She records that it takes 120 ns to go by. How fast is the car moving, as a fraction of c?
51. || A spaceship flies past an experimenter who measures its length to be one-half the length he had measured when the spaceship was at rest. An astronaut aboard the spaceship notes that his clock ticks at 1-second intervals. What is the time between ticks as measured by the experimenter?
52. ||| Marissa's spaceship approaches Joseph's at a speed of $0.99c$. As Marissa passes Joseph, they synchronize their clocks to both read $t = 0$ s. When Marissa's clock reads 100 s, she sends a light signal back to Joseph. According to his clock, when does he receive this signal?
53. ||| At a speed of $0.90c$, a spaceship travels to a star that is 9.0 ly distant.
    a. According to a scientist on earth, how long does the trip take?
    b. According to a scientist on the spaceship, how long does the trip take?
    c. According to the scientist on the spaceship, what is the distance traveled during the trip?
    d. At what speed do observers on the spaceship see the star approaching them?

Watch video solutions for problems: 27.61  27.63 

54. ‖ In an attempt to reduce the extraordinarily long travel times INT for voyaging to distant stars, some people have suggested traveling at close to the speed of light. Suppose you wish to visit the red giant star Betelgeuse, which is 430 ly away, and that you want your 20,000 kg rocket to move so fast that you age only 20 years during the round trip.
 a. How fast must the rocket travel relative to earth?
 b. How much energy is needed to accelerate the rocket to this speed?
 c. How many times larger is this energy than the total energy used by the United States in the year 2012, which was roughly 1.0×10^{20} J?
55. |||| A rocket traveling at $0.500c$ sets out for the nearest star, Alpha Centauri, which is 4.25 ly away from earth. It will return to earth immediately after reaching Alpha Centauri. What distance will the rocket travel and how long will the journey last according to (a) stay-at-home earthlings and (b) the rocket crew? (c) Which answers are the correct ones, those in part a or those in part b?
56. ‖ A distant quasar is found to be moving away from the earth at $0.80c$. A galaxy closer to the earth and along the same line of sight is moving away from us at $0.20c$. What is the recessional speed of the quasar as measured by astronomers in the other galaxy?

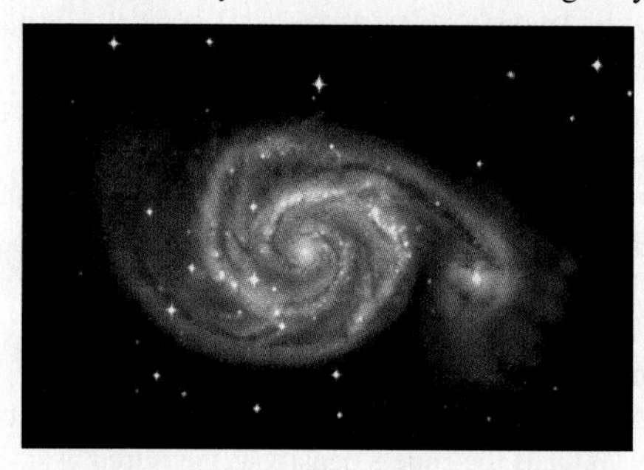

57. ‖ Two rockets approach each other. Each is traveling at $0.75c$ in the earth's reference frame. What is the speed of one rocket relative to the other?
58. ||| A railcar has a velocity of $0.5c$ relative to the ground. To another train traveling on the same track, the railcar's velocity is $-0.2c$. What is the velocity of the train relative to the ground?
59. |||| A military jet traveling at 1500 m/s has engine trouble and the pilot must bail out. Her ejection seat shoots her forward at 300 m/s relative to the jet. According to the Lorentz velocity transformation, by how much is her velocity relative to the ground less than the 1800 m/s predicted by Galilean relativity? **Hint:** Use the binomial approximation.
60. ||| James, Daniella, and Tara all possess identical clocks. As Daniella passes James in her rocket, James observes that her clock runs at 80% the rate of his clock. As Tara passes in her rocket, in the same direction as Daniella, James observes that her clock runs at 70% the rate of his clock. At what rate, relative to her clock, does Daniella observe Tara's clock to run?
61. ||||| A space beacon on Planet Karma emits a pulse of light every second. Spaceman Trevor flies directly toward the beacon at a speed of $0.95c$. According to Trevor, what is the time between one pulse reaching his ship and the next?
 Hint: First find the time between pulses reaching the ship in Planet Karma's reference frame.
62. ||| Two rockets, A and B, approach the earth from opposite directions at speed $0.800c$. The length of each rocket measured in its rest frame is 100 m. What is the length of rocket A as measured by the crew of rocket B?
63. ||| The highest-energy cosmic ray ever detected had an energy of about 3.0×10^{20} eV. Assume that this cosmic ray was a proton.
 a. What was the proton's speed as a fraction of c?
 b. If this proton started at the same time and place as a photon traveling at the speed of light, how far behind the photon would it be after traveling for 1 ly?
64. ||| What is the speed of an electron after being accelerated from INT rest through a 20×10^6 V potential difference?
65. ||| What is the speed of a proton after being accelerated from INT rest through a 50×10^6 V potential difference?
66. ‖ The half-life of a muon at rest is 1.5 μs. Muons that have been accelerated to a very high speed and are then held in a circular storage ring have a half-life of 7.5 μs.
 a. What is the speed of the muons in the storage ring?
 b. What is the total energy of a muon in the storage ring? The mass of a muon is 207 times the mass of an electron.
67. ‖ What is the momentum of a particle with speed $0.95c$ and total energy 2.0×10^{-10} J?
68. ||| What is the momentum of a particle whose total energy is four times its rest energy? Give your answer as a multiple of mc.
69. ‖ What is the total energy, in MeV, of
 INT a. A proton traveling at 99.0% of the speed of light?
 b. An electron traveling at 99.0% of the speed of light?
70. | What is the velocity, as a fraction of c, of
 INT a. A proton with 500 GeV total energy?
 b. An electron with 2.0 GeV total energy?
71. |||| At what speed is the kinetic energy of a particle twice its Newtonian value?
72. ‖ The factor γ appears in many relativistic expressions. A value $\gamma = 1.01$ implies that relativity changes the Newtonian values by approximately 1% and that relativistic effects can no longer be ignored. At what kinetic energy, in MeV, is $\gamma = 1.01$ for (a) an electron, and (b) a proton?
73. ‖ The chemical energy of gasoline is 46 MJ/kg. If gasoline's mass could be completely converted into energy, what mass of gasoline would be needed to equal the chemical energy content of 1.0 kg of gasoline?
74. |||| A standard nuclear power plant generates 3.0 GW of thermal INT power from the fission of ^{235}U. Experiments show that, on average, 0.19 u of mass is lost in each fission of a ^{235}U nucleus. How many kilograms of ^{235}U undergo fission each year in this power plant?
75. | The sun radiates energy at the rate 3.8×10^{26} W. The source INT of this energy is fusion, a nuclear reaction in which mass is transformed into energy. The mass of the sun is 2.0×10^{30} kg.
 a. How much mass does the sun lose each year?
 b. What percentage is this of the sun's total mass?
 c. Estimate the lifetime of the sun.
76. ‖ The radioactive element radium (Ra) decays by a process known as *alpha decay,* in which the nucleus emits a helium nucleus. (These high-speed helium nuclei were named alpha particles when radioactivity was first discovered, long before the identity of the particles was established.) The reaction is $^{226}\text{Ra} \rightarrow {}^{222}\text{Rn} + {}^{4}\text{He}$, where Rn is the element radon. The accurately measured atomic masses of the three atoms are 226.025, 222.017, and 4.003. How much energy is released in each decay? (The energy released in radioactive decay is what makes nuclear waste "hot.")

77. | The nuclear reaction that powers the sun is the fusion of four protons into a helium nucleus. The process involves several steps, but the net reaction is simply $4p \rightarrow {}^4He + \text{energy}$. The mass of a helium nucleus is known to be 6.64×10^{-27} kg.
   a. How much energy is released in each fusion?
   b. What fraction of the initial rest mass energy is this energy?

78. || When antimatter (which we'll learn more about in Chapter 30) INT interacts with an equal mass of ordinary matter, both matter and antimatter are converted completely into energy in the form of photons. In an antimatter-fueled spaceship, a staple of science fiction, the newly created photons are shot from the back of the ship, propelling it forward. Suppose such a ship has a mass of 2.0×10^6 kg, and carries a mass of fuel equal to 1% of its mass, or 1.0×10^4 kg of matter and an equal mass of antimatter.
   a. What is the final speed of the ship, assuming it starts from rest, if all energy released in the matter-antimatter annihilation is transformed into the kinetic energy of the ship?
   b. Not only do photons have energy, as you learned in Chapter 25, they also have momentum. Explain why, when energy and momentum conservation are both considered, the final speed of the ship will be less than you calculated in part a.

MCAT-Style Passage Problems

Pion Therapy BIO

Subatomic particles called *pions* are created when protons, accelerated to speeds very near c in a particle accelerator, smash into the nucleus of a target atom. Charged pions are unstable particles that decay into muons with a half-life of 1.8×10^{-8} s. Pions have been investigated for use in cancer treatment because they pass through tissue doing minimal damage until they decay, releasing significant energy at that point. The speed of the pions can be adjusted so that the most likely place for the decay is in a tumor.

Suppose pions are created in an accelerator, then directed into a medical bay 30 m away. The pions travel at the very high speed of $0.99995c$. Without time dilation, half of the pions would have decayed after traveling only 5.4 m, not far enough to make it to the medical bay. Time dilation allows them to survive long enough to reach the medical bay, enter tissue, slow down, and then decay where they are needed, in a tumor.

79. | What is the half-life of a pion in the reference frame of the patient undergoing pion therapy?
   A. 1.8×10^{-10} s B. 1.8×10^{-8} s
   C. 1.8×10^{-7} s D. 1.8×10^{-6} s

80. | According to the pion, what is the distance it travels from the accelerator to the medical bay?
   A. 0.30 m B. 3.0 m C. 30 m D. 3000 m

81. | The proton collision that creates the pion also creates a gamma-ray photon traveling in the same direction as the pion. The photon will get to the medical bay first because it is moving faster. What is the speed of the photon in the pion's reference frame?
   A. $0.00005c$ B. $0.5c$ C. $0.99995c$ D. c

82. | If the pion slows down to $0.99990c$, about what percentage of its kinetic energy is lost?
   A. 0.03% B. 0.3% C. 3% D. 30%

STOP TO THINK ANSWERS

Chapter Preview Stop to Think: C. The relative velocity $(v)_{CA}$ of the car (C) relative to Amy (A) is $(v)_{CA} = (v)_{CB} + (v)_{BA} =$ 10 m/s + 5 m/s = 15 m/s.

Stop to Think 27.1: A, C, and F. These move at constant velocity, or very nearly so. The others are accelerating.

Stop to Think 27.2: A. $u' = u - v = -10$ m/s $- 6$ m/s $= -16$ m/s. The *speed* is 16 m/s.

Stop to Think 27.3: C. Even the light has a slight travel time. The event is the hammer hitting the nail, not your seeing the hammer hit the nail.

Stop to Think 27.4: At the same time. Mark is halfway between the tree and the pole, so the fact that he *sees* the lightning bolts at the same time means they *happened* at the same time. It's true that Nancy *sees* event 1 before event 2, but the events actually occurred before she sees them. Mark and Nancy share a reference frame, because they are at rest relative to each other, and all experimenters in a reference frame, after correcting for any signal delays, *agree* on the spacetime coordinates of an event.

Stop to Think 27.5: After. This is the same as the case of Peggy and Ryan. In Mark's reference frame, as in Ryan's, the events are simultaneous. Nancy *sees* event 1 first, but the time when an event is seen is not when the event actually happens. Because all experimenters in a reference frame agree on the spacetime coordinates of an event, Nancy's position in her reference frame cannot affect the order of the events. If Nancy had been passing Mark at the instant the lightning strikes occurred in Mark's frame, then Nancy would be equivalent to Peggy. Event 2, like the firecracker at the front of Peggy's railroad car, occurs first in Nancy's reference frame.

Stop to Think 27.6: A. The two events—firecracker 1 explodes and firecracker 2 explodes—occur in the *same* place, in front of Martin. Thus the proper time is the time measured in Martin's frame, or 30 μs. The time interval measured by Lisa must be *longer* than the proper time.

Stop to Think 27.7: C. Nick measures proper time because Nick's clock is present at both the "nose passes Nick" event and the "tail passes Nick" event. Proper time is the smallest measured time interval between two events.

Stop to Think 27.8: A. The car is at rest in Peggy's frame, so she measures the proper length ℓ. The length measured in any other frame is shorter than the proper length, so Peggy measures the car to be longer than does Ryan.

Stop to Think 27.9: C. The bullet must obviously travel faster than Sam, eliminating answers D and E. And we know that no object can travel at or faster than the speed of light, eliminating A and B. The bullet travels faster than Sam, or faster than $0.75c$, but slower than c.

Stop to Think 27.10: B. Inspection of the graph in Figure 27.23a shows that the graph for relativistic momentum rapidly rises above the linear graph for Newtonian momentum. Thus if Lauren's speed doubles, her momentum must more than double.

Stop to Think 27.11: C. The kinetic energy is $(\gamma - 1)mc^2 = 0.5mc^2$, which is less than the rest energy mc^2.

28 Quantum Physics

This false-color image showing individual rod cells (green) and cone cells (blue) on the human retina was made with an electron microscope. Such exquisite detail would not be possible in an image created with a light microscope. Why is greater resolution possible in an image made with a beam of electrons?

LOOKING AHEAD ▸

Goal: To understand the quantization of energy for light and matter.

Waves Behave Like Particles

An interference pattern made with very low-intensity light shows that the light hits the screen in discrete "chunks" called **photons.**

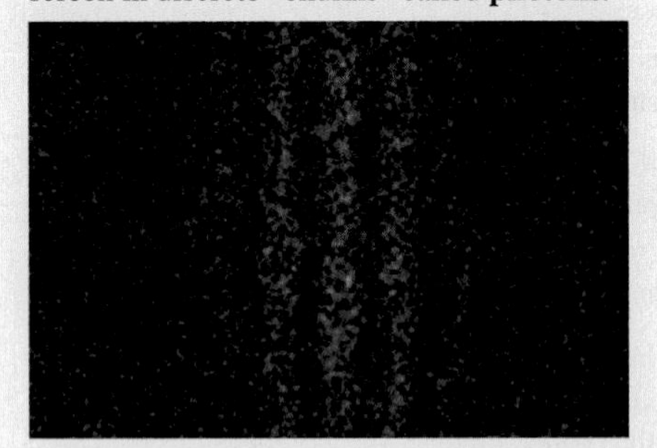

You'll learn how light sometimes behaves like a wave and sometimes like a particle.

Particles Behave Like Waves

This image of electrons diffracting from an aluminum target shows that, surprisingly, particles have a wave nature.

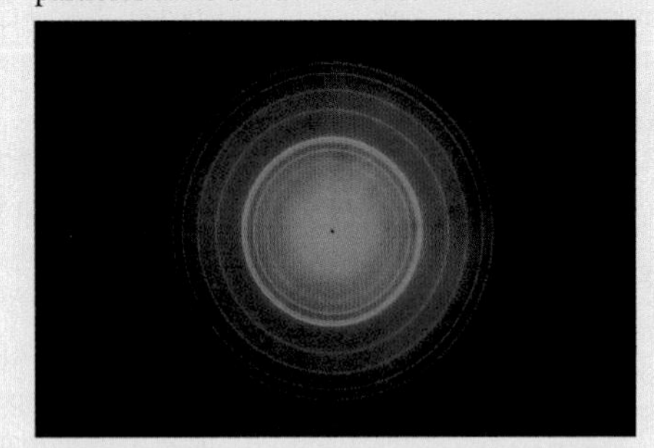

You'll learn that the wavelength of a particle is directly related to its momentum.

Quantization of Energy

These discrete colors are emitted by helium atoms as their electrons "jump" between quantized **energy levels.**

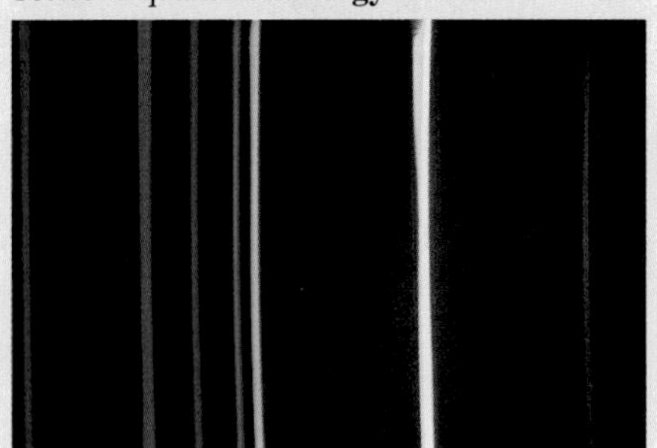

You'll learn how to calculate photon energies and wavelengths as quantum systems emit or absorb photons.

LOOKING BACK ◂

Double-Slit Interference

In Section 17.2 you learned that light waves spreading out from two narrow slits interfere, producing distinct fringes on a screen. This interference is clear evidence of the wave nature of light.

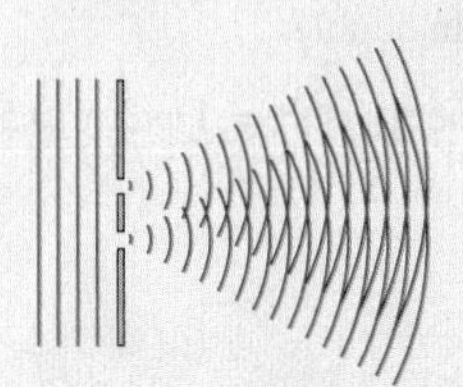

In this chapter, the double-slit interference experiment will reveal striking new properties of light and a surprising wave-like behavior of electrons.

STOP TO THINK

A laser illuminates two slits, leading to the interference pattern shown below. After the right-hand slit is covered up, what will the pattern look like?

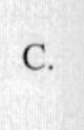

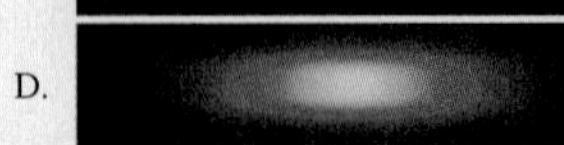

A.

B.

C.

D.

Except for relativity, everything we have studied until this point in the textbook was known by 1900. Newtonian mechanics, thermodynamics, and the theory of electromagnetism form what we call *classical physics*. It is an impressive body of knowledge with immense explanatory power and a vast number of applications.

But a spate of discoveries right around 1900 showed that classical physics, though remarkable, was incomplete. Investigations into the nature of light and matter led to many astonishing discoveries that classical physics simply could not explain. Sometimes, as you will see, light refuses to act like a wave and seems more like a collection of particles. Other experiments found that electrons sometimes behave like waves. These discoveries eventually led to a radical new theory of light and matter called *quantum physics*.

This chapter will introduce you to this strange but wonderful quantum world. We will take a more historical approach than in previous chapters. As we introduce new ideas, we will describe in some detail the key experiments and the evolution of theories to explain them.

28.1 X Rays and X-Ray Diffraction

The rules of quantum physics apply at the scale of atoms and electrons. Experiments to elucidate the nature of the atom and the physics of atomic particles produced results that defied explanation with classical theories. Investigators saw things no one had ever seen before, phenomena that needed new principles and theories to explain them.

In 1895, the German physicist Wilhelm Röntgen was studying how electrons travel through a vacuum. He sealed an electron-producing cathode and a metal target electrode into a vacuum tube. A high voltage pulled electrons from the cathode and accelerated them to very high speed before they struck the target electrode. One day, by chance, Röntgen left a sealed envelope containing film near the vacuum tube. He was later surprised to discover that the film had been exposed even though it had never been removed from the envelope. Some sort of penetrating radiation from the tube had exposed the film.

Röntgen had no idea what was coming from the tube, so he called them x rays, using the algebraic symbol x meaning "unknown." X rays were unlike anything, particle or wave, ever discovered before. Röntgen was not successful at reflecting the rays or at focusing them with a lens. He showed that they travel in straight lines, like particles, but they also pass right through most solid materials with very little absorption, something no known particle could do. The experiments of Röntgen and others led scientists to conclude that these mysterious rays were electromagnetic waves with very short wavelengths, as we learned in ◀◀ SECTION 25.7. These short-wavelength waves were produced in Röntgen's apparatus by the collision of fast electrons with a metal target. X rays are still produced this way, as shown in the illustration of the operation of a modern x-ray tube in FIGURE 28.1.

FIGURE 28.1 The operation of a modern x-ray tube.

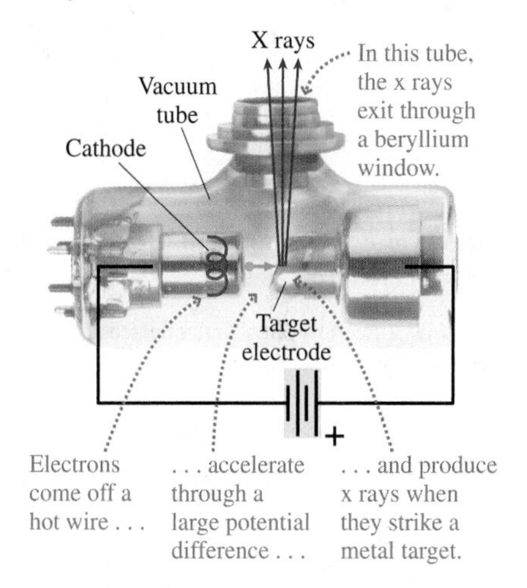

X-Ray Images BIO

X rays are penetrating, and Röntgen immediately realized that x rays could be used to create an image of the interior of the body. One of Röntgen's first images showed the bones in his wife's hand, dramatically demonstrating the medical potential of these newly discovered rays. Substances with high atomic numbers, such as lead or the minerals in bone, are effective at stopping the rays; materials with low atomic numbers, such as the beryllium window of the x-ray tube in Figure 28.1 or the water and organic compounds of soft tissues in the body, diminish them only slightly. As illustrated in FIGURE 28.2, an x-ray image is essentially a shadow of the bones and dense components of the body; where these tissues stop the x rays, the film is not exposed. The basic procedure for producing an x-ray image on film is little changed from Röntgen's day.

FIGURE 28.2 Creating an x-ray image.

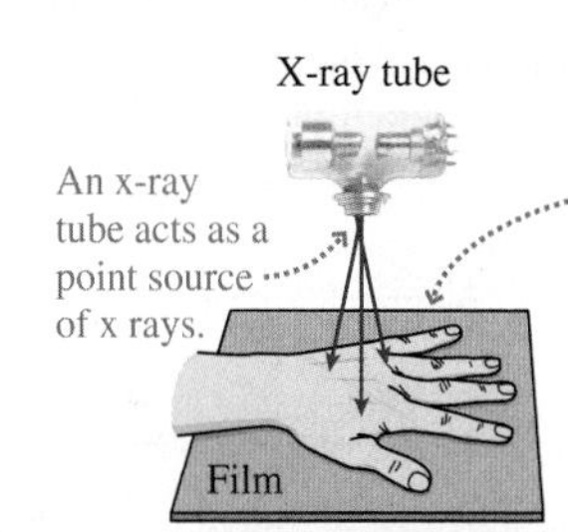

X-ray tube
An x-ray tube acts as a point source of x rays.
The part of the body to be imaged is on top of a piece of film. Dense tissues pass few x rays; the film is not exposed below these tissues.
Film

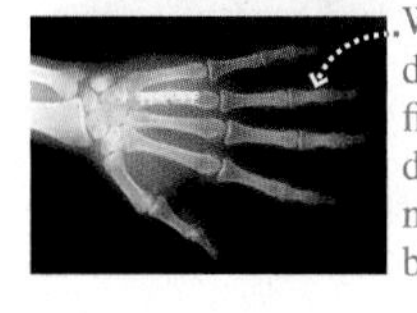

When the film is developed, the film is light where dense tissues or metal have blocked the x rays.

This use of x rays was of tremendous practical importance, but more important to the development of our story is the use of x rays to probe the structure of matter at an atomic scale.

X-Ray Diffraction

FIGURE 28.3 X rays incident on a simple cubic lattice crystal.

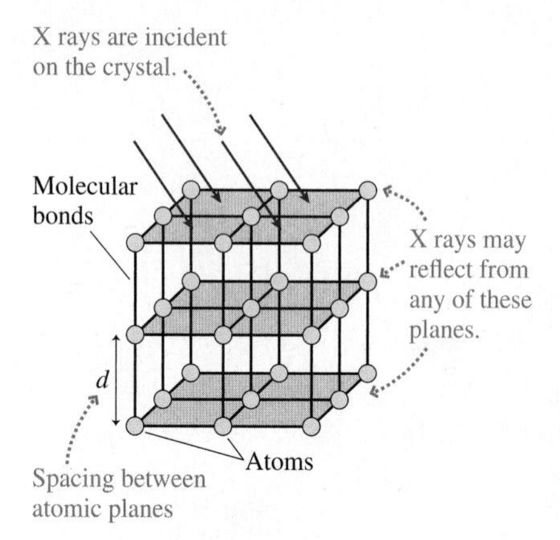
X rays are incident on the crystal.
Molecular bonds
X rays may reflect from any of these planes.
d
Spacing between atomic planes
Atoms

At about the same time scientists were first concluding that x rays were very-short-wavelength electromagnetic waves, researchers were also deducing that the size of an atom is ≈ 0.1 nm, and it was suggested that solids might consist of atoms arranged in a regular crystalline *lattice.* In 1912, the German scientist Max von Laue noted that x rays passing through a crystal ought to undergo diffraction from the "three-dimensional grating" of the crystal in much the same way that visible light diffracts from a diffraction grating. Such x-ray diffraction by crystals was soon confirmed experimentally, and measurements confirmed that x rays are indeed electromagnetic waves with wavelengths in the range 0.01 nm to 10 nm—a much shorter wavelength than visible light.

To understand x-ray diffraction, we begin by looking at the arrangement of atoms in a solid. FIGURE 28.3 shows x rays striking a crystal with a *simple cubic lattice.* This is a very straightforward arrangement, with the atoms in planes with spacing d between them.

FIGURE 28.4a shows a side view of the x rays striking the crystal, with the x rays incident at angle θ. Most of the x rays are transmitted through the plane, but a small fraction of the wave is reflected, much like the weak reflection of light from a sheet of glass. The reflected wave obeys the law of reflection—the angle of reflection equals the angle of incidence—and the figure has been drawn accordingly.

As we saw in Figure 28.3, a solid has not one single plane of atoms but many parallel planes. As x rays pass through a solid, a small fraction of the wave reflects from each of the parallel planes of atoms shown in FIGURE 28.4b. The *net* reflection from the solid is the *superposition* of the waves reflected by each atomic plane. For most angles of incidence, the reflected waves are out of phase and their superposition is very nearly zero. However, as in the thin-film interference we studied in ◀◀ SECTION 17.4, there are a few specific angles of incidence for which the reflected waves are in phase. For these angles of incidence, the reflected waves interfere constructively to produce a strong reflection. This strong x-ray reflection at a few specific angles of incidence is called **x-ray diffraction.**

FIGURE 28.4 X-ray reflections from parallel atomic planes.

(a) X rays are transmitted and reflected at one plane of atoms.

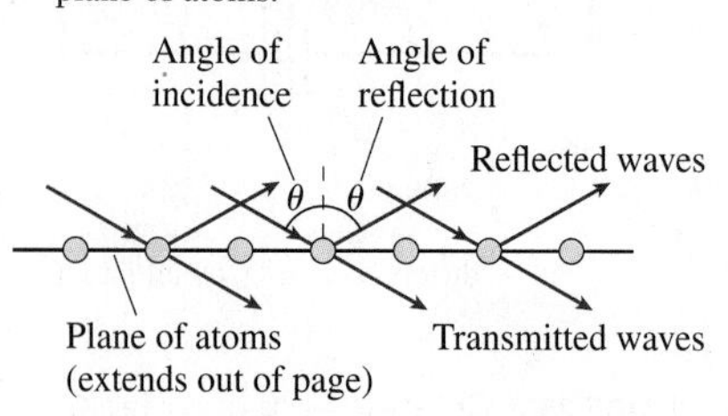
Angle of incidence
Angle of reflection
Reflected waves
θ
θ
Plane of atoms (extends out of page)
Transmitted waves

(b) The reflections from parallel planes interfere.

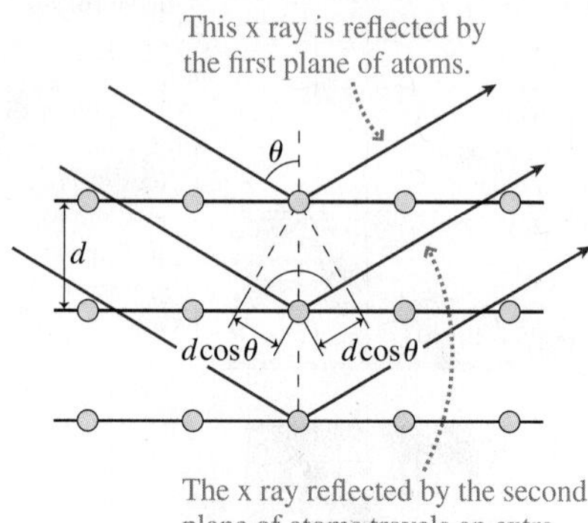
This x ray is reflected by the first plane of atoms.
θ
d
dcosθ
dcosθ
The x ray reflected by the second plane of atoms travels an extra distance Δr = 2dcosθ.

You can see from Figure 28.4b that the wave reflecting from any particular plane travels an extra distance $\Delta r = 2d\cos\theta$ before combining with the reflection from the plane immediately above it, where d is the spacing between the atomic planes. If Δr is a whole number of wavelengths, then these two waves will be in phase when they recombine. But if the reflections from two neighboring planes are in phase, then *all* the reflections from *all* the planes are in phase and will interfere constructively to produce a strong reflection. Consequently, x rays will reflect from the crystal when the angle of incidence θ_m satisfies the **Bragg condition:**

$$\Delta r = 2d\cos\theta_m = m\lambda \qquad m = 1, 2, 3, \ldots \tag{28.1}$$

The Bragg condition for constructive interference of x rays reflected from a solid

NOTE ▶ This formula is similar to that for constructive interference for light passed through a grating that we saw in Chapter 17. In both cases, we get constructive interference at only a few well-defined angles. ◀

EXAMPLE 28.1 **Analyzing x-ray diffraction**

X rays with a wavelength of 0.105 nm are diffracted by a crystal with a simple cubic lattice. Diffraction maxima are observed at angles 31.6° and 55.4° and at no angles between these two. What is the spacing between the atomic planes causing this diffraction?

PREPARE The angles must satisfy the Bragg condition. We don't know the values of m, but we know that they are two consecutive integers. In Equation 28.1 θ_m *decreases* as m increases, so 31.6° corresponds to the larger value of m. We will assume that 55.4° corresponds to m and 31.6° to $m+1$.

SOLVE The values of d and λ are the same for both diffractions, so we can use the Bragg condition to find

$$\frac{m+1}{m} = \frac{\cos 31.6°}{\cos 55.4°} = 1.50 = \frac{3}{2}$$

Thus 55.4° is the second-order diffraction and 31.6° is the third-order diffraction. With this information we can use the Bragg condition again to find

$$d = \frac{2\lambda}{2\cos\theta_2} = \frac{0.105 \text{ nm}}{\cos 55.4°} = 0.185 \text{ nm}$$

ASSESS We learned above that the size of atoms is ≈ 0.1 nm, so this is a reasonable value for the atomic spacing in a crystal.

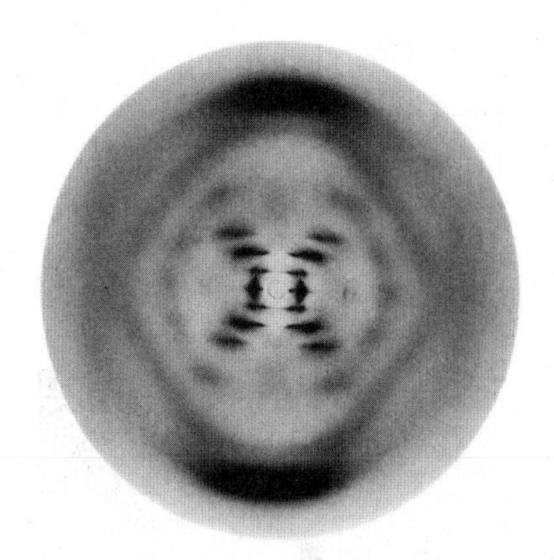

X marks the spot BIO Rosalind Franklin obtained this x-ray diffraction pattern for DNA in 1953. The cross of dark bands in the center of the diffraction pattern reveals something about the arrangement of atoms in the DNA molecule—that the molecule has the structure of a helix. This x-ray diffraction image was a key piece of information in the effort to unravel the structure of the DNA molecule.

Example 28.1 shows that an x-ray diffraction pattern reveals details of the crystal that produced it. The structure of the crystal was quite simple, so the example was straightforward. More complex crystals produce correspondingly complex patterns that can help reveal the structure of the crystals that produced them. As investigators developed theories of atoms and atomic structure, x rays were an invaluable tool—as they still are. X-ray diffraction is still widely used to decipher the three-dimensional structure of biological molecules such as proteins.

STOP TO THINK 28.1 The first-order diffraction of x rays from two crystals with simple cubic structure is measured. The first-order diffraction from crystal A occurs at an angle of 20°. The first-order diffraction of the same x rays from crystal B occurs at 30°. Which crystal has the larger atomic spacing?

28.2 The Photoelectric Effect

In ◀◀ SECTION 25.6, we introduced the idea that light can be thought of as *photons,* packets of a particular amount of energy. This is an idea that you have likely heard before, but when it was first introduced, it was truly revolutionary. For such an odd idea to find broad acceptance, compelling experimental evidence was needed. This evidence was provided by studies of the *photoelectric effect,* which we will explore in detail in this section to recognize the rationale for and the impact of this startling new concept.

The first hints about the photon nature of light came in the late 1800s with the discovery that a negatively charged electroscope could be discharged by shining ultraviolet light on it. The English physicist J. J. Thomson found that the ultraviolet light was causing the electroscope to emit electrons, as illustrated in **FIGURE 28.5**. The emission of electrons from a substance due to light striking its surface came to be called the **photoelectric effect.** This seemingly minor discovery became a pivotal event that opened the door to the new ideas we discuss in this chapter.

FIGURE 28.5 Ultraviolet light discharges a negatively charged electroscope.

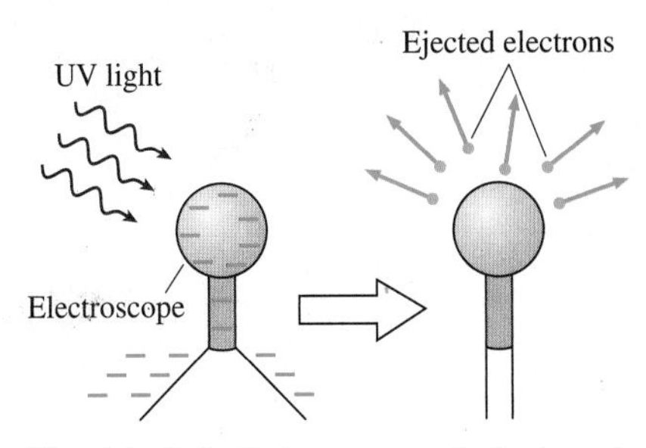

Characteristics of the Photoelectric Effect

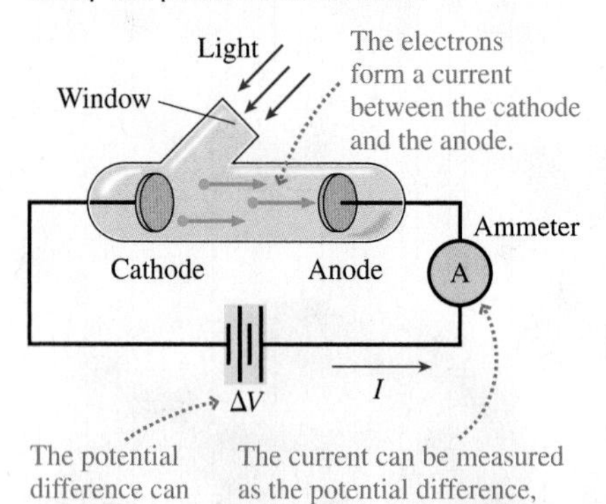

FIGURE 28.6 An experimental device to study the photoelectric effect.

Figure Video

Class Video

FIGURE 28.6 shows an evacuated glass tube with two facing electrodes and a window. When ultraviolet light shines on the cathode, a steady counterclockwise current (clockwise flow of electrons) passes through the ammeter. There are no junctions in this circuit, so the current must be the same all the way around the loop. The current in the space between the cathode and the anode consists of electrons moving freely through space (i.e., not inside a wire) at the *same rate* as the current in the wire. There is no current if the electrodes are in the dark, so electrons don't spontaneously leap off the cathode. Instead, the light causes electrons to be ejected from the cathode at a steady rate.

The battery in Figure 28.6 establishes an adjustable potential difference ΔV between the two electrodes. With it, we can study how the current I varies as the potential difference and the light's wavelength and intensity are changed. Doing so reveals the following characteristics of the photoelectric effect:

1. The current I is directly proportional to the light intensity. If the light intensity is doubled, the current also doubles.
2. The current appears without delay when the light is applied.
3. Electrons are emitted *only* if the light frequency f exceeds a **threshold frequency** f_0. This is shown in the graph of **FIGURE 28.7a**.
4. The value of the threshold frequency f_0 depends on the type of metal from which the cathode is made.
5. If the potential difference ΔV is more than about 1V positive (anode positive with respect to the cathode), the current changes very little as ΔV is increased. If ΔV is made negative (anode negative with respect to the cathode), by reversing the battery, the current decreases until at some voltage $\Delta V = -V_{stop}$ the current reaches zero. The value of V_{stop} is called the **stopping potential.** This behavior is shown in **FIGURE 28.7b**.
6. The value of V_{stop} is the same for both weak light and intense light. A more intense light causes a larger current, but in both cases the current ceases when $\Delta V = -V_{stop}$.

FIGURE 28.7 The photoelectric current dependence on the light frequency f and the battery potential difference ΔV.

(a)

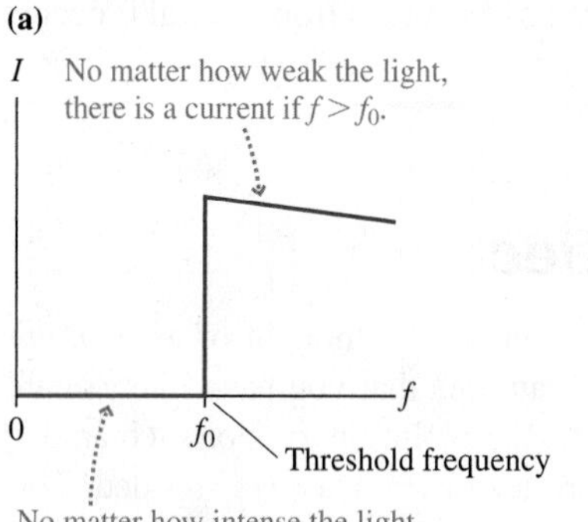

(b)

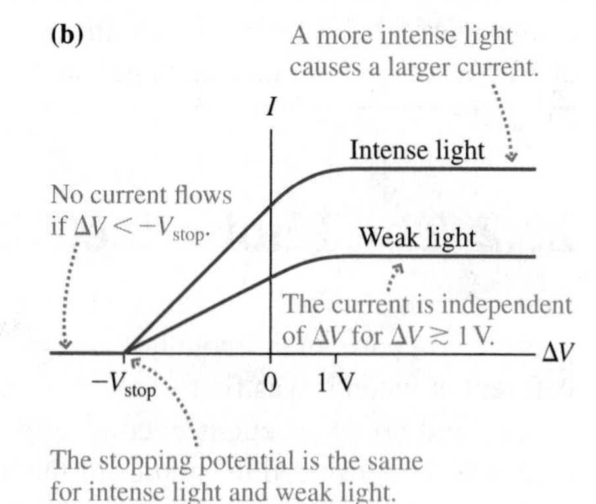

NOTE ► We're defining V_{stop} to be a *positive* number. The potential difference that stops the electrons is $\Delta V = -V_{stop}$, with an explicit minus sign. ◄

Understanding the Photoelectric Effect

You learned in Chapter 22 that electrons are the charge carriers in a metal and move around freely inside like a sea of negatively charged particles. The electrons are bound inside the metal and do not spontaneously spill out of an electrode at room temperature.

A useful analogy, shown in **FIGURE 28.8**, is the water in a swimming pool. Water molecules do not spontaneously leap out of the pool if the water is calm. To remove a water molecule, you must do *work* on it to lift it upward, against the force of gravity, to the edge of the pool. A minimum energy is needed to extract a water molecule—namely, the energy needed to lift a molecule that is right at the surface. Removing a water molecule that is deeper requires more than the minimum energy.

Similarly, a *minimum* energy is needed to free an electron from a metal. To extract an electron, you need to increase its energy until its speed is fast enough to escape. The minimum energy E_0 needed to free an electron is called the **work function** of the metal. Some electrons, like deeper water molecules, may require more energy than E_0 to escape, but all will require *at least* E_0. Table 28.1 lists the work functions in eV of some elements. (Recall that the conversion to joules is $1\ \text{eV} = 1.60 \times 10^{-19}$ J.)

FIGURE 28.8 A swimming pool analogy of electrons in a metal.

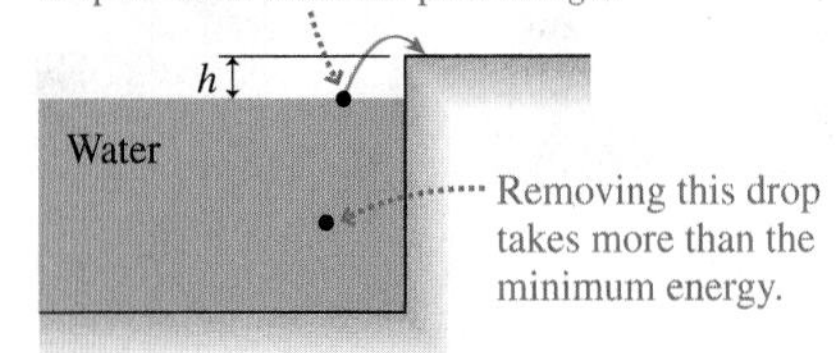
The *minimum* energy to remove a drop of water from the pool is *mgh*.
h
Water
Removing this drop takes more than the minimum energy.

Now, let's return to the photoelectric effect experiment of Figure 28.6. When ultraviolet light shines on the cathode, electrons leave with some kinetic energy. An electron with energy E_{elec} inside the metal loses energy ΔE as it escapes, so it emerges as an electron with kinetic energy $K = E_{elec} - \Delta E$. The work function energy E_0 is the *minimum* energy needed to remove an electron, so the *maximum* possible kinetic energy of an ejected electron is

$$K_{max} = E_{elec} - E_0$$

TABLE 28.1 The work functions for some metals

Element	E_0 (eV)
Potassium	2.30
Sodium	2.75
Aluminum	4.28
Tungsten	4.55
Copper	4.65
Iron	4.70
Gold	5.10

The electrons, after leaving the cathode, move out in all directions. **FIGURE 28.9** shows what happens as the potential difference ΔV between the cathode and the anode is varied:

- If the potential difference between the cathode and the anode is $\Delta V = 0$, there will be no electric field between the plates. Some electrons will reach the anode, creating a measurable current, but many do not.
- If the anode is positive, it attracts *all* of the electrons to the anode. A further increase in ΔV does not cause any more electrons to reach the anode and thus does not cause a further increase in the current I. This is why the curves in Figure 28.7b become horizontal for ΔV more than about 1V positive.
- If the anode is negative, it repels the electrons. However, an electron leaving the cathode with sufficient kinetic energy can still reach the anode, just as a ball hits the ceiling if you toss it upward with sufficient kinetic energy. A slightly negative anode voltage turns back only the slowest electrons. The current steadily decreases as the anode voltage becomes increasingly negative until, as the left side of Figure 28.7b shows, at the stopping potential, *all* electrons are turned back and the current ceases.

FIGURE 28.9 The effect of different voltages between the anode and cathode.

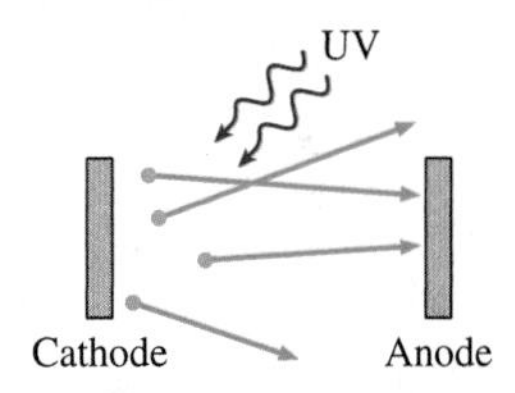
UV
Cathode
Anode

$\Delta V = 0$: The electrons leave the cathode in all directions. Only some reach the anode.

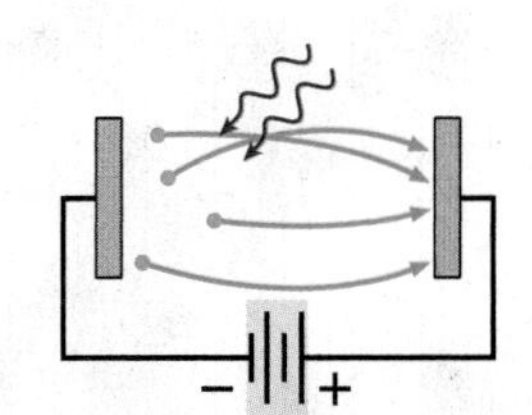
−
+

$\Delta V > 0$: Making the anode positive creates an electric field that pushes all the electrons to the anode.

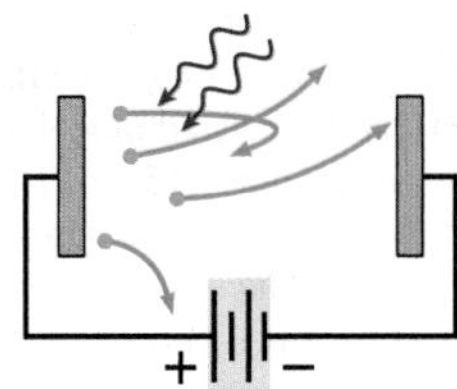
+
−

$\Delta V < 0$: Making the anode negative repels the electrons. Only the very fastest make it to the anode.

We can use conservation of energy to relate the maximum kinetic energy to the stopping potential. When ΔV is negative, as in the bottom panel of Figure 28.9, electrons are "going uphill," converting kinetic energy to potential energy as they slow down. That is, $\Delta U = -e\Delta V = -\Delta K$, where we've used $q = -e$ for electrons and ΔK is negative because the electrons are losing kinetic energy. When $\Delta V = -V_{stop}$, where the current ceases, the very fastest electrons, with K_{max}, are being turned back *just* as they reach the anode. They're converting 100% of their kinetic energy into potential energy, so $\Delta K = -K_{max}$. Thus $e\,V_{stop} = K_{max}$, or

$$V_{stop} = \frac{K_{max}}{e} \tag{28.2}$$

In other words, **measuring the stopping potential tells us the maximum kinetic energy of the electrons.**

Einstein's Explanation

When light shines on the cathode in a photoelectric effect experiment, why do electrons leave the metal at all? Early investigators suggested explanations based on classical physics. It was known that a heated electrode spontaneously emits electrons, so it was natural to suggest that the light falling on the cathode simply heated it, causing it to emit electrons. But this explanation is not consistent with experiment. For light to

Einstein's "Miracle Year" Albert Einstein was a little-known young man of 26 in 1905. In that single year, Einstein published three papers on three different topics, each of which would revolutionize physics. One was his initial paper on the theory of relativity. A second paper used statistical mechanics to explain a phenomenon called *Brownian motion,* the random motion of small particles suspended in water. It is Einstein's third paper of 1905, on the nature of light, in which we are most interested in this chapter.

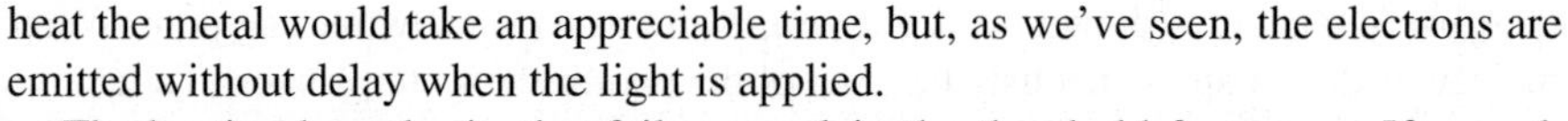

heat the metal would take an appreciable time, but, as we've seen, the electrons are emitted without delay when the light is applied.

The heating hypothesis also fails to explain the threshold frequency. If a weak intensity at a frequency just slightly above the threshold can generate a current, then certainly a strong intensity at a frequency just slightly below the threshold should be able to do so—it will heat the metal even more. There is no reason that a slight change in frequency should matter. Yet the experimental evidence shows a sharp frequency threshold, as we've seen.

A new physical theory was needed to fully explain the photoelectric-effect data. The currently accepted solution came in a 1905 paper by Albert Einstein in which he offered an exceedingly simple but amazingly bold idea that explained all of the noted features of the data.

Einstein's paper extended the work of the German physicist Max Planck, who had found that he could explain the form of the spectrum of a glowing, incandescent object that we saw in ◀◀ SECTION 25.7 only if he assumed that the oscillating atoms inside the heated solid vibrated in a particular way. The energy of an atom vibrating with frequency f had to be one of the specific energies $E = 0, hf, 2hf, 3hf, \ldots$, where h is a constant. That is, the vibration energies are **quantized.** The constant h, now called **Planck's constant,** is

$$h = 6.63 \times 10^{-34}\ \text{J} \cdot \text{s} = 4.14 \times 10^{-15}\ \text{eV} \cdot \text{s}$$

The first value, with SI units, is the proper one for most calculations, but you will find the second to be useful when energies are expressed in eV.

Einstein was the first to take Planck's idea seriously. Einstein went even further and suggested that **electromagnetic radiation itself is quantized!** That is, light is not really a continuous wave but, instead, arrives in small packets or bundles of energy. Einstein called each packet of energy a **light quantum,** and he postulated that the energy of one light quantum is directly proportional to the frequency of the light. That is, each quantum of light, which is now known as a **photon,** has energy

$$E = hf \tag{28.3}$$

The energy of a photon, a quantum of light, of frequency f

where h is Planck's constant. Higher-frequency light is composed of higher-energy photons—it is composed of bundles of greater energy. This seemingly simple assumption was completely at odds with the classical understanding of light as a wave, but it allowed Einstein to explain all of the properties of the photoelectric effect.

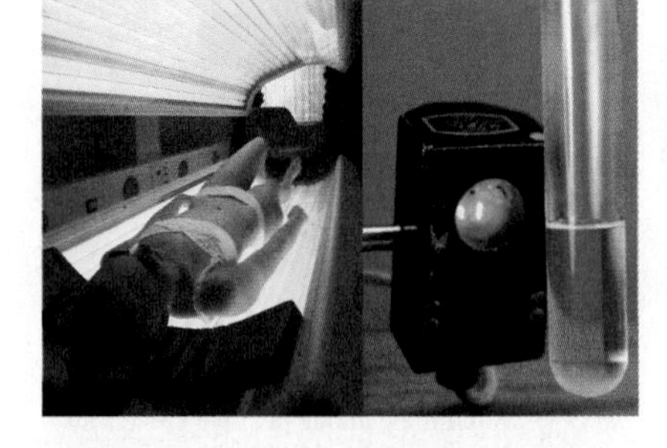

Not all ultraviolet is created equal BIO
In Example 28.2, you see that ultraviolet light sources with small differences in wavelength can have very different biological effects. Tanning beds emit nearly all of their energy at wavelengths greater than 315 nm. This light stimulates cells to produce melanin—resulting in a tan—but produces little short-term cell damage. Germicidal lamps use ultraviolet peaked at 254 nm, which will damage and even kill cells. Exposure to such a source will result in very painful sunburn.

EXAMPLE 28.2 Finding the energy of ultraviolet photons

Ultraviolet light at 290 nm does 250 times as much cellular damage as an equal intensity of ultraviolet at 310 nm; there is a clear threshold for damage at about 300 nm. What is the energy, in eV, of photons with a wavelength of 300 nm?

PREPARE The energy of a photon is related to its frequency by $E = hf$.

SOLVE The frequency at wavelength 300 nm is

$$f = \frac{c}{\lambda} = \frac{3.00 \times 10^{8}\ \text{m/s}}{300 \times 10^{-9}\ \text{m}} = 1.00 \times 10^{15}\ \text{Hz}$$

We can now use Equation 28.3 to calculate the energy, using the value of h in eV · s:

$$E = hf = (4.14 \times 10^{-15}\ \text{eV} \cdot \text{s})(1.00 \times 10^{15}\ \text{Hz}) = 4.14\ \text{eV}$$

ASSESS This number seems reasonable. We saw in Chapter 25 that splitting a bond in a water molecule requires an energy of 4.7 eV. We'd expect photons with energies in this range to be able to damage the complex organic molecules in a cell. As the problem notes, there is a sharp threshold for this damage. For energies larger than about 4.1 eV, photons can disrupt the genetic material of cells. Lower energies have little effect.

Einstein's Postulates and the Photoelectric Effect

The idea that light is quantized is now widely understood and accepted. But at the time of Einstein's paper, it was a truly revolutionary idea. Though we have used the photon model before, it is worthwhile to look at the theoretical underpinnings in more detail. In his 1905 paper, Einstein framed three postulates about light quanta and their interaction with matter:

1. Light of frequency f consists of discrete quanta, each of energy $E = hf$. Each photon travels at the speed of light c.
2. Light quanta are emitted or absorbed on an all-or-nothing basis. A substance can emit 1 or 2 or 3 quanta, but not 1.5. Similarly, an electron in a metal cannot absorb half a quantum but only an integer number.
3. A light quantum, when absorbed by a metal, delivers its entire energy to *one* electron.

NOTE ▸ These three postulates—that light comes in chunks, that the chunks cannot be divided, and that the energy of one chunk is delivered to one electron—are crucial for understanding the new ideas that will lead to quantum physics. ◂

Let's look at how Einstein's postulates apply to the photoelectric effect. We now think of the light shining on the metal as a torrent of photons, each of energy hf. Each photon is absorbed by *one* electron, giving that electron an energy $E_{elec} = hf$. This leads us to several interesting conclusions:

FIGURE 28.10 The ejection of an electron.

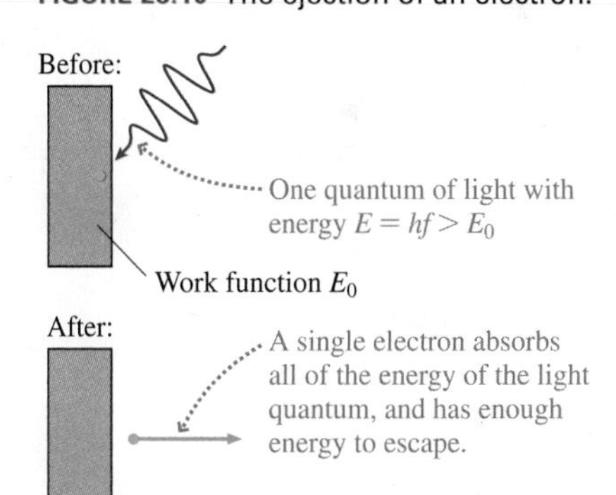
Before:
One quantum of light with energy $E = hf > E_0$
Work function E_0
After:
A single electron absorbs all of the energy of the light quantum, and has enough energy to escape.

1. An electron that has just absorbed a quantum of light energy has $E_{elec} = hf$. **FIGURE 28.10** shows that this electron can escape from the metal if its energy exceeds the work function E_0, or if

$$E_{elec} = hf \geq E_0 \tag{28.4}$$

In other words, there is a *threshold frequency*

$$f_0 = \frac{E_0}{h} \tag{28.5}$$

for the ejection of electrons. If f is less than f_0, even by just a small amount, none of the electrons will have sufficient energy to escape no matter how intense the light. But even very weak light with $f \geq f_0$ will give a few electrons sufficient energy to escape **because each photon delivers all of its energy to one electron.** This threshold behavior is exactly what the data show.

2. A more intense light delivers a larger number of photons to the surface. These eject a larger number of electrons and cause a larger current, exactly as observed.
3. There is a distribution of kinetic energies, because different electrons require different amounts of energy to escape, but the *maximum* kinetic energy is

$$K_{max} = E_{elec} - E_0 = hf - E_0 \tag{28.6}$$

As we noted in Equation 28.2, the stopping potential V_{stop} is a measure of K_{max}. Einstein's theory predicts that the stopping potential is related to the light frequency by

$$V_{stop} = \frac{K_{max}}{e} = \frac{hf - E_0}{e} \tag{28.7}$$

According to Equation 28.7, the stopping potential does *not* depend on the intensity of the light. Both weak light and intense light will have the same stopping potential. This agrees with the data.

4. If each photon transfers its energy hf to just one electron, that electron immediately has enough energy to escape. The current should begin instantly, with no delay, exactly as experiments had found.

Seeing the world in a different light BIO Plants use photosynthesis to convert the energy of light to chemical energy. Photons of visible light have sufficient energy to trigger the necessary molecular transitions, but infrared photons do not, so visible photons are absorbed while infrared photons are reflected. In this infrared photo, infrared photons strongly reflected from the trees' leaves make the trees appear a ghostly white.

FIGURE 28.11 A pebble transfers energy to the water.

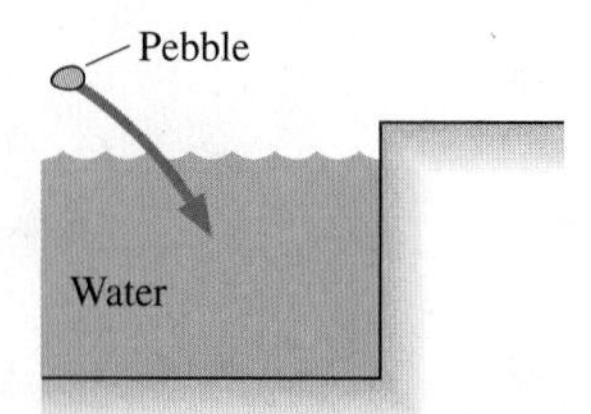

Classically, the energy of the pebble is shared by all the water molecules. One pebble causes only very small waves.

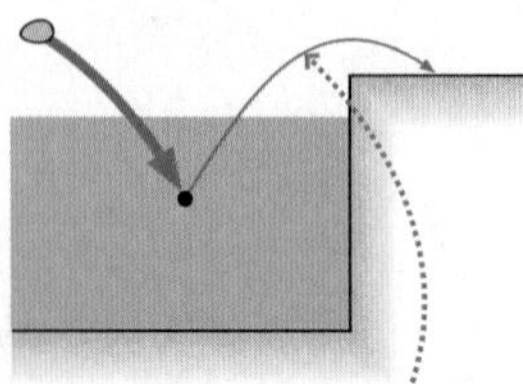

If the pebble could give *all* its energy to one drop, that drop could easily splash out of the pool.

Ultimately, Einstein's postulates are able to explain all of the observed features of the data for the photoelectric effect, though they require us to think of light in a very different way.

Let's use the swimming pool analogy again to help us visualize the photon model. **FIGURE 28.11** shows a pebble being thrown into the pool. The pebble increases the energy of the water, but the increase is shared among all the molecules in the pool. The increase in the water's energy is barely enough to make ripples, not nearly enough to splash water out of the pool. But suppose *all* the pebble's energy could go to *one drop* of water that didn't have to share it. That one drop of water would easily have enough energy to leap out of the pool. Einstein's hypothesis that a light quantum transfers all its energy to one electron is equivalent to the pebble transferring all its energy to one drop of water.

Einstein was awarded the Nobel Prize in 1921 not for his theory of relativity, as many would suppose, but for his explanation of the photoelectric effect. Einstein showed convincingly that energy is quantized and that light, even though it exhibits wave-like interference, comes in the particle-like packets of energy we now call photons. This was the first big step in the development of the theory of quantum physics.

EXAMPLE 28.3 Finding the photoelectric threshold frequency

What are the threshold frequencies and wavelengths for electron emission from sodium and from aluminum?

PREPARE Table 28.1 gives the work function for sodium as $E_0 = 2.75$ eV and that for aluminum as $E_0 = 4.28$ eV.

SOLVE We can use Equation 28.5, with h in units of eV · s, to calculate

$$f_0 = \frac{E_0}{h} = \begin{cases} 6.64 \times 10^{14}\ \text{Hz} & \text{sodium} \\ 10.34 \times 10^{14}\ \text{Hz} & \text{aluminum} \end{cases}$$

These frequencies are converted to wavelengths with $\lambda = c/f$, giving

$$\lambda = \begin{cases} 452\ \text{nm} & \text{sodium} \\ 290\ \text{nm} & \text{aluminum} \end{cases}$$

ASSESS The photoelectric effect can be observed with sodium for $\lambda < 452$ nm. This includes blue and violet visible light but not red, orange, yellow, or green. Aluminum, with a larger work function, needs ultraviolet wavelengths, $\lambda < 290$ nm.

EXAMPLE 28.4 Determining the maximum electron speed

What are the maximum electron speed and the stopping potential if sodium is illuminated with light of wavelength 300 nm?

PREPARE The kinetic energy of the emitted electrons—and the potential difference necessary to stop them—depends on the energy of the incoming photons, $E = hf$, and the work function of the metal from which they are emitted, $E_0 = 2.75$ eV.

SOLVE The light frequency is $f = c/\lambda = 1.00 \times 10^{15}$ Hz, so each light quantum has energy $hf = 4.14$ eV. The maximum kinetic energy of an electron is

$$K_{\text{max}} = hf - E_0 = 4.14\ \text{eV} - 2.75\ \text{eV} = 1.39\ \text{eV}$$
$$= 2.22 \times 10^{-19}\ \text{J}$$

Because $K = \frac{1}{2}mv^2$, where m is the electron's mass, not the mass of the sodium atom, the maximum speed of an electron leaving the cathode is

$$v_{\text{max}} = \sqrt{\frac{2K_{\text{max}}}{m}} = 6.99 \times 10^5\ \text{m/s}$$

Note that K_{max} must be in J, the SI unit of energy, in order to calculate a speed in m/s.

Now that we know the maximum kinetic energy of the electrons, we can use Equation 28.7 to calculate the stopping potential:

$$V_{\text{stop}} = \frac{K_{\text{max}}}{e} = 1.39\ \text{V}$$

An anode voltage of -1.39 V will be just sufficient to stop the fastest electrons and thus reduce the current to zero.

ASSESS The stopping potential has the *same numerical value* as K_{max} expressed in eV, which makes sense. An electron with a kinetic energy of 1.39 eV can go "uphill" against a potential difference of 1.39 V, but no more.

STOP TO THINK 28.2 The work functions of metals A, B, and C are 3.0 eV, 4.0 eV, and 5.0 eV, respectively. UV light shines on all three metals, causing electrons to be emitted. Rank in order, from largest to smallest, the stopping voltages for A, B, and C.

28.3 Photons

We've now seen compelling evidence for the photon nature of light, but this leaves an important question: Just what *are* photons? To begin our explanation, let's return to the experiment that showed most dramatically the wave nature of light—the double-slit interference experiment. We will make a change, though: We will dramatically lower the light intensity by inserting filters between the light source and the slits. The fringes will be too dim to see with the naked eye, so we will use a detector that can build up an image over time. (This is the same sort of detector we imagined using for the extremely low-light photograph in ◀◀ SECTION 25.6.)

FIGURE 28.12 A double-slit experiment performed with light of very low intensity.

(a) Image after a very short time

(b) Image after a slightly longer time

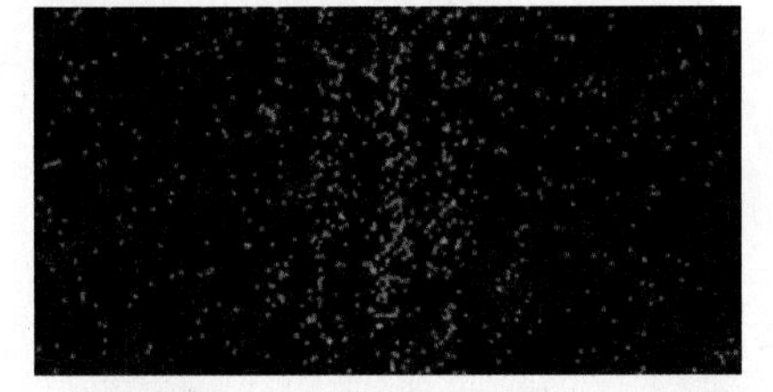

(c) Continuing to build up the image

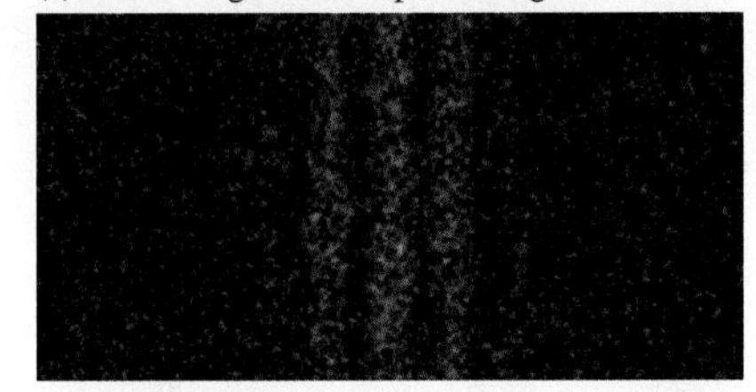

(d) Image after a very long time

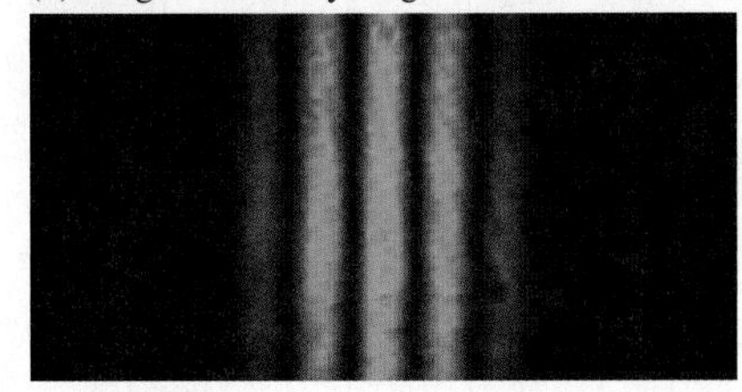

FIGURE 28.12 shows the outcome of such an experiment at four different times. At early times, very little light has reached the detector, and there are no discernible fringes. Instead, the detector shows dots; it is registering the arrival of particle-like objects.

As the detector builds up the image for a longer time, we see that the positions of the dots are not entirely random. They are grouped into bands at *exactly* the positions where we expect to see bright constructive-interference fringes. As the detector continues to gather light, the light and dark fringes become quite distinct. After a long time, the individual dots overlap and the image looks exactly like those we saw in Chapter 17.

The dots of light on the screen, which we'll attribute to the arrival of individual photons, are particle-like, but the overall picture clearly does not mesh with the classical idea of a particle. A classical particle, when faced with a double-slit apparatus, would go through one slit or the other. If light consisted of classical particles, we would see two bright areas on the screen, corresponding to light that has gone through one or the other slit. Instead, we see particle-like dots forming wave-like interference fringes.

This experiment was performed with a light level so low that only one photon at a time passed through the apparatus. If particle-like photons arrive at the detector in a banded pattern as a consequence of wave-like interference, as Figure 28.12 shows, but if only one photon at a time is passing through the experiment, what is it interfering with? The only possible answer is that the photon is somehow interfering *with itself.* Nothing else is present. But if each photon interferes with itself, rather than with other photons, then each photon, despite the fact that it is a particle-like object, must somehow go through *both* slits! This is something only a wave could do.

This all seems pretty crazy. But crazy or not, this is the way light behaves in real experiments. **Sometimes it exhibits particle-like behavior and sometimes it exhibits wave-like behavior.** The thing we call *light* is stranger and more complex than it first appeared, and there is no way to reconcile these seemingly contradictory behaviors. We have to accept nature as it is, rather than hoping that nature will conform to our expectations. Furthermore, as we will see, this half-wave/half-particle behavior is not restricted to light.

The Photon Rate

The photon nature of light isn't apparent in most cases. Most light sources with which you are familiar emit such vast numbers of photons that you are aware of only their wave-like superposition, just as you notice only the roar of a heavy rain on your roof and not the individual raindrops. Only at extremely low intensities does the light begin to appear as a stream of individual photons, like the random patter of raindrops when it is barely sprinkling.

EXAMPLE 28.5 How many photons per second does a laser emit?

The 1.0 mW light beam from a laser pointer ($\lambda = 670$ nm) shines on a screen. How many photons strike the screen each second?

PREPARE The power of the beam is 1.0 mW, or 1.0×10^{-3} J/s. Each second, 1.0×10^{-3} J of energy reaches the screen. It arrives as individual photons of energy given by Equation 28.3.

SOLVE The frequency of the photons is $f = c/\lambda = 4.48 \times 10^{14}$ Hz, so the energy of an individual photon is $E = hf = (6.63 \times 10^{-34}\ \text{J} \cdot \text{s})(4.48 \times 10^{14}\ \text{Hz}) = 2.97 \times 10^{-19}$ J. The number of photons reaching the screen each second is the total energy reaching the screen each second divided by the energy of an individual photon:

$$\frac{1.0 \times 10^{-3}\ \text{J/s}}{2.97 \times 10^{-19}\ \text{J/photon}} = 3.4 \times 10^{15}\ \text{photons per second}$$

ASSESS Each photon carries a small amount of energy, so there must be a huge number of photons per second to produce even this modest power.

CONCEPTUAL EXAMPLE 28.6 Comparing photon rates

A red laser pointer and a green laser pointer have the same power. Which one emits a larger number of photons per second?

REASON Red light has a longer wavelength and thus a lower frequency than green light, so the energy of a photon of red light is less than the energy of a photon of green light. The two pointers emit the same amount of light energy per second. Because the red laser emits light in smaller "chunks" of energy, it must emit more chunks per second to have the same power. The red laser emits more photons each second.

ASSESS This result can seem counterintuitive if you haven't thought hard about the implications of the photon model. Light of different wavelengths is made of photons of different energies, so these two lasers with different wavelengths—though they have the same power—must emit photons at different rates.

FIGURE 28.13 The operation of a solar cell.

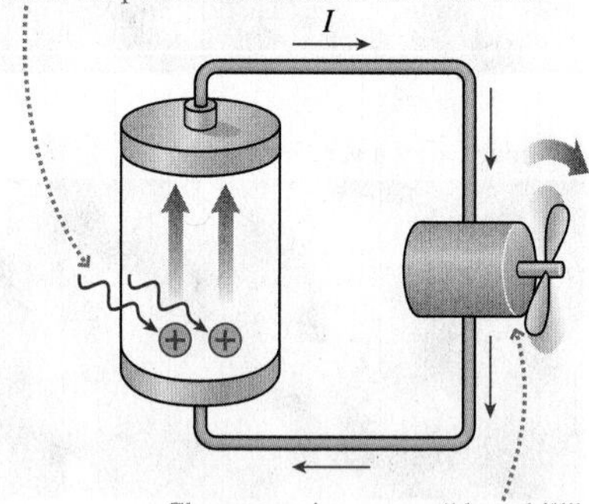

Photons with energy greater than the threshold give their energy to charge carriers, increasing their potential energy and lifting them to the positive terminal of the solar cell.
I
Charge carriers move "downhill" through the circuit. Their energy can be used to run useful devices.

Detecting Photons

Early light detectors, which used the photoelectric effect directly, consisted of a polished metal plate in a vacuum tube. When light fell on the plate, an electron current was generated that could trigger an action, such as sounding an alarm, or could provide a measurement of the light intensity.

Modern devices work on similar principles. In a *solar cell*, incoming photons give their energy to charge carriers, lifting them into higher-energy states. Recall the charge escalator model of a battery in ◀◀ SECTION 22.3. The solar cell works much like a battery, but the energy to lift charges to a higher potential comes from photons, not chemical reactions, as shown in FIGURE 28.13. The photon energy must exceed some minimum value to cause this transition, so solar cells have a threshold frequency, just like a device that uses the photoelectric effect directly. For a silicon-based solar cell, the most common type, the energy threshold is about 1.1 eV, corresponding to a wavelength of about 1200 nm, just beyond the range of the visible light spectrum, in the infrared.

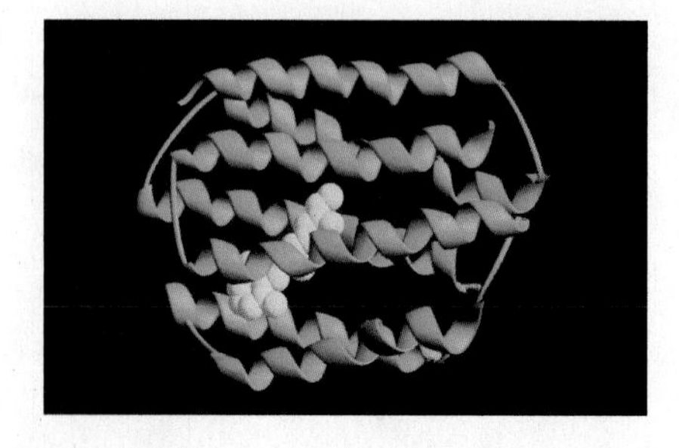

◀ **Seeing photons** BIO The basis of vision is the detection of single photons by specially adapted molecules in the rod and cone cells of the eye. This image shows a molecule of *rhodopsin* (blue) with a molecule called *retinal* (yellow) nested inside. A single photon of the right energy triggers a transition of the retinal molecule, changing its shape so that it no longer fits inside the rhodopsin "cage." The rhodopsin then changes shape to eject the retinal, and this motion leads to an electrical signal in a nerve fiber.

EXAMPLE 28.7 Finding the current from a solar cell

1.0 W of monochromatic light of wavelength 550 nm illuminates a silicon solar cell, driving a current in a circuit. What is the maximum possible current this light could produce?

PREPARE The wavelength is shorter than the 1200 nm threshold wavelength noted for a silicon solar cell, so the photons will have sufficient energy to cause charge carriers to flow. Each photon of the incident light will give its energy to a single charge carrier. The maximum number of charge carriers that can possibly flow in each second is thus equal to the number of photons that arrive each second.

SOLVE The power of the light is $P = 1.0\ \text{W} = 1.0\ \text{J/s}$. The frequency of the light is $f = c/\lambda = 5.45 \times 10^{14}\ \text{Hz}$, so the energy of individual photons is $E = hf = 3.61 \times 10^{-19}\ \text{J}$. The number of photons arriving per second is $(1.0\ \text{J/s})/(3.61 \times 10^{-19}\ \text{J/photon}) = 2.77 \times 10^{18}$. Each photon can set at most one charge carrier into motion, so the maximum current is 2.77×10^{18} electrons/s. The current in amps—coulombs per second—is the electron flow rate multiplied by the charge per electron:

$$I_{\text{max}} = (2.77 \times 10^{18}\ \text{electrons/s})(1.6 \times 10^{-19}\ \text{C})$$
$$= 0.44\ \text{C/s} = 0.44\ \text{A}$$

ASSESS The key concept underlying the solution is that one photon gives its energy to a single charge carrier. We've calculated the current if all photons give their energy to charge carriers. The current in a real solar cell will be less than this because some photons will be reflected or otherwise "lost" and will not transfer their energy to charge carriers.

The *charge-coupled device* (CCD) or *complementary metal oxide semiconductor* (CMOS) detector in a digital camera consists of millions of *pixels,* each a microscopic silicon-based photodetector. Each photon hitting a pixel (if its frequency exceeds the threshold frequency) liberates one electron. These electrons are stored inside the pixel, and the total accumulated charge is directly proportional to the light intensity—the number of photons—hitting the pixel. After the exposure, the charge in each pixel is read and the value stored in memory; then the pixel is reset to be ready for the next picture.

STOP TO THINK 28.3 The intensity of a beam of light is increased but the light's frequency is unchanged. Which one (or perhaps more than one) of the following is true?

A. The photons travel faster.
B. Each photon has more energy.
C. There are more photons per second.

28.4 Matter Waves

Prince Louis-Victor de Broglie was a French graduate student in 1924. It had been 19 years since Einstein had shaken the world of physics by introducing photons and thus blurring the distinction between a particle and a wave. As de Broglie thought about these issues, it seemed that nature should have some kind of symmetry. If light waves could have a particle-like nature, why shouldn't material particles have some kind of wave-like nature? In other words, could **matter waves** exist?

With no experimental evidence to go on, de Broglie reasoned by analogy with Einstein's equation $E = hf$ for the photon and with some of the ideas of his theory of relativity. De Broglie determined that *if* a material particle of momentum $p = mv$ has a wave-like nature, its wavelength must be given by

$$\lambda = \frac{h}{p} = \frac{h}{mv} \tag{28.8}$$

De Broglie wavelength for a moving particle

where h is Planck's constant. This wavelength is called the **de Broglie wavelength.**

EXAMPLE 28.8 Calculating the de Broglie wavelength of an electron

What is the de Broglie wavelength of an electron with a kinetic energy of 1.0 eV?

SOLVE An electron with kinetic energy $K = \frac{1}{2}mv^2 = 1.0\text{ eV} = 1.6 \times 10^{-19}\text{ J}$ has speed

$$v = \sqrt{\frac{2K}{m}} = 5.9 \times 10^5\text{ m/s}$$

Although fast by macroscopic standards, the electron gains this speed by accelerating through a potential difference of a mere 1 V. The de Broglie wavelength is

$$\lambda = \frac{h}{mv} = 1.2 \times 10^{-9}\text{ m} = 1.2\text{ nm}$$

ASSESS The electron's wavelength is small, but it is similar to the wavelengths of x rays and larger than the approximately 0.1 nm spacing of atoms in a crystal. We can observe x-ray diffraction, so if an electron has a wave nature, it should be easily observable.

FIGURE 28.14 A double-slit interference pattern created with electrons.

What would it mean for matter—an electron or a proton or a baseball—to have a wavelength? Would it obey the principle of superposition? Would it exhibit diffraction and interference? Surprisingly, **matter exhibits all of the properties that we associate with waves.** For example, **FIGURE 28.14** shows the intensity pattern recorded after 50 keV electrons passed through two narrow slits separated by 1.0 μm. The pattern is clearly a double-slit interference pattern, and the spacing of the fringes is exactly as the theory of Chapter 17 would predict for a wavelength given by de Broglie's formula. **The electrons are behaving like waves!**

But if matter waves are real, why don't we see baseballs and other macroscopic objects exhibiting wave-like behavior? The key is the wavelength. We found in Chapter 17 that diffraction, interference, and other wave-like phenomena are observed when the wavelength is comparable to or larger than the size of an opening a wave must pass through. As Example 28.8 just showed, a typical electron wavelength is somewhat larger than the spacing between atoms in a crystal, so we expect to see wave-like behavior as electrons pass through matter or through microscopic slits. But the de Broglie wavelength is inversely proportional to an object's mass, so the wavelengths of macroscopic objects are millions or billions of times smaller than the wavelengths of electrons—vastly smaller than the size of any openings these objects might pass through. The wave nature of macroscopic objects is unimportant and undetectable because their wavelengths are so incredibly small, as the following example shows.

EXAMPLE 28.9 Calculating the de Broglie wavelength of a smoke particle

One of the smallest macroscopic particles we could imagine using for an experiment would be a very small smoke or soot particle. These are $\approx 1\ \mu$m in diameter, too small to see with the naked eye and just barely at the limits of resolution of a microscope. A particle this size has mass $m \approx 10^{-18}$ kg. Estimate the de Broglie wavelength for a 1-μm-diameter particle moving at the very slow speed of 1 mm/s.

SOLVE The particle's momentum is $p = mv \approx 10^{-21}\text{ kg} \cdot \text{m/s}$. The de Broglie wavelength of a particle with this momentum is

$$\lambda = \frac{h}{p} \approx 7 \times 10^{-13}\text{ m}$$

ASSESS The wavelength is much, much smaller than the particle itself—much smaller than an individual atom! We don't expect to see this particle exhibiting wave-like behavior.

The preceding example shows that a very small particle moving at a very slow speed has a wavelength that is too small to be of consequence. For larger objects moving at higher speeds, the wavelength is even smaller. A pitched baseball will have a wavelength of about 10^{-34} m, so a batter cannot use the wave nature of the ball as an excuse for not getting a hit. With such unimaginably small wavelengths, it is little wonder that we do not see macroscopic objects exhibiting wave-like behavior.

The Interference and Diffraction of Matter

Though de Broglie made his hypothesis in the absence of experimental data, experimental evidence was soon forthcoming. **FIGURES 28.15a** and **b** show diffraction patterns produced by x rays and electrons passing through an aluminum-foil target. The primary observation to make from Figure 28.15 is that **electrons diffract and interfere exactly like x rays.**

FIGURE 28.15 The diffraction patterns produced by x rays, electrons, and neutrons passing through an aluminum-foil target.

(a) X-ray diffraction pattern

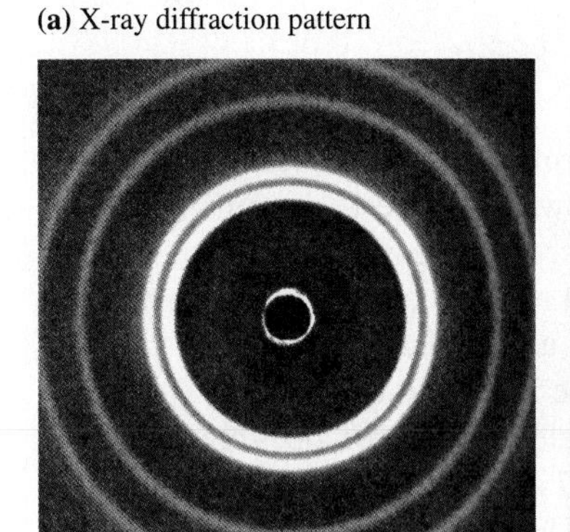

(b) Electron diffraction pattern

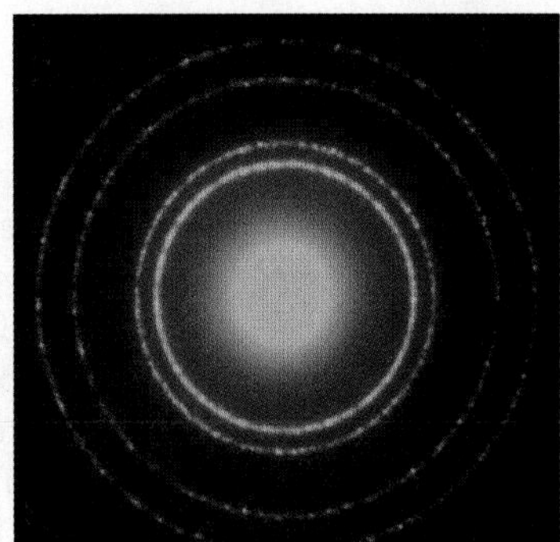

(c) Neutron diffraction pattern

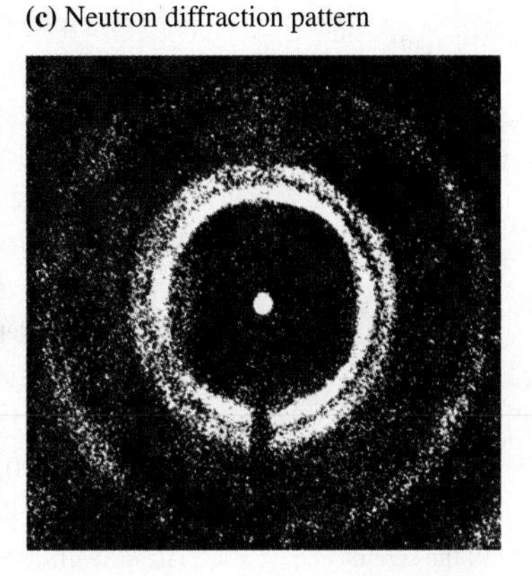

Later experiments demonstrated that de Broglie's hypothesis applies to other material particles as well. Neutrons have a much larger mass than electrons, which tends to decrease their de Broglie wavelength, but it is possible to generate very slow neutrons. The much smaller speed compensates for the heavier mass, so neutron wavelengths can be made comparable to electron wavelengths. **FIGURE 28.15c** shows a neutron diffraction pattern. It is similar to the x-ray and electron diffraction patterns, although of lower quality because neutrons are harder to detect. A neutron, too, is a matter wave. In recent years it has become possible to observe the interference and diffraction of atoms and even large molecules!

The Electron Microscope

As you learned in ◀◀ SECTION 19.7, the wave nature of light limits the ultimate resolution of an optical microscope—the smallest resolvable separation between two objects—to about half a wavelength of light. For visible light, the smallest feature that can be resolved, even with perfect lenses, is about 200 or 250 nm. But the picture of the retina at the start of the chapter can show details much finer than this because it wasn't made with light—it was made with a beam of electrons.

The electron microscope, invented in the 1930s, works much like a light microscope. In the absence of electric or magnetic fields, electrons travel through a vacuum in straight lines much like light rays. Electron trajectories can be bent with electric or magnetic fields. A coil of wire carrying a current can produce a magnetic field that bends parallel electron trajectories so that they all cross at a single point; we call this an *electron lens.* An electron lens focuses electrons in the same way a glass lens bends and focuses light rays.

FIGURE 28.16 shows how a *transmission electron microscope* (TEM) works. This is purely classical physics; the electrons experience electric and magnetic forces, and they follow trajectories given by Newton's second law. Our ability to control electron trajectories allows electron microscopes to have magnifications far exceeding those of light microscopes. But just as in a light microscope, the resolution is ultimately limited by wave effects. Electrons are not classical point particles; they have wave-like properties and a de Broglie wavelength $\lambda = h/p$.

FIGURE 28.16 The electron microscope.

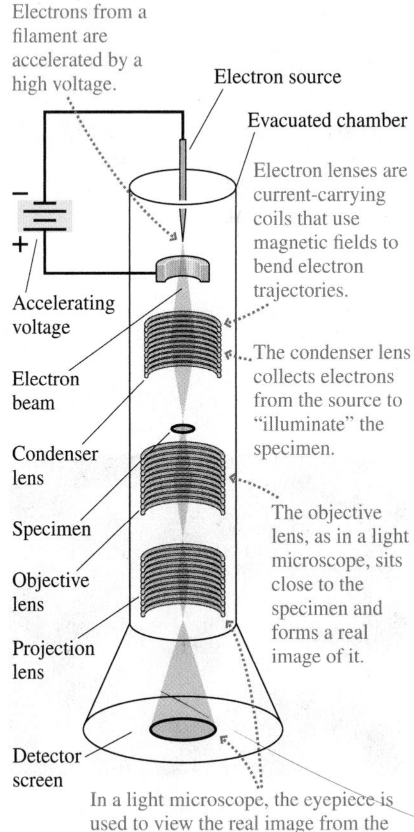

CONCEPTUAL EXAMPLE 28.10 Which wavelength is shorter?

An electron is accelerated through a potential difference ΔV. A second electron is accelerated through a potential difference that is twice as large. Which electron has a shorter de Broglie wavelength?

REASON The wavelength is inversely proportional to the speed. The electron that is accelerated through the larger potential difference will be moving faster and so will have a shorter de Broglie wavelength.

ASSESS Creating an electron micrograph requires high-speed electrons. Higher accelerating voltages mean higher speeds and shorter wavelengths, which would—in principle—allow for better resolution.

FIGURE 28.17 BIO TEM image of a pigment molecule from a crustacean shell.

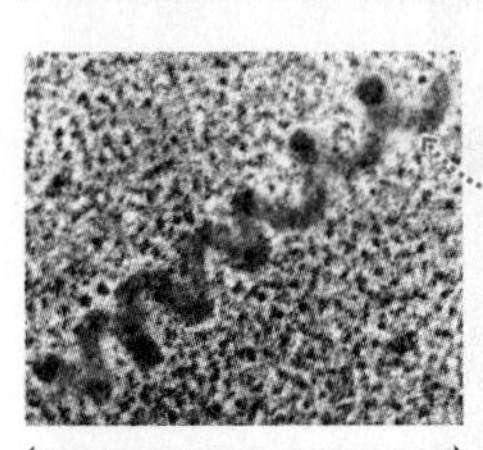

The great resolving power of the microscope allows the imaging of incredibly fine detail, in this case the actual structure of a molecule.

20 nm

The total image size is 1/10 the size of the smallest feature that can be resolved by a light microscope. A light microscope could not detect this individual molecule, let alone show its structure.

The reasoning used in Section 19.7 to determine maximum resolution applies equally well to electrons. Thus the resolving power is, at best, about half the electrons' de Broglie wavelength. For a 100 kV accelerating voltage, which is fairly typical, the de Broglie wavelength is $\lambda \approx 0.004$ nm (the electrons are moving fast enough that the momentum has to be calculated using relativity) and thus the theoretical resolving power of an electron microscope is about 0.002 nm.

In practice, the resolving powers of the best electron microscopes are limited by imperfections in the electron lenses to about 0.07 nm, just sufficient to resolve individual atoms with diameters of about 0.1 nm. This resolving power is about 3000 times smaller than can be achieved with light microscopes, as noted in **FIGURE 28.17**. Good light microscopes function at their theoretical limit, but there's still room to improve electron microscopes if a clever scientist or engineer can make a better electron lens.

STOP TO THINK 28.4 A beam of electrons, a beam of protons, and a beam of oxygen atoms each pass at the same speed through a 1-μm-wide slit. Which will produce the widest central maximum on a detector behind the slit?

A. The beam of electrons.
B. The beam of protons.
C. The beam of oxygen atoms.
D. All three patterns will be the same.
E. None of the beams will produce a diffraction pattern.

28.5 Energy Is Quantized

De Broglie hypothesized that material particles have wave-like properties, and you've now seen experimental evidence that this must be true. Not only is this bizarre, but the implications are profound.

You learned in ◀◀ SECTION 16.3 that the waves on a string fixed at both ends form standing waves. Wave reflections from both ends create waves traveling in both directions, and the superposition of two oppositely directed waves produces a standing wave. Could we do something like this with particles? Is there such a thing as a "standing matter wave"? In fact, you are probably already familiar with standing matter waves—the atomic electron orbitals that you learned about in chemistry.

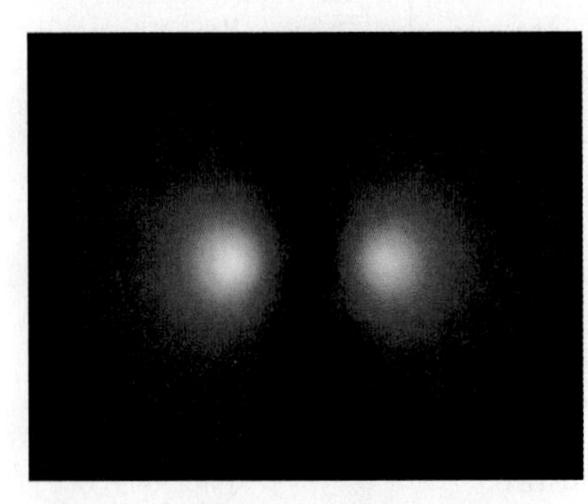

This computer simulation shows the p orbital of an atom. This orbital is an electron standing wave with a clear node at the center.

We'll have more to say about these orbitals in Chapter 29. For now, we'll start our discussion of standing matter waves with a simpler physical system called a "particle in a box." For simplicity, we'll consider one-dimensional motion, a particle that moves back and forth along the x-axis. The "box" is defined by two fixed ends, and the particle bounces back and forth between these boundaries as in **FIGURE 28.18**. We'll assume that collisions with the ends of the box are perfectly elastic, with no loss of kinetic energy.

Figure 28.18a shows a classical particle, such as a ball or a dust particle, in the box. This particle simply bounces back and forth at constant speed. But if particles have wave-like properties, perhaps we should consider a *wave* reflecting back and forth from the ends of the box. The reflections will create the standing wave shown in Figure 28.18b. This standing wave is analogous to the standing wave on a string that is tied at both ends.

What can we say about the properties of this standing matter wave? We can use what we know about matter waves and standing waves to make some deductions.

For waves on a string, we saw that there were only certain possible modes. The same will be true for the particle in a box. In Chapter 16, we found that the wavelength of a standing wave is related to the length L of the string by

$$\lambda_n = \frac{2L}{n} \qquad n = 1, 2, 3, 4, \ldots \tag{28.9}$$

The wavelength of the particle in a box will follow the same formula, but the wave describing the particle must also satisfy the de Broglie condition $\lambda = h/p$. Equating these two expressions for the wavelength gives

$$\frac{h}{p} = \frac{2L}{n} \tag{28.10}$$

Solving Equation 28.10 for the particle's momentum p, we find

$$p_n = n\left(\frac{h}{2L}\right) \qquad n = 1, 2, 3, 4, \ldots \tag{28.11}$$

This is a remarkable result; it is telling us that the momentum of the particle can have only certain values, the ones given by the equation. Other values simply aren't possible. The energy of the particle is related to its momentum by

$$E = \frac{1}{2}mv^2 = \frac{p^2}{2m} \tag{28.12}$$

If we use Equation 28.11 for the momentum, we find that the particle's energy is also restricted to a specific set of values given by

$$E_n = \frac{1}{2m}\left(\frac{hn}{2L}\right)^2$$

or

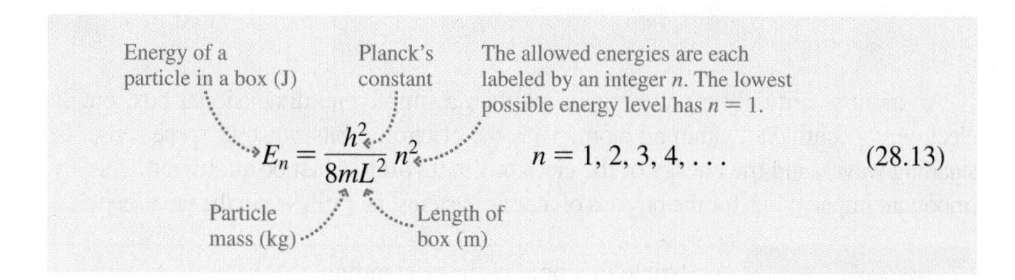

$$E_n = \frac{h^2}{8mL^2}n^2 \qquad n = 1, 2, 3, 4, \ldots \tag{28.13}$$

This conclusion is one of the most profound discoveries of physics. Because of the wave nature of matter, **a confined particle can have only certain energies.** This result—that a confined particle can have only discrete values of energy—is called the **quantization** of energy. More informally, we say that energy is *quantized.* The number n is called the **quantum number,** and each value of n characterizes one **energy level** of the particle in the box.

The lowest possible energy the particle in the box can have is

$$E_1 = \frac{h^2}{8mL^2} \tag{28.14}$$

FIGURE 28.18 A particle of mass m confined in a box of length L.

(a) A classical particle of mass m bounces back and forth between two boundaries.

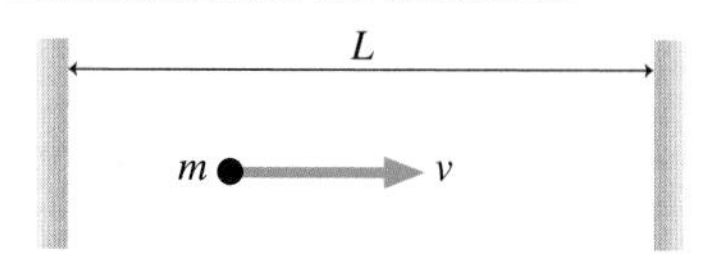

(b) Matter waves moving in opposite directions create standing waves.

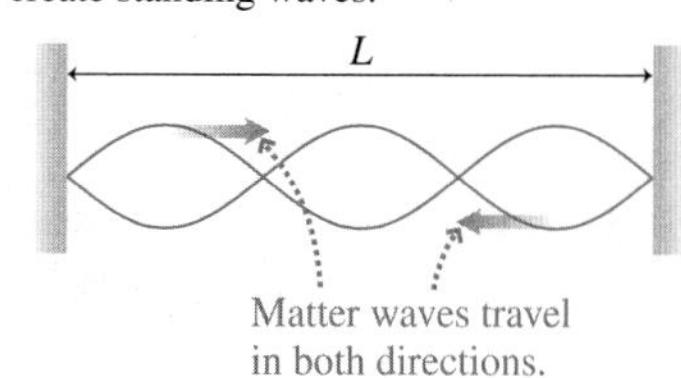

In terms of this lowest possible energy all other possible energies are

$$E_n = n^2 E_1 \tag{28.15}$$

This quantization is in stark contrast to the behavior of classical objects. It would be as if a baseball pitcher could throw a baseball at only 10 m/s, or 20 m/s, or 30 m/s, and so on, but at no speed in between. Baseball speeds aren't quantized, but the energy levels of a confined electron are—a result that has far-reaching implications.

EXAMPLE 28.11 Finding the allowed energies of a confined electron

An electron is confined to a region of space of length 0.19 nm—comparable in size to an atom. What are the first three allowed energies of the electron?

PREPARE We'll model this system as a particle in a box, with a box of length 0.19 nm. The possible energies are given by Equation 28.13.

SOLVE The mass of an electron is $m = 9.11 \times 10^{-31}$ kg. Thus the first allowed energy is

$$E_1 = \frac{h^2}{8mL^2} = 1.7 \times 10^{-18}\ \text{J} = 10\ \text{eV}$$

This is the lowest allowed energy. The next two allowed energies are

$$E_2 = 2^2 E_1 = 40\ \text{eV}$$

$$E_3 = 3^2 E_1 = 90\ \text{eV}$$

ASSESS These energies are significant; E_1 is larger than the work function of any metal in Table 28.1. Confining an electron to a region the size of an atom limits its energy to states separated by significant differences in energy. Clearly, our treatment of electrons in atoms must be a quantum treatment.

The energies allowed by Equation 28.13 are inversely proportional to both m and L^2. Both m and L have to be exceedingly small before energy quantization has any significance. Classical physics still works for baseballs! It is only at the atomic scale that quantization effects become important, as the following example shows.

EXAMPLE 28.12 The energy levels of a virus

A 30-nm-diameter virus is about the smallest imaginable macroscopic particle. What is the fundamental quantum of energy for this virus if it is confined in a one-dimensional cell of length 1.0 μm? The density of a virus is very close to that of water.

MODEL Model the virus as a particle in a box.

SOLVE The mass of a virus is $m = \rho V$, where the volume is $\frac{4}{3}\pi r^3$. A quick calculation shows that a 30-nm-diameter virus has mass $m = 1.4 \times 10^{-20}$ kg. The confinement length is $L = 1.0 \times 10^{-6}$ m. From Equation 28.14, the fundamental quantum of energy is

$$E_1 = \frac{h^2}{8mL^2} = \frac{(6.63 \times 10^{-34}\ \text{J}\cdot\text{s})^2}{8(1.4 \times 10^{-20}\ \text{kg})(1.0 \times 10^{-6}\ \text{m})^2}$$

$$= 3.9 \times 10^{-36}\ \text{J} = 2.5 \times 10^{-17}\ \text{eV}$$

ASSESS This is such an incredibly small amount of energy that there is no hope of distinguishing between energies of E_1 or $4E_1$ or $9E_1$. For any macroscopic particle, even one this tiny, the allowed energies will *seem* to be perfectly continuous. We will not observe the quantization.

An atom is certainly more complicated than a simple one-dimensional box, but an electron is "confined" within an atom. Thus the electron orbits must, in some sense, be standing waves, and **the energy of the electrons in an atom must be quantized.** This has important implications for the physics of atomic systems, as we'll see in the next section.

STOP TO THINK 28.5 A particle in a box, with the standing matter wave shown, has an energy of 8.0 eV. What is the lowest energy that this particle can have?

A. 1.0 eV B. 2.0 eV C. 4.0 eV D. 8.0 eV

28.6 Energy Levels and Quantum Jumps

Einstein and de Broglie introduced revolutionary new ideas—a blurring of the distinction between waves and particles, and the quantization of energy—but the first to develop a full-blown theory of quantum physics, in 1925, was the Austrian physicist

Erwin Schrödinger. Schrödinger's theory is now called *quantum mechanics*. It describes how to calculate the quantized energy levels of systems from a particle in a box to electrons in atoms. Quantum mechanics also describes another important piece of the puzzle: How does a quantized system gain or lose energy?

Erwin Schrödinger, one of the early architects of quantum mechanics.

Energy-Level Diagrams

The full theory of quantum mechanics shows that, just as for a particle in a box, the energy of a real physical system, such as an atom, is quantized: Only certain energies are allowed while all other energies are forbidden.

An **energy-level diagram** is a useful visual representation of the quantized energies. As an example, FIGURE 28.19a is the energy-level diagram for an electron in a 0.19-nm-long box. We computed these energies in Example 28.11. In an energy-level diagram, the vertical axis represents energy, but the horizontal axis is not a scale. Think of it as a ladder in which the energies are the rungs of the ladder. The lowest rung, with energy E_1, is called the **ground state.** Higher rungs, called **excited states,** are labeled by their quantum numbers, $n = 2, 3, 4, \ldots$. Whether it is a particle in a box, an atom, or the nucleus of an atom, quantum physics requires the system to be on one of the rungs of the ladder.

If a quantum system changes from one state to another, its energy changes. One thing that has not changed in quantum physics is the conservation of energy: If a system drops from a higher energy level to a lower, the excess energy ΔE_{system} must go somewhere. In the systems we will consider, this energy generally ends up in the form of an emitted photon. As FIGURE 28.19b shows, a quantum system in energy level E_i that "jumps down" to energy level E_f loses an energy $\Delta E_{\text{system}} = |E_f - E_i|$. This jump corresponds to the emission of a photon of frequency

$$f_{\text{photon}} = \frac{\Delta E_{\text{system}}}{h} \tag{28.16}$$

Conversely, if the system absorbs a photon, it can "jump up" to a higher energy level, as shown in FIGURE 28.19c. In this case, the frequency of the absorbed photon must follow Equation 28.16 as well. Such jumps are called **transitions** or **quantum jumps.**

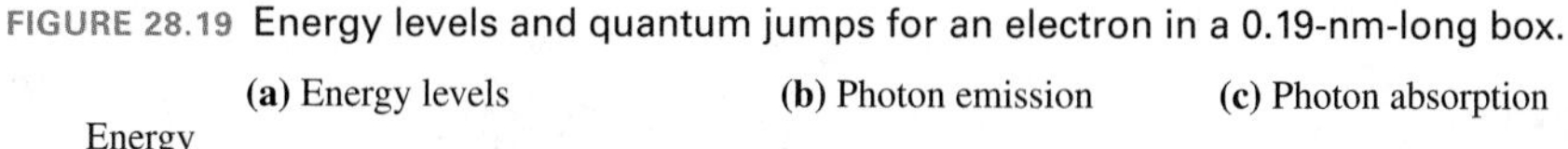

FIGURE 28.19 Energy levels and quantum jumps for an electron in a 0.19-nm-long box.

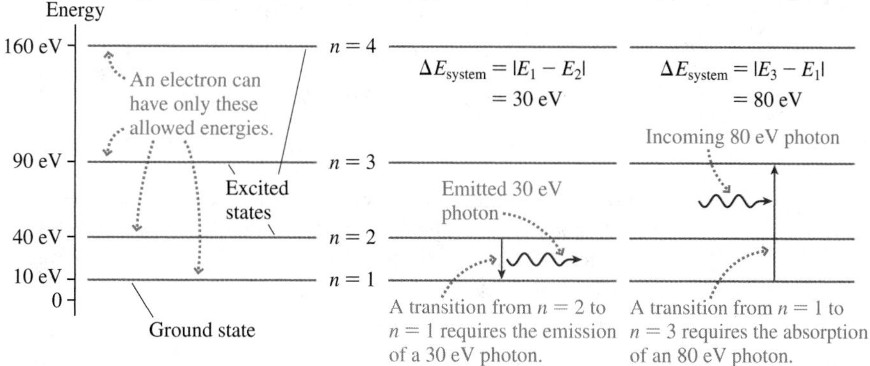

STOP TO THINK 28.6 For the system shown in Figure 28.19, which are possible energies of emitted photons? There may be more than one correct answer.

A. 40 eV B. 50 eV C. 60 eV D. 70 eV E. 80 eV

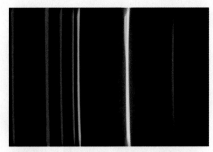

The spectrum of helium gas shows a discrete set of wavelengths, corresponding to the energies of possible transitions.

Notice that Equation 28.16 links Schrödinger's quantum theory to Einstein's earlier idea about the quantization of light energy. According to Einstein, a photon of frequency f has energy $E_{\text{photon}} = hf$. If a system jumps from an initial state with energy E_i to a final state with *lower* energy E_f, energy will be conserved if the system emits a photon with $E_{\text{photon}} = \Delta E_{\text{system}}$. The photon must have exactly the frequency given by Equation 28.16 if it is to carry away exactly the right amount of energy. As we'll see in the next chapter, these photons form the *emission spectrum* of the quantum system.

Similarly, a system can conserve energy while jumping to a higher-energy state, for which additional energy is needed, by absorbing a photon of frequency $f_{\text{photon}} = \Delta E_{\text{system}}/h$. The photon will not be absorbed unless it has exactly this frequency. The frequencies absorbed in these upward transitions form the system's *absorption spectrum.*

Let's summarize what quantum physics has to say about the properties of atomic-level systems:

1. **The energies are quantized.** Only certain energies are allowed; all others are forbidden. This is a consequence of the wave-like properties of matter.
2. **The ground state is stable.** Quantum systems seek the lowest possible energy state. A particle in an excited state, if left alone, will jump to lower and lower energy states until it reaches the ground state. Once in its ground state, there are no states of any lower energy to which a particle can jump.
3. **Quantum systems emit and absorb a *discrete spectrum* of light.** Only those photons whose frequencies match the energy *intervals* between the allowed energy levels can be emitted or absorbed. Photons of other frequencies cannot be emitted or absorbed without violating energy conservation.

We'll use these ideas in the next two chapters to understand the properties of atoms and nuclei.

EXAMPLE 28.13 Determining an emission spectrum from quantum states

An electron in a quantum system has allowed energies $E_1 = 1.0$ eV, $E_2 = 4.0$ eV, and $E_3 = 6.0$ eV. What wavelengths are observed in the emission spectrum of this system?

PREPARE FIGURE 28.20 shows the energy-level diagram for this system. Photons are emitted when the system undergoes a quantum jump from a higher energy level to a lower energy level. There are three possible transitions.

FIGURE 28.20 The system's energy-level diagram and quantum jumps.

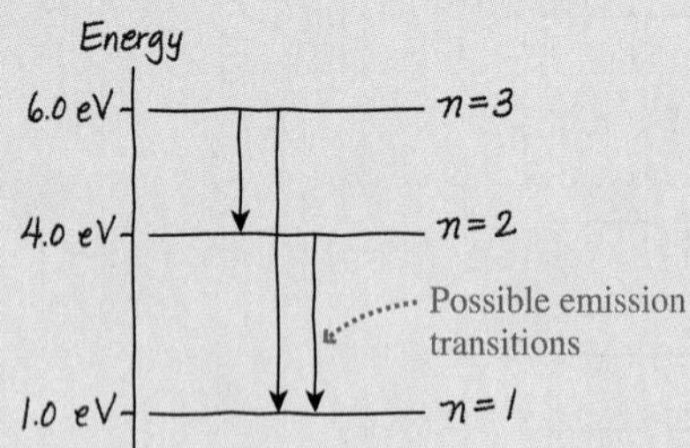

SOLVE This system will emit photons on the $3 \rightarrow 1$, $2 \rightarrow 1$, and $3 \rightarrow 2$ transitions, with $\Delta E_{3\rightarrow1} = 5.0$ eV, $\Delta E_{2\rightarrow1} = 3.0$ eV, and $\Delta E_{3\rightarrow2} = 2.0$ eV. From $f_{\text{photon}} = \Delta E_{\text{system}}/h$ and $\lambda = c/f$, we find that the wavelengths in the emission spectrum are

$3 \rightarrow 1$ $\quad f = 5.0\text{ eV}/h = 1.21 \times 10^{15}\text{ Hz}$

$\lambda = 250$ nm (ultraviolet)

$2 \rightarrow 1$ $\quad f = 3.0\text{ eV}/h = 7.25 \times 10^{14}\text{ Hz}$

$\lambda = 410$ nm (blue)

$3 \rightarrow 2$ $\quad f = 2.0\text{ eV}/h = 4.83 \times 10^{14}\text{ Hz}$

$\lambda = 620$ nm (orange)

ASSESS Transitions with a small energy difference, like $3 \rightarrow 2$, correspond to lower photon energies and thus longer wavelengths than transitions with a large energy difference like $3 \rightarrow 1$, as we would expect.

STOP TO THINK 28.7 A photon with a wavelength of 410 nm has energy $E_{\text{photon}} = 3.0$ eV. Do you expect to see a spectral line with $\lambda = 410$ nm in the emission spectrum of the system represented by this energy-level diagram?

Energy
5.0 eV — $n = 3$
3.0 eV — $n = 2$
1.0 eV — $n = 1$

28.7 The Uncertainty Principle

One of the strangest aspects of the quantum view of the world is an inherent limitation on our knowledge: **For a particle such as an electron, if you know where it is, you can't know exactly how fast it is moving.** This very counterintuitive notion is a result of the wave nature of matter and is worth a bit of explanation.

FIGURE 28.21 shows an experiment in which electrons moving along the y-axis pass through a slit of width a. Because of the wave nature of electrons, the slit causes them to spread out and produce a diffraction pattern.

But we can think of the experiment in a different way—as making a measurement of the position of the electrons. As an electron goes through the slit, we know something about its horizontal position. Our knowledge isn't perfect; we just know it is somewhere within the slit. We can establish an *uncertainty*, a limit on our knowledge. The uncertainty in the horizontal position is $\Delta x = a$, the width of the slit.

But, after passing through the slit, the electrons' waves spread out as they produce the diffraction pattern, causing the electrons to strike the screen over a range of positions. For this to happen, the electrons must have acquired a *horizontal* component of velocity v_x that varies from electron to electron. By sending the electrons through a slit—by trying to pin down their horizontal position—we've created an uncertainty in their horizontal velocity. Gaining knowledge of the *position* of the electrons has introduced uncertainty into our knowledge of the *velocity* of the electrons.

FIGURE 28.21 An experiment to illustrate the uncertainty principle.

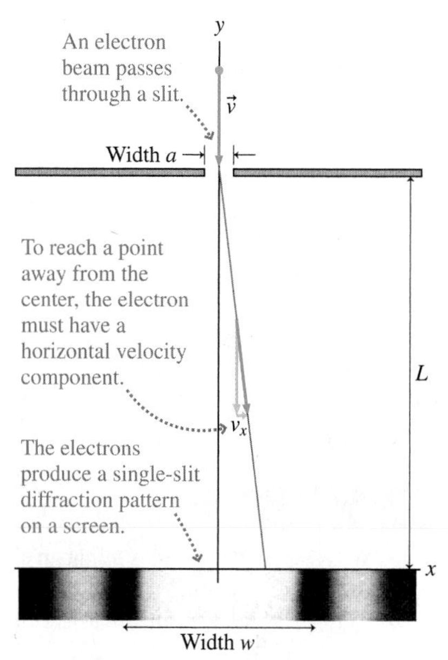

CONCEPTUAL EXAMPLE 28.14 Changing the uncertainty

Suppose we narrow the slit in the above experiment, allowing us to determine the electron's horizontal position more precisely. How does this affect the diffraction pattern? How does this change in the diffraction pattern affect the uncertainty in the velocity?

REASON We learned in Chapter 17 that the width of the central maximum of the single-slit diffraction pattern is $w = 2\lambda L/a$. Making the slit narrower—decreasing the value of a—increases the value of w, making the central fringe wider. If the fringe is wider, the spread of horizontal velocities must be greater, so there is a greater uncertainty in the horizontal velocity.

ASSESS Improving our knowledge of the position decreases our knowledge of the velocity. This is the hallmark of the *uncertainty principle*.

We've made this argument by considering a particular experiment, but it is an example of a general principle. In 1927, the German physicist Werner Heisenberg proved that, for any particle, the product of the uncertainty Δx in its position and the uncertainty Δp_x in its x-momentum has a lower limit fixed by the expression

$$\Delta x \, \Delta p_x \geq \frac{h}{4\pi} \tag{28.17}$$

Heisenberg uncertainty principle for position and momentum

A decreased uncertainty in position—knowing more precisely where a particle is—comes at the expense of an increased uncertainty in velocity and thus in momentum. But the relationship also goes the other way: Knowing a particle's velocity or momentum more precisely requires an increase in the uncertainty about its position.

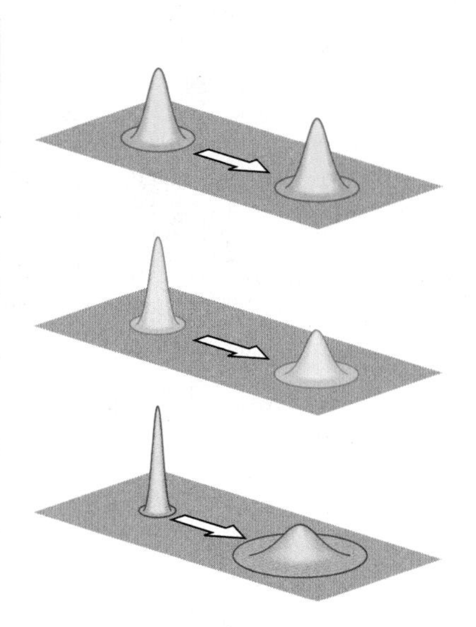

If I know where you are, I don't know where you're going In quantum physics, we represent particles by a *wave function* that describes their wave nature. This series of diagrams shows simulations of the evolution of three traveling wave functions. The top diagram shows the broad wave function of a particle whose position is not precisely defined. The uncertainty in momentum (and thus velocity) is small, so the wave function doesn't spread out much as it travels. The lower graphs show particles with more sharply peaked wave functions, implying less uncertainty in their initial positions. A reduced uncertainty in position means a larger uncertainty in velocity, so the wave functions spread out more quickly.

Uncertainties are associated with all experimental measurements, but better procedures and techniques can reduce those uncertainties. Classical physics places no limits on how small the uncertainties can be. A classical particle at any instant of time has an exact position x and an exact momentum p_x, and with sufficient care we can measure both x and p_x with such precision that we can make the product $\Delta x\,\Delta p_x$ as small as we like. There are no inherent limits to our knowledge.

In the quantum world, it's not so simple. No matter how clever you are, and no matter how good your experiment, you *cannot* measure both x and p_x simultaneously with arbitrarily good precision. Any measurements you make are limited by the condition that $\Delta x\,\Delta p_x \geq h/4\pi$. **The position and the momentum of a particle are *inherently* uncertain.**

Why? Because of the wave-like nature of matter! The "particle" is spread out in space, so there simply is not a precise value of its position x. Our belief that position and momentum have precise values is tied to our classical concept of a particle. As we revise our ideas of what atomic particles are like, we must also revise our ideas about position and momentum.

Let's revisit particles in a one-dimensional "box," now looking at uncertainties.

EXAMPLE 28.15 Determining uncertainties

a. What range of velocities might an electron have if confined to a 0.30-nm-wide region, about the size of an atom?

b. A 1.0-μm-diameter dust particle ($m \approx 10^{-15}$ kg) is confined within a 5-μm-long box. Can we know with certainty if the particle is at rest? If not, within what range is its velocity likely to be found?

PREPARE Localizing a particle means specifying its position with some accuracy—so there must be an uncertainty in the velocity. We can estimate the uncertainty by using Heisenberg's uncertainty principle.

SOLVE a. We aren't given the exact position of the particle, only that it is within a 0.30-nm-wide region. This means that we have specified the electron's position within a range $\Delta x = 3.0 \times 10^{-10}$ m. The uncertainty principle thus specifies that the least possible uncertainty in the momentum is

$$\Delta p_x = \frac{h}{4\pi\,\Delta x}$$

The uncertainty in the velocity is thus approximately

$$\Delta v_x = \frac{\Delta p_x}{m} \approx \frac{h}{4\pi m\,\Delta x} \approx 2 \times 10^5 \text{ m/s}$$

Because the *average* velocity is zero, (the particle is equally likely to be moving right or left) the best we can do is to say that the electron's velocity is somewhere in the interval $-1 \times 10^5 \text{ m/s} \leq v_x \leq 1 \times 10^5$ m/s. **It is simply not possible to specify the electron's velocity more precisely than this.**

b. We know the particle is somewhere in the box, so the uncertainty in our knowledge of its position is at most $\Delta x = L = 5\ \mu$m. With a finite Δx, the uncertainty Δp_x *cannot* be zero. **We cannot know with certainty if the particle is at rest inside the box.** No matter how hard we try to bring the particle to rest, the uncertainty in our knowledge of the particle's momentum will be approximately $\Delta p_x \approx h/(4\pi\,\Delta x) = h/(4\pi L)$. Consequently, the range of possible velocities is

$$\Delta v_x = \frac{\Delta p_x}{m} \approx \frac{h}{4\pi m L} \approx 1.0 \times 10^{-14} \text{ m/s}$$

This range of possible velocities will be centered on $v_x = 0$ m/s if we have done our best to have the particle be at rest. Therefore all we can know with certainty is that the particle's velocity is somewhere within the interval $-5 \times 10^{-15} \text{ m/s} \leq v_x \leq 5 \times 10^{-15}$ m/s.

ASSESS Our uncertainty about the electron's velocity is enormous. For an electron confined to a region of this size, the best we can do is to state that its speed is less than one million miles per hour! The uncertainty principle clearly sets real, practical limits on our ability to describe electrons. The situation for the dust particle is different. We can't say for certain that the particle is absolutely at rest. But knowing that its speed is less than 5×10^{-15} m/s means that the particle is at rest for all practical purposes. At this speed, the dust particle would require nearly 6 hours to travel the width of one atom! Again we see that the quantum view has profound implications at the atomic scale but need not affect the way we think of macroscopic objects.

NOTE ► In this text we will use the following convention for uncertainties: If the uncertainty calculated from Heisenberg's uncertainty principle is, say, Δx, this means that the possible values of x are in the range of $\pm\Delta x/2$ centered on the average value of x. ◄

STOP TO THINK 28.8 The speeds of an electron and a proton have been measured to the same uncertainty. Which one has a larger uncertainty in position?

A. The proton, because it's more massive. B. The electron, because it's less massive.
C. The uncertainty in position is the same, because the uncertainty in velocity is the same.

28.8 Applications and Implications of Quantum Theory

Quantum theory seems bizarre to those of us living at a scale where the rather different rules of classical physics apply. In this section we consider some of the implications of quantum theory and some applications that confirm these unusual notions.

Tunneling and the Scanning Tunneling Microscope

The fact that particles have a wave nature allows for imaging at remarkably small scales—the scale of single atoms! The *scanning tunneling microscope* doesn't work like other microscopes you have seen, but instead builds an image of a solid surface by scanning a probe near the surface.

FIGURE 28.22 shows the tip of a very, very thin metal needle, called the *probe tip,* positioned above the surface of a solid sample. The space between the tip and the surface is about 0.5 nm, only a few atomic diameters. Electrons in the sample are attracted to the positive probe tip, but no current should flow, according to classical physics, because the electrons cannot cross the gap between the sample and the probe; it is an incomplete circuit. As we found with the photoelectric effect, the electrons are bound inside the sample and not free to leave.

However, electrons are not classical particles. The electron has a wave nature, and waves don't have sharp edges. One startling prediction of Schrödinger's quantum mechanics is that the electrons' wave functions extend very slightly beyond the edge of the sample. When the probe tip comes close enough to the surface, close enough to poke into an electron's wave function, an electron that had been in the sample might suddenly find itself in the probe tip. In other words, a quantum electron *can* cross the gap between the sample and the probe tip, thus causing a current to flow in the circuit. This process is called **tunneling** because it is rather like tunneling through an uncrossable mountain barrier to get to the other side. Tunneling is completely forbidden by the laws of classical physics, so the fact that it occurs is a testament to the reality of quantum ideas.

The probability that an electron will tunnel across the gap is very sensitive to the size of the gap, which makes the **scanning tunneling microscope,** or STM, possible. When the probe tip passes over an atom or over an atomic-level bump on the surface, the gap narrows and the current increases as more electrons are able to tunnel across. Similarly, the tunneling probability decreases when the probe tip passes across an atomic-size valley, and the current falls. The current-versus-position data are used to construct an image of the surface.

The STM was the first technology that allowed imaging of individual atoms, and it was one of a handful of inventions in the 1980s that jump-started the current interest in nanotechnology. STM images offer a remarkable view of the world at an atomic scale. The STM image of **FIGURE 28.23a** clearly shows the hexagonal arrangement of the individual atoms on the surface of pyrolytic graphite. The image of a DNA molecule in **FIGURE 28.23b** shows the actual twists of the double-helix structure. Current research efforts aim to develop methods for sequencing DNA with scanning tunneling microscopes and other nanoprobes—to directly "read" a single strand of DNA!

FIGURE 28.22 The scanning tunneling microscope.

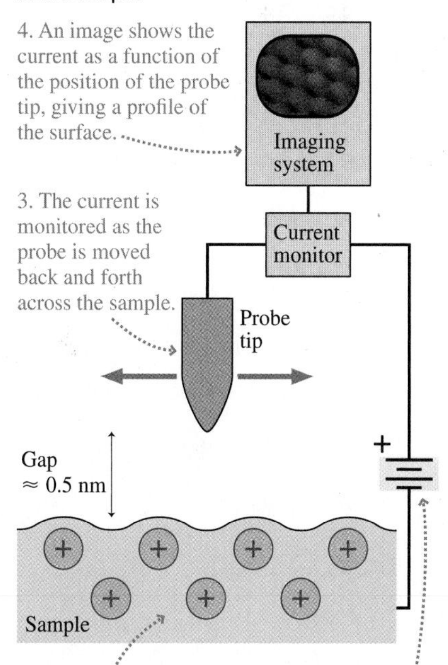
4. An image shows the current as a function of the position of the probe tip, giving a profile of the surface.
Imaging system
3. The current is monitored as the probe is moved back and forth across the sample.
Current monitor
Probe tip
Gap ≈ 0.5 nm
Sample
1. The sample can be modeled as positive ion cores in an electron "sea."
2. The small positive voltage of the probe causes electrons to tunnel across the narrow gap between the probe tip and the sample.

FIGURE 28.23 STM images.

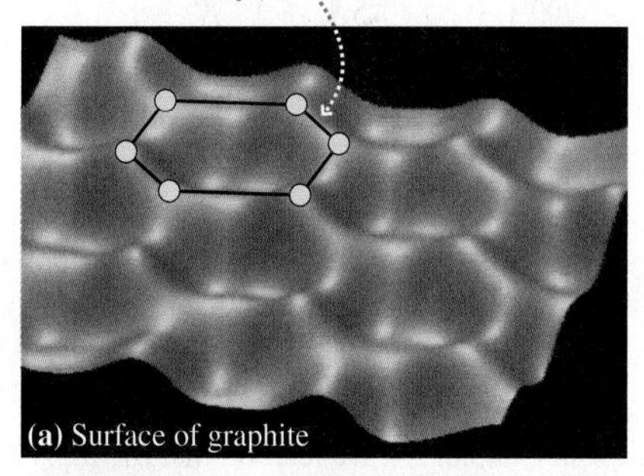
The hexagonal arrangement of atoms is clearly visible.
(a) Surface of graphite

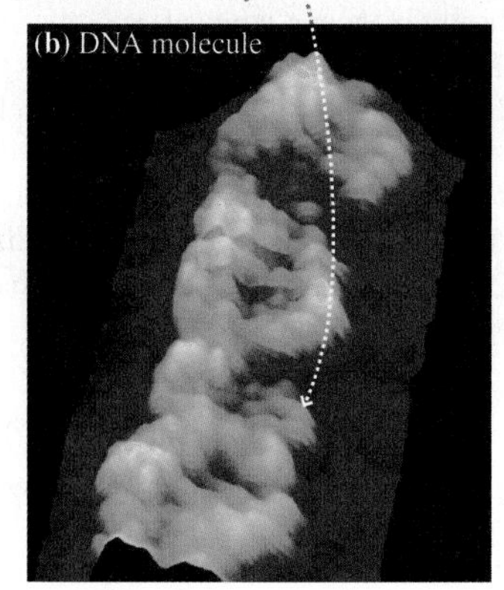
The double-helix structure of the DNA molecule is clearly visible.
(b) DNA molecule

Wave–Particle Duality

One common theme that has run through this chapter is the idea that, in quantum theory, things we think of as being waves have a particle nature, while things we think of as being particles have a wave nature. What is the true nature of light, or an electron? Are they particles or waves?

The various objects of classical physics are *either* particles *or* waves. There's no middle ground. Planets and baseballs are particles or collections of particles, while sound and light are clearly waves. Particles follow trajectories given by Newton's laws; waves obey the principle of superposition and exhibit interference. This wave–particle dichotomy seemed obvious until physicists encountered irrefutable evidence that light sometimes acts like a particle and, even stranger, that matter sometimes acts like a wave.

You might at first think that light and matter are *both* a wave *and* a particle, but that idea doesn't quite work. The basic definitions of particleness and waviness are mutually exclusive. Two sound waves can pass through each other and can overlap to produce a larger-amplitude sound wave; two baseballs can't. It is more profitable to conclude that light and matter are *neither* a wave *nor* a particle. At the microscopic scale of atoms and their constituents—a physical scale not directly accessible to our five senses—the classical concepts of particles and waves turn out to be simply too limited to explain the subtleties of nature.

Although matter and light have both wave-like aspects and particle-like aspects, they show us only one face at a time. If we arrange an experiment to measure a wave-like property, such as interference, we find photons and electrons acting like waves, not particles. An experiment to look for particles will find photons and electrons acting like particles, not waves. These two aspects of light and matter are *complementary* to each other, like a two-piece jigsaw puzzle. Neither the wave nor the particle model alone provides an adequate picture of light or matter, but taken together they provide us with a basis for understanding these elusive but most fundamental constituents of nature. This two-sided point of view is called *wave–particle duality.*

For over two hundred years, scientists and nonscientists alike felt that the clockwork universe of Newtonian physics was a fundamental description of reality. But wave–particle duality, along with Einstein's relativity, undermines the basic assumptions of the Newtonian worldview. The certainty and predictability of classical physics have given way to a new understanding of the universe in which chance and uncertainty play key roles—the universe of quantum physics.

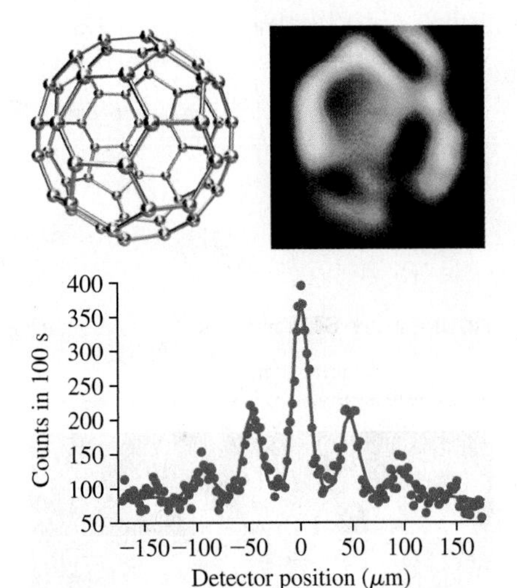

◂ The dual nature of a buckyball Treating atomic-level structures involves frequent shifts between particle and wave views. 60 carbon atoms can create the molecule diagrammed in the top-left image, known as C_{60}, or *buckminsterfullerene.* The scanning electron microscope image of a C_{60} molecule shown in the top-right image is a particle-like view of the molecule with individual carbon atoms clearly visible. The C_{60} molecule, though we can make a picture of it—showing the atoms that make it up—also has a wave nature. A beam of C_{60} sent through a grating will produce a diffraction pattern similar to those you studied for light waves, as the bottom graph shows.

INTEGRATED EXAMPLE 28.16 Magnetic resonance imaging BIO

In Chapter 24, we learned that the magnetism of permanent magnets arises because the inherent magnetic moment of electrons causes them to act like little compass needles. Protons also have an inherent magnetic moment, and this is the basis for magnetic resonance imaging (MRI) in medicine.

Although a compass needle would prefer to align with a magnetic field, the needle can point in *any* direction. This isn't the case for the magnetic moment of a proton. Quantum physics tells us that the proton's energy must be quantized. There are only two possible energy levels—and thus two possible orientations—for protons in a magnetic field:

$E_1 = -\mu B$ magnetic moment aligned with the field

$E_2 = +\mu B$ magnetic moment aligned opposite the field

where $\mu = 1.41 \times 10^{-26}$ J/T is the known value of the proton's magnetic moment. **FIGURE 28.24** shows the two possible energy states. The magnetic moment, like a compass needle, "wants" to align with the field, so that is the lower-energy state.

FIGURE 28.24 Energy levels for a proton in a magnetic field.

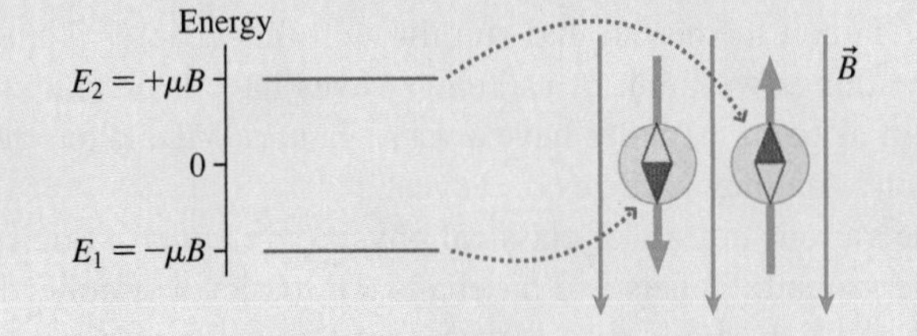

Human tissue is mostly water. Each water molecule has two hydrogen atoms whose nuclei are single protons. In a magnetic field, the protons go into one or the other quantum state. A photon of just the right energy can "flip" the orientation of a proton's magnetic moment by causing a quantum jump from one state to the other. The energy difference between the states is small, so the relatively low-frequency photons are in the radio portion of the electromagnetic spectrum. These photons are provided by a *probe coil* that emits radio waves. When the probe is tuned to just the right frequency, the waves are *in resonance* with the energy levels of the protons, thus giving us the name magnetic *resonance* imaging.

The rate of absorption of these low-energy photons is proportional to the density of hydrogen atoms. Hydrogen density varies with tissue type, so an MRI image—showing different tissues—is formed by measuring the variation across the body of the rate at which photons cause quantum jumps between the two proton energy levels. A figure showing the absorption rate versus position in the body is an image of a "slice" through a patient's body, as in **FIGURE 28.25**.

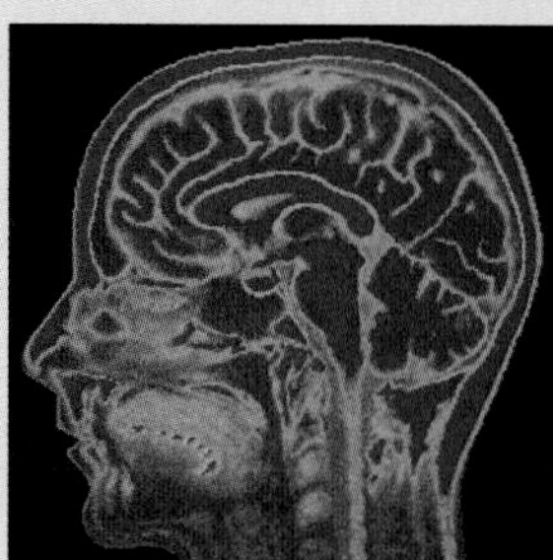

FIGURE 28.25 An MRI image shows the cross section of a patient's head.

a. An MRI patient is placed inside a solenoid that creates a strong magnetic field. If the field strength is 2.00 T, to what frequency must the probe coil be set? What is the wavelength of the photons produced?

b. In a uniform magnetic field, all protons in the body would absorb photons of the same frequency. To form an image of the body, the magnetic field is designed to vary from point to point in a known way. Because the field is different at each point in the body, each point has a unique frequency of photons that will be absorbed. The actual procedure is complex, but consider a simple model in which the field strength varies only along the axis of the patient's body, which we will call the x-axis. In particular, suppose that the magnetic field strength in tesla is given by $B = 2.00 + 1.60x$, where x, measured from a known reference point, is in meters. The probe coil is first tuned to the resonance frequency at the reference point. As the frequency is increased, a strong signal is observed at a frequency 4.7 MHz above the starting frequency. What is the location in the body, relative to the reference point, of the tissue creating this strong signal?

PREPARE If a photon has energy equal to the energy difference between the high- and low-energy states, it will be able to cause a quantum jump to the higher state—it will be absorbed. The photon energy E_{photon} must be equal to the energy difference between the two states: $\Delta E_{\text{system}} = 2\mu B$.

SOLVE a. At 2.00 T, the energy difference between the two proton states is

$$E_{\text{system}} = E_2 - E_1 = 2\mu B = 2(1.41 \times 10^{-26}\ \text{J/T})(2.00\ \text{T})$$
$$= 5.64 \times 10^{-26}\ \text{J}$$

This is a very low energy—only 3.5×10^{-7} eV. A photon will be absorbed if $E_{\text{photon}} = hf = E_{\text{system}}$. Thus the photon frequency must be

$$f = \frac{\Delta E_{\text{system}}}{h} = \frac{5.64 \times 10^{-26}\ \text{J}}{6.63 \times 10^{-34}\ \text{J} \cdot \text{s}}$$
$$= 85.1 \times 10^6\ \text{Hz} = 85.1\ \text{MHz}$$

This corresponds to a wavelength of $\lambda = c/f = 3.52$ m.

b. The magnetic field at the reference point ($x = 0$ m) is 2.00 T, so the probe frequency at this point is the 85.1 MHz we found in part a. The strong signal is 4.7 MHz above this, or 89.8 MHz. We can solve $E_{\text{photon}} = hf = E_{\text{system}} = 2\mu B$ to find the magnetic field at the point creating this strong signal:

$$B = \frac{hf}{2\mu} = 2.11\ \text{T}$$

We can then use the field-versus-distance formula given in the problem to find the position of this signal:

$$B = 2.11\ \text{T} = 2.00 + 1.60x$$
$$x = 0.069\ \text{m}$$

Thus there is a high density of protons 6.9 cm from the reference point.

ASSESS The frequency of the probe coil is in the radio portion of the electromagnetic spectrum, as we expected. The strong signal of part b is at a higher frequency, so this corresponds to a higher field and a positive value of x, as we found. The frequency is only slightly different from the original frequency, so we expect the point to be close to the reference position, as we found.

This is a simplified model of MRI, but the key features are present: A magnetic field that varies with position creates different energy levels for protons at different positions in the body, then tuned radio-wave photons measure the proton density at these different positions by causing and detecting quantum jumps between the two proton energy levels.

SUMMARY

Goal: To understand the quantization of energy for light and matter.

GENERAL PRINCIPLES

Light has particle-like properties

- The energy of a light wave comes in discrete packets (light quanta) we call **photons.**
- For light of frequency f, the energy of each photon is $E = hf$, where $h = 6.63 \times 10^{-34}\ \text{J}\cdot\text{s}$ is **Planck's constant**.
- When light strikes a metal surface, all of the energy of a single photon is given to a single electron.

Matter has wave-like properties

- The **de Broglie wavelength** of a particle of mass m is $\lambda = h/mv$.
- The wave-like nature of matter is seen in the interference patterns of electrons, protons, and other particles.

Quantization of energy

When a particle is confined, it sets up a de Broglie standing wave.

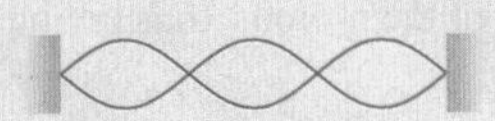

The fact that standing waves can have only certain allowed wavelengths leads to the conclusion that a confined particle can have only certain allowed energies.

Wave–particle duality

- Experiments designed to measure wave properties will show the wave nature of light and matter.
- Experiments designed to measure particle properties will show the particle nature of light and matter.

Heisenberg uncertainty principle

A particle with wave-like characteristics does not have a precise value of position x or a precise value of momentum p_x. Both are uncertain. The position uncertainty Δx and momentum uncertainty Δp_x are related by

$$\Delta x\, \Delta p_x \geq \frac{h}{4\pi}$$

The more you pin down the value of one, the less precisely the other can be known.

IMPORTANT CONCEPTS

Photoelectric effect

Light with frequency f can eject electrons from a metal only if $f \geq f_0 = E_0/h$, where E_0 is the metal's **work function.** Electrons will be ejected even if the intensity of the light is very small.

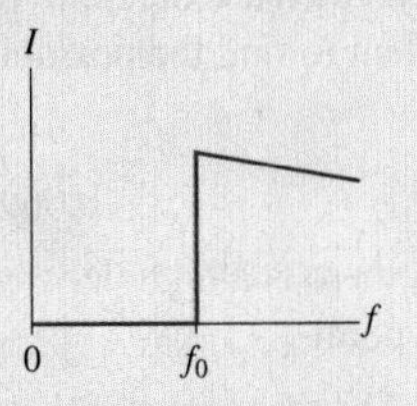

The **stopping potential** that stops even the fastest electrons is

$$V_{\text{stop}} = \frac{K_{\text{max}}}{e} = \frac{hf - E_0}{e}$$

The details of the photoelectric effect could not be explained with classical physics. New models were needed.

X-ray diffraction

X rays with wavelength λ undergo strong reflections from atomic planes spaced by d when the angle of incidence satisfies the **Bragg condition:**

$$2d\cos\theta = m\lambda$$

$$m = 1, 2, 3, \ldots$$

Energy levels and quantum jumps

The localization of electrons leads to quantized energy levels. An electron can exist only in certain energy states. An electron can jump to a higher level if a photon is absorbed, or to a lower level if a photon is emitted. The energy difference between the levels equals the photon energy.

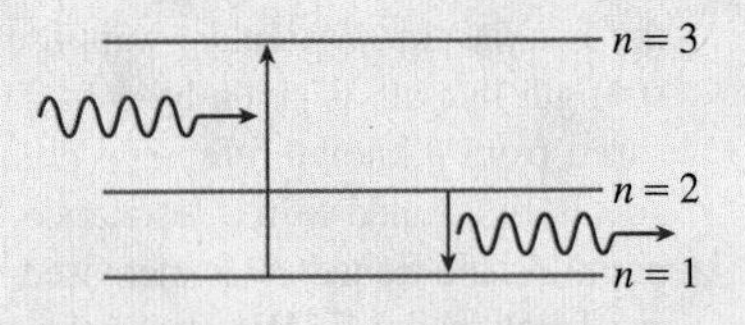

APPLICATIONS

The wave nature of light limits the resolution of a light microscope. A more detailed image may be made with an **electron microscope** because of the very small de Broglie wavelength of fast electrons.

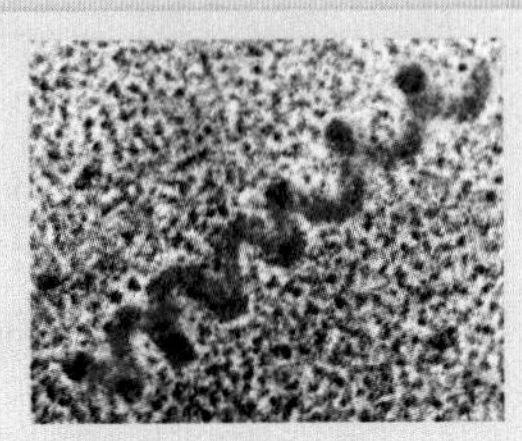

The wave nature of electrons allows them to **tunnel** across an insulating gap to the tip of a **scanning tunneling microscope,** revealing details of the atoms on a surface.

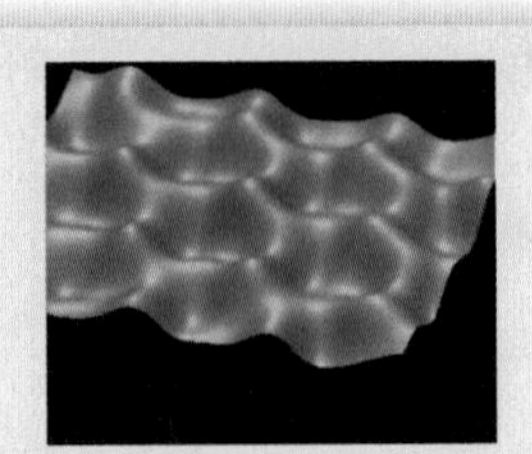

Problem difficulty is labeled as | (straightforward) to ||||| (challenging). Problems labeled INT integrate significant material from earlier chapters; BIO are of biological or medical interest.

For homework assigned on MasteringPhysics, go to MasteringPhysics®

Scan this QR code to launch a Video Tutor Solution that will help you solve problems for this chapter.

QUESTIONS

Conceptual Questions

1. The first-order x-ray diffraction of monochromatic x rays from a crystal occurs at angle θ_1. The crystal is then compressed, causing a slight reduction in its volume. Does θ_1 increase, decrease, or stay the same? Explain.
2. Explain the reasoning by which we claim that the stopping potential V_{stop} measures the maximum kinetic energy of the electrons in a photoelectric-effect experiment.
3. How does Einstein's explanation account for each of these characteristics of the photoelectric effect?
   a. The photoelectric current is zero for frequencies below some threshold.
   b. The photoelectric current increases with increasing light intensity.
   c. The photoelectric current is independent of ΔV for $\Delta V \gtrsim 1$V.
   d. The photoelectric current decreases slowly as ΔV becomes more negative.
   e. The stopping potential is independent of the light intensity.
   Which of these *cannot* be explained by classical physics? Explain.
4. A current is detected in a photoelectric-effect experiment when the cathode is illuminated with green light. Will a current necessarily be detected if the cathode is illuminated with blue light? With red light?
5. Figure Q28.5 shows the typical photoelectric behavior of a metal as the anode-cathode potential difference ΔV is varied.
   a. Why do the curves become horizontal for $\Delta V \gtrsim 1$V? Shouldn't the current increase as the potential difference increases? Explain.
   b. Why doesn't the current immediately drop to zero for $\Delta V < 0$ V? Shouldn't $\Delta V < 0$ V prevent the electrons from reaching the anode? Explain.
   c. The current is zero for $\Delta V < -2.0$ V. Where do the electrons go? Are no electrons emitted if $\Delta V < -2.0$ V? Or if they are, why is there no current? Explain.

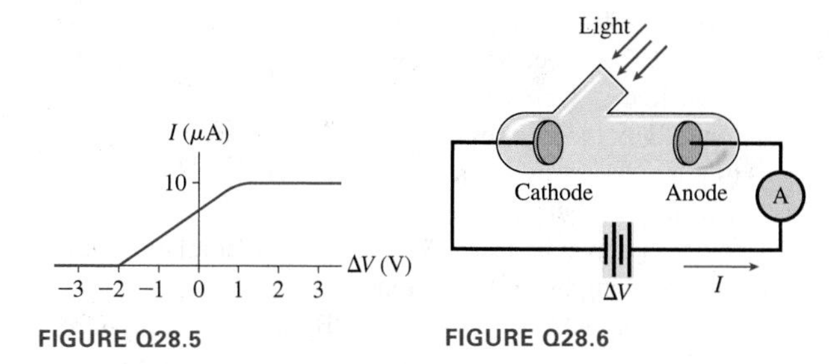

FIGURE Q28.5

FIGURE Q28.6

6. In the photoelectric effect experiment, as illustrated by Figure Q28.6, a current is measured while light is shining on the cathode. But this does not appear to be a complete circuit, so how can there be a current? Explain.
7. Metal surfaces on spacecraft in bright sunlight develop a net electric charge. Do they develop a negative or a positive charge? Explain.
8. Metal 1 has a larger work function than metal 2. Both are illuminated with the same short-wavelength ultraviolet light. Do electrons from metal 1 have a higher speed, a lower speed, or the same speed as electrons from metal 2? Explain.
9. A gold cathode is illuminated with light of wavelength 250 nm. It is found that the current is zero when $\Delta V = 1.0$ V. Would the current change if
   a. The light intensity is doubled?
   b. The anode-cathode potential difference is increased to $\Delta V = 5.5$ V?
10. Three laser beams have wavelengths $\lambda_1 = 400$ nm, $\lambda_2 = 600$ nm, and $\lambda_3 = 800$ nm. The power of each laser beam is 1 W.
    a. Rank in order, from largest to smallest, the photon energies E_1, E_2, and E_3 in these three laser beams. Explain.
    b. Rank in order, from largest to smallest, the number of photons per second N_1, N_2, and N_3 delivered by the three laser beams. Explain.
11. When we say that a photon is a "quantum of light," what does that mean? What is quantized?
12. A solar cell is illuminated with 2.0 W of light. Is the maximum possible current greater with 2.0 W of red light or 2.0 W of blue light?
13. An investigator is measuring the current in a photoelectric effect experiment. The cathode is illuminated by light of a single wavelength. What happens to the current if the wavelength of the light is reduced by a factor of two while keeping the intensity constant?
14. To have the best resolution, should an electron microscope use very fast electrons or very slow electrons? Explain.
15. An electron and a proton are accelerated from rest through potential differences of the same magnitude. Afterward, which particle has the larger de Broglie wavelength? Explain.
16. A neutron is shot straight up with an initial speed of 100 m/s. As it rises, does its de Broglie wavelength increase, decrease, or not change? Explain.
17. Electrons are accelerated from rest through an 8000 V potential difference. By what factor would their de Broglie wavelength increase if they were instead accelerated through a 2000 V potential?
18. Can an electron with a de Broglie wavelength of 2 μm pass through a slit that is 1 μm wide? Explain.
19. a. For the allowed energies of a particle in a box to be large, should the box be very big or very small? Explain.
    b. Which is likely to have larger values for the allowed energies: an atom in a molecule, an electron in an atom, or a proton in a nucleus? Explain.

20. Figure Q28.20 shows the standing de Broglie wave of a particle in a box.
    a. What is the quantum number?
    b. Can you determine from this picture whether the "classical" particle is moving to the right or to the left? If so, which is it? If not, why not?

FIGURE Q28.20

21. A particle in a box of length L_a has $E_1 = 2$ eV. The same particle in a box of length L_b has $E_2 = 50$ eV. What is the ratio L_a/L_b?
22. Imagine that the horizontal box of Figure 28.18 is instead oriented vertically. Also imagine the box to be on a neutron star where the gravitational field is so strong that the particle in the box slows significantly, nearly stopping, before it hits the top of the box. Make a *qualitative* sketch of the $n = 3$ de Broglie standing wave of a particle in this box.
    Hint: The nodes are *not* uniformly spaced.
23. Figure Q28.23 shows a standing de Broglie wave.
    a. Does this standing wave represent a particle that travels back and forth between the boundaries with a constant speed or a changing speed? Explain.
    b. If the speed is changing, at which end is the particle moving faster and at which end is it moving slower?

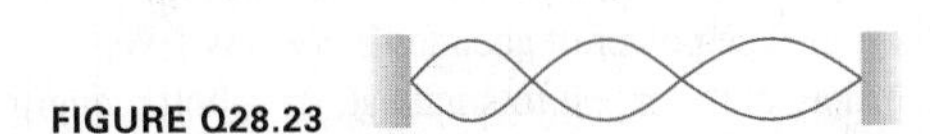
FIGURE Q28.23

24. The molecules in the rods and cones in the eye are tuned to BIO absorb photons of particular energies. The retinal molecule, like many molecules, is a long chain. Electrons can freely move along one stretch of the chain but are reflected at the ends, thus behaving like a particle in a one-dimensional box. The absorption of a photon lifts an electron from the ground state into the first excited state. Do the molecules in a red cone (which are tuned to absorb red light) or the molecules in a blue cone (tuned to absorb blue light) have a longer "box"?
25. Science fiction movies often use devices that transport people and objects rapidly from one position to another. To "beam" people in this fashion means taking them apart atom by atom, carefully measuring each position, and then sending the atoms in a beam to the desired final location where they reassemble. How do the principles of quantum mechanics pose problems for this futuristic means of transportation?

Multiple-Choice Questions

26. | A light sensor is based on a photodiode that requires a minimum photon energy of 1.7 eV to create mobile electrons. What is the longest wavelength of electromagnetic radiation that the sensor can detect?
    A. 500 nm B. 730 nm
    C. 1200 nm D. 2000 nm
27. | In a photoelectric effect experiment, the frequency of the light is increased while the intensity is held constant. As a result,
    A. There are more electrons. B. The electrons are faster.
    C. Both A and B. D. Neither A nor B.
28. | In a photoelectric effect experiment, the intensity of the light is increased while the frequency, which is above the threshold frequency, is held constant. As a result,
    A. There are more electrons. B. The electrons are faster.
    C. Both A and B. D. Neither A nor B.
29. | In the photoelectric effect, electrons are never emitted from a metal if the frequency of the incoming light is below a certain threshold value. This is because
    A. Photons of lower-frequency light don't have enough energy to eject an electron.
    B. The electric field of low-frequency light does not vibrate the electrons rapidly enough to eject them.
    C. The number of photons in low-frequency light is too small to eject electrons.
    D. Low-frequency light does not penetrate far enough into the metal to eject electrons.
30. || Visible light has a wavelength of about 500 nm. A typical radio wave has a wavelength of about 1.0 m. How many photons of the radio wave are needed to equal the energy of one photon of visible light?
    A. 2,000 B. 20,000
    C. 200,000 D. 2,000,000
31. | Two radio stations have the same power output from their antennas. One broadcasts AM at a frequency of 1000 kHz and one broadcasts FM at a frequency of 100 MHz. Which statement is true?
    A. The FM station emits more photons per second.
    B. The AM station emits more photons per second.
    C. The two stations emit the same number of photons per second.
32. | Light consisting of 2.7 eV photons is incident on a piece of potassium, which has a work function of 2.3 eV. What is the maximum kinetic energy of the ejected electrons?
    A. 2.3 eV B. 2.7 eV
    C. 5.0 eV D. 0.4 eV
33. || How many photons does a 5.0 mW helium-neon laser ($\lambda = 633$ nm) emit in 1 second?
    A. 1.2×10^{19} B. 4.0×10^{18}
    C. 8.0×10^{16} D. 1.6×10^{16}
34. | You shoot a beam of electrons through a double slit to make an interference pattern. After noting the properties of the pattern, you then double the speed of the electrons. What effect would this have?
    A. The fringes would get closer together.
    B. The fringes would get farther apart.
    C. The positions of the fringes would not change.
35. | Photon P in Figure Q28.35 moves an electron from energy level $n = 1$ to energy level $n = 3$. The electron jumps down to $n = 2$, emitting photon Q, and then jumps down to $n = 1$, emitting photon R. The spacing between energy levels is drawn to scale. What is the correct relationship among the wavelengths of the photons?

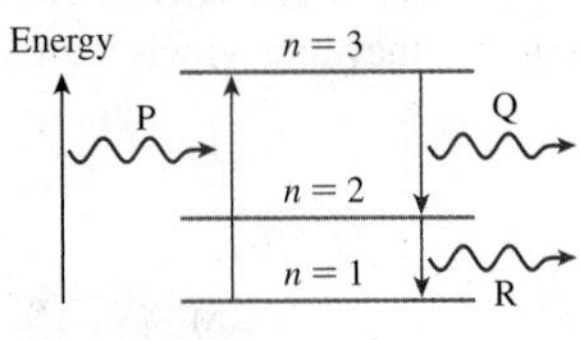

FIGURE Q28.35

    A. $\lambda_P < \lambda_Q < \lambda_R$ B. $\lambda_R < \lambda_P < \lambda_Q$
    C. $\lambda_Q < \lambda_P < \lambda_R$ D. $\lambda_R < \lambda_Q < \lambda_P$

PROBLEMS

Section 28.1 X Rays and X-Ray Diffraction

1. | X rays with a wavelength of 0.12 nm undergo first-order diffraction from a crystal at a 68° angle of incidence. What is the angle of second-order diffraction?
2. | X rays with a wavelength of 0.20 nm undergo first-order diffraction from a crystal at a 54° angle of incidence. At what angle does first-order diffraction occur for x rays with a wavelength of 0.15 nm?
3. | X rays diffract from a crystal in which the spacing between atomic planes is 0.175 nm. The second-order diffraction occurs at 45.0°. What is the angle of the first-order diffraction?
4. || The spacing between atomic planes in a crystal is 0.110 nm. If 12.0 keV x rays are diffracted by this crystal, what are the angles of (a) first-order and (b) second-order diffraction?
5. || X rays with a wavelength of 0.085 nm diffract from a crystal in which the spacing between atomic planes is 0.18 nm. How many diffraction orders are observed?

Section 28.2 The Photoelectric Effect

6. | Which metals in Table 28.1 exhibit the photoelectric effect for (a) light with $\lambda = 400$ nm and (b) light with $\lambda = 250$ nm?
7. | Electrons are emitted when a metal is illuminated by light with a wavelength less than 388 nm but for no greater wavelength. What is the metal's work function?
8. || Electrons in a photoelectric-effect experiment emerge from a copper surface with a maximum kinetic energy of 1.10 eV. What is the wavelength of the light?
9. || A metal surface is illuminated by light with a wavelength of 350 nm. The maximum kinetic energy of the emitted electrons is found to be 1.50 eV. What is the maximum electron kinetic energy if the same metal is illuminated by light with a wavelength of 250 nm?
10. ||| Light with a wavelength of 375 nm illuminates a metal cathode. The maximum kinetic energy of the emitted electrons is 0.76 eV. What is the longest wavelength of light that will cause electrons to be emitted from this cathode?
11. || You need to design a photodetector that can respond to the entire range of visible light. What is the maximum possible work function of the cathode?
12. || A photoelectric-effect experiment finds a stopping potential of 1.93 V when light of 200 nm wavelength is used to illuminate the cathode.
    a. From what metal is the cathode made?
    b. What is the stopping potential if the intensity of the light is doubled?
13. || Zinc has a work function of 4.3 eV.
    a. What is the longest wavelength of light that will release an electron from a zinc surface?
    b. A 4.7 eV photon strikes the surface and an electron is emitted. What is the maximum possible speed of the electron?
14. || Image intensifiers used in night-vision devices create a bright image from dim light by letting the light first fall on a *photocathode*. Electrons emitted by the photoelectric effect are accelerated and then strike a phosphorescent screen, causing it to glow more brightly than the original scene. Recent devices are sensitive to wavelengths as long as 900 nm, in the infrared.

    a. If the threshold wavelength is 900 nm, what is the work function of the photocathode?
    b. If light of wavelength 700 nm strikes such a photocathode, what will be the maximum kinetic energy, in eV, of the emitted electrons?
15. || Light with a wavelength of 350 nm shines on a metal surface, which emits electrons. The stopping potential is measured to be 1.25 V.
    a. What is the maximum speed of emitted electrons?
    b. Calculate the work function and identify the metal.

Section 28.3 Photons

16. | When an ultraviolet photon is absorbed by a molecule of DNA, the photon's energy can be converted into vibrational energy of the molecular bonds. Excessive vibration damages the molecule by causing the bonds to break. Ultraviolet light of wavelength less than 290 nm causes significant damage to DNA; ultraviolet light of longer wavelength causes minimal damage. What is the threshold photon energy, in eV, for DNA damage? BIO
17. | The spacing between atoms in graphite is approximately 0.25 nm. What is the energy of an x-ray photon with this wavelength?
18. || A firefly glows by the direct conversion of chemical energy to light. The light emitted by a firefly has peak intensity at a wavelength of 550 nm. BIO

    a. What is the minimum chemical energy, in eV, required to generate each photon?
    b. One molecule of ATP provides 0.30 eV of energy when it is metabolized in a cell. What is the minimum number of ATP molecules that must be consumed in the reactions that lead to the emission of one photon of 550 nm light?
19. | Your eyes have three different types of cones with maximum absorption at 437 nm, 533 nm, and 564 nm. What photon energies correspond to these wavelengths? BIO
20. | What is the wavelength, in nm, of a photon with energy (a) 0.30 eV, (b) 3.0 eV, and (c) 30 eV? For each, is this wavelength visible light, ultraviolet, or infrared?
21. | The laser in a DVD player emits light at a wavelength of 640 nm. The laser in a Blu-Ray player emits light at 405 nm. What are the photon energies of each laser, in eV?
22. || The wavelengths of light emitted by a firefly span the visible spectrum but have maximum intensity near 550 nm. A typical flash lasts for 100 ms and has a power of 1.2 mW. If we assume that all of the light is emitted at the peak-intensity wavelength of 550 nm, how many photons are emitted in one flash? BIO INT
23. || Station KAIM in Hawaii broadcasts on the AM dial at 870 kHz, with a maximum power of 50,000 W. At maximum power, how many photons does the transmitting antenna emit each second?

Watch video solutions for problems: 28.27 28.43 

24. ||| At 510 nm, the wavelength of maximum sensitivity of the human eye, the dark-adapted eye can sense a 100-ms-long flash of light of total energy 4.0×10^{-17} J. (Weaker flashes of light may be detected, but not reliably.) If 60% of the incident light is lost to reflection and absorption by tissues of the eye, how many photons reach the retina from this flash?
BIO
25. | 550 nm is the average wavelength of visible light.
    a. What is the energy of a photon with a wavelength of 550 nm?
    b. A typical incandescent lightbulb emits about 1 J of visible light energy every second. Estimate the number of visible photons emitted per second.
26. || *Dinoflagellates* are single-cell creatures that float in the world's oceans; many types are bioluminescent. When disturbed by motion in the water, a typical bioluminescent dinoflagellate emits 100,000,000 photons in a 0.10-s-long flash of light of wavelength 460 nm. What is the power of the flash in watts?
BIO

27. || A circuit employs a silicon solar cell to detect flashes of light lasting 0.25 s. The smallest current the circuit can detect reliably is 0.42 μA. Assuming that all photons reaching the solar cell give their energy to a charge carrier, what is the minimum power of a flash of light of wavelength 550 nm that can be detected?

Section 28.4 Matter Waves

28. | Estimate your de Broglie wavelength while walking at a speed of 1 m/s.
29. | a. What is the de Broglie wavelength of a 200 g baseball with a speed of 30 m/s?
    b. What is the speed of a 200 g baseball with a de Broglie wavelength of 0.20 nm?
30. | a. What is the speed of an electron with a de Broglie wavelength of 0.20 nm?
    b. What is the speed of a proton with a de Broglie wavelength of 0.20 nm?
31. || What is the kinetic energy, in eV, of an electron with a de Broglie wavelength of 1.0 nm?
32. || A paramecium is covered with motile hairs called cilia that propel it at a speed of 1 mm/s. If the paramecium has a volume of 2×10^{-13} m^3 and a density equal to that of water, what is its de Broglie wavelength when in motion? What fraction of the paramecium's 150 μm length does this wavelength represent?
BIO INT

33. || The diameter of an atomic nucleus is about 10 fm (1 fm $= 10^{-15}$ m). What is the kinetic energy, in MeV, of a proton with a de Broglie wavelength of 10 fm?
34. || Rubidium atoms are cooled to 0.10 μK in an atom trap. What is their de Broglie wavelength? How many times larger is this than the 0.25 nm diameter of the atoms?
INT
35. || Through what potential difference must an electron be accelerated from rest to have a de Broglie wavelength of 500 nm?

Section 28.5 Energy Is Quantized

36. || What is the length of a box in which the minimum energy of an electron is 1.5×10^{-18} J?
37. ||| What is the length of a one-dimensional box in which an electron in the $n = 1$ state has the same energy as a photon with a wavelength of 600 nm?
38. ||| An electron confined in a one-dimensional box is observed, at different times, to have energies of 12 eV, 27 eV, and 48 eV. What is the length of the box?
39. | The nucleus of a typical atom is 5.0 fm (1 fm $= 10^{-15}$ m) in diameter. A very simple model of the nucleus is a one-dimensional box in which protons are confined. Estimate the energy of a proton in the nucleus by finding the first three allowed energies of a proton in a 5.0-fm-long box.

Section 28.6 Energy Levels and Quantum Jumps

40. || The allowed energies of a quantum system are 1.0 eV, 2.0 eV, 4.0 eV, and 7.0 eV. What wavelengths appear in the system's emission spectrum?
41. || Figure P28.41 is an energy-level diagram for a quantum system. What wavelengths appear in the system's emission spectrum?

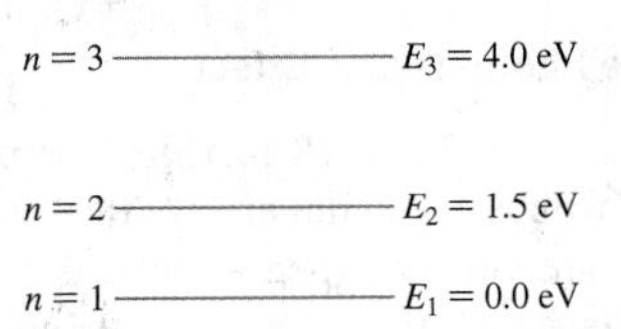

FIGURE P28.41

42. ||| The allowed energies of a quantum system are 0.0 eV, 4.0 eV, and 6.0 eV.
    a. Draw the system's energy-level diagram. Label each level with the energy and the quantum number.
    b. What wavelengths appear in the system's emission spectrum?
43. || A quantum system has three energy levels, so three wavelengths appear in its emission spectrum. The shortest observed wavelength is 248 nm; light with a 414 nm wavelength is also observed. What is the third wavelength?
44. ||| The allowed energies of a quantum system are 0.0 eV, 1.5 eV, 3.0 eV, and 6.0 eV. How many different wavelengths appear in the emission spectrum?

Section 28.7 The Uncertainty Principle

45. || The speed of an electron is known to be between 3.0×10^6 m/s and 3.2×10^6 m/s. Estimate the smallest possible uncertainty in its position.
46. || What is the smallest box in which you can confine an electron if you want to know for certain that the electron's speed is no more than 10 m/s?
47. ||| A spherical virus has a diameter of 50 nm. It is contained inside a long, narrow cell of length 1×10^{-4} m. What uncertainty does this imply for the velocity of the virus along the length of the cell? Assume the virus has a density equal to that of water.
BIO INT
48. ||| A thin solid barrier in the xy-plane has a 10-μm-diameter circular hole. An electron traveling in the z-direction with $v_x = 0$ m/s passes through the hole. Afterward, is v_x still zero? If not, within what range is v_x likely to be?
49. ||| A proton is confined within an atomic nucleus of diameter 4 fm (1 fm $= 10^{-15}$ m). Estimate the smallest range of speeds you might find for a proton in the nucleus.

General Problems

50. || X rays with a wavelength of 0.0700 nm diffract from a crystal. Two adjacent angles of x-ray diffraction are 45.6° and 21.0°. What is the distance in nm between the atomic planes responsible for the diffraction?

51. || Potassium and gold cathodes are used in a photoelectric-effect experiment. For each cathode, find:
    a. The threshold frequency
    b. The threshold wavelength
    c. The maximum electron ejection speed if the light has a wavelength of 220 nm
    d. The stopping potential if the wavelength is 220 nm

52. |||| In a photoelectric-effect experiment, the maximum kinetic energy of electrons is 2.8 eV. When the wavelength of the light is increased by 50%, the maximum energy decreases to 1.1 eV. What are (a) the work function of the cathode and (b) the initial wavelength?

53. |||| In a photoelectric-effect experiment, the stopping potential at a wavelength of 400 nm is 25.7% of the stopping potential at a wavelength of 300 nm. Of what metal is the cathode made?

54. || Light of constant intensity but varying wavelength was used to illuminate the cathode in a photoelectric-effect experiment. The graph of Figure P28.54 shows how the stopping potential depended on the frequency of the light. What is the work function, in eV, of the cathode?

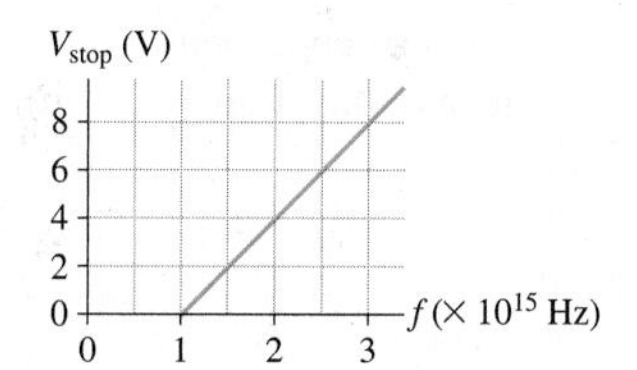

FIGURE P28.54

55. || What is the de Broglie wavelength of a red blood cell with a mass of 1.00×10^{-11} g that is moving with a speed of 0.400 cm/s? Do we need to be concerned with the wave nature of the blood cells when we describe the flow of blood in the body? BIO

56. || Suppose you need to image the structure of a virus with a diameter of 50 nm. For a sharp image, the wavelength of the probing wave must be 5.0 nm or less. We have seen that, for imaging such small objects, this short wavelength is obtained by using an electron beam in an electron microscope. Why don't we simply use short-wavelength electromagnetic waves? There's a problem with this approach: As the wavelength gets shorter, the energy of a photon of light gets greater and could damage or destroy the object being studied. Let's compare the energy of a photon and an electron that can provide the same resolution. BIO
    a. For light of wavelength 5.0 nm, what is the energy (in eV) of a single photon? In what part of the electromagnetic spectrum is this?
    b. For an electron with a de Broglie wavelength of 5.0 nm, what is the kinetic energy (in eV)?

57. | Gamma rays are photons with very high energy.
    a. What is the wavelength of a gamma-ray photon with energy 625 keV?
    b. How many visible-light photons with a wavelength of 500 nm would you need to match the energy of this one gamma-ray photon?

58. | A red laser with a wavelength of 650 nm and a blue laser with a wavelength of 450 nm emit laser beams with the same light power. What is the ratio of the red laser's photon emission rate (photons per second) to the blue laser's photon emission rate?

59. || A typical incandescent lightbulb emits approximately 3×10^{18} visible-light photons per second. Your eye, when it is fully dark adapted, can barely see the light from an incandescent lightbulb 10 km away. How many photons per second are incident at the image point on your retina? The diameter of a dark-adapted pupil is 6 mm. BIO INT

60. || The intensity of sunlight hitting the surface of the earth on a cloudy day is about 0.50 kW/m^2. Assuming your pupil can close down to a diameter of 2.0 mm and that the average wavelength of visible light is 550 nm, how many photons per second of visible light enter your eye if you look up at the sky on a cloudy day? BIO

61. ||| A red LED (light emitting diode) is connected to a battery; it carries a current. As electrons move through the diode, they jump between states, emitting photons in the process. Assume that each electron that travels through the diode causes the emission of a single 630 nm photon. What current is necessary to produce 5.0 mW of emitted light? INT

62. || A ruby laser emits an intense pulse of light that lasts a mere 10 ns. The light has a wavelength of 690 nm, and each pulse has an energy of 500 mJ.
    a. How many photons are emitted in each pulse?
    b. What is the *rate* of photon emission, in photons per second, during the 10 ns that the laser is "on"?

63. ||| In a laser range-finding experiment, a pulse of laser light is fired toward an array of reflecting mirrors left on the moon by Apollo astronauts. By measuring the time it takes for the pulse to travel to the moon, reflect off the mirrors, and return to earth, scientists can measure the distance to the moon to an accuracy of a few centimeters. The light pulses are 100 ps long, and the laser wavelength is 532 nm. When the pulse reaches the moon, it has an intensity of 300 W/m^2. How many photons strike a single 4.0-cm-diameter mirror from one laser pulse? INT

64. ||| The wavelength of the radiation in a microwave oven is 12 cm. How many photons are absorbed by 200 g of water as it's heated from 20°C to 90°C? INT

65. ||| Exposure to a sufficient quantity of ultraviolet light will redden the skin, producing *erythema*—a sunburn. The amount of exposure necessary to produce this reddening depends on the wavelength. For a 1.0 cm^2 patch of skin, 3.7 mJ of ultraviolet light at a wavelength of 254 nm will produce reddening; at 300 nm wavelength, 13 mJ are required. BIO
    a. What is the photon energy corresponding to each of these wavelengths?
    b. How many total photons does each of these exposures correspond to?
    c. Explain why there is a difference in the number of photons needed to provoke a response in the two cases.

66. ||| A silicon solar cell behaves like a battery with a 0.50 V terminal voltage. Suppose that 1.0 W of light of wavelength 600 nm falls on a solar cell and that 50% of the photons give their energy to charge carriers, creating a current. What is the solar cell's efficiency—that is, what percentage of the energy incident on the cell is converted to electric energy? INT

67. |||| What is the kinetic energy in eV of an electron whose de Broglie wavelength is 10% of the wavelength of a photon of the same energy?

68. ||| Electrons with a speed of 2.0×10^6 m/s pass through a double-slit apparatus. Interference fringes are detected with a fringe spacing of 1.5 mm.
INT
    a. What will the fringe spacing be if the electrons are replaced by neutrons with the same speed?
    b. What speed must neutrons have to produce interference fringes with a fringe spacing of 1.5 mm?
69. ||| Electrons pass through a 200-nm-diameter circular aperture, forming a diffraction pattern on a detector 1.2 m behind the aperture. What is the kinetic energy of the electrons if the width of the central maximum is 0.37 mm?
INT
70. | The electron interference pattern of Figure 28.14 was made by shooting electrons with 50 keV of kinetic energy through two slits spaced 1.0 μm apart. The fringes were recorded on a detector 1.0 m behind the slits.
INT
    a. What was the speed of the electrons? (The speed is large enough to justify using relativity, but for simplicity do this as a nonrelativistic calculation.)
    b. Figure 28.14 is greatly magnified. What was the actual spacing on the detector between adjacent bright fringes?
71. |||| It is stated in the text that special relativity must be used to calculate the de Broglie wavelength of electrons in an electron microscope. Let us discover how much of an effect relativity has. Consider an electron accelerated through a potential difference of 1.00×10^5 V.
INT
    a. Using the Newtonian (nonrelativistic) expressions for kinetic energy and momentum, what is the electron's de Broglie wavelength?
    b. The de Broglie wavelength is $\lambda = h/p$, but the momentum of a relativistic particle is not mv. Using the relativistic expressions for kinetic energy and momentum, what is the electron's de Broglie wavelength?
72. || An electron confined to a one-dimensional box of length 0.70 nm jumps from the $n = 2$ level to the ground state. What is the wavelength (in nm) of the emitted photon?
73. | a. What is the minimum energy of a 2.7 g Ping-Pong ball in a 10-cm-long box?
    b. What speed corresponds to this kinetic energy?
74. ||| The color of dyes results from the preferential absorption of certain wavelengths of light. Certain dye molecules consist of symmetric pairs of rings joined at the center by a chain of carbon atoms, as shown in Figure P28.74. Electrons of the bonds along the chain of carbon atoms are shared among the atoms in the chain, but are repelled by the nitrogen-containing rings at the end of the chain. These electrons are thus free to move along the chain but not beyond its ends. They look very much like a particle in a one-dimensional box. For the molecule shown, the effective length of the "box" is 0.85 nm. Assuming that the electrons start in the lowest energy state, what are the three longest wavelengths this molecule will absorb?

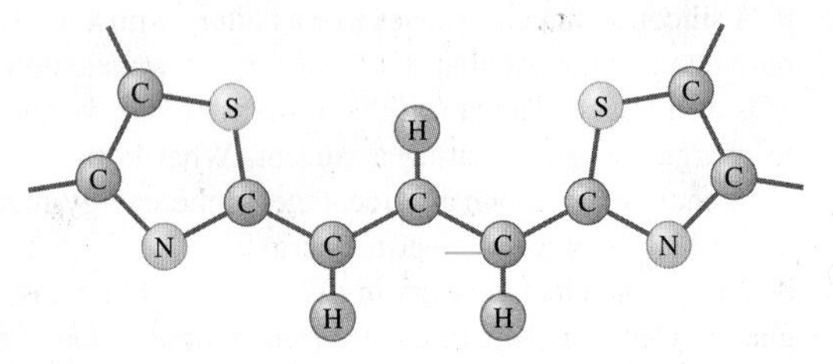

FIGURE P28.74

75. || What is the length of a box in which the difference between an electron's first and second allowed energies is 1.0×10^{-19} J?
76. || Two adjacent allowed energies of an electron in a one-dimensional box are 2.0 eV and 4.5 eV. What is the length of the box?
77. ||| An electron confined to a box has an energy of 1.28 eV. Another electron confined to an identical box has an energy of 2.88 eV. What is the smallest possible length for those boxes?
78. || Consider a small virus having a diameter of 10 nm. The atoms of the intracellular fluid are confined within this "box." Suppose we model the virus as a one-dimensional box of length 10 nm. What is the ground-state energy (in eV) of a sodium ion confined in such a box?
BIO INT
79. ||| It can be shown that the allowed energies of a particle of mass m in a two-dimensional square box of side L are

$$E_{nl} = \frac{h^2}{8mL^2}(n^2 + l^2)$$

The energy depends on two quantum numbers, n and l, both of which must have an integer value 1, 2, 3, ...
    a. What is the minimum energy for a particle in a two-dimensional square box of side L?
    b. What are the five lowest allowed energies? Give your values as multiples of E_{min}.
80. ||| An electron confined in a one-dimensional box emits a 200 nm photon in a quantum jump from $n = 2$ to $n = 1$. What is the length of the box?
81. ||| A proton confined in a one-dimensional box emits a 2.0 MeV gamma-ray photon in a quantum jump from $n = 2$ to $n = 1$. What is the length of the box?
82. ||| As an electron in a one-dimensional box of length 0.600 nm jumps between two energy levels, a photon of energy 8.36 eV is emitted. What are the quantum numbers of the two levels?
83. || Magnetic resonance is used in imaging; it is also a useful tool for analyzing chemical samples. Magnets for magnetic resonance experiments are often characterized by the proton resonance frequency they create. What is the field strength of an 800 MHz magnet?
84. ||| The electron has a magnetic moment, so you can do magnetic resonance measurements on substances with unpaired electron spins. The electron has a magnetic moment $\mu = 9.3 \times 10^{-24}$ J/T. A sample is placed in a solenoid of length 15 cm with 1200 turns of wire carrying a current of 3.5 A. A probe coil provides radio waves to "flip" the spins. What is the necessary frequency for the probe coil?
INT

MCAT-Style Passage Problems

Compton Scattering

Further support for the photon model of electromagnetic waves comes from *Compton scattering,* in which x rays scatter from electrons, changing direction and frequency in the process. Classical electromagnetic wave theory cannot explain the change in frequency of the x rays on scattering, but the photon model can.

Suppose an x-ray photon is moving to the right. It has a collision with a slow-moving electron, as in Figure P28.85. The photon transfers energy and momentum to the electron, which recoils at a high speed. The x-ray photon loses energy, and the photon energy formula $E = hf$ tells us that its frequency must decrease. The collision looks very much like the collision between two particles.

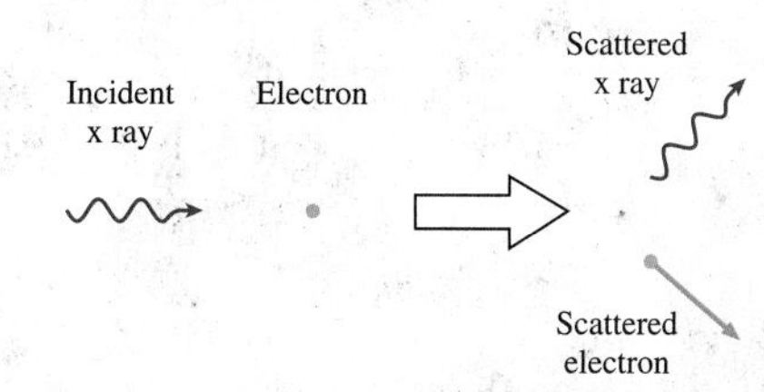

FIGURE P28.85

85. | When the x-ray photon scatters from the electron,
A. Its speed increases.
B. Its speed decreases.
C. Its speed stays the same.

86. | When the x-ray photon scatters from the electron,
A. Its wavelength increases.
B. Its wavelength decreases.
C. Its wavelength stays the same.

87. | When the electron is struck by the x-ray photon,
A. Its de Broglie wavelength increases.
B. Its de Broglie wavelength decreases.
C. Its de Broglie wavelength stays the same.

88. | X-ray diffraction can also change the direction of a beam of x rays. Which statement offers the best comparison between Compton scattering and x-ray diffraction?
A. X-ray diffraction changes the wavelength of x rays; Compton scattering does not.
B. Compton scattering changes the speed of x rays; x-ray diffraction does not.
C. X-ray diffraction relies on the particle nature of the x rays; Compton scattering relies on the wave nature.
D. X-ray diffraction relies on the wave nature of the x rays; Compton scattering relies on the particle nature.

STOP TO THINK ANSWERS

Chapter Preview Stop to Think: D. The fringes result from the interference of the waves spreading out from each of the two slits. With only one slit there is no interference, and only a broad, spread-out pattern is observed.

Stop to Think 28.1: B. The Bragg condition $2d\cos\theta_1 = \lambda$ tells us that larger values of d go with larger values of θ_1.

Stop to Think 28.2: $V_A > V_B > V_C$. For a given wavelength of light, electrons are ejected faster from metals with smaller work functions because it takes less energy to remove an electron. Faster electrons need a larger negative voltage to stop them.

Stop to Think 28.3: C. Photons always travel at c, and a photon's energy depends on only the light's frequency, not its intensity. Greater intensity means more energy each second, which means more photons.

Stop to Think 28.4: A. The widest diffraction pattern occurs for the largest wavelength. The de Broglie wavelength is inversely proportional to the particle's mass, and so will be largest for the least massive particle.

Stop to Think 28.5: B. The quantum number n is related to the wavelength as $\lambda_n = 2L/n$. The wave shown in the figure has a wavelength equal to L, so for this wave $n = 2$. The lowest energy is the ground state with $n = 1$. According to Equation 28.15, the $n = 1$ state has 1/4 the energy of the $n = 2$ state, or 2.0 eV.

Stop to Think 28.6: B, D, and E. The emitted photons' energies must be the difference in energies between two allowed energy levels. Thus $E_3 - E_2 = 50$ eV, $E_4 - E_3 = 70$ eV, and $E_3 - E_1 = 80$ eV are possible photon energies. 40 eV and 60 eV do not correspond to the difference between any two levels.

Stop to Think 28.7: No. The energy of an emitted photon is the energy *difference* between two allowed energies. The three possible quantum jumps have energy differences of 2.0 eV, 2.0 eV, and 4.0 eV.

Stop to Think 28.8: B. Because $\Delta p_x = m\,\Delta v_x$, the uncertainty in position is $\Delta x = \dfrac{h}{\Delta p_x} = \dfrac{h}{m\,\Delta v_x}$. A more massive particle has a smaller position uncertainty.

29 Atoms and Molecules

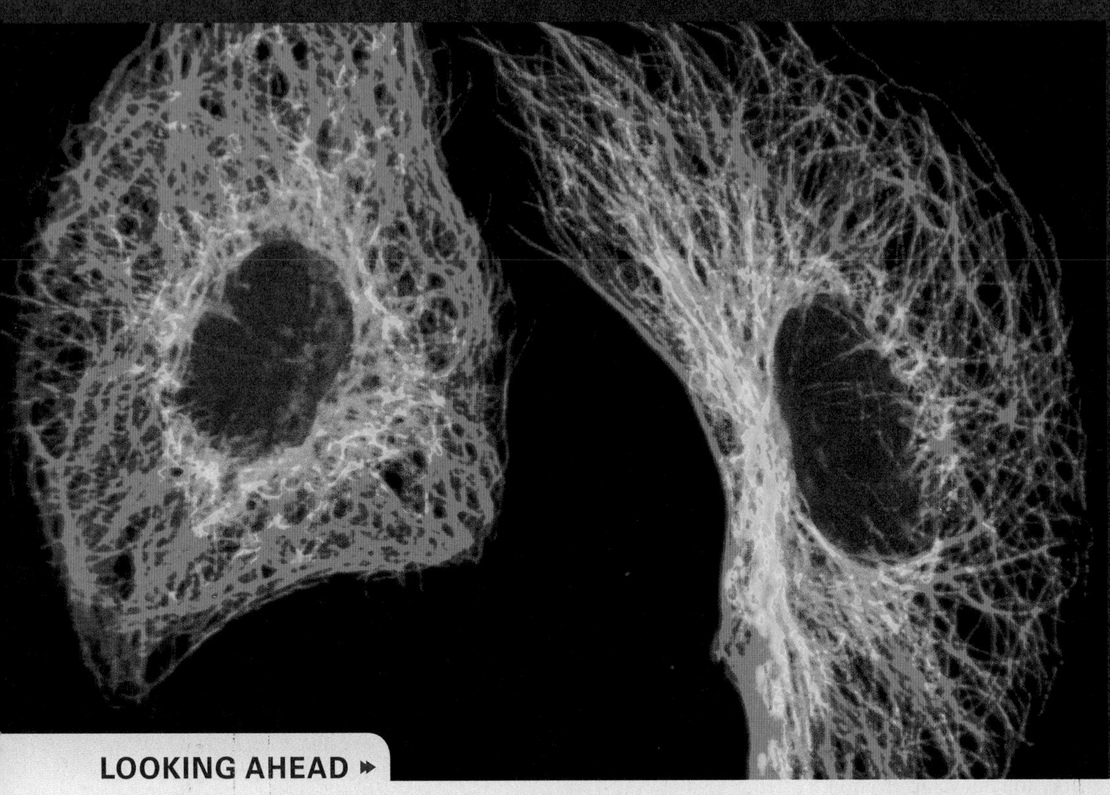

This microscopic image is a kidney cell from an African green monkey. The blue and green show, respectively, nuclear DNA and microtubules. What makes the different parts of this cell so brightly colored?

LOOKING AHEAD ▸▸

Goal: To use quantum physics to understand the properties of atoms, molecules, and their spectra.

Atomic Models

When light from a glowing gas is dispersed into colors by a diffraction grating, each element emits a unique set of wavelengths.

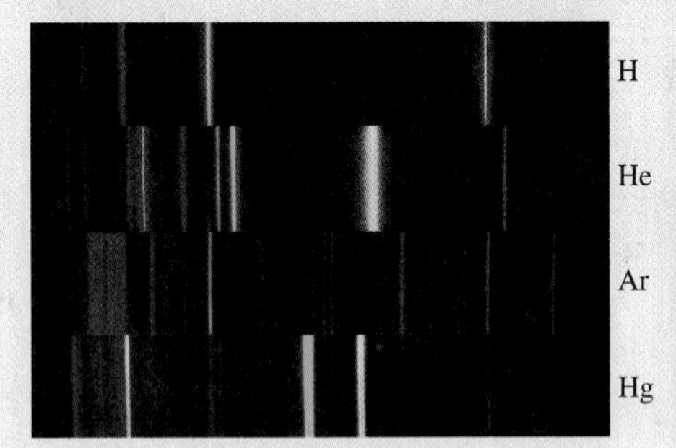

You'll learn how these **spectral lines** are important clues into the nature of the atom.

Molecules

Specialized molecules in certain microscopic plankton can emit light. These glowing waves are caused by such *bioluminescent* creatures.

You'll learn why, unlike atoms, the light given off by molecules contains a broad *band* of wavelengths.

Lasers

The intense beams of light at this laser light show result from the **stimulated emission** of photons by atoms.

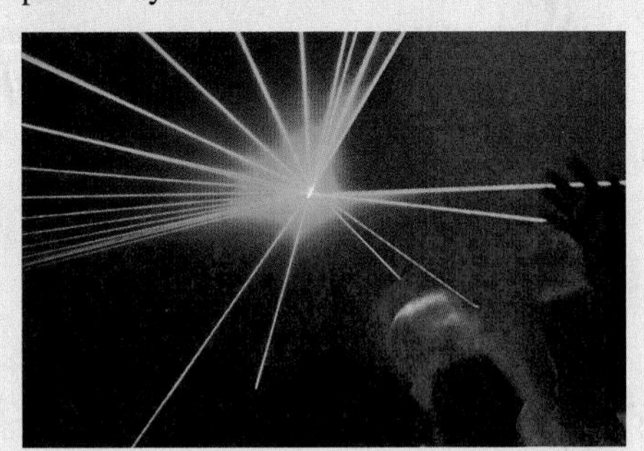

You'll learn that when atoms give off light, they can stimulate other atoms to emit light of exactly the same wavelength and phase.

LOOKING BACK ◂◂

Energy-Level Diagrams

In Sections 28.5 and 28.6 you learned that the energies of atomic-sized systems are *quantized*. The allowed energies of such a system can be represented in an energy-level diagram.

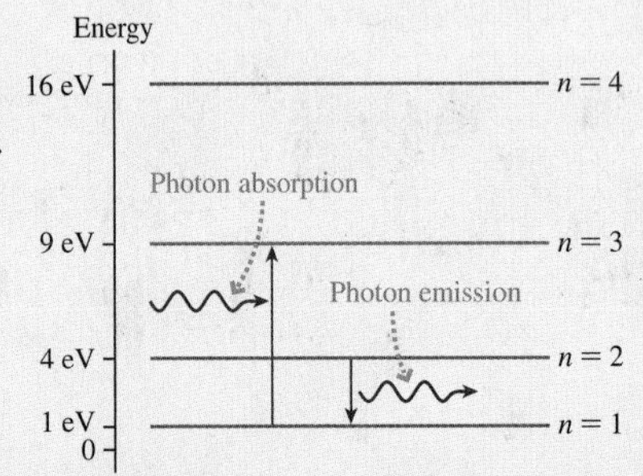

In this chapter, we'll use the ideas of quantization to understand in detail the energy levels and spectra of atoms and molecules.

STOP TO THINK

After the $n = 1$ to $n = 3$ absorption shown in the energy-level diagram to the left, which is a possible energy of an emitted photon?

A. 1 eV
B. 4 eV
C. 5 eV
D. 9 eV

29.1 Spectroscopy

The interference and diffraction of light were well understood by the end of the 19th century, and the knowledge was used to design practical tools for measuring wavelengths with great accuracy. The primary instrument for measuring the wavelengths of light is a **spectrometer,** such as the one shown in FIGURE 29.1. The heart of a spectrometer is a diffraction grating that causes different wavelengths of light to diffract at different angles.

Each wavelength is focused to a different position on the photographic plate or, more likely today, a CCD detector like the one in your digital camera. The distinctive pattern of wavelengths emitted by a source of light and recorded on the detector is called the **spectrum** of the light.

As you learned in ◀◀SECTION 25.7, hot, self-luminous objects, such as the sun or the filament of an incandescent lightbulb, emit a **continuous spectrum** in which a rainbow is formed by light being emitted at every possible wavelength. As you learned, this spectrum depends on only the object's temperature and thus contains no information about the atoms that the object is made of.

To learn about the light emitted and absorbed by individual atoms, we need to investigate them in the form of a low-pressure gas, where the atoms are far apart and isolated from one another. If a high voltage is applied to two electrodes sealed in a glass tube filled with a low-pressure gas, the gas begins to glow. In contrast to a hot solid object like a lightbulb filament, the light emitted by such a *gas discharge tube* contains only certain discrete, individual wavelengths. Such a spectrum is called a **discrete spectrum.**

FIGURE 29.2 shows examples of discrete spectra as they would appear on the detector of a spectrometer. Each bright line in a discrete spectrum, called a **spectral line,** represents *one* specific wavelength present in the light emitted by the source. A discrete spectrum is sometimes called a *line spectrum* because of its appearance on the detector. The familiar neon sign is actually a gas discharge tube containing neon, and you can see that a neon sign has its reddish-orange color because, as Figure 29.2 shows, nearly all of the wavelengths emitted by neon atoms fall within the wavelength range 600–700 nm that we perceive as orange and red.

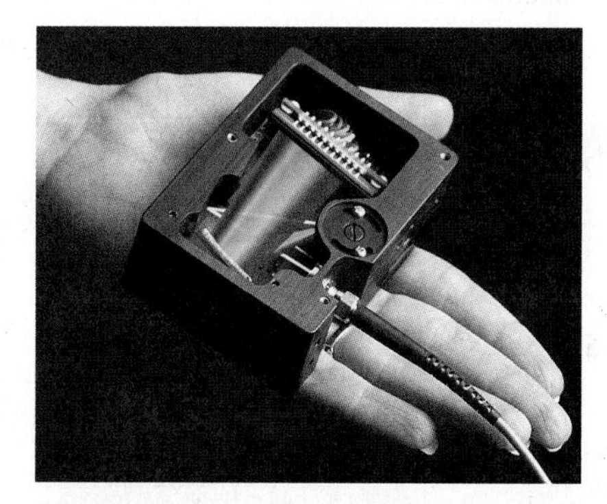

Some modern spectrometers are small enough to hold in your hand. (The rainbow has been added to show the paths that different colors take.)

FIGURE 29.1 A diffraction spectrometer.

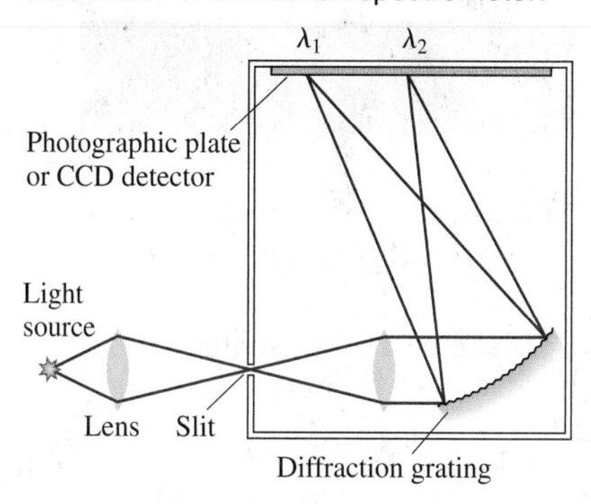

FIGURE 29.2 Examples of spectra in the visible wavelength range 400–700 nm.

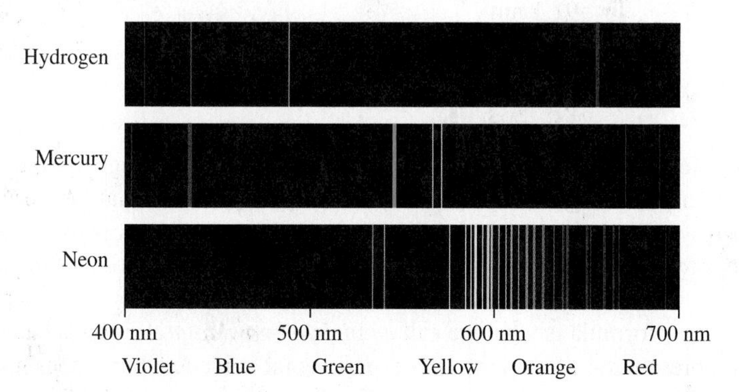

Figure Video

Figure 29.2 shows that the *atomic spectra* of hydrogen, mercury, and neon look very different from one another. In fact, every element in the periodic table has its own, unique discrete spectrum. The fact that each element emits a unique spectrum means that atomic spectra can be used as "fingerprints" to identify elements. Consequently, atomic spectroscopy is the basis of many contemporary technologies for analyzing the composition of unknown materials, monitoring air pollutants, and studying the atmospheres of the earth and other planets.

FIGURE 29.3 Measuring an absorption spectrum.

(a) Measuring an absorption spectrum

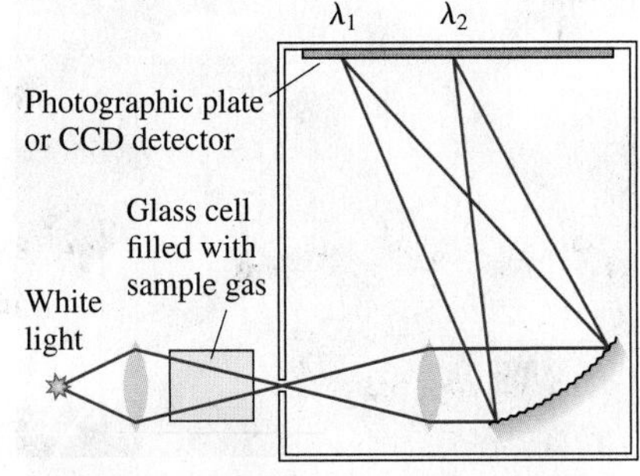

(b) Absorption and emission spectra of sodium

Absorption

Emission

300 nm 400 nm 500 nm 600 nm 700 nm

Ultraviolet Visible

TABLE 29.1 Wavelengths of visible lines in the hydrogen spectrum

656 nm
486 nm
434 nm
410 nm

Gases can also *absorb* light. In an absorption experiment, as shown in **FIGURE 29.3a**, a white-light source emits a continuous spectrum that, in the absence of a gas, uniformly illuminates the detector. When a sample of gas is placed in the light's path, any wavelengths absorbed by the gas are missing and the detector is dark at that wavelength.

Gases not only emit discrete wavelengths, but also absorb discrete wavelengths. But there is an important difference between the emission spectrum and the absorption spectrum of a gas: **Every wavelength that is absorbed by the gas is also emitted, but *not* every emitted wavelength is absorbed.** The wavelengths in the absorption spectrum appear as a *subset* of the wavelengths in the emission spectrum. As an example, **FIGURE 29.3b** shows both the absorption and the emission spectra of sodium atoms. All of the absorption wavelengths are prominent in the emission spectrum, but there are many emission lines for which no absorption occurs.

What causes atoms to emit or absorb light? Why a discrete spectrum? Why are some wavelengths emitted but not absorbed? Why is each element different? Nineteenth-century physicists struggled with these questions but could not answer them. Ultimately, classical physics was simply incapable of providing an understanding of atoms.

The only encouraging sign came from an unlikely source. While the spectra of other atoms have dozens or even hundreds of wavelengths, the visible spectrum of hydrogen, between 400 nm and 700 nm, consists of a mere four spectral lines (see Figure 29.2 and Table 29.1). If any spectrum could be understood, it should be that of the first element in the periodic table. The breakthrough came in 1885, not by an established and recognized scientist but by a Swiss school teacher, Johann Balmer. Balmer showed that the wavelengths in the hydrogen spectrum could be represented by the simple formula

$$\lambda = \frac{91.1 \text{ nm}}{\left(\dfrac{1}{2^2} - \dfrac{1}{n^2}\right)} \qquad n = 3, 4, 5, \ldots \tag{29.1}$$

Later experimental evidence, as ultraviolet and infrared spectroscopy developed, showed that Balmer's result could be generalized to

$$\lambda = \frac{91.1 \text{ nm}}{\left(\dfrac{1}{m^2} - \dfrac{1}{n^2}\right)} \tag{29.2}$$

m can be 1, 2, 3, . . .
n can be any integer greater than *m*.

Astronomical colors The red color of this nebula is due to the emission of light from hydrogen atoms. The atoms are excited by intense ultraviolet light from the star in the center. They then emit red light, with $\lambda = 656$ nm, as predicted by the Balmer formula with $m = 2$ and $n = 3$.

We now refer to Equation 29.2 as the **Balmer formula,** although Balmer himself suggested only the original version in which $m = 2$. Other than at the very highest levels of resolution, where new details appear that need not concern us in this text, the Balmer formula accurately describes *every* wavelength in the emission spectrum of hydrogen.

The Balmer formula is what we call *empirical knowledge.* It is an accurate mathematical representation found through experimental evidence, but it does not rest on any physical principles or physical laws. Yet the formula was so simple that it must, everyone agreed, have a simple explanation. It would take 30 years to find it.

STOP TO THINK 29.1 The black lines show the emission or absorption lines observed in two spectra of the same element. Which one is an emission spectrum and which is an absorption spectrum?

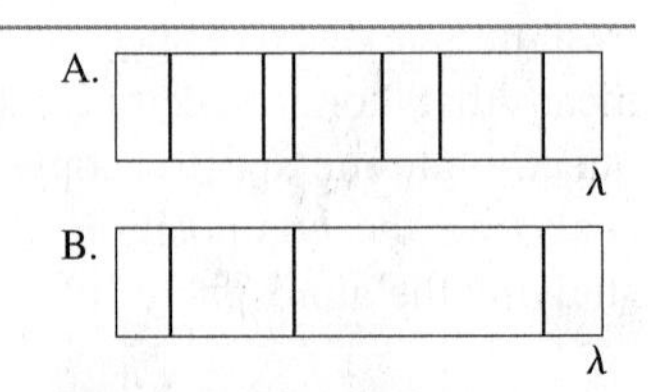

29.2 Atoms

It was the ancient Greeks who first had the idea of atoms as indivisible units of matter, but experimental evidence for atoms didn't appear until Dalton, Avogadro, and others began to formulate the laws of chemistry in the early 19th century. The existence of atoms with diameters of approximately 10^{-10} m was widely accepted by 1890, but it was still unknown if atoms were indivisible little spheres or if they had some kind of internal structure.

The 1897 discovery of the electron by J. J. Thomson had two important implications. First, atoms are not indivisible; they are built of smaller pieces. The electron was the first *subatomic* particle to be discovered. And second, the constituents of the atom are *charged particles*. Hence it seems plausible that the atom must be held together by electric forces.

Within a few years, measurements of the electron's mass m_e revealed that the electron is much less massive than even the smallest atom. Because the electrons are very small and light compared to the whole atom, it seemed reasonable to think that the positively charged part (protons were not yet known) had most of the mass and would take up most of the space. Thomson suggested that the atom consists of a spherical "cloud" of positive charge, roughly 10^{-10} m in diameter, in which the smaller negative electrons are embedded. The positive charge exactly balances the negative, so the atom as a whole has no net charge. This model of the atom has been called the "plum-pudding model" or the "raisin-cake model" for reasons that should be clear from the picture of **FIGURE 29.4**. However, Thomson's model of the atom did not stand the tests of time.

FIGURE 29.4 Thomson's raisin-cake model of the atom.

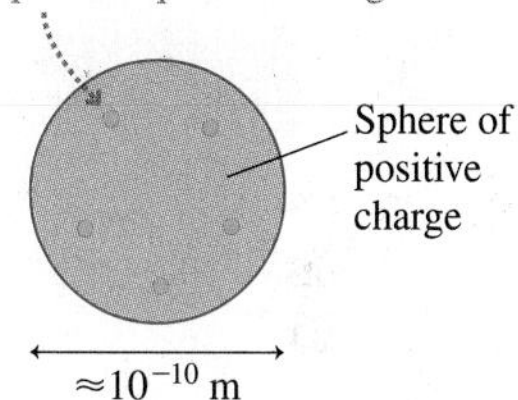
Thomson proposed that small negative electrons are embedded in a sphere of positive charge.
Sphere of positive charge
≈10^{-10} m

Almost simultaneously with Thomson's discovery of the electron, the French physicist Henri Becquerel announced his discovery that some new form of "rays" were emitted by crystals of uranium. These rays, like x rays, could expose film and pass through objects, but they were emitted continuously from the uranium without having to "do" anything to it.

One of Thomson's former students, Ernest Rutherford, began a study of these new rays and quickly discovered that a uranium crystal actually emits two *different* rays. **Beta rays** were eventually found to be high-speed electrons emitted by the uranium. **Alpha rays** (or alpha particles, as we now call them) consist of helium nuclei, with mass $m = 6.64 \times 10^{-27}$ kg, emitted at high speed from the sample.

It had been a shock to discover that atoms are not indivisible—they have an inner structure. Now, with the discovery of radioactivity, it appeared that some atoms were not even stable but could spit out various kinds of charged particles!

EXAMPLE 29.1 The speed of an alpha particle

Alpha particles are usually characterized by their kinetic energy in MeV. What is the speed of an 8.3 MeV alpha particle?

SOLVE Recall that 1 eV is the energy acquired by an electron accelerating through a 1 V potential difference, with the conversion $1.00 \text{ eV} = 1.60 \times 10^{-19}$ J. First, we convert the energy to joules:

$$K = 8.3 \times 10^6 \text{ eV} \times \frac{1.60 \times 10^{-19} \text{ J}}{1.00 \text{ eV}} = 1.3 \times 10^{-12} \text{ J}$$

Now, using the alpha-particle mass $m = 6.64 \times 10^{-27}$ kg given above, we can find the speed:

$$K = \frac{1}{2}mv^2 = 1.3 \times 10^{-12} \text{ J}$$

$$v = \sqrt{\frac{2K}{m}} = 2.0 \times 10^7 \text{ m/s}$$

ASSESS This is quite fast, about 7% of the speed of light.

The First Nuclear Physics Experiment

Rutherford soon realized that he could use these high-speed alpha particles as projectiles to probe inside other atoms. In 1909, Rutherford and his students set up the experiment shown in **FIGURE 29.5** to shoot alpha particles through very thin metal

FIGURE 29.5 Rutherford's experiment to shoot high-speed alpha particles through a thin gold foil.

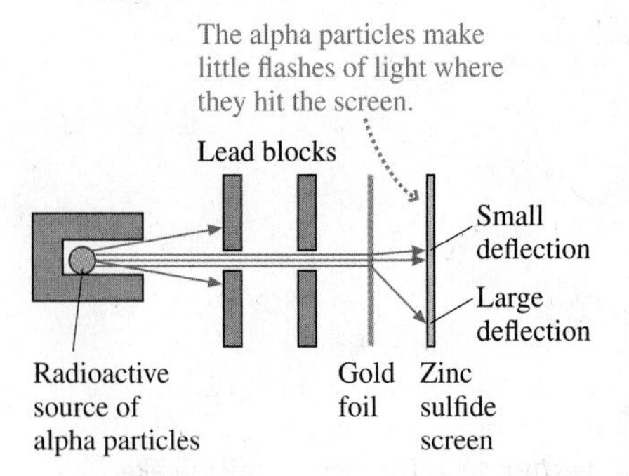

foils. The alpha particle is charged, and it experiences electric forces from the positive and negative charges of the atoms as it passes through the foil. According to Thomson's raisin-cake model of the atom, the forces exerted on the alpha particle by the positive atomic charges should roughly cancel the forces from the negative electrons, causing the alpha particles to experience only slight deflections. Indeed, this was the experimenters' initial observation.

At Rutherford's suggestion, his students then set up the apparatus to see if any alpha particles were deflected at *large* angles. It took only a few days to find the answer. Not only were alpha particles deflected at large angles, but a very few were reflected almost straight backward toward the source!

How can we understand this result? **FIGURE 29.6a** shows that an alpha particle passing through a Thomson atom would experience only a small deflection. But if an atom has a small positive core, such as the one in **FIGURE 29.6b**, a few of the alpha particles can come very close to the core. Because the electric force varies with the inverse square of the distance, the very large force of this very close approach can cause a large-angle scattering or even a backward deflection of the alpha particle.

FIGURE 29.6 Alpha particles interact differently with a concentrated positive nucleus than they would with the spread-out charges in Thomson's model.

(a) Thomson model

Alpha

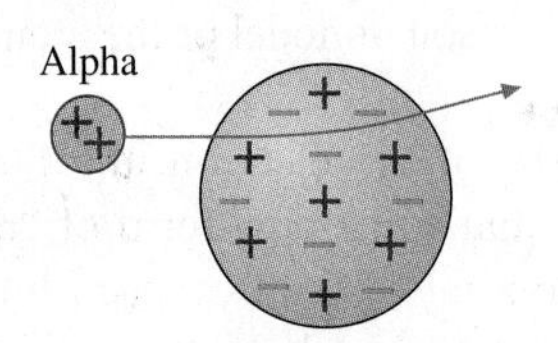

The alpha particle is only slightly deflected by a Thomson atom because forces from the spread-out positive and negative charges nearly cancel.

(b) Nuclear model

Alpha

If the atom has a concentrated positive nucleus, some alpha particles will be able to come very close to the nucleus and thus feel a very strong repulsive force.

The discovery of large-angle scattering of alpha particles quickly led Rutherford to envision an atom in which negative electrons orbit an unbelievably small, massive, positive **nucleus,** rather like a miniature solar system. This is the **nuclear model of the atom.** Further experiments showed that the diameter of the atomic nucleus is $\approx 1 \times 10^{-14}$ m $= 10$ fm (1 fm $= 1$ femtometer $= 10^{-15}$ m), a mere 0.01% the diameter of the atom itself. Thus nearly all of the atom is merely empty space—the void!

EXAMPLE 29.2 Going for the gold!

An 8.3 MeV alpha particle is shot directly toward the nucleus of a gold atom (atomic number 79). What is the distance of closest approach of the alpha particle to the nucleus?

PREPARE Energy is conserved in electric interactions. Assume that the gold nucleus, which is much more massive than the alpha particle, does not move. Also recall that the electric field and potential of a sphere of charge can be found by treating the total charge as a point charge. **FIGURE 29.7** is a before-and-after visual overview; the "before" situation is when the alpha is very far from the gold nucleus, and the "after" situation is when the alpha is at its distance of closest approach to the nucleus. The motion is in and out along a straight line.

FIGURE 29.7 A before-and-after visual overview of an alpha particle colliding with a nucleus.

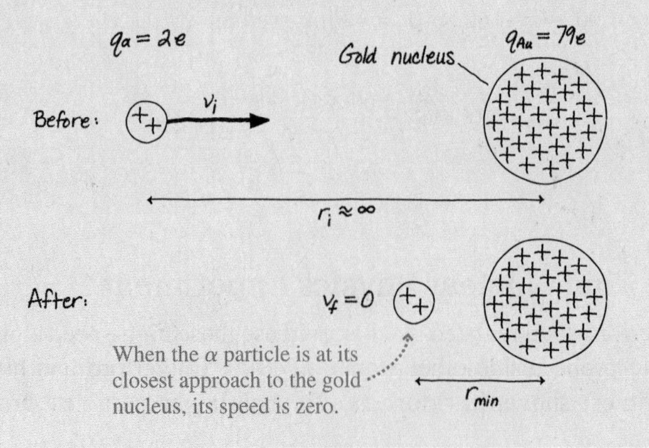

SOLVE Electric potential energy decreases rapidly with increasing separation, so initially, when the alpha particle is very far away, the system has only the initial kinetic energy of the alpha particle.

At the moment of closest approach the alpha particle has just come to rest and so the system has only potential energy. The conservation of energy statement $K_f + U_f = K_i + U_i$ is

$$0 + \frac{1}{4\pi\epsilon_0}\frac{q_\alpha q_{Au}}{r_{min}} = \frac{1}{2}mv_i^2 + 0$$

where q_α is the alpha-particle charge and we've treated the gold nucleus as a point charge q_{Au}. The solution for r_{min} is

$$r_{min} = \frac{1}{4\pi\epsilon_0}\frac{2q_\alpha q_{Au}}{mv_i^2}$$

The mass of the alpha particle is $m = 6.64 \times 10^{-27}$ kg and its charge is $q_\alpha = 2e = 3.20 \times 10^{-19}$ C. Gold has atomic number 79, so $q_{Au} = 79e = 1.26 \times 10^{-17}$ C. In Example 29.1 we found that an 8.3 MeV alpha particle has speed $v = 2.0 \times 10^7$ m/s. With this information, we can calculate

$$r_{min} = 2.7 \times 10^{-14}\ \text{m}$$

This is only about 1/10,000 the size of the atom itself!

ASSESS We ignored the gold atom's electrons in this example. In fact, they make almost no contribution to the alpha particle's trajectory. The alpha particle is exceedingly massive compared to the electrons, and the electrons are spread out over a distance very large compared to the size of the nucleus. Hence the alpha particle easily pushes them aside without any noticeable change in its velocity.

Using the Nuclear Model

The nuclear model of the atom makes it easy to picture atoms and understand such processes as ionization. For example, the **atomic number** of an element, its position in the periodic table of the elements, is the number of orbiting electrons (of a neutral atom) and the number of units of positive charge in the nucleus. The atomic number is represented by Z. Hydrogen, with $Z = 1$, has one electron orbiting a nucleus with charge $+1e$. Helium, with $Z = 2$, has two orbiting electrons and a nucleus with charge $+2e$. Because the orbiting electrons are very light, an x-ray photon or a rapidly moving particle, such as another electron, can knock one of the electrons away, creating a positive *ion.* Removing one electron makes a singly charged ion, with $q_{ion} = +e$. Removing two electrons creates a doubly charged ion, with $q_{ion} = +2e$. This is shown for lithium ($Z = 3$) in **FIGURE 29.8.**

FIGURE 29.8 Different ionization stages of the lithium atom ($Z = 3$).

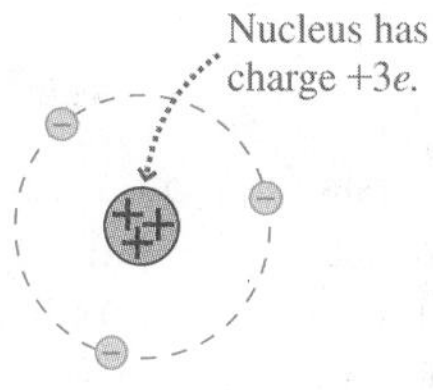

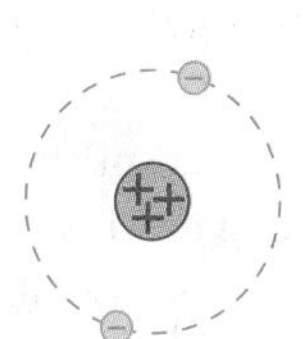

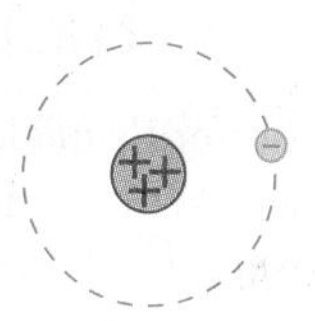

Experiments soon led to the recognition that the positive charge of the nucleus is associated with a positive subatomic particle called the **proton.** The proton's charge is $+e$, equal in magnitude but opposite in sign to the electron's charge. Further, with nearly all the atomic mass associated with the nucleus, the proton is about 1800 times more massive than the electron: $m_p = 1.67 \times 10^{-27}$ kg. Atoms with atomic number Z have Z protons in the nucleus, giving the nucleus charge $+Ze$.

This nuclear model of the atom allows us to understand why, during chemical reactions and when an object is charged by rubbing, electrons are easily transferred but protons are not. The protons are tightly bound in the nucleus, shielded by all the electrons, but outer electrons are easily stripped away. Rutherford's nuclear model has explanatory power that was lacking in Thomson's model.

But there was a problem. Helium, with atomic number 2, has twice as many electrons and protons as hydrogen. Lithium, $Z = 3$, has three electrons and protons. If a nucleus contains Z protons to balance the Z orbiting electrons, and if nearly all the atomic mass is contained in the nucleus, then helium should be twice as massive as hydrogen and lithium three times as massive. But it was known from chemistry measurements that helium is *four times* as massive as hydrogen and lithium is *seven times* as massive.

This difficulty was not resolved until the discovery, in 1932, of a third subatomic particle. This particle has essentially the same mass as a proton but *no* electric charge. It is called the **neutron.** Neutrons reside in the nucleus, with the protons, where they contribute to the mass of the atom but not to its charge. As you'll see in Chapter 30, neutrons help provide the "glue" that holds the nucleus together.

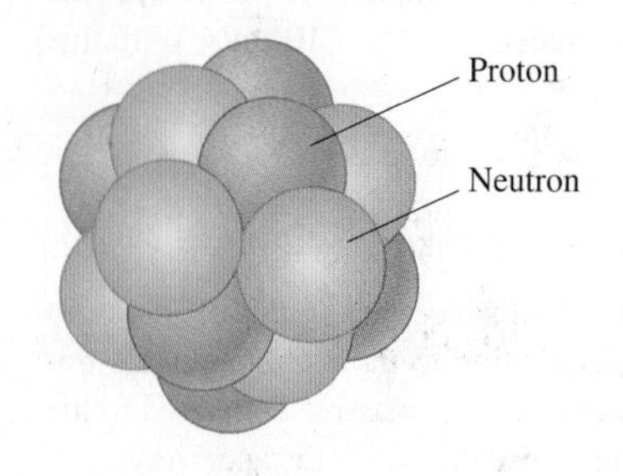

FIGURE 29.9 The nucleus of an atom contains protons and neutrons.

We now know that a nucleus contains Z protons plus N neutrons, as shown in FIGURE 29.9, giving the atom a **mass number** $A = Z + N$. The mass number, which is a dimensionless integer, is *not* the same thing as the atomic mass m. But because the proton and neutron masses are both ≈ 1 u, where

$$1 \text{ u} = 1 \text{ atomic mass unit} = 1.66 \times 10^{-27} \text{ kg}$$

the mass number A is *approximately* the mass in atomic mass units. For example, helium, with two protons and two neutrons ($A = 4$), has atomic mass $m = 6.646 \times 10^{-27}$ kg $= 4.003$ u ≈ 4 u.

There is a *range* of neutron numbers that happily form a nucleus with Z protons, creating a series of nuclei with the same Z-value (i.e., they are all the same chemical element) but different masses. Such a series of nuclei are called **isotopes.** The notation used to label isotopes is ^{A}Z, where the mass number A is given as a *leading* superscript. The proton number Z is not specified by an actual number but, equivalently, by the chemical symbol for that element. The most common isotope of neon has $Z = 10$ protons and $N = 10$ neutrons. Thus it has mass number $A = 20$ and is labeled ^{20}Ne. The neon isotope ^{22}Ne has $Z = 10$ protons (that's what makes it neon) and $N = 12$ neutrons. Helium has the two isotopes shown in FIGURE 29.10. The rare ^{3}He is only 0.0001% abundant, but it can be isolated and has important uses in scientific research.

FIGURE 29.10 The two isotopes of helium. ^{3}He is only 0.0001% abundant.

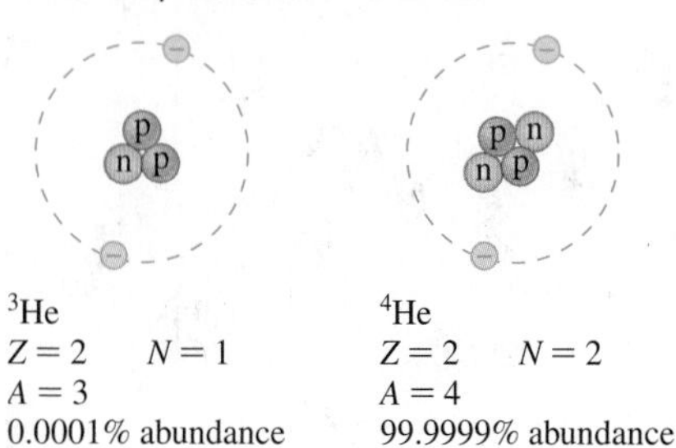

^{3}He
$Z = 2$ $N = 1$
$A = 3$
0.0001% abundance

^{4}He
$Z = 2$ $N = 2$
$A = 4$
99.9999% abundance

STOP TO THINK 29.2 Carbon is the sixth element in the periodic table. How many protons and how many neutrons are there in a nucleus of the isotope ^{14}C?

29.3 Bohr's Model of Atomic Quantization

Rutherford's nuclear model was an important step toward understanding atoms, but it had a serious shortcoming. According to Maxwell's theory of electricity and magnetism, the electrons orbiting the nucleus should act as small antennas and radiate electromagnetic waves. As FIGURE 29.11 shows, the energy carried off by the waves would cause the electrons to quickly—in less than a microsecond—spiral into the nucleus! In other words, classical Newtonian mechanics and electromagnetism predict that an atom with electrons orbiting a nucleus would be highly unstable and would immediately self-destruct. This clearly does not happen.

FIGURE 29.11 The fate of a Rutherford atom.

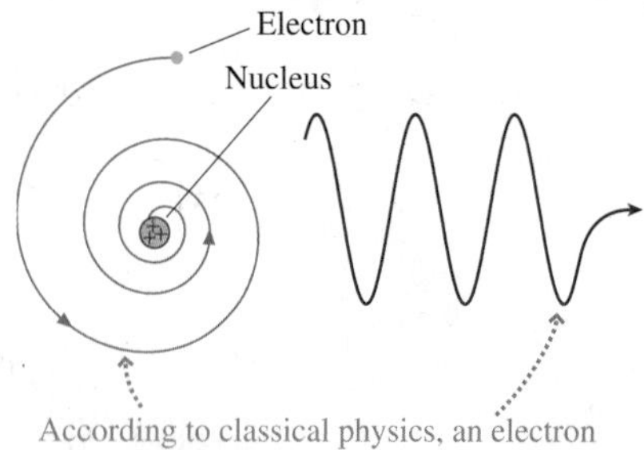

Electron
Nucleus
According to classical physics, an electron would spiral into the nucleus while radiating energy as an electromagnetic wave.

The experimental efforts of the late 19th and early 20th centuries had been impressive, and there could be no doubt about the existence of electrons, about the small positive nucleus, and about the unique discrete spectrum emitted by each atom. But the theoretical framework for understanding such observations had lagged behind. As the new century dawned, physicists could not explain the structure of atoms, could not explain the stability of matter, could not explain discrete spectra or why an element's absorption spectrum differs from its emission spectrum, and could not explain the origin of x rays or radioactivity.

A missing piece of the puzzle, although not recognized as such for a few years, was Einstein's 1905 introduction of light quanta. If light comes in discrete packets of energy, which we now call photons, and if atoms emit and absorb light, what does that imply about the structure of the atoms? This was the question posed by the Danish physicist Niels Bohr.

Bohr wanted to understand how a solar-system-like atom could be stable and not radiate away all its energy. He soon recognized that Einstein's light quanta had profound implications about the structure of atoms, and in 1913 Bohr proposed a radically new model of the atom in which he added quantization to Rutherford's nuclear atom. The basic assumptions of the **Bohr model of the atom** are as follows:

Understanding Bohr's model

Electrons can exist in only certain allowed orbits.

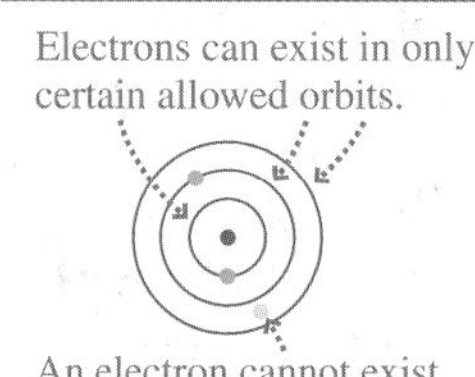

An electron cannot exist here, where there is no allowed orbit.

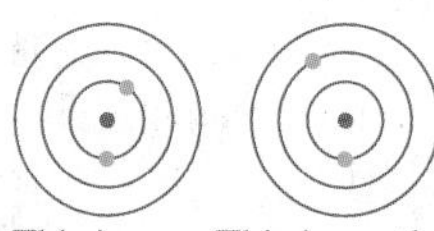

This is one stationary state. This is another stationary state.

1. The electrons in an atom can exist in only certain *allowed orbits*. A particular arrangement of electrons in these orbits is called a **stationary state.**

Energy-level diagram **Stationary states**

E_3 $n = 3$

These other states are *excited states*.

E_2 $n = 2$

E_1 $n = 1$

This state, with the lowest energy E_1, is the *ground state*. It is stable and can persist indefinitely.

2. Each stationary state has a discrete, well-defined energy E_n, as the energy-level diagram shows. That is, atomic energies are *quantized*. The stationary states are labeled by the *quantum number n* in order of increasing energy: $E_1 < E_2 < E_3 < \cdots$.

Photon emission

Excited-state electron

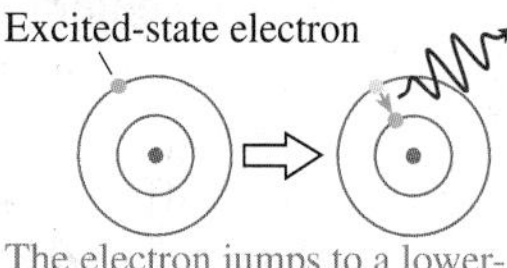

The electron jumps to a lower-energy stationary state and emits a photon.

Photon absorption

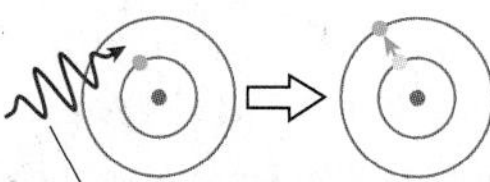

Approaching photon

The electron absorbs the photon and jumps to a higher-energy stationary state.

3. An atom can undergo a *transition* or *quantum jump* from one stationary state to another by emitting or absorbing a photon whose energy is exactly equal to the energy difference between the two stationary states.

Collisional excitation

Approaching particle

Particle loses energy.

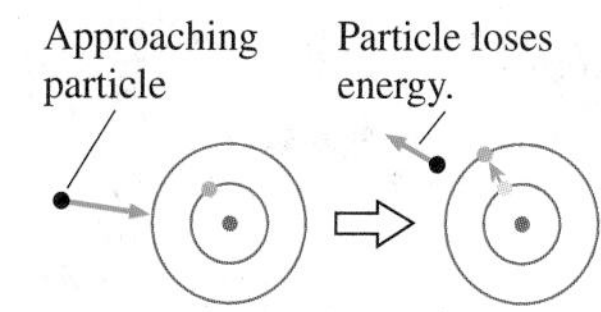

The particle transfers energy to the atom in the collision and excites the atom.

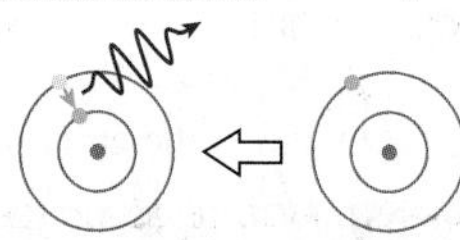

An atom in an excited state soon jumps to lower states, emitting a photon at each jump.

4. Atoms can also move from a lower-energy state to a higher-energy state by absorbing energy in a collision with an electron or other atom in a process called **collisional excitation.**

The excited atoms soon jump down to lower states, eventually ending in the stable ground state.

STOP TO THINK 29.3 3.0 eV photons excite a gas of atoms from their ground state to the $n = 3$ excited state. The atoms are then observed emitting 2.0 eV photons. Using what you learned in the table above, what other emitted photon energies would be observed?

A. 3.0 eV and 5.0 eV B. 3.0 eV and 1.0 eV C. 1.0 eV and 4.0 eV
D. 3.0 eV only E. 1.0 eV only

The implications of Bohr's model are profound. In particular:

1. **Matter is stable.** Once an atom is in its ground state, there are no states of any lower energy to which it can jump. It can remain in the ground state forever.
2. **Atoms emit and absorb a *discrete spectrum*.** When an atom jumps from an initial state with energy E_i to a final state with energy E_f, conservation of energy requires that it emit or absorb a photon with energy $E_{photon} = \Delta E_{atom} = |E_f - E_i|$. Because $E_{photon} = hf$, this photon must have frequency $f_{photon} = \Delta E_{atom}/h$. Photons of other frequencies cannot be emitted or absorbed without violating energy conservation.
3. **Emission spectra can be produced by collisions.** Energy from collisions can kick an atom up to an excited state. The atom then emits photons in a discrete emission spectrum as it jumps down to lower-energy states.
4. **Absorption wavelengths are a subset of the wavelengths in the emission spectrum.** Recall that all the lines seen in an absorption spectrum are also seen in emission, but many emission lines are *not* seen in absorption. According to Bohr's model, **most atoms, most of the time, are in their lowest energy state,** the $n = 1$ ground state. Thus the absorption spectrum consists of *only* those transitions such as $1 \rightarrow 2$, $1 \rightarrow 3$, ... in which the atom jumps from $n = 1$ to a higher value of n by absorbing a photon. Transitions such as $2 \rightarrow 3$ are not observed because there are essentially no atoms in $n = 2$ at any instant of time to do the absorbing. On the other hand, atoms that have been excited to the $n = 3$ state by collisions can emit photons corresponding to transitions $3 \rightarrow 1$ *and* $3 \rightarrow 2$. Thus

the wavelength corresponding to $\Delta E_{atom} = E_3 - E_1$ is seen in both emission and absorption, but photons with $\Delta E_{atom} = E_3 - E_2$ occur in emission only.

Video Tutor Demo

5. **Each element in the periodic table has a unique spectrum.** The energies of the stationary states are just the energies of the orbiting electrons. Different elements, with different numbers of electrons, have different stable orbits and thus different stationary states. States with different energies will emit and absorb photons of different wavelengths.

EXAMPLE 29.3 Wavelengths in emission and absorption spectra

An atom has stationary states $E_1 = 0.0$ eV, $E_2 = 2.0$ eV, and $E_3 = 5.0$ eV. What wavelengths are observed in the absorption spectrum and in the emission spectrum of this atom?

PREPARE FIGURE 29.12 shows an energy-level diagram for the atom. Photons are emitted when an atom undergoes a quantum jump from a higher energy level to a lower energy level. Photons are absorbed in a quantum jump from a lower energy level to a higher energy level. However, most of the atoms are in the $n = 1$ ground state, so the only quantum jumps seen in the absorption spectrum start from the $n = 1$ state.

FIGURE 29.12 The atom's energy-level diagram.

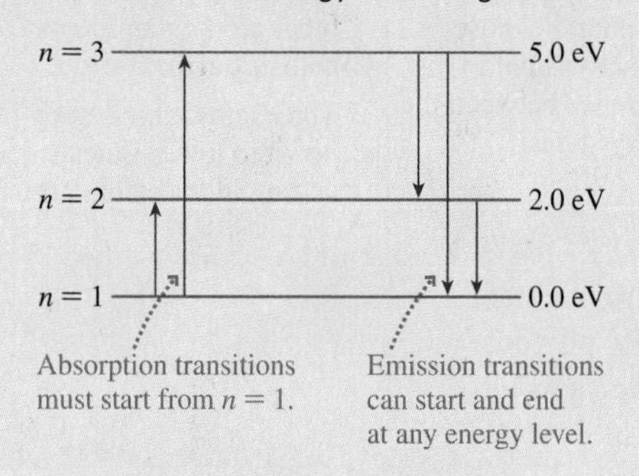
n = 3
5.0 eV
n = 2
2.0 eV
n = 1
0.0 eV
Absorption transitions must start from n = 1.
Emission transitions can start and end at any energy level.

SOLVE This atom absorbs photons on the $1 \rightarrow 2$ and $1 \rightarrow 3$ transitions, with $\Delta E_{1\rightarrow 2} = 2.0$ eV and $\Delta E_{1\rightarrow 3} = 5.0$ eV. From $f_{photon} = \Delta E_{atom}/h$ and $\lambda = c/f$, we find that the wavelengths in the absorption spectrum are

$1 \rightarrow 3$ $\quad f_{photon} = 5.0 \text{ eV}/h = 1.2 \times 10^{15}$ Hz

$\lambda = 250$ nm (ultraviolet)

$1 \rightarrow 2$ $\quad f_{photon} = 2.0 \text{ eV}/h = 4.8 \times 10^{14}$ Hz

$\lambda = 620$ nm (orange)

The emission spectrum also has the 620 nm and 250 nm wavelengths due to the $2 \rightarrow 1$ and $3 \rightarrow 1$ quantum jumps. In addition, the emission spectrum contains the $3 \rightarrow 2$ quantum jump with $\Delta E_{3\rightarrow 2} = 3.0$ eV that is *not* seen in absorption because there are too few atoms in the $n = 2$ state to absorb. A similar calculation finds $f_{photon} = 7.3 \times 10^{14}$ Hz and $\lambda = c/f = 410$ nm. Thus the emission wavelengths are

$2 \rightarrow 1$ $\quad \lambda = 620$ nm (orange)

$3 \rightarrow 2$ $\quad \lambda = 410$ nm (blue)

$3 \rightarrow 1$ $\quad \lambda = 250$ nm (ultraviolet)

STOP TO THINK 29.4 A photon with a wave-length of 410 nm has energy $E_{photon} = 3.0$ eV. Do you expect to see a spectral line with $\lambda = 410$ nm in the emission spectrum of the atom represented by this energy-level diagram? If so, what transition or transitions will emit it? Do you expect to see a spectral line with $\lambda = 410$ nm in the absorption spectrum? If so, what transition or transitions will absorb it?

$n = 4$ — 6.0 eV
$n = 3$ — 5.0 eV
$n = 2$ — 2.0 eV
$n = 1$ — 0.0 eV

29.4 The Bohr Hydrogen Atom

Bohr's hypothesis was a bold new idea, yet there was still one enormous stumbling block: What *are* the stationary states of an atom? Everything in Bohr's model hinges on the existence of these stationary states, of there being only certain electron orbits that are allowed. But nothing in classical physics provides any basis for such orbits. And Bohr's model describes only the *consequences* of having stationary states, not how to find them. If such states really exist, we will have to go beyond classical physics to find them.

To address this problem, Bohr did an explicit analysis of the hydrogen atom. The hydrogen atom, with only a single electron, was known to be the simplest atom. Furthermore, as we discussed in ◀◀ SECTION 29.1, Balmer had discovered a fairly simple formula that characterizes the wavelengths in the hydrogen emission spectrum. Anyone with a successful model of an atom was going to have to *predict*, from theory, Balmer's formula for the hydrogen atom.

Bohr's paper followed a rather circuitous line of reasoning. That is not surprising, because he had little to go on at the time. But our goal is a clear explanation of the ideas, not a historical study of Bohr's methods, so we are going to follow a different analysis using de Broglie's matter waves. De Broglie did not propose matter waves until 1924, 11 years after Bohr, but with the clarity of hindsight we can see that treating the electron as a wave provides a more straightforward analysis of the hydrogen atom. Although our route will be different from Bohr's, we will arrive at the same point, and, in addition, we will be in a much better position to understand the work that came after Bohr.

NOTE ▸ Bohr's analysis of the hydrogen atom is sometimes called the *Bohr atom.* It's important not to confuse this analysis, which applies only to hydrogen, with the more general postulates of the *Bohr model of the atom.* Those postulates, which we looked at in the previous section, apply to any atom. To make the distinction clear, we'll call Bohr's analysis of hydrogen the *Bohr hydrogen atom.* ◂

The Stationary States of the Hydrogen Atom

FIGURE 29.13 shows a Rutherford hydrogen atom, with a single electron orbiting a proton. We will assume a circular orbit of radius r and speed v. We will also assume the proton remains stationary while the electron revolves around it, a reasonable assumption because the proton is roughly 1800 times as massive as the electron.

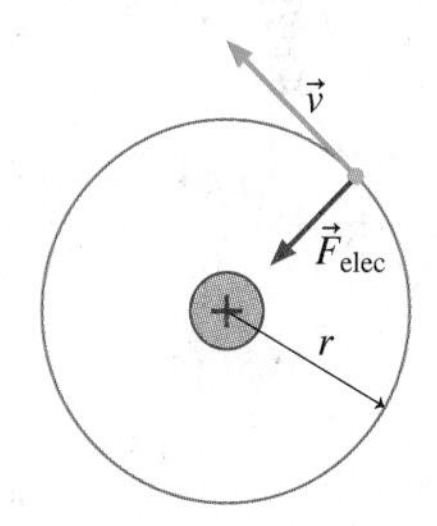

FIGURE 29.13 A Rutherford hydrogen atom. The size of the nucleus is greatly exaggerated.

Now the electron, as we are coming to understand it, has both particle-like and wave-like properties. First, let us treat the electron as a charged particle. The proton exerts a Coulomb electric force on the electron:

$$\vec{F}_{elec} = \left(\frac{1}{4\pi\epsilon_0}\frac{e^2}{r^2}, \text{ toward center}\right) \tag{29.3}$$

This force gives the electron an acceleration $\vec{a}_{elec} = \vec{F}_{elec}/m$ that also points to the center. This is a centripetal acceleration, causing the particle to move in its circular orbit. The centripetal acceleration of a particle moving in a circle of radius r at speed v is v^2/r, so that

$$a_{elec} = \frac{F_{elec}}{m} = \frac{e^2}{4\pi\epsilon_0 m r^2} = \frac{v^2}{r}$$

Rearranging, we find

$$v^2 = \frac{e^2}{4\pi\epsilon_0 m r} \tag{29.4}$$

NOTE ▸ m is the mass of the electron, *not* the mass of the entire atom. ◂

Equation 29.4 is a *constraint* on the motion. The speed v and radius r must satisfy Equation 29.4 if the electron is to move in a circular orbit. This constraint is not unique to atoms; we earlier found a similar relationship between v and r for orbiting satellites.

Now let's treat the electron as a de Broglie wave. In ◂◂ SECTION 28.5 we found that a particle confined to a one-dimensional box sets up a standing wave as it reflects back and forth. A standing wave, you will recall, consists of two traveling waves moving in opposite directions. When the round-trip distance in the box is equal to an integer number of wavelengths ($2L = n\lambda$), the two oppositely traveling waves interfere constructively to set up the standing wave.

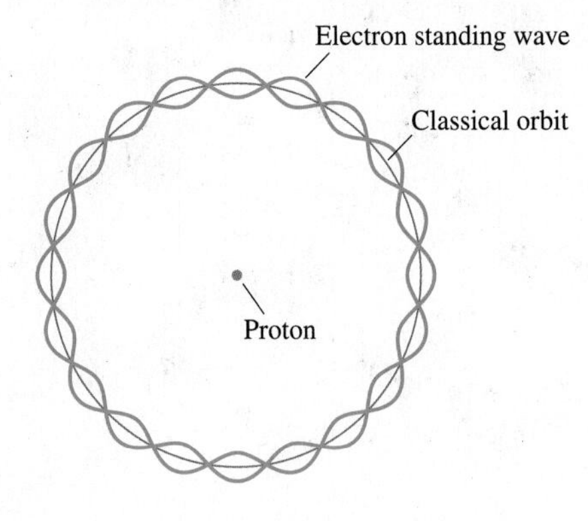

FIGURE 29.14 An $n = 10$ electron standing wave around the orbit's circumference.

Suppose that, instead of traveling back and forth along a line, our wave-like particle travels around the circumference of a circle. The particle will set up a standing wave, just like the particle in the box, if there are waves traveling in both directions and if the round-trip distance is an integer number of wavelengths. This is the idea we want to carry over from the particle in a box. As an example, **FIGURE 29.14** shows a standing wave around a circle with $n = 10$ wavelengths.

The mathematical condition for a circular standing wave is found by replacing the round-trip distance $2L$ in a box with the round-trip distance $2\pi r$ on a circle. Thus a circular standing wave will occur when

$$2\pi r = n\lambda \qquad n = 1, 2, 3, \ldots \tag{29.5}$$

But the de Broglie wavelength for a particle *has* to be $\lambda = h/p = h/mv$. Thus the standing-wave condition for a de Broglie wave is

$$2\pi r = n\frac{h}{mv}$$

This condition is true only if the electron's speed is

$$v = \frac{nh}{2\pi mr} \qquad n = 1, 2, 3, \ldots \tag{29.6}$$

In other words, the electron cannot have just any speed, only the discrete values given by Equation 29.6.

The quantity $h/2\pi$ occurs so often in quantum physics that it is customary to give it a special name. We define the quantity $\hbar$, pronounced "h bar," as

$$\hbar = \frac{h}{2\pi} = 1.05 \times 10^{-34}\ \text{J} \cdot \text{s} = 6.58 \times 10^{-16}\ \text{eV} \cdot \text{s}$$

With this definition, we can write Equation 29.6 as

$$v = \frac{n\hbar}{mr} \qquad n = 1, 2, 3, \ldots \tag{29.7}$$

This, like Equation 29.4, is another relationship between v and r. This is the constraint that arises from treating the electron as a wave.

Now if the electron can act as both a particle *and* a wave, then both the Equation 29.4 *and* Equation 29.7 constraints have to be obeyed. That is, v^2 as given by the Equation 29.4 particle constraint has to equal v^2 of the Equation 29.7 wave constraint. Equating these gives

$$v^2 = \frac{e^2}{4\pi\epsilon_0 mr} = \frac{n^2\hbar^2}{m^2r^2}$$

We can solve this equation to find that the radius r is

$$r_n = n^2\frac{4\pi\epsilon_0\hbar^2}{me^2} \qquad n = 1, 2, 3, \ldots \tag{29.8}$$

where we have added a subscript n to the radius r to indicate that it depends on the integer n.

The right-hand side of Equation 29.8, except for the n^2, is just a collection of constants. Let's group them all together and define the **Bohr radius** a_B to be

$$a_B = \text{Bohr radius} = \frac{4\pi\epsilon_0\hbar^2}{me^2} = 5.29 \times 10^{-11}\ \text{m} = 0.0529\ \text{nm}$$

With this definition, Equation 29.8 for the radius of the electron's orbit becomes

$$r_n = n^2 a_B \qquad n = 1, 2, 3, \ldots \tag{29.9}$$

Allowed radii of the Bohr hydrogen atom

r_n — n — p.44 — QUADRATIC

For example, $r_1 = 0.053$ nm, $r_2 = 0.212$ nm, and $r_3 = 0.476$ nm.

We have discovered stationary states! That is, **a hydrogen atom can exist *only* if the radius of the electron's orbit is one of the values given by Equation 29.9.** Intermediate values of the radius, such as $r = 0.100$ nm, cannot exist because the electron cannot set up a standing wave around the circumference. The possible orbits are *quantized,* and integer n is the quantum number.

Hydrogen Atom Energy Levels

The energy of a hydrogen atom is the kinetic energy of its electron plus the potential energy of the electron's interaction with the proton. This is

$$E = K + U = \frac{1}{2}mv^2 + \frac{1}{4\pi\epsilon_0}\frac{q_{\text{elec}}q_{\text{proton}}}{r} = \frac{1}{2}mv^2 - \frac{e^2}{4\pi\epsilon_0 r} \tag{29.10}$$

where we used $q_{\text{elec}} = -e$ and $q_{\text{proton}} = +e$.

Now we can proceed to find the energies of the stationary states. Knowing the possible radii, we can return to Equation 29.7 and find the possible electron speeds:

$$v_n = \frac{n\hbar}{mr_n} = \frac{1}{n}\frac{\hbar}{ma_B} = \frac{v_1}{n} \qquad n = 1, 2, 3, \ldots \tag{29.11}$$

where $v_1 = \hbar/ma_B = 2.19 \times 10^6$ m/s is the electron's speed in the $n = 1$ orbit. The speed decreases as n increases.

Finally, we can determine the energies of the stationary states by using Equations 29.9 and 29.11 for r and v in Equation 29.10 for the energy. The algebra is rather messy, but the result simplifies to

$$E_n = \frac{1}{2}mv_n^2 - \frac{e^2}{4\pi\epsilon_0 r_n} = -\frac{1}{n^2}\left(\frac{1}{4\pi\epsilon_0}\frac{e^2}{2a_B}\right) \tag{29.12}$$

Let's define

$$E_1 = \frac{1}{4\pi\epsilon_0}\frac{e^2}{2a_B} = 13.60 \text{ eV}$$

We can then write the energy levels of the stationary states of the hydrogen atom as

$$E_n = -\frac{E_1}{n^2} = -\frac{13.60 \text{ eV}}{n^2} \qquad n = 1, 2, 3, \ldots \tag{29.13}$$

Allowed energies of the Bohr hydrogen atom

This has been a lot of math, so we need to see where we are and what we have learned. Table 29.2 shows values of r_n, v_n, and E_n for quantum numbers $n = 1$ to 4. We do indeed seem to have discovered stationary states of the hydrogen atom. Each state, characterized by its quantum number n, has a unique radius, speed, and energy. These are displayed graphically in **FIGURE 29.15**, in which the orbits are drawn to scale. Notice how the atom's diameter increases very rapidly as n increases. At the same time, the electron's speed decreases.

TABLE 29.2 Radii, speeds, and energies for the first four states of the Bohr hydrogen atom

n	r_n (nm)	v_n (m/s)	E_n (eV)
1	0.053	2.19×10^6	−13.60
2	0.212	1.09×10^6	−3.40
3	0.476	0.73×10^6	−1.51
4	0.847	0.55×10^6	−0.85

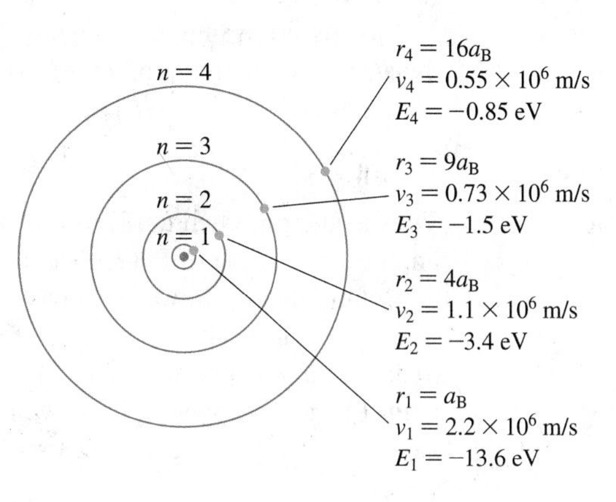

◀ **FIGURE 29.15** The first four stationary states, or allowed orbits, of the Bohr hydrogen atom drawn to scale.

EXAMPLE 29.4 Possible electron speeds in a hydrogen atom

Can an electron in a hydrogen atom have a speed of 3.60×10^5 m/s? If so, what are its energy and the radius of its orbit? What about a speed of 3.65×10^5 m/s?

PREPARE To be in a stationary state, the electron must have speed $v_n = v_1/n$, with $n = v_1/v$ an integer. Only if v_1/v is an integer is v an allowed electron speed.

SOLVE A speed of 3.60×10^5 m/s would require quantum number

$$n = \frac{v_1}{v} = \frac{2.19 \times 10^6 \text{ m/s}}{3.60 \times 10^5 \text{ m/s}} = 6.08$$

This is not an integer, so the electron can *not* have this speed. But if $v = 3.65 \times 10^5$ m/s, then

$$n = \frac{2.19 \times 10^6 \text{ m/s}}{3.65 \times 10^5 \text{ m/s}} = 6$$

This is the speed of an electron in the $n = 6$ excited state. An electron in this state has energy

$$E_6 = -\frac{13.60 \text{ eV}}{6^2} = -0.378 \text{ eV}$$

and the radius of its orbit is

$$r_6 = 6^2\, a_B = 6^2\,(0.0529 \text{ nm}) = 1.90 \text{ nm}$$

It is important to understand why the energies of the stationary states are negative. An electron and a proton bound into an atom have *less* energy than they do when they're separated. We know this because we would have to do work (i.e., *add* energy) to pull the electron and proton apart.

When the electron and proton are completely separated ($r \rightarrow \infty$) and at rest ($v = 0$), their potential energy $U = q_1q_2/4\pi\epsilon_0 r$ and kinetic energy $K = mv^2/2$ are zero. As the electron moves closer to the proton to form a hydrogen atom, its potential energy *decreases,* becoming negative, while the kinetic energy of the orbiting electron increases. Equation 29.12, however, shows that the potential energy decreases faster than the kinetic energy increases, leading to an overall negative energy for the atom.

Quantization of Angular Momentum

The angular momentum of a particle in circular motion, whether it is a planet or an electron, is

$$L = mvr$$

You will recall that angular momentum is conserved in orbital motion because a force directed toward a central point exerts no torque on the particle, so we expect the angular momentum of the electron in a hydrogen atom to be conserved as well.

The condition that a de Broglie wave for the electron set up a standing wave around the circumference was given, in Equation 29.5, as

$$2\pi r = n\lambda = n\frac{h}{mv}$$

Multiplying by mv and dividing by 2π, we have

$$mvr = n\frac{h}{2\pi} = n\hbar \qquad (29.14)$$

But mvr is the angular momentum L for a particle in a circular orbit. It appears that the angular momentum of an orbiting electron cannot have just any value. Instead, it must satisfy

$$L = n\hbar \qquad n = 1, 2, 3, \ldots \qquad (29.15)$$

Thus angular momentum is also quantized! The atom's angular momentum must be an integer multiple of Planck's constant $\hbar$.

Bohr correctly showed that the angular momentum of the hydrogen atom is quantized. As it turns out, however, the exact values of the angular momentum in the Bohr hydrogen atom (Equation 29.15) are incorrect. The full quantum-mechanical treatment of the hydrogen atom, given in Section 29.5, yields the correct values. But Bohr's analysis does give the correct values for the *energies* of the hydrogen atom.

STOP TO THINK 29.5 What is the quantum number of this hydrogen atom?

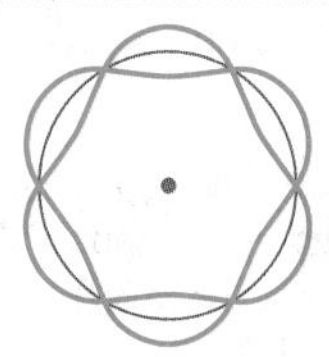

The Hydrogen Spectrum

The most important experimental evidence that we have about the hydrogen atom is its spectrum, so the primary test of the Bohr hydrogen atom is whether it correctly predicts the spectrum. **FIGURE 29.16** is an energy-level diagram for the hydrogen atom. As we noted in ◂◂ SECTION 28.6, the energies are like the rungs of a ladder. The lowest rung is the ground state, with $E_1 = -13.60$ eV. The top rung, with $E = 0$ eV, corresponds to a hydrogen ion in the limit $n \rightarrow \infty$. This top rung is called the **ionization limit.** In principle there are an infinite number of rungs, but only the lowest few are shown. The energy levels with higher values of n are all crowded together just below the ionization limit at $n = \infty$.

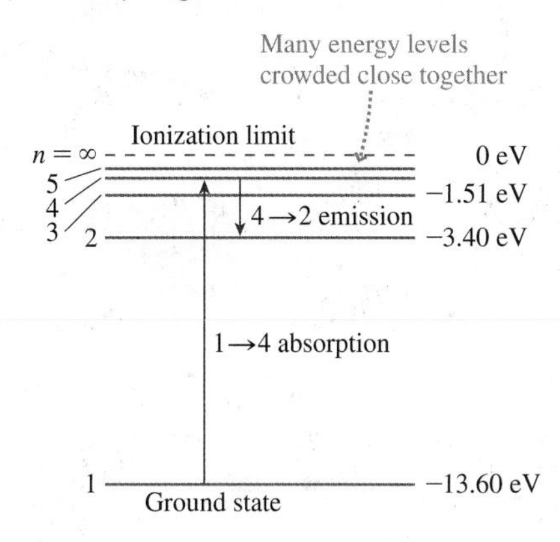

FIGURE 29.16 The energy-level diagram of the hydrogen atom.

The figure shows a $1 \rightarrow 4$ transition in which a photon is absorbed and a $4 \rightarrow 2$ transition in which a photon is emitted. For two quantum states m and n, where $n > m$ and E_n is the higher-energy state, an atom can *emit* a photon in an $n \rightarrow m$ transition or *absorb* a photon in an $m \rightarrow n$ transition.

According to the fifth assumption of Bohr's model of atomic quantization, the frequency of the photon emitted in an $n \rightarrow m$ transition is

$$f_{\text{photon}} = \frac{\Delta E_{\text{atom}}}{h} = \frac{E_n - E_m}{h} \qquad (29.16)$$

We can use Equation 29.13 for the energies E_n and E_m to predict that the emitted photon has frequency

$$f_{\text{photon}} = \frac{1}{h}\left(\frac{-13.60\text{ eV}}{n^2} - \frac{-13.60\text{ eV}}{m^2}\right) = \frac{13.60\text{ eV}}{h}\left(\frac{1}{m^2} - \frac{1}{n^2}\right)$$

The frequency is a positive number because $m < n$ and thus $1/m^2 > 1/n^2$.

We are more interested in wavelength than frequency, because wavelengths are the quantity measured by experiment. The wavelength of the photon emitted in an $n \rightarrow m$ quantum jump is

$$\lambda_{n \rightarrow m} = \frac{c}{f_{\text{photon}}} = \frac{\lambda_0}{\left(\dfrac{1}{m^2} - \dfrac{1}{n^2}\right)} \qquad \begin{array}{l} m = 1, 2, 3, \ldots \\ n = m+1, m+2, \ldots \end{array} \qquad (29.17)$$

with $\lambda_0 = 91.1$ nm. This should look familiar. It is the Balmer formula, Equation 29.2.

It works! Unlike previous atomic models, **the Bohr hydrogen atom correctly predicts the discrete spectrum of the hydrogen atom. FIGURE 29.17** shows two series of transitions that give rise to wavelengths in the spectrum. The *Balmer series,* consisting of transitions ending on the $m = 2$ state, gives visible wavelengths, and this is the series that Balmer initially analyzed. The *Lyman series,* ending on the $m = 1$ ground state, is in the ultraviolet region of the spectrum and was not measured until later. These series, as well as others in the infrared, are observed in a discharge tube where collisions with electrons excite the atoms upward from the ground state to state n. They then decay downward by emitting photons. Only the Lyman series is observed in the absorption spectrum because, as noted previously, essentially all the atoms in a quiescent gas are in the ground state.

FIGURE 29.17 Transitions producing the Lyman series and the Balmer series of lines in the hydrogen spectrum.

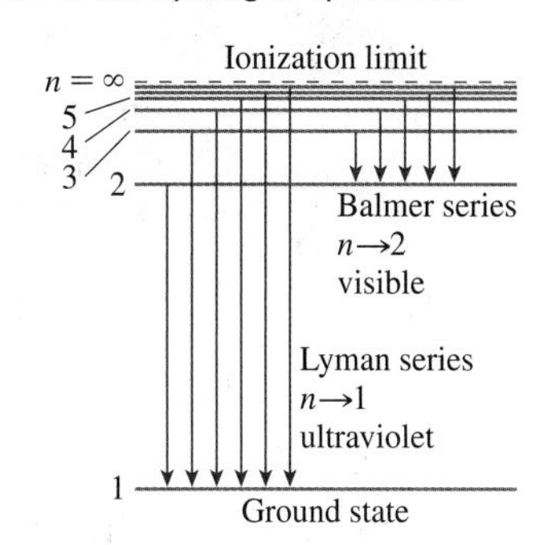

EXAMPLE 29.5 Wavelengths in galactic hydrogen absorption

Whenever astronomers look at distant galaxies, they find that the light has been strongly absorbed at the wavelength of the $1 \rightarrow 2$ transition in the Lyman series of hydrogen. This absorption tells us that interstellar space is filled with vast clouds of hydrogen left over from the Big Bang. What is the wavelength of the $1 \rightarrow 2$ absorption in hydrogen?

SOLVE Equation 29.17 predicts the *absorption* spectrum of hydrogen if we let $m = 1$. The absorption seen by astronomers is from the ground state of hydrogen ($m = 1$) to its first excited state ($n = 2$). The wavelength is

$$\lambda_{1 \rightarrow 2} = \frac{91.1 \text{ nm}}{\left(\frac{1}{1^2} - \frac{1}{2^2}\right)} = 122 \text{ nm}$$

ASSESS This wavelength is far into the ultraviolet. Ground-based astronomy cannot observe this region of the spectrum because the wavelengths are strongly absorbed by the atmosphere, but with space-based telescopes, first widely used in the 1970s, astronomers see 122 nm absorption in nearly every direction they look.

29.5 The Quantum-Mechanical Hydrogen Atom

Bohr's analysis of the hydrogen atom seemed to be a resounding success. By introducing stationary states, together with Einstein's ideas about light quanta, Bohr was able to provide the first solid understanding of discrete spectra and, in particular, to predict the Balmer formula for the wavelengths in the hydrogen spectrum. And the Bohr hydrogen atom, unlike Rutherford's model, was stable. There was clearly some validity to the idea of stationary states.

But Bohr was completely unsuccessful at explaining the spectra of any other atom. His method did not work even for helium, the second element in the periodic table with a mere two electrons. Although Bohr was clearly on the right track, his inability to extend the Bohr hydrogen atom to more complex atoms made it equally clear that the complete and correct theory remained to be discovered.

In 1925, Erwin Schrödinger introduced his general theory of *quantum* mechanics, a theory capable of calculating the allowed energy levels (i.e., the stationary states) of any system. The calculations are mathematically difficult, even for a system as simple as the hydrogen atom, and we will present results without proof.

The Bohr hydrogen atom was characterized by a single quantum number n. In contrast, Schrödinger's quantum-mechanical analysis of the hydrogen atom found that it must be described by *four* quantum numbers.

1. Schrödinger found that the *energy* of the hydrogen atom is given by the same expression found by Bohr, or

$$E_n = -\frac{13.60 \text{ eV}}{n^2} \qquad n = 1, 2, 3, \ldots \qquad (29.18)$$

   The integer n is called the **principal quantum number.**

2. The angular momentum L of the electron's orbit must be one of the values

$$L = \sqrt{l(l+1)}\,\hbar \qquad l = 0, 1, 2, 3, \ldots, n-1 \qquad (29.19)$$

   The integer l is called the **orbital quantum number.**

3. The plane of the electron's orbit can be tilted, but only at certain discrete angles. Each allowed angle is characterized by a quantum number m, which must be one of the values

$$m = -l, -l+1, \ldots, 0, \ldots, l-1, l \qquad (29.20)$$

   The integer m is called the **magnetic quantum number** because it becomes important when the atom is placed in a magnetic field.

4. The electron's *spin*—discussed below—can point only up or down. These two orientations are described by the **spin quantum number** m_s, which must be one of the values

$$m_s = -\frac{1}{2} \text{ or } +\frac{1}{2} \qquad (29.21)$$

In other words, each stationary state of the hydrogen atom is identified by a quartet of quantum numbers (n, l, m, m_s), and each quantum number is associated with a physical property of the atom.

NOTE ► For a hydrogen atom, the energy of a stationary state depends on only the principal quantum number n, not on l, m, or m_s. ◄

EXAMPLE 29.6 Listing quantum numbers

List all possible states of a hydrogen atom that have energy $E = -3.40$ eV.

SOLVE Energy depends on only the principal quantum number n. From Equation 29.18, states with $E = -3.40$ eV have

$$n = \sqrt{\frac{-13.60\text{ eV}}{-3.40\text{ eV}}} = 2$$

An atom with principal quantum number $n = 2$ could have either $l = 0$ or $l = 1$, but $l \geq 2$ is ruled out. If $l = 0$, the only possible value for the magnetic quantum number m is $m = 0$. If $l = 1$, then the atom could have $m = -1$, $m = 0$, or $m = +1$. For each of these, the spin quantum number could be $m_s = +\frac{1}{2}$ or $m_s = -\frac{1}{2}$. Thus the possible sets of quantum numbers are

n	l	m	m_s	n	l	m	m_s
2	0	0	$+\frac{1}{2}$	2	0	0	$-\frac{1}{2}$
2	1	1	$+\frac{1}{2}$	2	1	1	$-\frac{1}{2}$
2	1	0	$+\frac{1}{2}$	2	1	0	$-\frac{1}{2}$
2	1	-1	$+\frac{1}{2}$	2	1	-1	$-\frac{1}{2}$

These eight states all have the same energy.

Energy and Angular Momentum Are Quantized

The energy of the hydrogen atom depends on only the principal quantum number n. For other atoms, however, the allowed energies depend on both n and l. Consequently, it is useful to label the stationary states of an atom by their values of n and l. The lowercase letters shown in Table 29.3 are customarily used to represent the various values of quantum number l. Using these symbols, we call the ground state of the hydrogen atom, with $n = 1$ and $l = 0$, the $1s$ state; the $3d$ state has $n = 3$, $l = 2$.

FIGURE 29.18 is an energy-level diagram for the hydrogen atom in which the rows are labeled by n and the columns by l. The left column contains all of the $l = 0$ (or s) states, the next column is the $l = 1$ (or p) states, and so on.

As Equation 29.19 shows, the orbital quantum number l of an allowed state must be less than that state's principal quantum number n. For the ground state, with $n = 1$, only $l = 0$ is possible, so that the only $n = 1$ state is the $1s$ state. When $n = 2$, l can be 0 or 1, leading to both a $2s$ and a $2p$ state. For $n = 3$, there are $3s$, $3p$, and $3d$ states; and so on. Figure 29.18 shows only the first few energy levels for each value of l, but there really are an infinite number of levels, as $n \to \infty$, crowding together beneath $E = 0$. The dotted line at $E = 0$ is the atom's *ionization limit*, the energy of a hydrogen atom in which the electron has been moved infinitely far away to form an H^+ ion.

Classically, the angular momentum L of an orbiting electron can have any value. Not so in quantum mechanics. Equation 29.19 tells us that **the electron's orbital angular momentum is quantized.** The magnitude of the orbital angular momentum must be one of the discrete values

$$L = \sqrt{l(l+1)}\hbar = 0, \sqrt{2}\hbar, \sqrt{6}\hbar, \sqrt{12}\hbar, \ldots$$

where l is an integer. The Bohr atom also predicted quantized angular momentum, but the precise values of that prediction turned out to be in error. The quantum-mechanical prediction for L is more complex, but it agrees with experimental observations.

A particularly interesting prediction is that the ground state of hydrogen, with $l = 0$, has *no* angular momentum. A classical particle cannot orbit unless it has angular momentum, but apparently a quantum particle does not have this requirement.

STOP TO THINK 29.6 What are the quantum numbers n and l for a hydrogen atom with $E = -(13.60/9)$ eV and $L = \sqrt{2}\hbar$?

TABLE 29.3 Symbols used to represent quantum number l

l	Symbol
0	s
1	p
2	d
3	f

FIGURE 29.18 The energy-level diagram for the hydrogen atom.

n	E	l = 0 (Symbol s)	1 (p)	2 (d)	3 (f)
	$E = 0$ eV	Ionization limit			
4	-0.85 eV	$4s$	$4p$	$4d$	$4f$
3	-1.51 eV	$3s$	$3p$	$3d$	
2	-3.40 eV	$2s$	$2p$		
1	-13.60 eV	$1s$ (Ground state)			

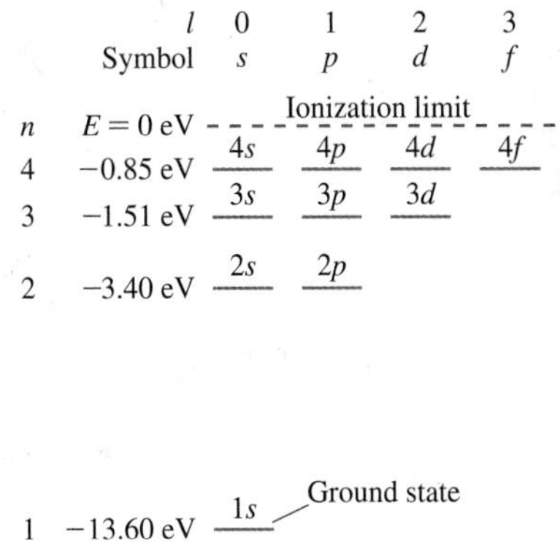

The Electron Spin

You learned in ◀◀ SECTION 24.8 that an electron has an inherent magnetic dipole moment—it acts as a tiny bar magnet with a north and a south pole. In association with its magnetic moment, an electron also has an intrinsic *angular momentum* called the *electron spin.* In the early years of quantum mechanics, it was thought that the electron was a very tiny ball of negative charge spinning on its axis, which would give the electron both a magnetic dipole moment and spin angular momentum. However, a spinning ball of charge would violate the laws of relativity and other physical laws. As far as we know today, the electron is truly a point particle that happens to have an intrinsic magnetic dipole moment and angular momentum.

The two possible spin quantum numbers $m_s = \pm\frac{1}{2}$ mean that the electron's intrinsic magnetic dipole points in the $+z$-direction or the $-z$-direction. These two orientations are called *spin up* and *spin down.* It is convenient to picture a little vector that can be drawn $\uparrow$ for a spin-up state and $\downarrow$ for a spin-down state. We will use this notation in the next section.

29.6 Multielectron Atoms

The quantum-mechanical solution for the hydrogen atom matches the experimental evidence, but so did the Bohr hydrogen atom. One of the first big successes of Schrödinger's quantum mechanics was an ability to calculate the stationary states and energy levels of *multielectron atoms*, atoms in which Z electrons orbit a nucleus with Z protons. As we've seen, a major difference between multielectron atoms and the simple one-electron hydrogen is that the energy of an electron in a multielectron atom depends on both quantum numbers *n and l.* Whereas the 2*s* and 2*p* states in hydrogen have the same energy, their energies are different in a multielectron atom. The difference arises from the electron-electron interactions that do not exist in a single-electron hydrogen atom.

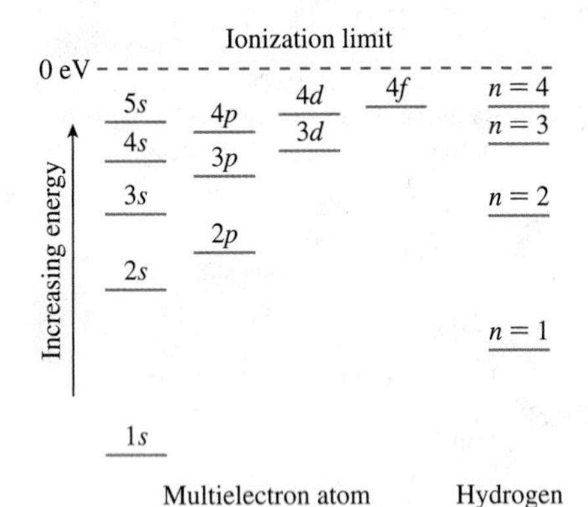

FIGURE 29.19 An energy-level diagram for electrons in a multielectron atom.

FIGURE 29.19 shows an energy-level diagram for the electrons in a multielectron atom. (Compare this to the hydrogen energy-level diagram in Figure 29.18.) For comparison, the hydrogen-atom energies are shown on the right edge of the figure. Two features of this diagram are of particular interest:

1. For each n, the energy increases as l increases until the maximum-l state has an energy very nearly that of the same n in hydrogen. States with small values of l are significantly lower in energy than the corresponding state in hydrogen.
2. As the energy increases, states with different n begin to alternate in energy. For example, the 3*s* and 3*p* states have lower energy than a 4*s* state, but the energy of an electron in a 3*d* state is slightly higher. This will have important implications for the structure of the periodic table of the elements.

The Pauli Exclusion Principle

By definition, the ground state of a quantum system is the state of lowest energy. What is the ground state of an atom having Z electrons and Z protons? Because the 1*s* state is the lowest energy state, it seems that the ground state should be one in which all Z electrons are in the 1*s* state. However, this idea is not consistent with the experimental evidence.

In 1925, the Austrian physicist Wolfgang Pauli hypothesized that no two electrons in a quantum system can be in the same quantum state. That is, **no two electrons can have exactly the same set of quantum numbers (n, l, m, m_s).** If one electron is present in a state, it *excludes* all others. This statement is called the **Pauli exclusion principle.** It turns out to be an extremely profound statement about the nature of matter.

The exclusion principle is not applicable to hydrogen, where there is only a single electron, but in helium, with $Z = 2$ electrons, we must make sure that the two electrons are in different quantum states. This is not difficult. For a $1s$ state, with $l = 0$, the only possible value of the magnetic quantum number is $m = 0$. But there are *two* possible values of m_s—namely, $-\frac{1}{2}$ and $+\frac{1}{2}$. If a first electron is in the spin-down $1s$ state $\left(1, 0, 0, -\frac{1}{2}\right)$, a second $1s$ electron can still be added to the atom as long as it is in the spin-up state $\left(1, 0, 0, +\frac{1}{2}\right)$. This is shown schematically in **FIGURE 29.20**, where the dots represent electrons on the rungs of the "energy ladder" and the arrows represent spin down or spin up.

The Pauli exclusion principle does not prevent both electrons of helium from being in the $1s$ state as long as they have opposite values of m_s, so we predict this to be the ground state. A list of an atom's occupied energy levels is called its **electron configuration.** The electron configuration of the helium ground state is written $1s^2$, where the superscript 2 indicates two electrons in the $1s$ energy level.

The states $\left(1, 0, 0, -\frac{1}{2}\right)$ and $\left(1, 0, 0, +\frac{1}{2}\right)$ are the only two states with $n = 1$. The ground state of helium has one electron in each of these states, so all the possible $n = 1$ states are filled. Consequently, the electron configuration $1s^2$ is called a **closed shell.**

The next element, lithium, has $Z = 3$ electrons. The first two electrons can go into $1s$ states, with opposite values of m_s, but what about the third electron? The $1s^2$ shell is closed, and there are no additional quantum states having $n = 1$. The only option for the third electron is the next energy state, $n = 2$. Figure 29.19 showed that for a multielectron atom, the next level above the $1s$ level is the $2s$ state, so lithium's third ground-state electron will be $2s$. **FIGURE 29.21** shows the electron configuration with the $2s$ electron being spin up, but it could equally well be spin down. The electron configuration for the lithium ground state is written $1s^2\,2s$. This indicates two $1s$ electrons and a single $2s$ electron.

FIGURE 29.20 The ground state of helium.

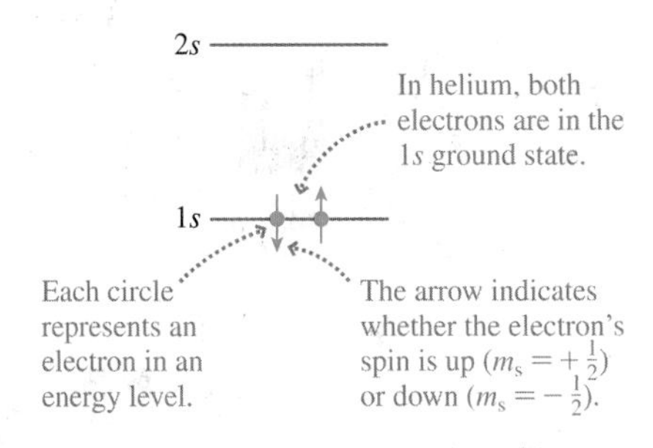

FIGURE 29.21 The ground state of lithium.

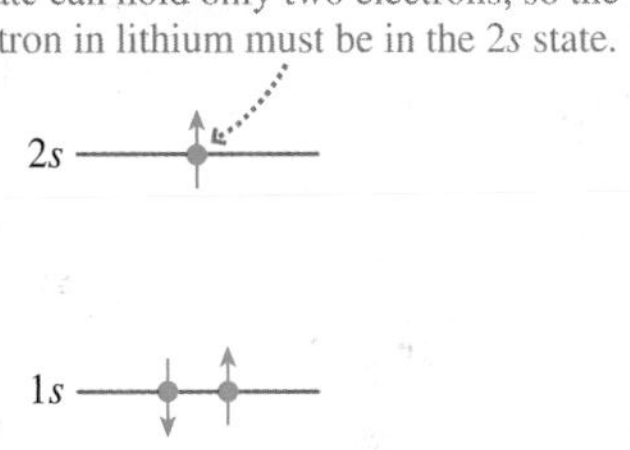

The 1s state can hold only two electrons, so the third electron in lithium must be in the 2s state.
2s
1s

The Periodic Table of the Elements

The 19th century was a time when chemists were discovering new elements and studying their chemical properties. The century opened with the atomic model still not completely validated, and with no one having any idea how many elements there might be. But chemistry developed quickly, and by mid-century it was clear that there were dozens of elements, but not hundreds.

The Russian chemist Dmitri Mendeléev was the first to propose, in 1867, a *periodic* arrangement of the elements based on the regular recurrence of chemical properties. He did so by explicitly pointing out "gaps" where, according to his hypothesis, undiscovered elements should exist. He could then predict the expected properties of the missing elements. The subsequent discovery of these elements verified Mendeléev's organizational scheme, which came to be known as the *periodic table of the elements.*

One of the great triumphs of the quantum-mechanical theory of multielectron atoms is that it explains the structure of the periodic table. We can understand this structure by looking at the energy-level diagram of **FIGURE 29.22**, which is an expanded version of the energy-level diagram of Figure 29.19. Just as for helium and lithium, atoms with larger values of Z are constructed by placing Z electrons into the lowest-energy levels that are consistent with the Pauli exclusion principle.

The s states of helium and lithium can each hold two electrons—one spin up and the other spin down—but the higher-angular-momentum states that will become filled for higher-Z atoms can hold more than two electrons. For each value l of the orbital quantum number, there are $2l + 1$ possible values of the magnetic quantum number m and, for each of these, two possible values of the spin quantum number m_s. Consequently, each energy *level* in Figure 29.22 is actually $2(2l + 1)$ different *states* that, taken together, are called a *subshell.* Table 29.4 lists the number of states in each subshell. Each state in a subshell is represented in Figure 29.22 by a colored dot. The dots' colors correspond to the periodic table in **FIGURE 29.23** on the next page, which is color coded to show which subshells are being filled as Z increases.

FIGURE 29.22 An energy-level diagram showing how many electrons can occupy each subshell.

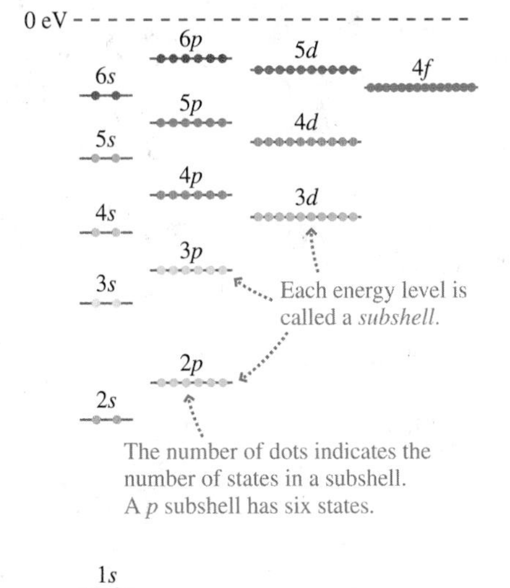

TABLE 29.4 Number of states in each subshell of an atom

Subshell	l	Number of states
s	0	2
p	1	6
d	2	10
f	3	14

FIGURE 29.23 The periodic table of the elements. The elements are color coded to the states in the energy-level diagram of Figure 29.22.

		Transition elements															
1 H (1s)																	2 He (1s)
3 Li	4 Be (2s)											5 B	6 C	7 N	8 O (2p)	9 F	10 Ne
11 Na	12 Mg (3s)											13 Al	14 Si	15 P	16 S (3p)	17 Cl	18 Ar
19 K	20 Ca (4s)	21 Sc	22 Ti	23 V	24 Cr	25 Mn	26 Fe (3d)	27 Co	28 Ni	29 Cu	30 Zn	31 Ga	32 Ge	33 As	34 Se (4p)	35 Br	36 Kr
37 Rb	38 Sr (5s)	39 Y	40 Zr	41 Nb	42 Mo	43 Tc	44 Ru (4d)	45 Rh	46 Pd	47 Ag	48 Cd	49 In	50 Sn	51 Sb	52 Te (5p)	53 I	54 Xe
55 Cs	56 Ba (6s)	71 Lu	72 Hf	73 Ta	74 W	75 Re	76 Os (5d)	77 Ir	78 Pt	79 Au	80 Hg	81 Tl	82 Pb	83 Bi	84 Po (6p)	85 At	86 Rn
87 Fr	88 Ra (7s)	103 Lr	104 Rf	105 Db	106 Sg	107 Bh	108 Hs (6d)	109 Mt	110 Ds	111 Rg	112 Cn						

Lanthanides	57 La	58 Ce	59 Pr	60 Nd	61 Pm	62 Sm	63 Eu	64 Gd (4f)	65 Tb	66 Dy	67 Ho	68 Er	69 Tm	70 Yb
Actinides	89 Ac	90 Th	91 Pa	92 U	93 Np	94 Pu	95 Am	96 Cm (5f)	97 Bk	98 Cf	99 Es	100 Fm	101 Md	102 No

Inner transition elements

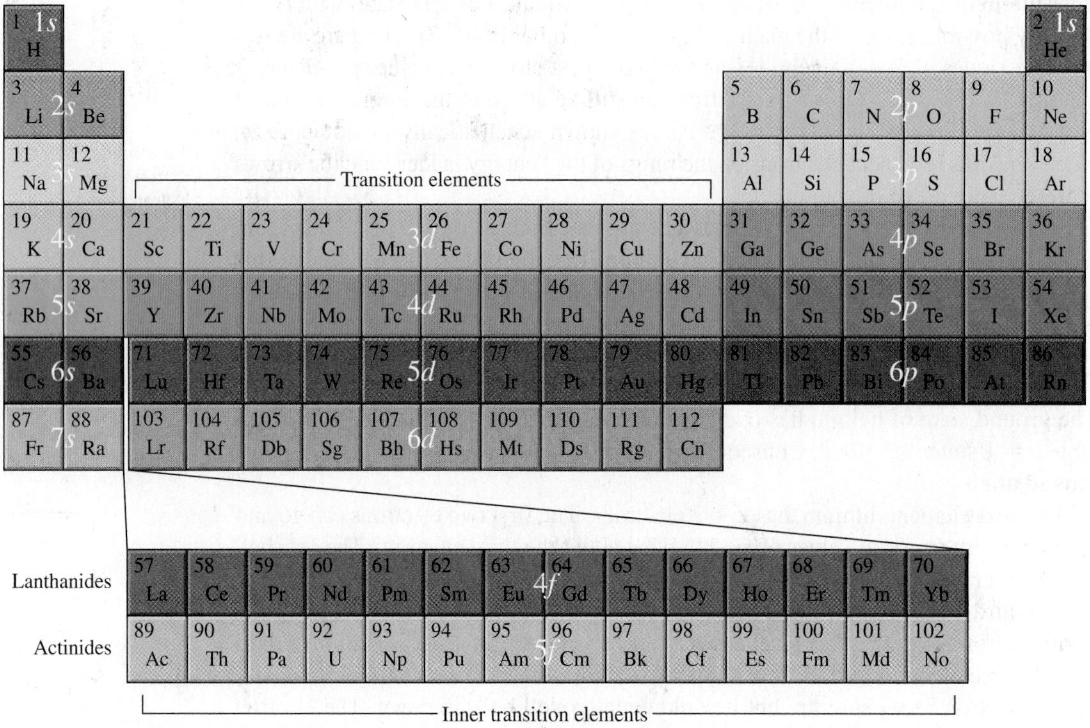

We can now use Figure 29.22 to construct the periodic table in Figure 29.23. We've already seen that lithium has two electrons in the $1s$ state and one electron in the $2s$ state. Four-electron beryllium ($Z = 4$) comes next. We see that there is still an empty state in the $2s$ subshell for this fourth electron to occupy, so beryllium closes the $2s$ subshell and has electron configuration $1s^2\,2s^2$.

As Z increases further, the next six electrons can each occupy states in the $2p$ subshell. These are the elements boron (B) through neon (Ne), completing the second row of the periodic table. Neon, which completes the $2p$ subshell, has ground-state configuration $1s^2\,2s^2\,2p^6$.

The $3s$ subshell is the next to be filled, leading to the elements sodium and magnesium. Filling the $3p$ subshell gives aluminum through argon, completing the third row of the table.

The fourth row is where the periodic table begins to get complicated. You might expect that once the $3p$ subshell in argon was filled, the $3d$ subshell would start to fill, starting with potassium. But if you look back at the energy-level diagram of Figure 29.22, you can see that the $3d$ state is slightly *higher* in energy than the $4s$ state. Because the ground state is the *lowest energy state* consistent with the Pauli exclusion principle, potassium finds it more favorable to fill a $4s$ state than to fill a $3d$ state. After the $4s$ state is filled (at calcium), the ten *transition elements* from scandium (Sc) through zinc (Zn) fill the 10 states of the $3d$ subshell.

The same pattern applies to the fifth row, where the $5s$, $4d$, and $5p$ subshells fill in succession. In the sixth row, however, after the initial $6s$ states are filled, the $4f$ subshell has the lowest energy, so it begins to fill *before* the $5d$ states. The elements corresponding to the $4f$ subshell, lanthanum through ytterbium, are known as the lanthanides, and they are traditionally drawn as a row separated from the rest of the table. The seventh row follows this same pattern.

Thus the entire periodic table can be built up using our knowledge of the energy-level diagram of a multielectron atom along with the Pauli exclusion principle.

EXAMPLE 29.7 The ground state of arsenic

Predict the ground-state electron configuration of arsenic.

SOLVE The periodic table shows that arsenic (As) has $Z = 33$, so we must identify the states of 33 electrons. Arsenic is in the fourth row, following the first group of transition elements. Argon ($Z = 18$) filled the $3p$ subshell, then calcium ($Z = 20$) filled the $4s$ subshell. The next 10 elements, through zinc ($Z = 30$), filled the $3d$ subshell. The $4p$ subshell starts filling with gallium ($Z = 31$), and arsenic is the third element in this group, so it will have three $4p$ electrons. Thus the ground-state configuration of arsenic is

$$1s^2 2s^2 2p^6 3s^2 3p^6 4s^2 3d^{10} 4p^3$$

STOP TO THINK 29.7 Which element has the ground-state electron configuration $1s^2 2s^2 2p^6 3s^2 3p^3$?

A. P B. Al C. B D. Ge

29.7 Excited States and Spectra

The periodic table organizes information about the *ground states* of the elements. These states are chemically most important because most atoms spend most of the time in their ground states. All the chemical ideas of valence, bonding, reactivity, and so on are consequences of these ground-state atomic structures. But the periodic table does not tell us anything about the excited states of atoms. It is the excited states that hold the key to understanding atomic spectra, and that is the topic to which we turn next.

Sodium ($Z = 11$) is a multielectron atom that we will use to illustrate excited states. The ground-state electron configuration of sodium is $1s^2 2s^2 2p^6 3s$. The first 10 electrons completely fill the $1s$, $2s$, and $2p$ subshells, creating a *neon core* whose electrons are tightly bound together. The $3s$ electron, however, is a *valence electron* that can be easily excited to higher energy levels. If this electron were excited to the $3p$ state, for instance, then we would write the electron configuration as $1s^2 2s^2 2p^6 3p$.

The excited states of sodium are produced by raising the valence electron to a higher energy level. The electrons in the neon core are unchanged. **FIGURE 29.24** is an energy-level diagram showing the ground state and some of the excited states of sodium. The $1s$, $2s$, and $2p$ states of the neon core are not shown on the diagram. These states are filled and unchanging, so only the states available to the valence electron are shown. Notice that the zero of energy has been shifted to the ground state. As we have discovered before, the zero of energy can be located where it is most convenient. With this choice, the excited-state energies tell us how far each state is above the ground state. The ionization limit now occurs at the value of the atom's ionization energy, which is 5.14 eV for sodium.

FIGURE 29.24 The $3s$ ground state of the sodium atom and some of the excited states.

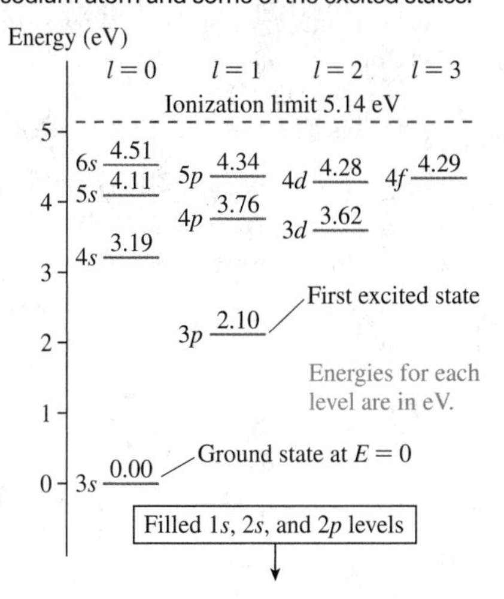

Excitation by Absorption

Left to itself, an atom will be in its lowest-energy ground state. How does an atom get into an excited state? The process of getting it there is called **excitation**, and there are two basic mechanisms: absorption and collision. We'll begin by looking at excitation by absorption.

One of the postulates of the basic Bohr model is that an atom can jump from one stationary state, of energy E_1, to a higher-energy state E_2 by absorbing a photon of frequency $f_{\text{photon}} = \Delta E_{\text{atom}}/h$. This process is shown in **FIGURE 29.25**. Because we are interested in spectra, it is more useful to write this in terms of the wavelength:

$$\lambda = \frac{c}{f_{\text{photon}}} = \frac{hc}{\Delta E_{\text{atom}}} = \frac{1240\ \text{eV} \cdot \text{nm}}{\Delta E_{\text{atom}}} \tag{29.22}$$

FIGURE 29.25 Excitation by photon absorption.

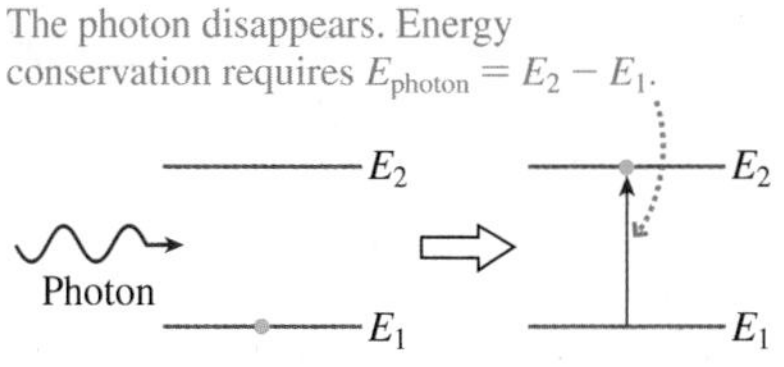

The photon disappears. Energy conservation requires $E_{\text{photon}} = E_2 - E_1$.

E_2 E_2

Photon

E_1 E_1

The final expression, which uses the value $hc = 1240\ \text{eV}\cdot\text{nm}$, gives the wavelength in nanometers if ΔE_{atom} is in electron volts.

Not every quantum jump allowed by Equation 29.22 can actually occur in an atom. A quantum-mechanical analysis of how the electrons in an atom interact with a light wave shows that transitions must also satisfy the following **selection rule:** Transitions (either absorption or emission) from a state with orbital quantum number l can occur to only another state whose orbital quantum number differs from the original state by ± 1, or

$$\Delta l = l_2 - l_1 = \pm 1 \qquad (29.23)$$

Selection rule for emission and absorption

EXAMPLE 29.8 Analyzing absorption in sodium

What are the two longest wavelengths in the absorption spectrum of sodium? What are the transitions?

PREPARE Absorption transitions always start from the ground state. Only transitions for which l changes by ± 1 are allowed.

SOLVE As Figure 29.24 shows, the sodium ground state is 3*s*. Starting from an *s* state ($l = 0$), the selection rule permits quantum jumps only to *p* states ($l = 1$). The lowest excited state is the 3*p* state and $3s \rightarrow 3p$ is an allowed transition ($\Delta l = 1$), so this will be the longest wavelength. You can see from the data in Figure 29.24 that $\Delta E_{\text{atom}} = 2.10\ \text{eV} - 0.00\ \text{eV} = 2.10\ \text{eV}$ for this transition. The corresponding wavelength is

$$\lambda = \frac{1240\ \text{eV}\cdot\text{nm}}{2.10\ \text{eV}} = 590\ \text{nm}$$

(Because of rounding, the calculation gives $\lambda = 590$ nm. The experimental value is actually 589 nm.)

The next excited state is 4*s*, but a $3s \rightarrow 4s$ transition is not allowed by the selection rule. The next allowed transition is $3s \rightarrow 4p$, with $\Delta E_{\text{atom}} = 3.76$ eV. The wavelength of this transition is

$$\lambda = \frac{1240\ \text{eV}\cdot\text{nm}}{3.76\ \text{eV}} = 330\ \text{nm}$$

ASSESS If you look at the sodium spectrum shown earlier in Figure 29.3b, you will see that 589 nm and 330 nm are, indeed, the two longest wavelengths in the absorption spectrum.

Collisional Excitation

An electron traveling with a speed of 1.0×10^6 m/s has a kinetic energy of 2.85 eV. If this electron collides with a ground-state sodium atom, a portion of its energy can be used to excite the atom to a higher-energy state, such as its 3*p* state. This process is called **collisional excitation** of the atom.

Collisional excitation differs from excitation by absorption in one very fundamental way. In absorption, the photon disappears. Consequently, *all* of the photon's energy must be transferred to the atom. Conservation of energy then requires $E_{\text{photon}} = \Delta E_{\text{atom}}$. In contrast, the electron is still present after collisional excitation and can still have some kinetic energy. That is, the electron does *not* have to transfer its entire energy to the atom. If the electron has an incident kinetic energy of 2.85 eV, it could transfer 2.10 eV to the sodium atom, thereby exciting it to the 3*p* state, and still depart the collision with a speed of 5.1×10^5 m/s and 0.75 eV of kinetic energy.

To excite the atom, the incident energy of the electron (or any other matter particle) merely has to *exceed* ΔE_{atom}; that is $E_{\text{particle}} \geq \Delta E_{\text{atom}}$. There's a threshold energy for exciting the atom, but no upper limit. It is all a matter of energy conservation. **FIGURE 29.26** shows the idea graphically.

FIGURE 29.26 Excitation by electron collision.

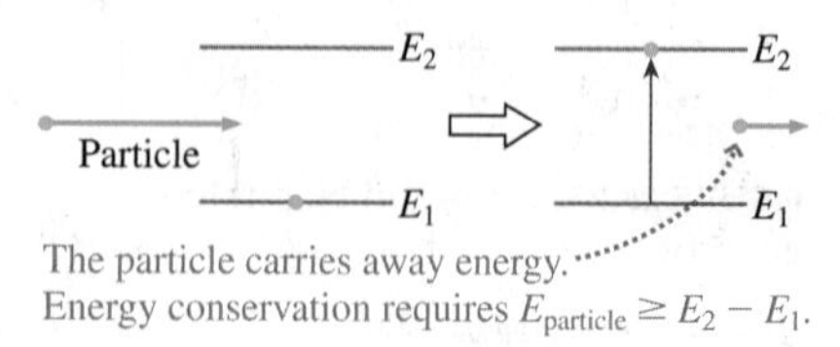

Collisional excitation by electrons is the predominant method of excitation in electrical discharges such as fluorescent lights, street lights, and neon signs. A gas is sealed in a tube at reduced pressure (≈ 1 mm Hg), then a fairly high voltage (≈ 1000 V) between electrodes at the ends of the tube causes the gas to ionize, creating a current in which both ions and electrons are charge carriers. The electrons accelerate in the electric field, gaining several eV of kinetic energy, then transfer some of this energy to the gas atoms upon collision.

NOTE ► In contrast to photon absorption, there are no selection rules for collisional excitation. Any state can be excited if the colliding particle has sufficient energy. ◄

CONCEPTUAL EXAMPLE 29.9 Possible excitation of hydrogen?

Can an electron with a kinetic energy of 11.4 eV cause a hydrogen atom to emit the prominent red spectral line (λ= 656 nm, $E_{\text{photon}} = 1.89$ eV) in the Balmer series?

REASON The electron must have sufficient energy to excite the upper state of the transition. The electron's energy of 11.4 eV is significantly greater than the 1.89 eV energy of a photon with wavelength 656 nm, but don't confuse the energy of the photon with the energy of the excitation. The red spectral line in the Balmer series is emitted in an $n = 3 \rightarrow 2$ quantum jump with $\Delta E_{\text{atom}} = 1.89$ eV, but to cause this emission, the electron must excite an atom from its *ground state,* with $n = 1$, up to the $n = 3$ level. From Figure 29.16, the necessary excitation energy is

$$\Delta E_{\text{atom}} = E_3 - E_1 = (-1.51 \text{ eV}) - (-13.60 \text{ eV}) = 12.09 \text{ eV}$$

The electron does *not* have sufficient energy to excite the atom to the state from which the emission would occur.

ASSESS As our discussion of absorption spectra showed, almost all excitations of atoms begin from the ground state. Quantum jumps down in energy, however, can begin and end at any two states allowed by selection rules.

Emission Spectra

The absorption of light is an important process, but it is the emission of light that really gets our attention. Understanding emission hinges upon the three ideas shown in **FIGURE 29.27**. Once we have determined the energy levels of an atom, from quantum mechanics, we can immediately predict its emission spectrum. We might also ask *how long* an atom remains in an excited state before undergoing a quantum jump to a lower-energy state and emitting a photon. Just as the uncertainty principle prevents us from knowing exactly where an electron is, we also can't determine exactly how long any particular atom spends in the excited state. However, we can determine the *average* time an atom spends in the excited state before emitting a photon. This average time, the **lifetime** of the state, is typically just a few nanoseconds.

FIGURE 29.27 Generation of an emission spectrum.

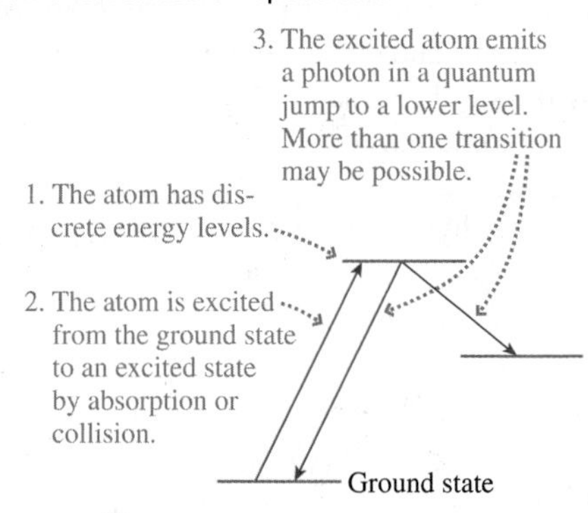

3. The excited atom emits a photon in a quantum jump to a lower level. More than one transition may be possible.
1. The atom has discrete energy levels.
2. The atom is excited from the ground state to an excited state by absorption or collision.
Ground state

As an example, **FIGURE 29.28a** shows some of the transitions and wavelengths observed in the emission spectrum of sodium. This diagram makes the point that each wavelength represents a quantum jump between two well-defined energy levels. Notice that the selection rule $\Delta l = \pm 1$ is obeyed in the sodium spectrum. The $5p$ levels can undergo quantum jumps to $3s$, $4s$, or $3d$ but *not* to $3p$ or $4p$.

FIGURE 29.28b shows the emission spectrum of sodium as it would be recorded in a spectrometer. (Many of the lines seen in this spectrum start from higher excited states that are not seen in the rather limited energy-level diagram of Figure 29.28a.) By comparing the spectrum to the energy-level diagram, you can recognize that the spectral lines at 589 nm, 330 nm, 286 nm, and 268 nm form a *series* of lines due to all the possible $np \rightarrow 3s$ transitions. They are the dominant features in the sodium spectrum.

The most obvious visual feature of sodium emission is its bright yellow color, produced by the 589 nm photons emitted in the $3p \rightarrow 3s$ transition. This is the basis of the *flame test* used in chemistry to test for sodium: A sample is held in a Bunsen burner, and a bright yellow glow indicates the presence of sodium. The 589 nm emission is also prominent in the pinkish-yellow glow of the common sodium-vapor street lights. These operate by creating an electrical discharge in sodium vapor. Most sodium-vapor lights use high-pressure lamps to increase their light output. The high pressure, however, causes the formation of Na_2 molecules, and these molecules emit the pinkish portion of the light.

FIGURE 29.28 The emission spectrum of sodium.

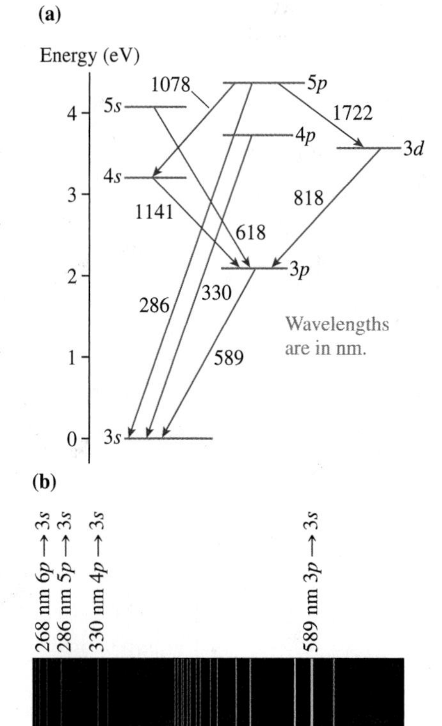

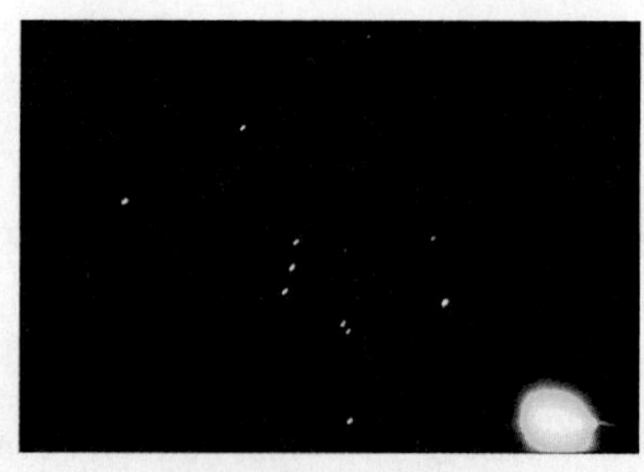

▲ **Seeing the light** Some cities close to astronomical observatories use low-pressure sodium lights, and these emit the distinctively yellow 589 nm light of sodium. The glow of city lights is a severe problem for astronomers, but the very specific 589 nm emission from sodium is easily removed with a *sodium filter,* an interference filter that lets all colors pass except the 589 nm yellow sodium light. The photos show the sky near a sodium streetlight without (left) and with (right) a sodium filter. The constellation of Orion, nearly obscured in the left photo, is clearly visible in the right photo.

X Rays

FIGURE 29.29 The generation of x rays from copper atoms.

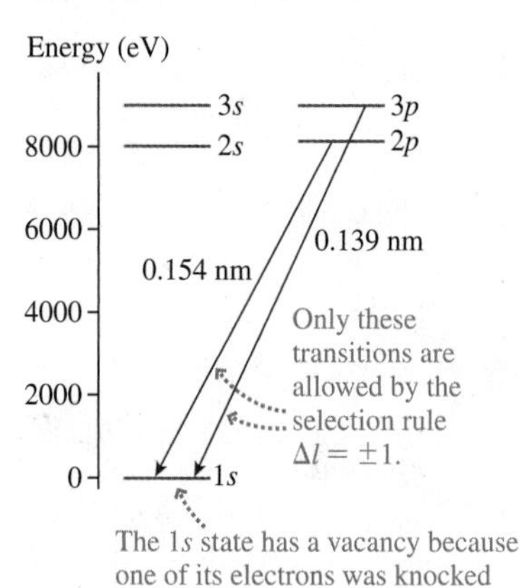

Chapter 28 noted that x rays are produced by causing very-high-speed electrons, accelerated with potential differences of many thousands of volts, to crash into metal targets. Rather than exciting the atom's valence electrons, such as happens in a gas discharge tube, these high-speed projectiles are capable of knocking inner-shell electrons out of the target atoms, producing an *inner-shell vacancy.* As **FIGURE 29.29** shows for copper atoms, this vacancy is filled when an electron from a higher shell undergoes a quantum jump into the vacancy, emitting a photon in the process.

In heavy elements, such as copper or iron, the energy difference between the inner and outer shells is very large—typically 10 keV. Consequently, the photon has energy $E_{\text{photon}} \approx 10$ keV and wavelength $\lambda \approx 0.1$ nm. These high-energy photons are the x rays discovered by Röntgen. X-ray photons are about 10,000 times more energetic than visible-light photons, and the wavelengths are about 10,000 times smaller. Even so, the underlying physics is the same: A photon is emitted when an electron in an atom undergoes a quantum jump.

STOP TO THINK 29.8 In this hypothetical atom, what is the photon energy E_{photon} of the longest-wavelength photons emitted by atoms in the 5*p* state?

A. 1.0 eV
B. 2.0 eV
C. 3.0 eV
D. 4.0 eV

Energy (eV)
5 — 5*s* — 4*d*
4 — 5*p*
3
2 — 4*p*
1 — 3*d*
0 — 4*s*

29.8 Molecules

Quantum mechanics applies to molecules just as it does to atoms, but molecules are more complex because they have internal modes of storing energy. In particular, molecules can *rotate* about their center of mass, and the atoms can *vibrate* back and forth as if the molecular bonds holding them together were little springs. For the most part, we'll overlook this internal motion and focus our attention on the electrons and the electron energy levels.

A quantum-mechanical analysis of molecules finds the following general results:

- The energy is quantized. Electrons can exist in only certain allowed energy levels.
- The number of energy levels is so extraordinarily high, and the energy levels are jammed so close together, that for all practical purposes the energy levels group into *bands* of allowed energy.
- In thermal equilibrium, nearly all molecules are in the very lowest energy levels.

FIGURE 29.30 shows a generic molecular energy-level diagram for a medium-size molecule. Whereas an atom has a well-defined ground state, a molecule has a broad band of lower energy levels. Similarly, a single excited state, such as the 2*s* state of the hydrogen atom, has been replaced by a band of excited energy levels. Despite the vast number of allowed energy levels, nearly all molecules spend nearly all their time in the very lowest energy levels.

FIGURE 29.30 The molecular energy-level diagram for a medium-size molecule.

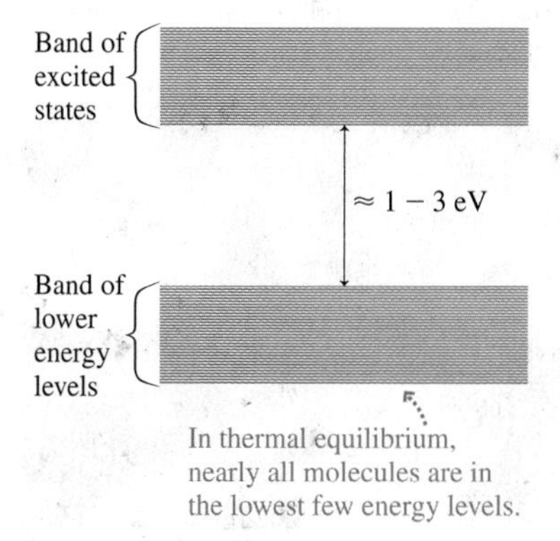

FIGURE 29.31 uses the energy-level diagram to explain two important phenomena of molecular spectroscopy: absorption and fluorescence. Whereas the absorption spectrum of an atom consists of discrete spectral lines, a molecule has a continuous *absorption band.* The absorption of light at a higher frequency (shorter wavelength) followed by the emission of light at a lower frequency (longer wavelength) is called **fluorescence.** Fluorescence occurs in molecules, but not atoms, because molecules can transform some of the absorbed energy into the vibrational energy of the atoms and thus increase the thermal energy of the molecules.

FIGURE 29.31 Molecular absorption and fluorescence.

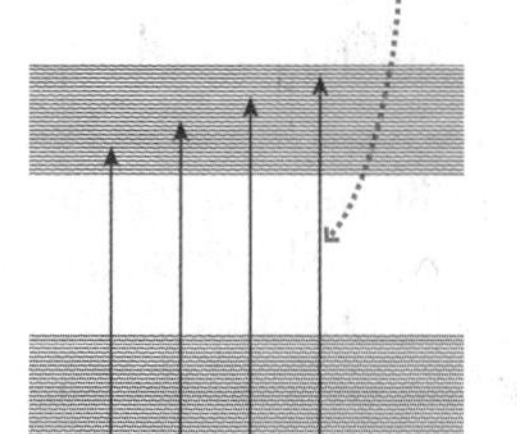

The molecules can absorb light over a range of wavelengths. This makes an *absorption band.*

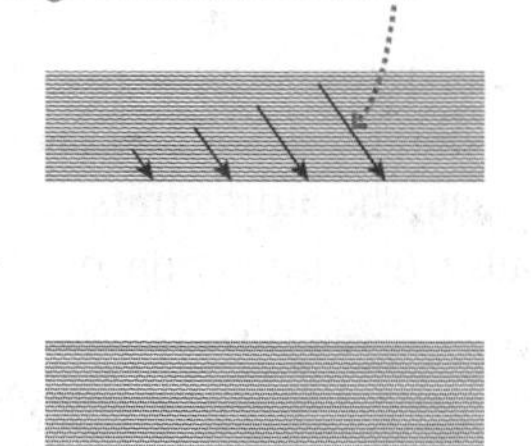

The molecules rapidly transform some of the absorbed energy into molecular vibrations, causing the molecules to fall to the bottom edge of the excited band.

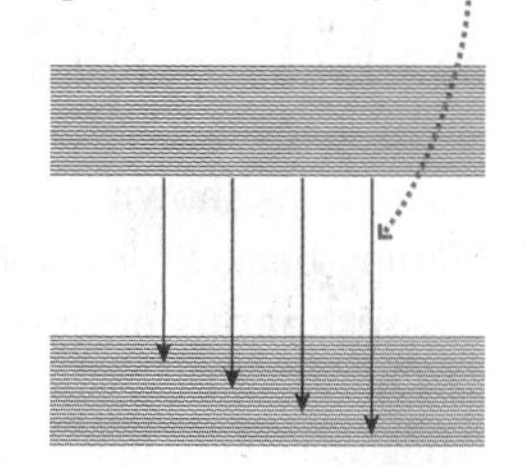

Quantum jumps back to the lower band have less energy than the original jumps up. Thus the *emission band* is at longer wavelength than the absorption band. This is *fluorescence.*

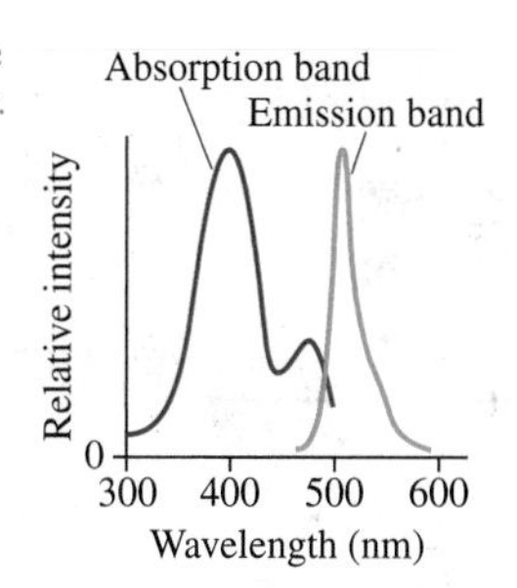

Fluorescence has many important applications. In biology, fluorescent dyes are used to stain tissue samples for microscopy. In the brilliantly colored photo that opened this chapter, the cells were stained with three different dyes, one of which (blue) was preferentially absorbed in the nucleus and another (green) by actin. The red dye was absorbed by other cell structures. When illuminated with ultraviolet light, which we can't see, each dye fluoresces with its own characteristic color.

▶ **Whiter than white** Laundry detergents often contain fluorescent dyes that absorb invisible ultraviolet light and then fluoresce in a broad band of visible wavelengths; that is, they give off white light. Sunlight contains lots of UV, so outdoors your white clothes not only reflect the white visible sunlight, but fluoresce even more white light—"whiter than white." In the photo, you can see how brightly the girl's shirt glows when illuminated with UV light (and, evidently, hair and teeth fluoresce this way as well).

Many biological molecules are fluorescent. A good example is chlorophyll, the green pigment in plants that allows photosynthesis to happen. When illuminated by blue or ultraviolet light, chlorophyll fluoresces an unexpected dark red color. This property of chlorophyll is used by marine biologists to measure the concentration of phytoplankton—microscopic plants—in seawater.

Quite recently, scientists isolated a fluorescent protein from a species of jellyfish. This protein fluoresces green when illuminated with ultraviolet light, so it's been dubbed GFP, for *green fluorescent protein.* GFP has become an important tool in genetics because it can be used to identify when particular genes are being expressed in living cells. This is done by fusing the jellyfish GFP gene to the gene being studied. When the gene is active, the cell manufactures GFP in addition to the usual protein coded by the gene. If the cells are observed under ultraviolet light, a bright

Class Video

FIGURE 29.32 BIO Green fluorescent protein shows the locations in mosquito larvae at which a particular gene is being expressed.

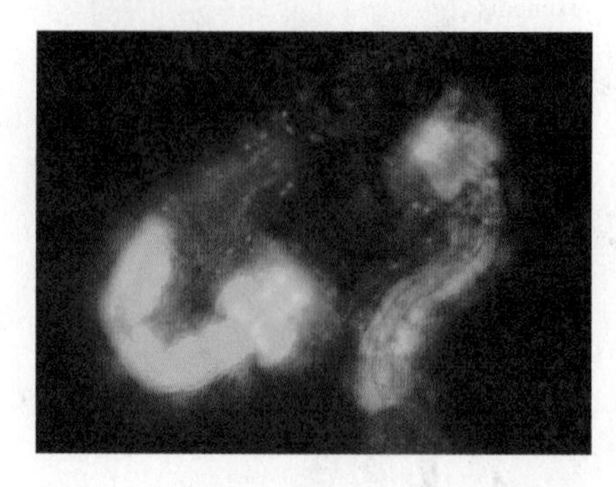

green glow from the GFP indicates that the gene is turned on. The cells are dark where the gene is not active. FIGURE 29.32 shows an example.

The color of GFP may remind you of *bioluminescence*—the summer flashes of fireflies or the green glow of various deep-sea fish—and, indeed, there is a connection. Bioluminescence is actually a form of *chemiluminescence,* the production of light in chemical reactions. Some chemical reactions create reactant molecules in an excited state. These molecules then emit light as they jump to lower energy levels, just as if they had first absorbed shorter-wavelength light. The emitted light has exactly the same spectrum as fluorescence, but the method by which the molecules are excited is different. Light sticks, which come in many colors, are an example of chemiluminescence.

Bioluminescence is just chemiluminescence in a biological organism. Some biochemical reaction—often a catalyzed reaction in an organism—produces a molecule in an excited state, and light is emitted as the molecule jumps to a lower state. In fireflies, the reaction involves an enzyme with the intriguing name *luciferase.* In the jellyfish *Aequoria victoria,* the source of GFP, bioluminescent reactions actually create blue light. However, the blue light is absorbed by the green fluorescent protein—just as predicted by the absorption curve in Figure 29.31—and re-emitted as green fluorescence, giving the jellyfish an eerie green glow when seen in the dark ocean.

29.9 Stimulated Emission and Lasers

FIGURE 29.33 Three types of radiative transitions.

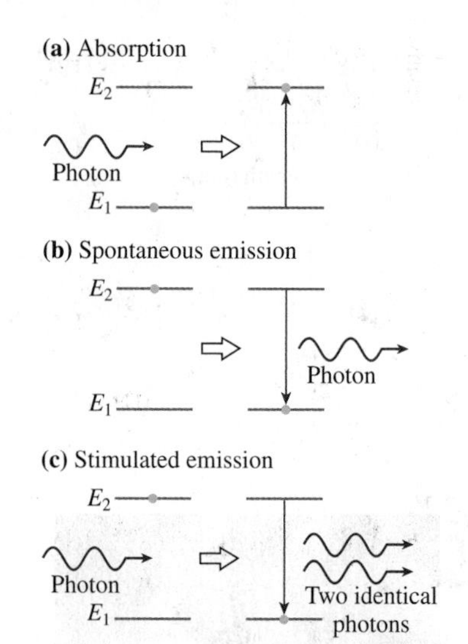

We have seen that an atom can jump from a lower-energy level E_1 to a higher-energy level E_2 by absorbing a photon. FIGURE 29.33a illustrates the process. Once in level 2, as shown in FIGURE 29.33b, the atom emits a photon of the same energy as it jumps back to level 1. Because this transition occurs spontaneously, it is called **spontaneous emission.**

In 1917, Einstein proposed a second mechanism by which an atom in state 2 can make a transition to state 1. The left part of FIGURE 29.33c shows a photon approaching an atom in its excited state 2. According to Einstein, if the energy of the photon is exactly equal to the energy difference $E_2 - E_1$ between the two states, this incoming photon can *induce* the atom to make the $2 \rightarrow 1$ transition, emitting a photon in the process. This process is called **stimulated emission.**

The incident photon is *not* absorbed in the process, so now there are *two* photons. And, interestingly, the emitted photon is *identical* to the incident photon. This means that as the two photons leave the atom they have exactly the same frequency and wavelength, are traveling in exactly the same direction, and are exactly in phase with each other. In other words, **stimulated emission produces a second photon that is an exact clone of the first.**

Stimulated emission is of no importance in most practical situations. Atoms spend almost all their time in their ground states, which makes stimulated emission extremely unlikely under ordinary conditions. In 1960, however, scientists discovered how to use stimulated emission to make a revolutionary new device: the laser.

Lasers

A spectacular laser light show depends on three key properties of coherent laser light: It can be very intense, its color is extremely pure, and the laser beam is narrow with little divergence.

The word **laser** is an acronym for the phrase *light amplification by the stimulated emission of radiation.* But what *is* a laser? Basically it is a device that produces a beam of highly *coherent* and essentially monochromatic (single-color) light as a result of stimulated emission. **Coherent** light is light in which all the electromagnetic waves have the same phase, direction, and amplitude. It is the coherence of a laser beam that allows it to be very tightly focused or to be rapidly modulated for communications.

Let's take a brief look at how a laser works. FIGURE 29.34 represents a system of atoms that have a lower energy level E_1 and a higher energy level E_2. Suppose that there are N_1 atoms in level 1 and N_2 atoms in level 2. Left to themselves, all the atoms would soon end up in level 1 because of the spontaneous emission $2 \rightarrow 1$. To prevent this, we can imagine that some type of excitation mechanism, perhaps an electrical discharge, continuously produces new excited atoms in level 2.

Let a photon of frequency $f = (E_2 - E_1)/h$ be incident on this group of atoms. Because it has the correct frequency, it could be absorbed by one of the atoms in level 1. Another possibility is that it could cause stimulated emission from one of the level 2 atoms. Ordinarily $N_2 \ll N_1$, so absorption events far outnumber stimulated emission events. Even if a few photons were generated by stimulated emission, they would quickly be absorbed by the vastly larger group of atoms in level 1.

But what if we could somehow arrange to place *every* atom in level 2, making $N_1 = 0$? Then the incident photon, upon encountering its first atom, will cause stimulated emission. Where there was initially one photon of frequency f, now there are two. These will strike two additional excited-state atoms, again causing stimulated emission. Then there will be four photons. As **FIGURE 29.35** shows, there will be a *chain reaction* of stimulated emission until all N_2 atoms emit a photon of frequency f.

In stimulated emission, each emitted photon is *identical* to the incident photon. The chain reaction of Figure 29.35 will lead not just to N_2 photons of frequency f, but to N_2 *identical* photons, all traveling together in the same direction with the same phase. If N_2 is a large number, as would be the case in any practical device, the one initial photon will have been *amplified* into a gigantic, coherent pulse of light!

FIGURE 29.34 Energy levels 1 and 2, with populations N_1 and N_2.

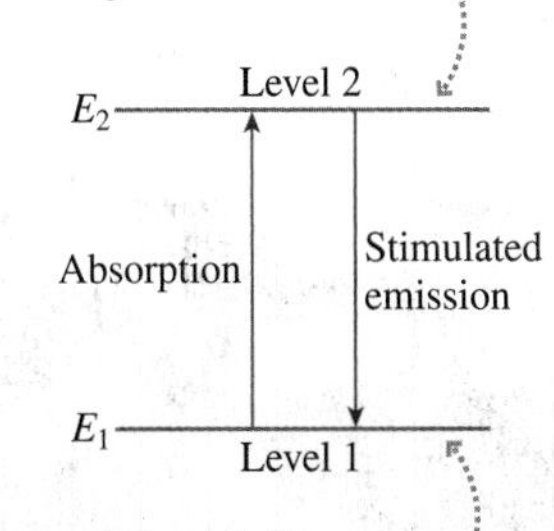

FIGURE 29.35 Stimulated emission creates a chain reaction of photon production in a population of excited atoms.

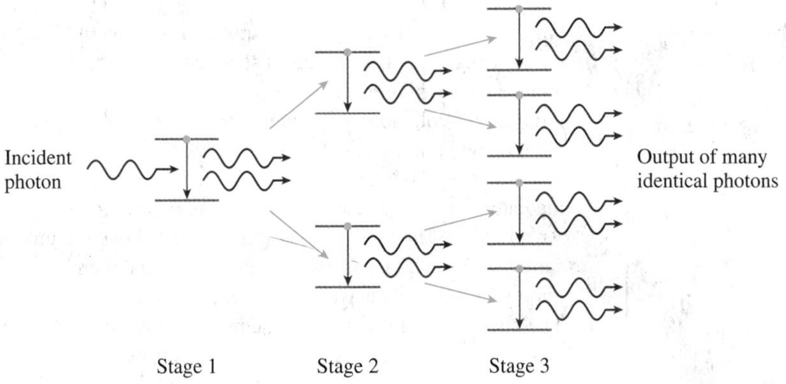

Although the chain reaction of Figure 29.35 illustrates the idea most clearly, it is not necessary for every atom to be in level 2 for amplification to occur. All that is needed is to have $N_2 > N_1$ so that stimulated emission exceeds absorption. Such a situation is called a **population inversion.** The stimulated emission is sustained by placing the *lasing medium*—the sample of atoms that emits the light—in an **optical cavity** consisting of two facing mirrors. As **FIGURE 29.36** shows, the photons interact repeatedly with the atoms in the medium as they bounce back and forth. This repeated interaction is necessary for the light intensity to build up to a high level. If one of the mirrors is partially transmitting, some of the light emerges as the *laser beam.*

FIGURE 29.36 Lasing takes place in an optical cavity.

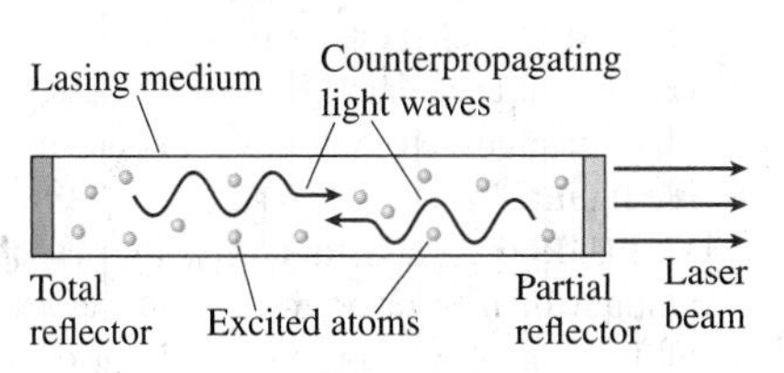

Lasers in Medicine

The invention of lasers was followed almost immediately by medical applications. Even a small-power laser beam can produce a significant amount of very localized *heating* if focused with a lens. More powerful lasers can easily *cut* through tissue by literally vaporizing it, replacing a stainless steel scalpel with a beam of light. Not only can laser surgery be very precise but it generally has less blood loss than conventional surgery because the heat of the laser seals the blood vessels and capillaries.

One common medical use of lasers is to remove plaque from artery walls, thus reducing the risk of stroke or heart attack. In this procedure, an optical fiber is threaded through arteries to reach the site. A powerful laser beam is then fired through the fiber to carefully vaporize the plaque. Laser beams traveling through optical fibers are also used to kill cancer cells in *photodynamic therapy*. In this case, light-sensitive chemicals are injected into the bloodstream and are preferentially taken up by cancer cells. The optical fiber is positioned next to the tumor and illuminates it with just the right wavelength to activate the light-sensitive chemicals and kill the cells. These procedures are minimally invasive, and they can reach areas of the body not readily accessible by conventional surgery.

◄ **Laser vision** BIO Laser-based LASIK surgery can correct for vision defects, such as near- or farsightedness, that result from an incorrect refractive power of the eye. You learned in Chapter 19 that the majority of your eye's focusing power occurs at the surface of the cornea. In LASIK, a special knife first cuts a small, thin flap in the cornea, and this flap is folded out of the way. A computer-controlled ultraviolet laser very carefully vaporizes the underlying corneal tissue to give it the desired shape; then the flap is folded back into place. The procedure takes only a few minutes and requires only a few numbing drops in the eye.

INTEGRATED EXAMPLE 29.10 Compact fluorescent lighting

You learned in Chapters 22 and 25 how an ordinary incandescent bulb works: Current passes through a filament, heating it until it glows white hot. But such bulbs are very inefficient, giving off only a few watts of visible light for every 100 W of electric power supplied to the bulb. Most of the power is, instead, converted into thermal energy.

For many years, offices and commercial buildings have used a very different type of lighting, the *fluorescent lamp*. Recently, compact fluorescent bulbs that can screw into ordinary lamp sockets have begun to make significant inroads into residential use. These bulbs are about four times more efficient than incandescent bulbs at transforming electric energy to visible light.

Inside the glass tube of a fluorescent bulb is a very small amount of mercury, which is in the form of a vapor when the bulb is on. Producing visible light occurs by a three-step process. First, a voltage of about 100 V is applied between electrodes at each end of the tube. This imparts kinetic energy to free electrons in the vapor, causing them to slam into mercury atoms and excite them by collisional excitation. Second, the excited atoms then jump to lower-energy states, emitting UV photons. Finally, the UV photons strike a *phosphor* that coats the inside of the tube, causing it to fluoresce with visible light. This is the light you see.

a. A mercury atom, after being collisionally excited by an electron, emits a photon with a wavelength of 185 nm in a quantum jump back to the ground state. If the electron starts from rest, what minimum distance must it travel to gain enough kinetic energy to cause this excitation? The 60-cm-long tube has 120 V applied between its ends.
b. After being collisionally excited, atoms sometimes emit two photons by jumping first from a high energy level to an intermediate level, giving off one photon, and then from this intermediate level to the ground state, giving off a second photon. An atom is excited to a state that is 7.79 eV above the ground state. It emits a 254-nm-wavelength photon and then a second photon. What is the wavelength of the second photon?
c. The energy-level diagram of the molecules in the phosphor is shown in **FIGURE 29.37**. After excitation by UV photons, what range of wavelengths can the phosphor emit by fluorescence?

FIGURE 29.37 Energy-level diagram of the phosphor molecules.

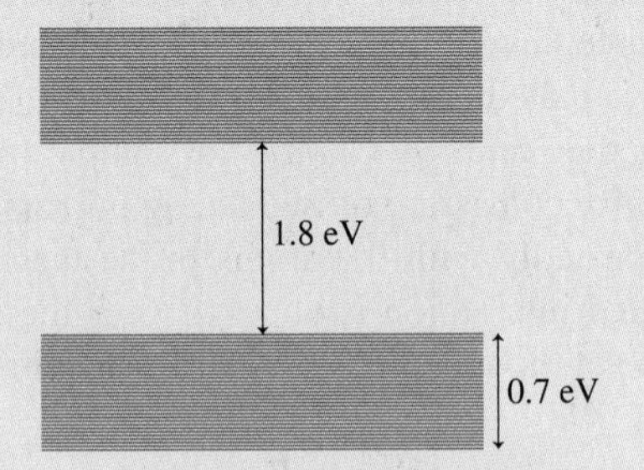

PREPARE An electron collisionally excites the mercury atom from its ground state to an excited state, increasing the atom's energy by ΔE_{atom}. Then the atom decays back to the ground state, giving off a 185-nm-wavelength photon. In order for collisional excitation to work, the kinetic energy of the incident electron must equal or exceed ΔE_{atom}; that is, it must equal or exceed the energy of a 185-nm-wavelength photon. The free electrons gain kinetic energy by accelerating through the potential difference inside the tube. We can use conservation of energy to find the distance an electron must travel to gain ΔE_{atom}, the minimum kinetic energy needed to cause an excitation.

In part b, the excited energy of the atom, 7.79 eV, is converted into photon energies. If we can find the energy of the first 254-nm-wavelength photon, the remaining energy must be that of the second photon, from which we can find its wavelength.

An inspection of Figure 29.31 shows that quantum jumps during fluorescence all begin at the bottom of the band of excited states but can end anywhere in the lower energy band. This range of energies will give us the range of wavelengths of the emitted photons.

SOLVE a. The minimum kinetic energy of the electron equals the energy of the 185-nm-wavelength photon that is subsequently emitted. This energy is

$$K_{\text{min}} = \Delta E_{\text{min}} = E_{\text{photon}} = \frac{hc}{\lambda} = \frac{1240 \text{ eV} \cdot \text{nm}}{185 \text{ nm}} = 6.7 \text{ eV}$$

where we have used the value of hc from Equation 29.22.

Recall that 1 eV is the kinetic energy gained by an electron as it accelerates through a 1 V potential difference. Here, the electron must gain a kinetic energy of 6.7 eV, so it must accelerate through a potential difference of 6.7 V. The fluorescent tube has a total potential drop of 120 V in 60 cm, or 2.0 V/cm. Thus to accelerate through a 6.7 V potential difference, the electron must travel a distance

$$\Delta x = \frac{6.7 \text{ V}}{2.0 \text{ V/cm}} = 3.4 \text{ cm}$$

b. The first emitted photon has energy

$$E_{\text{photon}} = \frac{hc}{\lambda} = \frac{1240 \text{ eV} \cdot \text{nm}}{254 \text{ nm}} = 4.88 \text{ eV}$$

The energy remaining to the second photon is then $7.79 \text{ eV} - 4.88 \text{ eV} = 2.91 \text{ eV}$; the wavelength of this photon is

$$\lambda = \frac{hc}{E_{\text{photon}}} = \frac{1240 \text{ eV} \cdot \text{nm}}{2.91 \text{ eV}} = 426 \text{ nm}$$

c. The energy of the photon emitted during a quantum jump from the bottom of the upper energy band to the top of the lower energy band is 1.8 eV, corresponding to a wavelength of

$$\lambda = \frac{hc}{E_{\text{photon}}} = \frac{1240 \text{ eV} \cdot \text{nm}}{1.8 \text{ eV}} = 690 \text{ nm}$$

The photon energy for a jump to the bottom of the lower energy band is $1.8 \text{ eV} + 0.7 \text{ eV} = 2.5 \text{ eV}$, with a wavelength of

$$\lambda = \frac{hc}{E_{\text{photon}}} = \frac{1240 \text{ eV} \cdot \text{nm}}{2.5 \text{ eV}} = 500 \text{ nm}$$

Thus this phosphor, after absorbing UV photons, emits visible light with a wavelength range of 500–690 nm.

ASSESS For part a, it seems reasonable that the electron travels only a small fraction of the tube length before gaining enough energy to collisionally excite an atom. The bulb would be very inefficient if electrons had to travel the full length before colliding with a mercury atom. The photon wavelengths in parts b and c also seem reasonable. In particular, the range of wavelengths emitted by the phosphor is a little more than half the visible spectrum, from green through red but missing blue and violet. Compact fluorescent tubes have three different phosphors, each with a somewhat different range of emission wavelengths, to give the full spectrum of white light. Slightly altering the balance between these phosphors distinguishes a "warm white" bulb from a "cool white" bulb.

SUMMARY

Goal: To use quantum physics to understand the properties of atoms, molecules, and their spectra.

IMPORTANT CONCEPTS

The Structure of an Atom

An atom consists of a very small, positively charged nucleus, surrounded by orbiting electrons.

- The number of protons is the atom's **atomic number** Z.
- The **atomic mass number** A is the number of protons + the number of neutrons.

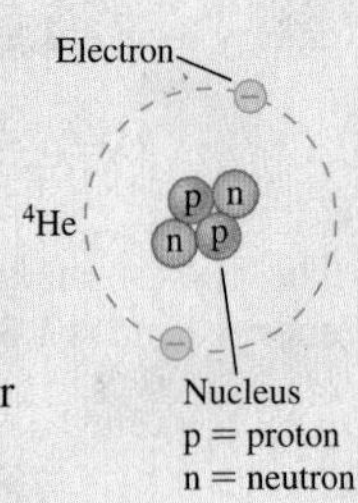

The Bohr Atom

In Bohr's model,

- The atom can exist in only certain **stationary states.** These states correspond to different electron orbits. Each state is numbered by **quantum number** $n = 1, 2, 3, \ldots$.
- Each state has a discrete, well-defined energy E_n.
- The atom can change its energy by undergoing a **quantum jump** between two states by emitting or absorbing a photon of energy $E_{\text{photon}} = \Delta E_{\text{atom}} = |E_f - E_i|$.

The Hydrogen Atom

In **Bohr's model of the hydrogen atom** the stationary states are found by requiring an integer number of de Broglie wavelengths to fit around the circumference of the electron's orbit: $2\pi r = n\lambda$. The integer n is the *principal quantum number.*

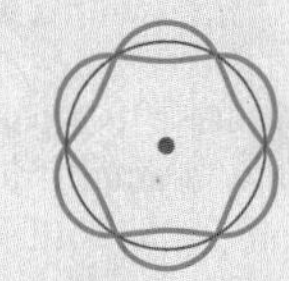

This leads to energy quantization with

$$E_n = -\frac{13.60 \text{ eV}}{n^2}$$

$n = 3$, $r_3 = 9a_B$, $E_3 = -1.5$ eV
$n = 2$, $r_2 = 4a_B$, $E_2 = -3.4$ eV
$n = 1$, $r_1 = a_B$, $E_1 = -13.6$ eV

and orbit radii $r_n = n^2 a_B$, where $a_B = 0.053$ nm is the **Bohr radius.**

The wavelengths of light in the hydrogen atom spectrum are given by the **Balmer formula:**

$$\lambda_{n \to m} = \frac{91.1 \text{ nm}}{\left(\frac{1}{m^2} - \frac{1}{n^2}\right)} \qquad \begin{array}{l} m = 1, 2, 3, \ldots \\ n = m + 1, m + 2, \ldots \end{array}$$

Beyond the Bohr model, quantum mechanics adds other quantized parameters, each with its own quantum number:

- The *orbital angular momentum,* quantum number l:

$$L = \sqrt{l(l+1)}\hbar \qquad l = 0, 1, 2, 3, \ldots, n - 1$$

- The *angle of the electron's orbit,* quantum number m:

$$m = -l, -l + 1, \ldots, 0, \ldots, l - 1, l$$

- The *direction of the electron* **spin,** quantum number m_s:

$$m_s = -\tfrac{1}{2} \text{ or } +\tfrac{1}{2}$$

The energy of a hydrogen atom depends only on n:

0 eV
3s 3p 3d
2s 2p
All states with the same n have the same energy.
−13.60 eV 1s

Multielectron atoms

Each electron is described by the same quantum numbers (n, l, m, m_s) used for the hydrogen atom, but the energy now depends on l as well as n.

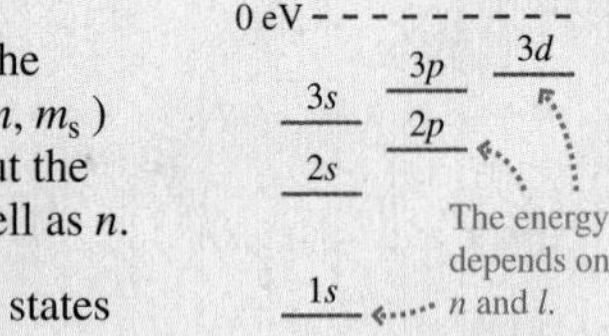

The **Pauli exclusion principle** states that no more than one electron can occupy each quantum state.

Molecules

In molecules, the states are spaced very closely into **bands** of states. Because electrons can be excited to and from many states, the spectra of molecules are broad, not discrete.

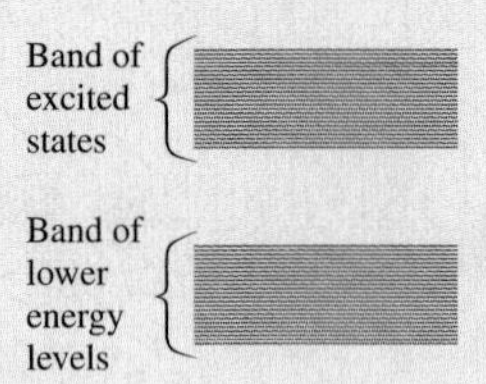

APPLICATIONS

Atomic emission spectra are generated by excitation followed by a photon-emitting quantum jump.

- **Excitation** occurs by absorption of a photon or by collision.
- A quantum jump can occur only if $\Delta l = \pm 1$.
- Quantized energies give rise to a **discrete spectrum.**

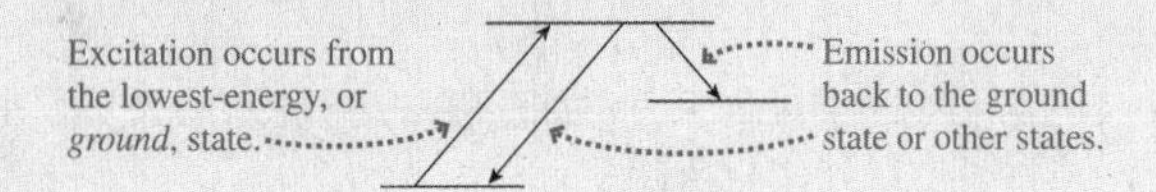

Excitation occurs from the lowest-energy, or ground, state.
Emission occurs back to the ground state or other states.

Lasers

A photon with energy $E_{\text{photon}} = E_2 - E_1$ can induce **stimulated emission** of a second photon identical to the first. These photons can then induce more atoms to emit photons. If more atoms are in state 2 than in state 1, this process can rapidly build up an intense beam of identical photons. This is the principle behind the laser.

Problem difficulty is labeled as | (straightforward) to ||||| (challenging). Problems labeled INT integrate significant material from earlier chapters; BIO are of biological or medical interest.

For assigned homework and other learning materials, go to MasteringPhysics®

Scan this QR code to launch a Video Tutor Solution that will help you solve problems for this chapter.

QUESTIONS

Conceptual Questions

1. A neon discharge emits a bright reddish-orange spectrum. But a glass tube filled with neon is completely transparent. Why doesn't the neon in the tube absorb orange and red wavelengths?
2. The two spectra shown in Figure Q29.2 belong to the same element, a fictional Element X. Explain why they are different.

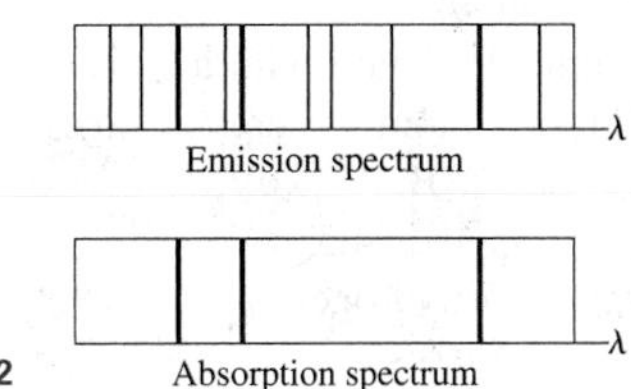

FIGURE Q29.2

3. Is a spectral line with wavelength 656 nm seen in the absorption spectrum of hydrogen atoms? Why or why not?
4. Can nuclei of the same element have different values of Z? Of N? Of A? Can nuclei of different elements have the same values of Z? Of N? Of A?
5. An atom has four energy levels. How many spectral lines are seen in its emission spectrum? In its absorption spectrum? How many absorption lines?
6. If an electron is in a *stationary state* of an atom, is the electron at rest? If not, what does the term mean?
7. The $n = 3$ state of hydrogen has $E_3 = -1.51$ eV.
   a. Why is the energy negative?
   b. What is the physical significance of the specific number 1.51 eV?
8. For a hydrogen atom, list all possible states (n, l, m, m_s) that have $E = -1.51$ eV.
9. What are the n and l values of the following states of a hydrogen atom: (a) $4d$, (b) $5f$, (c) $6s$?
10. How would you label the hydrogen-atom states with the following (n, l, m) quantum numbers: (a) (4, 3, 0), (b) (3, 2, 1), (c) (3, 2, −1)?
11. A hydrogen atom is in a state with principal quantum number $n = 5$. What possible values of the orbital quantum number l could this atom have?
12. Consider the two hydrogen-atom states $5d$ and $4f$. Which has the higher energy? Explain.
13. In a multielectron atom, does a $5s$ electron have higher or lower energy than a $4f$ electron? Explain.
14. Do the following electron configurations represent a possible state of an element? If so, (i) identify the element and (ii) determine if this is the ground state or an excited state. If not, why not?
   a. $1s^2 2s^2 2p^6 3s^2$
   b. $1s^2 2s^2 2p^7 3s$
   c. $1s^2 2s^2 2p^4 3s^2 3p^2$
15. Why is the section of the periodic table labeled as "transition elements" exactly 10 elements wide in all rows?
16. An electron is in an f state. Can it undergo a quantum jump to an s state? A p state? A d state? Explain.
17. Figure Q29.17 shows the energy-level diagram of Element X.
   a. What is the ionization energy of Element X?
   b. An atom in the ground state absorbs a photon, then emits a photon with a wavelength of 1240 nm. What conclusion can you draw about the energy of the photon that was absorbed?
   c. An atom in the ground state has a collision with an electron, then emits a photon with a wavelength of 1240 nm. What conclusion can you draw about the initial kinetic energy of the electron?

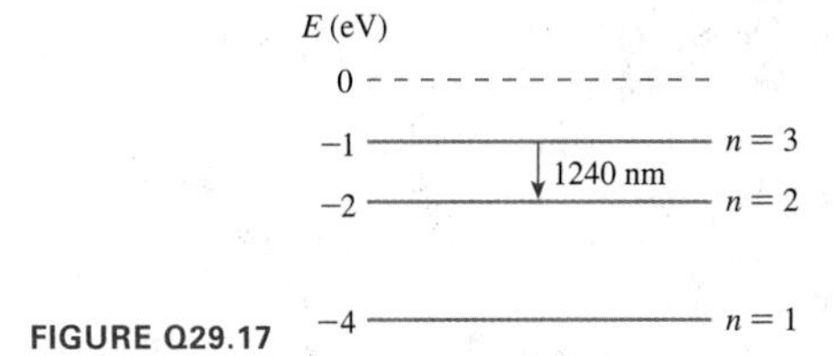

FIGURE Q29.17

18. a. Which states of a hydrogen atom can be excited by a collision with an electron with kinetic energy $K = 12.5$ eV? Explain.
   b. After the collision the atom is not in its ground state. What happens to the electron? (i) It bounces off with $K > 12.5$ eV, (ii) It bounces off with $K = 12.5$ eV, (iii) It bounces off with $K < 12.5$ eV, (iv) It is absorbed by the atom. Explain your choice.
   c. After the collision, the atom emits a photon. List all the possible $n \rightarrow m$ transitions that might occur as a result of this collision.
19. What *is* an atom's ionization energy? In other words, if you know the ionization energy of an atom, what is it that you know about the atom?
20. Figure Q29.20 shows the energy levels of a hypothetical atom.
   a. What *minimum* kinetic energy (in eV) must an electron have to collisionally excite this atom and cause the emission of a 620 nm photon? Explain.
   b. Can an electron with $K = 6$ eV cause the emission of 620 nm light? If so, what is the final kinetic energy of the electron? If not, why not?
   c. Can a 6 eV photon cause the emission of 620 nm light from this atom? Why or why not?
   d. Can a 7 eV photon cause the emission of 620 nm light from this atom? Why or why not?

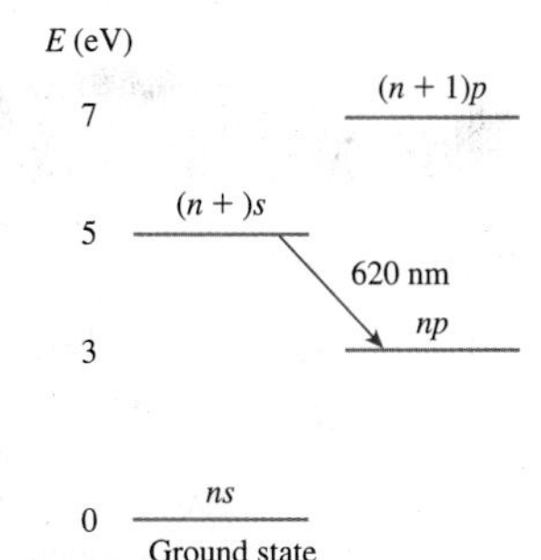

FIGURE Q29.20

21. Seven possible transitions are identified on the energy-level diagram in Figure Q29.21. For each, is this an allowed transition? If allowed, is it an emission or an absorption transition, and is the photon infrared, visible, or ultraviolet? If not allowed, why not?

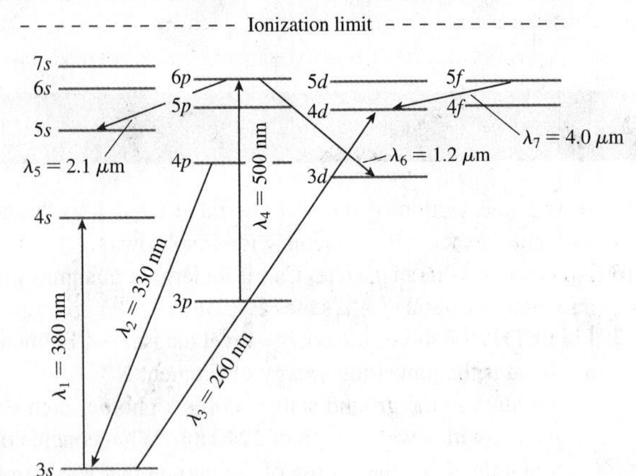

FIGURE Q29.21

22. A 2.0 eV photon is incident on an atom in the p state, as shown in the energy-level diagram in Figure Q29.22. Does the atom undergo an absorption transition, a stimulated emission transition, or neither? Explain.

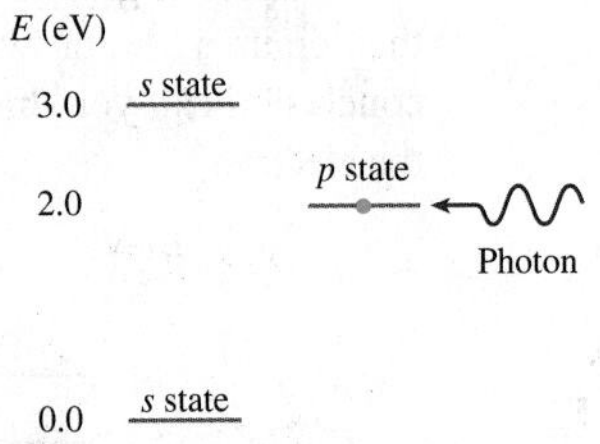

E (eV)
3.0 s state
2.0 p state — Photon
0.0 s state

FIGURE Q29.22

23. A glass tube contains 2×10^{11} atoms, some of which are in the ground state and some of which are excited. Figure Q29.23 shows the populations for the atoms' three energy levels. Is it possible for these atoms to be a laser? If so, on which transition would laser action occur? If not, why not?

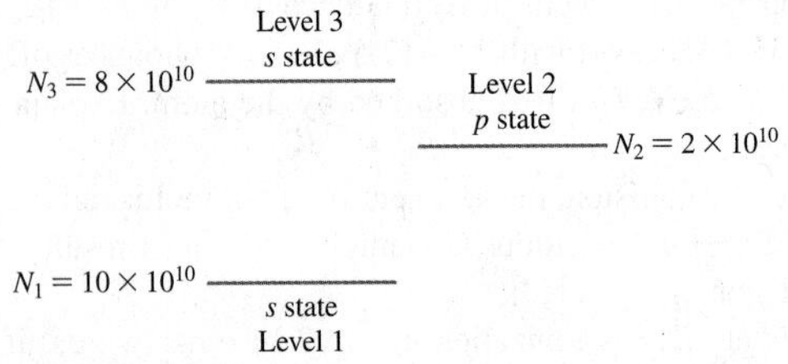

Level 3, s state: $N_3 = 8 \times 10^{10}$
Level 2, p state: $N_2 = 2 \times 10^{10}$
Level 1, s state: $N_1 = 10 \times 10^{10}$

FIGURE Q29.23

Multiple-Choice Questions

24. | An electron collides with an atom in its ground state. The atom then emits a photon of energy E_{photon}. In this process the *change* ΔE_{elec} in the electron's energy is
    A. Greater than E_{photon}.
    B. Greater than or equal to E_{photon}.
    C. Equal to E_{photon}.
    D. Less than or equal to E_{photon}.
    E. Less than E_{photon}.
25. || How many states are in the $l = 4$ subshell?
    A. 8 B. 9 C. 16 D. 18 E. 22
26. | What is the ground-state electron configuration of calcium ($Z = 20$)?
    A. $1s^2 2s^2 2p^6 3s^2 3p^8$
    B. $1s^2 2s^2 2p^6 3s^2 3p^6 4s^1 4p^1$
    C. $1s^2 2s^2 2p^6 3s^2 3p^6 4s^2$
    D. $1s^2 2s^2 2p^6 3s^2 3p^6 4p^2$
27. | An atom emits a photon with a wavelength of 275 nm. By how much does the atom's energy change?
    A. 0.72 eV B. 1.06 eV C. 2.29 eV
    D. 3.06 eV E. 4.51 eV
28. || The energy of a hydrogen atom is -3.40 eV. What is the electron's kinetic energy?
    A. 1.70 eV B. 2.62 eV C. 3.40 eV
    D. 5.73 eV E. 6.80 eV
29. | The angular momentum of an electron in a Bohr hydrogen atom is 3.18×10^{-34} kg · m^2/s. What is the atom's energy?
    A. -13.60 eV
    B. -6.73 eV
    C. -3.40 eV
    D. -1.51 eV
    E. -0.47 eV
30. | A "soft x-ray" photon with an energy of 41.8 eV is absorbed by a hydrogen atom in its ground state, knocking the atom's electron out. What is the speed of the electron as it leaves the atom?
    A. 1.84×10^5 m/s
    B. 3.08×10^5 m/s
    C. 8.16×10^5 m/s
    D. 3.15×10^6 m/s
    E. 3.83×10^6 m/s

PROBLEMS

Section 29.1 Spectroscopy

1. | Figure 29.2 and Table 29.1 showed the wavelengths of the first four lines in the visible spectrum of hydrogen.
    a. Determine the Balmer formula n and m values for these wavelengths.
    b. Predict the wavelength of the fifth line in the spectrum.
2. | The wavelengths in the hydrogen spectrum with $m = 1$ form a series of spectral lines called the Lyman series. Calculate the wavelengths of the first four members of the series.
3. | The Paschen series is analogous to the Balmer series, but with $m = 3$. Calculate the wavelengths of the first three members in the Paschen series. What part(s) of the electromagnetic spectrum are these in?

Section 29.2 Atoms

4. | How many electrons, protons, and neutrons are contained in the following atoms or ions: (a) ^{6}Li, (b) ^{13}C$^+$, and (c) ^{18}O^{++}?
5. | How many electrons, protons, and neutrons are contained in the following atoms or ions: (a) ^{9}Be$^+$, (b) ^{12}C, and (c) ^{15}N^{+++}?
6. | Write the symbol for an atom or ion with:
    a. four electrons, four protons, and five neutrons.
    b. six electrons, seven protons, and eight neutrons.
7. | Write the symbol for an atom or ion with:
    a. three electrons, three protons, and five neutrons.
    b. five electrons, six protons, and eight neutrons.

Watch video solution for problem: 29.33

Section 29.3 Bohr's Model of Atomic Quantization

8. | Figure P29.8 is an energy-level diagram for a simple atom. What wavelengths appear in the atom's (a) emission spectrum and (b) absorption spectrum?

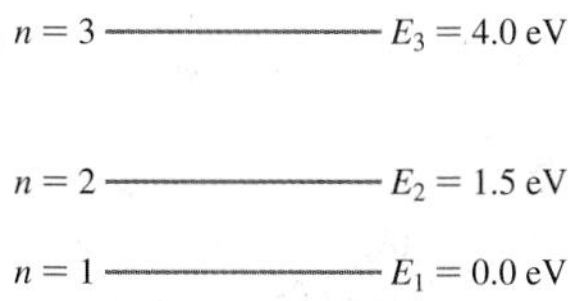

$n = 3$ ——— $E_3 = 4.0$ eV

$n = 2$ ——— $E_2 = 1.5$ eV

$n = 1$ ——— $E_1 = 0.0$ eV

FIGURE P29.8

9. | An electron with 2.0 eV of kinetic energy collides with the atom whose energy-level diagram is shown in Figure P29.8.
   a. Is the electron able to kick the atom to an excited state? Why or why not?
   b. If your answer to part a was yes, what is the electron's kinetic energy after the collision?
10. | The allowed energies of a simple atom are 0.0 eV, 4.0 eV, and 6.0 eV.
    a. Draw the atom's energy-level diagram. Label each level with the energy and the principal quantum number.
    b. What wavelengths appear in the atom's emission spectrum?
    c. What wavelengths appear in the atom's absorption spectrum?
11. ||| The allowed energies of a simple atom are 0.0 eV, 4.0 eV, and 6.0 eV. An electron traveling at a speed of 1.6×10^6 m/s collisionally excites the atom. What are the minimum and maximum speeds the electron could have after the collision?

Section 29.4 The Bohr Hydrogen Atom

12. || A researcher observes hydrogen emitting photons of energy 1.89 eV. What are the quantum numbers of the two states involved in the transition that emits these photons?
13. | A hydrogen atom is in the $n = 3$ state. In the Bohr model, how many electron wavelengths fit around this orbit?
14. || A hydrogen atom is in its $n = 1$ state. In the Bohr model, what is the ratio of its kinetic energy to its potential energy?
15. || Infrared light with a wavelength of 1870 nm is emitted from hydrogen. What are the quantum numbers of the two states involved in the transition that emits this light?
16. | a. What quantum number of the hydrogen atom comes closest to giving a 500-nm-diameter electron orbit?
    b. What are the electron's speed and energy in this state?
17. | a. Calculate the de Broglie wavelength of the electron in the $n = 1$, 2, and 3 states of the hydrogen atom. Use the information in Table 29.2.
    b. Show numerically that the circumference of the orbit for each of these stationary states is exactly equal to n de Broglie wavelengths.
    c. Sketch the de Broglie standing wave for the $n = 3$ orbit.
18. ||| A hydrogen atom is excited from its ground state to the $n = 4$ state. The atom subsequently emits three photons, one of which has a wavelength of 656 nm. What are the wavelengths of the other two photons?
19. | Determine all possible wavelengths of photons that can be emitted from the $n = 4$ state of a hydrogen atom.

Section 29.5 The Quantum-Mechanical Hydrogen Atom

20. | List the quantum numbers of (a) all possible $3p$ states and (b) all possible $3d$ states.
21. || When all quantum numbers are considered, how many different quantum states are there for a hydrogen atom with $n = 1$? With $n = 2$? With $n = 3$? List the quantum numbers of each state.
22. | What is the angular momentum of a hydrogen atom in (a) a $4p$ state and (b) a $5f$ state? Give your answers as a multiple of $\hbar$.
23. || The energy of a hydrogen atom is 12.09 eV above its ground-state energy. As a multiple of $\hbar$, what is the largest angular momentum that this atom could have?
24. || A hydrogen atom is in the $5p$ state. Determine (a) its energy, (b) its angular momentum, (c) its quantum number l, and (d) the possible values of its magnetic quantum number m.
25. ||| The angular momentum of a hydrogen atom is 4.70×10^{-34} J · s. What is the minimum energy, in eV, that this atom could have?

Section 29.6 Multielectron Atoms

26. | Predict the ground-state electron configurations of Mg, Sr, and Ba.
27. || Predict the ground-state electron configurations of Si, Ge, and Pb.
28. | Identify the element for each of these electron configurations. Then determine whether this configuration is the ground state or an excited state.
    a. $1s^22s^22p^5$
    b. $1s^22s^22p^63s^23p^63d^{10}4s^24p$
29. || a. With what element is the $3s$ subshell first completely filled?
    b. With what element is the $5d$ subshell first half filled?
30. | Identify the element for each of these electron configurations. Then determine whether this configuration is the ground state or an excited state.
    a. $1s^22s^22p^63s^23p^64s^23d^9$
    b. $1s^22s^22p^63s^23p^64s^23d^{10}4p^65s^24d^{10}5p^66s^24f^{14}5d^7$
31. | Explain what is wrong with these electron configurations:
    a. $1s^22s^22p^83s^23p^4$
    b. $1s^22s^32p^4$
32. || Which has higher energy: an electron in the $4f$ state or an electron in the $7s$ state? Explain.

Section 29.7 Excited States and Spectra

33. | An electron with a speed of 5.00×10^6 m/s collides with an atom. The collision excites the atom from its ground state (0 eV) to a state with an energy of 3.80 eV. What is the speed of the electron after the collision?
34. | Hydrogen gas absorbs light of wavelength 103 nm. Afterward, what wavelengths are seen in the emission spectrum?
35. || What is the longest wavelength of light that can excite the $4s$ state of sodium?
36. || An electron with a kinetic energy of 3.90 eV collides with a sodium atom. What possible wavelengths of light are subsequently emitted?
37. | a. Is a $4p \rightarrow 4s$ transition allowed in sodium? If so, what is its wavelength? If not, why not?
    b. Is a $3d \rightarrow 4s$ transition allowed in sodium? If so, what is its wavelength? If not, why not?

Section 29.8 Molecules

38. || Figure P29.38 shows a molecular energy-level diagram. What are the longest and shortest wavelengths in (a) the molecule's absorption spectrum and (b) the molecule's fluorescence spectrum?

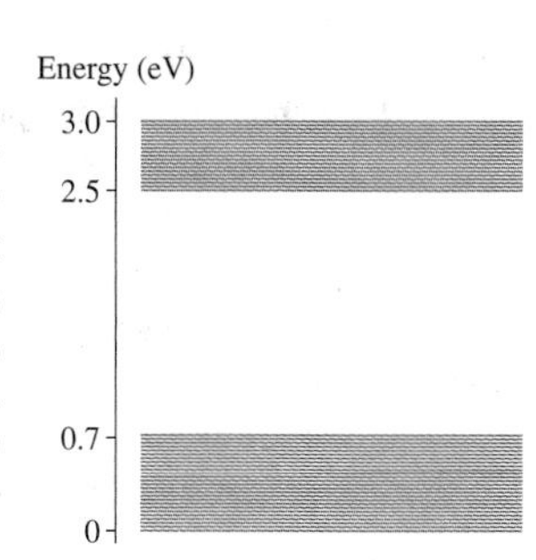

FIGURE P29.38

Watch video solutions for problems: 29.47 29.55

39. | The molecule whose energy-level diagram is shown in Figure P29.38 is illuminated by 2.7 eV photons. What is the longest wavelength of light that the molecule can emit?

Section 29.9 Stimulated Emission and Lasers

40. | A 1000 W carbon dioxide laser emits an infrared laser beam with a wavelength of 10.6 μm. How many photons are emitted per second?
41. || A 1.00 mW helium-neon laser emits a visible laser beam with a wavelength of 633 nm. How many photons are emitted per second?
42. || In LASIK surgery, a laser is used to reshape the cornea of the eye to improve vision. The laser produces extremely short pulses of light, each containing 1.0 mJ of energy. BIO INT
    a. In each pulse there are 9.7×10^{14} photons. What is the wavelength of the laser?
    b. Each pulse lasts only 20 ns. What is the average power delivered to the eye during a pulse?
43. || A ruby laser emits an intense pulse of light that lasts a mere 10 ns. The light has a wavelength of 690 nm, and each pulse has an energy of 500 mJ.
    a. How many photons are emitted in each pulse?
    b. What is the *rate* of photon emission, in photons per second, during the 10 ns that the laser is "on"?

General Problems

44. ||| A 2.55 eV photon is emitted from a hydrogen atom. What are the Balmer formula n and m values corresponding to this emission?
45. | Two of the wavelengths emitted by a hydrogen atom are 102.6 nm and 1876 nm.
    a. What are the Balmer formula n and m values for each of these wavelengths?
    b. For each of these wavelengths, is the light infrared, visible, or ultraviolet?
46. |||| In Example 29.2 it was assumed that the initially stationary gold nucleus would remain motionless during a head-on collision with an 8.3 MeV alpha particle. What is the actual recoil speed of the gold nucleus after that elastic collision? Assume that the mass of a gold nucleus is exactly 50 times the mass of an alpha particle. INT
    Hint: Review the discussion of perfectly elastic collisions in Chapter 10.
47. | Consider the gold isotope ^{197}Au. INT
    a. How many electrons, protons, and neutrons are in a neutral ^{197}Au atom?
    b. The gold nucleus has a diameter of 14.0 fm. What is the density of matter in a gold nucleus?
    c. The density of lead is 11,400 kg/m^3. How many times the density of lead is your answer to part b?
48. || Consider the lead isotope ^{207}Pb. INT
    a. How many electrons, protons, and neutrons are in a neutral ^{207}Pb atom?
    b. The lead nucleus has a diameter of 14.2 fm. What are the electric potential and the electric field strength at the surface of a lead nucleus?
49. ||| The diameter of an atom is 1.2×10^{-10} m and the diameter of its nucleus is 1.0×10^{-14} m. What percent of the atom's volume is occupied by mass and what percent is empty space?
50. | The charge-to-mass ratio of a nucleus, in units of e/u, is $q/m = Z/A$. For example, a hydrogen nucleus has $q/m = 1/1 = 1$.
    a. Make a graph of charge-to-mass ratio versus proton number Z for nuclei with $Z = 5, 10, 15, 20, \ldots, 90$. For A, use the average atomic mass shown on the periodic table of elements in Appendix B. Show each of these 18 nuclei as a dot, but don't connect the dots together as a curve.
    b. Describe any trend that you notice in your graph.
    c. What's happening in the nuclei that is responsible for this trend?
51. | If the nucleus is a few fm in diameter, the distance between the centers of two protons must be ≈ 2 fm. INT
    a. Calculate the repulsive electric force between two protons that are 2.0 fm apart.
    b. Calculate the attractive gravitational force between two protons that are 2.0 fm apart. Could gravity be the force that holds the nucleus together?
52. | In a head-on collision, the closest approach of a 6.24 MeV alpha particle to the center of a nucleus is 6.00 fm. The nucleus is in an atom of what element? Assume that the nucleus is heavy enough to remain stationary during the collision. INT
53. |||| A 20 MeV alpha particle is fired toward a ^{238}U nucleus. It follows the path shown in Figure P29.53. What is the alpha particle's speed when it is closest to the nucleus, 20 fm from its center? Assume that the nucleus doesn't move. INT

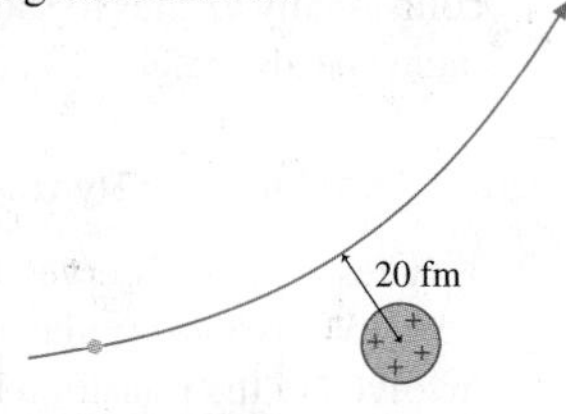

FIGURE P29.53

54. || The oxygen nucleus ^{16}O has a radius of 3.0 fm. INT
    a. With what speed must a proton be fired toward an oxygen nucleus to have a turning point 1.0 fm from the surface? Assume that the nucleus is heavy enough to remain stationary during the collision.
    b. What is the proton's kinetic energy in MeV?
55. || The absorption spectrum of an atom consists of the wavelengths 200 nm, 300 nm, and 500 nm.
    a. Draw the atom's energy-level diagram.
    b. What wavelengths are seen in the atom's emission spectrum?
56. || The first three energy levels of the fictitious element X are shown in Figure P29.56.
    a. What wavelengths are observed in the absorption spectrum of element X? Give your answers in nm.
    b. State whether each of your wavelengths in part a corresponds to ultraviolet, visible, or infrared light.
    c. An electron with a speed of 1.4×10^6 m/s collides with an atom of element X. Shortly afterward, the atom emits a 1240 nm photon. What was the electron's speed after the collision? Assume that, because the atom is so much more massive than the electron, the recoil of the atom is negligible.

    Hint: The energy of the photon is *not* the energy transferred to the atom in the collision.

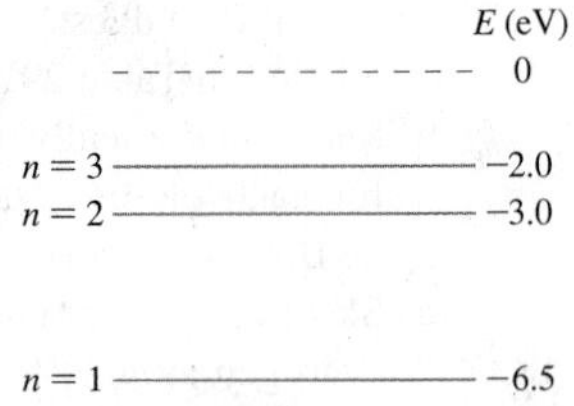

	E (eV)
	0
$n = 3$	−2.0
$n = 2$	−3.0
$n = 1$	−6.5

FIGURE P29.56

57. || A simple atom has four lines in its absorption spectrum. Ignoring any selection rules, how many lines will it have in its emission spectrum?

58. ||| A simple atom has only two absorption lines, at 250 nm and 600 nm. What is the wavelength of the one line in the emission spectrum that does not appear in the absorption spectrum?

59. | What is the wavelength of the series limit (i.e., the shortest possible wavelength) of the Lyman series in hydrogen?

60. || What is the energy of a Bohr hydrogen atom with a 5.18 nm diameter?

61. | A hydrogen atom in the ground state absorbs a 12.75 eV photon. Immediately after the absorption, the atom undergoes a quantum jump to the next-lowest energy level. What is the wavelength of the photon emitted in this quantum jump?

62. ||| Potassium atom 1, with a mass of 6.47×10^{-26} kg, has a speed of 1.30×10^4 m/s. It collides with potassium atom 2 that is at rest, exciting atom 2 to a state from which it emits a 3.06 eV photon as it returns to the ground state. Atom 2 leaves the collision with a speed of 1.17×10^4 m/s. What is the final speed of atom 1?

63. | INT
a. Calculate the orbital radius and the speed of an electron in both the $n = 99$ and the $n = 100$ states of hydrogen.
b. Determine the orbital frequency of the electron in each of these states.
c. Calculate the frequency of a photon emitted in a $100 \rightarrow 99$ transition.
d. Compare the photon frequency of part c to the *average* of your two orbital frequencies from part b. By what percent do they differ?

64. ||| INT Two hydrogen atoms collide head-on. The collision brings both atoms to a halt. Immediately after the collision, both atoms emit a 121.6 nm photon. What was the speed of each atom just before the collision?

65. ||| INT A beam of electrons is incident on a gas of hydrogen atoms.
a. What minimum speed must the electrons have to cause the emission of 656 nm light from the $3 \rightarrow 2$ transition of hydrogen?
b. Through what potential difference must the electrons be accelerated to have this speed?

66. ||| A hydrogen atom in its fourth excited state emits a photon with a wavelength of 1282 nm. What is the atom's maximum possible orbital angular momentum after the emission? Give your answer as a multiple of $\hbar$.

67. || A particular emission line in the hydrogen spectrum has a wavelength of 656.5 nm. What are all possible transitions (e.g., $6d \rightarrow 2s$) that could give rise to this emission?

68. ||| BIO INT Germicidal lamps are used to sterilize tools in biological and medical facilities. One type of germicidal lamp is a low-pressure mercury discharge tube, similar to a fluorescent lightbulb, that has been optimized to emit ultraviolet light with a wavelength of 254 nm. A 15-mm-diameter, 25-cm-long tube emits 4.5 W of ultraviolet light. The mercury vapor pressure inside the tube is 1.0 Pa at the operating temperature of 40°C.
a. On average, how many ultraviolet photons does each mercury atom emit per second?
b. The lifetime of the excited state in mercury is a rather long 120 ns. On average, what fraction of its time does a mercury atom spend in the excited state?

69. | a. What downward transitions are possible for a sodium atom in the $6s$ state? (See Figure 29.24.)
b. What are the wavelengths of the photons emitted in each of these transitions?

70. || The $5d \rightarrow 3p$ transition in the emission spectrum of sodium (see Figure 29.24) has a wavelength of 499 nm. What is the energy of the $5d$ state?

71. | A sodium atom (see Figure 29.28) in the $3p$ state emits a photon, decreasing the energy of the atom by 2.10 eV. By how much does the orbital angular momentum of the atom change? Does it increase or decrease?

72. || Figure P29.72 shows a few energy levels of the mercury atom. One valence electron is always in the $6s$ state; the other electron changes states.
a. Make a table showing all the allowed transitions in the emission spectrum. For each transition, indicate the photon wavelength, in nm.
b. What minimum speed must an electron have to excite the 492-nm-wavelength blue emission line in the Hg spectrum?

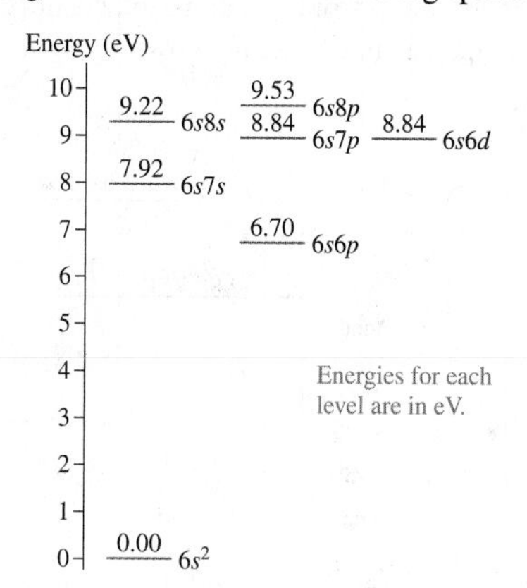

FIGURE P29.72

73. | Figure P29.73 shows the first few energy levels of the lithium atom. Make a table showing all the allowed transitions in the emission spectrum. For each transition, indicate
a. The wavelength, in nm.
b. Whether the transition is in the infrared, the visible, or the ultraviolet spectral region.
c. Whether or not the transition would be observed in the lithium absorption spectrum.

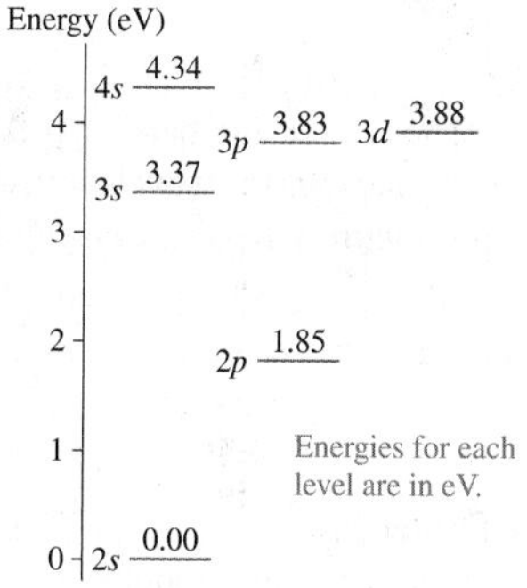

Energy (eV): $4s$ 4.34; $3p$ 3.83; $3d$ 3.88; $3s$ 3.37; $2p$ 1.85; $2s$ 0.00. Energies for each level are in eV.

FIGURE P29.73

74. | INT A laser emits 1.00×10^{19} photons per second from an excited state with energy $E_2 = 1.17$ eV. The lower energy level is $E_1 = 0$ eV.
a. What is the wavelength of this laser?
b. What is the power output of this laser?

75. | BIO INT Fluorescence microscopy, discussed in Section 29.8, is an important tool in modern cell biology. A variation on this technique is called *two-photon excitation*. If two photons are absorbed simultaneously their energies can add. For example, molecule that is normally excited by a 350-nm-wavelength photon can be excited by two photons each having half as much energy. For this process to be useful, photons must illuminate the sample at the very high rate of at least 1.0×10^{29} photons/m² · s. This is achieved by focusing a laser beam to a small spot and by concentrating the energy of the laser into very short ($<10^{-12}$ s) pulses that are fired 10^8 times each second. Suppose a biologist wants to use two-photon excitation to excite a dye molecule that would be excited by 500-nm-wavelength light in normal one-photon fluorescence microscopy. What minimum intensity (W/m²) must the laser beam have during each pulse?

MCAT-Style Passage Problems

Light-Emitting Diodes

Light-emitting diodes, known by the acronym LED, produce the familiar green and red indicator lights used in a wide variety of consumer electronics. LEDs are *semiconductor devices* in which the electrons can exist only in certain energy levels. Much like molecules, the energy levels are packed together close enough to form what appears to be a continuous band of possible energies. Energy supplied to an LED in a circuit excites electrons from a *valence band* into a *conduction band.* An electron can emit a photon by undergoing a quantum jump from a state in the conduction band into an empty state in the valence band, as shown in Figure P29.76.

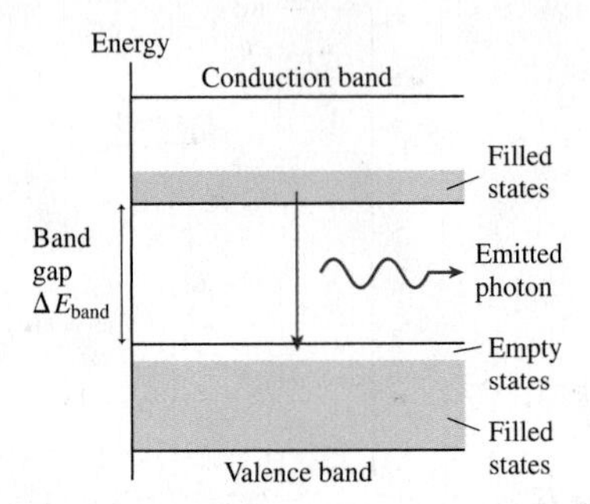

FIGURE P29.76 Energy-level diagram of an LED.

The size of the band gap ΔE_{band} determines the possible energies—and thus the wavelengths—of the emitted photons. Most LEDs emit a narrow range of wavelengths and thus have a distinct color. This makes them well-suited for traffic lights and other applications where a certain color is desired, but it makes them less desirable for general illumination. One way to make a "white" LED is to combine a blue LED with a substance that fluoresces yellow when illuminated with the blue light. The combination of the two colors makes light that appears reasonably white.

76. | An LED emits green light. Increasing the size of the band gap could change the color of the emitted light to
A. Red B. Orange
C. Yellow D. Blue

77. | Suppose the LED band gap is 2.5 eV, which corresponds to a wavelength of 500 nm. Consider the possible electron transitions in Figure P29.76. 500 nm is the
A. Maximum wavelength of the LED.
B. Average wavelength of the LED.
C. Minimum wavelength of the LED.

78. | The same kind of semiconducting material used to make an LED can also be used to convert absorbed light into electrical energy, essentially operating as an LED in reverse. In this case, the absorption of a photon causes an electron transition from a filled state in the valence band to an unfilled state in the conduction band. If $\Delta E_{band} = 1.4$ eV, what is the minimum wavelength of electromagnetic radiation that could lead to electric energy output?
A. 140 nm B. 890 nm
C. 1400 nm D. 8900 nm

79. | The efficiency of a light source is the percentage of its energy input that gets radiated as visible light. If some of the blue light in an LED is used to cause a fluorescent material to glow,
A. The overall efficiency of the LED is increased.
B. The overall efficiency of the LED does not change.
C. The overall efficiency of the LED decreases.

STOP TO THINK ANSWERS

Chapter Preview Stop to Think: C. Once the system is in the $n = 3$ state, it can emit a photon as it jumps directly to the $n = 1$ ground state, or two photons as it jumps first to the $n = 2$ state and then to the $n = 1$ state. Energy conservation requires that the energy of the emitted photon equal the *difference* between the energies of the two states. These differences are $E_3 - E_1 = 8$ eV, $E_3 - E_2 = 5$ eV, and $E_2 - E_1 = 3$ eV.

Stop to Think 29.1: A is emission, B is absorption. All wavelengths in the absorption spectrum are seen in the emission spectrum, but not all wavelengths in the emission spectrum are seen in the absorption spectrum.

Stop to Think 29.2: 6 protons and 8 neutrons. The number of protons is the atomic number, which is 6. That leaves $14 - 6 = 8$ neutrons.

Stop to Think 29.3: B. Because the atoms absorb 3.0 eV photons, the $n = 3$ state must be 3.0 eV above the ground state. The excited atom can then jump directly to the ground state, emitting a 3.0 eV photon in the process. But the atom could also undergo a $3 \rightarrow 2$ transition followed by a $2 \rightarrow 1$ transition. Suppose it is the $2 \rightarrow 1$ transition that emits the 2.0 eV photon. Then the $n = 2$ state must have an energy of 2.0 eV, and so the $3 \rightarrow 2$ transition must emit a 1.0 eV photon. (The same result is found if it is the $3 \rightarrow 2$ transition that emits the 2.0 eV photon.)

Stop to Think 29.4: In emission from the $n = 3$ to $n = 2$ transition, but not in absorption. The photon energy has to match the energy *difference* between two energy levels. Absorption is from the ground state, at $E_1 = 0$ eV. There's no energy level at 3 eV to which the atom could jump.

Stop to Think 29.5: $n = 3$. Each antinode is half a wavelength, so this standing wave has three full wavelengths in one circumference.

Stop to Think 29.6: $n = 3$, $l = 1$, or a $3p$ state.

Stop to Think 29.7: A. An inspection of the periodic table in Figure 29.23 shows that the element that has three of the possible six $3p$ states filled is phosphorus (P).

Stop to Think 29.8: C. Emission is a quantum jump to a lower-energy state. The $5p \rightarrow 4p$ transition is not allowed because $\Delta l = 0$ violates the selection rule. The lowest-energy allowed transition is $5p \rightarrow 3d$, with $E_{photon} = \Delta E_{atom} = 3.0$ eV.

30 Nuclear Physics

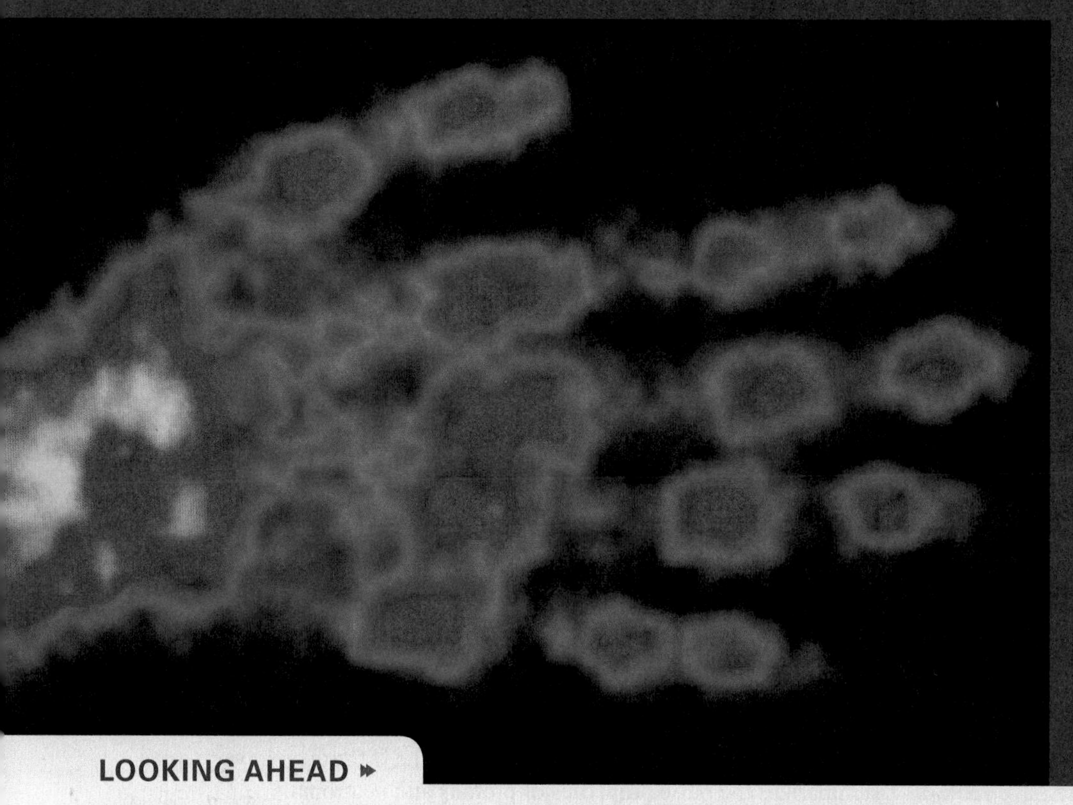

This is a bone scan, not an x ray. It was created using the radioactive decay of a particular type of nucleus inside the body. How can a process that occurs in the nucleus of an atom allow us to create an image of tissues of the human body?

LOOKING AHEAD ▸▸

Goal: To understand the physics of the nucleus and some of the applications of nuclear physics.

Nuclei and Isotopes

The ratio of two stable **isotopes** of oxygen in arctic ice gives us a record of past temperatures.

You'll learn about nuclear structure. The number of protons determines the element; the number of neutrons, the isotope.

Radioactivity and Radiation

These **radioactive** nuclei in this tank are unstable. They decay, emitting high-energy particles—**radiation**—that ionize the water.

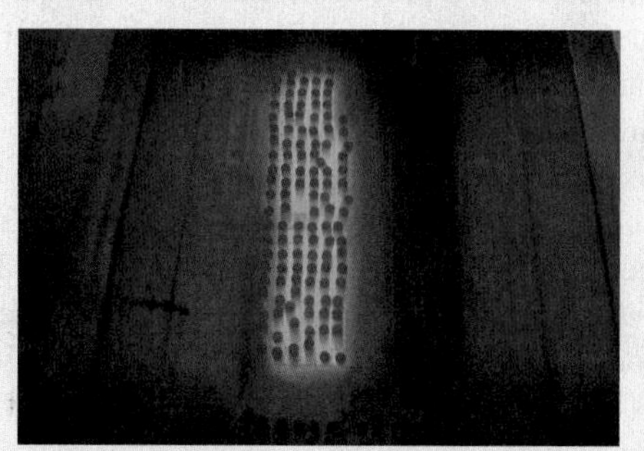

You'll learn about different nuclear decay modes (alpha, beta, and gamma) and the resulting radiation for each.

Decay and Half-Life

Measurements of carbon isotopes in these cave drawings show that they are 30,000 years old.

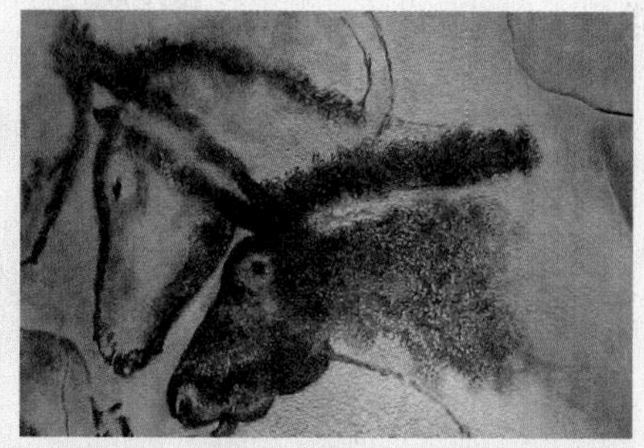

In any sample of ^{14}C, half the nuclei decay in 5700 years. You'll see how to use this half-life to calculate an object's age.

LOOKING BACK ◂◂

Energy levels in atoms

In Chapter 29, you learned how the periodic table of the elements is based on the energy levels of multielectron atoms.

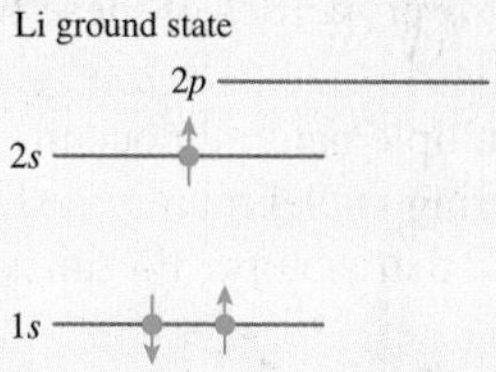

Li ground state

2p

2s

1s

The protons and neutrons in nuclei also have energy levels. Understanding these energy levels will allow you to understand nuclear decay modes.

STOP TO THINK

This energy-level diagram represents an atom with four electrons. What element is this? And is this the ground state of the atom or an excited state?

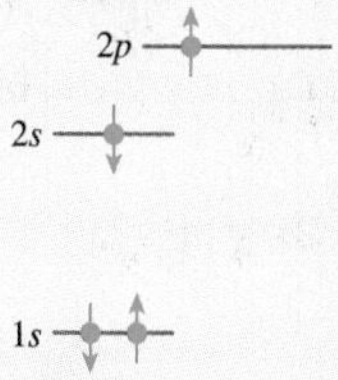

A. Lithium, ground state
B. Lithium, excited state
C. Beryllium, ground state
D. Beryllium, excited state
E. Boron, ground state
F. Boron, excited state

2p

2s

1s

FIGURE 30.1 The nucleus is a tiny speck within an atom.

This picture of an atom would need to be 10 m in diameter if it were drawn to the same scale as the dot representing the nucleus.

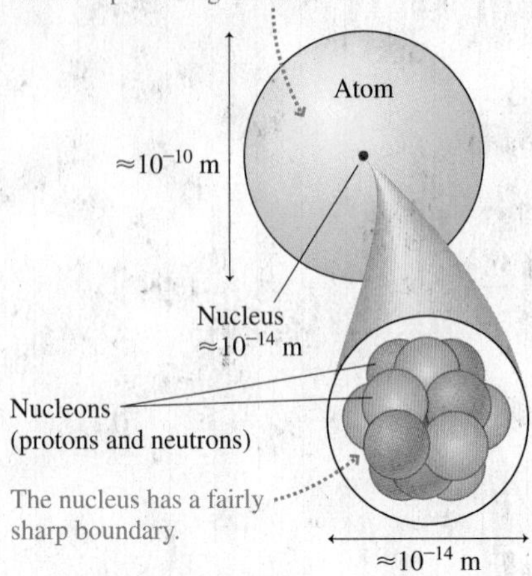

30.1 Nuclear Structure

For 29 chapters, we've made frequent references to properties of atoms that are due to the negative electrons surrounding the nucleus. In this final chapter it's time to dig deeper, to talk about the nucleus itself. In particular:

- What is nuclear matter? What are its properties?
- What holds the nucleus together? Why doesn't the electrostatic force blow it apart?
- What is the connection between the nucleus and radioactivity?

These are questions of **nuclear physics.**

The nucleus is a tiny speck in the center of a vastly larger atom. As **FIGURE 30.1** shows, the nuclear diameter of roughly 10^{-14} m is only about 1/10,000 the diameter of the atom. What we call *matter* is overwhelmingly empty space!

The nucleus is composed of two types of particles: protons and neutrons. Together, these are referred to as **nucleons.** The role of the neutrons, which have nothing to do with keeping electrons in orbit, is an important issue that we'll address in this chapter. The number of protons Z is the element's atomic number. An element is identified by the number of protons in the nucleus, not by the number of orbiting electrons. Electrons are easily added and removed, forming negative and positive ions, but doing so doesn't change the element. The mass number A is defined to be $A = Z + N$, where N is the neutron number. The mass number is the total number of nucleons in a nucleus.

NOTE ▶ The mass number, which is dimensionless, is *not* the same thing as the atomic mass m. We'll look at actual atomic masses later. ◀

TABLE 30.1 Protons and neutrons

	Proton	Neutron
Number	Z	N
Charge q	$+e$	0
Mass, in u	1.00728	1.00866

Protons and neutrons are virtually identical other than the fact that the proton has one unit of the fundamental charge e whereas the neutron is electrically neutral. The neutron is slightly more massive than the proton, but the difference is very small. Table 30.1 summarizes the basic properties of protons and neutrons.

Isotopes

Class Video

As we learned in Chapter 29, not all atoms of the same element (and thus the same Z) have the same mass. There is a *range* of neutron numbers that happily form a nucleus with Z protons, creating a series of nuclei having the same Z-value (i.e., they are all the same chemical element) but different A-values. Each A-value in a series of nuclei with the same Z-value is called an *isotope.* Isotopes for some of the elements are given in a table in Appendix D.

FIGURE 30.2 Three isotopes of carbon.

The leading superscript gives the total number of nucleons, which is the mass number A.

$^{12}_{6}C$ $^{13}_{6}C$ $^{14}_{6}C$

The leading subscript (if included) gives the number of protons.

The three nuclei all have the same number of protons, so they are isotopes of the same element, carbon.

The notation used to label isotopes uses the mass number A as a *leading* superscript, as shown in **FIGURE 30.2.** Hence ordinary carbon, which has six protons and six neutrons in the nucleus (and thus has $A = 12$), is written ^{12}C and pronounced "carbon twelve." The radioactive form of carbon used in carbon dating is ^{14}C. It has six protons, making it carbon, and eight neutrons, for a total of 14 nucleons. The isotope ^{2}H is a hydrogen atom in which the nucleus is not simply a proton but a proton and a neutron. Although the isotope is a form of hydrogen, it is called **deuterium.** Sometimes, for clarity, we will find it useful to include the atomic number as a leading *subscript.* Ordinary carbon is then written as $^{12}_{6}C$; deuterium as $^{2}_{1}H$.

The chemical behavior of an atom is largely determined by the orbiting electrons. Different isotopes of the same element have very similar *chemical* properties. ^{14}C will form the same chemical compounds as ^{12}C and will generally be treated the same by the body, a fact that permits the use of ^{14}C to determine the age of a sample. But the *nuclear* properties of these two isotopes are quite different, as we will see.

Most elements have multiple naturally occurring isotopes. For each element, the fraction of naturally occurring nuclei represented by one particular isotope is called the **natural abundance** of that isotope. For instance, oxygen has two primary isotopes, ^{16}O and ^{18}O. The data in Appendix D show that the natural abundance of ^{16}O is 99.76%, meaning that 9976 out of every 10,000 naturally occurring oxygen atoms are the isotope ^{16}O. Most of the remaining 0.24% of naturally occurring oxygen is the isotope ^{18}O, which has two extra neutrons. These two isotopes of oxygen are chemically equivalent, but the mass difference can lead to subtle differences in macroscopic behaviors. Atmospheric water vapor is always slightly deficient in water molecules that contain ^{18}O compared to water in the ocean because the lighter molecules that contain ^{16}O evaporate more readily.

More than 3000 isotopes are known. The majority of these are **radioactive,** meaning that the nucleus is not stable but, after some period of time, will either fragment or emit some kind of subatomic particle to reach a more stable state. Many of these radioactive isotopes are created by nuclear reactions in the laboratory and have only a fleeting existence. Only 266 isotopes are **stable** (i.e., nonradioactive) and occur in nature. In addition, there are a handful of radioactive isotopes with such long decay times, measured in billions of years, that they also occur naturally.

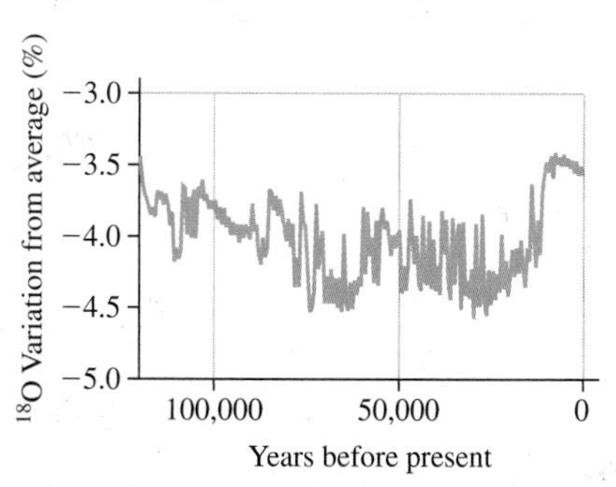

Taking the earth's temperature A core sample of Greenland's ice pack is composed of yearly snowfall compacted into ice, and the layers can be "read" to give a record of past temperatures. Evaporation from the ocean's surface favors water molecules that contain the lighter ^{16}O isotope over those that contain ^{18}O, but the difference decreases with increasing temperature. The fraction of water molecules that contain ^{18}O is greater in snow that forms at higher temperatures. The graph shows the isotopic composition of an ice core that reached back more than 100,000 years; higher numbers correspond to higher temperatures. Broad trends, such as the increase in temperatures at the end of the last ice age, are clearly seen.

Atomic Mass

You learned in ◀◀ SECTION 12.1 that atomic masses are specified in terms of the *atomic mass unit* u, defined such that the atomic mass of the isotope ^{12}C is exactly 12 u. The conversion to SI units is

$$1 \text{ u} = 1.6605 \times 10^{-27} \text{ kg}$$

Alternatively, as we saw in ◀◀ SECTION 27.10, we can use Einstein's $E_0 = mc^2$ to express masses in terms of their energy equivalent. The energy equivalent of 1 u of mass is

$$\begin{aligned} E_0 &= (1.6605 \times 10^{-27} \text{ kg})(2.9979 \times 10^8 \text{ m/s})^2 \\ &= 1.4924 \times 10^{-10} \text{ J} = 931.49 \text{ MeV} \end{aligned} \quad (30.1)$$

A mass of 1 u has an energy equivalent of 931.49 MeV, so we can use the following equation to find the energy equivalent of any atom or particle whose mass is given in atomic mass units:

$$E_0 \text{ (in MeV)} = m\text{(in u)} \times (931.49 \text{ MeV/u}) \quad (30.2)$$

By noting that Einstein's formula implies $m = E_0/c^2$, we can also write 1 u in the following form:

$$1 \text{ u} = \frac{E_0}{c^2} = 931.49\left(\frac{\text{MeV}}{c^2}\right) \quad (30.3)$$

It may seem unusual, but the units MeV/c^2 are units of mass. This will be a useful unit for us when we need to compute energy equivalents. The energy equivalent of mass 1 MeV/c^2 is simply 1 MeV.

Table 30.2 shows some important atomic mass values. Notice that the mass of a hydrogen atom is equal to the sum of the masses of a proton and an electron. But a quick calculation shows that the mass of a helium atom (2 protons, 2 neutrons, and 2 electrons) is 0.03038 u *less* than the sum of the masses of its constituents. The difference is due to the *binding energy* of the nucleus, a topic we'll look at in Section 30.2.

TABLE 30.2 Some atomic masses

Particle	Symbol	Mass (u)	Mass (MeV/c^2)
Electron	e	0.000549	0.51
Proton	p	1.007276	938.27
Neutron	n	1.008665	939.56
Hydrogen	^1H	1.007825	938.78
Helium	^4He	4.002602	3728.38

NOTE ▶ The atomic masses of the neutron and the proton are both ≈1 u. In earlier chapters, we often used the approximation that the atomic mass in u is equal to the mass number A. This approximation is sufficient in many contexts, such as when we calculated the speeds of gas molecules in Chapter 12. But many nuclear calculations involve the small difference between two masses that are almost the same. The two masses must be calculated or specified to four or five significant figures if their difference is to be meaningful. In calculations of nuclear energies, you should use more significant figures than usual, and you should start with the accurate values for nuclear masses given in Table 30.2 or Appendix D. ◀

The *chemical* atomic mass shown on the periodic table of the elements is the *weighted average* of the atomic masses of all naturally occurring isotopes. For example, chlorine has two stable isotopes: ^{35}Cl, with atomic mass $m = 34.97$ u, has an abundance of 75.8% and ^{37}Cl, at 36.97 u, has an abundance of 24.2%. The average, weighted by abundance, is $(0.758 \times 34.97\text{ u}) + (0.242 \times 36.97\text{ u}) = 35.45$ u. This is the value shown on the periodic table and is the correct value for most chemical calculations, but it is not the mass of any particular isotope of chlorine.

NOTE ▸ Nuclear physics calculations involve the masses of specific isotopes. The mass values for specific isotopes are given in Appendix D; these are the values you'll need for calculations in this chapter. Don't use the chemical atomic masses given in the periodic table! ◂

STOP TO THINK 30.1 Three electrons orbit a neutral ^{6}Li atom. How many electrons orbit a neutral ^{7}Li atom?

30.2 Nuclear Stability

Because nuclei are characterized by two independent numbers, N and Z, it is useful to show the known nuclei on a plot of neutron number N versus proton number Z. FIGURE 30.3 shows such a plot. Stable nuclei are represented by blue diamonds and unstable, radioactive nuclei by red dots.

FIGURE 30.3 Stable and unstable nuclei shown on a plot of neutron number N versus proton number Z.

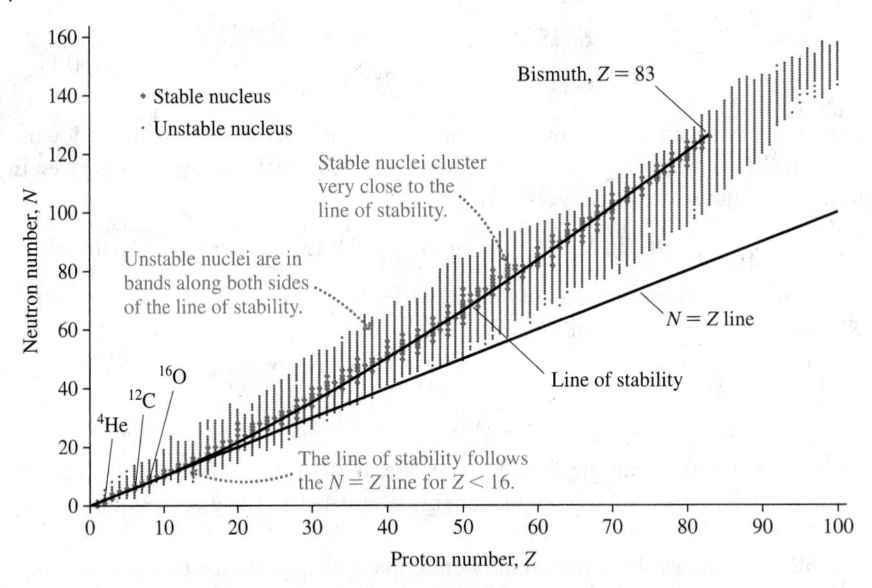

We can make several observations from this graph:

- The stable nuclei cluster very close to the curve called the **line of stability.**
- There are no stable nuclei with $Z > 83$ (bismuth). Heavier elements (up to $Z = 92$ (uranium)) are found in nature, but they are radioactive.
- Unstable nuclei are in bands along both sides of the line of stability.
- The lightest elements, with $Z < 16$, are stable when $N \approx Z$. The familiar isotopes ^{4}He, ^{12}C, and ^{16}O all have equal numbers of protons and neutrons.
- As Z increases, the number of neutrons needed for stability grows increasingly larger than the number of protons. The N/Z ratio is ≈ 1.2 at $Z = 40$ but has grown to ≈ 1.5 at $Z = 80$.

These observations—especially the fact that $N \approx Z$ for small Z but $N > Z$ for large Z—will be explained by the model of the nucleus that we'll explore in Section 30.3.

Unstable but ubiquitous uranium All the isotopes of uranium are unstable, but one is quite long-lived. Half the ^{238}U that was present at the formation of the earth is still around—and it is found all around you at low concentrations in nearly all of the rocks and soil on the earth's surface. Much of the radiation that you are exposed to comes from this naturally occurring and widely distributed unstable element.

Binding Energy

A nucleus is a *bound system*. That is, you would need to supply energy to disperse the nucleons by breaking the nuclear bonds between them. **FIGURE 30.4** shows this idea schematically.

You learned a similar idea in atomic physics. The energy levels of the hydrogen atom are negative numbers because the bound system has less energy than a free proton and electron. The energy you must supply to an atom to remove an electron is called the *ionization energy*.

The energy you would need to supply to a nucleus to disassemble it into individual protons and neutrons is called the **binding energy**. Whereas ionization energies of atoms are only a few eV, the binding energies of nuclei are tens or hundreds of MeV, energies large enough that their mass equivalent is not negligible.

Suppose we break a helium atom into two hydrogen atoms (taking account of the two protons and the two electrons) and two free neutrons as shown in **FIGURE 30.5**. The mass of the separated components is greater than that of the helium atom. The difference in mass $\Delta m = 0.03038$ u arises from the energy that was put into the system to separate the tightly bound nucleons. We can use the conversion of Equation 30.2 to find the energy equivalent of this mass difference; this energy is the binding energy B:

$$B = (0.03038 \text{ u})(931.49 \text{ MeV/u}) = 28.30 \text{ MeV}$$

Generally, the nuclear binding energy is computed by considering the mass difference between the atom and its separated components, Z hydrogen atoms and N neutrons:

$$B = (Zm_{\text{H}} + Nm_{\text{n}} - m_{\text{atom}}) \times (931.49 \text{ MeV/u}) \quad (30.4)$$

Nuclear binding energy for an atom of mass m_{atom} with Z protons and N neutrons

FIGURE 30.4 The nuclear binding energy.

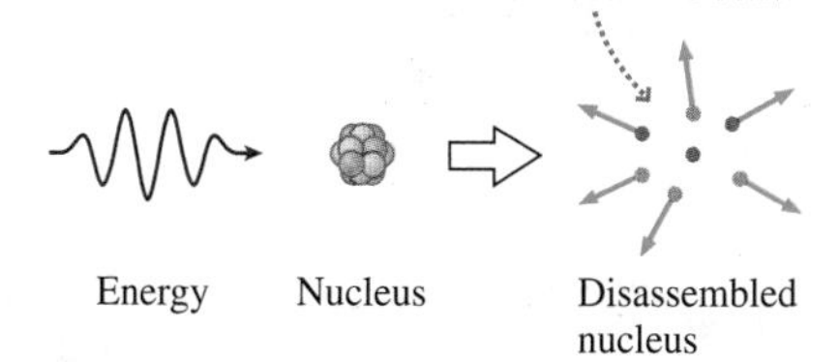

The binding energy is the energy that would be needed to disassemble a nucleus into individual nucleons.
Energy
Nucleus
Disassembled nucleus

FIGURE 30.5 The binding energy of the helium nucleus.

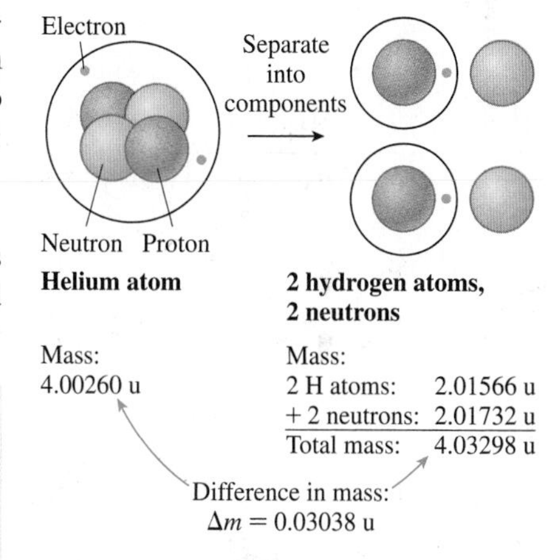

EXAMPLE 30.1 **Finding the binding energy of iron**

What is the nuclear binding energy of ^{56}Fe to the nearest MeV?

PREPARE Appendix D gives the atomic mass of ^{56}Fe as 55.934940 u. Iron has atomic number 26, so an atom of ^{56}Fe could be separated into 26 hydrogen atoms and 30 neutrons. The mass of the separated components is more than that of the iron nucleus; the difference gives us the binding energy.

SOLVE We solve for the binding energy using Equation 30.4. The masses of the hydrogen atom and the neutron are given in Table 30.2. We find

$$B = (26(1.007825 \text{ u}) + 30(1.008665 \text{ u}) - 55.934940 \text{ u})(931.49 \text{ MeV/u})$$
$$= (0.52846 \text{ u})(931.49 \text{ MeV/u}) = 492.26 \text{ MeV} \simeq 492 \text{ MeV}$$

ASSESS The difference in mass between the nucleus and its components is a small fraction of the mass of the nucleus, so we must use several significant figures in our mass values. The mass difference is small—about half that of a proton—but the energy equivalent, the binding energy, is enormous.

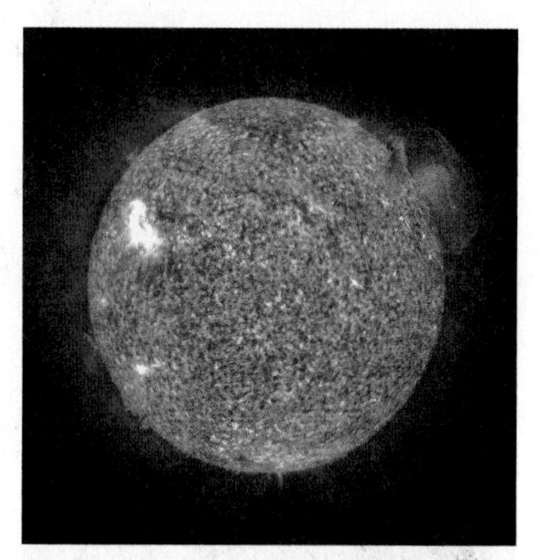

A nuclear fusion weight-loss plan The sun's energy comes from reactions that combine four hydrogen atoms to create a single atom of helium—a process called **nuclear fusion.** Because energy is released, the mass of the helium atom is less than that of the four hydrogen atoms. As the fusion reactions continue, the mass of the sun decreases—by 130 trillion tons per year! That's a lot of mass, but given the sun's enormous size, this change will amount to only a few hundredths of a percent of the sun's mass over its 10-billion-year lifetime.

How much energy is 492 MeV? To make a comparison with another energy value we have seen, the binding energy of a single iron nucleus is equivalent to the energy released in the metabolism of nearly *2 billion* molecules of ATP! The energy scale of nuclear processes is clearly quite different from that of chemical processes.

As A increases, the nuclear binding energy increases, simply because there are more nuclear bonds. A more useful measure for comparing one nucleus to another is the quantity B/A, called the *binding energy per nucleon*. Iron, with B= 492 MeV and $A = 56$, has 8.79 MeV per nucleon. This is the amount of energy, on average, you would need to supply in order to remove *one* nucleon from the nucleus. Nuclei with larger values of B/A are more tightly held together than nuclei with smaller values of B/A.

FIGURE 30.6 The curve of binding energy for the stable nuclei.

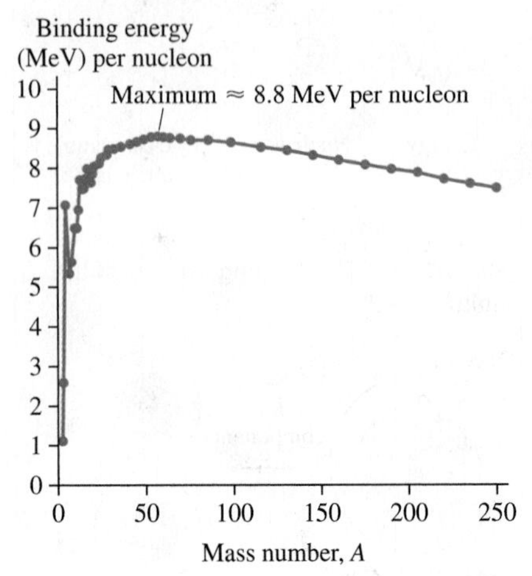

FIGURE 30.6 is a graph of the binding energy per nucleon versus mass number A. The line connecting the points is often called the **curve of binding energy.**

For small values of A, adding more nucleons increases the binding energy per nucleon. If two light nuclei can be joined together to make a single, larger nucleus, the final nucleus will have a higher binding energy per nucleon. Because the final nucleus is more tightly bound, energy will be released in this *nuclear fusion* process. Nuclear fusion of hydrogen to helium is the basic reaction that powers the sun.

The curve of binding energy has a broad maximum at $A \approx 60$. Nuclei with $A > 60$ become less stable as their mass increases because adding nucleons *decreases* the binding energy per nucleon. *Alpha decay*, one of the three basic types of radioactive decay that we'll examine later in this chapter, occurs when a heavy nucleus becomes more stable by ejecting a small group of nucleons in order to decrease its mass, releasing energy in the process. The decrease in binding energy per nucleon as mass increases helps explain why there are no stable nuclei beyond $Z = 83$.

A few very heavy nuclei, especially some isotopes of uranium and plutonium, are so unstable that they can be induced to fragment into two lighter nuclei in the process known as *nuclear fission*. For example, the collision of a slow-moving neutron with a ^{235}U nucleus causes the reaction

$$\text{n} + {}^{235}\text{U} \rightarrow {}^{236}\text{U} \rightarrow {}^{90}\text{Sr} + {}^{144}\text{Xe} + 2\text{n} \tag{30.5}$$

The ^{235}U nucleus absorbs the neutron to become ^{236}U, but ^{236}U is so unstable that it immediately fragments—in this case into a ^{90}Sr nucleus, a ^{144}Xe nucleus, and two neutrons. The less massive ^{90}Sr and ^{144}Xe nuclei are more tightly bound than the original ^{235}U nucleus, so a great deal of energy is released in this reaction. As we've seen, nuclei with lower values of Z have relatively smaller numbers of neutrons, meaning there will be neutrons "left over" after the reaction. Equation 30.5 shows some free neutrons among the reaction products, but the two nuclear fragments have "extra" neutrons as well—they have too many neutrons and will be unstable. This is generally true for the products of a fission reaction. The fact that the waste products of nuclear fission are radioactive has important consequences for the use of nuclear fission as a source of energy.

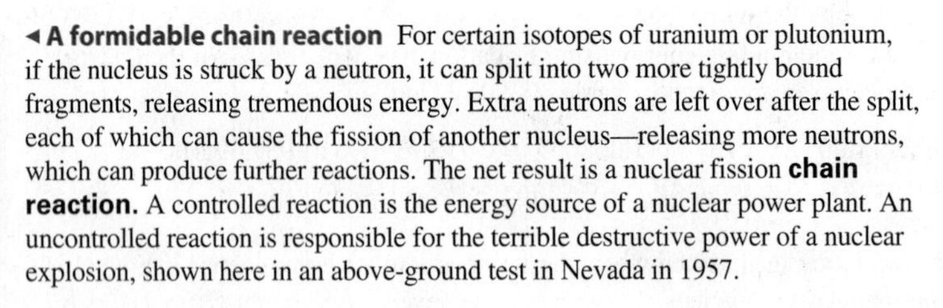

◀ **A formidable chain reaction** For certain isotopes of uranium or plutonium, if the nucleus is struck by a neutron, it can split into two more tightly bound fragments, releasing tremendous energy. Extra neutrons are left over after the split, each of which can cause the fission of another nucleus—releasing more neutrons, which can produce further reactions. The net result is a nuclear fission **chain reaction.** A controlled reaction is the energy source of a nuclear power plant. An uncontrolled reaction is responsible for the terrible destructive power of a nuclear explosion, shown here in an above-ground test in Nevada in 1957.

STOP TO THINK 30.2 ^{238}U is long-lived but ultimately unstable; it will eventually spontaneously break into two fragments, a ^{4}He nucleus and a ^{234}Th nucleus, in a process called alpha decay, which we'll learn about in the next section. A great deal of energy is released in the process. What must be true about the masses of the nuclei involved?

A. $m_U > m_{Th} + m_{He}$
B. $m_U = m_{Th} + m_{He}$
C. $m_U < m_{Th} + m_{He}$

30.3 Forces and Energy in the Nucleus

The nucleus of the atom is made up of protons, which are positively charged, and neutrons, which have no charge. Why doesn't the repulsive electrostatic force between the protons simply cause the nucleus to fly apart? There must be another force at work within the nucleus that holds the nucleons together. This force is called the **strong nuclear force,** or just the strong force.

The strong force has four important properties:

1. It is an *attractive* force between any two nucleons.
2. It does not act on electrons.
3. It is a *short-range* force, acting only over nuclear distances. We see no evidence for nuclear forces outside the nucleus.
4. Over the range where it acts, it is *stronger* than the electrostatic force that tries to push two protons apart.

FIGURE 30.7 summarizes the three types of interactions that take place within the nucleus. Many decades of research have shown that the strong force between two nucleons is independent of whether they are protons or neutrons. Charge is the basis for electromagnetic interactions, but it is of no relevance to the strong force. Protons and neutrons are identical as far as nuclear forces are concerned.

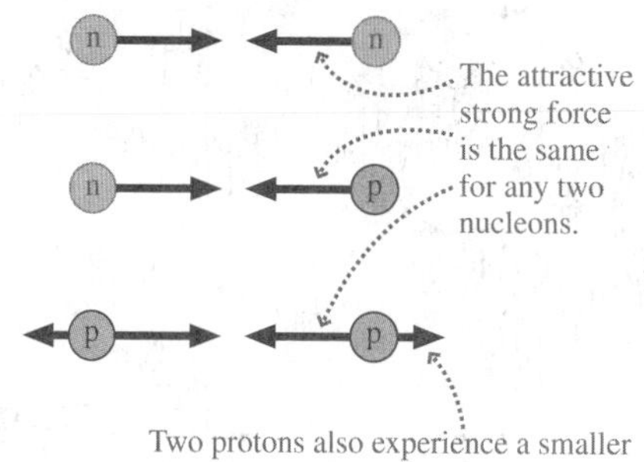

n
n
The attractive strong force is the same for any two nucleons.
n
p
p
p
Two protons also experience a smaller electrostatic repulsive force.

FIGURE 30.7 Forces between pairs of particles in the nucleus.

Protons throughout the nucleus exert repulsive electrostatic forces on each other, but, because of the short range of the strong force, a proton feels an attractive force from only the very few other protons with which it is in close contact. A nucleus with too many protons will be unstable because the repulsive electrostatic forces will overcome the attractive strong forces. Because neutrons participate in the strong force but exert no repulsive forces, **the neutrons provide the extra "glue" that holds the nucleus together.** In small nuclei, one neutron per proton is sufficient for stability. Hence small nuclei have $N \approx Z$. But as the nucleus grows, the repulsive force increases faster than the binding energy. More neutrons are needed for stability, so heavy nuclei have $N > Z$.

In ◀◀ SECTION 28.5 we learned that confining a particle to a region of space results in the quantization of energy. In Chapter 29 you saw the consequences of this quantization for electrons bound in atoms; in this chapter we will look at the quantized energy levels for protons and neutrons bound in the nucleus. Just as we did with electron energy levels in atoms, we can "build up" the nuclear state by adding nucleons to successively higher energy levels. Neutrons and protons have spin and follow the Pauli exclusion principle, which applies to nucleons just as it did to electrons. Each energy level can hold only a certain number of spin-up particles and spin-down particles, depending on the quantum numbers; additional nucleons must go into higher energy levels.

Although there is a great deal of similarity between the descriptions of nucleon energy levels in the nucleus and electron energy levels in the atom, there is one important difference: the energy scale. The electron energy levels are typically separated by a few eV, but the proton and neutron energy levels are separated by a few MeV—a million times as much.

Low-Z Nuclei

As our first example of an energy-level description of the nucleus, we'll consider the energy levels of low-Z nuclei ($Z < 8$). Because these nuclei have so few protons, we can neglect the electrostatic potential energy due to proton-proton repulsion and consider only the much larger nuclear potential energy. In that case, the energy levels of the protons and neutrons are essentially identical.

FIGURE 30.8 shows the three lowest allowed energy levels for protons and neutrons and the maximum number of protons and neutrons the Pauli principle allows in each. Energy values vary from nucleus to nucleus, but the spacing between these levels is several MeV.

FIGURE 30.8 The three lowest energy levels of neutrons and protons in a low-Z nucleus.

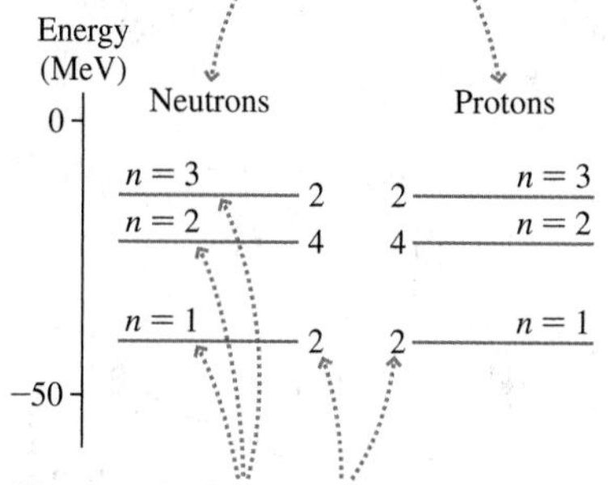

The proton potential energy is nearly identical to the neutron potential energy when Z is small.
Energy (MeV)
Neutrons
Protons
0
n = 3
2
2
n = 3
n = 2
4
4
n = 2
n = 1
2
2
n = 1
−50
These are the first three allowed energy levels. They are spaced several MeV apart.
These are the maximum numbers of protons and neutrons allowed by the Pauli principle.

Suppose we look at a series of nuclei, all with $A = 12$ but with different numbers of protons and neutrons: ^{12}B, ^{12}C, ^{12}N. All have 12 nucleons, but only ^{12}C is a stable isotope; the other two are not. Why do we see this difference in stability?

FIGURE 30.9 Nuclear energy-level diagrams of ^{12}B, ^{12}C, and ^{12}N.

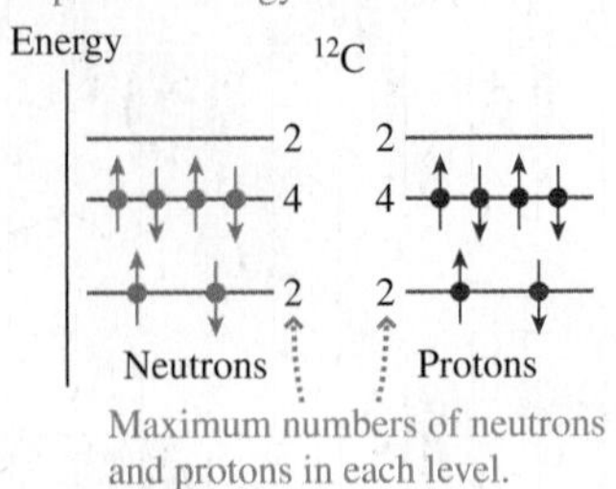
A ^{12}C nucleus is in its lowest possible energy state.
Energy
^{12}C
2
4
2
2
4
2
Neutrons
Protons
Maximum numbers of neutrons and protons in each level.

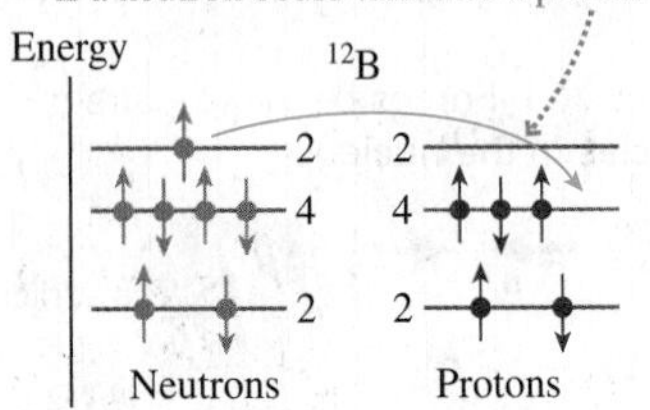
A ^{12}B nucleus could lower its energy if a neutron could turn into a proton.
Energy
^{12}B
2
4
2
2
4
2
Neutrons
Protons

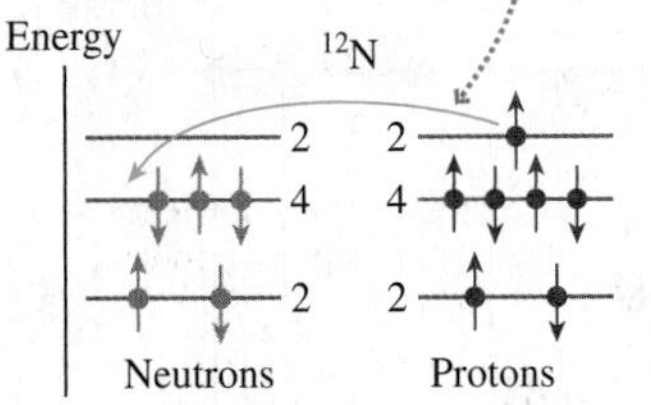
A ^{12}N nucleus could lower its energy if a proton could turn into a neutron.
Energy
^{12}N
2
4
2
2
4
2
Neutrons
Protons

FIGURE 30.9 shows the energy-level diagrams of ^{12}B, ^{12}C, and ^{12}N. Look first at ^{12}C, a nucleus with six protons and six neutrons. You can see that exactly six protons are allowed in the $n = 1$ and $n = 2$ proton energy levels. The same is true for the six neutrons. No other arrangement of the nucleons would lower the total energy, so this nucleus is stable.

^{12}B has five protons and seven neutrons. The sixth neutron fills the $n = 2$ neutron energy level, so the seventh neutron has to go into the $n = 3$ energy level. (This is just like the third electron in Li having to go into the $n = 2$ energy level because the first two electrons have filled the $n = 1$ energy level.) The $n = 2$ proton energy level has one vacancy because there are only five protons.

In atoms, electrons in higher energy levels move to lower energy levels by emitting a photon as the electron undergoes a quantum jump. That can't happen here because the higher-energy nucleon in ^{12}B is a neutron whereas the vacant lower energy level is that of a proton. But an analogous process could occur *if* a neutron could somehow turn into a proton, allowing it to move to a lower energy level. ^{12}N is just the opposite, with the seventh proton in the $n = 3$ energy level. If a proton could somehow turn into a neutron, it could move to a lower energy level, lowering the energy of the nucleus.

You can see from the diagrams that the ^{12}B and ^{12}N nuclei have significantly more energy—by several MeV—than ^{12}C. If a neutron could turn into a proton, and vice versa, these nuclei could move to a lower-energy state—that of ^{12}C. In fact, that's exactly what happens! These nuclei are not stable. We'll explore the details in Section 30.4. Both ^{12}B and ^{12}N decay into the more stable ^{12}C in the process known as *beta decay.*

High-Z Nuclei

We can use the energy levels of protons and neutrons in the nucleus to give a qualitative explanation for one more observation. FIGURE 30.10 shows the neutron and proton energy levels of a high-Z nucleus. In a nucleus with many protons, the increasing electrostatic potential energy raises the proton energy levels but not the neutron energy levels. Protons and neutrons now have a different set of energy levels.

FIGURE 30.10 The proton energy levels are displaced upward in a high-Z nucleus because of the electric potential energy.

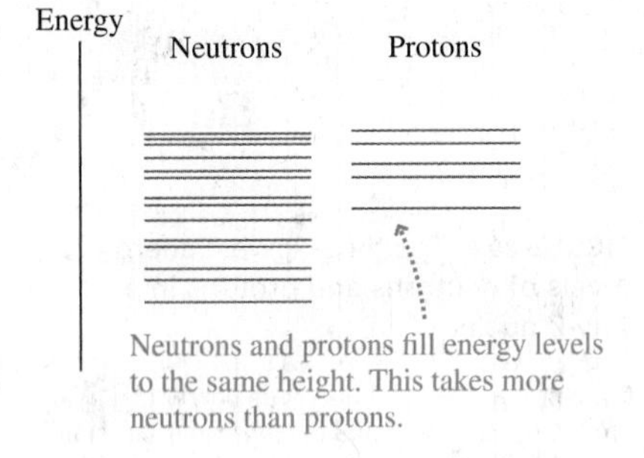
Energy
Neutrons
Protons
Neutrons and protons fill energy levels to the same height. This takes more neutrons than protons.

As a nucleus is "built," by the adding of protons and neutrons, the proton energy levels and the neutron energy levels must fill to just about the same height. If there were neutrons in energy levels above vacant proton levels, the nucleus would lower its energy by changing neutrons into protons. Similarly, a proton would turn into a neutron if there were a vacant neutron energy level beneath a filled proton level. **The net result is that the filled levels for protons and neutrons are at just about the same height.**

Because the neutron energy levels start at a lower energy, *more neutron states* are available than proton states. Consequently, a high-Z nucleus will have more neutrons than protons. This conclusion is consistent with our observation in Figure 30.3 that $N > Z$ for heavy nuclei.

STOP TO THINK 30.3 Based on the model of nuclear energy levels and transitions you have seen, would you expect ^{13}C to be stable?

30.4 Radiation and Radioactivity

When a nucleus is unstable, it will decay. But early investigators didn't observe the decay directly; instead, they observed the high-energy particles or rays emitted in the decay of unstable nuclei. By measuring the properties of the emitted radiation, they identified three different types of decay, which they called alpha, beta, and gamma. We now define *radioactivity* or *radioactive* decay as the spontaneous emission of particles or high-energy photons from unstable nuclei as they decay from higher-energy to lower-energy states. In this section, we will explore the decay mechanisms that result in the three different types of radiation, considering the changes in the nuclei that produce these decays.

Class Video

Alpha Decay

Many large nuclei are unstable; it is energetically favorable for them to spontaneously break apart into smaller fragments. A combination of two neutrons and two protons, a ^{4}He nucleus, is an especially stable nuclear combination. When a large nucleus spontaneously decays by breaking into two smaller fragments, one of these fragments is almost always a helium nucleus—an alpha particle, symbolized by α.

An unstable nucleus that ejects an alpha particle loses two protons and two neutrons, so we can write this decay as

$$^{A}_{Z}\text{X} \rightarrow {}^{A-4}_{Z-2}\text{Y} + \alpha + \text{energy} \tag{30.6}$$

Alpha decay of a nucleus

FIGURE 30.11 shows the alpha-decay process. The original nucleus X is called the **parent nucleus,** and the decay-product nucleus Y is the **daughter nucleus.**

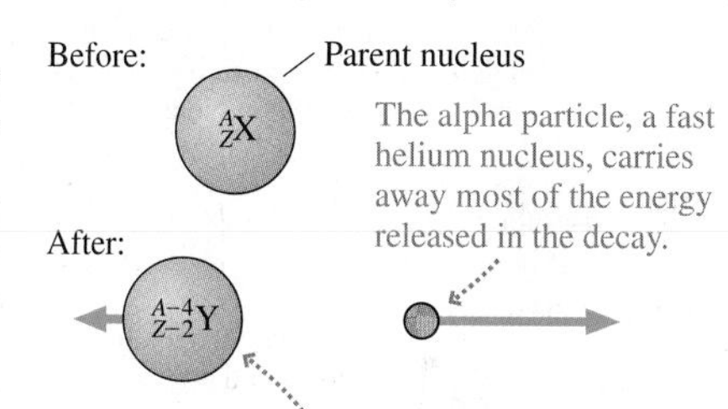
Before:
Parent nucleus
$^{A}_{Z}$X
The alpha particle, a fast helium nucleus, carries away most of the energy released in the decay.
After:
$^{A-4}_{Z-2}$Y
The daughter nucleus has two fewer protons and four fewer nucleons. It has a small recoil.

FIGURE 30.11 Alpha decay.

Energy conservation tells us that an alpha decay can occur only when the mass of the parent nucleus is greater than the mass of the daughter nucleus plus the mass of the alpha particle. This requirement is often met for heavy, high-Z nuclei beyond the maximum of the curve of binding energy of Figure 30.6. It is energetically favorable for these nuclei to eject an alpha particle because the daughter nucleus is more tightly bound than the parent nucleus.

The daughter nucleus, which is much more massive than an alpha particle, undergoes only a slight recoil, as we see in Figure 30.11. Consequently, **the energy released in an alpha decay ends up mostly as the kinetic energy of the alpha particle.** This energy is approximately equal to the mass energy difference between the initial and final states:

$$K_\alpha \approx \Delta E = (m_X - m_Y - m_{He})c^2 \tag{30.7}$$

EXAMPLE 30.2 Analyzing alpha decay in a smoke detector

Americium, atomic number 95, doesn't exist in nature; it is produced in nuclear reactors. An isotope of americium, ^{241}Am, is part of the sensing circuit in most smoke detectors. ^{241}Am decays by emitting an alpha particle. What is the daughter nucleus?

SOLVE Equation 30.6 shows that an alpha decay causes the atomic number to decrease by 2 and the atomic weight by 4. Let's write an equation for the decay showing the alpha particle as a helium nucleus, including the atomic weight superscript and the atomic number subscript for each element. There is no change in the total number of neutrons or protons, so the subscripts and superscripts must "balance" in the reaction:

$$^{241}_{95}\text{Am} \rightarrow {}^{237}_{93}? + {}^{4}_{2}\text{He} + \text{energy}$$

A quick glance at the periodic table reveals the unknown element in this equation, the daughter nucleus, to be an isotope of neptunium, $^{237}_{93}$Np.

ASSESS Balancing the two sides of the above reaction is similar to balancing the equation for a chemical reaction.

EXAMPLE 30.3 Finding the energy of an emitted alpha particle

The uranium isotope ^{238}U undergoes alpha decay to an isotope of thorium, ^{234}Th. What is the kinetic energy, in MeV, of the alpha particle?

PREPARE The decay products have less mass than the initial nucleus. This difference in mass is released as energy, most of which goes to the kinetic energy of the alpha particle. Because the energy of the alpha particle is only approximately equal to the reaction energy, we needn't use the full accuracy that the values in Appendix D provide. To 4 decimal places, the atomic mass of ^{238}U is 238.0508 u, that of ^{234}Th is 234.0436 u, and that of ^{4}He—the alpha particle—is 4.0026 u.

SOLVE We can calculate the kinetic energy of the alpha particle using Equation 30.7:

$$K_\alpha = (238.0508\text{ u} - 234.0436\text{ u} - 4.0026\text{ u})c^2$$
$$= (0.0046\text{ u})c^2$$

If we convert from u to MeV/c^2, using the conversion factor 1 u = 931.49 MeV/c^2, the c^2 cancels, and we end up with

$$K_\alpha = \left(0.0046\text{ u} \times \frac{931.49\text{ MeV}/c^2}{1\text{ u}}\right)c^2 = 4.3\text{ MeV}$$

ASSESS This is a typical alpha-particle energy, corresponding to a speed of about 5% of the speed of light. Notice that with a careful use of conversion factors we never had to evaluate c^2.

Beta Decay

In beta decay, a nucleus decays by emitting a very light particle, which investigators determined to be an electron, e^-. A typical example of beta decay occurs in the carbon isotope ^{14}C, which undergoes the beta-decay process

$$^{14}\text{C} \rightarrow {}^{14}\text{N} + e^-$$

Carbon has $Z = 6$ and nitrogen has $Z = 7$. Because Z increases by 1 but A doesn't change, it appears that a neutron within the nucleus has changed itself into a proton by emitting an electron. That is, the basic decay process appears to be

$$\text{n} \rightarrow \text{p} + e^- \qquad (30.8)$$

FIGURE 30.12 Beta decay.

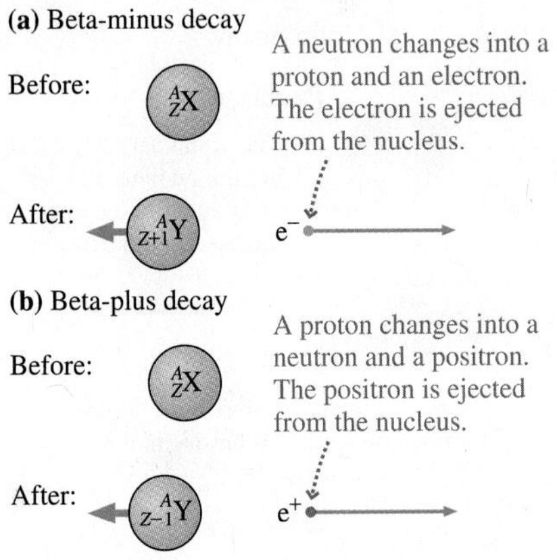

(a) Beta-minus decay

Before:

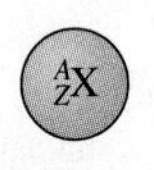

A neutron changes into a proton and an electron. The electron is ejected from the nucleus.

After:

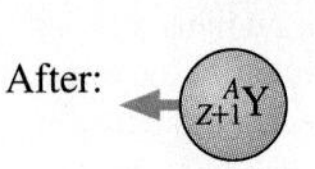

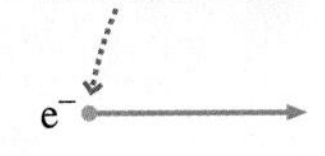

(b) Beta-plus decay

Before:

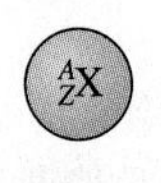

A proton changes into a neutron and a positron. The positron is ejected from the nucleus.

After:

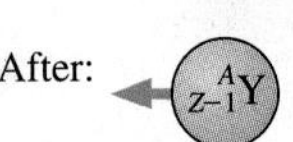

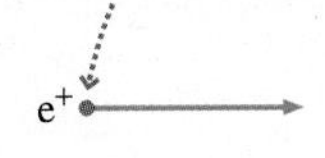

The electron is ejected from the nucleus but the proton is not. Thus the beta-decay process, shown in **FIGURE 30.12a**, is

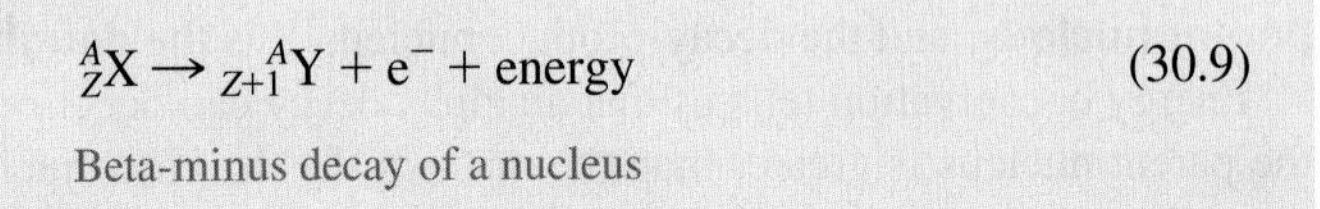

$$^A_Z\text{X} \rightarrow {}^{A}_{Z+1}\text{Y} + e^- + \text{energy} \qquad (30.9)$$

Beta-minus decay of a nucleus

Do neutrons *really* turn into protons? It turns out that a free neutron—one not bound in a nucleus—is *not* a stable particle. It decays into a proton and an electron, with a half-life of approximately 10 minutes. This decay conserves energy because $m_n > m_p + m_e$. Furthermore, it conserves charge.

Whether a neutron *within* a nucleus can decay depends not only on the masses of the neutron and proton but also on the masses of the parent and daughter nuclei, because energy has to be conserved for the entire nuclear system. **Beta decay occurs only if $m_X > m_Y$.** ^{14}C can undergo beta decay to ^{14}N because $m(^{14}C) > m(^{14}N)$. But $m(^{12}C) < m(^{12}N)$, so ^{12}C is stable and its neutrons will not decay.

A few nuclei undergo a slightly different form of beta decay by emitting a *positron.* A positron, for which we use the symbol e^+, is identical to an electron except that it has a positive charge. As we saw in ◀◀ SECTION 27.10, the positron is the *antiparticle* of the electron. To distinguish between these two forms of decay, we call the emission of an electron *beta-minus decay* and the emission of a positron *beta-plus decay.*

Inside a nucleus undergoing beta-plus decay, a proton changes into a neutron and a positron:

$$\text{p}^+ \rightarrow \text{n} + e^+ \qquad (30.10)$$

The full decay process, shown in **FIGURE 30.12b**, is

$$^A_Z\text{X} \rightarrow {}^{A}_{Z-1}\text{Y} + e^+ + \text{energy} \qquad (30.11)$$

Beta-plus decay of a nucleus

Beta-plus decay does *not* happen for a free proton because $m_p < m_n$. It *can* happen within a nucleus as long as energy is conserved for the entire nuclear system, but it is far less common than beta-minus decay.

In our earlier discussion of Section 30.3 we noted that the ^{12}B and ^{12}N nuclei could reach a lower-energy state if a proton could change into a neutron, and vice versa. Now we see that such changes can occur if the energy conditions are favorable. And, indeed, ^{12}B undergoes beta-minus decay to ^{12}C, while ^{12}N undergoes beta-plus decay to ^{12}C.

In general, beta decay is a process of nuclei with too many neutrons or too many protons that moves them closer to the line of stability in Figure 30.3.

NOTE ▶ The electron emitted in beta decay has nothing to do with the atom's valence electrons. The beta particle is created in the nucleus and ejected directly from the nucleus when a neutron is transformed into a proton and an electron. ◀

EXAMPLE 30.4 Analyzing beta decay in the human body

Your body contains several radioactive isotopes. Approximately 20% of the radiation dose you receive each year comes from the radioactive decay of these atoms. Most of this dose comes from one potassium isotope, ^{40}K, which most commonly decays by beta-minus emission. What is the daughter nucleus?

SOLVE Rewriting Equation 30.9 as

$$^{40}_{19}K \rightarrow ^{40}_{20}? + e^- + \text{energy}$$

we see that the daughter nucleus must be the calcium isotope $^{40}_{20}Ca$.

Gamma Decay

Gamma decay is similar to quantum processes you saw in earlier chapters. In Chapter 28, you learned that an atomic system can emit a photon with $E_{photon} = \Delta E_{atom}$ when an electron undergoes a quantum jump from an excited energy level to a lower energy level. Nuclei are no different. A proton or a neutron in an excited nuclear state, such as the one shown in **FIGURE 30.13**, can undergo a quantum jump to a lower-energy state by emitting a high-energy photon. This is the gamma-decay process.

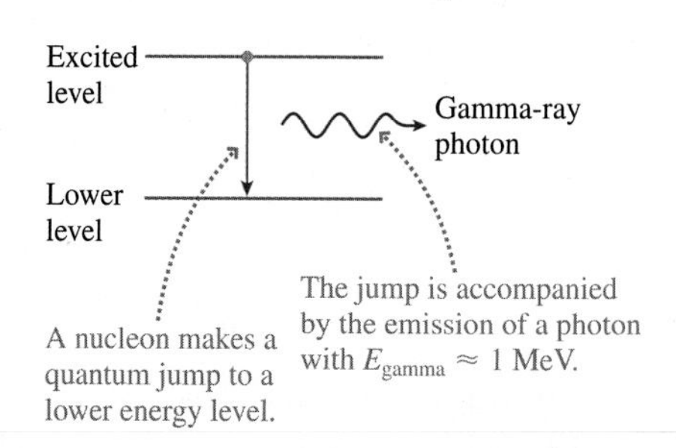
Excited level
Gamma-ray photon
Lower level
The jump is accompanied by the emission of a photon with $E_{gamma} \approx 1$ MeV.
A nucleon makes a quantum jump to a lower energy level.

FIGURE 30.13 Gamma decay.

The spacing between atomic energy levels is only a few eV. Nuclear energy levels, by contrast, are on the order of 1 MeV apart, meaning gamma-ray photons will have energies $E_{gamma} \approx 1$ MeV. Photons with this much energy are quite penetrating and deposit an extremely large amount of energy at the point where they are finally absorbed.

Nuclei left to themselves are usually in their ground states and thus cannot emit gamma-ray photons. However, alpha and beta decay often leave the daughter nucleus in an excited nuclear state, so gamma emission is often found to accompany alpha and beta emission.

Let's look at an example. One of the most important isotopes for medical imaging is ^{99}Tc, an isotope of the element technetium. An excited state of ^{99}Tc is produced in the beta decay of the molybdenum isotope ^{99}Mo:

$$^{99}Mo \rightarrow ^{99}Tc^* + e^- + \text{energy}$$

The asterisk signifies that the technetium nucleus is in an excited state. The excited nucleus then makes a transition to a lower-energy state via the emission of a 140 keV gamma ray:

$$^{99}Tc^* \rightarrow ^{99}Tc + \gamma$$

The full decay process is shown in **FIGURE 30.14**. The final state of the technetium nucleus is much more stable than the excited state.

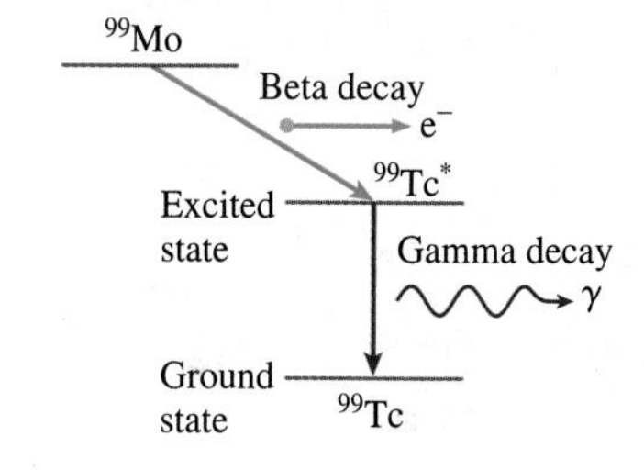
^{99}Mo
Beta decay
e^-
$^{99}Tc^*$
Excited state
Gamma decay
γ
Ground state
^{99}Tc

FIGURE 30.14 $^{99}Tc^*$, a gamma emitter, is produced in the beta decay of ^{99}Mo.

It's useful to collect information about the different possible decays that we've seen, so that we can compare and contrast the different decay modes.

SYNTHESIS 30.1 Nuclear decay modes

Alpha decay

Binding energy

Nuclei with large mass numbers can become more stable by shedding mass through alpha decay.

0 100 200

Mass number, A

The nucleus loses 4 nucleons . . .

$$^{A}_{Z}X \rightarrow ^{A-4}_{Z-2}Y + \alpha + \text{energy}$$

. . . of which 2 are protons.

Radium is a heavy element that is produced by decay of uranium.

$$^{226}_{88}Ra \rightarrow ^{222}_{86}Rn + \alpha + \text{energy}$$

Beta decay

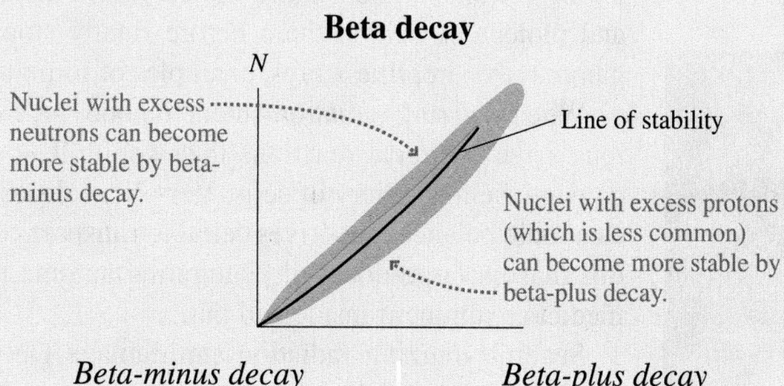
N
Nuclei with excess neutrons can become more stable by beta-minus decay.
Line of stability
Nuclei with excess protons (which is less common) can become more stable by beta-plus decay.

Beta-minus decay

A neutron turns into a proton, so:

The number of nucleons stays the same . . .

$$^{A}_{Z}X \rightarrow ^{A}_{Z+1}Y + e^- + \text{energy}$$

. . . and the nucleus gains a proton.

This chlorine isotope is present at low levels in the environment.

$$^{36}_{17}Cl \rightarrow ^{36}_{18}Ar + e^- + \text{energy}$$

Beta-plus decay

A proton turns into a neutron, so:

The number of nucleons stays the same . . .

$$^{A}_{Z}X \rightarrow ^{A}_{Z-1}Y + e^+ + \text{energy}$$

. . . and the nucleus loses a proton.

This carbon isotope is produced in cyclotrons.

$$^{11}_{6}C \rightarrow ^{11}_{5}B + e^+ + \text{energy}$$

Gamma decay

Nuclei left in an excited state by an alpha or beta decay can drop into a lower energy state through gamma decay

Excited level

Lower level

Gamma-ray photon

No change in the number of neutrons or protons means no change of element or isotope.

$$^{A}_{Z}X^* \rightarrow ^{A}_{Z}X + \gamma$$

This excited form of nickel is produced in the beta-minus decay of ^{60}Co.

$$^{60}_{28}Ni^* \rightarrow ^{60}_{28}Ni + \gamma$$

EXAMPLE 30.5 **What type of decay?**

Phosphorus has one stable isotope, ^{31}P. The isotope ^{32}P is a neutron-rich radioactive isotope that is used in nuclear medicine. What is the likely daughter nucleus of ^{32}P decay?

PREPARE Synthesis 30.1 contains details of when different decay modes are expected. $A = 32$ isn't an especially large mass number, so alpha decay is unlikely. But ^{32}P is neutron rich; it has one more neutron than the only stable phosphorus isotope, so it is likely to undergo beta-minus decay.

SOLVE Phosphorus has atomic number 15. When $^{32}_{15}$P undergoes beta-minus decay, the daughter nucleus has the same number of nucleons and one more proton. Following the details in Synthesis 30.1, we can write the decay as

$$^{32}_{15}\text{P} \rightarrow {}^{32}_{16}\text{S} + \text{e}^-$$

The daughter nucleus of this decay is ^{32}S.

ASSESS $^{32}_{16}$S is a stable isotope of sulfur, so our solution has a radioactive isotope decaying to a more stable state. This gives us confidence in our result.

Decay Series

A radioactive nucleus decays into a daughter nucleus. In many cases, the daughter nucleus is also radioactive and decays to produce its own daughter nucleus. The process continues until reaching a daughter nucleus that is stable. The sequence of isotopes, starting with the original unstable isotope and ending with the stable isotope, is called a **decay series.**

The elements with $Z > 83$ present in the earth's crust are part of the decay series of a few long-lived isotopes of uranium and thorium. As an example, **FIGURE 30.15** shows the decay series of ^{235}U, an isotope of uranium with a 700-million-year half-life. There are many unstable nuclei in the decay series. Ultimately, all ^{235}U nuclei end as the ^{207}Pb isotope of lead, a stable nucleus.

Notice that some nuclei can decay by either alpha *or* beta decay. Thus there are a variety of paths that a decay can follow, but they all end at the same point.

FIGURE 30.15 ^{235}U decay series.

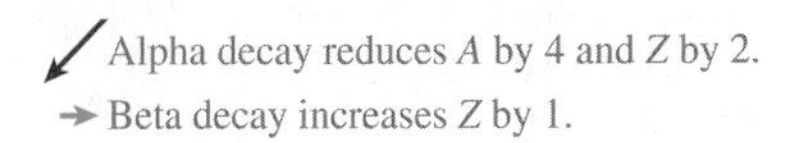

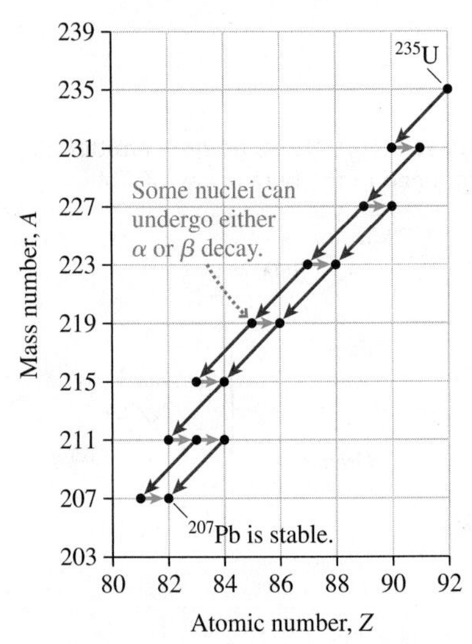

Nuclear Radiation Is a Form of Ionizing Radiation

The energies of the alpha and beta particles and the gamma-ray photons of nuclear decay are typically in the range 0.1–10 MeV. These energies are much higher than the ionization energies of atoms and molecules, which, we've seen, are a few eV. When the sun shines on your skin, it warms it; the low-energy photons are absorbed, their energy converted to thermal energy. The much higher energies of alpha and beta particles and gamma rays cause them to interact very differently with matter, *ionizing* atoms and *breaking* molecular bonds, leaving long trails of ionized atoms and molecules behind them before finally stopping. Alpha and beta particles and gamma rays are, like x rays, examples of ionizing radiation.

When ionizing radiation enters the body, it causes damage in two ways. First, the ions drive chemical reactions that wouldn't otherwise occur. These reactions may damage the machinery of cells. Very large doses of ionizing radiation upset the delicate ionic balance that drives cellular transport, and can rapidly lead to cell death. For this reason, large doses of penetrating gamma rays are sometimes used to sterilize medical equipment inside and out.

Second, ionizing radiation can damage DNA molecules by ionizing them and breaking bonds. If the damage is extensive, cellular repair mechanisms will not be able to cope, and the DNA will be permanently damaged, possibly creating a mutation or a tumor. Tissues with rapidly proliferating cells, such as bone marrow, are quite sensitive to ionizing radiation. Those with less-active cell reproduction, such as the nervous system, are much less sensitive.

A speck of radium placed on a photographic plate emits alpha particles that leave clearly visible ionization trails.

NOTE ▸ Ionizing radiation causes damage to materials and tissues, but **objects irradiated with alpha, beta, or gamma radiation do not become radioactive.** Ionization drives chemical processes involving the electrons. An object could become radioactive only if its nuclei were somehow changed, and that does not happen. ◂

STOP TO THINK 30.4 The cobalt isotope ^{60}Co ($Z = 27$) decays to the nickel isotope ^{60}Ni ($Z = 28$). The decay process is

A. Alpha decay.
B. Beta-minus decay.
C. Beta-plus decay.
D. Gamma decay.

30.5 Nuclear Decay and Half-Lives

The decay of nuclei is different from other types of decay you are familiar with. A tree branch that falls to the forest floor decays. It darkens, becomes soft, and crumbles. You might be able to tell, just by looking at it, about how long it had been decaying. Nuclear decay is different. The nucleus doesn't "age" in any sense. Instead, a nucleus has a certain probability that, within the next second, it will spontaneously turn into a different nucleus and, in the process, eject an alpha or beta particle or a gamma ray.

We can use an analogy here: If you toss a coin, it always has a 50% probability of showing tails, no matter what previous tosses might have been. You might "expect" heads if you've tossed 10 tails in a row, but the 11th toss still has a 50% chance of coming up tails. Likewise with nuclei. If a nucleus doesn't decay in this second, it is no more or less likely to decay in the next. The nucleus remains just as it was, without any change, until the decay finally occurs.

In fact, the mathematics of radioactive decay is the same as that of tossing coins. Suppose you have a large number N_0 of coins. You toss them all and then keep those that come up heads while setting aside those that come up tails. Probability dictates that about half the coins will show tails and be set aside. Now you repeat the process over and over. With each subsequent toss, about half the coins are set aside—they "decay."

After the first toss, the number of coins you have left is about $(1/2)N_0$ because you set aside about half the coins. After the second toss, when you set aside about half of that half, the number of remaining coins is about $(1/2) \times (1/2)N_0$, or $(1/2)^2N_0 = N_0/4$. Half of these coins will be set aside in the third toss, leaving you with $(1/2)^3N_0$, or 1/8 of what you started with. After m tosses—assuming you started with a very large number of coins—the number of coins left is $N = (1/2)^m N_0$.

Something similar happens with radioactive nuclei. If you start with N_0 unstable nuclei, after an interval of time we call one *half-life,* you'll have $N = (1/2)N_0$ nuclei remaining. The **half-life** $t_{1/2}$ is the average time required for one-half the nuclei to decay. This process continues, with one-half the remaining nuclei decaying in each successive half-life. The number of nuclei N remaining at time t is

$$N = N_0\left(\frac{1}{2}\right)^{t/t_{1/2}} \qquad (30.12)$$

Number of atoms remaining after time t → N
Number of atoms at the start, $t = 0$ → N_0
The units for t and $t_{1/2}$ must be the same.

Thus $N = N_0/2$ at $t = t_{1/2}$, $N = N_0/4$ at $t = 2t_{1/2}$, $N = N_0/8$ at $t = 3t_{1/2}$, and so on, with the ratio $t/t_{1/2}$ playing the role of the "number of tosses." **No matter how many nuclei there are at any point in time, the number decays by half during the next half-life.**

NOTE ▶ Each isotope that is unstable and decays has a characteristic half-life, which can range from a fraction of a second to billions of years. Appendix D provides nuclear data and half-lives for the isotopes referred to in this textbook. ◀

The nature of the decay process is shown graphically in FIGURE 30.16.

Figure Video

FIGURE 30.16 The decay of a sample of radioactive nuclei.

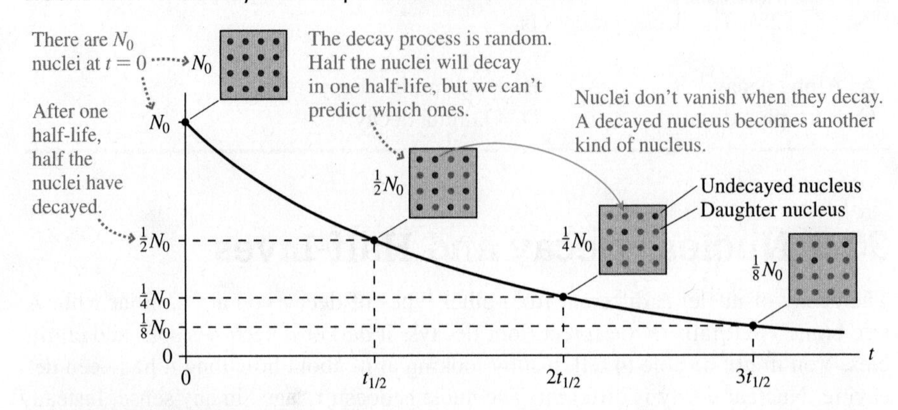

STOP TO THINK 30.5 Nucleus A decays to nucleus B, with a half-life of 10 minutes. A pure sample of nucleus A contains 24,000 particles. Using what you've learned in Figure 30.16, after 30 minutes the sample will contain

A. 3000 particles of nucleus B.
B. 4000 particles of nucleus B.
C. 12,000 particles of nucleus B.
D. 20,000 particles of nucleus B.
E. 21,000 particles of nucleus B.

The graph in Figure 30.16 has a form we have seen before: exponential decay. The exponential decay of the number of nuclei is analogous to the exponential decay of the capacitor voltage in an *RC* circuit or the amplitude of a damped harmonic oscillator. We can write Equation 30.12 in the form of an exponential decay in terms of a *time constant* τ that is related to the half-life:

The time constant τ is proportional to the half-life $t_{1/2}$

$$N = N_0 e^{-t/\tau}$$
$$\tau = \frac{t_{1/2}}{\ln 2} = (1.44)t_{1/2} \qquad (30.13)$$

N p. 456 EXPONENTIAL

We can demonstrate the relationship between the half-life and the time constant because we know, by definition, that $N = N_0/2$ at $t = t_{1/2}$. Thus, according to Equation 30.13,

$$\frac{N_0}{2} = N_0 e^{-t_{1/2}/\tau}$$

The N_0 cancels, and we can then take the natural logarithm of both sides to find

$$\ln\left(\frac{1}{2}\right) = -\ln 2 = -\frac{t_{1/2}}{\tau}$$

We can make two different rearrangements of this equation, finding the time constant in terms of the half-life, or the half-life in terms of the time constant:

$$\tau = \frac{t_{1/2}}{\ln 2} = 1.44t_{1/2}$$
$$t_{1/2} = \tau \ln 2 = 0.693\tau \qquad (30.14)$$

Whether we use Equation 30.12 and the half-life or Equation 30.13 and the time constant is a matter of convenience; both equations describe the same decay.

FIGURE 30.17 is a graphical representation of Equation 30.13; it is the same graph as that of Figure 30.16, simply written in a different mathematical form. The number of radioactive nuclei decreases from N_0 at $t = 0$ to $N = N_0 e^{-1} = 0.37N_0$ at time $t = \tau$. In practical terms, the number decreases by roughly two-thirds during one time constant.

FIGURE 30.17 The number of radioactive atoms decreases exponentially with time.

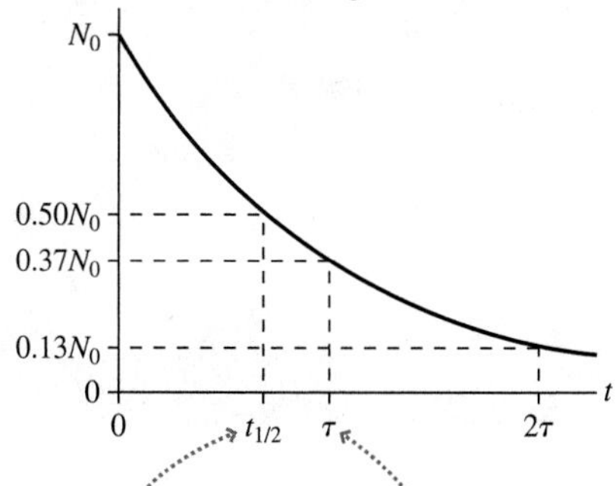

NOTE ▶ There is no natural "starting time" for an exponential decay; you can choose any instant you wish to be $t = 0$. The number of radioactive nuclei present at that instant is N_0. If at one instant you have 10,000 radioactive nuclei whose time constant is $\tau = 10$ min, you'll have roughly 3700 nuclei 10 min later. The fact that you may have had more than 10,000 nuclei at an earlier time isn't relevant. ◀

EXAMPLE 30.6 Determining the decay of radioactive iodine BIO

Patients with Graves disease have an overactive thyroid gland. A common treatment uses radioactive iodine, which is taken up by the thyroid. The radiation emitted in its decay will damage the tissues of the gland. A single pill is produced with 4.0×10^{14} atoms of the isotope ^{131}I, which has a half-life of 8.0 days.

a. How many atoms remain 24 hours after the pill's creation, when the pill is delivered to a hospital?
b. Although the iodine in the pill is constantly decaying, it is still usable as long as it contains at least 1.1×10^{14} ^{131}I atoms. What is the maximum delay before the pill is no longer usable?

PREPARE The atoms in the sample undergo exponential decay, decreasing steadily in number.

SOLVE a. The half-life is $t_{1/2} = 8.0$ days $= 192$ h. Using Equation 30.12, we can find the number of atoms remaining after 24 h have elapsed:

$$N = (4.0 \times 10^{14})\left(\frac{1}{2}\right)^{24/192} = 3.7 \times 10^{14} \text{ atoms}$$

b. The time after which 1.1×10^{14} atoms remain is given by

$$1.1 \times 10^{14} = (4.0 \times 10^{14})\left(\frac{1}{2}\right)^{t/192}$$

To solve for t, we write this as

$$\frac{1.1 \times 10^{14}}{4.0 \times 10^{14}} = \left(\frac{1}{2}\right)^{t/192}$$

or

$$0.275 = \left(\frac{1}{2}\right)^{t/192}$$

Now, we take the natural logarithm of both sides:

$$\ln(0.275) = \ln\left(\left(\frac{1}{2}\right)^{t/192}\right)$$

We can solve for t by using the fact that $\ln(a^x) = x\ln(a)$. This allows us to "pull out" the $t/192$ exponent to find

$$\ln(0.275) = \left(\frac{t}{192}\right)\ln\left(\frac{1}{2}\right)$$

Solving for t, we find that the pill ceases to be useful after

$$t = 192\frac{\ln(0.275)}{\ln(1/2)} = 360 \text{ h} = 15 \text{ days}$$

ASSESS The weakest usable concentration of iodine is approximately one-fourth of the initial concentration. This means that the decay time should be approximately equal to two half-lives, which is what we found.

Activity

The **activity** R of a radioactive sample is the number of decays per second. Each decay corresponds to an alpha, beta, or gamma emission, so the activity is a measure of how much radiation is being given off. A detailed treatment of the mathematics of decay shows that the activity of a sample of N nuclei having time constant τ (and half-life $t_{1/2}$) is

$$R = \frac{N}{\tau} = \frac{0.693N}{t_{1/2}} \tag{30.15}$$

A sample with $N = 1.0 \times 10^{10}$ nuclei decaying with time constant $\tau = 100$ s would, at that instant, have activity $R = 1.0 \times 10^8$ decays/s—in each second, 1.0×10^8 atoms would decay.

We see from Equation 30.15 that **activity is inversely proportional to the half-life.** If two samples have the same number of nuclei, the sample with the shorter half-life

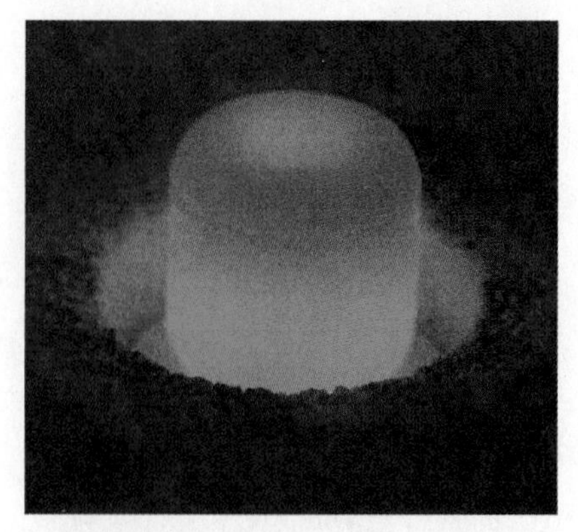

Powered by decay This pellet is made of a short-lived isotope of plutonium. The short half-life means that this isotope has a very high activity. The high decay rate warms the pellet enough to make it glow. This pellet is for a thermoelectric generator, which uses the heat of the pellet to produce electricity. The radioactive decay of short-lived isotopes of plutonium is used to provide power for spacecraft on voyages far from the sun. For many years, plutonium "batteries" were used to power heart pacemakers as well.

has the larger activity. We can combine Equation 30.15 with Equations 30.12 and 30.13 to obtain expressions for the variation of activity with time:

$$R = \frac{N}{\tau} = \frac{N_0}{\tau}\left(\frac{1}{2}\right)^{t/t_{1/2}} = \frac{N_0}{\tau}e^{-t/\tau}$$

N_0/τ is the initial activity R_0, so we find the following expression for the decay of activity:

Activity after time t

Activity at the start, $t = 0$

$$R = R_0\left(\frac{1}{2}\right)^{t/t_{1/2}} = R_0 e^{-t/\tau} \quad (30.16)$$

EXPONENTIAL p. 456

This equation has the same form as that for the decay of the sample. The activity of a sample decreases exponentially along with the number of remaining nuclei.

The SI unit of activity is the **becquerel,** defined as

$$1 \text{ becquerel} = 1 \text{ Bq} = 1 \text{ decay/s or } 1 \text{ s}^{-1}$$

An older unit of activity, but one that continues in widespread use, is the **curie.** The conversion factor is

$$1 \text{ curie} = 1 \text{ Ci} = 3.7 \times 10^{10} \text{ Bq}$$

1 Ci is a substantial amount of radiation. The radioactive samples used in laboratory experiments are typically $\approx 1\ \mu$Ci or, equivalently, $\approx 40{,}000$ Bq. These samples can be handled with only minor precautions. Larger sources of activity require thick shielding and other special precautions to prevent exposure to high levels of radiation.

CONCEPTUAL EXAMPLE 30.7 Relative activities of isotopes in the body

^{40}K ($t_{1/2} = 1.3 \times 10^9$ yr) and ^{14}C ($t_{1/2} = 5.7 \times 10^3$ yr) are two radioactive isotopes found in measurable quantities in your body. Suppose you have 1 mole of each. Which is more radioactive—that is, which has a greater activity?

REASON Equation 30.15 shows that the activity of a sample is proportional to the number of atoms and inversely proportional to the half-life. Because both samples have the same number of atoms, the sample of ^{14}C, with its much shorter half-life, has a much greater activity.

EXAMPLE 30.8 Determining the decay of activity BIO

A ^{60}Co (half-life 5.3 yr) source used to provide gamma rays to irradiate tumors has an activity of 0.43 Ci.

a. How many ^{60}Co atoms are in the source?
b. What will be the activity of the source after 10 yr?

PREPARE The activity of the source depends on the half-life and the number of atoms. If we know the activity and the half-life, we can compute the number of atoms. The number of atoms will undergo exponential decay, and so will the activity.

SOLVE a. Equation 30.15 relates the activity and the number of atoms. Rewriting the equation, we can relate the initial activity and the initial number of atoms as $N_0 = t_{1/2}R_0/0.693$. To use this equation, we need numbers in SI units. In Bq, the initial activity of the source is

$$R_0 = (0.43 \text{ Ci})\left(\frac{3.7 \times 10^{10} \text{ Bq}}{1 \text{ Ci}}\right) = 1.59 \times 10^{10} \text{ Bq}$$

The half-life $t_{1/2}$ in s is

$$t_{1/2} = (5.3 \text{ yr})\left(\frac{3.16 \times 10^7 \text{ s}}{1 \text{ yr}}\right) = 1.67 \times 10^8 \text{ s}$$

Thus the initial number of ^{60}Co atoms in the source is

$$N_0 = \frac{t_{1/2}R_0}{0.693} = \frac{(1.67 \times 10^8 \text{ s})(1.59 \times 10^{10} \text{ Bq})}{0.693} = 3.8 \times 10^{18} \text{ atoms}$$

b. The variation of activity with time is given by Equation 30.16. After 10 yr, the activity is

$$R = R_0\left(\frac{1}{2}\right)^{t/t_{1/2}} = (1.59 \times 10^{10} \text{ Bq})\left(\frac{1}{2}\right)^{10/5.3} = 4.3 \times 10^9 \text{ Bq} = 0.12 \text{ Ci}$$

ASSESS Although N_0 is a very large number, it is a very small fraction of a mole. The sample contains only about 400 μg of ^{60}Co. In the first part of the question we needed to convert the half-life to s. The second part of the question used a ratio of two times, so we can use any units we like as long as the units of both times are the same.

Radioactive Dating

Many geological and archaeological samples can be dated by measuring the decays of naturally occurring radioactive isotopes.

The most well-known dating technique uses the radioactive carbon isotope ^{14}C and is known as carbon dating or **radiocarbon dating.** ^{14}C has a half-life of 5730 years, so any ^{14}C present when the earth formed 4.5 billion years ago has long since decayed away. Nonetheless, ^{14}C is present in atmospheric carbon dioxide because high-energy cosmic rays collide with gas molecules high in the atmosphere to create ^{14}C nuclei from nuclear reactions with nitrogen and oxygen nuclei. The creation and decay of ^{14}C have reached a steady state in which the $^{14}C/^{12}C$ ratio is relatively stable at 1.3×10^{-12}.

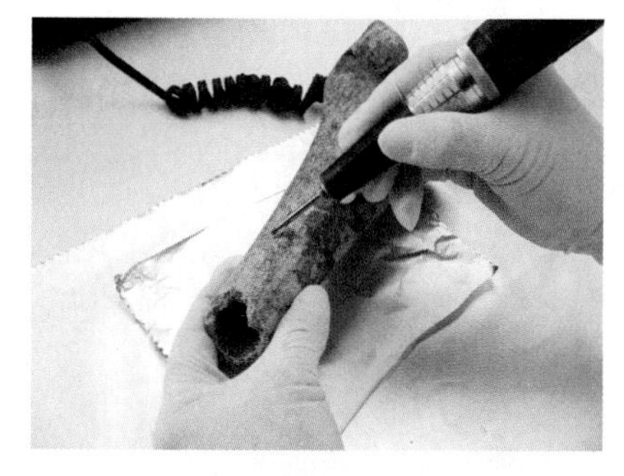

A researcher is extracting a small sample of an ancient bone. By measuring the ratio of carbon isotopes present in the sample she will determine the age of the bone.

All living organisms constantly exchange carbon dioxide with the environment, so the $^{14}C/^{12}C$ ratio in living organisms is also 1.3×10^{-12}. As soon as an organism dies, the ^{14}C in its tissue begins to decay and no new ^{14}C is added. As time goes on, the ^{14}C decays at a well-known rate. Thus, a measurement of the activity of an ancient organic sample permits a determination of the age. "New" samples have a higher fraction of ^{14}C than "old" samples.

The first step in radiocarbon dating is to extract and purify carbon from the sample. The carbon is then placed in a shielded chamber and its activity measured. This activity is then compared to the activity of an identical modern sample. Equation 30.16 relates the activity of a sample at a time t to its initial activity. If we assume that the original activity R_0 was the same as the activity of the modern sample, we can determine the time since the decay began—and thus the age of the sample.

EXAMPLE 30.9 Carbon dating a tooth BIO

A rear molar from a mammoth skeleton is dated using a measurement of its ^{14}C content. Carbon from the tooth is chemically extracted and formed into benzene. The benzene sample is placed in a shielded chamber. Decays from the sample come at an average rate of 11.5 counts per minute. A modern benzene sample of the exact same size gives 54.9 counts per minute. What is the age of the skeleton?

PREPARE We can assume that, thousands of years ago, the sample had an initial activity of 54.9 counts per minute— identical to the activity of a modern sample. The present activity is lower due to the decay of the ^{14}C since the death of the mammoth.

SOLVE Equation 30.16 gives the decrease of the activity as a function of time as $R = R_0(1/2)^{t/t_{1/2}}$. The current activity is $R = 11.5$ counts per minute, and we assume that the initial activity was $R_0 = 54.9$. t is the time since the mammoth stopped growing—the age of the skeleton. We solve for t by rearranging terms and computing a natural logarithm, as in Example 30.6:

$$\frac{R}{R_0} = \left(\frac{1}{2}\right)^{t/t_{1/2}}$$

$$\ln\left(\frac{R}{R_0}\right) = \left(\frac{t}{t_{1/2}}\right)\ln\left(\frac{1}{2}\right)$$

We then solve for the time t:

$$t = \frac{t_{1/2}}{\ln(1/2)}\ln\left(\frac{R}{R_0}\right) = \frac{5730 \text{ yr}}{\ln(1/2)}\ln\left(\frac{11.5}{54.9}\right) = 12{,}900 \text{ yr}$$

ASSESS The final time is in years, the same unit we used for the half-life. This is a realistic example of how such radiocarbon dating is done; the numbers and details used in this example come from an actual experimental measurement. The age of the sample places it at the end of the last ice age, when mammoths last roamed the earth, so our result seems reasonable.

CONCEPTUAL EXAMPLE 30.10 Source contamination

One possible problem with carbon dating is contamination with modern carbon sources. Suppose an archaeologist has unearthed and carbon-dated a fragment of wood that has absorbed carbon of recent vintage from organic molecules in groundwater. Does he underestimate or overestimate the age of the wood?

REASON Because the wood has absorbed modern carbon, it will have more ^{14}C than it would had it decayed undisturbed. The present activity is higher than it would otherwise be. This will lead to an underestimate of the age of the wood.

▶ **Responsible dating** BIO Measuring the activity of the carbon in a sample to determine the fraction of ^{14}C can require a significant amount of organic material—perhaps 25 g. For an artifact of great historical importance, such as this parchment fragment from the Dead Sea scrolls, this would be unacceptable. Instead, dates are obtained by using a mass spectrometer to directly measure the ratio of carbon isotopes, from which the age can be determined. This can be done with excellent accuracy on as little as 0.1 g of material.

Carbon dating can be used to date skeletons, wood, paper, fur, food material, and anything else made of organic matter. It is quite accurate for ages to about 15,000 years, about three half-lives. Items are dated to about 50,000 years with a fair degree of reliability.

Isotopes with longer half-lives are used to date geological samples. Potassium-argon dating, using ^{40}K with a half-life of 1.25 billion years, is especially useful for dating rocks of volcanic origin.

STOP TO THINK 30.6 A sample of 1000 radioactive atoms has a 10 minute half-life. How old is the sample when 750 atoms have decayed?

A. 10 minutes B. 15 minutes C. 20 minutes D. 30 minutes

30.6 Medical Applications of Nuclear Physics BIO

Nuclear physics has brought both peril and promise to society. Radioactivity can cause tumors. At the same time, radiation can be used to diagnose and cure some cancers. This section is a brief survey of medical applications of nuclear physics.

Radiation Dose

Ionizing radiation damages cells of the body, but it also damages bacteria and other pathogens. This gamma source is used for sterilizing medical equipment. The blue glow is due to the ionization of the air around the source.

Nuclear radiation disrupts a cell's machinery by altering and damaging biological molecules, as we saw in Section 30.4. The biological effects of radiation depend on two factors. The first is the physical factor of how much energy is absorbed by the body. The second is the biological factor of how tissue reacts to different forms of radiation.

Suppose a beta particle travels through tissue, losing kinetic energy as it ionizes atoms it passes. The energy lost by the beta particle is a good measure of the number of ions produced and thus the amount of damage done. In a certain volume of tissue, more ionization means more damage. For this reason, we define the radiation **dose** as the energy from ionizing radiation absorbed by 1 kg of tissue. The SI unit for the dose is the **gray,** abbreviated Gy. The Gy is defined as

$$1 \text{ Gy} = 1.00 \text{ J/kg of absorbed energy}$$

The number of Gy depends only on the energy absorbed, not on the type of radiation or on what the absorbing material is. Another common unit for dose is the *rad;* 1 rad = 0.01 Gy.

A 1 Gy dose of gamma rays and a 1 Gy dose of alpha particles have different biological consequences. To account for such differences, the **relative biological effectiveness** (RBE) is defined as the biological effect of a given dose relative to the biological effect of an equal dose of x rays. Table 30.3 lists the relative biological effectiveness of different forms of radiation. Larger values correspond to larger biological effects.

TABLE 30.3 Relative biological effectiveness of radiation

Radiation type	RBE
X rays	1
Gamma rays	1
Beta particles	1
Protons	5
Neutrons	5–20
Alpha particles	20

The radiation **dose equivalent** is the product of the energy dose in Gy and the relative biological effectiveness. Dose equivalent is measured in **sieverts,** abbreviated Sv. To be precise,

$$\text{dose equivalent in Sv} = \text{dose in Gy} \times \text{RBE}$$

One Sv of radiation produces the same biological damage regardless of the type of radiation. Another common unit of dose equivalent (also called biologically equivalent dose) is the *rem;* 1 rem = 0.01 Sv.

NOTE ► In practice, the term "dose" is often used for both dose and dose equivalent. Use the units as a guide. If the unit is Sv or rem, it is a dose equivalent; if Gy or rad, a dose. ◄

EXAMPLE 30.11 Finding energy deposited in radiation exposure BIO

A 75 kg patient is given a bone scan. A phosphorus compound containing the gamma-emitter ^{99}Tc is injected into the patient. It is taken up by the bones, and the emitted gamma rays are measured. The procedure exposes the patient to 3.6 mSv (360 mrem) of radiation. What is the total energy deposited in the patient's body, in J and in eV?

PREPARE The exposure is given in Sv, so it is a dose equivalent, a combination of deposited energy and biological effectiveness. The RBE for gamma rays is 1. Gamma rays are penetrating, and the source is distributed throughout the body, so this is a whole-body exposure. Each kg of the patient's body will receive approximately the same energy.

SOLVE The dose in Gy is the dose equivalent in Sv divided by the RBE. In this case, because RBE = 1, the dose in Gy is numerically equal to the equivalent dose in Sv. The dose is thus 3.6 mGy = 3.6×10^{-3} J/kg. The radiation energy absorbed in the patient's body is

$$\text{absorbed energy} = (3.6 \times 10^{-3}\ \text{J/kg})(75\ \text{kg}) = 0.27\ \text{J}$$

In eV, this is

$$\text{absorbed energy} = (0.27\ \text{J})(1\ \text{eV}/1.6 \times 10^{-19}\ \text{J}) = 1.7 \times 10^{18}\ \text{eV}$$

ASSESS The total energy deposited, 0.27 J, is quite small; there will be negligible heating of tissue. But radiation produces its effects in other ways, as we have seen. Because it takes only ≈ 10 eV to ionize an atom, this dose is enough energy to ionize over 10^{17} atoms, meaning it can cause significant disruption to the cells of the body.

TABLE 30.4 Radiation exposure

Radiation source	Typical exposure (mSv)
PET scan	7.0
Natural background (1 year)	3.0
Mammogram	0.70
Chest x ray	0.30
Transatlantic airplane flight	0.050
Dental x ray	0.030

The question inevitably arises: What is a safe dose? Unfortunately, there is no simple or clear definition of a safe dose. A prudent policy is to avoid unnecessary exposure, and to weigh the significance of an exposure in relation to the *natural background.* We are all exposed to continual radiation from cosmic rays and from naturally occurring radioactive atoms in the ground, the atmosphere, and even the food we eat. This background averages about 3 mSv (300 mrem) per year, although there are wide regional variations depending on the soil type and the elevation.

Table 30.4 lists the expected exposure from several different sources. A dental x ray subjects a person to approximately 1% of the yearly natural background that he or she would normally receive and is likely not a cause for significant worry. Mammograms involve a much larger dose, concentrated in a small region of the body. A nuclear medicine procedure, like a PET scan (which is discussed below), may involve an exposure that is larger than the typical yearly background dose. This significant dose must be weighed against the medical benefits of the procedure.

Nuclear Medicine

The tissues in the body most susceptible to radiation are those that are rapidly proliferating—including tumors. The goal in *radiation therapy* is to apply a large enough dose of radiation to destroy or shrink a tumor while producing minimal damage to surrounding healthy tissue.

In **FIGURE 30.18a**, a patient with a brain tumor is fitted with a metal *collimator* that absorbs gamma rays except for those traveling along desired paths. The collimator is fashioned so that gamma rays from an external source will be concentrated on the tumor, as shown in **FIGURE 30.18b**. Because the rapidly dividing cells of a tumor are much more sensitive to radiation than the tissues of the brain, and because surgical options carry risks of significant complications, radiation is a common means of treating tumors of the brain.

Other tumors are treated by surgically implanting radioactive "seeds" within the tumor. One common type of seed contains ^{125}I, which undergoes a nuclear decay followed by emission of a 27.5 keV photon. The relatively low-energy photon has a very short range so that it will damage only tissue close to the seed.

FIGURE 30.18 The use of gamma rays to treat a tumor in the brain.

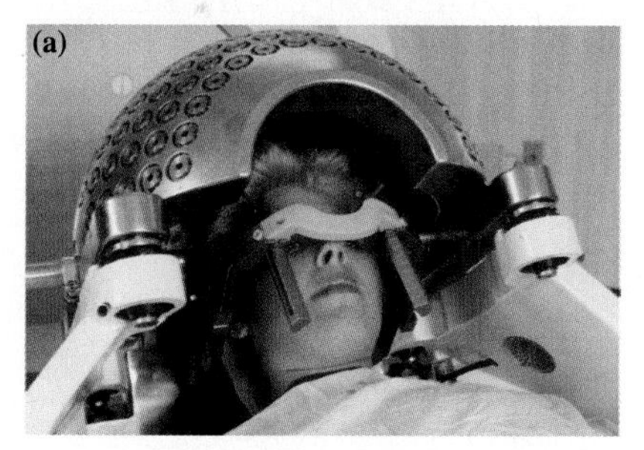
(a)

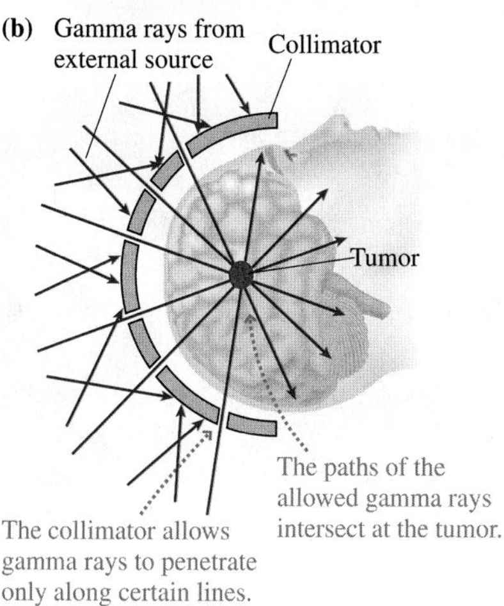
(b) Gamma rays from external source
Collimator
Tumor
The paths of the allowed gamma rays intersect at the tumor.
The collimator allows gamma rays to penetrate only along certain lines.

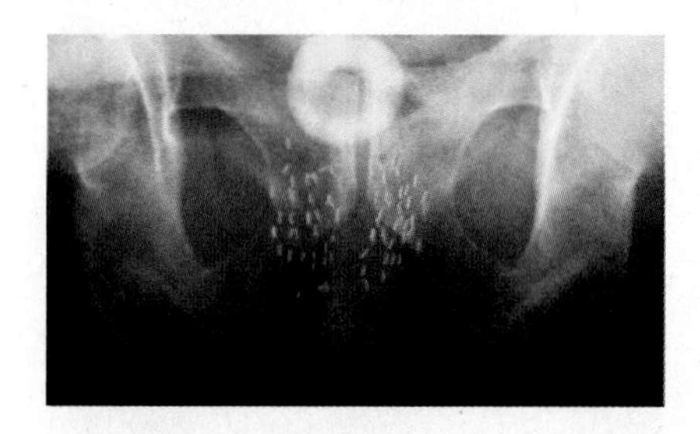

Some tissues in the body will preferentially take up certain isotopes, allowing for treatment by isotope ingestion. A common treatment for hyperthyroidism, in which the thyroid gland is overactive, is to damage the gland with the isotope ^{131}I, a beta-emitter with a half-life of 8.0 days. A patient is given a tablet containing ^{131}I. The iodine in the blood is taken up and retained by the thyroid gland, resulting in a reduction of the gland's activity with minimal disruption of surrounding tissue.

◀Metal seeds containing radioactive isotopes implanted in the prostate gland to shrink a tumor.

EXAMPLE 30.12 Using radiation to treat disease BIO

A patient with prostate cancer is treated with implanted seeds that contain the samarium isotope ^{153}Sm which decays with a half-life of 1.9 days by emitting a 0.81 MeV beta particle. On average, 50% of the decay energy is absorbed by the 24 g prostate gland. If the seeds have an activity of 100 MBq when first implanted, what dose equivalent in Sv does the gland receive during the first day?

PREPARE We can't use the activity to compute the dose directly because the activity will change with time. Instead, we'll figure out how many nuclei decay during the first day. We know the initial activity and the half-life, so we can find the initial number of nuclei. To do this, we'll need the half-life in seconds:

$$t_{1/2} = (1.9 \text{ days})\left(\frac{8.64 \times 10^4 \text{ s}}{1 \text{ day}}\right) = 1.64 \times 10^5 \text{ s}$$

To complete the dose calculation, we'll need the absorbed energy in joules. Each atom that decays emits a 0.81 MeV beta particle, and half of this energy is deposited in the gland. In joules, this is

$$\text{energy deposited per decay} = (0.5)(0.81 \text{ MeV})(1.60 \times 10^{-13} \text{ J/MeV}) = 6.48 \times 10^{-14} \text{ J}$$

SOLVE We can find the initial number of radioactive nuclei as we did in Example 30.8:

$$N_0 = \frac{t_{1/2}R_0}{0.693} = \frac{(1.64 \times 10^5 \text{ s})(100 \times 10^6 \text{ decays/s})}{0.693} = 2.37 \times 10^{13}$$

After 1.0 day, the number of radioactive nuclei in the sample has decreased to

$$N = N_0\left(\frac{1}{2}\right)^{t/t_{1/2}} = (2.37 \times 10^{13})\left(\frac{1}{2}\right)^{1.0 \text{ day}/1.9 \text{ days}} = 1.65 \times 10^{13}$$

The number of decays during the first day is the difference between these numbers:

$$N_0 - N = 2.37 \times 10^{13} - 1.65 \times 10^{13} = 7.2 \times 10^{12} \text{ decays}$$

The energy deposited in the gland by these decays is

$$\text{total energy deposited} = (6.48 \times 10^{-14} \text{ J/decay})(7.2 \times 10^{12} \text{ decays}) = 0.467 \text{ J}$$

The radiation dose to the 0.024 kg gland is thus

$$\text{dose} = \frac{0.467 \text{ J}}{0.024 \text{ kg}} = 19 \text{ Gy}$$

Beta particles have RBE = 1, so the dose equivalent in Sv is the same:

$$\text{dose equivalent} = 19 \text{ Sv}$$

ASSESS This is a very large dose equivalent, but the goal is to destroy the tumor, so this is a reasonable result.

Nuclear Imaging

X rays from an external source may be used to make an image of the body, as described in Chapter 28. *Nuclear imaging* uses an internal source—radiation from isotopes in the body—to produce an image of tissues in the body. The bone-scan image that opened the chapter is an example of nuclear imaging.

There is a key difference between x rays and nuclear imaging procedures. **An x ray is an image of anatomical structure;** it is excellent for identifying structural problems like broken bones. **Nuclear imaging creates an image of the biological activity of tissues in the body.** For example, nuclear imaging can detect reduced metabolic activity of brain tissue after a stroke.

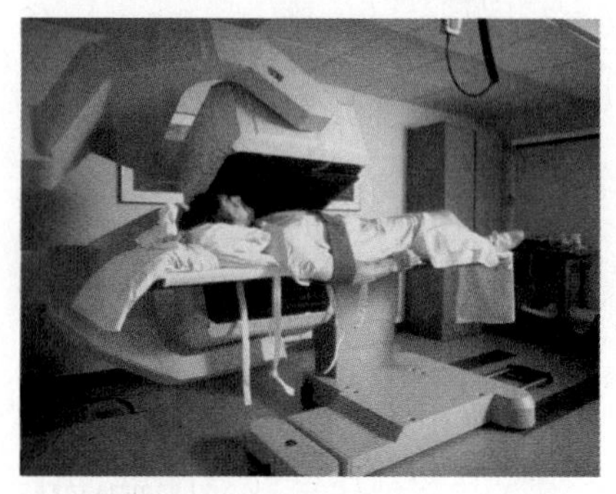

The detector is measuring gamma radiation emitted by isotopes taken up by tissues in the woman's head and neck.

Let's look at an example that illustrates the difference between a conventional x-ray image or CAT scan and an image made with a nuclear imaging technique. Suppose a doctor suspects a patient has cancerous tissue in the bones. An x ray does not show anything out of the ordinary; the tumors may be too small or may appear similar to normal bone. The doctor then orders a scan with a **gamma camera,** a device that can measure and produce an image from gamma rays emitted within the body.

The patient is given a dose of a phosphorus compound labeled with the gamma-emitter ^{99}Tc. This compound is taken up and retained in bone tissue where active growth is occurring. The ^{99}Tc will be concentrated in the bones where there has been recent injury or inflammation—or where a tumor is growing. The patient is then scanned with a gamma camera. **FIGURE 30.19a** shows how the gamma camera can pinpoint the location of the gamma-emitting isotopes in the body and produce an image that reveals their location and intensity. A typical image is shown in **FIGURE 30.19b**. The bright spots show high concentrations of ^{99}Tc, revealing areas of tumor growth. The tumors may be too small to show up on an x ray, but their activity is easily detected with the gamma camera. With such early detection, the patient's chance of a cure is greatly improved.

FIGURE 30.19 The operation of a gamma camera.

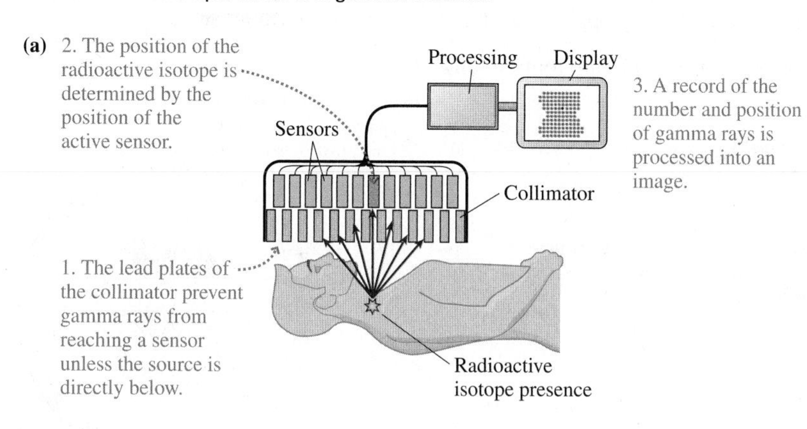
(a)
2. The position of the radioactive isotope is determined by the position of the active sensor.
Processing
Display
3. A record of the number and position of gamma rays is processed into an image.
Sensors
Collimator
1. The lead plates of the collimator prevent gamma rays from reaching a sensor unless the source is directly below.
Radioactive isotope presence

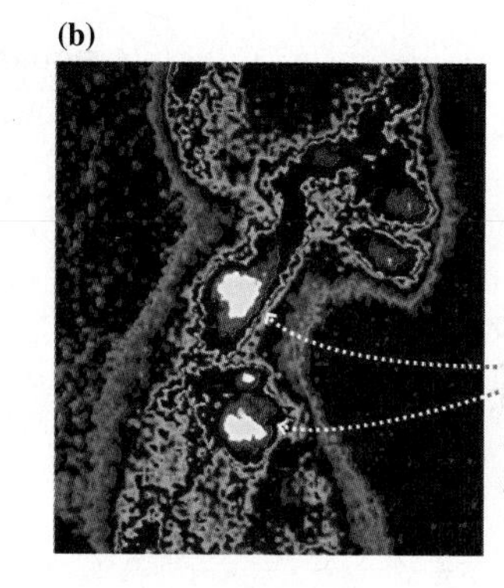
(b)
The bright spots show areas of active tumor growth.

CONCEPTUAL EXAMPLE 30.13 **Using radiation to diagnose disease** BIO

A patient suspected of having kidney disease is injected with a solution containing molecules that are taken up by healthy kidney tissue. The molecules have been "tagged" with radioactive ^{99}Tc. A gamma camera scan of the patient's abdomen gives the image in **FIGURE 30.20**. In this image, blue corresponds to the areas of highest activity. Which of the patient's kidneys has reduced function?

FIGURE 30.20 A gamma scan of a patient's kidneys.

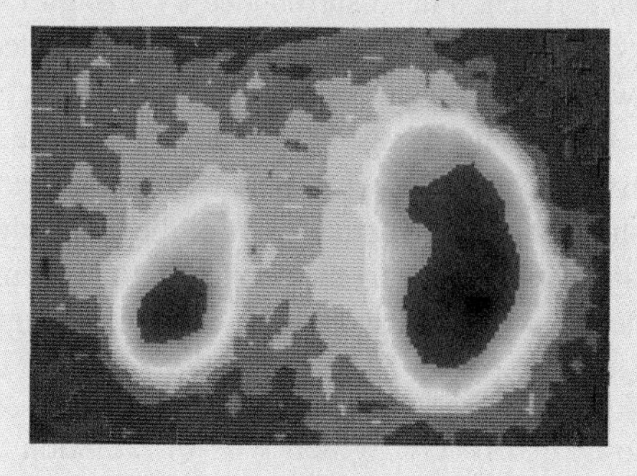

REASON Healthy tissue should show up in blue on the scan because healthy tissue will absorb molecules with the ^{99}Tc attached and will thus emit gamma rays. The kidney imaged on the right shows normal activity throughout; the kidney imaged on the left appears smaller, so it has a smaller volume of healthy tissue. The patient is ill; the problem is with the kidney imaged on the left.

ASSESS Depending on the isotope and how it is taken up by the body, either healthy tissue or damaged tissue could show up on a gamma camera scan.

Positron-Emission Tomography

Class Video

We have seen that a small number of radioactive isotopes decay by the emission of a positron. Such isotopes can be used for an imaging technique known as *positron-emission tomography,* or *PET.* PET is particularly important for imaging the brain.

The imaging process relies on the mass–energy conversion resulting from the combination of an electron and a positron. Suppose an electron and a positron are at rest, or nearly so; the combined momentum is nearly zero. The electron and positron have opposite charges and so will attract each other. When they meet, as we saw in ◀◀ SECTION 27.10, they completely annihilate—but energy and momentum are still conserved. The conservation of energy means that the annihilation will produce one or more high-energy photons—gamma rays. We learned in Chapter 27 that photons have momentum, so the annihilation can't produce a single photon because that photon would leave the scene of the annihilation with momentum. Instead the most likely result is a pair of photons directed exactly opposite each other, as shown in FIGURE 30.21a.

FIGURE 30.21 Positron-emission tomography.

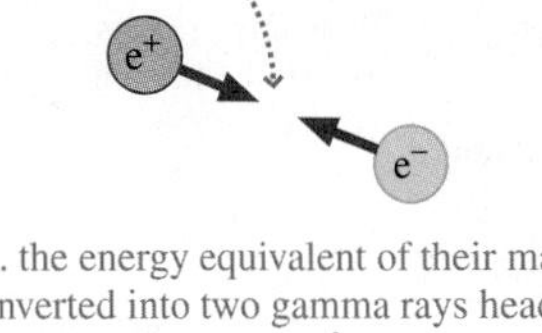
(a) When the electron and positron meet . . .
e^+
e^-
. . . the energy equivalent of their mass is converted into two gamma rays headed in opposite directions.
(b) Coincident detection of two gamma rays means that the positron source is along this line.

Most PET scans use the fluorine isotope ^{18}F, which emits a positron as it undergoes beta-plus decay to ^{18}O with a half-life of 110 minutes. ^{18}F is used to create an analog of glucose called fluorine-18 fluorodeoxyglucose (F-18 FDG). This compound is taken up by tissues in the brain. Areas that are more active are using more glucose, so the F-18 FDG is concentrated in active brain regions. When a fluorine atom in the F-18 FDG decays, the emitted positron immediately collides with a regular electron. The two annihilate to produce two gamma rays that travel out of the brain in opposite directions, as shown in Figure 30.21a.

FIGURE 30.21b shows a patient's head surrounded by a ring of gamma-ray detectors. Because the gamma rays from the positron's annihilation are emitted back to back, simultaneous detection of two gamma rays on opposite sides of the subject indicates that the annihilation occurred somewhere along the line between those detectors. Recording many such pairs of gamma rays shows with great accuracy where the decays are occurring. A full scan will show more activity in regions of the brain where metabolic activity is enhanced, less activity in regions where metabolic activity is depressed. An analysis of these scans can provide a conclusive diagnosis of stroke, injury, or Alzheimer's disease.

STOP TO THINK 30.7 A patient ingests a radioactive isotope to treat a tumor. The isotope provides a dose of 0.10 Gy. Which type of radiation will give the highest dose equivalent in Sv?

A. Alpha particles B. Beta particles C. Gamma rays

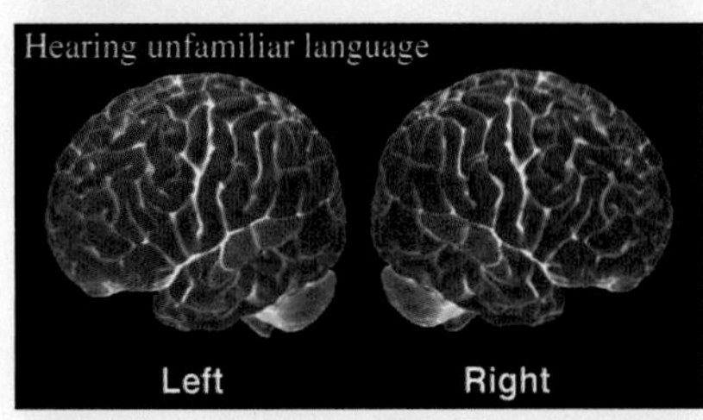
Hearing unfamiliar language
Left
Right

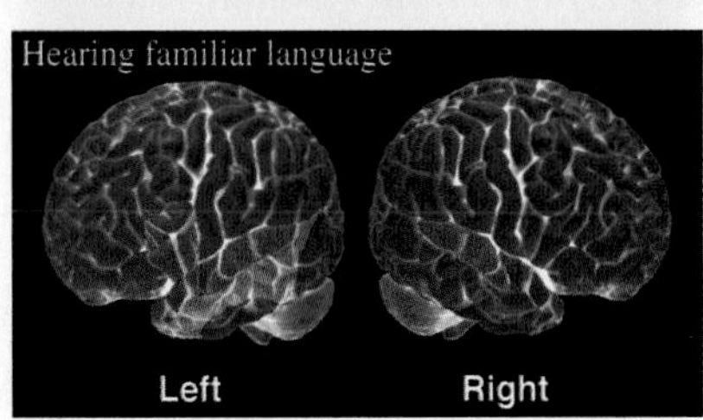
Hearing familiar language
Left
Right

◀ **This is your brain on PET** BIO The woman in the top photo is undergoing a PET scan not to diagnose disease but to probe the workings of the brain. While undergoing a PET scan, a subject is asked to perform different mental tasks. The lower panels show functional images from a PET scan super-imposed on anatomical images from a CAT scan. The subject first listened to speech in an unfamiliar language; the active areas of the brain were those responsible for hearing. Next, she listened to speech in a familiar language, resulting in activity in the parts of the brain responsible for speech and comprehension.

30.7 The Ultimate Building Blocks of Matter

As we've seen, modeling the nucleus as being made of protons and neutrons allows a description of all of the elements in the periodic table in terms of just three basic particles—protons, neutrons, and electrons. But are protons and neutrons *really* basic building blocks? Molecules are made of atoms. Atoms are made of a cloud of electrons surrounding a positively charged nucleus. The nucleus is composed of protons and neutrons. Where does this process end? Are electrons, protons, and neutrons the basic building blocks of matter, or are they made of still smaller subunits?

This question takes us into the domain of what is known as **particle physics**—the branch of physics that deals with the basic constituents of matter and the relationships among them. Particle physics starts with the constituents of the atom, the proton, neutron, and electron, but there are many other particles below the scale of the atom. We call these particles **subatomic particles.**

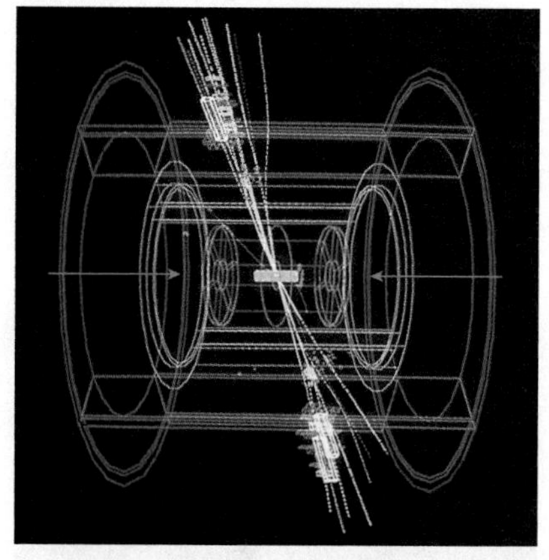

Subatomic crash tests The above picture is the record of a collision between an electron and a positron (paths represented by the green arrows) brought to high speeds in a collider. The particles annihilated in the center of a detector that measured the paths of the particles produced in the collision. In this case, the annihilation of the electron and positron created a particle known as a Z boson. The Z boson then quickly decayed into the two jets of particles seen coming out of the detector.

Antiparticles

We've described the positron as the *antiparticle* to the electron. In what sense is a positron an *anti*electron? As we've seen, when a positron and an electron meet, they annihilate each other, turning the energy equivalent of their masses into the pure energy of two photons. Mass disappears and light appears in one of the most spectacular confirmations of Einstein's relativity.

Every subatomic particle that has been discovered has an antiparticle twin that has the same mass and the same spin but opposite charge. In addition to positrons, there are antiprotons (with $q = -e$), antineutrons (also neutral, but not the same as regular neutrons), and antimatter versions of all the various subatomic particles we will see. The notation to represent an antiparticle is a bar over the top of the symbol. A proton is represented as p, an antiproton as $\bar{\text{p}}$.

Antiparticles provide interesting opportunities for creating "exotic" subatomic particles. When a particle meets its associated antiparticle, the two annihilate, leaving nothing but their energy behind. This energy must go somewhere. Sometimes it is emitted as gamma-ray photons, but this energy can also be used to create other particles.

The major tool for creating and studying subatomic particles is the *particle collider.* These machines use electric and magnetic fields to accelerate particles and their antiparticles, such as e and $\bar{\text{e}}$, or p and $\bar{\text{p}}$, to speeds very close to the speed of light. These particles then collide head-on. As they collide and annihilate, their mass-energy and kinetic energy combine to produce exotic particles that are not part of ordinary matter. These particles come in a dizzying variety—pions, kaons, lambda particles, sigma particles, and dozens of others—each with its own antiparticle. Most live no more than a trillionth of a second.

EXAMPLE 30.14 Determining a possible outcome of a proton-antiproton collision

When a proton and an antiproton annihilate, the resulting energy can be used to create new particles. One possibility is the creation of electrically neutral particles called *neutral pions.* A neutral pion has a rest mass of 135 MeV/c^2. How many neutral pions could be produced in the annihilation of a proton and an antiproton? Assume the proton and antiproton are moving very slowly as they collide.

PREPARE The mass of a proton is given in Table 30.2 as 938 MeV/c^2. The mass of an antiproton is the same. Because the proton and antiproton are moving slowly, with essentially no kinetic energy, the total energy available for creating new particles is the energy equivalent of the masses of the proton and the antiproton.

SOLVE The total energy from the annihilation of a proton and an antiproton is the energy equivalent of their masses:

$$E = (m_{\text{proton}} + m_{\text{antiproton}})c^2 = (938\ \text{MeV}/c^2 + 938\ \text{MeV}/c^2)c^2 = 1876\ \text{MeV}$$

Continued

It takes 135 MeV to create a neutral pion. The ratio

$$\frac{\text{energy available}}{\text{energy required to create a pion}} = \frac{1876 \text{ MeV}}{135 \text{ MeV}} = 13.9$$

tells us that we have enough energy to produce 13 neutral pions from this process, but not quite enough to produce 14.

ASSESS Because the mass of a pion is much less than that of a proton or an antiproton, the annihilation of a proton and antiproton can produce many more pions than the number of particles at the start. Though the production of 13 neutral pions is a possible outcome of a proton-antiproton interaction, it is not a likely one. In addition to the conservation of energy, there are many other physical laws that determine what types of particles, and in what quantities, are likely to be produced.

A big detector for a small particle The rubber raft in the photo is floating inside a particle detector designed to measure neutrinos. Neutrinos are so weakly interacting that a neutrino produced in a nuclear reaction in the center of the sun will likely pass through the entire mass of the sun and escape. Of course, the neutrino's weakly interacting nature also means that it is likely to pass right through a detector. The Super Kamiokande experiment in Japan monitors interactions in an enormous volume of water in order to spot a very small number of neutrino interactions.

Neutrinos

The most abundant particle in the universe is not the electron, proton, or neutron; it is a particle you may have never heard of. Vast numbers of these particles pass through your body every day, only in extremely rare cases leaving any trace of their passage. This elusive particle is the **neutrino,** a neutral, nearly massless particle that interacts only weakly with matter.

The neutrino is represented by the symbol ν, a lowercase Greek nu. There are three types of neutrinos. The neutrino involved in beta decay is the *electron neutrino* ν_e; it shows up in processes involving electrons and positrons. The electron neutrino of course has an antiparticle, the antineutrino $\bar{\nu}_e$. The full descriptions of beta-minus and beta-plus decays, including the neutrinos, are

$$n \rightarrow p^+ + e^- + \bar{\nu}_e$$
$$p^+ \rightarrow n + e^+ + \nu_e$$

NOTE ► If we are concerned with the accurate balance of energy and momentum in a beta decay, we must include the neutrino or the antineutrino, but we can generally ignore the presence of this weakly interacting particle for the problems we solve in this chapter. ◄

It was initially thought that the neutrino, like the photon, was a *massless* particle. However, experiments in the last few years have shown that the neutrino mass, while tiny, is not zero. The best current evidence suggests a mass about one-millionth the mass of an electron. Experiments now under way will determine a more precise value. Because the neutrino is so abundant in the universe, this small mass may be of great cosmological significance.

Quarks

The process of beta decay, in which a neutron can change into a proton, and vice versa, gives a hint that the neutron and the proton are *not* fundamental units but are made of smaller subunits.

There is another reason to imagine such subunits: the existence of dozens of subatomic particles—muons, pions, kaons, omega particles, and so on. Just as the periodic table explains the many different atomic elements in terms of three basic particles, perhaps it is possible to do something similar for this "subatomic zoo."

We now understand protons and neutrons to be composed of smaller charged particles whimsically named **quarks.** The quarks that form protons and neutrons are called **up quarks** and **down quarks,** symbolized as u and d, respectively. The nature of these quarks and the composition of the neutron and the proton are shown in **FIGURE 30.22**.

FIGURE 30.22 The quark content of the proton and neutron.

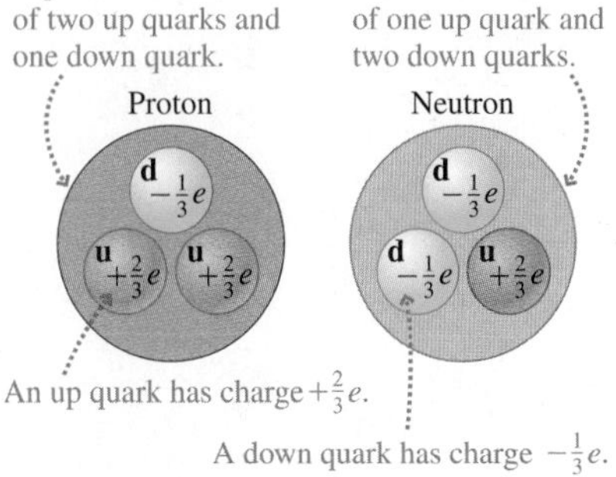

A proton is made of two up quarks and one down quark.
A neutron is made of one up quark and two down quarks.
Proton
Neutron
d $-\frac{1}{3}e$
u $+\frac{2}{3}e$
An up quark has charge $+\frac{2}{3}e$.
A down quark has charge $-\frac{1}{3}e$.

NOTE ▶ It seems surprising that the charges of quarks are *fractions* of e. Don't charges have to be integer multiples of e? It's true that atoms, molecules, and all macroscopic matter must have $q = Ne$ because these entities are constructed from electrons and protons. But no law of nature prevents other types of matter from having other amounts of charge. ◀

A neutron and a proton differ by one quark. Beta decay can now be understood as a process in which a down quark changes to an up quark, or vice versa. Beta-minus decay of a neutron can be written as

$$\mathrm{d} \rightarrow \mathrm{u} + \mathrm{e}^- + \bar{\nu}_\mathrm{e}$$

The existence of quarks thus provides an explanation of how a neutron can turn into a proton.

CONCEPTUAL EXAMPLE 30.15 Quarks and beta-plus decay

What is the quark description of beta-plus decay?

REASON In beta-plus decay, a proton turns into a neutron, with the emission of a positron and an electron neutrino. To turn a proton into a neutron requires the conversion of an up quark into a down quark; the total reaction is thus

$$\mathrm{u} \rightarrow \mathrm{d} + \mathrm{e}^+ + \nu_\mathrm{e}$$

Fundamental Particles

Our current understanding of the truly *fundamental* particles—the ones that cannot be broken down into smaller subunits—is that they come in two basic types: **leptons** (particles like the electron and the neutrino) and quarks (which combine to form particles like the proton and the neutron). The leptons and quarks are listed in Table 30.5. A few points are worthy of note:

- Each particle has an associated antiparticle.
- There are three *families* of leptons. The first is the electron and its associated neutrino, and their antiparticles. The other families are based on the muon and the tau, heavier siblings to the electron. Only the electron and positron are stable.
- There are also three families of quarks. The first is the up-down family that makes all "normal" matter. The other families are pairs of heavier quarks that form more exotic particles.

TABLE 30.5 Leptons and quarks

Leptons		Antileptons	
Electron	e^-	Positron	e^+
Electron neutrino	ν_e	Electron antineutrino	$\bar{\nu}_\mathrm{e}$
Muon	μ^-	Antimuon	μ^+
Muon neutrino	ν_μ	Muon antineutrino	$\bar{\nu}_\mu$
Tau	τ^-	Antitau	τ^+
Tau neutrino	ν_τ	Tau antineutrino	$\bar{\nu}_\tau$
Quarks		**Antiquarks**	
Up	u	Antiup	$\bar{\mathrm{u}}$
Down	d	Antidown	$\bar{\mathrm{d}}$
Strange	s	Antistrange	$\bar{\mathrm{s}}$
Charm	c	Anticharm	$\bar{\mathrm{c}}$
Bottom	b	Antibottom	$\bar{\mathrm{b}}$
Top	t	Antitop	$\bar{\mathrm{t}}$

As far as we know, this is where the trail ends. Matter is made of molecules; molecules of atoms; atoms of protons, neutrons, and electrons; protons and neutrons of quarks. Quarks and electrons seem to be truly fundamental. But scientists of the early 20th century thought they were at a stopping point as well—they thought that they knew all of the physics that there was to know. As we've seen over the past few chapters, this was far from true. New tools such as the next generation of particle colliders will certainly provide new discoveries and new surprises.

The early chapters of this textbook, in which you learned about forces and motion, had very obvious applications to things in your daily life. But in these past few chapters we see that even modern discoveries—discoveries such as antimatter that may seem like science fiction—can be put to very practical use. As we come to the close of this book, we hope that you have gained an appreciation not only for what physics tells us about the world, but also for the wide range of problems it can be used to solve.

INTEGRATED EXAMPLE 30.16 Čerenkov radiation

We know that nothing can travel faster than c, the speed of light in a vacuum. But we also know that light itself goes slower as it travels through a medium. Consequently particles moving at high speeds through a medium can be traveling faster than a light wave in the medium.

FIGURE 30.23 is a photo of the core of a nuclear reactor. The core is immersed in water, which carries away the thermal energy produced in the reactor. As high-energy electrons emerge with a speed very close to c from nuclear reactions in the core, they move through the water faster than the speed of light in water.

FIGURE 30.23 Čerenkov light illuminates the water surrounding a nuclear reactor core.

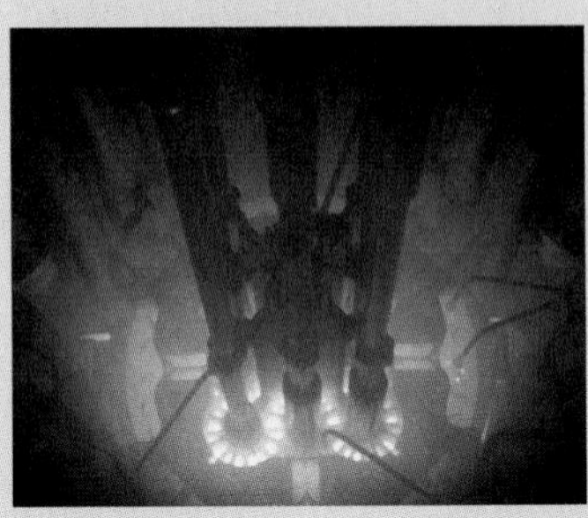

Recall that a shock wave—a sonic boom—is produced when an airplane moves faster than the speed of sound. An electron moving faster than the speed of light in water makes an electromagnetic shock wave—a light pulse analogous to a sonic boom. This particular type of electromagnetic radiation is known as **Čerenkov radiation,** or Čerenkov light, and it is responsible for the blue glow around the reactor core in Figure 30.23.

One source of high-speed electrons is the beta-minus decay of ^{133}Xe, a radioactive isotope of xenon that is produced in the fission of uranium and accumulates in the reactor core.

a. What is the daughter nucleus of this decay?
b. Assume that all of the energy released in the decay goes to kinetic energy of the emitted electron. What is the electron's kinetic energy?
c. Use the equations of special relativity to determine the electron's speed.
d. Would the emitted electron be moving at a speed high enough to cause Čerenkov light in water?
e. Based on the color of the Čerenkov light you can see in Figure 30.23, which of the following describes the spectrum of Čerenkov light?
   - The intensity of Čerenkov light is uniform at all frequencies.
   - The intensity of Čerenkov light is proportional to frequency.
   - The intensity of Čerenkov light is proportional to wavelength.

   Explain.

PREPARE ^{133}Xe undergoes beta decay, so the mass of this nucleus must be greater than that of the daughter nucleus. The "lost" mass is converted to energy, which we assume goes to the kinetic energy of the electron, so the kinetic energy of the electron will be $K = \Delta m \cdot c^2$. Once we know the electron's kinetic energy, we can determine the speed to see if it exceeds the speed of light in water. The speed of light in a medium is given by $v = c/n$, where n is the index of refraction. Water has $n = 1.33$.

SOLVE a. Equation 30.9 tells us that beta-minus decay increases Z by 1 while leaving A unchanged. Xe has $Z = 54$; $Z = 55$ is Cs (cesium), so the daughter nucleus, still with $A = 133$, is ^{133}Cs.

b. We use the data in Appendix D to find the mass difference between the parent and daughter nuclei:

$$\Delta m = m(^{133}\text{Xe}) - m(^{133}\text{Cs})$$
$$= 132.905906 \text{ u} - 132.905436 \text{ u} = 0.00047 \text{ u}$$

We assume that the energy corresponding to this mass difference is the kinetic energy of the emitted electron, so the electron's kinetic energy is

$$K = \Delta m \cdot c^2 = (0.00047 \text{ u})(931.49 \text{ MeV}/c^2)c^2 = 0.44 \text{ MeV}$$

c. The kinetic energy of the electron is large enough that we'll need to consider relativity—a classical treatment won't be sufficient. Equation 27.22 gives the relationship between an object's kinetic energy and its rest energy E_0 as $K = (\gamma - 1)E_0$. We can rearrange this to give γ in terms of a ratio of two energies that we know: the electron's rest energy, 0.51 MeV, and the electron's kinetic energy, 0.44 MeV. Because it's a ratio, we need not convert units:

$$\gamma = 1 + \frac{K}{E_0} = 1 + \frac{0.44 \text{ MeV}}{0.51 \text{ MeV}} = 1.9$$

We can now use the definition of γ to solve for the electron's speed:

$$\gamma = \frac{1}{\sqrt{1 - (v/c)^2}}$$
$$v = c\sqrt{1 - 1/\gamma^2} = 2.5 \times 10^8 \text{ m/s}$$

d. Čerenkov light will be emitted if the speed of the emitted electron is greater than the speed of a light wave in the water. The speed of light in water is

$$v = \frac{3.00 \times 10^8 \text{ m/s}}{1.33} = 2.3 \times 10^8 \text{ m/s}$$

The electron is moving faster than this, and so it will emit Čerenkov light.

e. The photo reveals that Čerenkov light appears blue, so more high-frequency blue light is emitted than low-frequency red light. The intensity is greater at higher frequencies, so the intensity is proportional to frequency—at least for visible light. The index of refraction decreases for very high frequencies, returning to $n = 1$ for x rays. Light speed at these very high frequencies is no longer slower than the particle speed, so Čerenkov light "cuts off" at very high frequencies.

ASSESS The energy of the beta particle is reasonably typical for particles emitted by nuclear decays, and we know that Čerenkov light is observed around reactor cores, so it's reasonable to expect the beta particle to be moving fast enough to emit Čerenkov light.

SUMMARY

Goal: To understand the physics of the nucleus and some of the applications of nuclear physics.

GENERAL PRINCIPLES

The Nucleus

The nucleus is a small, dense, positive core at the center of an atom.

Z protons, charge $+e$, spin $\frac{1}{2}$
N neutrons, charge 0, spin $\frac{1}{2}$

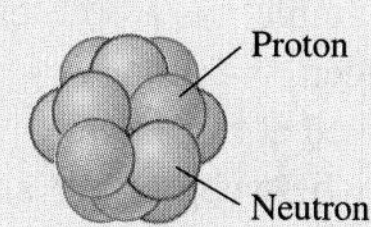

The **mass number** is

$$A = Z + N$$

Isotopes of an element have the same value of Z but different values of N.

The strong force holds nuclei together:

- It acts between any two nucleons.
- It is short range.

Adding neutrons to a nucleus allows the strong force to overcome the repulsive Coulomb force between protons.

The **binding energy** B of a nucleus depends on the mass difference between an atom and its constituents:

$$B = (Zm_H + Nm_n - m_{atom}) \times (931.49 \text{ MeV/u})$$

Nuclear Stability

Most nuclei are not **stable.** Unstable nuclei undergo **radioactive decay.** Stable nuclei cluster along the **line of stability** in a plot of the isotopes.

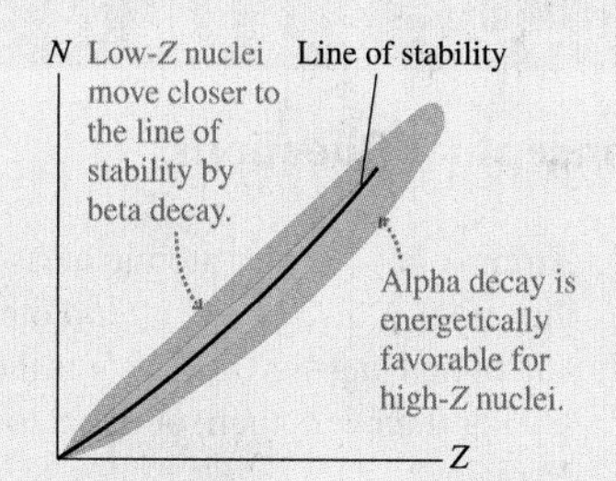

N
Low-Z nuclei move closer to the line of stability by beta decay.
Line of stability
Alpha decay is energetically favorable for high-Z nuclei.
Z

Mechanisms by which unstable nuclei decay:

Decay	Particle
alpha	^{4}He nucleus
beta-minus	e^-
beta-plus	e^+
gamma	photon

Alpha and beta decays change the nucleus; the daughter nucleus is a different element.

Alpha decay:

$${}^A_Z\text{X} \rightarrow {}^{A-4}_{Z-2}\text{Y} + \alpha + \text{energy}$$

Beta-minus decay:

$${}^A_Z\text{X} \rightarrow {}_{Z+1}^{A}\text{Y} + \beta + \text{energy}$$

IMPORTANT CONCEPTS

Energy levels

Nucleons fill nuclear energy levels, similar to filling electron energy levels in atoms. Nucleons can often jump to lower energy levels by emitting beta particles or gamma photons.

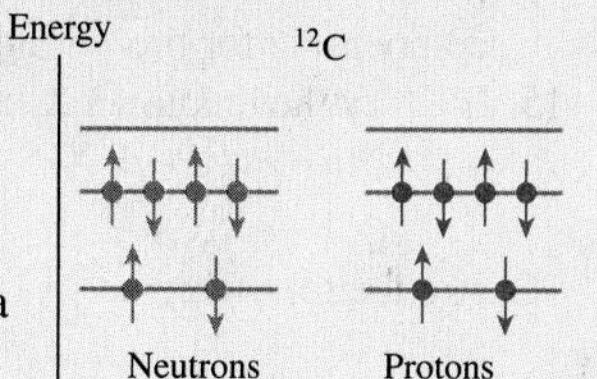

Quarks

Nucleons (and other particles) are made of quarks. Quarks and leptons are fundamental particles.

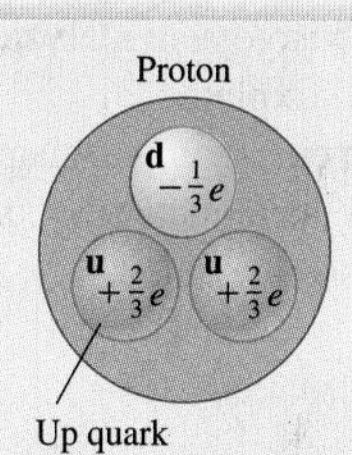

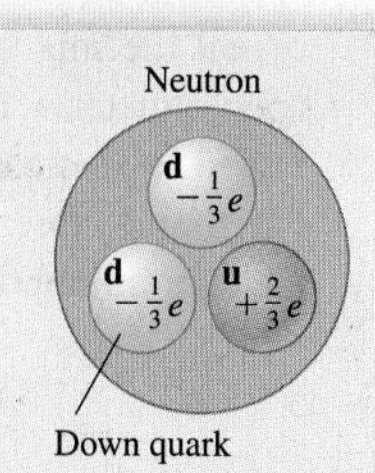

APPLICATIONS

Radioactive decay

The number of undecayed nuclei decreases exponentially with time t:

$$N = N_0 e^{-t/\tau}$$

$$N = N_0\left(\frac{1}{2}\right)^{t/t_{1/2}}$$

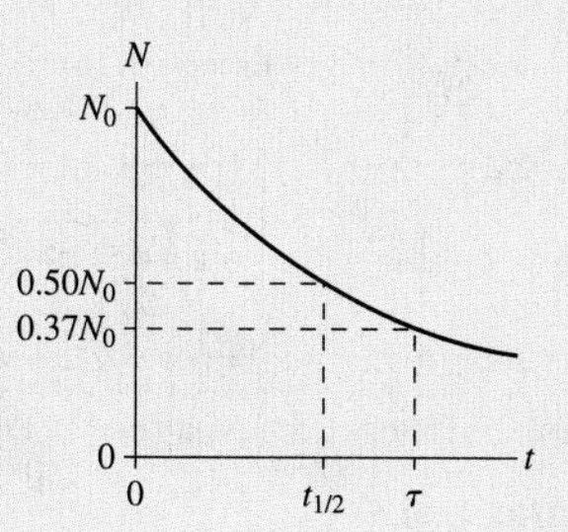

The **half-life**

$$t_{1/2} = \tau \ln 2 = 0.693\tau$$

is the time in which half of any sample decays.

Measuring radiation

The **activity** of a radioactive sample is the number of decays per second. Activity is related to the half-life as

$$R = \frac{0.693N}{t_{1/2}} = \frac{N}{\tau}$$

The radiation **dose** is measured in grays, where

$$1 \text{ Gy} = 1.00 \text{ J/kg of absorbed energy}$$

The **relative biological effectiveness** (RBE) is the biological effect of a dose relative to the biological effects of x rays. The **dose equivalent** is measured in sieverts, where

$$\text{dose equivalent in Sv} = \text{dose in Gy} \times \text{RBE}$$

Problem difficulty is labeled as | (straightforward) to ||||| (challenging). Problems labeled INT integrate significant material from earlier chapters; BIO are of biological or medical interest.

For assigned homework and other learning materials, go to MasteringPhysics®

Scan this QR code to launch a **Video Tutor Solution** that will help you solve problems for this chapter.

QUESTIONS

Conceptual Questions

1. Atom A has a larger atomic mass than atom B. Does this mean that atom A also has a larger atomic number? Explain.
2. Given that $m_H = 1.007825$ u, is the mass of a hydrogen atom ^{1}H greater than, less than, or equal to 1/12 the mass of a ^{12}C atom? Explain.
3. a. Is there a stable $^{30}_{3}$Li nucleus? Explain how you made your determination.
   b. Is there a stable $^{184}_{92}$U nucleus? Explain how you made your determination.
4. Rounding slightly, the nucleus ^{3}He has a binding energy of 2.5 MeV/nucleon and the nucleus ^{6}Li has a binding energy of 5 MeV/nucleon.
   a. What is the binding energy of ^{3}He?
   b. What is the binding energy of ^{6}Li?
   c. Is it energetically possible for two ^{3}He nuclei to join or fuse together into a ^{6}Li nucleus? Explain.
   d. Is it energetically possible for a ^{6}Li nucleus to split or fission into two ^{3}He nuclei? Explain.
5. A sample contains a mix of isotopes of an element. Using a spectrometer to measure the spectrum of emitted light will not reveal the mix of isotopes; analyzing the sample with a mass spectrometer will. Explain.
6. For each nuclear energy-level diagram in Figure Q30.6, state whether it represents a nuclear ground state, an excited nuclear state, or an impossible nucleus.

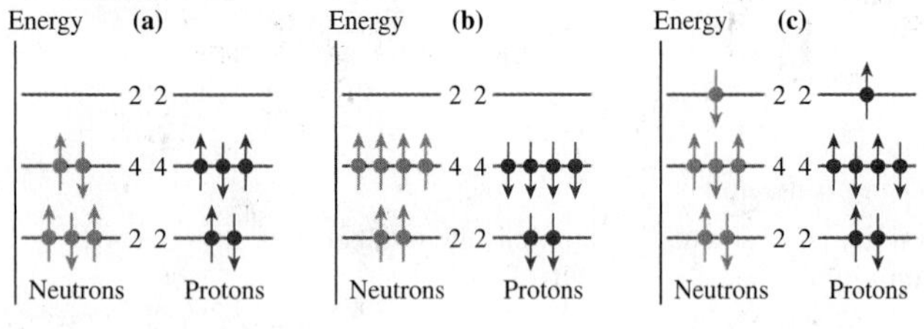

FIGURE Q30.6

7. Figure Q30.7 shows how the number of nuclei of one particular isotope varies with time. What is the half-life of the nucleus?

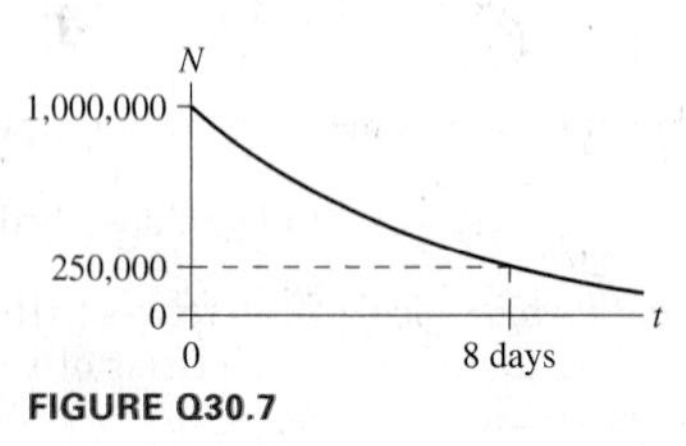

FIGURE Q30.7

8. A radioactive sample has a half-life of 10 s. 10,000 nuclei are present at $t = 20$ s.
   a. How many nuclei were there at $t = 0$ s?
   b. How many nuclei will there be at $t = 40$ s?
9. Nucleus A decays into the stable nucleus B with a half-life of 10 s. At $t = 0$ s there are 1000 A nuclei and no B nuclei. At what time will there be 750 B nuclei?
10. A radioactive sample's half-life is 1.0 min, so each nucleus in the sample has a 50% chance of undergoing a decay sometime between $t = 0$ and $t = 1$ min. One particular nucleus has not decayed at $t = 15$ min. What is the probability this nucleus will decay between $t = 15$ and $t = 16$ min?
11. The material that formed the earth was created in a supernova explosion approximately 6 billion years ago. The two most common isotopes of uranium, ^{238}U and ^{235}U, were created in roughly equal amounts. Today, the earth contains more than 100 times as much ^{238}U as ^{235}U. Which isotope has a longer half-life?
12. An investigator receives a sample of a short-lived isotope that was created using a cyclotron. Explain why it is not possible to tell the age of an individual nucleus, but it is possible to determine the age of the sample.
13. Oil and coal generally contain no measurable ^{14}C. What does this tell us about how long they have been buried?
14. Radiocarbon dating assumes that the abundance of ^{14}C in the environment has been constant. Suppose ^{14}C was less abundant 10,000 years ago than it is today. Would this cause a lab using radiocarbon dating to overestimate or underestimate the age of a 10,000-year-old artifact? (In fact, the abundance of ^{14}C in the environment does vary slightly with time. But the issue has been well studied, and the ages of artifacts are adjusted to compensate for this variation.)
15. Identify the unknown X in the following decays:
    a. $^{222}_{86}\text{Rn} \rightarrow {}^{218}_{84}\text{Po} + \text{X}$ b. $^{228}_{88}\text{Ra} \rightarrow {}^{228}_{89}\text{Ac} + \text{X}$
    c. $^{140}_{54}\text{Xe} \rightarrow {}^{140}_{55}\text{Cs} + \text{X}$ d. $^{64}_{29}\text{Cu} \rightarrow {}^{64}_{28}\text{Ni} + \text{X}$
16. Are the following decays possible? If not, why not?
    a. $^{232}_{90}\text{Th} \rightarrow {}^{236}_{92}\text{U} + \alpha$ b. $^{238}_{94}\text{Pu} \rightarrow {}^{236}_{92}\text{U} + \alpha$
    c. $^{33}_{15}\text{P} \rightarrow {}^{32}_{16}\text{S} + e^-$
17. What kind of decay, if any, would you expect for the nuclei with the energy-level diagrams shown in Figure Q30.17?

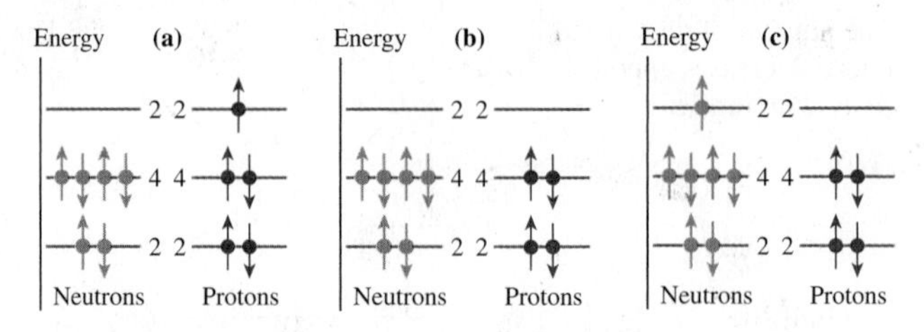

FIGURE Q30.17

18. The nuclei of ^{4}He and ^{16}O are very stable and are often referred to as "doubly magic" nuclei. Use what you know about energy levels to explain what is special about these particular nuclei.
19. BIO A and B are fresh apples. Apple A is strongly irradiated by nuclear radiation for 1 hour. Apple B is not irradiated. Afterward, in what ways are apples A and B different?
20. BIO A patient's tumor is irradiated with gamma rays from an external source. Afterward, is his body radioactive? Explain.

21. It's possible that a bone tumor will not show up on an x-ray BIO image but will show up in a gamma scan. Explain why this is so.
22. Four radiation doses are as follows: Dose A is 10 rad with an BIO RBE of 1, dose B is 20 rad with an RBE of 1, dose C is 10 rad with an RBE of 2, and dose D is 20 rad with an RBE of 2.
    a. Rank in order, from largest to smallest, the amount of energy delivered by these four doses.
    b. Rank in order, from largest to smallest, the biological damage caused by these four doses.
23. Two different sources of radiation give the same dose equiva- BIO lent in Sv. Does this mean that the radiation from each source has the same RBE? Explain.
24. Some types of MRI can produce images of resolution and detail BIO similar to PET. Though the images are similar, MRI is generally preferred over PET for studies of brain function involving healthy subjects. Why?
25. Sulfur colloid particles tagged with ^{99}Tc are taken up and BIO retained by cells in the liver and spleen. A patient is suspected of having a liver tumor that would destroy these cells. Explain how a gamma camera scan could be used to confirm or rule out the existence of a tumor.
26. The first two letters in the acronym SPECT, which describes BIO a nuclear imaging technique, stand for "single photon." Is a SPECT done with a gamma emitter or a positron emitter?

The following two questions concern an uncommon nuclear decay mode known as *electron capture*. Certain nuclei that are proton-rich but energetically prohibited from undergoing beta-plus decay can capture an electron from the 1*s* shell, which then combines with a proton to make a neutron. The basic reaction is

$$\mathrm{p} + \mathrm{e}^- \rightarrow \mathrm{n} + \nu_\mathrm{e}$$

27. Give a description of the electron capture process in terms of quarks.
28. Electron capture is usually followed by the emission of an x ray. Why?

Multiple-Choice Questions

29. | ^{74}As is a beta-plus emitter used for locating tumors with PET. BIO What is the daughter nucleus?
    A. $^{73}_{33}$As B. $^{74}_{32}$Ge C. $^{74}_{34}$Se D. $^{75}_{33}$As
30. || When uranium fissions, the fission products are radioactive because the nuclei are neutron-rich. What is the most likely decay mode for these nuclei?
    A. Alpha decay B. Beta-minus decay
    C. Beta-plus decay D. Gamma decay
31. | A certain watch's luminous glow is due to zinc sulfide paint that is energized by beta particles given off by *tritium*, the radioactive hydrogen isotope ^{3}H, which has a half-life of 12.3 years. This glow has about 1/10 of its initial brightness. How many years old is the watch?

    A. 20 yr B. 30 yr C. 40 yr D. 50 yr
32. | What is the unknown isotope in the following fission reaction: n + ^{235}U → ^{131}I + ? + 3n
    A. ^{86}Rb B. ^{102}Rb C. ^{89}Y D. ^{102}Y
33. | An investigator has 0.010 μg samples of two isotopes of strontium, ^{89}Sr ($t_{1/2}$ = 51 days) and ^{90}Sr ($t_{1/2}$ = 28 years). The samples contain approximately the same number of atoms. What can you say about the activity of the two samples?
    A. The ^{89}Sr sample has a higher activity.
    B. The ^{90}Sr sample has a higher activity.
    C. The two samples have about the same activity.
34. || The uranium in the earth's crust is 0.7% ^{235}U and 99.3% ^{238}U. Two billion years ago, ^{235}U comprised approximately 3% of the uranium in the earth's crust. This tells you something about the relative half-lives of the two isotopes. Suppose you have a sample of ^{235}U and a sample of ^{238}U, each with exactly the same number of atoms.
    A. The sample of ^{235}U has a higher activity.
    B. The sample of ^{238}U has a higher activity.
    C. The two samples have the same activity.
35. | Suppose you have a 1 g sample of ^{226}Ra, half-life 1600 years. How long will it be until only 0.1 g of radium is left?
    A. 1600 yr B. 3200 yr C. 5300 yr D. 16,000 yr
36. | A sample of ^{131}I, half-life 8.0 days, is registering 100 decays per second. How long will be it before the sample registers only 1 decay per second?
    A. 8 days B. 53 days C. 80 days D. 800 days
37. | The complete expression for the decay of the radioactive hydrogen isotope *tritium* may be written as ^{3}H → ^{3}He + X + Y. The symbols X and Y represent
    A. X = e^+, Y = $\bar{\nu}_e$ B. X = e^-, Y = ν_e
    C. X = e^+, Y = ν_e D. X = e^-, Y = $\bar{\nu}_e$
38. | The quark compositions of the proton and neutron are, respectively, uud and udd, where u is an up quark (charge $+\frac{2}{3}e$) and d is a down quark (charge $-\frac{1}{3}e$). There are also antiup $\bar{u}$ (charge $-\frac{2}{3}e$) and antidown $\bar{d}$ (charge $+\frac{1}{3}e$) quarks. The combination of a quark and an antiquark is called a *meson*. The mesons known as *pions* have the composition $\pi^+ = u\bar{d}$ and $\pi^- = \bar{u}d$. Suppose a proton collides with an antineutron. During such collisions, the various quarks and antiquarks annihilate whenever possible. When the remaining quarks combine to form a single particle, it is a
    A. Proton B. Neutron C. π^+ D. π^-

PROBLEMS

Section 30.1 Nuclear Structure

1. | How many protons and how many neutrons are in (a) ^{3}H, (b) ^{40}Ar, (c) ^{40}Ca, and (d) ^{239}Pu?
2. | How many protons and how many neutrons are in (a) ^{3}He, (b) ^{20}Ne, (c) ^{60}Co, and (d) ^{226}Ra?
3. || Use the data in Appendix D to calculate the chemical atomic mass of lithium, to two decimal places.
4. | Use the data in Appendix D to calculate the chemical atomic mass of neon, to two decimal places.
5. | What are the energy equivalents (in MeV) of the mass of (a) an electron, (b) a proton, (c) a hydrogen atom, and (d) a helium atom?
6. | The chemical atomic mass of hydrogen, with the two stable isotopes ^{1}H and ^{2}H (deuterium), is 1.00798 u. Use this value to determine the natural abundance of these two isotopes.

Watch video solutions for problems: 30.13  30.23  30.39 

Section 30.2 Nuclear Stability

7. | Calculate (in MeV) the total binding energy and the binding energy per nucleon (a) for ^{3}H and (b) for ^{3}He.
8. | Calculate (in MeV) the total binding energy and the binding energy per nucleon (a) for ^{40}Ar and (b) for ^{40}K.
9. | Calculate (in MeV) the binding energy per nucleon for ^{3}He and ^{4}He. Which is more tightly bound?
10. || Calculate (in MeV) the binding energy per nucleon for ^{12}C and ^{13}C. Which is more tightly bound?
11. | Calculate (in MeV) the binding energy per nucleon for (a) ^{14}N, (b) ^{56}Fe, and (c) ^{207}Pb.
12. || When a nucleus of ^{235}U undergoes fission, it breaks into two smaller, more tightly bound fragments. Calculate the binding energy per nucleon for ^{235}U and for the fission product ^{137}Cs.
13. || When a nucleus of ^{240}Pu undergoes fission, it breaks into two smaller, more tightly bound fragments. Calculate the binding energy per nucleon for ^{240}Pu and for the fission product ^{133}Xe.
14. || a. Compute the binding energy of the reactants and of the products in the nuclear fusion reaction

$$^2\text{H} + {}^6\text{Li} \rightarrow {}^4\text{He} + {}^4\text{He}$$

b. What is the change in binding energy per nucleon in this reaction?

15. || a. Compute the binding energy of the reactants and of the products in the nuclear fusion reaction

$$^3\text{H} + {}^3\text{He} \rightarrow {}^2\text{H} + {}^4\text{He}$$

b. What is the change in binding energy per nucleon in this reaction?

Section 30.3 Forces and Energy in the Nucleus

16. || Draw an energy-level diagram, similar to Figure 30.9, for the protons and neutrons in ^{11}Be. Do you expect this nucleus to be stable?
17. || Draw energy-level diagrams, similar to Figure 30.9, for all $A = 10$ nuclei listed in Appendix D. Show all the occupied neutron and proton levels. Given the diagrams, are any of the nuclei obviously unstable?
18. || Draw energy-level diagrams, similar to Figure 30.9, for all $A = 14$ nuclei listed in Appendix D. Show all the occupied neutron and proton levels. Given the diagrams, are any of the nuclei obviously unstable?
19. ||| You have seen that filled electron energy levels correspond to chemically stable atoms. A similar principle holds for nuclear energy levels; nuclei with equally filled proton and neutron energy levels are especially stable. What are the three lightest isotopes whose proton and neutron energy levels are both filled, and filled equally?

Section 30.4 Radiation and Radioactivity

20. | ^{15}O and ^{131}I are isotopes used in medical imaging. ^{15}O is a beta-plus emitter, ^{131}I a beta-minus emitter. What are the daughter nuclei of the two decays? BIO
21. | Spacecraft have been powered with energy from the alpha decay of ^{238}Pu. What is the daughter nucleus?
22. | Identify the unknown isotope X in the following decays.
    a. $^{234}\text{U} \rightarrow \text{X} + \alpha$
    b. $^{32}\text{P} \rightarrow \text{X} + e^-$
    c. $\text{X} \rightarrow {}^{30}\text{Si} + e^+$
    d. $^{24}\text{Mg} \rightarrow \text{X} + \gamma$
23. | Identify the unknown isotope X in the following decays.
    a. $\text{X} \rightarrow {}^{224}\text{Ra} + \alpha$
    b. $\text{X} \rightarrow {}^{207}\text{Pb} + e^-$
    c. $^7\text{Be} + e^- \rightarrow \text{X}$
    d. $\text{X} \rightarrow {}^{60}\text{Ni} + \gamma$
24. | What is the energy (in MeV) released in the alpha decay of ^{239}Pu?
25. | What is the energy (in MeV) released in the alpha decay of ^{228}Th?
26. | Medical gamma imaging is generally done with the technetium isotope ^{99}Tc*, which decays by emitting a gamma-ray photon with energy 140 keV. What is the mass loss of the nucleus, in u, upon emission of this gamma ray? BIO
27. || Cobalt has one stable isotope, ^{59}Co. What are the likely decay modes and daughter nuclei for (a) ^{56}Co and (b) ^{62}Co?
28. || Manganese has one stable isotope, ^{55}Mn. What are the likely decay modes and daughter nuclei for (a) ^{51}Mn and (b) ^{59}Mn?

Section 30.5 Nuclear Decay and Half-Lives

29. | The radioactive hydrogen isotope ^{3}H is called tritium. It decays by beta-minus decay with a half-life of 12.3 years.
    a. What is the daughter nucleus of tritium?
    b. A watch uses the decay of tritium to energize its glowing dial. What fraction of the tritium remains 20 years after the watch was created?
30. | The barium isotope ^{133}Ba has a half-life of 10.5 years. A sample begins with 1.0×10^{10} ^{133}Ba atoms. How many are left after (a) 2 years, (b) 20 years, and (c) 200 years?
31. | The cadmium isotope ^{109}Cd has a half-life of 462 days. A sample begins with 1.0×10^{12} ^{109}Cd atoms. How many are left after (a) 50 days, (b) 500 days, and (c) 5000 days?
32. || How many half-lives must elapse until (a) 90% and (b) 99% of a radioactive sample of atoms has decayed?
33. || The Chernobyl reactor accident in what is now Ukraine was the worst nuclear disaster of all time. Fission products from the reactor core spread over a wide area. The primary radiation exposure to people in western Europe was due to the short-lived (half-life 8.0 days) isotope ^{131}I, which fell across the landscape and was ingested by grazing cows that concentrated the isotope in their milk. Farmers couldn't sell the contaminated milk, so many opted to use the milk to make cheese, aging it until the radioactivity decayed to acceptable levels. How much time must elapse for the activity of a block of cheese containing ^{131}I to drop to 1.0% of its initial value? BIO
34. |||| What is the age in years of a bone in which the ^{14}C/^{12}C ratio is measured to be 1.65×10^{-13}?
35. || ^{85}Sr is a short-lived (half-life 65 days) isotope used in bone scans. A typical patient receives a dose of ^{85}Sr with an activity of 0.10 mCi. If all of the ^{85}Sr is retained by the body, what will be its activity in the patient's body after one year has passed? BIO
36. || What is the half-life in days of a radioactive sample with 5.0×10^{15} atoms and an activity of 5.0×10^8 Bq?
37. ||| What is the activity, in Bq and Ci, of 1.0 g of ^{226}Ra? Marie Curie was the discoverer of radium; can you see where the unit of activity named after her came from? INT
38. || Many medical PET scans use the isotope ^{18}F, which has a half-life of 1.8 h. A sample prepared at 10:00 A.M. has an activity of 20 mCi. What is the activity at 1:00 P.M., when the patient is injected? BIO
39. ||| An investigator collects a sample of a radioactive isotope with an activity of 370,000 Bq. 48 hours later, the activity is 120,000 Bq. What is the half-life of the sample?

40. | ^{235}U decays to ^{207}Pb via the decay series shown in Figure 30.15. The first decay in the chain, that of ^{235}U, has a half-life of 7.0×10^8 years. The subsequent decays are much more rapid, so we can take this as the half-life for the complete decay of ^{235}U to ^{207}Pb. Certain minerals exclude lead but not uranium from their crystal structure, so when the minerals form they have no lead, only uranium. As time goes on, the uranium decays to lead, so measuring the ratio of lead atoms to uranium atoms allows investigators to determine the ages of the minerals. If a sample of a mineral contains 3 atoms of ^{207}Pb for every 1 atom of ^{235}U, how many years ago was it formed?
41. || A sample of 1.0×10^{10} atoms that decay by alpha emission has a half-life of 100 min. How many alpha particles are emitted between $t = 50$ min and $t = 200$ min?
42. || BIO The technetium isotope ^{99}Tc is useful in medical imaging, but its short 6.0 h half-life means that shipping it from a source won't work; it must be created where it will be used. Hospitals extract the ^{99}Tc daughter product from the decay of the molybdenum isotope ^{99}Mo. The ^{99}Mo source has a half-life of 2.7 days, so it must be replaced weekly. If a hospital receives a 20 GBq ^{99}Mo source, what is its activity one week later?

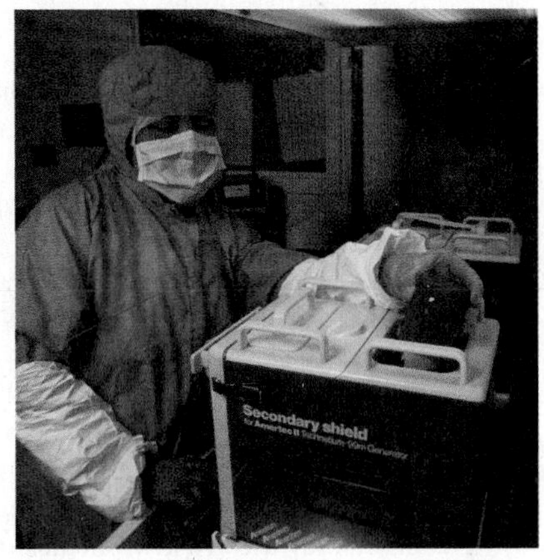

Section 30.6 Medical Applications of Nuclear Physics

43. | BIO A passenger on an airplane flying across the Atlantic receives an extra radiation dose of about 5μSv per hour from cosmic rays. How many hours of flying would it take in one year for a person to double his or her yearly radiation dose? Assume there are no other significant radiation sources besides natural background.
44. || BIO A 50 kg nuclear plant worker is exposed to 20 mJ of neutron radiation with an RBE of 10. What is the dose in mSv?
45. || BIO The decay chain of uranium includes radon, a noble gas. When uranium in the soil decays to radon, it may seep into houses; this can be a significant source of radiation exposure. Most of the exposure comes from the decay products of radon, but some comes from alpha decay of the radon itself. If radon in the air in your home is at the maximum permissible level, the gas in your lungs will have an activity of about 0.22 Bq. Each decay generates an alpha particle with 5.5 MeV of energy, and essentially all that energy is deposited in lung tissue. Over the course of 1 year, what will be the dose equivalent in Sv to the approximately 0.90 kg mass of your lungs?
46. || BIO A gamma scan showing the active volume of a patient's lungs can be created by having a patient breathe the radioactive isotope ^{133}Xe, which undergoes beta-minus decay with a subsequent gamma emission from the daughter nucleus. A typical procedure gives a dose of 0.30 rem to the lungs. How much energy is deposited in the 1.2 kg mass of a patient's lungs?

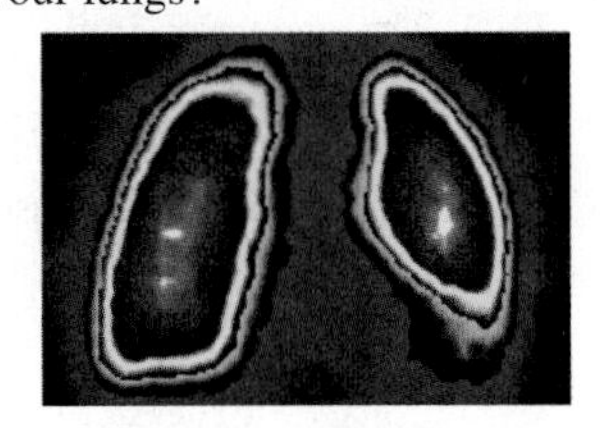

47. || BIO How many rad of gamma-ray photons cause the same biological damage as 30 rad of alpha radiation?
48. | BIO 150 rad of gamma radiation are directed into a 150 g tumor. How much energy does the tumor absorb?
49. || BIO During the 1950s, nuclear bombs were tested on islands in the South Pacific. In one test, personnel on a nearby island received 10 mGy per hour of beta and gamma radiation. At this rate, how long would it take to receive a potentially lethal dose equivalent of 4.5 Sv?
50. ||| BIO ^{131}I undergoes beta-minus decay with a subsequent gamma emission from the daughter nucleus. Iodine in the body is almost entirely taken up by the thyroid gland, so a gamma scan using this isotope will show a bright area corresponding to the thyroid gland with the surrounding tissue appearing dark. Because the isotope is concentrated in the gland, so is the radiation dose, most of which results from the beta emission. In a typical procedure, a patient receives 0.050 mCi of ^{131}I. Assume that all of the iodine is absorbed by the 0.15 kg thyroid gland. Each ^{131}I decay produces a 0.97 MeV beta particle. Assume that half the energy of each beta particle is deposited in the gland. What dose equivalent in Sv will the gland receive in the first hour?

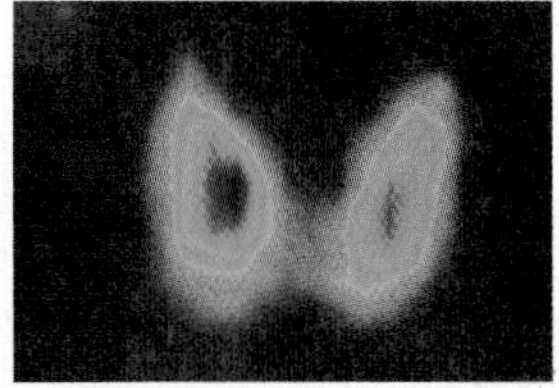

51. ||| BIO ^{90}Sr decays with the emission of a 2.8 MeV beta particle. Strontium is chemically similar to calcium and is taken up by bone. A 75 kg person exposed to waste from a nuclear accident absorbs ^{90}Sr with an activity of 370,000 Bq. Assume that all of this ^{90}Sr ends up in the skeleton. The skeleton forms 17% of the person's body mass. If 50% of the decay energy is absorbed by the skeleton, what dose equivalent in Sv will be received by the person's skeleton in the first month?
52. ||| BIO A patient receives a gamma scan of his liver. He ingests 3.7 MBq of ^{198}Au, which decays with a 2.7 day half-life by emitting a 1.4 MeV beta particle. Medical tests show that 60% of this isotope is absorbed and retained by the liver. If all of the radioactive decay energy is deposited in the liver, what is the total dose equivalent received by the patient's 1.5 kg liver once all of the atoms have decayed?
53. ||| BIO A 75 kg patient swallows a 30 μCi beta emitter with a half-life of 5.0 days, and the radioactive nuclei are quickly distributed throughout his body. The beta particles are emitted with an average energy of 0.35 MeV, 90% of which is absorbed by the body. What dose equivalent does the patient receive in the first week?

Section 30.7 The Ultimate Building Blocks of Matter

54. || BIO What are the minimum energies of the two oppositely directed gamma rays in a PET procedure?
55. |||| Positive and negative pions, denoted π^+ and π^-, are antiparticles of each other. Each has a rest mass of 140 MeV/c^2. Suppose a collision between an electron and positron, each with kinetic energy K, produces a π^+, π^- pair. What is the smallest possible value for K?
56. || In a particular beta-minus decay of a free neutron (that is, one not part of an atomic nucleus), the emitted electron has exactly the same kinetic energy as the emitted electron antineutrino. What is the value, in MeV, of that kinetic energy? Assume that the recoiling proton has negligible kinetic energy.

Watch video solution for problem: 30.65

General Problems

57. ||| Use the graph of binding energy of Figure 30.6 to estimate the total energy released if a nucleus with mass number 240 fissions into two nuclei with mass number 120.
58. ||| What is the activity in Bq and in Ci of a 2.0 mg sample of ^{3}H?
59. || The activity of a sample of the cesium isotope ^{137}Cs is 2.0×10^8 Bq. Many years later, after the sample has fully decayed, how many beta particles will have been emitted?
60. ||| You are assisting in an anthropology lab over the summer BIO by carrying out ^{14}C dating. A graduate student found a bone he believes to be 20,000 years old. You extract the carbon from the bone and prepare an equal-mass sample of carbon from modern organic material. To determine the activity of a sample with the accuracy your supervisor demands, you need to measure the time it takes for 10,000 decays to occur.
    a. The activity of the modern sample is 1.06 Bq. How long does that measurement take?
    b. It turns out that the graduate student's estimate of the bone's age was accurate. How long does it take to measure the activity of the ancient carbon?
61. |||| A sample of wood from an archaeological excavation is dated BIO by using a mass spectrometer to measure the fraction of ^{14}C atoms. Suppose 100 atoms of ^{14}C are found for every 1.0×10^{15} atoms of ^{12}C in the sample. What is the wood's age?
62. || A sample contains radioactive atoms of two types, A and B. Initially there are five times as many A atoms as there are B atoms. Two hours later, the numbers of the two atoms are equal. The half-life of A is 0.50 hours. What is the half-life of B?
63. |||| The technique known as potassium-argon dating is used to date volcanic rock and ash, and thus establish dates for nearby fossils, like this 1.8-million-year-old hominid skull. The potassium isotope ^{40}K decays with a 1.28-billion-year half-life and is naturally present at very low levels. The most common decay mode is beta-minus decay into the stable isotope ^{40}Ca, but 10.9% of decays result in the stable isotope ^{40}Ar. The high temperatures in volcanoes drive argon out of solidifying rock and ash, so there is no argon in newly formed material. After formation, argon produced in the decay of ^{40}K is trapped, so ^{40}Ar builds up steadily over time. Accurate dating is possible by measuring the ratio of the number of atoms of ^{40}Ar and ^{40}K. 1.8 million years after its formation,
    a. What fraction of the ^{40}K initially present in a sample has decayed?
    b. What is the ^{40}Ar/^{40}K ratio of the sample?

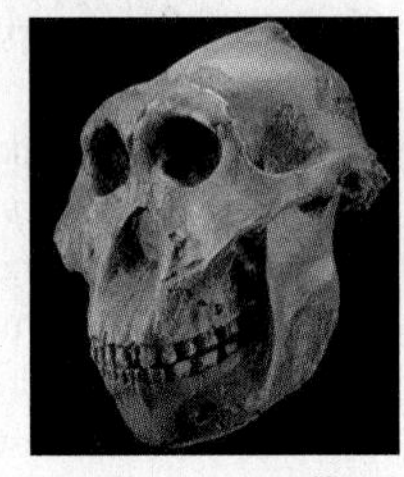

64. |||| Corals take up certain elements from seawater, including uranium but not thorium. BIO After the corals die, the uranium isotopes slowly decay into thorium isotopes. A measurement of the relative fraction of certain isotopes therefore provides a determination of the coral's age. A complicating factor is that the thorium isotopes decay as well. One scheme uses the alpha decay of ^{234}U to ^{230}Th. After a long time, the two species reach an equilibrium in which the number of ^{234}U decays per second (each producing an atom of ^{230}Th) is exactly equal to the number of ^{230}Th decays per second. What is the relative concentration of the two isotopes—the ratio of ^{234}U to ^{230}Th—when this equilibrium is reached?
65. ||| All the very heavy atoms found in the earth were created long ago by nuclear fusion reactions in a supernova, an exploding star. The debris spewed out by the supernova later coalesced to form the sun and the planets of our solar system. Nuclear physics suggests that the uranium isotopes ^{235}U ($t_{1/2} = 7.04 \times 10^8$ yr) and ^{238}U ($t_{1/2} = 4.47 \times 10^9$ yr) should have been created in roughly equal amounts. Today, 99.28% of uranium is ^{238}U and 0.72% is ^{235}U. How long ago did the supernova occur?
66. |||| About 12% of your body mass is carbon; some of this is BIO INT radioactive ^{14}C, a beta-emitter. If you absorb 100% of the 49 keV energy of each ^{14}C decay, what dose equivalent in Sv do you receive each year from the ^{14}C in your body?
67. ||||| Ground beef may be irradiated with high-energy electrons BIO INT from a linear accelerator to kill pathogens. In a standard treatment, 1.0 kg of beef receives 4.5 kGy of radiation in 40 s.
    a. How much energy is deposited in the beef?
    b. What is the average rate (in W) of energy deposition?
    c. Estimate the temperature increase of the beef due to this procedure. The specific heat of beef is approximately 3/4 of that of water.
68. ||| A 70 kg human body typically contains 140 g of potassium. BIO Potassium has a chemical atomic mass of 39.1 u and has three naturally occurring isotopes. One of those isotopes, ^{40}K, is radioactive with a half-life of 1.3 billion years and a natural abundance of 0.012%. Each ^{40}K decay deposits, on average, 1.0 MeV of energy into the body. What yearly dose in Gy does the typical person receive from the decay of ^{40}K in the body?
69. || A chest x ray uses 10 keV photons. A 60 kg person receives BIO a 30 mrem dose from one x ray that exposes 25% of the patient's body. How many x-ray photons are absorbed in the patient's body?

MCAT-Style Passage Problems

Plutonium-Powered Exploration

The Curiosity rover sent to explore the surface of Mars has an electric generator powered by heat from the radioactive decay of ^{238}Pu, a plutonium isotope that decays by alpha emission with a half-life of 88 years. At the start of the mission, the generator contained 9.6×10^{24} nuclei of ^{238}Pu.

70. | What is the daughter nucleus of the decay?
    A. ^{238}Am B. ^{238}Pu C. ^{238}Np D. ^{236}Th E. ^{234}U
71. || What was the approximate activity of the plutonium source at the start of the mission?
    A. 2×10^{21} Bq
    B. 2×10^{19} Bq
    C. 2×10^{17} Bq
    D. 2×10^{15} Bq
    E. 2×10^{13} Bq
72. || The generator initially provided 125 W of power. If you assume that the power of the generator is proportional to the activity of the plutonium, by approximately what percent did the power output decrease over the first two years of the rover's mission?
    A. 1.5% B. 2.0% C. 2.5% D. 3.0% E. 3.5%

Nuclear Fission

The uranium isotope ^{235}U can *fission*—break into two smaller-mass components and free neutrons—if it is struck by a free neutron. A typical reaction is

$$^1_0n + ^{235}_{92}U \rightarrow ^{141}_{56}Ba + ^{92}_{36}Kr + 3^1_0n$$

As you can see, the subscripts (the number of protons) and the superscripts (the number of nucleons) "balance" before and after the fission event; there is no change in the number of protons or neutrons. Significant energy is released in this reaction. If a fission event happens in a large chunk of ^{235}U, the neutrons released may induce the fission of other ^{235}U atoms, resulting in a chain reaction. This is how a nuclear reactor works.

The number of neutrons required to create a stable nucleus increases with atomic number. When the heavy ^{235}U nucleus fissions, the lighter reaction products are thus neutron rich and are likely unstable. Many of the short-lived radioactive nuclei used in medicine are produced in fission reactions in nuclear reactors.

73. | What statement can be made about the masses of atoms in the above reaction?
A. $m(^{235}_{92}U) > m(^{141}_{56}Ba) + m(^{92}_{36}Kr) + 2m(^1_0n)$
B. $m(^{235}_{92}U) < m(^{141}_{56}Ba) + m(^{92}_{36}Kr) + 2m(^1_0n)$
C. $m(^{235}_{92}U) = m(^{141}_{56}Ba) + m(^{92}_{36}Kr) + 2m(^1_0n)$
D. $m(^{235}_{92}U) = m(^{141}_{56}Ba) + m(^{92}_{36}Kr) + 3m(^1_0n)$
74. | Because the decay products in the above fission reaction are neutron rich, they will likely decay by what process?
A. Alpha decay B. Beta decay C. Gamma decay
75. || ^{235}U is radioactive, with a long half-life of 704 million years. The decay products of a ^{235}U fission reaction typically have half-lives of a few minutes. This means that the decay products of a fission reaction have
A. Much higher activity than the original uranium.
B. Much lower activity than the original uranium.
C. The same activity as the original uranium.
76. | If a $^{238}_{92}U$ nucleus is struck by a neutron, it may absorb the neutron. The resulting nucleus then rapidly undergoes beta-minus decay. The daughter nucleus of that decay is
A. $^{239}_{91}Pa$ B. $^{239}_{92}U$ C. $^{239}_{93}Np$ D. $^{239}_{94}Pu$

STOP TO THINK ANSWERS

Chapter Preview Stop to Think: D. There are four electrons, so this is atomic number 4—beryllium. There is one electron in the 2*s* level and one in the 2*p* level. The 2*s* level could hold two electrons, so this is an excited state; if the electron drops from 2*p* to 2*s*, the energy is decreased.

Stop to Think 30.1: Three. Different isotopes of an element have different numbers of neutrons but the same number of protons. The number of electrons in a neutral atom matches the number of protons.

Stop to Think 30.2: A. Energy is released in the decay process, so the mass must decrease.

Stop to Think 30.3: Yes. ^{12}C has filled levels of protons and neutrons; the neutron we add to make ^{13}C will be in a higher energy level, but there is no "hole" in a lower level for it to move to, so we expect this nucleus to be stable.

Stop to Think 30.4: B. An increase of Z with no change in A occurs when a neutron changes to a proton and an electron, ejecting the electron.

Stop to Think 30.5: E. 30 minutes is $3t_{1/2}$ so the number of nuclei A decreases by a factor of $2^3 = 8$ to 3000. This means that 21,000 A nuclei have decayed, and each became nucleus B.

Stop to Think 30.6: C. One-quarter of the atoms are left. This is one-half of one-half, or $(1/2)^2$, so two half-lives, 20 minutes, have elapsed.

Stop to Think 30.7: A. Dose equivalent is the product of dose in Gy (the same for each) and RBE (highest for alpha particles).

Modern Physics

A common theme runs through the final chapters of this textbook: Nature is stranger than we thought it was. From the bizarre paradoxes of special relativity to the dizzying array of subatomic particles, the physical description of the world around us has taken on an almost science-fiction air in Part VII. But this material isn't science fiction; it's real, the product of decades of careful experiments.

Relativity requires us to stretch our notions of space and time. Time really does slow down for particles moving at high speeds, as decades of experiments have shown. Classical Newtonian physics has a comforting predictability, but this is left behind in quantum theory. We simply can't know an electron's position and velocity at the same time. Many decades of clever experiments have shown conclusively that *no* underlying laws can restore the predictability of classical physics at the atomic scale. And nuclear physics has shown that the alchemist's dream is true—you *can* turn one element into another!

This new physics is surprisingly practical. Relativistic corrections allow GPS systems to give extraordinarily accurate measurements of your position anywhere on the earth. Quantum mechanics is the theory underlying the development of computer chips and other modern electronics. And your smoke detector probably contains a small amount of an element not found in nature, an element created in a nuclear reactor.

As we conclude our journey, the knowledge structure for Part VII summarizes the key ideas of relativity, quantum physics, and nuclear physics. These are the theories behind the emerging technologies of the 21st century.

KNOWLEDGE STRUCTURE VII Modern Physics

BASIC GOALS

What are the properties and characteristics of space and time?
What do we know about the nature of light and atoms?
How are atomic and nuclear phenomena explained by energy levels, wave functions, and photons?

GENERAL PRINCIPLES

Principle of relativity	All the laws of physics are the same in all inertial reference frames.
Quantization of energy	Particles of matter and photons of light have only certain allowed energies.
Uncertainty principle	$\Delta x\, \Delta p \geq h/4\pi$
Pauli exclusion principle	No more than one electron can occupy the same quantum state.

Relativity

- The speed of light c is the same in all inertial reference frames.
- No particle or causal influence can travel faster than c.
- Length contraction: The length of an object in a reference frame in which the object moves with speed v is

$$L = \sqrt{1-\beta^2}\,\ell \leq \ell$$

where ℓ is the proper length.
- Time dilation: The proper time interval $\Delta\tau$ between two events is measured in a reference frame in which the two events occur at the same position. The time interval Δt in a frame moving with relative speed v is

$$\Delta t = \Delta\tau/\sqrt{1-\beta^2} \geq \Delta\tau$$

- Particles have energy even when at rest. Mass can be transformed into energy and vice versa: $E_0 = mc^2$.

Quantum physics

- Matter has wave-like properties. A particle has a de Broglie wavelength:

$$\lambda = \frac{h}{mv}$$

- Light has particle-like properties. A photon of light of frequency f has energy:

$$E_{\text{photon}} = hf = \frac{hc}{\lambda}$$

- The wave nature of matter leads to quantized energy levels in atoms and nuclei. A transition between quantized energy levels involves the emission or absorption of a photon.

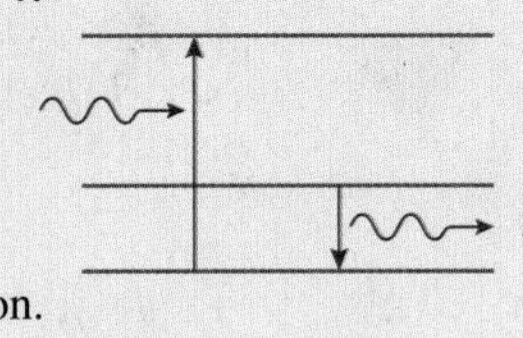

Properties of atoms

- Quantized energy levels depend on quantum numbers n and l.
- An atom can jump from one state to another by emitting or absorbing a photon of energy $E_{\text{photon}} = \Delta E_{\text{atom}}$.
- The ground-state electron configuration is the lowest-energy configuration consistent with the Pauli principle.

Properties of nuclei

- The nucleus is the small, dense, positive core at the center of an atom. The nucleus is held together by the strong force, an attractive short-range force between any two nucleons.

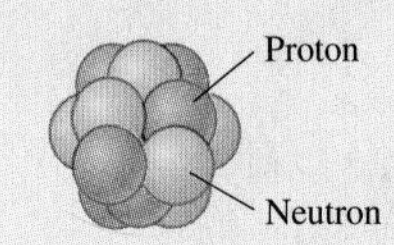

- Unstable nuclei decay by alpha, beta, or gamma decay. The number of undecayed nuclei decreases exponentially with time:

$$N = N_0\left(\frac{1}{2}\right)^{t/t_{1/2}}$$

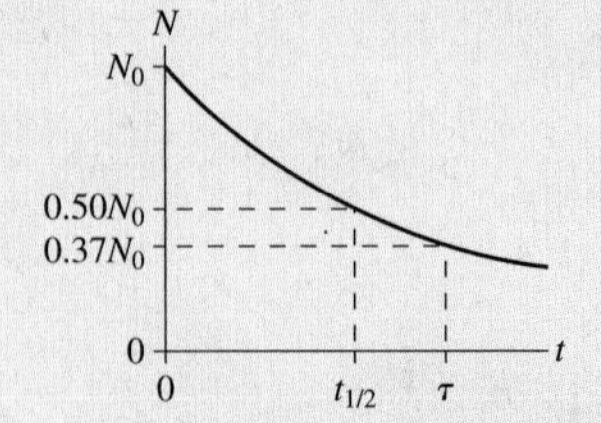

The Physics of Very Cold Atoms

Modern physics is a study of extremes. Relativity deals with the physics of objects traveling at near-light speeds. Quantum mechanics is about the physics of matter and energy at very small scales. Nuclear physics involves energies that dwarf anything dreamed of in previous centuries.

Some of the most remarkable discoveries of recent years are at another extreme—very low temperatures, mere billionths of a degree above absolute zero. Let's look at how such temperatures are achieved and some new physics that emerges.

You learned in Part III that the temperature of a gas depends on the speeds of the atoms in the gas. Suppose we start with atoms at or above room temperature. Cooling the gas means slowing the atoms down. How can we drastically reduce their speeds, bringing them nearly to a halt? The trick is to slow them, thus cooling them, using the interactions between light and atoms that we explored in Chapters 28 and 29.

Photons have momentum, and that momentum is transferred to an atom when a photon is absorbed. Part a of the figure shows an atom moving "upstream" against a laser beam tuned to an atomic transition. Photon absorptions transfer momentum, slowing the atom. Subsequent photon emissions give the atom a "kick," but in random directions, so on average the emissions won't speed up the atom in the same way that the absorptions slow it. A beam of atoms moving "upstream" against a correctly tuned laser beam is slowed down—the "hot" beam of atoms is cooled.

Once a laser cools the atoms, a different configuration of laser beams can trap them. Part b of the figure shows six overlapped laser beams, each tuned slightly *below* the frequency of an atomic absorption line. If an atom tries to leave the overlap region, it will be moving "upstream" against one of the laser beams. The atom will see that laser beam Doppler-shifted to a higher frequency, matching the transition frequency. The atom will then absorb photons from this laser beam, and the resulting kick will nudge it back into the overlap region. The atoms are trapped in what is known as *optical molasses* or, more generally, an *atom trap*. More effective traps can be made by adding magnetic fields. The final cooling of the atoms is by evaporation—letting the more energetic atoms leave the trap.

Ultimately, these techniques produce a diffuse gas of atoms moving at only ≈ 1 mm/s. This corresponds to a nearly unbelievable temperature of just a few nanokelvin—billionths of a degree above absolute zero. This is colder than outer space. The coldest spot in the universe is inside an atom trap in a physics lab.

Once the atoms are cooled, some very remarkable things happen. As we saw in Part VII, all particles, including atoms, have a wave nature. As the atoms slow, their wavelengths increase. In a correctly prepared gas at a low enough temperature, the de Broglie wavelength of an individual atom is larger than the spacing between atoms, and the wave functions of multiple atoms overlap. When this happens, some atoms undergo *Bose-Einstein condensation,* coalescing into one "super atom," with thousands of atoms occupying the same quantum state. An example of the resulting Bose-Einstein condensate is shown in part c of the figure. In the condensate, the atoms—that is, their wave functions—are all in the same place at the same time! This truly bizarre state of matter is a remarkable example of the counterintuitive nature of the quantum world.

Are there applications for Bose-Einstein condensation? Current talk of atom lasers and other futuristic concepts aside, no one really knows, just as the early architects of quantum mechanics didn't know that their theory would be used to design the chips that power personal computers.

At the start of the 20th century, there was a worry that everything in physics had been discovered. There is no such worry at the start of the 21st century, which promises to be full of wonderful discoveries and remarkable applications. What do you imagine the final chapter of a physics textbook will look like 100 years from now?

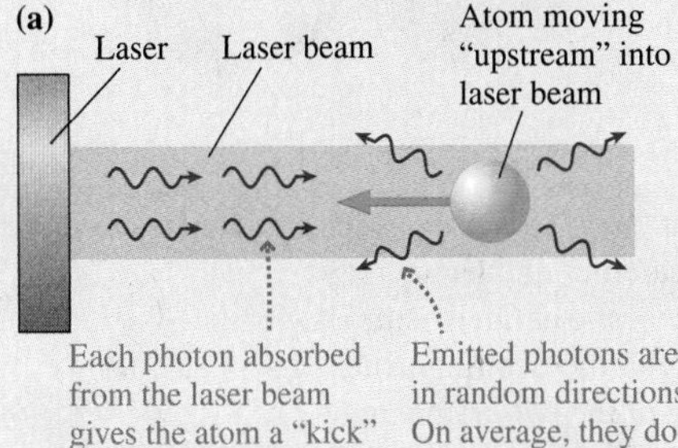
(a)
Laser
Laser beam
Atom moving "upstream" into laser beam
Each photon absorbed from the laser beam gives the atom a "kick" opposite its direction of travel. Thus, each absorption slightly slows the atom.
Emitted photons are in random directions. On average, they do not affect the atom's momentum.

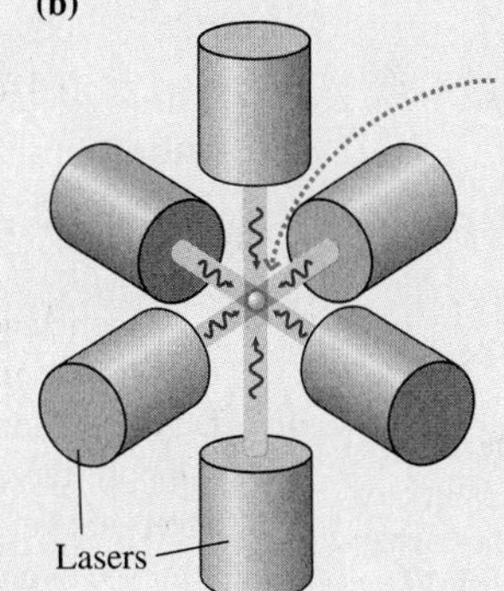
(b)
The intersection of laser beams creates an atom trap. An atom moving out of the trap will be moving upstream into one of the laser beams and will be pushed back.
Lasers

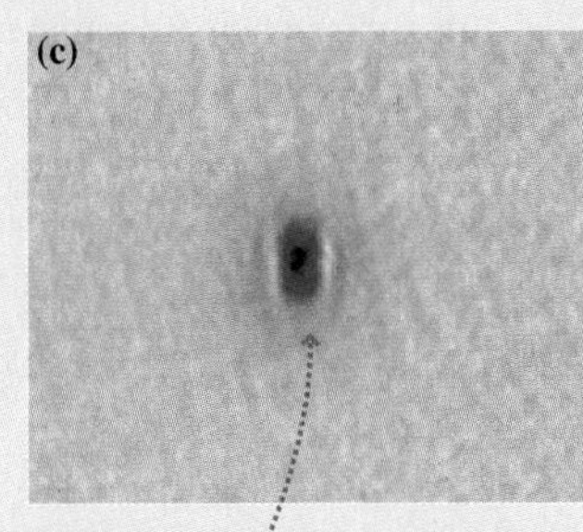
(c)
The dense collection of atoms in the center of the trap is a Bose-Einstein condensate.

Laser cooling and trapping.

The following questions are related to the passage "The Physics of Very Cold Atoms" on the preceding page.

1. Why is it useful to create an assembly of very slow-moving cold atoms
   A. The atoms can be more easily observed at slow speeds.
   B. Lowering the temperature this way permits isotopes that normally decay in very short times to persist long enough to be studied.
   C. At low speeds the quantum nature of the atoms becomes more apparent, and new forms of matter emerge.
   D. At low speeds the quantum nature of the atoms becomes less important, and they appear more like classical particles.
2. The momentum of a photon is given by $p = h/\lambda$. Suppose an atom emits a photon. Which of the following photons will give the atom the biggest "kick"—the highest recoil speed?
   A. An infrared photon B. A red-light photon
   C. A blue-light photon D. An ultraviolet photon
3. When an atom moves "upstream" against the photons in a laser beam, the energy of the photons appears to be _______ if the atom were at rest.
   A. Greater than B. Less than C. The same as
4. A gas of cold atoms strongly absorbs light of a specific wavelength. Warming the gas causes the absorption to decrease. Which of the following is the best explanation for this reduction?
   A. Warming the gas changes the atomic energy levels.
   B. Warming the gas causes the atoms to move at higher speeds, so the atoms "see" the photons at larger Doppler shifts.
   C. Warming the gas causes more collisions between the atoms, which affects the absorption of photons.
   D. Warming the gas makes it more opaque to the photons, so fewer enter the gas.
5. A gas of cold atoms starts at a temperature of 100 nK. The average speed of the atoms is then reduced by half. What is the new temperature?
   A. 71 nK B. 50 nK C. 37 nK D. 25 nK
6. Rubidium is often used for the type of experiments noted in the passage. At a speed of 1.0 mm/s, what is the approximate de Broglie wavelength of an atom of ^{87}Rb?
   A. 5 nm B. 50 nm C. 500 nm D. 5000 nm
7. A gas of rubidium atoms and a gas of sodium atoms have been cooled to the same very low temperature. What can we say about the de Broglie wavelengths of typical atoms in the two gases?
   A. The sodium atoms have the longer wavelength.
   B. The wavelengths are the same.
   C. The rubidium atoms have the longer wavelength.

The following passages and associated questions are based on the material of Part VII.

Splitting the Atom

"Splitting" an atom in the process of nuclear fission releases a great deal of energy. If all the atoms in 1 kg of ^{235}U undergo nuclear fission, 8.0×10^{13} J will be released, equal to the energy from burning 2.3×10^{6} kg of coal. What is the source of this energy? Surprisingly, the energy from this nuclear disintegration ultimately comes from the electric potential of the positive charges that make up the nucleus.

The protons in a nucleus exert repulsive forces on each other, but this force is less than the short-range attractive nuclear force. If a nucleus breaks into two smaller nuclei, the nuclear force will hold each of the fragments together, but it won't bind the two positively charged fragments to each other. This is illustrated in Figure VII.1. The two fragments feel a strong repulsive electrostatic force. The charges are large and the distance is small (roughly equal to the sum of the radius of each of the fragments), so the force—and thus the potential energy—is quite large.

In a fission reaction, a neutron causes a nucleus of ^{235}U to split into two smaller nuclei; a typical reaction is

$$n + {}^{235}\text{U} \rightarrow {}^{87}\text{Br} + {}^{147}\text{La} + \text{neutrons} + \approx 200\text{ MeV}$$

Right after the nucleus splits, with only the electric force now acting on the two fragments, the electrostatic potential energy of the two fragments is

$$U = \frac{kq_1q_2}{r_1 + r_2}$$

This is the energy that will be released, transformed into kinetic energy, when the fragments fly apart. If we use reasonable estimates for the radii of the two fragments, we compute a value for the energy that is close to the experimentally observed value of 200 MeV for the energy released in the fission reaction. The energy released in this *nuclear* reaction is actually *electric* potential energy.

8. How many neutrons are "left over" in the noted fission reaction?
   A. 1 B. 2 C. 3 D. 4

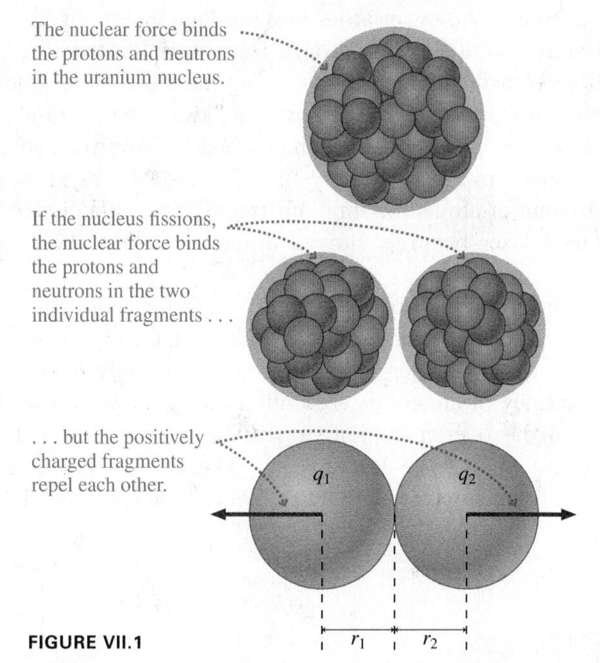

FIGURE VII.1

9. After a fission event, most of the energy released is in the form of
   A. Emitted beta particles and gamma rays.
   B. Kinetic energy of the emitted neutrons.
   C. Nuclear energy of the two fragments.
   D. Kinetic energy of the two fragments.
10. Suppose the original nucleus is at rest in the fission reaction noted above. If we neglect the momentum of the neutrons, after the two fragments fly apart,
    A. The Br nucleus has more momentum.
    B. The La nucleus has more momentum.
    C. The momentum of the Br nucleus equals that of the La nucleus.

11. Suppose the original nucleus is at rest in the fission reaction noted above. If we neglect the kinetic energy of the neutrons, after the two fragments fly apart,
A. The Br nucleus has more kinetic energy.
B. The La nucleus has more kinetic energy.
C. The kinetic energy of the Br nucleus equals that of the La nucleus.
12. 200 MeV is a typical energy released in a fission reaction. To get a sense for the scale of the energy, if we were to use this energy to create electron-positron pairs, approximately how many pairs could we create?
A. 50 B. 100 C. 200 D. 400
13. The two fragments of a fission reaction are isotopes that are neutron-rich; each has more neutrons than the stable isotopes for their nuclear species. They will quickly decay to more stable isotopes. What is the most likely decay mode?
A. Alpha decay B. Beta decay C. Gamma decay

Detecting and Deciphering Radiation

A researcher has placed a sample of radioactive material in an enclosure and blocked all emissions except those that travel in a particular direction, creating a beam of radiation. The beam then passes through a uniform magnetic field, as shown in Figure VII.2, before reaching a bank of detectors. Only three of the detectors record significant signals, so the researcher deduces that the particles coming from the source have taken three different paths, illustrated in the figure, and concludes that the sample is emitting three different kinds of radiation. Assume that the emitted particles move with similar speeds.

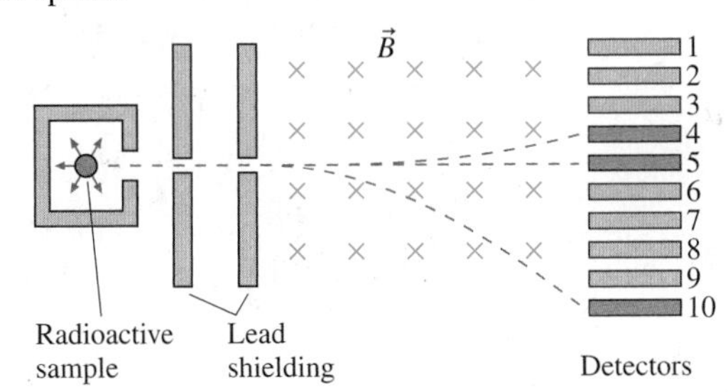

FIGURE VII.2

14. What type of radiation is detected by detector 4?
A. Alpha B. Beta-minus C. Beta-plus D. Gamma
15. What type of radiation is detected by detector 5?
A. Alpha B. Beta-minus C. Beta-plus D. Gamma
16. What type of radiation is detected by detector 10?
A. Alpha B. Beta-minus C. Beta-plus D. Gamma

Additional Integrated Problems

17. The glow-in-the-dark dials on some watches and some keychain lights shine with energy provided by the decay of radioactive tritium, $^3_1$H. Tritium is a radioactive isotope of hydrogen with a half-life of 12 years. Each decay emits an electron with an energy of 19 keV. A typical new watch has tritium with a total activity of 15 MBq.

A keychain light powered by the decay of tritium.

a. What is the speed of the emitted electron? (This speed is high enough that you'll need to do a relativistic calculation.)
b. What is the power, in watts, provided by the radioactive decay process?
c. What will be the activity of the tritium in a watch after 5 years, assuming none escapes?
18. BIO An x-ray tube is powered by a high-voltage supply that delivers 700 W to the tube. The tube converts 1% of this power into x rays of wavelength 0.030 nm.
a. Approximately how many x-ray photons are emitted per second?
b. If a 75 kg technician is accidentally exposed to the full power of the x-ray beam for 1.0 s, what dose equivalent in Sv does he receive? Assume that the x-ray energy is distributed over the body, and that 80% of the energy is absorbed.
19. Many speculative plans for spaceships capable of interstellar travel have been developed over the years. Nearly all are powered by the fusion of light nuclei, one of a very few power sources capable of providing the incredibly large energies required. A typical design for a fusion-powered craft has a 1.7×10^6 kg ship brought up to a speed of $0.12c$ using the energy from the fusion of $^2_1$H and $^3_2$He. Each fusion reaction produces a daughter nucleus and one free proton with a combined kinetic energy of 18 MeV; these high-speed particles are directed backward to create thrust.
a. What is the kinetic energy of the ship at the noted top speed? For the purposes of this problem you can do a nonrelativistic calculation.
b. If we assume that 50% of the energy of the fusion reactions goes into the kinetic energy of the ship (a very generous assumption), how many fusion reactions are required to get the ship up to speed?
c. How many kilograms of $^2_1$H and of $^3_2$He are required to produce the required number of reactions?
20. A muon is a lepton that is a higher-mass (rest mass 105 MeV/c^2) sibling to the electron. Muons are produced in the upper atmosphere when incoming cosmic rays collide with the nuclei of gas molecules. As the muons travel toward the surface of the earth, they lose energy. A muon that travels from the upper atmosphere to the surface of the earth typically begins with kinetic energy 6.0 GeV and reaches the surface of the earth with kinetic energy 4.0 GeV. The energy decreases by one-third of its initial value. By what fraction does the speed of the muon decrease?
21. A muon is a lepton that is a higher-mass (rest mass 105 MeV/c^2) sibling to the electron. Muons are produced in the upper atmosphere when incoming cosmic rays collide with the nuclei of gas molecules. The muon half-life is 1.5 μs, but atmospheric muons typically live much longer than this because of time dilation, as we saw in Chapter 27. Suppose 100,000 muons are created 120 km above the surface of the earth, each with kinetic energy 10 GeV. Assume that the muons don't lose energy but move at a constant velocity directed straight down toward the surface of the earth. How many muons survive to reach the surface?

Mathematics Review

Algebra

Using exponents: $a^{-x} = \frac{1}{a^x}$ $\quad a^x a^y = a^{(x+y)}$ $\quad \frac{a^x}{a^y} = a^{(x-y)}$ $\quad (a^x)^y = a^{xy}$

$a^0 = 1$ $\quad a^1 = a$ $\quad a^{1/n} = \sqrt[n]{a}$

Fractions: $\left(\frac{a}{b}\right)\left(\frac{c}{d}\right) = \frac{ac}{bd}$ $\quad \frac{a/b}{c/d} = \frac{ad}{bc}$ $\quad \frac{1}{1/a} = a$

Logarithms: Natural (base e) logarithms: If $a = e^x$, then $\ln(a) = x$ $\quad \ln(e^x) = x$ $\quad e^{\ln(x)} = x$

Base 10 logarithms: If $a = 10^x$, then $\log_{10}(a) = x$ $\quad \log_{10}(10^x) = x$ $\quad 10^{\log_{10}(x)} = x$

The following rules hold for both natural and base 10 algorithms:

$\ln(ab) = \ln(a) + \ln(b)$ $\quad \ln\left(\frac{a}{b}\right) = \ln(a) - \ln(b)$ $\quad \ln(a^n) = n\ln(a)$

The expression $\ln(a + b)$ cannot be simplified.

Linear equations: The graph of the equation $y = ax + b$ is a straight line. a is the slope of the graph. b is the y-intercept.

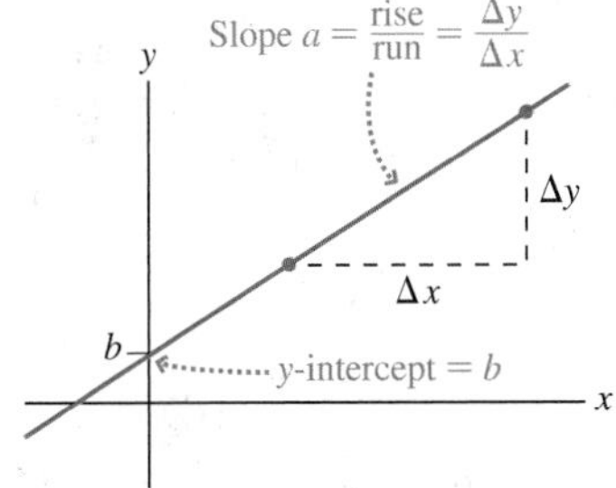

Proportionality: To say that y is proportional to x, written $y \propto x$, means that $y = ax$, where a is a constant. Proportionality is a special case of linearity. A graph of a proportional relationship is a straight line that passes through the origin. If $y \propto x$, then

$$\frac{y_1}{y_2} = \frac{x_1}{x_2}$$

Quadratic equation: The quadratic equation $ax^2 + bx + c = 0$ has the two solutions $x = \frac{-b \pm \sqrt{b^2 - 4ac}}{2a}$.

Geometry and Trigonometry

Area and volume:

Rectangle

$A = ab$

Rectangular box

$V = abc$

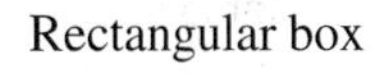

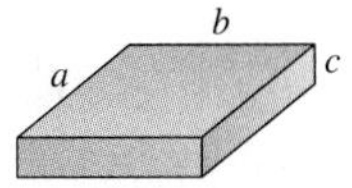

Triangle

$A = \frac{1}{2}ab$

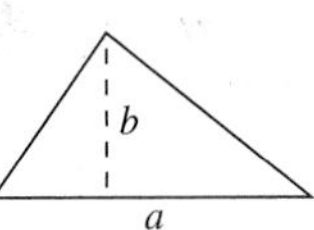

Right circular cylinder

$V = \pi r^2 l$

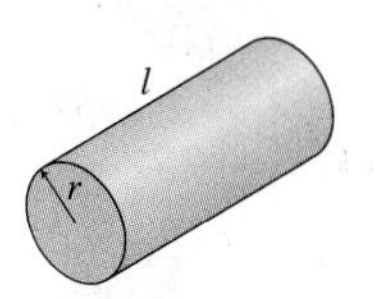

Circle

$C = 2\pi r$

$A = \pi r^2$

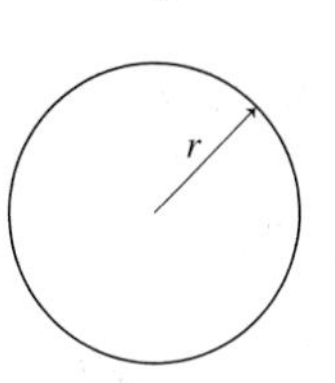

Sphere

$A = 4\pi r^2$

$V = \frac{4}{3}\pi r^3$

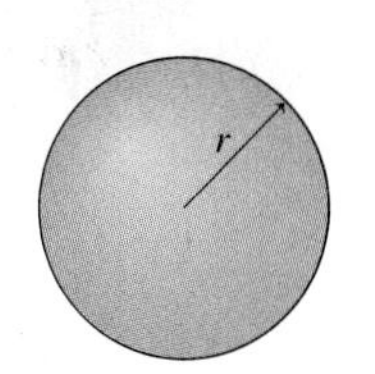

Arc length and angle: The angle θ in radians is defined as $\theta = s/r$.

The arc length that spans angle θ is $s = r\theta$.

2π rad $= 360°$

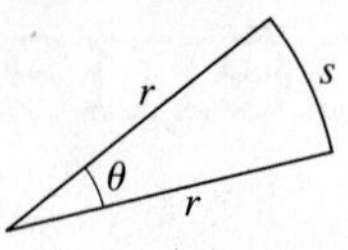

Right triangle: Pythagorean theorem $c = \sqrt{a^2 + b^2}$ or $a^2 + b^2 = c^2$

$$\sin\theta = \frac{b}{c} = \frac{\text{far side}}{\text{hypotenuse}} \qquad \theta = \sin^{-1}\left(\frac{b}{c}\right)$$

$$\cos\theta = \frac{a}{c} = \frac{\text{adjacent side}}{\text{hypotenuse}} \qquad \theta = \cos^{-1}\left(\frac{a}{c}\right)$$

$$\tan\theta = \frac{b}{a} = \frac{\text{far side}}{\text{adjacent side}} \qquad \theta = \tan^{-1}\left(\frac{b}{a}\right)$$

In general, if it is known that sine of an angle θ is x, so $x = \sin\theta$, then we can find θ by taking the *inverse sine* of x, denoted $\sin^{-1}x$. Thus $\theta = \sin^{-1}x$. Similar relations apply for cosines and tangents.

General triangle: $\alpha + \beta + \gamma = 180° = \pi$ rad

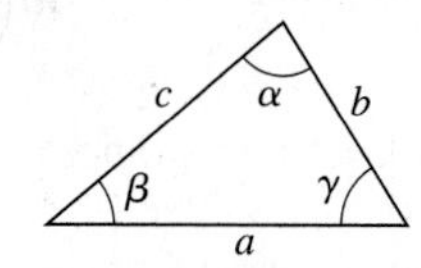

Identities:

$$\tan\alpha = \frac{\sin\alpha}{\cos\alpha} \qquad \sin^2\alpha + \cos^2\alpha = 1$$

$$\sin(-\alpha) = -\sin\alpha \qquad \cos(-\alpha) = \cos\alpha$$

$$\sin(2\alpha) = 2\sin\alpha\cos\alpha \qquad \cos(2\alpha) = \cos^2\alpha - \sin^2\alpha$$

Expansions and Approximations

Binomial approximation: $(1 + x)^n \approx 1 + nx \quad \text{if} \quad x \ll 1$

Small-angle approximation: If $\alpha \ll 1$ rad, then $\sin\alpha \approx \tan\alpha \approx \alpha$ and $\cos\alpha \approx 1$.

The small-angle approximation is excellent for $\alpha < 5°$ (≈ 0.1 rad) and generally acceptable up to $\alpha \approx 10°$.

Periodic Table of Elements

Atomic number — 27; Symbol — Co; Atomic mass — 58.9

Period																		
1	1 H 1.0																	2 He 4.0
2	3 Li 6.9	4 Be 9.0											5 B 10.8	6 C 12.0	7 N 14.0	8 O 16.0	9 F 19.0	10 Ne 20.2
3	11 Na 23.0	12 Mg 24.3											13 Al 27.0	14 Si 28.1	15 P 31.0	16 S 32.1	17 Cl 35.5	18 Ar 39.9
4	19 K 39.1	20 Ca 40.1	21 Sc 45.0	22 Ti 47.9	23 V 50.9	24 Cr 52.0	25 Mn 54.9	26 Fe 55.8	27 Co 58.9	28 Ni 58.7	29 Cu 63.5	30 Zn 65.4	31 Ga 69.7	32 Ge 72.6	33 As 74.9	34 Se 79.0	35 Br 79.9	36 Kr 83.8
5	37 Rb 85.5	38 Sr 87.6	39 Y 88.9	40 Zr 91.2	41 Nb 92.9	42 Mo 95.9	43 Tc [98]	44 Ru 101.1	45 Rh 102.9	46 Pd 106.4	47 Ag 107.9	48 Cd 112.4	49 In 114.8	50 Sn 118.7	51 Sb 121.8	52 Te 127.6	53 I 126.9	54 Xe 131.3
6	55 Cs 132.9	56 Ba 137.3	71 Lu 175.0	72 Hf 178.5	73 Ta 180.9	74 W 183.9	75 Re 186.2	76 Os 190.2	77 Ir 192.2	78 Pt 195.1	79 Au 197.0	80 Hg 200.6	81 Tl 204.4	82 Pb 207.2	83 Bi 209.0	84 Po [209]	85 At [210]	86 Rn [222]
7	87 Fr [223]	88 Ra [226]	103 Lr [262]	104 Rf [265]	105 Db [268]	106 Sg [271]	107 Bh [272]	108 Hs [270]	109 Mt [276]	110 Ds [281]	111 Rg [280]	112 Cn [285]	113	114 Fl [289]	115	116 Lv [293]	117	118

Transition elements: groups from 21 Sc through 30 Zn.

Inner transition elements

Lanthanides 6	57 La 138.9	58 Ce 140.1	59 Pr 140.9	60 Nd 144.2	61 Pm 144.9	62 Sm 150.4	63 Eu 152.0	64 Gd 157.3	65 Tb 158.9	66 Dy 162.5	67 Ho 164.9	68 Er 167.3	69 Tm 168.9	70 Yb 173.0
Actinides 7	89 Ac [227]	90 Th 232.0	91 Pa 231.0	92 U 238.0	93 Np [237]	94 Pu [244]	95 Am [243]	96 Cm [247]	97 Bk [247]	98 Cf [251]	99 Es [252]	100 Fm [257]	101 Md [258]	102 No [259]

An atomic mass in brackets is that of the longest-lived isotope of an element with no stable isotopes.

Video Resources MP

The following lists describe the videos available in the Pearson eText and in the Study Area of MasteringPhysics, with the corresponding textbook section and page references.

Class Videos

Video Tutor Solutions

Video Tutor Demonstrations

APPENDIX D

Atomic and Nuclear Data

Atomic Number (Z)	Element	Symbol	Mass Number (A)	Atomic Mass (u)	Percent Abundance	Decay Mode	Half-Life $t_{1/2}$
0	(Neutron)	n	1	1.008 665		β^-	10.4 min
1	Hydrogen	H	1	1.007 825	99.985	stable	
	Deuterium	D	2	2.014 102	0.015	stable	
	Tritium	T	3	3.016 049		β^-	12.33 yr
2	Helium	He	3	3.016 029	0.000 1	stable	
			4	4.002 602	99.999 9	stable	
			6	6.018 886		β^-	0.81 s
3	Lithium	Li	6	6.015 121	7.50	stable	
			7	7.016 003	92.50	stable	
			8	8.022 486		β^-	0.84 s
4	Beryllium	Be	9	9.012 174	100	stable	
			10	10.013 534		β^-	1.5×10^6 yr
5	Boron	B	10	10.012 936	19.90	stable	
			11	11.009 305	80.10	stable	
			12	12.014 352		β^-	0.020 2 s
6	Carbon	C	10	10.016 854		β^+	19.3 s
			11	11.011 433		β^+	20.4 min
			12	12.000 000	98.90	stable	
			13	13.003 355	1.10	stable	
			14	14.003 242		β^-	5 730 yr
			15	15.010 599		β^-	2.45 s
7	Nitrogen	N	12	12.018 613		β^+	0.011 0 s
			13	13.005 738		β^+	9.96 min
			14	14.003 074	99.63	stable	
			15	15.000 108	0.37	stable	
			16	16.006 100		β^-	7.13 s
			17	17.008 450		β^-	4.17 s
8	Oxygen	O	15	15.003 065		β^+	122 s
			16	15.994 915	99.76	stable	
			17	16.999 132	0.04	stable	
			18	17.999 160	0.20	stable	
			19	19.003 577		β^-	26.9 s
9	Fluorine	F	18	18.000 937		β^+	109.8 min
			19	18.998 404	100	stable	
			20	19.999 982		β^-	11.0 s
10	Neon	Ne	19	19.001 880		β^+	17.2 s
			20	19.992 435	90.48	stable	
			21	20.993 841	0.27	stable	
			22	21.991 383	9.25	stable	
17	Chlorine	Cl	35	34.968 853	75.77	stable	
			36	35.968 307		β^-	3.0×10^5 yr
			37	36.965 903	24.23	stable	

Atomic Number (Z)	Element	Symbol	Mass Number (A)	Atomic Mass (u)	Percent Abundance	Decay Mode	Half-Life $t_{1/2}$
18	Argon	Ar	36	35.967 547	0.34	stable	
			38	37.962 732	0.06	stable	
			39	38.964 314		β^-	269 yr
			40	39.962 384	99.60	stable	
			42	41.963 049		β^-	33 yr
19	Potassium	K	39	38.963 708	93.26	stable	
			40	39.964 000	0.01	β^-	1.28×10^9 yr
			41	40.961 827	6.73	stable	
26	Iron	Fe	54	54.939 613	5.9	stable	
			56	55.934 940	91.72	stable	
			57	56.935 396	2.1	stable	
			58	57.933 278	0.28	stable	
			60	59.934 072		β^-	1.5×10^6 yr
27	Cobalt	Co	59	58.933 198	100	stable	
			60	59.933 820		β^-	5.27 yr
38	Strontium	Sr	84	83.913 425	0.56%	stable	
			86	85.909 262	9.86%	stable	
			87	86.908 879	7.00%	stable	
			88	87.905 614	82.58%	stable	
			89	88.907 450		β^-	50.53 days
			90	89.907 738		β^-	27.78 yr
53	Iodine	I	127	126.904 474	100	stable	
			129	128.904 984		β^-	1.6×10^7 yr
			131	130.906 124		β^-	8 days
54	Xenon	Xe	128	127.903 531	1.9	stable	
			129	128.904 779	26.4	stable	
			130	129.903 509	4.1	stable	
			131	130.905 069	21.2	stable	
			132	131.904 141	26.9	stable	
			133	132.905 906		β^-	5.4 days
			134	133.905 394	10.4	stable	
			136	135.907 215	8.9	stable	
55	Cesium	Cs	133	132.905 436	100	stable	
			137	136.907 078		β^-	30 yr
82	Lead	Pb	204	203.973 020	1.4	stable	
			206	205.974 440	24.1	stable	
			207	206.975 871	22.1	stable	
			208	207.976 627	52.4	stable	
			210	209.984 163		α, β^-	22.3 yr
			211	210.988 734		β^-	36.1 min
83	Bismuth	Bi	209	208.980 374	100	stable	
			211	210.987 254		α	2.14 min
			215	215.001 836		β^-	7.4 min
86	Radon	Rn	219	219.009 477		α	3.96 s
			220	220.011 369		α	55.6 s
			222	222.017 571		α, β^-	3.823 days

Atomic Number (Z)	Element	Symbol	Mass Number (A)	Atomic Mass (u)	Percent Abundance	Decay Mode	Half-Life $t_{1/2}$
88	Radium	Ra	223	223.018 499		α	11.43 days
			224	224.020 187		α	3.66 days
			226	226.025 402		α	1 600 yr
			228	228.031 064		β^-	5.75 yr
90	Thorium	Th	227	227.027 701		α	18.72 days
			228	228.028 716		α	1.913 yr
			229	229.031 757		α	7 340 yr
			230	230.033 127		α	7.54×10^4 yr
			231	231.036 299		α, β^-	25.52 h
			232	232.038 051	100	α	1.40×10^{10} yr
			234	234.043 593		β^-	24.1 days
92	Uranium	U	233	233.039 630		α	1.59×10^5 yr
			234	234.040 946		α	2.45×10^5 yr
			235	235.043 924	0.72	α	7.04×10^8 yr
			236	236.045 562		α	2.34×10^7 yr
			238	238.050 784	99.28	α	4.47×10^9 yr
93	Neptunium	Np	237	237.048 168		α	2.14×10^6 yr
			238	238.050 946		β^-	2.12 days
			239	239.052 939		β^-	2.36 days
94	Plutonium	Pu	238	238.049 555		α	87.7 yr
			239	239.052 157		α	2.412×10^4 yr
			240	240.053 808		α	6560 yr
			242	242.058 737		α	3.73×10^6 yr
			244	244.064 200		α	8.1×10^7 yr
95	Americium	Am	241	241.056 823		α	432.21 yr
			243	243.061 375		α	7 370 yr

Answers

Chapter 17

Answers to odd-numbered multiple-choice questions

19. B
21. C
23. B
25. D

Answers to odd-numbered problems

1. a. 1.5×10^{-11} s b. 3.4 mm
3. 0.40 ns
5. 670 ns
7. 0.020 rad, 1.1°
9. 0.050 rad
11. 0.36 mm
13. 2,000, 201.2λ
15. 3.2°, 6.3°
17. 221 lines/mm
19. a. 1.3 m b. 7
21. 217 nm
23. 85.9 nm
25. 121 nm
27. 97.8 nm
29. 94 μm
31. 610 nm
33. 4.9 mm
35. 78 cm
37. 6.4 cm
39. 500 nm
41. 450 nm
43. Grating

Screen y

$y_1 = 0.58$ m $m = 1$

$\theta_1 = 0.53$ rad

$y_0 = 0$ m $m = 0$

$\theta_1 = 0.53$ rad

$y_1 = 0.58$ m $m = 1$

1.0 m

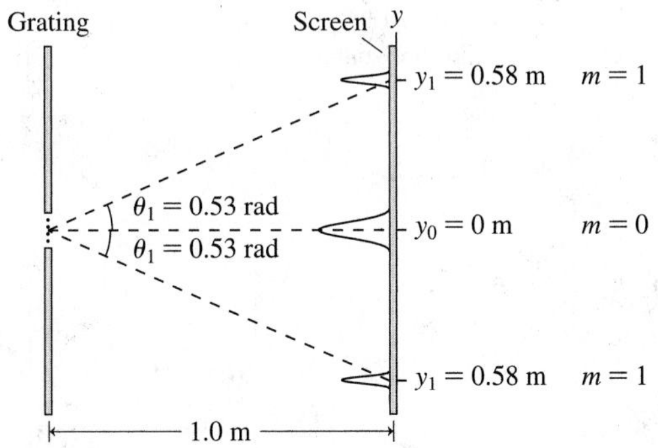

45. 43.2°
47. 43 cm
49. 500 nm
51. 25 cm
53. 1.8 μm
55. 500 nm
57. 477 nm
59. 410 nm, 690 nm, purple
61. 2.8°
63. 7.6 m
65. 0.10 mm
67. a. Double b. 0.15 mm
69. 0.1 mm
71. a. Diffraction b. 0.044° c. 4.6 mm d. 1.5 m
73. D
75. A

Chapter 18

Answers to odd-numbered multiple-choice questions

17. A
19. C
21. D
23. A
25. A
27. A

Answers to odd-numbered problems

1. 11 ft
3. 8.00 cm
5. B and C
7. 433 cm
9. 31°
11. 35°
13. 65°
15. 1.30
17. 3.2 cm
19. a. Closer b. Farther
21. $s' = -15$ cm, upright and virtual
23. $s' = -6$ cm, upright and virtual
25. $s' = 30$ cm, inverted and real
27. Concave, $f = 2.4$ cm
29. $f = -36$ cm
31. $s' = -15$ cm, $h' = 1.5$ cm
33. $s' = 50$ cm, $h' = 0.67$ cm
35. $s' = -20$ cm, $h' = 0.33$ cm
37. $s' = -16$ cm, $h' = 1.1$ cm
39. $s' = 56$ cm, $h' = 3.7$ cm
41. 1.0 m
43. 42°
45. 60°
47. a. 32° from the normal b. No
49. 82.8°
53. 35°
55. 48.8°
57. a. $\beta = \sin^{-1}\left(n \sin\left(\frac{1}{2}\alpha\right)\right)$ b. 1.58
59. 42°
61. 110 cm
63. Raised by 0.10 mm
65. $s' = -40$ cm, $h' = 10$ cm, upright and virtual
67. $s' = -40$ cm, $h' = 2.5$ cm
69. $s' = -30$ cm, $h' = 3.0$ cm
71. $s' = -30$ cm, $h' = 0.50$ cm
73. $s' = 160$ cm, $h' = 0.50$ cm; $s' = 40$ cm, $h' = 8.0$ cm
75. 67 cm
77. a. 180 cm b. 26 m c. 4.7 cm d. 1.5 m
    e. So you won't stop suddenly with someone behind you
79. C

Chapter 19

Answers to odd-numbered multiple-choice questions

17. C
19. C
21. C
23. C

Answers to odd-numbered problems

1. a. 5.8 mm b. The object must be getting closer.
3. 0.29 m
5. 17 mm
7. a. 2.5 cm b. 36 cm c. 2.3 cm
9. −0.33 D
11. 2.0 D
13. a. Farsighted b. ∞
15. 2.1 m
17. 5.0 cm
19. a. 1.20 cm b. 1.67 cm
21. 11 mm
23. 2.5 cm
25. $s' = -6.4$ cm left of the diverging lens, $h' = 1.0$ cm, same orientation as the object
27. a. 0.058 D b. 17 mm
29. a. 14° b. 6.4 mm
31. 0.28 mm, violet
33. 55 km
35. 4.9 mm
37. 3.1
39. 0.70 cm
41. b. $h'' = 2.0$ cm, $s'' = -20$ cm
43. 5.0 mm
45. 250
47. −4.0
49. 26 cm
51. 24 cm
53. 2.8×10^{-7} rad
55. 15 km
57. No
59. A
61. B

Part V Problems

Answers to odd-numbered problems

1. C
3. A
5. A
7. B
9. D
11. A
13. A
15. Aberrations are reduced.
17. 7.2 mm, 0.069 W/m^2

Chapter 20

Answers to odd-numbered multiple-choice questions

23. A
25. D
27. D
29. A

Answers to odd-numbered problems

1. a. Electrons have been removed. b. 3.1×10^{10}
3. 1.5×10^6 C
5. a. Electrons from sphere to rod b. 2.5×10^{10}
7. A has −160 nC; B has 0 C
9. a. 9.0×10^{-3} N b. 9.0×10^{-3} m/s^2
11. 2.6 cm
13. −10 nC
15. 8.0 nC
17. a. 1.7×10^{14} m/s^2, away from bead
    b. 3.2×10^{17} m/s^2, toward bead
19. 1.4×10^5 N/C, away from bead
21. a. 1.4×10^{-3} N/C, away from proton
    b. 1.4×10^{-3} N/C, toward electron
23. a. 3.6×10^4 N/C, 1.8×10^4 N/C, 1.8×10^4 N/C
    b.

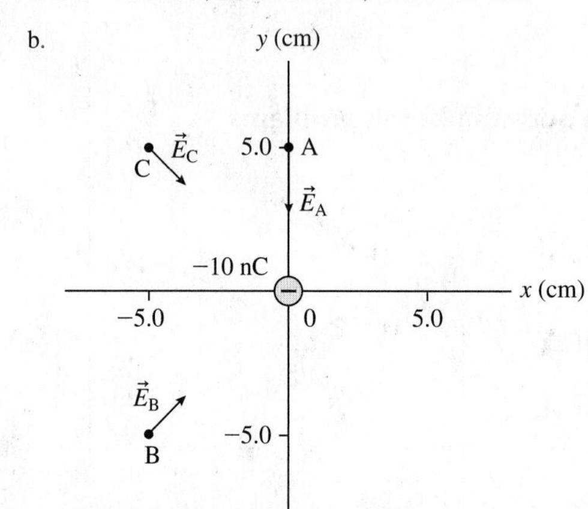

25. 2500 N/C, downward
27. 6.1×10^5 N/C, downward
29. ±14 nC
31. a. Upper b. 1.6×10^{-7} C
35. 7.2×10^{-15} N
37. a. 0 b. 5.6×10^{-27} N·m
39. 1.4×10^5 C
41. a. 500 N b. 3.0×10^{29} m/s^2
43. a. 0.45 N b. $q_A = 1.0 \times 10^{-6}$ C, $q_B = 0.50 \times 10^{-6}$ C
45. 4000 N/C, 9.3° above the horizontal
47. 3.1×10^{-4} N, upward
49. 0 N
51. 1.1×10^{-5} N, upward
53. 0.68 nC
55. 28 nC
57. 6.6×10^{15} rev/s
59. (−6.0 cm, 3.0 cm)
61. 5.9×10^5 N/C
63. 5.7 cm
65. 180 nC
67. 150 nC
69. 1.3 cm
71. B
73. D

Chapter 21

Answers to odd-numbered multiple-choice questions

21. A
23. A
25. C
27. D
29. B
31. A

Answers to odd-numbered problems

1. $-3.0\ \mu C$
3. 530 V
5. a. 3000 V b. $75\ \mu J$
7. a. 1000 eV b. 1.60×10^{-16} J c. 4.4×10^5 m/s
9. a. Lower b. -0.71 V c. 0.71 eV
11. 400 V
13. a. $E = 200$ kV/m, $\Delta V_C = 200$ V b. $E = 200$ kV/m, $\Delta V_C = 400$ V
15. 3.8 cm
17. 3.8 nC
19. a. 1.5 V b. 8.3 pC
21. -352 V
23. 10 kV/m, left
25. a. Positive b. Negative c. Positive
27. a. 7.1 pF b. 0.71 nC
29. 4.8 cm
31. 24 V
33. $15\ \mu C$
35. 13 nF
37. a. $Q = 5.0 \times 10^{-10}$ C, $\Delta V_C = 9.0$ V, $E = 9.0 \times 10^4$ V/m
    b. $Q = 2.5 \times 10^{-10}$ C, $\Delta V_C = 9.0$ V, $E = 9.0 \times 10^4$ V/m
39. a. 12.0 V b. 24 V
41. a. 1.7×10^{-2} C b. 14 J
43. $3.00\ \mu J$
45. a. 1.1×10^{-7} J b. $0.71\ J/m^3$
47. -140 V
49. a. 1.5 cm b. 6000 V, 8.0×10^5 V/m
51. -2.2×10^{-19} J
53. 1.4 mN
55. ± 12 cm
57. a. 100 V b. 5.4×10^4 N/C, along the $+x$-axis
59. a. 27 V b. -4.3×10^{-18} J
61. a. 3100 V b. 5.0×10^{-16} J
63. 1.5×10^6 m/s
65. 9 MV/m
67. 2.2×10^{-11} N
69. 7.5×10^5 m/s
71. 2.7×10^6 m/s
73. a. 1000 V b. 1.4×10^{-9} C c. 7.0×10^6 m/s
75. a. 9.6×10^7 m/s b. 0.82 mm c. $3.0 \times 10^{18}\ m/s^2$
77. a. $Q = 2.1 \times 10^{-10}$ C, $E = 3000$ V/m, $\Delta V = 15$ V
    b. $Q = 2.1 \times 10^{-10}$ C, $E = 3000$ V/m, $\Delta V = 30$ V
79. $\vec{E}_1 =$ (2500 V/m, down); $\vec{E}_2 =$ (5000 V/m, up)
81. b. $\vec{E}_A =$ (500 V/m, left), $\vec{E}_B =$ (1000 V/m, right),
    $\vec{E}_C =$ (710 V/m, 45° above straight left),
    $\vec{E}_D =$ (710 V/m, 45° below straight left)
83. a. 4.5×10^{-4} J b. 0.36 J
85. $22\ \mu F$
87. A
89. D

Chapter 22

Answers to odd-numbered multiple-choice questions

23. C
25. C
27. D
29. C

Answers to odd-numbered problems

1. 3000 C, 1.9×10^{22}
3. a. 1.6×10^{18} electrons/s b. Into
5. 9.4×10^{13} electrons
7. 13 A
9. 120 C
11. $I_B = 5$ A, $I_C = -2$ A
13. $1.5\ \mu J$
15. a. 45 C b. 4.1×10^2 J
17. 3200
19. a. 1.0 b. 0.50
21. $50\ \Omega \cdot m$
23. 8.7 m
25. 2.0 A
27. $1 \times 10^5\ \Omega$
29. a. $2.0\ \Omega$
    b.

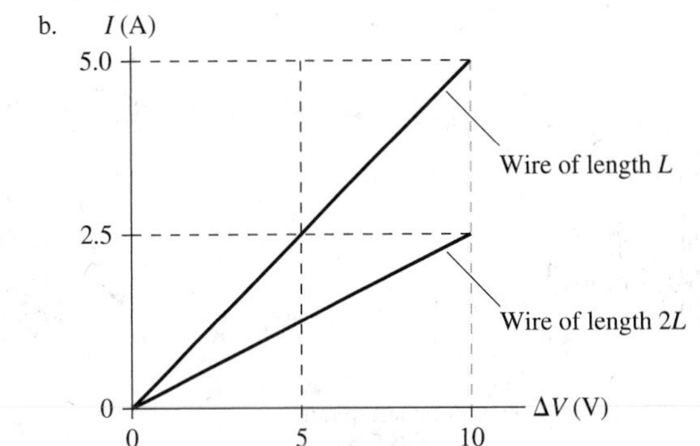

31. a. 200 V/m b. 100 V/m c. 0.25 A d. $480\ \Omega$
33. 0.43 V/m
35. 3000 J
37. 6.0 V
39. a. 0.38 A b. $320\ \Omega$
41. 15 kJ
43. 450 C
45. 28 min
47. 20%
49. $42\ G\Omega$
51. 33 h
53. a. $15\ \Omega$ b. $2.1 \times 10^{-6}\ \Omega \cdot m$
55. a. 0.90 m b. $40\ \Omega$
57. 3.6 mm
59. 1/2
61. Iron
63. 2.6 m, 0.24 mm
65. a. $48\ \Omega$ b. 7.6 min
67. a. 25/26 b. 1/26
69. a. $1.0\ k\Omega$ b. 100 V
71. A
73. B

Chapter 23

Answers to odd-numbered multiple-choice questions

31. B
33. A
35. B
37. B

Answers to odd-numbered problems

1.

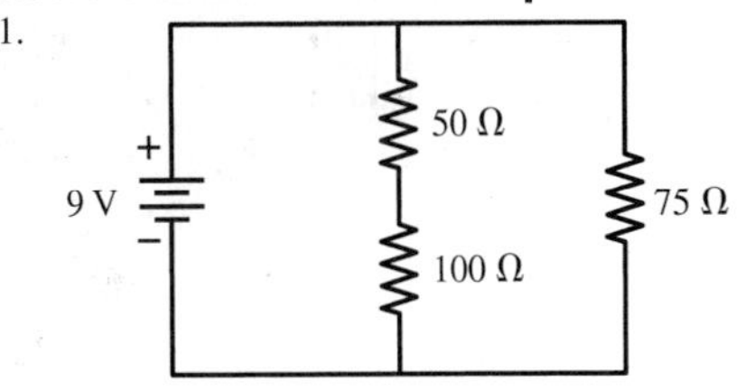

3\.

5\. a. $\Delta V_{12} = 2.0$ V, $\Delta V_{23} = 1.0$ V, $\Delta V_{34} = 0$ V
b. $\Delta V_{12} = 0$ V, $\Delta V_{23} = 3.0$ V, $\Delta V_{34} = 0$ V
7\. a. 0.17 A, to the left
11\. a. 1.0 Ω b. 1.0 Ω c. 0.50 Ω
13\. 270 Ω
15\. 170 Ω
17\. 24 Ω
19\. 300 Ω
21\. a. 2.0 A b. 5.0 A
23\. 20 Ω, 60 V
25\. Numbering the resistors from left to right:

R	*I* (A)	Δ*V* (V)
R_1	1.5	7.5
R_2	0.50	2.5
R_3	0.50	2.5
R_4	0.50	2.5

27\.

R	*I* (A)	Δ*V* (V)
6.0 Ω (left)	2.0	12
15 Ω	0.8	12
6.0 Ω (top)	1.2	7.2
4.0 Ω	1.2	4.8

29\.

R	*I* (A)	Δ*V* (V)
3 Ω	2.0	6.0
4 Ω	1.5	6.0
48 Ω	0.13	6.0
16 Ω	0.38	6.0

31\. 20 Ω
33\. 3.0 μF
35\. 150 μF in series
37\. 8.9 μF
39\. a. 0.57 μF b. 6.9 μC
41\. 2.0 ms
43\. 6.9 ms
45\. a. 18 μC, 180 mA b. 11 μC, 110 mA c. 2.4 μC, 24 mA
47\. 2.9 ms
49\. 33 m/s
51\. 20 m/s
53\. 19 W
55\. a. 1.0 V b. 1/3
57\. a. 2.3 V b. 1.0 kJ, 340 J
59\. $P_5 = 45$ W, $P_{20} = 20$ W
61\. 55 Ω
63\. 4
65\. a. Counterclockwise b. 6.0 Ω c. 0.38 W
67\. 10 kΩ
69\.

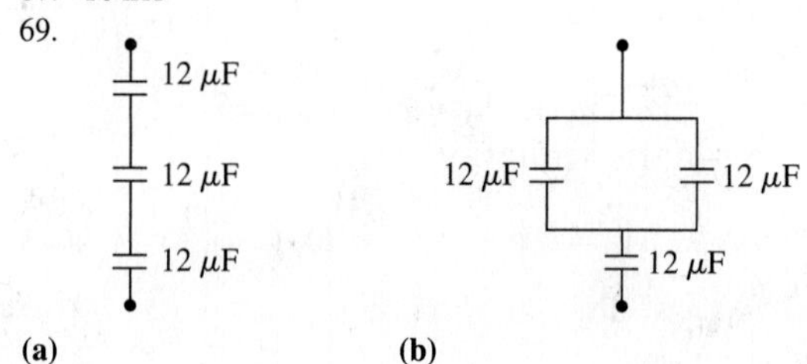

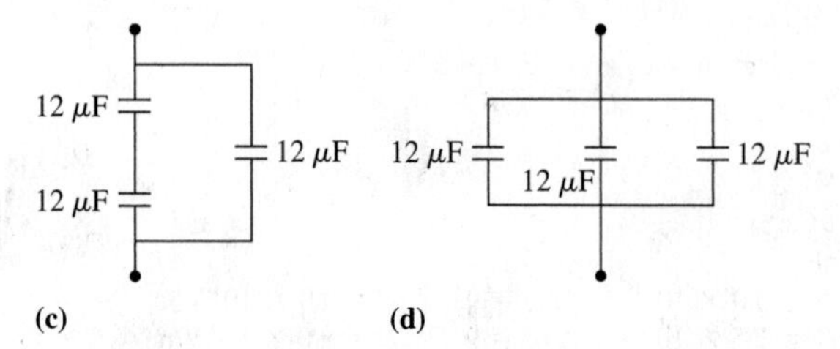

71\. a. 10 ms b. 5.2 ms
73\. 870 Ω
75\. 18 μF
77\. a. 18 μC b. 0.23 ms
79\. 6100
81\. a. 4.8 pC b. 3.0×10^7
83\. C
85\. B
87\. A
89\. B

Chapter 24

Answers to odd-numbered multiple-choice questions

29\. A
31\. A
33\. D
35\. C

Answers to odd-numbered problems

1\. 20 μT
3\. a. 20 A b. 1.6 mm
5\. Point 1: 6.0 mT into the page; point 2: 2.0 mT out of the page
7\. 5
9\. a. 2 μT, 4% b. The two fields nearly cancel
11\. 750 A
13\. 1.6 kA
15\. 3.8×10^{-7} A
17\. 4.1×10^{-4} T
19\. 8.6 μA
21\. 13 T
23\. a. (8.0×10^{-13} N, $-z$-direction)
b. (8.0×10^{-13} N, 45° clockwise from $-z$-axis in *yz* plane)
25\. a. 11 cm b. 10 m
27\. 58 kHz
29\. 0.086 T
31\. a. 0.61 m b. 18,000 orbits
33\. 0.13 T, out of the page
35\. 2.0×10^{-7} N
37\. a. 3.3 A b. 2.0 N c. 170 m/s^2
39\. 7.5×10^{-4} N·m
41\. a. 1.3×10^{-11} N·m b. Rotated clockwise 90°

43. a.

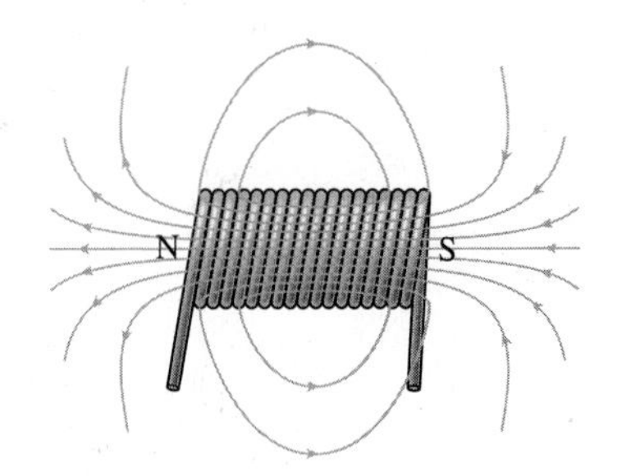

b. Attractive c. To the left, toward the solenoid
45. 3.0 Ω
47. 2.0 A
49. a. 0.11 mT b. North c. 2.5×10^{-23} N
51. The force of the earth's field
53. 0.20 A, axis of coil north-south
55. Rapid switching doesn't allow the moments to align.
57. 1.0 T
59. A
61. B
63. A
65. B
67. A
69. A

Chapter 25

Answers to odd-numbered multiple-choice questions

27. C
29. A
31. C
33. D
35. D

Answers to odd-numbered problems

1. 0.10 T, out of the page
3. a. 4.0 m/s b. 2.2 T
5. 9500 m/s
7. a. 2.9 mV b. Negative
9. For $\theta = 0°$, 3.9×10^{-4} Wb; for $\theta = 30°$, 3.4×10^{-4} Wb; for $\theta = 60°$, 2.0×10^{-4} Wb; for $\theta = 90°$, 0 Wb
11. a. 6.3×10^{-5} Wb b. 6.3×10^{-5} Wb
13. a. 8.7×10^{-4} Wb b. Clockwise
15. 3.1 V
17. a. 3.9 mV, 39 mA, clockwise b. 3.9 mV, 39 mA, clockwise c. 0 V, 0 A
19. 2.5 mV
21. 0.55 A
23. 0.97 A
25. 6.0×10^5 V/m
27. $E_0 = 1.4 \times 10^3$ V/m, $B_0 = 4.6 \times 10^{-6}$ T
29. $E_0 = 980$ V/m, $B_0 = 3.3 \times 10^{-6}$ T
31. 77 m
33. a. 2.2×10^{-6} W/m^2 b. 0.041 V/m
35. a. 10 W/m^2 b. 7.5 W/m^2 c. 5.0 W/m^2 d. 2.5 W/m^2 e. 0 W/m^2
37. 60°
39. 67%
41. 1.2 keV
43. a. 250 keV b. 1.0×10^6 bonds
45. 2.1×10^{-20} J
47. 4.68×10^{21} photons
49. 1700 photons/s
51. 2100°C
53. a. 200°C b. 1.6
55. $\lambda_{dental} = 50$ pm; $\lambda_{microtomography} = 20$ pm
57. 1.1 mm
59. 42 mV
61. a. 5.2×10^{-9} J b. 6.6×10^{-11} K
63. a. 0.20 A b. 4.0×10^{-3} N c. 11°C
65. a. 6.3×10^{-4} N b. 3.1×10^{-4} W c. 1.3×10^{-2} A, counterclockwise d. 3.1×10^{-4} W
67. a. 250 kW b. 1.8×10^7 V/m
69. 0.016 V/m
71. $I_0/4$
73. 3.0×10^{31}
75. A
77. D
79. C

Chapter 26

Answers to odd-numbered multiple-choice questions

23. D
25. A
27. A
29. C

Answers to odd-numbered problems

1. a. 50 mA b. 50 mA
3. 22.4 V
5. a. 9.0 Ω b. 19 A c. 3200 W
7. 530 A
9. 0.50 A
11. 17 mA
13. a. 15 turns b. 32 mA
15. a. 18 kW b. 1.0 kW
17. 12 A
19. Yes
21. $26.30
23. No
25. 34 V
27. 190 mA
29. 59 mA, yes
31. Yes
33. 270 Hz
35. a. 80 kHz b. 0 V
37. 130 Ω
39. a. 0.80 A b. 0.80 mA
41. a. 3.2×10^4 Hz b. 0 V
43. 80 Ω
45. 0.25 μH
47. 1.0 Ω
49. 10 mH
51. 7.1 Ω, 2.5 mH
53. 25 A
55. 5.8 mm
57. a. 24 V b. 4.5 mΩ c. 5600 A
59. a. 500 Hz b. 1.0 A
61. 35 mH
63. D
65. C
67. A
69. B

Part VI Problems

Answers to odd-numbered problems

1. B
3. C
5. B

7. A
9. B
11. C
13. C
15. D
17. B
19. C
21. 20 A, clockwise

Chapter 27

Answers to odd-numbered multiple-choice questions

21. B
23. B
25. C

Answers to odd-numbered problems

1. $v_{sound} = 345$ m/s, $v_{sprinter} = 10$ m/s
3. a. 15 m/s b. 5 m/s
5. 30 s
7. 3.0×10^8 m/s
9. a. 3.0×10^6 m b. 3.0×10^8 m
11. $t_1 = 1.0\ \mu s$, $t_2 = 2.0\ \mu s$
13. She sees flash 2 40 μs after flash 1.
15. Barn 1.0 μs before the tree.
17. a. Simultaneous b. Event 2 occurs first
19. $0.87c$
21. a. $0.80c$ b. 15 yr
23. Yes
25. 0.78 m
27. 0.35 nm
29. $0.60c$
31. a. 1.12×10^{-17} kg · m/s b. 22.4
33. $0.87c$
35. 6.0×10^{13} J, 9.0×10^{13} J, 1.5×10^{14} J
37. $0.42c$
39. 1.7 μs
41. 1.0 ps
43. 12 m
45. 0.50 s
47. 6.2 s
49. 4600 kg/m^3
51. 2 s
53. a. 10 yr b. 4.4 yr c. 3.9 ly d. $0.90c$
55. a. 170 yr b. 14.7 yr c. Both
57. $0.96c$
59. 9.00 nm/s
61. 0.16 s
63. a. $(1 - 4.9 \times 10^{-24})c$ b. 46 nm
65. $0.31c$
67. 6.3×10^{-19} kg · m/s
69. a. 6660 MeV b. 3.63 MeV
71. $0.79c$
73. 5.1×10^{-10} kg
75. a. 1.3×10^{17} kg b. 6.7×10^{-12}% c. 1.5×10^{13} yr
77. a. 4×10^{-12} J b. 0.6%
79. D
81. D

Chapter 28

Answers to odd-numbered multiple-choice questions

27. B
29. A
31. B
33. D
35. A

Answers to odd-numbered problems

1. 41°
3. 69.3°
5. 4
7. 3.20 eV
9. 3.47 eV
11. 1.8 eV
13. a. 290 nm b. 3.7×10^5 m/s
15. a. 6.63×10^5 m/s b. 2.30 eV, potassium
17. 5.0 keV
19. 2.84 eV, 2.33 eV, 2.20 eV
21. 1.94 eV, 3.07 eV
23. 8.7×10^{31}
25. a. 2.3 eV b. 3×10^{18}
27. 9.5×10^{-7} W
29. a. 1.1×10^{-34} m b. 1.7×10^{-23} m/s
31. 1.5 eV
33. 8.2 MeV
35. 6.0 μV
37. 0.43 nm
39. $E_1 = 1.3 \times 10^{-12}$ J, $E_2 = 5.3 \times 10^{-12}$ J, $E_3 = 1.2 \times 10^{-11}$ J
41. $\lambda = 830$ nm for the $2 \rightarrow 1$ transition
    $\lambda = 500$ nm for the $3 \rightarrow 2$ transition
    $\lambda = 310$ nm for the $3 \rightarrow 1$ transition
43. 618 nm
45. 5.8×10^{-10} m
47. 8.1×10^{-12}
49. 0 m/s to 1×10^7 m/s
51.
53. Sodium
55. No
57. a. 2.0×10^{-12} m b. 2.51×10^5
59. 7×10^4 s^{-1}
61. 2.3 mA
63. 1.0×10^8 photons
65. a. 7.83×10^{-19} J, 6.63×10^{-19} J b. 4.7×10^{15} photons, 2.0×10^{16} photons c. A sunburn will require more of the less energetic photons
67. 10 keV
69. 2.4 keV
71. a. 3.88×10^{-12} m b. 3.71×10^{-12} m
73. a. 2.0×10^{-63} J b. 1.2×10^{-30} m/s
75. 1.4 nm
77. 10.9×10^{-10} m
79. a. $\dfrac{h^2}{4mL^2}$ b. E_{min}, $\frac{5}{2}E_{min}$, $4E_{min}$, $5E_{min}$, $\frac{13}{2}E_{min}$
81. 18 fm
83. 19 T
85. C
87. B

Chapter 29

Answers to odd-numbered multiple-choice questions

25. D
27. E
29. D

Answers to odd-numbered problems

1. a.

Wavelength (nm)	m	n
656.6	2	3
486.3	2	4
434.2	2	5
410.3	2	6

   b. 397 nm
3. 1870 nm, 1280 nm, 1090 nm, infrared
5. a. 3 electrons, 4 protons, 5 neutrons
   b. 6 electrons, 6 protons, 6 neutrons
   c. 4 electrons, 7 protons, 8 neutrons
7. a. ^{8}Li b. $^{14}C^{+}$
9. a. Yes b. 0.50 eV
11. 6.7×10^5 m/s, 1.1×10^6 m/s
13. 3
15. $4 \rightarrow 3$
17. a. 0.332 nm, 0.665 nm, 0.997 nm b. 0.332 nm = $(2\pi(0.053 \text{ nm}))/1$, 0.665 nm = $(2\pi(0.232 \text{ nm}))/2$, 0.997 nm = $(2\pi(0.476 \text{ nm}))/3$
   c.

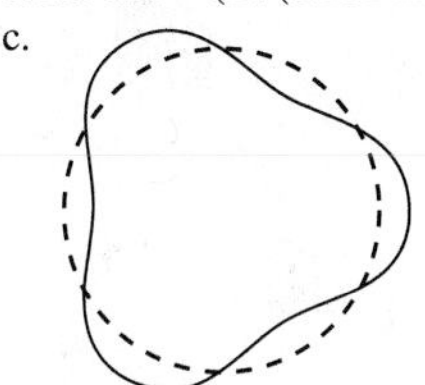

19. 97.2 nm, 486 nm, 1870 nm
21. $n = 1$: $(1, 0, 0, \pm\frac{1}{2})$ $n = 2$: $(2, 0, 0, \pm\frac{1}{2})$, $(2, 1, -1, \pm\frac{1}{2})$, $(2, 1, 0, \pm\frac{1}{2})$, $(2, 1, 1, \pm\frac{1}{2})$$n = 3$: $(3, 0, 0, \pm\frac{1}{2})$, $(3, 1, -1, \pm\frac{1}{2})$, $(3, 1, 0, \pm\frac{1}{2})$, $(3, 1, 1, \pm\frac{1}{2})$, $(3, 2, -2, \pm\frac{1}{2})$, $(3, 2, -1, \pm\frac{1}{2})$, $(3, 2, 0, \pm\frac{1}{2})$, $(3, 2, 1, \pm\frac{1}{2})$, $(3, 2, 2, \pm\frac{1}{2})$
23. $3\hbar$
25. −0.544 eV
27. $1s^2 2s^2 2p^6 3s^2 3p^2$, $1s^2 2s^2 2p^6 3s^2 3p^6 4s^2 3d^{10} 4p^2$, $1s^2 2s^2 2p^6 3s^2 3p^6 4s^2 3d^{10} 4p^6 5s^2 4d^{10} 5p^6 6s^2 4f^{14} 5d^{10} 6p^2$
29. a. Magnesium b. Molybdenum
31. a. Can't have 8 electrons in a p state b. Can't have 3 electrons in an s state
33. 4.86×10^6 m/s
35. 330 nm
37. a. Yes, 2180 nm b. No, violates $\Delta l = 1$
39. 690 nm
41. 3.18×10^{15} s^{-1}
43. a. 1.7×10^{18} b. 1.7×10^{26} photons/s
45. a. 102.6 nm: $n = 3$, $m = 1$; 1876 nm: $n = 4$, $m = 3$ b. Ultraviolet, infrared
47. a. 79 electrons, 79 protons, 118 neutrons b. 2.29×10^{17} kg/m^3 c. 2.01×10^{13}
49. 0.000000000058%, 99.999999999942%
51. a. 58 N b. 4.7×10^{-35} N
53. 1.8×10^7 m/s
55. b. $\lambda_{41} = 200$ nm $\lambda_{31} = 300$ nm $\lambda_{21} = 500$ nm $\lambda_{42} = 333$ nm $\lambda_{43} = 600$ nm $\lambda_{32} = 750$ nm
57. 10
59. 91.1 nm
61. 1870 nm
63. a. $r_{99} = 519$ nm, $v_{99} = 2.21 \times 10^4$ m/s, $r_{100} = 529$ nm, $v_{100} = 2.19 \times 10^4$ m/s b. $f_{99} = 6.79 \times 10^9$ Hz, $f_{100} = 6.59 \times 10^9$ Hz c. 6.68×10^9 Hz d. 0.15%
65. a. 2.06×10^6 m/s b. 12.1 V
67. $3s \rightarrow 2p$, $3p \rightarrow 2s$, $3d \rightarrow 2p$
69. a. $6s \rightarrow 5p$, $6s \rightarrow 4p$, $6s \rightarrow 3p$ b. 7300 nm, 1700 nm, 515 nm
71. $\sqrt{2}\hbar$, decrease
73.

Transition	a. Wavelength	b. Type	c. Absorption
$2p \rightarrow 2s$	670 nm	VIS	Yes
$3s \rightarrow 2p$	816 nm	IR	No
$3p \rightarrow 2s$	324 nm	UV	Yes
$3p \rightarrow 3s$	2696 nm	IR	No
$3d \rightarrow 2p$	611 nm	VIS	No
$3d \rightarrow 3p$	25 μm	IR	No
$4s \rightarrow 2p$	498 nm	VIS	No
$4s \rightarrow 3p$	2430 nm	IR	No

75. 2.0×10^{10} W/m^2
77. A
79. C

Chapter 30

Answers to odd-numbered multiple-choice questions

29. B
31. C
33. A
35. C
37. D

Answers to odd-numbered problems

1. a. 1 proton, 2 neutrons b. 18 protons, 22 neutrons c. 20 protons, 20 neutrons d. 94 protons, 145 neutrons
3. 6.94 u
5. a. 0.511 MeV b. 938.27 MeV c. 938.78 MeV d. 3728.4 MeV
7. a. Total 8.48 MeV, per nucleon 2.38 MeV b. Total 7.72 MeV, per nucleon 2.57 MeV
9. ^{3}He: 2.57 MeV, ^{4}He: 7.07 MeV; ^{4}He is more tightly bound
11. a. 7.58 MeV b. 8.79 MeV c. 7.87 MeV
13. ^{240}Pu: 7.56 MeV, ^{133}Xe: 8.41 MeV
15. a. ^{3}H: 8.482 MeV, ^{3}He: 7.718 MeV; ^{2}H: 2.224 MeV, ^{4}He: 28.297 MeV b. 2.389 MeV
17. None are obviously unstable
19. ^{4}He, ^{12}C, ^{16}O
21. ^{234}U
23. a. ^{228}Th b. ^{207}Tl c. ^{7}Li d. ^{60}Ni
25. 5.52 MeV
27. a. Beta-plus b. Beta-minus
29. a. ^{3}He b. 32%
31. a. 9.3×10^{11} b. 4.7×10^{11} c. 5.5×10^8
33. 53 d
35. 2.0 μCi
37. 3.7×10^{10} Bq, 1.0 Ci
39. 30 h
41. 4.6×10^9
43. 600 h
45. 0.14 mSv
47. 600 rad
49. 19 d
51. 17 mSv
53. 0.29 mSv
55. 139 MeV
57. 200 MeV
59. 2.7×10^{17}
61. 21,000 yr
63. a. 0.10% b. 0.00098
65. 5.9×10^9 yr
67. a. 4.5×10^3 J b. 110 W c. 1.4°C
69. 2.8×10^{12}

71. D
73. A
75. C

Part VII Problems

Answers to odd-numbered problems

1. C
3. A
5. D
7. A
9. D
11. A
13. B
15. D
17. a. 8.0×10^7 m/s b. 4.6×10^{-8} W c. 11 MBq
19. a. 1.1×10^{21} J b. 7.7×10^{32} reactions c. 2.6×10^6 kg of ${}^2_1H$, 3.8×10^6 kg of ${}^3_2H$
21. 15,000

Credits

PART IV Summary p. 531 WL I Delft Hydraulics; **p. 533** Zoonar GmbH/Alamy.

PART V Opener p. 534: Eric Isselée/Shutterstock (composite image of many eyes); Tomatito/Shutterstock (large inset of insect eye, top left); Angelo Gampiccolo/Fotolia (medium inset of yellow owl eye, upper middle); Eric Isselée/Fotolia (medium inset of fish eye, bottom left); filipesso/Fotolia (medium inset of butterfly wing "eyes", bottom right).

CHAPTER 17 P. 536 top Dave Watts/Nature Picture Library; **bottom left** Vasilius/Shutterstock; **bottom center** Tim Gainey/Alamy; **bottom right** Science Source; **p. 537 top** Sigitas Baltramaitis/Alamy; **bottom** Richard Megna/Fundamental Photographs; **p. 538** Richard Megna/Fundamental Photographs; **p. 547** Chris Collins/Corbis; **inset** Andrew Syred/Science Source; **p. 548** Stuart Field; **p. 549** Jian Zi; **p. 550 top** Pearson Education/Eric Schrader; **bottom** Edmund Optics, Inc.; **p. 551 top** Edmund Optics, Inc.; **middle** imagebroker / Alamy; **bottom** Richard Megna/Fundamental Photographs; **p. 553** Sabina Zigman; **p. 554** Ken Kay/Fundamental Photographs; **p. 557** NASA; **p. 562** Kim Taylor/Nature Picture Library; **p. 563** Tom Zagwodzki/Goddard Space Flight Center/NASA; **p. 564 both** Peter Vukusic.

CHAPTER 18 P. 565 top Michael Kappeler/Getty Images; **middle left** kjekol/Fotolia; **middle center** Anya Levinson/AAPT; **middle right** Image Source Alamy; **bottom** Richard Megna/Fundamental Photographs; **p. 567 top left** Onpoint Lasers, Inc.; **top second to left** Dynamic Graphics Group/IT Stock Free/Alamy; **top second to right** Mike Stone/Alamy; **top right** Paul Rapson/Alamy; **bottom** Stuart Field; **p. 568** Stuart Field; **p. 569** Sylvester Allred/Fundamental Photographs; **p. 570** Branko Čeak, The Slovenian Science Center; **p. 571** Pearson Education; **p. 574** Ken Kay/Fundamental Photographs; **p. 575** Pete Atkinson/Getty Images; **p. 576 top** Custom Medical Stock Photo, Inc.; **bottom** Pearson Education; **p. 578 left** Richard Megna/Fundamental Photographs; **right** Richard Megna/Fundamental Photographs; **p. 580** Pearson Education; **p. 582** Getty Images; **p. 586 top** Dr. Tamara Frank, Harbor Branch Oceanographic Institution, Inc.; **bottom** Alper Çuğun; **p. 587** Paul Silverman/Fundamental Photographs; **p. 592** Bajande/Science Source; **p. 594 top left** NASA; **bottom left** Erich Lessing/Art Resource; **right** Edouard Manet 1832-1883, "Bar at the Folies-Bergere". 1881/82. Oil on Canvas. 37 13/16" × 51" (90 × 130 cm). Courtauld Institute Galleries, London. AKG-Images; **p. 596 top** Vstock/Alamy; **bottom** Nano Calvo/Alamy; **p. 599 top** Jason Horowitz/Corbis; **bottom** David Hancock/Alamy.

CHAPTER 19 P. 600 top Juniors Bildarchiv/Alamy; **bottom left** Jupiter Images/Alamy; **bottom middle** auremar/Fotolia; **bottom right** Biophoto Associates/Science Source; **p. 601** Doug Perrine/Nature Picture Library; **p. 602 top** Alexander Raths/Fotolia; **bottom** NASA; **p. 604 top** Dynamics Graphics Group/Alamy; **bottom** Tetra Images/Alamy; **p. 607** all Stuart Field; **p. 609** Tetra Images/Alamy; **p. 610** Biophoto Associates/Science Source; **p. 611** NASA; **p. 613** Pearson Education/Eric Schrader; **p. 615** NASA; **p. 607 left** James Cavallini/Science Source; **right** STEVE GSCHMEISSNER/Science Photo Library/Alamy; **p. 619** Daniel Hebert/Shutterstock; **p. 621 left** Stuart Field; **right** Sam Ogden/Science Source; **p. 623** NASA; **p. 624 left** Alfred Pasieka/Science Source; **right** NASA.

PART V Summary p. 627 both Rob Meyer/Meyer Instruments; **p. 629** Jessica Riser.

PART VI Opener p. 630 Nonoy Tan/Alamy.

CHAPTER 20 P. 632 top Pacific Northwest National Laboratory; **bottom left** Charles D. Winters/Science Source; **bottom center** yevgeniy11/Fotolia; **bottom right** fergregory/Fotolia; **p. 633** Pearson Science; **p. 636** Allan Lee; **p. 638** Steve P. Hopkin/Getty Images; **p. 642** Science Pictures Limited/Science Source; **p. 645** Richard Megna/Fundamental Photographs; **p. 653 top** Biran Jones; **bottom** ARCO/Reinhard, H/age fotostock; **p. 658** David Lassman/The Image Works.

CHAPTER 21 P. 665 top Deborah Zemek; **middle left** steheap/Fotolia; **middle right** Kokhanchikov/Shutterstock; **bottom** Marianthi Lainas/Alamy; **p. 669 top** Aaron Horowitz/Corbis; **bottom** Ingram Publishing/Alamy; **p. 670 left** Stuart Field; **right** Juerg Streit; **p. 673 top** Pearson Education; **bottom** Varian Medical Systems, Inc.; **p. 682** Stephen Frink Collection/Alamy; **p. 685 top** Stocklite/Shutterstock; **bottom** Creative Digital Vision/Pearson Science; **p. 687** Helmut Foll; **p. 689** Andrew Lambert/Alamy; **p. 690 top** Dimitri Vervits/Getty Images; **bottom** Adam Hart-Davis/Science Source; **p. 692** NASA.

CHAPTER 22 P. 702 top May/Science Source; **bottom left** Pearson Science; **bottom center** Pearson Science; **bottom right** Roy McMahon/Corbis; **p. 708** Peter Scoones/Nature Picture Library; **p. 711** Dennis O'Clair/Getty Images; **p. 713 left** pixel shepherd/Alamy; **center** Igor S./Shutterstock; **right** Brian Jones; **p. 716** Brian Jones; **p. 717** FLIR Systems, Inc.; **p. 723** Oxford Scientific/Getty images.

CHAPTER 23 P. 727 top Alamy; **bottom left** Pavlo Sachek/Fotolia; **bottom center** Patrick Batchelder/Alamy; **bottom right** CNRI/Science Source; **p. 733** Dynamic Graphics Group/IT Stock Free/Alamy; **p. 735** Abelstock/Alamy; **p. 737** Simon Fraser/Science Source; **p. 739** Nick Veasey/Getty Images; **p. 742** Pearson Education; **p. 745top** Eric Schrader/Pearson Education; **bottom** Don W. Fawcett/Science Source; **p. 748** Stephane DeSakutin/AFP/Getty Images; **p. 749** SPL/Science Source; **p. 750** STEVE GSCHMEISSNER/Science Photo Library/Alamy; **p. 751** SPL/Science Source.

CHAPTER 24 P. 764 top Science Source; **bottom left** Daniel Dempster/Alamy; **bottom center** John Eisele; **bottom right** Romko_chuk/iStockphoto; **p. 767** John Eisele; **p. 768 all** John Eisele; **p. 769** Atsushi Arakaki et al., "Formation of magnetite by bacteria and its application" J R Soc Interface. 2008 September 6; 5(26): 977–999. doi: 10.1098/rsif.2008.0170; **p. 770 all** John Eisele; **p. 775** Kathleen Gustafson; **p. 776** Charles Thatcher/Getty Images; **p. 781** Chris Madeley/Science Source; **p. 782** Photo courtesy of GE Healthcare; **p. 783** March-McBirney/Hach Company; **p. 786** Brian James; **p. 789** UHB Trust/Getty Images; **p. 792 left** John Eisele; **right** IBM Corporation; **p. 793** Andrew Lambert/Science Source; **p. 796** Barrington Gradiometry.

CHAPTER 25 P. 804 top, both Ted Kinsman/Science Source; **bottom left** Yegor Korzh/Shutterstock; **bottom center** Scott McAlpine/ British Columbia Institute of Technology; **bottom right** Science Source; **p. 806** Science Source; **p. 807** James Watt/Stephen Frink Collection/Alamy; **p. 809** Pearson Education/Eric Schrader; **p. 814** James Leynse/Corbis; **p. 818** SSPL/The Image Works; **p. 820** Tony Craddock/Science Source; **p. 822 top three** Richard Megna/Fundamental Photographs; **bottom** Innerspace Imaging/Science Source; **p. 823 both** Brian Jones; **p. 824** Pearson Education/Pearson Science; **p. 827top three** Brian Jones; **bottom left** Ashok Captain/Alamy; **bottom right** Edward Kinsman/Science Source; **p. 829 top left** NOAA; **top right** NASA; **bottom left** Dr. Jack O. Burns; **bottom right** NASA; **p. 834** Dennis Nett/The Image Works.

CHAPTER 26 **P. 840 top** Duncan Usher/Alamy; **bottom left** Biehler Michael/Shutterstock; **bottom center** Tom Carter/Alamy; **bottom right** Eric Audras/Getty Images; **p. 841** Pearson Education; **p. 843** iStockphoto; **p. 844** nrt/Shutterstock; **p. 845** Pearson Education; **p. 848 top** Brian Jones; **bottom** 1125089601/Shutterstock; **p. 849** Paul A. Souders/Corbis; **p. 850** Pearson Education; **p. 851**Photos.com; **p. 852** ViennaPhoto/Alamy; **p. 854** Pearson Education/Eric Schrader; **p. 855** Galyna Andrushko/Shutterstock; **p. 858** David Burton / Alamy; **p. 867** Steve Sant/Alamy.

PART VII **Opener p. 872** Gregory G. Dimijian/Science Source.

CHAPTER 27 **P. 874 top** Chris Johnson/Alamy; **bottom left** iStockphoto; **bottom center** NASA; **bottom right** Sadeq/Fotolia; **p. 875 top** Topham/The Image Works; **bottom** Kaz Chiba/Getty Images; **p. 877** eStock Photo/Alamy; **p. 879** Jean Saint-Martin; **p. 879** Jim Corwin/Alamy; **p. 890 top** U.S. Department of Defense Visual Information Center; **bottom** Eckhard Slawik/Science Source; **p. 897** SSPL/The Image Works; **p. 898** LBNL/Science Source; **p. 904** Dynamic Graphics/age fotostock; **p. 905** NASA; **p. 906** John Chumack/Science Source.

CHAPTER 28 **P. 908 top** Omikron/Science Source; **middle left** Dr. Antoine Weis; **middle center** Science Source; **middle right** Ted Kinsman/Science Source; **bottom two** Science Source; **p. 909 top** Oxford Instruments; **bottom** Neil Borden/Science Source; **p. 911** M.H.F. Wilkins; **p. 914 top** Bettmann/Corbis; **bottom** Neil McAllister/Alamy; **p. 915** William Anthony Milberry; **p. 917** Dr. Antoine Weis; **p. 918** Kenneth Eward/BioGrafx/Science Source; **p. 920** Pearson Education; **p. 921 left** Educational Development Center, Inc.; **center** Educational Development Center, Inc.; **right** W.H. Freeman and Company; **p. 922 top** Science Source; **bottom** Brian Jones; **p. 925** Bettmann/Corbis; **p. 926** Ted Kinsman/Science Source; **p. 929 top** Colin Cuthbert/Science Source; **middle** Veeco Instruments, Inc.; **bottom** LLNL/Science Source; **p. 930 left** Ogwen/Shutterstock; **right** Hou et al., in the 18 October 1999 issue of Physical review Letters 83, 3001-3004 (1999); **p. 931** Mehau Kulyk/Science Source; **p. 932 left** Science Source; **right** Veeco Instruments, Inc.; **p. 935left** Bill Corwin; **right** Darwin Dale/Science Source; **p. 936 top** Solent News/Splash News/Newscom; **bottom** SPL.

CHAPTER 29 **P. 940 top** Jennifer Waters/Science Source; **bottom left** Stuart Field; **bottom middle** Philip Hart/Stocktrek Images/Getty Images; **bottom right** G.K./Shutterstock; **p. 941** Ocean Optics, Inc.; **p. 942** NASA; **p. 962 both** International Dark Sky Association; **p. 963** Edward Kinsman/Science Source; **p. 964 top** Sinclair Stammers/Science Source; **bottom** Frankwalker.de/Fotolia; **p. 966 top** U.S. Air Force photo/Tech. Sgt. Larry A. Simmons; **bottom** iStockphoto.

CHAPTER 30 **P. 975 top** CNRI/Science Source; **bottom left** British Antarctic Survey/Science Source; **bottom center** Earl Roberge/Science Source; **bottom right** HTO/Pearson Education; **p. 977** The National Ice Core Laboratory; NSF and USGS; **p. 978** Roberto De Gugliemo/Science Source; **p. 979** NASA; **p. 980** U.S. Department of Energy; **p. 986** C. Powell, P. Fowler & D. Perkin/Science Source; **p. 990** Hank Morgan/Science Source; **p. 991 top** James King-Holmes/Science Source; **bottom** Bruce E. Zuckerman/Corbis; **p. 992** Hank Morgan/Science Source; **p. 993** Custom Medical Stock Photo, Inc.; **p. 994 top** Chauvaint-Chapon/Phanie/SuperStock; **bottom** Custom Medical Stock Photo, Inc.; **p. 995 both** CNRI/Zephyr/Science Source; **p. 996 top** Hank Morgan/Science Source; **middle** Zephyr/Science Source; **bottom** Zephyr/Science Source; **p. 997** CERN/Science Source; **p. 998** The Sudbury Neutrino Observatory Institute SNOI; **p. 1000** U. S. Department of Energy; **p. 1003** Brian Jones; **p. 1005 top left** David Parker/Science Source; **bottom left** Jean-Perrin/CNRI/Science Source; **right** Science Source; **p. 1006** National Museum of Natural History.

PART VII **Summary p. 1009** Peter Hannaford; **p. 1011** Bgran, Used under a Creative Commons License, http://creativecommons.org/licenses/by/2.0/deed.en.

Index

For users of the two-volume edition, pages 1–533 are in Volume 1 and pages 534–1011 are in Volume 2.

A

B

C

D

E

M

Q

R

W

X

Y

Z